BIG MECHANISMS IN SYSTEMS BIOLOGY

BIG MECHANISMS IN SYSTEMS BIOLOGY

Big Data Mining, Network Modeling, and Genome-Wide Data Identification

BOR-SEN CHEN
National Tsing Hua University, Hsinchu, Taiwan

CHENG-WEI LI
National Tsing Hua University, Hsinchu, Taiwan

ELSEVIER

AMSTERDAM • BOSTON • HEIDELBERG • LONDON
NEW YORK • OXFORD • PARIS • SAN DIEGO
SAN FRANCISCO • SINGAPORE • SYDNEY • TOKYO
Academic Press is an imprint of Elsevier

Academic Press is an imprint of Elsevier
125 London Wall, London EC2Y 5AS, United Kingdom
525 B Street, Suite 1800, San Diego, CA 92101-4495, United States
50 Hampshire Street, 5th Floor, Cambridge, MA 02139, United States
The Boulevard, Langford Lane, Kidlington, Oxford OX5 1GB, United Kingdom

British Library Cataloguing-in-Publication Data
A catalogue record for this book is available from the British Library

Library of Congress Cataloging-in-Publication Data
A catalog record for this book is available from the Library of Congress

ISBN: 978-0-12-809479-2

Publisher: Mica Haley
Acquisition Editor: Rafael Teixeira
Editorial Project Manager: Mariana Kuhl
Production Project Manager: Karen East and Kirsty Halterman
Designer: Victoria Pearson

Typeset by MPS Limited, Chennai, India

CONTENTS

CHAPTER 1

Introduction to Big Mechanisms in Systems Biology

INTRODUCTION

Currently, the reductionist approach in science leads to a causal model that has small fragments from complicated systems. However, constructing causal models of entire systems is difficult because the required information is distributed across the exhaustive literature. In the United States, The Defense Advanced Research Projects Agency (DARPA) created the Big Mechanism program (BMP) to develop technology for constructing, understanding, and reasoning of large, complicated systems such as climatic, economic, ecological, and biological systems. At present, BMP focuses on cancer signaling pathways, but the technology is intended to be applied for general purposes [1,2]. In the future, the BMP aims to produce machines that can read literature and assemble causal fragments found in individual articles into larger causal models. For example, the computer can gather information from the literature on cancer biology, extract fragments of causal mechanisms from publications, assemble the mechanisms into executable models of unprecedented scale and fidelity, use these models to explain and predict aspects of cancer biology, and even test these predictions in vitro [1,2]. BMP aims to use existing online sources of biological knowledge as well as existing machine reading and information extraction methods, such as big data and big data mining methods, to develop representation and inference methods for mechanistic biology models and methods for the machine to understand the claims and evidence in papers, algorithms, or system modeling for inferring causal relationships through system identification via big data and big data mining [2].

In this chapter, we introduce Big Mechanisms, which are mechanistic models of complicated systems with too many elements and relationships or too many possible current or future states to be easily comprehended by humans. Next, we introduce Big Mechanisms in systems biology via system modeling and identification through genome-wide data and big data mining methods. Since the focus of the BMP is on systems biology, in this book, the scope of Big Mechanisms in systems biology is on the system mechanisms of the cell cycle, signal flow, immunity, regeneration, infection, aging, evolution, carcinogenesis, and medicine based on system models, big database mining, and genome-wide high-throughput experimental data. This book is

Big Mechanisms in Systems Biology
DOI: http://dx.doi.org/10.1016/B978-0-12-809479-2.00001-9

designed to develop systematic methods to help scientists understand the big mechanisms of very complicated biological systems.

1.1 INTRODUCTION TO BIG MECHANISMS

The DARPA has recently introduced a search for technologies based on a new kind of science in which research is integrated automatically or semi-automatically into causal, explanatory models [1]. It emphasizes the need for solutions to complex systems such as ecosystems, brains, economics, and social systems that have parts and processes that are currently studied piecemeal, with literature and data that are fragmented, distributed, and inconsistent. It addresses the problem of information overload that we are all currently experiencing in "Big Data" environments. In recent years, "Big Data" has transformed science, engineering, medicine, healthcare, finance, business, military, and ultimately society itself. For example, analyses of command information flows during military crises have been suggested as an approach to ontogenetic learning that avoids both the problem of describing the subject covered in a document, and the problem of integrating new subject matter into a predetermined classification code. We need to make computer-readable documents and data, and assemble the fragments into explanatory pattern. This would be part of the Big Mechanisms that explain causes and effects within systems.

Therefore, Big Mechanisms are made up of many small mechanisms, which are dispersed and need to be integrated into a knowledge base to understand the mechanisms of big and complex systems. However, our knowledge of these mechanisms is increasingly fragmented, voluminous, and inconsistent. It is of central importance to plan and follow the big database mining process to create a knowledge base, and investigate how to put this knowledge base into both a relational and graph format to perform analyses and visualizations for the development of knowledge and for application. DARPA's goal is to develop technologies for a new kind of science in which many areas of research are integrated into causal, explanatory, or mechanistic models, with deeper semantics to represent the causal and often kinetic models in response to the challenge of big data that science and industry are focus today.

At present, the domain of the BMP is systems biology [1,3,4]. The amount of data produced by the study of biological systems in different species and at different scales, from molecules to ecosystems, is growing exponentially. As this increase in information presents challenges to some areas of biology, systems biology researchers employ bioinformatics technologies to handle large amounts of data, from local production through to storage in publicly accessible and integrative depositories. Systems biology is a new area of study in biology that seeks to incorporate bioinformatics, statistics, mathematics, physics, chemistry, biology, and engineering, and that promises to advance and transform our ability to increase biological understanding, program biological functions, and render the engineering of biological circuits or pathways faster and more predictable. A large-scale, integrative, and multidisciplinary approach is needed for systems biology to flourish.

1.2 BIG MECHANISMS IN SYSTEMS BIOLOGY

This book addresses Big Mechanisms of systems biology by system identification and big data mining methods using models of biological systems [2,5]. Faced with a large volume of available literature, complicated mechanisms, small prior knowledge, few classes on the topic, and causal and mechanistic language, biological research is currently undergoing revolutionary changes in response to the integration of powerful technologies. Immunity, regeneration, infection, aging, evolution, and carcinogenesis are complicated biological systems, and related biological knowledge is fragmented, voluminous, and even inconsistent. The authors have observed remarkably poor agreement among different databases regarding components of these biological systems. These inconsistencies may reflect the underlying biology with time-varying systems and signal transduction events are often context-dependent. This raises a significant problem for mechanistic modeling since it is not clear which genes/proteins to include in models or experimental measurements.

In this book, we first construct candidate biological networks from omics data and big data mining that contain many false positives, interaction inconsistencies, and contradictions (see Fig. 1.1). System models are then employed to describe the relationship between components in the biological network. System identification (reverse engineering) methods are used to identify the interaction parameters of candidate biological network by experimental measurement via microarray data, NGS (next-generation sequencing) data, or real-time PCR data in a specific biological condition, such as lung cancer. Finally, the system order detection method is employed to determine the system order (the number of interactions) of the biological network so we can delete the insignificant interactions from the system order and prune false positives, inconsistencies, or contradictory interactions in the candidate biological network to obtain the real biological network of a specific biological condition.

After detection and removing inconsistencies between the experimental data and candidate biological network, biological networks in different biological conditions are obtained. We can compare two biological networks in different conditions to find significant changes in interactions or evolution of biological conditions (e.g., from a normal to cancer condition) as the network markers of these conditions change. From some related biological websites, like GO Enrichment Analysis, we can find biological functions or pathways to interpret system mechanisms of these network markers and identify adequate drug targets by methods to analyze sensitivity and robustness for the therapeutic treatment.

In this book, the Big Mechanisms of biological systems, like the immune system, regeneration process, photosynthetic system, and cancer are constructed by systems biology and big database mining methods. Systems biology sits at the heart of a new integrative 21st century paradigm of biology. It has introduced a search for a new

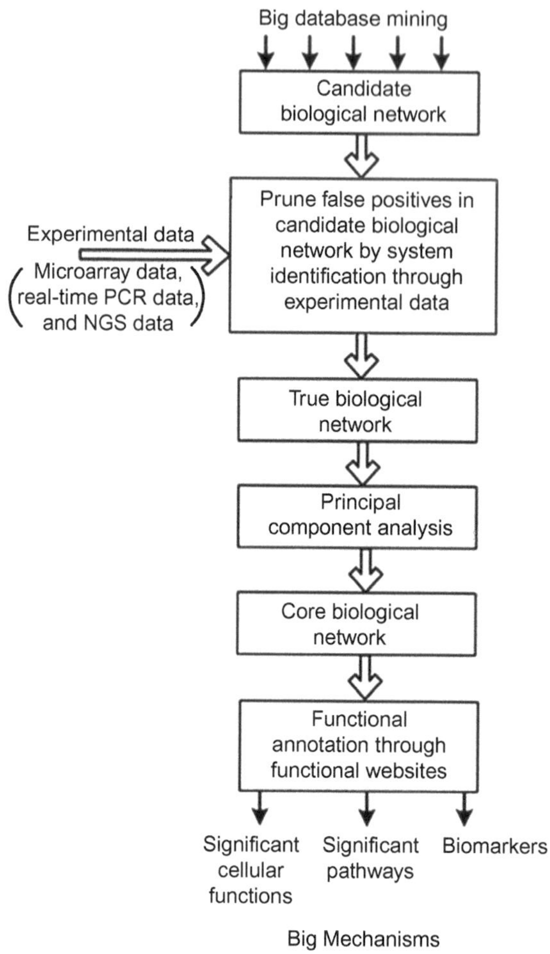

Figure 1.1 A block diagram of Big Mechanisms of systems biology via system identification and big data mining.

kind of scientific research based on big data mining and causal and explanatory models of experimental data. If we could resolve Big Mechanisms of systems biology by system modeling, system identification, and big data, the results could provide a paradigm for solutions to Big Mechanisms of complex systems, such as ecosystems, brains, economics, and social systems in the future.

1.3 THE SCOPE OF BIG MECHANISMS OF SYSTEMS BIOLOGY IN THIS BOOK

In order to describe the system mechanisms and behaviors, system modeling and system identification methods are introduced with big data mining in Chapter 2, System

Modeling and System Identification Methods for Big Mechanisms in Biological Systems. Linear and nonlinear models and related systematic and mathematic analyses for biological networks are introduced to investigate the Big Mechanisms in Systems Biology. Then, least square parameter estimation and maximum likelihood parameter estimation methods are introduced to estimate the system parameters of biological networks using experimental data or high-throughput data [5]. Some stochastic testing, systems theory, and system order detection methods are provided for statistical inference and hypothesis testing of biological mechanisms in bioinformatics and systems biology [6,7].

In Chapter 3, Procedure for Exploring Big Mechanisms of Systems Biology Through System Identification and Big Database Mining, how to construct Big Mechanisms in systems biology by integrating many smaller mechanisms from the dispersed heterogeneous omics data via system models and big database mining is introduced. Since the main domain of Big Mechanisms is systems biology, several examples, i.e., how to construct a gene regulatory network (GRN), protein–protein interaction network (PPIN), and integrated genetic and epigenetic networks (IGEN) from omics data, are given to illustrate the procedure for following big data mining processes and system identification procedures to elucidate system-based Big Mechanisms.

The most remarkable features of cells and entire organisms are their ability to reproduce, and the cell cycle entails an ordered series of macromolecular events that lead to cell division and the production of two daughter cells. As a consequence, we choose Chapter 4, Big Cellular Mechanisms in the Cell Cycle by System Identification and Big Data Mining, to look at the regulation of transcription in the progression of the yeast cell cycle as an example for analyzing Big Cellular Mechanisms using big data mining processes and system identification procedures in yeast [8–10].

Since the control of transcription factors (TFs) on their target genes is one of the most important cellular mechanisms, it is more appealing to identify the regulatory trans/cis mechanisms of TFs via dynamic regulatory models and genome-wide microarray data. Therefore, in Chapter 5, Big Regulatory Mechanisms in the Transcriptional Regulation Control of Gene Expression Using a Stochastic System Model and Genome-Wide Experimental Data, stochastic system models and experimental data are used to analyze Big Regulatory Mechanisms in transcription control [11–13].

The cells sense extracellular signals through intercellular communication or stress responses, cellular responses to sudden environmental stresses or physiological changes that provide living organisms with the opportunity for survival and further development. Therefore, defending against environmental stresses with protective mechanisms is an important topic. In Chapter 6, Big Mechanisms of Information Flow in Cellular Systems in Response to Environmental Stress Signals via System Identification and Data Mining, cellular systems under environmental stresses are given as examples to identify Big Mechanisms of information flow and protective mechanisms in response to environmental stresses using system identification and database mining [14–17].

The major task of the immune system is to defend the host against infections. Recently, a large variety of experimental techniques and high-throughput NGS data has been created and provides a sound scientific basis for integrative approaches to advanced, quantitative, and qualitative systems biology, using computation and modeling that will allow us to understand Big Mechanisms of complex immune responses. In Chapter 7, Big Offensive and Defensive Mechanisms in Systems Immunity From System Modeling and Big Data Mining, system modeling and big data mining are used to identify the big offensive and defensive mechanisms in the immune system [18–20].

Regeneration is one of the most intriguing and fascinating biological phenomena, but the molecular and cellular bases of regeneration are still not fully understood. Unlike the vertebrates with high regenerative capacity such as amphibians and fish, mammals have a restricted ability to regenerate lost cells, tissues, and organs at the adult stage. Therefore, understanding the molecular mechanism of regeneration in organisms is not only valuable in shedding light on the longstanding question of regeneration mechanisms, but it is also useful for stem cell-based therapies in the future. In Chapter 8, Big Regeneration Mechanisms via Systems Biology and Big Database Mining Methods, Big Regeneration Mechanisms are given as an example of using systems biology and big database mining methods [21,22].

Cancer is a genetic disease. Many screening arrays for finding new cancer genes have been constructed. Cancer is also a systemic disease. Therefore, tumorigenesis mechanisms should be investigated from a systems biology perspective. In Chapter 9, Big Tumorigenesis Mechanisms in Systems Cancer Biology via Big Database Mining and Network Modeling, cancer systems biology is used to identify Big Tumorigenesis Mechanisms through big database mining and network modeling [20,23–25].

Living organisms are complex systems characterized by emergent properties. A ubiquitous property in complex systems is robustness. Robustness is not a trivial biological mechanism to be studied using a reductionism approach. Robust mechanisms are frequently and widely distributed, guaranteeing organization of the system and cooperation between various parts. Signal transductivity, a system property, is complementary to robustness, i.e., a biological system with greater network robustness will show greater loss of signal transductivity and vice versa. Since aging and cancer are both systems diseases of somatic cell evolution, it is appealing to investigate the network robustness and signal transductivity of somatic cells in the aging and carcinogenic processes. In Chapter 10, Big Evolutionary Mechanisms of Network Robustness and Signaling Transductivity in Aging and Carcinogenic Process by System Modeling and Database Mining, the Big Mechanisms in aging processes and carcinogenesis are given as examples of examining the mechanisms of network robustness and signaling transductivity through system modeling and database mining [26,27].

Aging, an extremely complex and system-level process, has attracted much attention in medical research, especially as chronic diseases are quite prevalent in the elderly population. These may be the result of both accumulated genetic and epigenetic variations that lead to intrinsic perturbations and environmental changes that may stimulate signaling in the body. Decades of research have demonstrated that higher incidence of many pathophysiological phenotypes, such as cancer, inflammation, diabetes, cardiovascular diseases, and neurodegenerative disorders, are associated with the process of aging. In Chapter 11, Big Mechanisms of Aging via System Identification and Big Database Mining, we first investigate the systematic mechanism of GRN in the aging process by systems biology method via microarray data. Then we will investigate specific core genetic and epigenetic network for cellular mechanisms of human aging [28].

In the past few decades, medicine has sought a "magic bullet" that targets a simple disease-causing molecule and that could cure cancer, diabetes, and other complex diseases. Although some drugs have proven successful, many others have been found to be ineffective or the cause of significant side effects. This disappointing outcome highlights the limitation of single target drug paradigms and is considered to be the underlying cause of stagnation in the productivity of pharmaceutical industry. However, compound efficacy and safety in humans, including toxicity and pharmacokinetic profiles, rather than target selection, are the criteria that determine which drug candidates enter the clinic. Therefore, drug discovery using database mining and systems biology will become the most important approach to drug discovery. Finally, in Chapter 12, Big Drug Design Mechanisms via Systems Biology and Big Database Mining, Big Drug Design Mechanisms are introduced as therapeutic treatments using systems biology and big database mining. At first, we will give an overview of drug discovery using systems biology, and then introduce how to investigate core and specific network markers for multiple drug targets from the perspective of systems biology. Finally, a cocktail drug design is also introduced for multiple drug targets [29,30].

REFERENCES

[1] A data science Big Mechanism for DARPA, <http://semanticommunity.info/>.

[2] Cohen PR. DARPA's Big Mechanism program. Phys Biol 2015;12(4):1–9.

[3] Chen B-S, Wu C-C. Systems biology: an integrated platform for bioinformatics, systems synthetic biology and systems metabolic engineering. New York, NY: Nova Science Pub Inc.; 2014.

[4] Klipp E, Liebermeister W, Wierling C, Kowald A, Lehrach H, Herwig R. Systems biology. Weinheim: Wiley; 2011.

[5] Johansson R. System modeling and identification. Englewood Cliffs, NJ: Prentice Hall; 1993.

[6] Chen BS, Li CW. Constructing an integrated genetic and epigenetic cellular network for whole cellular mechanism using high-throughput next-generation sequencing data. BMC Syst Biol 2016;10:18.

[7] Wang YC, Chen BS. Integrated cellular network of transcription regulations and protein-protein interactions. BMC Syst Biol 2010;4:20.

[8] Chen HC, Lee HC, Lin TY, Li WH, Chen BS. Quantitative characterization of the transcriptional regulatory network in the yeast cell cycle. Bioinformatics 2004;20(12):1914–27.

[9] Wu WS, Li WH, Chen BS. Computational reconstruction of transcriptional regulatory modules of the yeast cell cycle. BMC Bioinformatics 2006;7:421.
[10] Wu WS, Li WH, Chen BS. Identifying regulatory targets of cell cycle transcription factors using gene expression and ChIP-chip data. BMC Bioinformatics 2007;8:188.
[11] Chang YH, Wang YC, Chen BS. Identification of transcription factor cooperativity via stochastic system model. Bioinformatics 2006;22(18):2276−82.
[12] Chang YH, Wang YC, Chen BS. Nonlinear dynamic trans/cis regulatory circuit for gene transcription via microarray data. Gene Regul Syst Bio 2007;1:151−66.
[13] Lin LH, Lee HC, Li WH, Chen BS. Dynamic modeling of cis-regulatory circuits and gene expression prediction via cross-gene identification. BMC Bioinformatics 2005;6:258.
[14] Li CW, Chen BS. Identifying functional mechanisms of gene and protein regulatory networks in response to a broader range of environmental stresses. Comp Funct Genomics 2010;Article ID:408705, 20 pages.
[15] Wu WS, Li WH, Chen BS. Reconstructing a network of stress-response regulators via dynamic system modeling of gene regulation. Gene Regul Syst Bio 2008;2:53−62.
[16] Yao CW, Hsu BD, Chen BS. Constructing gene regulatory networks for long term photosynthetic light acclimation in *Arabidopsis thaliana*. BMC Bioinformatics 2011;12:335.
[17] Przytycka TM, Kim YA. Network integration meets network dynamics. BMC Biol 2010;8:48.
[18] Chen BS, Yang SK, Lan CY, Chuang YJ. A systems biology approach to construct the gene regulatory network of systemic inflammation via microarray and databases mining. BMC Med Genomics 2008;1:46.
[19] Kuo ZY, Chuang YJ, Chao CC, Liu FC, Lan CY, Chen BS. Identification of infection- and defense-related genes via a dynamic host-pathogen interaction network Using a *Candida albicans*-zebrafish infection model. Journal of Innate Immunity 2013;5(2):137−52.
[20] Chen BS, Li CW. Constructing an integrated genetic and epigenetic cellular network for whole cellular mechanism using high-throughput next-generation sequencing data. BMC Syst Biol 2016;10(1):18.
[21] Wu CC, Lin C, Chen BS. Dynamic network-based relevance score reveals essential proteins and functional modules in directed differentiation. Stem Cells Int 2015;2015:792843.
[22] Wu CC, Tsai TH, Chang C, Lee TT, Lin C, Cheng IHJ, et al. On the crucial cerebellar wound healing-related pathways and their cross-talks after traumatic brain injury in *Danio rerio*. PLoS One 2014;9(6).
[23] Chu LH, Chen BS. Construction of a cancer-perturbed protein-protein interaction network for discovery of apoptosis drug targets. BMC Syst Biol 2008;2:56.
[24] Li CW, Chu YH, Chen BS. Construction and clarification of dynamic gene regulatory network of cancer cell cycle via microarray data. Cancer Inform 2006;2:223−41.
[25] Wang YC, Chen BS. A network-based biomarker approach for molecular investigation and diagnosis of lung cancer. BMC Med Genomics 2011;4:2.
[26] Chen BS, Tsai KW, Li CW. Using nonlinear stochastic evolutionary game strategy to model an evolutionary biological network of organ carcinogenesis under a natural selection scheme. Evol Bioinform 2015;11:155−78.
[27] Tu CT, Chen BS. On the increase in network robustness and decrease in network response ability during the aging process: a systems biology approach via microarray data. IEEE-ACM Trans Comput Biol Bioinform 2013;10(2):468−80.
[28] Li CW, Wang WH, Chen BS. Investigating the specific core genetic-and-epigenetic networks of cellular mechanisms involved in human aging in peripheral blood mononuclear cells. Oncotarget 2016;7(8):8556−79.
[29] Chen BS, Li CW. Analysing microarray data in drug discovery using systems biology. Expert Opin Drug Discov 2007;2(5):755−68.
[30] Wong YH, Lin CL, Chen TS, Chen CA, Jiang PS, Lai YH, et al. Multiple target drug cocktail design for attacking the core network markers of four cancers using ligand-based and structure-based virtual screening methods. BMC Med Genomics 2015;8(Suppl. 4):S4.

CHAPTER 2

System Modeling and System Identification Methods for Big Mechanisms in Biological Systems

INTRODUCTION

Consider the cellular system as shown in Fig. 2.1. Through intercellular communication and cellular stress responses, the cells sense extracellular signals. Different external changes or events may induce signaling in cells. Typical signals are hormones, pheromones, pathogens, heat, cold, light, osmotic pressure, and changes in the appearance or concentration of substances such as glucose, potassium ions, calcium ions, or cyclic adenosine monophosphate (cAMP) [1−7]. In the signal transduction pathway, extracellular signals are perceived by transmembrane receptors. The receptor changes its own state from susceptible to active and then it triggers subsequent cellular processes.

The active receptor stimulates an internal signaling cascade. This cascade frequently includes a series of changes in protein phosphorylation states. The sequence of state changes crosses the nuclear membrane. Eventually, some transcription factors (TFs) are activated or deactivated, which change their ability to bind a set of genes that produce the corresponding proteins in response to extracellular signals or stresses. In addition to genetic regulation by TFs, epigenetic regulation due to DNA methylation and microRNAs (miRNA) can also alter gene expression. However, owing to the complex nature of dynamic biological systems, knowledge of their components and interactions is not sufficient to interpret behaviors in the cellular system. In this book, as shown in Fig. 1.1, we employ big database mining to construct candidate-integrated genetic and epigenetic networks (GENs) to interpret some cellular mechanisms for some biological conditions. However, there exists a large amount of false positives in a big database mining process. Therefore, it is very important to prune these false positive regulations and interactions from the candidate GEN by real high-throughput experimental data. We need to model interactions and regulations in the candidate GEN and then prune these false positive interactions and regulations to obtain the real GEN through system modeling and identification methods by microarray data or next generation sequencing (NGS) data.

Therefore, in this chapter, we will describe the models for these cellular systems and use experimental data, such as microarray data, NGS data, or real-time polymerase

Big Mechanisms in Systems Biology
DOI: http://dx.doi.org/10.1016/B978-0-12-809479-2.00002-0

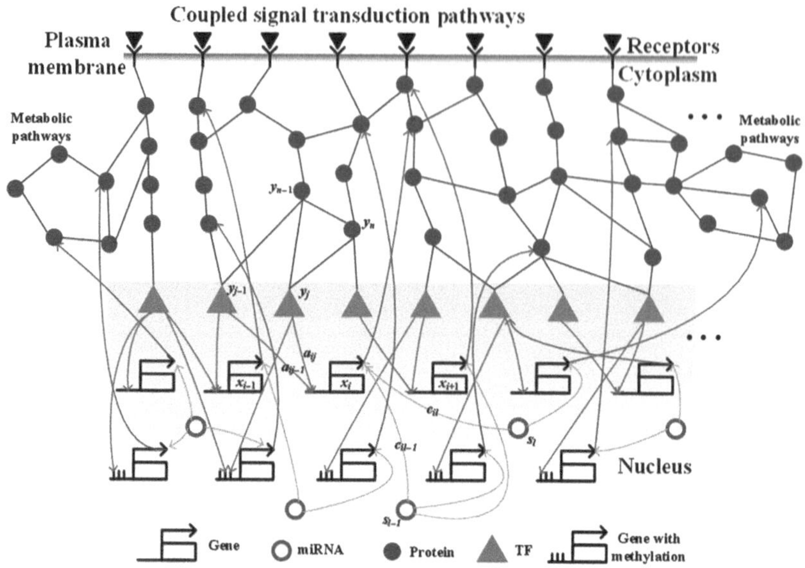

Figure 2.1 An integrated signal transduction pathway (i.e., protein–protein interaction network (PPIN), gene regulatory network (GRN), and miRNA regulatory network).

chain reaction (PCR) data, to identify the parameters of the model that will allow us to efficiently and accurately integrate this data into the Big Mechanism, and to interpret the behaviors of cellular systems by system modeling and identification methods. To begin, we will model and identify the protein–protein interaction network (PPIN) and gene regulatory network (GRN) by protein expression and microarray data, respectively. Then, these genetic networks will be integrated with epigenetic regulation by miRNA expression data to construct an integrated genetic and epigenetic cellular network. Further, core network extracted by applying principal network projection (PNP) to the integrated genetic and epigenetic cellular network is also introduced in this chapter to investigate significant network mechanism by principal component analysis (PCA) via singular value decomposition (SVA) method.

For the convenience of illustration, only linear system models are introduced for system identification of biological networks to investigate the Big Mechanisms through genome-wide high-throughput data. When necessary, the system identification methods based on nonlinear system models will be introduced in the following chapters.

2.1 DYNAMIC SYSTEM MODELS AND THEIR PARAMETER ESTIMATION BY TIME-PROFILE EXPERIMENTAL DATA

For the simplicity of analysis, we first discuss only the coupled signal transduction pathways shown in Fig. 2.2, which is extracted from Fig. 2.1. In Fig. 2.2, $y_i(t)$ denotes

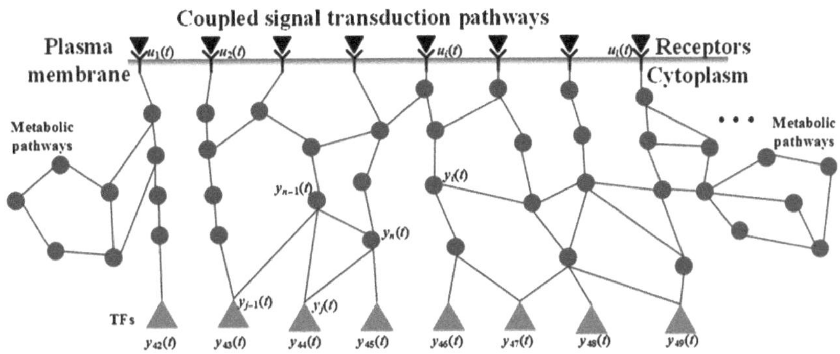

Figure 2.2 The coupled signal transduction pathways, i.e., protein–protein interaction network, or PPIN, extracted from Fig. 2.1.

the expression level of the *i*th protein in the coupled signal transduction pathways. $u_i(t)$ denotes extracellular signals. Different external changes or events outside the cell may stimulate signaling. $y_i(t)$ denotes the expression level of the *i*th protein in the transduction pathways. In the flow chart of the coupled signal transduction pathways in Fig. 2.2, $y_{42}(t)$, $y_{43}(t)$, $y_{44}(t)$, $y_{45}(t)$, $y_{46}(t)$, $y_{47}(t)$,... represent the expression levels of the terminal TFs in the pathways as a PPIN.

2.1.1 Dynamic model and parameter estimation for PPIN

For the purpose of system identification in the flow of signal transduction pathways in Fig. 2.2, the simple linear dynamic regression model for the expression level of the *i*th protein at time $t+1$ can be described as follows:

$$y_i(t+1) = y_i(t) + \sum_{\substack{j=1 \\ j\neq i}}^{n_i} c_{i,j}y_j(t) - \lambda_i y_i(t) + \sum_{j=1}^{l_i} b_{i,j}u_j(t) + h_i + w_i(t), \quad \text{for } i = 1,\ldots, N, \tag{2.1}$$

where $y_i(t)$ indicates the expression level of the *i*th protein at the present time t; $c_{i,j}$ denotes the interaction ability between protein i and protein j; λ_i denotes the degradation rate of protein i; h_i denotes the basal level of the *i*th protein; $b_{i,j}$ denotes the binding ability of extracellular signal u_j to the *i*th protein. In general, extracellular signals always bind to the receptor proteins on the membrane. If the *i*th protein is not bound by extracellular signal $u_j(t)$, then $b_{i,j}=0$. $w_i(t)$ denotes model residue or noise.

Let us denote the state vector and system matrices of the PPINs from Fig. 2.2 in Eq. (2.1) as follows:

$$y(t)=\begin{bmatrix} y_1(t) \\ \vdots \\ y_N(t) \end{bmatrix},\quad C=\begin{bmatrix} 1-\lambda_1 & c_{1,2} & \cdots & c_{1,N} \\ c_{2,1} & \ddots & \ddots & \vdots \\ \vdots & \ddots & \ddots & c_{N-1,N} \\ c_{N,1} & \cdots & c_{N,N-1} & 1-\lambda_N \end{bmatrix},\quad H=\begin{bmatrix} h_1 \\ \vdots \\ h_N \end{bmatrix},$$

$$B=\begin{bmatrix} b_{1,1} & \cdots & b_{1,l} \\ \vdots & \ddots & \vdots \\ b_{N,1} & \cdots & b_{N,l} \end{bmatrix},\quad u(t)=\begin{bmatrix} u_1(t) \\ \vdots \\ u_l(t) \end{bmatrix},\quad w(t)=\begin{bmatrix} w_1(t) \\ \vdots \\ w_N(t) \end{bmatrix},$$

where N is the total number of proteins and l is the total number of extracellular signals. Then, the PPI network of the coupled signal transduction pathways in Fig. 2.2 is represented by the following discrete-time dynamic equation

$$y(t+1)=Cy(t)+Bu(t)+H+w(t). \tag{2.2}$$

To exploit the effect of extracellular signals $u(t)$ on the coupled signal transduction pathways, the effect of basal level H and noise $w(t)$ should be extracted from the dynamic network in Eq. (2.2). Without consideration of the extracellular signal $u(t)$, the pathway network in Eq. (2.2) can be represented by

$$\hat{y}(t+1)=C\hat{y}(t)+H+w(t), \tag{2.3}$$

which is only the effect of basal level H and noise $w(t)$.

Let us denote $\tilde{y}(t+1)=y(t)-\hat{y}(t)$ and subtract Eq. (2.3) from (2.2). We then get

$$\tilde{y}(t+1)=C\tilde{y}(t)+Bu(t), \tag{2.4}$$

where $\tilde{y}(t)$ denotes the signal flow in the pathways due to extracellular signals $u(t)$ in Fig. 2.2.

In this case, the initial signal $\tilde{y}(0)=0$. The solution of the dynamic signal transduction system in Eq. (2.4) is given by the following signal flow equation [8]

$$\tilde{y}(t+1)=\sum_{j=0}^{t} C^{t-j}Bu(j). \tag{2.5}$$

If we want to know the signal flow from $u(t)$ to any protein, then the coupled signal transduction dynamic equation in Eq. (2.4) should be represented by

$$\begin{aligned} \tilde{y}(t+1) &= C\tilde{y}(t) + Bu(t), \\ \tilde{y}_i(t) &= [\underbrace{0\ldots0}_{1\sim(i-1)} \quad 1 \quad \underbrace{0\ldots0}_{(i+1)\sim M}]\,\tilde{y}(t) = D_i\tilde{y}(t), \end{aligned} \tag{2.6}$$

where $\tilde{y}_i(t)$ denotes the expression of the ith protein due to extracellular signal $u(t)$ and D_i is a row vector with all zeros except 1 at the ith element.

Therefore, the signal flow from $u(t)$ to the ith protein $\tilde{y}_i(t)$ in Eq. (2.6) in the signal transduction dynamic equation in Eq. (2.6) is given by

$$\tilde{y}_i(t+1) = \sum_{j=0}^{t} D_i C^{t-j} Bu(j). \tag{2.7}$$

Remark 2.1

1. Considering basal level and noise, the solution of the dynamic equation of coupled transduction pathways in Eq. (2.2) is given by

$$y(t+1) = C^{t+1}y(0) + \sum_{j=0}^{t} C^{t-j}(Bu(j) + H + w(j)), \tag{2.8}$$

where $y(0)$ denotes the initial condition. In this case, the expression of the ith protein in Eq. (2.7) should be modified as

$$y_i(t+1) = D_iC^{t+1}y(0) + \sum_{j=0}^{t} D_iC^{t-j}(Bu(j) + H + w(j)). \tag{2.9}$$

2. In the continuous dynamic case, Eq. (2.2) should be modified as

$$\dot{y}(t) = Cy(t) + Bu(t) + H + w(t), \tag{2.10}$$

with the solution as follows [9]:

$$y(t) = e^{ct}y(0) + \int_0^t e^{c(t-\tau)}(Bu(\tau) + H + w(\tau))d\tau, \tag{2.11}$$

in which the signal flow from $u(t)$ to the ith protein is given by

$$\tilde{y}(t) = \int_0^t D_i e^{c(t-\tau)} Bu(\tau)d\tau. \tag{2.12}$$

3. In nonlinear biological systems, Eqs. (2.2) and (2.10) should be modified as

$$y(t+1) = f(y(t)) + g(u(t)) + H + w(t), \tag{2.13}$$

and

$$\dot{y}(t) = f(y(t)) + g(u(t)) + H + w(t), \tag{2.14}$$

respectively, for nonlinear interaction functions $f(y(t))$ and $g(u(t))$.

4. In the metabolic pathways in Fig. 2.2, the stoichiometric description of the flux network based on a balanced equation can be also modeled as Eq. (2.2) or (2.10) in linear dynamic systems or Eq. (2.13) or (2.14) in nonlinear dynamic systems.

Based on the above analyses, if the system matrices C, H, and B in Eq. (2.2) or (2.10) can be identified from time-profile microarray data or protein microarray data, behavior and system characteristics like robustness, sensitivity, and transductivity of the biological network can be further analyzed. Therefore, we will follow this with an estimation of system matrices of the coupled signal transduction pathways in Eq. (2.2) from time-profile microarray data as an illustrative example of system parameter estimation of the PPIN. In general, we do not identify C, H, and B from Eq. (2.2) directly owing to the complex computation involved, with roundoff errors in the parameter estimation process. We want to estimate these parameters protein-by-protein from Eq. (2.1). From Eq. (2.1), they can be represented by the following linear regression form [10]

$$\begin{aligned} y_i(t+1) &= [y_1(t)\ldots y_i(t)\ldots y_{n_i}(t)u_1(t)\ldots u_{l_i}(t)1]\begin{bmatrix} c_{i,1} \\ \vdots \\ (1-\lambda_i) \\ \vdots \\ c_{i,n_i} \\ b_{i,1} \\ \vdots \\ b_{i,l_i} \\ h_i \end{bmatrix} + w_i(t) \\ &= \phi(t)\theta_i + w_i(t), \quad \text{for } i = 1,\ldots,N, \end{aligned} \tag{2.15}$$

with regression vector $\phi(t)$ and parameter vector θ_i to be estimated.

The parameter estimation problem is to find an estimate $\hat{\theta}_i$ of the parameter vector θ_i from the following M time-profile data $\{y_i(t)\}$ and $\{\phi(t)\}$ for $t = 1,\ldots,M$, i.e.,

$$\begin{cases} y_i(1) = \phi(0)\theta_i + w_i(0) \\ \quad\vdots \\ y_i(M) = \phi(M-1)\theta_i + w_i(M-1) \end{cases}. \tag{2.16}$$

For simplicity, we obtain the vector of observation $Y_i(M)$, regression matrix $\Phi(M)$, and noise vector $W_i(M)$, as follows:

$$Y_i(M) = \begin{bmatrix} y_i(1) \\ \vdots \\ y_i(M) \end{bmatrix}, \quad \Phi(M) = \begin{bmatrix} \phi(0) \\ \vdots \\ \phi(M-1) \end{bmatrix}, \quad W_i(M) = \begin{bmatrix} w_i(0) \\ \vdots \\ w_i(M-1) \end{bmatrix}. \tag{2.17}$$

Finally, the resulting parameter estimation model for linear regression model is

$$Y_i(M) = \Phi(M)\theta_i + W_i(M). \tag{2.18}$$

By the least squares parameter estimation method, we select parameter estimate $\hat{\theta}_i$ to minimize the following square error [8]

$$V(\hat{\theta}_i) = \frac{1}{2}(Y_i(M) - \Phi(M)\hat{\theta}_i)^T(Y_i(M) - \Phi(M)\hat{\theta}_i). \tag{2.19}$$

The optimal $\hat{\theta}_i$ to minimize the square error in Eq. (2.19) is obtained as [8]

$$\begin{aligned}\hat{\theta}_i &= (\Phi^T(M)\Phi(M))^{-1}\Phi^T(M)Y_i(M)\\ &= \Phi^{\dagger}(M)Y_i(M),\end{aligned} \tag{2.20}$$

where $\Phi^{\dagger}(M)$ denotes the pseudo-inverse of $\Phi(M)$.

If the constraints of parameter $\lambda_i \geq 0$, and $h_i \geq 0$, i.e., $\begin{bmatrix} 0 & 0 & 1 & 0 & 0 & \dots & \dots & \dots & 0 \\ 0 & \dots & \dots & \dots & 0 & \dots & \dots & 0 & -1 \end{bmatrix}\theta \leq \begin{bmatrix} 1 \\ 0 \end{bmatrix}$, need to be considered in the parameter estimation procedure, the constrained least squares parameter estimation method should be considered, which can be solved by using the optimization toolbox in Matlab [11,12].

The least squares parameter estimation algorithm in Eq. (2.20) for the dynamic model (2.1) of the ith protein is equivalent to the following recursive least squares parameter estimation algorithm [8]

$$\begin{aligned}\hat{\theta}_i(t+1) &= \hat{\theta}_i(t) + P_i(t)\phi(t)[y_i(t) - \phi(t)\theta_i(t-1)]\\ P_i(t) &= P_i(t-1) - \frac{P_i(t-1)\phi(t)\phi^T(t)P_i(t-1)}{1 + \phi^T(t)P_i(t-1)\phi(t)}, \quad \text{for } i = 1, \dots, N.\end{aligned} \tag{2.21}$$

The initials $\hat{\theta}_i(0)$ and $P_i(0)$ are given.

This recursive least square parameter estimation algorithm uses time-profile microarray data or real-time PCR data to update parameters protein-by-protein. Therefore, it can be used for real-time parameter estimation. If the number of time-profile data $y_i(t)$ is small, we can repeat several rounds of the recursive parameter estimation algorithm in Eq. (2.21), with the previous round result as initial parameter estimate $\hat{\theta}_i(0)$ and initial $P_i(0)$ of the next round to achieve the optimal parameter estimate. After parameter estimates $\hat{\theta}_i$ for $i = 1, \dots, N$ are obtained from the least square parameter estimation algorithm in Eq. (2.20) or (2.21) for all proteins in the coupled signal transduction pathways in Eq. (2.2) from the time-profile microarray data, we can obtain the system parameter estimates $\hat{C}$, $\hat{B}$, and $\hat{H}$ of the transduction pathways in Eq. (2.2) as follows:

$$[\hat{\theta}_1 \ \ \hat{\theta}_2 \ \ \dots \ \ \hat{\theta}_M] = [\hat{C} \ \ \hat{B} \ \ \hat{H}]^T, \tag{2.22}$$

where $\hat{C}$, $\hat{B}$, and $\hat{H}$ denote the estimates of C, B, and H in Eq. (2.2), respectively.

Remark 2.2

1. The parameter estimation problem in Eq. (2.15) or (2.18) can be also solved by the maximum likelihood estimation (MLE) method [8]. The MLE parameter estimation method will be introduced in the following chapters. However, in the case of Gaussian noise, the MLE method is equivalent to the constrained least square estimation method using the optimization toolbox in Matlab [11,12].
2. Since the coupled signal transduction network in Fig. 2.2 may be constructed by big database mining, there may be many false positive protein interactions in this candidate pathway. After the estimation by the least square parameter estimation algorithm in Eq. (2.20) or (2.21), we will prune these false positive protein interactions by the system order (interaction number in our case) detection method, i.e., the insignificant parameters in the system order will be removed from the coupled signal transduction pathways. The parameter estimation error variance and model complexity can be included in one statistic by the Akaike information criterion (AIC) as follows [8]:

$$AIC(n_i + l_i) = \log(\frac{1}{M}(Y_i(M) - \hat{Y}_i(M))^T(Y_i(M) - \hat{Y}_i(M))) + \frac{2(n_i + l_i)}{M}, \quad (2.23)$$

where the first term is related to the parameter estimation error variance with $\hat{Y}_i(M) = \Phi(M)\hat{\theta}_i$ and the second term is related to the model complexity; M denotes the number of time-profile data points and $(n_i + l_i)$ denotes the number of interaction parameters in Eq. (2.1) or (2.15) for the ith protein. AIC in Eq. (2.23) decreases as the estimation error variance decreases and increases as the number of parameters $(n_i + l_i)$ increases. AIC should have a minimum at the correct system order or interaction number [8,10]. Therefore, the true system order (protein interaction number) $n_i + l_i$ can minimize $AIC(n_i + l_i)$ in Eq. (2.23). In this situation, the insignificant parameters out of the true system order (the true interaction number) in $\hat{\theta}_i$ must be removed to prune the false positive protein interactions in Fig. 2.2 by the system identification method using microarray data.

Remark 2.3 If the protein interaction in Eq. (2.1) is of the following nonlinear dynamic model

$$\begin{aligned} y_i(t+1) &= \sum_{j=1}^{n_i} c_{i,j} f_{i,j}(y_j(t)) + \sum_{k=1}^{l_i} b_{i,k} g_{i,k}(u_k(t)) + h_i + w_i(t) \\ &= [f_{i,1}(y_1(t)) \ldots f_{i,n_i}(y_{n_i}(t)) g_{i,1}(u_1(t)) \ldots g_{i,l_i}(u_{l_i}(t)) 1] \begin{bmatrix} c_{i,1} \\ \vdots \\ c_{i,n_i} \\ b_{i,1} \\ \vdots \\ b_{i,l_i} \\ h_i \end{bmatrix} + w_i(t) \qquad (2.24) \\ &= \phi_i(t)\theta_i + w_i(t), \quad \text{for } i = 1, \ldots, N, \end{aligned}$$

where $f_{i,j}(y_j(t))$ and $g_{i,k}(u_k(t))$ denote the nonlinear interactions of protein i with other proteins $y_j(t)$ and extracellular signals $u_k(t)$, respectively, then the least square estimation algorithm in Eq. (2.20) or (2.21) can be also employed for parameter estimation.

2.1.2 Dynamic models and their parameter estimation by time-profile microarray data for GRNs

Suppose the microarray data is used for the system identification of candidate GRNs in Fig. 2.3, which is also extracted from Fig. 2.1 for the simplicity of analysis. The regulatory dynamics of the *i*th gene can be represented by the following regressive model

$$x_i(t+1) = x_i(t) + a_{i,1}x_1(t) + \ldots + a_{i,n}x_n(t) - \beta_i x_i(t) + k_i + v_i(t), \quad \text{for } i = 1, \ldots, n, \tag{2.25}$$

where $x_i(t)$ denotes the gene expression level of the *i*th gene at time *t*; $v_i(t)$ denotes the model residue and noise; $a_{i,j}$ denotes the regulatory ability of gene *j* on gene *i*; β_i denotes the degradation rate of mRNA of the *i*th gene; and k_i denotes the mRNA basal level, which represents regulations other than genetic regulations, such as epigenetic regulations.

In general, the regulation of the *j*th gene on the *i*th gene must be through its translational protein $y_i(t)$ (the TF $y_i(t)$ of the *i*th gene). However, if we only want to study the GRN among genes, the translational process of proteins to TFs is neglected and the regulatory ability $a_{i,j}$ has considered the effect of the translational process. Further, the epigenetic regulation of miRNAs on genes is also neglected here. The regulation of genes by TFs and miRNAs will be discussed in Section 2.3 in detail.

In general, a positive value of $a_{i,j}$ means the *j*th gene is an activating regulator, while a negative value of $a_{i,j}$ means the *j*th gene is an inhibiting regulator. The regulatory parameters $a_{i,j}$ and k_i, for $i = 1,\ldots,n$, in Eq. (2.25) can be identified with the constraint $k_i \geq 0$ by the constrained least square parameter estimation algorithm in Eq. (2.20) or (2.21) through the corresponding time-profile microarray data $x_i(t)$, for $i = 1,\ldots,n$ [7]. Similarly, after parameter estimation, the AIC in Eq. (2.23) can be also employed to prune the false positive regulators in Eq. (2.25) if the candidate GRN in Fig. 2.3 is constructed through big database mining.

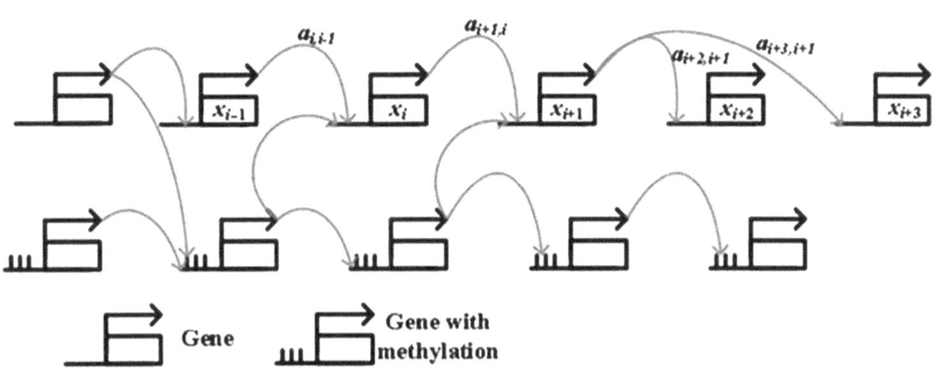

Figure 2.3 The candidate gene regulatory network (GRN) extracted from Fig. 2.1.

Remark 2.4 If the gene regulatory functions in Eq. (2.25) are of the following nonlinear form

$$\begin{aligned} x_i(t+1) &= a_{i,1}f_{i,1}(x_1(t)) + \ldots + a_{i,n}f_{i,n}(x_n(t)) + k_i + v_i(t) \\ &= [\, f_{i,1}(x_1(t)) \ldots f_{i,n}(x_n(t))\ 1] \begin{bmatrix} a_{i,1} \\ \vdots \\ a_{i,n} \\ k_i \end{bmatrix} \\ &= \phi_i(t)\theta_i + v_i(t), \quad \text{for } i = 1, \ldots, n, \end{aligned} \tag{2.26}$$

where $f_{i,j}(x_j(t))$ denotes the gene regulatory function from gene j to gene i, then the least square parameter estimation algorithm in Eq. (2.20) or (2.21) can be also employed for system identification of Eq. (2.26) through the corresponding time-profile microarray data. Similarly, after parameter estimation, the AIC in Eq. (2.23) can be used to prune the false positive regulations in the candidate GRN in Fig. 2.3 from big data mining to obtain the real GRN in a specific biological condition.

Therefore, based on the regulatory dynamic in Eq. (2.25), the GRN in Fig. 2.3 can be shown as follows:

$$\begin{bmatrix} x_1(t+1) \\ \vdots \\ x_i(t+1) \\ \vdots \\ x_n(t+1) \end{bmatrix} = \begin{bmatrix} (1-\beta_1) & \cdots & a_{1,i} & \cdots & a_{1,j} & \cdots & a_{1,n} \\ \vdots & \ddots & \cdots & \ddots & \vdots & \ddots & \vdots \\ a_{i,1} & \cdots & (1-\beta_i) & \cdots & a_{i,j} & \cdots & a_{i,n} \\ \vdots & \ddots & \cdots & \ddots & \vdots & \ddots & \vdots \\ a_{n,1} & \cdots & a_{n,i} & \cdots & a_{n,j} & \cdots & (1-\beta_n) \end{bmatrix} \begin{bmatrix} x_1(t) \\ \vdots \\ x_i(t) \\ \vdots \\ x_j(t) \\ \vdots \\ x_n(t) \end{bmatrix} + \begin{bmatrix} k_1 \\ \vdots \\ k_i \\ \vdots \\ k_n \end{bmatrix} + \begin{bmatrix} v_1(t) \\ \vdots \\ v_i(t) \\ \vdots \\ v_n(t) \end{bmatrix}, \tag{2.27}$$

which can be represented by

$$x(t+1) = Ax(t) + k + v(t). \tag{2.28}$$

The solution of the state dynamic equation in Eq. (2.28) is given by

$$x(t+1) = A^{t+1}(x(0)+k) + \sum_{j=1}^{t} A^{t-j} v(j).$$

Suppose we want to calculate the regulatory information flow from gene j to gene i. In general, it is difficult to solve the regulatory information problem for the GRN in Fig. 2.3 from the graph theory perspective [13], especially for a digraph (directed graph). In the GRN case, an input/output state space method is proposed to solve this difficult information flow problem of the diagraph as follows. First, the dynamic model of a GRN in Eq. (2.27) can be represented by the following input/output dynamic state space equation [14]

$$\begin{bmatrix} x_1(t+1) \\ \vdots \\ x_{i-1}(t+1) \\ x_i(t+1) \\ x_{i+1}(t+1) \\ \vdots \\ x_n(t+1) \end{bmatrix} = \begin{bmatrix} (1-\beta_1) & \cdots & a_{1,i-1} & a_{1,i} & a_{1,i+1} & \cdots & a_{1,j} & \cdots & a_{1,n} \\ \vdots & \ddots & \ddots & \vdots & \vdots & \ddots & \ddots & \ddots & \vdots \\ a_{i-1,1} & \ddots & (1-\beta_{i-1}) & a_{i-1,i} & a_{i-1,i+1} & \cdots & a_{i-1,j} & \cdots & a_{i-1,n} \\ a_{i,1} & \cdots & a_{i,i-1} & (1-\beta_i) & a_{i,i+1} & \cdots & (a_{i,j}-1) & \cdots & a_{i,n} \\ a_{i+1,1} & \cdots & a_{i+1,i-1} & a_{i+1,i} & (1-\beta_{i+1}) & \ddots & a_{i+1,j} & \cdots & a_{i+1,n} \\ \vdots & \ddots & \vdots & \vdots & \ddots & \ddots & \cdots & \ddots & \vdots \\ a_{n,1} & \cdots & a_{n,i-1} & a_{n,i} & a_{n,i+1} & \cdots & a_{n,j} & \cdots & (1-\beta_n) \end{bmatrix}$$

$$\bullet \begin{bmatrix} x_1(t) \\ \vdots \\ x_{i-1}(t) \\ x_i(t) \\ x_{i+1}(t) \\ \vdots \\ x_j(t) \\ \vdots \\ x_n(t) \end{bmatrix} + \begin{bmatrix} 0 \\ \vdots \\ 0 \\ 1 \\ 0 \\ \vdots \\ 0 \end{bmatrix} x_j(t) + \begin{bmatrix} k_1 \\ \vdots \\ k_{i-1} \\ k_i \\ k_{i+1} \\ \vdots \\ k_n \end{bmatrix} + \begin{bmatrix} v_1(t) \\ \vdots \\ v_{i-1}(t) \\ v_i(t) \\ v_{i+1}(t) \\ \vdots \\ v_n(t) \end{bmatrix},$$

$$x_i(t) = \begin{bmatrix} 0 & \cdots & 0 & 1 & 0 & \cdots & 0 \end{bmatrix} \begin{bmatrix} x_1(t) \\ \vdots \\ x_{i-1}(t) \\ x_i(t) \\ x_{i+1}(t) \\ \vdots \\ x_n(t) \end{bmatrix}. \qquad (2.29)$$

In the input/output dynamic state space system for the GRN in Eq. (2.29), $x_j(t)$ is considered as input signal and $x_i(t)$ as an output signal. Therefore, Eq. (2.29) is simply represented by

$$\begin{aligned} X(t+1) &= A_jX(t) + B_jx_j(t) + K + v(t) \\ x_i(t) &= D_iX(t) \end{aligned}, \tag{2.30}$$

which is similar to the signal transduction dynamic equation in Eq. (2.6), but with the gene expression level $x_j(t)$ replacing the extracellular signal $u(t)$.

Using systems theory [15], the regulatory signal flow from gene j to gene i in the GRN as shown in Fig. 2.3 is given by

$$x_i(t+1) = \sum_{k=0}^{t} D_iA_j^{\,t-k}B_jx_j(k). \tag{2.31}$$

2.2 STATIC MODELS AND THEIR PARAMETER ESTIMATION BY SAMPLE MICROARRAY DATA

2.2.1 Static modeling and parameter identification of PPIN by sample microarray data

If the available microarray data for measuring coupled signaling pathways in Fig. 2.2 are of one time point microarray from K different samples (e.g., K patients), the regression model for the protein expression of the ith protein cannot be represented by the discrete-time dynamic equation in Eq. (2.2), but can be represented by the following linear static regression form

$$y(k) = Cy(k) + Bu(k) + H + w(k), \quad \text{for } k = 1, \ldots, K, \tag{2.32}$$

where C and B are defined similarly in Eq. (2.2) but with $c_{i,i} = 0$, and $y(k)$, $u(k)$, and $w(k)$ given as follows:

$$y(k) = \begin{bmatrix} y_1(k) \\ \vdots \\ y_M(k) \end{bmatrix}, \quad u(k) = \begin{bmatrix} u_1(k) \\ \vdots \\ u_l(k) \end{bmatrix}, \quad w(k) = \begin{bmatrix} w_1(k) \\ \vdots \\ w_M(k) \end{bmatrix},$$

and $k = 1, \ldots, K$, denote the sample of microarray data. In this situation,

$$y(k) = (I-C)^{-1}Bu(k) + (I-C)^{-1}H + (I-C)^{-1}w(k). \tag{2.33}$$

From Eq. (2.33), it can be seen that the transduction function T from extracellular signal u to protein expression level y is

$$T = (I-C)^{-1}B. \tag{2.34}$$

If we want to estimate the signal transduction function T in Eq. (2.34), we need to estimate the system matrices C and B in Eq. (2.32) from the sample microarray data. From the linear regression model in Eq. (2.32), we get

$$\begin{aligned} y_i(k) &= [y_1(k) \ldots y_M(k) u_1(k) \ldots u_l(k) 1] \begin{bmatrix} c_{i,1} \\ \vdots \\ c_{i,M} \\ b_{i,1} \\ \vdots \\ b_{i,l} \\ h_i \end{bmatrix} + w_i(k) \\ &= \phi(k)\theta_i + w_i(k), \quad \text{for } k = 1, \ldots, K. \end{aligned} \tag{2.35}$$

Similarly, the least square parameter estimation algorithm with the constraint $h_i \geq 0$ can be employed to estimate θ_i as follows:

$$\begin{aligned} \hat{\theta}_i &= (\Phi^T(K)\Phi(K))^{-1}\Phi^T(K)Y_i(K) \\ &= \Phi^{\dagger}(K)Y_i(K), \quad \text{for } i = 1, \ldots, M, \end{aligned} \tag{2.36}$$

with the constraint $\begin{bmatrix} 0 & \ldots & 0 & -1 \end{bmatrix}\theta_i \leq 0$, where $\Phi(K) = \begin{bmatrix} \phi(1) \\ \vdots \\ \phi(K) \end{bmatrix}$, and $Y_i(K) = \begin{bmatrix} y_i(1) \\ \vdots \\ y_i(K) \end{bmatrix}$. The problem can be solved using the optimization toolbox in Matlab [11,12].

The recursive least square parameter identification for θ_i of the ith protein in Eq. (2.35) with K sample microarray is given by [8]

$$\begin{aligned} \hat{\theta}_i(k) &= \hat{\theta}_i(k-1) + P_i(k)\phi(k)\varepsilon_i(k), \\ \varepsilon_i(k) &= y_i(k) - \phi^T(k)\theta_i(k-1), \\ P_i(k) &= P_i(k-1) - \frac{P_i(k-1)\phi(k)\phi^T(k)P_i(k-1)}{1 + \phi^T(k)P_i(k-1)\phi(k)}, \\ &\phi(0) \text{ and } P_i(0) \text{ are given}, \quad \text{for } i = 1, \ldots, M, \; k = 1, \ldots, K. \end{aligned} \tag{2.37}$$

If the sample number K of the microarray data is small, we can repeat several rounds of the recursive parameter estimation algorithm in Eq. (2.37), with previous results as the initial conditions $\hat{\theta}_i(0)$ and $P_i(0)$ to achieve the optimal parameter estimate. After the parameters θ_i for $i = 1, \ldots, M$, are estimated by the recursive least square estimation algorithm in Eq. (2.37), we can use AIC in Eq. (2.23) for interaction number detection to prune the false positive interactions and obtain the true linear static model in Eq. (2.32). Then, the signal transduction function T in Eq. (2.34) can be calculated using the sample microarray data. If we only want to estimate the information transduced from all extracellular signals $u(k)$ to the ith protein (or TF) $y_i(k)$ in Fig. 2.2, then T in Eq. (2.34) should be modified as

$$\frac{y_i(k)}{u(k)} = T_i = D_i(I-C)^{-1}B, \tag{2.38}$$

where D_i is defined in Eqs. (2.29) and (2.30), i.e., a row vector with zero at all elements except 1 at the ith element. The information transduced from the jth extracellular signal $u_j(t)$ to the ith protein is given by

$$\frac{y_i(k)}{u_j(k)} = T_{i,j} = D_i(I-C)^{-1}B_j, \tag{2.39}$$

where B_j is defined in Eq. (2.30), i.e., a column vector with zero at all elements except 1 at the jth element.

2.2.2 Static modeling and parameter identification in the GRN by sample microarray data

If the microarray data measuring the regulatory information flow in the GRN in Fig. 2.3 are from one time point and K different samples, then the regression model for the regulation of the ith gene in Eq. (2.25) is modified to the following

$$\begin{aligned} x_i(k) &= a_{i,1}x_1(k) + \ldots + a_{i,n}x_n(k) + k_i + v_i(k), \\ &\text{for } k = 1,\ldots, K, \end{aligned} \tag{2.40}$$

where $x_1(k),\ldots, x_n(k)$ denote the gene expression levels of the GRN in Fig. 2.3 at the kth sample microarray. Therefore, the whole GRN in Fig. 2.3 can be represented by

$$X(k) = AX(k) + k + v(k), \quad \text{for } k = 1,\ldots, K, \tag{2.41}$$

where

$$X(k) = \begin{bmatrix} x_1(k) \\ \vdots \\ x_n(k) \end{bmatrix}, \quad v(k) = \begin{bmatrix} v_1(k) \\ \vdots \\ v_n(k) \end{bmatrix}, \quad \text{and } A = \begin{bmatrix} a_{1,1} & \cdots & a_{1,n} \\ \vdots & \ddots & \vdots \\ a_{n,1} & \cdots & a_{n,n} \end{bmatrix}, \quad k = \begin{bmatrix} k_1 \\ \vdots \\ k_n \end{bmatrix}.$$

Suppose we want to calculate the regulatory information flow from gene j to gene i in the GRN in Fig. 2.3. Then, Eq. (2.41) needs to be rearranged as

$$\begin{aligned} X(k) &= A_j X(k) + B_j x_j(k) + k + v(k), \\ x_i(k) &= D_i X(k), \end{aligned} \tag{2.42}$$

where A_j, B_j, and D_i are defined in Eqs. (2.29) and (2.30). From Eq. (2.42), we can obtain the regulatory information flow from gene j, $x_j(k)$, to gene i, $x_i(k)$, as follows

$$x_i(k) = D_i(I - A_j)^{-1} B_j x_j(k). \tag{2.43}$$

Hence, the regulatory information flow equation from gene j to gene i is given by

$$\frac{x_i(k)}{x_j(k)} = T_{i,j} = D_i(I - A_j)^{-1} B_j. \tag{2.44}$$

Therefore, if we can identify regulatory parameters $a_{1,1}, \ldots, a_{1,n}$, and k_i in Eq. (2.40) by the least square parameter estimation algorithm with the constraint $k_i \geq 0$ in Eq. (2.36) or (2.37) by microarray with K samples for all genes in the GRN, we can identify the regulatory parameter A in Eq. (2.41) and then determine A_j, B_j, and D_i in Eq. (2.42). In this situation, we can easily calculate the regulatory transfer function $T_{i,j}$ in Eq. (2.44) between any two genes in the GRN.

After the model and parameters of the PPIN in Fig. 2.2 and GRN in Fig. 2.3 are identified by microarray data, as described in the sections above, the PPIN and GRN should be integrated as a genetic and epigenetic cellular network as seen in Fig. 2.1.

2.3 MODELING AND IDENTIFICATION OF INTEGRATED GENETIC AND EPIGENETIC CELLULAR NETWORKS

2.3.1 Dynamic system model for integrated genetic and epigenetic cellular networks

The signal transduction pathways, GRNs, metabolic pathways, and epigenetic regulatory networks are always integrated as a genetic and epigenetic cellular network as shown in Fig. 2.1. In this situation, the translation from mRNA to protein needs to be considered at the protein expression level, and the protein interaction equation of the ith protein in Eq. (2.1) should be modified as

$$y_i(t+1) = \sum_{j=1}^{N_i} c_{i,j} y_j(t) + \sum_{n=1}^{l_i} b_{i,n} u_n(t) + (1 - \lambda_i) y_i(t) + \alpha_i x_i(t) + h_i + w_i(t), \tag{2.45}$$

where α_i denotes the translation effect from mRNA $x_i(t)$ to protein $y_i(t)$.

Similarly, if the gene i suffers from epigenetic inhibition by L_i miRNAs, then the gene regulatory equation of the ith gene in Eq. (2.25) should be modified as

$$x_i(t+1) = \sum_{j=1}^{N_i} a_{i,j}y_j(t) - \sum_{l=1}^{L_i} d_{l,i}m_l(t) - \beta_i x_i(t) + k_i + v_i(t), \tag{2.46}$$

where $a_{i,j}$ denotes the regulation from TF j and $d_{l,i}$ denotes the epigenetic regulation from miRNA $m_l(t)$, whose expression equation is described by the following

$$m_l(t+1) = -\sum_{i=1}^{N_l} d_{l,i}m_l(t) + (1-r_l)m_l(t) + \delta_i + n_l(t), \tag{2.47}$$

where r_l denotes the degradation rate of the lth miRNA $m_l(t)$. The term $\sum_{i=1}^{N_l} d_{l,i}m_l(t)$ denotes the loss of the lth miRNA for gene regulation in Eq. (2.46).

Remark 2.5

1. There are other epigenetic regulators such as DNA methylation in the gene transcription regulation process. Epigenetic DNA methylation will influence the regulation ability $a_{i,j}$ and mRNA basal level k_i in Eq. (2.46). Therefore, the effect of epigenetic methylation on gene regulation has been modeled into the values of $a_{i,j}$ and k_i in the gene regulation dynamic equation in Eq. (2.46).
2. After the integrated genetic and epigenetic cellular network is modeled as Eqs. (2.45)–(2.47), the parameter estimation procedures are similar to Eq. (2.20) or (2.21) if these interactive dynamic equations are formulated as the regression form in Eq. (2.15); e.g.,

$$x_i(t+1) = [y_1(t)\ldots y_{N_i}(t) - m_1(t)\ldots - m_{L_i}(t) - x_i(t)\ 1]\begin{bmatrix} a_{i,1} \\ \vdots \\ a_{i,N_i} \\ d_{1,i} \\ \vdots \\ d_{L_i,i} \\ \beta_i \\ k_i \end{bmatrix} + v_i(t) \tag{2.48}$$

$$= \phi(t)\theta_i + v_i(t), \quad \text{for } i = 1, \ldots, N_i.$$

If TF expression levels $y_i(t)$ and miRNA expression levels $m_l(t)$ can be obtained from high-throughput NGS data, then we can obtain $a_{i,j}$ and $d_{l,i}$ parameters from the least square estimation algorithm in Eq. (2.20) or (2.21). Similarly, the system order detection scheme AIC in Eq. (2.23) can also be employed to prune the false positives by removing the insignificant parameters out of the system order.

2.3.2 Static system model for integrated genetic and epigenetic cellular networks

If the experimental data for integrated genetic and epigenetic cellular networks are one time point data from different samples, then the dynamic genetic and epigenetic model in Eqs. (2.45)–(2.47) should be modified to the following static model for more suitable system parameter identification

$$y_i(k) = \sum_{\substack{j=1 \\ j \neq i}}^{N_i} c_{i,j} y_j(k) + \sum_{n=1}^{l_i} b_{i,n} u_n(k) + h_i + w_i(k),$$

$$x_i(k) = \sum_{\substack{j=1 \\ j \neq i}}^{N_i} a_{i,j} y_j(k) - \sum_{l=1}^{L_i} d_{l,i} m_l(k) + k_i + v_i(k), \tag{2.49}$$

$$m_l(k) = -\sum_{i=1}^{N_l} d_{l,i} x_i(k) + \delta_i + n_l(k), \quad \text{for } k = 1, \ldots, K,$$

where K denotes the number of sample data points.

These static models of integrated genetic and epigenetic regulation in Eq. (2.49) can be transformed to the regression form as seen in Eq. (2.35) for parameter estimation by the least square parameter estimation algorithm in Eq. (2.36) or (2.37) through the corresponding NGS data. Similarly, after parameter estimation, the system order detection scheme AIC in Eq. (2.23) can be employed to prune the false positives by removing the insignificant parameters out of the system order to obtain the real integrated genetic and epigenetic cellular network from candidate-integrated networks through big database mining and NGS data.

2.4 THE CORE NETWORK BY PNP OF THE INTEGRATED GENETIC AND EPIGENETIC CELLULAR NETWORK USING PCA

If the genetic and epigenetic cellular network is still very complex after database mining and system identification using NGS data have been employed to construct the real integrated network in Eqs. (2.45)–(2.47) for the dynamic model or Eq. (2.49) for the static model, it is still not easy to gain insight into the Big Mechanisms of systems biology. In this situation, the PNP is introduced to extract the most significant integrated genetic and epigenetic cellular network as the core network. Let the combined network matrix of the integrated genetic and epigenetic cellular network model in Eqs. (2.45)–(2.47) or in Eq. (2.49) be represented by:

$$H=\begin{bmatrix} c_{1,1} & \cdots & c_{1,j} & \cdots & c_{1,N} \\ \vdots & \ddots & \vdots & \ddots & \vdots \\ c_{i,1} & \cdots & c_{i,j} & \cdots & c_{i,N} \\ \vdots & \ddots & \vdots & \ddots & \vdots \\ c_{N,1} & \cdots & c_{N,j} & \cdots & c_{N,N} \\ a_{1,1} & \cdots & a_{1,j} & \cdots & a_{1,N} \\ \vdots & \ddots & \vdots & \ddots & \vdots \\ a_{i,1} & \cdots & a_{i,j} & \cdots & a_{i,N} \\ \vdots & \ddots & \vdots & \ddots & \vdots \\ a_{N,1} & \cdots & a_{N,j} & \cdots & a_{N,N} \\ -d_{1,1} & \cdots & -d_{1,j} & \cdots & -d_{1,N} \\ \vdots & \ddots & \vdots & \ddots & \vdots \\ -d_{l,1} & \cdots & -d_{l,j} & \cdots & -d_{l,N} \\ \vdots & \ddots & \vdots & \ddots & \vdots \\ -d_{L,1} & \cdots & -d_{L,j} & \cdots & -d_{L,N} \end{bmatrix}, \tag{2.50}$$

where $c_{i,j}$ denotes the interaction ability between proteins i and j, $a_{i,j}$ denotes the regulatory ability of gene j on gene i, and $-d_{l,j}$ denotes the inhibitive epigenetic regulation of miRNA l on gene j in Eq. (2.49). PCA is the linear transformation of H as the following SVD:

$$\begin{aligned} H=U\times D\times V^T &= \begin{bmatrix} u_1 & \cdots & u_N \end{bmatrix} \begin{bmatrix} d_1 & 0 & 0 \\ 0 & \ddots & 0 \\ 0 & 0 & d_N \end{bmatrix} \begin{bmatrix} v_1 & \cdots & v_N \end{bmatrix}^T \\ &= \sum_{i=1}^{N} u_i d_i {v_i}^T \end{aligned} \tag{2.51}$$

where $u_i \in \Re^{(2N+L)}$, and $v_i \in \Re^N$ are the ith left and right singular vectors of $H \in \Re^{(2N+L)\times N}$, respectively. The diagonal entries of the diagonal matrix, $D=\textbf{diag}(d_1, \ldots, d_m, \ldots, d_N)$, are the N singular values of H in descending order, i.e., $d_1 \geq \ldots \geq d_m \geq \ldots \geq d_N$. The eigen expression fraction (E_m) is defined as

$$E_m = \frac{d_m^2}{\sum_{m=1}^{N} d_m^2} \tag{2.52}$$

For example, we select the top M singular vectors of V such that $\sum_{m=1}^{M} E_m \geq 0.85$ with the minimal M, so that the M principal components contain 85% of the GEN from an energy point of view. Then, the principal projections of H to the top M singular vectors of V, or the similarities, are defined respectively as follows:

$$S(k,m) = h_k \bullet v_m{}^T \\ \text{for } k = 1,\ldots,(2N+L) \text{ and } m = 1,\ldots,M, \tag{2.53}$$

where h_k denotes the kth row vector of H in Eq. (2.50) and $v_m{}^T$ denotes the mth singular vector of V.

We further define the 2-norm distance from each gene to the top M singular vectors.

$$D(k) = \left[\sum_{m=1}^{M} [S(k,m)]^2\right]^{1/2}, \quad \text{for } k = 1,\ldots,(2N+L). \tag{2.54}$$

If $D(k)$ is close to zero, this implies that the kth node of the biological network is independent of the top M singular vectors. Thus, we define three thresholds, th_1, th_2, and th_3, to identify the respective core genes, $D(k) \geq th_1$ for $k = 1,\ldots, N$, and the core proteins, $D(k) \geq th_2$ for $k = N+1,\ldots, 2N$, and the core miRNAs $D(k) \geq th_3$ for $k = 2N+1,\ldots, 2N+L$, in the core network of the integrated genetic and epigenetic cellular network, which has the principal structure (or the so-called core) of the network.

REFERENCES

[1] Dibner C, Schibler U, Albrecht U. The mammalian circadian timing system: organization and coordination of central and peripheral clocks. Annu Rev Physiol 2010;72:517–49.

[2] Dodd AN, Kudla J, Sanders D. The language of calcium signaling. Annu Rev Plant Biol 2010;61:593–620.

[3] Kim EK, Choi EJ. Pathological roles of MAPK signaling pathways in human diseases. Biochim Biophys Acta 2010;1802(4):396–405.

[4] Kim TH, Bohmer M, Hu H, Nishimura N, Schroeder JI. Guard cell signal transduction network: advances in understanding abscisic acid, CO_2, and Ca^{2+} signaling. Annu Rev Plant Biol 2010;61:561–91.

[5] Leung AK, Sharp PA. MicroRNA functions in stress responses. Mol Cell 2010;40(2):205–15.

[6] Mosenden R, Tasken K. Cyclic AMP-mediated immune regulation—overview of mechanisms of action in T cells. Cell Signal 2011;23(6):1009–16.

[7] Klipp E. Systems biology in practice: concepts, implementation and application. Weinheim: Wiley-VCH; 2005.

[8] Johansson R. System modeling and identification. Englewood Cliffs, NJ: Prentice Hall; 1993.

[9] Boyd SP. Linear matrix inequalities in system and control theory. Philadelphia, PA: Society for Industrial and Applied Mathematics; 1994.

[10] Chen B-S, Wu C-C. Systems biology: an integrated platform for bioinformatics, systems synthetic biology and systems metabolic engineering. New York, NY: Nova Science Pub Inc.; 2014.
[11] Coleman TF, Li YY. A reflective Newton method for minimizing a quadratic function subject to bounds on some of the variables. Siam J Optim 1996;6(4):1040–58.
[12] Gill PE, Murray W, Wright MH. Practical optimization. San Diego, CA: Academic Press; 1986.
[13] Kreyszig E. Advanced engineering mathematics. 7th ed. New York, NY: Wiley; 1993.
[14] Chen B-S, Li C-W. Measuring information flow in cellular networks by the systems biology method through microarray data. Front Plant Sci. 2015;6:390.
[15] Ogata K. Discrete-time control systems. Englewood Cliffs, NJ: Prentice-Hall; 1987.

CHAPTER 3

Procedure for Exploring Big Mechanisms of Systems Biology Through System Identification and Big Database Mining

INTRODUCTION

After introducing system modeling and system identification methods of biological systems in Chapter 2, System Modeling and System Identification Methods for Big Mechanisms in Biological Systems, some examples and procedures on how to extract Big Mechanisms of biological systems by system identification and big database mining methods using genome-wide high-throughput data are introduced in this chapter. First, we introduce Big Mechanisms of gene regulatory networks (GRNs) and protein–protein interaction networks (PPINs) by dynamic interactive models and big data mining from microarray data. Then, we will introduce the Big Mechanisms of integrated GRNs and PPINs in cellular functions [1], and finally, the Big Mechanisms of integrated genetic and epigenetic cellular networks will be introduced using system identification, big database mining, and next generation sequencing (NGS) data [2–4] in Sections 3.1–3.4.

3.1 BIG MECHANISMS BASED ON GRNs BY SYSTEM IDENTIFICATION AND BIG DATABASE MINING

In this section, we will illustrate the big regulatory mechanisms of GRNs in yeast cells when they are faced with environmental stresses, e.g., osmotic stress. From these gene regulations and the corresponding GRN, we can investigate yeast's protective mechanisms in response to environmental stresses. The procedure for identifying Big Mechanisms of cellular functions in response to external environmental stresses is described in Fig. 3.1.

In the first step, using ChIP-Chip data, which provide information on all possible transcription factors (TFs) and the genes they regulate, and database mining for confirmed and predicted gene regulations, we can construct a candidate GRN, which may contain many false positive regulations to be deleted using the following system

Big Mechanisms in Systems Biology
DOI: http://dx.doi.org/10.1016/B978-0-12-809479-2.00003-2

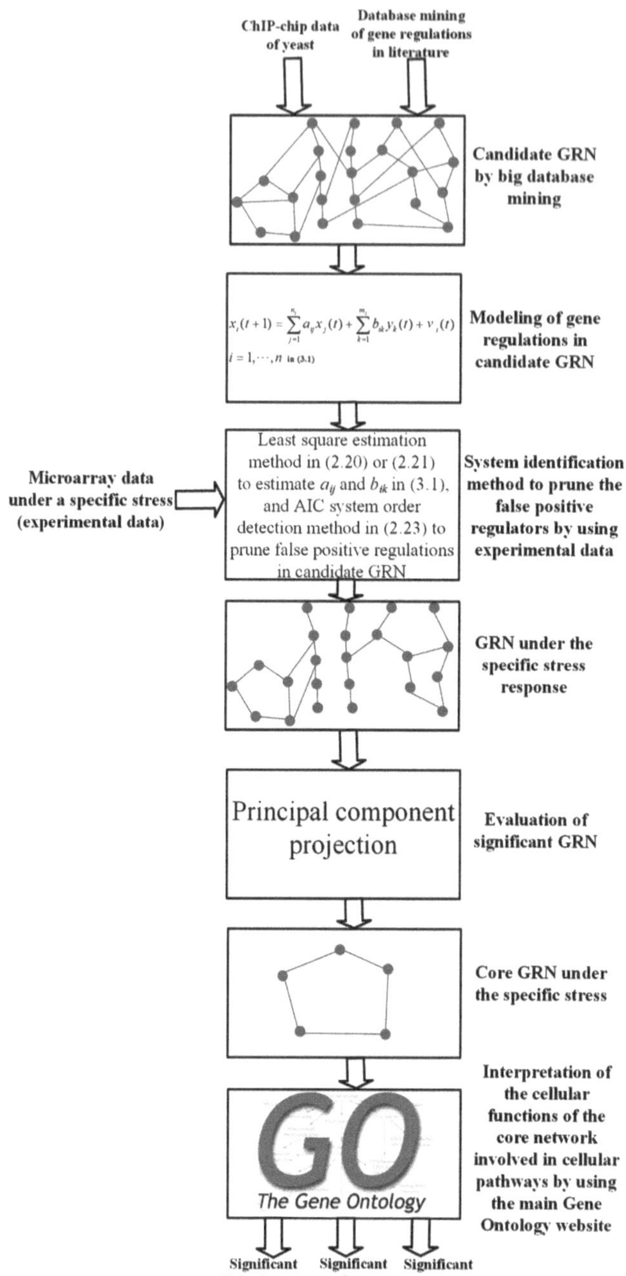

Figure 3.1 The procedure for extracting Big Regulatory Mechanisms based on the GRN involved in cellular functions in response to external environmental stresses using system identification and big database mining of microarray data.

identification method. In the second step, we can model the regulators for each gene in the candidate GRN as follows:

$$x_i(t+1) = \sum_{i=1}^{n_i} a_{ij} x_i(t) + \sum_{j=1}^{m_i} b_{ij} y_i(t) + v_i(t) \quad \text{for } i = 1, \ldots, n \tag{3.1}$$

where n_i denotes the number of regulators of gene i, including itself. m_i denotes the regulatory TFs on gene i.

In the third step, we use the least square parameter estimation algorithm of the system identification method in Eq. (2.20) or (2.21). Then, by AIC of the system order detected in Eq. (2.23), we delete the false positive regulations of the candidate GRN to obtain the real GRN under the specific stress response. If the constructed GRN is still complex, for simplicity, it is appealing to find the principal network projection of the GRN. With the principal network projection seen in in Section 2.4, we can obtain the core GRN under the specific stress. Finally, by GO annotation, we can find the significant cellular functions and pathways that are involved in the response to the specific stress and that protect the cell from that stress.

From Fig. 3.1, it is easy to determine the Big Mechanisms of a GRN involved in the cell cycles of other organisms or the GRN involved in cancers, which will be discussed in the following chapters.

3.2 BIG MECHANISMS BASED ON PPINs BY SYSTEM IDENTIFICATION AND BIG DATABASE MINING

In this section, we will illustrate how to extract the big response mechanisms of protein–protein interactions (PPIs) in yeast cells when they are faced with environmental stresses. From these PPIs and the related PPIN, we can investigate which protective mechanisms, at the protein level, are involved in the response to external environmental stresses. The procedure for how to extract the Big Mechanisms based on the PPIN of cellular responses to external environmental stresses through system identification and big database mining is illustrated in Fig. 3.2.

From Fig. 3.2, we first construct a candidate PPIN through using BioGRID and undertaking database mining of PPIs in the literature. Since the candidate PPIN is constructed from big database mining or bioinformatics predictions under different biological conditions, it will contain many false positive interactions. We will prune these false positive interactions by a system identification method using genome-wide high-throughput experimental data. Then, we construct the PPI dynamic model for each protein in the candidate PPIN as follows:

$$y_i(t+1) = \sum_{j=1}^{M_i} c_{ij} y_j(t) + (1 - \lambda_i) y_i(t) + h_i + w_i(t), \quad i = 1, \ldots, M \tag{3.2}$$

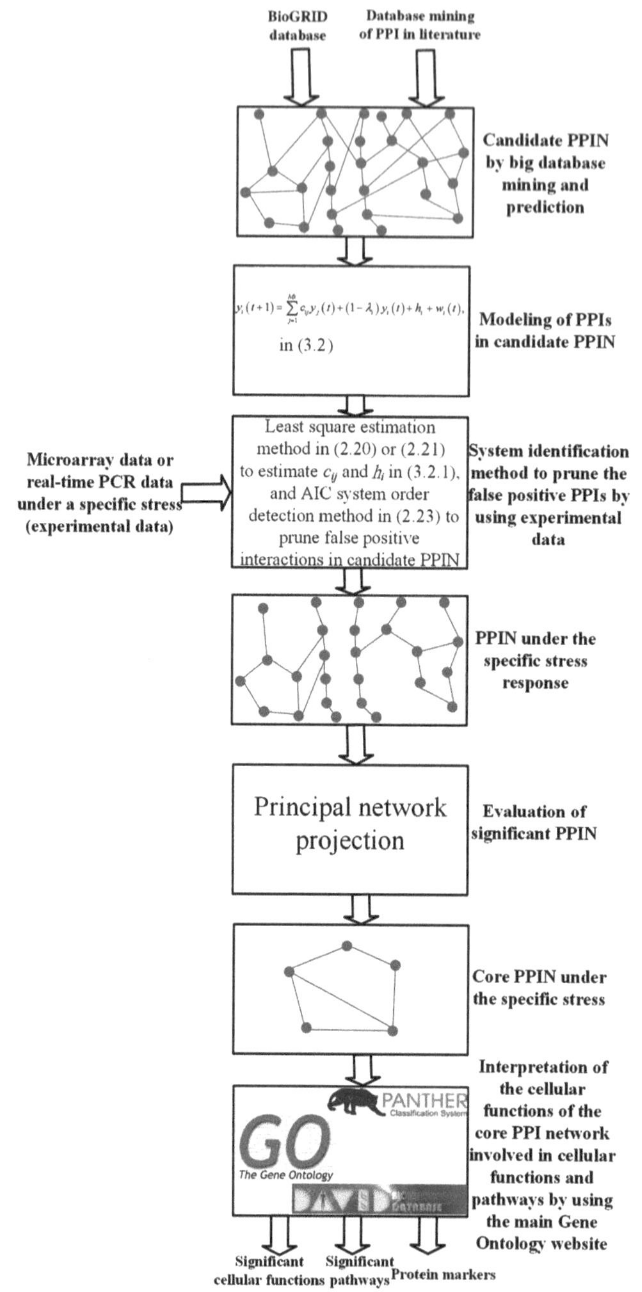

Figure 3.2 The procedure for exploring the Big Mechanisms of stress response based on PPIN-involved cellular functions and pathways in response to external environmental stresses by system identification and big database mining using microarray or real-time PCR data under a specific stress.

where $y_i(t)$ denotes the expression level of protein i, M_i is the number of proteins interacting with the protein i, c_{ij} denotes the interactivity between protein i and protein j, λ_i denotes the decay rate of protein i, h_i denotes the basal level of protein i, and $w_i(t)$ denotes the noise and model residue.

The third step in Fig. 3.2 is to employ experimental data (microarray or protein expression data) to identify the interactivities c_{ij} in Eq. (3.2) by the least square parameter estimation method as we saw in Eq. (2.20) or (2.21). Using the AIC system order detection method in Eq. (2.23), the insignificant interactivities c_{ij} out of the system order should be deleted to prune the false positive PPIs in the candidate PPIN. If the pruned PPIN is still large, we can use the principal network projection method in Section 2.4 to extract the most significant PPIN as the core PPIN under the specific stress. Finally, we can use GO annotation to interpret the Big Mechanisms in response to a specific environmental stress by the related cellular functions, pathways, or protein markers.

After illustrating the procedure for how to find the Big Mechanisms of PPINs in response to a specific environmental stress, we can apply a similar procedure to regeneration, infection, and cancer, which will be discussed in the following chapters.

3.3 BIG MECHANISMS BASED ON THE INTEGRATED GRN AND PPIN BY SYSTEM IDENTIFICATION AND BIG DATABASE MINING

In this section, we will illustrate how to extract the big stress-response mechanism for cells responding to external environmental stresses based on an integrated GRN and PPIN. The extraction procedure of the Big Mechanisms is given in Fig. 3.3. First, from ChIP-chip data, BioGRID, and big data in the literature, we construct a candidate-integrated GRN and PPIN using a big database mining method. We want to prune the false positives in the candidate GRN and PPIN using a system identification method and experimental data. Therefore, we can construct the integrated dynamic regulations and interactions in the candidate-integrated GRN and PPIN in response to a specific environmental stress as follows [1]:

$$
\begin{aligned}
y_i(t+1) &= \sum_{\substack{j=1 \\ j\neq i}}^{M_i} c_{ij}y_j(t) + (1-\lambda_i)y_i(t) + \sum_{j=1}^{l} b_{ij}u_i(t) + \alpha_i x_i(t) + h_i + w_i(t) \\
x_i(t+1) &= \sum\nolimits_{\substack{j=1 \\ j\neq i}} a_{ij}x_i(t) + (1-\beta_i)x_i(t) + k_i + \sum\nolimits_{k=1} f_{ik}y_k(t) + v_i(t)
\end{aligned}
\tag{3.3}
$$

where c_{ij} denotes the interaction between protein i and protein j, a_{ij} denotes the ability of gene j to regulate gene i, and f_{ik} denotes the ability of gene i to regulate TF k The third step is to employ experimental data (microarray data and omics data) to identify the parameters in the integrated genetic and protein interaction dynamic equation in Eq. (3.3) by

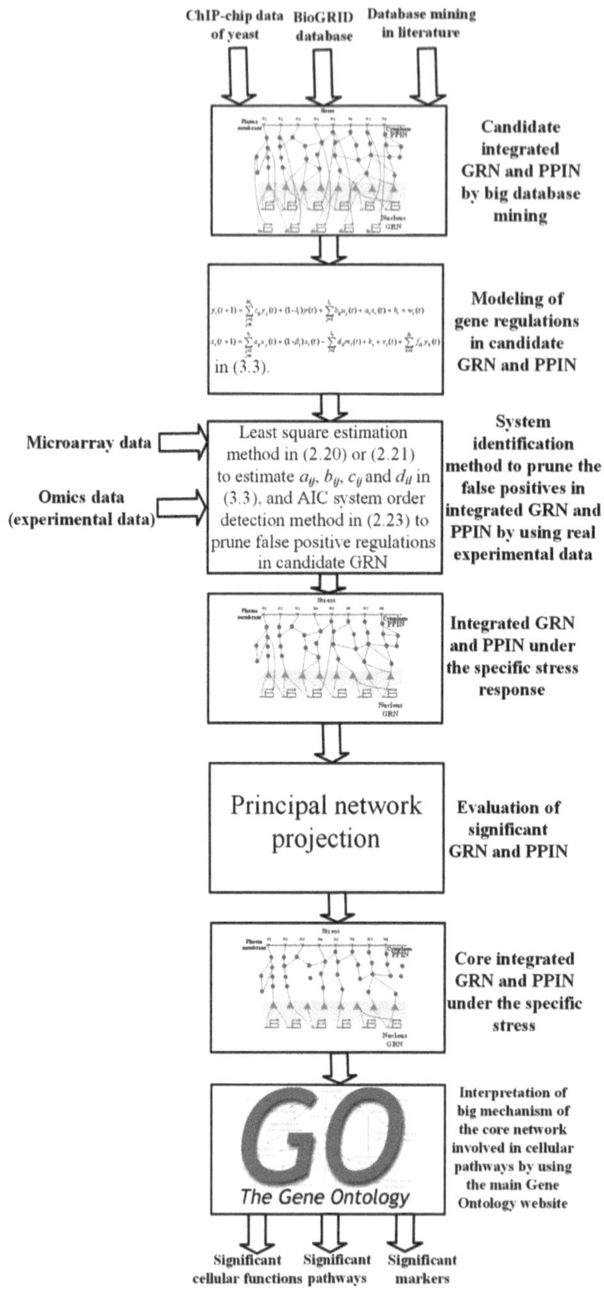

Figure 3.3 The procedure for exploring the Big Mechanisms of stress response based on integrated GRN and PPIN-involved cellular functions in response to external environmental stresses by system identification and big database mining.

the least square parameter estimation method in Eq. (2.20) or (2.21). Then, we prune the insignificant interactions out of the system order detected by the AIC method in Eq. (2.23) to delete the false positive regulations and interactions in the candidate-integrated GRN and PPIN to obtain the real integrated GRN and PPIN under the specific stress. If the integrated GRN and PPIN is still too complex to analyze, we can use the principal network projection method in Section 2.4 to extract a more significant integrated GRN and PPIN. Finally, we use GO annotation to find the Big Mechanisms based on the integrated GRN and PPIN of cells in response to a specific environmental stress. Other big cellular protection mechanisms of cellular functions based on the integrated GRN and PPIN in response to other biological conditions will be given in the following chapters with a detailed construction procedure.

3.4 BIG MECHANISMS BASED ON THE INTEGRATED GENETIC AND EPIGENETIC CELLULAR NETWORK BY SYSTEM IDENTIFICATION AND BIG DATABASE MINING

Recently, an enormous amount of genetic and epigenetic data has been generated by NGS technology. Therefore, we can investigate the Big Mechanisms based on integrated genetic and epigenetic cellular networks from the corresponding genetic and epigenetic data to gain greater insight into the complex cellular mechanisms of biological systems under different biological conditions.

In this section, the procedure to explore big cellular mechanisms in cancer, based on genetic and epigenetic cellular networks, is given in Fig. 3.4. This example illustrates how to extract the big cellular mechanism of carcinogenesis from genetic and epigenetic data from NGS by system identification and big database mining methods.

First, from BioGRID, ChIP-chip data from human samples, miRNA data, and data mining from the literature, we can construct a candidate-integrated genetic and epigenetic network (GEN) by the big database mining method. In order to prune false positive regulations and interactions in the candidate-integrated GENs in the third step of Fig. 3.4 by the system identification method, the dynamic model in the integrated network of cancer is given as follows [2]:

$$\begin{aligned}
y_i(t+1) &= \sum_{\substack{j=1\\ j\neq i}}^{M_i} c_{ij}y_j(t) + (1-\lambda_i)y_i(t) + \sum_{j=1}^{l_i} b_{ij}u_i(t) + \alpha_i x_i(t) + h_i + w_i(t) \\
x_i(t+1) &= \sum_{\substack{j=1\\ j\neq i}}^{n_i} a_{ij}x_i(t) + (1-\beta_i)x_i(t) - \sum_{l=1}^{L_i} d_{il}m_l(t) + k_i + \sum_{k=1}^{q_i} f_{ik}y_k(t) + v_i(t) \qquad (3.4) \\
m_l(t+1) &= (1-r_l)m_l(t) - \sum_{l=1}^{k_l} d_{il}m_l(t) + n_l(t)
\end{aligned}$$

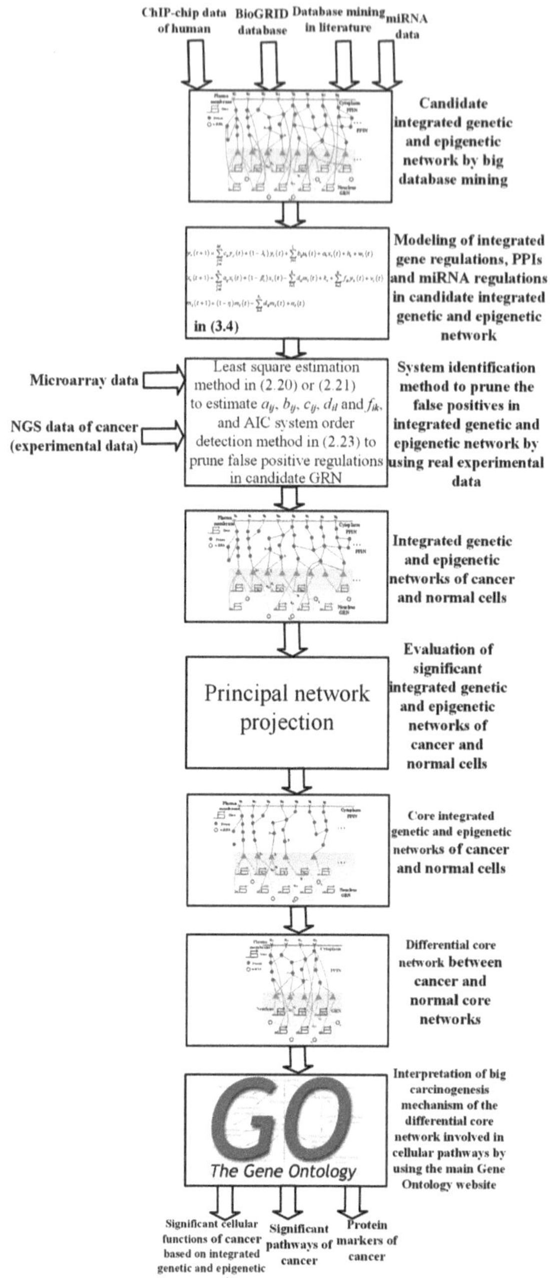

Figure 3.4 The procedure for exploring the Big Mechanisms of carcinogenesis based on integrated genetic and epigenetic cellular network-involved cellular functions in carcinogenesis by system identification and big database mining.

where $y_j(t)$, $x_i(t)$, and $m_l(t)$ denote the expression level of protein, mRNA, and miRNA, respectively. In Eq. (3.4), the methylation will influence the regulatory ability a_{ij} and the basal level k_i of gene expression.

Using genome-wide experimental data, including microarray, omics, and NGS of cancer, we identified the network parameters c_{ij}, b_{ij}, a_{ij}, d_{il}, and f_{ik} in Eq. (3.4), which represent the kinetic interactions and regulations in the candidate-integrated GEN of cancer in the third step of Fig. 3.4, determined by the least square parameter estimation method in Eq. (2.20) or (2.21). Then, by system order detection and minimizing the AIC in Eq. (2.23), the false positive interactions and regulations are pruned by deleting the insignificant network kinetic parameters out of the system order, to obtain the true integrated GEN of cancer. By a similar procedure, but with genome-wide experimental data (i.e., microarray, omics, and NGS) from normal cells, we can obtain the integrated GEN of normal cells. From the difference between the integrated GENs of cancer and normal cells, we can also obtain the differential integrated GENs. By the principal network projection method for this differential core integrated GEN, we can find significant pathways, cellular functions, and protein markers for the Big Mechanisms of carcinogenesis by GO enrichment analysis. Further, the core network-based biomarker can be obtained by principal component projection from the differential integrated GEN of cancer and normal cells. The Big Mechanisms of carcinogenesis, determined by system identification, omics data, and big database mining and based on the integrated GEN, will be explored in more detail in the subsequent chapters.

REFERENCES

[1] Wang YC, Chen BS. Integrated cellular network of transcription regulations and protein-protein interactions. BMC Syst Biol 2010;4:20.

[2] Chen BS, Li CW. Constructing an integrated genetic and epigenetic cellular network for whole cellular mechanism using high-throughput next-generation sequencing data. BMC Syst Biol 2016; 10(1):18.

[3] Li CW, Wang WH, Chen BS. Investigating the specific core genetic-and-epigenetic networks of cellular mechanisms involved in human aging in peripheral blood mononuclear cells. Oncotarget 2016;7(8):8556–79.

[4] Li C-W, Chen B-S. Network biomarkers of bladder cancer based on a genome-wide genetic and epigenetic network derived from next-generation sequencing data. Dis Markers 2016;2016:18.

CHAPTER 4

Big Cellular Mechanisms in the Cell Cycle by System Identification and Big Data Mining

INTRODUCTION

The most remarkable feature of cells and entire organisms is their ability to reproduce. The cell cycle entails an ordered series of macromolecular events that lead to cell division and the production of two daughter cells, each containing chromosomes identical to those of the parental cell. As we know from biology textbooks, duplication of parental chromosomes occurs during the S phase of the cell cycle, and one of the resulting daughter chromosomes is distributed to each daughter cell during mitosis. Precise temporal control of the events of the cell cycle ensures that the replication of chromosomes and their segregation to daughter cells occur in the proper order and with extremely high fidelity. Regulation of the cell cycle is critical for the normal development of multicellular organisms, and loss of control ultimately leads to cancer.

Recently, biochemical and genetic techniques, as well as recombinant DNA technology, have been employed to study various aspects of the eukaryotic cell cycle. These studies have revealed that the cell cycle is primarily controlled by regulation of nuclear DNA replication and mitosis. However, it is not easy to study the whole gene regulatory network (GRN) involved in the cell cycle, especially for various organisms. Fortunately, high-throughput microarray data and databases exist for different organisms. In this chapter (Section 4.1), we will show how to construct the gene regularity network involved in the cell cycle of yeast based on a dynamic gene regulatory model through system identification and big database mining to reveal the gene regulatory mechanisms. The abilities and timings of regulators for a specific target gene can also be characterized by the GRN of the yeast cell cycle. Then, we also introduce (in Section 4.2) a Module Finding Algorithm (MOFA) to construct a transcriptional regulatory module (TRM; a set of genes) regulated by a common set of transcription factors (TFs) that provides further insight into crosstalks between cell cycle and other cellular processes. The more complex GRNs of the cell cycle in plant and cancer cells will be further discussed in the following chapters.

Big Mechanisms in Systems Biology
DOI: http://dx.doi.org/10.1016/B978-0-12-809479-2.00004-4

4.1 CONSTRUCTING TRANSCRIPTIONAL REGULATORY NETWORK TO INVESTIGATE THE BIG MECHANISMS IN THE YEAST CELL CYCLE BY SYSTEM IDENTIFICATION AND BIG DATA MINING

Living cells are produced by a set of specific gene expression programs of transcription of thousands of genes under a coordinated arrangement. The regulation of gene expression programs in eukaryotes depends on the recognition of specific promoter sequences by transcriptional regulatory proteins [1,2] and often occurs through the coordinated action of multiple transcriptional regulators [3,4]. Combinational regulation of transcription has the advantages of including the control of gene expression in response to environmental changes and the use of limited number of TFs. Many combinations of regulators are produced and their regulatory activities are modified through diverse sets of conditions. Thus, the expression of each gene relies on the specific processes of a number of regulatory inputs, which are still unknown in most cases.

The molecular machinery of cell cycle control has been described in detail in the budding yeast *Saccharomyces cerevisiae*. The periodic alternation of gene transcript levels in *S. cerevisiae* is a major driving force of the cell cycle. Expression levels of hundreds of genes fluctuate throughout the cell cycle [5,6] and the self-degradation rates of mRNA have been precisely measured [7]. The functional classes of genes that show periodic transcription reflect the events that are occurring throughout the cell cycle, including bud emergence and growth, DNA synthesis, and spindle pole body duplication and migration. Therefore, it is of interest to decipher how these periodic recurring patterns are regulated.

Genome-wide microarray gene expression data relevant to the yeast cell cycle have been collected in a parallel manner [5–7]. Since the gene expression profile data are the network interactions resulting from the regulators and the target genes, it is reasonable to trace the interaction network from the microarray gene expression data. This treasure trove of data has been analyzed with a variety of clustering methods to group together genes with similar patterns of expression [8–10]. As mRNA levels are the outputs of gene expression regulatory networks, it is theoretically possible to use expression data to reverse-engineer the network architecture. A number of groups have coped with this problem using singular value decomposition analysis [11–13]. Instead of grouping genes together by their expression pattern, one can group genes by promoter sequence motif and then check whether this group of genes is coexpressed [14] or find combinations of TFs or binding sites [15,16]. To understand how genes are controlled by TFs, a genome-wide location analysis has been developed to identify which TFs bind to which promoters [17–19]. The information above provides us a clue as to how to decipher the gene expression regulatory network.

As an alternative approach, a method is devised for quantitatively characterizing transcriptional regulatory network (TRN) of the yeast cell cycle from genome-wide time-course expression data. In this section, we have not only identified the upstream suspected regulatory genes but also characterized the regulation significance by quantifying the regulatory ability. The first step is to construct a dynamic model of gene expression using the maximum likelihood method together with the Akaike Information Criterion (AIC) that we showed in Chapter 2, System Modeling and System Identification Methods for Big Mechanisms in Biological Systems. This allows us to measure the transcriptional regulation that controls the synthesis rate of a gene-specific mRNA level. The second step is to select a group of genes with regulatory ability through the cell cycle as regulator candidates and then deduce the relationship between the gene expression profiles of regulators and the transcriptional regulations of specific target genes. This algorithm iteratively picks out the genes with high correlations to the transcriptional regulation of target genes, and then identifies the associated parameters by minimizing the difference between the combined effects of regulators and the transcriptional regulation of target genes.

We applied this method to a publicly available data set of yeast from microarray experiments on the cell cycle [6], and used the measured mRNA decay rates [7] to identify the transcriptional regulators of the cell cycle and characterize their regulatory abilities on specific target genes. Through the quantitative analysis of the yeast TRN of the cell cycle–regulated genes, several TFs were identified and their regulatory abilities were determined. In addition, a few genes known to regulate kinase activity are predicted to be synergistic in coordinating gene expression. The TRN depicts the combined interactions of a small number of transcriptional regulators that are responsible for the gene expression program under certain conditions. This section provides a new approach for revealing GRN from gene expression data using a mathematical dynamic model and system identification techniques. The proposed approach is rather different from most existing statistical clustering methods for analyzing gene expression data, and these results show that this novel method is suitable for deciphering the complex interactions that regulate gene expression.

4.1.1 Methods to construct GRN of yeast cell cycle

The proposed method for quantitatively characterizing the TRN can be divided into two steps. In the first step, based on a dynamic modeling of gene expression, the transcriptional regulation and expression level of each gene are extracted from the raw microarray data. In the second step, based on the transcriptional regulation of the target gene, the upstream regulators will be deduced by correlating the transcriptional regulation of the target gene with the expression level of regulators.

Dynamic modeling of gene expression

The dynamics of gene expression can be modeled by a simple first-order differential equation, which is well established [20] and analyzed [21]. Let $X_i(t)$ denote the expression level of the ith gene at time t in a particular experiment. The differential equation has the form

$$\dot{X}_i(t) = G_i(t) - \lambda_i X_i(t) + \xi_i(t), \tag{4.1}$$

where $G_i(t)$ and λ_i are the transcriptional regulation and the self-degradation rate, respectively, of a specific gene i, and the last term $\xi_i(t)$ represents noise to stand for the data uncertainty and the model residuals. There is a time constant delay τ_i, $\tau_i = 1/\lambda_i$, from input $G_i(t)$ to output $X_i(t)$. The dynamic Equation (4.1) defines the changing rate of the mRNA level as the difference between the transcription regulation $G_i(t)$ controlling the synthesis rate and the natural degradation of the mRNA molecule, $\lambda_i X_i(t)$. The transcriptional regulation $G_i(t)$ is, in general, a complex function of combined effects of all upstream relevant regulators which determine the transcript level of a specific target gene.

To dissect the machinery controlling the dynamic model, the appropriate extraction of transcriptional regulation $G_i(t)$ plays a critical role. In the following, we want to estimate $G_i(t)$ from microarray data based on its basis functions (switch function). After the estimation of transcriptional regulation $G_i(t)$ from microarray data, we will identify the upstream regulators by correlating $G_i(t)$ with the possible regulatory genes. We assume that the transcriptional regulation is a linear combination of time-dependent switch functions:

$$G_i(t) = \sum_{j=1}^{m} a_{ij} S(t, \theta_j), \tag{4.2}$$

where the basis function $S(t,\theta_j)$ is a switch function of time specified by the parameter vector $\theta_j = \{\alpha,\beta_j,\gamma,\delta\}$ [22] which determines the shape and transition rate of the switch function and a_{ij} is an associated parameter that controls the weight of the specific switch function.

$$S(t, \theta_j) = \frac{1}{1 + \exp\{-\alpha[u(t, \beta_j, \delta) - \gamma]\}},$$

and

$$u(t, \beta_j, \delta) = \begin{cases} 0 & t \le \beta_j - \delta \\ \dfrac{t - (\beta_j - \delta)}{\delta} & \beta_j - \delta \le t \le \beta_j \\ \dfrac{(\beta_j + \delta) - t}{\delta} & \beta_j \le t \le \beta_j + \delta \\ 0 & t \ge \beta_j + \delta \end{cases},$$

i.e., the different values of the parameter a_{ij} will determine the different forms of the transcriptional regulation $G_i(t)$. The parameter vector θ_j determines the distribution of the switch function $S(t,\theta_j)$, where α and γ represent the transition slope and the cross-over threshold of the switch function, respectively, $u(t,\beta_j,\delta)$ represents a triangular function and the parameters $\{\beta_j-\delta,\delta,\beta_j+\delta\}$ determine the three corners of the underlying triangular function. Fig. 4.1 illustrates the distribution of different switch functions $S(t,\theta_j)$. These switch functions are employed to synthesize the transcriptional regulation $G_i(t)$ in Eq. (4.2) by specifying appropriate parameters a_{ij}. If the number of switch functions is increased, the resolution of synthesizing $G_i(t)$ in Eq. (4.2) will be increased simultaneously.

Reconsider Eqs. (4.1) and (4.2). If we have a set of data points $(\dot{X}_i(t_k), X_i(t_k))$, $k\in\{1,\ldots, N\}$ and the measured decay rates λ_i of mRNA for the ith gene expression, the parameters a_{ij} can be estimated for $j=1,\ldots, m$ by the following approach. Using a matrix notation, we can rewrite these two equations in a concise form:

$$\begin{bmatrix} \dot{X}_i(t_1) \\ \vdots \\ \dot{X}_i(t_N) \end{bmatrix} = \begin{bmatrix} S_1(t_1) & \cdots & S_m(t_1) \\ \vdots & \ddots & \vdots \\ S_1(t_N) & \cdots & S_m(t_N) \end{bmatrix} \begin{bmatrix} a_{i1} \\ \vdots \\ a_{im} \end{bmatrix} - \begin{bmatrix} \lambda_i X_i(t_1) \\ \vdots \\ \lambda_i X_i(t_N) \end{bmatrix} + \begin{bmatrix} \xi_i(t_1) \\ \vdots \\ \xi_i(t_N) \end{bmatrix}, \tag{4.3}$$

where $S_j(t_k) = S(t_k,\theta_j)$ and λ_i is the measured degradation rate from the study by Wang et al. [7]. If λ_i is unavailable, it should be estimated together with the parameters a_{ij}. For simplicity, we can further define the notations, Y_i, A_i, and Φ_i, to represent Eq. (4.3) as

$$Y_i = A_i\Phi_i + E_i, \tag{4.4}$$

where Y_i denotes an output vector,

$$Y_i = \begin{bmatrix} \dot{X}_i(t_1) + \lambda_i X_i(t_1) \\ \vdots \\ \dot{X}_i(t_N) + \lambda_i X_i(t_N) \end{bmatrix} = \begin{bmatrix} y_i(t_1) \\ \vdots \\ y_i(t_N) \end{bmatrix},$$

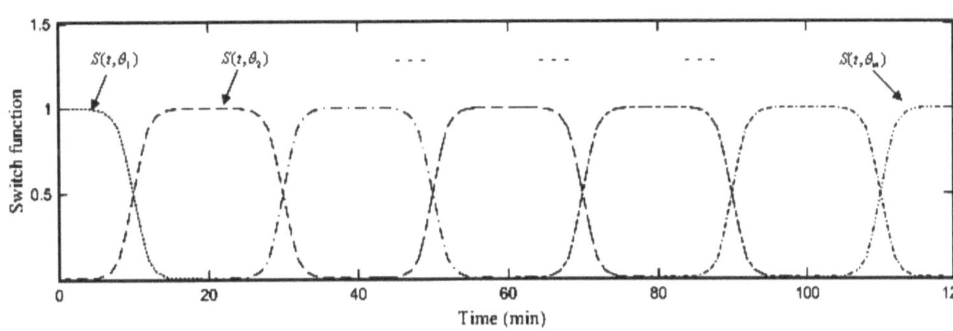

Figure 4.1 Illustration of switch functions for transcriptional regulation $G_i(t)$ in Eqs. (4.1) and (4.2), different dot lines indicate different switch functions $S(t,\theta_j)$.

A_i denotes a design matrix,

$$A_i = \begin{bmatrix} S_1(t_1) & \dots & S_m(t_1) \\ \vdots & \ddots & \vdots \\ S_1(t_N) & \dots & S_m(t_N) \end{bmatrix},$$

Φ_i denotes the parameter vector,

$$\Phi_i = \begin{bmatrix} a_{i1} \\ \vdots \\ a_{im} \end{bmatrix},$$

and E_i denotes the error vector to account for measurement noise and modeling errors

$$E_i = \begin{bmatrix} \xi_i(t_1) \\ \vdots \\ \xi_i(t_N) \end{bmatrix}.$$

If λ_i is unavailable, Eq. (4.3) should be modified to include λ_i in the parameter vector Φ_i, e.g., $Y_i = [\dot{X}_i(t_1)\dots\dot{X}_i(t_N)]^T$, $\Phi_i = [a_{i1}, \dots, a_{im}, \lambda_i]^T$ and A_i should include the data set $[X_i(t_1), \dots, X_i(t_N)]^T$. We assume that each element in the error vector $\xi_i(t_k)$, $k \in \{1, \dots, N\}$, is an independent random variable with a normal distribution with zero mean and variance σ^2, and estimate the parameter $\hat{\Phi}_i$ using the maximum likelihood method. The probability distribution of one data point $y_i(t_k)$ in Eq. (4.4), given t_k, is

$$p\left[y_i(t_k)\middle|t_k; \Phi_i, \sigma^2\right] = \frac{1}{\sqrt{2\pi\sigma^2}} \exp\left\{-\frac{[y_i(t_k) - z_i(t_k)]^2}{2\sigma^2}\right\}, \tag{4.5}$$

where $z_i(t_k) = R_{Ai}(t_k)\Phi_i$ and $R_{Ai}(t_k)$ is the row vector of A_i at time point t_k. The log-likelihood function for N data points given is then given by

$$L(\Phi_i, \sigma^2) = -\frac{N}{2}\ln\left(2\pi\sigma^2\right) - \frac{1}{2\sigma^2}\sum_{k=1}^{N}[y_i(t_k) - z_i(t_k)]^2. \tag{4.6}$$

The necessary condition for the maximum likelihood estimator of variance σ^2 is

$$\frac{\partial L(\Phi_i, \sigma^2)}{\partial \sigma^2} = 0.$$

By solving the preceding equation, we obtain

$$\hat{\sigma}^2 = \frac{1}{N}\sum_{k=1}^{N}[y_i(t_k) - z_i(t_k)]^2. \tag{4.7}$$

Substitution of Eq. (4.7) into Eq. (4.6) gives us the following form:

$$L(\Phi_i, \hat{\sigma}^2) = -\frac{N}{2}\ln(2\pi\hat{\sigma}^2) - \frac{N}{2}.$$

This means that we can find the maximum likelihood estimation of $\hat{\Phi}_i$ by minimizing the value of $\hat{\sigma}^2$. It can be achieved by the best choice of the parameter vector $\hat{\Phi}_i$ in Eq. (4.4) using the least squares method in Eq. (2.20), such that

$$\hat{\Phi}_i = (A_i^T A_i)^{-1} A_i^T Y_i, \tag{4.8}$$

i.e., in Gaussian noise case, the maximum likelihood estimation method is equivalent to the least squares method. It was shown that the maximum log likelihood has a tendency to overestimate the true value of the expected log likelihood. This tendency becomes more potent for models with a large number of parameters. Consequently, the number of free parameters we choose for Φ_i determines the goodness of fitting the true model. Here, we take the well-known system order detection scheme AIC of Eq. (2.23) into account for determining the number of parameter vectors (the system order) which were used in Eq. (4.3).

$$\begin{aligned} \text{AIC} = & -2(\text{maximum log likelihood of the estimated model}) \\ & +2(\text{number of free parameters of the estimated model}) \end{aligned}. \tag{4.9}$$

The number of parameters, in this case the value of m in Eq. (4.3) that minimizes the AIC, is chosen to create a specific dynamic expression model that is considered to be the most appropriate model order (i.e., the number m of a_{ij} in Eq. (4.2)) for characterizing the expression profile in this analysis.

After the parameters a_{ij} have been estimated, the transcriptional regulation $G_i(t)$ and expression level $X_i(t)$ can be specified from the microarray data by solving Eq. (4.1). The next step in this work is to identify the upstream regulators by correlating the transcriptional regulation $G_i(t)$ of target gene i with the expression level $X_j(t)$ of other genes via their regulatory functions, which will be defined in "Iterative algorithm for finding regulators in yeast cell cycle" section.

Iterative algorithm for finding regulators in yeast cell cycle

It is practical to consider that specific biochemical reactions are influenced by the concentration of relevant products [23,24]. For this purpose, we describe the regulatory function of a specific gene as a sigmoid function:

$$f_j\left[X_j(t), \theta_j\right] = \frac{1}{1 + \exp\{-r[X_j(t) - M_j]\}}, \tag{4.10}$$

where $X_j(t)$ is the expression level of the jth gene defined in Eq. (4.1), and $\theta_j \triangleq (r, M_j)$ indicates the transition rate(r) of the sigmoid function and the mean expression level (M_j) of gene j throughout the experiment, respectively. To further dissect the TRN from the dynamics of gene expression profiles, we derive upstream regulatory functions from the transcriptional regulation $G_i(t)$ of the target gene in Eq. (4.1), which has been estimated in the previous step. For simplicity, we match the transcriptional regulation of target gene i by a combination of a set of regulatory functions from regulators:

$$G_i(t) = c_{i0} + \sum_{j \in L_i} c_{ij} f_j[X_j(t), \theta_j], \tag{4.11}$$

where c_{ij} is the regulatory capability from regulator j to target gene i, $f_j[X_j(t),\theta_j]$ is a regulatory function defined in Eq. (4.10) and L_i is a possible set of indices of regulator genes for target gene i. The regulatory parameters c_{ij} are used to mimic modifications such as phosphorylation and dephosphorylation of proteins, and proteolysis through the dynamics of expression profiles, i.e., the activation and repression are characterized by these kinetic parameters c_{ij} similar to Eq. (4.3), we can rewrite Eq. (4.11) in the following matrix form:

$$\begin{bmatrix} G_i(t_1) \\ \vdots \\ G_i(t_N) \end{bmatrix} = \begin{bmatrix} 1 & f_1(t_1) & \cdots & f_{L_i}(t_1) \\ \vdots & \vdots & \ddots & \vdots \\ 1 & f_1(t_N) & \cdots & f_{L_i}(t_N) \end{bmatrix} \begin{bmatrix} c_{i0} \\ c_{i1} \\ \vdots \\ c_{iL_i} \end{bmatrix}. \tag{4.12}$$

Thus, the regulatory parameters c_{ij} can be directly estimated through the least squares method according to Eq. (4.8).

If we integrate Eqs. (4.1) and (4.11) into one equation, the dynamics of the TRN can be represented as the following nonlinear differential equation

$$\dot{X}_i(t) = c_{i0} + \sum_{j \in L_i} c_{ij} f_j[X_j(t), \theta_j] - \lambda_i X_i(t) + \xi_i(t). \tag{4.13}$$

In fact, Eq. (4.13) contains the rich information from exploring the regulatory network of each specific target gene. The genes that belong to a specific set L_i represent the potential upstream regulators of target gene i. The regulatory parameter c_{ij} characterizes the type and intensity of the influence of the jth gene on the ith target gene, with a positive sign indicating activation and a negative sign indicating repression, which will be defined as the regulatory ability hereafter. Apart from this, the first term in Eq. (4.13), c_{i0}, refers to the basal expression of a target gene. The goal of reverse engineering the regulatory network is to deduce the possible set of regulators and to identify their associated regulation abilities from the available data set. For this purpose, we devise an iterative algorithm based on the dynamic gene expression model for searching possible upstream regulators and then identifying the relevant regulatory abilities according to Eq. (4.12). This iterative algorithm can be constructed stepwise as follows:

Step 0. Based on the dynamic gene expression model described in Eq. (4.1), estimate the transcriptional regulation $G_i(t)$ and reconstruct the gene expression profile $X_i(t)$ for any interested target gene i.

Step 1. Select around 50 genes that have been confirmed to be TFs, cell cycle controlled genes and kinase activity regulators in the literature, and then calculate their regulatory function by Eq. (4.10). Consider these genes as the primary regulator candidates having the putative potential to regulate the periodic expressions of cell cycle–related genes.

Step 2. For any specific target gene i, examine the correlation between the transcriptional regulation $G_i(t)$ and the regulatory function $f_j[(X_j(t),\theta_j]$ stored in the candidate pool using a matched filter (see Appendix A for details). Select a few genes from the candidate pool with higher correlation as possible regulators. If there are TFs among the selected genes, then consider the TFs as putative regulators, and identify the regulatory ability parameters $c_{ij}S$ associated with the inferred regulator and the basal expression c_{i0} by the least square method via Eq. (4.12). Finally, an estimated transcriptional regulation $\hat{G}_i(t)$ is generated as described in Eq. (4.11). If there are no TFs found in the selected genes, generate the estimated transcriptional regulation $\hat{G}_i(t)$ for each possible regulator and choose the gene with the minimum root mean square error (RMSE) between the estimated $\hat{G}_i(t)$ and $G_i(t)$. The L_i set in Eq. (4.12) contains either the TFs or the chosen gene as putative regulators at the moment.

Step 3. In order to find regulators that might play the role of synergistic partners for regulating the target gene, define a new term $\Delta G_i(t) = G_i(t) - \hat{G}_i(t)$, which is used for correlating the remaining genes in the candidate pool instead of the original transcriptional regulation $G_i(t)$ used in Step 2. Use the output scores of the matched filter in Appendix A to help recognize the higher correlation between $\Delta G_i(t)$ and possible regulators. Select the genes with the five highest and five lowest correlations in turn to put into Eq. (4.11) and identify all the parameters again by the least squares method to form a new estimated transcriptional regulation $\hat{G}_i(t)$. Among the 10 estimated transcriptional regulation $\hat{G}_i(t)$ results, choose the gene with the minimum RMSE between the estimated $\hat{G}_i(t)$ and $G_i(t)$, and add the newly found gene into the regulator set L_i defined in Eq. (4.11).

Step 4. Repeat Step 3 until the regulator set L_i contains a predetermined number of regulators (for ease in interpretation, the algorithm identifies five possible regulators). Tabulate all the final regulatory ability parameters, the basal expression level and the possible regulator set L_i. For a new interested target gene, go to Step 1 to start a new program.

After analyzing all the interested target genes according to the above iterative algorithm, we obtained useful information for characterizing the TRN in the yeast cell cycle.

The algorithm aims to investigate the transcriptional regulation in the cell cycle by inferring the relationship between the transcriptional regulation $G_i(t)$ of the target gene and the transformed regulatory function $f_j[X_j(t),\theta_j]$ from the expression pattern of regulators. In addition, for easy interpretation, the algorithm only finds five possible regulators for each target gene and the target gene could be any interested gene in the whole genome, i.e., this algorithm can be applied on a genome-wide scale.

4.1.2 Cellular mechanisms of GRN in yeast cell cycle

Yeast cell cycle data

To identify the regulators and to characterize the regulatory abilities of yeast cell cycle−regulated genes, we applied this approach to a publicly available data set of Spellman et al. [6], which contained the expression of 6178 open reading frames in *S. cerevisiae* during the cell cycle (original data sets are available at http://cellcycle-www.standford.edu/) [6]. In Spellman et al.'s experiments, cell cultures were synchronized with different methods including α factors arrest, temperature arrest of a temperature-sensitive mutant, and elutriation synchronization. This analysis was applied on the α factors arrested data set. The raw data were transformed into a linear scale from the original log ratio given by Spellman et al. [6]. The raw data certainly contain uncertainty. This section employs the following method to eliminate the effect of uncertainty in the estimation of the transcriptional regulation profiles. The cubic spline was used first for data interpolation and smoothing to obtain a less sensitive first derivative of the expression pattern and the uncertainty is considered in the noise term $\xi_i(t)$ in Eq. (4.1). Then we used AIC combined with the least squares method to minimize the effect of uncertainty in parameter estimation. The cubic spline method has been used for a long time by engineers to fit curves through given points. A cubic spline interpolation defined and proved that any partition between each given data point is a continuous function that has continuous first and second derivatives everywhere in that corresponding interval and is represented by a polynomial of degree not exceeding three. This well-developed scheme allows us to determine a smooth curve representing the original limited experimental data and having a continuous first derivative function, which is indispensable information for this analysis.

For efficiency, the raw data was preprocessed by filtering out those genes with a flat expression profile (the difference between the maximum and the minimum expression level is less than twofold). After filtering the original raw data of 6178 genes, 2119 genes were considered as input expression profiles $X_i(t)$ to be examined in this assay. The input to this analysis required at every stage and the entire flowchart are depicted in Fig. 4.2. Fig. 4.3 shows the computed results of the dynamic expression model (Eq. 4.1) with the experimental expression patterns for some typical cell cycle−related genes. The transcriptional regulations $G_i(t)$s, defined in Eq. (4.1), change with time

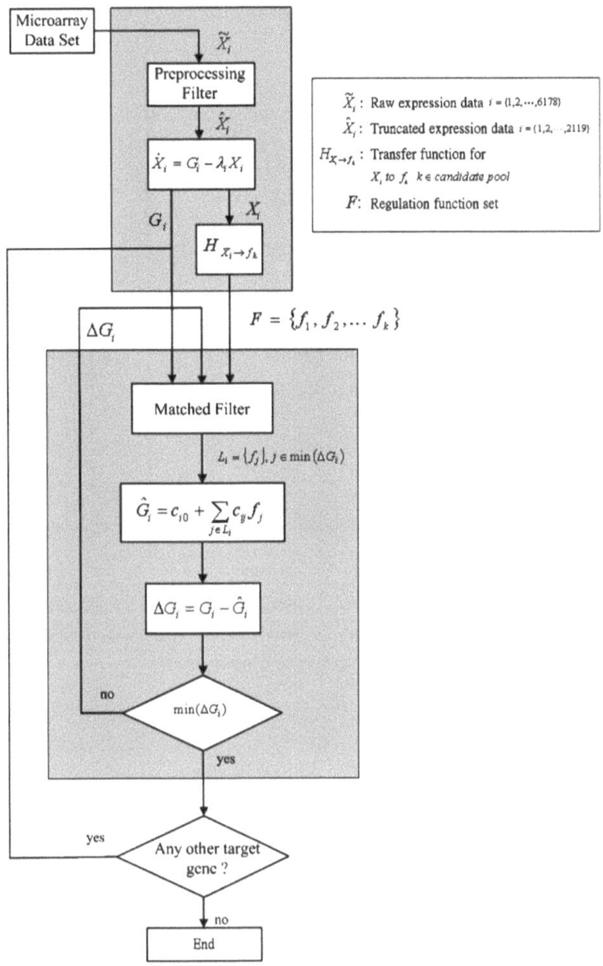

Figure 4.2 The overall flowchart of the analysis and data required in Eq. (4.11) to investigate the upstream regulators of GRNs at every stage of yeast cell cycle.

(Fig. 4.4). According to Eq. (4.1), the transcriptional regulation $G_i(t)$ is the major driving force for regulating the expression profile. The good fit of these computed results and experimental data points (Fig. 4.3) indicates that the dynamic expression model is suitable for describing the character of gene expression and the transcriptional regulation $G_i(t)$ (Fig. 4.4). We established the dynamic expression profiles of a group of genes that were previously denoted as cell cycle regulators [6,19] and characterized their regulatory function according to Eq. (4.10). This iterative algorithm can find the most likely genes that participate in regulating the expression of any target gene of interest on a genome-wide scale.

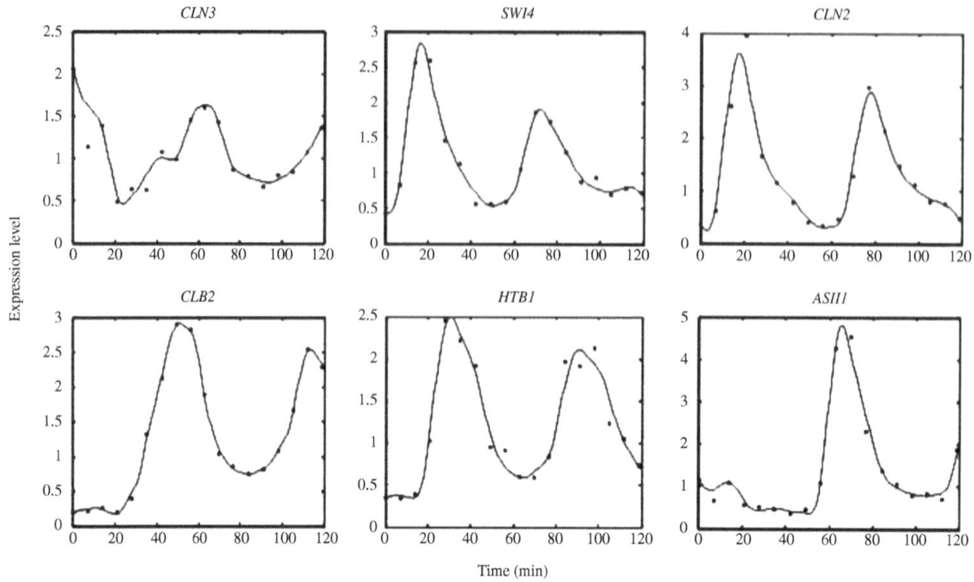

Figure 4.3 The expression profiles with corresponding dynamic model fitting. The experimental data are shown as dots and the dynamic modeling results are shown as *solid lines*. The six cell cycle–related genes are *CLN3*, *SWI4*, *CLN2*, *CLB2*, *HTB1*, and *ASH1*.

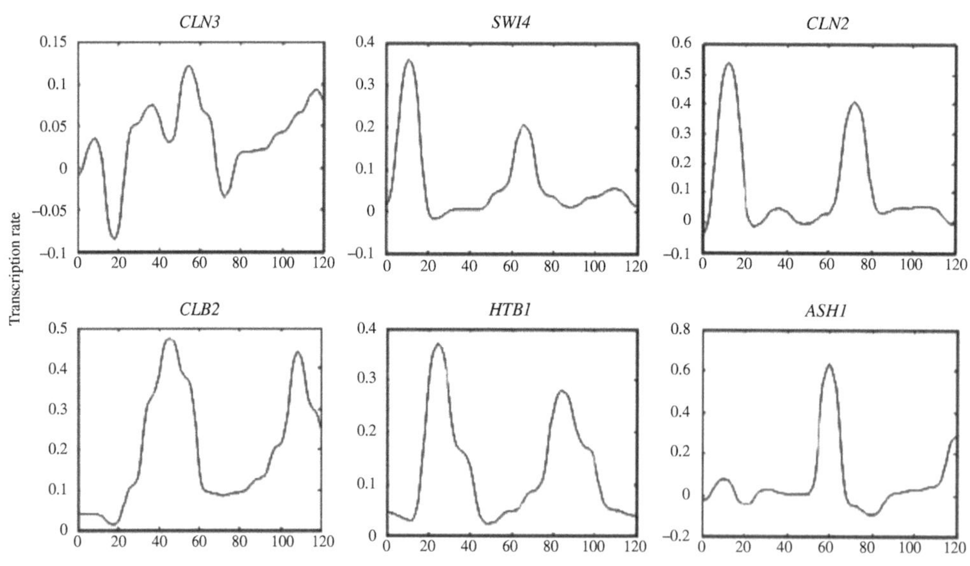

Figure 4.4 The identified transcriptional regulations, $G_i(t)$s, for the six cell cycle–related genes *CLN3*, *SWI4*, *CLN2*, *CLB2*, *HTB1*, and *ASH1*.

Logical inference from regulatory synergy

An attractive feature of combined synergy between regulatory proteins, such as TFs, kinase activity regulators, and cyclins, is that it accommodates complex mechanisms of gene regulation. Fig. 4.5 shows an example of the combinatorial gene regulation. Consider the following scenario that regulates the expression program of a specific target gene: two activators are capable of causing transcription initiation and a repressor is responsible for reducing the expression level. Although each of the regulators might work at different stages, the overall effect of transcriptional activity caused by these regulators does periodically fluctuate with time and peaks at a specific time period.

Fig. 4.5B illustrates how the expression program is operated under cooperative gene regulation in which two activators induce expression and one repressor reduces it. The top row shows the regulatory functions of three regulators x, y, and z; and the bottom row shows the overall transcriptional regulation of the target gene expression. If we assume that the relationship of the regulatory function with the expression level is described by Eq. (4.10), then the action period of regulators (i.e., the time period of regulatory function with the magnitude equal to 1) will determine the times of maximum and minimum transcriptional regulations. For example, repressor z acts at three time periods (0−15, 55−75, and 115−120 min) and subsequently causes the

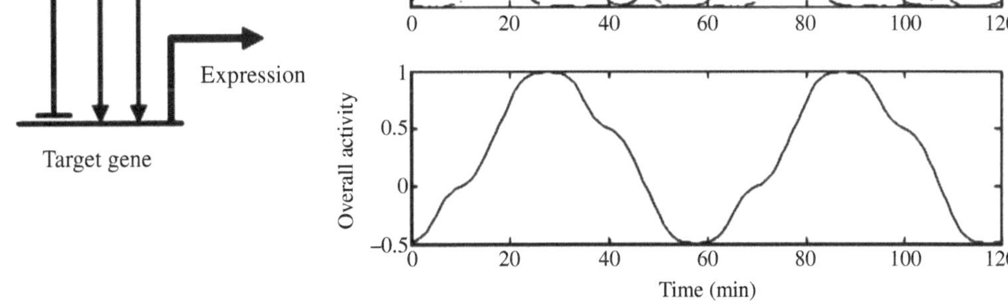

Figure 4.5 A simple model of a hypothetic target gene with two activators and one repressor, and three regulators synergistically coordinating the expression program of the target gene. (A) A schematic diagram of the gene expression program: activators x and y induce the expression of the target gene and a repressor z reduces its expression later. (B) Regulatory function of the three transcriptional regulators (top row) and the overall transcriptional regulation caused by the combinatorial regulation of the three different regulators (bottom row).

valley in the transcriptional regulation that leads to a low level of expression of the target gene. On the other hand, activators x and y work at later time periods; a possible explanation is that activator x decreases the synthesis of repressor z and subsequently promotes the synthesis of another activator y, which leads to an abrupt peak in the transcriptional regulation within the overlapping time period of activators x and y. Thus, the periodically fluctuating mRNA levels or regulatory functions of specific regulators indeed control the expression program of their target genes in a periodic way and usually in a synergetic way. Based on this simple example and the dynamic gene expression model, we were able to examine the periodically expressed genes observed in yeast during the cell division cycle and to dissect the specific expression program temporally mediated by a small number of regulators.

Transcriptional regulatory network

Our iterative scheme for finding regulators is conveniently illustrated by tracking between the transcriptional regulation $G_i(t)$ and the estimated transcriptional regulation $\hat{G}_i(t)$ for the target gene. This system provides the information of putative upstream regulators with their regulatory abilities, and depicts the overall transcriptional regulation controlled by the five putative regulators as a function of time for a given target gene (Table 4.1; Fig. 4.6). The extensive studies on the yeast cell cycle permit us to explore the features of the final regulatory network result [25–27]. For convenience in illustration, we only studied 20 target genes that are cell cycle–regulated and are highly expressed at different phases. Five possible regulators of each specific target gene were identified and the regulatory abilities were also characterized (Table 4.1). For example, the important cell cycle control gene *CLN2*, encoding cyclin protein in conjunction with cyclin-dependent kinase, CDK (Cdc28), forms a complex Cln2/Cdc28 that plays a major role in budding and spindle pole body duplication.

Our plots revealed that the peak of the transcriptional regulation of *CLN2* (~15 min; Fig. 4.4) precedes its expression peak at the G1 phase (~20 min; Fig. 4.3). The five regulators of the target *CLN2* include the positive regulators Cln3, Swi4, and Yhp1, and the negative regulators Clb4 and Clb2. According to the identified regulatory abilities, we propose that Cln3, Swi4, and Yhp1 are the major activators for *CLN2* transcription due to the significant values of their regulatory abilities, and thus Clb4 and Clb2 are the major repressors. These results indicate that the expression of *CLN2* was very significantly regulated by Cln3, Swi4, and Yhp1 (0.3211, 0.2745, and 0.2020, respectively; Table 4.1). The highest regulatory abilities of Cln3 and Swi4 imply that the association of Cln3 and Swi4 mediates *CLN2* transcription. SBF (Swi4Swi6 cell cycle box binding factor) is a TF binding on the promoters of many genes induced at the G1/S transition. SBF contains protein Swi6 as a subcomponent and Swi4 as the DNA-binding protein. Therefore, the periodic

Table 4.1 Five possible regulators and their regulatory abilities for 20 selected cell cycle–related genes

Target genes	Regulators (regulatory abilities)					Basal level C_0
SW14	Cln3 (0.2469)	Swi4[a] (0.0733)	Mbpl[a] (0.0216)	Swel (− 0.0839)	Spo12 (− 0.1909)	0.0437
CLN2	Cln3 (0.3211)	Swi4[a] (0.2745)	Yhp1 (0.2020)	Clb4 (− 0.1450)	Clb2 (− 0.1843)	− 0.0943
SWE1	Cln3 (0.0855)	Swi4[a] (0.0375)	Pcl2 (0.0222)	Spol2 (− 0.0463)	Pog1 (− 0.0678)	0.0463
HCM1	Swi4[a] (0.0708)	Cln3 (0.0598)	Rmel (0.0253)	Mfa1 (− 0.0355)	Clb2 (− 0.0899)	0.1368
NDD1	Skn7 (0.0525)	Far1 (0.0273)	Swi4[a] (0.0088)	Rmel (− 0.0126)	Dbf2 (− 0.433)	0.0322
SVS1	Yhp1 (0.1778)	Clb6 (0.1691)	Swi4[a] (0.0971)	Pc17 (− 0.1209)	Fkh1 (− 0.1735)	0.0869
MNN1	Yhp1 (0.1913)	Cln3 (0.1668)	Mfa2 (0.1425)	Swi4[a] (0.1087)	Clb2 (− 0.0543)	− 0.2223
CLB6	Cln3 (0.6335)	Yhp1 (0.2449)	Swi4[a] (0.1851)	Fkh1 (− 0.1607)	Spo12 (− 0.3443)	− 0.0877
HTA1	Clb4 (0.1065)	Swe1 (0.0264)	Swi4[a] (0.0263)	Yhp1 (− 0.1415)	Cln3 (− 0.2994)	0.2815
HTB1	Clb4 (0.0803)	Swi4[a] (0.0524)	Swe1 (0.0135)	Yhp1 (− 0.1252)	Cln3 (− 0.2218)	0.2176
HHT1	Swe1 (0.1080)	Clb4 (0.0413)	Swi4 (0.0144)	Yhp1 (− 0.1425)	Cln3 (− 0.2280)	0.2449
CLB4	Ndd1 (0.0368)	Cln2 (0.0168)	Clb1 (− 0.0225)	Pc12 (− 0.0303)	Yhp1 (− 0.0580)	0.0544
CLB2	Fkh2[a] (0.1488)	Clb1 (0.1341)	Ndd1 (0.1229)	Cln1 (− 0.269)	Sic1 (− 0.0552)	0.0307
CDC20	Spo12 (0.0864)	Ndd1 (0.0487)	Fkh2[a] (0.0347)	Ash1 (− 0.0199)	Mcm1[a] (− 0.0447)	0.0247
SPO12	Fkh2[a] (0.1079)	Ste12 (0.0990)	Mfa2 (− 0.0347)	Mcm1[a] (− 0.0470)	Ndd1[a] (− 0.0941)	0.0461
SW15	Clb2 (0.0927)	Fkh2[a] (0.0574)	Ash1 (− 0.0232)	Yox1 (− 0.0567)	Mob1 (− 0.0971)	0.1031
ASH1	Swi5[a] (0.1207)	Ash1 (0.1135)	Cln2 (− 0.1113)	Yhp1 (− 0.1337)	Rmel (− 0.1909)	0.1534
SIC1	Pc19 (0.1193)	Swi5[a] (0.0261)	Rme1 (− 0.0422)	Cln2 (− 0.0581)	Yhp1 (− 0.0954)	0.1277
CLN3	Sim1 (0.1149)	Swi5 (0.0220)	Mcm1[a] (0.0177)	Swe1 (− 0.0470)	Yhp1 (− 0.0624)	0.0075
CDC46	Cln3 (0.1114)	Clb2 (0.0141)	Rmel (− 0.0213)	Clb6 ((− 0.0269)	Opy2 (− 0.0441)	0.0348

The first column indicates the specific target genes, the second to sixth columns outline the identified regulators which were sorted in descending order according to the value of their associated regulation abilities and the last column is the basal level of transcription rate.

[a]Transcription factors that were previously reported to bind to the promoters of target genes (Lee et al., 2002), *p*-value $< .001$.

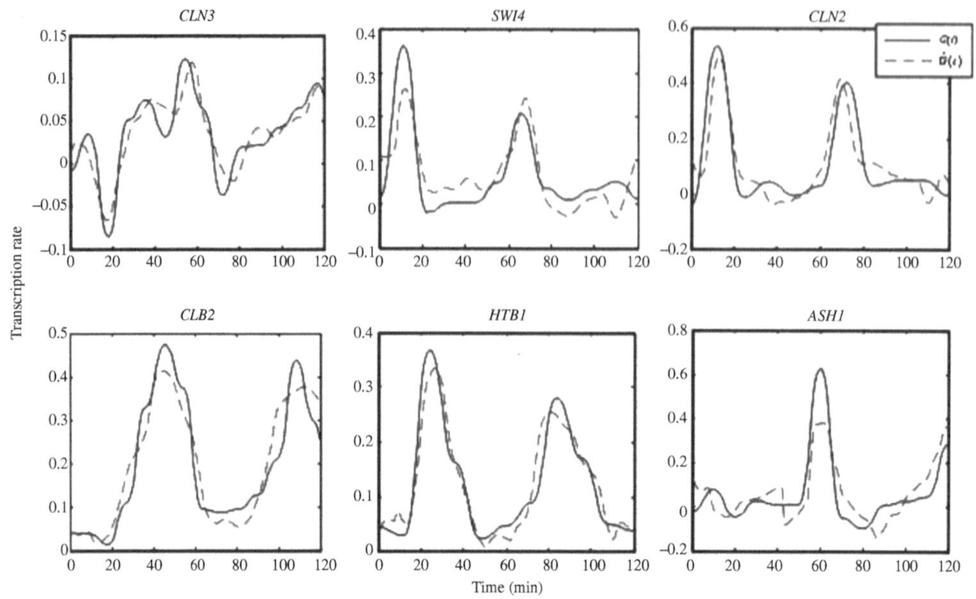

Figure 4.6 The comparison between identified transcriptional regulations $G_i(t)$s and the estimated transcriptional regulations generated by five possible regulators $\hat{G}_i(t)$s (broken lines) for the six cell cycle–related genes *CLN3, SWI4, CLN2, CLB2, HTB1*, and *ASH1*.

expression of *SWI4* and the putative activator identified by this method suggest that Swi4 plays an essential role in regulating the expression programs of genes peaked at Start. In addition to *CLN2*, Swi4 was identified as a putative activator associated with several G1 phase genes, including *SWI4, SWE1, NDD1, SVS1*, and *HCM1* (Table 4.1). All these genes contain an SBF-binding site in their promoter regions [17,19]. In this analysis, the regulatory ability of Swi4 was lower than that of Cln3, suggesting that Swi4 conducts activation succeeding the Cln3 activator. This result coincides with the biological evidence that Swi4 is a DNA-binding subunit of the TF SBF [28]. Cln3 was identified as a positive regulator of target genes *SWI4* and *CLN2* (Table 4.1). *CLN3* is expressed at the late M phase, and Cln3 might consequently activate the expression of *SWI4* and turn on the activity of SBF. Similar to the immense amount of evidence showing that Cln3 is the most prominent activator of SBF [29–31], this result displayed that Cln3 significantly regulated the expression of *SWI4* and several G1 phase genes, *CLN2, SWE1*, and *HCM1* (Table 4.2).

We selected a set of target genes with peak expression levels at various stages of the cell cycle (Table 4.1) and diagrammatically depict the regulatory abilities of the regulators on their target genes to illustrate the TRN of the yeast cell cycle (Fig. 4.7). It is well known that Clb2 is primarily responsible for inactivating SBF and turning off CLN2 synthesis as cells enter mitosis. This analysis also identified that Clb2 is a

Table 4.2 Five possible regulators and their influenced target genes involved in yeast cell cycle

Regulators	Target genes: Upregulation	Target genes: Downregulation
Cln3	*SWI4, CLN2, SWE1, HCM1, MNN1, CLB6, CDC46*	*HTA1, HTB1, HHT1*
Swe1	*HTA1, HTB1, HHT1*	*SWI4, CLN3*
Spo12	*CDC20*	*SWI4, SWE1, CLB6*
Yhp1	*CLN2, SVS1, MNN1, CLB6*	*HTA1, HTB1, HHT1, CLB4, SIC1, ASH1, CLN3*
Clb2	*SWI5, CDC46*	*CLN2, HCM, MNN1*
Rme1	*HCM1*	*NDD1, ASH1, SIC1,CDC46*
Swi4	*SWI4, CLN2, SWE1, HCM1, NDD1, SVS1, MNN1, CLB6, HTA1, HTB1, HHT1*	
Swi5	*ASH1, SIS1, CLN3*	
Ndd1	*CLB4, CLB2, CDC20*	*SPO12*
Fkh2	*CLB2, CDC20, SPO12, SWI5*	

The upstream regulators and their relevant target genes. The first column indicates the specific regulators; the second and third columns outline the upregulating target genes and downregulating target genes, respectively.

negative regulator of *CLN2* (−0.1843; Table 4.1), the negative correlation between the transcriptional regulation of *CLN2* and the expression profile of regulator Clb2 explains the relationship of this target−regulator pair. This network reveals that Spo12 had a significantly negative regulatory ability to coordinate the expression of several G1 genes, *SWI4* and *SWE1*. The molecular function of Spo12 is unclear. This protein plays the role of a positive regulator of exit from the M phase in mitosis and meiosis [32−34]. SPO12 transcripts are expressed in a cell cycle−regulated manner and peaked at the M phase [6]. The strong expression of *SPO12* at this period might cause the downregulation of its target genes through direct or indirect molecular interactions in a temporal manner. Swe1 is a protein having the molecular function of regulating CDK activity and is expressed abruptly during the G1 phase [35]. Swe1 represses the transcription of *CLN3* (6.2-fold of the basal level) and *SWI4* (1.8-fold of the basal level) while the expression of SWE1 is activated by Cln3 and Swi4 to form a feedback loop chain (Fig. 4.7; Table 4.1). This feedback loop chain might play a sophisticated role in regulating the expression of *CLN3* and *SWI4* and then mediating hundreds of genes whose expression peak at the G1 phase through binding to SBF subsequently [17,19].

Clbs are cyclins crucial for completing the G_2/M transition of mitotic cell cycle and the most typical one is Clb2. Mcm1, Fkh1/2, and Ndd1 are identified as TFs binding to the promoter sequence of CLB2 [17,19,36,37]. In this analysis, the TFs Fkh2 and Ndd1 are recognized as activators with significant regulatory abilities (4.8-fold and 4.1-fold of the basal level, respectively), while the cyclin Clb1 that

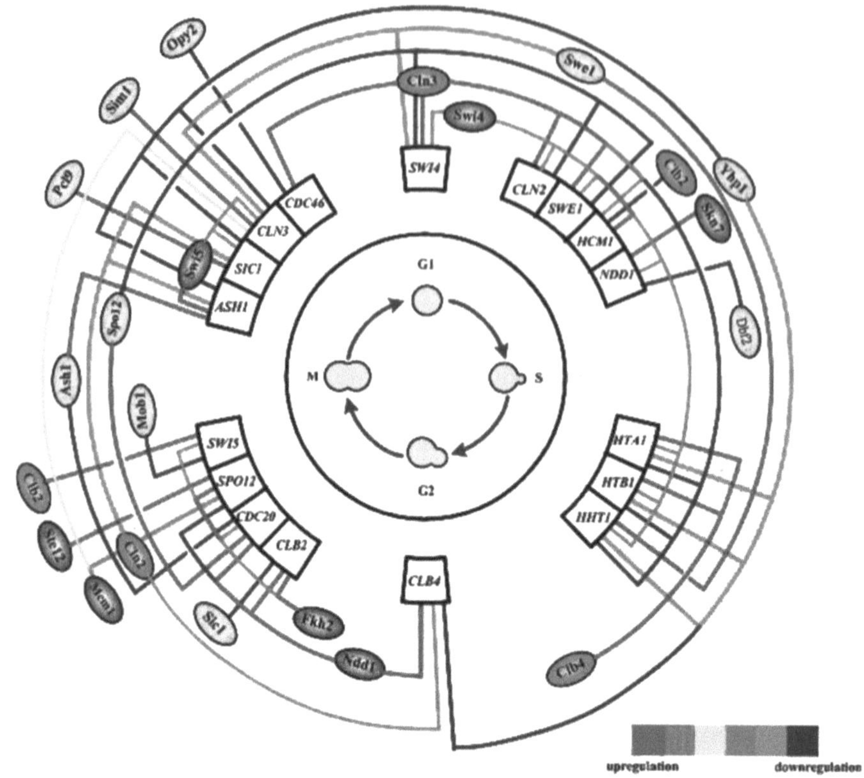

Figure 4.7 A transcriptional regulatory network of 17 selected target genes involved in the Big Cellular Mechanisms of yeast cell cycle. Each *white box* represents a target gene and the target gene is allocated in the cell cycle according to the time of its expression peak. Regulators are indicated by ovals of different colors. *Blue ovals* (white in print versions) denote transcription factors, *green ovals* (light gray in print versions) denote cyclins, and *orange ovals* (dark gray in print versions) denote cell cycle control proteins. The target gene and regulators are connected by *solid lines*. The regulatory abilities of the regulators for a specific target are characterized by colored lines: *red* (black in print versions) for upregulation and *blue* (gray in print versions) for downregulation as indicated.

might activate the transcription by itself is also recognized as another activator (Table 4.1). The significant regulatory abilities suggest that the regulators might play the primary role in controlling the expression of *CLB2*, concurring with the experimental evidence that Fkh2 and Ndd1 were recruited to the chromatin of the *CLB2* promoter [36]. In addition to the activators, a CDK inhibitor Sic1 that binds and inhibits the Clb2–Cdc28 complex was found to downregulate *CLB2* [38]. Sic1 did not show a high level of regulation (Table 4.1) and may act through indirect interaction. Cdc20 is a subcomponent of the anaphase promoting complex (APC), which is required for the onset of anaphase [39]. Similar to *CLB2*, the expression of *CDC20* showed a cell cycle–regulated manner and peaked at the M phase. The two TFs Ndd1 and Fkh2 were also identified as positive regulators of *CDC20*. In addition, Spo12 showed a significant regulatory ability, suggesting an activation role of Spo12 in concert with Ndd1 to mediate the expression of *CDC20* (Table 4.1). As mentioned earlier, Spo12 is thought to be a positive regulator of mitotic exit [32–34]. Spo12 may cooperate with Ndd1 and Fkh2 to turn on the transcription of *CDC20* for subsequently promoting the mitotic metaphase/anaphase transition.

Another crucial regulator for cell cycle control is Sic1. According to Verma et al. [40], when Cln2-associated kinase activity arises at Start, Sic1 is phosphorylated and Sic1p is rapidly presented by Cdc4 to the SCF (SCF is a complex of Skp1, Cdc34, Cdc53, and an F box-containing protein) for ubiquitination and subsequent proteolysis. The expression of *SIC1* reached a peak at the M/G1 phase; two putative activators Pcl9 and Swi5 were found to regulate its transcription (Table 4.1). Pcl9 is a cyclin-dependent protein kinase whose expression peaked at the M/G1 phase [41] and Swi5 is a TF whose expression fluctuated in a cell cycle–regulated manner and peaked at the M phase [6]. Cln2 and Yhp1 were identified as negative regulators for *SIC1* (Table 4.1). Yhp1 is a negative regulator for transcription and its expression peaked at the S phase [42,43]. These results agreed with previous reports that *SIC1* transcription is upregulated by Swi5 and downregulated by Cln2 [19,40]. The basal level of *SIC1* exceeds the regulatory ability of its regulators (Table 4.1), suggesting the constitutive expression of *SIC1*. Mutation of Fkh2 decreased Swi5 transcripts and consequently reduced the *SIC1* expression [37]. This scenario can be identified by this network. Swi5 bound to the promoter of a cyclin gene *CLN3*, which plays a critical role for the cell cycle control [19]. Moreover, Sim1 is identified as the primary activator together with Swi5 and Mcm1 for upregulating the transcription of *CLN3* (Table 4.1). A possible molecular function of Sim1 is the involvement in control of DNA replication [44]. The expression of *SIM1* peaked at the G2 phase might cooperate with TFs Swi5 and Mcm1 for the onset of *CLN3* transcription and subsequently mediate the cell cycle process.

The three target genes *HTA1*, *HTB1*, and *HHT1* (Table 4.1) are classified as the "histone" cluster whose expressions are highly conserved and peaked at the S phase [6]. Histones play an important role in chromatin assembly and disassembly, and thus are strongly cell cycle-regulated and peaked at the S phase. The three histone genes shared identical putative regulators and had relatively high levels of basal expression (Table 4.1). The negative regulators Cln3 and Yhp1 had regulatory abilities beyond the basal level and might directly or indirectly inhibit the function of activators at the basal level; thus, histone genes are expressed in a cell cycle–regulated manner. In addition, the regulatory abilities of the positive regulators Clb4, Swe1, and Swi4 were below the basal level. These positive regulators may help the unknown ubiquitous activators when the expression levels of *CLN3* and *YHP1* are comparatively low.

The fact that Swi4 was identified as a positive regulator in this system explains the reason why, through genome-wide binding location analysis, certain histone gene promoters were found to have SBF binding sites [19].

Table 4.1 lists 20 target genes that are expressed strongly at different phases throughout the cell cycle. Other target genes possessing similar expression patterns with these 20 genes might have the same regulators. In order to specify the target genes mediated by specific regulators that were identified in this analysis, we organized Table 4.2 to list the regulators and their target genes. The newly found pathways in Table 4.2 might be useful for dissecting the regulatory mechanism in the yeast cell cycle and could be examined in the future.

4.1.3 Discussion of big cellular mechanisms in yeast cell cycle

As a vast accumulation of gene expression microarray data becomes available, there is a growing need for computational approaches to study genome-wide transcriptional regulation. Although the potential of the information contained in the large and diverse databases of genome-wide expression profiles is amply appreciated, the radicalization of biological transcriptional regulation from such data remains a challenging task. This section presents a systematically iterative approach to discern and characterize the TRN in yeast from the raw expression profiles focused on the cell cycle [6]. This method provides a useful tool for analyzing gene expression data and validates a few regulators with periodic expression that are enough to coordinate the cell cycle program (Fig. 4.7). To date, the most widely used method for organizing and visualizing expression data is clustering, which determines a number of gene groups, each holding a similar, particular pattern across the whole data set. There is no doubt that the adequate and accurate clustering of expression profiles is the first step to dissecting the transcription network. However, this section provides another aspect of view for inferring the intertwined relationship between different genes that organize the genome-wide expression program.

Our approach also offers the following advantages. First, we adopt a limited number of sample points and transform the raw data into a mathematical kinetic model that describes the dynamics of the expression profile of any gene of interest and provides valuable information for further analysis. Second, the well-constructed dynamic model allows us to extract the transcriptional regulation that controls the expression of any gene of interest. This calculated term can be used as a touchstone for deducing the TRN. Third, we characterize the relationship between gene expression pattern and its regulatory function as a sigmoid function and quantifying the transcriptional regulation of each specific target gene that is composed of the combinations of particular regulatory functions contributed by upstream regulators. This is the first time we have used the dynamic model to describe TRNs based on microarray data.

This model explicitly describes a superposition of a smaller number of regulatory effects, including the genes responding to different combinations of common input variables and the regulatory functions that captures the biologically relevant combinatorial regulation. Last, the identified regulatory ability for each specific regulator evaluates the contribution of this regulator; the positive sign stands for activation and the negative sign stands for repression, and the magnitude represents the activity. These advantages of this approach will ameliorate analysis to cope with rapidly growing data and can be extended to systematic analysis for higher eukaryotes.

A few studies have focused on the GRN recently using different methods. For example, a fuzzy logic [45] approach was proposed to generate a connected network of genes using gene expression data; the fuzzy logic algorithm provides a way to transform expression values into qualitative descriptions that can be evaluated by using heuristic rules, and constructs a model to find triplets of activators, repressors and targets [45]. A smooth surface response algorithm has been developed to improve and extend the fuzzy logic algorithm for identifying the regulations of activators, repressors and targets from gene expression data, and constructing a GRN [46]. All of the above studies have done a great job of uncovering TRN from expression data. Compared with these approaches, this model not only qualitatively reveals the regulatory relationship between regulators and target genes but also quantitatively reveals the capabilities of regulators to the specific target genes. Moreover, this regulatory network dissects the synergistic nature of multiregulation caused by a set of regulators against a simple triplet model of activator, repressor and target. The systems analysis of this section provides an independent method for uncovering a TRN using a mathematical dynamic model and quantitatively characterizing the regulatory ability between regulators and target genes from mRNA transcript data.

The weakness of this model is that the transcriptional control network is oversimplified. In reality, the cell cycle is controlled by a much more complex network that includes not only transcription, but also proteolysis, phosphorylation, compartmentalization, and other regulatory machinery. To uncover the real mechanism for regulating the cell cycle, more deliberate experiments and sophisticated computational approaches need to be developed and examined. The limitation of this approach is that all the analyzed information only came from the expression data for cell cycle–regulated genes of yeast. For example, several crucial proteins promoting the events of cell cycle, such as Cdc28, Swi6, Cdh1, and Cdc14 [47] were not analyzed and characterized in this work, because the expression levels of these proteins are relatively consistent throughout the cell cycle [6]. It is simplistic to infer the real TRN from data of mRNA transcripts alone, because the real machinery involves more complicated biological processes, for instance, the posttranscriptional regulation and the protein–protein interactions. Apart from this, the real regulation for transcription might not be a simple combination as we proposed. A more detailed regulatory mechanism in transcription control will be given in Chapter 5, Big Regulatory Mechanisms in the Transcriptional Regulation Control of Gene Expression Using a Stochastic System Model and Genome-Wide Experimental Data. At present, this section is the first one undertaken with the application of quantitative dynamic modeling on the assay of the transcription network. The dynamic analysis unraveled gene transcription as well as the contribution of their activators and suppressors. As more detailed worthwhile experimental results like protein microarray data become available and as a more exhaustive knowledge of transcriptional mechanism is gained, a more accurate model can be developed to augment this approach for deciphering the real TRN. It is believed that such a systematic approach will be critical for dissecting the complex architecture of transcriptional networks in more complex eukaryotes, in anticipation of an avalanche of microarray data from the human and mouse genome.

APPENDIX A: MATCHED FILTER FOR SELECTING MORE CORRELATED REGULATORS IN YEAST CELL CYCLE

To fully specify the correlation or matching between the transcriptional regulation $G_i(t)$ and regulation function $f_j(t)$, we introduce a powerful technique, matched filter, which was widely used for detecting signals in the communication field and for the correlation classifier in the pattern recognition field. In general, let us suppose that the impulse response of the filter is

$$h_k(t) = s_k(T - t), \quad 0 \leq t \leq T$$

where $s_k(t)$ is the basis function and $h_k(t) = 0$ outside of the interval $0 \leq t \leq T$. If we let the input signal to the filter be $r(t)$, then the output of this filter is [48]

$$y_k(t) = \begin{cases} \int_0^t r(\tau)h_k(t-\tau)d\tau \\ \int_0^t r(\tau)s_k(T-t+\tau)d\tau \end{cases} \quad \text{for } k = 1, \ldots, N.$$

Now, if we sample the output of the filter at $t = T$, we obtain

$$y_k(T) = \int_0^T r(\tau)s_k(\tau)d\tau, \quad \text{for } k = 1, \ldots, N.$$

Hence, the sampled output of the filter at time $t = T$ is exactly the correlation between $r(t)$ and $s_k(t)$. A filter whose impulse response $h(t) = s(T - t)$, where $s(t)$ is assumed to be confined to the time interval $0 \leq t \leq T$, is called the matched filter to the signal $s(t)$. Based on this concept, we let the input signal as $G_i(t)$ and the impulse response of the filter as $f_k(t)$, where $G_i(t)$ and $f_k(t)$ are transcriptional regulation and regulation function, respectively. Therefore, the sampled output of the filter can be represented as the score that stands for the correlation between $G_i(t)$ and $f_k(t)$.

After investigating the TRN of 17 selected target genes involved in the cellular mechanism of yeast cell cycle, we will identify more target genes for more general TRN of cell cycle. In the following section, we will develop a TRIA algorithm to identify TFs' target genes through microarray data to understand how cells change their transcription program including cell cycle to adapt to their environmental stimuli.

4.2 CONSTRUCTING TRMs FOR BIG REGULATORY MECHANISMS OF THE YEAST CELL CYCLE

A TRM is a set of genes that is regulated by a common set of TFs. By organizing the genome into TRMs, a living cell can coordinate the activities of many genes and carry out complex functions. Therefore, identifying TRMs is helpful for understanding gene regulation. Integrating gene expression and ChIP-chip data, a method called MOFA has been developed, for reconstructing TRMs of the yeast cell cycle. MOFA identified 87 TRMs, which together contain 336 distinct genes regulated by 40 TFs. Using various kinds of data, we validated the biological relevance of the identified TRMs. The analysis in this section shows that different combinations of a fairly small number of TFs are responsible for regulating a large number of genes involved in different cell cycle phases and that there may exist crosstalk between the cell cycle and other cellular processes. MOFA is capable of finding many novel TF-target gene relationships and can determine whether a TF is an activator or/and a repressor in the yeast cell cycle. Finally, MOFA refines some clusters proposed by previous studies and provides a better understanding of how the complex expression program of the cell cycle is regulated. MOFA was developed to reconstruct TRMs of the yeast cell cycle. Many of these TRMs are in agreement with previous studies. Further, MOFA inferred many interesting modules and novel TF combinations in the TRN of yeast cell cycle. We believe that computational analysis of multiple types of data will be a powerful approach to studying Big Mechanisms of complex biological systems when more and more genomic resources such as genome-wide protein activity data and protein–protein interaction data become available.

4.2.1 Background of MOFA

A TRM is a set of genes that is regulated by a common set of TFs. By organizing the genome into TRMs, a living cell can coordinate the activities of many genes and carry out complex cellular functions. Therefore, identifying TRMs is useful for understanding cellular mechanisms in response to internal and external signals. The advances of high-throughput genomic tools such as DNA microarray [49,50] and chromatin immunoprecipitation-DNA chip (ChIP-chip) [17,51] have made the computational reconstruction of TRMs of an eukaryotic cell possible.

Genome-wide gene expression analysis has been used to investigate TRMs controlling a variety of cellular processes in yeast [5,6,52–54]. Clustering and motif-discovering algorithms have been applied to gene expression data to find sets of coregulated genes and have identified plausible binding motifs of their TFs [6,8,10]. Such approaches have also been expanded to incorporate previous knowledge about the genes, such as cellular functions [55] or promoter sequence motifs [16]. Moreover, some researchers used model-based approaches such as random Boolean networks [56] and Bayesian networks [57,58] to infer regulatory network architectures. However, this approach provides only indirect evidence of genetic regulatory interactions and does not identify the relevant TFs. On the other hand, the ChIP-chip technique was developed to identify physical interactions between TFs and DNA regions. Using ChIP-chip data, Simon et al. [19] investigated how the yeast cell-cycle gene-expression program is regulated by each of the nine major transcriptional activators. Lee et al. [18] constructed a network of TF-gene interactions and Harbison et al. [59] constructed an initial map of yeast's transcriptional regulatory code. However, ChIP-chip data alone cannot tell whether a TF is an activator or a repressor and, most importantly, ChIP-chip data are noisy and, depending on the chosen *p*-value cutoff, include many false positive or false negative TF-DNA binding relationships.

As gene expression and ChIP-chip data provide complementary information, some researchers [60–62] have integrated both types of data in their studies. However, most previous studies except the GRAM algorithm [60] assumed that a gene is regulated by a TF only if the *p*-value of TF-gene binding in the ChIP-chip data is $\leq .001$, thus suffering a false negative rate of $\sim 24\%$ in determining TF-gene binding [59].

In order to reduce the high false negative rate, the Temporal Relationship Identification Algorithm (TRIA) was developed, to use the information provided by gene expression data to alleviate the effect of using a stringent threshold in determining TF-gene binding. A TF-gene pair is said to have a positively (negatively) temporal relationship if the gene's expression profile is positively (negatively) correlated with the TF's regulatory profile possibly with time lags (see Section 4.2.5). TRIA identifies TF-gene pairs with a temporal relationship. We define that a TF binds to a specific gene if: (1) the *p*-value for the TF to bind the gene is $\leq .001$ in the ChIP-chip data, or (2) $.001 < p \leq .01$ and the TF-gene pair has a temporal relationship. That is, we allow the *p*-value cutoff to be relaxed to .01 if the TF-gene pair has a temporal relationship. This approach is different from the GRAM algorithm [60], which relied on sets of coexpressed gene to relax the stringent *p*-value cutoff.

From the above procedure, we derive a binding score matrix. Then we develop the MOFA that combines this binding score matrix with the gene expression matrix to reconstruct TRMs of the yeast cell cycle (see Section 4.2.5). For each of the five cell cycle phases (M/G1, G1, S, S/G2, and G2/M), MOFA exhaustively searches for all possible TF combinations and find their target genes. Once the set of target genes to which a common set of TFs bind is inferred, MOFA identifies a subset of these target genes whose gene expression profiles are positively correlated possibly with time lags. That is, the genes of a module not only share a common set of TFs but also have positively (time-shifted) correlated expression profiles. This gene module is more general than that of GRAM algorithm [60], which only searched coexpressed genes to form a module. MOFA reconstructs 87 TRMs. We then validate the biological relevance of each inferred TRM using existing experimental data, enrichment for genes in the same MIPS functional category [65], known DNA-binding motifs [6], etc.

4.2.2 MOFA algorithm in finding TRMs in yeast cell cycle

By integrating the gene expression and ChIP-chip data, MOFA identified 87 TRMs, which together contain 336 distinct genes regulated by 40 distinct TFs (see Fig. 4.8). In the literature [6,64–66], 139 of the 336 genes and 30 of the 40 TFs are known to be involved in the cell cycle.

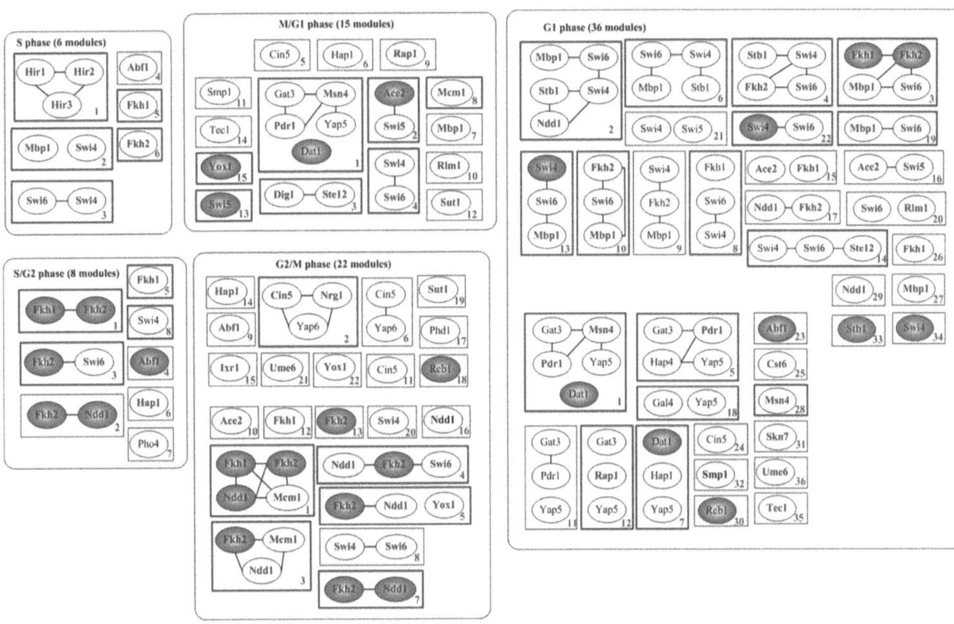

Figure 4.8 The 87 TRMs identified in this section. Each rectangle stands for a module and the ovals in a rectangle indicate the TFs that regulate the module. TF name is colored blue (gray in print versions) if its function is consistent with one of the module's over represented MIPS functional categories with adjusted *p*-value < .05 (after the Bonferroni correction for multiple tests) using the cumulative hypergeometric distribution or black if not. Two ovals are connected by a line if previous studies indicated that the two TFs interact with each other physically (forming a protein complex), functionally (in the same cellular process) or statistically (co-occurrence) [59,60,62–64]. An oval is colored red (light gray in print versions) (green (dark gray in print versions)) if the TF is identified as an activator (repressor). The periphery of a rectangle is colored purple (dark gray in print versions) if this module implicates important TF combinations or is discussed in the text.

Validation of the identified modules

Analysis of the identified modules in Fig. 4.8 suggests that MOFA identifies biologically relevant groups of genes. First, 83 of the 87 modules contain genes that are known to be involved in the cell cycle. Second, 51% (44/87) identified module includes groups of genes that function in the same cellular process: each of these modules contains at least one over represented MIPS functional category with adjusted p-value $<.05$ (after the Bonferroni correction for multiple tests) using the cumulative hypergeometric distribution. Third, the modules are generally accurate in assigning TFs to sets of genes whose functions are consistent with the TFs' known roles. We found that the regulatory functions of the 71% (120/169 counting multiplicity) TFs are consistent with one of their modules' over represented MIPS functional categories with adjusted p-value$<.05$ (see Fig. 4.8). As an example, Dig1 and Ste12 are known to regulate mating and pseudohyphal growth [67] and M/G1 is the critical phase for these processes. All five genes (*FUS1, GPA1, KAR4, SST2, TEC1*) of the {Ste12, Dig1} module are important for mating, pseudohyphal growth, or pheromone response. Fourth, 33% (188/568 counting multiplicity) genes are known by previous studies to be regulated by at least one of the TFs that we assigned to the module. Fifth, the genes of a module usually have the same binding motifs of the important cell cycle TFs such as SCB (bound by SBF), MCB (bound by MBF), SFF (bound by SFF), ECB (bound by Mcm1), and SWI5 (bound by Ace2 and Swi5). We found that in the majority of cases (36/45) in which a module is controlled by at least one of the important cell cycle TFs (SBF, MBF, SFF, Mcm1, Ace2, and Swi5), there always exist genes that have the known binding motifs of the corresponding TFs. Finally, in most cases in which a module is controlled by more than one TF, there is evidence that these TFs may interact physically or functionally (see Fig. 4.8). About 59% (70/118) of the TF interactions that we identified have been experimentally proven or identified by computational algorithms [59–62,65]. Taken together, these results provide evidence that MOFA identifies not only sets of biologically related genes, but also TFs that individually or cooperatively control these genes.

Identification of important TFs and their combinations for yeast cell cycle

MOFA identified 40 TFs that regulate genes of the yeast cell cycle and Fig. 4.9 shows the cell cycle phases in which these TFs carry out their regulatory functions. Table 4.3 lists these 40 TFs according to the number of target genes. The nine well-known cell cycle TFs (Ace2, Fkh1, Fkh2, Mbp1, Mcm1, Ndd1, Swi4, Swi5, and Swi6) are ranked within the top 14, suggesting the effectiveness of MOFA to find important cell cycle TFs. Moreover, we found another 21 TFs (Abf1, Cin5, Cst6, Dig1, Gal4, Gat3, Hap4, Hir1, Hir2, Hir3, Ixr1, Msn4, Rap1, Rlm1, Skn7, Stb1,

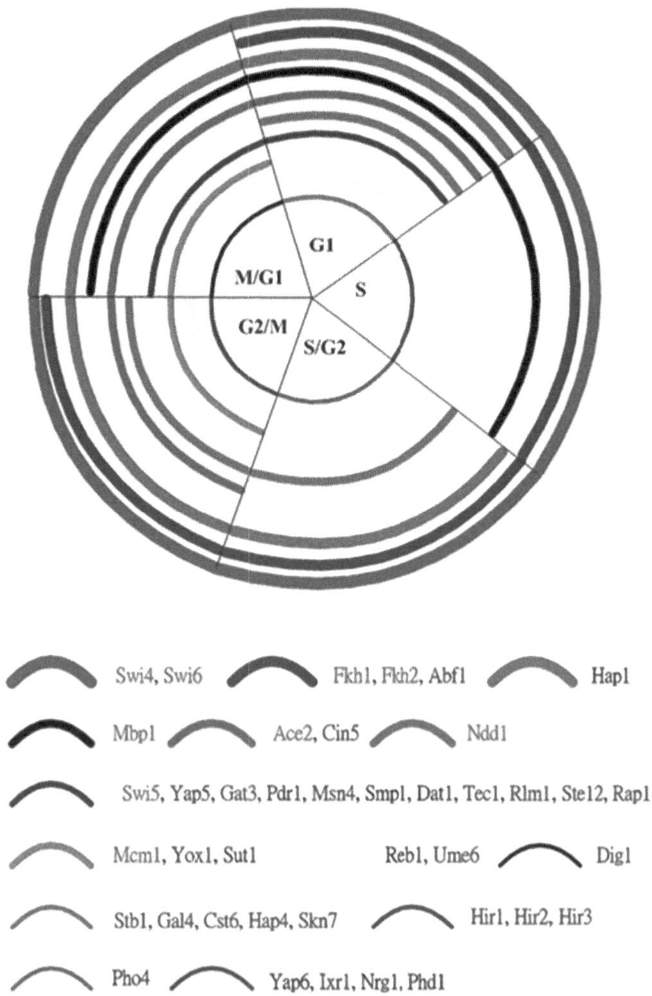

Figure 4.9 The cell cycle phases in which each of the 40 identified TFs carries out its regulatory function. Nine well-known cell cycle TFs are colored red (dark gray in print versions) and another 21 TFs that are also involved in the cell cycle [64–66] are colored blue (light gray in print versions).

Ste12, Tec1, Ume6, Yap5, and Yox1) that are relative to the cell cycle process, consistent with the previous studies [64–66]. The remaining 10 TFs (Dat1, Hap1, Nrg1, Pdr1, Phd1, Pho4, Reb1, Smp1, Sut1, and Yap6) are putative cell cycle related TFs. Among them, Hap1 is more plausible than the others to be related to the cell cycle process since the number of cell cycle genes that it regulates is much larger than that of the others (see Table 4.3). Actually, it has been shown that Hap1 (also called Ape1

Table 4.3 The number of genes regulated by each of the 40 TFs in each cell cycle phase

TF	All phases	M/G1	G1	S	S/G2	G2/M
Swi4	78	5	45	9	9	10
Swi6	74	5	50	7	5	7
Fkh2	67		14	5	15	33
Mbp1	57	5	47	5		
Fkh1	53		17	5	18	13
Ndd1	50		12		5	33
Abf1	34		10	5	10	9
Swi5	34	24	10			
Cin5	33	8	7			18
Hap1	30	9	5		5	11
Stb1	24		24			
Mcm1	22	7				15
Yap5	22	5	17			
Ace2	21	6	10			5
Gat3	20	5	15			
Pdr1	17	5	12			
Yox1	16	5				11
Msn4	15	5	10			
Reb1	15		8			7
Smp1	15	8	7			
Dat1	13	5	8			
Sut1	13	5				8
Tec1	13	5	8			
Rlm1	12	7	5			
Ste12	12	5	7			
Ume6	11		6			5
Yap6	11					11
Rap1	10	5	5			
Gal4	6		6			
Hir1	6			6		
Hir2	6			6		
Hir3	6			6		
Cst6	5		5			
Dig1	5	5				
Hap4	5		5			
Ixr1	5					5
Nrg1	5					5
Phd1	5					5
Pho4	5				5	
Skn7	5		5			

Nine well-known cell cycle TFs are colored red (dark gray in print versions) and another 21 TFs that are also involved in the cell cycle [64–66] are colored blue (light gray in print versions). The TFs are ordered by the number of their target genes in all phases.

AP endonuclease) regulates *APE1* [68]. Ape1 is a dual function enzyme and its cell cycle-dependent expression might affect both DNA repair and the activity of various TFs as a function of the cell cycle [68]. This evidence validates that MOFA has the ability to find novel TFs which may play a role in the cell cycle or are involved in other cellular processes that have crosstalk with the cell cycle process.

TF combinations and their target genes that are important for each cell cycle phase are also found. We found that different combinations of a fairly small number of TFs are responsible for regulating a large number of genes in different cell cycle phases.

Detailed discussions of the TF combinations and their target genes in each specific cell cycle phase are given below.

1. *The M/G1 Phase*

   Ace2 and Swi5 have been shown to control certain genes expressed in M/G1 [69]. We successfully found that {Ace2, Swi5} and {Swi5} regulate, respectively, modules 2 and 13 in M/G1. Both Ace2 and Swi5 were found to regulate *EGT2*, whose product is involved in cell wall biogenesis and cytokinesis. Swi5 also regulates *PCL9*, whose product is the only cyclin known to act in M/G1 [70] and *SIC1*, whose product is a cyclin regulator that inhibits the Cdc28-Clb kinase activity. Furthermore, Swi5 regulates several Y′ genes, which are a subgroup of a larger group of subtelomeric genes that share DNA sequence similarity and whose expression peaks in early G1 [6]. It is known that in the absence of Ndd1 and Fkh2, Mcm1 participates in the regulation of genes essential for cellular functions specific to late mitosis and early G1 [29,71]. Indeed, we found that {Mcm1} regulates module 8 in M/G1. It regulates *CDC46*, which encodes a protein involved in prereplication complex formation and *AGA2*, which involves in mating. In addition, Yox1 was recently characterized as a binding partner of Mcm1 in M/G1 [71]. We found that {Yox1}, acting as a repressor, regulates module 15 in M/G1. Three genes *CDC46*, *PIG1* and *YOR066W* are found to be regulated by both Mcm1 and Yox1, confirming that Yox1 and Mcm1 may co-regulate a group of genes.

   In addition, some cell-wall genes are known to be under the control of the M-phase regulator Mcm1 or the G1-phase regulator SBF. The M/G1 phase is a crucial time for cell wall synthesis because the bud separates from the mother right after the M/G1 phase. We successfully found TF combinations {Mcm1} and {Swi4, Swi6}, whose common target genes include *SWI4*, which encodes a late G1 TF, and *UTR2*, which is involved in cell-wall organization and polarized growth. The dual regulation of *SWI4* by Mcm1 and Swi4 has been shown previously [29].

   We identified {Dig1, Ste12} to regulate module 3 in M/G1. The genes of this module include *FUS1*, *GPA1*, *KAR4*, *SST2*, and *TEC1*, which are important for mating or pseudohyphal growth. Dig1 and Ste12 are known to regulate mating and pseudohyphal growth [67], supporting the biological relevance of this finding. We also found novel TF combinations. For example, {Dat1, Gat3, Msn4, Pdr1, Yap5} is identified to regulate a group of genes that are similar to subtelomerically encoded proteins.

2. *The G1 Phase*

   Previous molecular and genetic analysis suggested that SBF and MBF are important activators of genes essential for cellular functions specific to late G1 [19,72]. This result confirms this model: 10 out of the 36 modules in G1 are regulated by MBF or SBF. SBF regulates *BUD9*, *EXG1* (both of module 8), *GAS1*, *MNN1*, *OCH1*, and *PSA1* (all of module 22). These genes are involved in the morphological changes associated with cell budding. MBF controls *PDS5, RAD51, RNR1* (all of module 3), *DUN1, IRR1,* and *RAD27* (all of module 19). These genes are involved in DNA replication and repair. Moreover, the targets of SBF and MBF also include key cell cycle regulators. Both SBF and MBF were found to regulate *CLN1, CLB6* (both of module 2), and *PCL1* (of module 6). In addition, SBF regulates *PCL2* (of module 14) and MBF regulates *CLB6* (of module 2).

We found that Stb1 together with SBF (Swi4 + Swi6) or MBF (Mbp1 + Swi6) regulates modules 2, 4, and 6 in G1. It has been known that Stb1 binds to Swi6 in vitro and is thought to interact with Swi6, a subunit of both SBF and MBF, to regulate transcription in vivo [73]. Also, consistent with these results, Kato et al. [62] claimed the presence of the complexes Stb1 + Swi6 + Swi4 and Stb1 + Swi6 + Mbp1. Moreover, we found that {Ste12, Swi4, Swi6} regulates module 14 in G1, which is also consistent with the result of [62].

We found that Fkh1/Fkh2 combines with MBF/SBF to regulate modules 3, 4, 8, and 10 in G1. It is known that Fkh1 and Fkh2 regulate genes expressed in G2/M and also genes expressed in other cell cycle phases [19], supporting this result. We also found some novel TF combinations. For example, {Dat1, Gat3, Msn4, Pdr1, Yap5}, which is also found in M/G1, {Gat3, Hap4, Pdr1, Yap5}, {Dat1, Hap1, Yap5}, {Gat3, Rap1, Yap5}, {Gal4, Yap5}, and {Msn4} are all identified to regulate genes whose products are similar to subtelomerically encoded proteins. All these genes share DNA sequence similarity and are found in Y′ elements, which are located at chromosomes ends [6].

3. *The S Phase*

We found that {Fkh2} regulates various genes that encode proteins associated with chromatin structure including histone genes *HHF1* and *HHT1* (both of module 6). We found that {Fkh1} regulates *TEL2* (of module 5), a telomere length regulator, and *ARP7* (of module 5), a subunit of the chromatin remodeling Swi/Snf complex. Histone genes can be found in the {Fkh1}, {Fkh2}, {Swi4, Swi6}, and {Mbp1, Swi4} modules, suggesting that SBF, Fkh1, and Fkh2 probably regulate histone genes. This result is consistent with a few genomic studies [18,26] that indicated the involvement of SBF and Fkh1/Fkh2 in regulating S phase genes. In addition, we successfully identified {Hir1, Hir2, Hir3} to regulate six histone genes (*HTA1, HTB1, HHT1, HHF1, HHT2, HHF2*) of module 1 in the S phase, supported by existing experimental results [74]. In summary, we suggest that SBF and Fkh1/Fkh2 are activators and Hir1, Hir2, and Hir3 are repressors of histone genes.

4. *The S/G2 and G2/M Phases*

Simon et al. [19] and Lee et al. [18] indicated the involvement of SBF and Fkh1/Fkh2 in regulating S/G2 genes. We confirmed that Fkh1, Fkh2, Swi4, and Swi6 are important TFs in this phase since five out of the eight modules in S/G2 are regulated by at least one of these TFs. Fkh2, Swi4, and Swi6 are identified to regulate *SIM1*, which is involved in cell cycle control, and Fkh1 is identified to regulate *CLB4*, which encodes an S/G2 cyclin.

Previous studies have demonstrated that Mcm1 collaborates with Ndd1 and Fkh1/Fkh2 to regulate genes necessary for both entry into and exit from mitosis [36,75]. We successfully identified this TF combination to regulate module 1 in G2/M. Four of the seven genes identified in this module have an SFF (bound by Ndd1 + Fkh1/Fkh2) or ECB (bound by Mcm1) motif. The Mcm1 + Ndd1 + Fkh1/Fkh2 protein complex regulates transcription of *CLB2* (of module 1), whose product is necessary to enter mitosis. Furthermore, SBF and MBF regulate *SWE1* (of module 13 in G1) and *GIN4* (of module 13 in G1). Swe1 is a protein kinase that regulates the G2/M transition by inhibition of Cdc28-Clb2 kinase activity and Gin4 regulates Swe1 [76]. The Mcm1 + Ndd1 + Fkh1/Fkh2 protein complex also sets the stage for exit from mitosis at several levels [19]. First, they regulate two key M/G1 TFs: *SWI5* (of module 3) and *ACE2* (of module 1). Second, they regulate *CDC20* (of module 1), an activator of the APC. Finally, these activators regulate *SPO12* (of module 3), which encodes a protein that regulates the mitotic exit.

It has been suggested that Fkh2 has a more prominent role than Fkh1 in G2/M transcription [36]. This analysis agrees with this suggestion since the number of G2/M genes regulated by Fkh2 is much larger than that of Fkh1 (see Table 4.3). We also found novel TF combinations. For example, we found that SFF instead of combining with MCM1 can also combine with Swi6 or Yox1 to regulate G2/M genes and {Cin5, Nrg1, Yap6} is identified to regulate a group of genes with unknown functions.

4.2.3 Discussion of the MOFA algorithm

Relationships between two TFs of a module

The relationships between two TFs that regulate the same module fall into three categories. First, both TFs bind DNA in the same promoter region but do not interact with each other. Different TFs may regulate the target gene to execute different functions in different cellular processes. Indeed, we found that TFs in this category usually regulate genes that are required for multiple cellular processes. For example, we found that {Ste12, Swi4, Swi6} regulates module 14 in G1. Since Ste12 and SBF (Swi4 + Swi6) are both DNA-binding TFs and there is no evidence that Ste12 interacts with SBF, the relationship between Ste12 and SBF belongs to this category. Ste12 is a regulator of the mating or pseudohyphal growth pathway and SBF is an important regulator in the G1 phase. This indicates that there may exist crosstalk between these two cellular processes. That is, the TF combination {Ste12, Swi4, Swi6} probably regulates genes needed for the G1 phase and also independently needed for mating, confirming the results of [62]. Second, both TFs bind DNA and interact with each other. For example, we found that {Fkh2, Ndd1, Mcm1} regulates module 3 in G2/M. Both Mcm1 and Fkh2 bind DNA and these two proteins together recruit Ndd1 to form a protein complex to control the transcription of G2/M genes [36]. Third, only one TF binds DNA and the other TF regulates the target genes through binding to the DNA-binding TF. For example, {Mbp1, Swi6} and {Swi4, Swi6} are found to regulate, respectively, modules 19 and 22 in G1. MBF (Mbp1 + Swi6) functions in DNA replication, and SBF (Swi4 + Swi6) predominantly controls the expression of budding and cell-wall genes [17]. Since Swi6 is a non-DNA-binding cofactor of Swi4 and Mbp1, the relationship between Swi6 and Swi4/Mbp1 falls into the third category.

Advantages of MOFA

MOFA has two features that make it more powerful than current methods. First, it can reduce false negatives in determining binding events in the ChIP-chip data. Most researchers except for Bar-Joseph et al. [60] have chosen a relatively stringent *p*-value threshold (.001) to determine binding in order to reduce false positives at the expense of false negatives [18,59,62,63]. In comparison, MOFA allows the *p*-value cutoff to be relaxed to .01 if a TF-gene pair has a temporal relationship. (The approach is different from the GRAM algorithm [60], which relied on sets of coexpressed gene to relax the stringent *p*-value cutoff.) As an example, consider Swi5, a well-characterized cell cycle TF in M/G1. The {Swi5} module we inferred contains 18 genes that have similar expression patterns (see Fig. B.1). Four of these genes (*YOR264W, PST1, SIC1,* and *YHB1*) would not have been identified as Swi5 targets using the stringent *p*-value threshold (.001). Previous studies identified these four genes as true targets of Swi5 [6,18]. This attests to the ability of MOFA to lower the rate of false negatives without substantially increasing the rate of false positives. Overall, 87 of the 988 unique TF-gene interactions discovered by MOFA would not have been detected using the current ChIP-chip data with the stringent *p*-value cutoff (.001). In addition, 312 of the 988 unique TF-gene interactions are supported by gene expression data. That is, each of the 312 TF-gene pairs is identified to have a temporal relationship.

Second, MOFA can determine the role of a TF in regulating genes of a module. A TF is said to be an activator (repressor) of a module if the *p*-value of observing TF-gene pairs of the module having a positively (negatively) temporal relationship is $\leq .001$. The *p*-value is the probability that an observation would be made by chance, and is calculated using the cumulative binomial distribution [77]. We found nine activators (Abf1, Ace2, Fkh1, Fkh2, Ndd1, Reb1, Stb1, Swi4, and Swi5) and six repressors (Abf1, Dat1, Fkh1, Fkh2, Reb1, and Yox1), consistent with the results of previous studies [78–90]. Interestingly, four TFs (Abf1, Fkh1, Fkh2, and Reb1) are capable of being activators and repressors to regulate different modules. Table 4.4 provides the detailed discussion of how we assign the regulatory roles of TFs and the known experimental evidence that supports these findings.

MOFA is more powerful than the GRAM algorithm [60] in two ways. First, MOFA has the ability to assign a TF to be an activator or/and a repressor

Table 4.4 Identifying regulatory roles of TFs

TF	Phase (module number)	Regulatory role	*p*-Value	Evidence from literature
Abf1	S/G2 (4)	Activator	6×10^{-5}	[85]
Abf1	G1 (23)	Repressor	.001	[83]
Fkh1	S/G2 (1); G2/M (1)	Activator	3×10^{-8}; 1×10^{-7}	[75]
Fkh1	G1 (3)	Repressor	3×10^{-5}	[82]
Fkh2	S/G2 (1) (2) (3); G2/M (1) (3) (4) (5) (7) (13)	Activator	3×10^{-8}; 3×10^{-7}; 3×10^{-5}; 1×10^{-7}; 6×10^{-9}; 3×10^{-5}; 9×10^{-5}; 6×10^{-9}; 3×10^{-5}	[75]
Fkh2	G1 (3)	Repressor	3×10^{-5}	[82]
Reb1	G2/M (18)	Activator	2×10^{-4}	[87]
Reb1	G1 (30)	Repressor	2×10^{-5}	[78,86,90]
Ace2	M/G1 (2)	Activator	9×10^{-5}	[69,81]
Ndd1	S/G2 (2); G2/M (1) (7)	Activator	3×10^{-5}; 1×10^{-7}; 6×10^{-6}	[80,84]
Stb1	G1 (33)	Activator	2×10^{-11}	[73,79]
Swi4	G1 (13) (22) (34)	Activator	6×10^{-9}; 1×10^{-6}; 2×10^{-4}	[88]
Swi5	M/G1 (13)	Activator	6×10^{-8}	[69,81]
Dat1	M/G1 (1); G1 (1) (7)	Repressor	3×10^{-5}; 3×10^{-7}; 3×10^{-5}	[89]
Yox1	M/G1 (15)	Repressor	3×10^{-7}	[71]

MOFA can determine the regulatory role of a TF in regulating genes of a module. A TF is said to be an activator/repressor of a module if the *p*-value of observing TF-gene pairs of the module having positively/negatively (time-shifted) correlated profiles is $\leq .001$. The *p*-value is the probability that an observation would be made by chance, and is calculated using the cumulative binomial distribution [77]: $P(x \geq n_0) = \sum_{x=n_0}^{N} \left[\frac{N!}{x!(N-x)!}\right] p^x (1-p)^{N-x}$,
where N is the total number of genes in a module, n_0 is the number of genes that have temporal relationships with the TF, and p is the probability of observing an arbitrary gene in the genome that has a temporal relationship with the TF.

(see Table 4.4). On the contrary, the GRAM algorithm cannot find any repressors or activators that are correlated with its target genes with time lags as the GRAM algorithm regards a TF to be an activator only when the expression profiles of the TF and the genes in the corresponding module are coexpressed. For example, the GRAM algorithm found only two (Fkh1 and Fkh2) of the nine activators and none of the six repressors that are found by MOFA (see Table 4.4). Second, MOFA is more powerful than the GRAM algorithm to find out coregulated genes that are not coexpressed. Whilst the GRAM algorithm assumed that the genes of a module are coexpressed, MOFA allows the genes of a module to be positively correlated with time lags. Since it is known that coregulated genes may not be coexpressed [91,92], the relaxation of coexpressed assumption of GRAM algorithm makes MOFA have a better ability to reconstruct gene modules with biological relevance. For example, MOFA identified four genes (*YOR264W*, *PST1*, *SIC1*, and *YHB1*) as Swi5 targets ({Swi5} module in M/G1) which is supported by previous studies [6,18]. However, none of them was found by GRAM algorithm.

Parameter settings of MOFA

The choices of both the relaxed *p*-value and time-lag parameter have biological meanings. Two previous papers [18,59] used a statistical error model to assign a *p*-value of the binding relationship of a TF-gene pair. They found that if $p \leq .001$, the binding relationship of a TF-gene pair is of high confidence and can usually be confirmed by gene-specific PCR. If $p > .01$, the binding relationship of a TF-gene pair is of low confidence and cannot be confirmed by gene-specific PCR most of the time. However, if $.001 < p \leq .01$, the binding relationship of a TF-gene pair is ambiguous and can be confirmed by gene-specific PCR in some cases but not in the other cases. This aim is to solve this ambiguity. This is why we choose .01 to be the relaxed *p*-value. We say that an ambiguous binding relationship of a TF-gene pair is plausible if $.001 < p < .01$ and if this TF-gene pair has a temporal relationship. As to the time-lag parameter, its value is chosen to make the maximal time lag approximately equal to two consecutive cell cycle phases because Simon et al. [19] found cases where a cell cycle TF that expresses in one phase of the cell cycle can regulate genes that function in the next phase.

Increasing the value of the relaxed *p*-value or the time-lag parameter may introduce some false positive binding relationships of TF-gene pairs into the binding score matrix. On the other hand, decreasing the value of the relaxed *p*-value or the time-lag parameter may fail to rescue some false negative binding relationships of TF-gene pairs. A binding score matrix is used to construct an original TRM and MOFA refines the TRM by identifying a subset of these coregulated genes in a TRM whose gene expression profiles are highly positively correlated possibly with time lags. MOFA can filter out false positives to some extent because the expression profiles of false positives are unlikely by chance to be similar to those of the highly positively time-delayed correlated genes identified by MOFA. As to the false negative problem, MOFA cannot alleviate the harmful effect since these plausible binding relationships of TF-genes pairs are not included in the first place. That is, false negative problem is a more serious issue than the false positive problem in MOFA. Therefore, if users have no idea about the appropriate values of the relaxed *p*-value and the time-lag parameter, they should first try larger values since MOFA has the ability to reduce this kind of noises.

Refining clusters from Spellman et al.

Spellman et al. [6] used a hierarchical clustering algorithm to group together coexpressed genes and searched the promoters of these genes for consensus binding motifs. They tried to use these clusters to understand the transcriptional mechanisms of cell cycle regulation. Their approach has some drawbacks. First, coexpressed genes are not necessarily coregulated. Second, even if the genes in a cluster are coregulated, the relevant TFs still cannot be easily identified by the consensus binding motifs.

MOFA can refine clusters in Ref. [18] and provide a better understanding of how the cell regulates the complex expression program of the yeast cell cycle. For example, MOFA reassigned genes of the MCM cluster in Ref. [18] to several modules. As shown in Fig. 4.10A, these modules differ not only in the set of TFs regulating the modules, but also in the different cell cycle phases to which they belong. These results confirm previous findings that Mcm1 collaborates with Yox1 to regulate genes in M/G1 (e.g., *YOR066W* and *CDC46*) [71] and collaborates with Ndd1 and Fkh1/Fkh2 to regulate genes in G2/M (e.g., *SPO12* and *KIN3*) [36]. In addition, MOFA provides regulation information of the Y′ cluster in [18]. The Y′ cluster contains genes that share DNA sequence similarity and are found in Y′ elements, which are located at chromosome ends. Spellman et al. [6] did not figure out how these genes are regulated. As shown in Fig. 4.10B, MOFA reassigned genes of the Y′ cluster to three modules and identified several possible regulators (Dat1, Gal4, Gat3, Hap1, Hap4, Msn4, Pdr1, Rap1, and Yap5), providing information for future experiments.

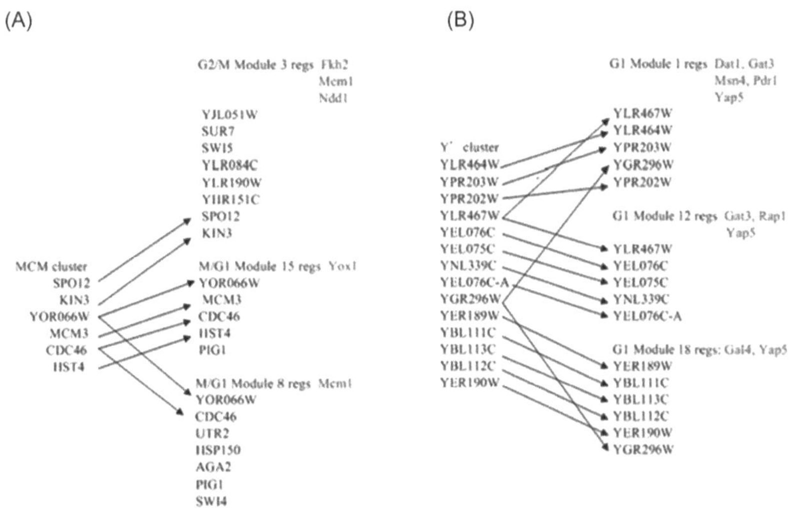

Figure 4.10 Refining clusters of TRM for the regulatory mechanism of cell cycle genes from Spellman et al. (A) Refining the MCM cluster in Ref. [6]. The modules identified by MOFA differ not only in the set of TFs regulating the modules, but also in the different cell cycle phases to which they belong, providing a better understanding of how the cell regulates the complex expression program of the yeast cell cycle. These results confirm previous findings that Mcm1 collaborates with Yox1 to regulate genes in M/G1 (e.g., YOR066W and CDC46) [71] and collaborates with Ndd1 and Fkh1/Fkh2 to regulate genes in G2/M (e.g., SPO12 and KIN3) [36]. (B) Refining the Y′ cluster in Ref. [6]. The Y′ cluster contains genes that share DNA sequence similarity and are found in Y′ elements, which are located at chromosome ends. Spellman et al. [6] did not figure out how these genes are regulated. MOFA reassigns genes in the Y′ cluster to three modules and identifies several possible regulators (Dat1, Gal4, Gat3, Hap1, Hap4, Msn4, Pdr1, Rap1, and Yap5), providing information for future experiments.

4.2.4 Conclusions of MOFA on identifying TRMs in yeast cell cycle

In this section, we discussed a method, called MOdule Finding Algorithm (MOFA), for reconstructing TRMs of the yeast cell cycle by integrating gene expression data and ChIP-chip data. MOFA identified 87 TRMs, which together contain 336 distinct genes regulated by 40 TFs. From the literature [6,64–66], 139 of the 336 genes and 30 of the 40 TFs are known to be involved in the cell cycle. The biological relevance of each inferred TRM was validated by using existing experimental data, enrichment for genes in the same MIPS functional category [65], known DNA-binding motifs [6], etc. This analysis shows that different combinations of a fairly small number of TFs are responsible for regulating a large number of genes involved in different cell cycle phases and that there may exist crosstalk between the cell cycle and other cellular processes. Besides, MOFA is capable of finding many novel TF-target gene relationships that could not be identified by using the current ChIP-chip data with the stringent p-value cutoff (.001) or the conventional correlation analysis that only checks the coexpressed relationship. In addition, MOFA can determine the relationships between TFs that regulating the same module and the regulatory roles of these TFs. We found nine activators and six repressors, consistent with the results of previous studies [78–90]. Finally, MOFA refines some clusters proposed by previous studies and provides a better understanding of how the complex expression program of the cell cycle is regulated.

We believe that computational analysis of multiple types of data will be a powerful approach to studying complex biological systems when more and more genomic resources such as genome-wide protein activity data and protein–protein interaction data become available.

4.2.5 Methods of MOFA for finding TRMs in yeast cell cycle

Data sets

We use the ChIP-chip data in Ref. [59] and the gene expression data (α factor) of the yeast cell cycle in Ref. [6]. Spellman et al. [6] used Fourier transform to identify 800 putative cell cycle genes (113 genes in M/G1, 300 in G1, 71 in S, 121 in S/G2, and 195 in G2/M). By integrating both types of data, this algorithm tries to reconstruct TRMs for each of the five cell cycle phases.

Identifying temporal relationships of TF-gene pairs

A cell cycle TF and its binding target are said to have a positively (negatively) temporal relationship if the target gene's expression profile is positively (negatively) correlated with the TF's regulatory profile possibly with time lags. It is known that TF binding affects gene expression in a nonlinear fashion: below some level it has no effect, and above some level the effect may saturate. This type of behavior can be modeled using a sigmoid function.

Therefore, we define the regulatory profile of a TF as a sigmoid function like previous studies [93–95].

The TRIA is developed to identify TF-gene pairs that have a temporal relationship. Let $\vec{x}=(x_1,\ldots,x_N)$ be the gene expression time profile of cell cycle TF x and $\vec{y}=(y_1,\ldots,y_N)$ be the expression profile of gene y. The regulatory profile $RP(\vec{x})=(f(x_1),\ldots,f(x_N))$ of TF x is defined as a sigmoid function, which is justified by some previous studies [93–95]

$$f(x_i)=\frac{1}{1+e^{-(x_i-\bar{x})/s)}}\quad \text{for } i=1,\ldots,N \tag{4.14}$$

where $\bar{x}$ is the sample mean and s is the sample standard deviation of $\vec{x}$. Compute the correlation between $\vec{y}$ and $RP(\vec{x})$ with a lag of k time points [96,97]:

$$r(k)=\left(\sum_{i=1}^{N-k}(y_{i+k}-\bar{y})(f(x_i)-\bar{m})\right)\Bigg/\left(\sqrt{\sum_{i=1}^{N-k}(y_{i+k}-\bar{y})^2}\sqrt{\sum_{i=1}^{N-k}(f(x_i)-\bar{m})^2}\right), \tag{4.15}$$

for $k=0,\ldots,L$

where $\bar{y}\triangleq\left(\sum_{i=1}^{N-k}y_{i+k}\right)\Big/(N-k)$, $\bar{m}\triangleq\left(\sum_{i=1}^{N-k}f(x_i)\right)\Big/(N-k)$ and L is the maximal time lag of the TF's regulatory profile considered. The value of L is chosen to make the maximal time lag approximately equal to two consecutive cell cycle phases because Simon et al. [19] found cases where a cell cycle TF that expresses in one phase of the cell cycle can regulate genes that function in the next phase.

Then we test the null hypothesis H_0: $r(k)=0$ and the alternative hypothesis H_1: $r(k)\neq 0$ by the bootstrap method (see Appendix B) and get a p-value $p(k)$. The time-lagged correlation (TlC) of $\vec{y}$ and $RP(\vec{x})$ is defined as $r(j)$ that has the smallest p-value (i.e., $TlC(\vec{y},RP(\vec{x}))=r(j)$ if $p(j)\leq p(k)\quad \forall k\neq j$). Note that $-1\leq TlC(\vec{y},RP(\vec{x}))\leq 1$. Two possible temporal relationships between $\vec{y}$ and $RP(\vec{x})$ can be identified by TRIA: $\vec{y}$ and $RP(\vec{x})$ are (1) positively correlated with a lag of j time points if $TlC(\vec{y},RP(\vec{x}))=r(j)>0$ and $p(j)\leq p_{\text{Threshold}}$ and (2) negatively correlated with a lag of j time points if $TlC(\vec{y},RP(\vec{x}))=r(j)<0$ and $p(j)\leq p_{\text{Threshold}}$. The $p_{\text{Threshold}}$ is chosen to ensure that we have at most a 5% false discovery rate [61].

Two observations motivated us to develop TRIA to detect the temporal relationship between a cell cycle TF and its regulatory targets. First, it has been shown that, at least in a few instances, the expression levels of TFs and their target genes were correlated [49,97–102]. Although this may not be true for TFs which are mainly regulated at the posttranscriptional level [14,103], it is not a serious problem for many cell cycle TFs whose expression levels significantly varies with times indicating that they are also under highly transcriptional control [77,93–95,100–102,104,105]. Second, the expression relationship between a TF and its regulatory targets may not be simultaneous but after some time lags [77,91,93,97,100–102,104,105].

The TRIA in the previous section was used to find regulatory targets of cell cycle TFs and its effectiveness was validated by statistically testing for the expression coherence, enrichment of functional groups and conserved binding motifs [106]. We found that when only cell cycle TFs are concerned, TRIA performed better than some pervious algorithms [107,108]. This may result from the fact that the previous algorithms are designed for all kinds of TFs but TRIA is specially designed for cell cycle TFs.

The Module finding algorithm (MOFA)

Before describing MOFA, we define some terms.

Definition 4.1: Let $E = [e_{ij}]$ be the gene expression matrix whose rows correspond to genes and columns correspond to time points, so that e_{ij} is the expression level of gene i at time point j.

Definition 4.2: Let $B = [b_{ij}]$ be the binding score matrix whose rows correspond to genes and columns correspond to TFs, so that b_{ij} denotes the binding score of TF j to bind gene i. We set $b_{ij} = 4$ if the p-value for TF j to bind gene i is $\leq .001$ in the ChIP-chip data and TF j and gene i are found to have a temporal relationship; $b_{ij} = 3$ if $p \leq .001$ but no temporal relationship; $b_{ij} = 2$ if $.001 < p \leq .01$ and a temporal relationship; $b_{ij} = 1$ if$.001 < p \leq .01$ but no temporal relationship; and $b_{ij} = 0$ if $p > .01$.

Definition 4.3: Let R be a set of TFs and $C(R,z)$ be the set of target genes to which all the TFs in R bind with a score $\geq z$. In addition, let SP be the set of all genes in a specific cell cycle phase (113 genes in M/G1, 300 in G1, 71 in S, 121 in S/G2, and 195 in G2/M).

Definition 4.4: The expression coherence score ($EC(A)$) for a set A is defined as the fraction of gene-gene pairs in A whose gene expression profiles are positively correlated possibly with time lags: $0 \leq EC(A) \leq 1$. Note that the higher the $EC(A)$ is, the more plausible the genes in A are coregulated.

Remark 4.1: The $EC(A)$ is a generalization of the expression correlation score used in [16,63]. Compared to theirs, this measure can in addition find coregulated genes whose gene expression profiles are positively correlated with time lags. As shown in Refs. [91,92], coregulated genes are not necessarily coexpressed. Since each gene may have a different response time to the same transcriptional regulatory mechanism in transcribing DNA to RNA, the RNA profiles of coregulated genes may not be conumber of members. Second, if $EC(C(\hat{R}, z)) \leq EC(SP)$, MOFA iteratively eliminates genes of the set $C(\hat{R}, z)$ starting from the one with the most dissimilar expression profile until $EC(\tilde{C}(\hat{R}, z)) > EC(SP)$, where $\tilde{C}(\hat{R}, z)$ is the set of the remaining genes and SP is the set of all genes in a specific cell cycle phase. That is, MOFA tries to identify a subset of coregulated genes whose gene expression profiles are highly positively correlated possibly with time lags compared to that of the set of all genes in the specific cell cycle phase. Finally, MOFA outputs a module $M(\hat{R}) \triangleq \tilde{C}(\hat{R}, z)$ if $\tilde{C}(\hat{R}, z)$ contains more than a certain number of genes, say five. The above procedure goes over all $\hat{R}'s$ in the specific cell cycle phase. We provide the pseudocode of MOFA in Fig. 4.11.

MOFA (E, B, z, min_size)
{
  For SP = M/G1, G1, S, S/G2 or G2/M
  {
    Search all possible R's in SP
    Record a particular R and the corresponding $C(R,z)$ if $\#(C(R,z)) \geq$ min_size
    Sort recorded R's according to their sizes and denote the sorting result as $\hat{R}'s$
    For $i = 1, 2, \cdots$
    {
      If $EC(C_i \triangleq C(\hat{R}_i, z)) \leq EC(SP)$, iteratively eliminate genes in C_i starting from the
        one with the most dissimilar expression profile, say h, until
        $EC(\tilde{C}_i \triangleq C_i \setminus \{h\}) > EC(SP)$
      Output the TF combination $\hat{R}_i$ and module $M(\hat{R}_i) \triangleq \tilde{C}_i$ if $\#(M(\hat{R}_i)) \geq$ min_size
      Mark all genes in $M(\hat{R}_i)$ so that they are not considered for $\hat{R}_j \subset \hat{R}_i$
    }
  }
}

Figure 4.11 The pseudocode of MOFA. In this section, we set the binding score $z = 2$. This means that a TF is regarded as binding to a gene if: (1) the p-value for the TF to bind the gene is $\leq .001$ in the ChIP-chip data, or (2) $.001 < p \leq .01$ and the TF-gene pair have a temporal relationship. Moreover, we require that the number of genes in a module must be ≥ 5. This value is the same as that in GRAM algorithm [60] for comparison purpose.

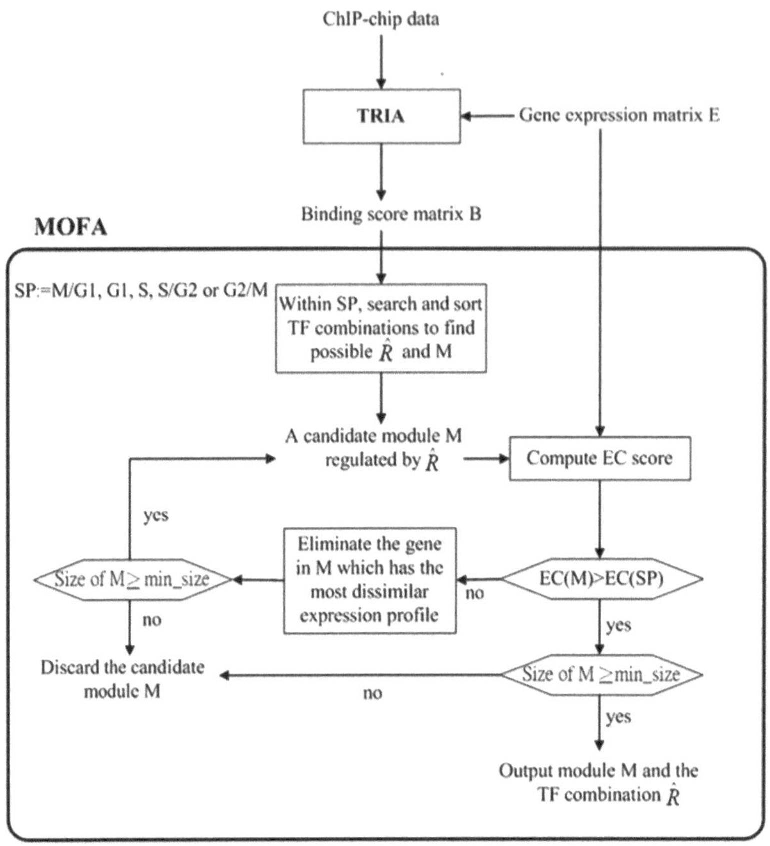

Figure 4.12 Flowchart of MOFA.

MOFA performs in two steps (see Fig. 4.12). First, for a specific cell cycle phase (M/G1, G1, S, S/G2, or G2/M), it exhaustively searches all possible $R's$ in order to find $C(R, z)'s$. A particular R and the corresponding $C(R,z)$ are recorded if $C(R,z)$ contains more than a certain number of genes.

APPENDIX B: METHODS AND FIGURES

The bootstrap method for testing the statistical significance of the difference between *r*(*k*) and 0

We observed $N-k$ pairs of observations, $Z = \{z_i: i = 1,\ldots,N-k \text{ and } \boldsymbol{z}_i = (f(x_i), y_{i+k})\}$. The correlation coefficient from the sample is calculated and denoted as

$$r(k) = \left(\sum_{i=1}^{N-k}(y_{i+k} - \overline{y})(f(x_i) - \overline{m})\right) \Bigg/ \left(\sqrt{\sum_{i=1}^{N-k}(y_{i+k} - \overline{y})^2}\sqrt{\sum_{i=1}^{N-k}(f(x_i) - \overline{m})^2}\right)$$

for $k = 0, 1, \ldots$

where $\overline{y} \triangleq \left(\sum_{i=1}^{N-k} y_{i+k}\right)/(N-k)$, $\overline{m} \triangleq \left(\sum_{i=1}^{N-k} f(x_i)\right)/(N-k)$ and $-1 \leq r(k) \leq 1$. It is aimed to use these observations to test if $r(k)$ is different from 0 significantly. Suppose the null hypothesis is H_0: $r(k) = 0$ and the alternative hypothesis is H_1: $r(k) \neq 0$. We will apply the bootstrap method to perform this hypothesis testing based on the observations. Keeping the pair relationship of these $N-k$ pairs to maintain the dependence between $(f(x_i), y_i + k)$, $\mathbf{z}_i$ are sampled with replacement $N-k$ times to form a bootstrap sample, $Z^* = \{\mathbf{z}_i^*: i = 1, \ldots, N-k$ and $\mathbf{z}_i^*$ belongs to $Z\}$. The correlation coefficient from the bootstrap sample Z^* is computed and denoted as $r^*(k)$, $-1 \leq r^*(k) \leq 1$. Repeat the resampling procedure B times, we will observed $r_1^*(k), \ldots, r_B^*(k)$. These bootstrap correlation coefficients are sorted to be $-1 \leq r_1^*(k) \leq \ldots \leq r_B^*(k) \leq 1$. Then, the $(1-\alpha)$ two-sided percentile interval is given by $[r^*_{(B \times \alpha/2)}(k), r^*_{(B \times (1-\alpha/2))}(k)]$ in this case [22]. If this percentile interval does not contain 0, then the null hypothesis is rejected at the significance level of α. Otherwise, the data fail to reject the null hypothesis at the significance level of α. Since the p-value is the smallest value of α for which the null hypothesis will be rejected based on the observation, the p-value for this test is estimated by the following:

$\widehat{p}(k) = 2min\{\widehat{p}_+(k), 1 - \widehat{p}_+(k)\}$, where $\widehat{p}_+(k) = \sum_{i=1}^{B} I\{r_i^*(k) \geq 0\}/B$, and $I\{\cdot\}$ is the indicator function whose value is one when the event is true and zero otherwise (see also Fig. B.1).

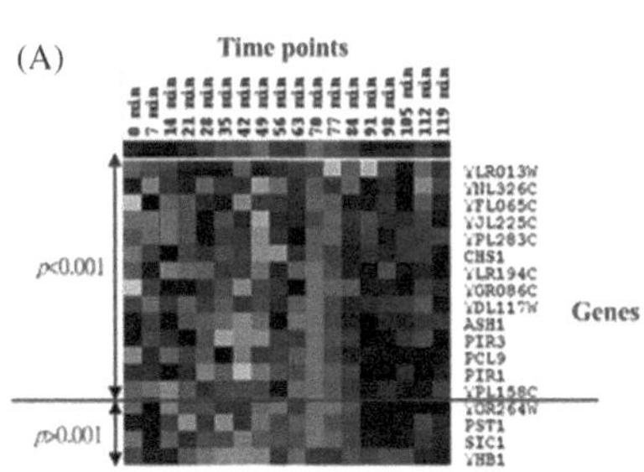

Figure B.1 (Continued)

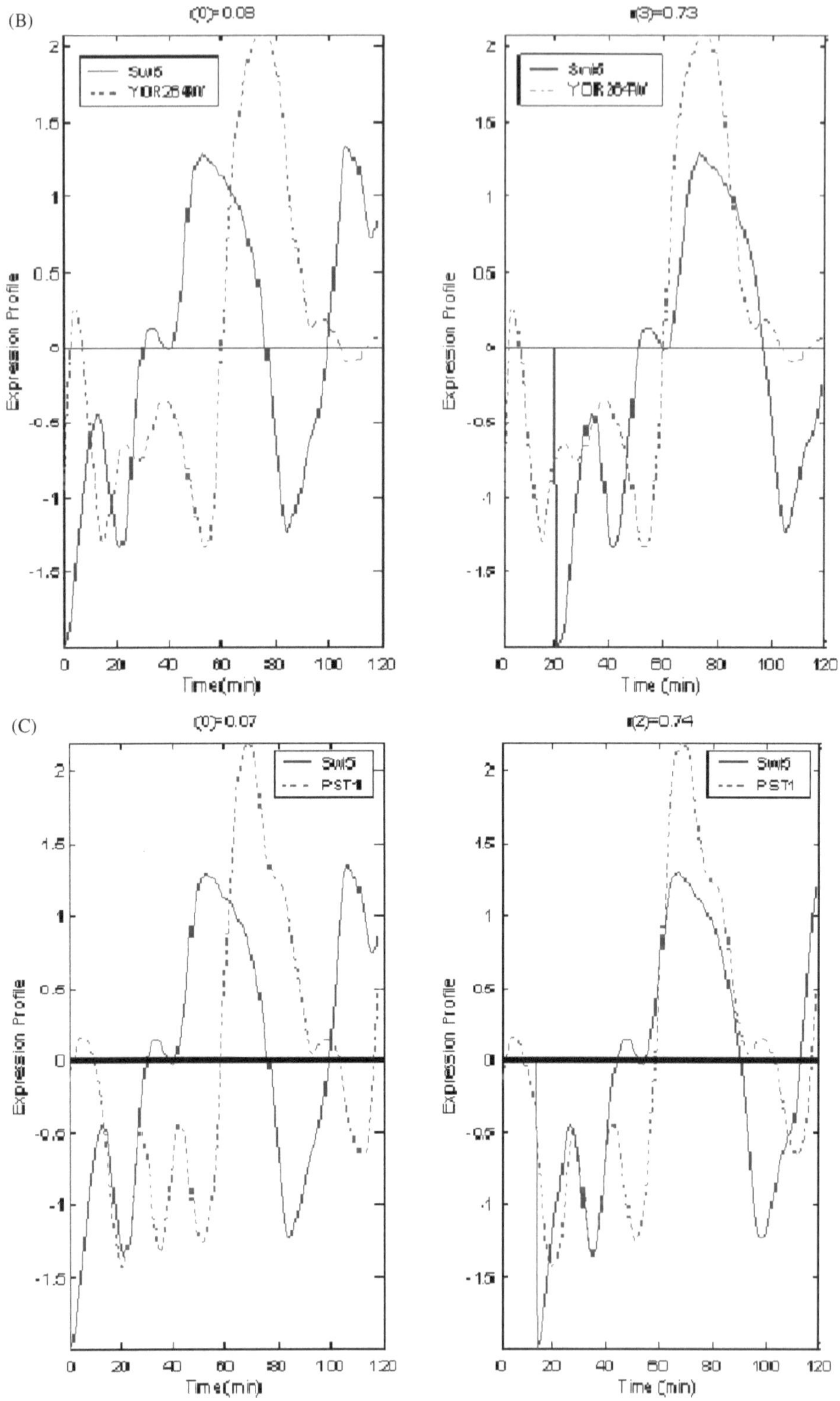

Figure B.1 (Continued)

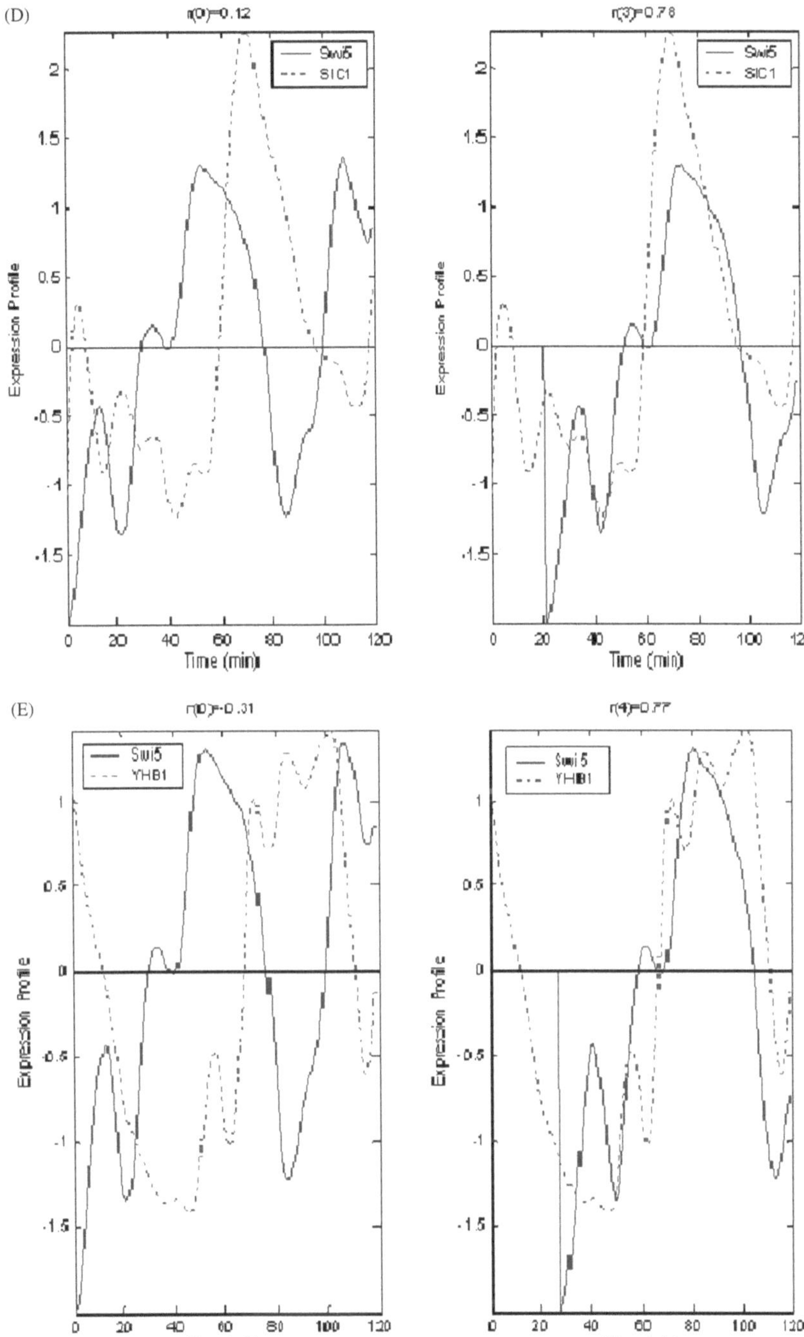

Figure B.1 MOFA can reduce false negatives in determining binding events in the ChIP-chip data. (A) Gene expression patterns of the 18 target genes of the {Swi5} module in M/G1. (B) TF-gene pair Swi5-*YOR264W* with binding *p*-value = .0053>.001. (C) TF-gene pair Swi5-*PST1* with binding *p*-value = .0085>.001. (D) TF-gene pair Swi5-*SIC1* with binding *p*-value = .0017>.001. (E) TF-gene pair Swi5-*YHB1* with binding *p*-value = .007>.001. These four TF-gene pairs all have positively time-shifted correlated relationships. Note that these four TF-gene pairs identified by MOFA could not be identified by using the current ChIP-chip data with the stringent *p*-value cutoff (.001) or the conventional correlation analysis that can only check coexpressed relationship.

REFERENCES

[1] Garvie CW, Wolberger C. Recognition of specific DNA sequences. Mol Cell 2001;8(5):937–46.

[2] Orphanides G, Reinberg D. A unified theory of gene expression. Cell 2002;108(4):439–51.

[3] Fickett JW, Wasserman WW. Discovery and modeling of transcriptional regulatory regions. Curr Opin Biotechnol 2000;11(1):19–24.

[4] Halfon MS, Carmena A, Gisselbrecht S, Sackerson CM, Jimenez F, Baylies MK, et al. Ras pathway specificity is determined by the integration of multiple signal-activated and tissue-restricted transcription factors. Cell 2000;103(1):63–74.

[5] Cho RJ, Campbell MJ, Winzeler EA, Steinmetz L, Conway A, Wodicka L, et al. A genome-wide transcriptional analysis of the mitotic cell cycle. Mol Cell 1998;2(1):65–73.

[6] Spellman PT, Sherlock G, Zhang MQ, Iyer VR, Anders K, Eisen MB, et al. Comprehensive identification of cell cycle-regulated genes of the yeast *Saccharomyces cerevisiae* by microarray hybridization. Mol Biol Cell 1998;9(12):3273–97.

[7] Wang YL, Liu CL, Storey JD, Tibshirani RJ, Herschlag D, Brown PO. Precision and functional specificity in mRNA decay. Proc Natl Acad Sci USA 2002;99(9):5860–5.

[8] Eisen MB, Spellman PT, Brown PO, Botstein D. Cluster analysis and display of genome-wide expression patterns. Proc Natl Acad Sci USA 1998;95(25):14863–8.

[9] Tamayo P, Slonim D, Mesirov J, Zhu Q, Kitareewan S, Dmitrovsky E, et al. Interpreting patterns of gene expression with self-organizing maps: methods and application to hematopoietic differentiation. Proc Natl Acad Sci USA 1999;96(6):2907–12.

[10] Tavazoie S, Hughes JD, Campbell MJ, Cho RJ, Church GM. Systematic determination of genetic network architecture. Nat Genet 1999;22(3):281–5.

[11] Alter O, Brown PO, Botstein D. Singular value decomposition for genome-wide expression data processing and modeling. Proc Natl Acad Sci USA 2000;97(18):10101–6.

[12] Holter NS, Maritan A, Cieplak M, Fedoroff NV, Banavar JR. Dynamic modeling of gene expression data. Proc Natl Acad Sci USA 2001;98(4):1693–8.

[13] Yeung MKS, Tegner J, Collins JJ. Reverse engineering gene networks using singular value decomposition and robust regression. Proc Natl Acad Sci USA 2002;99(9):6163–8.

[14] Bussemaker HJ, Li H, Siggia ED. Regulatory element detection using correlation with expression. Nat Genet 2001;27(2):167–71.

[15] GuhaThakurta D, Stormo GD. Identifying target sites for cooperatively binding factors. Bioinformatics 2001;17(7):608–21.

[16] Pilpel Y, Sudarsanam P, Church GM. Identifying regulatory networks by combinatorial analysis of promoter elements. Nat Genet 2001;29(2):153–9.

[17] Iyer VR, Horak CE, Scafe CS, Botstein D, Snyder M, Brown PO. Genomic binding sites of the yeast cell-cycle transcription factors SBF and MBF. Nature 2001;409(6819):533–8.

[18] Lee TI, Rinaldi NJ, Robert F, Odom DT, Bar-Joseph Z, Gerber GK, et al. Transcriptional regulatory networks in *Saccharomyces cerevisiae*. Science 2002;298(5594):799–804.

[19] Simon I, Barnett J, Hannett N, Harbison CT, Rinaldi NJ, Volkert TL, et al. Serial regulation of transcriptional regulators in the yeast cell cycle. Cell 2001;106(6):697–708.

[20] Singer RH, Penman S. Messenger RNA in HeLa cells: kinetics of formation and decay. J Mol Biol 1973;78(2):321–34.

[21] Sasik R, Iranfar N, Hwa T, Loomis WF. Extracting transcriptional events from temporal gene expression patterns during Dictyostelium development. Bioinformatics 2002;18(1):61–6.

[22] Efron B, Tibshirani R. An introduction to the bootstrap. New York, NY: Chapman & Hall; 1993.

[23] Goldbeter A, Koshland Jr DE. An amplified sensitivity arising from covalent modification in biological systems. Proc Natl Acad Sci USA 1981;78(11):6840–4.

[24] Mestl T, Plahte E, Omholt SW. A mathematical framework for describing and analyzing gene regulatory networks. J Theor Biol 1995;176(2):291–300.

[25] Banerjee N, Zhang MX. Functional genomics as applied to mapping transcription regulatory networks. Curr Opin Microbiol 2002;5(3):313–17.

[26] Futcher B. Transcriptional regulatory networks and the yeast cell cycle. Curr Opin Cell Biol 2002;14(6):676–83.

[27] Wyrick JJ, Young RA. Deciphering gene expression regulatory networks. Curr Opin Genet Dev 2002;12(2):130–6.
[28] Wijnen H, Landman A, Futcher B. The G(1) cyclin Cln3 promotes cell cycle entry via the transcription factor Swi6. Mol Cell Biol 2002;22(12):4402–18.
[29] McInerny CJ, Partridge JF, Mikesell GE, Creemer DP, Breeden LL. A novel Mcm1-dependent element in the SWI4, CLN3, CDC6, and CDC47 promoters activates M/G(1)-specific transcription. Genes Dev 1997;11(10):1277–88.
[30] Stuart D, Wittenberg C. CLN3, not positive feedback, determines the timing of CLN2 transcription in cycling cells. Genes Dev 1995;9(22):2780–94.
[31] Tyers M, Tokiwa G, Futcher B. Comparison of the *Saccharomyces cerevisiae* G1 Cyclins—CLN3 may be an upstream activator of CLN1, CLN2 and other cyclins. EMBO J 1993;12(5):1955–68.
[32] Buonomo SB, Rabitsch KP, Fuchs J, Gruber S, Sullivan M, Uhlmann F, et al. Division of the nucleolus and its release of CDC14 during anaphase of meiosis I depends on separase, SPO12, and SLK19. Dev Cell 2003;4(5):727–39.
[33] Marston AL, Lee BH, Amon A. The Cdc14 phosphatase and the FEAR network control meiotic spindle disassembly and chromosome segregation. Dev Cell 2003;4(5):711–26.
[34] Stegmeier F, Visintin R, Amon A. Separase, polo kinase, the kinetochore protein Slk19 and Spo12 function in a network that controls Cdc14 localization during early anaphase. Cell 2002;108 (2):207–20.
[35] Booher RN, Deshaies RJ, Kirschner MW. Properties of *Saccharomyces cerevisiae* wee1 and its differential regulation of p34CDC28 in response to G1 and G2 cyclins. EMBO J 1993;12(9):3417–26.
[36] Koranda M, Schleiffer A, Endler L, Ammerer G. Forkhead-like transcription factors recruit Ndd1 to the chromatin of G2/M-specific promoters. Nature 2000;406(6791):94–8.
[37] Zhu GF, Spellman PT, Volpe T, Brown PO, Botstein D, Davis TN, et al. Two yeast forkhead genes regulate the cell cycle and pseudohyphal growth. Nature 2000;406(6791):90–4.
[38] Schwob E, Bohm T, Mendenhall MD, Nasmyth K. The B-type cyclin kinase inhibitor p40SIC1 controls the G1 to S transition in *S. cerevisiae*. Cell 1994;79(2):233–44.
[39] Zachariae W, Nasmyth K. Whose end is destruction: cell division and the anaphase-promoting complex. Genes Dev 1999;13(16):2039–58.
[40] Verma R, Annan RS, Huddleston MJ, Carr SA, Reynard G, Deshaies RJ. Phosphorylation of Sic1p by G1 Cdk required for its degradation and entry into S phase. Science 1997;278 (5337):455–60.
[41] Wang Z, Wilson WA, Fujino MA, Roach PJ. The yeast cyclins Pc16p and Pc17p are involved in the control of glycogen storage by the cyclin-dependent protein kinase Pho85p. FEBS Lett 2001;506(3):277–80.
[42] Horak CE, Luscombe NM, Qian JA, Bertone P, Piccirrillo S, Gerstein M, et al. Complex transcriptional circuitry at the G1/S transition in *Saccharomyces cerevisiae*. Genes Dev 2002;16(23): 3017–33.
[43] Kunoh T, Kaneko Y, Harashima S. YHP1 encodes a new homeoprotein that binds to the IME1 promoter in *Saccharomyces cerevisiae*. Yeast 2000;16(5):439–49.
[44] Mouassite M, Guerin MG, Camougrand NM. The SUN family of *Saccharomyces cerevisiae*: the double knock-out of UTH1 and SIM1 promotes defects in nucleus migration and increased drug sensitivity. FEMS Microbiol Lett 2000;182(1):137–41.
[45] Woolf PJ, Wang YX. A fuzzy logic approach to analyzing gene expression data. Physiol Genomics 2000;3(1):9–15.
[46] Xu HQ, Wu PR, Wu CFJ, Tidwell C, Wang YX. A smooth response surface algorithm for constructing a gene regulatory network. Physiol Genomics 2002;11(1):11–20.
[47] Chen KC, Csikasz-Nagy A, Gyorffy B, Val J, Novak B, Tyson JJ. Kinetic analysis of a molecular model of the budding yeast cell cycle. Mol Biol Cell 2000;11(1):369–91.
[48] Proakis JG. Digital communications. 4th ed. Boston, MA: McGraw-Hill; 2001.
[49] DeRisi JL, Iyer VR, Brown PO. Exploring the metabolic and genetic control of gene expression on a genomic scale. Science 1997;278(5338):680–6.
[50] Schena M, Shalon D, Davis RW, Brown PO. Quantitative monitoring of gene-expression patterns with a complementary-DNA microarray. Science 1995;270(5235):467–70.

[51] Ren B, Robert F, Wyrick JJ, Aparicio O, Jennings EG, Simon I, et al. Genome-wide location and function of DNA binding proteins. Science 2000;290(5500):2306–9.
[52] Causton HC, Ren B, Koh SS, Harbison CT, Kanin E, Jennings EG, et al. Remodeling of yeast genome expression in response to environmental changes. Mol Biol Cell 2001;12(2):323–37.
[53] Chu S, DeRisi J, Eisen M, Mulholland J, Botstein D, Brown PO, et al. The transcriptional program of sporulation in budding yeast. Science 1998;282(5389):699–705.
[54] Gasch AP, Spellman PT, Kao CM, Carmel-Harel O, Eisen MB, Storz G, et al. Genomic expression programs in the response of yeast cells to environmental changes. Mol Biol Cell 2000;11 (12):4241–57.
[55] Ihmels J, Friedlander G, Bergmann S, Sarig O, Ziv Y, Barkai N. Revealing modular organization in the yeast transcriptional network. Nat Genet 2002;31(4):370–7.
[56] Liang S, Fuhrman S, Somogyi R. Reveal, a general reverse engineering algorithm for inference of genetic network architectures. Pac Symp Biocomput 1998;18–29.
[57] Friedman N, Linial M, Nachman I, Pe'er D. Using Bayesian networks to analyze expression data. J Comput Biol 2000;7(3–4):601–20.
[58] Segal E, Shapira M, Regev A, Pe'er D, Botstein D, Koller D, et al. Module networks: identifying regulatory modules and their condition-specific regulators from gene expression data. Nat Genet 2003;34(2):166–76.
[59] Harbison CT, Gordon DB, Lee TI, Rinaldi NJ, Macisaac KD, Danford TW, et al. Transcriptional regulatory code of a eukaryotic genome. Nature 2004;431(7004):99–104.
[60] Bar-Joseph Z, Gerber GK, Lee TI, Rinaldi NJ, Yoo JY, Robert F, et al. Computational discovery of gene modules and regulatory networks. Nat Biotechnol 2003;21(11):1337–42.
[61] Benjamini Y, Hochberg Y. Controlling the false discovery rate—a practical and powerful approach to multiple testing. J R Stat Soc B-Methodol 1995;57(1):289–300.
[62] Kato M, Hata N, Banerjee N, Futcher B, Zhang MQ. Identifying combinatorial regulation of transcription factors and binding motifs. Genome Biol 2004;5(8).
[63] Banerjee N, Zhang MQ. Identifying cooperativity among transcription factors controlling the cell cycle in yeast. Nucleic Acids Res 2003;31(23):7024–31.
[64] Tsai HK, Lu HHS, Li WH. Statistical methods for identifying yeast cell cycle transcription factors. Proc Natl Acad Sci USA 2005;102(38):13532–7.
[65] Mewes HW, Frishman D, Guldener U, Mannhaupt G, Mayer K, Mokrejs M, et al. MIPS: a database for genomes and protein sequences. Nucleic Acids Res 2002;30(1):31–4.
[66] Yang YL, Suen J, Brynildsen MP, Galbraith SJ, Liao JC. Inferring yeast cell cycle regulators and interactions using transcription factor activities. BMC Genomics 2005;6:90.
[67] Olson KA, Nelson C, Tai G, Hung W, Yong C, Astell C, et al. Two regulators of Ste12p inhibit pheromone-responsive transcription by separate mechanisms. Mol Cell Biol 2000;20(12): 4199–209.
[68] Fung H, Bennett RAO, Demple B. Key role of a downstream specificity protein 1 site in cell cycle-regulated transcription of the AP endonuclease gene APE1/APEX in NIH3T3 cells. J Biol Chem 2001;276(45):42011–17.
[69] McBride HJ, Yu YX, Stillman DJ. Distinct regions of the Swi5 and Ace2 transcription factors are required for specific gene activation. J Biol Chem 1999;274(30):21029–36.
[70] Aerne BL, Johnson AL, Toyn JH, Johnston LH. Swi5 controls a novel wave of cyclin synthesis in late mitosis. Mol Biol Cell 1998;9(4):945–56.
[71] Pramila T, Miles S, GuhaThakurta D, Jemiolo D, Breeden LL. Conserved homeodomain proteins interact with MADS box protein Mcm1 to restrict ECB-dependent transcription to the M/G1 phase of the cell cycle. Genes Dev 2002;16(23):3034–45.
[72] Koch C, Nasmyth K. Cell-cycle-regulated transcription in yeast. Curr Opin Cell Biol 1994;6 (3):451–9.
[73] Ho W, Costanzo M, Moore L, Kobayashi R, Andrews BJ. Regulation of transcription at the *Saccharomyces cerevisiae* start transition by Stb1, a Swi6-binding protein. Mol Cell Biol 1999;19 (8):5267–78.
[74] Dimova D, Nackerdien Z, Furgeson S, Eguchi S, Osley MA. A role for transcriptional repressors in targeting the yeast Swi/Snf complex. Mol Cell 1999;4(1):75–83.

[75] Kumar R, Reynolds DM, Shevchenko A, Shevchenko A, Goldstone SD, Dalton S. Forkhead transcription factors, Fkh1p and Fkh2p, collaborate with Mcm1p to control transcription required for M-phase. Curr Biol 2000;10(15):896–906.

[76] Barral Y, Parra M, Bidlingmaier S, Snyder M. Nim1-related kinases coordinate cell cycle progression with the organization of the peripheral cytoskeleton in yeast. Genes Dev 1999;13(2):176–87.

[77] Yu HY, Luscombe NM, Qian J, Gerstein M. Genomic analysis of gene expression relationships in transcriptional regulatory networks. Trends Genet 2003;19(8):422–7.

[78] Carmen AA, Holland MJ. The upstream repression sequence from the yeast enolase gene ENO1 is a complex regulatory element that binds multiple trans-acting factors including REB1. J Biol Chem 1994;269(13):9790–7.

[79] Costanzo M, Schub O, Andrews B. G(1) transcription factors are differentially regulated in *Saccharomyces cerevisiae* by the Swi6-binding protein Stb1. Mol Cell Biol 2003;23(14):5064–77.

[80] Darieva Z, Pic-Taylor A, Boros J, Spanos A, Geymonat M, Reece RJ, et al. Cell cycle-regulated transcription through the FHA domain of Fkh2p and the coactivator Ndd1p. Curr Biol 2003;13 (19):1740–5.

[81] Doolin MT, Johnson AL, Johnston LH, Butler G. Overlapping and distinct roles of the duplicated yeast transcription factors Ace2p and Swi5p. Mol Microbiol 2001;40(2):422–32.

[82] Hollenhorst PC, Bose ME, Mielke MR, Muller U, Fox CA. Forkhead genes in transcriptional silencing, cell morphology and the cell cycle: overlapping and distinct functions for FKH1 and FKH2 in *Saccharomyces cerevisiae*. Genetics 2000;154(4):1533–48.

[83] Loo S, Laurenson P, Foss M, Dillin A, Rine J. Roles of ABF1, NPL3, and YCL54 in silencing in *Saccharomyces cerevisiae*. Genetics 1995;141(3):889–902.

[84] Loy CJ, Lydall D, Surana U. NDD1, a high-dosage suppressor of cdc28-1N, is essential for expression of a subset of late-S-phase-specific genes in *Saccharomyces cerevisiae*. Mol Cell Biol 1999;19 (5):3312–27.

[85] Miyake T, Reese J, Loch CM, Auble DT, Li R. Genome-wide analysis of ARS (autonomously replicating sequence) binding factor 1 (Abf1p)-mediated transcriptional regulation in *Saccharomyces cerevisiae*. J Biol Chem 2004;279(33):34865–72.

[86] Morrow BE, Johnson SP, Warner JR. Proteins that bind to the yeast rDNA enhancer. J Biol Chem 1989;264(15):9061–8.

[87] Packham EA, Graham IR, Chambers A. The multifunctional transcription factors Abf1p, Rap1p and Reb1p are required for full transcriptional activation of the chromosomal PGK gene in *Saccharomyces cerevisiae*. Mol Gen Genet 1996;250(3):348–56.

[88] Partridge JF, Mikesell GE, Breeden LL. Cell cycle-dependent transcription of CLN1 involves Swi4 binding to MCB-like elements. J Biol Chem 1997;272(14):9071–7.

[89] Reardon BJ, Winters RS, Gordon D, Winter E. A peptide motif that recognizes a-center-dot-t tracts in DNA. Proc Natl Acad Sci USA 1993;90(23):11327–31.

[90] Wang KLC, Warner JR. Positive and negative autoregulation of REB1 transcription in *Saccharomyces cerevisiae*. Mol Cell Biol 1998;18(7):4368–76.

[91] Ji LP, Tan KL. Identifying time-lagged gene clusters using gene expression data. Bioinformatics 2005;21(4):509–16.

[92] Zhou XHJ, Kao MCJ, Huang HY, Wong A, Nunez-Iglesias J, Primig M, et al. Functional annotation and network reconstruction through cross-platform integration of microarray data. Nat Biotechnol 2005;23(2):238–43.

[93] Chang WC, Li CW, Chen BS. Quantitative inference of dynamic regulatory pathways via microarray data. BMC Bioinformatics 2005;6:44.

[94] Chen HC, Lee HC, Lin TY, Li WH, Chen BS. Quantitative characterization of the transcriptional regulatory network in the yeast cell cycle. Bioinformatics 2004;20(12):1914–27.

[95] Chen KC, Wang TY, Tseng HH, Huang CYF, Kao CY. A stochastic differential equation model for quantifying transcriptional regulatory network in *Saccharomyces cerevisiae*. Bioinformatics 2005;21 (12):2883–90.

[96] Arkin A, Shen PD, Ross J. A test case of correlation metric construction of a reaction pathway from measurements. Science 1997;277(5330):1275–9.

[97] Kato M, Tsunoda T, Takagi T. Lag analysis of genetic networks in the cell cycle of budding yeast. Genome Informatics 2001;12:266–7.
[98] Birnbaum K, Benfey PN, Shasha DE. cis element/transcription factor analysis (cis/TF): a method for discovering transcription factor/cis element relationships. Genome Res 2001;11(9):1567–73.
[99] Chu S, Herskowitz I. Gametogenesis in yeast is regulated by a transcriptional cascade dependent on Ndt80. Mol Cell 1998;1(5):685–96.
[100] Qian J, Dolled-Filhart M, Lin J, Yu HY, Gerstein M. Beyond synexpression relationships: local clustering of time-shifted and inverted gene expression profiles identifies new, biologically relevant interactions. J Mol Biol 2001;314(5):1053–66.
[101] Schmitt WA, Raab RM, Stephanopoulos G. Elucidation of gene interaction networks through time-lagged correlation analysis of transcriptional data. Genome Res 2004;14(8):1654–63.
[102] Zhu Z, Pilpel Y, Church GM. Computational identification of transcription factor binding sites via a transcription-factor-centric clustering (TFCC) algorithm. J Mol Biol 2002;318(1):71–81.
[103] Liao JC, Boscolo R, Yang YL, Tran LM, Sabatti C, Roychowdhury VP. Network component analysis: reconstruction of regulatory signals in biological systems. Proc Natl Acad Sci USA 2003;100(26):15522–7.
[104] Qian J, Lin J, Luscombe NM, Yu HY, Gerstein M. Prediction of regulatory networks: genome-wide identification of transcription factor targets from gene expression data. Bioinformatics 2003;19(15):1917–26.
[105] Reis BY, Butte AJ, Kohane IS. Approaching causality: discovering time-lag correlations in genetic expression data with static and dynamic relevance networks. In: Proc of RECOMB. Citeseer; 2000.
[106] Wu W-S, Li W-H, Chen B-S. Identifying regulatory targets of cell cycle transcription factors using gene expression and ChIP-chip data. BMC Bioinformatics 2007;8(1):188.
[107] Gao F, Foat BC, Bussemaker HJ. Defining transcriptional networks through integrative modeling of mRNA expression and transcription factor binding data. BMC Bioinformatics 2004;5:31.
[108] Yu TW, Li KC. Inference of transcriptional regulatory network by two-stage constrained space factor analysis. Bioinformatics 2005;21(21):4033–8.

CHAPTER 5

Big Regulatory Mechanisms in the Transcriptional Regulation Control of Gene Expression Using a Stochastic System Model and Genome-Wide Experimental Data

INTRODUCTION

After identifying the mechanisms for regulating the cell cycle basd on the gene regulatory networks (GRNs) and transcriptional regulatory modules in Chapter 4, Big Cellular Mechanisms in the Cell Cycle by System Identification and Big Data Mining, and also because the transcriptional control by transcription factors (TFs) of their target genes is one of the most important cellular mechanisms, this chapter will identify the regulatory trans/cis circuit (mechanism) of TFs and the cooperativity of these TFs to gain more insight into gene regulatory mechanisms via dynamic regulatory models and genome-wide microarray data.

TF binding sites are known to co-occur in the same gene owing to the cooperativity of the TFs that bind to them. Genome-wide location data can help us understand how an individual TF regulates its target genes. Nevertheless, how TFs cooperate to regulate their target genes still requires further study. In Section 5.1, genome-wide location data and expression profiles are integrated to reveal how TFs cooperate to regulate their target genes from a stochastic system model perspective. Based on the stochastic dynamic model, a new measure is introduced to indicate the potential cooperativity of TFs. Finally, a cooperative cell cycle TF network is constructed from significant cooperative TF pairs to investigate the big regulatory mechanisms in transcriptional control of gene expression.

In Section 5.2, dynamic regulatory mechanisms of cis-regulatory circuits for gene expression are investigated by cross-gene elements in the promoter region of target genes bound by TFs, but how TFs regulate gene expression via recognition of cis elements is still not clear. In this section, a novel cross-gene identification scheme is proposed to infer how multiple TFs coordinate to regulate gene transcription in

Big Mechanisms in Systems Biology
DOI: http://dx.doi.org/10.1016/B978-0-12-809479-2.00005-6

the yeast cell cycle and to uncover the regulatory function of a cis-regulatory circuit. In Section 5.3, we will introduce trans/cis-regulatory circuits for gene transcription using a nonlinear dynamic regulatory model and big database mining. The trans-regulatory circuit incorporates regulatory interactions between upstream regulatory genes and TF binding site motifs or cis elements, and the cis-regulatory circuit is viewed as a dynamic interactive circuit among binding site motifs with their effective action on target genes. In brief, gene transcription mechanism depends on trans/cis-regulatory circuits. In this section, nonlinear trans/cis-regulatory circuits for gene transcription regulation are constructed to gain insight into the transcriptional mechanism of yeast using microarray data, translation time delay, and information on TF binding sites. This method can be applied to not only yeast, but also to other species.

5.1 IDENTIFICATION OF TF COOPERATIVITY IN GENE REGULATION OF THE CELL CYCLE VIA THE STOCHASTIC SYSTEM MODEL

TF binding sites are known to co-occur with cooperativity to regulate genes. Genome-wide location data can help us understand how an individual TF regulates its target gene. Nevertheless, how TFs cooperate to regulate their target genes still needs further study. In this section, to explore the dynamic property of gene expression profiles, an approach using the stochastic system model and the maximum likelihood estimation method is proposed. Genome-wide location data and expression profiles are integrated to reveal how TFs cooperate to regulate their target genes from the stochastic system perspective. Based on a stochastic dynamic model, a new measurement of TF cooperativity is developed according to the regulatory abilities of cooperative TF pairs and the number of their occurrences. This method is employed to the yeast cell cycle and reveals successfully many cooperative TF pairs that had been confirmed by previous experiments. Other TF pairs with potential cooperativity mentioned in these results can provide new directions for future experiments. Finally, a cooperative TF network of cell cycle is constructed from significant cooperative TF pairs.

Precise transcriptional control is one of the major reasons for different gene expression and regulation. Owing to advances in DNA microarray technology and genome sequencing, measuring gene expression levels on a genomic scale has become possible. Measuring gene expression time profiles can help us understand mechanisms of transcriptional regulation, including functional significance of cell cycle regulation and response of environmental changes [1,2]. However, time profiles alone are not enough to identify precisely the whole transcriptional regulatory network. More precise methods also need the binding information of TFs to promoters in DNA. Genes are always regulated by a number of upstream regulatory genes through binding of TFs to specific sites in the DNA promoter region. To construct a gene regulatory network (i.e., trans-regulatory circuit), it is important to understand the binding relationship between target genes and TFs. In recent studies, the genome-wide location (ChIP-chip) analysis was employed to obtain the binding information of TFs to promoters in DNA [2–7]. However, these studies did not shed light on the interactions or cooperativity between TFs in a trans-regulatory circuit, which is very important to understand the Big Mechanisms of gene expression and regulation.

With advances in experimental approaches and abundant data sources, functional genomics has begun to investigate the more complex, cooperative TF interactions to regulate properly gene expression. In order to find cooperative TF interactions, a statistical technique was employed to identify significant homotypic or heterotypic TF binding site clusters [8]. Moreover, the correlation of genome-wide expression profiles was employed to uncover functional motif combinations in the promoters [9]. Furthermore, genome-wide location data (ChIP-chip) and gene expression profiles were integrated to rigorously assess TF cooperativity [10,11]. However, these studies only use expression correlation score in the view of statistics to determine TF cooperativity. They did not construct an overall transcriptional regulatory system to uncover TF cooperativity in the transcription regulation of gene expression from the dynamic system perspective. In this section, we assess TF cooperativity not only from a dynamic system perspective, but also in statistics. We could get more insight into the regulatory mechanism by constructing its dynamic system model and could also make more precise prediction of TF cooperativity.

A dynamic model is often employed to describe a mechanism of complex and kinetic systems in many fields. Systems biology and computational biology methods have recently been widely employed to describe the biological functions from the dynamic system point of view [12–16]. In this section, we exploit genome-wide location data [17] and gene expression profiles [2] to construct a dynamic regulatory model for the genes of interest. The dynamic regulatory model is developed to describe how the upstream regulatory genes control a target gene to produce the output mRNA expression through its transcriptional regulatory network (see Chapter 2: System Modeling and System Identification Methods for Big Mechanisms in Biological Systems). In order to reduce computational complexity, we estimate TF cooperativities one gene by one gene. In the dynamic regulatory model, there are two groups of input transcription regulations affecting expression profiles of the target gene. The first group is due to individual TF regulation, and the other is due to TF cooperativity regulation in the gene expression process. After estimating the regulatory abilities of these two groups, we can get all TF abilities for regulating the transcriptional process of the target gene. It is more possible to have TF cooperativity if binding sites for specific pairs of regulators co-occur more frequently within the same promoter regions of target genes [17]. Therefore, according to the estimated cooperative TF abilities and the number of TF pairs cobinding to the target genes in trans-regulatory circuit, a statistical measurement method is specified to assess TF cooperativity rigorously. The statistical measurement method is proposed to reduce the possible estimation error due to few data points. The new measurement of TF cooperativity is the multiplicative cooperative *p*-value, which can be obtained by multiplying all individual *p*-values of different target genes to estimate efficiently the cooperativity of any cooperative TF pairs on a genomic scale. Further, we use the shuffling method to avoid overfitting to confirm the reliability of the proposed method.

These results show that many cooperative TF pairs that were previously characterized for the Big Mechanisms of gene regulation and expression through experiments indeed have a high cooperativity in this analysis, thus validating the method we proposed as a TF cooperativity predictor and providing valuable information for further analysis. A cooperative network is also constructed with significant cooperative TF pairs for the yeast cell cycle. Finally, the results reveal several novel possible cooperative TF pairs not found in previous studies, thus providing new directions for future experiments.

5.1.1 Materials and measuring methods of TF cooperativity in trans-regulatory circuits

Selecting and processing experimental data

We used the genome-wide yeast microarray hybridization data of Spellman et al. (1998) as these mRNA expression profiles. They proposed many experimental methods for resetting the yeast cell cycle to measure mRNA expression profiles for the whole genome comprehensively. Here, we used the 768 cell cycle–related genes selected by Simon et al. (2001) from an "α factor" experimental cell cycle data as the target genes. The "α factor" is an experimental technique to synchronize cell cycle. The genome-wide location data are taken from Harbison et al. (2004), in which the genomic occupancy of 203 DNA-binding TFs is determined in yeast. We selected the significant binding set using binding *p*-value $p < .0015$ as the inputs in the dynamic gene transcriptional regulatory model.

In genome-wide expression profiles of Spellman et al. (1998), there are many missing data. We used the cubic spline interpolation method [18], which employs a fourth-order polynomial to fit data points, to compensate for these missing values in the profiles. Furthermore, in order to avoid the overfitting estimation of the profiles, we also use the cubic spline method to interpolate the data points when there are more than six TFs binding to the target gene, which implies that the number of parameters to be estimated is 22. Only 5% of the target genes are bound by more than six TFs (see Appendix A). In order to fit the gene dynamic model in the linear scale, the microarray data are returned from the $\log_2$ scale to the linear scale.

Dynamic model of trans-regulatory circuits

First, we consider a trans-regulatory circuit as a system block in Fig. 5.1 with several regulatory genes as inputs and a target gene as output. Owing to random noise and fluctuation at the molecular level, the transcriptional behavior of the gene regulatory network is described by a stochastic discrete dynamic equation, and the general form of transcriptional regulation for a target gene is written as follows:

$$y(t+1) = a\,y(t) + \sum_{i=1}^{N} b_i x_i(t) + G(t) + k + \varepsilon(t) \tag{5.1}$$

where $y(t)$ represents mRNA expression level of the target gene at time point t, and the parameter a indicates the effect of the present state value $y(t)$ to the next state value $y(t+1)$. $x_i(t)$, for $i \in \{1, \ldots, N\}$, represent the regulation functions of N TFs binding to the target gene, and b_i indicates the regulatory ability of the ith TF. $G(t)$ is the possible regulatory function of cooperative TFs.

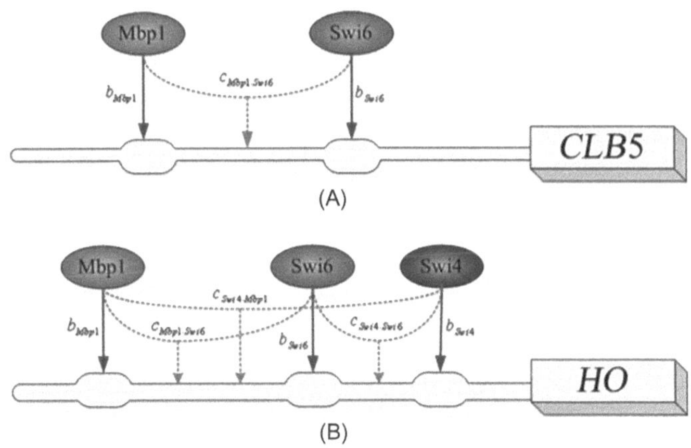

Figure 5.1 Examples of dynamic transcriptional regulatory models for trans-regulatory circuit of target genes *CLB5* and *HO*. The candidates of regulatory TFs of target gene were obtained from the genome-wide location data of Harbison et al. (2004) when *p*-value was chosen as $p < .0015$. In the examples, the regulatory TFs of *CLB5* are *Mbp1* and *Swi6*, and the regulatory TFs of *HO* are Mbp1, Swi6, and Swi4. The regulations of individual TF are represented by *solid lines* and the regulations of possible TF cooperativities are represented by *dotted lines*.

k represents the basal level from other factors, and $\varepsilon(t)$ denotes a stochastic noise owing to model uncertainty and fluctuation of mRNA microarray data in the target gene.

For system identification of trans-regulatory circuit, it is more practical to consider the biochemical reaction between the TF reaction functions $x_i(t)$ at the motif binding sites and their relevant mRNA expression profiles $y_i(t)$ of the upstream regulatory gene [19,20]. For this purpose, we describe the binding reaction function $x_i(t)$ of the TF as the following sigmoid function

$$x_i(t) = f_i(y_i) = \frac{1}{1 + \exp[-r(y_i(t) - M_i)]} \tag{5.2}$$

where r denotes the transition rate of the sigmoid function, and M_i denotes the mean of mRNA expression level of the regulatory gene i. $y_i(t)$ and $x_i(t)$ represent the mRNA expression profiles of the ith regulatory gene and the binding regulation function of the corresponding TF, respectively. The sigmoid function can also be considered as a method for normalizing the expression profiles of regulatory genes between 0 and 1, which has been successfully employed to describe the binding of regulatory genes [12].

The regulatory function $G(t)$ is combined by all possible TF cooperativities of the target gene. We describe the regulatory function $G(t)$ in Eq. (5.1) as the following

$$G(t) = \sum_{i=1}^{N} \sum_{j=i+1}^{N} c_{i\bullet j} x_{i\bullet j}(t) \tag{5.3}$$

where $x_{i\bullet j}(t) \equiv f_{i\bullet j}(y_i y_j)$ is a sigmoid function of product $y_i(t)y_j(t)$ to denote the binding function of cooperative TFs i and j, and $c_{i\bullet j}$ denotes the regulatory ability (or kinetic activity) of the cooperative TFs i and j. N denotes the number of cooperative TFs.

Substituting Eq. (5.3) into (5.1), we get the following dynamic transcriptional regulatory equation

$$y(t+1) = a\,y(t) + \sum_{i=1}^{N} b_i x_i(t) + \sum_{i=1}^{N} \sum_{j=i+1}^{N} c_{i\bullet j} x_{i\bullet j}(t) + k + \varepsilon(t) \tag{5.4}$$

In Fig. 5.1, there are examples of constructing the discrete dynamic models of trans-regulatory circuit for target genes. After describing the general stochastic dynamic model of transcriptional regulation, we can identify the models of gene expression from the available microarray data.

Identifying trans-regulatory circuits

After constructing the discrete stochastic dynamic model, we use the method of maximum likelihood to estimate the parameters. Eq. (5.4) can be written as the following regression form.

$$y(t+1) = [\,y(t) \quad x_1(t) \quad \ldots \quad x_N(t) \quad x_{1\bullet 2}(t) \quad \ldots \quad x_{(N-1)\bullet N}(t) \quad 1\,]\bullet \begin{bmatrix} a \\ b_1 \\ \vdots \\ b_N \\ c_{1\bullet 2} \\ \vdots \\ c_{(N-1)\bullet N} \\ k \end{bmatrix} + \varepsilon(t) \equiv \phi(t)\theta + \varepsilon(t) \tag{5.5}$$

where $\phi(t)$ denotes the regression vector which can be obtained from the processing above. θ is the parameter vector of the trans-regulatory circuit which is to be estimated.

If we have a lot of microarray data points obtained from the cubic spline method above, it is easy to get values of $\{\, y(t_l) \quad x_i(t_l) \quad x_{i\bullet j}(t_l)\,\}$ for $l \in \{1, \ldots, M\}$ and $i, j \in \{1, \ldots, N\}$, where M is the number of expression time points of a target gene, and N is the number of TFs binding to the target gene. The parameter vector θ can be estimated by the following approach. After a matrix notation, Eq. (5.5) at different time points is of the following form

$$\begin{bmatrix} y(t_2) \\ y(t_3) \\ \vdots \\ y(t_M) \end{bmatrix} = \begin{bmatrix} \phi(t_1) \\ \phi(t_2) \\ \vdots \\ \phi(t_{M-1}) \end{bmatrix} \theta + \begin{bmatrix} \varepsilon(t_1) \\ \varepsilon(t_2) \\ \vdots \\ \varepsilon(t_{M-1}) \end{bmatrix} \tag{5.6}$$

For simplicity, we can further define the notations Y, Φ, and e to represent Eq. (5.6) as follows

$$Y = \Phi\theta + e \tag{5.7}$$

In Eq. (5.6), we assume noises $\varepsilon(t_l)$ at different time points as independent random variables of normal distribution with zero mean and unknown variance σ^2, i.e., the variance of e is $\sum = E\{ee^T\} = \sigma^2 I$, where I is an identity matrix. In this section, a maximum likelihood parameter estimation method will be employed to estimate θ

and σ^2 from microarray data of regulatory genes and the target gene [21]. If e is assumed to be normally distributed with $M-1$ elements, its probability density function is of the following form

$$p(e) = \frac{1}{((2\pi)^{M-1}\det\Sigma)^{1/2}}\exp\left\{-\frac{e^T\Sigma^{-1}e}{2}\right\} \tag{5.8}$$

Under Eq. (5.7), we have the likelihood function

$$L(\theta,\sigma^2) = p(\theta,\sigma^2) = \frac{1}{(2\pi\sigma^2)^{(M-1)/2}}\exp\left\{-\frac{(Y-\Phi\theta)^T(Y-\Phi\theta)}{2\sigma^2}\right\} \tag{5.9}$$

Eq. (5.9) can be considered as a function of parameter θ and σ^2. We want to specify θ and σ^2 to maximize the likelihood function in Eq. (5.9). It is practical to take the logarithm of their likelihood function, and then we have the following log-likelihood function:

$$\log L(\theta,\sigma^2) = -\frac{M-1}{2}\log(2\pi\sigma^2) - \frac{1}{2\sigma^2}\sum_{k=1}^{M-1}[y(t_{k+1})-\phi(t_k)\theta]^2 \tag{5.10}$$

where $y(t_k)$ and $\phi(t_k)$ are the kth elements of Y and Φ, respectively.

Here we expect the log-likelihood function to have the maximum at $\theta=\hat{\theta}$ and $\sigma^2=\hat{\sigma}^2$. The necessary condition for the maximum likelihood estimates $\hat{\theta}$ and $\hat{\sigma}^2$ is as follows [21]

$$\left.\frac{\partial \log L(\theta,\sigma^2)}{\partial\theta}\right|_{\theta=\hat{\theta}} = 0$$
$$\left.\frac{\partial \log L(\theta,\sigma^2)}{\partial\sigma^2}\right|_{\sigma^2=\hat{\sigma}^2} = 0 \tag{5.11}$$

The estimated parameters $\hat{\theta}$ and $\hat{\sigma}^2$ are shown in Eqs. (5.12) and (5.13)

$$\hat{\theta} = (\Phi^T\Phi)^{-1}\Phi^T Y \tag{5.12}$$

$$\hat{\sigma}^2 = \frac{1}{M-1}\sum_{k=1}^{M-1}\left[y(t_{k+1}) - \phi(t_k)\hat{\theta}\right] = \frac{1}{M-1}(Y-\Phi\hat{\theta})^T(Y-\Phi\hat{\theta}) \tag{5.13}$$

where Y and Φ can be obtained from the microarray data of regulatory genes and the target gene. After obtaining the estimate $\hat{\theta}$ from Eq. (5.12) via microarray data, the transcriptional regulatory circuit of target genes could be modeled by the following dynamic equation

$$y(t+1) = \hat{a}y(t) + \sum_{i=1}^{N}\hat{b}_i x_i(t) + \sum_{i=1}^{N}\sum_{j=i+1}^{N}\hat{c}_{i\bullet j}x_{i\bullet j}(t) + \hat{k} \tag{5.14}$$

Therefore, the TF cooperativity at each target gene could be evaluated by $\hat{c}_{i\bullet j}$, the regulatory ability of the cooperative TF pair, in Eq. (5.14). However, $\hat{c}_{i\bullet j}$ in Eq. (5.14) is only from one target gene point of view. A statistical method should be developed to measure the cooperativity of TFs to combine all $\hat{c}_{i\bullet j}$ of the TF i and j for all target genes (i.e., on a genomic scale). It will be discussed in "Measurement of TF cooperativity in gene regulatory mechanism" section.

Measurement of TF cooperativity in gene regulatory mechanism

In order to measure the possible TF cooperativity, we define a new measurement, the multiplicative cooperative p-value P_C, according to the statistics of regulatory abilities $\hat{c}_{i\bullet j}$ of all target genes in the dynamic model of Eq. (5.14) and the number of TF pairs appearing in all target genes. For any TF pair, if the binding sites of specific TF pairs co-occur more frequently in the target genes, then it is more likely to have cooperativity between these two TFs [17]. In addition, the magnitude of the regulatory ability $\hat{c}_{i\bullet j}$ of TF cooperation in this dynamic model can be also useful to help us determine the significant TF cooperativity because the magnitude represents its importance in the transcriptional regulation of the target gene.

To define the multiplicative cooperative p-value P_C on a genomic scale, we first calculate the individual p-value for each regulatory ability of a cooperative TF pair. The p-value is defined as $(P_B)_m$ when a cooperative TF pair binds to the mth target gene. For the mth target gene bound by the cooperative TF pair i and j, we rewrote its regulatory ability $\hat{c}_{i\bullet j}$ as $\hat{c}^m_{i\bullet j}$. After constructing 1000 random permutations of $x_{i\bullet j}[t]$ in Eq. (5.4) for the mth target gene bound by TFs pair i and j, we used these random permutations to estimate 1000 different $\hat{c}^m_{i\bullet j}$ rewritten as $\tilde{c}^m_{i\bullet j}$ and then computed a probability density $p_{\left|\tilde{c}^m_{i\bullet j}\right|}(x)$ of all absolute values $\left|\tilde{c}^m_{i\bullet j}\right|$ according to their magnitudes by normalization. The probability density distribution $p_{\left|\tilde{c}^m_{i\bullet j}\right|}(x)$ of cooperativity is shown in Fig. 5.2. Then, using $p_{\left|\tilde{c}^m_{i\bullet j}\right|}(x)$, we calculated the individual p-value $(P_B)_m$ of TFs i and j at the mth target gene by

$$(P_B)_m = \int_{\left|\hat{c}^m_{i\bullet j}\right|}^{\infty} p_{\left|\tilde{c}^m_{i\bullet j}\right|}(x)dx \tag{5.15}$$

where $\left|\hat{c}^m_{i\bullet j}\right|$ denotes the regulatory ability of the cooperative TF pair i and j at the mth target gene without random permutation.

A TF pair may co-occur to bind many target genes, so we could obtain many individual p-values $(P_B)_m$, $m \in \{1, 2, \ldots, L\}$, where L is the number of the target genes bound by the possible cooperative TF pair i and j. In Fig. 5.1, e.g., both TFs

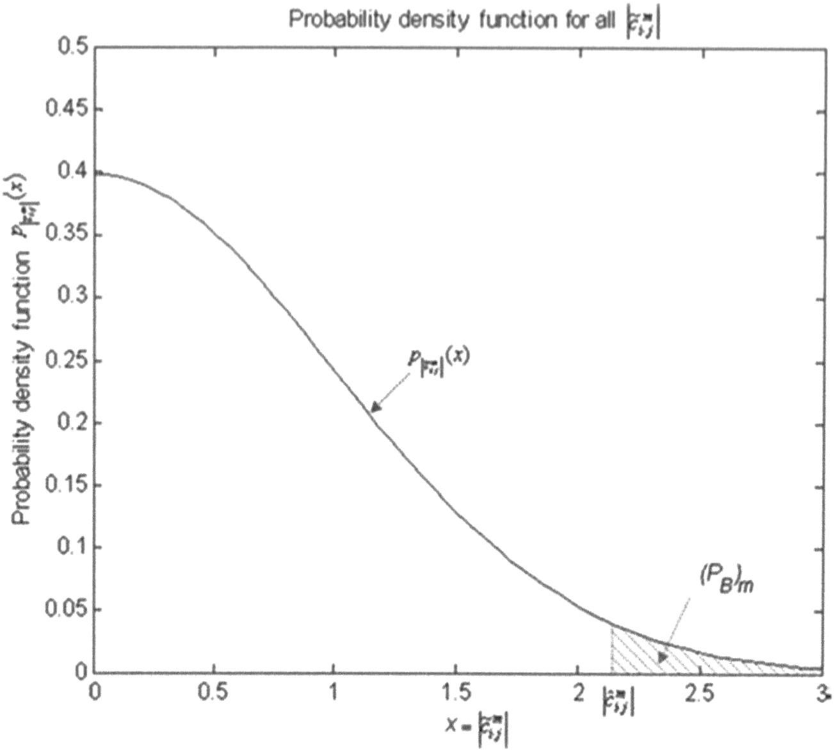

Figure 5.2 Probability density function $p_{|\tilde{c}^m_{i \bullet j}|}(x)$ for $|\tilde{c}^m_{i \bullet j}|$. The pdf $p_{|\tilde{c}^m_{i \bullet j}|}(x)$ of all $|\tilde{c}^m_{i \bullet j}|$ is according to the number of $|\tilde{c}^m_{i \bullet j}|$ occurrence estimated by random permuting $x_{i \bullet j}[t]$ for the mth target gene bound by TF pair i and j. The individual p-value $(P_B)_m$ of the mth target gene is the area of the pdf above its regulatory ability $|\tilde{c}^m_{i \bullet j}|$.

Mbp1 and Swi6 both bind to HO and CLB5. Therefore, the gene HO has its $c_{Mbp1 \bullet Swi6}$ and the corresponding $(P_B)_{HO}$, and so does CLB5 with its $c_{Mbp1 \bullet Swi6}$ and the corresponding $(P_B)_{CLB5}$. After computing all the individual p-value $(P_B)_m$ for the TF pair i and j for all target genes, i.e., $(P_B)_m$ for $m \in \{1, 2, \ldots, L\}$, we define the multiplicative cooperative p-value P_C for the TF pair i and j by multiplying all of its $(P_B)_m$ as follows

$$P_C = \prod_{m=1}^{L} (P_B)_m \tag{5.16}$$

where L is the number of the target genes bound by the possible cooperative TF pair i and j, and $(P_B)_m$ denotes the individual p-value of the cooperative TF pair i and j binding to the mth target gene.

In Eqs. (5.15) and (5.16), the more likely the TF cooperativity is, the less P_C of the TF pair is. Therefore, there are two situations that make P_C smaller. One is the larger number L of the TF pairs co-occurring in all target genes, i.e., with a larger number of individual p-values for the TF pair. This situation is the same with the result that the TF pair has a greater possibility of cooperation if binding sites of specific TF pairs co-occur more frequently in target genes [17]. Another situation is that the regulatory ability $|\hat{c}_{i\bullet j}|$ of the TF pair i and j is so significant at some target genes that the individual p-value $(P_B)_m$ of the TF pair is small at the target genes. In this way, the multiplicative cooperative p-value P_C also becomes small. Therefore, it is reasonable to use the multiplicative cooperative p-value P_C to evaluate the TF cooperativity.

5.1.2 Identifying TF cooperativity in cell cycle gene regulatory mechanism

Analysis of TF cooperativity

We integrated the expression profiles of the yeast cell cycle taken from Spellman et al. and genome-wide location data of Harbison et al. to identify the cooperativity of all possible TF interaction pairs among 203 TFs. After constructing and identifying parameters of gene transcriptional regulatory networks of yeast, we calculated the multiplicative cooperative p-value P_C of each possible TF interaction pair. Sorted by P_C values, all possible TF interaction pairs are also ranked according to their cooperative possibilities, and the results are shown in Table 5.1. According to these results, the cooperative network of significant cooperative TF pairs is given in Fig. 5.3. Of the results listed, 60% are confirmed by evidence in many studies. However, most of these TF interaction pairs confirmed are related to the cell cycle of yeast. It may be because these source data are expression profiles of the yeast cell cycle so that most of these TF interaction pairs found in the list are related to cell cycle. Moreover, the TF interaction pairs that have not yet been proved but are found by the method we proposed can provide a direction for future experiments. The details of the TF cooperativities we found are discussed below.

Cell cycle

1. G_1/S Phase (Swi4-Swi6, Mbp1-Swi6, Mbp1-Swi4, Stb1-Swi6)

   The G_1 to S phase transition of the eukaryotic cell cycle is a critical point for the coordination of cell cycle progression with cellular growth. SBF and MBF are sequence-specific TFs that activate gene expression to mediate the G_1/S transition of the cell cycle in yeast [30]. SBF is a protein complex of *Swi4* and *Swi6*, and MBF is a protein complex of *Mbp1* and *Swi6* [22]. The related *Swi4* and *Mbp1* proteins are the DNA-binding components of the respective factors, and *Swi6* may have a regulatory function [31,32]. The cooperativity between *Swi4* and *Swi6* is

Table 5.1 Significant cooperative TF pairs of cell cycle target genes

Ranking	TF1	TF2	P_C[a]	Literature evidence	Results of Banerjee and Zhang[b] method
1	Swi4	Swi6	3.91E-163	[22]	Yes
2	Mbp1	Swi6	2.78E-125	[22]	Yes
3	Fkh2	Ndd1	1.84E-88	[23]	Yes
4	Mbp1	Swi4	2.05E-84	[22,24]	
5	Gat3	Yap5	1.00E-76	[24]	
6	Fkh2	Swi6	7.80E-59		
7	Fkh1	Fkh2	2.31E-58	[23,25]	Yes
8	Pdr1	Yap5	2.82E-49	[24]	
9	Fkh2	Swi4	1.00E-48		
10	Mcm1	Ndd1	2.53E-46	[23]	Yes
11	Fkh2	Mbp1	1.36E-45		
12	Fkh1	Swi6	7.00E-45		
13	Gat3	Pdr1	3.97E-44	[24]	Yes
14	Fkh1	Ndd1	2.93E-43	[23]	Yes
15	Ndd1	Swi6	1.00E-42		
16	Msn4	Yap5	8.35E-42		Yes
17	Fkh2	Mcm1	3.72E-40	[23]	Yes
18	Gat3	Msn4	1.00E-39		Yes
19	Stb1	Swi6	3.02E-39	[26,27]	
20	Cin5	Yap6	5.03E-37	[24]	Yes
21	Ndd1	Swi4	7.30E-37	[24]	
22	Gat3	Hap4	1.00E-36		
23	Hap4	Yap5	1.00E-36		
24	Fkh1	Mbp1	3.30E-36	[27]	Yes
25	Stb1	Swi4	5.40E-35		Yes
26	Ste12	Swi6	3.20E-34		
27	Msn4	Pdr1	7.00E-33		
28	Dig1	Ste12	9.66E-33	[24,28]	
29	Mbp1	Ndd1	2.28E-32	[24]	
30	Ace2	Swi5	5.54E-31	[29]	Yes

The TF pairs are sorted by P_C value.
[a]List of the results with P_C smaller than 10^{-30}.
[b]The results confirmed by Banerjee and Zhang's study, too [10].

included in these results, and so is the cooperativity between *Mbp1* and *Swi6*. In addition, *Mbp1* and *Swi4* sharing 50% identity in their DNA-binding domains are found with the probable cooperativity in these results [22].

In the late G_1 phase, Stb1 is an important regulator controlling the timing of start transcription that is revealed in the absence of the G_1 regulator Cln3 and binds to *Swi6* in vivo [27]. The interaction of Stb1 and Swi6 confirmed by Ho et al. [27] and Costanzo et al. [26] is also found in these results.

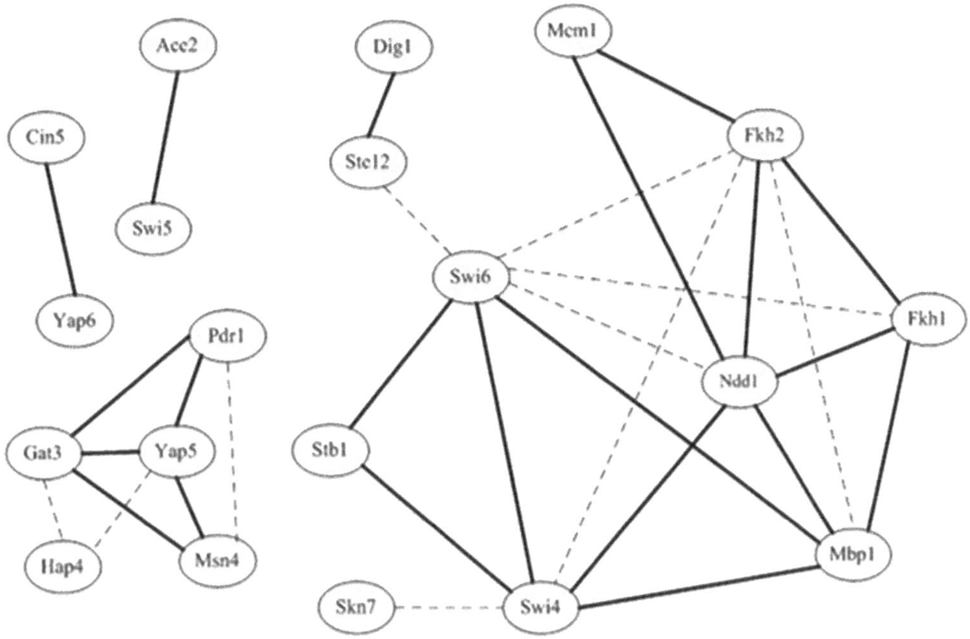

Figure 5.3 The significant cooperative TF network of cell cycle target genes. The cooperative TF pairs confirmed by literature evidences or the results of Banerjee and Zhang [10] are shown in *solid lines*, and those still to be confirmed are expressed in *dotted lines*.

2. G_2/M Phase (Ndd1-Fkh1, Ndd1-Fkh2, Fkh1-Fkh2, Mcm1-Fkh2, Ndd1-Mcm1)

   In the G_2/M phase, the important regulator is the TF SFF containing Ndd1 and Fkh1 or Fkh2 [23,33]. These results strongly indicate the cooperativity between Ndd1 and Fkh2, which are components of SFF. Another cooperativity between Ndd1 and Fkh1, which are also components of SFF, is found in these results, too. In addition, Fkh1 and Fkh2 are found with cooperativity in these results. Fkh1 and Fkh2 share 72% identical DNA-binding domain, and the double mutant of Fkh1 and Fkh2 displays obvious morphological change [23].

   SFF is thought to regulate a program of mitotic transcription in conjunction with the TF Mcm1. Moreover, Fkh2, a component of SFF, assembles into a ternary complex with Mcm1 on both the SWI5 and CLB2 cell cycle genes [23]. The fact that Fkh2 is cooperative with Mcm1 is also listed in this result. In addition to Ndd1 and Fkh2, Mcm1 and Fkh2, another cooperative TF pair in the G_2/M phase, Ndd1 and Mcm1, is found in these results. For the above cooperative TF pairs found in the G_2/M phase, it is not surprising to find experimentally that Mcm1, Fkh2, and Ndd1 also form a complex to regulate the CLB2 gene and other genes [23,25].

3. M/G_1 Phase (Ace2-Swi5)

In the M/G_1 phase, we found that the TF pair, Ace2 and Swi5, is cooperative. Ace2 and Swi5 is a pair of TFs of yeast that regulates the expression of many cell cycle-specific genes [29]. In recent studies, Ace2 and Swi5 cooperate to induce the expressions of a subset of genes, but the antagonistic interaction between Ace2 and Swi5 has been found [29]. With 82% identical DNA-binding domains, Ace2 and Swi5 bind to the same DNA sequence [34], and it is possible that proteins compete for access to these promoters, but only one activates transcription [29]. Therefore, one partner of Swi5 and Ace2 sometimes can have a stronger contribution toward regulation, and the finding of antagonistic interaction of Ace2 and Swi5 is not surprising.

Mating (Ste12-Dig1)

The TF Ste12 is responsible for activating genes in response to MAP kinase cascades controlling mating and filamentous growth. Two inhibitors Dig1 and Dig2 regulate Ste12 negatively [28]. It was found that Dig1 and Dig2 do not function through redundant mechanisms, but rather inhibit pheromone-responsive transcription through interactions with separate regions of Ste12 [28]. In these results, we indeed found that Ste12 is cooperative with Dig1. However, in the location data of Harbison et al., Dig2 is not listed in the 203 TF. That is why another protein Dig2 was not found to be cooperative with Ste12 in these results. If the genome-wide location data are comprehensive for more TFs, we believe that the method can predict more cooperative TF pairs, like Dig2 and Ste12.

Comparison with results of other methods

Comparing the TF cooperativities we found with Banerjee and Zhang's [10] results (see Table 5.1) shows that many cooperative TF pairs confirmed by literature evidences are found in both results. Several pairs, which are still not confirmed by experiments, are also found by both methods. Furthermore, these results indicate more cooperative TF pairs confirmed by literature evidences but not found by Banerjee and Zhang's methods. As seen in Fig. 5.4, this method is better and finds more real cooperative TF pairs than Banerjee and Zhang's methods.

However, there are still three pairs, namely Arg80-Arg81, Ace2-Hsf1, and Hsf1-Skn7, found in Banerjee and Zhang's results but not in these results. These three TF pairs could be considered as false negative TF pairs of this method. After checking these three cooperative TF pairs in detail, we found that Ace2 and Hsf1 did not co-occur to bind any target gene of the 768 cell cycle genes from the location data of Harbison et al. at $p < .0015$, and so did Hsf1 and Skn7. This method fails to find a cooperative TF pair that does not co-occur to bind any target gene from the information of location data. Another pair, Arg80 and Arg81, co-occurred to bind only one

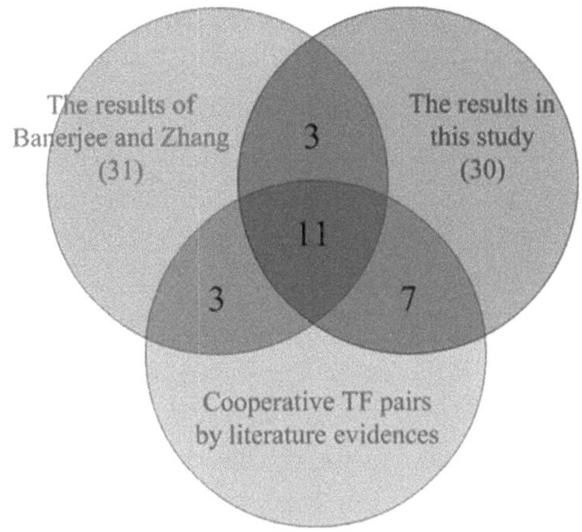

Figure 5.4 Overlap of these results, Banerjee and Zhang's results, and cooperative TF pairs supported by literature evidences. By comparing both results, we can easily find that the confirmed rate of this method is better than that obtained by Banerjee and Zhang's method [10].

target gene of the 768 cell cycle genes from the location data of Harbison et al. at $p < .0015$. For this cooperative TF pair, the number L in Eq. (5.16) is only 1, and that is why its P_C is not small enough to be considered significantly. Finally, we calculated the confirmed rates, which are the rates of both results compared with literature evidences, of the TF pairs listed in these findings and Banerjee and Zhang's results [10]. The confirmed rate of this method, 60%, is higher than that of Banerjee and Zhang's, 45.16%.

Pilpel et al. [9] uncovered functional motif combinations in the promoters of *Saccharomyces cerevisiae* using microarray data. They uncovered not only cell cycle-related motifs, but also sporulation and various stress responses. Focusing on cell cycle, they identified only 10 motif pairs. Comparing these results with the cell cycle results from Pilpel et al. [9], there are only three TF pairs in common (see Appendix A). This may be because they only use microarray data but not use ChIP-chip data to infer combinatorial motifs.

5.1.3 Discussion of TF cooperativity in gene transcriptional regulation by dynamic regulatory model and microarray data

In this section, we successfully identified cooperative TF pairs by integrating genome-wide location data and expression profiles. This method is based on transcriptional regulatory mechanisms of TFs and their cooperativities at binding sites from dynamic system perspective. Unlike others' studies using statistical correlation [9,10], this method provides the view of dynamic regulatory systems to mimic the transcriptional procedure. The measure of TF cooperativity of TFs is also developed by a multiplicative cooperative p-value according to the statistics of estimated regulatory ability of cooperative genes in the dynamic regulatory model. These results showed that many cooperative TF pairs identified by this method are confirmed by literature evidence. We also indicated several possible TF pairs with cooperativity that are listed in these results but not confirmed yet. These provide new directions for future experiments. Moreover, it was shown that this method is better than other methods using statistical correlation [9,10].

From the cooperative network of significant TF cooperative pairs shown in Fig. 5.3, we can find that the interaction lines among several TFs (Swi4, Swi6, Fkh1, Fkh2, Ndd1, and Mbp1) are more compact than others. All of these TFs have important roles in the yeast cell cycle, so it is likely that there exist TF cooperativities among these TFs. However, some of these TFs may cobind to many target genes because of their importance for the yeast cell cycle but no cooperativity among them. In this analysis, this situation will lead to a false positive. It may be another possible reason why the interaction lines among these important TFs are more compact.

In order to confirm the reliability of the proposed method, the shuffling method was used to test the overfitting. In the expression data shuffling case, only 36.67% of TF cooperativities are found, which may be due to the correct ChIP-chip data and the uncertainty reduction by p-value detection method. When comparing the results, we found that there is no overfitting by the proposed method (see Appendix A).

With the microarray data of the yeast cell cycle of Spellman et al. and the ChIP-chip data of Harbison et al., it is shown that this method can successfully predict the cooperative TF pairs of the yeast cell cycle. If the gene expression profiles and genome-wide location data of other process for yeast or other species, such as the heat shock stress, are available, this method could be employed to identify the cooperative TF pairs of different stress conditions or other species (see Appendix A). Furthermore, this method can be extended to identify the cooperativities among more than two TFs by adding extra terms generated by multiplications of their expression profiles through the sigmoid function. Moreover, we can integrate protein–protein interaction data to determine interactions not only among TFs but also among TFs and proteins in this model (see Appendix A).That is, by modifying the method we proposed, we can construct an integrated cellular network of transcription regulation and protein–protein interaction.

APPENDIX A: METHODS IN IDENTIFYING THE TF COOPERATIVITY

Avoiding overfitting

We have some methods to avoid overfitting in this section. First, we use the cubic spline method to interpolate the data points while the number of TFs binding to the target gene exceeds 6, which implied the number of parameters to be estimated is 22. Only 5% of target genes have more than 6 TFs. The more TFs binding to a target gene, the more data points we interpolated by the cubic spline method to avoid overfitting. Second, in the method we proposed, we used p-value to determine the cooperativity among TFs, which could reduce uncertainty due to small data to avoid overfitting. Third, as data shuffling technique is a good way to test the overfitting, we randomly shuffled the microarray expression data or genome-wide location data to prove that there is no overfitting in the method we proposed. By comparing the results of expression data shuffling and genome-wide location data shuffling with the result of normal data in Table 5.1, we found that there exists much difference in these results, showing that there is no overfitting. In the expression data shuffling case, only 36.67% of cooperativities of TFs are found, which come from the correct ChIP-chip data and the uncertainty reduction by p-value detection. In the location data shuffling, no cooperativities of TFs are found due to the same cutoff P_C as normal data, 10^{-30}. The shuffling results show that the normal genome-wide location data save the ratio of cooperativities to be found in the expression data shuffling case, and the cooperativites we found are more sensitive to genome-wide location data. The shuffling results are shown as follows

Expression data shuffling case

	TF1	TF2	P_C	In normal condition
1	Swi4	Swi6	3.24E-153	Yes
2	Mbp1	Swi6	3.43E-121	Yes
3	Mbp1	Swi4	1.44E-63	Yes
4	Gat3	Yap5	2.07E-52	Yes
5	Fkh2	Ndd1	7.74E-52	Yes
6	Fkh1	Fkh2	1.20E-42	Yes
7	Pdr1	Yap5	1.85E-40	Yes
8	Cin5	Yap6	6.15E-35	Yes
9	Stb1	Swi6	1.44E-34	Yes
10	Fkh1	Swi6	3.34E-34	Yes
11	Fkh2	Swi6	5.97E-33	Yes

Genome-wide location data shuffling case

	TF1	TF2	P_C
1	Gal3	Hir3	1.00E-12
2	Fhl1	Tec1	1.00E-09
3	Opi1	Spt2	1.67E-09
4	Mss11	Yap3	9.50E-08
5	Gln3	Rds1	1.17E-07
6	Ykr046w	Ynr063w	6.64E-07
7	Ifh1	Swi5	8.35E-07
8	Gal3	Stp4	9.00E-07

Note that when the cutoff P_C is as normal data (normal data means the data without shuffling), 10^{-30}, we could not find any cooperative pairs among TFs when shuffling genome-wide location data. We just list top eight pairs of the genome-wide location data shuffling case. By the three remedy methods presented above, we showed that there exists no overfitting in this section and the cooperativities we inferred are reliable.

Comparison with Pilpel et al.'s results

Pilpel et al. (2001) uncovered functional motif combinations in the promoters of *S. cerevisiae* using microarray data. They uncovered not only cell cycle–related motifs, but also sporulation and various stress responses. Focusing on cell cycle, they identified only 10 motif pairs. Comparing these results with the cell cycle results from Pilpel et al., there are only three TF pairs in common. This may be because they only use microarray data but not use ChIP-chip data to infer combinatorial motifs. The comparison is shown below.

Motif 1	Motif 2	TF 1	TF 2	This method
RAP1	mRPE6	Rap1		
RAP1	PDR	Rap1	Pdr1	
mRRPE	PAC			
ECB	SFF	Yox1,Yhp1	Ndd1, Fkh1/Fkh2	
ECB	SFF′	Yox1,Yhp1	Ndd1, Fkh1/Fkh2	
MCM1′	Deoxyribonucleotide_metabolism_ orfnum2SD_n4	Mcm1		
MCM1′	SFF	Mcm1	Ndd1, Fkh1/Fkh2	Yes
MCM1	SFF	Mcm1	Ndd1, Fkh1/Fkh2	Yes
MCM1	SFF′	Mcm1	Ndd1, Fkh1/Fkh2	Yes
ALPHA2	Deoxyribonucleotide_metabolism_ orfnum2SD_n8	MATalpha2		

Note that in the table above, motif pairs (motif 1 and motif 2) are from Pilpel et al. (2001). We find the TFs which bind to those motifs from database on the internet and compare the TF pairs with these results.

Other than cell cycle condition

Our method combines microarray expression data and ChIP-chip (genome-wide location) data to infer cooperativities among TFs. Thus, as long as we have both kinds of data, we could determine the cooperativity among TFs by the method we proposed. Here, we applied the method to *S. cerevisiae* under H_2O_2 stress (by combining microarray data from Gasch et al. and ChIP-chip data from Harbison et al.) to prove that this method is not only applicable to yeast cell cycle case. The result of TF cooperative interactions is as follows

	TF1	TF2	P_C
1	YAP1	YAP7	1.00E-51
2	MSN2	SKN7	2.00E-42
3	MSN2	MSN4	7.78E-28
4	CIN5	YAP7	1.00E-27
5	AFT2	PUT3	1.00E-24
6	HSF1	MSN2	1.00E-24
7	MSN2	PUT3	1.00E-24
8	MSN2	YAP7	1.00E-24
9	MSN2	YAP1	1.00E-21

The result comes from 856 H_2O_2 stress-related genes and 27 TFs. These 27 TFs are selected by Harbison et al., which are all related to H_2O_2 stress. We infer 9 cooperativities among 27 TFs when the cutoff P_C is 10^{-20}. These cooperativities need to be proved by biological experiments.

Integrating protein–protein interaction data

In addition to the interactions among TFs, we could integrate the protein–protein interaction data by including the proteins which interact with the TFs selected by Harbison et al. into the model we proposed. In this way, we could infer the cooperativities not only among the TFs but also among the TFs and proteins we include in the model. Thus, we could construct transcriptional regulatory interaction network other than TFs. However, in this situation, more parameters need to be estimated and more expression data points are needed.

Gene expression programs depend on regulation of cis elements in promoter region of target genes by TFs, but how TFs regulate gene expression via recognition of cis elements is still not clear. In the following section, a novel cross-gene identification scheme is proposed to infer how multiple TFs coordinate to regulate gene transcription in cis-regulatory circuit of a cell cycle gene and to uncover hidden regulatory functions of a cis-regulatory circuit.

5.2 CIS-REGULATORY MECHANISMS FOR GENE EXPRESSION VIA CROSS-GENE IDENTIFICATION AND DATA MINING

After the introduction of the cooperativity of TFs in the trans-regulatory circuit of gene regulation, we will investigate the cis-regulatory mechanisms of gene expression in this section. Cis-regulatory circuits have been applied to model cell cycle control and other developmental processes [35,36]. Recently, the genetic regulatory and transcriptional networks of yeast have been studied [5,7,30,37–39]. However, in order to understand the regulation of any particular gene in the network, one should study carefully how the genes in the network collectively operate with remarkable precision in response to environmental cues and the structure and function of the cis-regulatory circuit of the gene expression [39]. The cis-regulatory circuit of a gene consists of its cis elements, i.e., binding motifs of TFs, and the interactions among their recognizing TFs. The cis elements of a gene can be considered as the information processing units in the regulatory circuit; they receive multiple inputs from the TFs that bind the cis elements of the gene. The output is the instruction for the transcription apparatus to determine whether the gene is to be expressed at a specific rate or to be repressed [11,40].

A cis-regulatory circuit may be regarded as a control device that is called into play by the TFs that have target sites inside the promoter [2,10,12,41–44]. A well-known example is the promoter region of the developmentally regulated endo 16 gene of the sea urchin [44–46]. It is about ~2300 bp in length and consists of several clusters of target sites for distinct functions. Yuh et al. [44–46] have explored the function of each subregion of the endo 16 system and every target site within each subregion, using a cis-regulatory logic model.

However, a drawback of most current methods for inferring cis-regulatory circuits is that they rely on changing or deleting some binding site sequences (e.g., Refs. [44–46]), which may not provide intact functional information for reconstructing the cis-regulatory circuit. The deletion or mutation experiments may change or destroy the original cis-regulatory circuit structure. Using such data, one may lose significant interactions among TFs. Obviously, it is appealing to develop a method that can infer intact cis-regulatory circuits. Recently, there are some statistical and system level approaches to study the genome-wide transcription regulation and address cooperativity among TFs [10,39,41,43]. Important advances have been made toward understanding the control mechanisms of transcriptional regulatory networks. One strategy infers global networks directly from whole genome microarray data, and another strategy focuses on the identification of shared cis elements in the promoters of coregulated genes, signified by similar expression profiles [38]. In this section we discuss and develop a new method to combine microarray data and TF-binding location data by chromatin immunoprecipitation [2,5] to study the regulatory and interaction functions of various cis elements with regard to the target gene. The data of Lee et al. [5] can reveal the in vivo physical interactions of TFs with their cis elements on the promoter and therefore can provide a more reliable view of functional interactions between TFs and cis elements. Combining these types of data and microarray data, we propose a novel cross-gene identification scheme to infer how multiple TFs coordinate to regulate gene transcription. This approach is rather different from most existing statistical and system level methods for analyzing gene expression data. These results show that this novel method is suitable for deciphering the control mechanisms of complex TFs interactions and for predicting the gene expression. In addition, we also identify the dynamic regulatory functions of TFs interaction in the yeast cell cycle, which cannot be achieved by current methods.

5.2.1 Investigating cis-regulatory mechanisms in gene expression

Characterizing the cis-regulatory circuit of a gene

In this section, we show that there are two steps for characterizing the cis-regulatory circuit of a gene. The first step is to find a cluster of genes that includes the gene of interest and a number of other genes each of which shares a subset of cis elements with the gene of interest. Assuming a certain regulatory function for each of the TFs that recognize the cis elements of genes in the cluster and certain interaction functions among the TFs of a circuit, we set up dynamic equations for the cis-regulatory circuits of the genes in the cluster to describe their expression profiles. Since each gene in the cluster shares some cis elements with the gene of interest, a matrix of cis elements for the cluster of genes can be constructed. In this model, the regulatory functions of individual TFs and the interactions among TFs can be estimated from microarray data. In the second step, a cross-gene identification scheme is developed with an array of expression profiles of genes in the cluster (e.g., Ref. [2]) to identify regulatory functions of the TFs and their possible interactions for each gene in the cluster; the parameters are estimated by the least square estimation algorithm. Finally, plugging these estimated regulatory functions and interactions into the dynamic equations, one can explicitly describe the cis-regulatory circuit of the gene of interest.

Choice of a cluster of genes

As an illustration, suppose some genes of the yeast cell cycle are of interest. We find a cluster of genes for each gene of interest according to their cis elements found in Simon et al. [7]. To simplify the analysis, we consider only the nine TFs that are currently known to be important cell cycle TFs of the yeast (Mbp1, Swi4, Swi6, Mcm1, Fkh1, Fkh2, Ndd1, Swi5, and Ace2). The cluster of genes for the gene of interest is called the reference gene cluster (RGC). In an RGC, we assume that each gene shares some cis elements of the gene of interest. Furthermore, the regulatory functions and the interactions of the TFs recognizing the same cis elements are assumed to be the same for all genes in the RGC. For example, in Fig. 5.5A, *MFA2* is the gene of interest; it causes cell cycle arrest and is essential for mating in yeast [42]. This gene has three main cis elements: Ndd1, Mcm1, and Swi5 [44], from which we want to reconstruct the cis-regulatory circuit of MFA2 from yeast microarray data. The cis elements of MFA2 are denoted as follows:

$$MFA2{:}\{\text{Ndd}, \text{Mcm1}, \text{Swi5}\} \tag{5.17}$$

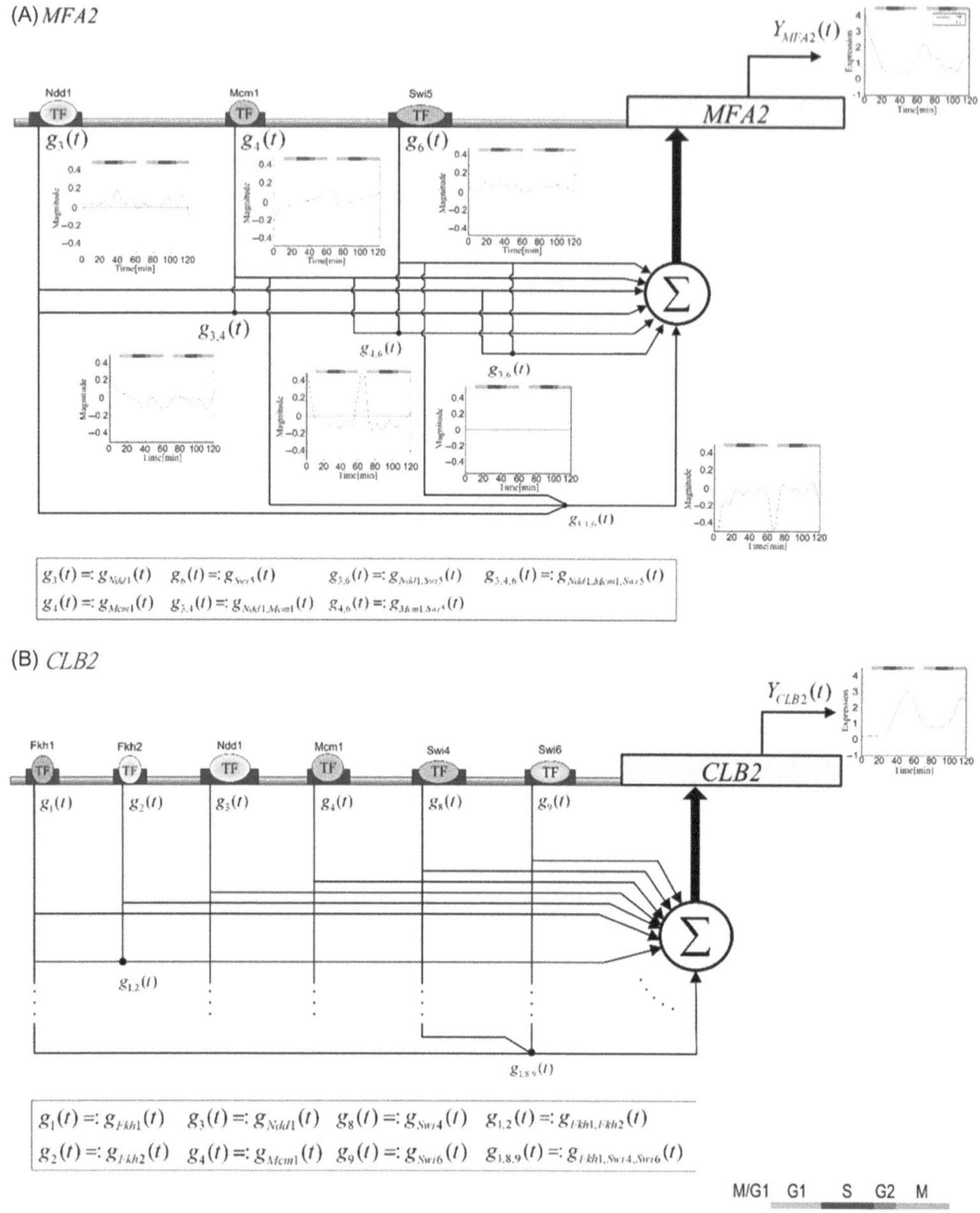

Figure 5.5 Dynamic model of the cis-regulatory circuit in the regulatory mechanism of gene *MFA2*, (A) and of gene *CLB2* (B) The genome-wide TF-binding location data obtained using chromatin immunoprecipitation [7] is used to identify the transcription factor binding motifs (cis elements). A binding transcription factor p has a regulatory function $g_p(t)$ and interacts with other recognizing TFs to produce the regulatory functions $g_{p,q}(t)$ and $g_{p,q,r}(t)$. These regulatory functions are the inputs of the cis-regulatory circuit and generate the dynamic output (i.e., the expression profile) of the target gene. Different phases of the cell cycle are indicated by the colored bar at the right lower corner.

Some genes chosen for the cluster and their cis elements are:

$$YAL022C{:}\{Swi5\}, YAR018C{:}\{Ndd1, Mcm1\}, YKL163W{:}\{Mcm1, \mathrm{Swi5}\}\ldots$$

The cluster of genes can be represented by a connectivity matrix of cis elements as shown in Table 5.2 in which “1” denotes the connection of a cis element with a gene, while “0” means no connection. Similarly, the RGC for gene *CLB2* can be represented by the connectivity matrix shown in Table 5.3.

Dynamic modeling of cis-regulatory circuits in gene regulatory mechanism

Fig. 5.5 illustrates the leaky integrator-based dynamic models of the cis-regulatory circuits of two yeast genes (*MFA2* and *CLB2*) [35,36]. The dynamics of gene expression can be modeled by a simple first-order nonlinear differential equation that is well established and analyzed in Ref. [12]; each model includes the possible regulatory functions of the individual TFs and possible interactions among the TFs. For the target gene *MFA2* (Fig. 5.5A), the cis-regulatory circuit is modeled by the following dynamic equation

$$\begin{aligned}\dot{Y}_{MFA2}(t) = {} & g_{Ndd1}(t) + g_{Mcm1}(t) + g_{Swi5}(t) + g_{Ndd1,Mcm1}(t) + g_{Ndd1,Swi5}(t) \\ & + g_{Mcm1,Swi5}(t) + g_{Ndd1,Mcm1,Swi5}(t) - \lambda_{MFA2} Y_{MFA2}(t) + \varepsilon_{MFA2}(t)\end{aligned} \tag{5.18}$$

where $\varepsilon_{MFA2}(t)$ denotes the noise (data uncertainty), $g_{Ndd1}(t)$, $g_{Mcm1}(t)$, and $g_{Swi5}(t)$ are the regulatory functions of TFs Ndd1, Mcm1, and Swi5 or the incident transcriptional regulations at the Ndd1, Mcm1, and Swi5 cis elements, respectively. λ_{MFA2} represents the mRNA decay rate of the target gene and we used the degradation rate measured by Wang et al. [43]. $g_{Ndd1,Mcm1}(t)$, $g_{Ndd1,Swi5}(t)$, $g_{Mcm1,Swi5}(t)$, and $g_{Ndd1,Mcm1,Swi5}(t)$ denote the following nonlinear interactions among the three TFs:

$$\begin{aligned} g_{Ndd1,Mcm1}(t) &= :f(g_{Ndd1}(t), g_{Mcm1}(t)) \\ g_{Ndd1,Swi5}(t) &= :f(g_{Ndd1}(t), g_{Swi5}(t)) \\ g_{Mcm1,Swi5}(t) &= :f(g_{Mcm1}(t) g_{Swi5}(t)) \end{aligned} \tag{5.19}$$

and

$$g_{Ndd1,Mcm1,Swi5}(t) = :f(g_{Ndd1}(t), g_{Mcm1}(t), g_{Swi5}(t)) \tag{5.20}$$

The biological meaning of Eq. (5.18) is that the change in the mRNA expression level of gene *MFA2* is due to the productions of regulatory functions of individual TFs and interactions among the TFs, i.e., $g_{Ndd1}(t) + g_{Mcm1}(t) + g_{Swi5}(t) + g_{Ndd1,Mcm1}(t) + g_{Ndd1,Swi5}(t) + g_{Mcm1,Swi5}(t) + g_{Ndd1,Mcm1,Swi5}(t)$, and $-\lambda_{MFA2YMFA2}(t)$, which is the degradation of mRNA. Similarly, the cis-regulatory circuit of the target gene *CLB2* in Fig. 5.5B is modeled by

$$\begin{aligned}\dot{Y}_{CLB2}(t) = {} & g_{Fkh1}(t) + g_{Fkh2}(t) + g_{Fkh1,Fkh2}(t) \\ & + g_{Fkh1,Fkh2,Ndd1}(t) - \lambda_{CLB2} Y_{CLB2}(t) + \varepsilon_{CLB2}(t)\end{aligned} \tag{5.21}$$

Table 5.2 The reference gene clusters (RGCs) of *MFA2*

	Fkh1	Fkh2	Ndd1	Mcm1	Ace2	Swi5	Mbp1	Swi4	Swi6
MFA2 (YNL145W)	0	0	1	1	0	1	0	0	0

Reference genes cluster (RGC) and their connectivities to cis elements

ORF	Fkh1	Fkh2	Ndd1	Mcm1	Ace2	Swi5	Mbp1	Swi4	Swi6	ORF	Fkh1	Fkh2	Ndd1	Mcm1	Ace2	Swi5	Mbp1	Swi4	Swi6
YAL022C	0	0	0	0	0	1	0	0	0	YKL209C	0	0	0	1	0	0	0	0	0
YAR018C	0	0	1	1	0	0	0	0	0	YLR274W	0	0	0	1	0	0	0	0	0
YDR150W	0	0	1	0	0	0	0	0	0	YML050W	0	0	1	1	0	0	0	0	0
YFL026W	0	0	0	1	0	0	0	0	0	YML125C	0	0	0	0	0	1	0	0	0
YGL032C	0	0	0	1	0	0	0	0	0	YMR001C	0	0	1	0	0	0	0	0	0
YIL050W	0	0	0	0	0	1	0	0	0	YMR002W	0	0	1	0	0	0	0	0	0
YIL129C	0	0	0	0	0	1	0	0	0	YMR253C	0	0	0	1	0	0	0	0	0
YJL079C	0	0	1	1	0	0	0	0	0	YNL056W	0	0	1	1	0	0	0	0	0
YJL157C	0	0	0	1	0	0	0	0	0	YNL058C	0	0	1	1	0	0	0	0	0
YKL163W	0	0	0	1	0	1	0	0	0	YNL145W	0	0	1	1	0	1	0	0	0
YKL164C	0	0	0	1	0	1	0	0	0	YOR066W	0	0	0	1	0	0	0	0	0

Target gene *MFA2* and the connectivities to cis elements.

Table 5.3 The reference gene clusters (RGCs) of *CLB2*

	Fkh1	Fkh2	Ndd1	Mcm1	Ace2	Swi5	Mbp1	Swi4	Swi6
CLB2 (YPR119W)	1	1	1	1	0	0	0	1	1

Reference genes cluster (RGC) and their connectivities to cis elements

ORF	Fkh1	Fkh2	Ndd1	Mcm1	Ace2	Swi5	Mbp1	Swi4	Swi6	ORF	Fkh1	Fkh2	Ndd1	Mcm1	Ace2	Swi5	Mbp1	Swi4	Swi6
YAR018C	0	0	1	1	0	0	0	0	0	YKR013W	0	0	0	0	0	0	0	1	1
YBR133C	0	1	0	0	0	0	0	0	0	YLR056W	0	0	0	0	0	0	0	1	1
YBR138C	1	0	1	0	0	0	0	0	0	YLR084C	0	1	1	1	0	0	0	1	0
YBR139W	1	0	1	0	0	0	0	0	0	YLR131C	0	1	1	1	0	0	0	0	0
YCL063W	1	1	0	0	0	0	0	0	0	YLR190W	0	1	1	1	0	0	0	0	0
YDL227C	0	0	0	0	0	0	0	1	1	YLR209C	1	0	0	0	0	0	0	0	0
YDR033W	0	1	1	0	0	0	0	0	0	YLR210W	1	0	0	0	0	0	0	0	0
YDR146C	0	1	1	1	0	0	0	0	0	YLR274W	0	0	0	1	0	0	0	0	0
YDR150W	0	0	1	0	0	0	0	0	0	YLR342W	0	0	0	0	0	0	0	1	1
YDR224C	0	0	0	0	0	0	0	1	1	YML050W	0	0	1	1	0	0	0	0	0
YDR225W	0	0	0	0	0	0	0	1	1	YML064C	1	1	0	0	0	0	0	0	0
YDR507C	0	0	0	1	0	0	0	1	1	YMR001C	0	0	1	0	0	0	0	0	0
YEL017W	1	0	0	0	0	0	0	0	0	YMR002W	0	0	1	0	0	0	0	0	0
YEL040W	0	1	0	1	0	0	0	1	1	YMR015C	1	0	0	0	0	0	0	1	0
YER001W	0	0	0	0	0	0	0	1	1	YMR183C	1	0	0	0	0	0	0	0	0
YFL026W	0	0	0	1	0	0	0	0	0	YMR253C	0	0	0	1	0	0	0	0	0
YGL032C	0	0	0	1	0	0	0	0	0	YMR305C	0	0	0	0	0	0	0	1	1
YGL038C	0	0	0	0	0	0	0	1	1	YMR307W	0	0	0	0	0	0	0	1	1
YGL116W	0	1	1	1	0	0	0	0	0	YNL056W	0	0	1	1	0	0	0	0	0
YGR014W	0	0	0	0	0	0	0	1	1	YNL058C	0	0	1	1	0	0	0	0	0
YGR099W	1	0	0	0	0	0	0	0	0	YNL231C	0	1	0	0	0	0	0	1	1
YGR151C	0	0	0	0	0	0	0	1	0	YNL300W	0	0	0	0	0	0	0	1	1
YGR152C	0	0	0	0	0	0	0	1	0	YOL011W	0	0	0	0	0	0	0	1	0
YGR153W	0	0	0	0	0	0	0	1	0	YOL030W	1	0	0	0	0	0	0	0	0
YGR221C	0	0	0	0	0	0	0	1	1	YOL114C	0	0	0	0	0	0	0	1	1

(*Continued*)

Table 5.3 (Continued)

	Fkh1	Fkh2	Ndd1	Mcm1	Ace2	Swi5	Mbp1	Swi4	Swi6										
YGR279C	0	0	0	0	0	0	0	1	1	YOR066W	0	0	0	1	0	0	0	0	0
YHR061C	0	1	0	0	0	0	0	1	1	YOR073W	0	1	0	0	0	0	0	0	0
YIL056W	0	1	1	0	0	0	0	1	1	YOR372C	0	0	0	0	0	0	0	1	1
YIL121W	0	0	0	0	0	0	0	1	0	YPL032C	1	0	0	0	0	0	0	0	0
YIL123W	0	1	0	1	0	0	0	1	1	YPL116W	1	0	0	0	0	0	0	0	0
YIL158W	0	1	1	1	0	0	0	0	0	YPL127C	0	0	0	0	0	0	0	1	1
YJL051W	0	1	1	1	0	0	0	0	0	YPL141C	1	0	0	0	0	0	0	0	0
YJL079C	0	0	1	1	0	0	0	0	0	YPL155C	0	1	0	0	0	0	0	0	0
YJL057C	0	0	0	1	0	0	0	0	0	YPL163C	0	0	0	0	0	0	0	1	1
YJL158C	0	1	1	0	0	0	0	1	1	YPL255W	0	0	0	0	0	0	0	0	1
YJR054W	0	0	0	0	0	0	0	1	1	YPL256C	0	0	0	0	0	0	0	0	1
YJR092W	1	1	1	1	0	0	0	0	0	YPR013C	1	0	0	0	0	0	0	1	0
YJR110W	0	1	0	0	0	0	0	0	0	YPR119W	1	1	1	1	0	0	0	1	1
YKL096W	0	1	0	0	0	0	0	1	1	YPR149W	0	1	1	0	0	0	0	1	0
YKL103C	0	0	0	0	0	0	0	1	0	YPR159W	0	0	0	0	0	0	0	1	1
YKL209C	0	0	0	1	0	0	0	0	0										

Target gene *CLB2* and the connectivities to cis elements.

For simplicity, the indices of target genes and cis elements are denoted by numerical notation, so that the cis-regulatory circuit of gene i can be written as

$$\dot{Y}_i(t) = \sum_{p}^{v} g_p(t) + \sum_{pq} g_{p,q}(t) + \sum_{pqr} g_{p,q,r}(t) - \lambda_i Y_i(t) + \varepsilon_i(t) \tag{5.22}$$

where v is the total number of cis elements in gene i and its corresponding degradation rate λ_i measured by Wang et al. [43]. If λ_i is unavailable, it should be estimated together with the parameters $g_p(t)$, $g_{p,q}(t)$, $g_{p,q,r}(t)$ (see Section 5.2.4). For the cis-regulatory circuits in Eqs. (5.18), (5.21), and (5.22), one obviously cannot estimate the multiple unknowns $g_p(t)$, $g_{p,q}(t)$, $g_{p,q,r}(t)$, ... by only the expression profile (i.e., $Y_i(t)$) of the ith target gene. However, since the functions $g_p(t)$, $g_{p,q}(t)$, $g_{p,q,r}(t)$, ... are assumed to be the same for all genes in the RGC and as there are overlaps of cis elements among genes in this RGC, one can estimate these functions from an array of expression profiles $Y_1(i)$, ..., $Y_N(t)$ of the genes in the RGC simultaneously, taking advantage of cross-information enhancement. The RGCs for *MFA2* and *CLB2* are shown in Tables 5.2 and 5.3, respectively. In this situation, a cross-gene identification method is proposed as follows. By integrating the dynamic equations of cis-regulatory circuits for N genes in the RGC of the gene of interest, we obtain the following array of dynamic equations

$$\begin{bmatrix} \dot{Y}_1(t) \\ \dot{Y}_2(t) \\ \vdots \\ \dot{Y}_i(t) \\ \vdots \\ \dot{Y}_N(t) \end{bmatrix} = \begin{bmatrix} 1 & 0 & \dots & 0 & 1 & \dots & 0 & \dots & 1 \\ 1 & 1 & \dots & 1 & 1 & \dots & 1 & \dots & 0 \\ \vdots & \vdots & \ddots & \vdots & \vdots & \ddots & \vdots & \ddots & \vdots \\ 0 & 1 & \dots & 0 & 0 & \dots & 0 & \dots & 1 \\ \vdots & \vdots & \ddots & \vdots & \vdots & \ddots & \vdots & \ddots & \vdots \\ 1 & 1 & \dots & 1 & 0 & \dots & 0 & \dots & 1 \end{bmatrix} \begin{bmatrix} g_1(t) \\ g_2(t) \\ \vdots \\ g_{1,2}(t) \\ g_{1,3}(t) \\ \vdots \\ g_{1,2,3}(t) \\ \vdots \\ g_{p,q,r}(t) \end{bmatrix} - \begin{bmatrix} \lambda_1 Y_1(t) \\ \lambda_2 Y_2(t) \\ \vdots \\ \lambda_i Y_i(t) \\ \vdots \\ \lambda_N Y_N(t) \end{bmatrix} + \begin{bmatrix} \varepsilon_1(t) \\ \varepsilon_1(t) \\ \vdots \\ \varepsilon_i(t) \\ \vdots \\ \varepsilon_N(t) \end{bmatrix} \tag{5.23}$$

In the cross-gene dynamic equations in Eq. (5.23), the process to identify regulatory functions $g_p(t)$, $g_{p,q}(t)$, and $g_{p,q,r}(t)$ from microarray data $Y_i(t)$, $i = 1, \ldots, N$ is called the cross-gene identification approach, in which the regulatory functions $g_p(t)$, $g_{p,q}(t)$, and $g_{p,q,r}(t)$ are shared by genes in the RGC. Therefore, the estimation of the regulatory functions of one gene can also use the information from the profiles $Y_1(t), \ldots,$ $Y_N(t)$ of other genes in RGC to improve the identification ability of the regulatory functions to reconstruct the cis-regulatory circuit of the gene of interest, which is called cross-information enhancement.

Remark 5.1: Suppose that the gene of interest in Eq. (5.22) has cis elements $p = 1,...,v$. Then all genes whose cis elements are subsets of these v cis elements are included in the same RGC of the gene of interest.

Cross-gene identification scheme for gene regulatory mechanism

Since the number of functions $g_p(t)$, $g_{p,q}(t)$, $g_{p,q,r}(t)$, ... is finite, we can estimate these functions if the number N of dynamic equations in Eq. (5.23) is large enough. Eq. (5.23) can be rewritten in an algebraic form

$$X(t) = AG(t) + E(t)$$

$$\text{where } X(t) = \begin{bmatrix} \dot{Y}_1(t) + \lambda_1 Y_1(t) \\ \dot{Y}_2(t) + \lambda_2 Y_2(t) \\ \vdots \\ \dot{Y}_i(t) + \lambda_i Y_i(t) \\ \vdots \\ \dot{Y}_N(t) + \lambda_N Y_N(t) \end{bmatrix}, \quad G(t) = \begin{bmatrix} g_1(t) \\ g_2(t) \\ \vdots \\ g_{1,2}(t) \\ g_{1,3}(t) \\ \vdots \\ g_{1,2,3}(t) \\ \vdots \\ g_{p,q,r}(t) \end{bmatrix}, \text{ and } E(t) = \begin{bmatrix} \varepsilon_1(t) \\ \varepsilon_1(t) \\ \vdots \\ \varepsilon_i(t) \\ \vdots \\ \varepsilon_N(t) \end{bmatrix}. \tag{5.24}$$

Remark 5.2: In order to calculate the derivatives in $X(t)$ from undersampled data, a cubic spline interpolation method [47,48] is employed for curve fitting to obtain more accurate differential values and to learn more reliable models.

In Eq. (5.24), $X(t)$ can be calculated from microarray data for the RGC of the gene of interest, which can then be used to estimate the vector $G(t)$ by the least squares method, leading to the following solution:

$$\hat{G}(t) = (A^T A)^{-1} A^T X(t) \tag{5.25}$$

for all t. After $\hat{G}(t)$ is estimated from Eq. (5.25), the regulatory functions $g_1(t)$, ..., $g_{1,2}(t)$, ... and $g_{p,q,r}(t)$ in Eq. (5.24) can be reconstructed for the genes in the RGC at every time point. If a cis-regulatory circuit is free of any function $g_p(t)$, $g_{p,q}(t)$, or $g_{p,q,r}(t)$, the value of the estimated function should be very small or zero. After the functions are estimated, they can be plugged into Eq. (5.22) and the reconstruction of the cis-regulatory circuit of the gene of interest is completed.

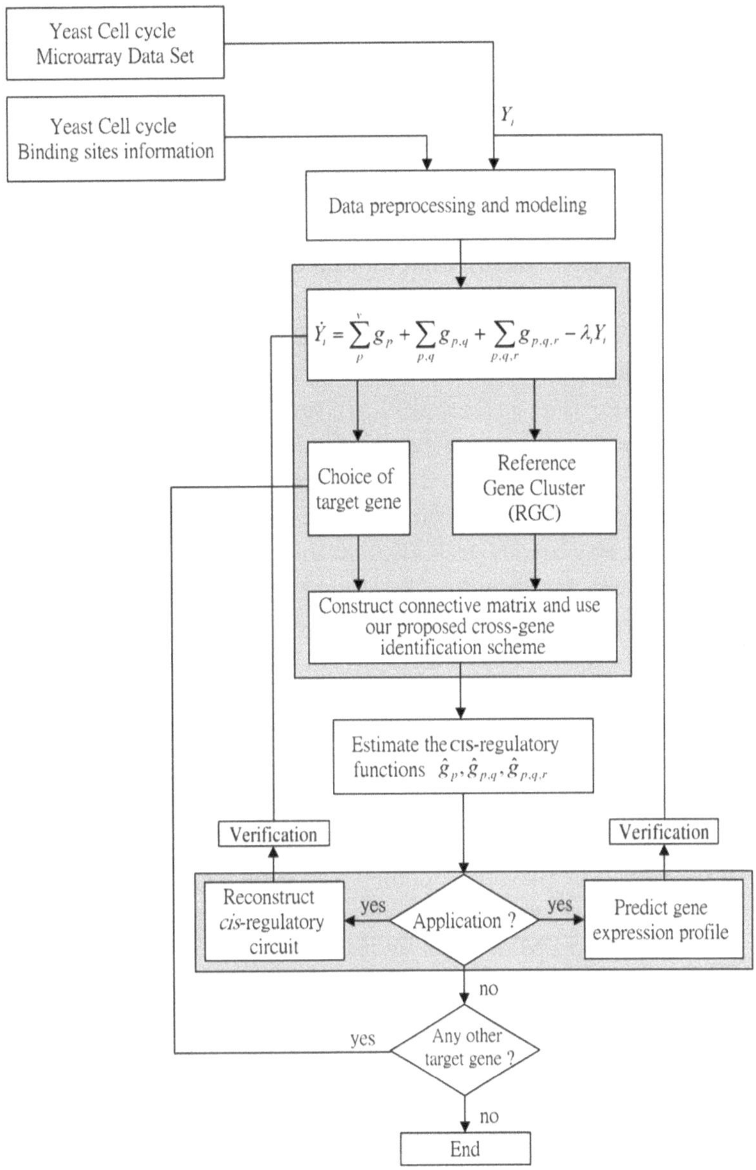

Figure 5.6 The overall flowchart of the modeling, identification, and prediction of a cis-regulatory circuit.

The flowchart for modeling, identifying, and predicting a cis-regulatory circuit is shown in Fig. 5.6.

In order to obtain more accurate cis-regulatory circuits, the model should include the triple interactions among the recognizing TFs; i.e., the vector $G(t)$ in Eq. (5.24) should include $g_{p,q,r}(t)$.

Remark 5.3: If the degradation parameters λ_i in Eq. (5.22) are unavailable, the estimation procedure of $G(t)$ and λ_i from Eq. (5.23) to (5.25) should be modified as in Section 5.2.4.

Two examples of identifying cis-regulatory mechanism in gene expression

1. The cis-regulatory circuit of MFA2

Suppose that the cis-regulatory circuit of the MFA2 gene is of interest. We construct a dynamic model of the cis-regulatory circuits of the genes in the RGC of MFA2 in Table 5.2 and then estimate the regulatory functions and interactions by the cross-gene identification scheme. The estimated transcriptional regulatory functions $g_p(t)$ and interactions $g_{p,q}(t)$ and $g_{p,q,r}(t)$ are shown as the insets in Fig. 5.5A. These insets indicate that cis elements Ndd1, Mcm1, and Swi5 in MFA2 all have cell cycle regulatory abilities in the late G1 phase. In addition, every individual cis element has a positive regulatory function on the MFA2 gene; e.g., the function $g_4(t)$ for Mcm1 has an obvious peak value in the transition phase late G1 of the cell cycle. Michaelis and Herskowitz [42] found that the MFA2 gene causes the cell cycle arrest at the G1 phase and is required for mating in yeast. Note that the interaction $g_{3,6}(t)$ between TFs Ndd1 and Swi5 is very weak or absent in the cell cycle. In contrast, the interaction $g_{4,6}(t)$ between TFs Mcm1 and Swi5 is dynamic; it has a high positive peak value in the late G1 phase, which coincides with MFA2's activity phase. This interaction seems to play an important role of positive regulation in this cis-regulatory circuit. On the other hand, the regulatory function $g_{3,4,6}(t)$ of the interaction among TFs Ndd1, Mcm1, and Swi5 is negative on gene MFA2. This regulation may repress the expression of gene MFA2 to make the expression decay to the steady state. If there is no repression function such as $g_{3,4,6}(t)$, the expression of MFA2 will increase and may disrupt in the cell cycle.

2. The cis-regulatory circuit of CLB2

Clb proteins are crucial cyclins for completing the G2/M transition of the mitotic cell cycle and the most typical one is the B-type mitotic cyclin Clb2, which is required for entry into mitosis [49]. Suppose that the cis-regulatory circuit of gene CLB2 is of interest. Fkh1, Fkh2, Ndd1, Mcm1, Swi4, and Swi6 have been identified as the TFs that bind to the promoter sequence of CLB2 [7,25,30,50]. As shown in Fig. 5.5B, the possible cis-regulatory circuit of CLB2 is very complex. Using the cross-gene identification scheme, we reconstructed the cis-regulatory functions as shown in Fig. 5.7. Although there are 41 possible regulatory functions, including $g_p(t)$, $g_{p,q}(t)$, and $g_{p,q,r}(t)$, only 20 regulatory functions are found to have nonzero values (Fig. 5.7).

The two cis elements *Fkh1* and *Fkh2* are found to have very similar regulatory functions $g_1(t)$ and $g_2(t)$, in agreement with the experimental evidence that the forkhead family members Fkh1 and Fkh2 of TFs have overlapping roles in the control of

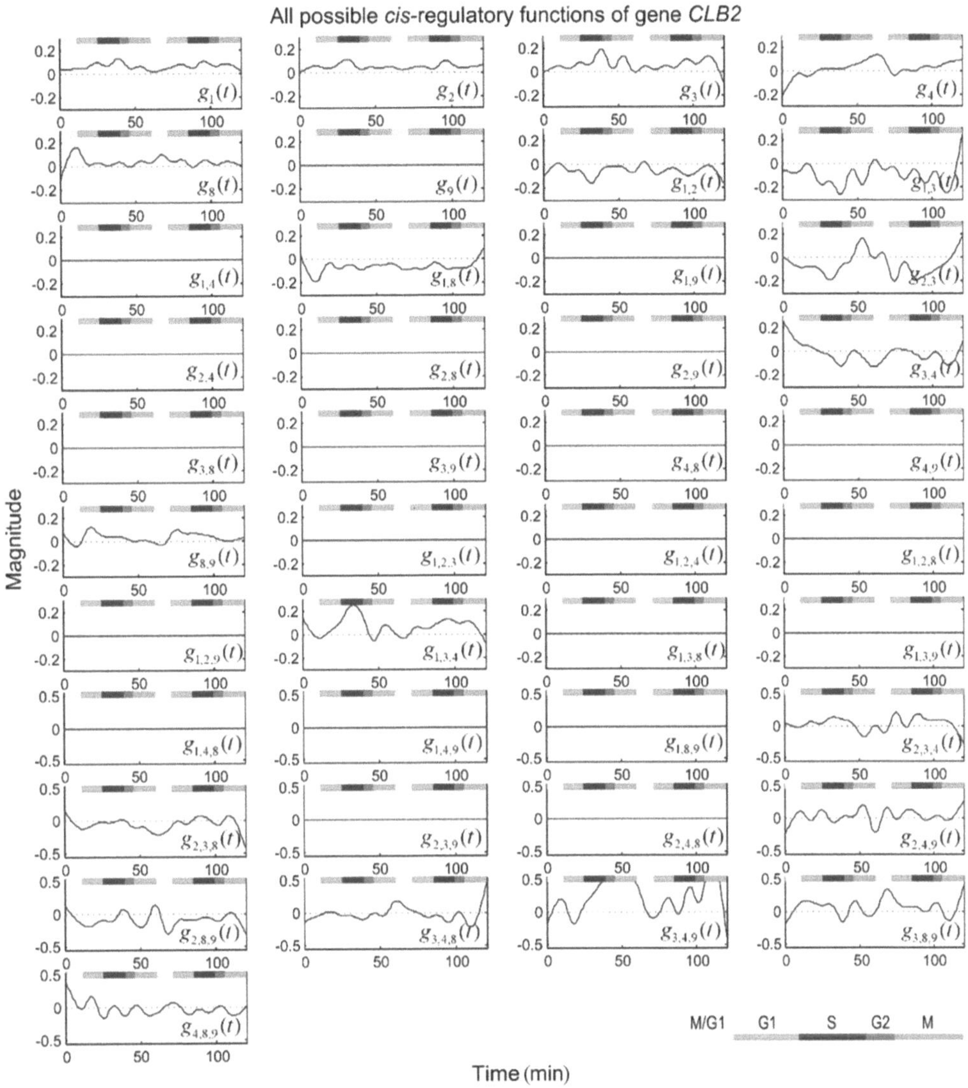

$g_1(t) =: g_{\text{Fkh1}}(t)$ $g_3(t) =: g_{\text{Ndd1}}(t)$ $g_8(t) =: g_{\text{Swi4}}(t)$ $g_{1,2}(t) =: g_{\text{Fkh1,Fkh2}}(t)$

$g_2(t) =: g_{\text{Fkh2}}(t)$ $g_4(t) =: g_{\text{Mcm1}}(t)$ $g_9(t) =: g_{\text{Swi6}}(t)$ $g_{1,3,4}(t) =: g_{\text{Fkh1,Ndd1,Mcm1}}(t)$

Figure 5.7 All estimated cis-regulatory functions, including the regulatory function $g_p(t)$ of each individual TF and the interactions $g_{p,q}(t)$ and $g_{p,q,r}(t)$ among the TFs that recognize the cis elements of the *CLB2* gene. The numerical notation of regulatory functions is given in the box at the bottom of the figure. Different phases in the cell cycle are indicated by the colored bar near the right lower corner.

the G2/M transition [23,25]. The regulatory function $g_2(t)$ of Fkh2 has a distinct cell cycle regulatory ability (Fig. 5.7) and especially, the interaction function $g_{2,3}(t)$ between Fkh2 and Ndd1 has a strong regulatory contribution to the gene expression profile in the M/G1 phase. The regulatory function $g_4(t)$ of Mcm1 has two peaks. There is experimental evidence that Mcm1 is a member of an evolutionarily conserved class of TFs that have related to DNA-binding sequences and dimerization domains. In addition, Mcm1 binds the early cell cycle box (ECB) that contains a Mcm1 cis element in the SWI4, CLN3, CDC6, and CDC47 promoters and activates M/G1-specific transcription [51].

The cell cycle genes that are activated during the late G1 or S phase have SBF or MBF sequence-specific TFs that bind the cis elements in their promoter region. SBF (the Swi4-Swi6 cell cycle box binding factor) is a heterodimer of Swi4 and Swi6 [27,30,31]. The regulatory functions $g_8(t)$ and $g_9(t)$ of Swi4 and Swi6 and their interaction function $g_{8,9}(t)$ are estimated using the dynamic expression model (Fig. 5.7). It is well known that neither Swi4 nor Swi6 alone has obvious cell cycle regulation ability, and indeed we found that $g_8(t)$ has only one peak and so shows no cycle and that $g_9(t)$ shows no capability of cell cycle regulation (Fig. 5.7). In contrast, the combination of Swi6 and Swi4 to make the complex SBF enables the cis elements Swi6 and Swi4 to provide cell cycle regulation capacity; that is, the interaction function $g_{8,9}(t)$ of Swi6 and Swi4 has a peak during the G1/S phase of the cell cycle. Ndd1 and Fkh2 are bound to identical promoters throughout the cell cycle and their interaction $g_{2,3}(t)$ is an important transcriptional process targeted by the Cdk activity [50,52]. In addition, there is another obvious positive interaction $g_{3,4,9}(t)$ among Ndd1, Mcm1, and Swi6 (Fig. 5.7). It has a large regulatory ability in the G1/S phase, which almost dominates the expression profile of CLB2. In contrast, $g_{3,4}(t)$ has a negative regulation contribution. We therefore propose that the key factor Swi6 in the interaction $g_{3,4,9}(t)$ is similar to its role in SBF and MBF. At any rate, this model suggests that Swi6 plays a key role in the interaction among Ndd1, Mcm1, and Swi6. This is a new finding in the cis-regulatory circuit of CLB2.

We also confirm the well-known interaction among Ndd1, Fkh2, and Mcm1 in the cis-regulatory circuits of the CLB2 and SWI5 genes because the interaction function $g_{2,3,4}(t)$ has a distinct regulatory ability in the G2/M phase of the cell cycle (Fig. 5.7). Interestingly, the interaction function $g_{3,4,9}(t)$ among Ndd1, Mcm1, and Swi6 is about two times higher than any of the other functions in Fig. 5.7. In addition, the interaction $g_{1,3,4}(t)$ among Fkh1, Ndd1, and Mcm1 is highly positive, the interaction $g_{3,4,8}(t)$ among Ndd1, Mcm1, and Swi4 is mildly positive, while the interaction $g_{3,8,9}(t)$ among Ndd1, Swi4, and Swi6 is negative. These observations are in agreement with the fact that the regulatory ability of an interaction among TFs is usually much stronger than that of an individual TF; in other words, there is synergy among TFs.

In summary, there are many experimental observations that support the cis-regulatory functions identified by the dynamic model and this model provides novel insights into the quantitative regulation of the cis-regulatory circuit of a gene of interest.

Support from expression phases of TFs in the cell cycle

In this section, we investigate the question of why the strengths of regulatory functions in the cis-regulatory circuits are different in different phases of the cell cycle. Based on the mRNA expression profiles of TF genes from experiments, the distribution of the expressions of the nine TF genes in the different phases of the cell cycle is shown in Fig. 5.8. In support of these results, the large positive interaction functions (peaks) among a set of TFs always occur during the expression phases of the genes of the interacting TFs. For example, for the cis-regulatory circuit of MFA2 (Fig. 5.5A), there is an obvious peak for the function $g_{4,6}(t)$ of the interaction between TFs Mcm1 and Swi5 during the M phase, and in Fig. 5.8, this peak indeed occurs during the expression phases of the two TF genes [2,51]. As another example, for the cis-regulatory circuit of CLB2, there is a strong interaction ($g_{3,4,9}(t)$) among Ndd1, Mcm1, and Swi6 starting from the G2 phase (Fig. 5.7). We can therefore infer that the expression of a cell cycle gene in a specific phase of the cell cycle needs a specific inducing signal, which is mainly from the interactions of certain specific TFs that bind the cis elements of the gene.

Accuracy of reconstructed cis-regulatory circuits in transcriptional regulation mechanism

The accuracy of the reconstructed cis-regulatory circuit of a gene can be evaluated by reconstructing the expression profile of the gene using the reconstructed cis-regulatory circuit

$$\dot{\hat{Y}}_i(t) = \sum_p^v \hat{g}_p(t) + \sum_{pq} \hat{g}_{p,q}(t) + \sum_{pqr} \hat{g}_{p,q,r}(t) - \lambda_i \hat{Y}_i(t) \tag{5.26}$$

where $\hat{g}_p(t)$, $\hat{g}_{p,q}(t)$, and $\hat{g}_{p,q,r}(t)$ have been estimated by the cross-gene identification scheme. The reconstructed profile and the observed profile are compared in Fig. 5.9. We find that if the number of the cis elements of a gene is large enough, the reconstructed expression profile is very accurate. Fortunately, although the reconstructed expression profile is not accurate in some genes, the trend of the expression profile for a gene is always correct.

Prediction of gene expression profile by the corresponding cis-regulatory circuits

In the above investigation, each regulatory circuit was identified using 95% of genes in its RGC, and the remaining 5% of genes in the RGC will now be used for predicting expression profiles, i.e., for cross validation. This cross-gene identification scheme assumes that the regulatory functions of TFs are universal in the cluster of genes with

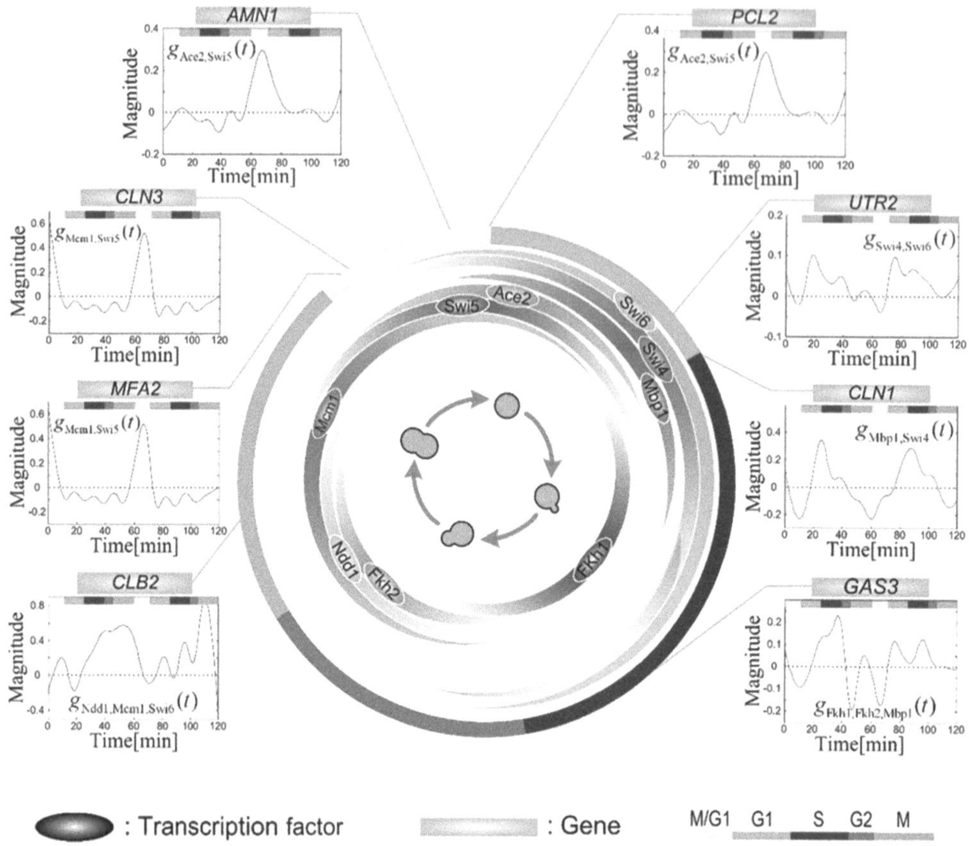

Figure 5.8 The gene expression phases match the main regulatory functions of TFs. It is seen that the main interaction functions of TFs have a peak value and always occur during or soon after the mRNA expression phases of the corresponding genes. For example, the gene *MFA2* has the main interaction regulatory function $g_{Mcm1,Swi5}(t)$ which has a peak during the expression phases between the TFs Mcm1 and Swi5 by identifying the cis-regulatory circuit. As another example, the gene *UTR2* has the main interaction regulatory function $g_{Swi4,Swi6}(t)$, which has a peak during the expression phases between the TFs Swi4 and Swi6 by identifying the cis-regulatory circuit. These results indicate that the main regulatory functions have a peak value phase to match the gene expression phase. Therefore, we can estimate the gene expression phase by identifying the main regulatory function. The width of a colored band in the inner circle is approximately proportional to the expression level of the TF gene of interest in the cell cycle. A *pink line* (black in print versions) points to the main expression phase of a target gene in the *pink box* (gray in print versions). Different phases in the cell cycle are indicated by the colored bar at the right lower corner.

similar functions. Under this assumption, this method should be able to predict the expression profiles of other genes in the same cluster that have not been employed in Eq. (5.23) to reconstruct the cis-regulatory circuits. This is one way to validate this model. In Fig. 5.10, we randomly chose 100 RGCs to test the prediction accuracy;

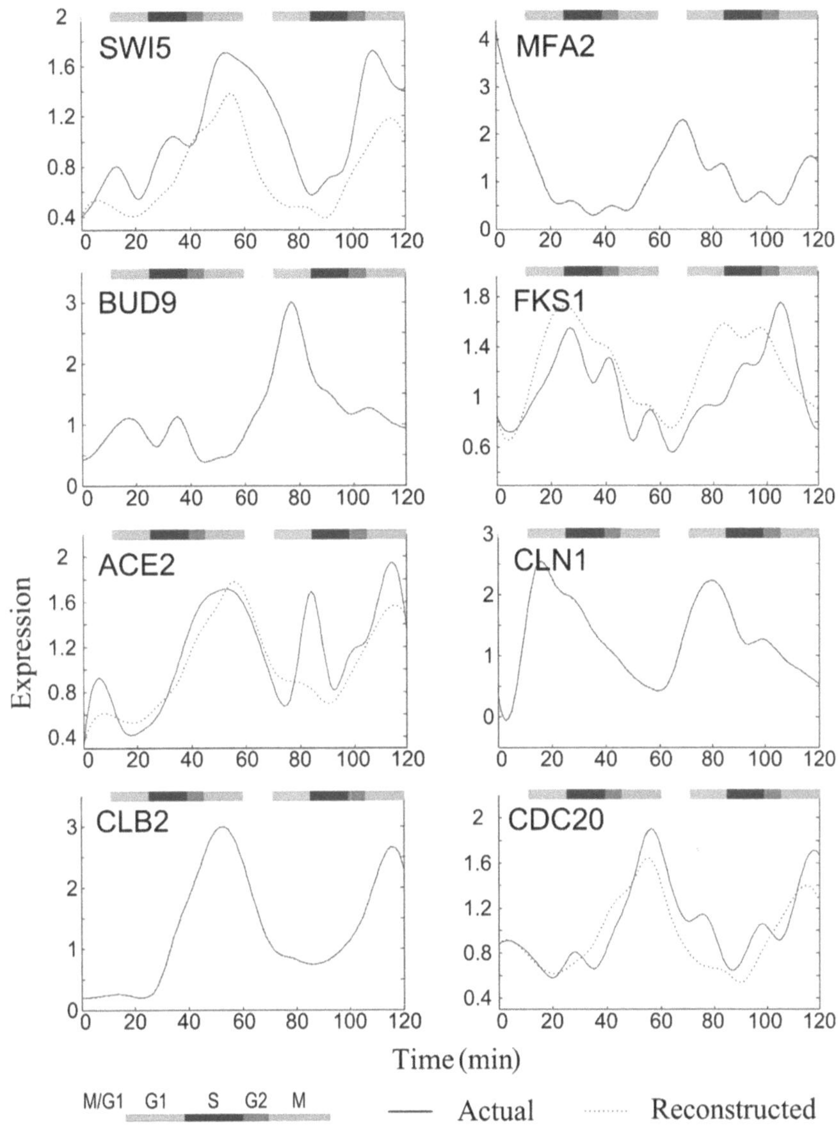

Figure 5.9 Comparison between the actual gene expression profiles (with cubic spline) and the reconstructed gene expression profiles. The examples shown were randomly chosen. The reconstructed gene expression profiles were obtained by integrating the estimated cis-regulatory functions and the chromatin immunoprecipitation data. When there is only one *blue line* (black in print versions) in a figure it means that the reconstructed function is very close to the actual gene expression profile. Different phases in the cell cycle are indicated by the colored bar.

(A)

Number of RGC	Predicted gene accession number	Mean square error (MSE)
RGC_1	YGL028C	1.2694
RGC_2	YJL157C	0.3482
RGC_3	YHR152W	0.1259
RGC_4	YHR143W	0.695
RGC_5	YMR305C	0.0855
RGC_6	YIL121W	0.1153
RGC_7	YOL011W	0.0869
RGC_8	YHR149C	0.311
RGC_9	YGR152C	0.4089
RGC_10	YPR120C	0.4474
RGC_11	YNR009W	0.4067
RGC_12	YPL255W	0.3916
RGC_13	YML064C	0.0736
RGC_14	YML125C	0.0939
RGC_15	YPR013C	0.0926
RGC_16	YKL151C	0.1541
RGC_17	YKL044W	0.9212
RGC_18	YNL339C	2.0549
RGC_19	YMR145C	0.1306
RGC_20	YMR015C	0.1359
RGC_21	YPL155C	0.6567
RGC_22	YJL051W	0.7856
RGC_23	YOL030W	0.4845
RGC_24	YJR110W	0.2617
RGC_25	YIL056W	0.8052
RGC_26	YNL056W	0.4925
RGC_27	YLR013W	0.4122
RGC_28	YLR209C	0.5594
RGC_29	YLR084C	0.68
RGC_30	YIL009W	0.2891
RGC_31	YGR086C	0.3596
RGC_32	YAL022C	0.1352
RGC_33	YNL058C	0.41
RGC_34	YIL158W	0.0466
RGC_35	YLR190W	0.3882
RGC_36	YPL242C	0.1649
RGC_37	YCR024C-A	0.1783
RGC_38	YHR151C	0.1891
RGC_39	YGL008C	0.7552
RGC_40	YIL050W	0.1061
RGC_41	YOR073W	0.2387
RGC_42	YEL017W	0.1188
RGC_43	YPL116W	0.1665
RGC_44	YCL063W	0.113
RGC_45	YMR183C	0.0296
RGC_46	YIL129C	0.1597
RGC_47	YPL032C	0.0551
RGC_48	YKL008C	0.117
RGC_49	YLR342W	0.0814
RGC_50	YDL055C	0.2077
RGC_51	YBL002W	0.316
RGC_52	YPL127C	0.2238
RGC_53	YDL018C	0.4313
RGC_54	YJR030C	0.3781
RGC_55	YMR078C	0.5847
RGC_56	YNL300W	0.7581
RGC_57	YGR153W	0.4001
RGC_58	YPR075C	0.1955
RGC_59	YGR189C	0.5323
RGC_60	YGR234W	0.8625
RGC_61	YJL159W	0.8619
RGC_62	YDL127W	1.5604
RGC_63	YER190W	0.0661
RGC_64	YOR074C	0.1065
RGC_65	YML027W	0.1176
RGC_66	YDL101C	0.0321
RGC_67	YMR179W	0.1823
RGC_68	YER189W	0.0655
RGC_69	YGR109C	1.2195
RGC_70	YER070W	0.3758
RGC_71	YDL003W	0.429
RGC_72	YBR070C	0.4805
RGC_73	YLR463C	0.5441
RGC_74	YLR465C	0.3695
RGC_75	YPR202W	0.3572
RGC_76	YLR464W	0.3935
RGC_77	YFL065C	0.369
RGC_78	YLR462W	0.3527
RGC_79	YBL111C	0.2549
RGC_80	YLR466W	0.253
RGC_81	YOR372C	0.3655
RGC_82	YJL115W	0.3465
RGC_83	YJL079C	0.4326
RGC_84	YER145C	0.253
RGC_85	YOL012C	0.4529
RGC_86	YDR225W	1.2079
RGC_87	YDR224C	0.7289
RGC_88	YMR215W	0.6068
RGC_89	YLR056W	0.2487
RGC_90	YLR131C	0.1322
RGC_91	YMR001C	0.3859
RGC_92	YMR002W	0.0965
RGC_93	YBL003C	0.1892
RGC_94	YMR307W	0.0919
RGC_95	YOR247W	0.0256
RGC_96	YOR248W	0.0249
RGC_97	YCR065W	0.492
RGC_98	YPL163C	1.1383
RGC_99	YPL256C	1.0978
RGC_100	YGR014W	0.1084

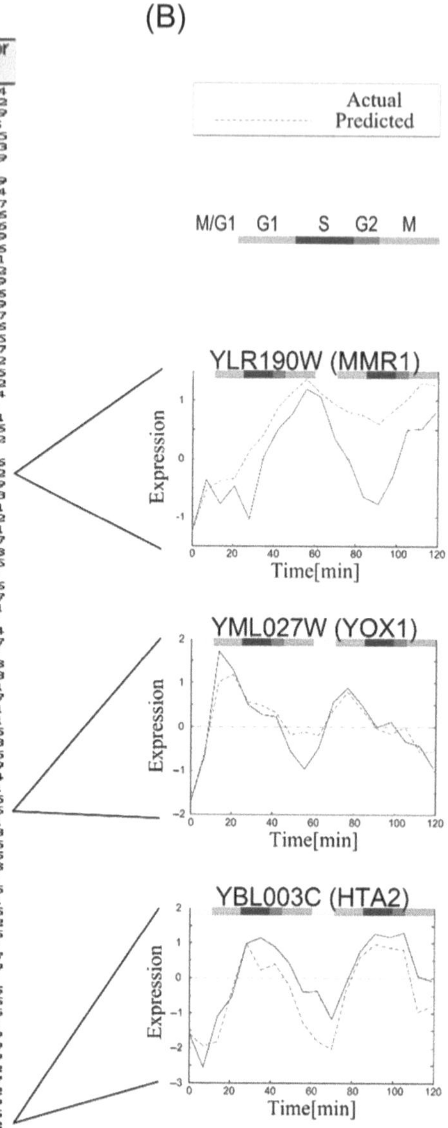

Figure 5.10 The regulatory functions integrated with ChIP-chip data to predict gene expression profiles. To test the prediction performance of this model, (A) 100 yeast cell cycle genes that have not been employed in the reconstruction of the cis-regulatory circuits are randomly chosen from their corresponding reference gene clusters (RGCs). The maximum mean square error (MSE) of prediction results is 2.055 and the minimum is 0.025. (B) Three examples of the comparison between the actual (*blue* (black in print versions)) and the predicted (*red* (gray in print versions)) gene expression profiles. Different phases in the cell cycle are indicated by the colored bar.

that is, for each RGC we show the prediction result for one of the unused genes. Fig. 5.10A shows the predicted target gene and the mean square error (MSE) of prediction results which has the maximum of 2.055 and the minimum of 0.025. In addition, three examples of the detailed comparison between the actual and the predicted gene expression profiles are shown in Fig. 5.10B. In general, the predicted profiles are satisfactory approximations of the observed expression profiles. We found that the smaller the number of the cis elements, the less accurate the prediction results. However, if some cis elements of a gene have strong regulatory functions, the expression profiles of this gene can be predicted accurately even when the number of cis elements is small. If some genes in the RGC have the same cis elements but have different observed gene expression profiles, these expression profiles will lead to poor estimation of parameters. This is the main cause of prediction error. Why does this situation arise? It may be that some cis elements of these genes have not yet been identified or there are some errors in the inference of the cis elements. For example, MMR1 may have another cis element such as Gcr2 [5] and this may be why the predicted profile is quite different from the observed profile (Fig. 5.10B).

5.2.2 Discussion of gene regulatory mechanism via identifying corresponding cis-regulatory circuits

In contrast to current methods, this method uses all possible expression profile information from the cluster of genes to reconstruct the cis-regulatory circuit of a target gene. In particular, this method is capable of extracting dynamic interactions among TFs. For this reason, the analysis and interpretation of output expression profiles become straightforward. Therefore, this method has a high potential for applications such as studying variations of the cis-regulatory circuit of the same gene in different yeast strains to investigate the regulatory evolution of the gene.

The novelties of this section include: (1) a nonlinear dynamical model is developed for cis-regulatory circuits of gene regulatory mechanism in terms of regulatory functions and interactions among TFs, (2) a cross-gene identification scheme is proposed to estimate many parameters involved in the dynamical model of cis-regulatory circuits from the expression profiles of genes in the RGC, (3) a detailed identification of the dynamic cis-regulatory abilities of TFs, which vary with time, and (4) a gene expression prediction method is developed by the proposed dynamic cis-regulatory circuit, assuming that the cis-regulatory functions of the same TFs in different circuits are the same. Three advantages of this method over current methods are that the cis-regulatory circuit is constructed with the circuit structure intact, that it uses the expression profiles of many genes simultaneously to obtain extra information, and that it is dynamic and quantitative.

Significantly, this model not only can confirm known regulations but also can provide conjectures for experimental verification. Consider the key positive cis-regulatory function $g_{4,6}(t)$ in Fig. 5.5A. We propose that during the expression of gene MFA2, the TF Ndd1 (the G2/M phase) communicates with the TF Mcm1 (the M phase) to transmit a specific signal to induce the expression of the MFA2 gene. Such conjectures from the reconstructed cis-regulatory circuits may be useful for studying the regulatory evolution of genes by comparing the cis-regulatory functions of different strains, or for predicting the gene expression behavior before conducting an experiment.

However, we found poor results in some cases. For example, in Fig. 5.7, we were unable to find the basal regulatory function $g_9(t)$ of individual Swi6 or the interaction $g_{2,4}(t)$ between Fkh2 and Mcm1 for gene CLB2 [53]. These regulatory functions have not been identified by this scheme because they have no obvious specific phase regulatory ability. Besides, from the cis-regulatory functions in Figs. 5.5 and 5.6, several cis-regulatory function profiles did not show a clear periodicity. A possible reason may be that the original microarray data were noisy and the use of cubic spline interpolation and linear transformation of microarray data in this scheme had introduced new noise and distortions. Most probably, the cis element information used to construct the cis-regulatory circuits under yeast cell cycle is not complete; only nine significant cis elements were considered in this section to reduce the complexity of the mathematical model. Another possible source of error is that we have not considered the order of the cis elements on the promoter region, which may affect the strength of the interaction between TFs [26]. Such differences, however, can be incorporated by putting, say both $g_{p,q}(t)$ and $g_{p,q,r}(t)$, into the model.

In view of the facts that there are uncertainties about the cis elements of some of the genes studied and that microarray data are noisy, it is remarkable that this method gave accurate results for the expression profiles of the majority of the cell cycle genes studied and also gave fairly accurate predictions of the expression profiles of other cell cycle genes. In the future, if better cis elements data and more accurate and longer gene expression profile data become available, we should be able to improve the reconstruction of cis-regulatory circuits. Also, this approach may be extended to reconstruct cis-regulatory circuits in more diverse conditions and more complex eukaryotes. After cis-regulatory circuits are accurately described by explicit dynamical equations, some applications will be straightforward.

5.2.3 Conclusion of cis-regulatory mechanisms of gene expression

In this section, we assume that the regulatory functions of the same cis elements and the interaction functions among their TFs are similar across genes within the cluster of genes with overlapping cis elements; i.e., the regulatory functions and interaction functions are universal in this cluster of genes. Under this assumption, the cross-gene identification scheme takes advantage of cross-information enhancement to improve the accuracy of parameter estimation. The number of genes used should be large enough, so that we have a large number of outputs (i.e., their microarray data) for parameter estimation.

After the parameters of the cis-regulatory circuits of interest have been estimated, the cis-regulatory mechanisms can be explicitly described by plugging these parameters into their corresponding dynamic equations of cis-regulatory circuits. Moreover, these estimated functions and interactions can be used to predict the expression profiles of other genes that share the same cis-regulatory elements but have not been used to identify the cis-regulatory circuits. In this manner, we can evaluate the performance of the proposed dynamic model of cis-regulatory circuits. From a number of examples, we have found that the predicted results are in most cases satisfactory, confirming the validity of the proposed dynamic model of cis-regulatory circuits. This modeling represents a new approach to studying cis-regulatory circuits from cross-gene expression data. It is a systems biology approach because we consider the regulatory circuits of many genes and many TFs at the same time and we use system identification techniques to estimate the parameters of the circuits. The results of expression prediction from experimental data suggest that this novel approach is suitable for deciphering the regulatory functions and the cooperativity of the TFs that regulate the expression of a gene.

5.2.4 Systems biology and identification methods of constructing cis-regulatory circuits

Experimental data

To identify the cis-regulatory circuit of a gene of interest in the yeast cell cycle, we apply this approach to the data of Spellman et al. [2], which contains expression profiles of 6178 open reading frames (ORFs) in the yeast *S. cerevisiae* during the cell cycle. This analysis was applied to the α-factor arrest data set. The raw data were transformed into a linear scale from the original log ratio carried out by Spellman et al. [2]. To reduce the effect of noise and to overcome undersampled microarray data in the estimation of cis-regulatory circuits, the cubic spline was used for data interpolation and smoothing to obtain a less sensitive first derivative of the expression pattern and to learn a more reliable model. Furthermore, the noise is modeled in the noise term $\varepsilon_i(t)$ in Eq. (5.23).

From the RGC of the gene of interest, the cross-gene identification method from Eqs. (5.24) to (5.25) is employed to reconstruct the cis-regulatory circuit. The connectivity information between TFs and their target genes was obtained from the yeast cell cycle analysis [7]. We focused on the nine TFs that have been identified to play important roles in the transcription regulation of a set of yeast genes whose expressions are cell–cycle dependent; these nine TFs are Mbp1, Swi4, Swi6, Mcm1, Fkh1, Fkh2, Ndd1, Swi5, and Ace2 [23,34,50,51].

Estimation of degradation rate

If the mRNA degradation rate λ_i in Eq. (5.22) has not been estimated experimentally, it should be estimated together with the parameters $g_p(t)$, $g_{p,q}(t)$, $g_{p,q,r}(t)$. The algorithm to estimate the λ_i is described as follows. First, Eq. (5.22) is changed to

$$\dot{Y}_i(t) = \sum_{p}^{v} g_p(t) + \sum_{pq} g_{p,q}(t) + \sum_{pqr} g_{p,q,r}(t) - \lambda_i Y_i(t) + \varepsilon_i(t) \tag{5.27}$$

where v is the total number of cis elements in gene i, and the degradation rate is substituted as $\lambda_i(t)$. Similarly, since the functions $g_p(t)$, $g_{p,q}(t)$, $g_{p,q,r}(t)$, ... are assumed to be the same for all genes in the RGC and since there are overlaps of cis elements among genes in this RGC, one can estimate these functions from an array of expression profiles $Y_1(t),\ldots, Y_N(t)$ of the genes in the RGC simultaneously, taking advantage of cross-information enhancement. The RGCs for MFA2 and CLB2 are shown in Tables 5.2 and 5.3, respectively.

Second, by integrating the dynamic equations of cis-regulatory circuits for N genes in the RGC of the gene of interest, we obtain the following array of dynamic equations

$$\begin{bmatrix} \dot{Y}_1(t) \\ \dot{Y}_2(t) \\ \vdots \\ \dot{Y}_i(t) \\ \vdots \\ \dot{Y}_N(t) \end{bmatrix} = \left[\begin{array}{ccccccccccc|cccccc} 1 & 0 & \cdots & 0 & 1 & \cdots & 0 & \cdots & 1 & -Y_1(t) & 0 & \cdots & \cdots & \cdots & 0 \\ 1 & 1 & \cdots & 1 & 1 & \cdots & 1 & \cdots & 0 & 0 & -Y_2(t) & \ddots & \ddots & \ddots & \vdots \\ \vdots & \vdots & \ddots & \vdots & \vdots & \ddots & \vdots & \ddots & \vdots & \vdots & \ddots & \ddots & \ddots & \ddots & \vdots \\ 0 & 1 & \cdots & 0 & 0 & \cdots & 0 & \cdots & 1 & \vdots & \ddots & \ddots & -Y_i(t) & \ddots & \vdots \\ \vdots & \vdots & \ddots & \vdots & \vdots & \ddots & \vdots & \ddots & \vdots & \vdots & \ddots & \ddots & \ddots & \ddots & 0 \\ 1 & 1 & \cdots & 1 & 0 & \cdots & 0 & \cdots & 1 & 0 & \cdots & \cdots & \cdots & 0 & -Y_N(t) \end{array}\right]$$

$$\bullet \begin{bmatrix} g_1(t) \\ g_2(t) \\ \\ g_{1,2}(t) \\ g_{1,3}(t) \\ \vdots \\ g_{1,2,3}(t) \\ \vdots \\ g_{p,q,r}(t) \\ \lambda_1 \\ \lambda_2 \\ \vdots \\ \lambda_i \\ \vdots \\ \lambda_N \end{bmatrix} + \begin{bmatrix} \varepsilon_1(t) \\ \varepsilon_1(t) \\ \vdots \\ \varepsilon_i(t) \\ \vdots \\ \varepsilon_N(t) \end{bmatrix} \tag{5.28}$$

In the dynamic equations in Eq. (5.28), the regulatory functions $g_p(t)$, $g_{p,q}(t)$, and $g_{p,q,r}(t)$ are shared by genes in the RGC. Therefore, the estimation of these functions from expression profiles $Y_1(t)$, $Y_2(\mathrm{t})$, ..., $Y_N(t)$ can use also information from other genes to enhance this ability to reconstruct the cis-regulatory circuit of the gene of interest.

Finally, as the number of functions $g_p(t)$, $g_{p,q}(t)$, $g_{p,q,r}(t)$, ... is finite, we can estimate these functions and the decay rates $\lambda_1(t)$, $\lambda_2(t)$, ..., $\lambda_N(t)$ if the number N of dynamic equations in Eq. (5.28) is large enough. Eq. (5.28) can be rewritten in an algebraic form

$$\tilde{X}(t) = \tilde{A}(t) \bullet \tilde{G}(t) + E(t) \tag{5.29}$$

where $\tilde{X}(t) = \begin{bmatrix} \dot{Y}_1(t) \\ \dot{Y}_2(t) \\ \vdots \\ \dot{Y}_i(t) \\ \vdots \\ \dot{Y}_N(t) \end{bmatrix}$, $E(t) = \begin{bmatrix} \varepsilon_1(t) \\ \varepsilon_1(t) \\ \vdots \\ \varepsilon_i(t) \\ \vdots \\ \varepsilon_N(t) \end{bmatrix}$, and $\tilde{G}(t) = \begin{bmatrix} g_1(t) \\ g_2(t) \\ \vdots \\ g_{1,2}(t) \\ g_{1,3}(t) \\ \vdots \\ g_{1,2,3}(t) \\ \vdots \\ g_{p,q,r}(t) \\ \lambda_1 \\ \lambda_2 \\ \vdots \\ \lambda_i \\ \vdots \\ \lambda_N \end{bmatrix}$.

Then, we have the following dynamic equations for all time profiles

$$\begin{bmatrix} \tilde{X}(t_1) \\ \vdots \\ \vdots \\ \tilde{X}(t_m) \end{bmatrix} = \begin{bmatrix} \tilde{A}(t_1) & 0 & \cdots & 0 \\ 0 & \ddots & \ddots & \vdots \\ \vdots & \ddots & \ddots & 0 \\ 0 & \cdots & 0 & \tilde{A}(t_m) \end{bmatrix} \begin{bmatrix} \tilde{G}(t_1) \\ \vdots \\ \vdots \\ \tilde{G}(t_m) \end{bmatrix} + \begin{bmatrix} E(t_1) \\ \vdots \\ \vdots \\ E(t_m) \end{bmatrix} \tag{5.30}$$

Let us denote the above equations in the following simple algebraic form

$$\tilde{X} = \tilde{\Phi}\tilde{\Theta} + M \tag{5.31}$$

In Eq. (5.31), $\tilde{X}$ and $\tilde{\Phi}$ can be calculated from microarray data for the RGC of the gene of interest, which can then be used to estimate the vector $\hat{\Theta}$ by the least squares method in Eq. (2.20), leading to the following solution:

$$\hat{\tilde{\Theta}} = (\tilde{\Phi}^T \tilde{\Phi})^{-1} \tilde{\Phi} \tilde{X} \tag{5.32}$$

After $\hat{\tilde{\Theta}}$ is estimated from Eq. (5.32), the regulatory functions $\tilde{G}(t)$ in Eq. (5.29) can be reconstructed for the genes in the RGC at every time point. However, in Eq. (5.28), the degradation rate $\lambda(t)$ is a time-varying function and is affected by both the error terms and experimental data. Therefore, in order to reduce the influence on degradation rate, we simplify Eq. (5.28) and average the negative gradients of $\lambda(t)$ to obtain the constant value $\hat{\lambda}$. Then the estimated $\hat{\lambda}$ is substituted into Eq. (5.23) to reidentify the cis-regulatory functions to derive the final regulatory functions. Using this procedure, we can avoid the effects of the time-varying function $\lambda(t)$ on the identification process and reduce the influence by different experimental data. Hence, the degradation rate can be estimated. After the functions are estimated, they can be plugged into Eq. (5.27) and then the reconstruction of the cis-regulatory circuit of the gene of interest is completed.

Gene transcription depends on trans/cis-regulatory circuits of a gene. After introducing trans-regulatory circuit and cis-regulatory circuit in the previous section, a trans/cis-regulatory circuits will be constructed by nonlinear dynamic model and microarray data to understand the regulatory mechanism of cell cycle genes in Section 5.3.

5.3 NONLINEAR DYNAMIC TRANS/CIS-REGULATORY MECHANISMS FOR GENE TRANSCRIPTION VIA MICROARRAY DATA

The trans-regulatory circuit is considered as the significant regulatory interactions between upstream regulatory genes and TF binding site motifs or cis elements as shown in Fig. 5.1. The cis-regulatory circuit is viewed as a dynamic interactive circuit among binding site motifs with their effective action on the expression mechanism of target gene as shown in Fig. 5.5. In brief, gene transcription depends on the trans/cis-regulatory circuits. In this section, nonlinear trans/cis-regulatory circuits for gene transcription in yeast are constructed using microarray data, translation time delay, and information of TF binding sites. We provide a useful nonlinear dynamic modeling and develop a parameter estimating method for the construction of trans/cis-regulatory circuits as shown in Fig. 5.11, which is powerful for understanding gene transcription mechanism. We apply this method to construct trans/cis-regulatory circuits of yeast cell cycle–related genes and successfully quantify their regulatory abilities and find possible cis-element interactions. Not only could the data of yeast be applied by this method, but those of other species could be used. The proposed method can provide a quantitative basis for system analysis of gene circuits, which is potential for gene regulatory circuit design with a desired gene expression.

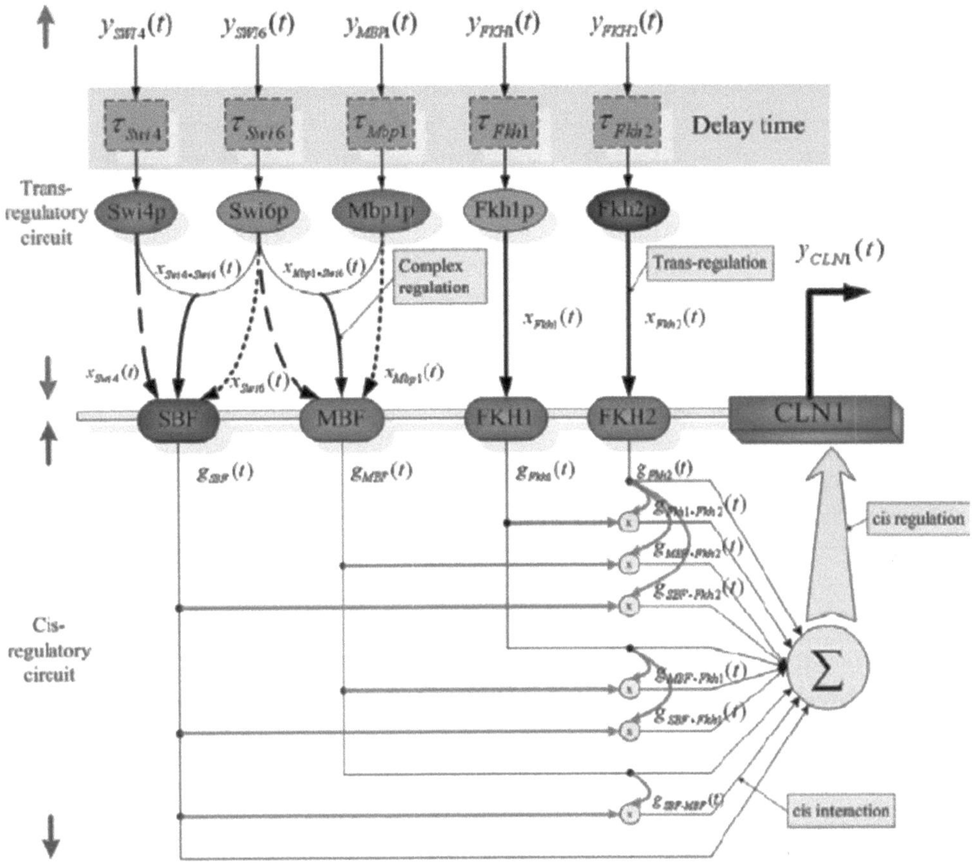

Figure 5.11 The dynamic trans/cis regulatory model of the gene transcription of target gene *CLN1*. $x_{ij}(t)$ denotes trans-regulatory function of the complex of the transcription factors *i* and *j* and $g_{ij}(t)$ denotes the interaction between the cis elements *i* and *j*.

Due to advances in DNA microarray technology and genome sequencing, it has become possible to measure gene expression levels on a genomic scale. Microarray technology provides insight into the transcriptional state of a cell. In recent studies, the expression profiles and motif binding sites of genes in yeast have been revealed, and regulatory networks have been proposed to explain their regulatory functions [2,5,7,17,30,54]. However, in order to gain more insight into the infrastructure of the regulatory scheme inside the gene transcription, trans/cis-regulatory circuit constructed by the view of the systems biology is an important topic.

Genes are always regulated by a number of upstream regulatory genes through their products, for instance, TFs bind to specific sites (i.e., cis elements) in the DNA promoter region. The interactions between upstream regulatory genes and cis elements are described by a trans-regulatory circuit (see Fig. 5.11). The DNA-binding motifs, serving as anchors to TFs, play the role of staying platforms for the assembly of multiproteins at the output. This mechanism specifies the cis-regulatory circuit, which is viewed as a dynamic interactive circuit among binding site motifs with their effective action on the expression scheme of target gene (Fig. 5.11). In brief, gene transcription depends on the trans/cis-regulatory circuits. Therefore, the comprehension of gene transcription mechanism needs to recognize the corresponding trans/cis-regulatory circuits. Unfortunately, although the cis-regulatory circuit has been widely discussed in *Drosophila* [55] and the sea urchin [45,46,56], not much is known about trans/cis-regulatory circuits because of their complicated interactive mechanisms, which are not easily detected directly by conventional experiments. That is why we want to propose a method to construct trans/cis-regulatory circuits to understand gene transcription process thoroughly.

Given a processed data set, one expects to be ready to tackle the biological data interpretation problem. It is popular to use clustering, classification, and projection methods to analyze the data set. However, most analysis methods only use microarray data, not both dynamic expression time profiles of microarray data and the information of motif binding sites [12,57,58]. Furthermore, the interactions among proteins and delay from transcription to translation are not considered in their models of gene transcription. Recently, systems biology and computational biology methods have been widely considered to describe the biological functions from the dynamic system perspective [12,13,16,59–61]. Engineering theory has also been used to know more about biological complexity [14,16,58,62–64]. Here, we provide a nonlinear dynamic model to gain more insight into trans/cis-regulatory circuits for gene transcription using the knowledge of microarray data, the information of motif binding sites, translation delays and protein complexes from the systems biology perspective. In this section, a nonlinear dynamic model is proposed to describe the kinetics of trans/cis-regulatory circuits for gene transcription mechanism of yeast genes. In the trans circuit part (see Fig. 5.11), the transcription regulatory functions on cis elements are described by the regulatory activities of TFs and reaction of complexes among TFs. The binding from upstream regulatory genes to the TF binding sites is described by a biological sigmoid function to model the binding (activating the binding function) and no binding (inactivating the binding function) signals beyond or below some threshold of mRNA concentrations of regulatory genes. The translation time delay from the mRNA of a regulatory gene to TF is also considered in this model. The Big Mechanisms (interactions and cooperations) of complexes of TFs are represented by nonlinear interaction system [59].

In this section, nonlinear interaction system between two proteins and among three proteins are also considered to mimic the effect of protein complexes on a gene transcription regulation process. In the cis-regulatory circuit part, the interactions of cis elements are also modeled by a nonlinear dynamic equation. The regulatory functions of double and triple nonlinear interactions among the cis elements are considered in this dynamic model. Furthermore, the decay rate is also considered to describe how the output mRNA is degraded in the target gene.

The nonlinear dynamic model is useful to describe how the upstream regulatory genes control a target gene to produce the output expression of mRNA through its trans/cis-regulatory circuit. Using the information on TFs [7] and the experimental profiles on upstream regulatory genes and target gene [2], the model of trans/cis-regulatory circuit of target gene can be transformed to an algebraic regression equation for parameter estimation of the dynamic transcription model. With this study, the stochastic noises of microarray data are also considered to describe the uncertainties of the data. This is capable of improving the accuracy of parameter estimations in the dynamic model. The kinetic coefficients of the stochastic dynamic model, the decay rate of target gene's mRNA, and the variance of noises are estimated via the maximum likelihood estimation method and the downhill simplex search method. After estimating these parameters via expression profiles of the target gene and its regulatory genes, they will be substituted into the nonlinear dynamic model to confirm the validity of trans/cis-regulatory circuits. For illustration, the trans/cis-regulatory circuits of the gene *CLN1* are constructed in detail by the proposed method from expression profiles in the microarray data [2] and knowledge of binding site motifs of yeast cell cycle [7]. After constructing the model, we are able to use the mRNA expression profiles of regulatory genes to predict the expression profiles of the target gene and analyze the characteristics of the nonlinear trans/cis-regulatory circuits to gain more insight into gene transcription, which also provides a potential method for gene regulatory circuit design with a desired gene expression.

5.3.1 Modeling and identification of trans/cis circuit for gene regulatory mechanism

First of all, we construct a dynamic trans/cis-regulatory circuit with two parts. One simulates the binding effects of regulatory proteins and their complexes, and the other considers interactions of cis elements in the target gene. After that, we apply the downhill simplex search method and the maximum likelihood method to estimate the dynamic parameters of the trans/cis-regulatory circuit. Finally, we show the experimental data that have been used and how to preprocess it before system identification of regulatory circuits. The proposed method will provide a way for system analysis of gene regulatory circuits and a method for gene circuit design with a desired gene expression.

5.3.2 Dynamic model of trans/cis-regulatory circuits

First, we consider a trans/cis-regulatory network as a system block with several regulatory genes as inputs and a target gene as output. In this model, we use the mRNA expression data of upstream regulatory genes corresponding to TFs as the system input and the mRNA expression data of the target gene as the system output. For the illustration of nonlinear dynamic trans/cis-regulatory circuit, an example of the target gene *CLN1* is shown in Fig. 5.11. From the binding site motif data [7], we know that the gene *CLN1* has five TFs (Swi4, Swi6, Mbp1, Fkh1, and Fkh2). Some previous investigations [22,65,66] show that some proteins could form a protein complex via protein–protein interaction, and then the protein complex binds to the motif of the target gene. The binding of TFs on binding site motifs is described by the sigmoid function to model the binding (ON) and nonbinding (OFF) through some threshold [19,20,58,67]. In the trans-regulatory circuit, we also consider this protein complex effect. The translated proteins Swi4 and Swi6 will form a protein complex (SBF) as a TF to bind to the motif in the target gene, and proteins Mbp1 and Swi6 form another complex, MBF [30]. Therefore, in Fig. 5.11, we use linear functions and nonlinear functions to express the trans-regulatory functions of the TFs and their complexes, respectively. The trans-regulatory functions $g_{SBF}(t)$, $g_{MBF}(t)$, $g_{Fkh1}(t)$, and $g_{Fkh2}(t)$ of the respective cis elements SBF, MBF, Fkh1, and Fkh2 are the regulatory mechanism of the expression profiles of TFs and their complexes interactions. In the trans-regulatory circuit, these trans-regulatory functions on cis elements, which are generated from upstream TFs and their interactive complexes, are shown as follows

$$\begin{aligned}
g_{SBF}(t) &= a_{Swi4}x_{Swi4}(t) + a_{Swi6}x_{Swi6}(t) + a_{Swi4\bullet Swi6}x_{Swi4\bullet Swi6}(t) \\
g_{MBF}(t) &= a_{Mbp1}x_{Mbp1}(t) + a_{Swi6}x_{Swi6}(t) + a_{Mbp1\bullet Swi6}x_{Mbp1\bullet Swi6}(t) \\
g_{Fkh1}(t) &= a_{Fkh1}x_{Fkh1}(t) \\
g_{Fkh2}(t) &= a_{Fkh2}x_{Fkh2}(t)
\end{aligned} \tag{5.33}$$

where $g_i(t)$ denotes the trans-regulatory function to the cis element i, $x_i(t)$ denotes the reaction of the TF i on the binding site, $x_{ij}(t)$ denotes the complex interaction between TFs $x_i(t)$ and $x_j(t)$, a_i denotes the regulatory ability of the TF $x_i(t)$ to the cis elements, and a_{ij} denotes the regulatory ability of the complex due to TFs x_i and x_j.

After considering the trans effects of regulatory proteins and their protein complexes on cis elements, we also take into account the cis effects of regulatory interactions of cis elements on the transcription of coding region of the target gene. We suppose that the interaction function of two cis elements could be represented by the product of trans-regulatory functions from their TFs. From Fig. 5.11, the interactions between cis elements are represented as follows:

$$\begin{aligned}
g_{SBF\bullet MBF}(t) &= b_{SBF\bullet MBF}x_{Swi4}(t)x_{Swi6}(t)x_{Mbp1}(t) \\
g_{SBF\bullet Fkh1}(t) &= b_{SBF\bullet Fkh1}x_{Swi4}(t)x_{Swi6}(t)x_{Fkh1}(t) \\
g_{SBF\bullet Fkh2}(t) &= b_{SBF\bullet Fkh2}x_{Swi4}(t)x_{Swi6}(t)x_{Fkh2}(t) \\
g_{MBF\bullet Fkh1}(t) &= b_{MBF\bullet Fkh1}x_{Swi6}(t)x_{Mbp1}(t)x_{Fkh1}(t) \\
g_{MBF\bullet Fkh2}(t) &= b_{MBF\bullet Fkh2}x_{Swi6}(t)x_{Mbp1}(t)x_{Fkh2}(t) \\
g_{Fkh1\bullet Fkh2}(t) &= b_{Fkh1\bullet Fkh2}x_{Fkh1}(t)x_{Fkh2}(t)
\end{aligned} \tag{5.34}$$

where $g_{ij}(t)$ denotes the interaction between cis elements i and j. The coefficient b_{ij} denotes the corresponding interactive ability.

The trans-regulatory functions in Eq. (5.33) and their cis interactions in Eq. (5.34) will lead to the transcription mechanism of gene *CLN1* with mRNA expression profiles $y_{CLN1}(t)$ in Fig. 5.11. The transcriptional behavior of cis-regulatory circuit will be described by the following stochastic dynamic equation.

$$\dot{y}_{CLN1}(t) = G(t) - \lambda y_{CLN1}(t) + \varepsilon(t) \tag{5.35}$$

where the nonlinear transcriptional regulatory function $G(t)$ is denoted as

$$\begin{aligned} G(t) = & g_{SBF}(t) + g_{MBF}(t) + g_{Fkh1}(t) + g_{Fkh2}(t) + g_{SBF \bullet MBF}(t) + g_{SBF \bullet Fkh1}(t) \\ & + g_{SBF \bullet Fkh2}(t) + g_{MBF \bullet Fkh1}(t) + g_{MBF \bullet Fkh2}(t) + c \end{aligned}$$

and λ denotes the decay rate of mRNA expression profiles, which represent the degradation of mRNA. The constant c denotes the basal level of regulation, which comes from factors other than TFs. The gene regulatory function $G(t)$ in Eq. (5.35) denotes the whole transcriptional regulation from the cis elements due to the binding of TFs of the gene *CLN1* and the basal level from other factors. $\varepsilon(t)$ denotes a stochastic noise due to the uncertainty and the fluctuation of mRNA microarray data.

The biological meaning of Eq. (5.35) is that the change of mRNA level in the gene CLN1 is the synthesis due to the transcriptional regulatory function $G(t)$ and the degradation $-\lambda y_{CLN1}(t)$. The mRNA expression of the gene *CLN1* will increase if the synthesis regulatory function $G(t)$ is greater than the degradation $\lambda y_{CLN1}(t)$. Otherwise, it will decrease. Substituting Eqs. (5.33) and (5.34) into Eq. (5.35), we get the dynamic equation of trans/cis-regulatory circuit in the gene transcription mechanism of the gene *CLN1* as follows

$$\begin{aligned} \dot{y}_{CLN1}(t) = & a_{Swi4}x_{Swi4}(t) + 2a_{Swi6}x_{Swi6}(t) + a_{Mbp1}x_{Mbp1}(t) + a_{Fkh1}x_{Fkh1}(t) \\ & + a_{Fkh2}x_{Fkh2}(t) + a_{Swi4 \bullet Swi6}x_{Swi4 \bullet Swi6}(t) + a_{Mbp1 \bullet Swi6}x_{Mbp1 \bullet Swi6}(t) \\ & + b_{SBF \bullet MBF}x_{Swi4}(t) \bullet x_{Swi6}(t) \bullet x_{Mbp1}(t) \\ & + b_{SBF \bullet Fkh1}x_{Swi4}(t) \bullet x_{Swi6}(t) \bullet x_{Fkh1}(t) \\ & + b_{SBF \bullet Fkh2}x_{Swi4}(t) \bullet x_{Swi6}(t) \bullet x_{Fkh2}(t) \\ & + b_{MBF \bullet Fkh1}x_{Swi6}(t) \bullet x_{Mbp1}(t) \bullet x_{Fkh1}(t) \\ & + b_{MBF \bullet Fkh2}x_{Swi6}(t) \bullet x_{Mbp1}(t) \bullet x_{Fkh2}(t) \\ & + b_{Fkh1 \bullet Fkh2}x_{Fkh1}(t) \bullet x_{Fkh2}(t) + c - \lambda y_{CLN1}(t) + \varepsilon(t) \end{aligned} \tag{5.36}$$

In the above dynamic equation of gene transcription mechanism, $x_i(t)$ denotes the expression profiles of the binding i trans/cis-regulatory circuit. The nonlinear term $\prod_i x_i(t)$ denotes the interactions of TFs in the trans/cis-regulatory circuit. The coefficients in the dynamic gene circuit denote the corresponding regulatory abilities of the corresponding regulatory mechanisms and will be identified later by the corresponding microarray data.

However, at present, it is still not easy to measure directly the expression profiles of TFs $x_i(t)$ in Eq. (5.36). Because the expression profiles of yeast mRNA are available, in this section, the expression levels of these TFs will be replaced by expression levels of mRNA microarray data of their upstream regulatory genes but with some translation process delay in the cell. Now, for system identification of trans/cis-regulatory circuit, it is more practical to consider the biochemical reactive relation between the TF profiles $x_i(t)$ at the motif binding sites and their relevant mRNA expression profiles $y_i(t)$ of the upstream regulatory gene [19,20]. For this purpose, we describe the binding reaction function $x_i(t)$ of the TF on its motif binding site as a sigmoid function of mRNA expression profiles of the corresponding regulatory gene [67]. Further, it takes time for mRNA to translate into proteins as TFs and move to motif binding sites of the following target gene [68]. We should consider this translation delay time to the sigmoid binding function when using mRNA expression profiles $y_i(t)$ to replace expression levels $x_i(t)$ of TFs in the trans/cis dynamic model (Table 5.4).

$$x_i(t) = f_i(y_i(t-\tau_i)) = \frac{1}{1+\exp\{-r[y_i(t-\tau_i)-M_i]\}} \tag{5.37}$$

where r denotes the transition rate of the sigmoid function, M_i denotes the mean of mRNA expression level of the regulatory gene i, and τ_i denotes the time delay of the translation time from mRNA to TF (protein) for the regulatory gene i. In Fig. 5.11, $y_i(t)$ and $x_i(t)$ represent the mRNA expression profile and the binding regulation function of the corresponding TF on its motif binding site, respectively. The biochemical meaning of Eq. (5.37) is that the regulation of TF $x_i(t)$ on the binding site motifs is between ON (binding) and OFF (no binding) signal with some transition region dependent on beyond or below some threshold of mRNA expression level of the regulatory gene after a time delay τ_i, which is available in Arava et al. [68]. The sigmoid function can also be considered as a method to normalize the expression profiles of

Table 5.4 Translation delay time of nine major transcription factors

Transcription factor	Translation delay time (min)
Fkh1	Unavailable
Fkh2	1.056818384
Mcm1	0.050641356
Ace2	0.20647784
Swi5	0.223001589
Ndd1	0.638633059
Swi4	0.495504558
Mbp1	0.408482192
Swi6	0.71182446

regulatory genes between 0 and 1 to model the binding and no binding, which has been successfully employed to describe the binding of regulatory gene [12,59,67].

Hence, based on the mRNA expression time profiles and the translation delay time, the dynamic equation of trans/cis-regulatory circuit for the gene transcription mechanism of CLN1 is described by

$$\begin{aligned}
\dot{y}_{CLN1}(t) = {} & a_{Swi4} f_{Swi4}(y_{Swi4}(t-\tau_{Swi4})) + 2a_{Swi6} f_{Swi6}(y_{Swi6}(t-\tau_{Swi6})) \\
& + a_{Mbp1} f_{Mbp1}(y_{Mbp1}(t-\tau_{Mbp1})) + a_{Fkh1} f_{Fkh1}(y_{Fkh1}(t-\tau_{Fkh1})) \\
& + a_{Fkh2} f_{Fkh2}(y_{Fkh2}(t-\tau_{Fkh2})) \\
& + a_{Swi4 \bullet Swi6} f_{Swi4 \bullet Swi6}(y_{Swi4}(t-\tau_{Swi4}) y_{Swi6}(t-\tau_{Swi6})) \\
& + a_{Mbp1 \bullet Swi6} f_{Mbp1 \bullet Swi6}(y_{Mbp1}(t-\tau_{Mbp1}) y_{Swi6}(t-\tau_{Swi6})) \\
& + b_{SBF \bullet MBF} f_{Swi4}(y_{Swi4}(t-\tau_{Swi4})) f_{Swi6}(y_{Swi6}(t-\tau_{Swi6})) \\
& \bullet f_{Mbp1}(y_{Mbp1}(t-\tau_{Mbp1})) \\
& + b_{SBF \bullet Fkh1} f_{Swi4}(y_{Swi4}(t-\tau_{Swi4})) f_{Swi6}(y_{Swi6}(t-\tau_{Swi6})) \\
& \bullet f_{Fkh1}(y_{Fkh1}(t-\tau_{Fkh1})) \\
& + b_{SBF \bullet Fkh2} f_{Swi4}(y_{Swi4}(t-\tau_{Swi4})) f_{Swi6}(y_{Swi6}(t-\tau_{Swi6})) \\
& \bullet f_{Fkh2}(y_{Fkh2}(t-\tau_{Fkh2})) \\
& + b_{MBF \bullet Fkh1} f_{Swi6}(y_{Swi6}(t-\tau_{Swi6})) f_{Mbp1}(y_{Mbp1}(t-\tau_{Mbp1})) \\
& \bullet f_{Fkh1}(y_{Fkh1}(t-\tau_{Fkh1})) \\
& + b_{MBF \bullet Fkh2} f_{Swi6}(y_{Swi6}(t-\tau_{Swi6})) f_{Mbp1}(y_{Mbp1}(t-\tau_{Mbp1})) \\
& \bullet f_{Fkh2}(y_{Fkh2}(t-\tau_{Fkh2})) \\
& + b_{Fkh1 \bullet Fkh2} f_{Fkh1}(y_{Fkh1}(t-\tau_{Fkh1})) f_{Fkh2}(y_{Fkh2}(t-\tau_{Fkh2})) \\
& + c - \lambda_{CLN1}(t) + \varepsilon(t)
\end{aligned} \tag{5.38}$$

Based on the above analysis, in general, a block diagram to construct trans/cis-regulatory circuit of gene transcription is shown in Fig. 5.12, and the nonlinear dynamic model of trans/cis-regulatory circuit of gene transcription mechanism of any gene of interest in yeast can be described as follows

$$\begin{aligned}
\dot{y}(t) = {} & \sum_{i=1}^{N} a_i f_i(y_i(t-\tau_i)) + \sum_{i,j} a_{i,j} f_i(y_i(t-\tau_i) y_j(t-\tau_j)) \\
& + \sum_{i,j,k} a_{i,j,k} f_{i,j,k}(y_i(t-\tau_i) y_j(t-\tau_j) y_k(t-\tau_k)) \\
& + \sum_{p,q} b_{p,q} \prod_r f_r(y_r(t-\tau_r)) + c - \lambda y(t) + \varepsilon(t)
\end{aligned} \tag{5.39}$$

Unlike the previous dynamic gene regulatory models only with linear regulatory terms [12,57,58], in this model, nonlinear interactions among two proteins and three

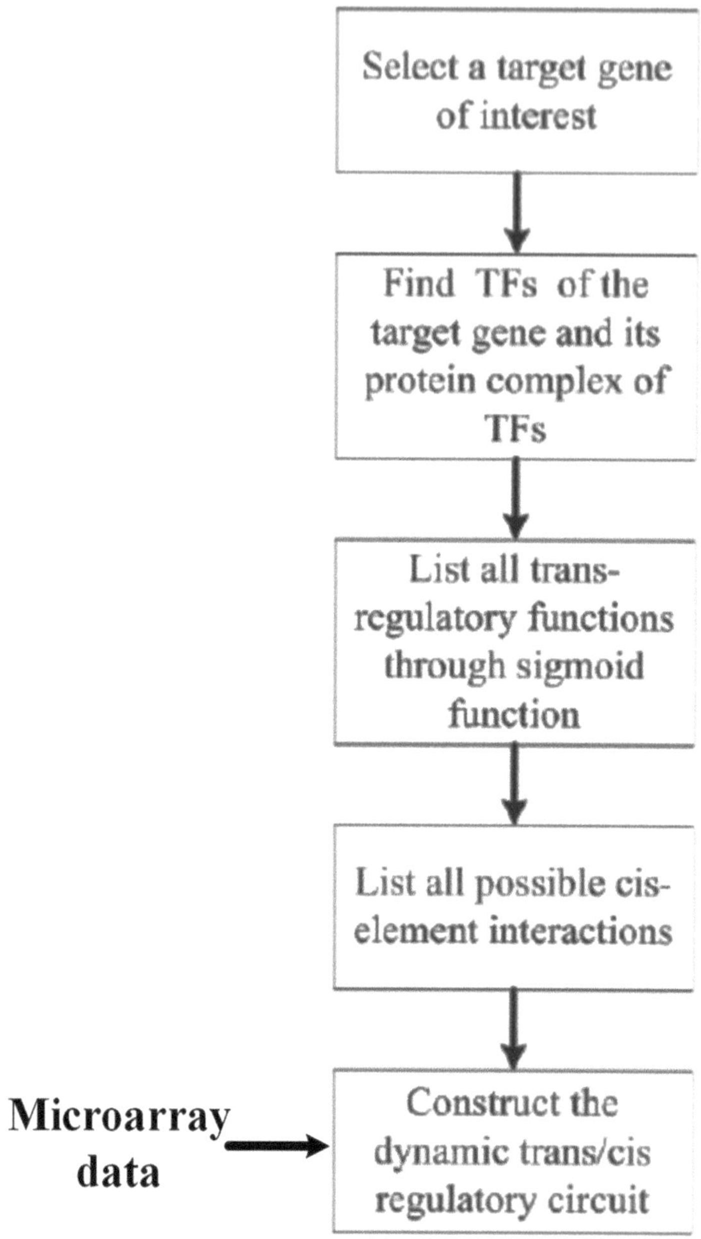

Figure 5.12 Block diagram to construct a dynamic trans/cis-regulatory circuit for gene transcription mechanism.

proteins are also considered in this gene expression dynamic model. Furthermore, the delay from transcription to translation is also considered in this dynamic model. According to the information on protein complexes and motif binding sites in the promoter region of the gene, $y(t)$ denotes the mRNA expression profiles of a gene of interest and $y_i(t)$, for $i\in\{1,\ldots,N\}$, denote the mRNA expression profiles of upstream regulatory genes. With sampling expression, in the next section, Eq. (5.39) can be rewritten as

$$\dot{y}(t) + \lambda y(t) = \sum_{i=1}^{L} \theta_i \xi_i(t) + c + \varepsilon(t) \tag{5.40}$$

where $\xi_i(t)$ for $i\in\{1,\ldots,L\}$ denotes all possible regulatory functions, θ_i for $i\in\{1,\ldots,L\}$ denotes the corresponding regulatory abilities (coefficients), and L is the total number of input functions.

The next step is to identify the parameters θ_i, λ, c, and the covariance σ^2 of noise from mRNA microarray data. The parameter estimations of θ_i,λ,c, and σ^2 are achieved by the combination of downhill simplex search algorithm and maximum likelihood parameter method using the Methods outlined below. After the parameter estimation is finished, we could construct a trans/cis-regulatory circuit for any gene of interest via mRNA microarray data and the motif binding site information of TFs and their complexes.

5.3.3 Experimental data for identifying trans/cis circuits for gene transcription mechanisms

We use the yeast microarray hybridization data of Spellman et al. [2] as these mRNA expression profiles. They have many experimental methods to reset the yeast cell cycle to measure mRNA expression profiles for the whole genome comprehensively. Here, we use the experimental cell cycle data of the "α factor." The information of motif binding sites in a target gene is from Simon et al. [7], Lee et al. [5], and Harbison et al. [17], and then the protein complex data are obtained from Simon et al. [7] and the MIPS database. In the study of Simon et al. [7], there are 9 major TFs of cell cycle genes in yeast, so we use these 9 major TFs as the regulatory input of this dynamic model, which could also avoid overfitting in parameter estimation due to only 18 time points in expression profiles. Using the above information, we could construct dynamic models as Eq. (5.39) for cell cycle genes and then identify the parameters from mRNA microarray data by the proposed parameter estimation algorithm methods. We also include the translation delay time to this dynamic model. The data of translation delay time is available from Arava et al. [68].

5.3.4 Investigating gene transcription mechanism using microarray data

Reconstruction of microarray raw data by the gene regulatory circuits

We use the "α factor" microarray data of Spellman et al. [2] and information on nine major TFs from Simon et al. [7] to construct this nonlinear dynamic trans/cis model. In the study of Simon et al. [7], there are 769 cell cycle genes, but only 189 cell cycle genes have at least one motif binding site of 9 major TFs (p-value $< .0015$). Therefore, we could construct dynamic trans/cis-regulatory circuits for these 189 cell cycle genes.

After estimating the parameters of dynamic trans/cis-regulatory circuit for each gene, we compare the actual expression profiles with the constructed expression profiles of 189 genes by the cluster analysis and visualization tool (details can be found in Section 5.3.6). The comparison can be seen in Fig. 5.13 and the correlation coefficient of these two data is 0.7276. However, we found that 80 cell cycle genes have only one motif binding site of 9 major TFs. These genes may have other motif binding sites which are not of the 9 major TFs, or they have posttranscription to dominate their expression profiles. For example, *FUN26*, *SHL7*, *STE2*, *AGA2*, *YBL111C*, *YBL112C*, and *YBL113C* have many other motif binding sites, which are not in the 9 major TFs [5,17]. Thus these 9 TFs may not play dominating roles in these genes, and the performances of predicted expression profiles of these genes by dynamic trans/cis-regulatory circuits may not be very satisfactory. After removing 80 genes with one motif binding site from 189 genes, the dynamic trans/cis-regulatory circuits of 109 cell cycle genes with at least two motif binding sites of these 9 TFs are considered to be more accurate. The actual profiles and the predicted profiles by the dynamic trans/cis-regulatory circuits are shown in Fig. 5.14, and their correlation coefficient is 0.8502. We also use the Chi-square method to test the hypothesis of trans/cis circuits by comparing these predicted data and real microarray data. We found that 176 of 189 target genes cannot be rejected by the proposed trans/cis circuit by 95% significance, and other 13 target genes cannot be tested by the Chi-square method because the number of their parameters is more than the number of their expression profiles data.

Test of shuffled microarray data

In order to test the accuracy of this model, we shuffled Spellman et al.'s "α factor" microarray data [2] of expression profiles by random. After shuffling the microarray data, we applied this method to construct another new trans/cis-regulatory circuit and then used the new regulatory circuit to generate profiles. For the 189 target genes with at least one motif of 9 major TFs, the correlation coefficient between shuffled profiles and their corresponding profiles predicted by new regulatory circuits via shuffled data is 0.1143 (shown in Fig. B.1), but the correlation coefficient between

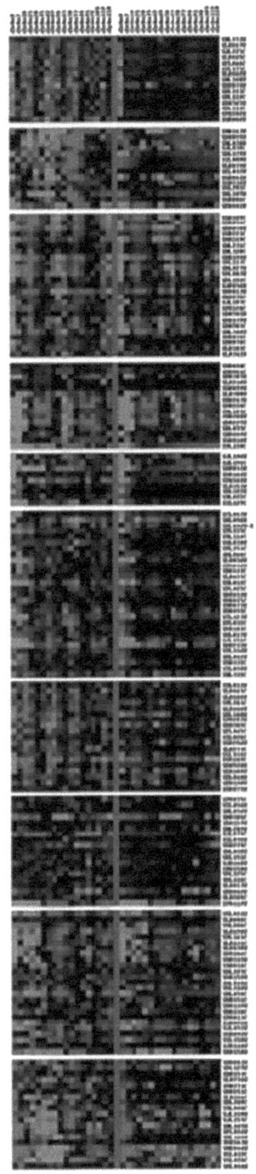

Figure 5.13 Comparison between the experimental mRNA expression profiles and those predicted by the proposed model. The experimental mRNA expression profiles of 189 cell cycle genes are at left side, and the profiles which are generated by the predicted dynamic regulatory circuits are at right side. Genes represented by *red* (dark gray in print versions) tonalities are overexpressed and those represented by *green* (light gray in print versions) ones are downregulated. The correlation coefficient of both profiles is 0.7276.

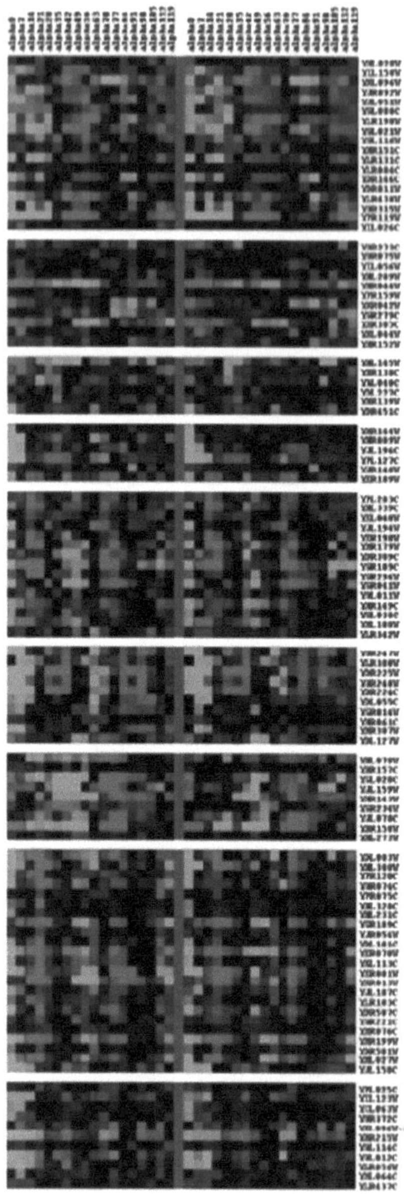

Figure 5.14 Comparison between the experimental mRNA expression profiles and those predicted by the proposed model. The experimental mRNA expression profiles of 109 cell cycle genes are at the left side, and the profiles predicted by the dynamic regulatory circuits are at the right side. The correlation coefficient of both profiles is 0.8502.

actual profiles and their corresponding predicted profiles is 0.7276. Furthermore, for the 109 target genes with at least two more motifs of 9 major TFs, the correlation coefficient between shuffled profiles and their corresponding profiles predicted by shuffled data is 0.5939 (shown in Fig. B2), but the correlation coefficient between actual profiles and their corresponding predicted profiles is 0.8502. Obviously, only data generated by real biological systems could be identified well by the proposed model. From the view of system identification, a proper system model will lead to a good system identification if overfitting could be avoided. In this case, the proposed model is applicable to the trans/cis-regulatory circuit because the correlation coefficient of real microarray data is much larger than that of shuffled microarray data. So we can say that this dynamic model is proper for the trans/cis-regulatory circuit of gene expression mechanism of yeast.

Regulatory activities of TFs in the target gene

From the estimated parameters, we can find which TF activates or represses its target genes and quantify its activity. Therefore, we can see which TF can greatly affect a certain target gene, and why the same TF may cause different effects on other target genes. In Table 5.5, we can see there are many TFs regulating the target gene CLN1. CLN1, a G1/S-specific cyclin, is regulated by SBF, MBF, Fkh1, and Fkh2 [7]. By the estimated kinetic parameters of MBF (Swi6, Mbp1, and Mbp1.Swi6), we found that Swi6 dominates the effect of MBF on target gene CLN1, and it may play an important regulatory role in CLN1. The kinetic parameters of Mbp1 and Mbp1.Swi6 are of smaller scale, which means that Mbp1 may play a certain kind of role but not an important regulatory role in CLN1. Mbp1 may be necessary for CLN1, but has no more effect on CLN1 after the expression of Mbp1 exceeds certain level. This result could reflect that Mbp1 is a DNA-binding component of MBF, and Swi6 may have a regulatory function [30]. For parameters of SBF (Swi4, Swi6, and Swi4 · Swi6), Swi4 and Swi6 play major roles of SBF in CLN1, but the role of Swi4 · Swi6 is minor. It shows that the linear combination of Swi4 and Swi6 has significant effect on target gene CLN1. Hence, we consider that Swi6 may have a regulatory function of SBF in CLN1, and Swi4 is also significant on target gene CLN1. From Table 5.5, Fkh1 and Fkh2 contribute negative regulation to target gene CLN1. Comparing these kinetic parameters, we know that Fkh1, Swi4, and Swi6 affect target gene CLN1 more than Fkh2 and Mbp1 do.

Our model also provides another important estimation of interactions among the cis elements. All possible interactions among the cis elements are considered, and then their kinetic parameters are estimated by expression profiles of microarray data. These estimated kinetic parameters could not only tell us the possibilities of these interactions of cis elements, but also the significance of these interactions. For example, we consider 5 cis-element interactions in target gene CLN1. From the kinetic parameters

Table 5.5 Parameters of dynamic trans/cis regulatory models of CLN1, UTR2, and MNN1 based on Eq. (5.36)

CLN1	Terms due to MBF	$-0.009x_{Mbp1}(t-0.4)+0.025x_{Swi6}(t-0.7)-0.0027f_{Mbp1\bullet Swi6}(y_{Mbp1}(t-0.4)\bullet y_{Swi6}(t-0.7))$
$\dot{y}_{CLN1}(t)=$	Terms due to SBF	$-0.01_{Swi4}(t-0.5)+0.02516x_{Swi6}(t-0.7)+0.003435f_{Swi4\bullet Swi6}(y_{Swi4}(t-0.5)\bullet y_{Swi6}(t-0.6))$
	Terms of Fkh1 and Fkh2	$-0.36396x_{Fkh1}(t)-0.024962x_{Fkh2}(t-1)$
	Interaction between MBF and SBF	$-0.0027745x_{Swi4}(t-0.5)x_{Swi6}(t-0.7)x_{Mbp1}(t-0.4)$
	Interaction between SBF and Fkh1	$-0.011154x_{Swi4}(t-0.5)x_{Swi6}(t-0.7)x_{Fkh1}(t)$
	Interaction between SBF and Fkh2	$-0.014026x_{Swi4}(t-0.5)x_{Swi6}(t-0.7)x_{Fkh2}(t-1)$
	Interaction between MBF and Fkh1	$-0.0057164x_{Swi6}(t-0.7)x_{Mbp1}(t-0.4)x_{Fkh1}(t)$
	Interaction between MBF and Fkh2	$-0.046408x_{Swi6}(t-0.7)x_{Mbp1}(t-0.4)x_{Fkh2}(t-1)$
	Interaction between Fkh1 and Fkh2	$-0.0023141x_{Fkh1}(t)x_{Fkh2}(t-1)$
	Decay rate and basal level	$-4.0839-0.20719y_{CLN1}(t)$
PCL2	Terms due to SBF	$0.0167x_{Swi4}(t-0.5)-0.0384x_{Swi6}(t-0.7)-0.0464f_{Swi4\bullet Swi6}(y_{Swi4}(t-0.5)\bullet y_{Swi6}(t-0.6))$
$\dot{y}_{PCL2}(t)=$	Terms due to Ace2 and Swi5	$+0.070897x_{Ace2}(t-0.21)+0.1044x_{Swi5}(t-0.22)$
	Interaction between Ace2 and Swi5	$+0.045873x_{Ace2}(t-0.21)x_{Swi5}(t-0.22)$
	Interaction between SBF and Ace2	$+0.17035x_{Swi4}(t-0.5)x_{Swi6}(t-0.7)x_{Ace2}(t-0.21)$
	Interaction between SBF and Swi5	$+0.093195x_{Swi4}(t-0.5)x_{Swi6}(t-0.7)x_{Swi5}(t-0.22)$
	Decay rate and basal level	$+1.2225-0.43022y_{PCL2}(t)$
MNN1	Terms due to SBF	$-0.0432x_{Swi4}(t-0.5)-0.394x_{Swi6}(t-0.7)-0.142f_{Swi4\bullet Swi6}(y_{Swi4}(t-0.5)\bullet y_{Swi6}(t-0.6))$
$\dot{y}_{MNN1}(t)=$	Decay rate and basal level	$+0.81055-0.45975y_{MNN1}(t)$

estimated, there are three parameters larger than the others. We may conclude that these three terms, MBF · Fkh2, SBF · Fkh1, and SBF · Fkh2, could be the three possible interactions among real cis elements.

With the same TF binding to different target genes, different kinetic parameters of this TF determine different regulatory effects on the target genes. A larger parameter of the same TF binding to different target genes means that it plays a more important role and is more sensitive than those binding to another target gene. From these estimated kinetic parameters, we can compare these effects on different target genes with the same TF.

The G1 cyclin PCL2, which associates with Pho85p cyclin-dependent kinase (Cdk) to contribute to entry into the mitotic cell cycle, is regulated by SBF, Ace2, and Swi5 [7].

On the kinetic parameters of PCL2 (see Table 5.5), the effect of TF SBF on PCL2 is dominated by Swi6 and Swi4 · Swi6. For another target gene MNN1, required for addition of alpha1,3-mannose linkages to N-linked and O-linked oligosaccharides, the effect of TF SBF on MNN1 is also dominated by Swi6 and Swi4 · Swi6, which can be seen in SGD database. However, the kinetic parameters of SBF on PCL2 [7] are all smaller than those of SBF on MNN1. We can consider that SBF has more regulatory effect on MNN1 than PCL2 and the sensitivity of SBF on MNN1 is more than that on PCL2. After simulation by changing the expression profiles of Swi4 or Swi6, of which SBF is composed, we find that the result is the same as we discussed above.

Possibilities of cis-element Interactions

In order to find the most possible cis-element interactions, we sorted all possible kinetic parameters of cis-element interactions calculated by this method as a distribution and then found those within a two-sided 90% confidence interval of the distribution. There are 314 possible cis-element interactions within the 189 cell cycle genes selected before, and then we found 15 cis-element interaction terms in each side of this two-sided interval. Some of these 30 interaction terms appear more frequently, which include complex interactions of Swi4/Mbp1/Swi6, Fkh2/Swi4/Swi6, and Ace2/Swi5. In Table 5.6, we count the frequencies of these cis-element interactions, which occur more than two times within the two-sided 90% confidence interval.

For the complex interactions of Swi4/Mbp1/Swi6, both SBF, the complex of Swi4 and Swi6, and MBF, the complex of Mbp1 and Swi6, control the transcription of G1/S cyclin genes, and many genes such as Cdc6, Swi4, Clb6, Swe1, and Cln1, are both bound by SBF and MBF at the same time. Therefore, it is possible that

Table 5.6 The possible cis element interactions which appear within the two-sided 90% confidence interval (there are 30 cis-element interactions within the two-sided 90% confidence interval)

Swi4/Mbp1/Swi6	5
Fkh2/Swi4/Swi6	4
Ace2/Swi5	3
Mcm1/Swi5	3
Fkh1/Fkh2	2
Fkh2/Mcm1	2
Fkh2/Mbp1/Swi6	2
Mcm1/Swi4/Swi6	2
Swi4/Swi5/Swi6	2

the activators (Swi4/Mbp1/Swi6) have a cis-element interaction, and the possibility of their relationship has been reported by Kato et al. [11]. The other two possible cis-element interaction terms (Fkh2/Swi4/Swi6 and Ace2/Swi5) calculated by this method may also exist. A few genomic analyses have indicated the involvement of SBF and Fkh1/Fkh2 in S phase [5], and SBF and Fkh2 most probably regulate budding, cell-wall synthesis, and spindle-related genes in S phase [11]. Ace2 and Swi5 are a pair of TFs of yeast that regulate the expression of many cell cycle-specific genes, including Sic1, an inhibitor of Cdc28 protein kinase, Rme1, a regulator of meiosis, and Ash1, a regulator of meiosis [29]. In recent studies, Ace2 and Swi5 cooperate to induce the expressions of a subset of genes, but the antagonistic interaction of Ace2 and Swi5 was found [29]. With 82% identical DNA-binding domains, Ace2 and Swi5 bind to the same DNA sequence [34], and it is possible that proteins compete for access on these promoters, but only one activates transcription [29]. Therefore, one partner of Swi5 and Ace2 sometimes can have a stronger contribution toward regulation, and the antagonistic interaction of Ace2 and Swi5 found is not surprising.

5.3.5 Discussion of gene transcription mechanism by identifying trans/cis circuits

Our method uses mRNA expression profiles, protein complexes, translation time delay, and the information of binding site motif to construct a dynamic trans/cis-regulatory circuit to gain more insight into the gene expression of yeast. It is more precise to use these kinds of data to construct a nonlinear stochastic regulatory model for the gene transcription. Furthermore, a stochastic regulatory model can easily describe the properties of change, interaction, and uncertainty in mRNA expression profiles. Recently, Vu and Vohradsky also used nonlinear dynamic model to infer the transcriptional regulators of target genes [69]. They successfully identified possible transcriptional factors of yeast cell cycle, which shows the power of nonlinear dynamic models.

Constructing the trans/cis-regulatory circuits of the cell cycle genes of yeast is useful to quantify the influence of TFs on their target genes, and then the possibilities of cis element interactions could be found from the statistical perspective. The confirmation of these possible cis element interactions would be a direction of further research. In addition, the proposed method can provide a quantitative basis for system analysis of gene circuit and give a scheme for gene circuit design with a desired gene expression in the future. Not only could the data of *S. cerevisiae* be applied by this method but those of other species also could. Any species with their mRNA expression profiles and the information of binding site motif can be applied using this method to construct their trans/cis-regulatory circuits.

However, we found that the mRNA expression profiles of several genes generated by the predicted dynamic trans/cis regulatory-circuit have a little difference from their original mRNA expression profiles. It may be because nine transcriptional factors are not enough in the cases in which the genes may not be controlled dominantly by nine TFs. We may use ChIP-chip data [17] to replace the nine major TFs from Simon et al. [7] to construct the transcriptional regulatory circuit. 204 DNA-binding transcriptional factors are found in yeast [17], which may be useful for this elaboration of the transcriptional regulatory circuit. Furthermore, only 18 time points of mRNA expression profiles are used to estimate the parameters of dynamic model in each gene, which are not enough to get precise parameter estimation because of the large number of kinetic parameters to be estimated, which will easily lead to overfitting. If we include all possible 204 TFs into this model, the number of parameters that we want to estimate may be so large that overfitting would happen. Consequently, we only choose the most important nine TFs in this model to avoid the overfitting in parameter estimation. If there are more time point profiles of mRNA expression and information of binding site motifs, the accuracy of the dynamic trans/cis-regulatory circuit could be improved by the proposed method.

In conclusion, unlike the convention linear dynamic models, this section provides a nonlinear stochastic dynamic trans/cis-regulatory circuit for any species via mRNA expression profiles, the information of binding site motif, protein complexes, and translation time delay to gain more insight into transcriptional regulatory infrastructures of gene expression program of yeast. The results are confirmed by a statistical hypothesis test.

5.3.6 Systems biology and identification methods of constructing dynamic trans/cis circuits

Parameter estimation of dynamic trans/cis-regulatory circuit

The dynamic trans/cis-regulatory circuit of a gene of interest can be described by the dynamic Eq. (5.39) according to the mRNA expression profiles of upstream regulatory genes and their nonlinear interactions. The parameters of the dynamic trans/cis circuit are then specified to meet the practical input/output microarray data of the regulatory genes and the target gene.

In order to avoid the interruption of high frequency noise with the use of the derivative information $y(t)$, the solution of the dynamic Eq. (5.40) is expressed by the following integration equation.

$$\begin{aligned} y(t) &= y(0)e^{-\lambda t} + \sum_i e^{-\lambda t}\int_0^t e^{\lambda\tau}\xi_i(\tau)d\tau + ce^{-\lambda t}\int_0^t e^{\lambda\tau}d\tau + ce^{-\lambda t}\int_0^t \varepsilon(\tau)e^{\lambda\tau}d\tau \\ &= Ke^{-\lambda t} + \sum_i \theta_i e^{-\lambda t}\int_0^t e^{\lambda\tau}\xi_i(\tau)d\tau + k + e(t) \end{aligned} \tag{5.41}$$

where $k \equiv c/\lambda$ and $K \equiv y(0) - c/\lambda$ are constants. Noise of $y(t)$ is written with $e(t) \equiv e^{-\lambda t}\int_0^t \varepsilon(t)e^{\lambda\tau}d\tau$, i.e., the mRNA concentration could be generated by dynamic Eq. (5.41) through upstream regulatory mRNA $y_1(t)\ldots y_N(t)$ if their parameters could be estimated.

In Eq. (5.41), $\xi_i(t)$ is a combinative regulatory function of the regulatory genes. We use a third order polynomial cubic spline method to approximate $\xi_i(t)$ through microarray data with 18 time points, and then use the partial integration method to calculate $e^{-\lambda t}\int_0^t e^{\lambda\tau}\xi_i(\tau)d\tau$ [18].

For $\xi_i(t)$ function, we use the cubic spline method to approach it. Cubic spline is a method of using a third order polynomial to approach every four data points [18]. So we can rewrite the function $\xi_i(t)$ as follows.

$$\xi_i(t) = \alpha_i t^3 + \beta_i t^2 + \gamma_i t + \delta_i \tag{5.42}$$

From Eqs. (5.41) and (5.42), we can use partial integration to refine Eq. (5.41) as follows:

$$y(t) = \Lambda e^{-\lambda t} + \sum_i \theta_i \bullet \left(\frac{\xi_i(t)}{\lambda} - \frac{\xi_i'(t)}{\lambda^2} + \frac{\xi_i''(t)}{\lambda^3} - \frac{\xi_i'''(t)}{\lambda^4}\right) + k + e(t) \tag{5.43}$$

where $\xi_i'(t) = 3\alpha_i t^2 + 2\beta_i t + \gamma_i$, $\xi_i''(t) = 6\alpha_i t + 2\beta_i$, $\xi_i'''(t) = 12\alpha_i$, and $\Lambda = K - \sum_i \theta_i \bullet \left(\frac{\xi_i(0)}{\lambda} - \frac{\xi_i'(0)}{\lambda^2} + \frac{\xi_i''(0)}{\lambda^3} - \frac{\xi_i'''(0)}{\lambda^4}\right)$.

After applying the cubic spline method to solve and rewrite the dynamic Eq. (5.41), we use a search algorithm, downhill simplex search method, for parameter estimations [4]. However, this search method is not very efficient. Therefore, we combine the downhill simplex search method and maximum likelihood method to estimate the parameters. We use the downhill simplex search method to estimate the nonlinear parameter λ at first and then use the maximum likelihood method to estimate linear parameters K, θ_i, and k.

After estimating λ by downhill simplex research method, we use the method of maximum likelihood to estimate the other linear parameters. Eq. (5.43) can be written as the following regression form.

$$y(t) = \left[e^{-\lambda t} \quad \left(\frac{\xi_1(t)}{\lambda} - \frac{\xi_1'(t)}{\lambda^2} + \frac{\xi_1''(t)}{\lambda^3} - \frac{\xi_1'''(t)}{\lambda^4} \right) \quad \cdots \quad \left(\frac{\xi_i(t)}{\lambda} - \frac{\xi_i'(t)}{\lambda^2} + \frac{\xi_i''(t)}{\lambda^3} - \frac{\xi_i'''(t)}{\lambda^4} \right) \quad 1 \right] \begin{bmatrix} \Lambda \\ \theta_1 \\ \vdots \\ \theta_i \\ k \end{bmatrix} + e(t) \equiv \phi(t)\theta + e(t) \tag{5.44}$$

where $\phi(t)$ denotes the regression vector which can be obtained from the above processing. θ is the parameter vector of trans/cis-regulatory circuit which is to be estimated to make the transcriptional expression of trans/cis circuit meet the microarray data. If we have a lot of microarray data points to process by the method above to get values of $\left\{ y(t_l) \quad \left(\frac{\xi_i(t_l)}{\lambda} - \frac{\xi_i'(t_l)}{\lambda^2} + \frac{\xi_i''(t_l)}{\lambda^3} - \frac{\xi_i'''(t_l)}{\lambda^4} \right) \right\}$ for $l \in \{1, \ldots, 18\}$ and $i \in \{1, \ldots, L\}$, the parameter vector can be estimated by the following approach. Using a matrix notation, Eq. (5.44) at different time points is of the following form

$$\begin{bmatrix} y(t_1) \\ \vdots \\ y(t_{18}) \end{bmatrix} = \begin{bmatrix} \phi(t_1) \\ \vdots \\ \phi(t_{18}) \end{bmatrix} \theta + \begin{bmatrix} e(t_1) \\ \vdots \\ e(t_{18}) \end{bmatrix} \tag{5.45}$$

For simplicity, we can further define the notations Y, Φ, and E to represent Eq. (5.45) as follows.

$$Y = \Phi\theta + E \tag{5.46}$$

In Eq. (5.45), we assume noises $e(t_i)$ at different time points as independent random variables of normal distribution with zero mean and unknown variance σ^2, i.e., the variance of E is $\Sigma = E[EE^T] = \sigma^2 I$, where I is a unit matrix. In this section, a maximum likelihood parameter estimation method will be used to estimate θ and σ^2 from microarray data of regulatory genes and the target gene [21]. If E is assumed to be normally distributed with N elements, its probability density function is of the following form

$$p(E) = \frac{1}{\left((2\pi)^N \det\Sigma\right)^{-1/2}} \exp\left\{-\frac{E^T\Sigma^{-1}E}{2}\right\} \tag{5.47}$$

Under Eq. (5.46) we have the likelihood function

$$p(\theta,\sigma^2) = \frac{1}{(2\pi\sigma^2)^{-N/2}} \exp\left\{-\frac{(Y-\Phi\theta)^T(Y-\Phi\theta)}{2\sigma^2}\right\} \tag{5.48}$$

Eq. (5.48) can be considered as a function of parameters θ and σ^2. We want to specify θ and σ^2 to maximize the likelihood function in Eq. (5.48). It is practical to take the logarithm of their likelihood function, and then we have the following log-likelihood function as follows.

$$\log L(\theta,\sigma^2) = -\frac{N}{2}\log(2\pi\sigma^2) - \frac{1}{2\sigma^2}\sum_{k=1}^{N}[y(t_k) - \phi(t_k)\theta]^2 \tag{5.49}$$

where $y(t_k)$ and $\phi(t_k)$ are kth elements of Y and Φ, respectively.

Here we expect the log-likelihood function to have the maximum at $\theta = \hat{\theta}$ and $\sigma^2 = \hat{\sigma}^2$. The necessary condition for maximum likelihood estimates and as follows [21]

$$\begin{aligned} \left.\frac{\partial \log L(\theta,\sigma^2)}{\partial\theta}\right|_{\theta=\hat{\theta}} &= 0 \\ \left.\frac{\partial \log L(\theta,\sigma^2)}{\partial\sigma^2}\right|_{\sigma=\hat{\sigma}} &= 0 \end{aligned} \tag{5.50}$$

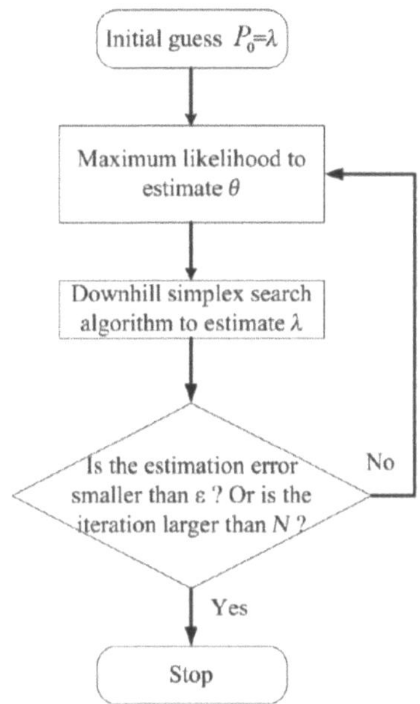

Figure 5.15 The combined method of the downhill simplex search and the maximum likelihood estimation for λ and θ, iteratively.

The estimated parameters $\hat{\theta}$ and $\hat{\sigma}^2$are shown in Eqs. (5.51) and (5.52).

$$\hat{\theta} = (\Phi^T\Phi)^{-1}\Phi^T Y \tag{5.51}$$

$$\begin{aligned}\hat{\sigma}^2 &= \frac{1}{N}\sum_{k=1}^{N}[y(t_k) - \phi(t_k)\hat{\theta}] \\ &= \frac{1}{N}(Y-\Phi\hat{\theta})^T(Y-\Phi\hat{\theta})\end{aligned} \tag{5.52}$$

where $Y$and Φ can be obtained from the microarray data of regulatory genes and the target gene.

After estimating $\hat{\theta}$ by maximum likelihood method, we take $\hat{\theta}$ into downhill simplex search method to estimate λ again. We iterate these two methods until the differences of real data and estimated data are as small as possible. All iteration processes are shown in Fig. 5.15.

Comparison between actual/constructed expression profiles

We use the cluster analysis and visualization tool (Cluster and Treeview) written by Michael Eisen (http://rana.lbl.gov/EisenSoftware.htm) to compare the actual expression profiles with the constructed expression profiles. When clustering the expression profiles, we use hierarchical clustering methods [70].

APPENDIX B: FIGURES

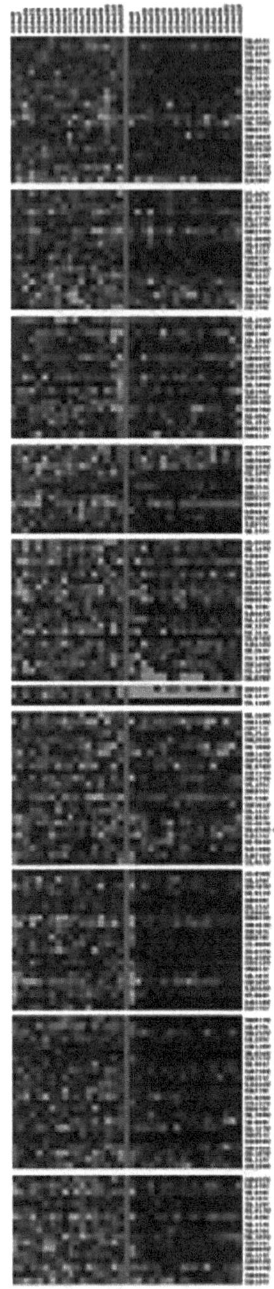

Figure B.1 Comparison between the shuffled experimental mRNA expression profiles and those predicted by the proposed model. The shuffled experimental mRNA expression profiles of 189 cell cycle genes are at the left side, and the profiles predicted by the dynamic regulatory circuits are at the right side. And the correlation coefficient of both profiles is 0.1143.

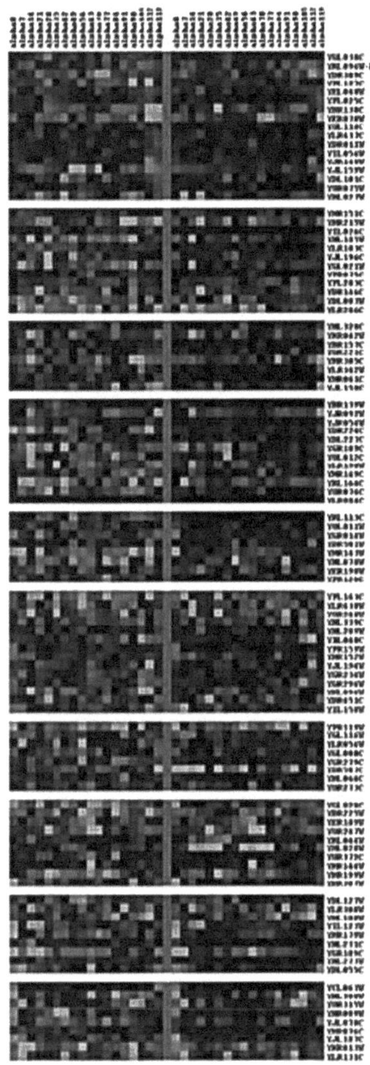

Figure B.2 Comparison between the shuffled experimental mRNA expression profiles and those predicted by the proposed model. The shuffled experimental mRNA expression profiles of 109 cell cycle genes are at the left side, and the profiles predicted by the dynamic regulatory circuits are at the right side. And the correlation coefficient of both profiles is 0.5939.

REFERENCES

[1] Gasch AP, Spellman PT, Kao CM, Carmel-Harel O, Eisen MB, Storz G, et al. Genomic expression programs in the response of yeast cells to environmental changes. Mol Biol Cell 2000;11 (12):4241–57.
[2] Spellman PT, Sherlock G, Zhang MQ, Iyer VR, Anders K, Eisen MB, et al. Comprehensive identification of cell cycle-regulated genes of the yeast *Saccharomyces cerevisiae* by microarray hybridization. Mol Biol Cell 1998;9(12):3273–97.
[3] Chang WC, Li CW, Chen BS. Quantitative inference of dynamic regulatory pathways via microarray data. BMC Bioinformatics 2005;6:44.
[4] Jang J-SR, Sun C-T, Mizutani E. Neuro-fuzzy and soft computing: a computational approach to learning and machine intelligence. Upper Saddle River, NJ: Prentice Hall; 1997.
[5] Lee TI, Rinaldi NJ, Robert F, Odom DT, Bar-Joseph Z, Gerber GK, et al. Transcriptional regulatory networks in *Saccharomyces cerevisiae*. Science 2002;298(5594):799–804.
[6] Lin LH, Lee HC, Li WH, Chen BS. Dynamic modeling of cis-regulatory circuits and gene expression prediction via cross-gene identification. BMC Bioinformatics 2005;6:258.
[7] Simon I, Barnett J, Hannett N, Harbison CT, Rinaldi NJ, Volkert TL, et al. Serial regulation of transcriptional regulators in the yeast cell cycle. Cell 2001;106(6):697–708.
[8] Wagner A. Genes regulated cooperatively by one or more transcription factors and their identification in whole eukaryotic genomes. Bioinformatics 1999;15(10):776–84.
[9] Pilpel Y, Sudarsanam P, Church GM. Identifying regulatory networks by combinatorial analysis of promoter elements. Nat Genet 2001;29(2):153–9.
[10] Banerjee N, Zhang MQ. Identifying cooperativity among transcription factors controlling the cell cycle in yeast. Nucleic Acids Res 2003;31(23):7024–31.
[11] Kato M, Hata N, Banerjee N, Futcher B, Zhang MQ. Identifying combinatorial regulation of transcription factors and binding motifs. Genome Biol 2004;5(8):R56.
[12] Chen HC, Lee HC, Lin TY, Li WH, Chen BS. Quantitative characterization of the transcriptional regulatory network in the yeast cell cycle. Bioinformatics 2004;20(12):1914–27.
[13] Davidson EH, McCay DR, Hood L. Regulatory gene networks and the properties of the developmental process. Proc Natl Acad Sci USA 2003;100(4):1475–80.
[14] Hasty J, McMillen D, Collins JJ. Engineered gene circuits. Nature 2002;420(6912):224–30.
[15] Hood L. Systems biology: integrating technology, biology, and computation. Mech Ageing Dev 2003;124(1):9–16.
[16] Tegner J, Yeung MKS, Hasty J, Collins JJ. Reverse engineering gene networks: Integrating genetic perturbations with dynamical modeling. Proc Natl Acad Sci USA 2003;100(10):5944–9.
[17] Harbison CT, Gordon DB, Lee TI, Rinaldi NJ, Macisaac KD, Danford TW, et al. Transcriptional regulatory code of a eukaryotic genome. Nature 2004;431(7004):99–104.
[18] Faires JD, Burden RL. Numerical methods. 2nd ed. Pacific Grove, CA: Brooks/Cole Pub. Co. 1998.
[19] Goldbeter A, Koshland DE. An amplified sensitivity arising from covalent modification in biological-systems. Proc Natl Acad Sci USA Biol Sci 1981;78(11):6840–4.
[20] Mestl T, Plahte E, Omholt SW. A mathematical framework for describing and analyzing gene regulatory networks. J Theor Biol 1995;176(2):291–300.
[21] Johansson R. System modeling and identification. Englewood Cliffs, NJ: Prentice Hall; 1993.
[22] Koch C, Moll T, Neuberg M, Ahorn H, Nasmyth K. A role for the transcription factors MBP1 and SWI4 in progression from G1 to S-Phase. Science 1993;261(5128):1551–7.
[23] Kumar R, Reynolds DM, Shevchenko A, Shevchenko A, Goldstone SD, Dalton S. Forkhead transcription factors, Fkh1p and Fkh2p, collaborate with Mcm1p to control transcription required for M-phase. Curr Biol 2000;10(15):896–906.
[24] Manke T, Bringas R, Vingron M. Correlating protein-DNA and protein-protein interaction networks. J Mol Biol 2003;333(1):75–85.
[25] Zhu GF, Spellman PT, Volpe T, Brown PO, Botstein D, Davis TN, et al. Two yeast forkhead genes regulate the cell cycle and pseudohyphal growth. Nature 2000;406(6791):90–4.

[26] Costanzo M, Schub O, Andrews B. G(1) transcription factors are differentially regulated in *Saccharomyces cerevisiae* by the Swi6-binding protein Stb1. Mol Cell Biol 2003;23(14):5064–77.
[27] Ho W, Costanzo M, Moore L, Kobayashi R, Andrews BJ. Regulation of transcription at the *Saccharomyces cerevisiae* start transition by Stb1, a Swi6-binding protein. Mol Cell Biol 1999;19 (8):5267–78.
[28] Olson KA, Nelson C, Tai G, Hung W, Yong C, Astell C, et al. Two regulators of Ste12p inhibit pheromone-responsive transcription by separate mechanisms. Mol Cell Biol 2000;20(12):4199–209.
[29] Doolin MT, Johnson AL, Johnston LH, Butler G. Overlapping and distinct roles of the duplicated yeast transcription factors Ace2p and Swi5p. Mol Microbiol 2001;40(2):422–32.
[30] Iyer VR, Horak CE, Scafe CS, Botstein D, Snyder M, Brown PO. Genomic binding sites of the yeast cell-cycle transcription factors SBF and MBF. Nature 2001;409(6819):533–8.
[31] Dirick L, Moll T, Auer H, Nasmyth K. A central role for SWI6 In modulating cell-cycle start-specific transcription in yeast. Nature 1992;357(6378):508–13.
[32] Primig M, Sockanathan S, Auer H, Nasmyth K. Anatomy of a transcription factor important for the start of the cell-cycle in *Saccharomyces cerevisiae.* Nature 1992;358(6387):593–7.
[33] Futcher B. Transcriptional regulatory networks and the yeast cell cycle. Curr Opin Cell Biol 2002;14(6):676–83.
[34] McBride HJ, Yu YX, Stillman DJ. Distinct regions of the Swi5 and Ace2 transcription factors are required for specific gene activation. J Biol Chem 1999;274(30):21029–36.
[35] Mjolsness E, Sharp DH, Reinitz J. A connectionist model of development. J Theor Biol 1991;152 (4):429–53.
[36] Shea MA, Ackers GK. The OR control system of bacteriophage lambda—a physical-chemical model for gene-regulation. J Mol Biol 1985;181(2):211–30.
[37] Gao F, Foat BC, Bussemaker HJ. Defining transcriptional networks through integrative modeling of mRNA expression and transcription factor binding data. BMC Bioinformatics 2004;5:31.
[38] Haverty PM, Hansen U, Weng ZP. Computational inference of transcriptional regulatory networks from expression profiling and transcription factor binding site identification. Nucleic Acids Res 2004;32(1):179–88.
[39] Wei HR, Kaznessis Y. Inferring gene regulatory relationships by combining target-target pattern recognition and regulator-specific motif examination. Biotechnol Bioeng 2005;89(1):53–77.
[40] Nachman I, Regev A, Friedman N. Inferring quantitative models of regulatory networks from expression data. Bioinformatics 2004;20:248–56.
[41] Das D, Banerjee N, Zhang MQ. Interacting models of cooperative gene regulation. Proc Natl Acad Sci USA 2004;101(46):16234–9.
[42] Michaelis S, Herskowitz I. The A factor pheromone of *Saccharomyces cerevisiae* is essential for mating. Mol Cell Biol 1988;8(3):1309–18.
[43] Wang YL, Liu CL, Storey JD, Tibshirani RJ, Herschlag D, Brown PO. Precision and functional specificity in mRNA decay. Proc Natl Acad Sci USA 2002;99(9):5860–5.
[44] Yuh CH, Bolouri H, Davidson EH. Cis-regulatory logic in the endo16 gene: switching from a specification to a differentiation mode of control. Development 2001;128(5):617–29.
[45] Yuh CH, Bolouri H, Davidson EH. Genomic cis-regulatory logic: experimental and computational analysis of a sea urchin gene. Science 1998;279(5358):1896–902.
[46] Yuh CH, Moore JG, Davidson EH. Quantitative functional interrelations within the cis-regulatory system S-purpuratus Endo16 gene. Development 1996;122(12):4045–56.
[47] Bar-Joseph Z, Gerber G, Gifford DK, Jaakkola TS, Simon I. A new approach to analyzing gene expression time series data. In: Proceedings of the sixth annual international conference on computational biology: 2002. ACM; 2002. p. 39–48.
[48] De Boor C. A practical guide to splines. New York, NY: Springer-Verlag; 1978.
[49] Maher M, Cong F, Kindelberger D, Nasmyth K, Dalton S. Cell-cycle-regulated transcription of the CLB2 gene is dependent on MCM1 and a ternary complex factor. Mol Cell Biol 1995;15 (6):3129–37.
[50] Koranda M, Schleiffer A, Endler L, Ammerer G. Forkhead-like transcription factors recruit Ndd1 to the chromatin of G2/M-specific promoters. Nature 2000;406(6791):94–8.

[51] McInerny CJ, Partridge JF, Mikesell GE, Creemer DP, Breeden LL. A novel Mcm1-dependent element in the SWI4, CLN3, CDC6, and CDC47 promoters activates M/G(1)-specific transcription. Genes Dev 1997;11(10):1277−88.

[52] Loy CJ, Lydall D, Surana U. NDD1, a high-dosage suppressor of cdc28-1N, is essential for expression of a subset of late-S-phase-specific genes in *Saccharomyces cerevisiae.* Mol Cell Biol 1999;19 (5):3312−27.

[53] Lim FL, Hayes A, West AG, Pic-Taylor A, Darieva Z, Morgan BA, et al. Mcm1p-induced DNA bending regulates the formation of ternary transcription factor complexes. Mol Cell Biol 2003;23 (2):450−61.

[54] Hartemink AJ, Gifford DK, Jaakkola TS, Young RA. Combining location and expression data for principled discovery of genetic regulatory network models. Pac Symp Biocomput 2002;437−49.

[55] Berman BP, Nibu Y, Pfeiffer BD, Tomancak P, Celniker SE, Levine M, et al. Exploiting transcription factor binding site clustering to identify cis-regulatory modules involved in pattern formation in the Drosophila genome. Proc Natl Acad Sci USA 2002;99(2):757−62.

[56] Yuh CH, Davidson EH. Modular cis-regulatory organization of Endo16, a gut-specific gene of the sea urchin embryo. Development 1996;122(4):1069−82.

[57] Chen KC, Wang TY, Tseng HH, Huang CYF, Kao CY. A stochastic differential equation model for quantifying transcriptional regulatory network in *Saccharomyces cerevisiae.* Bioinformatics 2005;21 (12):2883−90.

[58] Gardner TS, di Bernardo D, Lorenz D, Collins JJ. Inferring genetic networks and identifying compound mode of action via expression profiling. Science 2003;301(5629):102−5.

[59] Chang Y-H, Wang Y-C, Chen B-S. Identification of transcription factor cooperativity via stochastic system model. Bioinformatics 2006;22(18):2276−82.

[60] Sontag E, Kiyatkin A, Kholodenko BN. Inferring dynamic architecture of cellular networks using time series of gene expression, protein and metabolite data. Bioinformatics 2004;20(12):1877−86.

[61] Yeung MKS, Tegner J, Collins JJ. Reverse engineering gene networks using singular value decomposition and robust regression. Proc Natl Acad Sci USA 2002;99(9):6163−8.

[62] Carlson JM, Doyle J. Highly optimized tolerance: a mechanism for power laws in designed systems. Phys Rev E 1999;60(2):1412−27.

[63] Csete ME, Doyle JC. Reverse engineering of biological complexity. Science 2002;295(5560):1664−9.

[64] Yi TM, Huang Y, Simon MI, Doyle J. Robust perfect adaptation in bacterial chemotaxis through integral feedback control. Proc Natl Acad Sci USA 2000;97(9):4649−53.

[65] Ito T, Chiba T, Ozawa R, Yoshida M, Hattori M, Sakaki Y. A comprehensive two-hybrid analysis to explore the yeast protein interactome. Proc Natl Acad Sci USA 2001;98(8):4569−74.

[66] Uetz P, Giot L, Cagney G, Mansfield TA, Judson RS, Knight JR, et al. A comprehensive analysis of protein-protein interactions in *Saccharomyces cerevisiae.* Nature 2000;403(6770):623−7.

[67] Klipp E, Herwig R, Kowald A, Wierling C, Lehrach H. Systems biology in practice: concepts, implementation and application. Weinheim: John Wiley & Sons; 2005.

[68] Arava Y, Wang YL, Storey JD, Liu CL, Brown PO, Herschlag D. Genome-wide analysis of mRNA translation profiles in *Saccharomyces cerevisiae.* Proc Natl Acad Sci USA 2003;100(7):3889−94.

[69] Vu TT, Vohradsky J. Nonlinear differential equation model for quantification of transcriptional regulation applied to microarray data of *Saccharomyces cerevisiae.* Nucleic Acids Res 2007;35(1):279−87.

[70] Eisen MB, Spellman PT, Brown PO, Botstein D. Cluster analysis and display of genome-wide expression patterns. Proc Natl Acad Sci USA 1998;95(25):14863−8.

CHAPTER 6

Big Mechanisms of Information Flow in Cellular Systems in Response to Environmental Stress Signals via System Identification and Data Mining

INTRODUCTION

The cell senses extracellular signals through intercellular communication, resulting in cellular stress responses. Environmental changes or different events may stimulate different signals. Typical signals are hormones, pheromones, pathogens, heat, cold, light, and osmotic pressure. In general, these signals can be called environmental stresses (or stimuli). These signals are received by their corresponding receptors and transmitted through specific signal transduction pathways to the corresponding transcription factors (TFs) to regulate genes to produce the corresponding proteins in response to the stress. Cellular responses to sudden environmental stresses or physiological changes provide living organisms with the opportunity for survival and further development. Therefore, it is important to understand protective mechanisms against environmental stresses. The mechanisms for this information flow in signal transduction pathways in response to stress are very complex. In this chapter, we will investigate the cellular mechanisms behind information flow in cellular systems through system identification and data mining.

As unicellular organisms such as yeast have evolved mechanisms to respond to environmental stresses by rapidly reorganizing gene expression through stress-response TFs, we use (see Section 6.1) a dynamic system model of gene regulation to describe the information transmission mechanism of the control of TFs on gene expression in response to a specific environmental stress. Then, based on the dynamic system model, we introduce a Stress Regulator Identification Algorithm (SRIA) to identify stress-response TFs for six stresses. We have found that a small number of TFs are probably sufficient to control a wide variety of expression patterns in yeast under different stresses, which suggests that the response mechanisms to different stresses may have a bow-tie structure and that there is crosstalk among different stress responses. In Section 6.2, we identify the mechanisms of gene and protein regulatory network responses to a

Big Mechanisms in Systems Biology
DOI: http://dx.doi.org/10.1016/B978-0-12-809479-2.00006-8

broad range of environmental stress-activated genes and protein regulatory networks from database mining and microarray data in response to environmental stresses. According to the reconstructed gene/protein networks, possible mutual interactions, as well as feedforward and feedback loops, are found to accelerate the response and filter noise in these signaling transduction pathways. A bow-tie core network is specifically identified to coordinate mutual interactions and feedforward loops, feedback inhibition, feedback activation, and crosstalk as protective mechanisms to cope efficiently with a broad range of environmental stresses with limited proteins and pathways.

Light is considered as an external stimulus to plant cells. The photosynthetic light response is an important process that allows plants to optimize the efficiency of photosynthesis, which is the core technology for green energy. However, currently, little is known about the molecular mechanisms behind the regulation of the photosynthetic light response. In Section 6.3, a systematic method is proposed to investigate the response mechanism to light stimuli by constructing a gene regulatory network (GRN) from database mining and microarray data of *Arabidopsis thaliana*. If more experimental whole-genome Chromatin ImmunoPrecipitation (ChIP)-chip and next-generation sequencing (NGS) data from multiple sampling times become available in the future, the proposed method will be improved for constructing a whole-genome GRN to understand the cellular mechanisms of photosynthesis.

In Section 6.4, a systematic method is proposed to construct an integrated genetic and epigenetic cellular network (IGECN) using a dynamic model and high-throughput NGS data to investigate the entire cellular signal response mechanism. Three kinds of "omics" data, including genomics, transcriptomics, proteomics, as well as epigenetic microRNAs (miRNAs) and methylation data, are employed to provide a signal flow view of cellular mechanisms of cancer and infectious cells.

6.1 CONSTRUCTING STRESS-RESPONSE MECHANISMS VIA DYNAMIC GENE REGULATORY MODELING AND DATA MINING

Single-celled organisms such as yeasts constantly battle variable or even harsh external environments that threaten their survival or, at least, prevent them from performing optimally. Environmental changes can be of a physical or chemical nature such as temperature, oxidation, osmolarity, acidity, and nutrient availability [1]. The response mechanisms to environmental stresses are highly complex. One aspect of this cellular adaptation is the reorganization of gene expression. The gene expression program required for maintenance of the optimal cell physiology in one environment may be far from optimal in another environment. Thus, when the environmental condition changes abruptly, the cell must rapidly adjust its gene expression program to adapt to the new conditions [2].

Large-scale reprogramming of gene expression can be revealed as stress-response mechanism from genome-wide DNA microarrays, which measure the relative transcription levels of every gene in the yeast genome at any given moment, providing a snapshot of the genomic expression program [3]. Exploring the dynamic nature of the yeast genome through time-course experiments can illuminate yeast stress responses. For example, Gasch et al. [2] and Causton et al. [4] used genome-wide expression analysis to explore how gene expression in yeast is remodeled over time as cells respond to heat shock, oxidative shock, osmotic shock, acidic stress, nitrogen depletion, amino acid starvation as well as other environmental stresses. They discovered that more than half of the genome is involved in responding to at least one of the investigated environmental changes. A set of genes (~10% of yeast genes), termed the environmental stress response (ESR) genes or common environmental response (CER) genes, showed a similar drastic response to almost all of these environmental changes. Other gene expression responses appeared to be specific to particular environmental conditions. However, the regulators of these stress-response genes are not revealed by their studies. The complete network of stress-response regulators and the details of their actions remain to be investigated [2].

Computational and statistical methods have been developed to identify plausible regulators of many cellular processes in yeast [5–14]. When the time-course data of a system are available, using dynamic system modeling is more appropriate than using a statistical approach because it can model the dynamic behavior of the system [15–18]. Therefore, in this section we use a dynamic system model of gene regulation to describe the mechanism of how TFs may control a gene's expression. Then, based on the dynamic system model, the SRIA is developed to identify stress-response TFs for six different kinds of stresses. This goal is to reconstruct a network of stress-response regulators and study the details of their actions.

6.1.1 Systems biology methods of constructing big mechanisms of information flow

Datasets and data preprocessing

Two kinds of datasets are used. First, for each gene in the yeast genome, the TFs that may regulate its expression are retrieved from the YEASTRACT database [19]. The regulatory associations between the target gene and the TFs are included if they are supported by published data showing at least one of the following experimental evidences: (1) Change in the expression of the target gene due to a deletion (or mutation) in the gene encoding transcription factor; these evidences may come from detailed gene by gene analysis or genome-wide expression analysis; (2) Binding of the transcription factor to the promoter region of the target gene, as supported by band-shift, foot-printing or ChIP assays. However, the TF-gene regulatory association data are noisy. The genes whose expressions are affected by the mutation of a TF may not be the direct targets of that TF. Moreover, the genes that are bound by a TF identified by ChIP-chip experiments may not be regulated by that TF because TF binding does not necessarily mean regulation. Therefore, other independent data sources such as gene expression or proteomic data should be used to filter out the noise inherent in the TF-gene regulatory association data.

In this section, we incorporate the gene expression data into this analysis. The genome-wide gene expression time profiles under various stress conditions such as heat shock, oxidative shock, osmotic shock, acidic stress, nitrogen depletion, and amino acid starvation are from Gasch et al. (2000) and Causton et al. (2001) [3,4]. Under each stress condition, samples for all genes in the yeast genome are collected at multiple time points. In order to reconstruct the missing values and avoid overfitting, the cubic spline method [20] is applied to interpolate some extra data points. (The missing values of the microarray data are those data points whose values are questionable due to the experimental defects. Therefore, these data points are not reported in the microarray data and regarded as the missing values.) In order to fit the dynamic system model in the linear scale, the microarray data are transformed from the log2 scale to the linear scale.

Dynamic system model of gene regulation by TFs

We consider the transcriptional regulatory mechanism of a target gene as a system with the regulatory profiles of several TFs as the inputs and the gene expression profile of the target gene as the output. Owing to random noise and fluctuations at the molecular level, the transcriptional regulation of a target gene is described by the following stochastic dynamic equation:

$$y(t+1) = \left(\sum_{i=1}^{N} b_i x_i(t) + k \right) - \lambda y(t) + \varepsilon(t) \tag{6.1}$$

where $y(t)$ represents the target gene's expression profile at time point t, N denotes the number of TFs that may regulate the target gene's expression, b_i indicates the regulatory ability of TFi, $x_i(t)$ represents the regulatory profile of TFi at time point t, k represents the basal level induced by RNA polymerase II, λ indicates the degrading effect of the present state value $y(t)$ on the next state value $y(t+1)$ and $\varepsilon(t)$ denotes a stochastic noise due to the modeling error and measuring error of the target gene's expression profile. $\varepsilon(t)$ is assumed to be a Gaussian noise with mean zero and unknown standard deviation σ. The biological meaning of Eq. (6.1) is that $y(t+1)$ (the target gene's expression value at the next state) is determined by $\sum_{i=1}^{N} b_i x_i(t) + k$ (the production effect of the N TFs at the present state and RNA polymerase II) and $-\lambda y(t)$ (the degradation effect of the target gene at the present state).

It has been shown that TF binding usually affects gene expression in a nonlinear fashion: below some level it has no effect, while above a certain level the effect may become saturated. This type of binding behavior can be modeled using a sigmoid function [12,13,15–18]. Therefore, we define $x_i(t)$ (the regulatory profile of TFi at time point t) as a sigmoid function of $z_i(t)$ (the gene expression profile of TFi at time point t):

$$x_i(t) = f(z_i(t)) = \frac{1}{1 + \exp[-r(z_i(t) - A_i)]} \tag{6.2}$$

where r denotes the transition rate of the sigmoid function and A_i denotes the mean of the gene expression profile of TFi.

Estimating the parameters of the dynamic system model

Using the TF-gene regulatory association data from the YEASTRACT database and gene expression data from Gasch et al. [2] and Causton et al. [4], we can estimate the parameters of the dynamic system model in Eq. (6.1) [3,4]. We rewrite Eq. (6.1) to the following regression form:

$$\begin{aligned} y(t+1) &= \begin{bmatrix} x_1(t) & \ldots & x_N(t) & 1 & -y(t) \end{bmatrix} \begin{bmatrix} b_1 \\ \vdots \\ b_N \\ k \\ \lambda \end{bmatrix} + \varepsilon(t) \\ &= \phi(t)\theta + \varepsilon(t) \end{aligned} \tag{6.3}$$

where $\phi(t) = [x_1(t)\ldots x_N(t) 1 - y(t)]$ denotes the regression vector and $\theta = \begin{bmatrix} b_1 \\ \vdots \\ b_N \\ k \\ \lambda \end{bmatrix}$ is the parameter vector. N is the number of the TFs that may regulate the target gene's expression and can be inferred from the TF-gene regulatory association data.

From the gene expression data, it is easy to get the values of $\{x_i(t_v),\ y(t_v)\}$ for $i \in \{1,\ldots,N\}$, $v \in \{1,\ldots,M\}$, where M is the number of the time points of a target gene's expression profile. Eq. (6.3) at different time points can be put together as follows:

$$\begin{bmatrix} y(t_2) \\ \vdots \\ y(t_M) \end{bmatrix} = \begin{bmatrix} \phi(t_1) \\ \vdots \\ \phi(t_{M-1}) \end{bmatrix} \theta + \begin{bmatrix} \varepsilon(t_1) \\ \vdots \\ \varepsilon(t_{M-1}) \end{bmatrix} \tag{6.4}$$

For simplicity, we can further define the notations Y, Φ, and e to represent Eq. (6.4) as follows:

$$Y = \Phi\theta + e \tag{6.5}$$

The parameter vector θ can be estimated by the maximum likelihood (ML) method in Eqs. (5.7) – (5.12) as follows [21]:

$$\hat{\theta} = (\Phi^T\Phi)^{-1}\Phi^T Y \tag{6.6}$$

Stress regulator identification algorithm

After constructing the dynamic system model of gene regulation as in Eq. (6.1) and estimate the parameter vector as in Eq. (6.6), we are now ready to identify the stress-response TFs for six kinds of stresses. We develop the SRIA to do this task. SRIA can be divided into the following three steps (see the flowchart in Fig. 6.1).

Step 1: Identification of stress-response genes

The first step is to find out all the genes in the yeast genome that respond to a specific stress (e.g., heat shock, oxidative shock, osmotic shock, acidic stress, nitrogen depletion,

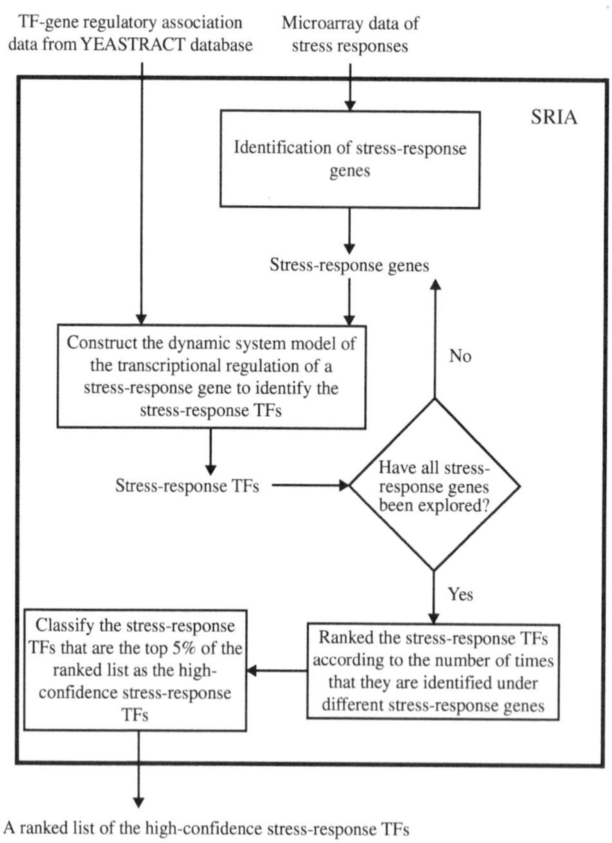

Figure 6.1 The flowchart of SRIA.

and amino acid starvation). A gene is said to respond to a specific stress if more than two time points of its gene expression profile measured under that specific stress are induced or repressed by more than threefold compared to that of the unstressed condition.

Step 2: Identification of stress-response TFs

For each stress-response gene identified in Step 1, we retrieve all TFs that may regulate its expression from the TF-gene regulatory association data. Knowing this information enables us to construct the dynamic system model of the transcriptional regulation of the stress-response gene as in Eq. (6.1). Then we apply the ML method to estimate the parameter vector $\hat{\theta} = \begin{bmatrix} \hat{b}_1 & \ldots & \hat{b}_N & \hat{k} & \hat{\lambda} \end{bmatrix}^T$ of the model and get the ML estimate as in Eq. (6.6). Since b_i stands for the regulatory ability of TFi, a small absolute value of b_i means that TFi only has a negligible effect on the stress-response gene's expression. Therefore, this TF-gene regulatory association may be a false positive and should be eliminated from this analysis. On the other hand, a large absolute value of b_i means that TFi has a large regulatory effect on the stress-response gene's expression. We regard TFi to be a stress-response TF if its regulatory ability b_i is statistically significantly different from zero (i.e., $|b_i| \gg 0$). The test statistic $t = \hat{b}_i/(s\sqrt{u_{ii}})$, a t-distribution with degree of freedom $(M-1)-(N+2)$ is used to assign a p-value for rejecting the null hypothesis H_0: $b_i = 0$, where u_{ii} is the ith diagonal element of the matrix $(\Phi^T\Phi)^{-1}$ and $s = \sqrt{(Y-\Phi\hat{\theta})^T(Y-\Phi\hat{\theta})/[(M-1)-(N+2)]}$ unbiased estimator of σ (the standard deviation of the stochastic noise $\varepsilon(t)$) [22]. The p-value computed by the t-distribution is then adjusted by Bonferroni correction to represent the true α level in the multiple hypothesis testing [23]. Then, TFi is said to be involved in response to a specific stress if the adjusted p-value $p_{\text{adjusted}} \leq .01$.

Step 3

The stress-response TFs identified in Step 2 are ranked according to the number of times that they are identified under different stress-response genes. The first TF in the list is the one that is identified with the largest number of times. The TFs that are at the top 5% of the ranked list are classified as the high-confidence stress-response TFs. Finally, we output a ranked list of the high-confidence stress-response TFs for each of the six different kinds of stresses. The flowchart of SRIA could be seen in Fig. 6.1.

6.1.2 Results of stress-response mechanisms by SRIA

Stress-response TFs

We identified the TFs that respond to heat shock, oxidative shock, osmotic shock, acidic stress, nitrogen depletion, and amino acid starvation, respectively. Table 6.1 shows the high-confidence stress-response TFs for each of the above six stresses.

Table 6.1 The high-confidence TFs in response to each of the six different stresses

Heat shock TFs	Oxidative shock TFs	Osmotic shock TFs	Acidic stress TFs	Nitrogen depletion TFs	Amino acid starvation TFs
Arr1	Sfp1	Arr1	Msn2	Sfp1	Arr1
Rpn4	Rpn4	Sfp1	Msn4	Rpn4	Gcn4
Msn2	Arr1	Rpn4	Arr1	Rap1	Rpn4
Msn4	Yap1	Yap1	Rpn4	Arr1	Sfp1
Yap1	Rap1	Msn2	Sok2	Fhl1	Rap1
Cad1	Ifh1	Msn4		Ifh1	

The high-confidence TFs in response to heat shock, oxidative shock, osmotic shock, acidic stress, nitrogen depletion, and amino acid starvation are shown. The TFs are in *red* (dark gray in print versions) (*blue* (light gray in print versions)) if there exist solid (partial) experimental evidence showing that they are involved in the same stress as we predicted.

The identified stress-response TFs can be divided into two categories. The first category is the well-known stress-response TFs with solid literature evidence that directly indicates involvement of these TFs in response to that specific stress. The second category is the novel stress-response TFs that have only partial or no literature support. (The partial literature support means that these TFs are predicted by previous studies as plausible stress-response TFs but still need further verification.) We found that 38% (13/34 counting multiplicity) of the identified stress-response TFs belongs to the first category, indicating the effectiveness of SRIA. In addition, 52% (11/21 counting multiplicity) of the second category has partial literature support, revealing the predictive power of SRIA. Therefore, the novel stress-response TFs (Arr1, Ifh1, Rpn4, and Sok2) that have no literature evidence yet are worthy of experimental verification.

Biological validation of predictions

Now we discuss in detail these predictions that are supported by experimental evidence in the literature.

1. Heat Shock

   The predicted heat shock TFs of Msn2, Msn4, Rpn4, Yap1, and Cad1 have solid or partial literature evidence. First, Msn2 and Msn4 bind DNA at stress response element (STRE) and activate many STRE-regulated genes in response to many stresses such as heat shock, oxidative shock, and osmotic shock [24]. Second, it has been demonstrated that the heat shock TF Hsf1 coordinates a feedforward gene regulatory circuit for *RPN4* heat induction [25]. Third, Yap1 is known to induce the expression of *GSH1* and *GSH2* to synthesize glutathione in heat shock response [26]. Fourth, overexpression of Cad1 is known to increase thermotolerance of a cell under starvation conditions [24].

2. Oxidative Shock

The predicted oxidative shock TFs of Sfp1, Yap1, Rpn4, and Rap1 have solid or partial literature evidence. First, Sfp1 is known to regulate ribosomal protein (RP) gene expression in response to various stresses such as oxidative shock and osmotic shock [27]. Second, Yap1 is known to regulate genes that respond to oxidative shock. For example, Yap1 regulates *TRX2*, a cytoplasmic thioredoxin isoenzyme of the thioredoxin system which protects cells against oxidative stress [28]. Third, *RPN4* promoter contains a Yap1 binding site (YRE) which is responsible for *RPN4* induction in response to oxidative stress [25]. Fourth, it is known that Rap1 signaling is required for suppression of Ras-generated reactive oxygen species and protection against oxidative stress [29].

3. Osmotic Shock

The predicted osmotic shock TFs of Msn2, Msn4, Sfp1, and Yap1 have solid or partial literature evidence. First, the msn2-msn4 double deletion mutants exhibit higher sensitivity to severe osmotic stress, indicating Msn2 and Msn4 are involved in response to osmotic stress [24]. Second, Sfp1 regulates RP gene expression in response to various stresses such as oxidative shock and osmotic shock [27]. Third, the *YAP4* gene, previously shown to play a role in response to hyper-osmotic stress, is regulated by the transactivators Yap1 and Msn2 [30].

4. Acidic Stress

The predicted acidic stress TFs of Msn2, Msn4, and Rpn4 have solid or partial literature evidence. First, it is known that *RGD1* is activated at low pH. The transcription level at low pH was demonstrated to depend on the STRE box located in the *RGD1* promoter. The general stress-activated TFs Msn2 and Msn4 were shown to act mainly on the basal *RGD1* transcriptional level in normal and stress conditions [31]. Second, under acute herbicide 2,4-dichlorophenoxyacetic acid (2,4-D) stress, 14% of the yeast transcripts suffered a greater than twofold change. Most of the upregulated genes in response to 2,4-D are known targets of Msn2, Msn4, and Rpn4 [32].

5. Nitrogen Depletion

The predicted nitrogen depletion TFs of Sfp1, Ifh1, Fhl1, Rpn4, and Rap1 have partial literature evidence. First, RP genes in eukaryotes are coordinately regulated in response to growth stimuli and environmental stress, thereby permitting cells to adjust ribosome number and overall protein synthetic capacity to physiological conditions. Sfp1, Fhl1, and Ifh1 are known to regulate RP gene expression in response to nutrient depletion [24,27,28]. Second, on solid growth media with limiting nitrogen source, diploid budding-yeast cells differentiate from the yeast form to a filamentous, adhesive and invasive form. Both low availability of nitrogen and a solid growth substrate are required to induce diploid filamentous-form growth. It is known that Rpn4 regulates filamentous growth, indicating that Rpn4 is involved in response to nitrogen depletion [33]. Third, Rap1 is known to be involved in the regulation of nitrogen, sulfur, and selenium metabolism [28].

6. Amino Acid Starvation

The predicted amino acid starvation TFs of Gcn4, Rap1, and Sfp1 have solid or partial literature evidence. First, Gcn4 is a well-known transcriptional activator of amino acid biosynthetic genes in response to amino acid starvation [24]. Second, *HIS4* encodes an enzyme that catalyzes the histidine biosynthesis and Rap1 is required for the rapid increase in the *HIS4* mRNA level following amino acid starvation [34]. Third, RP genes in eukaryotes are coordinately regulated in response to growth stimuli and environmental stress, thereby permitting cells to adjust ribosome number and overall protein synthetic capacity to physiological conditions. Sfp1 is known to regulate RP gene expression in response to nutrient depletion [24,27].

6.1.3 Discussions and conclusions of stress-response mechanism in the information flow of yeast cell

In this section, we use a dynamic system model of gene regulation to describe the mechanism of how the stress-response TFs may control a stress-response gene's expression. Based on the dynamic system model, we develop the SRIA to identify the stress-response TFs for each of six kinds of stresses. Some general stress-response TFs that respond to various stresses and some specific stress-response TFs that respond to a specific stress are identified. For example, we found the well-known general stress-response TFs of Msn2, Msn4, Yap1, Rpn4, and Sfp1, consistent with the results of *Saccharomyces* Genome Database (SGD) [24]. Besides, the well-known heat shock TF Cad1, nitrogen depletion TF Fhl1 and amino acid starvation TF Gcn4 are identified. The ability to identify these well-known stress-response TFs validates the effectiveness of SRIA.

SRIA identified 12 distinct TFs (Arr1, Cad1, Fhl1, Gcn4, Ifh1, Msn2, Msn4, Rap1, Rpn4, Sfp1, Sok2, and Yap1) to be in response to at least one of the six stresses under study (see Fig. 6.2). This indicates that a small number of TFs may be sufficient to control a wide variety of expression patterns in yeast under different stresses. Two implications can be inferred from this observation. First, the response mechanisms to different stresses may have a bow-tie structure [35]. As shown in Fig. 6.2, the core stress-response TFs make up the "knots" of a bow tie, facilitating the fan in of a large variety of environmental stresses through signal transduction pathways, and fan out of an even larger variety of stress-response proteins through activating stress-response target genes. Actually, approximate two-thirds of the yeast genome (about 3600 genes) are involved in responding to the changes in environment [4]. Second, there may exist regulatory crosstalks among different stress responses (see Fig. 6.3). We found that heat shock, osmotic shock and acidic stress all can trigger TFs Arr1, Msn2, Msn4, and Rpn4, indicating that these three stresses may share a similar stress-response mechanism. In addition, we found that oxidative shock, nitrogen depletion, and amino acid

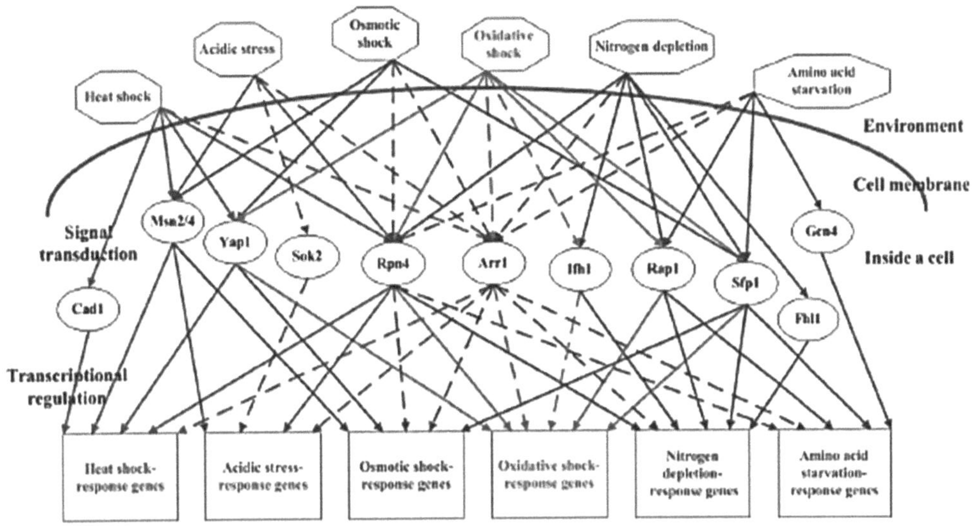

Figure 6.2 The information flow network of stress-response mechanism in yeast. Environmental stresses are represented by octagons, stress-response TFs are represented by ellipses and stress-response genes are represented by *rectangles*. *Solid (or dashed) lines* indicate the known (or predicted) regulatory relationships among environmental stresses, stress-response TFs and stress-response genes. Heat shock responses are in *red*, acidic stress responses are in *green*, osmotic shock responses are in *blue*, oxidative shock responses are in *orange*, nitrogen depletion responses are in *brown*, and amino acid starvation responses are in *purple*.

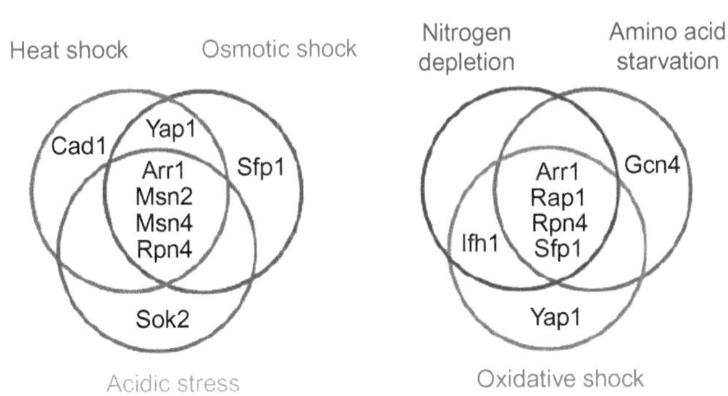

Figure 6.3 Regulatory crosstalks among different stress responses. The cellular response mechanisms to heat shock, osmotic shock, and acidic stress may have regulatory crosstalk. They all trigger TFs Arr1, Msn2, Msn4, and Rpn4. Moreover, the cellular response mechanisms to oxidative shock, nitrogen depletion, and amino acid starvation may have regulatory crosstalks. They all trigger TFs Arr1, Rap1, Rpn4, and Sfp1.

starvation all can trigger TFs of Arr1, Rap1, Rpn4, and Sfp1, indicating crosstalk among the cellular responses to these three stresses. The fact that different stress response mechanisms share some, but not all, of their regulators suggests a higher level of modularity of the yeast stress response network [11].

In Step 3 of SRIA, only those stress-response TFs that are at the top 5% of the ranked list are classified as the high-confidence stress-response TFs and reported as the final result. The reason for choosing only the top 5% of the ranked list is that when the cutoff threshold equals 5%, SRIA has the best performance in terms of the trade-off between maximizing the true positive rate and minimizing the false negative rate to find out the known amino acid starvation TFs (see Fig. 6.4 for details). The true positive and false negative rates of SRIA using different cutoff thresholds are shown. When the cutoff threshold equals 5%, SRIA has the best performance in terms of the tradeoff between maximizing the true positive rate and minimizing the false negative rate to find out the known amino acid starvation TFs. (The true positives are those known amino acid starvation TFs that are found by SRIA and the false negatives are those known amino acid starvation TFs that are not found by SRIA.)

The response mechanisms to stress are highly complex. They require a complex network of sensing and signal transduction leading to the adaptation of cell growth and proliferation along with the adjustments of the gene expression program, metabolic activities and other features of the cell [1]. This section focuses on the regulation of gene expression and proposed a network of stress-response regulators and their details of actions. Thus, it provides a starting point for understanding the adaptation mechanisms that yeast uses to survive some of the environmental stress conditions, experienced in the wild. We believe that as more gene expression data are accumulated, in combination with data from other whole-organism approaches, novel

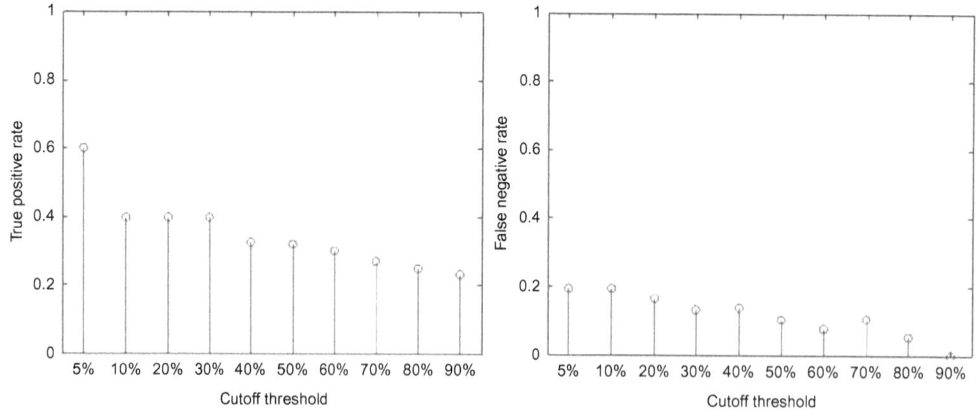

Figure 6.4 Statistics of the performance of SRIA using different cutoff thresholds.

computational algorithms such as SRIA have the potential to construct a dynamic picture of the integrated cellular response of yeast cells to environmental changes.

After constructing GRN to interpret stress-response mechanisms in this section, in order to understand protective mechanisms of yeast against environmental stresses, we need to construct integrated GRN and protein–protein interaction network (PPIN) to get more insight into the protective mechanisms of yeast in the following section.

6.2 IDENTIFYING PROTECTIVE MECHANISMS OF GENE AND PROTEIN NETWORKS IN RESPONSE TO A BROAD RANGE OF ENVIRONMENTAL STRESS SIGNALS

Cellular response mechanisms to sudden environmental stresses or physiological changes provide living organisms with the opportunity for survival and further development. Therefore, it is an importance to understand protective mechanisms against environmental stresses from the viewpoint of gene and protein networks. We propose two coupled nonlinear stochastic dynamic models to reconstruct stress-activated gene and protein regulatory networks to investigate protective mechanisms via microarray data in response to environmental stresses. According to the reconstructed gene/protein networks, some possible mutual interactions, feedforward and feedback loops are found for accelerating response and filtering noises in these signaling pathways. A bow-tie core network is also identified to coordinate mutual interactions, feedforward loops, feedback inhibitions, feedback activations and crosstalk as protective mechanisms to cope efficiently with a broader range of environmental stresses with limited proteins and pathways.

Eukaryotic cells have developed protective mechanisms in response to external environmental or physiological changes (stresses). For living organisms, cellular response to sudden environmental or physiological changes determines their fate: life or death. Survivors play an essential role in adjusting the adaptation of the whole organism to such changes or just remain uncorrelated with these changes. When unicellular organisms like *Saccharomyces cerevisiae* suffer from drastic environmental changes, cells may respond swiftly to variations. Therefore, many kinds of signaling pathways exist to construct a protective mechanism to transcriptionally regulate responsible target genes in response to a broader range of environmental stresses. Of these, the high osmolarity glycerol (HOG) pathway, which is the best understood osmoresponsive system among eukaryotes, is activated by high osmolarity, e.g., sorbitol osmotic stress. In contrast to the HOG pathway, the cell-wall integrity pathway is activated by low osmolarity, e.g., hypo-osmotic stress. In response to signaling osmotic changes, about 10% of genes are significantly affected in yeast *S. cerevisiae* [2,4,36]. The mitogen-activated protein kinase (MAPK) pathways play an essential role in response to several environmental changes, e.g., growth factors, hormones, cytokines, and environmental signals, control stress response, cell growth, morphogenesis, and proliferation. Although some nodes of MAPK pathways and the edges of their connections are suggested [37–39], the way they work and connect together as a protective mechanism in response to specific environmental change is still unclear. Therefore, in this section we try to reconstruct three different directional MAPK signaling pathways (MAPK protein regulatory pathways) in response to the hypo-osmotic stress, the sorbitol osmotic stress, and the pheromone stress, respectively, because pheromone-activated pathways, i.e., the pheromone response pathway, share common components such as *CDC24*, *STE20*, *STE50*, *STE11*, *STE7*, *KSS1*, and *STE12*, with the HOG pathway [38,40].

In recent studies, the systems biology method and computational systems biology schemes have been widely used to construct dynamic models for GRNs [41–46]. In order to reconstruct stress-activated gene and protein regulatory networks (see the flowchart of gene/protein network reconstruction in Fig. 6.5), we develop regulatory dynamic models which can reconstruct not only a directional protein regulatory pathway but also the responsible gene transcription regulatory network, simultaneously. When the time-course data of a dynamic system is available, the use of system dynamic modeling is more appropriate than the use of a statistical approach, because it can model the dynamic behavior of the system. Recently, the dynamic model of genetic network could be identified by microarray data via so-called reverse engineering methods [47,48]. By incorporating various types of genomic data, e.g., motif information, ChIP-chip data and microarray data, transcriptional modules have been explored [49,50]. Previous studies have examined dynamic system identifications of upstream regulators to target genes based on gene expression profiles under cell cycles [15,17]. However, the binding information of TFs to promoters of DNA sequences has not been fully utilized to seek possible regulations between TFs and target genes. Thus, in this section, microarray data [2,51], ChIP-chip data [52], and protein–protein interaction (PPI) data are used to investigate the interplay between internal transcriptional regulations and signaling transduction pathways to investigate the corresponding protective mechanism. Three environmental stresses, i.e., the hypo-osmotic stress, the sorbitol osmotic stress and the pheromone stress (α factor), are considered to identify the responsive mechanisms of gene and protein regulatory networks which activate the mostly concerned MAPK pathway in yeast.

The transcriptional regulatory dynamic model and protein interaction dynamic model are exploited to simultaneously reconstruct the stress-activated gene and protein regulatory networks via a combination of ChIP-chip data, genome-wide microarray data and PPI data (80,537 interacting pairs obtained from BioGRID (the Biological General Repository for Interaction Datasets, http://www.thebiogrid.org/). The significant upstream TFs of the target genes and the significant upstream interacting proteins were selected through Akaike Information Criterion (AIC) in Eq. (2.23)

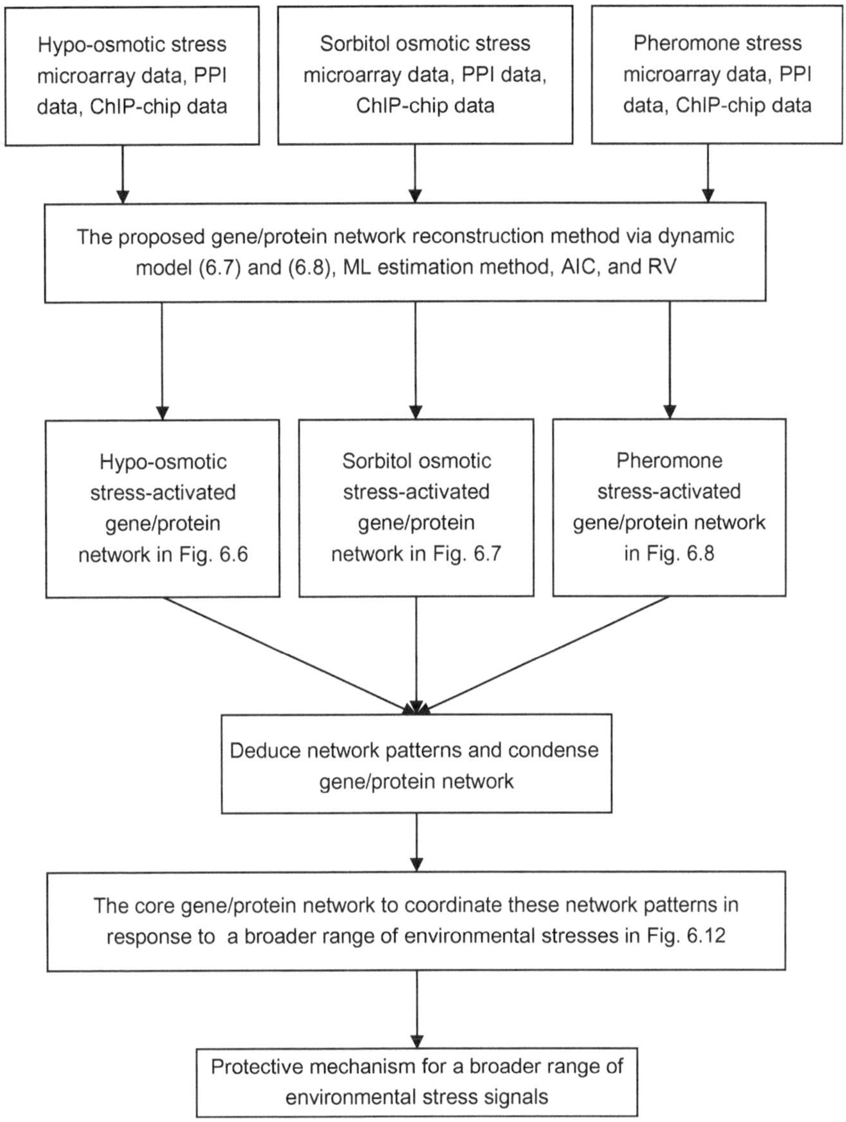

Figure 6.5 The flowchart of the proposed reconstruction method for gene/protein networks to investigate protective mechanisms in response to a broader range of environmental stresses. Based on the microarray data, protein/protein interaction data and ChIP-chip data, a gene/protein network is reconstructed in response to each specific stress in Figs. 6.6–6.8. Then these stress-activated gene/protein networks are condensed to find a bow-tie core gene/protein network in Fig. 6.12 to investigate protective mechanism in response to a broader range of environmental stresses.

via backstepping the selection procedure in the pruning of identified environmental stress-activated gene and protein regulatory networks [15,52]. Then we can reconstruct the significant stress-activated signaling pathways and GRNs under different environmental stresses. After investigating the network patterns of these gene/protein regulatory networks, we can deduce a bow-tie core network to coordinate these network patterns in response to a broader range of environmental stresses (see flowchart in Fig. 6.5).

Recently, transcription and translation represent structures where common machineries are used to trigger the corresponding biological functions in response to a broader range of environmental stresses, but versatile mechanisms make up the conserved core network. A bow-tie core network previously found for the web network and metabolic network has many inputs and outputs that are connected through a conserved core and versatile weak linkages with extensive signaling pathways governing network responses [35,53,54]. Various signaling pathways can be interfaced with a core network through versatile interfaces, so that different pathways can be recruited or removed easily without seriously affecting other parts of the network except the network connections in response to the environmental stresses. Therefore, a bow-tie architecture improves the robustness of the network against external perturbations by a robust core where numerous reactions are mediated [55,56]. In additional to the robustness of the conserved core, a bow-tie architecture also facilitates feedback regulation, because the whole network needs to respond to environmental stresses through a bow-tie core with a minimal effect on other networks [35].

Increased stability and switch-like response may exist in a MAPK pathway in response to corresponding stimuli in order to reduce response time [57,58]. In addition, a shutoff mechanism may exist in a MAPK pathway in response to noncorresponding stimuli in order to prevent proliferation of inappropriate pathway activation(s) [59]. In line with the investigations of the network patterns in previous studies [58,60–63], the patterns can be considered as the specific mechanisms for yeast in response to different environmental changes. Hence, the PPIs of MAPK pathways and transcription regulations are coordinated to make up for the corresponding mechanisms and adapt to environmental changes. Therefore, in the coordinate pathways we are concerned with two feedback loops, i.e., feedback activation and feedback inhibition, which provide the switch-like response and shutoff mechanism for the networks, respectively. In a MAPK pathway, feedback inhibition, which is used to decrease the protein concentration of the pathway by negative transcription regulation performed by the downstream TF of the pathway, acts like a shutoff mechanism to prevent the proliferation of inappropriate pathway activation [59,64]; while feedback activation, which is used to increase the protein concentration of the pathway by positive transcription regulation performed by the downstream TF of the pathway, can increase stability and reduce response time to environmental stimuli [57,58]. Moreover, signal accelerators may exist in a MAPK pathway in order to promote the efficiency of signaling. Since mutual interaction, i.e., the action of signaling proteins on each other, is expected to increase reactants, which can dramatically accelerate signaling [65], mutual interactions in gene and protein regulatory networks are concerned with to investigate which stress-activated pathways are likely accelerated by mutual interactions. If enough time profiles of microarray data and PPI data for the reconstruction of gene and protein regulatory networks are available, the proposed method can be applied to investigate some specific mechanisms of gene and protein networks for all species. Finally, the coordination of these specific mechanisms of gene and protein network via a bow-tie architecture will be also investigated in detail to provide us with in depth insight into the nature of this protective system against a broader range of environmental stresses on yeast.

6.2.1 Construction of gene transcriptional regulatory network to investigate protective mechanism in response to environmental stress signals

The flowchart of gene/protein network reconstruction algorithm is shown in Fig. 6.5. First, we need to develop a dynamic model for a gene/protein network. In general, biological regulatory systems can be described using simple mathematical models to reveal their biological characteristics. Two coupled stochastic differential equations are constructed in the description of the causality of gene and protein regulatory network, respectively [15].

Let $X_i(t)$ and $Y_i(t)$ respectively denote the ith gene expression profile and the corresponding protein level at time point t. The dynamic mechanisms lying in the transcript level $X_i(t)$ are formulated by the following dynamic model:

$$\dot{X}_i(t) + a_i X_i(t) = \sum_{j=1}^{J_i} b_{i,j} Y_j(t) + b_{i,0} + \varepsilon_i(t) \quad \text{for } i = 1, 2 \ldots O, \tag{6.7}$$

where $Y_j(t)$ are the protein levels. In Eq. (6.7), three dynamic parameters are used to describe the dynamic interplaying behavior of the transcription regulatory model: (1) Degradation role of mRNA expression, a_i ($a_i \geq 0$); (2) Transcription regulation of the ith gene by the jth regulatory protein (TF) with kinetic parameter $b_{i,j}$; and (3) Basal level of gene expression of the ith gene, $b_{i,0}$. Additionally, ε_i stands for the measurement noise of microarray data or residue of the dynamic model. J_i is the number of candidate TFs binding to the promoter regions of the ith target gene obtained from the ChIP-chip data [66]. We set the p-value threshold for .001 to choose the candidate TFs of each target gene among 203 TFs by the ChIP-chip data. Here, the protein level, $Y_j(t)$, is obtained by the sigmoid function (Hill function) as the translation from mRNA to protein [67,68], i.e., $Y_j(t) = (X_j(t)^n)/[p_j^n + X_j(t)^n]$, where p_j is defined as the mean of $X_j(t)$.

6.2.2 Protein regulatory pathway

The dynamic mechanisms of regulatory proteins lying in protein level $Y_i(t)$ are formulated by the following dynamic model:

$$\dot{Y}_i(t) + \lambda_i Y_i(t) = g_i X_i(t) - \sum_{k=1}^{K_i} d_{i,k} Y_i(t) Y_k(t) + d_{i,0} + \xi_i(t) \quad \text{for } i = 1, 2 \ldots O. \tag{6.8}$$

In Eq. (6.8), three dynamic parameters are used to describe the dynamic interplaying behavior of the regulatory protein system: (1) Degradation rate λ_i of protein. ($\lambda_i \geq 0$); (2) Interaction parameter $d_{i,k}$ between the ith protein and its K_i upstream proteins; and (3) The protein production rate g_i from the translation of mRNA $X_i(t)$ in Eq. (6.7) and the basal level $d_{i,0}$. Additionally, ξ_i is the measurement noise of microarray data or residue of the dynamic model. K_i is the number of upstream regulatory protein candidates interacting directionally with the ith protein which are obtained from BioGRID. In BioGRID, two directions of each protein interacting pair are both considered as interaction candidates in the protein interaction dynamic model. Here, the protein data, $Y_i(t)$, is also obtained from the mRNA level $X_i(t)$ through the sigmoid function (Hill function) as translation function [67].

As per the assumption of previous studies in the mathematical model of protein interactions [69,70], a protein level will be diminished by association of PPIs. Although PPIs are not always significant for the change of protein level, the term $-d_{i,k} Y_i(t) Y_k(t)$ exists in the general form (6.8) of the dynamic model. The insignificant interactions $d_{i,k}$ will be further pruned off by the proposed AIC backstepping method (see Section 6.2.6).

In the reconstructions of stress-activated gene and protein regulatory networks, we initially selected the proteins which have been well-considered in MAPK pathways. However, which proteins are selected is not crucial, because the proteins in each MAPK pathway under different environmental stresses are updated specifically based on the results of these dynamic models.

According to the reconstructed protein regulatory networks (or MAPK pathways) under different stresses, we can suggest which MAPK pathway would be accelerated or wired with the ability to avoid noise under a specific stress via the findings of mutual interactions and feedforward loops. Moreover, according to the reconstructed combinative gene and protein regulatory networks, we can suggest which pathway response is likely to be rapidly and stably activated or which is probably mediated to prevent noises by a specific stress via the findings of feedback activations and feedback inhibitions.

6.2.3 Selection of MAPK pathway-related proteins

In this section, because three environmental stresses, i.e., hypo-osmotic stress, sorbitol osmotic stress, and pheromone stress (α factor), are investigated either for their oppositely activating stresses or for their highly overlapped signaling proteins, the preselected proteins are selected within three MAPK pathways of the pheromone response pathway, the HOG pathway and the cell-wall integrity pathway [37,38,40,64,65,71–74]. The 40 proteins in Figs. 6.6–6.8 are selected within three MAPK pathways as with the preselected proteins [37,40,75,76].

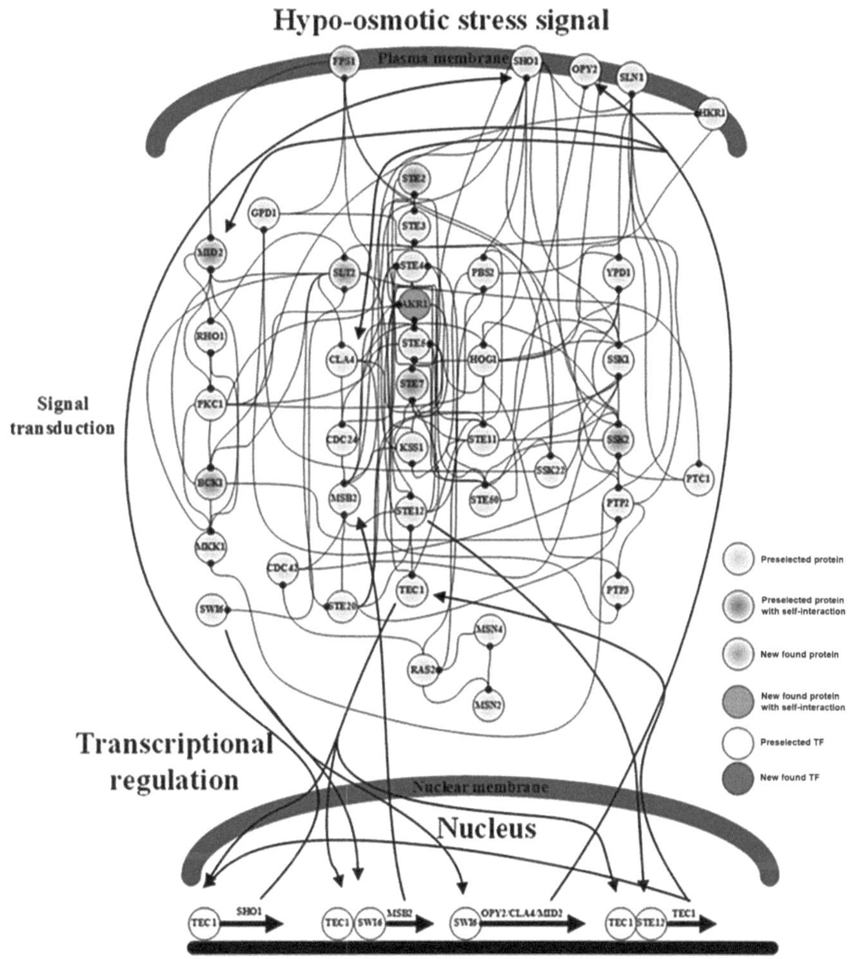

Figure 6.6 The interplaying between internal gene regulatory network and protein signal pathways as a specific protective mechanism in response to hypo-osmotic stress signal. A *black dot* stands for protein interactions.

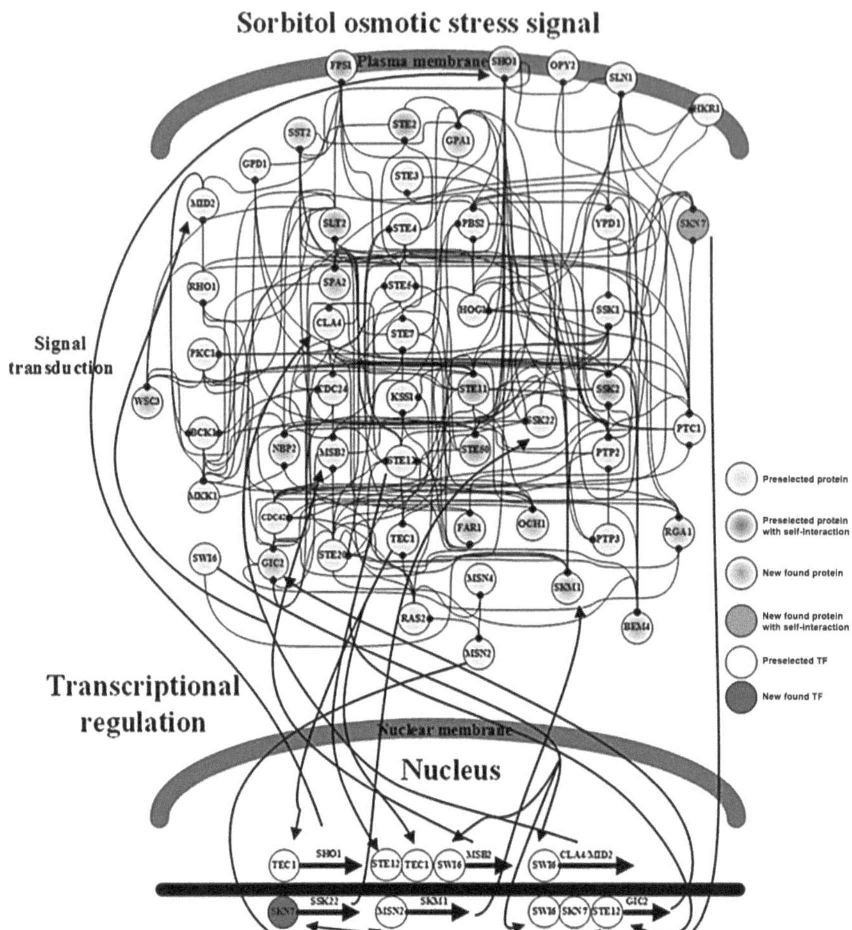

Figure 6.7 The interplaying between internal gene regulatory network and protein signal pathways as a specific protective mechanism in response to sorbitol osmotic stress signal.

6.2.4 Protective mechanisms under different environmental stress signals

According to ML method combined with the backstepping AIC, the identified models for the interplaying between the gene transcriptional regulatory network and protein regulatory pathway are given as:

$$\begin{aligned} \dot{X}_i(t) + \hat{a}_i X_i(t) &= \sum_{j=1}^{J'_i} \hat{b}_{i,j} Y_j(t) + \hat{b}_{i,0} + \hat{\varepsilon}_i(t) \quad \text{for } i = 1, 2 \ldots O, \\ \dot{Y}_i(t) + \hat{\lambda}_i Y_i(t) &= \hat{g}_i X_i(t) - \sum_{k=1}^{K'_i} \hat{d}_{i,k} Y_i(t) Y_k(t) + \hat{d}_{i,0} + \hat{\xi}_i(t), \end{aligned} \tag{6.9}$$

where $\hat{a}_i$, $\hat{b}_{i,j}$, $\hat{\lambda}_i$, $\hat{g}_i$, and $\hat{d}_{i,k}$ are obtained from the ML parameter estimation method (see Section 6.2.6). J_i candidate TFs and K_i upstream regulatory proteins are pruned

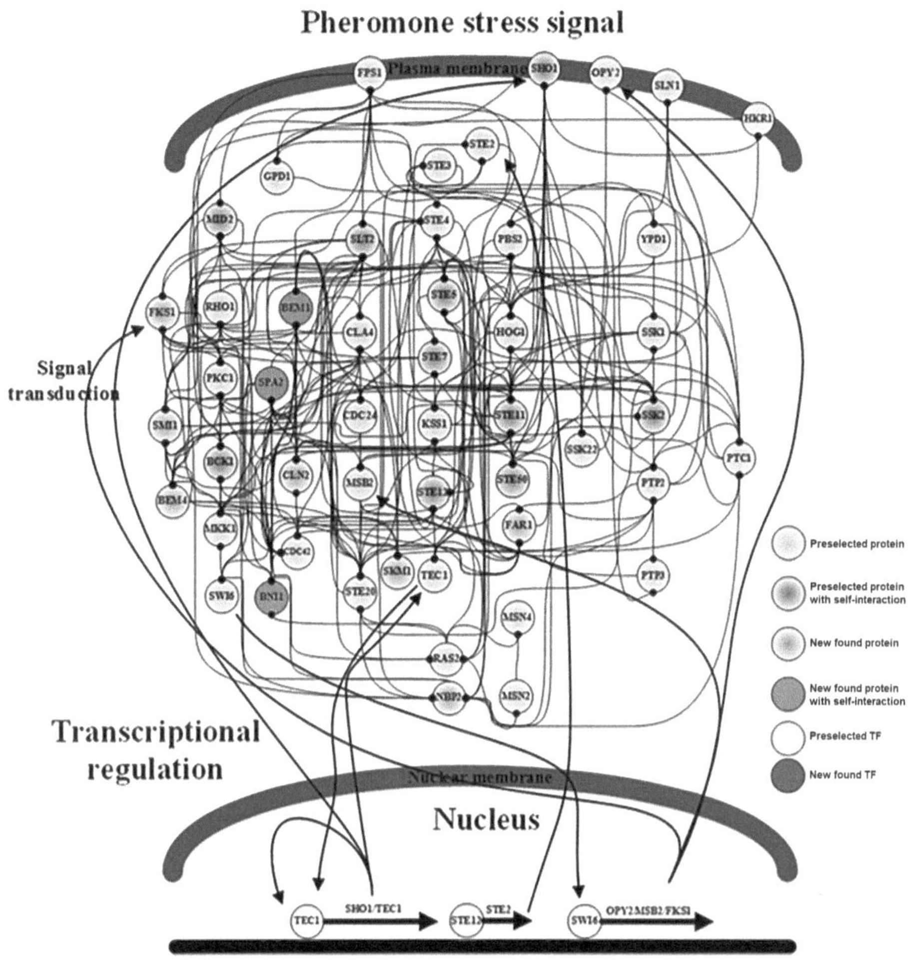

Figure 6.8 The interplaying between the internal gene regulatory network and outer protein signal pathways as a specific protective mechanism in response to pheromone stress signal.

down to J_i' and K_i' by backstepping AIC (see Section 6.2.6). The covariances of residuals $\hat{\varepsilon}_i$ and $\hat{\xi}_i$ are also estimated by the ML parameter estimation method (see Section 6.2.6). After identifying dynamic regulation function for each target gene and dynamic interaction function for each protein in Eq. (6.9), we could construct the whole gene/protein network by linking the related genes and proteins via the identified transcriptional regulation coefficients $\hat{b}_{i,j}$ and protein interaction parameter $\hat{d}_{i,k}$ in Eq. (6.9) one by one.

According to the identified models, the gene and protein regulatory networks within the 40 preselected proteins are reconnected for the three environmental stresses, i.e., hypo-osmotic stress (Fig. 6.6), sorbitol osmotic stress (Fig. 6.7), and pheromone stress (α factor) (Fig. 6.8) so that the cellular molecular mechanisms in response to environmental stresses will be identified more thoroughly. Furthermore, the protein regulatory pathways in Figs. 6.9–6.11 that contain the MAPK module [40,64] are respectively extracted from the whole protein signal pathways in Figs. 6.6–6.8.

Although the gene and protein regulatory networks are reconstructed according to the identified model, some specific target proteins, which are not included in the 40 preselected proteins, may exist in response to specific stresses. In order to improve the accuracy of the reconstructed gene and protein regulatory network and reveal the specific mechanisms of each MAPK pathway in response to each stress, we try to find the specific target proteins or TFs with a significant number of upstream interacting proteins (>5) in the gene and protein regulatory pathways under each stress (Table 6.2A–C). For the significant target proteins of the three stresses (Table 6.2A–C), we find 1, 12, and 10 new proteins respectively, which are not included in the 40 preselected proteins. The new found proteins probably play an important role in coordinating MAPK pathways in response to specific stresses. In accord with the evidence in Table 6.2, in order to investigate what possible roles the new found proteins play in response to a specific stress, the target proteins are grouped respectively into their possible functional pathways (MAPK pathways), so that we can discuss their possible functions in the sequel. After the grouping of the new found proteins into their specific MAPK pathways as a specific protective mechanism in

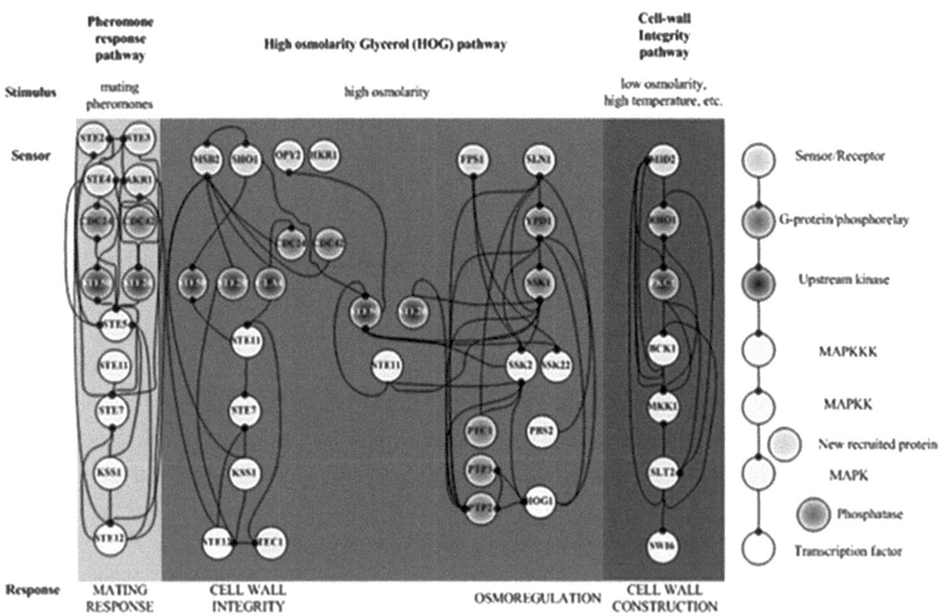

Figure 6.9 The MAPK module contained in protein regulatory pathways as a specific protective mechanism in response to hypo-osmotic stress signal. The new recruited protein, *AKR1*, is found and grouped based on Table 6.2A. The protein regulatory pathways are the same as those in Fig. 6.6 but with a MAPK module form according to the identified parameter, $\hat{d}_{i,k}$.

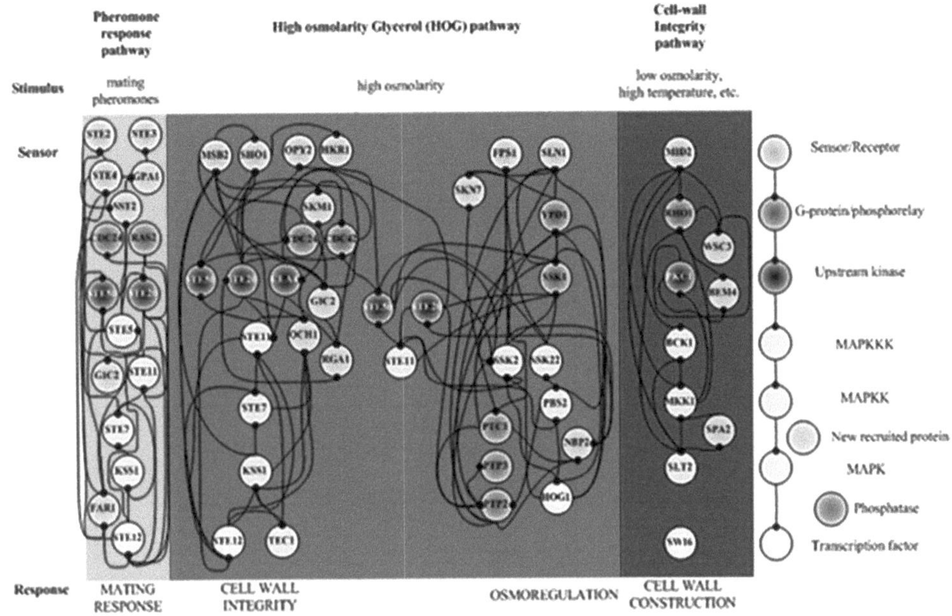

Figure 6.10 The MAPK module contained in protein regulatory pathways as a specific protective mechanism in response to sorbitol osmotic stress signal. The 12 new recruited proteins are identified and grouped based on Table 6.2B. The protein regulatory pathways are the same as those in Fig. 6.7 but with MAPK module form according to the identified parameter, $\hat{d}_{i,k}$.

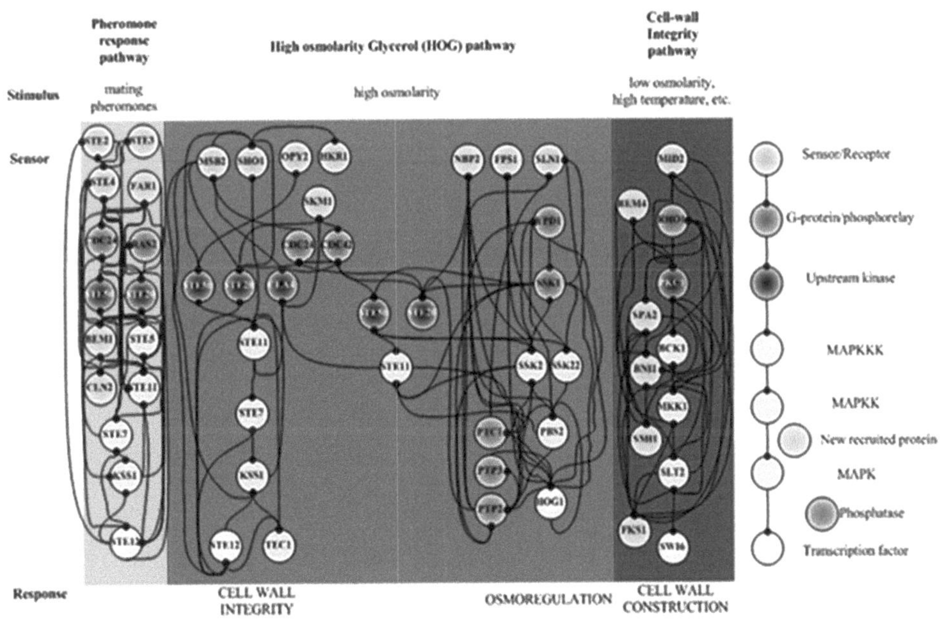

Figure 6.11 The MAPK module contained in protein regulatory pathways as a specific protective mechanism in response to pheromone stress signal. The 10 new recruited proteins are identified and grouped based on Table 6.2C. The protein regulatory pathways are the same as those in Fig. 6.8 but with MAPK module form according to the identified parameter, $\hat{d}_{i,k}$.

Table 6.2 Under hypo-osmotic stress (A), sorbitol osmotic stress (B), and pheromone stress (C), the listed target proteins are interacted by at least five preselected proteins

(A) Hypo-osmotic stress

Target protein	Literature evidence	Upstream proteins within 40 preselected proteins											
SSK1^{N}	[38–40,64]	SLN1 (0.050)	HOG1 (0.054)	STE50 (0.091)	CDC42 (0.095)	SSK2 (0.122)	STE20 (0.163)	OPY2 (0.181)	STE11 (0.272)	SHO1 (0.299)	PTP2 (0.56)	YPD1 (0.58)	SSK22 (0.905)
SSK2^{N}	[37–40,64,75]	STE11 (0.016)	SSK2 (0.018)	GPD1 (0.025)	PTP2 (0.041)	PTC1 (0.222)	YPD1 (0.249)	SSK1 (0.295)	FPS1 (0.531)	HOG1 (0.578)	SHO1 (0.819)		
PTP2^{N}	[37,38,40,64]	YPD1 (0.100)	SSK1 (0.140)	SSK2 (0.143)	SLN1 (0.160)	PTC1 (0.320)	STE20 (0.325)	HOG1 (0.330)	BCK1 (0.338)	PBS2 (0.500)			
STE50^{P,O}	[39,40,64]	STE5 (0.047)	STE4 (0.208)	SSK1 (0.271)	SSK22 (0.286)	STE50 (0.306)	SSK2 (0.443)	SHO1 (0.479)	STE11 (0.627)				
MSB2^{O}	[39]	CDC24 (0.015)	STE20 (0.028)	CLA4 (0.049)	CDC42 (0.056)	HOG1 (0.185)	SHO1 (0.432)	KSS1 (0.441)	STE12 (0.667)				
STE5^{P}	[64]	CDC24 (0.045)	KSS1 (0.079)	STE7 (0.095)	STE50 (0.431)	STE20 (0.556)	STE12 (0.739)	STE2 (0.958)					
STE7^{P,O}	[39,40,64]	STE11 (0.035)	KSS1 (0.125)	STE5 (0.180)	STE7 (0.187)	STE4 (0.412)	STE12 (0.630)						
YPD1^{N}	[39,40,64]	HOG1 (0.030)	SLN1 (0.060)	PTP2 (0.200)	PBS2 (0.200)	SSK2 (0.280)	SSK1 (0.640)						
AKR1^{P}	[77]	KSS1 (0.014)	STE3 (0.236)	STE5 (0.267)	STE20 (0.365)	STE12 (0.391)	CLA4 (0.396)						
SLN1^{N}	[38–40,64]	HOG1 (0.024)	PTC1 (0.111)	SSK1 (0.166)	PTP2 (0.291)	SHO1 (0.457)	YPD1 (0.571)						
PBS2^{N}	[37–40,64,75]	PKC1 (0.009)	RAS2 (0.067)	SSK22 (0.104)	HOG1 (0.124)	SHO1 (0.148)	PTP2 (0.260)						
MID2^{I}	[40,64]	SLT2 (0.080)	MID2 (0.125)	FPS1 (0.130)	PKC1 (0.158)	MKK1 (0.316)	BCK1 (0.656)						
MKK1^{I}	[37,40,64]	PKC1 (0.004)	MID2 (0.045)	PTP2 (0.221)	RHO1 (0.238)	SLT2 (0.416)	BCK1 (0.544)						

(B) Sorbitol osmotic stress

Target protein	Literature evidence	Upstream proteins within 40 preselected proteins												
STE4^{P}	[64]	STE12 (0.001)	TEC1 (0.005)	PKC1 (0.006)	HOG1 (0.014)	STE50 (0.015)	RHO1 (0.017)	STE11 (0.020)	STE7 (0.021)	STE20 (0.023)	CDC24 (0.024)	STE2 (0.037)	STE5 (0.209)	STE3 (0.236)
SHO1^{O}	[37–40,64,75]	SHO1 (0.001)	SSK1 (0.002)	MSB2 (0.003)	HOG1 (0.004)	SSK2 (0.006)	STE11 (0.007)	PKC1 (0.008)	STE50 (0.013)	STE20 (0.014)	PBS2 (0.017)	SLN1 (0.024)		
STE11^{P,O}	[37–40,64,75]	TEC1 (0.003)	STE20 (0.004)	KSS1 (0.007)	SSK1 (0.010)	STE7 (0.011)	STE11 (0.014)	CLA4 (0.014)	SHO1 (0.016)	GPD1 (0.017)	STE4 (0.025)	STE5 (0.516)		
PTP2^{N}	[37,40,64,78]	SSK2 (0.020)	PTC1 (0.045)	HOG1 (0.180)	PTP3 (0.195)	BCK1 (0.200)	SSK1 (0.213)	SLT2 (0.360)	STE20 (0.405)	PBS2 (0.413)	SLN1 (0.500)	YPD1 (0.525)		
STE50^{P,O}	[39,40,64,75]	STE50 (0.022)	SSK1 (0.074)	CDC42 (0.095)	SSK2 (0.111)	STE4 (0.118)	STE5 (0.231)	SSK22 (0.249)	STE11 (0.281)	SHO1 (0.391)	OPY2 (0.408)			
STE5^{P}	[64]	KSS1 (0.003)	CDC24 (0.011)	STE11 (0.013)	HOG1 (0.024)	STE7 (0.026)	STE12 (0.030)	STE50 (0.031)	STE5 (0.073)	STE4 (0.333)	STE2 (0.809)			
PBS2^{N}	[37–40,64,75]	KSS1 (0.002)	RAS2 (0.004)	STE11 (0.009)	PKC1 (0.010)	PTP2 (0.011)	YPD1 (0.012)	SSK22 (0.017)	HOG1 (0.019)	PTC1 (0.023)	SHO1 (0.026)			
SSK2^{N}	[37–40,64,75]	SSK2 (0.014)	STE50 (0.058)	GPD1 (0.080)	FPS1 (0.086)	HOG1 (0.144)	PTC1 (0.154)	PTP2 (0.168)	YPD1 (0.229)	SSK1 (0.288)	STE11 (0.736)			
SSK1^{N}	[37–40,64,75]	HOG1 (0.061)	PTP2 (0.092)	STE50 (0.092)	YPD1 (0.112)	SSK22 (0.163)	SSK2 (0.255)	STE11 (0.357)	SHO1 (0.643)	STE20 (0.796)				
STE12^{P,O}	[40,64]	STE7 (0.002)	STE11 (0.004)	STE20 (0.006)	MSB2 (0.012)	RAS2 (0.014)	STE5 (0.289)	STE4 (0.351)	STE2 (0.907)					
FPS1^{N}	[38]	STE11 (0.001)	GPD1 (0.003)	HOG1 (0.007)	MID2 (0.009)	SSK22 (0.011)	FPS1 (0.014)	SSK2 (0.016)	SLT2 (0.017)					
MSB2^{N}	[39]	HOG1 (0.048)	CDC42 (0.097)	STE12 (0.104)	CDC24 (0.135)	CLA4 (0.228)	KSS1 (0.249)	STE20 (0.824)						
FAR1^{P}	[76]	STE11 (0.007)	STE12 (0.011)	CDC42 (0.021)	CDC24 (0.022)	SSK2 (0.023)	STE5 (0.086)	STE4 (0.296)						
SPA2^{I}	[64]	SLT2 (0.001)	SSK2 (0.005)	STE7 (0.010)	CLA4 (0.012)	STE11 (0.015)	RAS2 (0.016)	STE12 (0.022)						

(Continued)

Table 6.2 (Continued)

(B) Sorbitol osmotic stress

Target protein	Literature evidence	Upstream proteins within 40 preselected proteins												
PTC1^{N}	[37,38,64,75]	CLA4 (0.001)	SSK2 (0.002)	PBS2 (0.004)	PKC1 (0.007)	PTP2 (0.010)	SLN1 (0.015)							
NBP2^{N}	[75]	SSK2 (0.025)	PTC1 (0.044)	BCK1 (0.127)	PTP2 (0.222)	STE20 (0.452)	SLT2 (0.645)							
GIC2^{P,O}	[79]	MSB2 (0.180)	STE50 (0.413)	STE20 (0.471)	CDC24 (0.645)	CLA4 (0.676)	CDC42 (0.712)							
OCH1^{O}	[64]	KSS1 (0.050)	CLA4 (0.223)	STE20 (0.331)	STE7 (0.331)	STE12 (0.347)	STE11 (0.744)							
GPA1^{P}	[64,80]	STE7 (0.051)	STE5 (0.112)	STE2 (0.147)	STE4 (0.357)	STE12 (0.388)	STE11 (0.520)							
SKN7^{N}	[37,64]	PKC1 (0.005)	YPD1 (0.009)	CDC42 (0.009)	SLN1 (0.011)	PTC1 (0.014)	RHO1 (0.016)							
SST2^{P}	[81]	STE50 (0.001)	KSS1 (0.015)	SHO1 (0.016)	STE4 (0.175)	STE5 (0.242)	STE2 (0.623)							
WSC3^{I}	[82]	RHO1 (0.032)	SSK22 (0.036)	STE11 (0.041)	PKC1 (0.100)	SSK2 (0.215)	MID2 (0.299)							
SKM1^{O}	[37]	CLA4 (0.001)	SWI6 (0.010)	STE20 (0.015)	CDC42 (0.020)	PTC1 (0.026)	STE3 (0.051)							
RGA1^{O}	[83]	CLA4 (0.001)	RHO1 (0.006)	CDC42 (0.007)	STE20 (0.019)	PBS2 (0.032)	STE4 (0.442)							
MKK1^{I}	[37,40,64]	PKC1 (0.028)	RHO1 (0.038)	BCK1 (0.061)	SLT2 (0.153)	MID2 (0.436)	PTP2 (0.670)							
BEM4^{I}	[78]	CLA4 (0.006)	HKR1 (0.011)	CDC24 (0.016)	CDC42 (0.019)	STE20 (0.021)	RHO1 (0.025)							

(C) Pheromone stress

Target protein	Literature evidence	Upstream proteins within 40 preselected proteins												
STE20^{P,O}	[39,40,64,75]	STE4 (0.001)	STE3 (0.002)	STE5 (0.008)	CLA4 (0.010)	MSB2 (0.016)	SSK1 (0.016)	STE12 (0.069)	RAS2 (0.080)	CDC24 (0.175)	PTP2 (0.342)	SLT2 (0.471)	CDC42 (0.535)	SHO1 (0.634)
HOG1^{N}	[37−40,64,75]	STE5 (0.001)	STE4 (0.002)	SSK2 (0.035)	YPD1 (0.099)	MSB2 (0.109)	PTP2 (0.202)	SSK1 (0.235)	FPS1 (0.265)	SHO1 (0.385)	PTC1 (0.393)	SLN1 (0.417)	PTP3 (0.420)	STE11 (0.718)
STE4^{P}	[64]	STE5 (0.012)	STE3 (0.019)	STE2 (0.022)	RHO1 (0.036)	STE50 (0.126)	PKC1 (0.128)	STE7 (0.201)	KSS1 (0.231)	STE20 (0.261)	HOG1 (0.365)	STE12 (0.370)	STE11 (0.373)	CDC24 (0.544)
STE50^{P,O}	[39,40,64,75]	SSK1 (0.011)	STE50 (0.023)	STE5 (0.106)	OPY2 (0.251)	SSK22 (0.318)	SHO1 (0.321)	SSK2 (0.416)	CDC42 (0.539)	STE4 (0.739)	STE11 (0.749)			
SHO1^{O}	[37−40,64,75]	PBS2 (0.029)	SSK1 (0.102)	STE11 (0.132)	MSB2 (0.165)	STE50 (0.217)	HKR1 (0.260)	SSK2 (0.263)	HOG1 (0.607)	SHO1 (0.640)	STE20 (0.725)			
SPA2^{I}	[64]	MKK1 (0.070)	STE7 (0.085)	PKC1 (0.162)	CLA4 (0.188)	RAS2 (0.234)	STE11 (0.304)	SSK2 (0.444)	RHO1 (0.609)	SLT2 (0.641)	STE12 (0.672)			
PKC1^{I}	[37,40,64]	STE4 (0.010)	PBS2 (0.063)	SLT2 (0.080)	SHO1 (0.187)	BCK1 (0.313)	RHO1 (0.340)	MKK1 (0.365)	PTC1 (0.644)	MID2 (0.819)				
BEM1^{P}	[37,77,84]	STE4 (0.004)	RHO1 (0.009)	STE5 (0.009)	CDC42 (0.033)	CDC24 (0.037)	STE7 (0.045)	CLA4 (0.065)	STE20 (0.245)	STE11 (0.411)				
STE5^{P}	[64]	STE5 (0.012)	STE4 (0.060)	STE12 (0.151)	HOG1 (0.173)	STE11 (0.230)	CDC24 (0.235)	STE2 (0.303)	STE50 (0.346)	KSS1 (0.560)				
STE12^{P,O}	[40,64]	STE4 (0.012)	STE5 (0.016)	STE7 (0.076)	STE12 (0.078)	RAS2 (0.109)	TEC1 (0.180)	STE20 (0.217)	MSB2 (0.431)	STE11 (0.657)				
PBS2^{N}	[37−40,64,75]	SHO1 (0.003)	YPD1 (0.005)	RAS2 (0.010)	PTC1 (0.011)	PKC1 (0.021)	KSS1 (0.022)	SSK2 (0.025)	HOG1 (0.026)	SSK22 (0.029)				
SSK2^{N}	[37−40,64,75]	SSK2 (0.009)	SSK1 (0.034)	PBS2 (0.106)	HOG1 (0.106)	STE50 (0.227)	SHO1 (0.386)	YPD1 (0.431)	FPS1 (0.484)	PTC1 (0.567)				
PTP2^{N}	[37−40,64,75]	PBS2 (0.039)	MKK1 (0.039)	HOG1 (0.047)	BCK1 (0.150)	PTP3 (0.249)	SSK2 (0.266)	PTC1 (0.288)	SSK1 (0.305)	SLT2 (0.798)				
CLN2^{P}	[37,64,85,86]	SWI6 (0.001)	STE4 (0.001)	STE11 (0.003)	CLA4 (0.007)	SLT2 (0.009)	CDC42 (0.074)	STE20 (0.096)	RAS2 (0.525)	PKC1 (0.552)				
NBP2^{N}	[75]	PTP2 (0.021)	STE20 (0.065)	PBS2 (0.120)	HOG1 (0.477)	SSK2 (0.529)	SLT2 (0.567)	BCK1 (0.645)	PTC1 (0.696)					

(Continued)

Table 6.2 (Continued)

(C) Pheromone stress

Target protein	Literature evidence	Upstream proteins within 40 preselected proteins												
BCK1^{I}	[37,40,64]	HKR1 (0.003)	MKK1 (0.065)	PTP2 (0.139)	BCK1 (0.164)	SLT2 (0.166)	PKC1 (0.250)	CLA4 (0.267)	MID2 (0.911)					
KSS1^{P,O}	[39,40,64]	STE4 (0.022)	STE12 (0.031)	HOG1 (0.179)	STE7 (0.190)	PBS2 (0.237)	TEC1 (0.305)	MSB2 (0.692)						
SMI1^{I}	[86,87]	SLT2 (0.029)	PKC1 (0.236)	CLA4 (0.251)	FPS1 (0.340)	BCK1 (0.380)	PTC1 (0.650)	MID2 (0.790)						
FPS1^{N}	[38]	MID2 (0.028)	STE11 (0.047)	SSK22 (0.053)	SSK2 (0.063)	GPD1 (0.202)	SLT2 (0.260)	HOG1 (0.709)						
BNI1^{I}	[64,88,89]	RHO1 (0.000)	CDC42 (0.000)	STE11 (0.004)	CLA4 (0.005)	STE12 (0.006)	PKC1 (0.009)	BCK1 (0.011)						
CLA4^{O}	[38,39]	MSB2 (0.001)	CDC24 (0.002)	SLT2 (0.003)	BCK1 (0.003)	STE20 (0.006)	STE11 (0.006)	CDC42 (0.007)						
PTC1^{N}	[37,38,64,75]	SSK2 (0.070)	PKC1 (0.107)	PTP2 (0.172)	PBS2 (0.342)	HOG1 (0.680)	SLT2 (0.736)							
STE7^{P,O}	[38–40,64]	STE5 (0.012)	STE4 (0.089)	STE12 (0.130)	STE7 (0.331)	STE11 (0.385)	KSS1 (0.568)							
SLT2^{I}	[37,40,64]	BCK1 (0.001)	PKC1 (0.001)	MID2 (0.003)	FPS1 (0.003)	PTC1 (0.006)	CLA4 (0.007)							
FAR1^{P}	[76]	SSK2 (0.008)	STE12 (0.010)	STE5 (0.011)	CDC42 (0.021)	CDC24 (0.029)	STE4 (0.035)							
MID2^{I}	[40,64]	SLT2 (0.019)	MID2 (0.019)	FPS1 (0.065)	MKK1 (0.083)	PKC1 (0.150)	RHO1 (0.166)							
FKS1^{I}	[37,64]	MKK1 (0.009)	SLT2 (0.030)	MID2 (0.050)	RHO1 (0.191)	PKC1 (0.332)	PTC1 (0.414)							
SKM1^{O}	[37]	STE3 (0.027)	PTC1 (0.125)	CLA4 (0.131)	SWI6 (0.150)	STE20 (0.197)	CDC42 (0.312)							
MKK1^{I}	[37,40,64]	BCK1 (0.022)	PTP2 (0.104)	SLT2 (0.196)	PKC1 (0.329)	RHO1 (0.678)	MID2 (0.726)							
BEM4^{I}	[78]	PKC1 (0.009)	RHO1 (0.015)	CDC24 (0.018)	CDC42 (0.020)	CLA4 (0.024)	HKR1 (0.027)							

According to the literature, target proteins can be grouped respectively into the functional pathways denoted by P (pheromone response pathway), O (SHO1 branch of the HOG pathway), N (SLN1 branch of the HOG pathway), and I (cell-wall integrity pathway). The *RV* of each PPI is also given inside the brackets. A protein interaction with smaller *RV* implies a more significant protein interaction. New found target proteins are highlighted with the color gray.

response to each stress, the reconstruction of the gene and protein regulatory network in each stress is improved, as shown in Figs. 6.6–6.8.

For the convenience of analysis below based on the investigated functional pathways (patterns), the gene and protein regulatory networks in Figs. 6.6–6.8 are redrawn in a condensed form as shown in Fig. 6.12. There are nine membrane proteins and four TFs in Fig. 6.12 which are the same as those in MAPK pathways

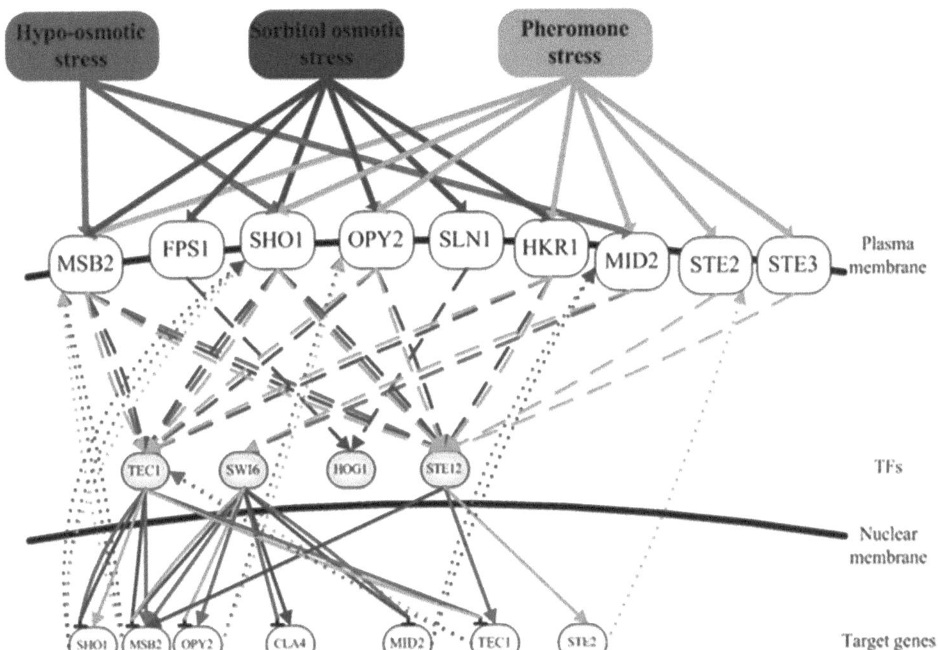

Figure 6.12 Feedforward activation, feedback inhibition, and activation and crosstalks of the condensed gene and protein regulatory networks are coordinated in response to environmental stress signal by a bow-tie core, i.e., TFs, of gene and protein network as a set of core protective mechanisms in response to a broader range of hypo-osmotic stress (*blue* (gray in print versions)), sorbitol osmotic stress (*purple* (dark gray in print versions)), and pheromone stress (*orange* (light gray in print versions)), simultaneously: The connections from stresses to TFs are reconstructed according to Figs. 6.9–6.11. For example, if a MAPK pathway under stress A, from membrane protein B to the downstream TF C of the pathway, is fully connected in Figs. 6.9–6.11, the relationship, from stress A to membrane protein B and then to TF C, is connected with the same color through the whole pathway. The transcription regulations from TFs to target genes (*solid thin lines*) are deduced according to Figs. 6.6–6.8. A target gene transcriptionally regulated under a certain stress may be transcribed to influence the concentration of its protein (TF) (*dotted lines*) in response to the stress. In order to provide as a set of protection mechanisms against a variety of environmental stresses, the bow-tie core network can detect a broader range of molecular signatures for stresses and invoke effective response under resource-limited condition.

(Figs. 6.9–6.11). The connections in Fig. 6.12 of stresses to target genes are based on Figs. 6.9–6.11. For example, if a MAPK pathway under stress A, from membrane protein (sensor/receptor) B to the downstream TF C of the pathway, is fully connected in Figs. 6.9–6.11, the connection, A→B→C, will be drawn in Fig. 6.12.

Obviously, there exists a bow-tie structure to coordinate these functional pathways (patterns) in the condensed stress-activated network. The bow-tie architecture can entail inherent trade-offs among robustness, resource limitations, and performance [53], i.e., the bow-tie core network can enhance the robustness of the stress responsive system by enabling it to cope efficiently with a broader range of environmental stresses with limited resources of proteins and pathways.

According to the coordination of feedback activation, feedback inhibition, and crosstalk by a bow-tie structure of the condensed stress-activated network in Fig. 6.12, we discuss what kinds of possible mechanisms the stress-activated networks possess to make fast or noiseless responses to a broader range of environmental stresses in the sequel.

Two MAPK pathways, the HOG pathway and the cell-wall integrity pathway, activated respectively by high and low osmotic stresses are the most concerned MAPK pathways in yeast *S. cerevisiae*. Additionally, the HOG pathway shares seven common components with the pheromone response pathway. Therefore, these methods are applied to the three stresses, i.e., hypo-osmotic stress, sorbitol osmotic stress, and pheromone stress (α factor), to suggest probable specific mechanisms and to deduce their coordinate bow-tie core network in response to a broader range of environmental stresses.

The specific protective mechanism in response to hypo-osmotic stress signal

When yeast cells are exposed to osmotic stress, several cellular responses, such as solute transporters, solute synthesis, stress resistance, and cell wall structure, are induced by MAPK pathways. Two MAPK pathways, the HOG pathway and the cell integrity pathway (or protein kinase C pathway), have opposing functions stimulated respectively by hyper-osmotic stress and hypo-osmotic stress. The consistent properties between the cell integrity pathway and the pathway for controlling cell wall metabolism lead to cells with a stronger cell wall at low osmolarity rather than cells at high osmolarity [64,90]. According to Fig. 6.9 (the hypo-osmotic stress) and Fig. 6.12, we suggest that a strong cell wall of yeast under hypo-osmotic stress may result from two fully connected pathways, i.e., the SHO1 branch of the HOG pathway and the cell-wall integrity pathway.

Under hypo-osmotic stress, the cell integrity pathway senses the osmotic changes at the cell wall and controls the production of enzyme involved in cell wall metabolism by transcription regulation to efficiently diminishing turgor pressure [64].

In the cell integrity pathway, we find that 12 genes, which probably exist in the function category of cell walls, i.e., *PSA1*, *UTR2*, *CIS3*, *GON7*, *MNN5*, *CWP1*, *EXG1*, *MID2*, *PAH1*, *GAS3*, *FKS3*, and *KRE6*, with *p*-value 4.50E-4 (evaluated by MIPS database) among 121 SWI6 transcriptionally regulating-genes identified by this model, may play an important role in efficiently diminishing turgor pressure.

As hypo-osmotic stress accompanied by other environmental changes, which stimulate the pathways mediated by the components just as with the hypo-osmotic stress, exist prevalently, feedforward loops are probably wired to the cell integrity pathway to filter out other signals (noises) [67]. Additionally, protein interactions are expected to increase reactants, which should dramatically accelerate the signaling pathway [65]. According to Fig. 6.9, although the HOG pathway is fully connected except PBS2 and most of the connections are mutually interacted, the SLN1 branch of the HOG pathway is not activated by hypo-osmotic stress. It has been proven that this is due to scaffold proteins, which can provide specificities to the pathway by different bindings and interactions [65,71,91]. Therefore, according to Figs. 6.9 and 6.10, we suggest that the scaffold protein, *PBS2* plays an important role in recognizing hypo- and sorbitol osmotic stress in the HOG pathway.

According to Table 6.2A, *AKR1*, which interacts with a significant number (>5) of preselected proteins, is a newly discovered protein. *AKR1* has been suggested to play a role as a negative regulator of the pheromone response pathway and is also involved in cell shape control [77]. Therefore, we suggest that *AKR1* is an important protein participating in the pheromone response pathway under hyper-osmotic stress.

The specific protective mechanism in response to sorbitol osmotic stress signal

At high osmolarity, two branches of the HOG pathway, the SHO1 branch and the SLN1 branch, are observed to sense osmotic changes and rapidly make internal adjustments. In Fig. 6.10, sorbitol osmotic stress is shown to have many more mutual interactions and feedforward loops in the HOG pathway than hypo-osmotic stress (Fig. 6.9). The connections may make pathways more rapid and more robust (acting against external noise) in response to sorbitol osmotic stress.

According to Table 6.2B, the new found proteins are the 12 proteins (highlighted by gray color) which interact with a significant number (>5) of preselected proteins. The 12 proteins are grouped and will be discussed in the following paragraphs based on the research shown in Table 6.2B.

Two proteins, namely *WSC3* and *SPA2*, are the new found proteins which are known as members of the cell-wall integrity pathway. *WSC3* is involved in the maintenance of cell wall integrity [82], while *SPA2* acts as a scaffold protein for *MKK*1 and *MPK1* [64]. In addition, *BEM4* is probably involved in the RHO1-mediating signaling pathway [78]. *BEM4* is functionally relevant to *RHO1* and should play a novel role in the signaling pathway mediated by *RHO1*. One possible role of *BEM4* is to act like chaperone in the stabilizing or folding of RHO1. According to the cell-wall integrity pathway shown in Fig. 6.10, we suggest that *RHO1*, *PKC1*, and *SLT2* may play important roles in the inactive cell-wall integrity pathway under sorbitol osmotic stress.

In the pheromone response pathway, four proteins, including *GPA1*, *SST2*, *FAR1*, and *GIC2*, are the new found proteins as shown in Table 6.2B. *GIC2*, whose function is still unknown, can interact with *CDC42*, and therefore GIC2 is grouped with the pheromone response pathway and the SHO1 branch of the HOG pathway [79]. *GPA1*, a G_α subunit, has been involved in mediating pheromone response pathway [64,80]. *SST2* is required to prevent receptor-independent signaling of the pheromone response pathway [81]. Additionally, Far1 is a cell cycle arrest mediator [76].

In the SLN1 branch of the HOG pathway, *SKN7* and *NBP2* are new found proteins participating in this important pathway under sorbitol osmotic stress. SLN1-YPD1-SKN7 has been proven to act as a phosphorelay system that turns on the HOG pathway until the yeast suffers from cell shrinking (Fig. 6.10) [37,64]. In addition, *SKN7* appears to have different functions, such as acting as a transcription factor or a protein in signaling systems, not only mediating different stresses but also linking the cell-wall integrity pathway to the HOG pathway mediated by interacting directly with *RHO1* (Fig. 6.7). During yeast adaptation, *NBP2* is predicted to act as an adapter, recruiting PTC1 to the PBS2-HOG1 complex in the *PTC1* inactivation of HOG1. We suggest that the activated HOG pathway under sorbitol osmotic stress is due to the unbound NBP2-PBS2 complex which results in HOG1 which cannot be inactivated by *PTC1* [75].

In the *SHO1* branch of the HOG pathway, the new found proteins *OCH1*, *SKM1*, and RGA1 are probably important in response to high osmolarity. The promoter of OCH1, which encodes a mannosyltransferase, responds to the presence of *SLN1*, and *KSS1* is activated by the mutation of *OCH1* [64]. Therefore, we suggest that *OCH1* participates in the SHO1 branch of the HOG pathway under sorbitol osmotic stress. SKM1, which is similar to *STE20* and *CLA4*, is probably a downstream effector of *CDC42*, but the function of *SKM1* is still unclear [37]. According to [92], *CDC42* may promote the phosphorylation of GIC2 by recruiting STE20 and SKM1. Therefore, we suggest that SKM1 is a member of the SHO1 branch of the HOG pathway under sorbitol osmotic stress. *RGA1* is suggested as a link between *CDC42* and pheromone pathway components [83] (Fig. 6.7). Although the 12 new found proteins are probably important in response to sorbitol osmotic stress, most of them, such as *BEM4*, *GIC2*, *FAR1*, *OCH1*, *SKM1*, and *RGA1*, are functionally unclear, and likely even participate in multiple pathways with complicated roles. We can only infer some possible protective mechanisms according to previous studies and these results.

The specific protective mechanism in response to pheromone stress signal

When yeast cells are stimulated by mating pheromones, i.e., **a**-factor and α-factor, haploid cells are matted to a diploid form. Cellular functions, i.e., polarized growth, cell cycle arrest in G1, cell adhesion, and cell fusion, are activated under the pheromone stress. Therefore, some genes must exist to link up cellular functions and the pheromone response pathway. According to [87], SMI1, suppressor of matrix-association region inhibition, probably coordinates cell wall synthesis and budding [87]. This coordination is also found in this section (Fig. 6.11). Further, it is also found that the three MAPK pathways, the pheromone response pathway, the SHO1 branch of the HOG pathway, and the cell-wall integrity pathway, are probably induced by pheromone stress (Figs. 6.8 and 6.12). According to Fig. 6.11, mutual interactions apparently exist in two pathways, namely the pheromone response pathway and the SHO1 branch of the HOG pathway. Therefore, once again we infer that the existence of mutual interactions probably leads to rapid response under environmental stresses. According to [37,77,84], because the deletion of *AKR1* or *BEM1* results in severe effects on cell morphology and viability, and both of them are required for efficient pheromone signaling, in the pheromone response pathway, *AKR1* and *BEM1* play an important role in cellular morphogenesis. *CLN2*, a G1 cyclin, is involved in the cell cycle regulation of MAP kinase signaling [37,64,85,86]. *FKS1* and *SMI1* are required for the synthesis of a major structural component of cell walls, 1,3-β-glucan [37,64,87,93]. *BNI1* is known to interact with some members of the Rho-GTPase family, and is probably involved in cell integrity signaling [64,88,89].

According to the above discussions, we conclude that the SHO1 branch of the HOG pathway under sorbitol osmotic stress (Fig. 6.10) and pheromone response pathway, the SHO1 branch of the HOG pathway and the cell-wall integrity pathway under pheromone stress (Fig. 6.11) are most likely accelerated by mutual interactions. There are a considerable number of feedforward loops wired to the cell-wall integrity pathway under hypo-osmotic stress as shown in Fig. 6.9, the HOG pathway under sorbitol osmotic stress as in Fig. 6.10, and the pheromone response pathway, the SLN1 branch of the HOG pathway and the cell-wall integrity pathway under pheromone stress as in Fig. 6.11, to avoid inappropriate stimulations.

Bow-tie protective mechanisms via coordination of mutual interaction, feedforward and feedback loop and crosstalk for a broader range of environmental stress signals

The response of yeast to osmotic stress is a transient protection event. According to the observation seen in Ref. [64], the raising time and falling time of HOG1 are respectively within about 1 and 30 min in response to osmotic shock. An intrinsic feedback mechanism therefore probably exists in the HOG pathway to maintain signaling competence for a fast response to osmotic shock and is required for a homeostatic process [64,94]. According to these results about the intrinsic feedback mechanism, we find that *TEC1* and *SWI6* are required respectively for feedback inhibition of SHO1 and MID2 under both osmotic stresses, i.e., hypo-osmotic stress and sorbitol osmotic stress (Figs. 6.9 and 6.12). Moreover, at high osmolarity, both *TEC1* and *SWI6* are required for feedback inhibition of MSB2, while at low osmolarity SWI6 is required for feedback inhibition of *CLA4* (Figs. 6.9 and 6.12). Feedback inhibition acts like a shutoff mechanism to prevent proliferation from inappropriate pathway activation [64]. In addition, feedback activation can increase stability and reduce response time to environmental stimuli [58]. We find that at high osmolarity the SHO1 branch of the HOG pathway rapidly responds to *STE12* via STE12-mediated upregulation of MSB2 (Figs. 6.9 and 6.12), while at low osmolarity the SHO1 branch of HOG pathway rapidly responds to the downstream TF of the cell-wall integrity pathway, i.e., SWI6, via SWI6-mediated upregulation of MSB2 and OPY2 (Figs. 6.9 and 6.12).

Moreover, it is evident that rapid response to the pheromone stresses via feedback activation probably exists in both the pheromone pathway and the SHO1 branch of the HOG pathway respectively via STE12-mediated upregulation of STE2 and via TEC1-mediated upregulation of SHO1 (Figs. 6.11 and 6.12). Additionally, we found that feedback inhibitions may mediate SWI6-mediated downregulation of both OPY2 and MSB2 in response to pheromone stress (Figs. 6.11 and 6.12). The feedback regulations implicate that a secondary pathway, the SHO1 branch of the HOG pathway, may be induced by pheromone stress mediated by the cell-wall integrity pathway combined with feedback inhibition to filter out inappropriate signals.

From the results in Table 6.2A–C, we find that many TFs are detected in a broader range of environmental stresses, including TEC1, STE12, ADR1, ARR1, CIN5, FKH2, GAT1, and HIR2. It implies that they may be crosstalk among the MAPK pathways, such as the pheromone response pathway, HOG pathway and cell-wall integrity pathway. Although these pathways respond to different stresses, they may induce different levels of the same TFs to perform cell protection. In Fig. 6.12, we find that TEC1 and STE12 are detected in hypo-osmotic stress, sorbitol osmotic stress and pheromone stress. A previous study [64] showed that TEC1 promoter has a regulatory element, called the FRE (filamentation and invasion responsive element), which is both necessary and sufficient for transcription regulation by upstream activating signals in the pheromone response pathway. Both STE12 and TEC1 regulate the genes containing one copy of a FRE in close proximity to a binding site and are required for the pheromone response. Thus, the FRE of TEC1 promoter provides a positive feedback mechanism for upregulation of TEC1, which can be detected in Fig. 6.8. Furthermore, from the established signaling pathways, Gat-Viks and Shamir [95] showed that STE12 and TEC1, which are the TFs in the downstream of the SHO1 branch of the HOG pathway, may respond to osmotic stress. Together, the SHO1 branch of the HOG pathway integrates the need for cell expansion in not only the pheromone stress but also the osmotic stress [64]. Therefore, from Table 6.2A–C, we can infer that several TFs may have crosstalks in response to a broader range of environmental stresses.

Finally, we conclude that in the cell-wall integrity pathway, RHO1, PKC1, and SLT2 may play an important role in the activation of the cell-wall integrity pathway under both hypo-osmotic stress and pheromone stress. Under hypo-osmotic stress the SHO1 branch of the HOG pathway acts like a secondary pathway, which is accelerated by the activated cell-wall integrity pathway, and both of the pathways, i.e., the SHO1 branch of the HOG pathway and the cell-wall integrity pathway, might combine with feedback inhibitions to prevent inappropriate activations (Figs. 6.9 and 6.12). In the SLN1 branch of the HOG pathway, a scaffold protein, PBS2, can provide the specificity to the activation of the SLN1 branch of the HOG pathway under sorbitol osmotic stress by different kinds of binding and interaction. Under sorbitol osmotic stress, the SHO1 branch of the HOG pathway can itself be accelerated and has the ability to prevent inappropriate activation by feedback inhibition and activation (Figs. 6.10 and 6.12). In the pheromone response pathway, once again we suggest that a scaffold protein, STE5, can provide the specific protective mechanism to the activation of the pheromone response pathway under pheromone stress [65,71,91]. Therefore, when the pheromone response pathway suffers from pheromone stress via membrane receptors, i.e., STE2 and STE3, the activated pheromone response pathway speeds its response by feedback activation via STE12-mediated upregulation of STE2 (Fig. 6.12). The activated pheromone response pathway accelerates itself via feedback activation (Fig. 6.12). Under pheromone stress (α factor), the SHO1 branch of the HOG pathway acts like a secondary pathway, that can sense the external signal mediated by the cell-wall integrity pathway, which has the ability to prevent inappropriate activation for the SHO1 branch of the HOG pathway. Additionally, the SHO1 branch of the HOG pathway can also accelerate itself via feedback activation (Figs. 6.11 and 6.12). Furthermore, we compare the results of Table 6.2A–C with Ref. [7] to find the overlap TFs between them. We find that the overlap TFs are participating significantly in not only protein regulatory network but also GRNs under a specific stress. Under hypo-osmotic stress, sorbitol osmotic stress and pheromone stress, there are respectively five TFs: SWI6^{1}, PHD1^{3}, SWI5^{4}, SKN7^{6}, and IXR1^{9}, five TFs: SKN7^{1}, FKH2^{3}, HSF1^{5}, MBP1^{6}, and PHD1^{7}, and six TFs: SWI4^{2}, PHD1^{3}, MBP1^{4}, SWI5^{5}, FKH2^{6}, and SKN7^{10}, found in both studies. The superscript of the TFs obtained from Table 2 of Ref. [7] shows order of the significance in response to specific stress. Therefore, the other TFs in Table 2 of Ref. [7] not found in this section such as INO4^{2} and BAS1^{5} under hypo-osmotic stress, SMP1^{2} and FKH1^{4}

under sorbitol osmotic stress and MCM1[1] under pheromone stress may only play an important role in transcription regulations as a specific protective mechanism in response to specific stress. Additionally, SWI6 under sorbitol osmotic stress and HOG1 under both hypo-osmotic stress and pheromone stress are not found in the significant interacting proteins (TFs) by both Ref. [7] and these results.

By comparing with previous studies [38–40,64,96–98], the false negative rate is 26.47% (9/34) for HOG pathway under sorbitol osmotic stress and 8.33% (1/12) for pheromone response pathway under pheromone stress. Owing to the insufficient studies in MAPK pathway under hypo-osmotic stress, we cannot compare these findings with others under hypo-osmotic stress.

From the core gene/protein network in Fig. 6.12, feedforward activation, feedback inhibition and activation, and crosstalks are coordinated by a bow-tie structure in response to three environmental stresses. In order to provide protection against a variety of environmental stresses, the protection system has to detect a broader range of molecular signatures for stresses and invoke effective responses. It is clear that this has to be performed under resource-limited conditions because the number of genes, proteins, and pathways that can involve in stress responses is not infinite. Since a bow-tie architecture comprises conserved and efficient core processes with diverse and redundant input and output processes and entails inherent trade-offs among robustness, fragility, resource limitation and performance, the bow-tie gene and protein core network has been investigated by the proposed method to coordinate the mutual interactions, feedforward loops, feedback inhibition and activation and crosstalks as a core protective mechanism in response to a boarder range of environmental stresses.

According to the bow-tie protective network under different stresses in Fig. 6.12, TFs exist as bow-tie cores to trigger appropriate biological responses via receptors (or sensors) in response to environmental stresses. In Fig. 6.12, there are five loops, which not only trigger the biological response of MAPK pathway but also regulate the expression of membranes proteins (sensors/receptors) via transcription regulations of TFs. For example, TEC1 triggers the response of cell wall integrity via the receptor (or sensor), SHO1, in response to all three stresses; TEC1 triggers the response of cell wall integrity via the receptor (or sensor), MSB2, in response to both osmotic stresses; STE12 triggers the response of cell wall integrity via the receptor (or sensor), MSB2, in response to sorbitol osmotic stress; STE12 triggers the downstream response of pheromone response pathway via the receptor (or sensor), STE2, in response to pheromone stress; and SWI6 triggers the response of cell wall construction via the receptor (or sensor), MID2, in response to hypo-osmotic stress. Therefore, these findings in Fig. 6.12 obviously show that the gene and protein regulatory networks display a set of core protective mechanisms just like the nested bow-tie architecture. Although there exist many protein interactions, the existence of a conserved core and versatile weak linkages provides an opportunity to alter network connections and further trigger appropriate biological responses without seriously affecting other parts of the network in response to environmental stresses. This investigation provides us with in-depth insight into the nature of the protection mechanism of yeast against a variety of environmental stresses.

6.2.5 Conclusion of protective mechanisms in response to a broad range of environmental stress signals

In this section, based on microarray data, ChIP-chip data and PPI data, gene/protein regulatory networks are constructed via dynamic model, ML estimation, AIC pruning methods, and ranking value (RV) to investigate their interplaying protection roles in response to environmental stresses on yeast. Then, a bow-tie core network structure is found to coordinate some functional pathways of these condensed gene/protein networks as a set of core protective mechanisms in response to a broader range of environmental stresses. The main works of this section are summarized as follows: (1) We construct two nonlinear stochastic coupled dynamic models to combine gene transcriptional regulatory networks and protein regulatory pathways (signaling pathway) in response to different stresses, i.e., hypo-osmotic stress, sorbitol osmotic stress, and pheromone stress (α-factor). (2) We use the AIC method combined with backstepping selection procedure to prune the insignificant regulations in the stress-activated gene and protein regulatory networks. Furthermore we construct the ranking value of the significant regulations and interactions, *RV*, for each upstream protein (TF) to the corresponding target based on both the pruning result of the backstepping AIC and the identified parameters, $\hat{b}_{i,j}$ and $\hat{d}_{i,k}$. (3) According to the significantly regulated proteins under each stress (Table 6.2), we find that some new found proteins probably participate in specific MAPK pathways in response to specific stress. (4) According to functional patterns of the reconstructed stress-activated gene and protein regulatory networks, we find that the coordination of mutual interaction, feedforward loop, feedback activation, feedback inhibition, and crosstalk by bow-tie core gene/protein network provides a set of core protection mechanisms in response to a broader range of environmental stress, i.e., with speedy signaling transduction, noiseless and fast response and network robustness under limited resources for preventing proliferation of inappropriate activation. Although the stress-activated gene and protein regulatory networks in response to different stresses are reconstructed by these models via microarray data, ChIP-chip data, and PPI data, a weakness of this section is that some ChIP-chip data is to date still unavailable for all 203 TFs to obtain whole protection mechanism for gene and protein regulatory networks in response to a broader range of environmental changes.

6.2.6 Methods of parameter estimation in protective mechanisms

System identification and AIC backstepping method

The coupled dynamic models in Eqs. (6.7) and (6.8) can be combined and rewritten in the following matrix forms:

$$Z_i = \Phi_i \Theta_i + \psi_i \tag{6.10}$$

where $Z_i = \begin{bmatrix} \dot{X}_i(t_1) & \dots & \dot{X}_i(t_N) & \dot{Y}_i(t_1) & \dots & \dot{Y}_i(t_N) \end{bmatrix}^T$,

$$\Phi_i = \begin{bmatrix} -X_i(t_1) & 1 & Y_1(t_1) & \dots & Y_{J_i}(t_1) & & & & & & \\ \vdots & \vdots & \vdots & \ddots & \vdots & & & & 0 & & \\ -X_i(t_{1N}) & 1 & Y_1(t_N) & \dots & Y_{J_i}(t_N) & & & & & & \\ & & & & & -Y_i(t_1) & 1 & X_i(t_1) & -Y_i(t_1)\bullet Y_1(t_1) & \dots & -Y_i(t_1)\bullet Y_{K_i}(t_1) \\ & & 0 & & & \vdots & \vdots & \vdots & \vdots & \ddots & \vdots \\ & & & & & -Y_i(t_N) & 1 & X_i(t_{1N}) & -Y_i(t_N)\bullet Y_1(t_N) & \dots & -Y_i(t_N)\bullet Y_{K_i}(t_N) \end{bmatrix},$$

$$\Theta_i = [\underbrace{a_i \quad b_{i,0} \quad b_{i,1} \quad \dots \quad b_{i,J_i}}_{p_i} \quad \underbrace{\lambda_i \quad d_{i,0} \quad g_i \quad d_{i,1} \quad \dots \quad d_{i,K_i}}_{q_i}\ ]^T \in \Re^{p_i+q_i},$$

$\psi_i = \begin{bmatrix} \varepsilon_i(t_1) & \dots & \varepsilon_i(t_N) & \xi_i(t_1) & \dots & \xi_i(t_N) \end{bmatrix}^T$ where $p_i = J_i + 2$ and $q_i = K_i + 3$.

Suppose the noise components ε_i and ξ_i are independently and normally distributed, and the noise vector ψ_i has zero mean and variances $\sigma_i^2 I$ [99]. Using ML method, we solve the parameter estimation problem with the optimum estimation $\hat{\Theta}_i$ and $\hat{\sigma}_i^2$. The likelihood function of Z_i is defined as follows:

$$p_i\left(Z_i|\Theta_i, \sigma_i^2\right) = (2\pi)^{-1/2}\det\left(\sigma_i^2\right)^{-N/2}\exp\left[\left(-1/2\sigma_i^2\right)\bullet(Z_i - \Phi_i\Theta_i)^T(Z_i - \Phi_i\Theta_i)\right] \tag{6.11}$$

The log-likelihood function for the given M data points in Z_i can be defined as follows:

$$L_i(\Theta_i, \sigma_i^2) = \text{constant} - \left(\frac{M}{2}\right)\ln\left[\det(\sigma_i^2)\right] - \left(\frac{1}{2\sigma_i^2}\right)\sum_{k=1}^{M} Z_i(t_k) - R_{\Phi_i}(t_k)\Theta_i \tag{6.12}$$

where R_{Φ_i} is the row vector of $\boldsymbol{\Phi}_i$. We then estimate unknown parameters Θ_i and σ_i^2 by maximizing $L_i(\Theta_i, \sigma_i^2)$, i.e., $\partial L_i(\Theta_i, \sigma_i^2)/\partial\Theta_i = 0$ and $\partial L_i(\Theta_i, \sigma_i^2)/\partial\sigma_i = 0$, as follows:

$$\hat{\sigma}_i^2 = \frac{1}{M}(Z_i - \Phi_i\Theta_i)^T(Z_i - \Phi_i\Theta_i) \tag{6.13}$$

$$\hat{\Theta}_i = (\Phi_i^T\Phi_i)^{-1}\Phi_i^T Z_i \tag{6.14}$$

Because all the significant transcription regulations and PPIs do not simultaneously occur in each environmental condition, AIC combined with the backstepping method [100] is used to select the significant regulations, i.e., $b_{i,j}$ and $d_{i,k}$, to prune the constructed networks to achieve the minimization of the AIC_i in Eq. (2.23), which is defined in this case as follows:

$$AIC_i(l) = \log\left[\frac{1}{M}(Z_i - \Phi_i\hat{\Theta}_i)^T(Z_i - \Phi_i\hat{\Theta}_i)\right] + \frac{2l}{M} \tag{6.15}$$

where l is the number of reserved parameters, i.e., $p_i + q_i$, in the backstepping AIC. The order of two parameters, $b_{i,j}$ and $d_{i,k}$, in Eq. (6.10) are separately treated by the backstepping AIC. The order of the parameters selected by the backstepping AIC implies the importance of each PPI and transcription regulation, i.e., the first selected protein (TF) is more important to the regulation of the target gene (or target protein) than the later selected proteins (TFs).

Ranking value

Because both the order of the parameters selected by the backstepping AIC and the absolute values of the parameters imply the importance of each PPI and transcription regulation, we create a RV to evaluate the importance of each upstream protein (TF) to its target protein (gene). According to the identified models (6.9), we define indices to represent respectively the order of the absolute value of the parameters (j,k), i.e., $|\hat{b}_{i,j}| < |\hat{b}_{i,j-1}|$ and $|\hat{d}_{i,k}| < |\hat{d}_{i,k-1}|$, and the order of the parameters selected by the backstepping AIC (m,n), i.e., $\hat{b}_{i,m}$ and $\hat{d}_{i,n}$ represent respectively the mth selected parameter and the nth selected parameter in the AIC selections. The RV is defined as $RV = j \bullet n / J_i'^2$ and $RV = k \bullet m / K_i'^2$, respectively, in the gene transcriptional regulatory network and the protein regulatory pathway, i.e., the regulation $\hat{b}_{i,j}$ (or $\hat{d}_{i,k}$) with smaller RV implies that the regulation $\hat{b}_{i,j}$ (or $\hat{d}_{i,k}$) is more significant in the regulation of gene i (or protein i).

p-Value evaluation of identified regulation relationships

We construct 1000 random permutations of $Y_j(t)$ in the first equation of (6.9) (or $Y_i(t)Y_k(t)$ in the second equation of (6.9)) and then we use the random permutations to estimate 1000 different $\hat{b}_{i,j}$ (or $\hat{d}_{i,k}$) with $AIC_{i,j}^b$ (or $AIC_{i,j}^d$). The individual p-value for each $\hat{b}_{i,j}$ (or $\hat{d}_{i,k}$) is $P_{i,j}^b = \int_{|\hat{b}_{i,j}|}^{\infty} p_{|\hat{b}_{i,j}|}(x)dx \bullet \int_{-\infty}^{AIC_{i,j}^b} p_{AIC_{i,j}^b}(x)dx$ (or $P_{i,j}^d = \int_{|\hat{d}_{i,j}|}^{\infty} p_{|\hat{d}_{i,j}|}(x)dx \bullet \int_{-\infty}^{AIC_{i,j}^d} p_{AIC_{i,j}^d}(x)dx$) where $p_{|\hat{b}_{i,j}|}(x)$, $p_{AIC_{i,j}^b}(x)$, $p_{|\hat{d}_{i,j}|}(x)$, and $p_{AIC_{i,j}^d}(x)$ are respectively the probability distributions of $|\hat{b}_{i,j}|$, $AIC_{i,j}^b$, $|\hat{d}_{i,j}|$, and $AIC_{i,j}^d$.

As plants must maintain high photosynthetic efficiency for survival and energy production, changes of light signals are considered environmental changes to affect photosynthetic light reactions of plants. Theoretically, we will construct GRN of *A. thaliana* to investigate control mechanisms of photosynthetic light acclimation in response to different light signals in the following section.

6.3 CONSTRUCTING GRNs FOR CONTROL MECHANISMS OF PHOTOSYNTHETIC LIGHT ACCLIMATION IN RESPONSE TO DIFFERENT LIGHT SIGNALS

Life on earth is dependent on energy derived from the sun, and photosynthesis is the only biological process able to harvest this energy. Plants must maintain high photosynthetic efficiency to ensure sufficient energy for survival and seed production for the next generation. As plants are sessile organisms and cannot escape environmental changes that directly affect photosynthetic light reactions, they have evolved regulatory mechanisms that optimize photosynthetic electron transport to acclimate the photosynthetic process to the prevailing environment [101,102]. Changes in the intensity and spectral quality of light received by plants beneath a tree canopy or within dense plant population contribute to imbalance in the excitation of energy distribution between photosystem II (PSII) and photosystem I (PSI). Because PSI and PSII work electrochemically in series, if either of the two photosystems is imbalanced and the redox state of the electron transport chain components is changed, there is a decrease in the efficiency of electron flow [103]. Such imbalances are counterbalanced by two different acclimation responses: state transition, which is a short-term response, and adjustment of photosystem stoichiometry, which is a long-term response (LTR) [104]. The focus of this section is on the regulatory mechanisms of the LTR to fluctuating light quality in the nucleus.

Adjustment of photosystem stoichiometry is a LTR that occurs on a time scale of hours to days. It redistributes excitation energy between PSII and PSI mainly by reconfiguring the relative amount of the two photosystems, enabling them to optimize light utilization [102,103,105–107]. Imbalances in components of the photosynthetic electron transport chain are sensed within the chloroplast. The change in redox state of the plastoquinone (PQ) pool of the electron transport chain is the critical regulatory signal source for transcriptional control. This change delivers a signal to the nucleus and chloroplast to modulate the expression of photosynthetic genes encoding the PSII and PSI proteins (Fig. 6.13). Such redox signals from chloroplast to nucleus are the so-called retrograde signals, in which photosynthetic efficiency is thought to be a sensor for fluctuations in the environment. The retrograde signals represent a functional feedback control which directs the expression of genes in the nucleus to respond to disturbances from the surrounding environment [105]. This LTR process has been relatively well characterized at the biophysical level, however the acclimation process remains poorly understood at the bio-molecular level. For instance, most subunits of photosystems are encoded in the nucleus and require special regulation, but these regulatory processes are not yet understood. The primary aim of this section is to use a systems biology approach to investigate the molecular control mechanisms of photosynthetic light acclimation by constructing nuclear transcriptional GRNs under different PQ pool redox states. This has been achieved using time series microarray

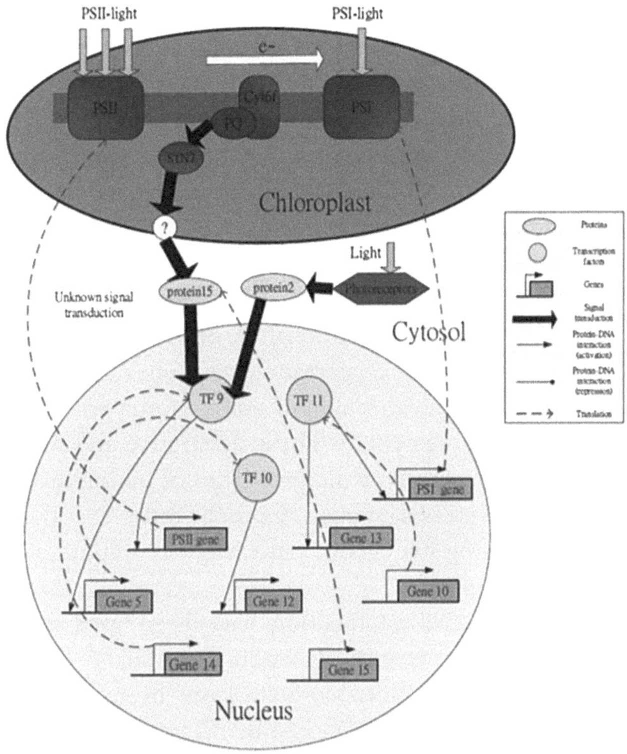

Figure 6.13 Effect of different light signals on gene expression. The three cell compartments are depicted schematically: chloroplast (*green* (dark gray in print versions)), cytosol (*white*), and nucleus (*yellow* (light gray in print versions)). The integrated cellular network consists of two subnetworks. One network is the signaling transduction pathways in the cytosol and the other network is the gene regulatory network for transcriptional regulation in nucleus. An important mediator for PQ pool redox states is the thylakoid kinase *STN7*. *STN7* kinase transmits the decisive signal to the nucleus, resulting in the ensuing regulation of the relative amount of each of the photosystems. However, the mechanism by which the redox signal is transmitted from the chloroplast double membrane into the cytosol is poorly understood. In this section, the gene regulatory network (the protein–DNA interaction) in the nucleus is constructed under different retrograde signals originating from different redox states in the PQ pool. The light environment is perceived by cytosolic photoreceptors. Although both PQ pool and photoreceptor systems report changes of ambient light environment by different signal transduction pathways to the nucleus, some common TFs in nucleus may be employed simultaneously in different light-related systems as a regulatory mechanism to respond to the prevailing environment.

data of *A. thaliana* under an artificial light system. The artificial light system mimics light conditions by preferentially exciting PSI or PSII to induce more reduced or oxidized states of the PQ pool by shifting from PSI light to PSII light, or from PSII light to PSI light.

The foremost purpose of the section of molecular biology is to decipher the mechanisms behind the biological processes of cells. Recently, the development of high-throughput genomic tools such as DNA microarray [108,109] and chromatin immunoprecipitation-DNA chip (ChIP-chip) data [110,111] have provided comprehensive information about gene expression activity and potential interactions between transcription factors and genes. Although this data catalogs significant patterns of the dynamic expression of thousands of genes, identifying the characteristics of a biological process under a specific condition among vast amounts of experimental data is still an extremely difficult task. These vast amounts of data emphasize the need for systematic approaches that enable the computational reconstruction of dynamic transcriptional regulatory networks [44,45]. Systematic approaches are based on integrating computational methods with high-throughput and genome-wide data to inspect the nodes and edges of gene networks [112]. These descriptive and mathematical models for GRNs can uncover significant dynamic properties of biological systems from their responses to internal and external signals. Consequently, the use of a systematic approach to reverse engineering has become an extremely important tool for identifying GRNs via gene expression data.

Various approaches of reverse engineering have been used to identify plausible gene networks, including clustering algorithms, ordinary differential equation models and Bayesian networks (BNs), which have been extended to construct dynamic Bayesian networks (DBNs) [113,114]. Clustering algorithms have been used to find coregulated genes based on gene coexpression properties. They are therefore suitable to find a set of genes that potentially have the same regulators or functional modules. Besides, both DBNs and ODE network models have previously been used successfully to model dynamic processes with time series microarray data. Furthermore, some studies [115] have incorporated many other types of heterogeneous data to reconstruct gene networks. However, in the plant model, *A. thaliana*, the types of heterogeneous data are limited, making it difficult to apply a systematic approach to construct gene networks. In a previous study [116], a novel BN-based algorithm was adapted to construct a GRN for *A. thaliana* from a large number of separate microarray experiments without heterogeneous data. Jiao et al. [117] constructed a GRN for *A. thaliana* by utilizing promoter analysis to predict plausible transcriptional relationship. They were able to identify large-scale GRNs that included photosynthetic genes but did not mention the central molecular mechanism of photosynthesis. In this section, we have analyzed the molecular mechanisms of photosynthetic light acclimation by adapting a systems biology approach to construct GRNs using prediction database and time series microarray data. We constructed two GRNs under different PS-light shifts to gain insight into the mechanism of photosynthetic light acclimation. Such gene networks based on this algorithm can find the TFs integral to regulating the photosynthetic light acclimation response.

In this section, we propose a systematic approach based on the dynamic gene regulatory model and time series microarray data to gradually refine the rough GRN. The first requirement is to identify the candidate TF-gene interactions and construct a rough GRN. Due to the limit of ChIP-chip data and network-related studies on *A. thaliana*, we have used Plant Promoter Analysis Navigator (PlantPAN) [118] to compensate for this insufficiency by inferring potential TF-gene interactions from data mining of the literature and databases. Pearson's correlation between each candidate TF-gene pair is then computed to gain a rough GRN. This rough GRN is added to using an efficient dynamic model of gene regulation [16,119] and time series microarray data [120], thus producing a dynamic gene regulation system. As a final step, ML system identification methods [21] and the AIC, a method to detect the order of the dynamic gene regulatory system [21], are employed to identify kinetic parameters in the dynamic gene regulation model and to delete regulatory genes that do not have a significant influence on the target gene in the rough GRN. This method of applying an algorithm which combines system identification and AIC to reduce the insignificant interactions of rough gene regulatory and PPINs has been used successfully in many studies, including the study of a putative GRN of systemic inflammation in humans [119], a biofilm-related GRN in *Candida albicans* [16], a cancer-perturbed PPI [121], and a PPI under several stresses in *S. cerevisiae* [122]. These results were validated by evidence within literature, thus demonstrating that such systems biology approaches provide a powerful and flexible tool which can be used for different species under different conditions.

The time series microarray data is the profile of each gene expression at several specific time points under a specific stress or condition. Using the dynamic model of gene regulation and system identification method alongside time series microarray data, false positive interactions between transcription factors and target genes can be removed to obtain a refined GRN. This refined GRN depicts a real GRN under photosynthetic light acclimation responses, as confirmed by real microarray data. Therefore, pruning the rough gene network based on a systematic approach and time series microarray data of *A. thaliana* can produce a refined GRN which deciphers photosynthetic light acclimation at both bio-molecular level and a system-wide level.

6.3.1 Systems biology methods of constructing GRN of photosynthetic light acclimation in response to different light signals

In this section, the construction of GRNs for long-term photosynthetic light acclimation in *A. thaliana* can be divided into two stages. During the first stage, the candidate TF-gene interactions are identified by data mining of the literature and PlantPAN [118] to develop a rough GRN. During the second stage, the rough GRN is pruned by the ML system identification and AIC system order detection.

These pruning methods have been supplemented with time series microarray data to obtain a refined GRN.

Data used for analysis

We have used previous microarray data [120] as this mRNA expression profile. Plants acclimated to PSI or PSII lights were shifted to the alternate condition to induce either a strong reduction (PSI-to-PSII light shift) signal or a strong oxidation (PSII-to-PSI light shift) signal. The LTR to such shifts is followed by collecting samples prior to the shift ($t = 0$ control) and at 0.5, 2, 8, and 48 h thereafter. Transcript profiles from those samples were obtained using an established DNA filter array with gene-specific tags (GSTs) for nuclear genes encoding chloroplast proteins. The filter array, which carries 3292 GSTs, including 2661 nuclear genes for chloroplast proteins and 631 genes encoding for nonchloroplast proteins, was used to test the impact of light-induced redox signals on their expression. PSI-to-PSII and PSII-to-PSI experiments were conducted separately in neighboring growth cabinets of the same climate chamber. Plants were acclimated to PSI and PSII light. Following this acclimation period the plants were moved to the alternate light. Three independent samples were harvested at each indicated time point for RNA preparation. Total RNA was isolated from 200 to 500 individual plants per sample. Three independent hybridization experiments were performed for each time point, with different filters and independent radioactively label cDNA probes, thus minimizing variation between individual plants, filters, and probes.

Stages for constructing GRN for photosynthetic light acclimation

Refer to Fig 6.14 in conjunction with these stages.

Stage I: Construction of a rough (candidate) GRN for photosynthetic light acclimation

At this stage, the construction of a rough GRN is performed in the following four steps:

Step 1: The target gene pool that is potentially related to the LTR of photosynthetic light acclimation is determined by data mining of the literature and database. According to previous studies [102], the LTR of photosynthetic acclimation can counteract an imbalance by adjusting the relative amount of the two photosystems. Therefore we explore the Kyoto Encyclopedia of Genes and Genomes (KEGG) database to retrieve the nuclear genes encoding photosystems or light-harvesting protein complexes. From this database, we also select the genes involved in the photosynthetic light reaction pathway. Moreover, the tentative target genes of the long-term photosynthetic light acclimation identified in Ref. [123] by analyzing photosynthetic impairment in varieties with double mutated gene are added into the target gene pool. This final aim is to select candidate regulators of the target gene pool to obtain the potential TF-gene interactions involved in photosynthetic light acclimation response to different light environments.

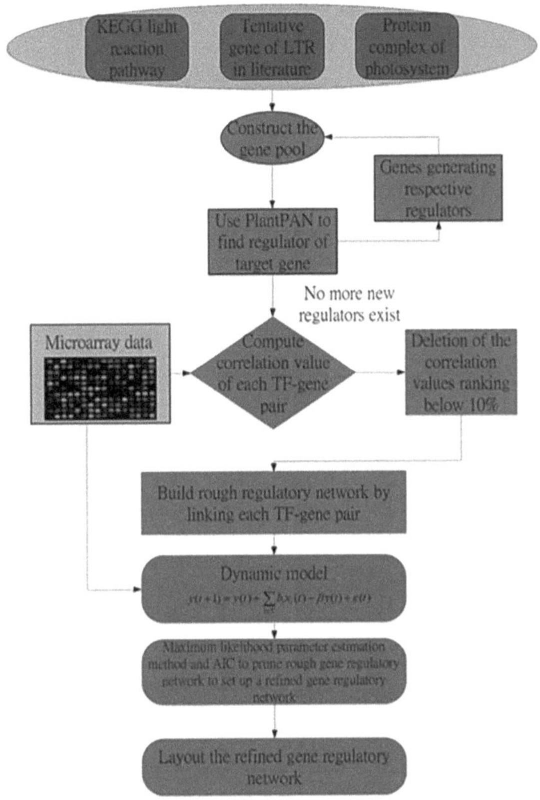

Figure 6.14 The flowchart for constructing the gene regulatory network for control mechanism of photosynthetic light acclimation response. The figure depicts the procedure for construction of the gene regulatory networks. *Green* (light gray in print versions) is representative of data that is needed while *red* (dark gray in print versions) represents the procedure to construct the rough gene regulatory network for photosynthetic light acclimation (Stage I). *Blue* (gray in print versions) is representative of the procedures which prune the rough gene regulatory network using the dynamic model with system identification method (Stage II).

Step 2: Each target gene is imported to PlantPAN, which recognizes candidate binding sites of transcription factors in the promoter sequence of target gene, from -1000 (upstream) to $+100$ (downstream). At this stage, the candidate TFs predicted by the PlantPAN database could be considered candidate regulators for the target genes. Any candidate TFs that are not present in the original target gene pool should be added to it in order to find their regulators iteratively. The iterative procedure is discontinued when no new interactions are generated from the PlantPAN database. Based on predictions of the PlantPAN database, it is possible to obtain the potential TF-gene interaction pairs involved in photosynthetic light acclimation response. However, the potential binding of a TF does not imply that the TF regulates the gene in photosynthetic light acclimation. In order to reduce the false positive errors, it is necessary to refine the TF-gene binding pairs. This is achieved by using time series microarray data of photosynthetic light acclimation to confirm the existence of these transcriptional bindings.

At this point, we have obtained an extended target gene pool coupled with its corresponding TF-gene pairs. However, due to the limited microarray data used in this section, genes and potential regulators that aren't present in the microarray data probe must be removed. Following this removal, a new target gene pool is obtained, with 65 genes containing 7 TFs.

Step 3: In a previous study [13], a TF and its target gene were said to have a positive or negative temporal relationship if the target gene's expression profile is similarly correlated with the TF's expression profile. The Pearson's correlation was computed between each TF-gene pair based on this assumption. The candidate TF-gene pairs were then ranked according to the absolute value of correlation. Finally, any TF-gene pair below 10% rank was deleted. Because the correlation value is only the first discrimination parameter in the overall procedure of gradual refinement, it is necessary to avoid missing any possible candidate TF-gene pair at this early stage. The primary aim in this step is to delete only the highly unlikely regulatory interactions.

Step 4: At this stage, the first selection of possible regulatory interactions is selected from the TF-gene pairs identified in Step 3. They are regarded as candidate TF-gene pairs with the potential to become transcriptional interactions in GRN under photosynthetic light acclimation. These candidate TF-gene pairs constitute a rough regulatory network. Refinement of this preliminary result requires more rigorous pruning methods.

Stage II: Pruning the rough GRN through the use of a dynamic gene regulation model with system identification methods

Step 1: Dynamic gene regulatory model: the rough GRN constructed in the previous stage using data mining and the Pearson's statistical inferences is expected to contain some false positive TF-gene pairs. Therefore the rough gene network should be confirmed by gene expressions of microarray data. Gene regulations that cannot be matched by time-profile microarray data are deleted. The dynamic gene regulation model is used to depict the transcriptional regulatory mechanism as a system with several regulatory genes as transcriptionary inputs and a target gene's expression as an output. This allows the dynamic transcriptional regulation of a target gene to mimic a subsystem in the rough GRN. The multi-input single-output stochastic dynamic gene regulation model is proposed as follows:

$$y(t+1) = y(t) + \sum_{i \in S} b_i x_i(t) - \beta y(t) + \varepsilon(t) + k \tag{6.16}$$

where $y(t)$ represents the target gene expression level at time t and S represents the set of potential regulators of the target gene in the rough gene network and could be set as $S = [L] = \{1,\ldots, L\}$. For each potential regulator $i \in S$, b_i explains how the expression of target gene is affected by the regulatory ability of TFi. β is attributable to the degradation effect of the target gene itself and k denotes the basal molecular level. In addition, $\varepsilon(t)$ represents a stochastic noise due to uncertainty within the model and measurement error of microarray data in the target gene. Here $\varepsilon(t)$ is assumed to be a Gaussian noise with zero mean and unknown standard deviation. Because biological systems seem to exhibit nonlinear characteristics, we assume a sigmoidal regulatory binding relationship between the regulator and the target gene. Furthermore, $x_i(t)$ is the transcriptional regulatory function of TFs on their motif binding sites, as described by following the sigmoid function of mRNA expression profiles on their corresponding regulatory genes [94], where the sigmoid function denotes the threshold of binding for a TF on its promoter binding site for the transcriptional regulation in Eq. (6.16). $x_i(t)$ can be expressed as follows:

$$x_i(t) = f(y_i) = \frac{1}{1 + \exp(-r(y_i(t) - m_i))} \tag{6.17}$$

where r denotes the transition rate of sigmoid function and m_i denotes the mean of the TF-gene expression profile.

Step 2: Refinement of the rough GRN: The model describes how upstream transcriptional factors modulate their target genes' mRNA expression through a transcriptional GRN. With the help of dynamic gene regulation model, the problem of network construction, which is difficult to decipher in biology, is transferred into a preidentified system in engineering. The rough regulatory network is a large system which contains several subsystems. Each subsystem consists of a target gene and its regulatory TFs. The state of each target gene at different time points depends on the time series microarray data. This produces a large network system in which each state of the subsystem is known and the parameters of the subsystem have been estimated. These difficult parameter estimations are conducted using system identification methods, working with one target gene at a time. This estimation method involves combining the ML parameter estimation method with the most parsimonious model order detection, AIC, to identify the unknown element of the network system. In short, the parameters of system can be estimated using the ML method and AIC can then delete any interactions without significant influence on this network system. These deletions are based on the results of the ML method and the degree detection of interactions in the network system. By approaching the network one target gene at a time, it is possible to identify all of the parameters of the rough GRN. Finally, to produce the pruned GRN, the overall gene regulatory system is deciphered dynamically and insignificant regulations are eliminated. This pruned GRN is close to real GRN in the cell. The overall flowchart of the proposed method is shown in Fig. 6.14.

In summary, the potential regulators of a target gene are first selected by the data mining of the literature and the PlantPAN database. Following this any highly unlikely regulatory relationship is removed via microarray with the help of Pearson's correlation. These possible relationships are finally pruned by time series microarray data of *A. thaliana* via ML system identification and Akaike's system order detection through a dynamic gene regulatory model. We have combined several algorithms and tools to improve the performance of the gene network construction of the photosynthetic light acclimation.

6.3.2 Investigating control mechanism of photosynthetic light acclimation at different light shift conditions

Based on data collected in Stage I, we can construct a rough GRN of the photosynthetic light acclimation system. We then establish a dynamic model for the rough gene network. By combining the system identification scheme and the parsimonious AIC method, the rough gene network was pruned using time series microarray data. According to the flowchart of Fig. 6.14, it is possible to construct two refined GRNs, representing the PSI-to-PSII (reduced state of plastoquinone) and the PSII-to-PSI (oxidized state of plastoquinone) light shift conditions. These two GRNs have been rearranged and depicted by Cytoscape [124] (Fig. 6.15A and B). Fig. 6.15A depicts the GRN under the reduced state of the PQ pool and identifies 55 nodes with 109 edges for the network. Alternatively, Fig. 6.15B depicts the GRN for the oxidized state of the PQ pool and identifies 55 nodes with 92 edges for the network. The numbers of nodes, edges, and highly connected hubs at different light conditions are shown in Table 6.3.

Once the refined gene networks under different light shift systems have been constructed, the topology of the network can be examined. The degree of connection of each node in the two different conditions has been summarized (Table 6.3). The results demonstrate that small-degree nodes are most abundant and high-degree nodes are relatively rare. This verifies that the networks constructed via this algorithm are scale-free, as opposed to random. Scale-free networks have smaller path lengths than random networks and can tolerate random removal of nodes; however, loss of the hubs may cause the network to collapse into clusters. Hence, smaller path lengths ensure efficient reaction and higher tolerance of random mutation against internal and external perturbations [125].

There are two hubs, *ATHB-1* and *Arabidopsis* Response Regulator 10 (*ARR10*), which can be regarded as highly connected hubs of signal transduction in the two different light conditions. Inactivation of these highly connected hubs by mutation may lead to collapse of the photosynthetic LTR system. Interestingly, the two hubs appear to have high importance for different light shift conditions. For the reduced state of the PQ pool, the gene *ATHB-1* is the top rank and the gene *ARR10* is the second

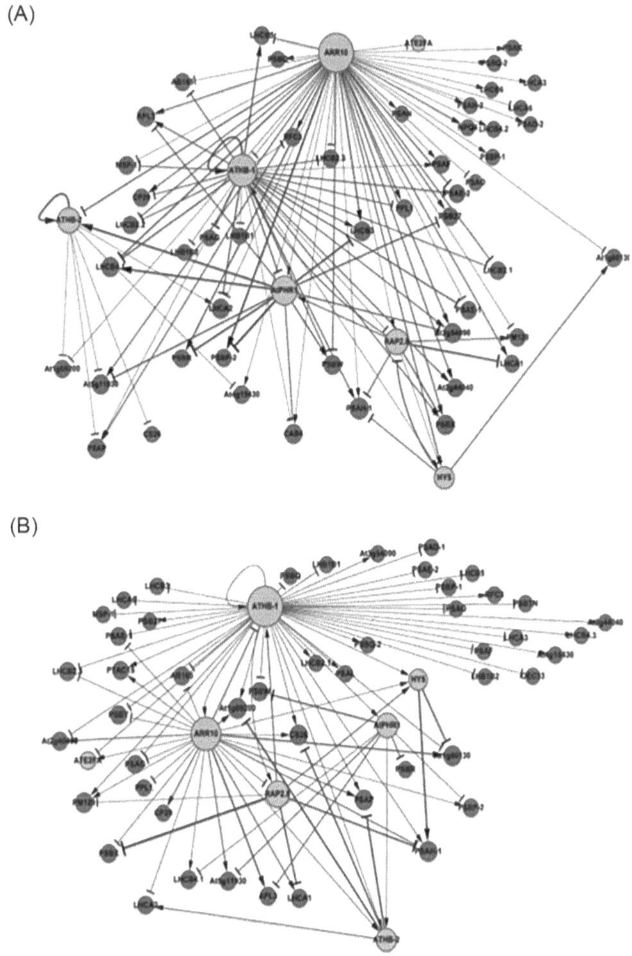

Figure 6.15 The gene regulatory networks for control mechanism of long-term light acclimation response to two light shift conditions. (A) Under reduced state of the PQ pool (PSI-to-PSII light shift). (B) Under oxidized state of the PQ pool (PSII-to-PSI light shift). The gene regulatory networks of long-term light acclimation response visualized in a hierarchical form. *Yellow* (light gray in print versions) represents the transcription factor and *green* (dark gray in print versions) represents the gene. The node size is proportional to the degree of each node and the edge width represents the magnitude of the regulatory ability in Eq. (6.16). *This figure has been created by Cytoscape (Shannon P, Markiel A, Ozier O, Baliga NS, Wang JT, Ramage D, et al. Cytoscape: a software environment for integrated models of biomolecular interaction networks. Genome Res 2003;13(11):2498–504).*

rank in Table 6.3. However, for the oxidized state of the PQ pool, the ranking of these two genes is reversed. Within the two gene regulatory systems, some interactive genes can easily be removed whilst others can be easily added under different external stimuli. This concept is known as "weak linkage" in network theory [125].

Table 6.3 Highly connected genes and their gene connectivities in two different light shift conditions

		PSI-to-PSII light shift	PSII-to-PSI light shift
	Node	55	55
	Edge	109	92
Gene name	AGI ID	Reduced state of the PQ	Oxidized state of the PQ
ATHB-1	At3g01470	33	48
ARR10	At4g31920	51	28
AtPHR1	At4g28610	14	8
RAP2.8	At1g68840	10	9
ATHB-2	At4g16780	10	7

"Weak linkage" structures enable the addition or removal of new or old processes to the existing core process using common versatile mechanisms that operate on diverse inputs and outputs [53]. As a consequence, "weak linkage" can improve the information exchange, signal transduction, and network robustness in response to external stimuli [53]. Moreover, the most important interactions and information exchanges sometimes occur via nodes from otherwise unrelated networks implying that the nonhubs may play a critical role in gene regulation [53,125]. These nonhubs appear to have evolved by natural selection to improve the robustness of biological networks.

The GRNs in Fig. 6.15A and B are too complex to distinguish which transcriptional interactions are present only in reduced state of the PQ pool or oxidized state of the PQ pool. Due to this, the differential GRNs (Fig. 6.16A and B) have been obtained by comparing the network under PSI-to-PSII light shift conditions with the network under PSII-to-PSI light shift conditions. By matching the interactions found in both light shift conditions, we can identify the specific connections present only in PSI-to-PSII or PSII-to-PSI light shift conditions. Common interactions found in both conditions are considered to be a common subnetwork in both light conditions, which could be regarded as an inherent regulation under normal light conditions. The differential GRNs produced on elimination of this common subnetwork provide a clearer representation of which regulatory interactions effect photosynthetic light acclimation under the different redox states of the PQ pool. In Fig. 6.16A, the differential regulatory network is found only in reduced state of the PQ pool, and it is clear that *ARR10* is the main hub with the highest connectivity. In Fig. 6.16B, the differential regulatory network is found only in oxidized state of the PQ pool, and *ATHB-1* is the main regulator in this differential network. These observation shown in Fig. 6.16A and B further confirm the importance of *ATHB-1* and *ARR10* in the opposite conditions. Principally, these observations indicate that *ARR10* is responsible for the signaling transduction and transcriptional regulation of acclimation response under the reduced state of the PQ pool and that *ATHB-1* is prominent under the

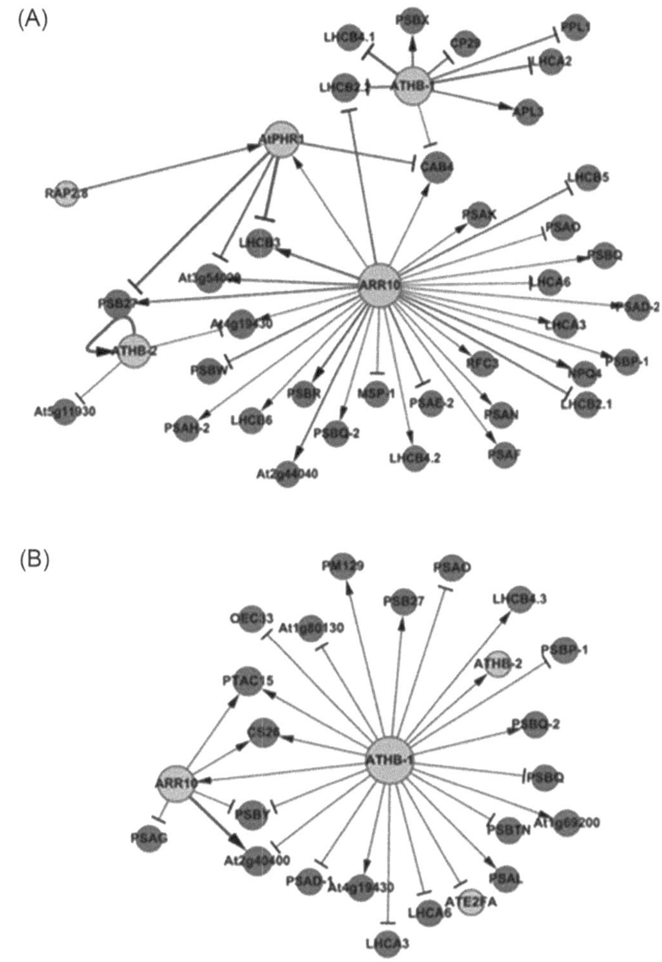

Figure 6.16 The differential gene regulatory networks to investigate the response mechanism to light stimulus environment. (A) Under reduced state of the PQ pool (PSI-to-PSII light shift). (B) Under oxidized state of the PQ pool (PSII-to-PSI light shift). *Yellow* (light gray in print versions) represents the transcription factor and *green* (dark gray in print versions) represents the gene. The node size is proportional to the degree of each node and the edge width represents the magnitude of the regulatory ability in Eq. (6.16). *This figure has been created by Cytoscape (Shannon P, Markiel A, Ozier O, Baliga NS, Wang JT, Ramage D, et al. Cytoscape: a software environment for integrated models of biomolecular interaction networks. Genome Res 2003;13(11):2498–504).*

oxidized state of the PQ pool. Rather than investigating the entire GRNs, we can easily obtain the important information behind it in two antagonistic conditions by examining the differential networks.

According to these results, the hubs *ATHB-1* and *ARR10* are identified with different significant status in opposite redox states of the PQ pool. As shown in

Ref. [120], the initial transcriptional response occurred in the nucleus 30 min after PSI-to-PSII light shift. Alternatively, following the PSII-to-PSI light shift experiment, the initial transcriptional response occurred in nucleus 2 h after the shift. Fig. 6.17 shows the time series expression profiles of both *ATHB-1* and *ARR10* in two different light shift conditions. Fig. 6.17A and B demonstrates that the gene expression of *ATHB-1* has an obvious change 8 h after the PSII-to-PSI light shift, but it does not appear to change following the PSI-to-PSII light shift, verifying that *ATHB-1* is involved in the regulatory mechanism in the oxidized state of the PQ pool. The transcriptional expression of ARR10 exhibits a more noticeable change following the PSI-to-PSII light shift than following the PSII-to-PSI light shift in Fig. 6.17C and D. Fig. 6.17C shows *ARR10* has significant change 2 h after the PSI-to-PSII light shift,

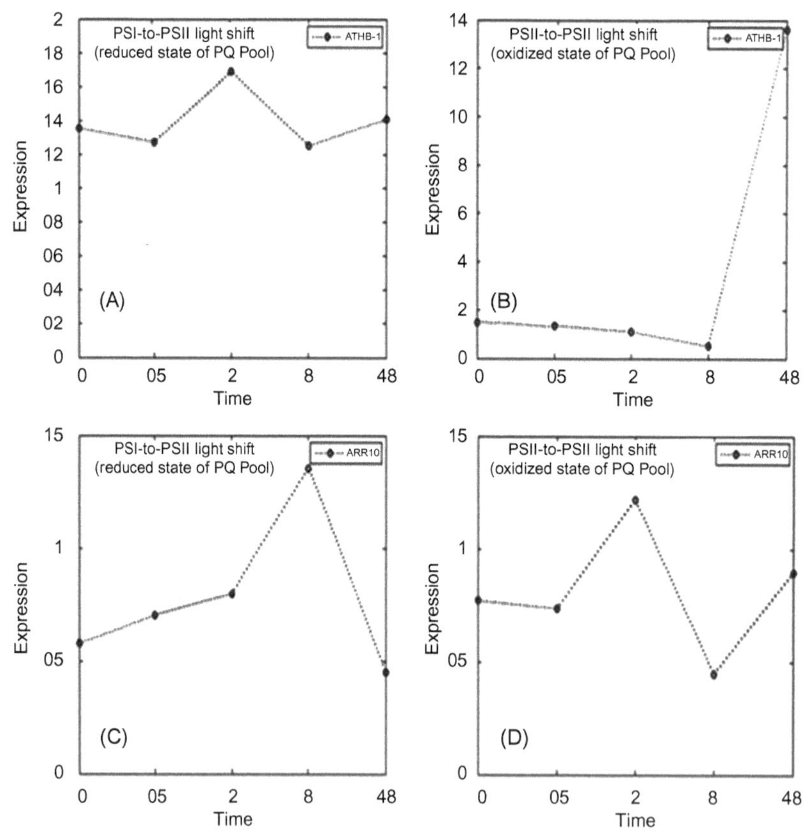

Figure 6.17 The gene expression profile of ATHB-1 and ARR10. (A) and (B) are the gene expression profiles of ATHB-1 after PSI-to-PSII (reduced state of PQ pool) and PSII-to-PSI (oxidized state of PQ pool) light shifts, respectively; (C) and (D) are the gene expression profiles of ARR10 after PSI-to-PSII (reduced state of PQ pool) and PSII-to-PSI (oxidized state of PQ pool) light shifts, respectively.

verifying that *ARR10* is involved in the regulatory mechanism under the reduced state of the PQ pool. Furthermore, transcription factor *ARR10* is involved in the regulation of the chlorophyll Biosynthetic process [126]. There are different types of chlorophyll in a chloroplast: chlorophyll a and chlorophyll b, both of which absorb different light spectra. Due to the treatment of PS-light shift, certain spectrum light is deficient which is not enough to excite with adequate energy to lead to the change of redox state of the PQ pool. Therefore *ARR10* plays a key role in maintaining homeostasis by regulating chlorophyll biosynthesis to increase the absorption of the portions of light spectrum that are deficient. By increasing the absorption of deficient light, any imbalance in the excitation of energy distribution between PSII and PSI will be counteracted. This further confirms that ARR10 is involved in the long-term photosynthetic acclimation response under different spectral qualities of light.

In Table 6.3, the connectivity between the TFs AtPHR1 and RAP2.8 is similar to the connectivity between *ATHB-1* and *ARR10*. The connectivity number of AtPHR1 is greater than that of RAP2.8 under the PSI-to-PSII light shift and this ordering is reversed under the PSII-to-PSI light shift. This demonstrates that AtPHR1 and RAP2.8 play different roles under different light shift conditions. Alternatively, one of the transcription factors in Table 6.3, HY5, does not demonstrate a change of connectivity between the networks of the different light shifts and consequently is not present in two differential networks (Fig. 6.16A and B). Since the connectivity is not influenced by the different light shifts, it appears that HY5 may be involved in the inherent process of photosynthesis during the normal light conditions. These results demonstrate that the transcription factors that have been identified play different roles under different light conditions, and that ATHB-1, ARR10, AtPHR1, and RAP2.8 participate in the long-term photosynthetic light acclimation response.

A total of seven regulators are considered as potential *A. thaliana* LTR-related TFs, which are predicted from this target gene pool via PlantPAN database. We seek evidence from the literature to validate their inferred function in regulation of the photosynthetic light acclimation response. *ATHB-1* is involved in response to blue light [127], while *ATHB-2* and *HY5* are strongly induced by far-red light [128,129] and involved in red or far-red light signaling pathways [130,131]. This suggests that they are light-related transcription factors and may be related to the photosynthetic light acclimation. Although red and far-red light signaling may be induced by the cytosolic photoreceptor phytochrome, it has not been determined whether signals from photoreceptors and the PQ pool converge to a molecular component under a long-term PSI or PSII light conditions. Fey et al. [132] identified photoreceptors that are not essential for adjustment of photosystem stoichiometry, however these results indicate that two signal sources coming from different light systems may share the same TFs to regulate gene expression in the nucleus regardless of whether their signal transductions are coupled or not in the cytosol.

According to the results, the transcription factors maintain homeostasis under two different qualities of light by altering regulation of two gene regulatory systems. Fig. 6.18 provides a visualized representation of the photosynthetic light acclimation mechanism as a feedback control system in which the transcription factors function in the role of mode switching. Based on this feedback control system, photosynthetic light acclimation under the two PS-lights can be regarded as a mode switch between two different gene regulatory patterns to improve the efficiency of energy conversion.

We explored The *Arabidopsis* Information Resource (TAIR) database to obtain annotations of seven TFs in the GRNs. Interestingly, five of the seven TFs (ARR10, AtPHR1, ATE2FA, RAP2.8, and ATHB-2) are involved in the leaf senescence stage of *Arabidopsis* [133]. ARR10 and ATHB-1 are involved in the cytokinin-mediated signaling pathway and respond to cytokinin stimulus [126,134]. *HY5* is involved in photomorphogenesis [135] and *ATHB-1* is involved in the leaf morphogenesis [136]. The diversity of these TFs demonstrates that the long-term acclimation response in plants is complicated and is connected to many biological processes. For instance, when a plant is in the shadow of a newly established building, the building can change

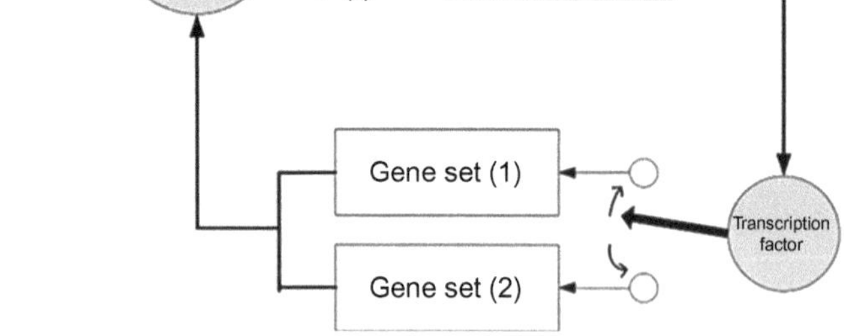

Figure 6.18 Feedback control system to mimic the photosynthetic light acclimation mechanism. Two PS-lights represent high energy input (PSII-light) and low energy input (PSI-light), respectively. $e^{-}(k)$ denotes the difference between the oxidized state of PQ pool and the homeostatic state. $e^{+}(k)$ denotes the difference between the reduced state of PQ pool and the homeostatic state. Chloroplast signals inform the transcription factors how to cope with changes in the surrounding environment via the signal transduction pathway in the cytosol. The transcription factors act as mode switch between different regulating gene sets to display different modes which counteract any change in light input.

the redox state of the PQ pool for an extended period of time. This phenomenon will cause the plant to attempt to maintain homeostasis by reconfiguring the relative amount of the two photosystems via this mode switch. If the mode switch is not sufficient to maintain homeostasis, the plant may respond to the change through alternate processes, such as leaf senescence, cytokinin stimulus, and altering leaf morphogenesis to construct a new leaf structure. For long-term changes in light quality, the plant will facilitate leaf senescence to remove old leaves and allow new buds to rapidly develop into the leaves with a structure more suitable for the current spectral quality of light. This response is a reconstruction as opposed to a mode switch and is also a light-induced LTR.

Through analysis of these results, it is possible to predict the important TFs corresponding to opposite redox states of the PQ pool. More biological insight into the gene networks in two different photosystem light shift conditions is given in the discussion section below.

6.3.3 Discussion of control mechanism in photosynthetic light acclimation

The LTR is initiated whenever a photosynthetic organism is subjected to a stable light gradient for an extended period of time. Because this response can be easily investigated by growing plants in the laboratory under an artificial light source that preferentially excite PSII or PSI (so-called PSII-light or PSI-light systems), the response has been well characterized at the biological level. However, the integration of gene regulation and metabolism during the acclimation process is currently poorly understood. Some studies have attempted to further this understanding of the LTR by identifying primary target genes under different light intensities and spectral qualities [120,123,137–139]. The number of these identified genes that respond to the photosynthetic redox signal in artificial light system can then be tested by observing microarray expression data following mutation of the specific gene or treatment with site-specific electron transport inhibitors. However, the identification of tentative target genes may help only in understanding single aspects of these processes. The full extent to which the LTR controls gene expression in higher plants is still poorly understood and will require systems biology approaches to assess large scale time series microarray data [140].

This section has employed a systems biology approach, combined with a prediction database, to construct GRNs using microarray data. The proposed systematic approach can depict the transcriptional regulatory mechanism of a photosynthesis network as a dynamic system. Furthermore, the time profile of mRNA of each target gene can be described by the dynamic gene regulatory equation in (6.16). Some researchers have applied similar dynamic equations to construct regulatory networks in yeast and *Escherichia coli* with high accuracy [141,142]. However, they did not consider that the potential regulator set S in Eq. (6.16) may include false positive regulators. In this section, we therefore employed ML method and the AIC [21] via system order detection. To approach the minimum AIC value in Eq. (2.23), iteratively removing insignificant regulators one by one, the correct regulators of the target gene can be obtained. The parameter estimation method and system order detection method are widely applied in systems biology have been shown to be an efficient tool of reverse engineering technology.

The adjustment of photosystem stoichiometry is a LTR to changed light quality by changing the relative number of the two photosystems [102,103,105–107]. Fig. 6.19 shows the average expression of genes encoding each photosystem and its corresponding light-harvesting protein complex present in two differential GRNs (Fig. 6.16A and B), respectively. In Fig. 6.19A, due to the lack of PSI-light, the plant must counteract the reduced state of the PQ pool by increasing the relative expression of PSI-related genes. The average expression of PSI-related genes is greater than that of the PSII-related genes at each time point. In contrast, Fig. 6.19B illustrates that a lack of PSII-light causes the plant to counteract the oxidized state of the PQ pool by increasing the relative expression level of PSII-related genes. In this situation, the average expression of PSI-related genes tends to decrease after 8 h. Moreover, the average expression of PSII-related genes has no significant change in both light shift conditions. This result is consistent with results gathered for mustard (*Sinapis alba* L.) [101] which found that the stoichiometry is regulated by only changing the number of PSI, while PSII remains constant. Although mRNA expression may not exactly

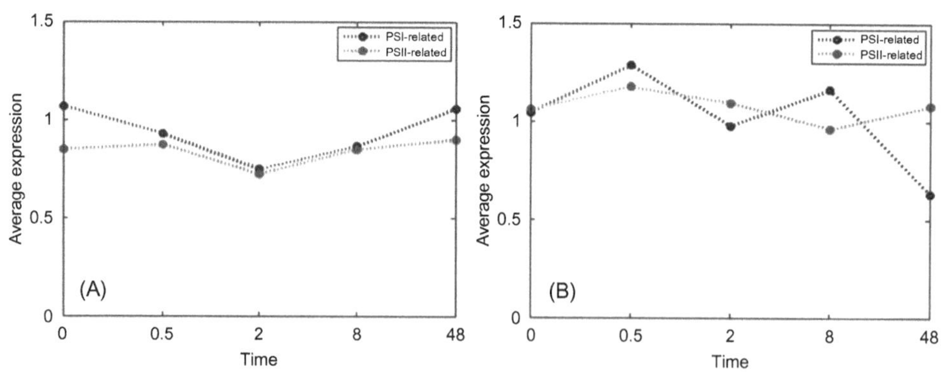

Figure 6.19 The average expression of PSI and PSII in two light shift conditions. (A) represents the average expression of PSI-related and PSII-related genes in reduced state of the PQ pool, (B) represents the average expression of PSI-related and PSII-related genes in oxidized state of the PQ pool. The PSI-related genes consist of genes encoded the photosystem I reaction center subunit and LHCI. The PSII-related genes consist of genes encoded the photosystem II reaction center subunit and LHCII. Every gene must be regulated in differential gene regulatory networks.

represent its corresponding protein concentration, mRNA expression is thought to at least partially reflect protein level [143,144]. Consequently, the results above demonstrate that the constructed networks have a high level of confidence.

Recent studies [104,145] have shown that the *STN7* kinase mediates the redox states of the PQ pool. STN7 kinase is regarded as a source of retrograde signals under changes in the spectral quality of light. These light-related retrograde signals are transported across the double membrane of the chloroplast and initiate an unknown signal transduction pathway in the cytosol which carries the signal into the nucleus to affect the expression of nuclear genes (Fig. 6.13). The mechanism by which expression of these genes is altered remains unknown. Furthermore, upstream TFs in the nucleus which are directly activated by a signal from cytosol are also poorly understood. These TFs serve as the interface between the signaling pathway and the GRN (Fig. 6.13). In step 2 of Stage I, the initial rough regulatory network is developed by iteratively searching for TF-binding sites via the PlantPAN database. Due to the limit of microarray data, the iterative procedure is stopped when no new regulatory interactions are generated from the PlantPAN database. This searching method can identify the TFs which are on the top of overall GRN and are directly activated by signals from the cytosol. In this section, the GRNs in Fig. 6.15A and B are visualized in a hierarchical form. This shows the stratum of each gene in the network. In Fig. 6.15A, ARR10 is clearly on the top of the GRN and thus can be considered as the source of the overall gene network for LTR in the nucleus. When the reduced state of the PQ pool is signaled to the nucleus, *ARR10* is directly induced. This transduces the signal to the downstream genes to counteract any imbalance in excitation of energy distribution between two photosystems. In Fig. 6.15B, *ARR10* is also on the top of GRN of the oxidized state of the PQ pool. However, it is noteworthy that *ARR10* represses *ATHB-1* and then *ATHB-1* activates *ARR10*. Therefore, we can hypothesize that both *ARR10* and *ATHB-1* may be directly induced by the signal of oxidized state of the PQ pool. Because the microarray data is not the genome-wide data, this is only a prediction. If the genome-wide microarray data becomes available, it will be possible to obtain more TFs by this proposed algorithm.

In this section, we use a multi-input/single-output regulatory model to dynamically describe this gene regulatory system. The model analyzes multiple regulators in respect to one target gene, combining time series microarray data to determine the regulatory relationship between this target gene and its upstream regulators. By using Pearson's correlation and the AIC to prune the complex GRN, an accurate GRN of the LTR system in the nucleus can be formed. While this method combines many algorithms to construct a GRN for the photosynthetic acclimation system, the primary obstacle faced in development of this network is the construction of a rough GRN. Due to the lack of complete ChIP-chip data for *A. thaliana*, the identification of transcriptional regulatory relationships relies on the prediction database. The candidate regulators are selected from the pool of potential regulators typically defined by computational prediction or sequence similarity analysis. If an important regulator is not included in the pool, it will inevitably escape identification by the proposed modeling approach.

6.3.4 Conclusions of control mechanism in response to different light acclimation environments

The photosynthetic light acclimation response is a fundamental process in plants which can optimize the efficiency of photosynthesis in fluctuating light quality. However, little is known about the molecular mechanisms that regulate these acclimation responses. In this section, we investigate this molecular mechanism from the physiological level by constructing a GRN from the limited data sources for *A. thaliana* using an efficient computational framework. This type of systems biology approach has become increasingly common in recent years and can provide insight into the underlying mechanisms of the photosynthetic light acclimation response. GRNs under PSI-to-PSII light shift and PSII-to-PSI light shift are compared to identify the important transcription factors involved in the regulation of photosynthetic LTR. The hubs and "weak linkages" are also analyzed for the robust gene network under different light conditions. Although the GRNs are a small part of photosynthetic light acclimation response in plants, they may provide a foundation on which the overall molecular mechanism underlying the photosynthetic process in organisms may be deciphered. Moreover, it is hoped that the results of this section will attract more attention to the topic of photosynthetic light acclimation with large-scale experiments. As experimental data for whole-genome ChIP-chip data and microarray data with more multiple sampling points become available in the future, the performance of the proposed method will be improved and enable more efficient construction of genome-wide regulatory networks of photosynthetic system. Such genome-wide GRNs may provide a better understanding for the molecular interaction mechanisms between the chloroplastic, nucleic, and mitochondrial components of the cell under the long-term photosynthetic acclimation response. For instance, the regulatory mechanisms for gene network of state transition, a short-term response, in the chloroplast. Furthermore, if more PPI data becomes available, the protein interactions of the signal transduction pathway may be identified and integrated with the gene transcriptional regulatory network to provide more insight into the overall photosynthetic light acclimation response.

As NGS omics data can involve to transcriptional regulation, protein interaction, and miRNA and methylation with more predictive power to investigate cellular mechanisms than other datasets, in the following section, we will construct integrated genetic and epigenetic network to investigate the whole cellular mechanisms in response to environmental stress signals via NGS data.

6.4 CONSTRUCTING IGECN FOR INVESTIGATING WHOLE CELLULAR SIGNAL FLOW MECHANISMS IN RESPONSE TO ENVIRONMENTAL STRESS SIGNALS USING HIGH-THROUGHPUT NGS

Epigenetics has been investigated in cancer initiation, and development, especially, since the appearance of epigenomics. Epigenetics may be defined as the mechanisms that lead to heritable changes in gene function without affecting the sequence of genome. These mechanisms explain how individuals with the same genotype produce phenotypic differences in response to environmental stimuli. Recently, with the accumulation of high-throughput NGS data, a key goal of systems biology is to construct networks for different cellular levels to explore whole cellular mechanisms. At present, there is no satisfactory method to construct an IGECN, which combines NGS omics data with GRNs, miRNAs regulatory networks, PPINs, and epigenetic regulatory networks of methylation using high-throughput NGS data. We investigated different kinds of NGS omics data to develop a systems biology method to construct an integrated cellular network for cellular genetic and epigenetic signal flow based on three coupling models that describe GRNs, PPINs, miRNA regulatory networks, and methylation regulation. The proposed method was applied to construct IGECNs of gastric cancer and the human immune response to human immunodeficiency virus (HIV) infection, to elucidate human defense response mechanisms from the perspective of genetic and epigenetic cellular signal flow. We successfully constructed an IGECN and validated it by using evidence from literature search. The integration of NGS omics data related to transcription regulation, PPIs, and miRNA and methylation regulation has more predictive power than independent datasets. We found that dysregulation of MIR7 contributes to the initiation and progression of inflammation-induced gastric cancer; dysregulation of MIR9 contributes to HIV-1 infection to hijack CD4 + T cells through dysfunction of the immune and hormone pathways; dysregulation of MIR139-5p, MIRLET7i, and MIR10a contributes to the HIV-1 integration/replication stage; dysregulation of MIR101, MIR141, and MIR152 contributes to the HIV-1 virus assembly and budding mechanisms; dysregulation of MIR302a contributes to not only microvesicle-mediated transfer of miRNAs but also dysfunction of NF-κB signaling pathway in hepatocarcinogenesis. The coupling dynamic systems of the whole IGECN can allow us to investigate genetic and epigenetic cellular mechanisms through whole cellular signal flow via omics data and big database mining, and are useful for further experiments in the field of systems and synthetic biology.

With advances in molecular biology technologies for whole genome sequencing, expression profiling, and high-throughput experiments, large amounts of biological data covering various biological levels have emerged [146−148]. These kinds of omics data include genetic sequences (genomics), microarray-based genome-wide expression profiles (transcriptomics), protein abundance data (proteomics), and miRNA and methylation data, and they provide an unprecedented view of cellular components and their cellular mechanisms in biological systems [149]. With the large amount of genomic/transcriptomic and epigenetic data that has accumulated, a more complex cellular function understanding of living organisms is possible. Computational systems biology techniques that are able to combine these large, heterogeneous genomic/transcriptomic and epigenetic datasets will provide useful tools to gain additional systems biology insights into cellular mechanism under a specific biological condition, e.g., cancer cells or infected cells [150].

Many studies have focused on GRNs and PPINs. Classic graph algorithms have been used to integrate cellular networks of protein–protein and protein–DNA interactions. Well known simple and complex regulatory circuits have been identified and many putative regulatory circuits have been discussed. In Ref. [151], this study includes searches for composite network motifs, which consist of both transcription regulation and PPIs that recur significantly more often than expected in random networks. In Ref. [152], integrated networks comprise transcriptional and PPI data. Interspecific analysis has shown that several types of network motifs are not subject to any particular evolutionary pressure to preserve the corresponding interaction patterns of important biological functions. Studies have also integrated protein–protein and protein–DNA interactions to infer signaling regulatory pathways, but they focus only on pathways that explain gene expression changes in response to gene knockouts [153]. Recently, an integrated cellular network of transcription regulation and PPIs was introduced by Wang and Chen [154,155]. They used gene-expression data at multiple time points to prune and combine candidate gene regulatory and signaling networks obtained from genome-scale data. An integrated and focused network for a specific condition of interest is then obtained. The transcriptional network is characterized as a dynamical system in which the expression of a target gene is modeled as a function of the regulatory effect of its corresponding TFs and mRNA degradation. The modeling of a signaling/protein interaction network accounts for the activity of neighboring loci in the network. Genomic/transcriptomic and high-throughput methods have successfully identified many GRNs and PPINs. However, to explore cellular mechanism, we need more genomic data, such as epigenetic regulation data.

In real cellular systems, the expression of protein-coding genes is controlled by a complex network of genetic and epigenetic regulatory interactions. In addition to the abovementioned genetic regulation mechanisms, epigenetic regulation via DNA methylation is also an important regulatory mechanism in several cellular processes. Recently, high-throughput NGS data containing information on mRNAs, miRNAs, and methylation has been generated. Therefore, many compelling biological questions center on how regulation and interactions among genes, proteins, and epigenetic regulators give rise to specific cellular mechanisms. To address this problem, we proposed a method to construct an integrated cellular network that can explain specific cellular mechanisms under genetic and epigenetic regulation in response to specific biological conditions, based on the coupling of stochastic dynamic models. Recently, systems biology and computational biology methods have been widely employed to develop stochastic dynamic models that describe biological functions from a dynamic systems perspective [156–160]. Dynamic models to construct an IGECN not only provide a quantitative description of the integrated cellular network, but also predict the cellular mechanism of the network in response to various conditions, gene knockouts, and treatments with external agents [161].

In this section, we integrated omics data, including NGS [162], mRNA and miRNA expression [163], RNA sequencing (RNA-seq) [164], PPIs [165], transcription regulation interaction [166–169], miRNA-target gene association [170–174], and gene ontology (GO) data (http://geneontology.org/) to construct a candidate IGECN. A schematic diagram of the candidate IGECN is shown in Fig. 6.20. The candidate IGECN mainly consisted of three subnetworks. The first was the candidate PPIN, which included candidate PPIs in signal transduction pathways and metabolic

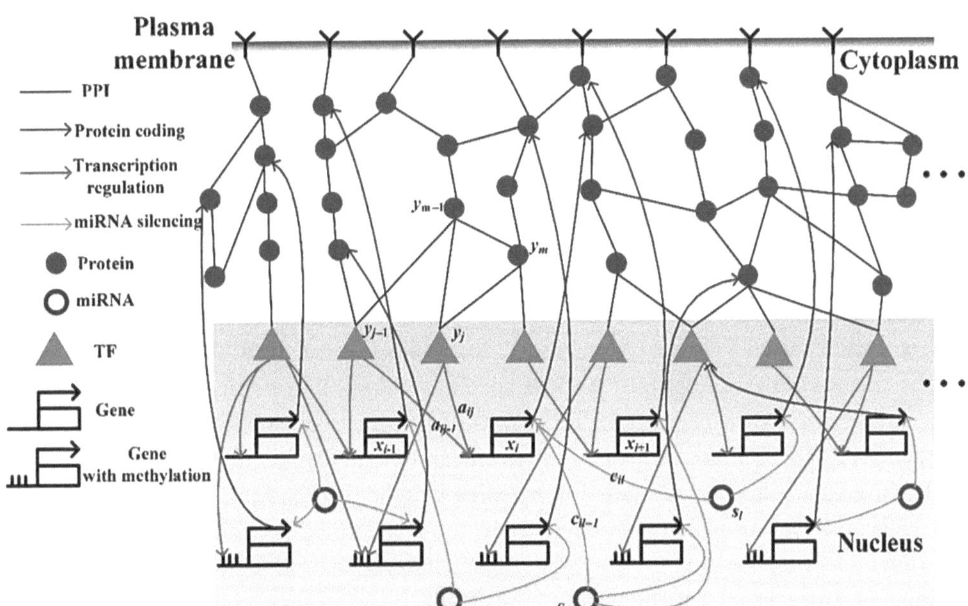

Figure 6.20 Schematic diagram of candidate integrated genetic and epigenetic cellular networks (IGECN). The candidate IGECN was constructed using omics data and database mining. The candidate IGECN mainly consisted of three subnetworks: the candidate PPIN including signal transduction pathways and metabolic pathways, the candidate GRN in which transcriptional regulation occurs, and the candidate miRNA regulatory network. The epigenetic regulation of DNA methylation was considered to influence gene transcription. TFs represent the interface between PPINs and GRNs, and genes represent the interface between GRNs, miRNA regulatory networks, PPINs, and methylation regulatory networks.

pathways; the second was the GRN, which described transcription regulation; the third was the candidate miRNA regulatory network. Epigenetic DNA methylation was also considered, which involves the modification of DNA to influence mRNA transcription. The TFs in Fig. 6.20 are at the interface between the PPIN and GRN, and genes are at the interface between the GRN, PPIN, and miRNA and methylation regulatory networks.

Since the candidate IGECN was constructed using omics data and big database mining, for data collected under various experimental conditions (including some that were gathered in vitro), it must contain many false positive interactions and regulations in the candidate PPIN and GRN, respectively. Therefore, it is necessary to prune these false positives, which may not explain real-world cellular mechanisms of biological organisms. In this situation, it is more appealing to evaluate these interactions and regulations in the candidate IGECN for investigating whole cellular mechanisms of a real organism under specific conditions (e.g., HIV infections or cancer) using gene expression profile data, protein expression profile data, and miRNA expression profile data via dynamic interaction and regulation models. The false positive interactions and regulations can be pruned based on these real-time data to construct an accurate IGECN for some specific cellular mechanisms.

In this section, three coupling stochastic dynamic models were proposed to develop a candidate IGECN that describes the interplay between the GRN, PPIN, and miRNA network and accounts for methylation regulation in the GRN. After system parameter estimation for coupling stochastic dynamic models of candidate IGECN by these time-profile data, the AIC and the *p*-value statistical method [155] were employed to detect the system order (the number of interactions and regulations) of coupling stochastic dynamic models, and then prune the false positive regulations and interactions, which are not significant and out of system order, to obtain the real IGECN for some specific cellular mechanisms in response to some specific biological condition. By the interplay of genetic transcription regulation, PPIs, miRNA regulation, and DNA methylation of the identified IGECN, the cellular mechanism can be elucidated comprehensively.

To investigate host–pathogen interspecies cellular interactions and regulations in host cells is a very important research topic for attempting to understand pathogenic and defensive mechanisms in the HIV infection process and help control the clinic pathogenic infections. In this study, the IGECN is constructed for investigating host–pathogen crosstalks in HIV-infected cells by gene expression profile and miRNA profile of NGS data. Some significant crosstalk hubs in the interspecies IGECN can be extracted as significant crosstalk network markers, from which we could get more insight into the pathogenic mechanism of HIV virus and the defense mechanism of human. Further, the IGECN is also constructed for cancer cells and compared with normal cells to find the significant network marker to investigate some significant cues from the genetic and epigenetic regulatory mechanisms using NGS data.

6.4.1 Materials for constructing IGECN

Data selection and processing in gastric cancer and liver cancer

Kim and colleagues [163] proposed microarray data, including gene expression data and miRNA expression data, 1 pair of data for normal and tumor tissues of stage I gastric cancer in humans, 10 pairs of data for normal and tumor tissues of stage II gastric cancer in humans, and 15 pairs of data for normal and tumor tissues of stage III gastric cancer in humans. According to The Cancer Genome Atlas (TCGA; https://tcga-data.nci.nih.gov/tcga/), the mRNA and miRNA expression data including 17 dataset for normal tissues and 163 dataset for tumor tissues of stage I liver cancer, 7 dataset for normal tissues and 82 dataset for tumor tissues of stage II liver cancer, 8 dataset for normal tissues and 80 dataset for tumor tissues of stage III liver cancer, and 1 dataset for normal tissues and 6 dataset for tumor tissues of stage IV liver cancer were available. The candidate associations for miRNA-gene regulatory associations are available at the TargetScan [175–178], TF-gene regulatory associations are available at the Human Transcriptional Regulation Interactions database (HTRIdb) [166] and Integrated Transcription Factor Platform (ITFP) [169], and PPIs can be found at BioGRID [165] and available in the manually curated human signaling network in Edwin Wang Lab website (http://www.cancer-systemsbiology.org/) [179]. In order to integrate the above databases, we use gene symbols in NCBI human gene database to denote the standard names of genes in this study. We used Matlab's text-file and string manipulation tools in text mining. The interaction candidates were integrated and unified from different databases.

Data selection and processing in CD4 + T cells of human with HIV-1 infection

Mohammadi and colleagues [162] reported gene expression data and miRNA expression data in HIV-1-infected cluster of differentiation 4 + (CD4 +) T cells and noninfected (normal) cells to analyze host–pathogen interactions at early (or reverse transcription), middle (integration/replication), and late (virus assembly/budding) infection stages.

Construction of the dynamic models in IGECN

Our goal was to construct an IGECN in which the transcription regulation mechanisms, miRNA regulations, PPIs, and epigenetic regulation via DNA methylation are integrated for investigating whole cellular mechanisms under specific biological conditions. A flowchart describing the proposed method to construct the IGECN is shown in Fig. 6.21. Several kinds of NSG omics data and information derived from databases are integrated, including microarray gene expression data, PPI data, miRNA expression data, and miRNA-target gene association data, as the input for the proposed IGECN construction method. From these data, the candidate GRN, PPIN, and miRNA regulatory network using the broad TF-gene regulation pool, PPI pool, and miRNA regulatory pool, respectively, under various biological conditions were

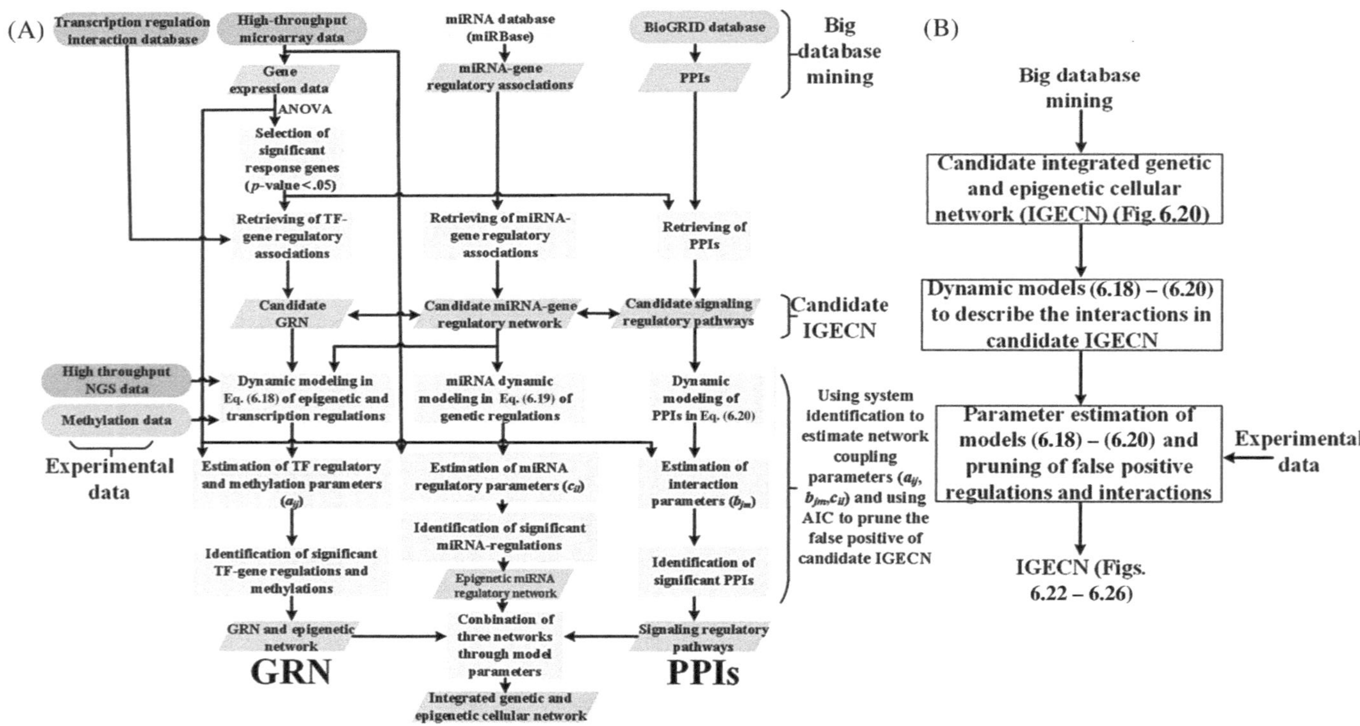

Figure 6.21 Flowchart (A) and block diagram (B) of the proposed method to integrate genetic and epigenetic cellular networks.

retrieved to construct a candidate IGECN. In this study, NGS data including gene expression and miRNA profiles were used to validate these integrated TF-gene regulation data, PPIs, and miRNA regulation data in the candidate IGECN. Therefore, stochastic dynamic models were used to identify interactions and regulatory mechanisms in the candidate IGCEN by gene expression and miRNA expression profiles using NGS data under a specific biological condition of interest, e.g., cancer or infections.

In the transcriptional regulation subnetwork of the candidate IGECN, the candidate GRN was depicted as a stochastic dynamic system in which the regulations of TFs and miRNAs are the inputs and gene expression levels of target genes are outputs (Fig. 6.20). For target gene i in the candidate GRN, the stochastic dynamic model is described by the following stochastic discrete dynamic equation:

$$x_i(t+1) = x_i(t) + \sum_{j=1}^{J_i} a_{ij} y_j(t) - \sum_{l=1}^{L_i} c_{il} x_i(t) s_l(t) - \lambda_i x_i(t) + k_i + w_i(t), \tag{6.18}$$

where $x_i(t)$ represents the mRNA expression level of target gene i at time t, a_{ij} denotes the regulatory ability of the jth TF to the ith target gene (with a positive value indicating activation and a negative value indicating repression), the regulatory ability a_{ij} is also influenced by the DNA methylations of the corresponding gene [180], $y_j(t)$ represents the regulation function of the jth TF binding to the target gene i, λ_i indicates the degradation effect from the present time t to time $t+1$, k_i represents the basal level (which is also dependent on the methylation of gene i), $s_l(t)$ denotes the expression level of the lth miRNA, the parameter c_{il} represents the rate of miRNA–mRNA coupling (and is dependent on the energy of RNA–RNA binding [181]), and $w_i(t)$ denotes the stochastic noise due to modeling residue and fluctuation in the expression profile of the target gene. The biological interpretation of Eq. (6.18) is that the mRNA expression $x_i(t+1)$ of the target gene i at time $t+1$ is a function of the present mRNA expression $x_i(t)$ and mRNA expression due to transcriptional regulation of J_i TFs that bind to the target gene at time t, minus the inhibitory regulation of L_i miRNAs at time t, minus the mRNA due to the degradation over time, plus the basal level of mRNA expression, and stochastic noise due to measurement and random fluctuations in the expression profile of target gene i.

In the candidate miRNA regulatory subnetwork, the stochastic dynamic model of the lth miRNA can be described by the following stochastic discrete time dynamic equation [182]:

$$s_l(t+1) = s_l(t) - \sum_{i=1}^{I_l} c_{il} s_l(t) x_i(t) - r_l s_l(t) + k_l + \varpi_l(t), \tag{6.19}$$

where the degradation of miRNA, i.e., $\sum_{i=1}^{I_l} c_{il} s_l(t) x_i(t)$, is accompanied by mRNA degradation events induced by the miRNA in Eq. (6.18). That is, the lth miRNA induces the degradation of I_l mRNAs at present. r_l denotes the self-degradation rate, k_l denotes the production rate of the lth miRNA, and $\varpi_l(t)$ denotes the stochastic noise.

In the PPI subnetwork, the candidate PPIs were depicted as a system in which expressions of mRNA and TFs are the input/output of the system (Fig. 6.20). For a target protein j in the candidate PPI subnetwork, the stochastic dynamic model of protein expression level is given as follows:

$$y_j(t+1) = y_j(t) - \sum_{m=1}^{M_j} b_{jm} y_j(t) y_m(t) + \alpha_j x_j(t) - \beta_j y_j(t) + h_j + \zeta_j(t) \tag{6.20}$$

where $y_j(t)$ represents the expression level at time t of the target protein j, b_{jm} denotes the interaction ability between the mth interactive protein $y_m(t)$ and the target protein $y_j(t)$, α_j denotes the effect of translation from mRNA to protein, $x_j(t)$ represents the mRNA expression level of the corresponding target protein j, h_j represents the basal expression level, and $\zeta_j(t)$ is the stochastic noise. The rate of formation of the protein complex $y_j(t)y_m(t)$ is proportional to the product of the concentration of each proteins [183,184], i.e., it is proportional to the probability of molecular collisions between two proteins; thus, the protein interaction complex was modeled as nonlinear multiplication scheme.

The biological interpretation of Eq. (6.20) is that the expression level of the target protein j at time $t+1$ is a function of the protein expression level at time t, minus protein complex interactions with M_j proteins, plus the translation effect, minus the present protein degradation, plus the basal protein level from other sources and stochastic interactions. It is noted that there is no directionality for interacting proteins in the PPI subnetwork.

The regulatory effects and interactions between genes, proteins, miRNA, and DNA methylation in the candidate IGECN are now described. Some TFs $y_j(t)$ at the end of candidate signal transduction pathways (PPINs) regulate target genes according to the regulation function $a_{ij}y_j(t)$ shown in Eq. (6.18), and mRNAs of target genes are also negatively regulated by some miRNAs $s_l(t)$. The regulated genes influence the corresponding protein expression level via translation from mRNA $x_j(t)$ to protein $\alpha_j x_j(t)$, as described in Eq. (6.20). An mRNA degradation event, $-c_{il}x_i(t)s_l(t)$, described in Eq. (6.18), induced by miRNA $s_l(t)$, also leads to the degradation of miRNA, $-c_{il}s_l(t)x_i(t)$, in Eq. (6.19). Epigenetic DNA methylation also influences the regulation parameter a_{ij} and basal level k_i in Eq. (6.18) and the translation rate α_j from mRNA to protein in Eq. (6.20). The interplay among genes, proteins, miRNAs, and DNA methylation can be seen in Eqs. (6.18–6.20) and Fig. 6.20, in which TFs are at the interface between PPINs and GRNs, and genes are at the interface between GRNs, miRNA, PPINs, and DNA methylation.

Based on the above stochastic dynamic models (6.18)−(6.20), the candidate GRN can be linked through the regulatory parameters a_{ij} in Eq. (6.18) between genes and their possible regulatory TFs (Fig. 6.20), and the candidate PPIN can be linked through the interaction parameters b_{jm} in Eq. (6.20) between possible interacting proteins. The candidate GRN and miRNA regulatory network are linked through the miRNA−mRNA coupling parameter c_{il} in Eqs. (6.18) and (6.19). Since the candidate IGECN only indicated potential TF-gene regulatory effects, PPIs, and miRNA−mRNA couplings based on data collected by database mining, they should be confirmed by gene expression profile, protein expression profile, and miRNA expression profile in NGS data. The regulatory effects a_{ij} and c_{il} in Eq. (6.18), interaction b_{jm} in Eq. (6.20) and regulation c_{il} in Eq. (6.19), which represent the edges of the candidate IGECN, should be identified and validated by empirical gene, protein, and miRNA expression data. These regulatory and interactive parameters were evaluated using temporal data by solving the constrained least square parameter estimation problem. Owing to the limited number of temporal data points, to avoid overfitting when identifying the parameters in Eqs. (6.18)−(6.20), the cubic spline method [155] was also used to interpolate extra time points for gene expression data. The gene regulatory parameters a_{ij} and c_{il} were identified gene by gene (and PPI parameters b_{jm} protein by protein), so that the candidate integrated network identification process was not limited by the number of genes, proteins, and miRNAs in the candidate IGECN. In other words, the regulatory parameters a_{ij} and c_{il} in Eq. (6.18) were first identified for target gene i and then for target gene $i+1$, $i+2$, etc. The models in Eqs. (6.18)−(6.20) only represent a small fraction of real biology. For example, many posttranscriptional regulators (RNA-binding proteins) as well as epigenetic regulations beyond methylation were involved. The model uncertainty of real biological system and the fluctuation of expression data were involved in the stochastic noises, $w_i(t)$ in Eq. (6.18), $\varpi_l(t)$ in Eq. (6.19), and $\zeta_j(t)$ in Eq. (6.20). A model order detection method such as AIC and statistical assessment like the Student's t-test [155] were used to prune the false-positive regulatory effects and interactions and then detect the true number of each in the candidate IGECN. AIC [21,155,185] was applied to detect the real regulations and interactions by pruning the false positive regulations and interactions in the candidate GRN (6.18), candidate miRNA regulatory network (6.19), and the candidate PPI network (6.20). AIC, which includes both estimated residual error and model complexity in one statistics, quantifies the relative goodness of fit of a model The minimization achieved in AIC will indicate the real network with the true model order. By applying AIC detection method, the insignificant parameters, a_{ij}, c_{il}, and b_{jm}, out of network order are considered as false positives and pruned from candidate IGECNs. We then used Student's t-test to evaluate the significance of each regulatory parameter in Eqs. (6.18)−(6.20) [155]. In this way, the candidate GRN, candidate PPIN, and candidate miRNA regulatory network based on large-scale data mining were pruned using the temporal data, leading to the construction of an accurate GRN, PPIN, and miRNA regulatory network. Based on the interactive effects between GRNs and signal transduction pathways (or PPINs) via TFs, and among GRNs, miRNA regulatory networks, and PPINs via coupling genes, these three networks were coupled to consist of the IGECN.

We introduced a general method to construct the IGECN using large-scale database mining and system identification for cellular functions. In the following, two examples are given to illustrate the application of the IGECN system to investigate cellular mechanisms under different biological conditions. In the first example, an IGECN was constructed to investigate cellular mechanisms under cancer conditions and in the second example, an IGECN was constructed to investigate pathogenic and defensive cellular mechanisms under HIV infection conditions.

Identification of the Regulatory Parameter a_{ij} from the ith TF to the jth Target Gene, Interaction b_{jm} between the jth and mth Proteins, and Coupling Rate c_{il} between the lth miRNA and the ith mRNA in IGECN.

After constructing the coupling dynamic models (6.18)–(6.20) in the candidate IGECN, the regulative and interactive parameters have to be identified using NGS data. The strategy is to identify the IGECN gene by gene, protein by protein, and miRNA by miRNA. Before applying the identification method, we first examine the dynamic models carefully. In Eq. (6.18), the basal level k_i should be always nonnegative, because the gene expressions are always nonnegative. Because the parameters in Eq. (6.18) have certain constraints, the regulatory parameters were identified by solving the constrained least square parameter estimation.

The mRNA model (6.18) was rewritten as the following linear regression form:

$$x_i(t+1) = \begin{bmatrix} y_1(t) & \ldots & y_{J_i}(t) & x_i(t) & -s_1(t)x_i(t) & \ldots & -s_{L_i}(t)x_i(t) & 1 \end{bmatrix} \begin{bmatrix} a_{i1} \\ \vdots \\ a_{iJ_i} \\ (1-\lambda_i) \\ c_{i1} \\ \vdots \\ c_{iL_i} \\ k_i \end{bmatrix} + w_i(t)$$
$$= \phi_i(t)\theta_i^1 + w_i(t) \tag{6.21}$$

where $\phi_i(t)$ denotes the regression vector which can be obtained from the processing above and θ_i^1 is the parameter vector of target gene i to be estimated. In order to avoid overfitting in the parameter estimation process, when identifying the regulatory parameters, the cubic spline method [20,186,187] was also used to interpolate extra time points for $y_i(t)$, $x_i(t)$, and $s_i(t)$ from NGS data.

The mRNA model (6.21) at different time points can be arranged as follows:

$$\begin{bmatrix} x_i(t_2) \\ x_i(t_3) \\ \vdots \\ x_i(t_L) \end{bmatrix} = \begin{bmatrix} \phi_i(t_1) \\ \phi_i(t_2) \\ \vdots \\ \phi_i(t_{L-1}) \end{bmatrix} \theta_i^1 + \begin{bmatrix} w_i(t_1) \\ w_i(t_3) \\ \vdots \\ w_i(t_{L-1}) \end{bmatrix} \tag{6.22}$$

where L is the number of expression time points of NGS data after cubic spline interpolation.

For simplicity, we define the notations X_i, Φ_i, and W_i to represent Eq. (6.22) as:

$$X_i = \Phi_i \theta_i^1 + W_i \tag{6.23}$$

The constrained least square parameter estimation problem of θ_i^1 is formulated as follows:

$$\min_{\theta_i^1} \left\| \Phi_i \theta_i^1 - X_i \right\|_2^2 \text{ subject to } A\theta_i^1 \leq b \tag{6.24}$$

where $A = \begin{bmatrix} \overbrace{0 & \cdots & 0}^{J_i+1} & -1 & \overset{L_i+1}{0} & 0 \\ \vdots & \ddots & \vdots & 0 & \ddots & 0 \\ 0 & \cdots & 0 & 0 & 0 & -1 \end{bmatrix}$, and $b = [0 \ldots 0]^T$ that give the constraints to force the miRNA inhibition $-c_{il}$ to be always nonpositive, and the basal level k_i to be always nonnegative in Eq. (6.18), i.e., $-c_{il} \leq 0$, and $k_i \geq 0$.The constrained least squares problem was solved using the active set method for quadratic programming [188,189].

Similarly, the miRNA model (6.19) was rewritten in the following regression form:

$$\begin{aligned} s_l(t+1) &= \begin{bmatrix} s_l(t) & -s_l(t)x_1(t) & \ldots & -s_l(t)x_{I_l}(t) & 1 \end{bmatrix} \begin{bmatrix} 1-r_l \\ c_{1l} \\ \vdots \\ c_{I_l l} \\ k_l \end{bmatrix} + \varpi_l(t) \\ &= \psi_l(t)\theta_l^2 + \varpi_l(t) \end{aligned} \tag{6.25}$$

where $\psi_l(t)$ indicates the regression vector and θ_l^2 is the parameter vector to be estimated. By cubic spline method, at different time points, Eq. (6.25) can be represented as the following equation:

$$S_l = \Psi_l \theta_l^2 + W_l \tag{6.26}$$

The parameter identification problem is then formulated as follows:

$$\min_{\theta_i^2} \left\| \Psi_l \theta_l^2 - S_l \right\|_2^2 \text{ subject to } B\theta_l^1 \leq b \tag{6.27}$$

where $B = \begin{bmatrix} 0 & \overbrace{-1 \quad 0 \quad 0}^{I_l+1} \\ 0 & 0 \quad \ddots \quad 0 \\ 0 & 0 \quad 0 \quad -1 \end{bmatrix}$, and $b = [0 \ldots 0]^T$ that give the constraint to force the miRNA degradation $-c_{il}$ to be always nonpositive, and the basal level k_l to be always nonnegative in Eq. (6.19), i.e., $-c_{il} \leq 0$, and $k_l \geq 0$. Finally, the protein model (6.20) uses the same way like above to make sure $-b_{jm} \leq 0$, and α_j, $h_j \geq 0$. Since large-scale measurement of protein activities has yet to be realized and 73% of the variance in protein abundance can be explained by mRNA abundance [190], mRNA expression profiles were always used to substitute for the protein expression profiles. If the simultaneously measured genome-wide protein expression data and mRNA expression data in each cancer stage are available, the general models in Eqs. (6.18)–(6.20) can also be applied to identify the real network in cancer. After the parameter identification problem was solved, we can identify the IGECN. For example, we identified the regulatory parameter $a_{\mathrm{RBM5,ETS1}} = -0.22$ from the TF ETS1 to the target gene RBM5, interaction parameter $b_{\mathrm{PGR,NCOA2}} = 0.06$ between the 2 proteins PGR and NCOA2, and coupling rate $c_{\mathrm{USP15,MIR381}} = 0.001$ between the miRNA MIR381 and the mRNA USP15 in gastric cancer cells.

By using Student's t-test [191], the p-values for the estimated parameters, including the regulatory ability a_{ij} in Eq. (6.18), the miRNA–mRNA coupling rate c_{il} in Eqs. (6.18) and (6.19), the protein interaction ability b_{jm} in Eq. (6.20), were calculated to determine the significance of the parameters. By using one-way ANOVA, we calculated the p-value for each gene to determine the significance of expression change between normal and tumor cells in cancers, and between normal and HIV-infected human cells. The human DNA methylation profiles of liver cancers (including 41 normal and 369 tumor cells), and gastric cancer (including 2 normal and 396 tumor cells) were available in TCGA. A one-way ANOVA was also used to calculate the p-value of each gene to determine the significance of DNA methylation change between normal and tumor cells in each cancer.

6.4.2 Results of IGECN by high-throughput NGS data

The construction of IGECN is very useful to investigate the whole cellular mechanisms of specific biological conditions by big database mining and NGS data. The following two examples are given to illustrate how to construct IGECN to investigate the roles of genetic and epigenetic methylation and miRNA dysregulations in gastric tumorigenesis and HIV infection. Based on these significant core network biomarkers in IGECN, potential multiple drug targets were also proposed for the development of therapeutic treatment.

(I) Construction of IGECN to Investigate Significant Cellular Mechanisms in Gastric Tumorigenesis and Hepatocarcinogenesis.

Cancer is a complex and heterogeneous disease that is highly robust against therapeutic interventions. It is a constellation of diverse and evolving disorders characterized by the uncontrolled proliferation of cells that may eventually lead to fatal dysfunction of the cellular system. Although some cancer subtypes can be cured by early diagnosis and specific treatment, no effective treatment has been established for many subtypes. Due to the complex, heterogeneous, and evolving nature of cancer, it is appealing to construct an IGECN to gain insight into the carcinogenesis process using large-scale data mining and systematic genetic modeling and various types of omics and temporal data.

Based on the importance of epigenetic modifications in cancer, we identified IGECNs using raw data collected at various stages of cancer as the temporal information for gastric tumorigenesis (Fig. 6.22) and hepatocarcinogenesis (Fig. 6.23),

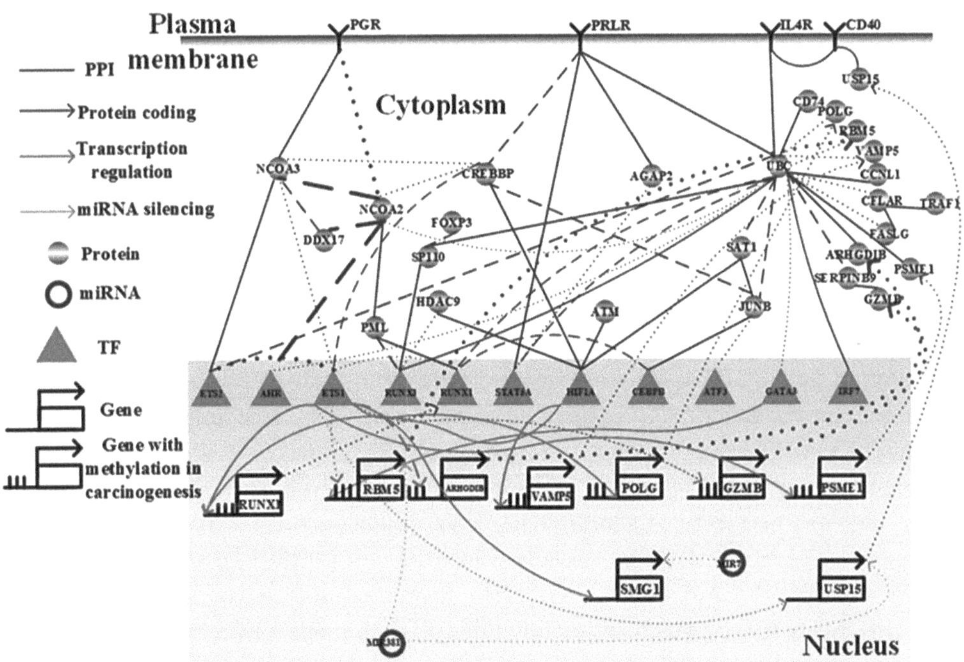

Figure 6.22 The integrated genetic and epigenetic cellular network for investigating the cellular mechanisms of normal and tumor tissues in gastric cancer through the change of genetic and epigenetic cellular information flow. The associations include the networks in tumor cells (*dotted lines*), normal cells (*dashed lines*), and both (*solid lines*). The *bold lines* indicate large genetic and epigenetic regulatory effects or genetic expression. Dysregulation of MIR7 contributes to initiation and progression of inflammation-induced gastric cancer, and PML, a potential drug target, overcomes platinum resistance in platinum-based chemotherapy in gastric cancer.

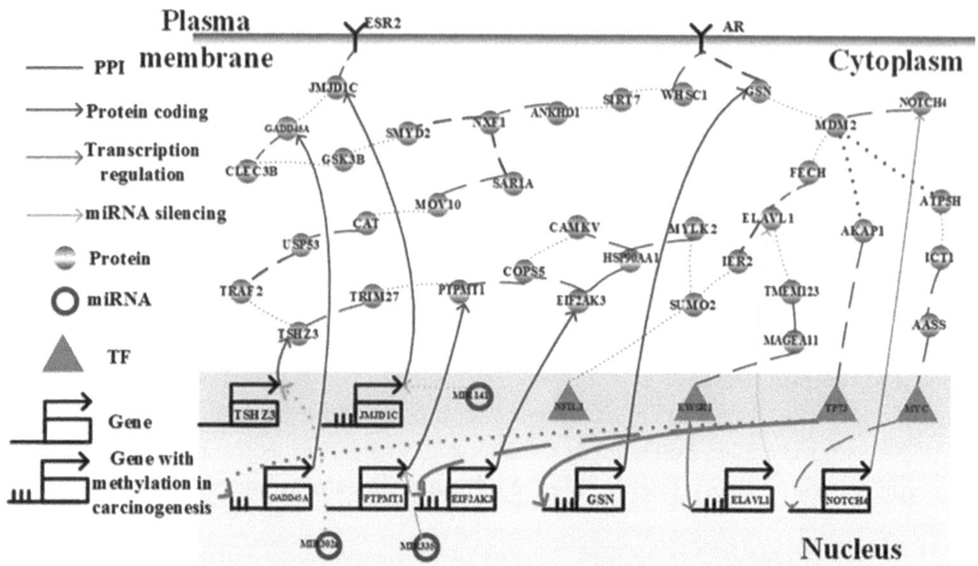

Figure 6.23 The integrated genetic and epigenetic cellular network for investigating the cellular mechanisms of normal and tumor tissues in liver cancer through the change of genetic and epigenetic cellular information flow. The associations include the networks in tumor cells (*dotted lines*), normal cells (*dashed lines*), and both (*solid lines*). The bold lines indicate large genetic and epigenetic regulatory effects or genetic expression. Dysregulations of MIR141 and MIR335 contribute to dysfunction of DNA damage response and cell proliferation in human HCC, respectively.

demonstrating large differences between miRNA regulations of normal and tumor tissues in gastric tumorigenesis. The results (Fig. 6.22) show that two cell cycle genes, ubiquitin C (*UBC*) and ataxia telangiectasia mutated (*ATM*), and one chromosomal translocation gene, CREB binding protein (*CREBBP*), play important roles in gastric tumorigenesis via genetic and epigenetic regulation. ATM is an important cell cycle checkpoint kinase [192], and chromosomal translocation plays a critical role in gastric tumorigenesis [193]. We identified that seven genes have basal level difference between cancer and normal cells, including *RUNX1* (p-value $< 1.4 \times 10^{-3}$), *ARHGDIB* (p-value $< 1.6 \times 10^{-2}$), *PSME1* (p-value $< 4 \times 10^{-2}$), *VAMP5* (p-value $< 2 \times 10^{-2}$), *POLG* (p-value $< 1.3 \times 10^{-2}$), *GZMB* (p-value $< 1.44 \times 10^{-2}$), and *RBM5* (p-value $< 8.2 \times 10^{-3}$), which are mainly due to DNA methylation of the corresponding genes. In order to validate our finding, we used the epigenomics data of human gastric cancer in TCGA. According to DNA methylation profiles of gastric cancer cells and normal gastric cells from TCGA, we found that five genes have methylation change, including *RUNX1* (p-value $< 3.56 \times 10^{-1}$), *ARHGDIB* (p-value $< 1.81 \times 10^{-1}$), *PSME1* (p-value $< 2.59 \times 10^{-1}$), *GZMB* (p-value $< 1.79 \times 10^{-1}$), and *RBM5* (p-value $< 3.94 \times 10^{-1}$), and two genes have small methylation change including *VAMP5* (p-value $< 8.61 \times 10^{-1}$), *POLG* (p-value $< 8.38 \times 10^{-1}$) between gastric cancer cells and normal gastric cells (Fig. 6.22).

Moreover, we also identified that five genes have basal level difference between hepatocellular carcinomas (HCCs) and normal cells, including *ELAVL1* (p-value $< 3.69 \times 10^{-5}$), *JMJD1C* (p-value $< 3.59 \times 10^{-11}$), *GADD45A* (p-value $< 7.32 \times 10^{-11}$), *EIF2AK3* (p-value $< 5 \times 10^{-5}$), and *GSN* (p-value $< 1.19 \times 10^{-2}$), which is mainly due to DNA methylation of the corresponding genes (Fig. 6.23). We used the epigenomics data of HCCs in TCGA to support the results. According to DNA methylation profiles of HCCs and normal liver cells from TCGA, we found that five genes have methylation changes, including *ELAVL1* (p-value $< 5.7 \times 10^{-2}$), *JMJD1C* (p-value $< 1.4 \times 10^{-2}$), *GADD45A* (p-value $< 9.6 \times 10^{-2}$), *EIF2AK3* (p-value $< 1.85 \times 10^{-6}$), and *GSN* (p-value $< 8.71 \times 10^{-4}$).

(II) Construction of an IGECN for Investigating Pathogenic and Defensive Mechanisms of HIV-1-infected Cells.

Human HIV was first identified in 1983 and rapidly emerged as a virus responsible for a pandemic. At the beginning of this century, the HIV epidemic received global attention, and methods to halt and reverse the AIDS epidemic became an important United National Millennium topic. Currently, approximately 35 million people are living with HIV worldwide. With the advent of highly active antiretroviral therapies, HIV now can be managed as a chronic disease, but developing a cure remains a significant challenge. Therefore, the investigation of interspecific host–pathogen regulatory mechanisms and interactions in cellular genetic and epigenetic systems is a very important research topic to improve understanding of pathogenic and defensive mechanisms in the HIV infection process and to help control clinical pathogenic infections. In this section, using two-sided time-profile HIV-human high-throughput sequence data (NGS), RT-PCR data, miRNA data, and other omic data, the interspecific host–pathogen IGECN was constructed for the HIV infection process. To construct this model, the coupling dynamic Eqs. (6.18)–(6.20) should be modified to consider the pathogen in addition to the host species as follows:

$$
\begin{aligned}
x_i(t+1) &= x_i(t) + \sum_{j=1}^{J_i} a_{ij} y_j(t) - \sum_{l=1}^{L_i} c_{il} x_i(t) s_l(t) + \sum_{o=1}^{O_i} d_{io} h_o(t) - \lambda_i x_i(t) + k_i + w_i(t), \\
y_j(t+1) &= y_j(t) - \sum_{m=1}^{M_j} b_{jm} y_j(t) y_m(t) - \sum_{o=1}^{O_j} g_{jo} y_j(t) h_o(t) + \alpha_j x_j(t) - \beta_j y_j(t) + h_j + \zeta_j(t), \\
h_o(t+1) &= h_o(t) - \sum_{j=1}^{J_n} g_{jo} h_o(t) y_j(t) - \sum_{l=1}^{L_o} m_{ol} h_o(t) s_l(t) - \kappa_o h_o(t) + \delta_o + \xi_o(t), \\
s_l(t+1) &= s_l(t) - \sum_{i=1}^{I_l} c_{il} s_l(t) x_i(t) - \sum_{o=1}^{O_l} m_{ol} s_l(t) h_o(t) - r_l s_l(t) + k_l + \varpi_l(t),
\end{aligned}
\tag{6.28}
$$

where $h_o(t)$ denotes expression of the oth pathogen protein, and the crosstalk effect of pathogen proteins $h_o(t)$ are considered. After system identification of coupling IGECN between host and pathogen in Eq. (6.28) by NGS data at different infection stages, we found that three genes had differences in their basal levels between cancer and normal cells at the early infection stage, *TAF5* (p-value $< 2.4 \times 10^{-1}$), *DDX3X* (p-value $< 2 \times 10^{-1}$), and *CELF1* (p-value $< 9.47 \times 10^{-2}$); two genes had basal level differences between cancer and normal cells at the middle stage, *TAF5* (p-value $< 5.8 \times 10^{-2}$) and *CELF1* (p-value $< 2.07 \times 10^{-7}$); two genes had basal level differences between cancer and normal cells at the late stage, *ZNF451* (p-value $< 9 \times 10^{-2}$) and *CELF1* (p-value $< 1.6 \times 10^{-3}$). These genes with different basal levels at different infection stages are mainly due to DNA methylation of stage-specific genes. Previous evidence for DNA methylation, including *TAF5* [194] and *DDX3X* [195] methylation in individuals with HIV infection, *ZNF451* methylation in individuals with Epstein–Barr virus (EBV) infection [196], and *CELF1* methylation in individuals with Moloney leukemia virus infection [197], supports our findings. However, DNA methylation of genes are highly tissue-specific. If the methylation profiles of genes in HIV-infected and mock cells are available in the future, the results can be specifically supported.

By applying the raw expression data and the candidate associations in cancer cells to our model (6.28), we identified the IGECNs at the three infection stages (Figs. 6.24–6.26). In the following section, we discuss the contribution of miRNA dysregulation to gastric cancer, HIV-1 infection progress, and liver cancer with respect to biological functions. The results have applications to the development of therapeutic treatments.

Briefly, dysregulation of MIR7 contributes to the initiation and progression of inflammation-induced gastric cancer; dysregulation of MIR9 contributes to HIV-1 infection to hijack CD4 + T cells through dysfunction of the immune and hormone pathways; dysregulation of MIR139-5p, MIRLET7i, and MIR10a contributes to the HIV-1 integration/replication stage through DNA hypermethylation and immune system dysfunction; dysregulation of MIR101, MIR141, and MIR152 contributes to the HIV-1 virus assembly/budding stage through DNA hypermethylation, ubiquitin transfer, and endoplasmic reticulum-associated degradation; dysregulation of MIR302a contributes to not only microvesicle-mediated transfer of miRNAs but also dysfunction of NF-κB signaling pathway in hepatocarcinogenesis.

6.4.3 Discussion

Gastric cancer

MIR7 contributes to the initiation and progression of inflammation-induced gastric cancer

Until now, it has been suggested that miRNAs play important roles in regulating proliferation and apoptosis to promote the invasion, metastasis, and angiogenesis of gastric cancer cells [198–200]. Although a few connections between genetic and epigenetic

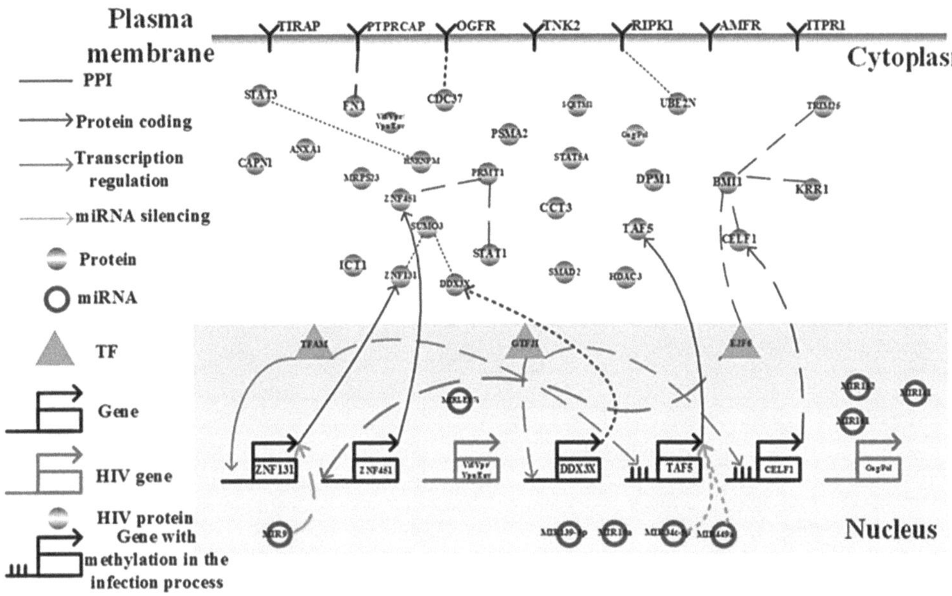

Figure 6.24 The integrated genetic and epigenetic cellular network for investigating the cellular mechanisms of normal and HIV-infected T cells at the first HIV-1 infection stage, i.e., the reverse transcription stage. The signal associations include those in HIV-infected cells (*dotted lines*), normal cells (*dashed lines*), and both cell types (*solid lines*). The *bold lines* indicate large genetic and epigenetic regulatory signals or genetic expression. Dysregulation of the signaling cascade from MIR9 to SUMO3 contributes to HIV-1 infection, hijacking CD4 + T cells via dysfunction of the immune and hormone pathways at the early infection stage.

regulatory mechanisms in gastric cancer cells can be considered independently using biological experiments [198–201], the precise biological functions of miRNAs in gastric cancer cells and miRNAs affecting gastric cancer progression need to be determined by comparing the IGECN with that of normal cells.

In this study, we identified the IGECN including the largest differences in miRNA regulation between normal and cancer cells (Fig. 6.22). We identified that MIR7 represses SMG1 (p-value $< 10^{-16}$) in cancer cells. It has been reported that both MIR7 and SMG1 are involved in the initiation and progression of inflammation-induced gastric cancer [202,203]. The significant expression changes of MIR7 (p-value $< .151$) and SMG1 (p-value $< .009$) between normal gastric cells and gastric cancer cells also supported the result that MIR7 and SMG1 play important role in Human gastric carcinogenesis.

We identified that the expression change of MIR381 (p-value $< .83$) contributes to the expression changes of RNA-binding protein 5 (RBM5) (p-value $< .008$) and ubiquitin specific peptidase 15 (USP15) (p-value $< 6 \times 10^{-4}$) via significant miRNA regulations (p-value $< 10^{-16}$) in cancer cells. The activities of RBM5 and USP15

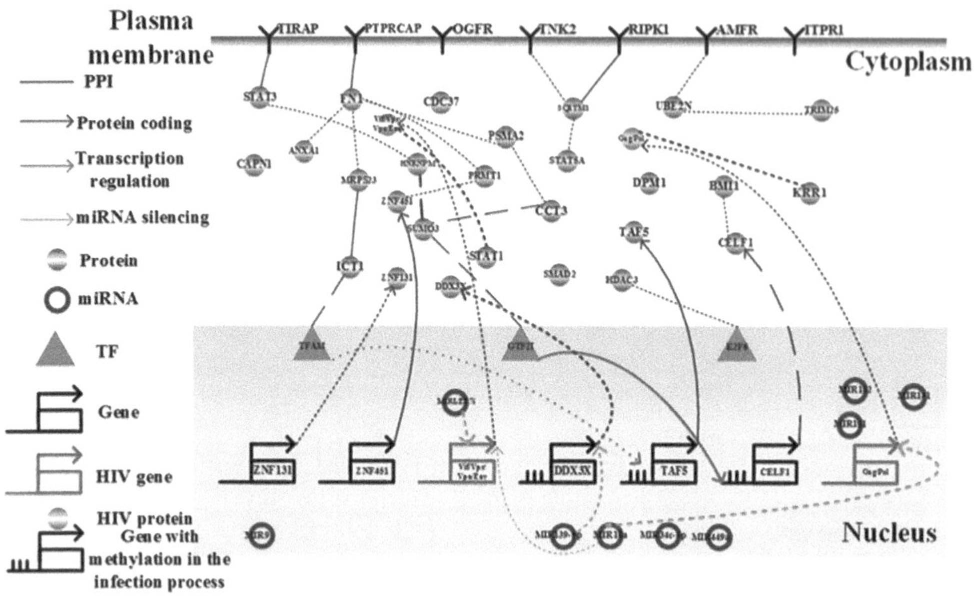

Figure 6.25 The integrated genetic and epigenetic cellular network for investigating the cellular mechanisms of normal and HIV-infected T cells at the intermediate HIV infection stage, i.e., the integration/replication stage. The signal associations include those in HIV-infected cells (*dotted lines*), normal cells (*dashed lines*), and both cell types (*solid lines*). The *bold lines* indicate large genetic and epigenetic regulatory signals or genetic expression. Dysregulation of MIR139-5p and MIRLET7i contributes to dysfunction of the immune response and viral replication via DNA hypermethylation.

were also significantly activated in gastric cancer. It has been reported that the reduced RBM5 and USP15 expressions result in tumor cell proliferation [204,205] and induce tumor cell apoptosis [206], respectively. It has also been suggested that dysregulation of two proteins, IL4 and CD40, leads to dysfunction of cell proliferation in mantle cell lymphoma [207], and dysregulation of MIR381 leads to dysfunction of proliferation in esophageal cancer [208]. Thus, we proposed that dysregulation of miRNAs in gastric cancer contributes substantially to the dysfunction of proliferation and apoptosis in gastric cancer via the ubiquitin-proteasome system [209]. TCGA offers a multi-layered view of genomics and epigenomics data of many human cancer types. According to DNA methylation profiles of gastric cancer cells and normal gastric cells from TCGA, the methylation changes of GZMB (p-value $< .18$), ARHGDIB (p-value $< .18$), PSME1 (p-value $< .26$), RUNX1 (p-value $< .36$), RBM5 (p-value $< .39$), POLG (p-value $< .84$), and VAMP5 (p-value $< .86$) were found by the IGECN. Although the methylation changes of POLG and VAMP5 were insignificant, the gene expression changes of POLG (p-value $< .01$) and VAMP5 (p-value $< .02$)

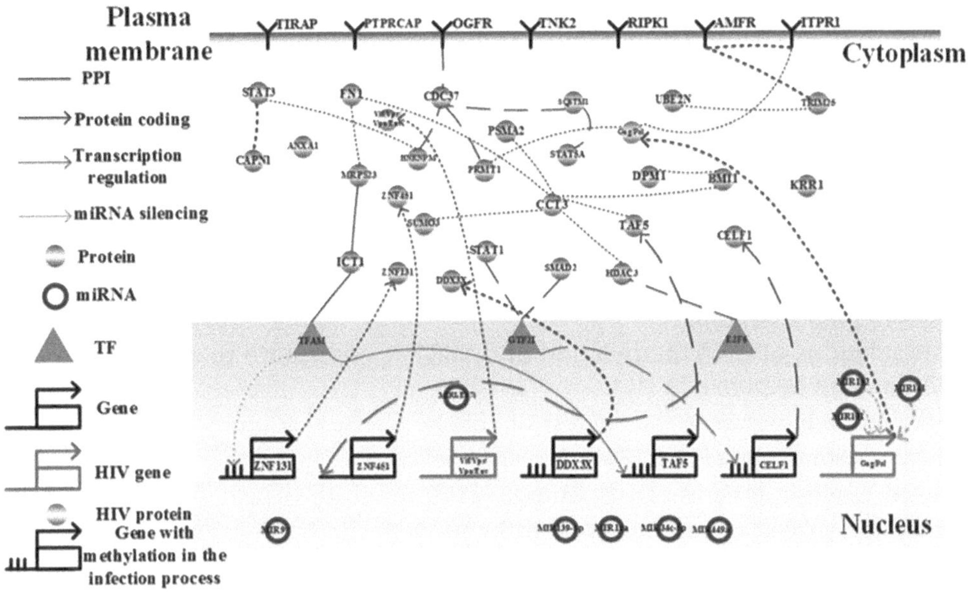

Figure 6.26 The integrated genetic and epigenetic cellular network for investigating the cellular mechanisms of normal and HIV-infected T cells at the late HIV infection stage, i.e., the virus assembly/budding stage. The signal associations include those in HIV-infected cells (*dotted lines*), normal cells (*dashed lines*), and both cell types (*solid lines*). The *bold lines* indicate large genetic and epigenetic regulatory signals or genetic expression. Dysregulation of MIR101, MIR141, and MIR152 to the HIV-1 Gag protein contributes to HIV-1 budding and release via DNA hypermethylation, ubiquitin transfer, and endoplasmic reticulum-associated degradation at the late infection stage.

were still significant. Therefore, we suggested that other regulators may exist in regulating the expressions of POLG and VAMP5 in gastric cancer.

PML, a potential drug target, overcomes platinum resistance in platinum-based chemotherapy in gastric cancer

Although platinum-based chemotherapy has been used to treat most malignant diseases, platinum resistance still limits its efficacy. The nuclear receptor coactivator family (NCOAs) and the p160 steroid receptor coactivator family (SRCs) in gastric cancer cells exhibit significant differences in interactions observed in the IGECN than in normal cells. It has been proposed that SRC3 plays an important role in intrinsic resistance to platinum therapy in cancers [210]. The results revealed that dysregulation of NCOA2 is transduced by signaling cascades via promyelocytic leukemia (PML) and RUNX1, which is transcriptionally regulated by v-ets erythroblastosis virus E26 oncogene homolog 1 (ETS1). It has been suggested that ETS1 plays an important role in regulating gastric cancer cell invasion, metastasis, and angiogenesis [201,211].

Because RUNX1 is essential for the development of human hematopoietic stem cells [212], we proposed that PML is a potential drug target that can overcome platinum resistance in the platinum-based chemotherapy in gastric cancer. It has also been suggested that PML mediates glioblastoma resistance to the mammalian target in rapamycin-targeted therapies [213]. Two pharmaceutical drugs, arsenic trioxide and retinoic acid, potentially target PML to overcome platinum resistance in gastric cancer.

Hepatocellular carcinoma

Dysregulations of DNA methylation and MIR141 contribute to dysfunction of DNA damage response in HCC

HCC frequently occurs in the context of chronic disease that promotes DNA damage and chromosomal aberrations [214]. Recently, it has been found that the overexpressed MIR141 results in extensive DNA damage [215], and JMJD1C and TP73 are the components of DNA-damage response pathway with implications for tumorigenesis [216,217]. The accumulated alterations in the DNA-damage response pathway could trigger hepatocarcinogenesis [218,219]. In this study, we identified that the three genetic and epigenetic regulations, including the gain of the miRNA regulation of MIR141 to JMJD1C (p-value $< .789$) and the genetic regulation of TP73 to GADD45A (p-value $< .037$) in cancer cells, and DNA methylations of JMJD1C and GADD45A, led to not only changes of gene expression profiles of JMJD1C (p-value $< 7.3 \times 10^{-11}$) and GADD45A (p-value $< 7.3 \times 10^{-11}$) but also changes of DNA methylation profiles of JMJD1C (p-value $< .014$) and GADD45A (p-value $< .096$) between human normal liver cells and liver cancer cells (Fig. 6.23). The changes also resulted in the gain of PPI between JMJD1C and GADD45A (p-value $< .48$) in hepatocarcinogenesis. Therefore, we suggested that the dysregulations of DNA methylation and MIR141 give rise to dysfunction of DNA damage response in HCC.

Dysregulation of MIR335 contributes human HCC cell proliferation

Recently, it has been proposed that downregulation of PTPMT1 is sufficient to trigger cancer cell death [220], and TRIM27 participates in regulating cell proliferation in the development of cancer [221]. We identified that the change in the miRNA expression profile of MIR335 (p-value $< .808$) results in the significant change in the gene expression profile of PTPMT1 (p-value $< 8.4 \times 10^{-4}$) between normal liver cells and liver cancer cells (Fig. 6.23). The expressions of TRIM27 and PTPMT1 were significantly activated in HCC. The significant genetic alterations of TRIM27 (p-value $< 8.4 \times 10^{-4}$) and PTPMT1, and their interaction (p-value $< .32$) in HCC promote aberrant proliferation. Therefore, we suggest that dysregulation of MIR335 contributes human HCC cell proliferation.

Dysregulation of MIR302a contributes to not only microvesicle-mediated transfer of miRNAs but also dysfunction of NF-κB signaling pathway in hepatocarcinogenesis

It has been identified that microvesicles transfer miRNA between cells and alter biological functions by affecting signaling pathways in hepatocarcinogenesis [222]. Microvesicles are enriched with maladjusted miRNAs, which could regulate zinc finger proteins. We identified that the change in the miRNA expression profile of MIR302a (p-value < .734) leads to the change in the gene expression profile of teashirt zinc finger homeobox 3 (TSHZ3) (p-value < .189), also known as zinc finger protein 537 (ZNF537), between normal liver cells and liver cancer cells (Fig. 6.23). The expression change of TSHZ3 further contributes to the change in the gene expression profile of TNF receptor-associated factor 2 (TRAF2) (p-value $< 1.73 \times 10^{-11}$), which mediates the dysfunction of the nuclear factor-kappa B (NF-κB) signaling pathway in hepatocarcinogenesis. The dysregulated NF-κB signaling pathway results in HCC angiogenesis and metastasis [223]. Suppression of TRAF2 inhibits the proliferation of several cancer cells [224], and the oncogene in epithelial cancers, TRAF2, contributing to the dysregulated NF-κB signaling pathway has been also identified [224]. Therefore, dysregulation of MIR302a contributes to not only microvesicle-mediated transfer of miRNAs but also dysfunction of NF-κB signaling pathway in hepatocarcinogenesis.

Dysregulation of the hormone receptors, ESR2 and AR, contributes to hepatocarcinogenesis

The most recent study proposed that menopause, which results from estrogen deficiency and physiological aging, increases risk of immune system disorders, mitochondrial dysfunction, cellular senescence, and imbalance between antioxidant formation and oxidative stress [225]. The physiologic and biochemical changes in those functions have a direct effect on liver function and mediate the development of liver disease. In this study, we identified two hormone receptors, estrogen receptor 2 (ESR2) and androgen receptor (AR), which were involved in multiple changes in intracellular signaling cascades. The gene expression profile of ESR2 showed the change (p-value < .091) between young and old normal liver cells, and the change (p-value < .063) between young normal liver cells and young liver cancer cells (Fig. 6.23). The relatively small change in the gene expression profile of ESR2 (p-value < .45) between old normal liver cells and old liver cancer cells implicates that the gene expression change of ESR2 could be necessary but not sufficient for hepatocarcinogenesis. The significant change in the gene expression profile of AR not only between young normal liver cells and young liver cancer cells (p-value $< 2.46 \times 10^{-6}$) but also between old normal liver cells and old liver cancer cells

(p-value $< 6.13 \times 10^{-7}$) implicates that the gene expression change of AR may be required for hepatocarcinogenesis. The result can be supported by the most recent study in hepatocarcinogenesis [226].

HIV-1 infection

The signaling cascade from MIR9 to SUMO3 contributes to HIV-1 infection to hijack CD4 + T cells through dysfunction of the immune and hormone pathways at early stage

Owing to the available HIV-1 mRNA gene expression data at middle and late infection stages and the lack of HIV-1 protein expression data, the impact of HIV-1 proteins on the host cells cannot be identified, directly. In this study, we identified the IGECNs at early, middle, and late infection stages (Figs. 6.24–6.26), including the largest differences in the miRNA regulation and HIV-1 protein interactions at the middle and late stages between normal and infected cells. At early infection stage (Fig. 6.24), we identified that the expression changes of MIR9 (p-value $< .448$) and general transcription factor IIi (GTF2I) (p-value $< .96$) contribute to the expression change of small ubiquitin-like modifier 3 (SUMO3) (p-value $< .016$) via respectively regulating zinc finger protein 131 (ZNF131) (p-value $< 9.3 \times 10^{-14}$) and DEAD (Asp-Glu-Ala-Asp) box helicase 3, X-linked (DDX3X) (p-value $< 1 \times 10^{-16}$). By comparing miRNA expression profiles between the wild-type HIV-1 infected human CEMx174 lymphocytes and Tat RNA silencing suppressor in HIV-1 infected human CEMx174 lymphocytes [227], Tat RNA silencing suppressor contributed to the expression change of MIR9 (p-value $< .035$). It has been also proposed that Tat can induce MIR9 to control inflammatory responses [228], and Tat can also control GTF2I expression during HIV infection [229]. Additionally, it has been also reported that SUMO modification of ZNF131 suppressed estrogen signaling [230], and HIV transcription was suppressed by inducing estrogen signaling [231]. Therefore, we suggested that the HIV infection leads to dysregulation of immune and hormone pathways to hijack CD4 + T cells.

Dysregulation of MIR34c-5p and MIR449a contributes to the frequency of latent HIV infection

Moreover, we also identified that the expression changes of MIR34c-5p (p-value $< .33$) and MIR449a (p-value $< .50$) contributed to the expression changes of TATA binding protein-associated factor 5 (TAF5) (p-value $< .25$) via miRNA regulations (p-value $< 1 \times 10^{-16}$). It has been suggested that the genetic and epigenetic regulations in TATA protein binding may control the frequency of latent HIV infection [232]. Therefore, we suggested that genetic and epigenetic regulations in TAF5 also contribute to control HIV infection at early stage.

Dysregulation of MIR139-5p contributes to HIV-1 replication via DNA hypermethylation

At the intermediate HIV infection stage, we identified that the expression change of MIR139-5p (p-value $< .23$) contributes to the expression change of DDX3X (p-value $< .14$) via miRNA regulation (p-value $< 1 \times 10^{-16}$). The expression of DDX3X was also activated at the intermediate HIV infection stage. Therefore, we suggested that DDX3X could regulate HIV replication through MIR139-5p regulations [233,234].

Dysregulation of MIRLET7i contributes to dysfunction of the immune response and viral replication

It has been reported that the HIV-1 Vif/Vpr protein is essential for viral replication [235,236]. We determined that the Vif/Vpr protein is regulated by MIRLET7i (p-value $< 1 \times 10^{-16}$) and interacts with the human STAT1 protein (p-value $< 1 \times 10^{-16}$) (Fig. 6.25). Therefore, the expression change of MIRLET7i (p-value $< .079$) leads to the expression change of Vif/Vpr protein (p-value $< .318$) during the intermediate HIV infection stage. It was supported by the previous observations that MIRLET7i has high homology to HIV-1 sequences (90–100%) [237] and the immune response of T cells induced by the Vpr-STAT1 interaction was involved in regulating HIV-1 replication [235]. Moreover, by comparing miRNA expression profiles between the wild-type HIV-1 infected human CEMx174 lymphocytes and Vif/Vpr RNA silencing suppressor in HIV-1 infected human CEMx174 lymphocytes [227], Vif/Vpr RNA silencing suppressor contributes to the expression change of MIR7i (p-value $< .107$). Therefore, we proposed that the miRNA regulation of MIRLET7i to the Vif/Vpr protein is involved in the induction of the immune response and viral replication (Fig. 6.25).

Dysregulation of MIR10a contributes to dysfunction of immune response in CD4 + T cells and viral replication

At the second infection stage, we determined that MIR10a regulates HIV-1 proteins including Gag/Pol (p-value $< 1 \times 10^{-16}$), which significantly interacts with small subunit (SSU) processome component, homolog (yeast) (KRR1) (p-value $< 1 \times 10^{-16}$), also known as HIV-1 Rev binding protein 2 (HRB2) (Fig. 6.25). Therefore, the expression change of MIR10a (p-value $< .86$) contributes to the expression change of Gag/Pol (p-value $< .36$). It has been reported that KRR1, an evolutionarily conserved gene, is essential for viral replication [238]. Additionally, MIR10a is involved in the regulation of immune response in CD4 + T cells [239] and coxsackievirus group B type 3 (CVB3) replication [240]. Moreover, it has been shown that the copy number of loci belonging to the MIRLET7 family and MIR10a is correlated with HIV-1 viral load during viral infection [241]. Therefore, we proposed that dysregulation of MIR10a is involved in HIV-1 replication via posttranscriptional regulation of HIV-1.

Dysregulation of MIR101, MIR141, and MIR152 to HIV-1 budding and releasing through DNA hypermethylation, ubiquitin transfer, and endoplasmic reticulum-associated degradation at the virus assembly/budding infection stage

The final steps of HIV-1 replication are virus assembly and budding. It has been proposed that the HIV-1 Gag protein is involved in the regulation of these steps [242,243]. At the late infection stage, we determined that regulation of MIR101, MIR141, and MIR152 to Gag (p-value $< 1 \times 10^{-16}$) results in dysfunction of the infected cells through a signaling cascade of 5 proteins, PRMT1, inositol 1,4,5-trisphosphate receptor type 1 (ITPR1), autocrine motility factor receptor (AMFR), tripartite motif containing 25 (TRIM25), and ubiquitin-conjugating enzyme E2N (UBE2N) (Fig. 6.26). The four genes were also highly expressed in HIV-1-infected T cells. Therefore, the expression changes of MIR101, MIR141, and MIR152 (p-value $< .037$, p-value $< .108$, and p-value $< .126$, respectively) contributed to the expression change of Gag (p-value $< .090$). We proposed that DNA methylation-associated proteins including PRMT1 and UBE2N contributes to virus release from HIV-1-infected cells to dysregulated uninfected cells through aberrant hypermethylation of host target genes [244]. It has been also suggested that MIR101 [245], and MIR152 [246] induce aberrant DNA hypermethylation in the hepatitis B virus infection. Moreover, ubiquitin transfer is required for efficient HIV-1 release during the final step of viral replication. Therefore, we suggested that the UBE2N-associated ubiquitin transfer is induced by a signaling cascade leading to dysregulation of MIR101, MIR141, and MIR152. Additionally, we also found that ITPR1 is induced to regulate HIV-1 infectivity through endoplasmic reticulum-associated degradation [247].

6.4.4 Conclusions

In this section, we proposed a new method to construct IGECNs for describing cellular mechanisms by system dynamic modeling and large-scale database mining of NGS data. Furthermore, we applied the method to identify differences in the IGECNs between normal and cancer cells to investigate carcinogenesis, and between HIV-1-infected and noninfected cells to investigate pathogenic and defensive mechanisms at the reverse transcription stage, integration/replication stage, and virus assembly/budding stage based on omics databases and NGS data. Database mining provided all possible candidates for genetic and epigenetic regulatory effects and interactions in the IGECNs related to cellular functions. We used AIC and statistical assessment to prune the false-positive regulatory effects and interactions by applying the dynamic coupling model to NGS data. Finally, we compared the major differences in IGECNs between different biological conditions to identify mechanistic differences to explore the evolution of cellular mechanisms. Based on these network comparisons, we examined how genetic and epigenetic regulation mechanism affects gastric tumorigenesis, hepatocarcinogenesis, and the progression of HIV-1 infections in CD4 + T cells. Therefore, the proposed IGECN construction method allowed us to unravel cellular network mechanisms from genetic and epigenetic mechanisms using omics databases and NGS data. A cancer hallmark network framework has been proposed 9 cancer hallmark networks, including survival network, mutating network, dedifferentiation network, angiogenesis-inducing network, immune-escaping network, EMT (epithelial-mesenchymal transition) network, genome duplication network, metabolic network, and Stroma-network [248]. By comparing the hallmark network difference between normal and cancer cells, cancer clonal evolution and clinical phenotypes, which will have impact on diagnosis and personalized treatment, can be predicted.

REFERENCES

[1] Hohmann S, Mager WH. Yeast stress responses. Rev. ed. Berlin, New York: Springer; 2003.
[2] Gasch AP, Spellman PT, Kao CM, Carmel-Harel O, Eisen MB, Storz G, et al. Genomic expression programs in the response of yeast cells to environmental changes. Mol Biol Cell 2000;11 (12):4241–57.
[3] Gasch AP, Werner-Washburne M. The genomics of yeast responses to environmental stress and starvation. Funct Integr Genomics 2002;2(4–5):181–92.
[4] Causton HC, Ren B, Koh SS, Harbison CT, Kanin E, Jennings EG, et al. Remodeling of yeast genome expression in response to environmental changes. Mol Biol Cell 2001;12(2):323–37.
[5] Bar-Joseph Z, Gerber GK, Lee TI, Rinaldi NJ, Yoo JY, Robert F, et al. Computational discovery of gene modules and regulatory networks. Nat Biotechnol 2003;21(11):1337–42.
[6] Kim H, Hu W, Kluger Y. Unraveling condition specific gene transcriptional regulatory networks in *Saccharomyces cerevisiae*. BMC Bioinformatics 2006;7.
[7] Lin L-H, Lee H-C, Li W-H, Chen B-S. A systematic approach to detecting transcription factors in response to environmental stresses. BMC Bioinformatics 2007;8.
[8] Middendorf M, Kundaje A, Wiggins C, Freund Y, Leslie C. Predicting genetic regulatory response using classification. Bioinformatics 2004;20:232–40.
[9] Pilpel Y, Sudarsanam P, Church GM. Identifying regulatory networks by combinatorial analysis of promoter elements. Nat Genet 2001;29(2):153–9.
[10] Rokhlenko O, Wexler Y, Yakhini Z. Similarities and differences of gene expression in yeast stress conditions. Bioinformatics 2007;23(2):E184–90.
[11] Segal E, Shapira M, Regev A, Pe'er D, Botstein D, Koller D, et al. Module networks: identifying regulatory modules and their condition-specific regulators from gene expression data. Nat Genet 2003;34(2):166–76.
[12] Wu WS, Li WH, Chen BS. Identifying regulatory targets of cell cycle transcription factors using gene expression and ChIP-chip data. BMC Bioinformatics 2007;8.
[13] Wu W-S, Li W-H, Chen B-S. Computational reconstruction of transcriptional regulatory modules of the yeast cell cycle. BMC Bioinformatics 2006;7.
[14] Yu TW, Li KC. Inference of transcriptional regulatory network by two-stage constrained space factor analysis. Bioinformatics 2005;21(21):4033–8.
[15] Chang WC, Li CW, Chen BS. Quantitative inference of dynamic regulatory pathways via microarray data. BMC Bioinformatics 2005;6:44.

[16] Chang Y-H, Wang Y-C, Chen B-S. Identification of transcription factor cooperativity via stochastic system model. Bioinformatics 2006;22(18):2276–82.
[17] Chen HC, Lee HC, Lin TY, Li WH, Chen BS. Quantitative characterization of the transcriptional regulatory network in the yeast cell cycle. Bioinformatics 2004;20(12):1914–27.
[18] Chen KC, Wang TY, Tseng HH, Huang CYF, Kao CY. A stochastic differential equation model for quantifying transcriptional regulatory network in *Saccharomyces cerevisiae*. Bioinformatics 2005;21 (12):2883–90.
[19] Teixeira MC, Monteiro P, Jain P, Tenreiro S, Fernandes AR, Mira NP, et al. The YEASTRACT database: a tool for the analysis of transcription regulatory associations in *Saccharomyces cerevisiae*. Nucleic Acids Res 2006;34:D446–51.
[20] Faires JD, Burden RL. Numerical methods. 2nd ed. Pacific Grove, CA: Brooks/Cole Pub. Co. 1998.
[21] Johansson R. System modeling and identification. Englewood Cliffs, NJ: Prentice Hall; 1993.
[22] Mendenhall W, Sincich T. Statistics for engineering and the sciences. 5th ed. Upper Saddle River, NJ: Pearson Prentice-Hall; 2007.
[23] Mendenhall W, Sincich T. Statistics for engineering and the sciences. 4th ed. San Francisco: Prentice-Hall, Inc. 1995.
[24] Cherry JM, Adler C, Ball C, Chervitz SA, Dwight SS, Hester ET, et al. SGD: *Saccharomyces* Genome Database. Nucleic Acids Res 1998;26(1):73–9.
[25] Hahn JS, Neef DW, Thiele DJ. A stress regulatory network for co-ordinated activation of proteasome expression mediated by yeast heat shock transcription factor. Mol Microbiol 2006;60 (1):240–51.
[26] Sugiyama K, Izawa S, Inoue Y. The Yap1p-dependent induction of glutathione synthesis in heat shock response of *Saccharomyces cerevisiae*. J Biol Chem 2000;275(20):15535–40.
[27] Marion RM, Regev A, Segal E, Barash Y, Koller D, Friedman N, et al. Sfp1 is a stress- and nutrient-sensitive regulator of ribosomal protein gene expression. Proc Natl Acad Sci USA 2004;101(40):14315–22.
[28] Guldener U, Munsterkotter M, Kastenmuller G, Strack N, van Helden J, Lemer C, et al. CYGD: the comprehensive Yeast Genome Database. Nucleic Acids Res 2005;33:D364–8.
[29] Remans PHJ, Gringhuis SI, van Laar JM, Sanders ME, Papendrecht-van der Voort EAM, Zwartkruis FJT, et al. Rap1 signaling is required for suppression of Ras-generated reactive oxygen species and protection against oxidative stress in T lymphocytes. J Immunol 2004;173 (2):920–31.
[30] Nevitt T, Pereira J, Rodrigues-Pousada C. YAP4 gene expression is induced in response to several forms of stress in *Saccharomyces cerevisiae*. Yeast 2004;21(16):1365–74.
[31] Gatti X, de Bettignies G, Claret S, Doignon F, Crouzet M, Thoraval D. RGD1, encoding a RhoGAP involved in low-pH survival, is an Msn2p/Msn4p regulated gene in *Saccharomyces cerevisiae*. Gene 2005;351:159–69.
[32] Teixeira MC, Fernandes AR, Mira NP, Becker JD, Sa-Correia I. Early transcriptional response of *Saccharomyces cerevisiae* to stress imposed by the herbicide 2,4-dichlorophenoxyacetic acid. FEMS Yeast Res 2006;6(2):230–48.
[33] Prinz S, Avila-Campillo I, Aldridge C, Srinivasan A, Dimitrov K, Siegel AF, et al. Control of yeast filamentous-form growth by modules in an integrated molecular network. Genome Res 2004;14 (3):380–90.
[34] Devlin C, Ticebaldwin K, Shore D, Arndt KT. RAP1 is required for BAS1/BAS2-dependent and GCN4-dependent transcription of the yeast HIS4 gene. Mol Cell Biol 1991;11(7):3642–51.
[35] Csete M, Doyle J. Bow ties, metabolism and disease. Trends Biotechnol 2004;22(9):446–50.
[36] Yale J, Bohnert HJ. Transcript expression in *Saccharomyces cerevisiae* at high salinity. J Biol Chem 2001;276(19):15996–6007.
[37] Gustin MC, Albertyn J, Alexander M, Davenport K. MAP kinase pathways in the yeast *Saccharomyces cerevisiae*. Microbiol Mol Biol Rev 1998;62(4):1264–300.
[38] Hohmann S, Krantz M, Nordlander B. Yeast osmoregulation. In: Haussinger D, Sies H, editors. Osmosensing and osmosignaling, vol. 428. San Diego: Elsevier Academic Press; 2007. p. 29–30.

[39] Tatebayashi K, Tanaka K, Yang H-Y, Yamamoto K, Matsushita Y, Tomida T, et al. Transmembrane mucins Hkr1 and Msb2 are putative osmosensors in the SHO1 branch of yeast HOG pathway. EMBO J 2007;26(15):3521–33.

[40] Saito H, Tatebayashi K. Regulation of the osmoregulatory HOG MAPK cascade in yeast. J Biochem 2004;136(3):267–72.

[41] Akutsu T, Kuhara S, Maruyama O, Miyano S. A system for identifying genetic networks from gene expression patterns produced by gene disruptions and overexpressions. Genome Inform Ser Workshop Genome Inform 1998;9:151–60.

[42] Akutsu T, Miyano S, Kuhara S. Identification of genetic networks from a small number of gene expression patterns under the Boolean network model. Pac Symp Biocomput 1999;17–28.

[43] Akutsu T, Miyano S, Kuhara S. Inferring qualitative relations in genetic networks and metabolic pathways. Bioinformatics 2000;16(8):727–34.

[44] Kitano H. Computational systems biology. Nature 2002;420(6912):206–10.

[45] Kitano H. Systems biology: a brief overview. Science 2002;295(5560):1662–4.

[46] Kitano H. Looking beyond the details: a rise in system-oriented approaches in genetics and molecular biology. Curr Genet 2002;41(1):1–10.

[47] Ko CL, Wang FS, Chao YP, Chen TW. S-system approach to modeling recombinant *Escherichia coli* growth by hybrid differential evolution with data collocation. Biochem Eng J 2006;28(1):10–16.

[48] Tsai KY, Wang FS. Evolutionary optimization with data collocation for reverse engineering of biological networks. Bioinformatics 2005;21(7):1180–8.

[49] Tsai HK, Huang GTW, Chou MY, Lu HHS, Li WH. Method for identifying transcription factor binding sites in yeast. Bioinformatics 2006;22(14):1675–81.

[50] Tsai HK, Lu HHS, Li WH. Statistical methods for identifying yeast cell cycle transcription factors. Proc Natl Acad Sci USA 2005;102(38):13532–7.

[51] Spellman PT, Sherlock G, Zhang MQ, Iyer VR, Anders K, Eisen MB, et al. Comprehensive identification of cell cycle-regulated genes of the yeast *Saccharomyces cerevisiae* by microarray hybridization. Mol Biol Cell 1998;9(12):3273–97.

[52] Weisberg S. Applied linear regression. 2nd ed. New York, NY: Wiley; 1985.

[53] Kitano H. Biological robustness. Nat Rev Genet 2004;5(11):826–37.

[54] Ma HW, Zeng AP. The connectivity structure, giant strong component and centrality of metabolic networks. Bioinformatics 2003;19(11):1423–30.

[55] Caporale LH. Natural selection and the emergence of a mutation phenotype: an update of the evolutionary synthesis considering mechanisms that affect genome variation. Annu Rev Microbiol 2003;57:467–85.

[56] Gerhart J, Kirschner M. Cells, embryos, and evolution: toward a cellular and developmental understanding of phenotypic variation and evolutionary adaptability. Malden, MA: Blackwell Scientific; 1997.

[57] Ferrell JE, Machleder EM. The biochemical basis of an all-or-none cell fate switch in Xenopus oocytes. Science 1998;280(5365):895–8.

[58] Shen-Orr SS, Milo R, Mangan S, Alon U. Network motifs in the transcriptional regulation network of *Escherichia coli*. Nat Genet 2002;31(1):64–8.

[59] Swain PS. Efficient attenuation of stochasticity in gene expression through post-transcriptional control. J Mol Biol 2004;344(4):965–76.

[60] Kalir S, Mangan S, Alon U. A coherent feed-forward loop with a SUM input function prolongs flagella expression in *Escherichia coli*. Mol Syst Biol 2005;1:2005.0006.

[61] Mangan S, Itzkovitz S, Zaslaver A, Alon U. The incoherent feed-forward loop accelerates the response-time of the gal system of *Escherichia coli*. J Mol Biol 2006;356(5):1073–81.

[62] Mangan S, Zaslaver A, Alon U. The coherent feedforward loop serves as a sign-sensitive delay element in transcription networks. J Mol Biol 2003;334(2):197–204.

[63] Milo R, Shen-Orr S, Itzkovitz S, Kashtan N, Chklovskii D, Alon U. Network motifs: simple building blocks of complex networks. Science 2002;298(5594):824–7.

[64] Hohmann S. Osmotic stress signaling and osmoadaptation in yeasts. Microbiol Mol Biol Rev 2002;66(2):300–72.

[65] Schwartz MA, Madhani HD. Principles of map kinase signaling specificity in *Saccharomyces cerevisiae*. Annu Rev Genet 2004;38:725–48.

[66] Harbison CT, Gordon DB, Lee TI, Rinaldi NJ, Macisaac KD, Danford TW, et al. Transcriptional regulatory code of a eukaryotic genome. Nature 2004;431(7004):99–104.

[67] Alon U. An introduction to systems biology: design principles of biological circuits. Boca Raton, FL: Chapman & Hall/CRC; 2007.

[68] Han Z, Vondriska TM, Yang L, MacLellan WR, Weiss JN, Qu Z. Signal transduction network motifs and biological memory. J Theor Biol 2007;246(4):755–61.

[69] Sabouri-Ghomi M, Ciliberto A, Kar S, Novak B, Tyson JJ. Antagonism and bistability in protein interaction networks. J Theor Biol 2008;250(1):209–18.

[70] Zhang L, Athale CA, Deisboeck TS. Development of a three-dimensional multiscale agent-based tumor model: simulating gene-protein interaction profiles, cell phenotypes and multicellular patterns in brain cancer. J Theor Biol 2007;244(1):96–107.

[71] Andersson J, Simpson DM, Qi MS, Wang YM, Elion EA. Differential input by Ste5 scaffold and Msg5 phosphatase route a MAPK cascade to multiple outcomes. EMBO J 2004;23 (13):2564–76.

[72] Elion EA. The Ste5p scaffold. J Cell Sci 2001;114(22):3967–78.

[73] Errede B, Cade RM, Yashar BM, Kamada Y, Levin DE, Irie K, et al. Dynamics and organization of Map kinase signal pathways. Mol Reprod Dev 1995;42(4):477–85.

[74] Treisman R. Regulation of transcription by MAP kinase cascades. Curr Opin Cell Biol 1996;8 (2):205–15.

[75] Mapes J, Ota IM. Nbp2 targets the Ptc1-type 2C Ser/Thr phosphatase to the HOG MAPK pathway. EMBO J 2004;23(2):302–11.

[76] Remenyi A, Good MC, Bhattacharyya RP, Lim WA. The role of docking interactions in mediating signaling input, output, and discrimination in the yeast MAPK network. Mol Cell 2005;20 (6):951–62.

[77] Pryciak PM, Hartwell LH. AKR1 encodes a candidate effector of the G beta gamma complex in the *Saccharomyces cerevisiae* pheromone response pathway and contributes to control of both cell shape and signal transduction. Mol Cell Biol 1996;16(6):2614–26.

[78] Hirano H, Tanaka K, Ozaki K, Imamura H, Kohno H, Hihara T, et al. ROM7/BEM4 encodes a novel protein that interacts with the Rho1p small GTP-binding protein in *Saccharomyces cerevisiae*. Mol Cell Biol 1996;16(8):4396–403.

[79] Brown JL, Jaquenoud M, Gulli MP, Chant J, Peter M. Novel Cdc42-binding proteins Gic1 and Gic2 control cell polarity in yeast. Genes Dev 1997;11(22):2972–82.

[80] Xue Y, Batlle M, Hirsch JP. GPR1 encodes a putative G protein-coupled receptor that associates with the Gpa2p G(alpha) subunit and functions in a Ras-independent pathway. EMBO J 1998;17 (7):1996–2007.

[81] Druey KM, Blumer KJ, Kang VH, Kehrl JH. Inhibition of C-protein-mediated MAP kinase activation by a new mammalian gene family. Nature 1996;379(6567):742–6.

[82] Wojda I, Alonso-Monge R, Bebelman JP, Mager WH, Siderius M. Response to high osmotic conditions and elevated temperature in *Saccharomyces cerevisiae* is controlled by intracellular glycerol and involves coordinate activity of MAP kinase pathways. Microbiology 2003;149: 1193–204.

[83] Stevenson BJ, Ferguson B, Devirgilio C, Bi E, Pringle JR, Ammerer G, et al. Mutation of RGA1, which encodes a putative GTPase-activating protein for the polarity-establishment protein CDC42P, activates the pheromone-response pathway in the yeast *Saccharomyces cerevisiae*. Genes Dev 1995;9(23):2949–63.

[84] Kao LR, Peterson J, Ji R, Bender L, Bender A. Interactions between the ankyrin repeat-containing protein Akr1p and the pheromone response pathway in *Saccharomyces cerevisiae*. Mol Cell Biol 1996;16(1):168–78.

[85] Ogas J, Andrews BJ, Herskowitz I. Transcriptional activation of CLN1, CLN2, and a putative new G1-cyclin (HCS26) by SWI4, a positive regulator of G1-specific transcription. Cell 1991;66 (5):1015–26.

[86] Strickfaden SC, Winters MJ, Ben-Ari G, Lamson RE, Tyers M, Pryciak PM. A mechanism for cell-cycle regulation of MAP kinase signaling in a yeast differentiation pathway. Cell 2007;128 (3):519–31.
[87] Basmaji F, Martin-Yken H, Durand F, Dagkessamanskaia A, Pichereaux C, Rossignol M, et al. The 'interactome' of the Knr4/Smi1, a protein implicated in coordinating cell wall synthesis with bud emergence in *Saccharomyces cerevisiae*. Mol Genet Genomics 2006;275(3):217–30.
[88] Fujiwara T, Tanaka K, Mino A, Kikyo M, Takahashi K, Shimizu K, et al. Rho1p-Bni1p-Spa2p interactions: implication in localization of Bni1p at the bud site and regulation of the actin cytoskeleton in *Saccharomyces cerevisiae*. Mol Biol Cell 1998;9(5):1221–33.
[89] Imamura H, Tanaka K, Hihara T, Umikawa M, Kamei T, Takahashi K, et al. Bni1p and Bnr1p: downstream targets of the Rho family small G-proteins which interact with profilin and regulate actin cytoskeleton in *Saccharomyces cerevisiae*. EMBO J 1997;16(10):2745–55.
[90] Davenport KR, Sohaskey M, Kamada Y, Levin DE, Gustin MC. A 2nd osmosensing signal-transduction pathway in yeast—hypotonic shock activates the PKC1 protein kinase-regulated cell integrity pathway. J Biol Chem 1995;270(50):30157–61.
[91] Dard N, Peter M. Scaffold proteins in MAP kinase signaling: more than simple passive activating platforms. Bioessays 2006;28(2):146–56.
[92] Zhu H, Klemic JF, Chang S, Bertone P, Casamayor A, Klemic KG, et al. Analysis of yeast protein kinases using protein chips. Nat Genet 2000;26(3):283–9.
[93] Green R, Lesage G, Sdicu AM, Menard P, Bussey H. A synthetic analysis of the *Saccharomyces cerevisiae* stress sensor Mid2p, and identification of a Mid2p-interacting protein, Zeo1p, that modulates the PKC1-MPK1 cell integrity pathway. Microbiology 2003;149:2487–99.
[94] Klipp E, Nordlander B, Kruger R, Gennemark P, Hohmann S. Integrative model of the response of yeast to osmotic shock. Nat Biotechnol 2005;23(8):975–82.
[95] Gat-Viks I, Shamir R. Refinement and expansion of signaling pathways: the osmotic response network in yeast. Genome Res 2007;17(3):358–67.
[96] Estruch F. Stress-controlled transcription factors, stress-induced genes and stress tolerance in budding yeast. FEMS Microbiol Rev 2000;24(4):469–86.
[97] Hersen P, McClean MN, Mahadevan L, Ramanathan S. Signal processing by the HOG MAP kinase pathway. Proc Natl Acad Sci USA 2008;105(20):7165–70.
[98] O'Rourke SM, Herskowitz I, O'Shea EK. Yeast go the whole HOG for the hyperosmotic response. Trends Genet 2002;18(8):405–12.
[99] Zhang WH, Chen BS, Tseng CS. Robust H-infinity filtering for nonlinear stochastic systems. IEEE Trans Signal Process 2005;53(2):589–98.
[100] Miller AJ. Subset selection in regression. 2nd ed. Boca Raton, FL: Chapman & Hall/CRC; 2002.
[101] Pfannschmidt T, Nilsson A, Allen JF. Photosynthetic control of chloroplast gene expression. Nature 1999;397(6720):625–8.
[102] Walters RG. Towards an understanding of photosynthetic acclimation. J Exp Bot 2005;56 (411):435–47.
[103] Wagner R, Dietzel L, Braeutigam K, Fischer W, Pfannschmidt T. The long-term response to fluctuating light quality is an important and distinct light acclimation mechanism that supports survival of *Arabidopsis thaliana* under low light conditions. Planta 2008;228(4):573–87.
[104] Pfannschmidt T, Braeutigam K, Wagner R, Dietzel L, Schroeter Y, Steiner S, et al. Potential regulation of gene expression in photosynthetic cells by redox and energy state: approaches towards better understanding. Ann Bot (Lond) 2009;103(4):599–607.
[105] Fey V, Wagner R, Brautigam K, Pfannschmidt T. Photosynthetic redox control of nuclear gene expression. J Exp Bot 2005;56(416):1491–8.
[106] Murakami A, Kim SJ, Fujita Y. Changes in photosystem stoichiometry in response to environmental conditions for cell growth observed with the cyanophyte Synechocystis PCC 6714. Plant Cell Physiol 1997;38(4):392–7.
[107] Pfannschmidt T, Schutze K, Fey V, Sherameti I, Oelmuller R. Chloroplast redox control of nuclear gene expression—a new class of plastid signals in interorganellar communication. Antioxid Redox Signal 2003;5(1):95–101.

[108] DeRisi JL, Iyer VR, Brown PO. Exploring the metabolic and genetic control of gene expression on a genomic scale. Science 1997;278(5338):680–6.
[109] Schena M, Shalon D, Davis RW, Brown PO. Quantitative monitoring of gene-expression patterns with a complementary-DNA microarray. Science 1995;270(5235):467–70.
[110] Iyer VR, Horak CE, Scafe CS, Botstein D, Snyder M, Brown PO. Genomic binding sites of the yeast cell-cycle transcription factors SBF and MBF. Nature 2001;409(6819):533–8.
[111] Ren B, Robert F, Wyrick JJ, Aparicio O, Jennings EG, Simon I, et al. Genome-wide location and function of DNA binding proteins. Science 2000;290(5500):2306–9.
[112] Yeung MKS, Tegner J, Collins JJ. Reverse engineering gene networks using singular value decomposition and robust regression. Proc Natl Acad Sci USA 2002;99(9):6163–8.
[113] Yu J, Smith VA, Wang PP, Hartemink AJ, Jarvis ED. Advances to Bayesian network inference for generating causal networks from observational biological data. Bioinformatics 2004;20 (18):3594–603.
[114] Zou M, Conzen SD. A new dynamic Bayesian network (DBN) approach for identifying gene regulatory networks from time course microarray data. Bioinformatics 2005;21(1):71–9.
[115] Geier F, Timmer J, Fleck C. Reconstructing gene-regulatory networks from time series, knockout data, and prior knowledge. BMC Syst Biol 2007;1:11.
[116] Needham CJ, Manfield IW, Bulpitt AJ, Gilmartin PM, Westhead DR. From gene expression to gene regulatory networks in *Arabidopsis thaliana*. BMC Syst Biol 2009;3:85.
[117] Jiao Q, Yang Z, Huang J. Construction of a gene regulatory network for *Arabidopsis* based on metabolic pathway data. Chinese Sci Bull 2010;55(2):158–62.
[118] Chang W-C, Lee T-Y, Huang H-D, Huang H-Y, Pan R-L. PlantPAN: Plant promoter analysis navigator, for identifying combinatorial cis-regulatory elements with distance constraint in plant gene groups. BMC Genomics 2008;9:561.
[119] Chen B-S, Yang S-K, Lan C-Y, Chuang Y-J. A systems biology approach to construct the gene regulatory network of systemic inflammation via microarray and databases mining. BMC Med Genomics 2008;1:46.
[120] Braeutigam K, Dietzel L, Kleine T, Stroeher E, Wormuth D, Dietz K-J, et al. Dynamic plastid redox signals integrate gene expression and metabolism to induce distinct metabolic states in photosynthetic acclimation in *Arabidopsis*. Plant Cell 2009;21(9):2715–32.
[121] Chu L-H, Chen B-S. Construction of a cancer-perturbed protein-protein interaction network for discovery of apoptosis drug targets. BMC Syst Biol 2008;2:56.
[122] Wang Y-C, Chen B-S. Integrated cellular network of transcription regulations and protein-protein interactions. BMC Syst Biol 2010;4:20.
[123] Pesaresi P, Hertle A, Pribil M, Kleine T, Wagner R, Strissel H, et al. Arabidopsis STN7 kinase provides a link between short- and long-term photosynthetic acclimation. Plant Cell 2009;21 (8):2402–23.
[124] Shannon P, Markiel A, Ozier O, Baliga NS, Wang JT, Ramage D, et al. Cytoscape: a software environment for integrated models of biomolecular interaction networks. Genome Res 2003;13 (11):2498–504.
[125] Albert R. Scale-free networks in cell biology. J Cell Sci 2005;118(21):4947–57.
[126] Argyros RD, Mathews DE, Chiang Y-H, Palmer CM, Thibault DM, Etheridge N, et al. Type B response regulators of *Arabidopsis* play key roles in cytokinin signaling and plant development. Plant Cell 2008;20(8):2102–16.
[127] Henriksson E, Olsson ASB, Johannesson H, Johansson H, Hanson J, Engstrom P, et al. Homeodomain leucine zipper class I genes in *Arabidopsis*. Expression patterns and phylogenetic relationships. Plant Physiol 2005;139(1):509–18.
[128] Carabelli M, Sessa G, Baima S, Morelli G, Ruberti I. The *Arabidopsis* ATHB-2 and ATHB-4 genes are strongly induced by far-red-rich light. Plant J 1993;4(3):469–79.
[129] Tepperman JM, Hwang Y-S, Quail PH. phyA dominates in transduction of red-light signals to rapidly responding genes at the initiation of *Arabidopsis* seedling de-etiolation. Plant J 2006;48(5):728–42.
[130] Franklin KA, Whitelam GC. Light-quality regulation of freezing tolerance in *Arabidopsis thaliana*. Nat Genet 2007;39(11):1410–13.

[131] Khanna R, Shen Y, Toledo-Ortiz G, Kikis EA, Johannesson H, Hwang YS, et al. Functional profiling reveals that only a small number of phytochrome-regulated early-response genes in *Arabidopsis* are necessary for optimal deetiolation. Plant Cell 2006;18(9):2157–71.

[132] Fey V, Wagner R, Brautigam K, Wirtz M, Hell R, Dietzmann A, et al. Retrograde plastid redox signals in the expression of nuclear genes for chloroplast proteins of *Arabidopsis thaliana*. J Biol Chem 2005;280(17):5318.

[133] Schmid M, Davison TS, Henz SR, Pape UJ, Demar M, Vingron M, et al. A gene expression map of *Arabidopsis thaliana* development. Nat Genet 2005;37(5):501–6.

[134] Brenner WG, Romanov GA, Kollmer I, Burkle L, Schmulling T. Immediate-early and delayed cytokinin response genes of *Arabidopsis thaliana* identified by genome-wide expression profiling reveal novel cytokinin-sensitive processes and suggest cytokinin action through transcriptional cascades. Plant J 2005;44(2):314–33.

[135] Schwechheimer C, Deng XW. The COP/DET/FUS proteins—regulators of eukaryotic growth and development. Semin Cell Dev Biol 2000;11(6):495–503.

[136] Aoyama T, Dong CH, Wu Y, Carabelli M, Sessa G, Ruberti I, et al. Ectopic expression of the *Arabidopsis* transcriptional activator ATHB-1 alters leaf cell fate in tobacco. Plant Cell 1995;7(11):1773–85.

[137] Adamiec M, Drath M, Jackowski G. Redox state of plastoquinone pool regulates expression of *Arabidopsis thaliana* genes in response to elevated irradiance. Acta Biochim Pol 2008;55(1):161–73.

[138] Kurth J, Varotto C, Pesaresi P, Biehl A, Richly E, Salamini F, et al. Gene-sequence-tag expression analyses of 1,800 genes related to chloroplast functions. Planta 2002;215(1):101–9.

[139] Richly E, Dietzmann A, Biehl A, Kurth J, Laloi C, Apel K, et al. Covariations in the nuclear chloroplast transcriptome reveal a regulatory master-switch. EMBO Rep 2003;4(5):491–8.

[140] Dietzel L, Braeutigam K, Pfannschmidt T. Photosynthetic acclimation: state transitions and adjustment of photosystem stoichiometry–functional relationships between short-term and long-term light quality acclimation in plants. FEBS J 2008;275(6):1080–8.

[141] Wang Y, Wang G, Yang B, Tao H, Yang JY, Deng Y, et al. Reconstruct gene regulatory network using slice pattern model. BMC Genomics 2009;10:S2.

[142] Yip KY, Alexander RP, Yan K-K, Gerstein M. Improved reconstruction of in silico gene regulatory networks by integrating knockout and perturbation data. PLoS One 2010;5(1):e8121.

[143] Newman JRS, Ghaemmaghami S, Ihmels J, Breslow DK, Noble M, DeRisi JL, et al. Single-cell proteomic analysis of *S. cerevisiae* reveals the architecture of biological noise. Nature 2006;441(7095):840–6.

[144] Orntoft TF, Thykjaer T, Waldman FM, Wolf H, Celis JE. Genome-wide study of gene copy numbers, transcripts, and protein levels in pairs of non-invasive and invasive human transitional cell carcinomas. Mol Cell Proteomics 2002;1(1):37–45.

[145] Bonardi V, Pesaresi P, Becker T, Schleiff E, Wagner R, Pfannschmidt T, et al. Photosystem II core phosphorylation and photosynthetic acclimation require two different protein kinases. Nature 2005;437(7062):1179–82.

[146] Blair JD, Price EM. Illuminating potential technical artifacts of DNA-methylation array probes. Am J Hum Genet 2012;91(4):760–2.

[147] Chatr-aryamontri A, Breitkreutz BJ, Heinicke S, Boucher L, Winter A, Stark C, et al. The BioGRID interaction database: 2013 update. Nucleic Acids Res 2013;41(D1):D816–23.

[148] Wang Z, Gerstein M, Snyder M. RNA-Seq: a revolutionary tool for transcriptomics. Nat Rev Genet 2009;10(1):57–63.

[149] Broen JCA, Radstake TRDJ, Rossato M. The role of genetics and epigenetics in the pathogenesis of systemic sclerosis. Nat Rev Rheumatol 2014;10(11):671–81.

[150] Ke XY, Cortina-Borja M, Silva BC, Lowe R, Rakyan V, Balding D. Integrated analysis of genome-wide genetic and epigenetic association data for identification of disease mechanisms. Epigenetics 2013;8(11):1236–44.

[151] Yeger-Lotem E, Sattath S, Kashtan N, Itzkovitz S, Milo R, Pinter RY, et al. Network motifs in integrated cellular networks of transcription-regulation and protein-protein interaction. Proc Natl Acad Sci USA 2004;101(16):5934–9.

[152] Lin CC, Chen YJ, Chen CY, Oyang YJ, Juan HF, Huang HC. Crosstalk between transcription factors and microRNAs in human protein interaction network. BMC Syst Biol 2012;6:18.
[153] Herrgard MJ, Lee BS, Portnoy V, Palsson BO. Integrated analysis of regulatory and metabolic networks reveals novel regulatory mechanisms in *Saccharomyces cerevisiae*. Genome Res 2006;16 (5):627−35.
[154] Przytycka TM, Kim YA. Network integration meets network dynamics. BMC Biol 2010;8:48.
[155] Wang YC, Chen BS. Integrated cellular network of transcription regulations and protein-protein interactions. BMC Syst Biol 2010;4:20.
[156] Hsu HJ, Lee CF, Kaunas R. A dynamic stochastic model of frequency-dependent stress fiber alignment induced by cyclic stretch. PLoS One 2009;4(3):e4853.
[157] Chen LZ, Nguang SK, Chen X. Modelling and optimization of biotechnological processes: artificial intelligence approaches. Berlin: Springer; 2006.
[158] Chen LZ, Nguang SK, Chen XD. On-line identification and optimization of feed rate profiles for high productivity fed-batch culture of hybridoma cells using genetic algorithms. ISA Trans 2002;41(4):409−19.
[159] Chuang CH, Lin CL, Chang YC, Jennawasin T, Chen PK. Design of synthetic biological logic circuits based on evolutionary algorithm. IET Syst Biol 2013;7(4):89−105.
[160] Huang CH, Jan HY, Lin CL, Lee CS. System identification: DNA computing approach. ISA Trans 2009;48(3):254−63.
[161] Guintivano J, Brown T, Newcomer A, Jones M, Cox O, Maher BS, et al. Identification and replication of a combined epigenetic and genetic biomarker predicting suicide and suicidal behaviors. Am J Psychiatry 2014;171(12):1287−96.
[162] Mohammadi P, Desfarges S, Bartha I, Joos B, Zangger N, Munoz M, et al. 24 Hours in the Life of HIV-1 in a T Cell Line. PLoS Pathog 2013;9(1):e1003161.
[163] Kim SY, Park C, Kim HJ, Park J, Hwang J, Kim JI, et al. Deregulation of immune response genes in patients with Epstein-Barr virus-associated gastric cancer and outcomes. Gastroenterology 2015;148(1):137−574.
[164] Trapnell C, Pachter L, Salzberg SL. TopHat: discovering splice junctions with RNA-Seq. Bioinformatics 2009;25(9):1105−11.
[165] Stark C, Breitkreutz BJ, Reguly T, Boucher L, Breitkreutz A, Tyers M. BioGRID: a general repository for interaction datasets. Nucleic Acids Res 2006;34:D535−9.
[166] Bovolenta LA, Acencio ML, Lemke N. HTRIdb: an open-access database for experimentally verified human transcriptional regulation interactions. BMC Genomics 2012;13:405.
[167] Mootha VK, Lindgren CM, Eriksson KF, Subramanian A, Sihag S, Lehar J, et al. PGC-1 alpha-responsive genes involved in oxidative phosphorylation are coordinately downregulated in human diabetes. Nat Genet 2003;34(3):267−73.
[168] Subramanian A, Tamayo P, Mootha VK, Mukherjee S, Ebert BL, Gillette MA, et al. Gene set enrichment analysis: a knowledge-based approach for interpreting genome-wide expression profiles. Proc Natl Acad Sci USA 2005;102(43):15545−50.
[169] Zheng GY, Tu K, Yang Q, Xiong Y, Wei CC, Xie L, et al. ITFP: an integrated platform of mammalian transcription factors. Bioinformatics 2008;24(20):2416−17.
[170] Griffiths-Jones S. The microRNA Registry. Nucleic Acids Res 2004;32:D109−11.
[171] Griffiths-Jones S, Saini HK, van Dongen S, Enright AJ. miRBase: tools for microRNA genomics. Nucleic Acids Res 2008;36:D154−8.
[172] Kertesz M, Iovino N, Unnerstall U, Gaul U, Segal E. The role of site accessibility in microRNA target recognition. Nat Genet 2007;39(10):1278−84.
[173] Kozomara A, Griffiths-Jones S. miRBase: integrating microRNA annotation and deep-sequencing data. Nucleic Acids Res 2011;39:D152−7.
[174] Kozomara A, Griffiths-Jones S. miRBase: annotating high confidence microRNAs using deep sequencing data. Nucleic Acids Res 2014;42(D1):D68−73.
[175] Agarwal V, Bell GW, Nam JW, Bartel DP. Predicting effective microRNA target sites in mammalian mRNAs. Elife 2015;4.

[176] Friedman RC, Farh KKH, Burge CB, Bartel DP. Most mammalian mRNAs are conserved targets of microRNAs. Genome Res 2009;19(1):92–105.
[177] Garcia DM, Baek D, Shin C, Bell GW, Grimson A, Bartel DP. Weak seed-pairing stability and high target-site abundance decrease the proficiency of lsy-6 and other microRNAs. Nat Struct Mol Biol 2011;18(10):1139–46.
[178] Grimson A, Farh KK, Johnston WK, Garrett-Engele P, Lim LP, Bartel DP. MicroRNA targeting specificity in mammals: determinants beyond seed pairing. Mol Cell 2007;27(1):91–105.
[179] Cui QH, Ma Y, Jaramillo M, Bari H, Awan A, Yang S, et al. A map of human cancer signaling. Mol Syst Biol 2007;3:152.
[180] Huang YZ, Sun JJ, Zhang LZ, Li CJ, Womack JE, Li ZJ, et al. Genome-wide DNA methylation profiles and their relationships with mRNA and the microRNA transcriptome in bovine muscle tissue (Bos taurine). Sci Rep 2014;4:1–17.
[181] Muckstein U, Tafer H, Hackermuller J, Bernhart SH, Stadler PF, Hofacker IL. Thermodynamics of RNA-RNA binding. Bioinformatics 2006;22(10):1177–82.
[182] Nandi A, Vaz C, Bhattacharya A, Ramaswamy R. miRNA-regulated dynamics in circadian oscillator models. BMC Syst Biol 2009;3:45.
[183] Ferreira AEN, Freire AMJP, Voit EO. A quantitative model of the generation of N-epsilon-(carboxymethyl)lysine in the Maillard reaction between collagen and glucose. Biochem J 2003;376:109–21.
[184] Voit EO, Ferreira AEN. Computational analysis of biochemical systems: a practical guide for biochemists and molecular biologists. Cambridge; New York: Cambridge University Press; 2000.
[185] Akaike H. New look at statistical-model identification. IEEE Trans Automatic Control 1974;Ac19 (6):716–23.
[186] Bar-Joseph Z, Gerber G, Gifford DK, Jaakkola TS, Simon I. A new approach to analyzing gene expression time series data. Proceedings of the sixth annual international conference on computational biology. Washington, DC: ACM; 2002. p. 39–48.
[187] De Boor C. with 32 figures, Rev. ed. A practical guide to splines. New York, NY: Springer; 2001
[188] Coleman TF, Hulbert LA. A direct active set algorithm for large sparse quadratic programs with simple bounds. Math Program 1989;45(3):373–406.
[189] Gill PE, Murray W, Wright MH. Practical optimization. London; New York: Academic Press; 1981.
[190] Lu P, Vogel C, Wang R, Yao X, Marcotte EM. Absolute protein expression profiling estimates the relative contributions of transcriptional and translational regulation. Nat Biotechnol 2007;25 (1):117–24.
[191] Seber GAF, Lee AJ. Linear regression analysis. 2nd ed. Hoboken, NJ: Wiley-Interscience; 2003.
[192] Abraham RT. Cell cycle checkpoint signaling through the ATM and ATR kinases. Genes Dev 2001;15(17):2177–96.
[193] McMahon M. RAF translocations expand cancer targets. Nat Med 2010;16(7):749–50.
[194] Wilhelm E, Doyle MC, Nzaramba I, Magdzinski A, Dumais N, Bell B. CTGC motifs within the HIV core promoter specify Tat-responsive pre-initiation complexes. Retrovirology 2012;9:62.
[195] Sharma D, Bhattacharya J. Evolutionary constraints acting on DDX3X protein potentially interferes with rev-mediated nuclear export of HIV-1 RNA. PLoS One 2010;5(3):e9613.
[196] Kang M-S, Kieff E. Epstein–Barr virus latent genes. Exp Mol Med 2015;47(1):e131.
[197] Liu L, Ouyang M, Rao JN, Zou T, Xiao L, Chung HK, et al. Competition between RNA-binding proteins CELF1 and HuR modulates MYC translation and intestinal epithelium renewal. Mol Biol Cell 2015;26(10):1797–810.
[198] Xie LM, Zhang ZW, Tan ZQ, He RF, Zeng X, Xie YJ, et al. microRNA-124 inhibits proliferation and induces apoptosis by directly repressing EZH2 in gastric cancer. Mol Cell Biochem 2014;392(1–2):153–9.
[199] Zhao XN, Li XW, Yuan HQ. microRNAs in gastric cancer invasion and metastasis. Front Biosci-Landmark 2013;18:803–10.
[200] Chang S, Gao L, Yang Y, Tong D, Guo B, Liu L, et al. miR-145 mediates the antiproliferative and gene regulatory effects of vitamin D3 by directly targeting E2F3 in gastric cancer cells. Oncotarget 2015;6(10):7675–85.

[201] Zheng LD, Pu JR, Qi T, Qi M, Li D, Xiang X, et al. miRNA-145 Targets v-ets erythroblastosis virus e26 oncogene homolog 1 to suppress the invasion, metastasis, and angiogenesis of gastric cancer cells. Mol Cancer Res 2013;11(2):182–93.
[202] Cao DH, Jiang J, Tsukamoto T, Liu RM, Ma L, Jia ZF, et al. Canolol inhibits gastric tumors initiation and progression through COX-2/PGE2 pathway in K19-C2mE transgenic mice. PLoS One 2015;10(3):e0120938.
[203] Roberts TL, Ho U, Luff J, Lee CS, Apte SH, MacDonald KPA, et al. Smg1 haploinsufficiency predisposes to tumor formation and inflammation. Proc Natl Acad Sci USA 2013;110(4):E285–94.
[204] Ishida J. Clinicopathological study of the amount of LUCA15 expressed in gastric cancer tissue. Hokkaido Igaku Zasshi 2006;81(5):451–9.
[205] Peng J, Valeshabad AK, Li QF, Wang Y. Differential expression of RBM5 and KRAS in pancreatic ductal adenocarcinoma and their association with clinicopathological features. Oncol Lett 2013;5(3):1000–4.
[206] Zou Q, Jin J, Hu HB, Li HYS, Romano S, Xiao YC, et al. USP15 stabilizes MDM2 to mediate cancer-cell survival and inhibit antitumor T cell responses. Nat Immunol 2014;15(6):562–70.
[207] Guidoboni M, Zancai P, Cariati R, Rizzo S, Dal Col J, Pavan A, et al. Retinoic acid inhibits the proliferative response induced by CD40 activation and interleukin-4 in mantle cell lymphoma. Cancer Res 2005;65(2):587–95.
[208] Zhou S, Ye WG, Ren J, Shao QJ, Qi YH, Liang J, et al. MicroRNA-381 increases radiosensitivity in esophageal squamous cell carcinoma. Am J Cancer Res 2015;5(1):267–77.
[209] Mani A, Gelmann EP. The ubiquitin-proteasome pathway and its role in cancer. J Clin Oncol 2005;23(21):4776–89.
[210] Palmieri C, Gojis O, Rudraraju B, Stamp-Vincent C, Wilson D, Langdon S, et al. Expression of steroid receptor coactivator 3 in ovarian epithelial cancer is a poor prognostic factor and a marker for platinum resistance. Br J Cancer 2013;108(10):2039–44.
[211] Nishida T, Terashima M, Fukami K, Yamada Y. PIASy controls ubiquitination-dependent proteasomal degradation of Ets-1. Biochem J 2007;405:481–8.
[212] Lacaud G, Gore L, Kennedy M, Kouskoff V, Kingsley P, Hogan C, et al. Runx1 is essential for hematopoietic commitment at the hemangioblast stage of development in vitro. Blood 2002;100(2):458–66.
[213] Iwanami A, Gini B, Zanca C, Matsutani T, Assuncao A, Nael A, et al. PML mediates glioblastoma resistance to mammalian target of rapamycin (mTOR)-targeted therapies. Proc Natl Acad Sci USA 2013;110(11):4339–44.
[214] Jemal A, Bray F, Center MM, Ferlay J, Ward E, Forman D. Global cancer statistics. CA Cancer J Clin 2011;61(2):69–90.
[215] Itahana Y, Neo SH, Itahana K. miR-141, a new player, joins the senescence orchestra. Cell Cycle 2013;12(23):3586–7.
[216] Watanabe S, Watanabe K, Akimov V, Bartkova J, Blagoev B, Lukas J, et al. JMJD1C demethylates MDC1 to regulate the RNF8 and BRCA1-mediated chromatin response to DNA breaks. Nat Struct Mol Biol 2013;20(12):1425–33.
[217] Zheng TS, Wang JB, Song XA, Meng XZ, Pan SH, Jiang HC, et al. Nutlin-3 cooperates with doxorubicin to induce apoptosis of human hepatocellular carcinoma cells through p53 or p73 signaling pathways. J Cancer Res Clin Oncol 2010;136(10):1597–604.
[218] Lamy E, Herz C, Lutz-Bonengel S, Hertrampf A, Marton MR, Mersch-Sundermann V. The MAPK pathway signals telomerase modulation in response to isothiocyanate-induced DNA damage of human liver cancer cells. PLoS One 2013;8(1):e53240.
[219] Yang SF, Chang CW, Wei RJ, Shiue YL, Wang SN, Yeh YT. Involvement of DNA damage response pathways in hepatocellular carcinoma. Biomed Res Int 2014;2014:153867.
[220] Niemi NM, Lanning NJ, Westrate LM, MacKeigan JP. Downregulation of the mitochondrial phosphatase PTPMT1 is sufficient to promote cancer cell death. PLoS One 2013;8(1):e53803.
[221] Hatakeyama S. TRIM proteins and cancer. Nat Rev Cancer 2011;11(11):792–804.
[222] Li H, Sun L, Chen X, Xiong W, Hu D, Jie S. Microvesicle microRNA profiles and functional roles between chronic hepatitis B and hepatocellular carcinoma. Clin Trans Oncol 2014;16(3):315–21.

[223] Song RP, Song HW, Liang YJ, Yin DL, Zhang H, Zheng TS, et al. Reciprocal activation between ATPase Inhibitory Factor 1 and NF-kappa B drives hepatocellular carcinoma angiogenesis and metastasis. Hepatology 2014;60(5):1659–73.
[224] Shen RR, Zhou AY, Kim E, O'Connell JT, Hagerstrand D, Beroukhim R, et al. TRAF2 is an NF-kappa B-activating oncogene in epithelial cancers. Oncogene 2015;34(2):209–16.
[225] Brady CW. Liver disease in menopause. World J Gastroenterol 2015;21(25):7613–20.
[226] Tian Y, Xie X, Lin Y, Tan G, Zhong W. Androgen receptor in hepatocarcinogenesis: recent developments and perspectives (Review). Oncol Lett 2015;9(5):1983–8.
[227] Hayes AM, Qian SM, Yu LB, Boris-Lawrie K. Tat RNA silencing suppressor activity contributes to perturbation of lymphocyte miRNA by HIV-1. Retrovirology 2011;8:36.
[228] Yao H, Ma R, Yang L, Hu G, Chen XF, Duan M, et al. MiR-9 promotes microglial activation by targeting MCPIP1. Nat Commun 2014;5:4386.
[229] Taylor HE, Linde ME, Khatua AK, Popik W, Hildreth JEK. Sterol regulatory element-binding protein 2 couples HIV-1 transcription to cholesterol homeostasis and t cell activation. J Virol 2011;85(15):7699–709.
[230] Oh Y, Chung KC. Small Ubiquitin-like Modifier (SUMO) modification of zinc finger protein 131 potentiates its negative effect on estrogen signaling. J Biol Chem 2012;287(21):17517–29.
[231] Szotek EL, Narasipura SD, Al-Harthi L. 17 beta-Estradiol inhibits HIV-1 by inducing a complex formation between beta-catenin and estrogen receptor alpha on the HIV promoter to suppress HIV transcription. Virology 2013;443(2):375–83.
[232] Miller-Jensen K, Skupsky R, Shah PS, Arkin AP, Schaffer DV. Genetic selection for context-dependent stochastic phenotypes: Sp1 and TATA mutations increase phenotypic noise in HIV-1 gene expression. PLoS Comput Biol 2013;9(7):e1003135.
[233] Mahboobi SH, Javanpour AA, Mofrad MRK. The interaction of RNA helicase DDX3 with HIV-1 Rev-CRM1-RanGTP complex during the HIV replication cycle. PLoS One 2015;10(2): e0112969.
[234] Yedavalli VSRK, Jeang KT. Rev-ing up post-transcriptional HIV-1 RNA expression. RNA Biol 2011;8(2):195–9.
[235] Zahoor MA, Xue GG, Sato H, Murakami T, Takeshima S, Aida Y. HIV-1 Vpr induces interferon-stimulated genes in human monocyte-derived macrophages. PLoS One 2014;9(8): e106418.
[236] Zhang H, Pomerantz RJ, Dornadula G, Sun Y. Human immunodeficiency virus type 1 Vif protein is an integral component of an mRNP complex of viral RNA and could be involved in the viral RNA folding and packaging process. J Virol 2000;74(18):8252–61.
[237] Kanak M, Alseiari M, Balasubramanian P, Addanki K, Aggarwal M, Noorali S, et al. Triplex-forming MicroRNAs form stable complexes with HIV-1 provirus and inhibit its replication. Appl Immunohistochem Mol Morphol 2010;18(6):532–45.
[238] Panaro MA, Mitolo V, Cianciulli A, Cavallo P, Mitolo CI, Acquafredda A. The HIV-1 rev binding family of proteins: the dog proteins as a study model. Endocr Metab Immune Disord Drug Targets 2008;8(1):30–46.
[239] Kelada S, Sethupathy P, Okoye IS, Kistasis E, Czieso S, White SD, et al. miR-182 and miR-10a are key regulators of Treg specialisation and stability during schistosome and leishmania-associated inflammation. PLoS Pathog 2013;9(6):e1003451.
[240] Tong L, Lin LX, Wu S, Guo ZW, Wang TY, Qin Y, et al. MiR-10a(star) up-regulates coxsackievirus B3 biosynthesis by targeting the 3D-coding sequence. Nucleic Acids Res 2013;41 (6):3760–71.
[241] Abdel-Mohsen M, Deng XT, Danesh A, Liegler T, Jacobs ES, Rauch A, et al. Role of MicroRNA modulation in the Interferon-alpha/Ribavirin suppression of HIV-1 in vivo. PLoS One 2014;9(10):e109220.
[242] Grigorov B, Arcanger F, Roingeard P, Darlix JL, Muriaux D. Assembly of infectious HIV-1 in human epithelial and T-lymphoblastic cell lines. J Mol Biol 2006;359(4):848–62.
[243] De Clercq E. Current lead natural products for the chemotherapy of human immunodeficiency virus (HIV) infection. Med Res Rev 2000;20(5):323–49.

[244] Luzzi A, Morettini F, Gazaneo S, Mundo L, Onnis A, Mannucci S, et al. HIV-1 Tat induces DNMT over-expression through microRNA dysregulation in HIV-related non Hodgkin lymphomas. Infect Agent Cancer 2014;9:41.
[245] Wei XF, Xiang TX, Ren GS, Tan C, Liu R, Xu X, et al. miR-101 is down-regulated by the hepatitis B virus x protein and induces aberrant DNA methylation by targeting DNA methyltransferase 3**A**. Cell Signal 2013;25(2):439–46.
[246] Huang JF, Wang Y, Guo YJ, Sun SH. Down-regulated MicroRNA-152 induces aberrant DNA methylation in Hepatitis B virus-related hepatocellular carcinoma by targeting DNA methyltransferase 1. Hepatology 2010;52(1):60–70.
[247] Casini A, Olivieri M, Vecchi L, Burrone OR, Cereseto A. Reduction of HIV-1 infectivity through endoplasmic reticulum-associated degradation-mediated Env depletion. J Virol 2015;89(5):2966–71.
[248] Wang E, Zaman N, Mcgee S, Milanese JS, Masoudi-Nejad A, O'Connor-McCourt M. Predictive genomics: a cancer hallmark network framework for predicting tumor clinical phenotypes using genome sequencing data. Semin Cancer Biol 2015;30:4–12.

CHAPTER 7

Big Offensive and Defensive Mechanisms in Systems Immunity From System Modeling and Big Data Mining

INTRODUCTION

The major role of an immune system is to defend the host against infections. Immunity has always been viewed as a defense system, in the sense that its genetic foundation is complex and based on a multitude of proteins in many pathways, which interact with each other to coordinate the mechanisms of defense against infection. Full-scale system models for the entire immune system, therefore, are not simple, but rely on the integration of many different components that deal with pathogenic organisms. In the future, integrative approaches will form the basis for advanced, quantitative, and qualitative types of systems biology using computation and modeling to understand the complex, dynamic mechanisms of entire cells and their functional, molecular modules. The immune system is likely to be high on this agenda. A large variety of experimental techniques are creating a sound scientific basis for systems biology, and many techniques are capable of generating big data at the levels of entire cells, tissues, organs, or organisms. The data obtained are more complete and will allow for the implementation of a new wave of large-scale computational analysis of immune systems. This contrasts with the situation from a decade ago, when lack of experimental data prevented the creation of many data-driven systems biology approaches.

To begin this chapter, in Section 7.1, we first introduce a systems biology approach to construct the gene regulatory network (GRN) of systemic inflammation via microarray data and database mining. Further, we examine susceptibility to infection and robustness of the inflammatory GRN by analyzing network hubs and weak ties in systemic inflammation. Inflammation is a hallmark of many human diseases, and elucidating the cellular mechanisms underlying systemic inflammation has long been an important subject of basic and clinical research. When primary pathogenic events remain unclear due to their immense complexity, construction and analysis of the GRN of inflammation become the best way to understand the detrimental effects of disease.

Big Mechanisms in Systems Biology
DOI: http://dx.doi.org/10.1016/B978-0-12-809479-2.00007-X

Then, we identify the infection and defense-related genes via a dynamic host–pathogen interaction network using a *Candida albicans*-zebrafish inflection model in Section 7.2. Based on two-sided, interactive time profile microarray data of *C. albicans* and zebrafish during infection, the infection-related protein–protein interaction (PPI) network of these two species and the intercellular PPI networks between a host and pathogen are simultaneously constructed. This integrated network is a dynamic interaction model, unifying intercellular invasion and cellular defense mechanisms during infection. The signal transduction pathways that regulate morphogenesis and hyphal growth of *C. albicans* are also investigated based on significant interactions found in the intercellular PPI network. Two cellular networks are also developed corresponding to different infection stages (adhesion and invasion), and then compared with each other to identify shared proteins. Further, important defense-related mechanisms in zebrafish can also be found using the same approach.

Finally, we investigate the shift in host–pathogen crosstalk network markers to reveal the molecular mechanisms of pathogenesis and defense in the HIV infection process by a systems biology approach via dynamic interaction modeling and big data mining in Section 7.3. With the help of simultaneous, two-sided time-course HIV-human high-throughput sequence data (next-generation sequencing, NGS), real-time polymerase chain reaction (RT-PCR) data, and other omics data, an interspecies protein–protein and miRNA interaction (PPMI) network is constructed for host and pathogen interactions in the HIV infection process. Principal interspecies PPMI networks are extracted from host–pathogen crosstalk networks at different stages in HIV infection. By comparing these crosstalk networks in HIV infection, host–pathogen crosstalk network markers at different infection stages, as well as in the entire infectious process, can be extracted. By further investigating this common core of host–pathogen crosstalk network markers, we can gain additional insight into the pathologic mechanism of the HIV virus and the defense mechanism of humans. This might offer new drug targets for effective Acquired Immune Deficiency Syndrome (AIDS) therapies.

7.1 A SYSTEMS BIOLOGY APPROACH TO CONSTRUCT THE GRN OF SYSTEMIC INFLAMMATION MECHANISMS VIA MICROARRAY AND DATABASES MINING

Recently, the employment of microarray technology has rapidly produced vast catalogs of gene expression activities. The immense amount of data highlights the need for a systematic tool to identify and analyze the underlying GRNs [1,2]. Several computational methods for the inference of transcriptional regulatory networks (TRNs) from experimental microarray data in *Saccharomyces cerevisiae* have been published [3,4]. The genome-wide transcriptional responses of inflammation are usually focused on the known functional interactions of the master switch proteins, such as Rel or NF-κB proteins [5–7]. The identification of NF-κB as a key player in the pathogenesis of inflammation suggests that NF-κB-targeted therapeutics might be effective in treating diseases like rheumatoid arthritis, which is a well-known disease where inflammatory response is causing the primary damage [8]. However, inflammation is usually a life-preserving response, as reflected by the increased risk of grave infections in people with genetic deficiencies in key components of the inflammatory signaling pathways [9].

Although inflammation is a hallmark of many human diseases [10,11], few studies have evaluated the genome-wide responses induced by systemic inflammation in humans. DNA microarray has allowed the semi-quantitative measurement of gene expression programming in great depth and on a broad scale. However, it is a challenge to overcome the difficulties of recognizing and evaluating relevant biological processes from vast quantities of experimental data. Recently, systems biology has gained much attention due to emerging experimental and computation methods [1,2]. Systems biology is the coordinated study of biological systems by: (1) investigating the components of networks and their interactions; (2) applying experimental high-throughput and whole-genome techniques; and (3) integrating computational methods with experimental efforts [12]. Therefore, it is more appealing to adapt a systems biology approach to study the mechanism of systemic inflammation via high-throughput transcriptomic studies of human disease. Such systematic approach can provide insights into the regulation of immune cell activities, tolerance of innate immune system, and the susceptibility of infection in human. Based on a structured network-based approach and a statistical likelihood method, a network-based analysis of systemic inflammation mechanism in human has been given to evaluate genome-wide transcriptional responses in the context of known functional relationships among proteins, small molecules, and phenotypes [10,11]. The genome-wide interaction network is probed to identify functional modules that are perturbed in response to endotoxin exposure. A dynamic Bayesian network (DBN) approach has also been developed to predict the GRNs from time-course expression data [13].

Gene expression is transcriptionally controlled by inducible transcription factors (TFs). The transcription factor NF-κB in particular is pivotal in the regulation of inflammation. For example, an unstimulated macrophage is kept under an inactivated condition, and its NF-κB is retained in the cytoplasm through interaction with inhibitory proteins known as IκB. Cell stimulation by bacterial endotoxin will trigger a signaling pathway which results in the degradation of IκB, leading to nuclear translocation of NF-κB and activation of the transcription of various pro-inflammatory cytokines [14] (IL1A, IL1B, TNFA, IL6, IL8, etc.). Many crosstalks among the signaling pathways are recognized. It is now known that the biological functions of IL1A and TNFA overlap and complement with each other [4,15].

Thus, blocking only one mediator may not effectively reduce the overall inflammatory responses. Both IL1B and TNFA produce effects at an early stage of inflammation and the use of their inhibitory reagents at the later stage may not be able to reverse the most damaging events that are initiated by them. As a result, IL1B and TNFA may not represent the best targets for intervention in systemic inflammatory response. In another study [16], TNFA and IL1 were shown to have positive feedback loops to TNFR and IL1R, respectively. On the other hand, the NF-κB also initiate the transcription of an inhibitory protein (A20) which can inactivate NF-κB by suppressive phosphorylation in IKK [17]. The other important receptors in the immune system, TLR family members (TLR2 and TLR4), which recognize pathogens by means of conserved structural features of the microbes such as LPS for Gram-negative bacteria, would involve in activating the MyD88/IRAK signaling cascade, which bifurcates and leads to NF-κB and c-Jun/ATF2/TCF activation [18].

Because microarray data contain vast cataloged patterns of dynamic expression of the activated genes, we need systematic tools to identify the interaction architecture and the dynamics of the underlying gene networks. Indeed, the system identification problem of the underlying dynamic gene networks falls naturally into the category of reverse engineering [13]; a complex genetic network underlies a mass set of gene expression data, and the task is to infer the connectivity of gene circuit through dynamic gene regulatory model [12]. Therefore, to understand complex gene networks requires the integration of microarray data and dynamic modeling by a systematic approach. The systematic approach has to include computational dynamic modeling coupled with microarray data, data mining, dynamic view of rapid responses and network structural view arising from high-throughput analysis of the interacting species [19]. To achieve this, a DBN method has been developed to predict GRNs from time series data [13]. However, this section has not combined with other network algorithms and knowledge-based databases. It carries two fundamental problems which greatly reduce the effectiveness of the DBN approach. The first problem is the relatively low accuracy of prediction inherently, and the second is the excessive computation time.

Since the identification of a perturbed biological networks under the effect of bacterial endotoxin is an important topic in basic and clinical research, it is imperative to conduct systematic analysis based on the expression profiles of microarray data. An approach of combining genome-wide expression analysis with a clustering method has been introduced to identify functional networks using a Genetic Regulatory Modules (GRAM) algorithm to provide biological insights into GRNs [20]. Because the clustering algorithms are employed to identify sets of coexpressed and potentially coregulated genes from gene expression data, it is more suitable to find a gene module as a set of coexpressed genes to which the same set of TFs will bind to their promoter regions. Therefore, it is not suitable to construct the TRNs as a dynamic model. It is hence essential to provide a new way to identify the perturbed biological networks. To achieve this, systems biology and computational biology methods will need to be employed to describe the biological mechanisms from a dynamic systems perspective [21,22].

In this section, a systems biology approach is proposed to achieve a gradual refinement of inflammatory regulatory network. A rough (candidate) GRN of inflammation mechanism is first constructed by information extracted from the Ensembl database (http://www.ensembl.org/index.html) and JASPAR (http://jaspar.genereg.net/) algorithms. We then build a dynamic regulatory model according to the rough gene network with consideration of time-delay between regulatory gene and target gene to describe the GRN. Based on the dynamic regulatory model and microarray data in Refs. [10,11], a maximum likelihood method is used to identify the regulatory parameters of upstream regulatory genes for each target gene. Finally, we prune away the insignificant regulatory genes by AIC model order detection method (Eq. 2.23) in system identification [23] to refine the GRN of inflammatory mechanism in response to bacterial endotoxin. By comparing with normal GRNs, we obtain the perturbed gene network to analyze the effect of inflammatory stimulus on the immune system. The hubs and "weak ties" are also discussed for the robust inflammatory gene network. This section is also based on database mining to construct a rough inflammatory regulatory network.

7.1.1 Investigating systemic inflammation mechanisms by constructing the corresponding GRN

Construction of rough GRN of inflammation

The construction procedure for a GRN of inflammatory system can be divided into 7 steps in this approach (see Fig. 7.1). The rough GRN of inflammation is set up from step 1 to step 5, and the refinement is then performed from step 5 to step 7. The step numbers are marked alongside the blocks in the flowchart.

Step 1: We first select 49 genes (see Table 7.1) that are associated with the inflammatory responses based on data mining in the published literature [10,11]. Next, we cross-reference the findings reported in other literatures [5–9], and select the candidate genes that we are interested in with biofunctions like cell–cell signaling (IL17C, etc.), leukocyte migration (SCYE1, etc.) or detection of abiotic stimulus (TACR1, etc.) as candidates. (The annotations of different biological processes from Gene Ontology (GO) database for these 49 genes are shown in Table A.1 in Appendix A.) In order to distill the essence from the complicated global inflammatory gene network, we choose not to classify its function modules like Calvano et al. [10]. Instead, we only select 49 significant genes as a core in the inflammatory network, it becomes much easier to identify the permutations between normal and inflammatory conditions. It can also enable us to give biological function interpretations and to perform literature validations, especially on the NF-κB subnetwork.

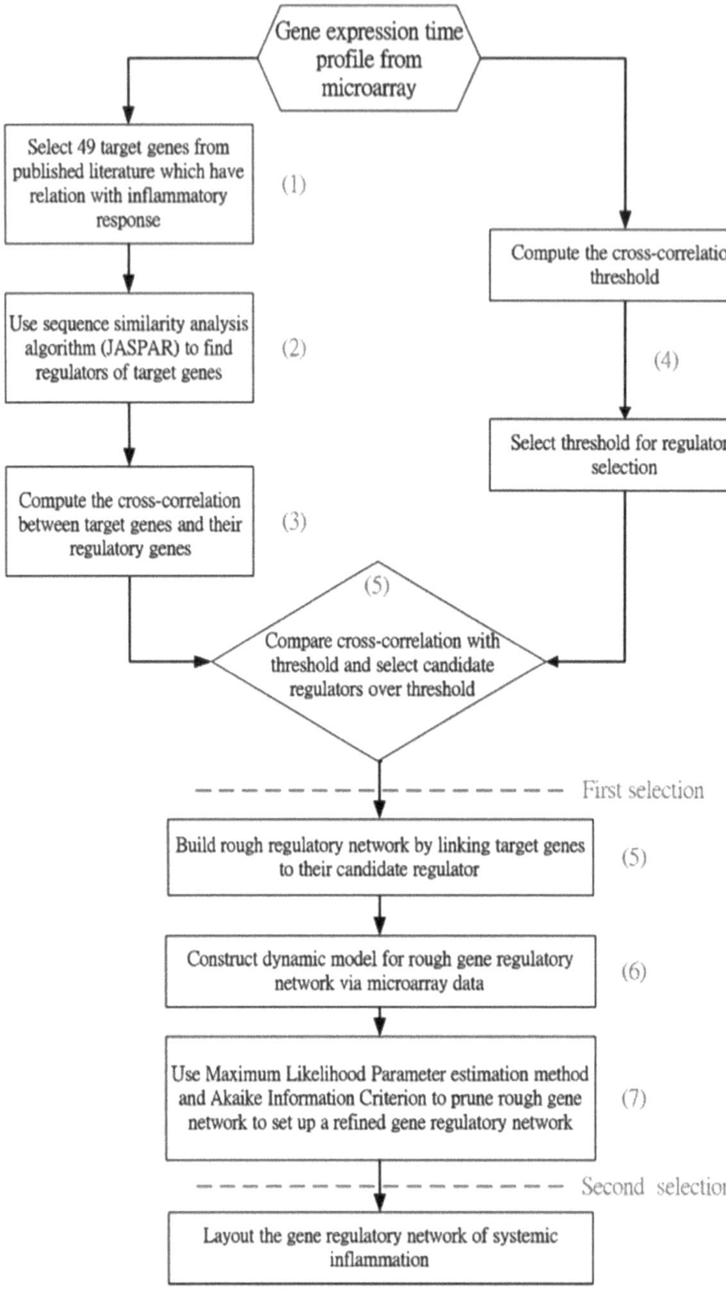

Figure 7.1 The flowchart for constructing the gene regulatory network of systemic inflammation mechanism. The left-hand-side path selects target genes and their potential regulatory genes, and the right-hand-side path generates a threshold of cross-correlation between each target gene and its upstream regulator to select possible regulatory genes from the left-hand-side path to construct a rough gene regulatory network of inflammatory response. Then the rough gene regulatory network is pruned by dynamic model and parsimonious Akaike Information Criterion to achieve a refined gene regulatory network of inflammation.

Table 7.1 Total 49 genes selected from published literatures

Gene name	Description	Gene name	Description
ABCFI	ATP-binding cassette, subfamily F (GCN20), member 1	*IL22*	Interleukin 22
ADORA2A	Adenosine A2a receptor	*IL6*	Interleukin 6
ADORA3	Adenosine A3 receptor	*IL8*	Interleukin 8
ALOX5	Arachidonate 5-lipoxygenase	*IRAK*	Interleukin 1 receptor-associated kinase 1
AMBP	Alpha-I-microglobulin/bikunin precursor	*ITGB2*	Integrin, beta 2
ANXAI	Annexin A1	*KNG*	Kininogen
AOAH	Acyloxyacyl hydrolase (neutrophil)	*MAPK10*	Mitogen-activated protein kinase 10
BLNK	B-cell linker	*NFATC3*	Nuclear factor of activated T cells, cytoplasmic, calcineurin-dependent 3
CCLI8	Chemokine (C-C motif) ligand 18 (pulmonary and activation-regulated)	*NFKBI*	Nuclear factor of kappa light polypeptide gene enhancer in B-cells I (p105)
CCR7	Chemokine (C-C motif) receptor 7	*NFKBIA*	Nuclear factor of kappa light polypeptide gene enhancer in B-cells inhibitor, alpha
CEBPD	CCAAT/enhancer binding protein (C/EBP), delta	*NFRKB*	Nuclear factor related to kappa B binding protein
CXCLI4	Chemokine (C-X-C motif) ligand 14	*NR3CI*	Nuclear receptor subfamily 3, group C, member 1 (glucocorticoid receptor)
CXCL2	Chemokine (C-X-C motif) ligand 2	*PLA2G4B*	Phospholipase A2, group IVB (cytosolic)
CYBB	Cytochrome b-245, beta polypeptide (chronic granulomatous disease)	*PLAA*	Phospholipase A2-activating protein
FOS	V-fos FBJ murine osteosarcoma viral oncogene homolog	*REG3A*	Pancreatitis-associated protein
GPRI32	G-protein coupled receptor 132	*SCCE*	Kallikrein 7 (chymotryptic, stratum corneum)
HDAC4	Histone deacetylase 4	*SCYEI*	Small inducible cytokine subfamily E, member I (endothelial monocyte-activating)
HDAC5	Histone deacetylase 5	*TACRI*	Tachykinin receptor I
HDAC7A	Histone deacetylase 7A	*TICAM2*	Toll-like receptor adaptor molecule 2
HDAC9	Histone deacetylase 9	*TLR4*	Toll-like receptor 4
HPSE	Heparanase	*TLR7*	Toll-like receptor 7
ILI7	Interleukin 17C	*TNFA*	Tumor necrosis factor
ILIA	Interleukin 1a	*TNFR*	Tumor necrosis factor receptor superfamily member 1A precursor
ILIB	Interleukin 1b	*TOLLIP*	Toil interacting protein
ILIR	Interleukin-1 receptor type I precursor		

Our goal is to select the candidate regulators (i.e., TFs) of 49 target genes in inflammatory response to construct the rough GRN of inflammation by linking these target genes to their regulators.

Step 2: We explore the Ensembl database (http://www.ensembl.org/index.html) to retrieve the promoter sequences of 49 target genes and then conduct sequence similarity analysis to identity candidate regulators of these target genes in JASPAR (http://asp.ii.uib.no:8090/cgi-bin/jaspar2005/jaspar_db.pl), which is a high-quality TF database. In this stage, we hypothesize that if some TFs are selected by the predictions of JASPAR using these criterions, the genes generating the respective TFs at the protein level could be considered as candidate regulators to the target genes.

After this step, we obtain a set of candidate regulators from the JASPAR analysis (Table A.2). However, there are still many false positive errors in these hits because the outcome has listed all possible regulators in conditions beyond inflammatory response. Some pruning methods based on microarray data of inflammatory response are necessary. The pruning procedure is described after step 5.

Step 3: We screen and select potential regulators from the JASPAR hits by cross-correlation threshold of gene expression data [24], which is based on the assumption that there are possible correlations between target gene and their upstream regulators, with or without time delays. We compute the cross-correlations between the target genes and their own regulatory genes separately, and the cross-correlation values is then used to identify the candidate regulators according to the assumption that the regulatory genes and target genes have a positively (or negatively) correlated temporal relationship if the target gene's expression profile is positively (or negatively) correlated with the regulatory genes profile, with or without time lags.

Step 4: A careful choice of proper threshold for correlation to discriminate the "by chance" associations is indeed important. In order to decide on a threshold of significant correlations between transcription regulators and target genes for selection of candidate transcription regulatory genes, we randomly choose 2000 genes from 22,577 genes and computed their correlations by the Pearson Correlation in Eq. (7.3) in Section 7.1.4, as ranked in Fig. 7.2. According to the ranking in Fig. 7.2, we select the 30% (i.e., 0.46451 in cross-correlation value) as this threshold. A lower threshold may recruit some "by chance" regulator genes, while a higher threshold may result in the increase of false-negative genes. However, we have learned from our own experience that high correlated genes may be associated due to the fact that they are coregulated by the same regulatory genes/TFs. In other word, those genes that are coregulated by a common set of genes but do not regulate each other always arise with high correlation. On the other hand, time delay in signaling pathways may mask the genes with co-regulatory

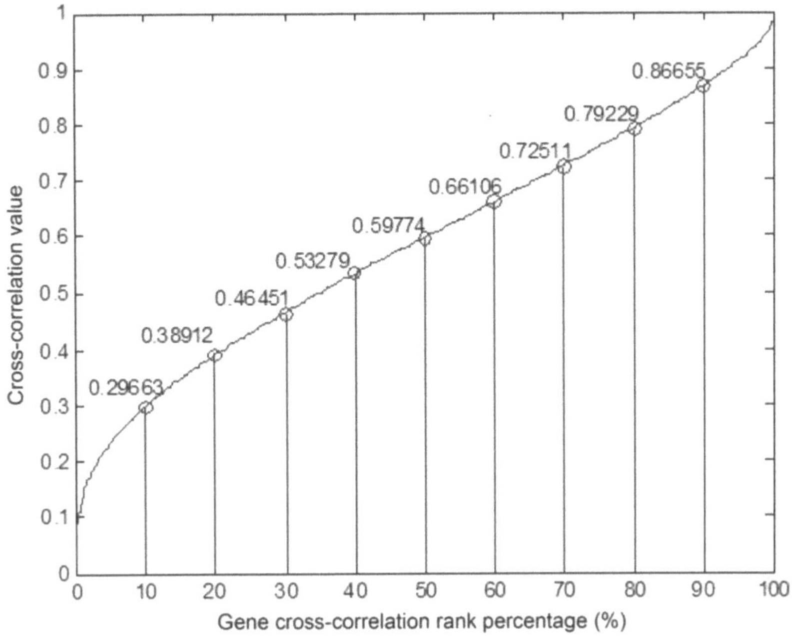

Figure 7.2 Distribution of a threshold for selecting candidate regulators by cross-correlation method. 2000 genes are randomly chosen from 22,577 genes to compute their correlation and then these correlations are ranked. A threshold 0.3 is specified to select possible candidate regulators from those based on DNA sequence similarity in the JASPAR database.

association with low correlation. Because the correlation is only the first discrimination parameter used in the gradual refinement of the inflammatory regulatory network, we do not want to miss all possible candidate regulatory genes at such early stage. The aim in the first stage of this algorithm is to delete those "impossible" regulatory associations that are truly false. Therefore, we cannot adapt a high threshold of correlation for discrimination. Furthermore, by using the permutation steps for data randomization, the probability density function of parameter estimation shows the parameter estimations of regulatory genes by the threshold of correlation has a *p*-value less than .001. So the genes selected by the proposed threshold (0.46451) are the significant candidates for regulatory genes.

Step 5: Here we make the first selection from the candidate regulators in Step 3. This implies that if the cross-correlation between a candidate regulator and the target gene is more than 0.46451, it will be considered as a candidate regulator for the target gene. After selecting potential regulators by cross-correlation threshold, these target genes and their candidate regulators are integrated to construct a preliminary GRN of inflammatory response. Results of the first selection are listed in Appendix A (Table A.2(B)).

1. Pruning the Preliminary GRN via a Dynamic Model

By this point we have constructed a preliminary network via the first five selection steps using statistical inferences. However, we have yet to consider the dynamic property of this network. To include the dynamic parameters, we apply the Akaike Information Criterion (AIC) in Chapter 2, to help us make a more comprehensive selection. The AIC algorithm is denoted as Step 7 in Fig. 7.1. A dynamic regulatory model is proposed to parsimoniously describe the gene regulatory genetic network of inflammation. It should be mentioned that the time delays from the regulators to their target genes, which have been detected by cross-correlation prediction algorithm via correlation, are considered in the dynamic regulatory model to mimic the delay phenomenon due to the transduction relay of the metabolic and signal pathways in the real transcriptional regulatory process. Details of the pruning are presented in the following paragraph and in the material and methods section. In this section, based on the possible interactions in a preliminary GRN of inflammation [10,11,25] obtained from the previous sections, a dynamic regulatory model for the transcription of an interested target gene of systemic inflammation is developed. This model describes how the upstream regulatory genes control their target genes to produce the output expression of mRNA through TRN. From the rough (or candidate) gene network through database-predicted information, we construct a dynamic regulatory model for each target gene of systemic inflammation in humans. Then, according to the microarray data of genetic expression, we identify the number of connections in the dynamic regulatory model of rough gene network in the inflammatory system. Based on the degree of interaction in the regulatory network, we prune the preliminary GRN of inflammation one target gene at a time via AIC. The pruning procedures to obtain a refined GRN (see Fig. 7.1) are given in the following steps.

Step 6: According to the rough gene network, the transcriptional regulation of a target gene in inflammatory system is dynamically modeled in the following multi-input/single-output stochastic process.

$$y(t+1) = ay(t) + \sum_{i=1}^{L} b_i x_i(t-\tau_i) + k + \varepsilon(t) \tag{7.1}$$

where $y(t)$ represents mRNA expression level of a target gene at time t, and parameter a indicates the effect of the present state $y(t)$ on the next state $y(t+1)$; $x_i(t-\tau_i)$, for $i=1,\ldots,L$, denotes the regulation functions of L upstream TFs in the rough gene network; and b_i, for $i=1,\ldots,L$, denotes their corresponding kinetic coefficients (or regulation abilities). In addition, τ_i denotes the expression delay from regulatory gene i to the target gene, which was detected via identifying the model by the fact that at the delay τ_i regulatory gene i has the highest correlation with the target gene. The value of τ_i will be iteratively detected from 0 to 2 h

(4 time points) by a minimum loss function based on AIC in the final pruning step. It can be ensured that the τ_i value we detect has the best model fitting, although it has a large amount of computations. k in Eq. (7.1) represents the basal molecular level to denote the regulation of unknown factors. $\varepsilon(t)$ denotes a stochastic noise due to model uncertainty and fluctuation of the mRNA microarray in the target gene. The binding transcriptional regulatory functions $x_i(t)$ of TFs on their motif binding sites are described by the following sigmoid functions of mRNA expression profiles of their corresponding regulatory genes, respectively [19]

$$x_i(t) = f_i(y_i) = \frac{1}{1 + \exp(-r(y_i(t) - m_i))} \tag{7.2}$$

i.e., the sigmoid functions in Eq. (7.2) denote the thresholds of bindings of TFs on motif binding sites for the transcriptional regulation in Eq. (7.1).

Step 7: By combining the maximum likelihood parameter estimation method with the most parsimonious model order detection using the AIC (see Section 7.1.4), we could prune the rough gene network to generate a more refined gene network through the most parsimonious gene transcription regulatory model in Eq. (7.1), i.e., the insignificant interactions (or small b_i) could be deleted by AIC. With the upstream regulatory genes as target genes, we can then trace back their upstream regulatory genes by a similar construction procedure. Iteratively, we could construct the whole GRN of systemic inflammation in the innate immune system. The results of selection are listed Table A.2(C).

2. Construction of Inflammatory Gene Network in Immune System

Based on the 49 target genes (see Table 7.1) and their candidate regulators (Table A.2(C)), we construct a rough GRN of the human inflammatory system. Then, according to the rough GRN, we set up the dynamic model to prune it once more, setting up a refined GRN by a system identification scheme and parsimonious AIC method via microarray data. At this point, we can construct two more refined GRNs for both the inflammatory/activated and the normal/resting conditions by the same construction flowchart shown in Fig. 7.1, and draw two GRNs by the Osprey tool [26] (see Figs. 7.3 and 7.4). In Fig. 7.3, there are 94 nodes with 336 edges for the inflammatory/activated gene network and in Fig. 7.4, there are 66 nodes with 264 edges for the normal/resting gene network.

By comparing the inflammatory network with the normal network, we obtain the differential/perturbed GRN (see Figs. 7.5 and 7.6). While some interactions can be found in both the normal and the inflammatory (LPS-treated) regulatory network of an inflamed system between the normal and inflammatory cells (Table A.3). This significant finding helps better understand the effect of inflammatory stimulus on the innate immune system. As noted, the perturbed inflammatory GRN in the immune system between normal and LPS-stress cells is the focus of this section.

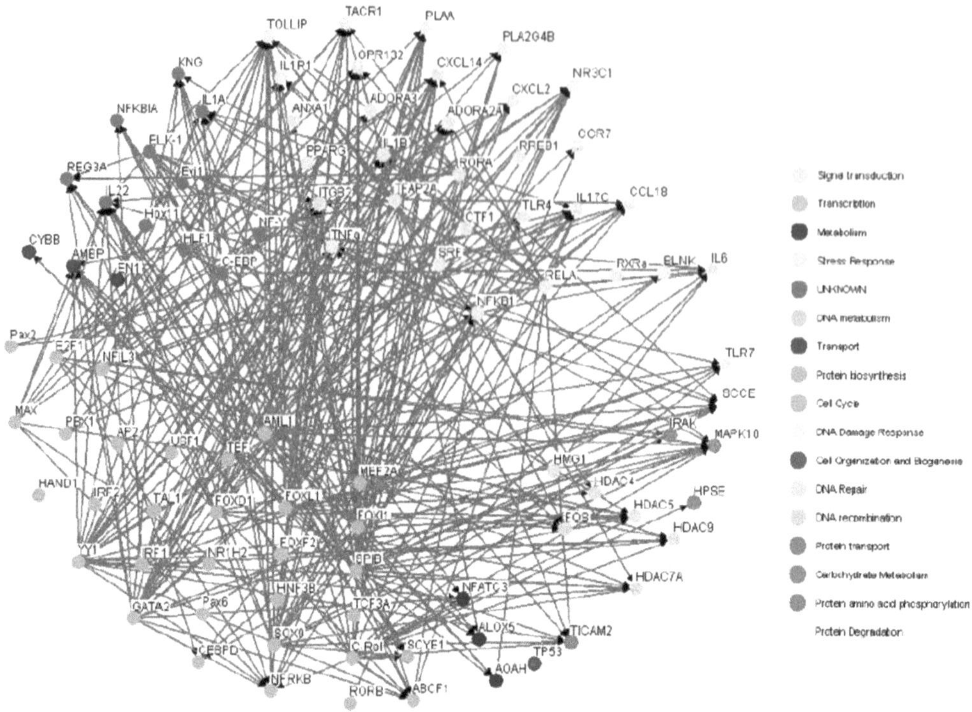

Figure 7.3 The inflammatory transcriptional gene network in immune system with LPS. The inflammatory gene network with LPS containing.

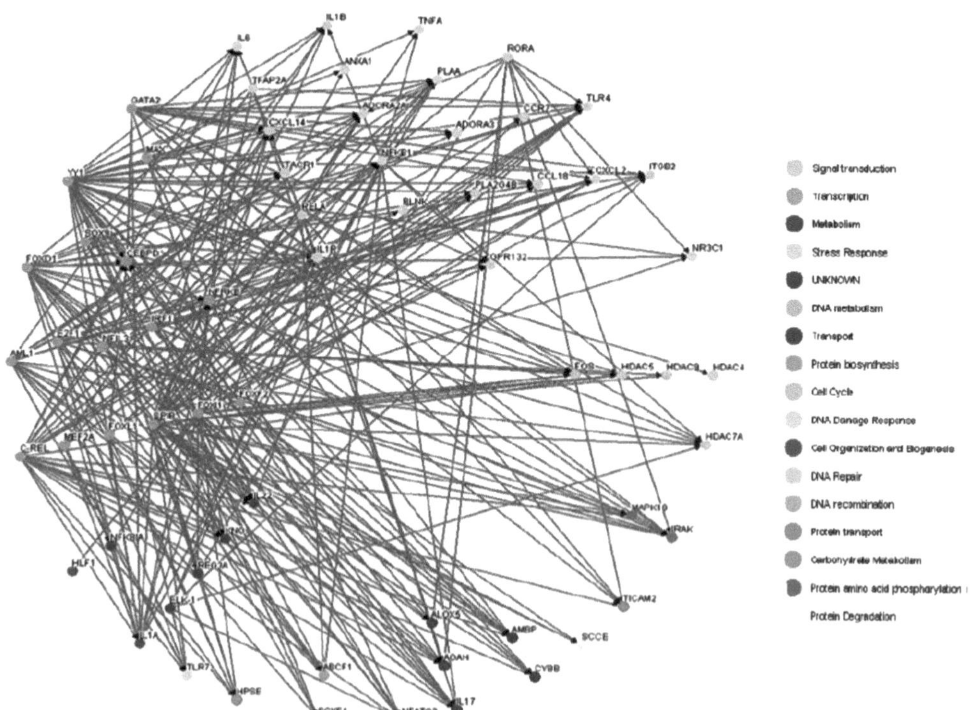

Figure 7.4 The inflammatory transcriptional gene network in immune system without LPS. The inflammatory gene network in normal condition.

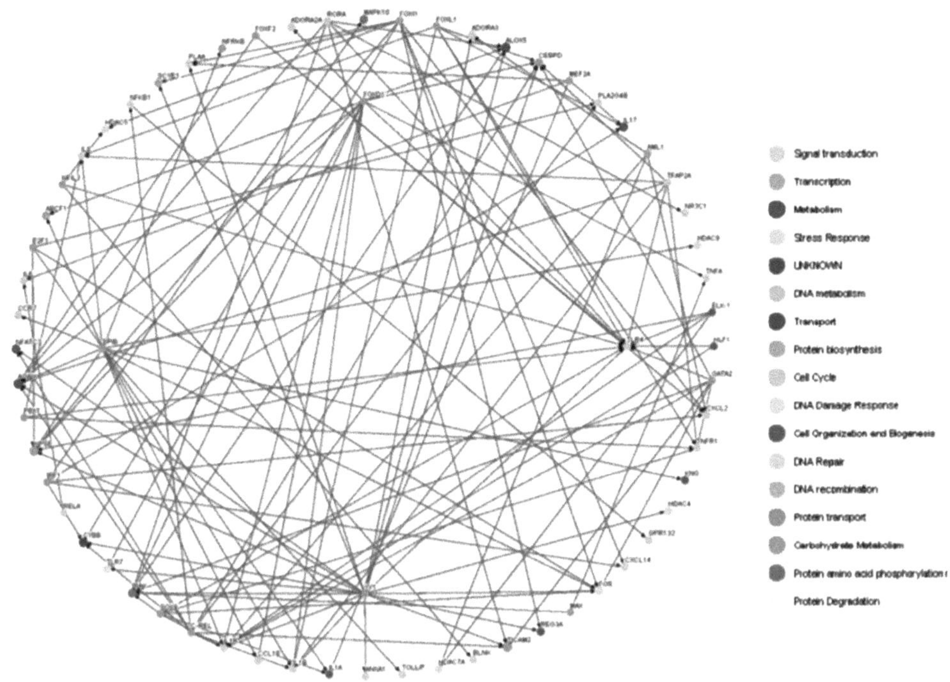

Figure 7.5 The perturbed transcriptional gene network. Gene network only in normal condition but not inflammatory condition.

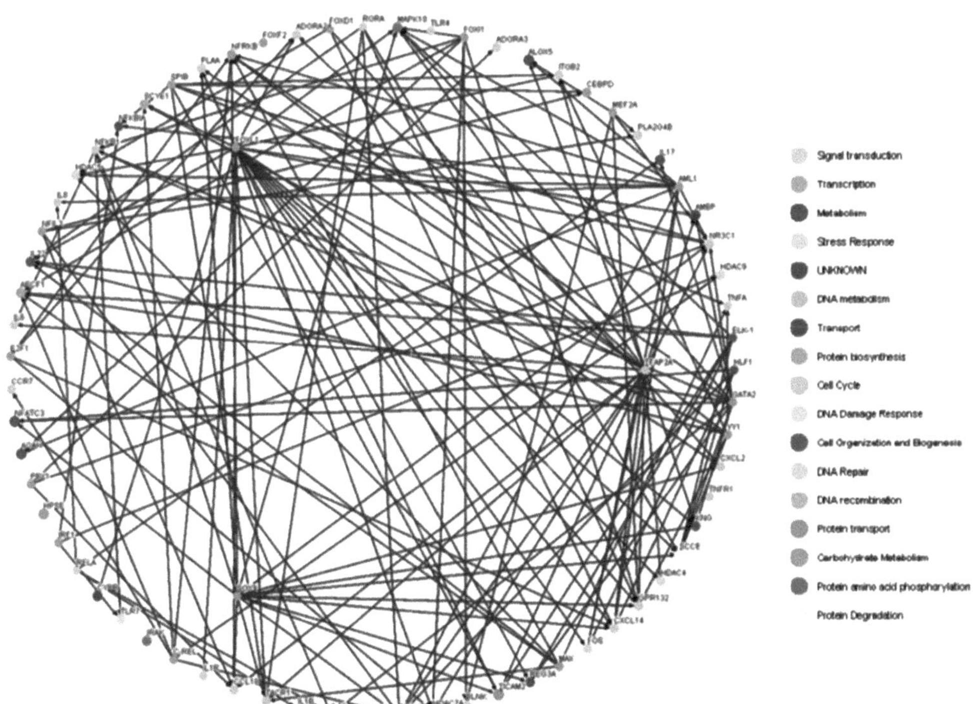

Figure 7.6 The perturbed transcriptional gene network. Gene network only in inflammatory condition but not in normal condition.

To create networks, we extracted the unique connections which only existed in one specific network but not in another. We showed the similarities and the differences in the gene. We further lay out the perturbed inflammatory GRN to locate the significance differential connection of the key components. We can observe many differences in normal and inflammatory conditions from the perturbed gene network. As shown in Fig. 7.5, the gene network contains 64 nodes with 131 edges found only in normal condition but not in inflammatory condition, and there are 4 hubs (FOXD1, SPIB, YY1, and TLR4) which appear to be highly connected. In Fig. 7.6, the gene network contains 70 nodes with 159 edges for gene network found only in inflammatory condition but not in normal condition, and there are clearly 3 hubs (FOXL1, TFAP2A, and SOX9) within this perturbed network. It is noteworthy that these highly connected hubs have been mentioned in several previous studies. For example, TFAP2A is inactivated [27] and SOX9 is inhibitive [28] in response to inflammation as shown in Fig. 7.6. Additionally, FOXL1 is dramatically induced during hepatic stellate cell activation [29] and preliminary experimental data indicates that FoxL1 is involved in the regulation of the adhesion molecule ICAM-1, an important mediator of neutrophil recruitment in liver injury. The current investigation is focused on delineating the mechanism by which FOXL1 regulates inflammatory signaling in the liver.

We summarize the connection degree (i.e., the number of connections) of each node of Fig. 7.6 and compile a list of regulators with connection degree of $\geqq 8$ (see Table 7.2) to identify perturbed hub proteins that induce differences between inflammatory and normal conditions. These proteins are possible target regulators for drug discovery investigation (such as antiinflammatory drugs [30–32]). Finally, we summarize the gene connectivity of six regulators (FOXL, TFAP2A, SOX9, GATA2, AML1, and NR3C1) with high degree of connectivity in Table 7.2, which are confirmed and in agreement with previous research findings [27–34].

It has been shown that the robust gene network can form a scale-free network, i.e., genes prefer to form links with other genes that already have the highest number of links [35,36]. Scale-free gene networks could tolerate random removal of nodes but are vulnerable to loss of highly interactive hubs [35,36]. This may result in the lethal

Table 7.2 Gene connectivity only in inflammatory condition but not in normal condition

Regulator	Connectivity	References
FOXL1	23	[18]
TFAP2A	19	[27]
SOX9	16	[28]
GATA2	12	[30,32]
AML1	11	[33]
NR3C1	8	[31,34]

outcome in a system's behavior when highly connected hubs are targeted. In the inflammatory gene network shown in Fig. 7.3, genes such as NF-κB, TNF-α, and RELA can be considered highly connected hubs of the signaling transduction. If they are inactivated by mutation or disease, the inflammatory gene network will lead to eventual collapse of the system. In order to overcome this lethal outcome, "weak linkage" architectures are evolved by natural selection to improve the robustness of GRNs. We argue that such versatile mechanisms underlie the essential regulatory process of robust gene networks. As a result, some connections can easily be removed and some connections can easily be added in the GRN. This concept is also known as "weak ties" in network theory [35]. "Weak ties" structures in biological networks enable remove of old processes and addition of new processes to the existing core processing to improve the information exchanges and signal transductions using common versatile mechanisms that operate on diverse inputs to various stimuli [36]. As a consequence, "weak ties" can improve a network's robustness against external stimuli. Obviously, the connections of the perturbed gene network in Fig. 7.5 are presented only in the normal condition. The perturbed gene network in Fig. 7.6 can hence be considered as additional connections in the inflammatory gene network. In response to bacterial endotoxin, the connections in Fig. 7.5 are removed and the connections in Fig. 7.6 are added. Apparently, this agrees with the concept of the so-called "strength of weak ties" in network theory, where the most important interactions and information exchanges sometimes occur via nodes from otherwise unrelated networks. This implies that non-hubs may play a pivotal role in the gene regulation [35,36]. Similarly, the "weak ties" architecture in NF-κB gene network in inflammatory condition is shown in the removal and addition of connections of gene network in Figs. 7.5 and 7.6.

In summary, the regulators of target genes are first selected by JASPAR, then truncated by the threshold of cross-correlation and finally pruned by AIC via microarray data and a dynamic model. We combine several algorithms and tools to improve the performance of the gene network construction of the target inflammatory system. All the data sources are independently produced by various research groups and the results are verified with more independent studies published previously. It is clear that the top-down procedures can predict the target genes and their candidate regulatory TFs well. More biological insight into the perturbed inflammatory network is given in Section 7.1.2 and details of the proposed GRN construction algorithm are shown later in Section 7.1.4.

7.1.2 Discussion of systemic inflammation mechanism via microarray data and database mining

The NF-κB pathway, which is an important modular inflammatory system, is illustrated as a trimmed down GRN in Figs. 7.7 and 7.8. This concise network includes

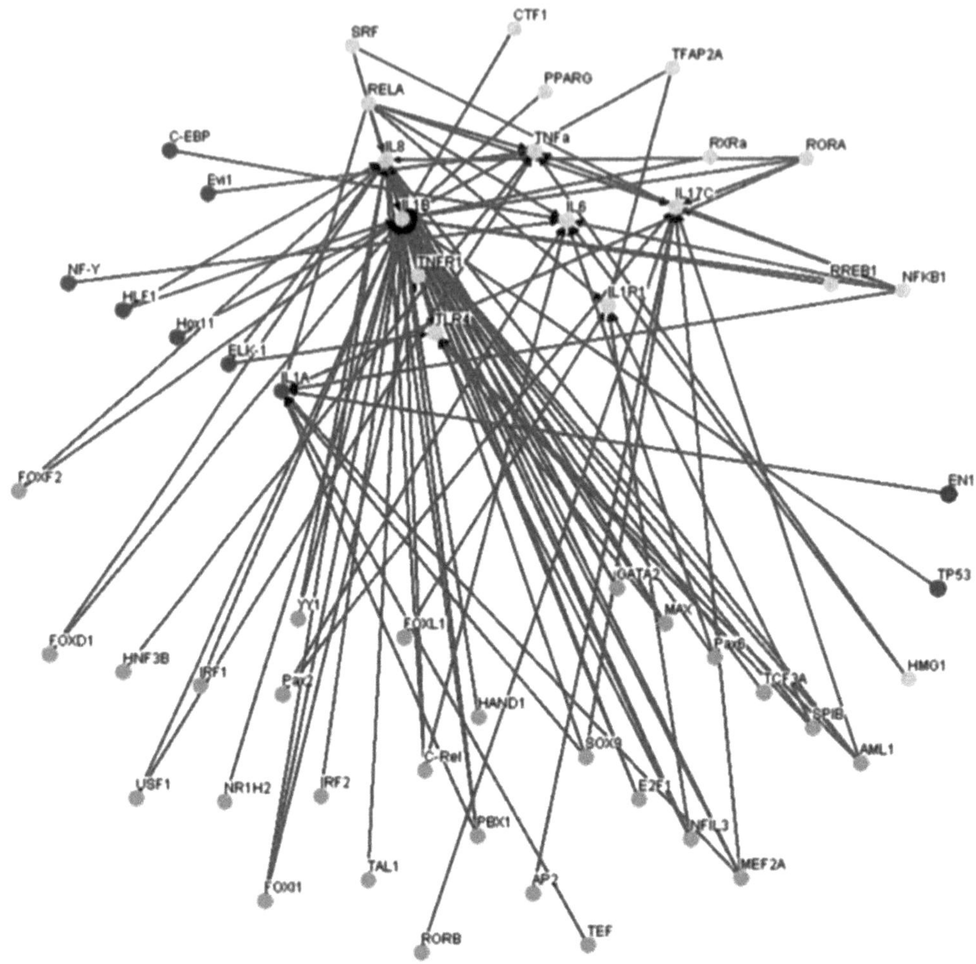

Figure 7.7 The important pro-inflammatory gene network induced or activated by NF-κB in immune system with LPS. The important pro-inflammatory gene network in inflammatory condition.

important pro-inflammatory cytokine genes: *IL1A*, *IL1B*, *IL1R*, *IL6*, *TNFA*, *IL17*, *IL8* and the receptor genes *IL1R*, *TLR4*, and *TNFR*, all of which have well-known roles in the NF-κB signaling pathway. This concise network can help us to monitor the performance of inflammatory responses under diverse conditions. By the proposed method shown in this section, we can predict the dynamic profiles of those cytokines. As expected, these results are comparable to the findings published in previous studies [8,25] discussed in the following paragraphs. These in silico findings confirm the wet-bench observation that many characterized genes in the common inflammation response are regulated by the transcription factor NF-κB [6].

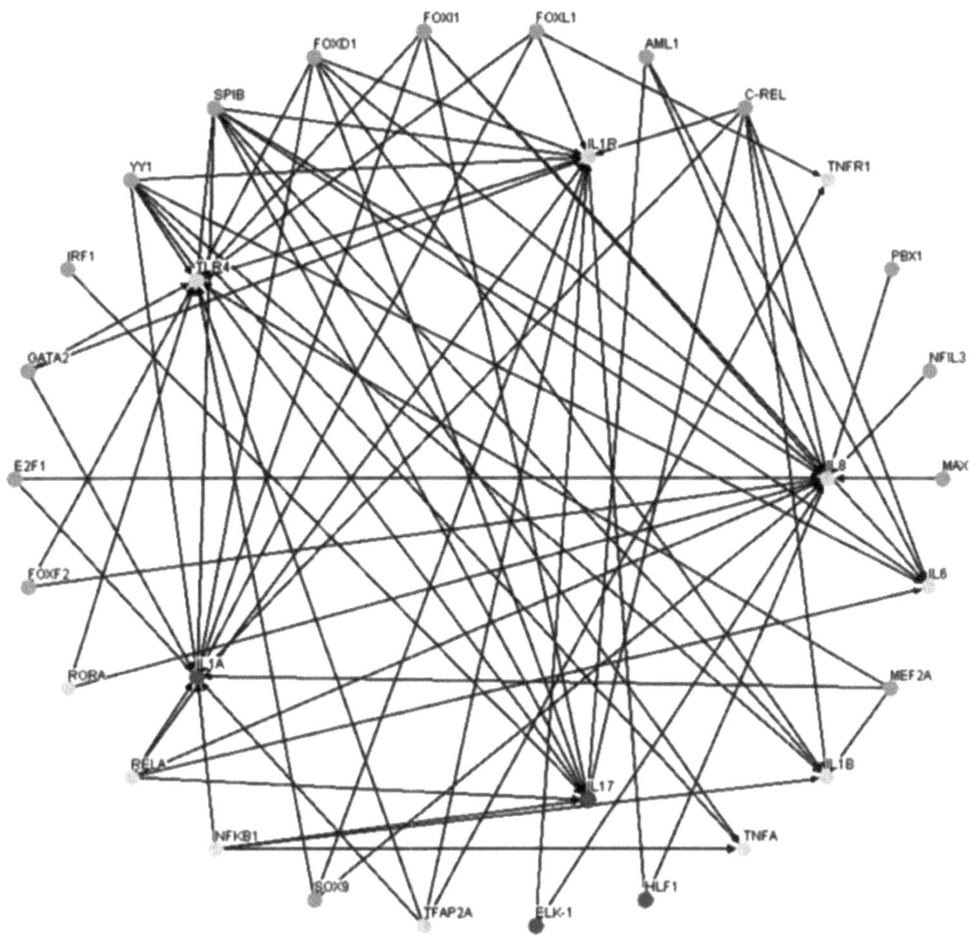

Figure 7.8 The important pro-inflammatory gene network induced or activated by NF-κB in immune system without LPS. The important pro-inflammatory gene network in normal condition.

On the other hand, the perturbed gene network of these pro-inflammatory genes in NF-κB signaling pathway is shown in Figs. 7.9 and 7.10. The perturbation in Fig. 7.9 is more complicated than the perturbation shown in Fig. 7.10 because in normal condition these genes have to fulfill other biochemical tasks other than inflammation. This analysis also reveals that the important genes (*IL1A*, *IL1B*, *IL1R*, *IL6*, *TNFA*, *IL17*, *IL8*, *IL1R*, *TLR4*, and *TNFR*) detected by these algorithms are vital for the inflammatory response because they are more connected during inflammation than in normal conditions. In inflamed conditions, they appear to work in accordance with each other to enhance their effects on the inflammatory responses. For example, there is strong evidence to support that NF-κB1 and RELA have to regulate the pro-inflammatory genes collectively when they are in inflammatory responses [7].

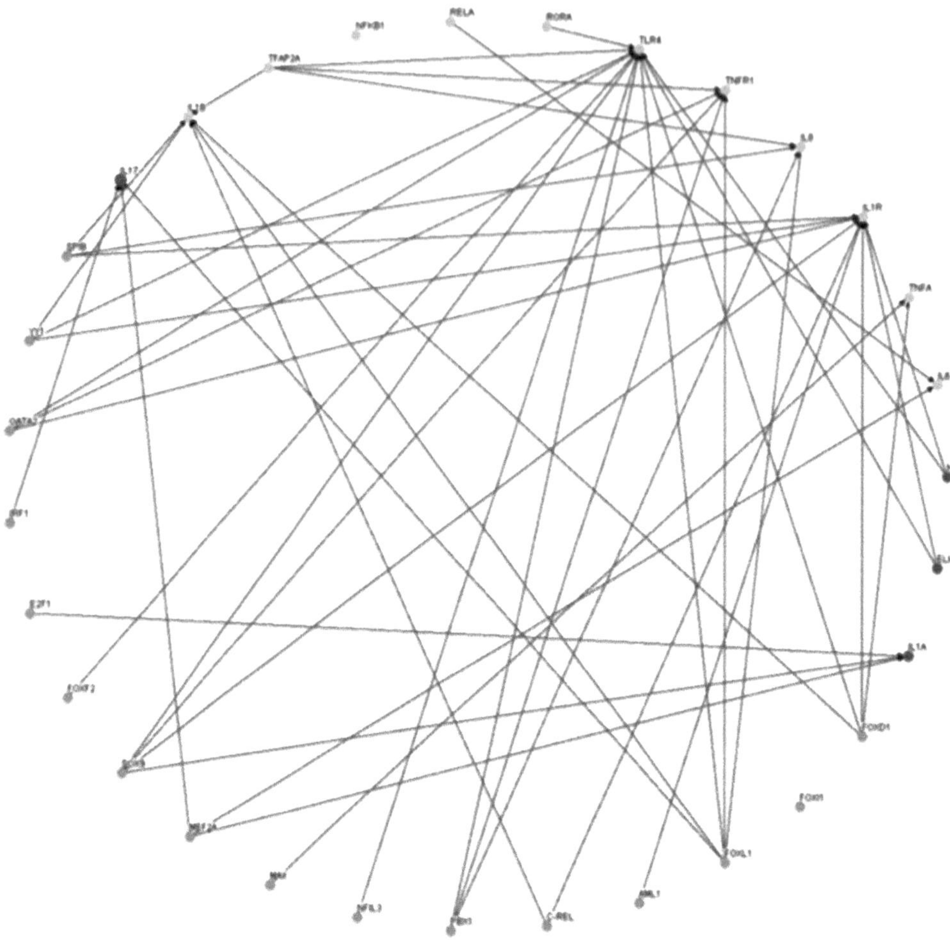

Figure 7.9 The important pro-inflammatory perturbed NF-κB gene network. Gene network only in normal condition but not inflammatory condition.

In recent studies [8,25], cytokine and chemokine networks have been shown to play a pivotal role in inflammation because they are involved either directly or indirectly in the innate and adaptive immune responses. It has been shown that Interleukin-1 alpha (IL1A) and Interleukin-1 beta (IL1B) act via their receptor (IL1R) to induce gene expressions which in turn mediate a feedback protein synthesis involved in the later wave of inflammatory responses [16,25]. This is in agreement with the dynamic profiles of the pro-inflammatory genes and their receptors (*IL1A*, *IL1B*, *IL1R*, *IL6*, *TNFA*, *IL17*, *TLR4*, *TNFR*, and *IL8*) which are simulated by this dynamic regulatory model (shown in Fig. 7.11). The accuracy of the curve fitting data has demonstrated the prediction power of the proposed method.

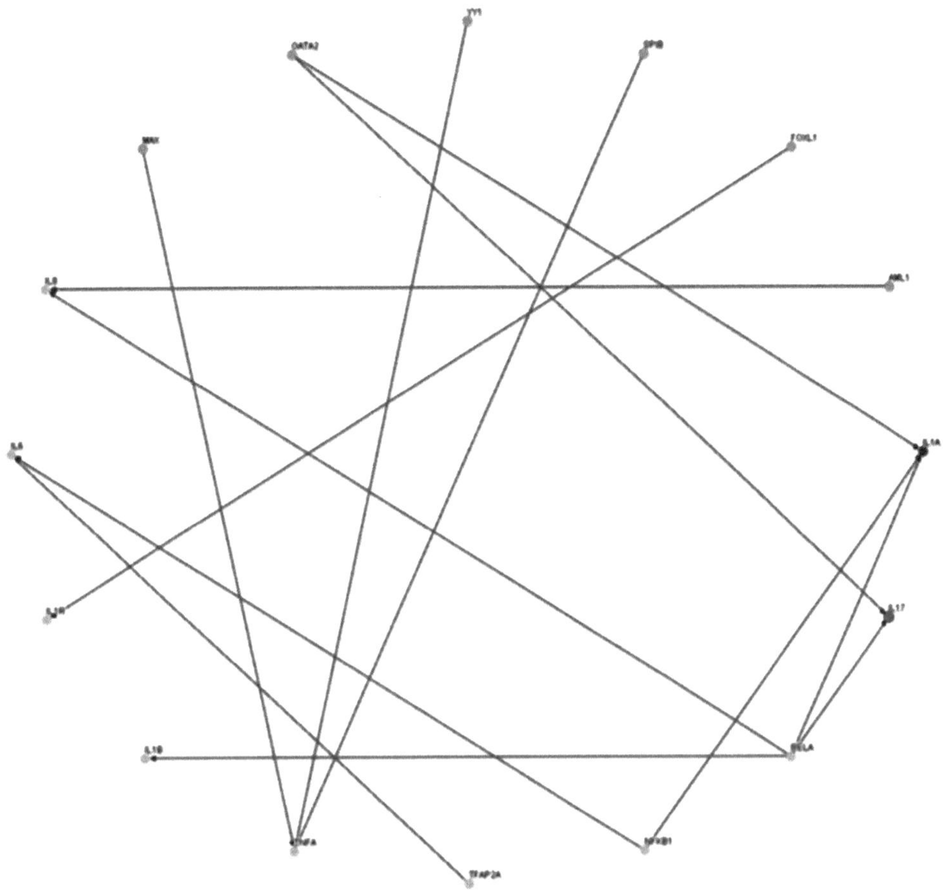

Figure 7.10 The important pro-inflammatory perturbed NF-κB gene network. Gene network only in inflammatory condition but not in normal condition.

Without a doubt, the performance of the proposed method is very satisfactory. In Fig. 7.11, the Interleukin-1 alpha (*IL1A*) and Interleukin-1 beta (*IL1B*) have a peak expression at 2~3 h poststimulation, and then gradually decays because of the removal of bacterial endotoxin. Interestingly, its receptor (*IL1R*) has a peak expression at 6~7 h poststimulation, while *IL1A* expression has reached another peak in about 8~9 h. This concerted changes in cytokine and receptor may be explained by the following mechanism in which *IL1A* has a positive feedback loop in NF-κB signaling pathway through *IL1R* when the affected signaling network suffers inflammatory stress [16]. The same situations occur as well in TNF-α and its designated receptor *TNFR*. The *TLR4*, which is in a growing TLR family structurally characterized by a cytoplasmic Toll/interleukin-1R (*TIR*) domain and by extracellular leucine-rich repeats

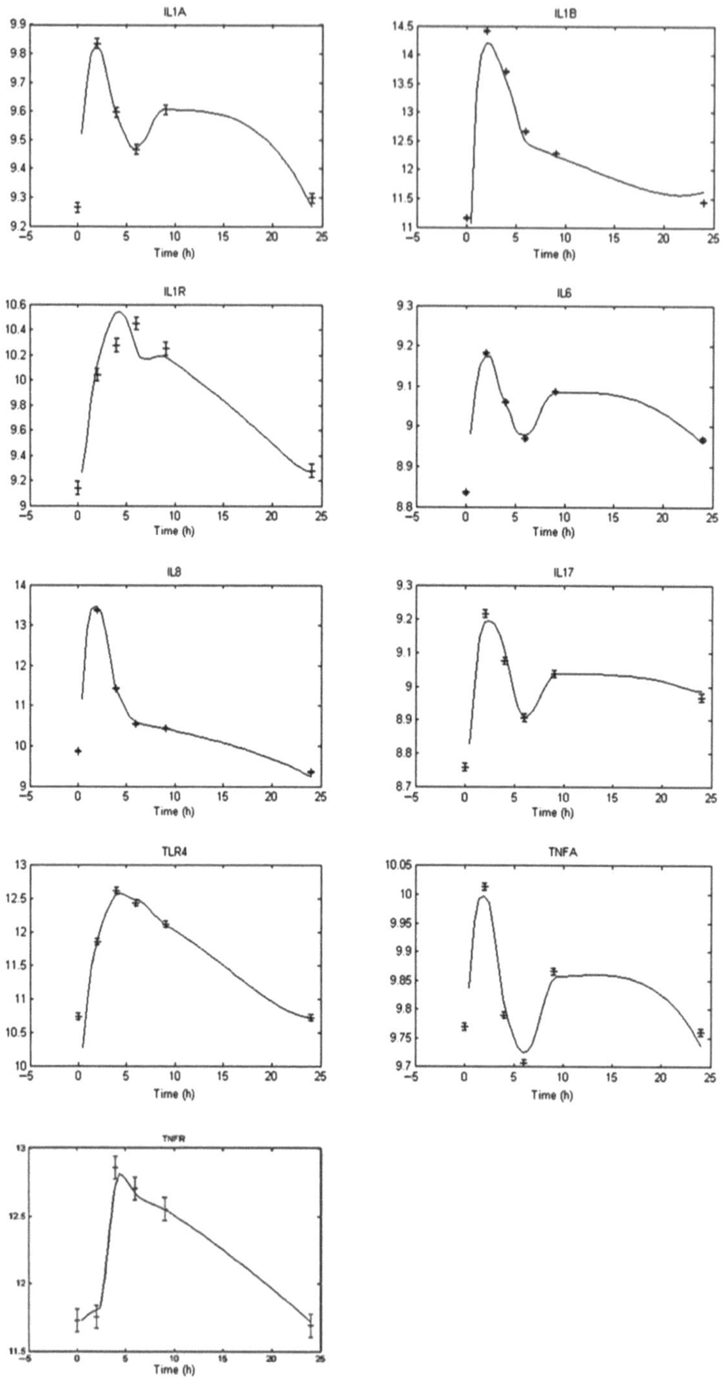

Figure 7.11 The curve fittings of dynamic regulatory model of pro-inflammatory gene and its regulators. The "o" is the data from microarray in 24 h and the *solid line* is the curve fitting by the proposed dynamic model in Eq. (7.1). The error bars for standard deviations have been marked. We denoted the curve fittings of nine target genes and their upstream regulators respectively, and the regulatory parameters for each dynamic model are presented.

[18], has the same dynamic fluctuation as seen on *IL1R* or *TNFR*. The other genes like *IL8*, *IL6*, *IL17* and their own receptors are all exhibiting similar behavior in this analyses (data not shown).

For Step 7, the identification of time delay and the estimated parameters are shown in Appendix A (Table A.4 and Figs. A.1–A.9). Although we consider the effect of time lag τ_i in this model, it is plausible that not all regulators have delay times on their transcription regulations.

It seems that the regulation in inflammation may act so swiftly that parameter τ_i cannot be detected (i.e., less than one time unit of microarray data or one half hour). However, there are several time lag regulations in IL8 and its regulators, such as *SOX9*, *MEF2A*, *NFIL3*, *ELK1*, *FOXF1*, *FOXD1*, *GATA2*, *FOXI1*, *REL*, and *RELA*. It is because IL8 has a more complicated regulatory mechanism through other pathways with considerable delay. The dynamic model assumes that the expression profile of a target gene results from the kinetic activity of one or more specific regulators, which bind to the downstream target gene's promoter site and initiate the transcription of that target gene to exert its effect on the inflammation network. In other words, it is possible to generate the target gene expression profile via the gene expression profiles of the upstream TFs using the dynamic regulatory model and its kinetic parameters in Eq. (7.1). The continuous gene expression profiles in Fig. 7.11 (see Figs. A.1–A.9) are generated by the identified dynamic model for all target genes and their corresponding regulators, which can fit the microarray data reasonably well. Dynamic modeling of biological systems including genetic networks and cell regulatory networks has been applied in functional analyses for a long time. However, very few of the other modelings has included the time delay parameter which is comprehensively factored into this stochastic dynamic model. The findings shown in this section successfully demonstrate that we can efficiently refine the GRN of systemic inflammation in human via microarray data, and to mimic the signaling transduction delay in the transcriptional regulatory process.

Combining the cross-correlation selection algorithm and the AIC, we created a novel dynamic modeling algorithm to trim down the tangled regulatory genetic network of human inflammatory system without any loss of biological meaning. The algorithm presented here can model all combinations of the target genes/regulators and produces the best predictions on gene expression by the dynamic regulatory model. Instead of attempting to model the whole complicated regulatory processes with the high risk of incorrect prediction, this dynamic model focuses only on a concise set of target genes with a more reliable outcome. Iteratively, we could eventually construct the whole GRN of systemic inflammation in response to bacterial endotoxin by this dynamic model through microarray data.

An essential problem with the application of the multivariate procedures to the microarray gene expression data, as expressed in recent publications, is associated with the reproducibility of the complex constructions resulting from such analyses. In order to confirm the reproducibility of the proposed method, we use this algorithm to rebuild the GRN via the microarray data published in Ref. [37]. In Ref. [37], the authors found 19 genes with significant inflammatory responses. In this situation, we reconstruct the inflammatory gene network based on these 19 genes. After comparing the reconstructed inflammatory GRN with the one in the text, we found some similarities and differences. The same highly connected hubs are *GATA2*, *AML1* (*RUNX1*), and *YY1*. There are more than five connections for these hubs in both perturbed inflammatory networks. However, due to the lack of some specific gene expression data in Ref. [37], we were unable to verify a part of highly interactive genes in the text (i.e., *FOXL1*, *TFAP2A*, and *SOX9*). Interestingly, we also found there are some hubs only present in the reconstructed network but not in the text like *GATA3* and *FPR*, which would be involved in host defense against bacterial infection and in the clearance of damaged cells [38]. The reason why these 19 candidate genes still discovered new hubs is because some of 19 candidate genes are not included in the previous 49 genes. For different experimental conditions, research topics, and technology platforms, the data pool from different literature may be different. Therefore, the candidates of target genes we chose here differed from the text, so the computational results would not be identical (Figs. A.10–A.12).

In this section, we use multi-input/single-output regulatory model to dynamically describe this gene regulatory system (i.e., multiple regulators and one target gene) that can mimic the real gene regulation in response to inflammation. The simulation can figure out the regulatory relationship and time lag value between upstream regulator and downstream target genes using time-series microarray data. In the research of Zou et al. [13], the authors used the concept of time delay just in a static state analysis of gene network, without applying it to dynamic modeling to mimic the bona fide gene regulatory behavior. Furthermore, the apparent shortcoming of the static state analysis is the limitation on a single-input single-output system (i.e., one regulator and one target gene). Such single-input single-output system is rarely existed in actual gene regulation.

While significant improvement in network construction has been achieved by this method, there are still two drawbacks in this section. First, although we present a multi-input/single-output system, it still cannot represent the actual biological conditions because they are multi-input/multi-output systems in most situations. This means when using AIC to trim the initial tangled GRN, we should prune down all data simultaneously rather than separately. However, such approach will increase the computational complexity in the combinatorial way and thus become computationally infeasible. The second drawback of all published algorithms for inference of TRNs in inflammation, including this study, is that the candidate regulators are selected from the pool of potential regulators typically defined by computational prediction, either by sequence similarity analysis, or by other genome annotation methods. If a true regulator is not included in the pool, it will inevitably escape identification by the modeling approach. This type of error will likely become a very significant problem in a poorly characterized genome of a model organism.

7.1.3 Conclusions of systemic inflammation mechanism

Our dynamic modeling represents a new approach to the study of GRN in inflammatory response. It is based on database mining to construct an inflammatory regulatory network. It is also a systems biology approach because we process the complex regulatory network of numerous genes and regulators from various data sources at the same time. The trimmed down algorithm presented here can also be extended for global GRN analysis other than the inflammatory system in the future. From the curve fitting data generated by the proposed method, it can be seen that the performance is very satisfactory. By comparing with normal GRNs, we obtain the perturbed gene network to analyze the effect of inflammatory stimulus on the immune system. The hubs and "weak ties" are also discussed for the robust inflammatory gene network. The proposed GRN is also confirmed by published evidence in the literature. In the future research, we will investigate the dynamic networks in a host–pathogen interaction mechanism on an animal model organism. We will also consider extending the algorithm to the identification and analysis of crosstalking TRNs between host and pathogen.

7.1.4 Materials and methods of constructing GRN for systemic inflammation mechanism

Dataset selection

We used previous microarray data [10,11] as these mRNA expression profiles. Gene expression in whole blood leukocytes was determined at 0, 2, 4, 6, 9, and 24 h after the intravenous administration of bacterial endotoxin to four healthy human subjects. In those experiments, four additional subjects were studied under identical conditions but without endotoxin administration. The infusion of endotoxin activates innate immune responses and presents physiological responses of brief duration. It should be noted that there is an initial pro-inflammatory phase and a subsequent counter-regulatory phase, with resolution of virtually all clinical perturbation within 24 h.

Construction of rough gene networks of systemic inflammation

The cross-correlation method was developed to identify target genes that are regulated by a common set TFs. The cross-correlation uses continuous gene expression with the assumption that the regulatory genes and target genes have a level of positively (negatively) temporal correlation relationship if the target gene's expression profile is positively (negatively) correlated with the regulatory gene's profile, possibly with time lags. The next procedure is to specify the threshold for the correlation between target genes and their regulators. In this section, there are 22,577 gene expression time profiles [10,11]. We choose 2000 gene expression profiles randomly and computed their correlations with different time lags or lead to evaluate a threshold for significant correlations for possible regulators of target genes, which are useful for selecting candidate regulators from those via JASPAR. Let $\vec{u} = (u_1, \ldots, u_N)$ be the expression profile of gene u and $\vec{v} = (v_1, \ldots v_N)$ be another expression profile of gene v. N is the time points of expression profile. Compute the correlation between $\vec{u}$ and $\vec{v}$ with the lag or lead of h time points as follows:

$$\begin{aligned} r(h) &\triangleq \left(\sum_{i=1}^{N-h} (u_{i+h} - \overline{u})(v_i - \overline{v})\right) \Big/ \left(\sqrt{\sum_{i=1}^{N-h} (u_{i+h} - \overline{u})^2} \cdot \sqrt{\sum_{i=1}^{N-h} (v_i - \overline{v})^2}\right) \\ \overline{u} &\triangleq \sum_{i=1}^{N-h} u_{i+h} \Big/ (N-h),\ \overline{v} \triangleq \sum_{i=1}^{N-h} v_i \Big/ (N-h), \quad \text{for } h = -M, \ldots, 0, \ldots, M \end{aligned} \tag{7.3}$$

Here M is the maximal time lead or lag between each two genes. Initially we do not know which are the target genes and which are the regulator genes. Since each time-interval in h is a half-hour, we allow 2 h lead and lag and compute the correlation between a gene and a TF with all possible time lags or leads that are less than 2 h for regulatory response.

Finally, we select the maximum correlation between two genes with different time delays or time leads as their correlation and rank them as in Fig. 7.2 for all regulatory genes. We can obtain the distribution of correlation based on their ranks. Then, we can decide a threshold for a possible regulatory relationship between regulators and their target genes (Fig. 7.2). In this section, a correlation larger than 30% (or 0.46451) is selected as a threshold for possible regulators, which is used to truncate all impossible regulators from the pool of regulators suggested by JASPAR via DNA sequence similarity analysis. Then, we link the remainder regulators selected by cross-correlation threshold with their target genes to construct a rough GRN. After the rough GRN of inflammation system is constructed by integrating target genes with their regulators selected by cross-correlation, the rough GRN is modeled by the dynamic Eq. (7.1) for further pruning. Therefore, the kinetic parameters of regulatory dynamic model are identified by the maximum likelihood parameter estimation via microarray data. After parameter identification, the insignificant interaction coefficients of dynamic model are pruned by the most parsimonious AIC to refine the GRN in the inflammatory condition. The possible regulators selected by JASPAR algorithm are pruned two times in this method, once by correlation threshold via cross-correlation and again by AIC via dynamic model and microarray data. The details are described below.

Constructing a dynamic model for GRN via microarray data

After constructing the stochastic dynamic equation in Eq. (7.1) to model the regulation of a target gene, we use the method of maximum likelihood to estimate the kinetic parameters of dynamic model. Eq. (7.1) can be rewritten in the following form.

$$\begin{aligned} y(t+1) &= \begin{bmatrix} y(t) & x_1(t-\tau_i) & \ldots & x_L(t-\tau_i) & 1 \end{bmatrix} \begin{bmatrix} a \\ b_1 \\ \vdots \\ b_L \\ k \end{bmatrix} + \varepsilon(t) \\ &= \phi(t)\theta + \varepsilon(t) \end{aligned} \tag{7.4}$$

where $\phi(t)$ denotes the regression vector which can be obtained from microarray data, and $\theta \in R_p$ denotes the parameter vector of dimension p in regression Eq. (7.4).

After applying the cubic spline method to interpolate the microarray data, we can obtain as many data points as we want. Then it is easy to obtain values of $\{y(t+l)\ x_i(t+l)\}$ for $l \in \{1,\ldots,m\}$ and $i \in \{1,\ldots,L\}$, where m is the number of expression time points of a target gene, and L is the number of TFs binding to the target gene in the rough gene network. By further computation of Eq. (7.4) at different time points we can obtain the following vector form equation by data point interpolation.

$$\begin{bmatrix} y(t+2) \\ \vdots \\ y(t+m) \end{bmatrix} = \begin{bmatrix} \phi(t+1) \\ \vdots \\ \phi(t+m-1) \end{bmatrix} \theta + \begin{bmatrix} \varepsilon(t+1) \\ \vdots \\ \varepsilon(t+m-1) \end{bmatrix} \tag{7.5}$$

For simplicity, it can be represented as follows:

$$Y = \Phi\theta + e \tag{7.6}$$

In Eq. (7.6), the random noise $\varepsilon(t_k)$ is regarded as a random variables of white Gaussian noise with zero mean and unknown variance σ^2, i.e., $E\{e\} = 0$, and $\Sigma_e = E[ee^T] = \sigma^2 I$, where I is an identity matrix. In this section, a maximum likelihood parameter estimation method is used to estimate θ and σ^2 by the regression data obtained from the microarray data of regulatory genes and the target gene [33]. Under the assumption of the Gaussian noise vector e with $m-1$ elements, its probability density function is given as follows:

$$p(e) = \frac{1}{((2\pi)^{m-1} \det \Sigma_e)^{1/2}} \exp\left(-\frac{1}{2} e^T \Sigma_e^{-1} e\right) \tag{7.7}$$

From Eq. (7.7), we can obtain the likelihood function

$$L(\theta,\sigma^2)=P(\theta,\sigma^2)=\frac{1}{(2\pi\sigma^2)^{(m-1)/2}}\exp\left\{-\frac{(X-\Phi\bullet\theta)^T(X-\Phi\bullet\theta)}{2\sigma^2}\right\} \tag{7.8}$$

Eq. (7.8) can be considered as a function of parameters θ and σ^2. In order to simplify the computation, it is practical to take the logarithm of Eq. (7.8), which yields the following log-likelihood function:

$$\log L(\theta,\sigma^2)=-\frac{m-1}{2}\log(2\pi\sigma^2)-\frac{1}{2\sigma^2}\sum_{k=1}^{m-1}\left|y(t+k+1)-\phi(t+k)\bullet\theta\right|^2 \tag{7.9}$$

where $y(t+k)$ and $\phi(t+k)$ are the kth elements of Y and Φ in Eq. (7.6), respectively. By the maximum likelihood parameter estimation method, we expect the log-likelihood function to have the maximum at $\theta=\hat{\theta}$ and $\hat{\sigma}^2=\sigma^2$. The necessary conditions for the maximum likelihood estimates $\hat{\theta}$ and $\hat{\sigma}^2$ are as follows [23]:

$$\begin{aligned}\frac{\partial\log L(\theta,\sigma^2)}{\partial\theta}&=0\\ \frac{\partial\log L(\theta,\sigma^2)}{\partial\sigma^2}&=0\end{aligned} \tag{7.10}$$

The estimated parameters $\hat{\theta}$ and $\hat{\sigma}^2$ are shown below,

$$\hat{\theta}=(\Phi^T\Phi)^{-1}\Phi^TY \tag{7.11}$$

$$\hat{\sigma}^2=\frac{1}{m-1}\sum_{k=1}^{m-1}(y(t+k+1)-\varphi(t+k)\hat{\theta})^2=\frac{1}{m-1}(Y-\Phi\hat{\theta})^T(Y-\Phi\hat{\theta}) \tag{7.12}$$

where Y and Φ can be obtained from the microarray data of regulatory genes and the target gene. After obtaining the estimated parameter $\hat{\theta}$, the dynamic equation of the target gene in the estimated TRN can be expressed as follows:

$$y(t+1)=\hat{a}y(t)+\sum_{i=1}^{L}\hat{b}_ix_i(t-\hat{\tau}_i)+\hat{k}+\varepsilon(t) \tag{7.13}$$

where $\hat{a}$, $\hat{b}_i$, and $\hat{k}$ are obtained from Eq. (7.11) and the variance of $\varepsilon(t)$ is obtained from Eq. (7.12). Iteratively, one target gene at a time, we can construct the overall dynamic equations of TRN of inflammation, which are interconnected through the regulations $\sum_{i=1}^{L} \hat{b}_i x_i(t)$ of TFs.

Since some interaction coefficients $\hat{b}_i$ of the GRN in Eq. (7.13) are insignificant, they should be pruned off by the parsimonious AIC criterion. This is discussed in the "Pruning the GRN" section.

Pruning the GRN

First, in this section, we use the JASPAR database to identify plausible binding motifs of their TFs roughly and select candidate regulators from the pool of DNA sequence similarity analysis. A rough GRN of inflammation is constructed by linking target genes and their regulators with a cross-correlation threshold larger than 30% (see Fig. 7.2). Then we use the maximum likelihood estimation method to estimate the parameters of the dynamic model for a preliminary GRN of the inflammatory system.

Although the maximum likelihood estimation method can help us quantify the regulatory abilities of all the possible interactive candidates of regulators on target genes, we still do not know exactly how significantly the regulatory ability can be regarded as a true regulator. In order to determine whether a regulator is significant or not, a statistical approach based on model validation is proposed for evaluating the significance of these model parameters to prune the preliminary gene network. In this section, AIC is employed to validate the model order (or the number of model parameters) to determine the significance of these dynamic model parameters [23]. The AIC, which attempts to include both the estimated residual variance and the model complexity in one statistic, decreases as the residual variance $\hat{\sigma}^2$ decreases and increases as the number p of parameters increases. As the expected residual variance decreases with increasing p for nonadequate model complexities, there should be a minimum around the correct number p of network parameters. For a transcriptional regulatory model with p regulatory parameters to fit with data from N samples, the AIC can be written as follows [23]:

$$AIC(p) = \log\left(\frac{1}{N}(Y-\hat{Y})^T(Y-\hat{Y})\right) + \frac{2p}{N} \tag{7.14}$$

where $\hat{Y}$ denotes the estimated expression profile of the target gene, i.e., $\hat{Y} = \Phi\hat{\theta}$.

This is a tradeoff between residual variance and model order. The minimization of Eq. (7.14) will achieve the true model order (i.e., the number of regulators of the target gene) of the gene regulatory system [23]. After the statistical selection of p parameters by minimizing the AIC, we can easily determine whether the regulatory TFs candidate is a significant or just a false positive and then construct a refined GRN model for inflammation. Finally, evidence from previous studies is an important validation to support this refined GRN.

APPENDIX A: TABLES AND FIGURES

Dataset selection

We used the microarray data of Boldrick et al. [37] as these mRNA expression profiles. Gene expression in whole blood leukocytes was determined at 0, 0.5, 1, 2, 4, 6, 12, and 24 h after the intravenous administration of bacterial endotoxin to Human peripheral blood mononuclear cells. In their experiments, eight additional stresses were studied under identical conditions, we adopt the MONK-treated sample as this normal condition and the LPS-treated sample as the inflammatory condition. The infusion of endotoxin activates innate immune responses and presents with physiological responses of brief duration. It should be noted that there is an initial pro-inflammatory phase and a subsequent counter-regulatory phase, with resolution of virtually all clinical perturbation within 24 h.

Network construction

In the first step, we pick up 9 genes which are common induction and 10 genes of the host response which are common repression (see Tables A.5 and A.6). These genes are discussed in the innate immune response from microarray of Boldrick et al. [37] and we want to select the candidate regulators of these 19 target genes in diversity pathogen infect in innate immune response. We would construct the GRN in both inflammatory and normal cases. The section of regulators needs to analyze changes in blood leukocyte gene expression patterns by Temporal Relationship Identification Algorithm (TRIA) prediction algorithm in Chapter 4 [24]. The TRIA prediction algorithm could identify the regulators by the correlations among continuous time gene expressions with the assumption that the regulatory genes and target genes have a positively (negatively) temporal relationship if the target gene's expression profile is positively (negatively) correlated with the regulatory genes profile, possibly with time lags.

We follow the steps of the flowchart in the main manuscript then rebuild the GRN in both inflammation condition and normal condition (see Figs. A.10 and A.11). Fig. A.12 shows the gene network only in inflammatory condition but not in normal condition. The results of the selection in inflammation condition are shown in Table A.7.

Here we show Tables A.1–A.7 and Figs A.1–A.12.

Table A.1 Short characteristics of 49 target genes

Gene name	Description	Gene ontology: biological process
IL17C	Interleukin 17C	GO:0007166: cell surface receptor linked signal transduction GO:0007267: cell–cell signaling GO:0006954: inflammatory response
IL1A	Interleukin 1a	GO:0006954: inflammatory response
TNFA	Tumor necrosis factor	GO:0006959: humoral immune response GO:0006954: inflammatory response GO:0043123: positive regulation of I-kappaB kinase/NF-kappaB cascade GO:0051092: positive regulation of NF-kappaB transcription factor activity GO:0051023: regulation of immunoglobulin secretion
IL6	Interleukin 6	GO:0007267: cell–cell signaling GO:0006959: humoral immune response GO:0006954: inflammatory response GO:0045727: positive regulation of translation
IL1B	Interleukin 1b	GO:0007267: cell–cell signaling GO:0006954: inflammatory response GO:0007165: signal transduction
TLR4	Toll-like receptor 4	GO:0007249: I-kappaB kinase/NF-kappaB cascade GO:0042116: macrophage activation GO:0007165: signal transduction GO:0042088: T-helper 1 type immune response
NFATC3	Nuclear factor of activated T cells, cytoplasmic, calcineurin-dependent 3	GO:0006954: inflammatory response
SCYE1	Small inducible cytokine subfamily E, member 1 (endothelial monocyte-activating)	GO:0007267: cell–cell signaling GO:0006954: inflammatory response GO:0050900: leukocyte migration GO:0007165: signal transduction
TICAM2	Toll-like receptor adaptor molecule 2	GO:0043123: positive regulation of I-kappaB kinase/NF-kappaB cascade
HDAC4	Histone deacetylase 4	GO:0030183: B-cell differentiation GO:0006954: inflammatory response
HDAC5	Histone deacetylase 5	GO:0030183: B-cell differentiation GO:0006954: inflammatory response GO:0006355: regulation of transcription, DNA-dependent

(*Continued*)

Table A.1 (Continued)

Gene name	Description	Gene ontology: biological process
HDAC7A	Histone deacetylase 7A	GO:0030183: B-cell differentiation GO:0006954: inflammatory response
HDAC9	Histone deacetylase 9	GO:0030183: B-cell differentiation GO:0006954: inflammatory response
ITGB2	Integrin, beta 2	GO:0007267: cell–cell signaling GO:0006954: inflammatory response GO:0007159: leukocyte adhesion GO:0008360: regulation of cell shape GO:0030593: neutrophil chemotaxis
CXCL2	Chemokine (C-X-C motif) ligand 2	GO:0006954: inflammatory response
ALOX5	Arachidonate 5-lipoxygenase	GO:0006954: inflammatory response GO:0019370: leukotriene biosynthetic process GO:0006691: leukotriene metabolic process
NFKBIA	Nuclear factor of kappa light polypeptide gene enhancer in B-cells inhibitor, alpha	GO:0007253: cytoplasmic sequestering of NF-kappaB GO:0042345: regulation of NF-kappaB import into nucleus
NR3C1	Nuclear receptor subfamily 3, group C, member 1 (glucocorticoid receptor)	GO:0007165: signal transduction
CEBPD	CCAAT/enhancer binding protein (C/EBP), delta	GO:0006366: transcription from RNA polymerase II promoter
ANXA1	Annexin A1	GO:0007166: cell surface receptor linked signal transduction GO:0006954: inflammatory response
CYBB	Cytochrome b-245, beta polypeptide (chronic granulomatous disease)	GO:0006954: inflammatory response GO:0045087: innate immune response
AOAH	Acyloxyacyl hydrolase (neutrophil)	
REG3A	Pancreatitis-associated protein	GO:0008283: cell proliferation
FOS	V-fos FBJ murine osteosarcoma viral oncogene homolog	GO:0006954: inflammatory response GO:0006357: regulation of transcription from RNA polymerase II promoter
IRAK	Interleukin-1 receptor-associated kinase 1	GO:0007250: activation of NF-kappaB-inducing kinase activity GO:0045941: positive regulation of transcription GO:0007165: signal transduction
PLAA	Phospholipase A2-activating protein	GO:0007165: signal transduction
CCR7	Chemokine (C-C motif) receptor 7	GO:0006954: inflammatory response

(*Continued*)

Table A.1 (Continued)

Gene name	Description	Gene ontology: biological process
CXCL14	Chemokine (C-X-C motif) ligand 14	GO:0007267: cell–cell signaling GO:0007165: signal transduction
PLA2G4B	Phospholipase A2, group IVB (cytosolic)	GO:0006954: inflammatory response
NFRKB	Nuclear factor related to kappa B binding protein	GO:0006954: inflammatory response GO:0006366: transcription from RNA polymerase II promoter
MAPK10	Mitogen-activated protein kinase 10	GO:0007165: signal transduction
ADORA2A	Adenosine A2a receptor	GO:0008015: blood circulation GO:0007596: blood coagulation GO:0007267: cell–cell signaling GO:0006968: cellular defense response GO:0006954: inflammatory response GO:0006909: phagocytosis GO:0007600: sensory perception
SCCE	Kallikrein 7 (chymotryptic, stratum corneum)	GO:0008544: epidermis development
ADORA3	Adenosine A3 receptor	GO:0006954: inflammatory response GO:0007165: signal transduction
NFKB1	Nuclear factor of kappa light polypeptide gene enhancer in B-cells 1 (p105)	GO:0006954: inflammatory response GO:0045941: positive regulation of transcription GO:0006366: transcription from RNA polymerase II promoter
CCL18	Chemokine (C-C motif) ligand 18 (pulmonary and activation-regulated)	GO:0007267: cell–cell signaling GO:0006955: immune response GO:0009607: response to biotic stimulus GO:0007165: signal transduction
AMBP	Alpha-1-microglobulin/bikunin precursor	GO:0007155: cell adhesion GO:0050777: negative regulation of immune response
TACR1	Tachykinin receptor 1	GO:0009582: detection of abiotic stimulus GO:0006954: inflammatory response GO:0007638: mechanosensory behavior
KNG	Kininogen	GO:0006954: inflammatory response GO:0030195: negative regulation of blood coagulation GO:0007162: negative regulation of cell adhesion GO:0006939: smooth muscle contraction

(*Continued*)

Table A.1 (Continued)

Gene name	Description	Gene ontology: biological process
BLNK	B-cell linker	GO:0030183: B-cell differentiation GO:0006959: humoral immune response GO:0006954: inflammatory response GO:0007242: intracellular signaling cascade
ABCF1	ATP-binding cassette, subfamily F (GCN20), member 1	GO:0006954: inflammatory response GO:0006412: translation
HPSE	Heparanase	
TLR7	Toll-like receptor 7	GO:0051607: defense response to virus GO:0007249: I-kappaB kinase/NF-kappaB cascade GO:0045416: positive regulation of interleukin 8 biosynthetic process
IL22	Interleukin 22	GO:0006953: acute-phase response GO:0007267: cell–cell signaling GO:0006954: inflammatory response
GPR132	G-protein coupled receptor 132	GO:0007186: G-protein coupled receptor protein signaling pathway
IL1R	Interleukin-1 receptor antagonist protein precursor	GO:0006955: immune response GO:0006954: inflammatory response
TOLLIP	Toll interacting protein	GO:0007267: cell–cell signaling GO:0006954: inflammatory response GO:0007242: intracellular signaling cascade GO:0045321: leukocyte activation
IL8	Interleukin 8	GO:0007267: cell–cell signaling GO:0007186: G-protein coupled receptor protein signaling pathway GO:0006954: inflammatory response GO:0007242: intracellular signaling cascade GO:0030155: regulation of cell adhesion GO:0045091: regulation of retroviral genome replication
TNFR	CD27 antigen precursor	GO:0016064: immunoglobulin mediated immune response GO:0045579: positive regulation of B-cell differentiation GO:0008588: release of cytoplasmic sequestered NF-kappaB

Table A.2 The inflammatory genes and their regulators

Gene name	(A) Possible regulators from JASPAR	(B) Candidate regulators from cross-correlation threshold	(C) Refined regulators from AIC
IL17	RUNX1,SOX9,RORA,MEF2A,HLF,TFAP2A, NFIL3,ELK1,FOXD1,FOXL1,GATA2, FOXI1,IRF1,YY1,REL,RELA,NFKB1,SPIB	RUNX1,SOX9,RORA,HLF,TFAP2A, NFIL3,ELK1,FOXD1,FOXL1,GATA2, FOXI1,YY1,REL,RELA,NFKB1,SPIB	RUNX1,SOX9,RORA,HLF,TFAP2A, NFIL3,ELK1,FOXD1,GATA2,FOXI1, YY1,REL,RELA,NFKB1,SPIB
IL1A	RUNX1,Pbx,SOX9,MEF2A,HLF,TFAP2A, E2F1,NFIL3,ELK1,FOXD1,FOXL1,GATA2, FOXI1,YY1,REL,RELA,NFKB1,SPIB	RUNX1,Pbx,SOX9,HLF,TFAP2A,E2F1, NFIL3,ELK1,FOXD1,FOXL1,GATA2, FOXI1,YY1,REL,RELA,NFKB1	RUNX1,Pbx,HLF,TFAP2A,NFIL3, ELK1,FOXD1,FOXL1,GATA2, FOXI1,YY1,REL,RELA,NFKB1
TNFA	TFAP2A,FOXD1,FOXL1,GATA2,FOXI1,MAX, YY1,NFKB1,SPIB	TFAP2A,FOXD1,FOXL1,GATA2,FOXI1, MAX,YY1,NFKB1,SPIB	TFAP2A,GATA2,FOXI1,MAX,YY1, NFKB1,SPIB
IL6	RUNX1,MEF2A,HLF,TFAP2A,ELK1,FOXL1, GATA2,FOXI1,YY1,REL,RELA,NFKB1,SPIB	RUNX1,HLF,TFAP2A,ELK1,FOXL1, GATA2,FOXI1,YY1,REL,NFKB1	RUNX1,HLF,TFAP2A,FOXL1,GATA2, FOXI1,YY1,REL,NFKB1
IL1B	SOX9,MEF2A,TFAP2A,ELK1,FOXD1,FOXL1, GATA2,FOXI1,YY1,REL,RELA,NFKB1,SPIB	SOX9,MEF2A,ELK1,GATA2,FOXI1, REL,RELA,NFKB1,SPIB	SOX9,MEF2A,ELK1,GATA2,FOXI1, RELA,NFKB1
TLR4	RUNX1,Pbx,SOX9,RORA,MEF2A,HLF, TFAP2A,NFIL3,ELK1,FOXF2,FOXD1, FOXL1,GATA2,FOXI1,IRF1,MAX,YY1, REL,RELA,SPIB	RUNX1,MEF2A,NFIL3,ELK1,FOXI1, IRF1,MAX,REL,RELA,SPIB	RUNX1,MEF2A,FOXI1,IRF1,MAX, REL,RELA,SPIB
NFATC3	SOX9,HLF,TFAP2A,NFIL3,ELK1,FOXD1, FOXL1,GATA2,FOXI1,MAX,YY1,REL	SOX9,HLF,NFIL3,ELK1,FOXD1,GATA2, FOXI1,MAX,YY1	SOX9,HLF,NFIL3,ELK1,FOXD1, GATA2,FOXI1,MAX
SCYE1	RUNX1,SOX9,TFAP2A,FOXD1,FOXL1, GATA2,FOXI1,YY1,REL,SPIB	RUNX1,TFAP2A,FOXL1,REL,SPIB	RUNX1,TFAP2A,FOXL1,REL,SPIB
TICAM2	SOX9,RORA,TFAP2A,ELK1,FOXL1,GATA2, YY1,REL,RELA,SPIB	SOX9,TFAP2A,ELK1,FOXL1,REL, RELA,SPIB	SOX9,TFAP2A,ELK1,FOXL1,RELA, SPIB
HDAC4	SOX9,HLF,TFAP2A,E2F1,NFIL3,ELK1,FOXL1, GATA2,YY1,REL	TFAP2A,FOXL1	TFAP2A,FOXL1
HDAC5	RUNX1,SOX9,HLF,TFAP2A,NFIL3,ELK1, FOXD1,FOXL1,GATA2,FOXI1,MAX,YY1, REL,SPIB	RUNX1,FOXD1,FOXL1,FOXI1,MAX, YY1,REL,SPIB	RUNX1,FOXD1,FOXL1,MAX,YY1, REL,SPIB
HDAC7A	RUNX1,SOX9,TFAP2A,E2F1,ELK1,FOXD1, FOXL1,GATA2,FOXI1,YY1,NFKB1,SPIB	SOX9,TFAP2A,E2F1,FOXD1,FOXL1, GATA2,FOXI1,YY1,NFKB1,SPIB	TFAP2A,E2F1,FOXD1,FOXL1, GATA2,FOXI1,YY1,NFKB1,SPIB
HDAC9	SOX9,RORA,MEF2A,TFAP2A,NFIL3,ELK1, FOXD1,FOXL1,GATA2,FOXI1,IRF1,YY1, REL,SPIB	MEF2A,NFIL3,IRF1,REL,SPIB	NFIL3,IRF1,REL

(*Continued*)

Table A.2 (Continued)

Gene name	(A) Possible regulators from JASPAR	(B) Candidate regulators from cross-correlation threshold	(C) Refined regulators from AIC
ITGB2	RUNX1,SOX9,TFAP2A,ELK1,FOXD1,FOXL1, GATA2,FOXI1,IRF1,MAX,YY1,SPIB	RUNX1,SOX9,TFAP2A,ELK1,FOXD1, FOXL1,GATA2,FOXI1,IRF1,MAX, YY1,SPIB	RUNX1,TFAP2A,ELK1,FOXD1, FOXL1,GATA2,FOXI1,IRF1,MAX, YY1,SPIB
CXCL2	RUNX1,Pbx,SOX9,RORA,E2F1,NFIL3, FOXF2,FOXD1,FOXL1,GATA2,FOXI1, IRF1,YY1,SPIB	RUNX1,Pbx,SOX9,RORA,E2F1,NFIL3, FOXF2,FOXD1,FOXL1,GATA2, FOXI1,IRF1,YY1	SOX9,RORA,E2F1,NFIL3,FOXD1, FOXL1,GATA2
ALOX5	RUNX1,Pbx,SOX9,RORA,MEF2A,TFAP2A, FOXD1,FOXL1,GATA2,FOXI1,MAX,YY1, SPIB	RUNX1,MEF2A,TFAP2A,FOXL1,SPIB	RUNX1,MEF2A,TFAP2A,SPIB
NFKBIA	RUNX1,SOX9,MEF2A,TFAP2A,FOXD1, FOXL1,GATA2,FOXI1,IRF1,YY1,SPIB	RUNX1,SOX9,MEF2A,GATA2,FOXI1, IRF1,SPIB	RUNX1,MEF2A,FOXI1,IRF1,SPIB
NR3C1	RUNX1,Pbx,SOX9,RORA,MEF2A,TFAP2A, FOXD1,FOXL1,GATA2,FOXI1,MAX,YY1, SPIB	RUNX1,Pbx,SOX9,RORA,MEF2A, TFAP2A,FOXD1,FOXL1,GATA2, FOXI1,YY1,SPIB	RUNX1,Pbx,SOX9,MEF2A,TFAP2A, FOXD1,FOXL1,GATA2,FOXI1, YY1,SPIB
CEBPD	TFAP2A,NFIL3,FOXF2,FOXD1,FOXL1, GATA2,FOXI1,YY1,REL,SPIB	NFIL3,FOXF2,FOXI1,REL,SPIB	NFIL3,FOXF2,FOXI1,SPIB
ANXA1	GATA2,YY1,SPIB	SPIB	SPIB
CYBB	RUNX1,SOX9,TFAP2A,FOXD1,FOXL1, GATA2,FOXI1,IRF1,YY1,REL,SPIB	FOXL1,IRF1,SPIB	FOXL1,SPIB
AOAH	RUNX1,SOX9,RORA,HLF,TFAP2A,ELK1, FOXF2,FOXD1,FOXL1,GATA2,FOXI1, YY1,REL,RELA,SPIB	RUNX1,TFAP2A,ELK1,FOXL1,FOXI1, RELA,SPIB	RUNX1,FOXL1,FOXI1,RELA
REG3A	SOX9,RORA,ELK1,FOXL1,GATA2,FOXI1, YY1,SPIB	SOX9,RORA,ELK1,FOXL1,GATA2, FOXI1,YY1,SPIB	SOX9,RORA,ELK1,FOXL1,GATA2, SPIB
FOS	RUNX1,SOX9,TFAP2A,E2F1,ELK1,FOXD1, FOXL1,GATA2,FOXI1,YY1,SPIB	RUNX1,TFAP2A,ELK1,FOXL1	RUNX1,TFAP2A,ELK1,FOXL1
IRAK	RUNX1,SOX9,RORA,TFAP2A,E2F1,ELK1, FOXD1,FOXL1,GATA2,MAX,YY1,SPIB	E2F1,MAX,SPIB	E2F1,SPIB
PLAA	RUNX1,Pbx,SOX9,RORA,MEF2A,HLF, TFAP2A,NFIL3,ELK1,FOXF2,FOXD1, FOXL1,GATA2,FOXI1,IRF1,MAX,YY1,SPIB	RUNX1,Pbx,SOX9,RORA,HLF,TFAP2A, ELK1,FOXF2,FOXD1,FOXL1,GATA2, FOXI1,MAX,YY1,SPIB	RUNX1,SOX9,TFAP2A,FOXF2, FOXD1,FOXL1,GATA2,MAX,SPIB

CCR7	RUNX1,SOX9,TFAP2A,ELK1,FOXD1, GATA2,IRF1,YY1,REL,SPIB	IRF1,REL,SPIB	IRF1,REL
CXCL14	RUNX1,SOX9,RORA,HLF,TFAP2A,E2F1, ELK1,FOXD1,FOXL1,GATA2,FOXI1,IRF1, YY1,REL,RELA,SPIB	RUNX1,SOX9,RORA,HLF,TFAP2A, E2F1,ELK1,FOXD1,FOXL1,GATA2, FOXI1,YY1,REL,RELA,SPIB	RUNX1,SOX9,HLF,TFAP2A,E2F1, FOXD1,FOXL1,GATA2,YY1,REL, RELA,SPIB
PLA2G4B	MEF2A,TFAP2A,E2F1,NFIL3,ELK1,FOXL1, GATA2,FOXI1,YY1,REL,SPIB	MEF2A,E2F1,NFIL3,GATA2,FOXI1, YY1,SPIB	MEF2A,NFIL3,FOXI1,YY1,SPIB
NFRKB	RUNX1,SOX9,RORA,MEF2A,TFAP2A,E2F1, FOXD1,FOXL1,GATA2,FOXI1,MAX,YY1, REL,SPIB	RUNX1,SOX9,RORA,E2F1,FOXD1, GATA2,FOXI1,MAX,YY1,REL,SPIB	RUNX1,SOX9,RORA,E2F1,FOXD1, REL,SPIB,GATA2,FOXI1,MAX, YY1
MAPK10	RUNX1,SOX9,MEF2A,HLF,TFAP2A,E2F1, ELK1,FOXL1GATA2,FOXI1,IRF1,YY1, REL,SPIB	RUNX1,SOX9,HLF,TFAP2A,E2F1, ELK1,FOXL1,GATA2FOXI1,YY1, REL,SPIB	RUNX1,SOX9,TFAP2A,E2F1,ELK1, FOXL1,GATA2,FOXI1,REL,SPIB
ADORA2A	RUNX1,SOX9,RORA,TFAP2A,E2F1,FOXD1, FOXL1,GATA2,IRF1,YY1,REL,RELA,SPIB	RUNX1,SOX9,RORA,TFAP2A,E2F1, FOXD1,FOXL1,GATA2,IRF1,YY1, REL,RELA,SPIB	RUNX1,RORA,TFAP2A,E2F1,REL, RELA,SPIB,FOXL1,GATA2,IRF1, YY1
SCCE	RUNX1,Pbx,SOX9,RORA,HLF,TFAP2A, ELK1,FOXD1,FOXL1,GATA2,FOXI1,IRF1, MAX,YY1,SPIB	RUNX1,Pbx,SOX9,RORA,HLF, TFAP2A,ELK1,FOXD1,FOXL1, GATA2,FOXI1,MAX,YY1,SPIB	RUNX1,SOX9,RORA,HLF,ELK1, FOXD1,FOXL1,GATA2,YY1,SPIB
ADORA3	RUNX1,Pbx,ELK1,FOXL1,GATA2,YY1	Pbx,GATA2	Pbx,GATA2
NFKB1	RUNX1,SOX9,TFAP2A,ELK1,FOXD1,FOXL1, GATA2,FOXI1,YY1,REL,SPIB	RUNX1,SOX9,ELK1,GATA2,FOXI1, YY1,REL,SPIB	RUNX1,SOX9,ELK1,GATA2,FOXI1, YY1,SPIB
CCL18	Pbx,SOX9,TFAP2A,NFIL3,FOXD1,FOXL1, GATA2,YY1,SPIB	Pbx,SOX9,TFAP2A,NFIL3,FOXD1, FOXL1,GATA2,YY1	Pbx,SOX9,TFAP2A,NFIL3,FOXL1, GATA2,YY1
AMBP	RUNX1,TFAP2A,ELK1,FOXF2,FOXD1, FOXL1,GATA2,FOXI1,MAX,YY1,REL, SPIB	RUNX1,TFAP2A,ELK1,FOXF2,FOXD1, FOXL1,GATA2,FOXI1,MAX,YY1, REL,SPIB	TFAP2A,FOXF2,FOXD1,FOXL1, GATA2,FOXI1,MAX,YY1,REL, SPIB
TACR1	RUNX1,SOX9,RORA,TFAP2A,E2F1,FOXD1, FOXL1,GATA2,MAX,YY1,SPIB	RUNX1,SOX9,RORA,TFAP2A,E2F1, FOXD1,FOXL1,GATA2,MAX,YY1, SPIB	RUNX1,SOX9,RORA,TFAP2A,E2F1, FOXD1,FOXL1,MAX,YY1,SPIB
KNG	RUNX1,SOX9,RORA,TFAP2A,FOXL1, GATA2,FOXI1,YY1,REL,SPIB	RUNX1,SOX9,RORA,TFAP2A,FOXL1, GATA2,FOXI1,YY1,REL,SPIB	RUNX1,SOX9,TFAP2A,FOXL1, GATA2,FOXI1,YY1,REL
BLNK	SOX9,MEF2A,NFIL3,ELK1,FOXF2,FOXD1, FOXL1,GATA2,FOXI1,YY1,SPIB	MEF2A,NFIL3,FOXI1,SPIB	MEF2A,FOXI1

(*Continued*)

Table A.2 (Continued)

Gene name	(A) Possible regulators from JASPAR	(B) Candidate regulators from cross-correlation threshold	(C) Refined regulators from AIC
ABCF1	RUNX1,SOX9,RORA,TFAP2A,E2F1,ELK1, FOXD1,FOXL1,GATA2,FOXI1,MAX,YY1, REL,RELA,NFKB1,SPIB	RUNX1,SOX9,E2F1,ELK1,GATA2, FOXI1,MAX,YY1,REL,RELA, NFKB1,SPIB	RUNX1,E2F1,ELK1,GATA2,FOXI1, MAX,YY1,REL,RELA,NFKB1
HPSE	RUNX1,SOX9,TFAP2A,E2F1,ELK1,FOXD1, FOXL1,GATA2,FOXI1,YY1,REL,SPIB	REL,SPIB	REL
TLR7	RUNX1,SOX9,MEF2A,HLFTFAP2A,FOXD1, FOXL1,GATA2,FOXI1,IRF1,YY1,REL, RELA,NFKB1,SPIB	RUNX1,MEF2A,IRF1,REL,RELA, NFKB1,SPIB	RUNX1,MEF2A,IRF1,REL,NFKB1, SPIB
IL22	RUNX1,SOX9,RORA,MEF2A,TFAP2A,E2F1, ELK1,FOXD1,FOXL1,GATA2,FOXI1,MAX, YY1,REL,NFKB1,SPIB	RUNX1,SOX9,RORA,TFAP2A,E2F1, ELK1,FOXD1,FOXL1,GATA2,FOXI1, MAX,YY1,REL,NFKB1,SPIB	RUNX1,SOX9,RORA,E2F1,MAX, YY1,REL,NFKB1,SPIB,ELK1, FOXD1,FOXL1,GATA2,FOXI1
GPR132	SOX9,RORA,MEF2A,TFAP2A,E2F1,FOXL1, GATA2,FOXI1,IRF1,YY1,NFKB1,SPIB	SOX9,RORA,MEF2A,TFAP2A,E2F1, FOXL1,GATA2,FOXI1,IRF1,YY1, NFKB1,SPIB	SOX9,RORA,MEF2A,TFAP2A,E2F1, FOXL1,GATA2,FOXI1,IRF1,YY1, SPIB
IL1R	RUNX1,SOX9,HLF,TFAP2A,NFIL3,ELK1, FOXD1,FOXL1,GATA2,FOXI1,YY1,REL, RELA,SPIB	TFAP2A,NFIL3,FOXL1,RELA,SPIB	TFAP2A,NFIL3,FOXL1,RELA
TOLLIP	RUNX1,SOX9,RORA,MEF2A,HLF,TFAP2A, NFIL3,ELK1,FOXD1,FOXL1,GATA2, FOXI1,MAX,YY1,REL,RELA,NFKB1,SPIB	RUNX1,SOX9,RORA,HLF,TFAP2A, NFIL3,ELK1,YY1,REL,RELA, NFKB1,SPIB,FOXD1,FOXL1,GATA2, FOXI1,MAX	RUNX1,SOX9,RORA,HLF,TFAP2A, NFIL3,ELK1,YY1,REL,RELA, NFKB1,FOXD1,FOXL1,GATA2, FOXI1,MAX
IL8	RUNX1,Pbx,SOX9,RORA,MEF2A,HLF, TFAP2A,E2F1,NFIL3,ELK1,FOXF2,FOXD1, FOXL1,GATA2,FOXI1,MAX,YY1,REL, RELA,SPIB	RUNX1,Pbx,SOX9,RORA,MEF2A,HLF, E2F1,NFIL3,ELK1,FOXF2,FOXD1, GATA2,FOXI1,MAX,YY1,REL,RELA	RUNX1,SOX9,RORA,MEF2A,HLF, E2F1,NFIL3,ELK1,FOXF2,FOXD1, GATA2,FOXI1,MAX,YY1,REL, RELA
TNFR	Pbx,SOX9,TFAP2A,FOXL1,GATA2,FOXI1	FOXI1	FOXI1

Table A.3 The gene regulatory network in immune system of unactivated and inflammatory cells

	In unactivated condition	In inflammatory condition
IL17	RUNX1,SOX9,RORA,MEF2A,HLF,TFAP2A,NFIL3,ELK1,FOXD1,FOXL1, FOXI1,IRF1,YY1,REL,NFKB1,SPIB	RUNX1,SOX9,RORA,HLF,TFAP2A,NFIL3,ELK1,FOXD1, GATA2,FOXI1,YY1,REL,RELA,NFKB1,SPIB
IL1A	RUNX1,Pbx,SOX9,MEF2A,HLF,TFAP2A,E2F1,NFIL3,ELK1,FOXD1, FOXL1,FOXI1,YY1,REL	RUNX1,Pbx,HLF,TFAP2A,NFIL3,ELK1,FOXD1,FOXL1, GATA2,FOXI1,YY1,REL,RELA,NFKB1
TNFa	TFAP2A,FOXD1,GATA2,FOXI1,MAX,NFKB1	TFAP2A,GATA2,FOXI1,MAX,YY1,NFKB1,SPIB
IL6	RUNX1,MEF2A,HLF,FOXL1,GATA2,FOXI1,YY1,REL,RELA,SPIB	RUNX1,HLF,TFAP2A,FOXL1,GATA2,FOXI1,YY1,REL, NFKB1
IL1B	SOX9,MEF2A,TFAP2A,ELK1,FOXD1,FOXL1,GATA2,FOXI1,YY1,REL, NFKB1,SPIB	SOX9,MEF2A,ELK1,GATA2,FOXI1,RELA,NFKB1
TLR4	RUNX1,Pbx,SOX9,RORA,MEF2A,HLF,TFAP2A,NFIL3,ELK1,FOXF2, FOXD1,FOXL1,GATA2,FOXI1,IRF1,MAX,YY1,REL,RELA,SPIB	RUNX1,MEF2A,FOXI1,IRF1,MAX,REL,RELA,SPIB
NFATC3	NFIL3,ELK1,FOXD1,FOXI1,MAX,YY1,REL	SOX9,HLF,NFIL3,ELK1,FOXD1,GATA2,FOXI1,MAX
SCYE1	RUNX1,FOXI1,YY1,SPIB	RUNX1,TFAP2A,FOXL1,REL,SPIB
TICAM2	RORA,GATA2,YY1,REL,RELA	SOX9,TFAP2A,ELK1,FOXL1,RELA,SPIB
HDAC4	YY1	TFAP2A,FOXL1
HDAC5	NFIL3,FOXI1,YY1,REL,SPIB	RUNX1,FOXD1,FOXL1,MAX,YY1,REL,SPIB
HDAC7A	RUNX1,ELK1,FOXD1,GATA2,YY1,NFKB1,SPIB	TFAP2A,E2F1,FOXD1,FOXL1,GATA2,FOXI1,YY1,NFKB1,SPIB
HDAC9	NFIL3,SPIB	NFIL3,IRF1,REL
ITGB2	RUNX1,ELK1,FOXD1,GATA2,FOXI1,IRF1,MAX	RUNX1,TFAP2A,ELK1,FOXD1,FOXL1,GATA2,FOXI1,IRF1, MAX,YY1,SPIB
CXCL2	RUNX1,RORA,E2F1,NFIL3,FOXF2,GATA2,FOXI1,IRF	SOX9,RORA,E2F1,NFIL3,FOXD1,FOXL1,GATA2
ALOX5	RORA,FOXD1,YY1,SPIB	RUNX1,MEF2A,TFAP2A,SPIB
NFKBIA	MEF2A,FOXI1	RUNX1,MEF2A,FOXI1,IRF1,SPIB
NR3C1	RORA,FOXD1,FOXI1,YY1	RUNX1,Pbx,SOX9,MEF2A,TFAP2A,FOXD1,FOXL1,GATA2, FOXI1,YY1,SPIB
CEBPD	TFAP2A,FOXF2,FOXD1,FOXL1,GATA2,FOXI1,YY1,REL	NFIL3,FOXF2,FOXI1,SPIB
ANXA1	YY1,SPIB	SPIB
CYBB	FOXI1,IRF1,YY1,REL,SPIB	FOXL1,SPIB
AOAH	RORA,TFAP2A,FOXF2,FOXD1,FOXL1,YY1,REL,RELA,SPIB	RUNX1,FOXL1,FOXI1,RELA
REG3A	RORA,FOXI1,YY1,SPIB	SOX9,RORA,ELK1,FOXL1,GATA2,SPIB
FOS	SOX9,E2F1,GATA2,FOXI1,YY1,SPIB	RUNX1,TFAP2A,ELK1,FOXL1
IRAK	RUNX1,SOX9,RORA,E2F1,FOXD1,GATA2,MAX, YY1,SPIB	E2F1,SPIB
PLAA	RUNX1,MEF2A,FOXF2,FOXD1,GATA2,FOXI1,IRF1,MAX,YY1,SPIB	RUNX1,SOX9,TFAP2A,FOXF2,FOXD1,FOXL1,GATA2, MAX,SPIB
CCR7	IRF1,SPIB	IRF1,REL

(Continued)

Table A.3 (Continued)

	In unactivated condition	In inflammatory condition
CXCL14	GATA2,FOXI1,IRF1,YY1,REL,RELA,SPIB	RUNX1,SOX9,HLF,TFAP2A,E2F1,FOXD1,FOXL1,GATA2, YY1,REL,RELA,SPIB
PLA2G4B	E2F1,FOXI1,YY1,REL,SPIB	MEF2A,NFIL3,FOXI1,YY1,SPIB
NFRKB	RORA,FOXD1,FOXI1,SPIB	RUNX1,SOX9,RORA,E2F1,FOXD1,REL,SPIB,GATA2, FOXI1,MAX,YY1
MAPK10	E2F1,IRF1,REL,SPIB	RUNX1,SOX9,TFAP2A,E2F1,ELK1,FOXL1,GATA2,FOXI1, REL,SPIB
ADORA2A	RUNX1,RORA,FOXD1,GATA2,IRF1,YY1,REL, RELA,SPIB	RUNX1,RORA,TFAP2A,E2F1,REL,RELA,SPIB,FOXL1, GATA2,IRF1,YY1
SCCE	RUNX1,FOXD1,SPIB	RUNX1,SOX9,RORA,HLF,ELK1,FOXD1,FOXL1,GATA2, YY1,SPIB
ADORA3	RUNX1,GATA2,YY1	Pbx,GATA2
NFKB1	RUNX1,ELK1,REL,SPIB	RUNX1,SOX9,ELK1,GATA2,FOXI1,YY1,SPIB
CCL18	FOXD1,GATA2,YY1,SPIB	Pbx,SOX9,TFAP2A,NFIL3,FOXL1,GATA2,YY1
AMBP	FOXF2,FOXD1,FOXI1,MAX,YY1,REL,SPIB	TFAP2A,FOXF2,FOXD1,FOXL1,GATA2,FOXI1,MAX,YY1, REL,SPIB
TACR1	RUNX1,E2F1,FOXD1,YY1	RUNX1,SOX9,RORA,TFAP2A,E2F1,FOXD1,FOXL1,MAX, YY1,SPIB
KNG	RUNX1,FOXI1,YY1,REL,SPIB	RUNX1,SOX9,TFAP2A,FOXL1,GATA2,FOXI1,YY1,REL
BLNK	NFIL3	MEF2A,FOXI1
ABCF1	RUNX1,E2F1,FOXD1,YY1,RELA,NFKB1,SPIB	RUNX1,E2F1,ELK1,GATA2,FOXI1,MAX,YY1,REL,RELA, NFKB1
HPSE	E2F1,ELK1,FOXD1,FOXI1,YY1,REL,SPIB,	REL
TLR7	RUNX1,MEF2A,FOXD1,YY1,REL,SPIB	RUNX1,MEF2A,IRF1,REL,NFKB1,SPIB
IL22	RUNX1,SOX9,E2F1,FOXD1,GATA2,FOXI1,REL, NFKB1,SPIB	RUNX1,SOX9,RORA,E2F1,MAX,YY1,REL,NFKB1,SPIB, ELK1,FOXD1,FOXL1,GATA2,FOXI1
GPR132	E2F1,GATA2,IRF1,NFKB1,SPIB	SOX9,RORA,MEF2A,TFAP2A,E2F1,FOXL1,GATA2,FOXI1, IRF1,YY1,SPIB
IL1R	RUNX1,SOX9,HLF,TFAP2A,NFIL3,ELK1,FOXD1, GATA2,YY1,REL, RELA,SPIB	TFAP2A,NFIL3,FOXL1,RELA
TOLLIP	ELK1,FOXD1,GATA2,FOXI1,MAX,YY1,RELA,NFKB1,SPIB	RUNX1,SOX9,RORA,HLF,TFAP2A,NFIL3,ELK1,YY1,REL, RELA,NFKB1,FOXD1,FOXL1,GATA2,FOXI1,MAX
IL8	Pbx,SOX9,RORA,MEF2A,HLF,TFAP2A,E2F1,NFIL3,ELK1,FOXF2, FOXD1,FOXL1,GATA2,FOXI1,MAX,YY1,REL,SPIB	RUNX1,SOX9,RORA,MEF2A,HLF,E2F1,NFIL3,ELK1, FOXF2,FOXD1,GATA2,FOXI1,MAX,YY1,REL,RELA
TNFR	Pbx,SOX9,TFAP2A,FOXL1,GATA2,FOXI1	FOXI1

Table A.4 The parameters of the inflammatory gene regulator models in Figs. A.1–A.9 (A) and standard deviations of them (B)

A. The parameters of the inflammatory gene regulator models

a_1	286.3	b_1	1.14	c_1	0.74	d_1	1.48	e_1	−3.19	f_1	−37.22	g_1	1.026	h_1	0.94	i_1	−0.07
a_2	−256.5	b_2	−7.90	c_2	17.12	d_2	−7.888	e_2	−5.68	f_2	−92.1	g_2	3.74	h_2	1.02	i_2	124.4
a_3	2678	b_3	−6.78	c_3	−13.89	d_3	18.35	e_3	186.2	f_3	157.6	g_3	−2.81			i_3	−601.6
a_4	−3107	b_4	30.71	c_4	−6.779	d_4	−15.36	e_4	−59.94	f_4	−490	g_4	10.97			i_4	1113
a_5	−756.8	b_5	7.04	c_5	8.94	d_5	−1.19	e_5	−7.90	f_5	310.2	g_5	5.27			i_5	−431
a_6	−33.95	b_6	−10.74			d_6	4.21	e_6	94.24	f_6	308.8	g_6	5.69			i_6	29.29
a_7	−1538	b_7	−32.87			d_7	−2.06	e_7	−164.2	f_{17}	−12.31	g_7	−3.60			i_7	−928.5
a_8	571.5	b_8	13.85			d_8	−3.28	e_8	38.94	f_8	164.1	g_8	−5.63			i_8	−378.9
a_9	−255.4							e_9	−32.11	F_9	−117.6	g_9	−13.6			i_9	−37.33
a_{10}	−1458							e_{10}	−0.47	f_{10}	95.22					i_{10}	636.3
a_{11}	312.5									f_{11}	215.3					i_{11}	708.3
a_{12}	456									f_{12}	−6.02					i_{12}	−211.7
a_{13}	−98.25									f_{13}	−106.6					i_{13}	226.4
a_{14}	−3.59									f_{14}	−4.48					i_{14}	237.1
a_{15}	4.88									f_{15}	4.98					i_{15}	−260.6
										f_{16}	1.57					i_{16}	−268.5
																i_{17}	34.03

B. Standard deviations for each gene

IL1A	0.017
IL1B	0.012
IL1R	0.051
IL6	0.0035
IL8	0.0219
IL17	0.011
TNFA	0.046
TLR4	0.0059
TNFR	0.0842

Table A.5 Features of the host response: common induction

Gene symbol	Description	GO bio-function
CD40	TNF receptor superfamily member 5	GO:0042100: B-cell proliferation GO:0006954: inflammatory response GO:0030168: platelet activation GO:0043123: positive regulation of I-kappaB kinase/NF-kappaB cascade GO:0006461: protein complex assembly
CSF	Combining with macrophage colony stimulating factor to initiate a change in cell activity	GO:0008283: cell proliferation GO:0007275: multicellular organismal development GO:0007165: signal transduction
IL-1α	Interleukin-1 alpha precursor	GO:0006954: inflammatory response
IL-1β	Interleukin-1 beta precursor	GO:0007267: cell–cell signaling GO:0006954: inflammatory response GO:0008285: negative regulation of cell proliferation GO:0007165: signal transduction
IL2RA	Interleukin-2 receptor alpha chain precursor	GO:0008283: cell proliferation GO:0007166: cell surface receptor linked signal transduction GO:0006955: immune response
IL-3	Interleukin-3 precursor	GO:0007267: cell–cell signaling GO:0008284: positive regulation of cell proliferation
IL-8	Monocyte-derived neutrophil chemotactic factor	GO:0007267: cell–cell signaling GO:0007186: G-protein coupled receptor protein signaling pathway GO:0006954: inflammatory response GO:0007242: intracellular signaling cascade GO:0030155: regulation of cell adhesion GO:0045091: regulation of retroviral genome replication
TNF-α	Tumor necrosis factor precursor	GO:0006959: humoral immune response GO:0006954: inflammatory response GO:0043123: positive regulation of I-kappaB kinase/NF-kappaB cascade GO:0051092: positive regulation of NF-kappaB transcription factor activity GO:0051023: regulation of immunoglobulin secretion
TNFSF1	Lymphotoxin-alpha precursor	GO:0007267: cell–cell signaling

Table A.6 Features of the host response: common repression

Gene symbol	Description	GO bio-function
ADAM8	A disintegrin and metalloproteinase domain 8	GO:0005887: integral to plasma membrane
CCR1	C-C chemokine receptor type 1	GO:0007155: cell adhesion GO:0007267: cell–cell signaling GO:0007187: G-protein signaling, coupled to cyclic nucleotide second messenger GO:0006955: immune response GO:0006954: inflammatory response
CD14	Monocyte differentiation antigen CD14 precursor	GO:0007166: cell surface receptor linked signal transduction GO:0006909: phagocytosis
CD31	Platelet endothelial cell adhesion molecule precursor	GO:0030334: regulation of cell migration GO:0042060: wound healing
CD64	High-affinity immunoglobulin gamma Fc receptor I precursor	GO:0001788: antibody-dependent cellular cytotoxicity GO:0019884: antigen processing and presentation of exogenous antigen GO:0007166: cell surface receptor linked signal transduction GO:0042742: defense response to bacterium GO:0006911: phagocytosis, engulfment
CYBB	Cytochrome b-245, beta polypeptide	GO:0006954: inflammatory response GO:0045087: innate immune response
FPR	fMet-Leu-Phe receptor	GO:0006928: cell motility GO:0007186: G-protein coupled receptor protein signaling pathway GO:0007188: G-protein signaling, coupled to cAMP nucleotide second messenger GO:0007165: signal transduction
ITGAX	Leukocyte adhesion receptor p150,95	GO:0007155: cell adhesion GO:0009887: organ morphogenesis
NCF1	Neutrophil cytosol factor 1	GO:0006968: cellular defense response GO:0005625: soluble fraction
WASP	Wiskott–Aldrich syndrome protein	GO:0007596: blood coagulation GO:0006952: defense response GO:0008544: epidermis development GO:0006955: immune response

Table A.7 The inflammatory genes and their regulators

Gene name	(A) Possible regulators from JASPAR	(B) Candidate regulators from cross-correlation threshold	(C) Refined regulators from AIC
ADAM8	RUNX1,TFAP2A,CREB1,ELK1, GATA2,GATA3,MAX,SP1,SPI1, SPIB,YY1,REL,NFKB1	CREB1,ELK1,GATA2,GATA3, MAX,SP1,SPI1,SPIB,YY1, NFKB1	CREB1,ELK1,GATA2,GATA3, MAX,SP1,SPI1,SPIB,YY1, NFKB1
CCR1	HLF,IRF1,MEF2A,REL,RELA, RORA,RUNX1,SP1,SPI1,SPIB, YY1	IRF1,MEF2A,RORA,SP1,SPI1, YY1	IRF1,MEF2A,RORA,SP1,SPI1, YY1
CD14	E2F1,ELK1,GATA2 GATA3,HLF, IRF1,MAX,NFIL3,NFKB1,REL, RUNX1,SP1,SPI1,SPIB,YY1	E2F1,ELK1,GATA2 GATA3,HLF, IRF1,MAX,NFIL3,NFKB1, RUNX1,SP1,SPI1,YY1	E2F1,ELK1,GATA2 GATA3,HLF, IRF1,MAX,NFIL3,NFKB1, RUNX1,SP1,SPI1,YY1
CD31	E2F1,ELK1,GATA2 GATA3,IRF1, MAX,NFIL3,REL,RUNX1 SP1, SPI1,SPIB,YY1	E2F1,ELK1,GATA2 GATA3,IRF1, MAX,NFIL3,RUNX1,SP1,SPI1, YY1	E2F1,ELK1,GATA3 IRF1,MAX, NFIL3,RUNX1,SP1,SPI1,YY1
CD64	E2F1,ELK1,GATA2 GATA3,HLF, IRF1,MAX,NFIL3,REL,RELA, RUNX1,SPI1,SPIB,YY1	E2F1,ELK1,GATA2 GATA3,HLF, IRF1,MAX,NFIL3,RELA, RUNX1,SPI1,SPIB,YY1	ELK1,GATA2,GATA3,HLF,IRF1, MAX,NFIL3,RELA,RUNX1, SPIB,YY1
CYBB	GATA2,GATA3,IRF1,REL,RUNX1, SP1,SPI1,SPIB,SRY,YY1	GATA2,IRF1,RUNX1,SP1,SPI1, SPIB,YY1	GATA2,IRF1,SP1,SPI1,SPIB,YY1
FPR	CREB1,E2F1,ELK1 GATA2,GATA3, HLF MAX,MEF2A,REL,RORA, RUNX1,SP1,SPI1,SPIB,SRY,YY1	CREB1,E2F1,ELK1 GATA2,HLF, MAX,MEF2A,RORA,RUNX1, SP1,SPI1,SPIB,YY1	CREB1,E2F1,ELK1 GATA2,HLF, MAX,MEF2A,RORA,RUNX1, SP1,SPI1,SPIB,YY1
ITGAX	CREB1,E2F1,ELK1 GATA2,GATA3, IRF1,NFKB1,RELA,RORA, RUNX1,SP1,SPI1,SPIB,SRY,YY1	CREB1,E2F1,ELK1 GATA2, GATA3,IRF1,NFKB1,RELA, RORA,RUNX1,SP1,SPI1,YY1	CREB1,E2F1,ELK1 GATA2, GATA3,IRF1,NFKB1,RELA, RUNX1,SP1,SPI1,YY1
NCF1	ELK1,GATA2,GATA3,HLF,MEF2A, SP1 SPI1,SPIB,SRY,YY1	ELK1,GATA2,GATA3,HLF,MEF2A, SP1 SPI1,SPIB,SRY,YY1	ELK1,GATA2,GATA3,HLF,MEF2A, SPI1,SPIB,SRY,YY1
WASP	CREB1,ELK1,GATA2,GATA3, MAX,NFKB1,REL,RELA, RUNX1,SP1,SPI1,SPIB,SRY,YY1	CREB1,ELK1,GATA2,GATA3, MAX,NFKB1,RELA,RUNX1 SP1,SPI1,SPIB,YY1	CREB1,ELK1,GATA3,MAX, RELA,RUNX1,SPI1,SPIB,YY1

CD40	E2F1,ELK1,GATA2 GATA3,IRF1, IRF2,MAX,MEF2A,NFIL3, NFKB1,Pbx,REL,RELA,RORA, RUNX1,SP1,SPI1,SPIB,SRY,YY1	E2F1,ELK1,GATA2 GATA3,IRF1, IRF2,MAX,MEF2A,NFIL3, NFKB1,Pbx,REL,RELA,RORA, RUNX1,SP1,SPI1,SPIB,YY1	E2F1,ELK1,GATA2 GATA3,IRF1, IRF2,MAX,MEF2A,NFIL3, NFKB1,Pbx,REL,RELA,RORA, RUNX1,SP1,SPI1,SPIB,YY1
CSF	E2F1,GATA2,GATA3,IRF1,IRF2, MAX,NFKB1,REL,RELA, RUNX1,SP1,SPI1,SPIB	E2F1,GATA2,GATA3,MAX, NFKB1,RELA,RUNX1,SP1, SPIB	E2F1,GATA2,GATA3,MAX, NFKB1,RELA,RUNX1,SP1, SPIB
IL1A	E2F1,ELK1,GATA2 GATA3,HLF, MEF2A,NFIL3,Pbx,RUNX1,SPI1, SPIB,SRY,YY1	E2F1,ELK1,GATA2 GATA3,HLF, MEF2A,NFIL3,Pbx,RUNX1, SPI1,SPIB,YY1	ELK1,GATA2,GATA3,HLF,MEF2A, NFIL3,Pbx,RUNX1,SPI1,SPIB, YY1
IL1B	ELK1,GATA2,GATA3,MEF2A,REL, RELA,SP1,SPI1,SPIB,SRY,YY1	ELK1,GATA2,GATA3,MEF2A, RELA,SP1,SPI1,YY1	ELK1,GATA2,GATA3,MEF2A, RELA,SP1,SPI1,YY1
IL2RA	CREB1,E2F1,ELK1 GATA2,GATA3, HLF IRF1,MEF2A,NFIL3 REL, RELA,RORA,RUNX1,SPI1,SPIB, SRY,YY1	CREB1,E2F1,ELK1 GATA2, GATA3,HLF IRF1,MEF2A, NFIL3 RELA,RORA,RUNX1, SPI1,YY1	CREB1,E2F1,ELK1 GATA2, GATA3,HLF IRF1,MEF2A, NFIL3 RELA,RORA,RUNX1, SPI1,YY1
IL3	CREB1,E2F1,ELK1 GATA2,GATA3, HLF IRF1,MEF2A,NFIL3 NFKB1,REL,RELA,RORA, RUNX1,SP1 SPI1,SPIB,SRF,SRY, YY1	CREB1,E2F1,ELK1 GATA2, GATA3,HLF IRF1,MEF2A, NFIL3 NFKB1,RELA,RORA, RUNX1,SP1,SPI1,SRF,YY1	CREB1,E2F1,ELK1 GATA2, GATA3,IRF1,MEF2A,NFIL3, NFKB1,RELA,RORA,RUNX1, SP1,SPI1,SRF,YY1
IL8	E2F1,ELK1,GATA2 GATA3,HLF, MAX,MEF2A,NFIL3,Pbx,REL, RELA,RORA,RUNX1,SPI1,SPIB, SRY,YY1	E2F1,ELK1,GATA2 GATA3,HLF, MAX,MEF2A,NFIL3,Pbx,RELA, RORA,RUNX1,SPI1,SPIB,YY1	E2F1,ELK1,GATA3 HLF,MAX, MEF2A,NFIL3,Pbx,RORA, RUNX1,SPI1,SPIB,YY1
TNFA	GATA2,GATA3,MAX,NFKB1,SPI1, SPIB,SRY,YY1	GATA2,GATA3,MAX,NFKB1, SPI1,SPIB,YY1	GATA2,GATA3,MAX,NFKB1, SPI1,YY1
TNFSF1	CREB1,E2F1,ELK1 GATA2,GATA3, HLF IRF1,MAX,MEF2A REL, RELA,RUNX1 SP1,SPI1,SPIB, YY1	CREB1,ELK1,GATA2,GATA3, IRF1,MAX,MEF2A,RELA,SP1, SPI1,YY1	CREB1,ELK1,GATA2,GATA3, IRF1,MAX,MEF2A,RELA,SP1, YY1

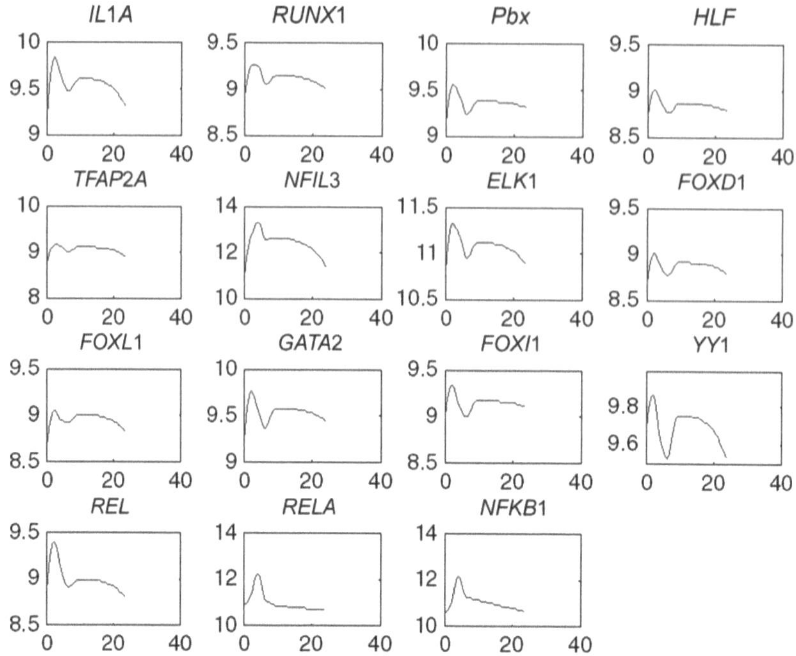

Figure A.1 The identified time delay of *IL1A* in Step 7.

$$
\begin{aligned}
X_{IL1A}(n+1) = {} & a_1 X_{IL1A}(n) + a_2 X_{RUNX1}(n) + a_3 X_{PBX}(n) \\
& + a_4 X_{HLF}(n) + a_5 X_{TFAP2A}(n) + a_6 X_{C-REL}(n) \\
& + a_7 X_{NFIL3}(n) + a_8 X_{ELK1}(n) + a_9 X_{FOXD1}(n) \\
& + a_{10} X_{FOXL1}(n) + a_{11} X_{GATA2}(n) + a_{12} X_{FOXI1}(n)
\end{aligned}
$$

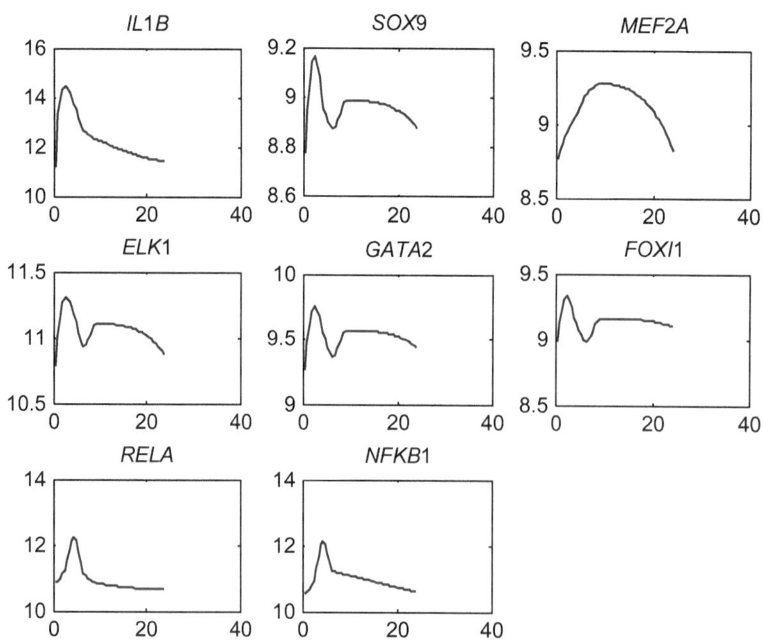

Figure A.2 The identified time delay of *IL1B* in Step 7.

$$
\begin{aligned}
X_{IL1B}(n+1) = {} & b_1 X_{IL1B}(n) + b_2 X_{SOX9}(n) + b_3 X_{MEF2A}(n-3) + b_4 X_{ELK1}(n) \\
& + b_5 X_{FOXI1}(n) + b_6 X_{GATA2}(n) + b_7 X_{NFKB1}(n) + b_8 X_{RELA}(n)
\end{aligned}
$$

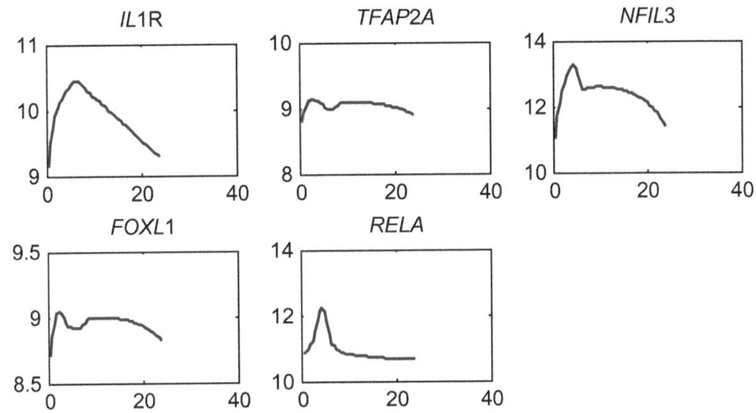

Figure A.3 The identified time delay of *IL1R* in Step 7.

$$X_{IL1R}(n+1) = c_1 X_{IL1R}(n) + c_2 X_{TFAP2A}(n) + c_3 X_{NFIL3}(n) + c_4 X_{FOXL1}(n) + c_5 X_{RELA}(n-4)$$

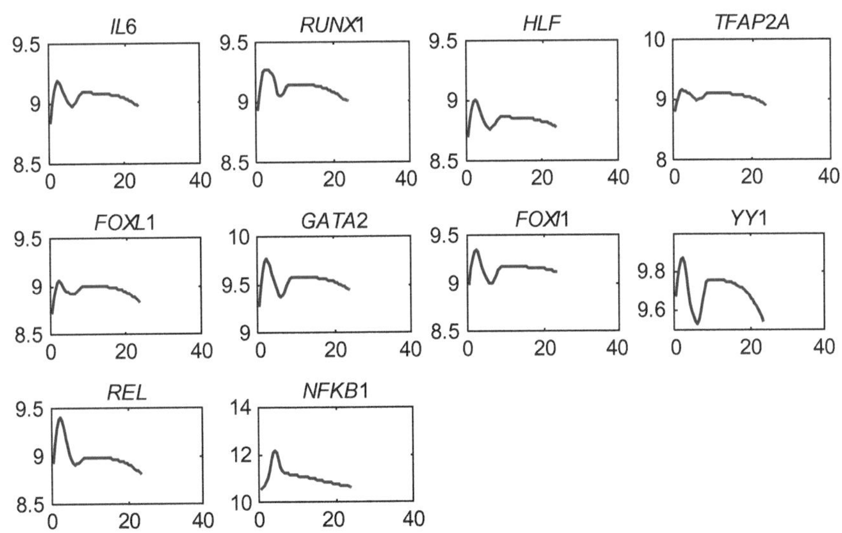

Figure A.4 The identified time delay of *IL6* in Step 7.

$$X_{IL6}(n+1) = e_1 X_{IL6}(n) + e_2 X_{RUNX1}(n) + e_3 X_{HLF}(n) + e_4 X_{TFAP2A}(n) + e_5 X_{FOXL2}(n) + e_6 X_{GATA2}(n) + e_7 X_{FOXI1}(n) + e_8 X_{YY1}(n) + e_9 X_{REL}(n) + e_{10} X_{NFKB1}(n)$$

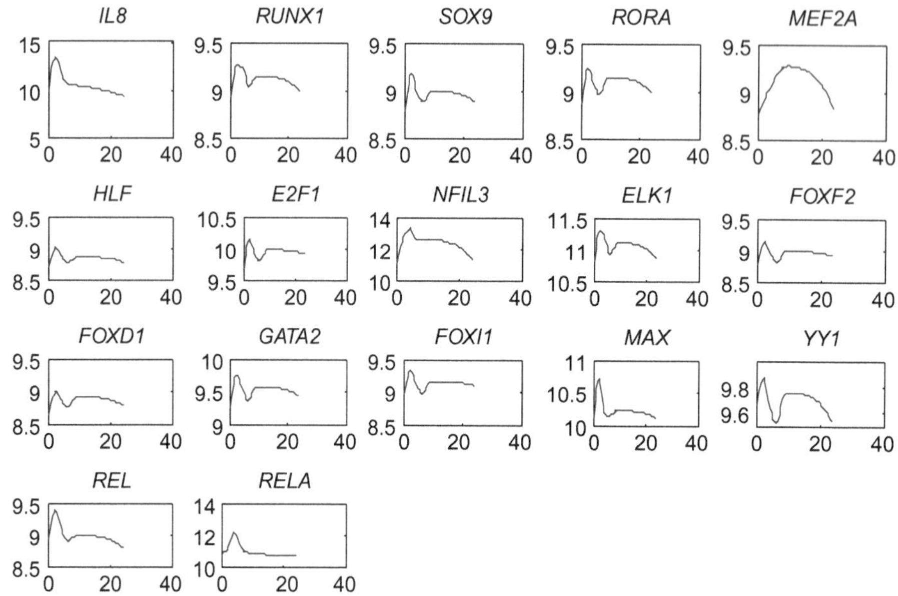

Figure A.5 The identified time delay of *IL8* in Step 7.

$$X_{IL8}(n+1) = i_1 X_{IL8}(n) + i_2 X_{PBX}(n) + i_3 X_{SOX9}(n-1) + i_4 X_{RORA}(n) + i_5 X_{MEF2A}(n-2) + i_6 X_{HLF}(n) + i_7 X_{E2F1}(n) + i_8 X_{NFIL3}(n-1) + i_9 X_{ELK1}(n-1) + i_{10} X_{FOXF1}(n-1)$$

7.1.5 Comments on systemic inflammatory mechanisms by constructing GRN

After comparing the reconstructed inflammatory GRN with the one in the text, we found some similarities and also some differences. The same highly connected hubs are GATA2, AML1 (RUNX1), and YY1. There are more than five connections for these hubs in both perturbed inflammatory networks. However, for the lack of some specific gene expression data in Ref. [37], we were unable to verify a part of highly interactive genes in the text (i.e., FOXL1, TFAP2A, and SOX9). Interestingly, we also found there are some hubs only present in the reconstructed network but not in the text like GATA3 and FPR, which would be involved in host defense against bacterial infection and in the clearance of damaged cells [38]. The reason that these 19 candidate genes still discovered new hubs is because some of 19 candidate genes are not included in the previous 49 genes. For different experimental conditions, research topics, and technology platforms, the data pool from different literature may be different. Therefore, the candidates of target genes we chose here differed from the text, so the computational results would not be identical.

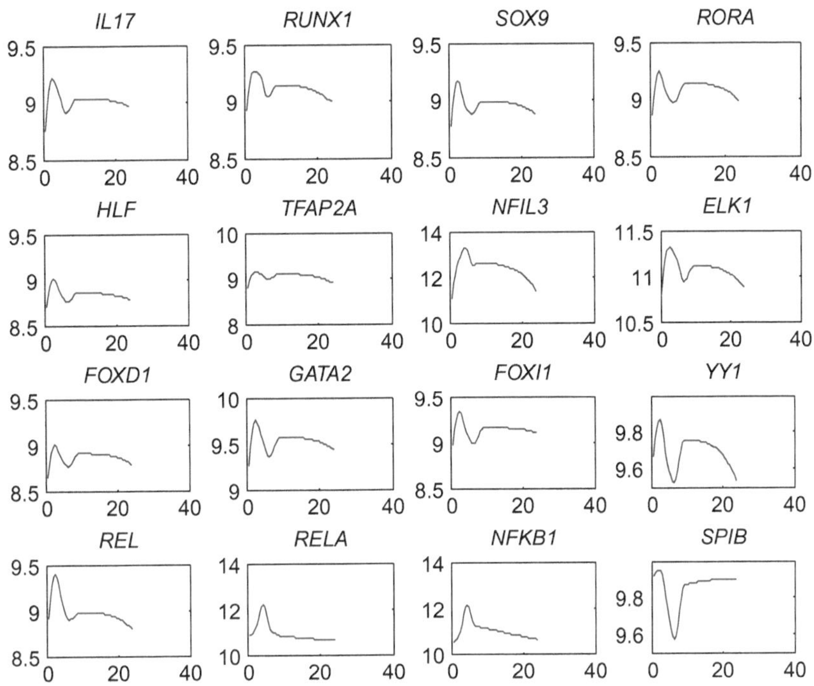

Figure A.6 The identified time delay of *IL17* in Step 7.

$$\begin{aligned}X_{IL17}(n+1) = & f_1 X_{IL17}(n) + f_2 X_{RUNX1}(n) + f_3 X_{RORA}(n) + f_4 X_{HLF}(n) \\ & + f_5 X_{TFAP2A}(n) + f_6 X_{NFIL3}(n) + f_7 X_{ELK1}(n) \\ & + f_8 X_{FOXD1}(n) + f_9 X_{FOXL1}(n) + f_{10} X_{GATA2}(n) \\ & + f_{11} X_{FOXI1}(n) + f_{12} X_{YY1}(n-1) + f_{13} X_{C-REL}(n) \\ & + f_{14} X_{RELA}(n) + f_{15} X_{NFKB1}(n) + f_{16} X_{SPIB}(n-2)\end{aligned}$$

7.2 IDENTIFICATION OF INFECTION AND DEFENSE-RELATED MECHANISMS VIA A DYNAMIC HOST–PATHOGEN INTERACTION NETWORK USING *C. ALBICANS*-ZEBRAFISH INFECTION MODEL

After introducing TRN of the systemic inflammation in response to bacteria endotoxin in the previous section, we are introducing cellular and intercellular infectious PPI networks between host and pathogen to investigate the infection and defense mechanism in this section. *C. albicans* is an opportunistic fungal pathogen responsible for various mucosal infections, such as candidiasis (e.g., oral thrush and vaginitis) and other potentially life-threatening diseases [39]. *C. albicans* is also the species most frequently responsible for hospital-acquired fungal infections. This pathogen can colonize various biomaterials, such as ventricular assist devices and urinary and vascular

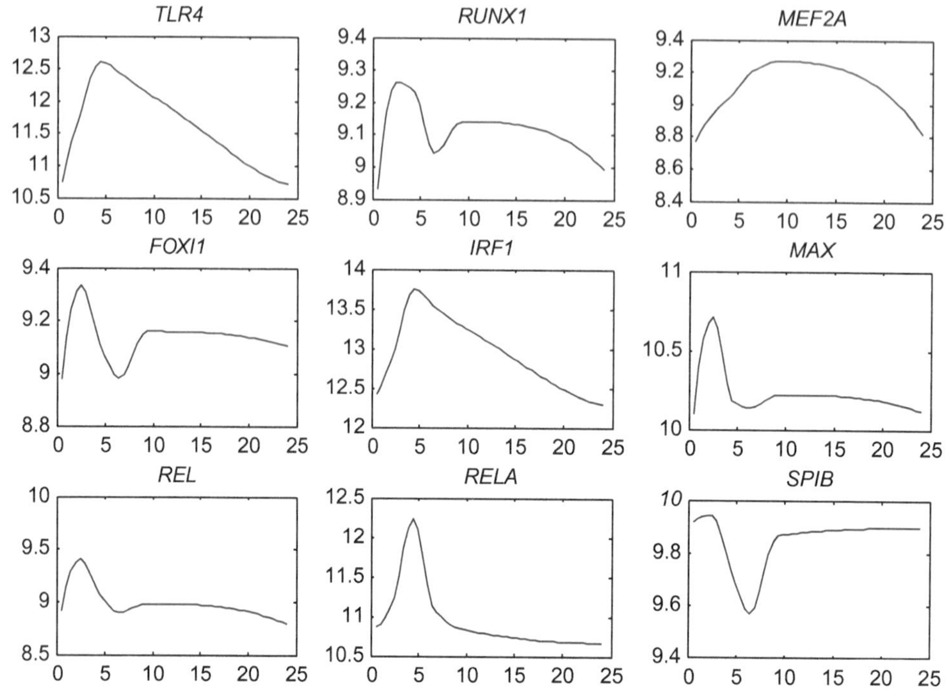

Figure A.7 The identified time delay of *TLR4* in Step 7.

$$\begin{aligned} X_{TLR4}(n+1) = & g_1 X_{TLR4}(n) + g_2 X_{RUNX1}(n-2) + g_3 X_{MEF2A}(n) \\ & + g_4 X_{FOXI1}(n) + g_5 X_{IRF1}(n) + g_6 X_{MAX}(n-4) \\ & + g_7 X_{C-REL}(n-4) + g_8 X_{RELA}(n-1) + g_9 X_{SPIB}(n) \end{aligned}$$

catheters, forming dense biofilms that are resistant to most antifungal drugs [40]. *C. albicans* infections and candidiasis are difficult to treat and create very serious therapeutic challenges. Mortality rates among patients with candidiasis can be as high as 40–60%, especially for those who have bloodstream infections [41]. Therefore, an understanding of the molecular mechanisms underlying the pathogenicity of *C. albicans* and of host defense systems could improve medical therapy and facilitate the development of new antifungal drugs.

Under normal circumstances, *C. albicans* lives in 80% of the human population with no harmful effects, although its overgrowth results in candidiasis, often observed in immunocompromised (e.g., HIV-positive) individuals [42,43]. *Candida albicans* can grow in a variety of morphological forms, ranging from yeast form to pseudohyphae form to true tubular hyphae form, depending on the growth conditions in the host environment [44]. A number of molecular factors have been implicated as associated with the virulence of *C. albicans*, such as host recognition biomolecules, secreted

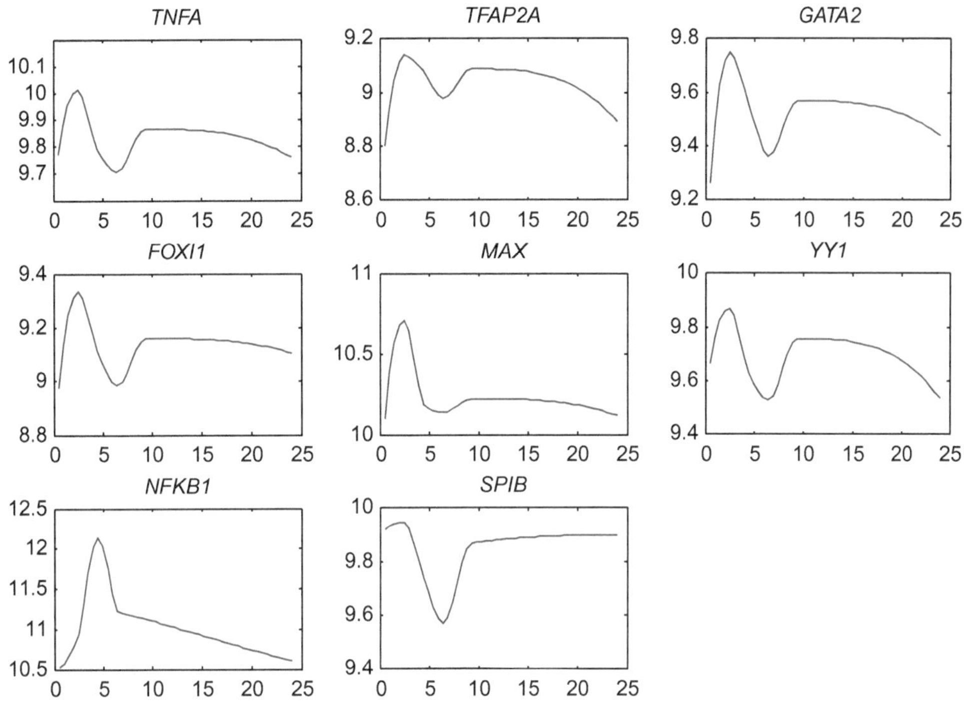

Figure A.8 The identified time delay of *TNFA* in Step 7.

$$X_{TNFA}(n+1) = d_1 X_{TNFA}(n) + d_2 X_{TFAP2A}(n) + d_3 X_{FOXD1}(n) + d_4 X_{GATA2}(n) + d_5 X_{FOXI1}(n) + d_6 X_{MAX}(n) + d_7 X_{NFKB1}(n) + d_8 X_{SPIB}(n-1)$$

aspartyl proteases and phospholipases, as well as life cycle factors like adhesion and morphogenesis [45]. Among those factors, the transition from yeast to hyphal form is considered to be critical for *C. albicans* pathogenesis. The ability of *C. albicans* to form hyphae has been proposed as a virulence factor, as these structures are often observed in invaded tissue and *C. albicans* strains unable to form hyphae (whether naturally or through introduced mutations) show defective infectivity [44,46]. Although previous studies have provided some insights, the details of the molecular mechanisms responsible for morphological forms still await elucidation. In this section, we utilized time series microarray data over 9 time points to construct two dynamic networks, which represent PPI in the adhesion stage (i.e., when hyphae are not growing) and the invasion stage (i.e., when they are). By comparing these two dynamic networks, we can investigate the details of molecular mechanisms responsible for changes in *C. albicans* infectivity across morphological forms.

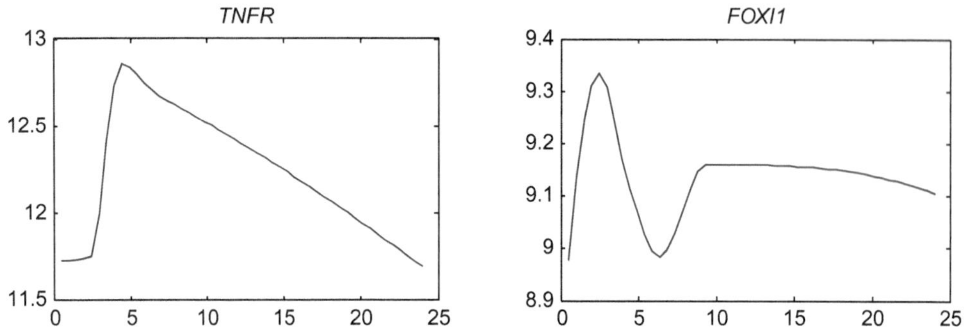

Figure A.9 The identified time delay of *TNFR* in Step 7.

$$X_{TNFR}(n+1) = h_1 X_{TNFR}(n) + h_2 X_{FOXI1}(n)$$

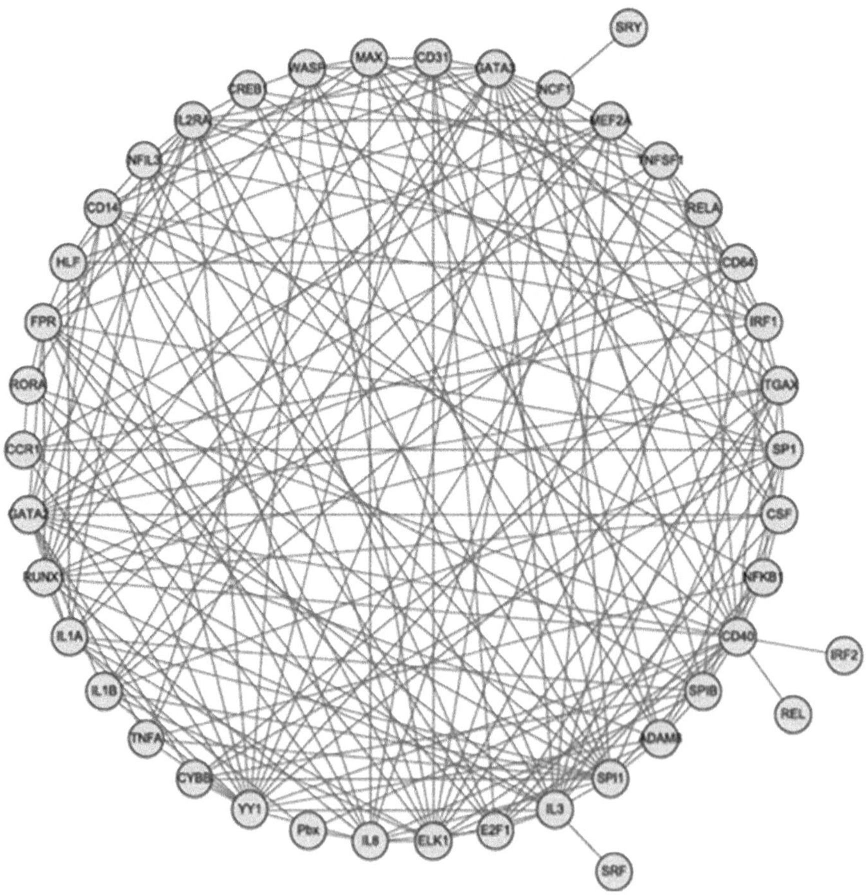

Figure A.10 The gene regulatory network in inflammatory condition.

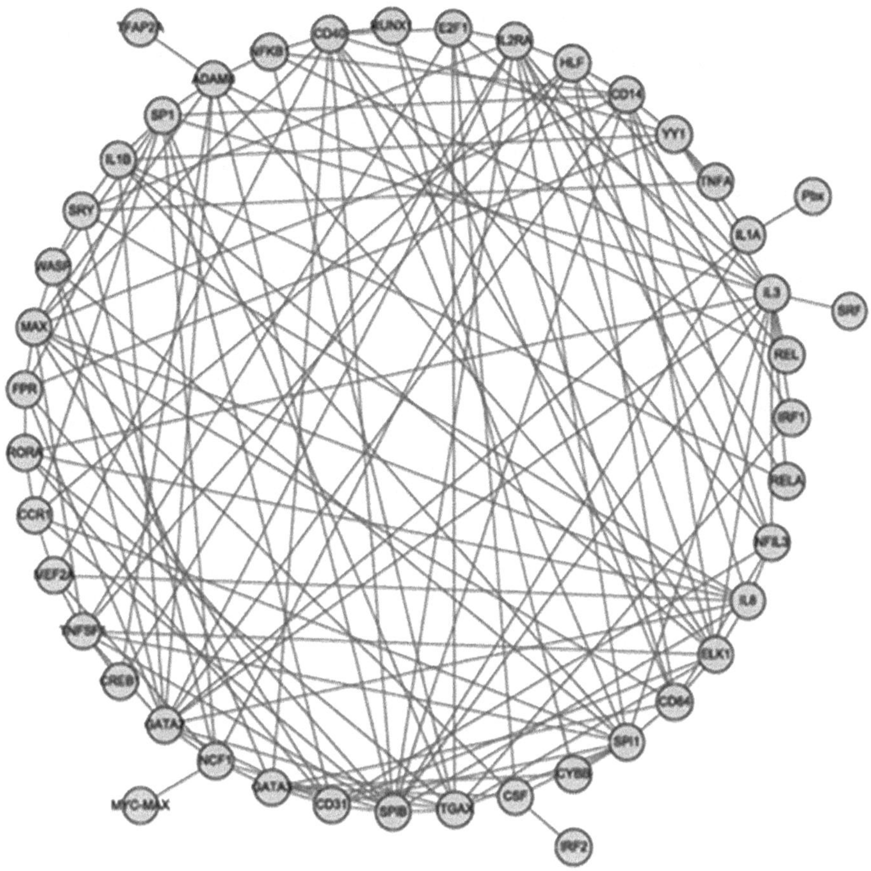

Figure A.11 The gene regulatory network in normal condition.

At present, we still lack sufficient high-throughput screening data for *C. albicans* such as PPI and ChIP-chip data, even though the genome for *C. albicans* has been identified, sequenced, and released to aid research on this significant pathogen [47]. The *C. albicans* genome for strain SC5314 has already been sequenced, revealing that almost two-thirds of its ~6000 open reading frames are orthologous to genes of *S. cerevisiae*, the most intensively studied eukaryotic model organism to have its entire genome sequenced [48,49]. The identification also revealed gene orthologs between *C. albicans* and *S. cerevisiae*. Unlike *C. albicans*, *S. cerevisiae* does not form true hyphae and is generally not considered a human pathogen. *S. cerevisiae* has abundant high-throughput screening data and it is closely related to *C. albicans* (i.e., both fall within the class *Hemiascomycetes*), the information from *S. cerevisiae* could be usefully adapted for our understanding of *C. albicans* biology and pathogenesis [49].

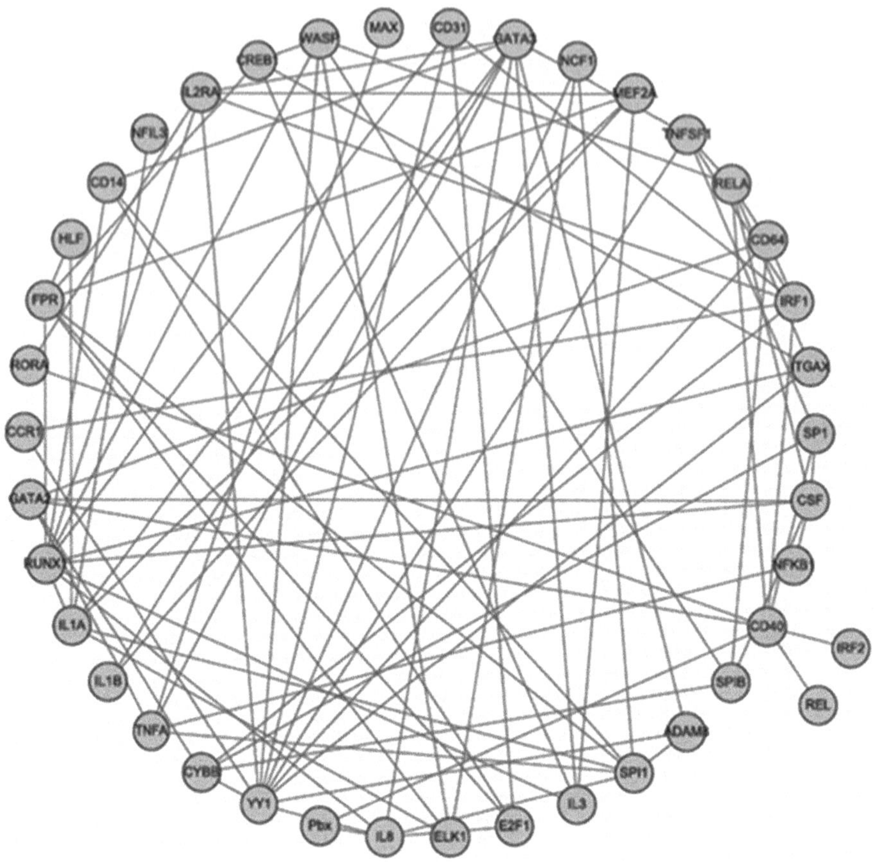

Figure A.12 Gene network only in inflammatory condition but not in normal condition.

The zebrafish (*Danio rerio*) has emerged as a powerful new vertebrate model for human disease. Numerous studies have already utilized the zebrafish system to study the pathogenesis of various human infectious diseases, including those caused by bacteria or viruses [50,51]. The zebrafish immune system displays remarkable similarities to mammalian counterparts. As a demonstration of the zebrafish's utility as a model organism for human disease, in 49 cases of a zebrafish mutant gene being cloned based on a forward genetic screen, the genes were found to have homologs in human diseases [52]. Overall, the zebrafish genetic map demonstrates highly conserved synteny with the human genome [53]. Chao et al. [54] have also demonstrated that *C. albicans* can colonize and invade the fish host at multiple anatomical sites and prove fatal in a dose-dependent manner. Therefore, a zebrafish infection model could be used to investigate the details of the *C. albicans* invasive process and infectious mechanisms.

In this section, we construct an infectious *C. albicans* and zebrafish intercellular PPI network by mining and integrating microarray data, PPI information and host−pathogen intercellular interactions in order to investigate how morphology regulates the infectious behavior of *C. albicans* on host tissue. Consequently, we discovered that all major hyphae-related pathways are visible in this hyphal PPI network, confirming the reliability and accuracy of these methods and results. From a systems perspective, we were able to predict the proteins with the largest changes in the number of interactions and the hub proteins for morphological switching processes. We identified several important hyphae growth-related proteins, e.g., Ubi4, Act1, Kex2, Hsl1, and Tsa1 and some proteins worth further exploration for pathogenicity research such as Hht21, Kre1, and Orf19.5438. Moreover, three noteworthy functions at work in *C. albicans* infection—cellular iron ion homeostasis, glucose transport and cell wall molecular biosynthesis—were named as pathogen invasion mechanisms from analysis of the integrated intercellular protein interaction networks. Furthermore, several functions, such as apoptosis and immune response, were also found to be involved in host defense mechanisms.

7.2.1 Materials and methods of identifying host and pathogen interaction networks for infection and defense mechanism

Simultaneous time-course microarray experiment during C. albicans *infection*

SC5314 strain *C. albicans* and adult AB strain zebrafish were used for the experiments to obtain two-side microarray data from host and pathogen, simultaneously. Their maintenance and preparation were performed according to procedures described previously [54]. Zebrafish were anesthetized by immersion in water containing 0.17 g/mL of Tricaine (Sigma) and then intraperitoneally injected with 1×10^8 CFU *C. albicans* cells suspended in 10 μL sterile phosphate-buffered saline. The infected fish were sacrificed by immersion in ice water at 0.5, 1, 2, 4, 6, 8, 12, 16, and 18 h postinfection and frozen in liquid nitrogen. *C. albicans*-infected zebrafish were treated with Trizol Reagent (Invitrogen, Carlsbad, CA, USA), pulverized in liquid nitrogen using a small mortar and pestle, and then disrupted using an MagNA Lyser System (Roche) with glass beads (cat. No. G8772-100G, Sigma) by shaking at 5000 rpm for 15 s. After phase separation by adding chloroform, the total RNA was purified using an RNeasy Mini Kit (Qiagen, Hilden, Germany). Purified RNA was quantified at OD260 nm using an ND-1000 spectrophotometer (NanoDrop Technology, Wilmington, DE, USA) and analyzed using a Bioanalyzer 2100 (Agilent Technologies, Santa Clara, CA, USA) with RNA 6000 Nano LabChip kit (Agilent Technologies). One microgram of the total RNA was amplified using a Quick-Amp Labeling kit (Agilent Technologies) and labeled with Cy3 (CyDye, PerkinElmer, Waltham, MA, USA) during the in vitro transcription process. 0.625 μg of Cy3 cRNA for the *C. albicans* array and 1.65 μg of Cy3 cRNA for the zebrafish array were fragmented to an average size of 50−100 nucleotides by incubation with fragmentation buffer at 60°C for 30 min. The fragmented labeled cRNA was then hybridized to both *C. albicans* and zebrafish oligo microarrays (Agilent Technologies) at 60°C for 17 h. After washing and drying using a nitrogen gun, microarrays were scanned using an Agilent microarray scanner (Agilent Technologies) at 535 nm for Cy3. For each time point, three biological replicates were done for both organisms. The raw data were processed with Loess normalization and the results have been deposited in the Gene Expression Omnibus (accession No. GSE32119; since the data are currently private, reviewers can access the data using the following link: http://www.ncbi.nlm.nih.gov/geo/query/acc.cgi?token=llqtnceackuwilu&acc=GSE32119).

Overview of the construction process of infection intercellular PPI network

The global screening method for infection-related proteins was divided into three key steps: (1) data selection and processing for proteins; (2) constructing dynamic hyphal growth networks for *C. albicans* and dynamic networks for *D. rerio*; and (3) constructing an intercellular PPI network between pathogen and host. The flowchart of the integrated cellular network construction is shown in Fig. 7.12. After constructing the overall network—which consists of the hyphal growth PPI network for *C. albicans*, the PPI network for *D. rerio* and the intercellular PPI network between pathogen and host—we searched for potential infection-related proteins and immune response pattern recognition molecules in both *C. albicans* and zebrafish.

Data selection and preprocessing

In this section, several types of data were mined and integrated to construct the integrated cellular network. In *C. albicans*, the required data included some of its microarray gene expression profiles, PPIs from *S. cerevisiae*, data on gene orthologs between *C. albicans* and *S. cerevisiae*, and gene annotations for *C. albicans*. There are 9 time points in the *C. albicans* microarray data spanning from 0.5 to 18 h postinfection (i.e., 0.5, 1, 2, 4, 6, 8, 12, 16, and 18 h). The gene ortholog data were acquired from the Candida Genome Database, and the *C. albicans* gene annotations were retrieved from the GO project. The PPI data for *S. cerevisiae* were extracted from the Biological General Repository for Interaction Datasets (BioGRID). In zebrafish, the data included microarray gene expression profiles, PPIs from *Homo sapiens*, data on human and zebrafish gene orthologs and functional gene annotations for zebrafish. There were 9 time points in the zebrafish microarray data spanning from 0.5 to 18 h (i.e., 0.5, 1, 2, 4, 6, 8, 12, 16, and 18 h). The gene ortholog data were acquired using the ZFIN, In Paranoid, and BLAST databases. The zebrafish gene annotations were retrieved from the GO project. The PPI data for *H. sapiens* were extracted from BioGRID and the Human Protein Reference Database.

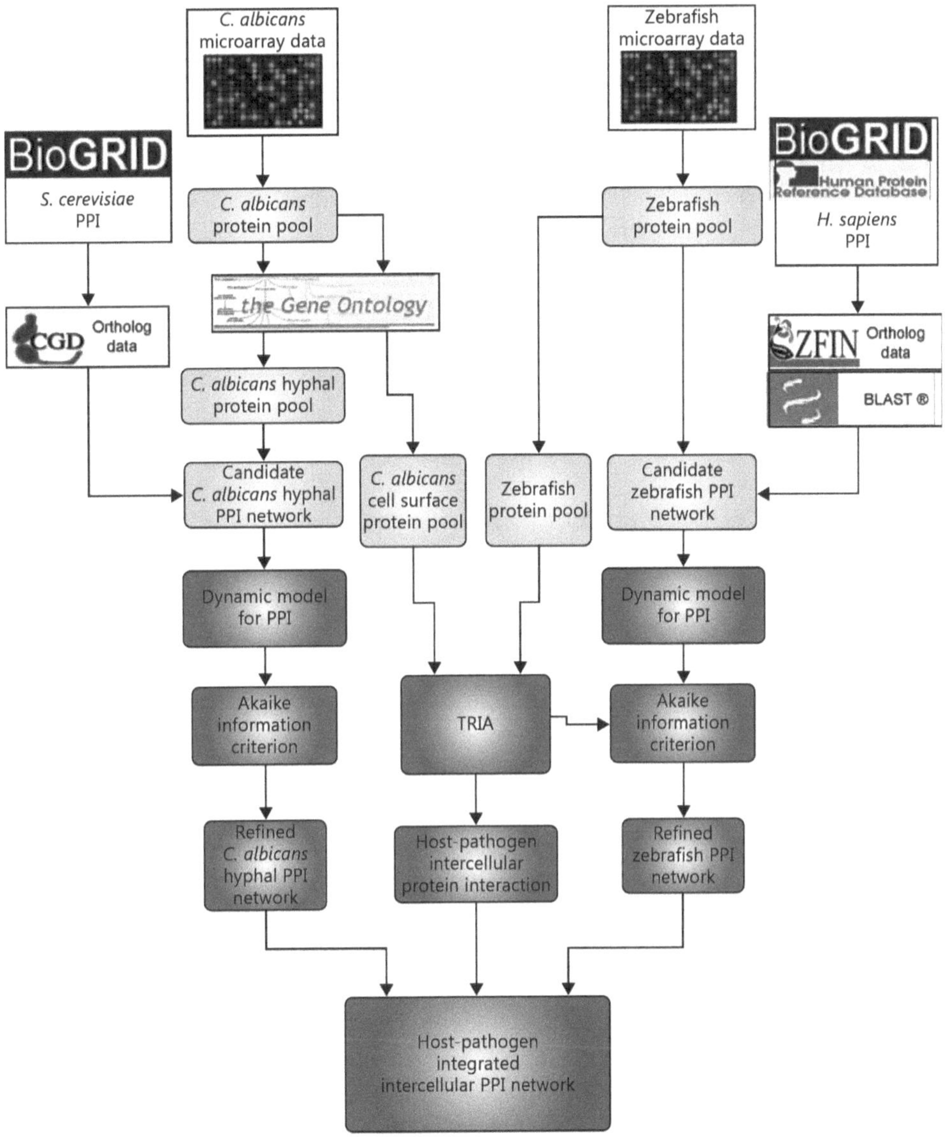

Figure 7.12 Flowchart of the construction of the integrated infection intercellular PPI network via database mining and integration. The construction of the integrated intercellular PPI network was performed by database mining and network identification. The network construction combines DNA microarray data with different types of information from various databases, as shown in the *white boxes*. The *blue boxes* (light gray in print versions) show the steps of candidate subnetwork construction. The bottom part (*orange boxes* (dark gray in print versions)) of the flowchart illustrates the steps of network identification and the subsequent construction of the integrated cellular network.

Selection of protein pool for rough (or candidate) PPI networks

Because of the lack of PPI databases between *C. albicans* and zebrafish at present, gene ortholog data between *C. albicans* and *S. cerevisiae* as well as between zebrafish and *H. sapiens* were utilized to set up protein data pools for these candidate *C. albicans* and zebrafish PPI networks, respectively. *C. albicans* PPIs can be inferred by applying ortholog data between *C. albicans* and *S. cerevisiae* to the latter's PPI data; similarly, zebrafish PPIs can be inferred by mapping *H. sapiens* PPI data to ortholog data between humans and zebrafish. Then, we set up the protein pool consisting of differentially expressed proteins. Since large-scale protein activity measurements are unobtainable, mRNA expression profiles were used as a substitute instead. Although the mRNA expression levels cannot be completely representative of the corresponding protein expression levels, they are at least partially and positively correlated [55,56]. The mRNA expression level for each protein was used to filter differentially expressed proteins using one-way analysis of variance (ANOVA), where the null hypothesis was the average mRNA expression levels at every time point being equivalent. In *C. albicans* and zebrafish alike, the proteins with p-values returned by ANOVA that were less than .01 were added to the protein pool. In this manner, we selected 4820 and 9665 proteins for inclusion in the protein pools of *C. albicans* and zebrafish, respectively. In this step, we found that a set of 4820 proteins is too large for constructing the PPI network for *C. albicans*. Because the high resultant PPI numbers exceed the size of the microarray data, we narrowed the protein pool of *C. albicans* to avoid overfitting in the parameter identification process for the PPI network construction. So, utilizing the GO database to further select a hyphal growth protein pool within the 4820 proteins set, we constructed a hyphal PPI network for *C. albicans* consisting of a subset of 403 proteins identified as related to hyphal growth. In addition, we were able to locate the beginning of hyphal growth in the body of the zebrafish at 2−4 h postinfection from microscopy images of the experiment (Fig. 7.13). Therefore, we selected 598 additional proteins of which mRNA levels changed by more than twofold in 1−6 h to another hyphal growth protein pool. Most of these 598 proteins had not yet been confirmed as associated with hyphal growth. Combining the 403 hyphae-related proteins with the 598 proteins having a more than twofold change in expression levels yielded 1001 proteins for the total hyphal growth protein pool. A candidate PPI network could be constructed based on this protein pool and PPI information. Since candidate PPI networks contain many false positive PPIs, the candidate PPI network was pruned using real-time series microarray data and based on a dynamic interaction model in "Dynamic model for organism protein interaction network during infection" section.

Dynamic model for organism protein interaction network during infection

The candidate PPI network can be depicted as a dynamic system in which interactive proteins and mRNA are considered as inputs of the system and protein activities as

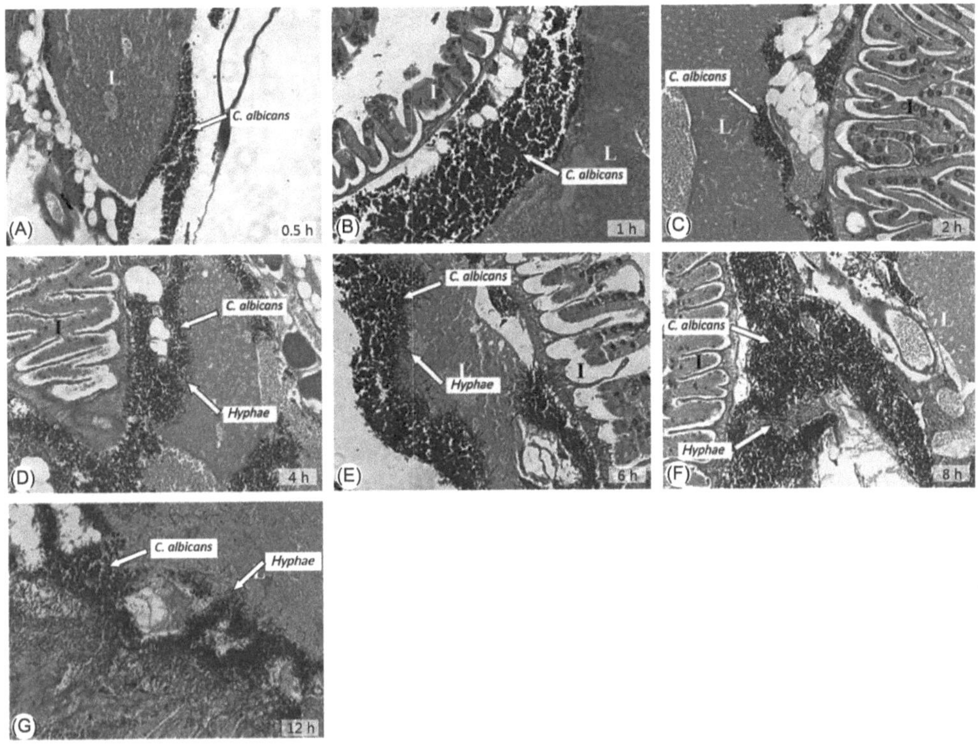

Figure 7.13 Experimental microscopy images of the infection process of *C. albicans* on zebrafish tissue: infection of zebrafish with *C. albicans*. Zebrafish sections were stained with HE and imaged by microscopy as described in Methods. The respective time points are 0.5 h (A), 1 h (B), 2 h (C), 4 h (D), 6 h (E), 8 h (F), and 12 h (G). "L" indicates liver and "I" indicates intestines. It is apparent that hyphae began to grow between the 2- and 4-h time points.

outputs of the system. All proteins in the candidate PPI network can be considered as target proteins. For a target protein p in the candidate PPI network with N interacting proteins, a dynamic model of the protein's activity can be represented as follows:

$$y_p(t+1) = y_p(t) + \sum_{q=1}^{Q_p} b_{pq} y_p(t) y_q(t) + \alpha_p x_p(t) - \beta_p y_p(t) + w_p(t), \quad \text{for } p = 1, \ldots, N \tag{7.15}$$

where $y_p(t)$ represents the activity level of p at time t, b_{pq} denotes the interaction ability of the qth interactive protein to p, $y_q(t)$ represents the protein activity level of the qth protein interacting with p, α_p denotes the translation rate from mRNA to protein, $x_p(t)$ represents the mRNA expression level of p, β_p indicates the decay rate of the protein, and $w_p(t)$ is stochastic noise. The PPI rate is proportional to the product of two proteins' concentrations [57] (i.e., proportional to the probability of molecular collisions between two proteins), and thus the interaction is modeled as a nonlinear multiplication scheme. The biological interpretation of Eq. (7.15) is that the protein activity level of target protein p at time $t+1$ is a function of the present protein activity level plus regulatory interactions with Q_p interactive proteins, plus additive translation effects from mRNA, minus present protein degradation effects and plus some stochastic noise. Because of the undirected nature of protein interactions, we did not assign direction for a two-protein interaction in the PPI subnetwork in Eq. (7.15). After the dynamic interaction model for the pth protein is constructed as in Eq. (7.15), the interaction parameters b_{pq}, translation parameter α_p and decay rate β_p can be estimated from microarray data in the following subsection (see Appendix B). Since the number of interactions in a candidate PPI network varies in the literature, dependent on the biological situation or condition targeted by a study, there exist many false positives and several interactions may not be relevant for these purposes. Therefore, the estimated interaction parameters $\hat{b}_{pq}$ should be pruned using the AIC model order selection method.

Determination of significant interaction pairings

When the regulatory interaction parameters $\hat{b}_{pq}$ have been identified, AIC [58] is then employed for both model order selection and determination of significant interactions in the infection PPI networks: i.e., to determine the number of interactions Q_p in Eq. (7.15). The AIC, which attempts to include both the estimated residual error and model complexity in one statistical measure, decreases as the residual variance decreases and increases as the number of interactions (i.e., complexity) increases [23].

$$AIC(Q_p) = 2\log \varepsilon_p + \frac{2Q_p}{L}, \quad \text{for } \varepsilon_p = Y_p - \Phi\hat{\theta}_p, \tag{7.16}$$

As the expected residual error decreases with increasing interactions for inadequate model complexities, there should be a minimum located near the correct interaction number [23,58]. Therefore, AIC can be used to select model order (i.e., the number of interactions) based on the protein interaction abilities $\hat{b}_{pq}$ identified above. In other words, we use the AIC model order selection method to reduce the likelihood of false positive PPIs from the candidate network using time profile microarray data to achieve a more realistic PPI network. After constructing the PPI networks for host and pathogen, we constructed a network for the protein interactions between pathogen and host to gain more insight into the offensive and defensive schemes of pathogen and host during the infection process.

Construction of an intercellular PPI network between pathogen and host

To identify the intercellular PPIs between pathogen and host during infection of zebrafish with *C. albicans*, we utilized the TRIA in Chapter 4, which uses gene expression data to identify a given TF's regulatory targets from its binding targets as inferred from ChIP-chip data [59]. The first step was to build a pool of *C. albicans* cell surface proteins. We used the GO database to select 195 cell surface proteins from the 4031-protein pool to build the resultant protein pool for *C. albicans*. Because host resistance against *C. albicans* infections is mediated predominantly by phagocytes, namely neutrophils and macrophages [60,61], we assumed that cell surface proteins of *C. albicans* may interact with any protein of zebrafish in the infectious process. So, we let $\vec{x} = (x_1, \ldots, x_N)$ denote the gene expression time profile of *C. albicans* cell surface protein x and $\vec{y} = (y_1, \ldots, y_N)$ denote the gene expression time profile of zebrafish protein y. We constructed the protein interactions between *C. albicans* and zebrafish via cross-correlation calculations of their time series microarray data.

We compute the cross-correlation between $\vec{x}$ and $\vec{y}$ with a lag of k time points as follows:

$$c(k) = \left(\sum_{i=1}^{N-k}(y_{i+k} - \overline{y})(x_i - \overline{x})\right) \Big/ \left(\sqrt{\sum_{i=1}^{N-k}(y_{i+k} - \overline{y})^2}\sqrt{\sum_{i=1}^{N-k}(x_i - \overline{x})^2}\right), \quad \text{for } k = 0, \ldots, T \tag{7.17}$$

where

$$\begin{aligned} \overline{x} &\triangleq \left(\sum_{i=1}^{N-k} x_i\right) / (N - K) \\ \overline{y} &\triangleq \left(\sum_{i=1}^{N-k} y_{i+k}\right) / (N - K) \end{aligned}, \tag{7.18}$$

and T is the maximal time lag after the *C. albicans* infection. We interpolated the 9 time points available for both *C. albicans* and zebrafish into 36 time points. The interval between each time point was 0.5 h. In this section, we set $T = 8$, meaning that we computed the cross-correlation between a *C. albicans* cell surface protein and a zebrafish protein for all possible time lags less than 4 h. Although the beginning of hyphal growth in the body of the zebrafish occurs at 2–4 h postinfection, we assumed the hyphae-related proteins of *C. albicans* might influence zebrafish proteins ahead of 4 h postinfection. Then, we tested the null hypothesis (H_0): $c(k) = 0$ (i.e., the cell surface proteins of *C. albicans* and zebrafish proteins are uncorrelated) and the alternative hypothesis (H_α): $c(k) \neq 0$ by the bootstrap method [62] to obtain a p-value $p(k)$ (see Appendix B). After all cross-correlations were calculated, we set the constraint that cross-correlation levels must be higher than 0.95. The PPIs satisfying this constraint were considered as potential intercellular PPIs between *C. albicans* and zebrafish.

7.2.2 Investigating infection and defense mechanisms by host and pathogen interaction network

Construction of the integrated intercellular PPI network during infection

This section aimed to construct the integrated intercellular interaction network between the hyphal proteins of *C. albicans* and zebrafish proteins during the infection process. The flowchart detailing its construction is shown in Fig. 7.12 and has three main routes, among which two separately construct the hyphal PPI network of *C. albicans* and the PPI network of zebrafish. The third constructs the host–pathogen intercellular PPI network. Based on the microarray data, we selected 4820 and 9665 proteins for inclusion in the source protein pools of *C. albicans* and zebrafish, respectively. In addition, we selected 1002 proteins for inclusion in the hyphal growth protein pool from the *C. albicans* protein pool due to the need to investigate what factors are behind the transition from yeast form to hyphal form in the infection process. In the candidate *C. albicans* hyphal PPI network, there were 3604 PPIs; in the candidate zebrafish PPI network, there were 1129.

We utilized the 9 time point *C. albicans* time series microarray data to construct two dynamic networks for different infection stages. Since hyphae appear to begin to grow in the zebrafish body from 2 to 4 h postinfection in the experimental microscopy images (Fig. 7.13), we collected two groups of data at different stages of infection to construct two separate networks. With the *C. albicans* microarray data spanning 0.5–4 h, we constructed a network called the "adhesive stage network," which represents *C. albicans* cells in the adhesion stage. Since cubic spline interpolation requires at least four data points to solve a cubic polynomial [63], we included the 4-h data point to construct this network. With the *C. albicans* microarray data spanning 2–12 h, we constructed another network called the "hyphal stage network," which represents *C. albicans* cells transitioning to the hyphal form. Similarly, we collected two groups of data at different stages of infection to construct two separate PPI networks for zebrafish as well: one for microarray data from 0.5 to 4 h, and another for data from 2 to 12 h, named the zebrafish stage 1 network and zebrafish stage 2 network, respectively. By estimating the system parameters using the time series microarray data and selecting model order using the AIC measurement [23,58], the likelihood of false positive interactions in the potential PPI network for the infection process was reduced. Network refinement yielded 550 proteins with 2725 PPIs in the adhesive stage network and 555 proteins with 3171 PPIs in the hyphal stage network: these two networks could then be combined into the *C. albicans* dynamic hyphal PPI network for the infection process (Fig. B.1). Similar refinements in the zebrafish data returned 1248 proteins with 2344 PPIs in the zebrafish stage 1 network and 1265 proteins with 2379 PPIs in the zebrafish stage 2 network, and these two networks could then be combined into the zebrafish dynamic PPI network for the defensive process (Fig. B.2). The *C. albicans* dynamic hyphal PPI network, the zebrafish dynamic PPI network and the host–pathogen intercellular PPI network could be merged into an integrated infection intercellular PPI network.

The global system view of the *C. albicans*- and zebra fish-integrated infection intercellular PPI network is illustrated in Fig. 7.14. The entire integrated infection intercellular network can be divided into eight levels according to the location of

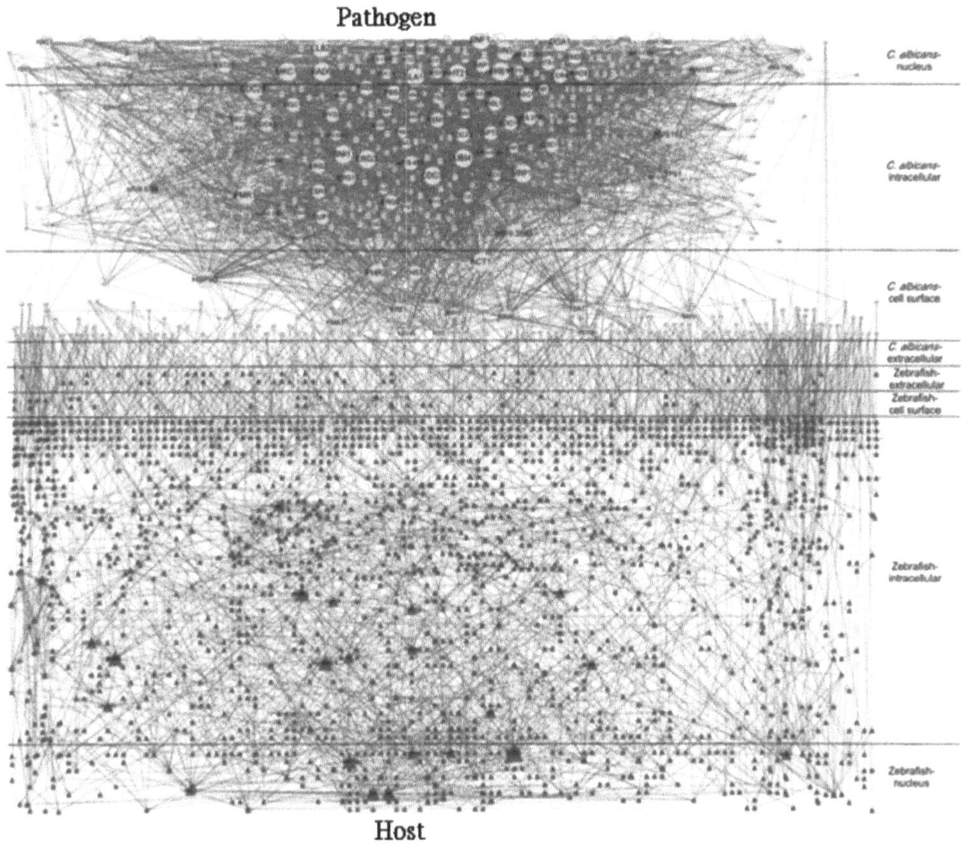

Figure 7.14 *C. albicans* and zebrafish integrated intercellular dynamic PPI network during *C. albicans* infection of zebrafish: the infectious intercellular network is composed of three subnetworks. The upper subnetwork is the dynamic hyphal PPI network of *C. albicans*. The middle subnetwork shows the host–pathogen intercellular interaction network. For simplicity, only the top five correlated interactions of the *C. albicans* cell surface proteins are listed. The bottom subnetwork is the dynamic defensive protein interaction network of zebrafish. This infectious intercellular PPI network contains lines and nodes of three different colors. The *red lines* (black in print versions) denote PPIs that did not appear in the stage 1 network but did in the stage 2 network. The *green lines* (dark gray in print versions) denote PPIs that appeared in the stage 1 network but did not in the stage 2 network. The *blue lines* (light gray in print versions) denote PPIs that appeared in both the stage 1 and 2 networks. The node size denotes connectivity degree. The drawing of the PPI network was created using Cytoscape.

protein action (i.e., nucleus, intracellular, cell surface, or extracellular) and species (i.e., *C. albicans* or zebrafish) and is composed of three subnetworks. The upper subnetwork is the dynamic hyphal PPI network of *C. albicans*. The middle subnetwork shows the host–pathogen intercellular interaction network. For simplicity, only the top five correlated interactions of the *C. albicans* cell surface proteins are listed. The bottom subnetwork is the dynamic defensive protein interaction network of zebrafish.

Inspection of the dynamic hyphal growth PPI network of C. albicans

In order to verify the accuracy of this dynamic hyphal growth protein interaction network, we investigated whether this network contains previously identified pathways related to hyphal growth, which are illustrated in Fig. B.3 [64]. This figure displays signal transduction pathways involved in regulating morphogenesis in *C. albicans*. An inspection of the integrated infectious intercellular network seen in Fig. 7.14 confirmed that this *C. albicans* dynamic hyphal PPI network includes the MAP kinase cascade, cyclic AMP/PKA pathway and other hyphae-associated pathways. We isolated these pathways from Fig. 7.14, and then constructed a new hyphae-related subnetwork as Fig. 7.15. It is apparent that this subnetwork is very similar to Fig. B.3. This new hyphae-related subnetwork contains almost all of the proteins and interactions of the already known hyphae-related pathways. The GTP-binding protein Ras1 was not contained in this dynamic hyphal growth PPI network because its *p*-value is greater than .01 in the original protein pool selection step for *C. albicans*; however, Ras2, which is in the same family as Ras1, appeared in the new subnetwork. Ras2 is similar to *S. cerevisiae* Ras2p, which can activate adenylate cyclase and is involved in *S. cerevisiae* pseudohyphal growth, and Ras2 mutants could have altered filamentous growth patterns [65]. Similar to Ras1 in Fig. B.3, Ras2 also stimulates Cyr1 (Cdc35), which in turn acts as an intracellular second messenger during morphological switching. Ras2 also stimulates Cdc42 through Ras-related protein (Rsr1), which is involved in budding, cell morphogenesis, and hyphal development processes [66]. In addition, an interaction between Cdc42 and Wal1 is represented with a *dotted line* in Fig. B.3, meaning this link is not completely known. However, we can see that Cdc42 links to Wal1 via Myo2 and Rho3 in this new hyphae-related subnetwork. Myo2 is required for polarized cell growth and dimorphic switching in *C. albicans* and is also involved in hyphal development [67]. Rho3 is required for polarized cell growth and cell separation and also involved in hyphal development [68]. Fortunately, the previously uncertain pathway between Cdc42 and Wal1 was more completely elucidated in this dynamic hyphal growth PPI network.

Aside from these three well-known hyphae-related signaling pathways—i.e., the MAP kinase and cyclic AMP signaling pathways and the polarized cell growth pathway—the pH-dependent Rim101 pathway was also identified from the dynamic integrated infection intercellular network. In this pathway, Nrg1 was not identified for

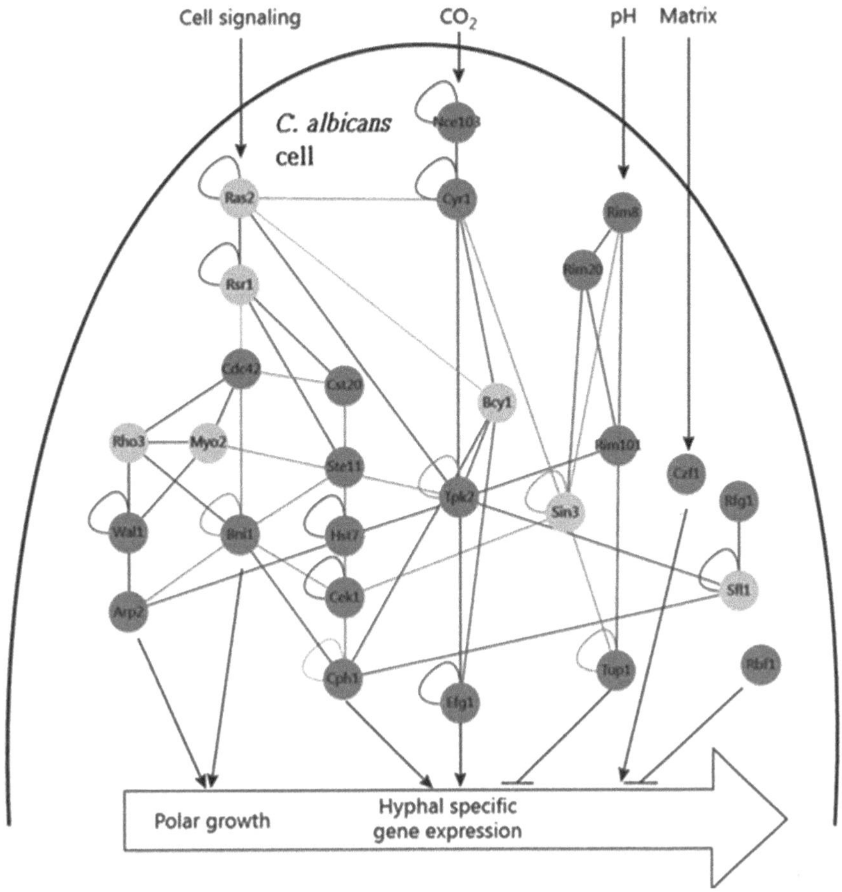

Figure 7.15 Signaling cascades involved in the dynamic hyphal growth protein interaction subnetwork of *C. albicans* at different stages of infection: the MAP kinase signaling pathway, cyclic AMP signaling pathway, polarized cell growth pathway, and Rim101 signaling pathway were identified in this dynamic hyphal growth PPI network of *C. albicans* as occurring during *C. albicans* infection of zebrafish. These signal transduction pathways are involved in the dynamic regulation of *C. albicans* morphological transitions. The *red lines* (dark gray in print versions) indicate PPIs that did not appear in the adhesive stage network but did in the hyphal stage network. The *green lines* (light gray in print versions) indicate PPIs that appeared in the adhesive stage network but did not appear in the hyphal stage network. The *blue lines* (gray in print versions) indicate PPIs that appeared in both the adhesive and hyphal stage networks. The *red nodes* (dark gray in print versions) indicate proteins that are also included in Fig. B.3; the *yellow nodes* (light gray in print versions) indicate proteins that are not included in Fig. B.3.

inclusion in Fig. 7.15 due to its *p*-value being greater than .01. However, Tup1, which has the same function as Nrg1, fits into the pathways shown in Fig. 7.15 and hence the pH-dependent pathway would seemingly be uninterrupted. In conclusion, 20 out of 22 proteins from already known pathways (i.e., those in Fig. B.3) are included in this *C. albicans* dynamic hyphal PPI network, in which the major hyphae-related pathways are all visible. These results verify the high accuracy of this infection intercellular PPI network. Moreover, this hyphal growth protein interaction subnetwork provides a dynamic and more complete hyphal network in comparison with Fig. B.3. Specifically, the *yellow nodes* (light gray in print versions) in Fig. 7.15 represent the proteins that are not contained in Fig. B.3. These proteins are all related to hyphal growth or filamentous growth, and the figure reflects the true pathways of these proteins.

Utilization of dynamic intracellular PPI networks to identify proteins with important roles in hyphal development mechanism

We utilized the dynamic intercellular PPI networks to investigate which proteins play important roles in hyphal development. Based on the dynamic hyphal PPI network of *C. albicans* during infection (Fig. B.1), we explored proteins of which interactions with other proteins displaying large variations between values in the adhesive and hyphal stage networks. In other words, the number of increased interactions and the number of reduced interactions for each protein would be summed to find out the largest interaction difference between the two stages. Table 7.3 lists the top

Table 7.3 Proteins with the largest changes in PPI number between the adhesive and hyphal stage networks during infection

C. albicans protein	Increased interactions	Reduced interactions	Total changed interactions	GO functional annotation	
				Hyphal growth	Filamentous growth
Ubi4	37	23	60		+
Act1	31	12	43	+	
Hsp90	20	15	35		+
Sla1	22	8	30	+	
Bnil	22	7	29	+	+
Sin3	22	6	28		+
Hht21	19	7	26		
Mkc1	16	10	26	+	+
Phr2	17	8	25		+
Pmr1	17	7	24	+	
Rvs161	14	9	23	+	
Rad6	16	6	22		+
Hgc1	15	6	21	+	
Vrp1	16	4	20	+	
Clb2	12	7	19		+

The top 15 *C. albicans* proteins in the hyphal growth PPI networks are ranked by change in the number of protein interactions. The number of increased PPIs of these proteins between the adhesive and hyphal stages is given, along with the number of reduced ones. Total changed interactions displays the sum of PPI changes. GO functional annotation provides the GO terms of proteins, which we have filtered to only list GO terms more concerned with hyphal growth and filamentous growth. The functional annotations are from the GO database (http://www.geneontology.org/).

15 proteins by magnitude of their PPI changes. In addition, these 15 proteins almost completely overlap with the hubs in the hyphal stage network (Table B.1). In general, the number of possible interactions for any given protein is paralleled by changes in protein interaction values. Therefore, the variations of the hub protein interactions should be larger than the conventional nodes in the infectious PPI network.

From Fig. B.4, it can be seen that this hyphal growth PPI network is scale free. In a scale-free network, the probability that a node is highly connected is significantly higher than in a random network, and the network's properties are often determined by a relatively small number of highly connected nodes known as hubs [69]. The scale-free networks are particularly resistant to random node removal but extremely sensitive to the targeted removal of hubs [70]. Hence, the hubs are believed to be essential to the robustness of information transmission in the morphological transition from yeast to hyphal form. In this hyphal growth PPI network, Ubi4 is the protein for which the number of interactions had the largest changes in Table 7.3 and it is also the biggest hub in the hyphal stage network (Table B.1). Ubi4 is involved in the negative control of morphological switching in *C. albicans*, as well as in maintaining yeast cell morphology [71]. From the time profile data of Ubi4, the expression of Ubi4 was reduced from the 0.5- to 4-h time points (Fig. B.5). It is clear that Ubi4 expression drops when hyphae began to grow. Act1 is the second-ranked protein in Table 7.3. Act1 is an actin that influences both cAMP synthesis and hyphal morphogenesis [72]. Consequently, Ubi4 and Act1 may play the most important roles in hyphal growth development, because of their high PPI variation between the adhesive and hyphal stage networks. Similarly, other highly ranked proteins in Table 7.3—Hsp90, Sla1, Bni1, etc.—have also been identified as potentially important in hyphal development [73–75]. In fact, Hht21 is the only protein in Table 7.3 that has not been verified as related to hyphal growth at present. We predict that Hht21 also plays an important role in hyphal development and was a worthwhile protein to identify in the present study.

Subsequently, we wanted to investigate which proteins had many increased interactions but fewer reduced interactions in the adhesive stage network than in the hyphal stage network. We supposed that these proteins would have a lot of influence on hyphal growth development. In order to discard the many proteins with only minor interaction variations between the networks, the minimum number of changed interactions for a protein was raised to 10. Column 6 of Table 7.4 indicates the ratio of increased interactions to total changed interactions. From Table 7.4, we can see that 13 of the top 15 proteins from the GO database are related to hyphal development—i.e., having GO terms of hyphal growth or filamentous growth. The first-ranked protein is Kex2, which influences the *C. albicans* proteinase secretion and hyphae formation. The disruption of Kex2 function in *C. albicans* has pleiotropic effects that may impinge on the ability of the organism to colonize and invade tissues [76].

Table 7.4 Proportion of increased PPIs to total changed PPIs between the adhesive and hyphal stage PPI network during infection

Ranking	*C. albicans* protein	Increased interactions	Reduced interactions	Total changed interactions	Ratio of increased interactions	GO functional annotation	
						Hyphal growth	Filamentous growth
1	Kex2	10	0	10	1	+	
2	Hsl1	15	1	16	0.9375		+
3	Tsa1	10	0	10	0.9167	+	
4	Cek1	9	1	10	0.9	+	
5	Chs3	9	1	10	0.9	+	
6	Top1	9	1	10	0.9	+	
7	Orf19.3843	15	2	17	0.8826		+
8	Kre1	10	2	12	0.8333		
9	Sec2	10	2	10	0.8333	+	
10	Cdc42	14	3	17	0.8235	+	
11	Cst20	14	3	17	0.8235	+	
12	Kem1	14	3	17	0.8235	+	
13	Erg5	9	2	11	0.8182		+
14	Gpi7	9	2	11	0.8182	+	
15	Orf19.5438	9	2	11	0.8182		

The top 15 *C. albicans* proteins in the hyphal growth PPI networks are ranked by the proportion of increased PPIs to reduced PPIs. The number of increased PPIs of these proteins between the adhesive and hyphal stages is given, along with the number of reduced PPIs. Total changed interactions displays the sum of PPI changes. The ratio of increased Interactions shows the proportion of increased PPIs to total changed PPIs. GO functional annotation provides the GO terms of proteins, which we have filtered to only list the GO terms more concerned with hyphal growth and filamentous growth. The functional annotations are from the GO database (http://www.geneontology.org/).

The second-ranked protein, Hsl1, is a probable protein kinase involved in morphological determination during the cell cycle of both yeast-form and hyphal cells via regulation of Swe1 and Cdc28 [77]. The third-ranked protein is Tsa1, indispensable for the yeast-to-hyphal transition when *C. albicans* is cultured under oxidative stress [78]. The fourth- to seventh-ranked proteins are Cek1, Chs3, Top1, and Orf19.3843, respectively, and are also all related to hyphal development [79–81].

The 8th- and 15th-ranked proteins are Kre1 and Orf19.5438, respectively. From the current literature, it is unknown whether Kre1 and Orf19.5438 are related to hyphal growth. However, we predict that Kre1 and Orf19.5438 may have some relationship to hyphal development, because the remaining 13 of the top 15 proteins in Table 7.4 have been confirmed to relate to this process.

From the dynamic PPI network of zebrafish during infection detailed in Fig. B.2, we also investigated which proteins have the largest interaction difference between zebrafish stage 1 and stage 2 networks. These protein interaction changes are related to molecular defensive mechanisms activated in response to *C. albicans* invasion. Table 7.5 lists the top 10 proteins with the largest changes in number of protein interactions. Moreover, we also listed the hub proteins of the zebrafish stage 2 network in Table B.2. Tp53 is the protein with the largest variation in interaction between

Table 7.5 Proteins with the most changes in PPI number between the zebrafish stage 1 and stage 2 networks during infection

D. rerio protein	Increased interactions	Reduced interactions	Total changed interactions	*D. rerio* GO annotation	*H. sapiens* GO annotation
Tp53	10	10	20	Apoptosis	Apoptosis
Esr1	8	5	13	Metal ion binding	Regulation of apoptosis
Traf6	6	5	11	Innate immune response Regulation of apoptosis Response to bacterium	Apoptosis Innate immune response
Jun	3	6	9	Canonical Wnt receptor signaling pathway	Innate immune response
Ar	5	4	9	Metal ion binding	Cell death Positive regulation of NF-kappaB Transcription factor activity
Usp14	5	2	7	Ubiquitin-dependent protein catabolic process	Ubiquitin thiolesterase activity
Hsp90a.1	5	2	7	Myofibril assembly	
Rb1	2	4	6	Myoblast differentiation Ubiquitin protein ligase binding	Negative regulation of cell growth
Mdm2	4	1	5	Negative regulation of apoptosis p53 binding	Fibroblast growth factor receptor Signaling pathway
Casp8	1	3	4	Apoptosis Proteolysis	Activation of innate immune response Apoptosis

The names of the top 10 zebrafish proteins in the PPI networks are ranked by change in the number of protein interactions. The number of increased PPIs of these proteins between stages 1 and 2 are given, along with the number of reduced PPIs. The final two columns provide the GO terms of proteins. *D. rerio* proteins were mapped to *H. sapiens* proteins by using ortholog data.

zebrafish stage 1 and stage 2 networks. Tp53 is already well known as an apoptosis protein [82]. Esr1 is the second-ranked protein in Table 7.5, which is identified as an apoptosis-related protein in the GO database. Traf6 is the third-ranked protein; a recent study using a zebrafish embryo model has analyzed the in vivo function of Traf6 in innate immune response without interference of adaptive immunity [83]. Traf6 can activate the NF-κB signal transduction pathway in zebrafish [84]. We can see that of the top 10 proteins in Table 7.5, many are related to innate immune response and apoptosis. It makes sense that the immunization-related proteins of a host would have a greater response to pathogen invasion. Additionally, previous studies have indicated that many bacteria are able to trigger apoptosis in the host cell [85]. Induction of apoptosis in epithelial or endothelial cells might break the epithelia/endothelial cell barrier and permit the bacteria to reach the submucosa. On the other hand, apoptosis might be beneficial for the infected organ, as apoptotic cell death of the infected target cell permits other cells to phagocytose the apoptotic bodies containing bacteria, possibly resulting in the rapid digestion of the pathogen. It is therefore also reasonable that apoptosis-associated protein interactions would have a greater defense response to pathogen invasion.

Intercellular PPI network between C. albicans *and zebrafish during infection*

After investigating the dynamic PPI networks of *C. albicans* hyphal growth and zebrafish defense, respectively, we investigated the intercellular PPI network between *C. albicans* and zebrafish shown in Fig. 7.14. The cross-correlation of each protein interaction between *C. albicans* and zebrafish was calculated using TRIA in Chapter 4, and the interactions higher than 0.95 were chosen as potential PPIs. In order to find out which *C. albicans* proteins have the greatest impact on zebrafish, we analyzed which cell surface proteins of *C. albicans* (taken from the 4031-protein set) had the most potential interactions with zebrafish. The top 20 *C. albicans* hub proteins ranked by number of interactions are listed in Table 7.6, with higher-ranked proteins deemed more important in the infection process. These hub proteins can be classified into seven major functional groups according to the GO database, including hyphal growth, cell adhesion, biofilm formation, cellular response to neutral pH, cellular iron ion homeostasis, glucose transport, and cell wall molecular biosynthesis. The first four are well-known functions that occur during *C. albicans* infection. When *C. albicans* infects a host, it adheres to the host at first, then grows hyphae to invade the host and further forms biofilms to parasitize in host. The ability to respond to ambient pH is also critical to the growth and virulence of *C. albicans*. It is well established that a near-neutral pH ($\sim$6.5) favors hyphal development of *C. albicans* in vitro, while a low pH ($<$6.5) blocks hyphal formation and stimulates growth of the yeast form [86]. It is interesting that iron and glucose and cell wall molecular biosynthesis appear in this important table. Notably, in the last several years iron ion uptake has been thought to be an important factor in

Table 7.6 Numbers of PPIs of *C. albicans* cell surface proteins in the host–pathogen intercellular PPI network

Interactions (*n*)	*C. albicans* protein	GO functional annotation						
		Cell adhesion	Hyphal growth	Biofilm formation	Cellular response to neutral pH	Iron ion homeostasis	Glucose transport	Cell wall molecular biosynthesis
1447	Csh1	+		+				
1089	Rbt5			+		+		
1078	Pga 10			+		+		
1027	Pga7							
964	Csa1			+		+		
747	Mp65	+	+	+	+		+	
728	Hgt2						+	
701	Chs3		+					+
643	Cht2		+					+
532	Chs2							+
532	Hgt9						+	
532	Mnt1	+	+					+
491	Hgt12		+		+		+	+
451	Phr1	+	+		+			+
435	Pga49							
419	Cdc48							
408	Hgt7						+	
367	Als1	+	+	+	+			
357	Tos1							
352	Mid1		+					

The top 20 *C. albicans* hub proteins with the most potential interactions with zebrafish are listed; their cross-correlation values between protein interactions were larger than 0.95. The number of PPIs of *C. albicans* cell surface proteins are sorted in descending order, along with the corresponding protein name.

pathogenesis. The ability to acquire iron from host tissues is a major contributing factor to the virulence of pathogenic microorganisms. *C. albicans*, like many pathogenic bacteria, is able to utilize hemin and hemoglobin as iron sources [87]. The availability of iron can have a significant impact on both pathogen virulence and host antimicrobial defenses. Some studies have shown that pretreatment of endothelial cells with an iron chelator can reduce the damage inflicted by *C. albicans* [88,89]. Glucose has previously been reported to induce germ-tube formation in *C. albicans* [90,91], and Paranjape and Datta [92] recognized it as critical for its pH-regulated pathways. Chitin, β-glucan, and mannose are essential molecules constituting most fungal cell walls. It is intriguing that the cell wall proteins of *C. albicans* are predicted by this modeling to have so many interactions with zebrafish proteins. Recent studies have shown that cell wall β-glucan is a key fungal signature molecule targeted by the innate immune system to clear fungal infection and that *C. albicans* masks β-glucan from immune recognition by using a mannoprotein coat [93,94]. The immune system may also be able to counteract this fungal defense by unmasking the signature components of the fungus during the course of infection. Wheeler et al. [94] showed that β-glucan is initially masked but subsequently exposed on the surface of *C. albicans* in the normal course of infection. Although it is unknown how this unmasking occurs, it is possible that with time immune cells accumulate in sufficient numbers to directly damage the cell wall and expose the β-glucan of *C. albicans*. The chitin of the cell wall may be similarly destroyed such that *C. albicans* needs to synthesize more essential cell wall molecules to protect itself. Csh1 (cell surface hydrophobicity 1) is the largest hub in the intercellular *C. albicans*-zebrafish PPI network. Knockout of the Csh1 gene has been undertaken to address the potential contribution of its antigen in mediating fungal cell adhesion to host tissue [95]. Another study has demonstrated that Csh1 contributes to virulence of *C. albicans* in mice [96]. The fourth hub in Table 7.6 is Pga7, and its function and those of two other proteins (Pga49 and Tos1) are unknown at present. As the top 20 hub-proteins in Table 7.6 have the most potential interactions with zebrafish, any one of these proteins should have a significant impact on zebrafish defense. It would be worthwhile to investigate the functions of these three unknown proteins.

In addition, we have also listed the top 25 zebrafish hub proteins that have the most intercellular interactions with *C. albicans* in Table B.3. There are several proteins related to cell proliferation, blood coagulation, and proteolysis. It makes sense that more cells would need to proliferate to defend and maintain the normal operation of the host body while *C. albicans* invades, and zebrafish may also secrete proteolytic compounds to damage *C. albicans*. Proteolytic cascades also play a crucial role in innate immune response because they can be triggered more quickly than the adaptive immune response, which requires that gene expression be altered [97]. To summarize these findings, Table 7.7 provides the biological functions of *C. albicans* and zebrafish that are observed in the intercellular infection PPI network.

Table 7.7 Important biological processes of the *C. albicans* invasion mechanism and zebrafish defense mechanism during infection

	***C. albicans* (invasion mechanism)**	**Zebrafish (defense mechanism)**
Biological process	Hyphal growth Cell adhesion Biofilm formation Cellular response to neutral pH Cellular iron ion homeostasis Glucose transport Cell wall molecular biosynthesis	Apoptosis Innate immune response Cell proliferation Blood coagulation Proteolysis

7.2.3 Discussion of infection and defense mechanism by host–pathogen infection network

To discover *C. albicans* proteins or pattern recognition molecules that play a critical role in the infection of zebrafish, we constructed an intercellular infection network consisting of *C. albicans* dynamic hyphal, zebrafish dynamic and host–pathogen intercellular PPI networks. To verify the reliability and accuracy of these methods for constructing the molecular networks and these results, we compared the *C. albicans* dynamic hyphal PPI network with already known pathways implicated in hyphal growth. Twenty out of 22 proteins in already known pathways (Fig. B.3) were included in the *C. albicans* dynamic hyphal PPI network, and all major hyphae-related pathways were visible in this hyphal PPI network. These results also provided a comprehensive and dynamic PPI network for hyphal development during the infection process, and we additionally found that Ras2, Myo2, Rho3, and Rsr1 may play significant roles in hyphal growth processes. Based on the time-course information of *C. albicans* microarray data, we were also able to construct two PPI networks that represented prehyphal growth (i.e., adhesion) mechanisms and hyphal growth (i.e., invasion) mechanisms for their respective stages. Using a similar approach for zebrafish, we were able to construct two stages of defense networks corresponding to these two invasion networks of *C. albicans*. The proteins with the largest PPI variation between the two networks were elucidated for both *C. albicans* and zebrafish. In the *C. albicans* hyphal growth network, Ubi4 was the protein with the most interaction changes between the adhesive stage and hyphal stage networks, and was also the biggest hub in the latter. Ubi4 in *C. albicans* is involved in the negative control of morphological switching, as well as in maintaining yeast cell morphology. We additionally listed other proteins with large variations—Act1, Hsp90, Sla1, Bni1, etc.—in Table 7.3, which also appear to be network hubs in Table B.1. We predict that deletion of these proteins in *C. albicans* could strongly impact the robustness of information transmission during the morphological transition from yeast to hyphal form. Hht21 is a noteworthy exception in Tables 7.3 and B.1 that has not been verified as related to hyphal growth.

In addition, we investigated some proteins with many increased and fewer reduced interactions in the hyphal stage network in comparison with the adhesive stage network. We found that 13 of these top 15 proteins were identified as related to hyphal growth development by the GO database. The top-ranked proteins, such as Kex2, Hsl1, Tsa1, Cek1, Chs3, and Top1, are all related to hyphal growth, and we predict they are very important in hyphal development. Moreover, whether the 8th- and 15th-ranked proteins—Kre1 and Orf19.5438, respectively—are related to hyphal growth is still unknown in the present literature; however, we might predict that Kre1 and Orf19.5438 are indeed related to hyphal development, because the other 13 of the top 15 proteins in Table 7.5 have all been confirmed as such. Interestingly, a recent study of deletion mutants in *S. cerevisiae* revealed that the Kre1-deleted strain significantly suppresses the hyperpseudohyphal phenotype [98]. This result implies that Kre1 may also be related to filamentous growth in *C. albicans*. Due to the limited relevant research on zebrafish PPIs, PPI mapping between zebrafish and *H. sapiens* still remains incompletely characterized, with the variations in protein interaction levels in zebrafish dynamic PPI networks appearing very small in comparison to the *C. albicans* hyphal PPI network. Nevertheless, we were able to list the top 10 proteins having the largest changes in the number of protein interactions, and found them to be mainly related to apoptosis and innate immunity.

The intercellular PPI network between *C. albicans* and zebrafish is also considered in this analysis. The top 20 *C. albicans* hub proteins having the most intercellular interactions with zebrafish proteins are listed in Table 7.6. These hub proteins can be classified into seven major functions according to their annotations in the GO database: hyphal growth, cell adhesion, biofilm formation, cellular response to neutral pH, cellular iron ion homeostasis, glucose transport, and cell wall molecular biosynthesis. The first four are the well-known functions that would occur when *C. albicans* infects its host. Ionic iron is necessary for hyphal growth and its availability has been considered as a factor connected with pathogenesis in the last several years. The ability to acquire iron from host tissues is a major factor affecting the pathogenicity of microorganisms. However, a detailed mechanism for the participation of iron in *C. albicans* infection is still unclear. Glucose has previously been reported to induce germ-tube formation in *C. albicans*. *Candida albicans* needs energy and nutrients while it infects its host. To obtain these nutrients it may acquire glucose from host tissues and so glucose transport functionality in *C. albicans* might have many intercellular interactions with zebrafish. Moreover, it would be worthwhile to investigate why cell wall molecular proteins have so many interactions with host proteins. Recent studies have shown that the immune system may be able to counteract this fungal defense by unmasking the signature molecular components of the fungus during the course of infection. Although it is not known how this unmasking occurs, it is possible that with time, immune cells accumulate in sufficient numbers to directly damage the cell wall and expose β-glucan of *C. albicans*. The chitin of the cell wall may also be destroyed during this accumulation as well, so *C. albicans* would need to synthesize more cell wall molecules to protect itself.

In this section, our aim was to construct a host−pathogen intercellular integrated PPI network by using the microarray data of *C. albicans* and zebrafish, and then utilize it to predict which proteins play critical roles in the infection mechanism of pathogens and the defense mechanism of hosts. In conclusion, we identified several important proteins related to *C. albicans* infection such as Ubi4, Act1, Kex2, Hsl1, and Tsa1, and some proteins whose contribution to pathogenicity warrants further investigation, such as Hht21, Kre1, and Orf19.5438. These proteins may exert a tremendous influence on morphological transition of *C. albicans* and hence they may provide useful drug targets for broad-spectrum treatments of *C. albicans* infection. Moreover, three noteworthy functions in *C. albicans* host infection—cellular iron ion homeostasis, glucose transport and cell wall molecular biosynthesis—were also discovered from the perspective of the intercellular PPI network. Several significant proteins related to innate immune and apoptotic function such as Tp53, Esr1, and Traf6 were found in the molecular defensive mechanisms of zebrafish responsive to *C. albicans* invasion. Furthermore, biological processes like apoptosis, innate immune response, cell proliferation, blood coagulation, and proteolysis were also found in systematic defensive mechanisms of zebrafish during *C. albicans* infection. We hope that the proposed intercellular protein interaction method implemented through this dynamic host−pathogen interaction scheme may ultimately help provide useful medical therapies and facilitate the development of new antifungal drugs.

APPENDIX B: METHODS, TABLES, AND FIGURES

Methods

1. Network Parameter Identification via Time Series Microarray Data

   After constructing the dynamic models of the candidate cellular protein interaction network, the interaction parameters in the models have to be identified using the collected microarray data. The identification strategy is to outline the cellular protein interaction network protein by protein, i.e., for $p = 1,\ldots,N$. The regulatory parameters are identified by solving a least square parameter estimation problem [99]. Eq. (7.15) can be rewritten as the following regression form:

$$\begin{aligned} y_p(t+1) &= \left[\, y_p(t)y_1(t) \quad \ldots \quad y_p(t)y_{Q_p}(t) \quad x_p(t) \quad y_p(t) \right] \begin{bmatrix} b_{p1} \\ \vdots \\ b_{pQ_p} \\ \alpha_p \\ (1-\beta_p) \end{bmatrix} + w_p(t) \\ &= \phi_p(t)\theta_p + w_p(t) \end{aligned} \tag{B.1}$$

where $\phi_p(t)$ indicates the regression vector and θ_p is the parameter vector to be estimated. In order to avoid overfitting, the cubic spline method [100,101] was used to interpolate extra time points for the gene expression data. For simplicity, at different time points i.e., $t = 1,\ldots,L$, Eq. (B.1) could be presented as follows:

$$Y_p = \Phi_p \theta_p + \Omega_p \tag{B.2}$$

where $Y_p = [y_p(2) \; \ldots \; y_p(L)]^T$, $\Phi_p = [\phi_p(1) \; \ldots \; \phi_p(L-1)]$ and L is the number of microarray data points after cubic spline interpolation. The parameter estimation problem can then be formulated as the following least square minimization problem

$$\min_{\theta_p} \frac{1}{2} \| Y_p - \Phi_p \theta_p \|_2^2 \tag{B.3}$$

The least square minimization problem in Eq. (B.3) can be solved for the optimal estimate θ_p^T by the quadratic programming in Eq. (2.20) [99]. We can estimate the regulatory interaction parameters b_{pq} protein by protein for the candidate PPI network via time series microarray data: i.e., through the estimated interaction abilities $\hat{b}_{pq}$, $p = 1,\ldots,N$, $q = 1,\ldots,Q$, we can estimate the regulatory interaction parameters of a potential PPI network.

2. Determination of Cross-Correlation Values

Keeping the pairwise relationship of these $N-k$ pairs to maintain dependence between $(x_i, y_{i+k})$is sampled with $N-k$ replacements to form a bootstrapped sample $Z^* = \{z_i^*: i = 1,\ldots, N-k$ and z_i^* belongs to $Z^*\}$. The correlation coefficient from the bootstrapped sample Z^* is computed and denoted as $c^*(k)$, $-1 \le c^*(k) \le 1$. Repeating the resampling procedure B times, we will observe $c_1^*(k),\ldots,c_B^*(k)$. These bootstrapped correlation coefficients are derived as $-1 \le c_1^*(k) \le \ldots \le c_B^*(k) \le 1$. In this case, the two-sided percentile interval of $(1-\alpha)$ is given by $[c^*_{(B\times -\alpha/2)}(k),\; c^*_{(B\times \alpha/2)}(k)]$ [62]. If this percentile interval does not contain 0, then the null hypothesis is rejected at significance level α. Otherwise, the data fails to reject the null hypothesis, again at significance level α. Since the p-value is the smallest value of α for which the null hypothesis will be rejected based on the observation, the p-value for this test is estimated as follows:

$$\hat{p}(k) = \min\{\hat{p}_+(k), 1 - \hat{p}_+(k)\}, \quad \text{where } \hat{p}_+(k) = \sum_{i=1}^{B} I\{c_i^*(k) \ge 0\}/B \tag{B.4}$$

where $I\{\cdot\}$ is the indicator function whose value is one when the event is true and zero otherwise.

The time-lagged correlation (TlC) of $\vec{x}$ and $\vec{y}$ is defined as $c(j)$ having the smallest p-value (i.e., $TlC(\vec{x},\vec{y}) = c(j)$ if $p(j) \le p(k)$, $\forall k \ne j$) . Note that $-1 \le TlC(\vec{x},\vec{y}) \le 1$. So that $TlC(\vec{x},\vec{y})$ is the cross-correlation between $\vec{x}$ and $\vec{y}$.

After investigating the infection and defense mechanisms between *C. albicans*-zebrafish using two-side microarray data in this section, we will investigate the pathogenic and defensive mechanisms between host and pathogen through crosstalk network marker in the HIV infection process by NGS data and database mining in Section 7.3.

Here we show Tables B.1–B.3 and Figs. B.1–B.5.

Table B.1 Hubs in the hyphal stage PPI network during infection

Hub	GO term	Number of interactions
Ubi4	GO:0000902: cell morphogenesis GO:0030447: filamentous growth GO:0009405: pathogenesis	115
Hsp90	GO:0030447: filamentous growth GO:0009405: pathogenesis	54
Sin3	GO:0030447: filamentous growth	54
Act1	GO:0030448: hyphal growth	53
Mkc1	GO:0030447: filamentous growth GO:0009405: pathogenesis GO:0030448: hyphal growth	52
Sla1	GO:0030448: hyphal growth	50
Cdc3	GO:0001411: hyphal tip	50
Pmr1	GO:0030448: hyphal growth	47
Phr2	GO:0030447: filamentous growth GO:0009405: pathogenesis	47
Cdc28	GO:0010570: regulation of filamentous growth GO:0051726: regulation of cell cycle	43
Bni1	GO:0030448: hyphal growth GO:0009405: pathogenesis GO:0030447: filamentous growth	41
Orf19.3843	GO:0030447: filamentous growth	40
Snf1	GO:0042710: biofilm formation GO:0007155: cell adhesion GO:0007124: pseudohyphal growth	39
Hht21	GO:0007094: mitotic cell cycle spindle assembly checkpoint GO:0009303: rRNA transcription	36
Erg3	GO:0030448: hyphal growth GO:0005506: iron ion binding	36

Column 1 gives the names of the top 17 hub proteins in the hyphal growth interaction network ranked by number of PPIs. Column 2 provides the GO terms of proteins, which we have filtered to only list the GO terms more concerned with hyphal growth. Column 3 shows the number of PPIs of these hub proteins.

Table B.2 Hubs in the zebrafish stage 2 PPI network during infection

Zebrafish protein	Zebrafish GO term	*H. sapiens* GO term	Number of interactions
Tp53	GO:0006915: apoptosis	GO:0006915: apoptosis	58
	GO:0043065: positive regulation of apoptosis	□	
Esr1	GO:0046872: metal ion binding	GO:0042981: regulation of apoptosis	33
	GO:0008270: zinc ion binding	□	
Hsp90a.1	GO:0014866: skeletal myofibril assembly	□	32
	GO:0045429: positive regulation of nitric oxide biosynthetic process	□	
	GO:0048769: sarcomerogenesis	□	
	GO:0030235: nitric-oxide synthase regulator activity	□	
Traf6	GO:0016567: protein ubiquitination	GO:0006915: apoptosis	31
	GO:0042981: regulation of apoptosis	GO:0045087: innate immune response	
	GO:0009617: response to bacterium	GO:0042088: T-helper 1 type immune response	
	GO:0009615: response to virus	□	
	GO:0046872: metal ion binding	□	
	GO:0008270: zinc ion binding	□	
Ikbkg	GO:0007252: I-kappaB phosphorylation	□	31
	GO:0051092: positive regulation of NF-kappaB transcription factor activity	□	
	GO:0006974: response to DNA damage stimulus	□	
Jun	GO:0060070: canonical Wnt receptor signaling pathway	GO:0045087: innate immune response	27
	GO:0006355: regulation of transcription, DNA-dependent	GO:0043525: positive regulation of neuron apoptosis	
	GO:0046686: response to cadmium ion	□	
Ube2i	GO:0007049: cell cycle	GO:0051301: cell division	27
	GO:0007088: regulation of mitosis	GO:0016567: protein ubiquitination	
Ar	GO:0046872: metal ion binding	GO:0008219: cell death	26
	GO:0008270: zinc ion binding	GO:0008283: cell proliferation	
Myca	GO:0006355: regulation of transcription, DNA-dependent	□	25
	GO:0006351: transcription, DNA-dependent	□	
Src	GO:0006468: protein phosphorylation	GO:0016337: cell–cell adhesion	24
	GO:0005524: ATP binding	□	

Column 1 gives the names of the top 10 hub proteins ranked by number of protein interactions in the zebrafish stage 2 PPI network. Columns 2 and 3 provide the GO terms of the proteins from zebrafish and *H. sapiens*, respectively. Column 4 indicates the number of PPIs of these hub proteins. □ denotes the blankness.

Table B.3 The number of PPIs of zebrafish proteins in host–pathogen intercellular PPI network in the infectious processing

Number of interaction	Zebrafish protein	GO term: Biological process	GO term: Molecular function
37	Igf1	GO:0009953: dorsal/ventral pattern formation	GO:0008083: growth factor activity
		GO:0008284: positive regulation of cell proliferation	GO:0005179: hormone activity
		□	GO:0005159: insulin-like growth factor receptor binding
37	Chia.2	GO:0005975: chitin catabolic process	GO:0008061: chitin binding
35	Nbl1	GO:0060872: semicircular canal development	□
34	F10	GO:0007596: blood coagulation	GO:0003824: catalytic activity
		GO:0006508: proteolysis	GO:0004252: serine-type endopeptidase activity
34	F7i	GO:0007596: blood coagulation	GO:0005509: calcium ion binding
		GO:0030195: negative regulation of blood coagulation	GO:0003824: catalytic activity
		□	GO:0004252: serine-type endopeptidase activity
33	Igfbp2a	GO:0001525: angiogenesis	□
		GO:0040007: growth	□
		GO:0007507: heart development	□
		GO:0007275: multicellular organismal development	□
		GO:0008285: negative regulation of cell proliferation	□
		GO:0008156: negative regulation of DNA replication	□
		GO:0001558: regulation of cell growth	□
		GO:0043567: regulation of insulin-like growth factor receptor signaling pathway	□
33	Mstnb	GO:0040007: growth	□
		GO:0007517: muscle organ development	□
		GO:0045926: negative regulation of growth	□
		GO:0007179: transforming growth factor beta receptor signaling pathway	□
33	Slc34a2a	GO:0006817: phosphate transport	GO:0015321: sodium-dependent phosphate transmembrane transporter activity
32	Cx30.9	GO:0007154: cell communication	□
32	Itgb1b.2	GO:0007155: cell adhesion	□
		GO:0007160: cell-matrix adhesion	□
		GO:0007229: integrin-mediated signaling pathway	□
		GO:0007275: multicellular organismal development	□
32	Chia.1	GO:0005975: chitin catabolic process	GO:0008061: chitin binding

(Continued)

Table B.3 (Continued)

Number of interaction	Zebrafish protein	GO term	
		Biological process	Molecular function
31	Asah2	GO:0006672: ceramide metabolic process	□
		GO:0006629: lipid metabolic process	□
		GO:0007275: multicellular organismal development	□
		GO:0006665: sphingolipid metabolic process	□
30	F9	GO:0007596: blood coagulation	□
		GO:0006508: proteolysis	□
30	Glra4b	GO:0006811: ion transport	□
30	Hbl4	□	GO:0005529: sugar binding
30	Hpx	GO:0042221: response to chemical stimulus	GO:0046872: metal ion binding
29	Cetp	GO:0043691: reverse cholesterol transport	GO:0017127: cholesterol transporter activity
		□	GO:0008289: lipid binding
29	F2	GO:0007596: blood coagulation	GO:0005509: calcium ion binding
		GO:0006508: proteolysis	GO:0003824: catalytic activity
		□	GO:0016787: hydrolase activity
		□	GO:0008233: peptidase activity
		□	GO:0004252: serine-type endopeptidase activity
		□	GO:0008236: serine-type peptidase activity
29	Si:ch211-140f21.1	□	GO:0004866: endopeptidase inhibitor activity
29	Zgc:163025	GO:0007596: blood coagulation	GO:0005509: calcium ion binding
		GO:0006508: proteolysis	GO:0003824: catalytic activity
		□	GO:0004252: serine-type endopeptidase activity
29	Kcna6	GO:0006811: ion transport	□
		GO:0006813: potassium ion transport	□
		GO:0055085: transmembrane transport	□
28	Colec11	□	GO:0005537: mannose binding
		□	GO:0005529: sugar binding
27	Chrnd	GO:0006811: ion transport	□
		GO:0030239: myofibril assembly	□
27	Glra3	GO:0006811: ion transport	□
27	Apoea	Unknown	□

The top 25 zebrafish hub proteins ranked by number of potential interactions with *C. albicans* proteins are listed in this table; their PPI cross-correlations are larger than 0.95. Column 1 indicates the number of PPIs of zebrafish proteins sorted in descending order. Column 2 provides the protein name in zebrafish; Columns 3 and 4 provide the corresponding GO terms. □ denotes the blankness.

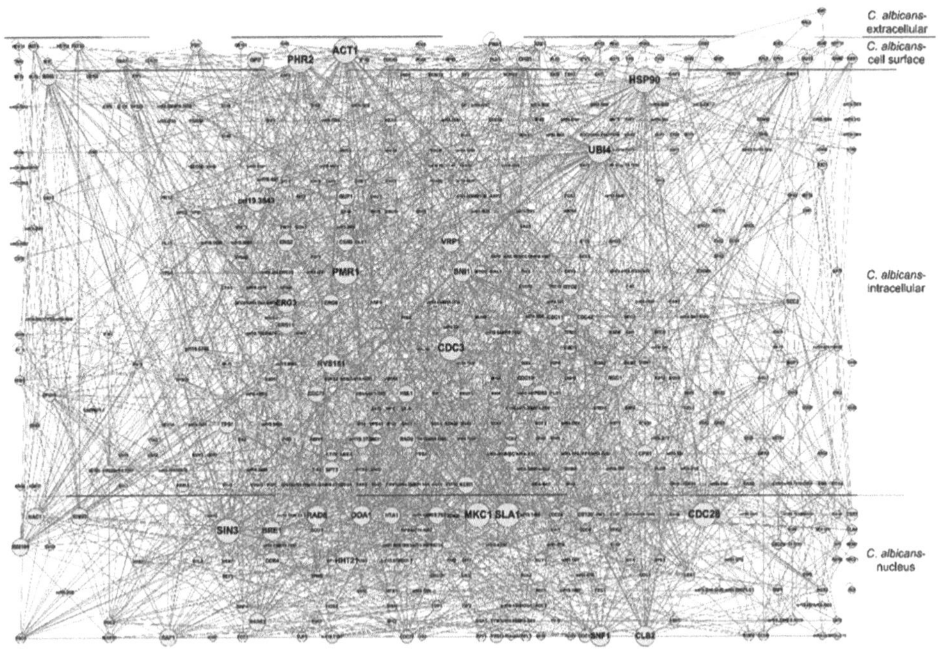

Figure B.1 The combined *C. albicans* PPI networks including 550 proteins with 2725 PPIs in the adhesive stage network and 555 proteins with 3171 PPIs in the hyphal stage network.

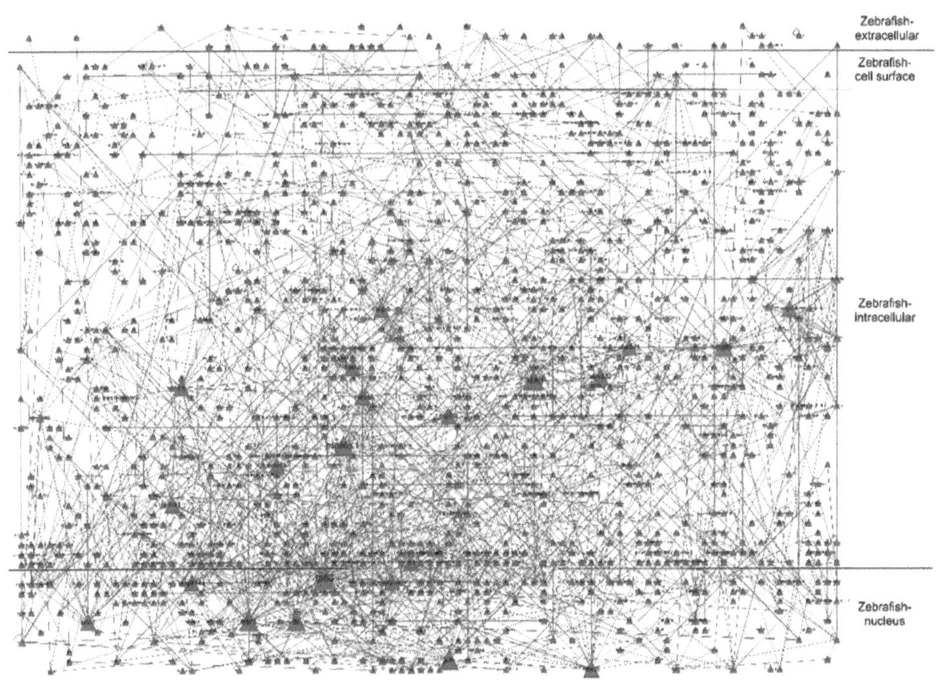

Figure B.2 The combined zebrafish PPI networks including 1248 proteins with 2344 PPIs in the zebrafish stage 1 network and 1265 proteins with 2379 PPIs in the zebrafish stage 2 network.

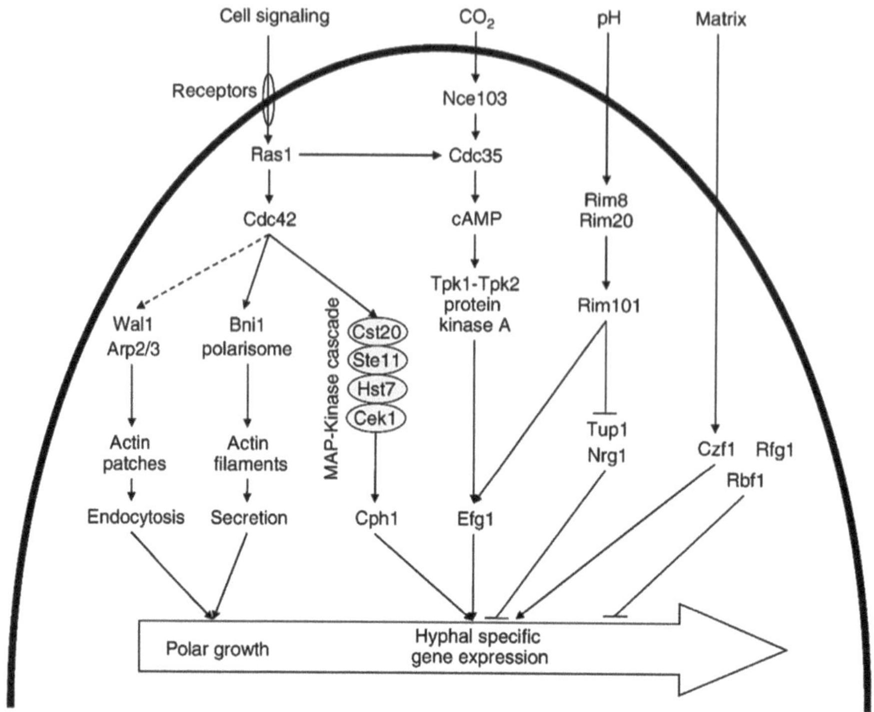

Figure B.3 The identified signal transduction pathways involved in regulating morphogenesis in *C. albicans*.

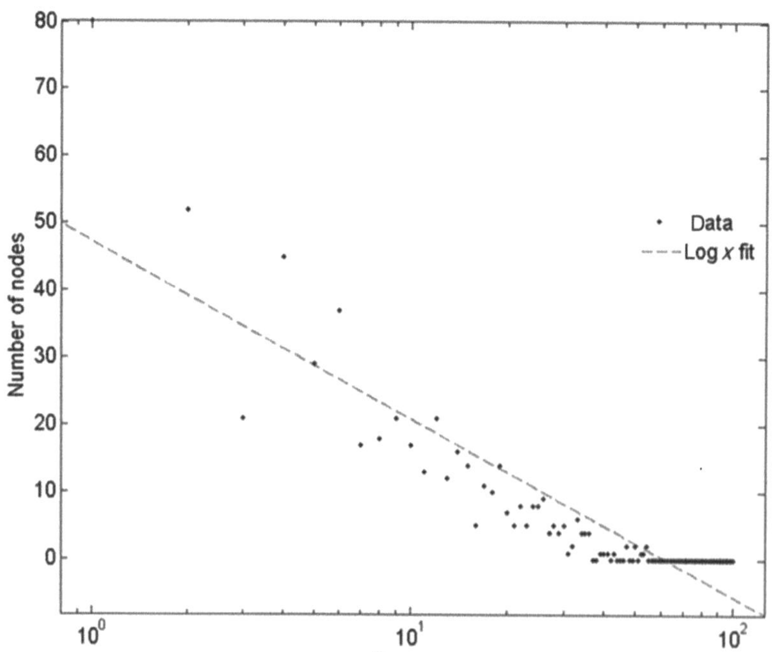

Figure B.4 The degree distribution of the identified hyphal growth PPI network, which follows a power law. In log–log scale, the power law function was close to a straight line.

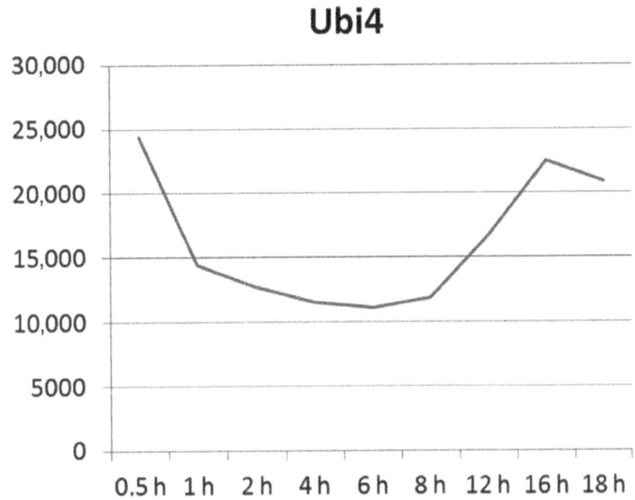

Figure B.5 The expression of Ubi4.

7.3 INVESTIGATING HOST–PATHOGEN INTERACTION NETWORKS TO REVEAL THE PATHOGENIC MECHANISM IN HIV INFECTION: A SYSTEMS BIOLOGY APPROACH

Two big issues in the study of pathogens are determining how pathogens attack their hosts and how the host defends itself against these attacks. Therefore, investigating host–pathogen interactions is important for understanding pathogenicity and host defensive mechanisms and controlling infections.

In this section, we used omics data, including time-course data from high-throughput sequencing, RT-PCR, and human miRNA and PPI to construct an interspecies PPMI network of human CD4+ T cells during HIV-1 infection through system modeling and identification.

By applying a functional annotation tool to the identified PPMI network at each stage of HIV infection, we found that regulation of three miRNAs: miR-140-5p, miR-320a, and miR-941, is involved in the development of autoimmune disorders, tumor proliferation, and the pathogenesis of T cells at the reverse transcription stage. Regulation of miR-331-3p and miR-320a is involved in HIV-1 replication, replicative spread, antiapoptosis, cell proliferation, and dysregulation of cell cycle control at the integration/replication stage. Regulation of miR-341-5p is involved in carcinogenesis at the late stage of HIV-1 infection.

By investigating the common core proteins and changes in specific proteins in the PPMI network during HIV-1 infection, we obtained insights into the functional core modules in HIV-infected CD4+ T cells. Based on the pathogenic mechanisms at each stage, we identified potential drug combinations for treating patients with HIV-1 infection, including thalidomide, oxaprozin, and metformin, at the reverse transcription stage; quercetin, nifedipine, and fenbendazole, at the integration/replication stage; and staurosporine, quercetin, prednisolone, and flufenamic acid, at the late stage.

HIV was first identified in 1983, and it quickly became a pandemic. According to the Joint United Nations Programme on HIV/AIDS, today, approximately 35 million people worldwide have HIV infection. Thus, HIV prevention and treatment became important at the beginning of this century. HIV infection depletes human CD4+ T cells, which are critical for immune responses. Thus, HIV infection impairs the immune system and causes AIDS. Although HIV can now be managed as a chronic disease through the use of highly active antiretroviral therapy (HAART) [102], curing the infection remains a significant challenge.

It has been demonstrated that a mutation in HIV-1 conferring drug resistance can occur within a single day [103]. Because of its high rates of mutation, methylation, and recombination, HIV can quickly evolve resistance to antiretroviral drugs. A clear understanding of the molecular mechanism of HIV infection is required for efficient antiretroviral drug design with minimal side effects; thus, our current lack of a complete understanding has led to unresolved disease in antiretroviral therapy. Avoiding the side effects of anti-HIV drugs requires uncovering the pathogenic mechanism of HIV infection from a genome-wide perspective.

Genome-wide expression data from human CD4+ T cells and HIV-1 gene expression data during HIV-1 infection have been reported [104]. Three stages of HIV-1 infection have been characterized by analyzing a time course of human gene/miRNA expression and viral gene expression of 24 h after infection. The first (or early) stage, also called the reverse transcription stage, occurs 2–12 h postinfection, during which the virus reverse transcribes its single-stranded RNA genome into double-stranded cDNA. The second stage, or the integration/replication stage, occurs 6–18 h after infection, during which the newly synthesized viral cDNA is integrated into the host genome. The late stage occurs 16–24 h after infection, at which point viral particles are assembled and packaged for release. The genome-wide data also shows the effect of miRNA regulation [105,106] and DNA methylation [107,108] on gene expression. Therefore, the molecular pathogenic mechanisms, including miRNA regulation and DNA methylation, were considered in the investigation of the pathogenic mechanism of HIV infection in human CD4+ T cells.

miRNAs, which are small noncoding RNA molecules (19–22 nucleotides in length), bind to target mRNAs through complementary base pairing to inhibit mRNA translation. miRNAs may play an important role in HIV infection [109,110], because they regulate many biological processes, including virus replication and translation [111]. Members of the same miRNA family that have distinct sequences and expression patterns contribute to the diversity in posttranscription modification and biological processes. However, miRNA-mediated regulation of HIV infection is complex and poorly understood. It has been reported that HIV-1 does not encode any miRNAs [112]. Thus, it is important to investigate the dynamic changes in human miRNA regulation in CD4+ T cells during HIV-1 infection from the reverse transcription stage to the late stage in addition to the changes in PPIs. Our results may provide insight into specific antiretroviral therapy for the different stages of infection from a genome-wide perspective.

Since Mohammadi and colleagues investigated the profound perturbation of cellular physiology during HIV infection in terms of several regulatory mechanisms and generated genome-wide time-course expression data for both HIV and the host SupT1 cell line [104], it is now possible to investigate the changes in the PPMI in mock-infected and HIV-1-infected CD4+ T cells. In these experiments, polybrene and HIV-1 spinoculation were added to increase the efficiency of infection [113]. Mock-infected cells were also treated with polybrene and mock spinoculation. PPMI has been used to investigate the topology and function of networks [114,115] and drug targets [116] in humans from a graph theory perspective. PPMI has also been used to analyze how miRNAs regulate biological processes through signaling pathways in human hepatocarcinogenesis [117] and renal carcinogenesis [118]. In this section, we analyzed genome-wide expression data [104] in human CD4+ T cells and HIV-1 to construct genome-wide PPMI networks at three stages of HIV-1 infection. As a control, we constructed three-stage PPMI networks in mock-infected cells. The identified genome-wide PPMI networks provide an opportunity to unravel the pathogenic mechanism of HIV infection to aid in the discovery of new anti-HIV drugs.

In a recent study, DNA methylation in HIV-1-infected CD4+ T cells was examined [119]. Another recent study reported that DNA methylation can directly affect the binding affinities of RNA polymerase and miRNAs in primary human somatic and germline cells [120] and indirectly affect PPIs [121]. In this section, we applied a dynamic system and system identification method to construct the genome-wide PPMI networks of mock- and HIV-infected CD4+ T cells. Methylation in CD4+ T cells was inferred by the identified parameters of the dynamic system. By comparing the genome-wide DNA methylation profiles of HIV-infected and uninfected human B cells, we determined the effect of DNA methylation on the pathogenic mechanism of the HIV-1-infection in CD4+ T cells. T cells (or thymus cells) and B cells (or bone marrow-derived cells) are the major cells in the human immune response.

In this section, we first used big data, including intraspecies and interspecies PPI [122,123] and miRNA regulation [124,125], to construct the candidate PPMI network for HIV-1-infected CD4+ T cells. We then applied a dynamic system and system identification method to identify the genome-wide real PPMI networks of HIV-infected and uninfected T cells during three stages of infection. To extract the core networks of the genome-wide PPMI networks, we defined an infection score based on the identified protein and miRNA interactions in the genome-wide networks. Finally, we applied big mechanism analysis to unravel the pathologic mechanisms of HIV-1 infection during these three stages using a GO tool and to identify the core proteins with an aim to discover new anti-HIV drugs. We used the drug response genome-wide microarray data [126] from a promyelocytic cell line to design three multiple drug combinations for the treatment of patients at the three stages of HIV-1 infection.

7.3.1 Methods of constructing the common core and specific interspecies PPMI networks

A flowchart of the strategy used to construct the common core and specific interspecies PPMI networks is shown in Fig. 7.16. The methods are described below.

Protein pool selection

Mohammadi and colleagues [104] generated genome-wide time-course expression data for HIV and human mRNA and human miRNA in HIV- and mock-infected T cells. The expression data at several time points was set to zero due to the low quality of the samples (according to Mohammadi and colleagues) [104]. The HIV expression data includes three transcript types: unspliced pre-mRNA, single-spliced RNA, and multiple-spliced RNA. Because the expression of viral and human genes was measured by RT-PCR and high-throughput RNA sequencing, respectively, it was necessary to normalize the mean expression of the three alternatively spliced viral mRNAs to a scale comparable to that of the human expression data. We normalized the mean viral gene expression to the mean human gene expression data in both HIV- and mock-infected cells.

Host proteins were selected by one-way ANOVA, where the null hypothesis was that the average expression of a protein is the same in HIV- and mock-infected cells. Host proteins with a p-value greater than .05 were pooled. A total of 3234 genes were identified, and an additional 159 miRNAs that potentially interact with these genes were added to the pool. All 16 HIV proteins are produced via alternative splicing of viral mRNA. Capsid, Gag, matrix, nucleocapsid, p1, p6, and pol are produced from unspliced pre-mRNA ($\sim$9 kb); Rev, Tat, and Nef are produced from multiple-spliced RNA ($\sim$2 kb); and Env, gp120, gp41, Vif, Vpr, and Vpu are produced from single-spliced RNA ($\sim$4 kb). The genome-wide expression data includes the expression of unspliced pre-mRNA, multiple-spliced RNA, and single-spliced RNA. Therefore, every protein from an RNA group (unspliced pre-mRNA, multiple-spliced RNA, or single-spliced RNA) has the same expression data. Expression data for the three HIV transcripts was used to represent the expression of the 16 HIV proteins and identify the real PPMI network.

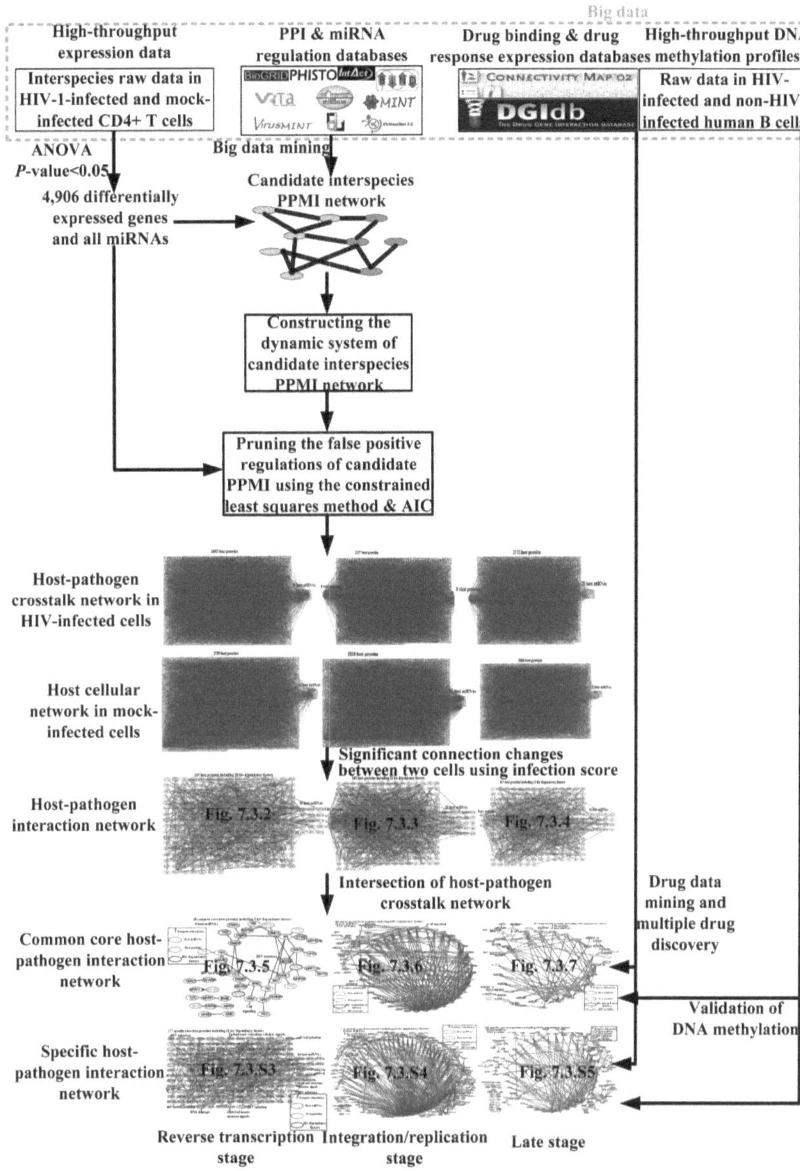

Figure 7.16 Flowchart for the construction of the common core and specific interspecies PPMI networks. Interspecies PPMI networks for each stage of HIV infection were constructed from protein–protein interaction, HIV–host interaction, miRNA regulation, and time-course gene expression data. Based on this information and the selected protein pool from big data mining, putative interspecies PPMIs were constructed as the candidate interspecies PPMI networks. Then, by using coupling dynamic models (7.19) and (7.20) and gene expression profiles, the interaction of edges in these networks were evaluated by the constrained least square parameter estimation method in Eq. (7.23). Candidate interspecies PPMI networks were pruned and refined by system order detection. By comparing HIV- and mock-infected cells, the infection indication matrix D in Eqs. (7.26) and (7.27) was obtained for each stage of infection. Infection scores were computed for each protein, and the top 10% of proteins with the highest infection scores were selected as the core proteins. Finally, intersecting networks from each stage of infection yielded the core proteins of the common core host–pathogen interaction networks at the three stages. The rest of the proteins were included in the specific host–pathogen interaction network at each infection stage.

The genome-wide time-course expression data for HIV-1 and HIV-infected and mock-infected CD4+ T cells showed three infection stages. The reverse transcription, integration/replication, and late stages occur 2–12, 6–18, and 16–24 h after infection, respectively. The three-stage expression data allowed us to identify the real PPMI network at each stage by applying the system identification method and system order detection scheme to the dynamic models described below.

To validate our DNA methylation findings in human immune cells during HIV-1 infection, we applied one-way ANOVA to the genome-wide DNA methylation profiles of HIV-infected and uninfected human B cells [127]. We found significant differences in the DNA methylation profiles between HIV-infected and non-HIV-infected human B cells ($p < .05$).

Reconstruction of the candidate PPMI network through big data mining

Human PPIs and miRNA regulation were collected from big datasets, including the Biological General Repository for Interaction Datasets (BioGRID; http://thebiogrid.org/), Database of Interacting Proteins (DIP; http://dip.doe-mbi.ucla.edu), Biomolecular Interaction Network Database (BIND; http://bind.ca), IntAct (http://www.ebi.ac.uk/intact/), molecular INTeraction database (MINT; http://mint.bio.uniroma2.it/mint/), TargetScanHuman (http://www.targetscan.org/), and CircuitsDB 2 (http://penelope.unito.it/circuitsdb2), which contain 8,578,807 protein interaction pairs and 492,550 regulatory miRNAs. HIV protein interaction data and interspecies HIV-host protein interaction data were obtained from the National Center for Biotechnology Information (NCBI; http://www.ncbi.nlm.nih.gov/genome/viruses/retroviruses/hiv-1/interactions/) [123], VirHostNet 2.0 (http://virhostnet.prabi.fr), VirusMINT database (http://mint.bio.uniroma2.it/virusmint/), and the pathogen–host interaction search tool (PHISTO; www.phisto.org/), which contain 14,437 interactions between human and HIV proteins and 21 HIV PPI pairs. These data include interactions deduced from both biological experiments and computational simulations. To identify the real interspecies PPMI network, we applied a dynamic model and system identification to the candidate interspecies PPMI network to prune false positive interactions using genome-wide time-course expression data. Therefore, although each HIV protein from an HIV RNA group, such as unspliced pre-mRNA, has the same expression, the big data provided different candidate interactions, which resulted in different real interactions after applying the system identification method.

Dynamic models of the host–pathogen interspecies PPMI network

To construct the real interspecies PPMI network from the candidate interspecies PPMI network, we applied the following dynamic models to characterize the molecular mechanisms, including PPIs and miRNA regulation, in the interspecies PPMI network.

For the host, the dynamic model of the ith host protein y_i in the interspecies PPMI network can be described by the following regression form:

$$y_i(t+1) = y_i(t) + \sum_{n=1}^{N} a_{in} y_n(t) + \sum_{k=1}^{K} b_{ik} h_k(t) - \sum_{z=1}^{Z} c_{iz} R_z(t) - \lambda_i y_i(t) + \delta_i + \omega_i(t) \tag{7.19}$$

where $y_i(t)$ and $y_n(t)$ are the expression of the ith and nth host proteins at time t for i, $n = 1,\ldots, N$, respectively; a_{in} and b_{ik} are the interactions between the ith and nth host proteins and between the kth viral protein and the ith host protein, respectively; $h_k(t)$ and $R_z(t)$ are the expression levels of the kth viral protein and the zth host miRNA at time t, respectively; $-c_{iz}$ is the negative regulatory ability of the zth miRNA on the ith host protein; $-\lambda_i$ and δ_i represent the degradation effect and the basal expression level of the ith host protein, respectively; $\omega_i(t)$ denotes stochastic noise of the ith host protein at time t; and N, K, and Z indicate the total number of host proteins, viral proteins, and host miRNAs in the candidate PPMI network, respectively.

For the virus, the dynamic model of the jth viral protein h_j in the interspecies PPMI network can be described by the following regression form:

$$h_j(t+1) = h_j(t) + \sum_{n=1}^{N} \alpha_{jn} y_n(t) + \sum_{k=1}^{K} \beta_{jk} h_k(t) - \sum_{z=1}^{Z} \gamma_{jz} R_z(t) - \varepsilon_j h_j(t) + \kappa_j + w_j(t) \tag{7.20}$$

where $h_j(t)$ and $w_j(t)$ are the expression level and stochastic noise of the jth viral protein at time t for $j = 1,\ldots, K$, respectively; α_{jn} and β_{jk} are the interaction between the jth viral protein and the nth host protein and between the jth and kth viral proteins, respectively; $-\gamma_{jz}$ indicates the negative regulatory ability of the zth host miRNA to the jth viral protein; and $-\varepsilon_j$ and κ_j are the degradation effect and the basal expression level of the jth viral protein, respectively.

In other words, the coupling dynamic models in Eqs. (7.19) and (7.20) characterize the molecular mechanisms in HIV-1 and CD4+ T cells, including the interspecies crosstalk between viral and host proteins using the terms $\sum_{k=1}^{K} b_{ik} h_k(t)$ in Eq. (7.19) and$\sum_{n=1}^{N} \alpha_{jn} y_n(t)$in Eq. (7.20), intraspecies interactions in viral and host cells using the terms $\sum_{n=1}^{N} a_{in} y_n(t)$ in Eq. (7.19) and $\sum_{k=1}^{K} \beta_{jk} h_k(t)$ in Eq. (7.20), respectively, the host miRNA negative regulation in host and viral proteins using the terms $-\sum_{z=1}^{Z} c_{iz} R_z(t)$ in Eq. (7.19) and $-\sum_{z=1}^{Z} \gamma_{jz} R_z(t)$ in Eq. (7.20), respectively, and the host and viral protein degradations using the terms $-\lambda_i y_i(t)$ in Eq. (7.19) and $-\varepsilon_j h_j(t)$ in Eq. (7.20), respectively. Furthermore, we applied the following criteria to the models in Eqs. (7.19) and (7.20) to identify the parameters and prune the candidate interspecies PPMI network using genome-wide temporal expression data.

Identification of the real PPMI network by pruning false positive interactions through system identification and system order detection

After constructing a candidate interspecies PPMI network through big data mining, the candidate interactions included many false positive interactions (edges). To prune false positive interactions and obtain the real interspecies PPMI networks at the three HIV infection stages, two coupling linear dynamic models in Eqs. (7.19) and (7.20) were proposed as a mathematical description for the interspecies PPIs between virus and host [110]. To identify the parameters, including a_{in}, b_{ik}, and c_{iz} in model (7.19) and α_{jn}, β_{jk}, and γ_{jz} in model (7.20), using genome-wide temporal expression data for host cells and RT-PCR temporal expression data for HIV, we rewrote Eq. (7.19) as the following regression form.

$$y_i(t+1) = \begin{bmatrix} y_1(t) & \cdots & y_N(t) & h_1(t) & \cdots & h_K(t) & -R_1(t) & \cdots & -R_Z(t) & y_i(t) & 1 \end{bmatrix} \begin{bmatrix} a_{i1} \\ \vdots \\ a_{iN} \\ b_{i1} \\ \vdots \\ b_{iK} \\ c_{i1} \\ \vdots \\ c_{iZ} \\ (1-\lambda_i) \\ \delta_i \end{bmatrix} + \omega_i(t) \tag{7.21}$$

$$\triangleq \phi_i(t)\theta_i + \omega_i(t)$$

where $\phi_i(t)$ is the regression vector that can be obtained from the expression data and θ_i is the parameter vector to be estimated for protein i of host cells. We defined N_i, K_i, and Z_i as the number of nonzero parameters in a_{in}, b_{ik}, and c_{iz}, respectively. In other words, we have to identify $N_i + K_i + Z_i + 2$ parameters in θ_i. For parameter estimation accuracy, the temporal expression data of the ith protein were interpolated to L_i data points, $L_i = 5(N_i + K_i + Z_i + 2)$ by applying the cubic spline method to avoid overfitting in the parameter estimation [128]. When the temporal expression data in $y_i(t)$, $h_k(t)$, and $R_z(t)$ of protein i is interpolated to L_i time points, we rewrote Eq. (7.21) in the following form for the ith protein in host cells:

$$Y_i = \Phi_i\theta_i + E_i \tag{7.22}$$

where

$$Y_i = \begin{bmatrix} y_i(2) \\ \vdots \\ y_i(L_i) \end{bmatrix}, \quad \Phi_i = \begin{bmatrix} \phi_i(1) \\ \vdots \\ \phi_i(L_i - 1) \end{bmatrix}, \quad \text{and} \quad E_i = \begin{bmatrix} \omega_i(1) \\ \vdots \\ \omega_i(L_i - 1) \end{bmatrix}.$$

Thus, the parameter estimation problem for $\hat{\theta}_i$ was solved by applying the following constrained least-squares minimization criteria [23,110]:

$$\hat{\theta}_i = \arg\min_{\theta_i} \frac{1}{2} \left\| \Phi_i\theta_i - Y_i \right\|_2^2$$

$$\text{Subject to} \begin{bmatrix} 0 & \cdots & 0 & 0 & \cdots & 0 & -1 & 0 & \cdots & \cdots & 0 \\ \vdots & \ddots & \vdots & \vdots & \ddots & \vdots & 0 & \ddots & \ddots & \cdots & \vdots \\ 0 & \cdots & 0 & 0 & \cdots & 0 & \vdots & \ddots & -1 & \ddots & \vdots \\ \vdots & \ddots & \vdots & \vdots & \ddots & \vdots & \vdots & \cdots & \ddots & 1 & 0 \\ 0 & \cdots & 0 & 0 & \cdots & 0 & 0 & \cdots & \cdots & 0 & -1 \end{bmatrix} \theta_i \leq \begin{bmatrix} 0 \\ \vdots \\ 0 \\ 1 \\ 0 \end{bmatrix} \tag{7.23}$$

The constraint set contained nonpositive miRNA regulation as $-c_{iz}$, nonpositive host protein degradation as $-\lambda_i$, and nonnegative host protein basal expression level as δ_i. The estimation problem was optimized using the constrained least-squares scheme based on a reflective Newton method for minimizing a quadratic function in the MATLAB optimization toolbox [129], and the optimal parameter $\hat{\theta}_i$ was then obtained. Furthermore, AIC in (2.23) [23] was applied to prune the false positive connections in the PPMI network at each infection stage. The AIC value decreases with the number of interaction parameters, and a minimum is achieved at the true number of interaction parameters. Therefore, the false positive PPMI interactions in the candidate PPMI network were pruned by using the AIC value to obtain the real PPMI network at each HIV infection stage. Because the genome-wide host and viral protein expression levels during HIV-1 infection have not yet be measured and 73% of the variance in protein abundance can be explained by mRNA abundance [130], mRNA expression profiles are frequently used as a substitute for the protein expression profiles. Furthermore, if the relationship between the protein expression $y(t)$ and mRNA expression $x(t)$ is linear, i.e., $y(t) = \alpha x(t)$, α could be canceled out on both sides of Eq. (7.21). Estimation of parameters a_{in}, b_{ik}, and c_{iz} is not influenced by replacing the protein expression with mRNA expression. By randomly introducing or removing five edges from the optimal parameter $\hat{\theta}_i$, we obtained the result with 30 trials shown in Fig. C.1, which showed that the optimal results may be able to identify biologically relevant signals consistent with what has been shown experimentally, i.e., the proposed model is able to treat them differently using the learning data.

Remark 7.1: To obtain a symmetric matrix of PPIs, we defined $a_{in} = \{a_{ni} | |a_{ni}| > |a_{in}|\}$. A Student's t-test was used to calculate the statistical significance (p-value) of the parameters in Eqs. (7.19) and (7.20) in the real PPMI network under the null hypothesis H_0: $a_{in} = 0$ and $b_{ik} = 0$ in Eq. (7.19) or $\alpha_{jn} = 0$ and $\beta_{jk} = 0$ in Eq. (7.20) [131]. The parameter identification procedures for the host proteins in Eq. (7.19) and the viral proteins in Eq. (7.20) were the same. We then obtained the pruned interspecies PPMI networks for both the host and virus at the three HIV infection stages. To compare the networks of the infected and mock-infected CD4+ T cells, we defined an infection score using the differential PPMI network between the infected and mock-infected host cells at each stage as follows.

Use of an infection score to determine the interaction changes in the interspecies PPMI networks of mock- and HIV-infected CD4+ T cells at each infection stage

After reconstructing the interspecies PPMI network, we obtained three pruned interspecies PPMI networks in HIV-infected cells at three infection stages, and three networks in mock-infected cells at the time points used for the three infection stages. In the host models (7.19) of the mock-infected cells, the interspecies interactions between HIV and T cells, b_{ik} for $i = 1, \ldots, N$, and $k = 1, \ldots, K$, are set to zero. In mock-infected cells, the HIV model shown in Eq. (7.20) does not exist.

To define an infection score based on the differences between the two interspecies PPMI networks of HIV-infected and mock-infected cells, we first defined the network matrix of HIV-infected cells, A_v, according to $\hat{\theta}_i$ in Eq. (7.23) as follows:

$$A_v \triangleq \begin{bmatrix} \hat{a}_{11,v} & \cdots & \hat{a}_{1N,v} & \hat{b}_{11,v} & \cdots & \hat{b}_{1K,v} & \hat{c}_{11,v} & \cdots & \hat{c}_{1Z,v} \\ \vdots & \ddots & \vdots & \vdots & \ddots & \vdots & \vdots & \ddots & \vdots \\ \hat{a}_{N1,v} & \cdots & \hat{a}_{NN,v} & \hat{b}_{N1,v} & \cdots & \hat{b}_{NK,v} & \hat{c}_{N1,v} & \cdots & \hat{c}_{NZ,v} \end{bmatrix} \tag{7.24}$$

Similarly, we defined the network matrix, A_o, of mock-infected cells as follows:

$$A_o \triangleq \begin{bmatrix} \hat{a}_{11,o} & \cdots & \hat{a}_{1N,o} & 0 & \cdots & 0 & \hat{c}_{11,o} & \cdots & \hat{c}_{1Z,o} \\ \vdots & \ddots & \vdots & \vdots & \ddots & \vdots & \vdots & \ddots & \vdots \\ \hat{a}_{N1,o} & \cdots & \hat{a}_{NN,o} & \underbrace{0 & \cdots & 0}_{K} & \hat{c}_{N1,o} & \cdots & \hat{c}_{NZ,o} \end{bmatrix} \tag{7.25}$$

Furthermore, we defined the infection indication matrix D based on the interaction differences between the networks of mock- and HIV-infected cells as follows:

$$D \triangleq |A_v - A_o|$$
$$= \begin{bmatrix} |d_{11}| & \cdots & |d_{1j}| & \cdots & |d_{1(N+K+Z)}| \\ \vdots & \ddots & \vdots & \ddots & \vdots \\ |d_{i1}| & \cdots & |d_{ij}| & \cdots & |d_{i(N+K+Z)}| \\ \vdots & \ddots & \vdots & \ddots & \vdots \\ |d_{N1}| & \cdots & |d_{Nj}| & \cdots & |d_{N(N+K+Z)}| \end{bmatrix}$$
$$= \begin{bmatrix} |\hat{a}_{11,v} - \hat{a}_{11,o}| & \cdots & |\hat{a}_{1N,v} - \hat{a}_{1N,o}| & |\hat{b}_{11,v}| & \cdots & |\hat{b}_{1K,v}| & |\hat{c}_{11,v} - \hat{c}_{11,o}| & \cdots & |\hat{c}_{1Z,v} - \hat{c}_{1Z,o}| \\ \vdots & \ddots & \vdots & \vdots & \ddots & \vdots & \vdots & \ddots & \vdots \\ |\hat{a}_{N1,v} - \hat{a}_{N1,o}| & \cdots & |\hat{a}_{NN,v} - \hat{a}_{NN,o}| & |\hat{b}_{N1,v}| & \cdots & |\hat{b}_{NK,v}| & |\hat{c}_{N1,v} - \hat{c}_{N1,o}| & \cdots & |\hat{c}_{NZ,v} - \hat{c}_{NZ,o}| \end{bmatrix} \quad (7.26)$$

where $|d_{ij}|$ denotes the difference in the connection from the jth node to the ith node between the PPMI networks of mock- and HIV-infected cells.

To determine the proteins with connection differences between mock- and HIV-infected cells, the infection score of the ith protein, I_i, was defined as follows.

$$I_i \triangleq \log_2 \text{link}_i + \frac{\sum_{j=1}^{N+K+Z} |d_{ij}|}{\text{link}_i} \quad (7.27)$$

where link_i indicates the number of the links to the ith protein; $\sum_{j=1}^{N+K+Z} |d_{ij}|/\text{link}_i$ represents the average weighting difference of the ith protein between the networks of mock- and HIV-infected cells; and $\log_2 \text{link}_i$ is a compensation term to weight the hubs in the differential PPMI network. High infection scores suggest differences in the corresponding biological processes between mock- and HIV-infected cells.

For the infection score in Eq. (7.27) of the ith protein in the interspecies PPMI network, we considered the average weighting difference of the ith protein between the networks of mock- and HIV-infected cells as a major criterion. However, numerous small connection differences in a hub protein lead to a decrease in the major criteria due to the average weighting, i.e., the second term of Eq. (7.27) (Fig. C.2). The link number is scaled by $\log_2$ as a compromise for compensation adjustment to enhance the significant node of the hub in the infection process. According to the distribution plots of I_i and $\sum_{j=1}^{N+K+Z} |d_{ij}|/\text{link}_i$ among the proteins (Fig. C.2) at each stage in the mock- and HIV-infected cells, the term $\log_2 \text{link}_i$ compensates for the effect of numerous small average connection differences on the infection scores of hub proteins. The compensation term brings the infection scores of the hubs closer to the center of the infection score distribution. In other words, we assumed that the infection score of the hub proteins should not be lower than the median infection score of all proteins. The infection score in Eq. (7.27) based on these two criteria was used to define I_i to determine which host proteins with high connection differences between the host–pathogen interaction networks of mock- and HIV-infected cells. In this section, the core crosstalk network defined at each infection stage comprises the host proteins with the highest I_i (top 10%) and their interacting host miRNAs and viral proteins (Figs. 7.17–7.19 and Figs. C.3–C.5).

Moreover, in order to unravel the pathogenic mechanisms during HIV infection, we extracted the common PPMI network and specific PPMI networks at each stage based on the crosstalk core networks as follows.

Extraction of the common and specific PPMI networks during HIV-1 infection

Based on the core crosstalk networks in the reverse transcription (Fig. 7.17), integration/replication (Fig. 7.18), and late infection (Fig. 7.19) stages, we found a set of 47 proteins that were common to all three networks (Table 7.8), which were designated as the common PPMI network of the host–pathogen interaction along with their interacting host miRNAs and viral proteins (Figs. 7.20–7.22). All other proteins, including the interacting host miRNAs and viral proteins, were members of the specific PPMI

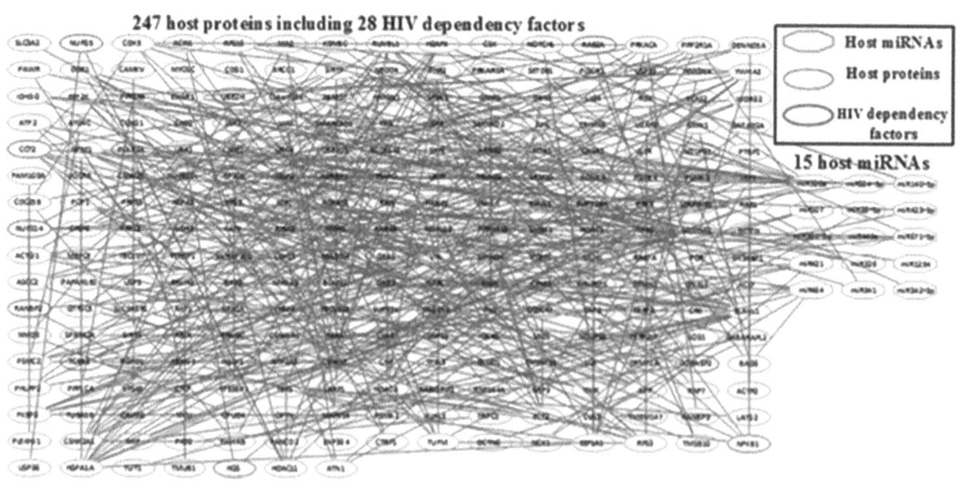

Figure 7.17 The interspecies host–pathogen interaction network at the reverse transcription stage. The top 10% of host proteins with the highest infection scores at the reverse transcription stage were selected as the core proteins of this early host–pathogen interaction network.

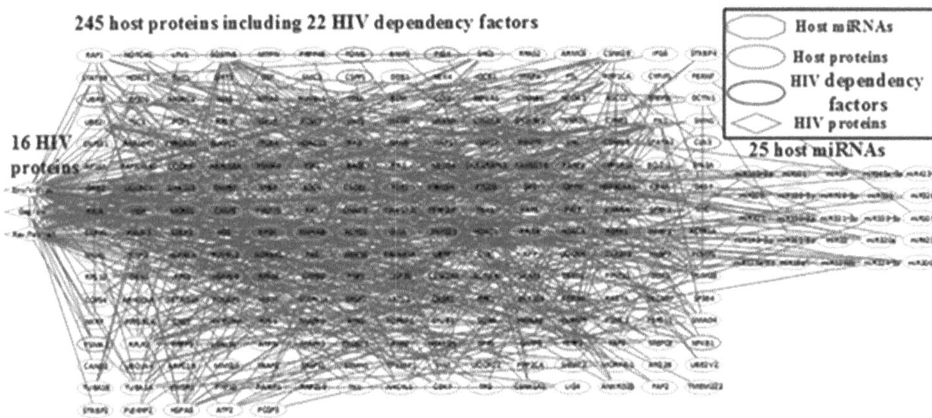

Figure 7.18 The interspecies host–pathogen interaction network during the integration/replication stage. The top 10% of host proteins with the highest infection scores during the integration/replication stage were selected as nodes of this host–pathogen interaction network.

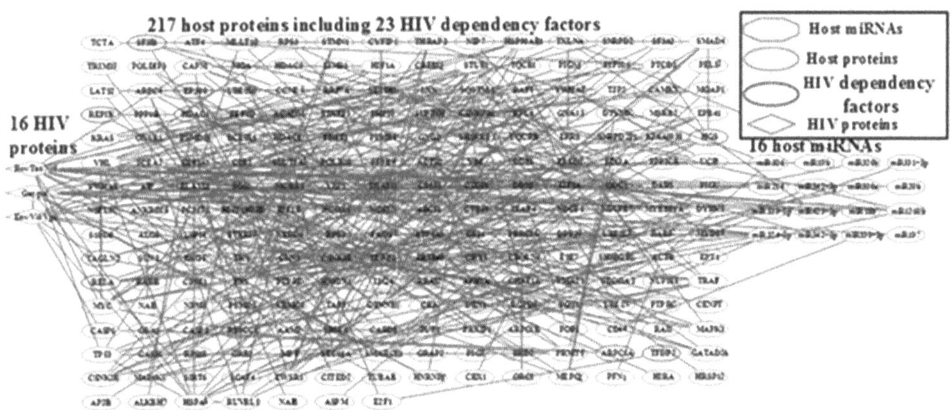

Figure 7.19 The interspecies host–pathogen interaction network in the late stage of HIV infection. Interactions of the top 10% of host proteins with the highest infection scores in the late stage of infection are shown.

networks at each infection stage (Figs. C.3–C.5). In a genome-wide siRNA screen to knock down more than 20,000 human genes, 932 proteins were found to be essential for HIV replication [132–134]. These factors were called HIV dependency factors and are highlighted in the results (Figs. 7.17–7.24 and Figs. C.3–C.5).

Design strategy for determining drug combinations for treating each stage of HIV-1 infection in CD4+ T cells

We applied Pearson correlation to the drug response genome-wide microarray data [126] to evaluate the effects of 1327 drugs on 14,825 genes/miRNAs.

Table 7.8 The 47 proteins in the core host–pathogen crosstalk network and the infection scores at different stages of HIV infection

	Infection score				Infection score		
Core proteins	**Reverse transcription stage**	**Integration/ replication stage**	**Late stage**	**Core proteins**	**Reverse transcription stage**	**Integration/ replication stage**	**Late stage**
APP	8.453	8.909	8.714	MYC	6.936	7.082	7.226
BAG6	6.307	6.068	6.016	NEDD4	7.454	7.014	8.436
CLN3	6.901	6.118	6.830	NPM1	6.006	6.701	6.059
CSNK1E	6.552	7.028	7.176	OXSR1	5.951	6.083	6.117
CSNK2A1	6.660	6.711	6.413	PAXIP1	6.506	6.184	6.984
CSNK2B	6.505	6.717	6.912	POT1	6.912	7.045	6.579
DBN1	5.804	6.293	6.472	PPP1CA	6.271	6.578	6.061
DDB1	5.948	6.492	6.348	RELA	5.592	6.726	6.238
EEF1A1	6.277	5.719	6.804	RUVBL1	6.676	5.986	6.514
ELAVL1	10.092	9.862	11.208	SH3KBP1	5.772	6.134	5.989
EP300	6.616	6.947	6.082	SMAD4	6.412	5.739	6.858
EWSR1	5.825	5.792	6.256	SQSTM1	8.372	8.363	7.937
FN1	9.521	8.667	9.609	STMN1	5.623	6.896	9.061
GATAD2A	6.277	6.463	7.805	TAF9	6.169	6.295	6.543
GRB2	6.737	7.200	6.857	TERF2	7.003	7.539	7.536
HDAC1	7.207	7.329	8.144	TP53	7.142	7.398	6.845
HDAC11	6.884	6.576	5.952	TUBA1B	6.678	6.490	6.052
HDAC3	6.059	6.730	7.587	TUT1	7.961	10.285	9.482
HGS	6.017	6.372	6.520	UQCRB	5.921	6.045	6.767
HSPA8	5.611	5.701	6.976	VHL	8.921	9.145	10.007
IQCB1	6.325	6.225	6.170	VIM	6.111	6.338	6.768
LATS2	5.902	8.290	6.222	YWHAB	6.100	6.340	6.395
LIG4	5.952	5.744	6.329	YWHAZ	6.228	6.700	6.084
MEPCE	5.644	5.786	6.845				

The core set of 47 proteins common to the interaction networks throughout the viral life cycle. The names of the core proteins and the corresponding infection scores at different stages of infection are shown.

If the correlation coefficient between a drug–gene (or drug–miRNA) pair is close to 1, gene expression can be activated by an increased concentration of the drug. Oppositely, if the correlation coefficient between a drug–gene (or drug–miRNA) pair is close to -1, gene expression can be inhibited by an increased concentration of the drug. Therefore, we can determine multiple drug combinations that activate the highly repressed genes and inhibit the highly expressed genes at a particular infection stage as a potential treatment for HIV patients.

7.3.2 Results and discussion

Functional analysis of the core PPMI networks at three infection stages

By applying a system identification method and system order detection scheme to dynamic models of the PPMI network at each stage of HIV-1 infection using

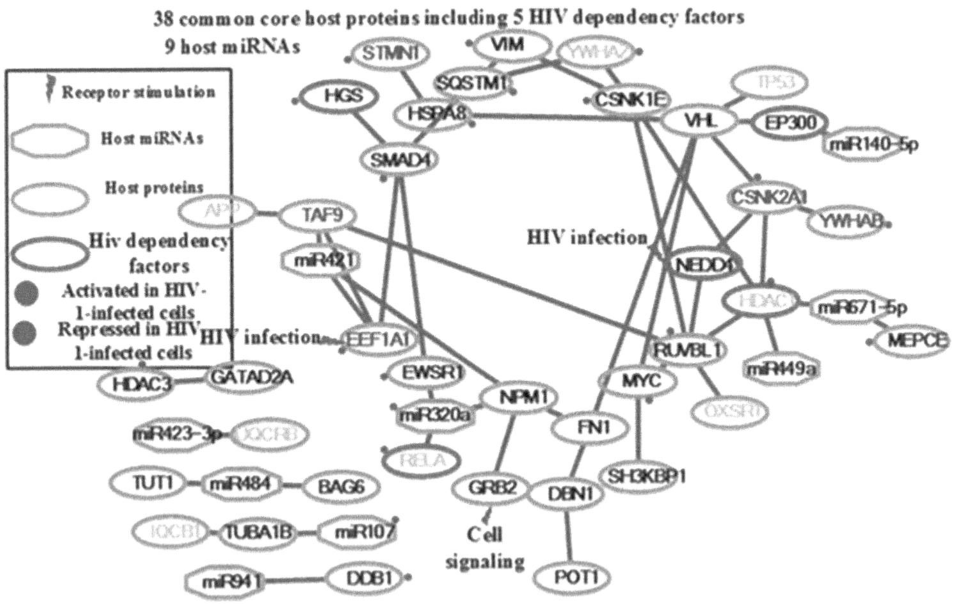

Figure 7.20 The core host–pathogen interaction network during the reverse transcription stage. By selecting the intersecting core host proteins from all three stages of infection, we identified 47 host proteins as core proteins (Table 7.8). Interactions among these core proteins during reverse transcription are depicted. *Pink* (light gray in print versions) and *red* (dark gray in print versions) labels denote the hypo- and hypermethylated proteins, respectively, i.e., *pink* and *red* labels denote that the fold-changes from the basal level, δ_i, are respectively less of greater than 35 and 100 in Eq. (7.19) between mock- and HIV-infected cells.

high-throughput gene expression data from HIV-1 and CD4+ T cells (see Section 7.3.1), we obtained the real PPMI network at the three HIV-1 infection stages. Based on the identified PPMI networks, we used an infection score to identify the core PPMI network with the top 10% of host proteins, which have specific biological functions at each stage of HIV-1 infection. The core PPMI networks include 247 host proteins and 15 host miRNAs at the reverse transcription stage; 245 host proteins, 25 host miRNAs, and 16 viral proteins at the integration/replication stage; and 217 host proteins, 16 host miRNAs, and 16 viral proteins at the late stage. We applied the GO tool DAVID (https://david.ncifcrf.gov/) to the host core proteins to identify the top categories based on *p*-value in Swiss-Prot and Protein Information Resource (SP-PIR) keywords, Biological Process, and Online Mendelian Inheritance in Man (OMIM) Disease. The SP-PIR keywords include molecular mechanisms, such as methylation, acetylation, and ubiquitin-like (Ubl) conjugation, and participating in the epigenetic alteration of gene expression. Epigenetic alteration of gene expression in CD4+ T cells has been observed during HIV-1 infection [135]. We assumed that the molecular mechanisms of epigenetic alterations, such as DNA

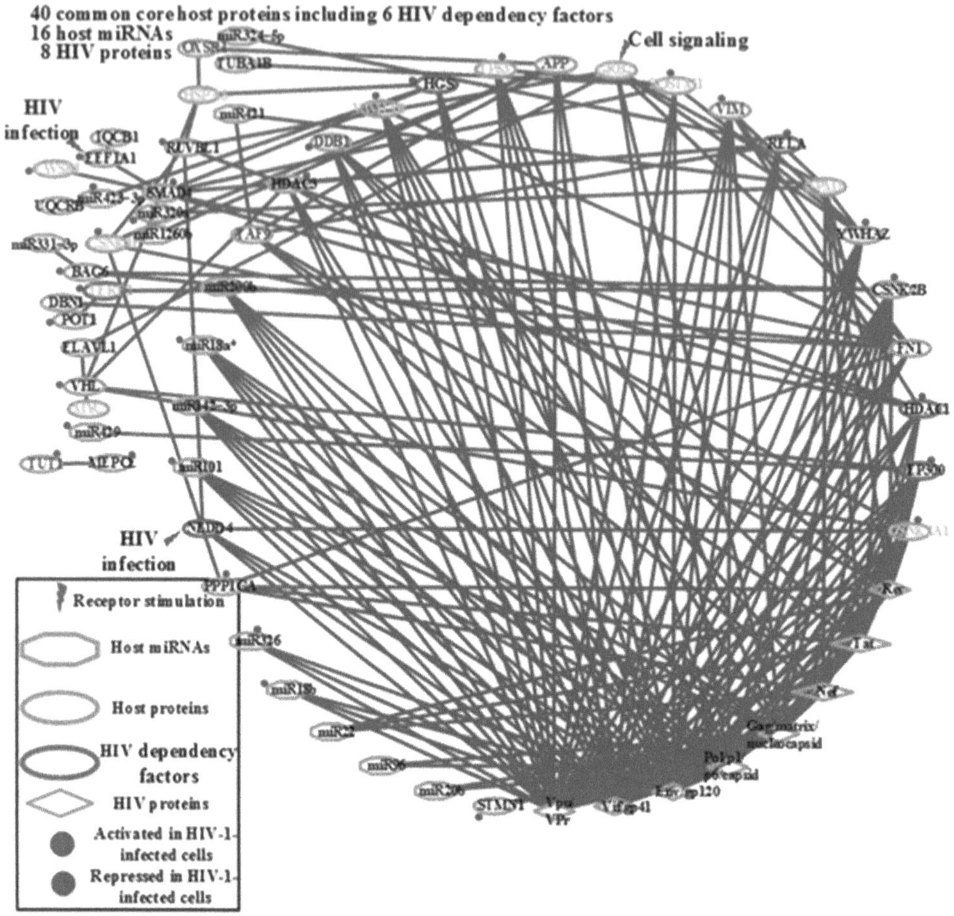

Figure 7.21 The core host–pathogen interaction network during the integration/replication stage. The symbols are the same as those shown in Fig. 7.20.

methylation and histone modification, could lead to basal changes in gene expression during HIV-1 infection in Eq. (7.19). The functional categories of the core proteins can help us determine the pathogenic and defensive mechanisms in HIV-1 infection.

We obtained the top functional categories of the core proteins at the reverse transcription stage (Fig. 7.17), which were acetylation ($p < 7.56 \times 10^{-45}$), negative regulation of macromolecule metabolic process ($p < 2.87 \times 10^{-11}$), Ubl conjugation ($p < 6.96 \times 10^{-11}$), and host–virus interaction ($p < 3.64 \times 10^{-9}$). One hundred and thirty seven core proteins were associated with acetylation and Ubl conjugation in the molecular mechanisms of epigenetics. The negative regulation of macromolecule metabolic processes could inhibit the degradation of host proteins. It has been reported that several human proteins, such as HDAC1, Gemin2, HuR, AKAP149, and DHX9,

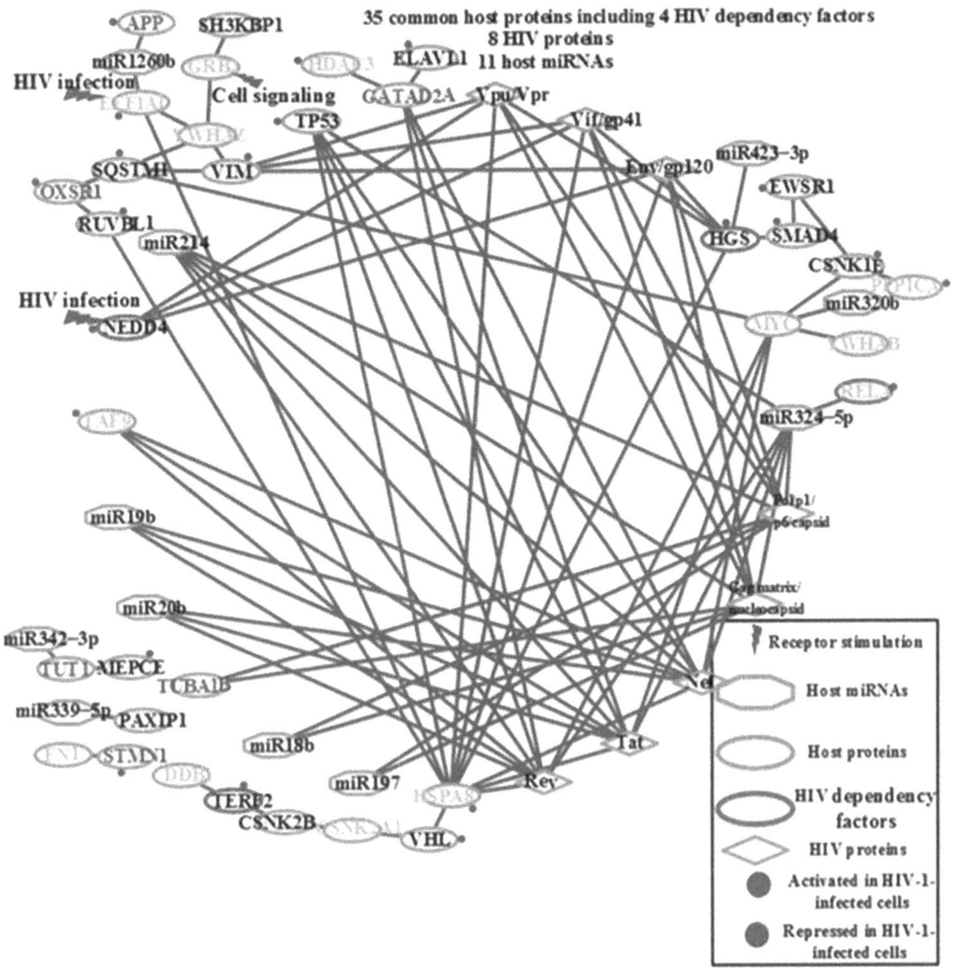

Figure 7.22 The core host–pathogen interaction network in the late-stage infection stage. The symbols are the same as those shown in Fig. 7.20.

are required for reverse transcription of HIV-1 [136]. Therefore, the core proteins in the PPMI network at the reverse transcription stage implicate epigenetic alteration of human genes during viral infection, and inhibition of human protein degradation may help increase the efficiency of the reverse transcription of HIV-1.

The top categories of the core proteins at the integration/replication stage (Fig. 7.18) include acetylation ($p < 2.68 \times 10^{-52}$), negative regulation of macromolecule metabolic process ($p < 4.08 \times 10^{-14}$), negative regulation of apoptosis ($p < 1.29 \times 10^{-10}$), and host–virus interaction ($p < 2.07 \times 10^{-9}$). It has been also observed that HIV-1 replication induces the proliferation of CD4+ T cells, which is

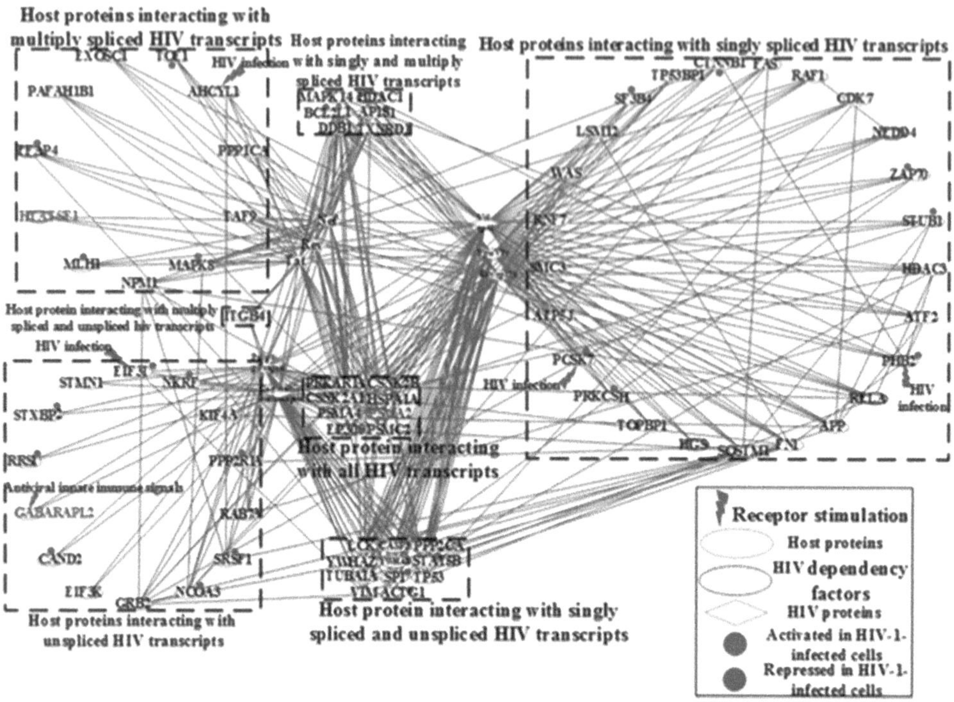

Figure 7.23 Interspecies interactions between viral and host proteins at the integration/replication stage. The symbols are the same as those shown in Fig. 7.20.

mediated by the effect of HIV-1 Vpr on apoptosis [137,138]. The host proteins identified in the core PPMI network at integration/replication suggest that epigenetic alteration of human genes, the effects of HIV-1 proteins, and the reduced degradation of human proteins lead to HIV-1 integration and replication and host cell proliferation.

The top categories of the core proteins at the late stage (Fig. 7.19) included retroviral genome replication ($p < 0.027$), hepatocellular carcinoma (HCC) ($p < 0.040$), telomere capping ($p < 0.053$), and protein repair ($p < 0.053$). Interestingly, the roles of the identified core host proteins at the late stage included retroviral genome replication.

Using confocal microscopy of live HIV-infected cells, we observed HIV replication in antigen-presenting cells during the late stage of infection before spreading to T cells [139]. The core proteins of the PPMI network at the late stage also suggest that the accumulated epigenetic alterations in human genes could cause incorrect folding of human proteins, which induces crucial repair functions. The accumulated misfolded proteins lead to endoplasmic reticulum stress, which has a profound effect on cancer cell proliferation [140]. Telomere capping is then induced to stabilize cell proliferation. The HCC-associated proteins in the core PPMI network at the late stage could cause mortality in HIV-infected patients [141].

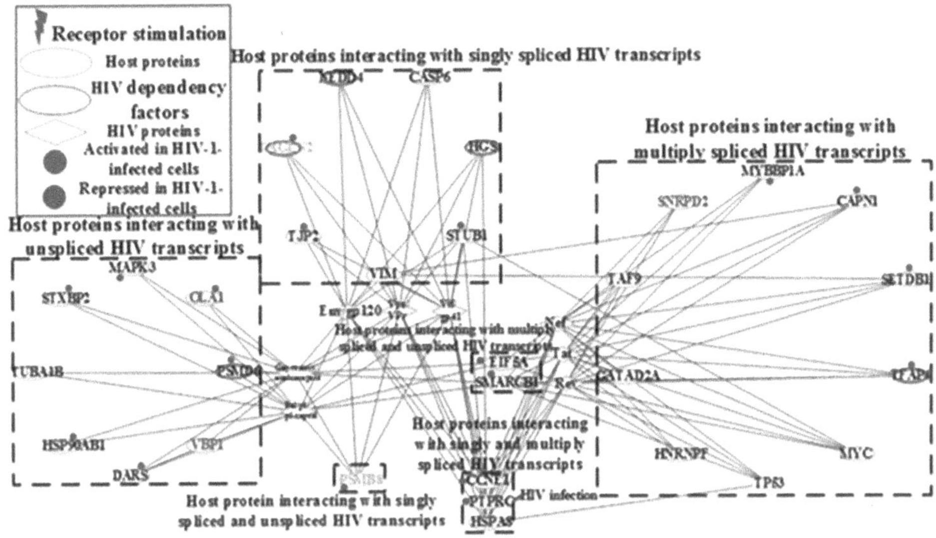

Figure 7.24 Late-stage interactions between viral and host proteins. The symbols are the same as those shown in Fig. 7.20.

Functional analysis of the common core PPMI network at the three infection stages

We extracted the common and unique functions at the three infection stages, and by comparing 247 host proteins at the reverse transcription stage, 245 host proteins at the integration/replication stage, and 217 host proteins at the late stage, we obtained 47 common proteins. We also obtained 200 specific proteins at the reverse transcription stage, 198 specific proteins at the integration/replication stage, and 170 specific proteins at the late stage. We removed some specific proteins that did not connect to any other protein or miRNA in the corresponding specific PPMI networks, and then we obtained the specific PPMI networks at the reverse transcription (Fig. C.3), integration/replication (Fig. C.4), and late (Fig. C.5) stages by connecting specific proteins to interacting host miRNAs and viral proteins. Similarly, the connections between common proteins and their interacting host miRNAs and viral proteins constitute the specific core PPMI network at each infection stage (Figs. 7.20–7.22).

The top categories for the 47 common proteins are acetylation ($p < 4.59 \times 10^{-11}$), host–virus interaction ($p < 1.69 \times 10^{-8}$), negative regulation of macromolecule metabolic process ($p < 3.05 \times 10^{-8}$), and Ubl conjugation ($p < 7.27 \times 10^{-6}$). Therefore, the functional categories in the common core PPMI networks show that epigenetic control of human gene expression and viral infection and inhibition of human protein degradation contribute to the progression of HIV-1 infection in CD4+ T cells. In other words, we found that the main functions of the core PPMI network at the reverse transcription stage are the same as those of the common core PPMI network. This result is supported by the previous suggestion that the progression of HIV-1 infection has a common mechanism that is largely independent of human and viral diversity [142].

Because no viral expression data is available at the reverse transcription stage, we cannot identify the direct target of the viral proteins identified based on system modeling at the early infection stage. Previous reports have shown that in chronic untreated HIV infection, the more Gag epitopes targeted by the immune response the lower the viral load [143]. It has also been reported that Gag directly interacts with EEF1A1 and NEDD4 in T cells [144,145]. We identified EEF1A1 and NEDD4 as direct targets of HIV-1 infection in the core PPMI network at the reverse transcription stage (Fig. 7.20). CSNK2A1 and RUVBL1, which interact with NEDD4, are involved in acetylation, while SMAD4 and TAF9, which interact with EEF1A1, are involved in the negative regulation of macromolecule metabolic process and Ubl conjugation. Hypermethylation of STMN1, CSNK2A1, YWHAB, and EEF1A1 and the miRNAs, miR-421, miR-320a, miR-449a, miR-671-5p, and miR-140-5p mainly regulate cellular functions, including epigenetic regulation, protein degradation, and HIV reverse transcription during early HIV-1 infection. By comparing the methylation profiles of HIV-infected and non-HIV-infected human B cells (see Section 7.3.1), we observed significant methylation differences in STMN1 ($p < 2.99 \times 10^{-6}$), CSNK2A1 ($p < 5.59 \times 10^{-7}$), YWHAB ($p < 1.33 \times 10^{-14}$), and EEF1A1 ($p < 6.77 \times 10^{-9}$).

For the common core PPMI network at the integration/replication stage (Fig. 7.21), we found several miRNAs, including miR-18a*, miR-18b, miR-20b, miR-22, miR-96, miR-101, miR-142-3p, miR-200b, and miR-326, that directly interact with HIV proteins. Among them, only miR-326 is significantly expressed in HIV-infected T cells at the integration/replication stage. It has been observed that antagomirs, which knocked down miR-326, increased HIV-1 replication in CD4+ T cells [146]. Therefore, we suggest that these HIV-interacting miRNAs modulate HIV-1 replication in human T cells. Hypermethylation of APP, VIM, and OXSR1 was also observed in the core PPMI network at the integration/replication stage (Fig. 7.21). Significant methylation differences were observed in the DNA methylation profiles of APP ($p < 1.46 \times 10^{-6}$), VIM ($p < 7.75 \times 10^{-7}$), and OXSR1 ($p < 9.83 \times 10^{-8}$). Methylated VIM interacts with HDAC1, which can participate in four of the top functional categories, including acetylation, host–virus interaction, negative regulation of macromolecule metabolic process, and apoptosis, at the integration/replication stage (Fig. 7.21). Methylated OXSR1 interacts with HSPA8, which can participate in acetylation and host–virus interaction. Methylated APP interacts with TAF9, which can participate in the negative regulation of apoptosis. It has been reported that HDAC1 is associated with acetylation and viral replication in

HIV-1-infected T cells [147]. It has also been shown that HSPA8 is associated with promoting viral replication in HIV-1-infected T cells [148]. Additionally, a review study proposed that TAF9 may be involved in the transcription and replication of HIV genes in HIV-infected T cells [149]. TAF9 is also the output of the Notch signaling pathway, which controls cell apoptosis [150]. A recent study has suggested that HIV proteins alter normal apoptotic signaling, resulting in an increased viral load in T cells and the formation of viral reservoirs that ultimately increase infectivity [151]. Based on these results, we suggest that epigenetic alterations in APP, VIM, and OXSR1 and miR-326 repression play an important role in negatively regulating cell apoptosis in the common core PPMI network at the integration/replication stage.

In the common core PPMI network at the late stage (Fig. 7.22), we identified numerous miRNAs, including miR-18b, miR-19b, miR-20b, miR-197, miR-214, and miR-324-5p, that directly regulate viral proteins. Among these miRNAs, miR-214 and miR-324-5p interact with the most viral proteins and have been proposed to target HIV-1 and inhibit virion production [152–155]. Hypermethylation of APP, GATAD2A, STMN1, and TUT1 was identified in the common core PPMI network at the late stage. Comparison of the methylation profiles in uninfected and HIV-infected B cells showed significant differences in the methylation profiles of APP ($p < 1.46 \times 10^{-6}$), GATAD2A ($p < 1.71 \times 10^{-10}$), STMN1 ($p < 2.99 \times 10^{-6}$), and TUT1 ($p < 1.07 \times 10^{-9}$). Interestingly, six methylation-associated host proteins/miRNAs, including the APP-interacting miRNA miR-1260b, the GATAD2A-interacting proteins ELAVL1 and HDAC3, the STMN1-interacting protein FN1, the TUT1-interacting protein/miRNA MEPCE, and miR-342-3p, are closely related to human carcinogenesis [156–163]. The common core proteins APP, OXSR1, TAF9, and ELAVL1 are significantly upregulated at the late stage of infection. APP and OXSR1 have been shown to be overexpressed in neuropathogenesis [164,165], and late HIV infection can also induce neuropathogenesis [166]. Additionally, because TAF9 depletion leads to apoptotic cell death [167] and TAF9 activity is essential for cell viability [168], we suggest that TAF9 overexpression is involved in regulating carcinogenesis at the late stage of HIV-1 infection.

Functional analysis of host–pathogen interaction networks

Before conducting the functional analysis of the specific core PPMI networks at the three infection stages, we investigated the effect of the direct interactions between host and viral proteins on host functions at the integration/replication and late stages. Because we only have the expression data for unspliced pre-mRNA, multiple-spliced RNA, and single-spliced RNA, we analyzed the impact of these three viral RNAs on host cellular functions. Based on the identified PPMI networks at the integration/replication (Fig. 7.18) and late (Fig. 7.19) stages, we obtained the interactions between host proteins and the viral RNAs encoded by these three viral RNAs.

The results show that most host proteins only interact with the viral proteins encoded by a single viral RNA at the integration/replication (Fig. 7.23) and late (Fig. 7.24) stages. At the integration/replication stage (Fig. 7.23), the host proteins that directly interact with viral proteins were categorized into the top four cellular functions, acetylation ($p < 4.75 \times 10^{-21}$), host–virus interaction ($p < 8.61 \times 10^{-10}$), regulation of apoptosis ($p < 3.94 \times 10^{-9}$), and negative regulation of macromolecule metabolic process ($p < 2.07 \times 10^{-7}$), based on the GO tool DAVID. The viral proteins encoded by unspliced pre-mRNA interact with 9, 1, 1, and 2 host proteins involved in acetylation, host–virus interaction, regulation of apoptosis, and negative regulation of macromolecule metabolic process, respectively. The viral proteins encoded by single-spliced RNA interact with 9, 2, 2, and 3 host proteins involved in acetylation, host–virus interaction, regulation of apoptosis, and negative regulation of macromolecule metabolic process, respectively. The viral proteins encoded by multiple-spliced RNA interact with 5, 3, and 2 host proteins involved in acetylation, regulation of apoptosis, and negative regulation of macromolecule metabolic process, respectively. Therefore, we suggest that most HIV proteins have indirect effects on epigenetically regulated host genes. Additionally, the proteins encoded by single-spliced RNA and multiple-spliced RNA increasingly contribute to the regulation of apoptosis and negative regulation of macromolecule metabolic process, leading to high carcinogenic risk [169].

At the late infection stage (Fig. 7.24), the host proteins that directly interact with viral proteins were categorized into the top four cellular functions, acetylation ($p < 3.17 \times 10^{-7}$), host–virus interaction ($p < 4.42 \times 10^{-6}$), regulation of protein catabolic process ($p < 4.94 \times 10^{-6}$), and response to radiation ($p < 9.4 \times 10^{-4}$). The viral proteins encoded by unspliced pre-mRNA interact with 5 and 1 host proteins involved in acetylation and host–virus interaction, respectively. The viral proteins encoded by single-spliced RNA interact with 3, 2, 3, and 2 host proteins involved in acetylation, host–virus interaction, regulation of protein catabolic process, and response to radiation, respectively. The viral proteins encoded by multiple-spliced RNA interact with 7, 1, 1, and 2 host proteins involved in acetylation, host–virus interaction, regulation of protein catabolic process, and response to radiation, respectively. Because a number of tumors occur at higher rates in HIV-positive individuals than in HIV-negative individuals [170,171], studies of HIV-positive individuals treated by radiation therapy have been reported. HIV-infected patients may be more sensitive to ionizing radiation than noninfected patients [172]. These results show that the viral proteins encoded by the single-spliced and multiple-spliced RNAs could contribute to the radiation sensitization of HIV-positive individuals. Additionally, the data also show a lesser effect of the viral proteins encoded by unspliced pre-mRNA on human proteins from the integration/replication stage to the late stage. This may be because the HIV protein Gag traffics and targets to the plasma membrane for retroviral assembly and release at the late stage of infection [173].

Functional analysis of the specific PPMI networks at the three infection stages

After extracting the biological functions of the host that directly interact with viral proteins at each stage of HIV-1 infection, we investigate the biological functions of the specific PPMI networks at the reverse transcription (Fig. C.3), integration/replication (Fig. C.4), and late (Fig. C.5) stages. We applied the GO tool DAVID to the specific PPMI networks at the three infection stages to identify the functional core module of the PPMI network at each infection stage. We then obtained the top three functional categories at each stage, which were protein kinase activity ($p < 2.56 \times 10^{-4}$), RNA splicing ($p < 4.53 \times 10^{-4}$), and macromolecular complex subunit organization ($p < 5.29 \times 10^{-4}$) at the reverse transcription stage (Fig. 7.25); steroid hormone receptor binding ($p < 4.05 \times 10^{-6}$), response to DNA damage stimulus ($p < 4.19 \times 10^{-6}$), and regulation of DNA replication ($p < 6.43 \times 10^{-5}$) at the integration/replication stage (Fig. 7.26); and prostate cancer ($p < 1.43 \times 10^{-5}$), protein biosynthesis ($p < 3.18 \times 10^{-5}$), and chronic myeloid leukemia ($p < 4.02 \times 10^{-5}$) at the late stage (Fig. 7.27).

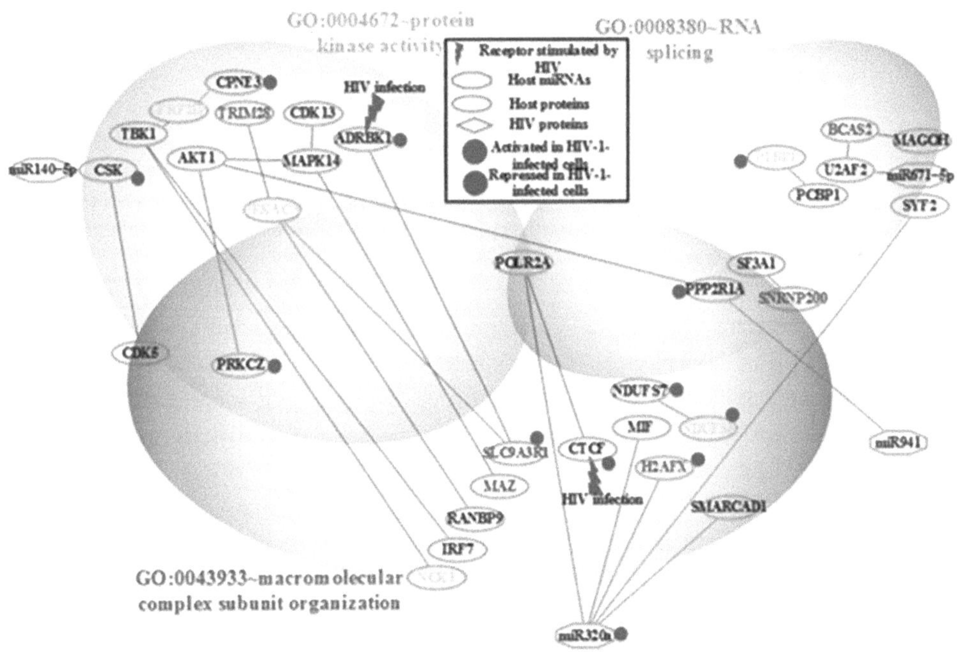

Figure 7.25 The identified functional core modules of HIV-1-infected CD4+ T cells at the reverse transcription stage. We applied DAVID to the specific PPMI network in Fig. C.3 to identify the three main cellular functions of the proteins in the network, which are protein kinase, RNA splicing, and macromolecular complex subunit organization, that are induced by these proteins in HIV-1-infected T cells compared with mock-infected T cells.

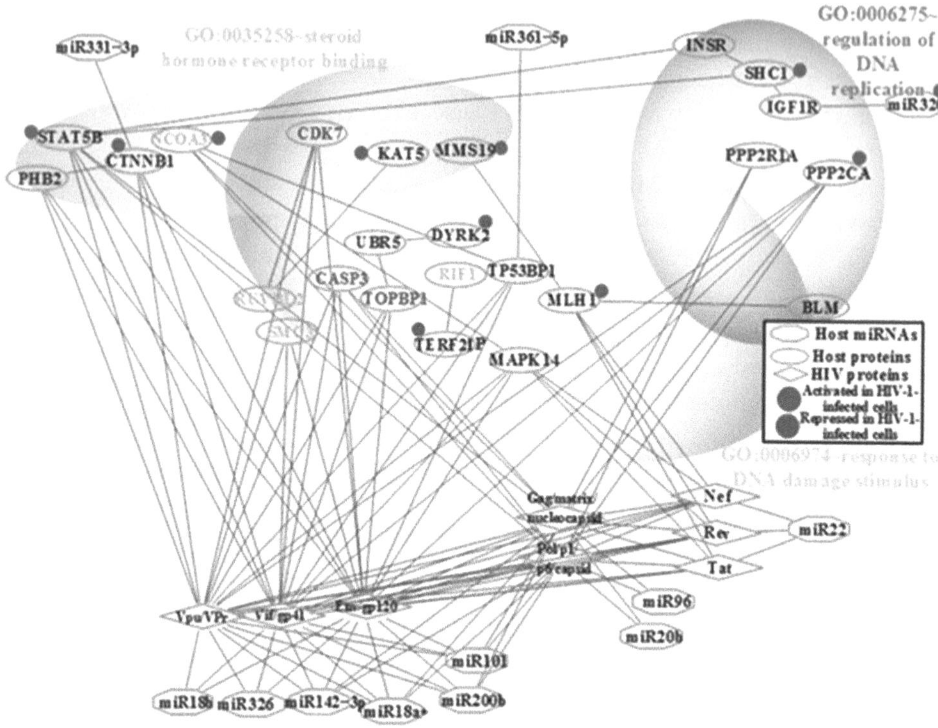

Figure 7.26 The identified functional core modules of HIV-1-infected CD4 + T cells at the integration/replication stage. We applied DAVID to the specific PPMI network in Fig. C.4 to identify the three main functions, steroid hormone receptor binding, response to DNA damage stimulus, and regulation of DNA replication, that are induced by the specific proteins in HIV-1-infected T cells compared with mock-infected T cells.

At the reverse transcription stage, several HIV proteins associated with the mature reverse transcription complex [174] and host protein kinases, such as the cAMP-dependent protein kinases (PKAs) [175] and cyclin-dependent kinases (CDKs) [176], play an important role in the reverse transcription of full-length viral RNA and spliced RNAs of HIV-1 [177]. The identified functional core modules shown in Fig. 7.25 characterize the molecular mechanisms that occur at the reverse transcription stage. Comparison of the methylation profiles in uninfected and HIV-infected cells identified several hypermethylated proteins, including TRIM28 ($p < 1.17 \times 10^{-11}$), MAZ ($p < 1.34 \times 10^{-9}$), SLC9A3R1 ($p < 2.56 \times 10^{-9}$), H2AFX ($p < 4.11 \times 10^{-8}$), SNRNP200 ($p < 5.38 \times 10^{-6}$), and BCAS2 ($p < 1.89 \times 10^{-5}$), at the reverse transcription stage (Fig. 7.25). The hypermethylated proteins TRIM28, MAZ, and SLC9A3R1 interact with ADRBK1, MAPK14, CDK13, and PRKACA, which are involved in protein kinase activity. It has been reported that CDK13 plays an important role in regulating both HIV replication [178] and mRNA splicing [179]

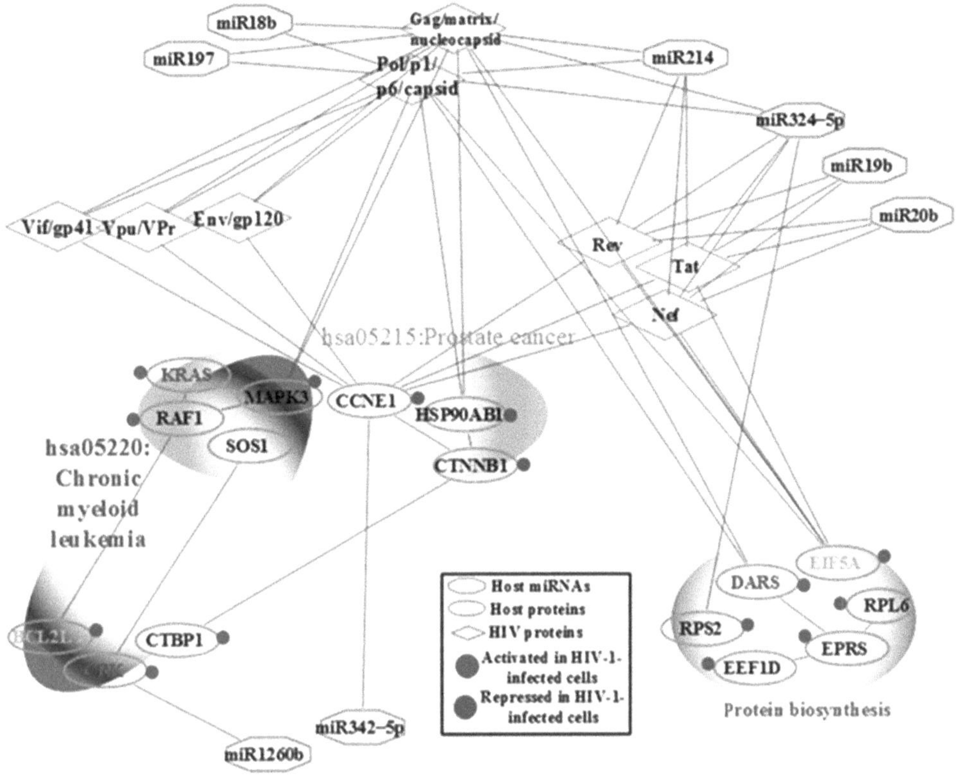

Figure 7.27 The identified functional core modules of the HIV-1-infected CD4 + T-cell line at the late stage. We applied DAVID to the specific PPMI network shown in Fig. C.5 to identify the three main functions, prostate cancer, protein biosynthesis, and chronic myeloid leukemia, that are induced by the specific proteins in HIV-1-infected T cells compared with mock-infected T cells.

at the reverse transcription stage. MAPK14 and PRKACA have been shown to induce T-cell immune responses in HIV-infected patients [180]. In the functional core modules of the specific PPMI network at the reverse transcription stage, we identified epigenetic regulation of TRIM28, which has also been shown to epigenetically regulate HIV-1 transcription [181]. We showed that miR-140-5p inhibits the expression of CSK. Like LYP [182], repressed CDK5 could inhibit T-cell activation and play an important role in development of autoimmune disorders [183] when dissociated from CSK at the reverse transcription stage. Therefore, we propose that miR-140-5p-mediated repression of CSK translation contributes to autoimmune disorders [184] and the repression of T-cell activation in HIV-1-infected T cells. Additionally, we showed that inhibition of miR-320a derepresses POLR2A, macrophage migration inhibitory factor (MIF), SYF2, and SMARCAD1 (Fig. 7.25). POLR2A is essential for the survival of lymphocytes [185]. MIF contributes to tumorigenesis in T cells by facilitating tumor proliferation [186]. HIV-induced production of SYF2 could be involved in the impaired proliferation capacity of T cells at the early stage of HIV infection [187]. SMARCAD1 plays a critical role in the efficient transmission of epigenetic information in stem cells [188]. Therefore, we propose that inhibition of miR-320a derepresses abnormal cell proliferation and epigenetic regulation in T cells. Moreover, we also showed that miR-941 inhibits PPP2R1A, leading to disruption of the PPI between PPP2R1A and AKT1. HIV-activated AKT1 has been implicated in the pathogenesis of T cells, which are the primary targets of HIV and respond to infection by providing several critical immune signals [189]. It has been also reported that disruption of the PPI between PPP2R1A and AKT1 contributes to cell survival [190]. Therefore, we propose that miR-941 regulation induces pathogenesis, activates critical immune signals, and contributes to cell survival.

At the integration/replication stage, it has been inferred that modifications of steroid hormones, particularly glucocorticoids and androgens, is associated with the loss of fat-free mass and fat mass observed in untreated HIV-infected patients [191–194]. The effects of steroid hormone modifications and DNA repair factors on HIV-1 long terminal repeat activity [195] and viral DNA [196], respectively, may contribute to viral replication. The identified functional core modules in Fig. 7.26 also characterize the molecular mechanisms at the integration/replication stage. Comparison of the methylation profiles in uninfected and HIV-infected cells also identified several hypermethylated proteins, including CASP3 ($p < 1.02 \times 10^{-7}$), CDK7 ($p < 4.09 \times 10^{-9}$), and TOPBP1 ($p < 9.88 \times 10^{-6}$), at the integration/replication stage (Fig. 7.26). The hypermethylated proteins CDK7 and TOPBP1 interact with RUVBL2 and UBR5, respectively. The use of genome-wide screening technologies in HIV-infected human cells showed that UBR5 and RUVBL2 are involved in the folding and assembly of proteins [197] and DNA repair factors [196], respectively, at the integration/replication stage. The effects of methylation on DNA repair [198,199] and posttranslation modification [200,201] in lymphocytes has also been discussed. We showed that miR-331-3p inhibits CTNNB1, which leads to an interaction between PHB2 and HIV proteins. It has been reported that PHB2 interacts with HIV-1 Env to promote replicative spread at the integration/replication stage [202]. Thus, inhibition of CTNNB1 can also promote HIV-1 replication [203]. A negative relationship or inverse relationship between PHB2 and CTNNB1 was also identified in cell migration [204]. Therefore, we propose that inhibition of CTNNB1 by miR-331-3p promotes HIV-1 replication and replicative spread. We observed that the interaction of IGF1R with miR-320a and SHC1 is disrupted by inhibition of miR-320a and SHC1 expression. Inhibition of SHC1 has been shown to lead to more sustained HIV-1 replication and reduced apoptosis of the infected T cells [205]. It has been shown that IGF1R is essential for cell growth and mediating antiapoptotic signals in AIDS/HIV-related Kaposi's sarcoma [206]. Activated SHC1 has also been shown to induce the Ras/MAP kinase pathway, which can trigger cell cycle arrest through IGF1R [207].

Therefore, we propose that inhibition of miR-320a and SHC1 contributes to cell proliferation and dysregulation of the cell cycle in HIV-1-infected cells at the integration/replication stage.

At the late stage of infection, several malignancies, especially prostate carcinoma in men [208] and chronic myeloid leukemia [209], have been known to occur more frequently. The increased carcinogenesis in HIV-infected patients is thought to be due to prolonged survival in the era of HAART. It has been also observed that in the prostate, the number of CD4+ T cells in the tumor is significantly greater than that in the surrounding benign tissue [210]. Based on the identified functional core modules of the HIV-1-infected CD4+ T cells at the late infection stage (Fig. 7..27), we propose that prostate carcinoma and chronic myeloid leukemia in HIV-infected patients are due to the pathogenesis of HIV infection, leading to a progressive depletion of CD4+ T-cell populations. We compared the synthesis of CD4 protein products in uninfected and HIV-infected CD4+ T cells [211]. The increased levels of cellular proteins have been suggested to interact with HIV-1 proteins participating in HIV infection [211]. The identified functional core modules shown in Fig. 7.27 characterize the molecular mechanisms occurring at the late stage of HOV infection. Comparison of the methylation profiles in uninfected and HIV-infected cells also identified several hypermethylated proteins, including CRK ($p < 1.81 \times 10^{-4}$), DARS ($p < 1.96 \times 10^{-8}$), and KRAS ($p < 1.69 \times 10^{-11}$), at the integration/replication stage (Fig. 7.27). The interaction of the hypermethylated and activated host protein CRK with SOS1 has been reported to promote cell proliferation and increase cell transformation and tumorigenicity in lymphocytes [212,213]. Methylated and activated DARS has been reported to be involved in the translation of HIV proteins in HIV particle assembly and budding at the late infection stage [214,215]. DARS and EPRS activation plays a critical role in formation of a protein complex called GAIT (IFN-gamma-activated inhibitor of translation) involved translation silencing and inhibiting tumor angiogenesis [216]. We propose that inhibition of EPRS could lead to angiogenesis associated with cancer development. The result shows that methylated and activated KRAS interacts with activated RAF1, which has been reported to enhance HIV-1 infectivity and function in virus assembly and release from infected cells [217,218]. Additionally, we showed that miR-341-5p inhibits CCNE1. It has been reported that the activated CCNE1 is an essential downstream effector of the retinoblastoma tumor suppressor (RB) pathway [219,220]. Disruption of the RB pathway has been reported in human cancers [221]. Therefore, the result suggests that inhibition of CCNE1 could result in carcinogenesis at the late stage of HIV-1 infection.

The pathogenic mechanisms shown in Figs. 7.25–7.27 during the three stages of HIV-1 infection are summarized in Fig. 7.28. At the reverse transcription stage, we identified the three main cellular functions, protein kinase, RNA splicing, and macromolecular complex subunit organization, that are induced by the specific proteins in HIV-1-infected T cells compared to mock-infected T cells (Fig. 7.28A).

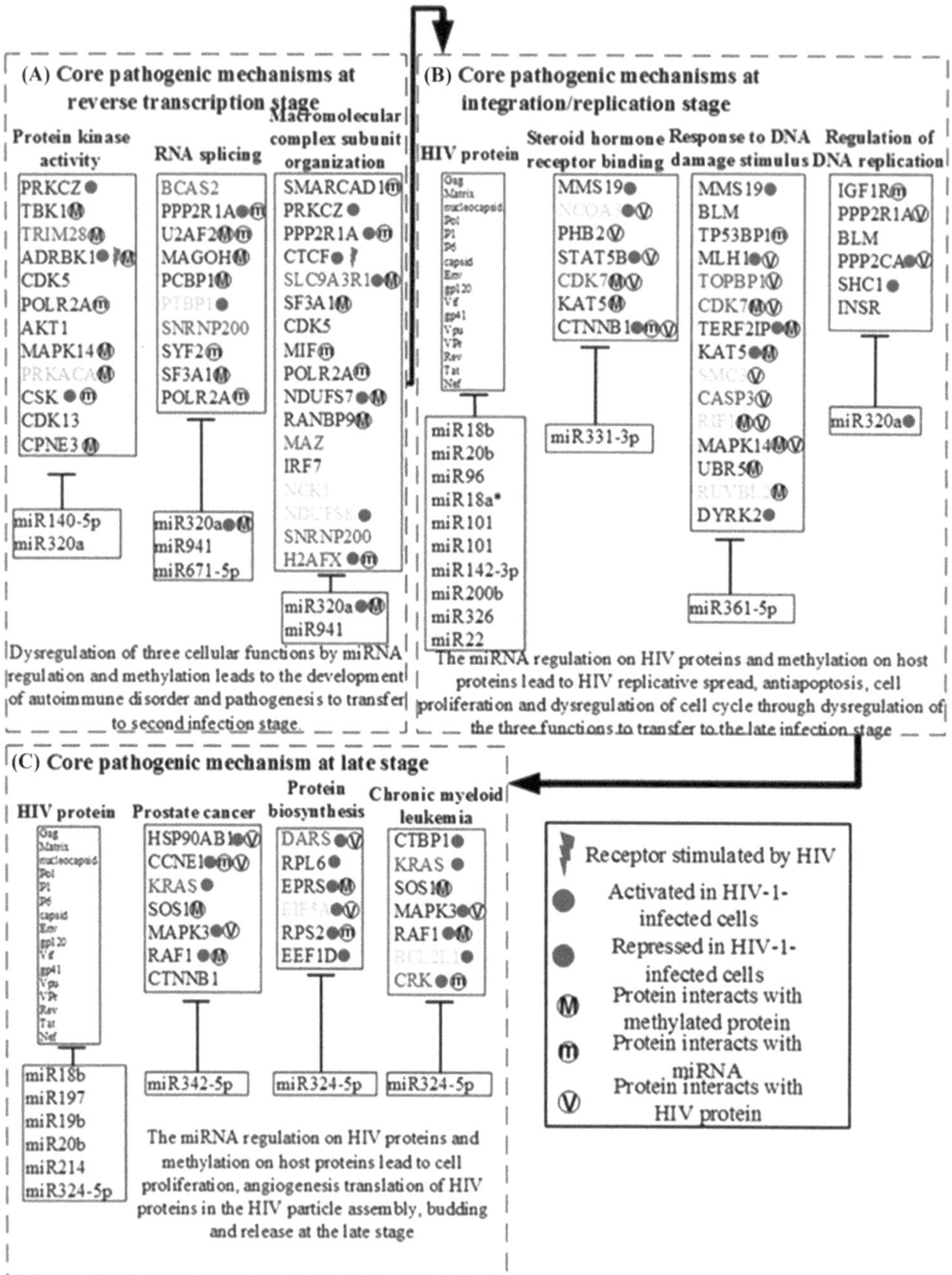

Figure 7.28 Pathogenic mechanisms based on the specific PPMI networks at the three stages of HIV infection. Based on the pathogenic mechanisms shown in Figs. 7.25–7.27, the identified core pathogenic mechanisms in the reverse transcription (A), integration/replication (B), and late (C) stages were determined. The *pink* (light gray in print versions) and *red* (dark gray in print versions) labels denote the identified hypo- and hypermethylated proteins, respectively.

Epigenetic regulation of TRIM28, MAZ, and SLC9A3R1 contributes to HIV replication, mRNA splicing, and the immune responses of T cells. Regulation of miR-140-5p, miR-320a, and miR-941, are involved in the development of autoimmune disorders, tumor proliferation, and the pathogenesis of T cells at the reverse transcription stage. At the integration/replication stage, we identified three main functions, steroid hormone receptor binding, response to DNA damage stimulus, and regulation of DNA replication, that are induced by the specific proteins in HIV-1-infected T cells when compared with mock-infected T cells (Fig. 7.28B). Epigenetic regulation of CDK7 and TOPBP contribute to DNA repair and the posttranslation modification, folding, and assembly of host proteins. The regulation of miR-331-3p, miR-320a, and miR-320a are involved in HIV-1 replication, replicative spread, antiapoptosis, cell proliferation, and dysregulation of the cell cycle at the integration/replication stage. At the late stage, we identified prostate cancer, protein biosynthesis, and chronic myeloid leukemia, as the three top functional categories of the specific proteins in HIV-1-infected T cells when compared with mock-infected T cells (Fig. 7.28C). Epigenetic regulation of CRK, DARS, and KRAS contribute to cell proliferation, angiogenesis, translation of HIV proteins in HIV particle assembly and budding, increased cell transformation and tumorigenicity, and enhanced HIV-1 infectivity, and functions in virus assembly and release. miR-341-5p is involved in carcinogenesis at the late stage of HIV-1 infection.

Multiple drug combinations for treatment at each stage of HIV-1 infection based on the specific PPMI networks

Finally, to identify drug combinations for the treatment of patients at each stage of HIV-1 infection, we used a strategy to discover a drug combination that could activate the highly repressed genes/miRNAs and inhibit the highly expressed genes/miRNAs in the functional core modules of the specific PPMI network (Figs. 7.25–7.27) at each HIV infection stage (see Section 7.3.1). At the reverse transcription stage, the gene *CPNE3* is highly expressed and ADRBK1, CSK, PRKCZ, SLC9A3R1, CTCF, H2AFX, NDUFS8, PTBP1, and miR-320a are highly repressed in HIV-1-infected T cells. Therefore, based on the functional core modules shown in Fig. 7.25, we propose a combination of thalidomide, oxaprozin, and metformin to treat patients with HIV-1 infection at the reverse transcription stage (Table 7.9). At the integration/replication stage, NCOA3 is highly expressed and STAT5B, CTNNB1, KAT5, MMS19, DYRK2, TERF2IP, MLH1, SHC1, PPP2CA, and miR-320a are highly repressed in HIV-1-infected T cells. Therefore, according to the functional core modules in Fig. 7.26, we propose a combination of quercetin, nifedipine, and fenbendazole to treat patients with HIV-1 infection at the integration/replication stage (Table 7.9). At the late stage, KRAS, RAF1, CRK, CTBP1, and DARS are highly expressed and MAPK3, CCNE1, HSP90AB1, CTNNB1, BCL2L1, EIF5A, RPL6, EPRS, RPS2, and EEF1D are highly repressed in HIV-1-infected T cells. Based on the functional core modules shown in Fig. 2.27, we propose a

Table 7.9 The strategy to identify potential drug combinations for treating HIV-1-infected patients at each stage of HIV-1 infection based on the drug response genome-wide microarray data

	Reverse transcription stage	Integration/ replication stage	Late stage
The highly expressed genes for potential inhibition strategy of multiple drug design	CPNE3	NCOA3	KRAS, RAF1, CRK, CTBP1, DARS
The suppressed genes for potential activation strategy of multiple drug design	ADRBK1, CSK, PRKCZ, SLC9A3R1, CTCF, H2AFX, NDUFS8, PTBP1, miR-320a	STAT5B, CTNNB1, KAT5, MMS19, DYRK2, TERF2IP, MLH1, SHC1, PPP2CA, miR-320a	MAPK3, CCNE1, HSP90AB1, CTNNB1, BCL2L1, EIF5A, RPL6, EPRS, RPS2, EEF1D
The potential multiple drug combination	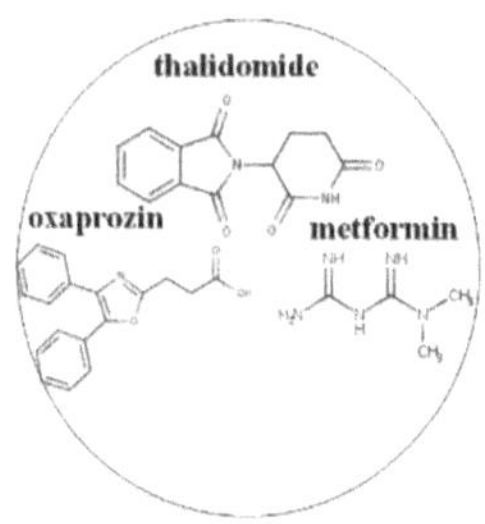	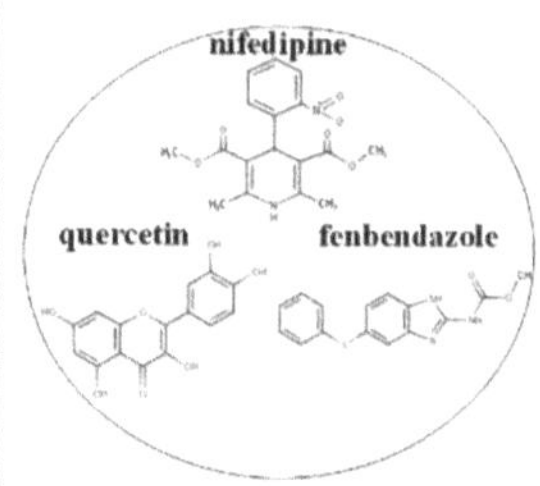	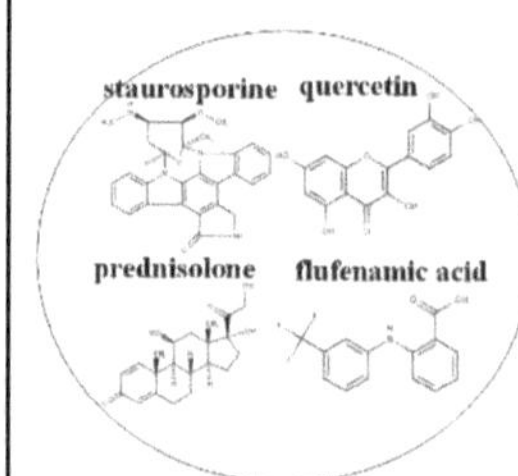

We used the correlation coefficients between drug–gene pairs (see Section 7.3.1) to determine their suitability for the treatment of HIV. If the correlation coefficient of a drug–gene pair was larger than 0.3, expression of the gene can be activated by the drug, whereas if the correlation coefficient of a drug–gene pair was less than 0.1, expression of the gene can be inhibited by the drug. Based on these criteria, we obtained a drug combination of thalidomide, oxaprozin, and metformin that could be used to treat patients with HIV-1 infection at the reverse transcription stage, a combination of quercetin, nifedipine, and fenbendazole that could be used to treat patients at the integration/replication stage, and a combination of staurosporine, quercetin, prednisolone, and flufenamic acid that could be used to treat patients at the late stage.

combination of staurosporine, quercetin, prednisolone, and flufenamic acid to treat patients with HIV-1 infection at the late stage (Table 7.9).

7.3.3 Conclusions

The mechanisms underlying HIV invasion and the host response are complex. HIV replication requires host machinery, and how the virus takes over and repurposes host resources has been extensively investigated for many years. Although some HIV therapies increase the life spans of HIV-infected patients, curing HIV infection is still a big challenge. Thus, understanding the interactions between HIV and host cells may identify new therapeutic targets. In this section, we provided a global view of PPMI using a systems biology approach. By comparing the PPMIs between HIV- and mock-infected cells, the common core PPMI network was extracted for each stage of infection, the intersection of which comprises the core set of proteins that define the entire infection cycle. Analysis of the common core PPMI network yielded new insight into the race between virus invasion and host defense. We also analyzed the role of each viral protein during infection, as well as the role of epigenetic and miRNA regulation. Moreover, we used the GO tool DAVID to unravel the pathogenic mechanism at each infection stage. Elucidation of the cellular processes associated with each viral protein may reveal how HIV hijacks host resources to facilitate infection, and investigation of miRNA and epigenetic regulation illustrates how miRNAs and methylation affect infection. Finally, we proposed multiple drug combinations for treating HIV-1-infected patients with at each stage of infection. Thus, the proposed systems biology methods can identify the functional core modules of host–pathogen networks as candidate drug targets for HIV therapy.

APPENDIX C: FIGURES

Here we show Figs. C.1–C.5.

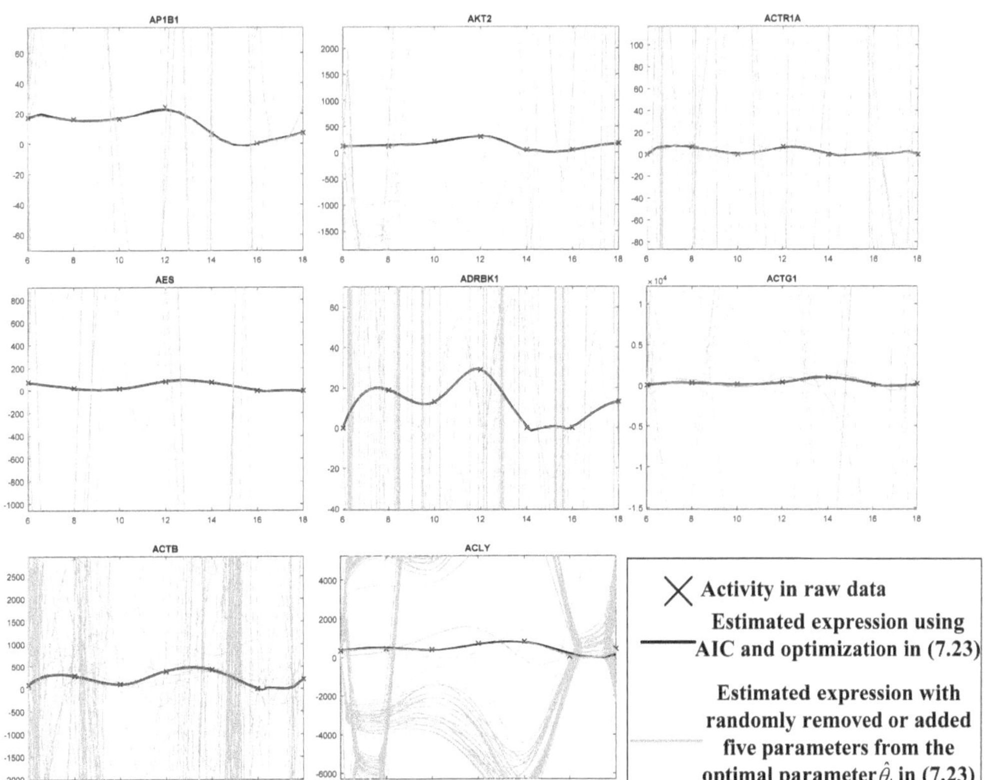

Figure C.1 Comparing the optimal results and the results with randomly removed and introduced five edges in 30 trials.

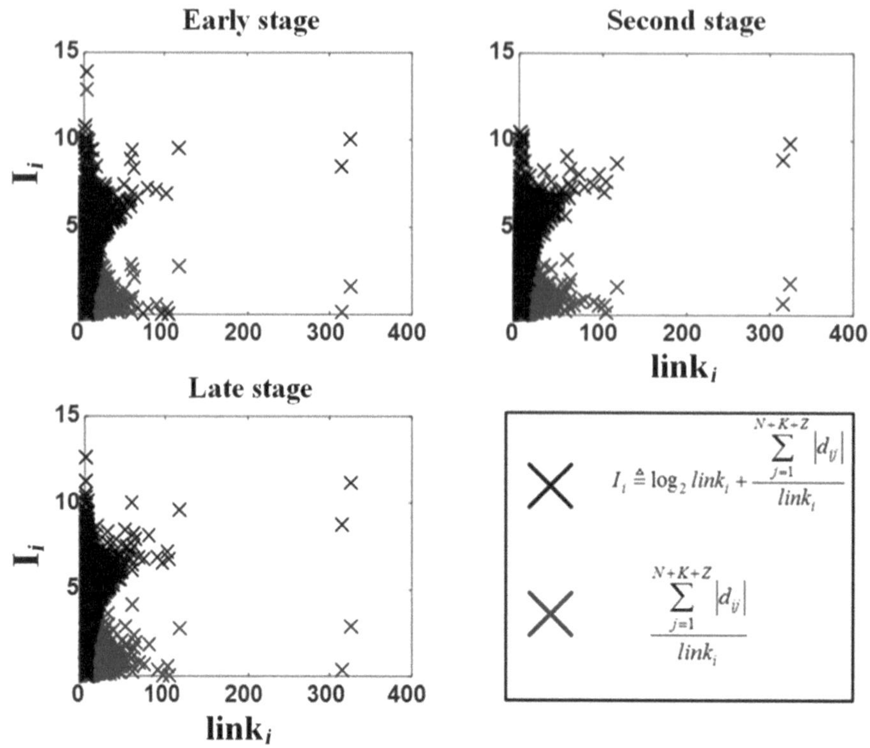

Figure C.2 The distribution plots of I_i and $\sum_{j=1}^{N+K+Z} |d_{ij}|/\text{link}_i$ among proteins.

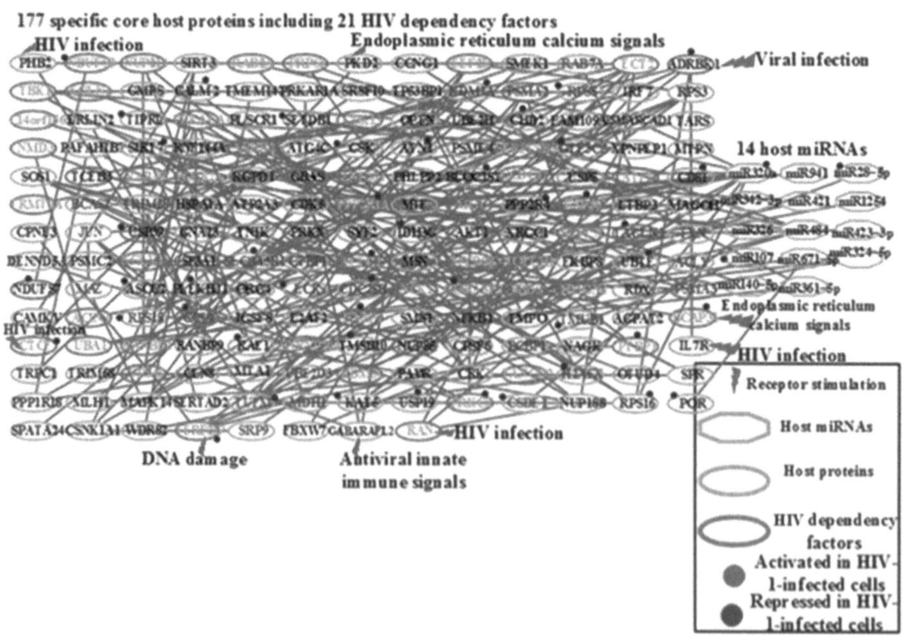

Figure C.3 The specific host–pathogen crosstalk network at the reverse transcription stage. The proteins are those proteins with the highest infection scores that are not common to all three stages of infection. The symbols are the same with those in Fig. 7.20.

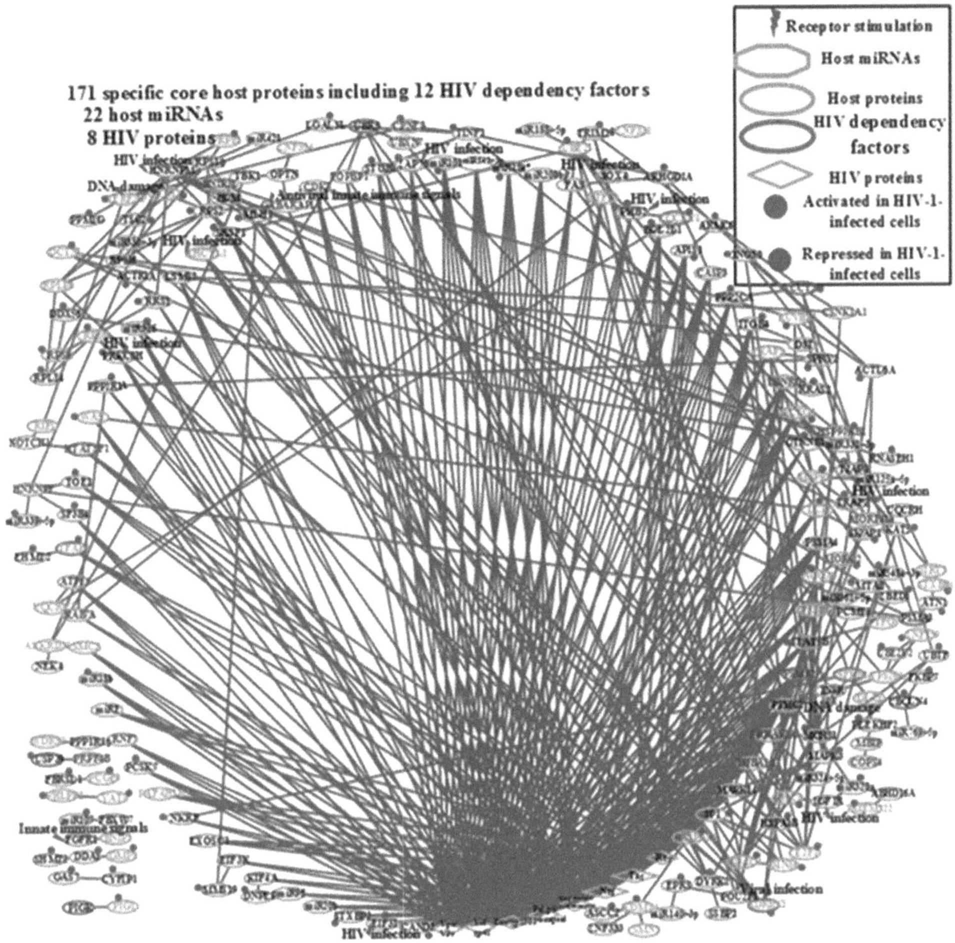

Figure C.4 The specific host–pathogen crosstalk network marker at the integration/replication stage. The symbols are the same with those in Fig. 7.20.

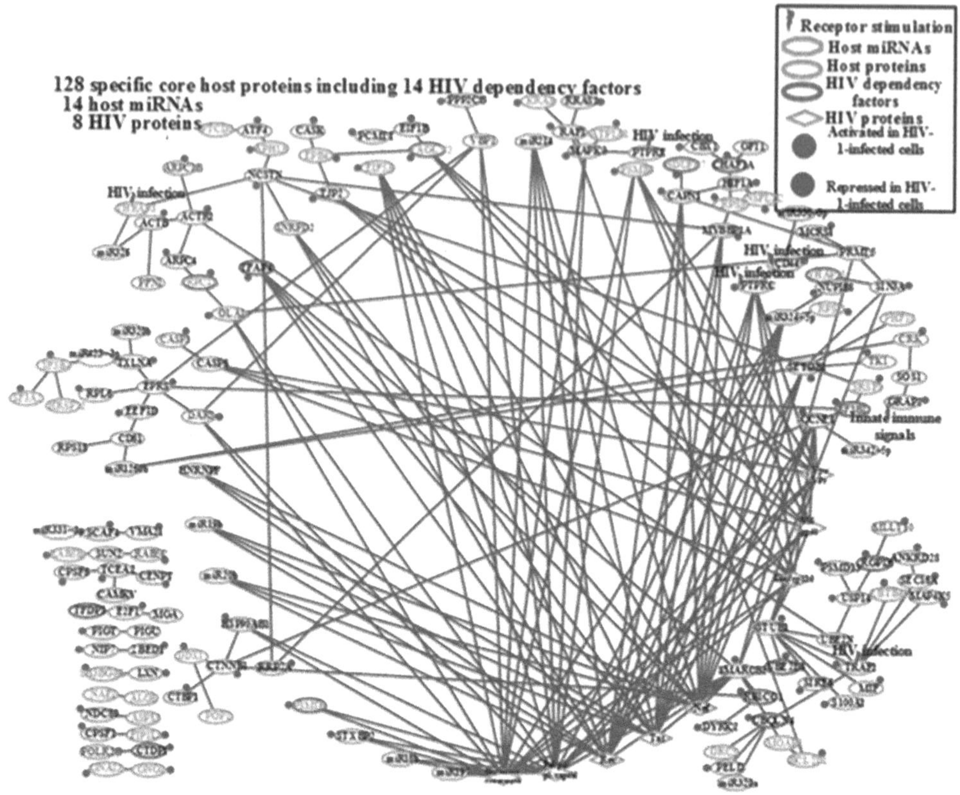

Figure C.5 The specific host–pathogen crosstalk network marker at the late-stage infection stage. The symbols are the same with those in Fig. 7.20.

REFERENCES

[1] Kitano H. Computational systems biology. Nature 2002;420(6912):206–10.

[2] Kitano H. Systems biology: a brief overview. Science 2002;295(5560):1662–4.

[3] Lin LH, Lee HC, Li WH, Chen BS. Dynamic modeling of cis-regulatory circuits and gene expression prediction via cross-gene identification. BMC Bioinformatics 2005;6:258.

[4] Vu TT, Vohradsky J. Nonlinear differential equation model for quantification of transcriptional regulation applied to microarray data of *Saccharomyces cerevisiae*. Nucleic Acids Res 2007;35(1):279–87.

[5] Baldwin AS. The NF-kappa B and I kappa B proteins: new discoveries and insights. Annu Rev Immunol 1996;14:649–83.

[6] Ghosh S, May MJ, Kopp EB. NF-kappa B and rel proteins: evolutionarily conserved mediators of immune responses. Annu Rev Immunol 1998;16:225–60.

[7] Hayden MS, Ghosh S. Signaling to NF-kappa B. Genes Dev 2004;18(18):2195–224.

[8] Makarov SS. NF-kappa B in rheumatoid arthritis: a pivotal regulator of inflammation, hyperplasia, and tissue destruction. Arthritis Res 2001;3(4):200–6.

[9] Coussens LM, Werb Z. Inflammation and cancer. Nature 2002;420(6917):860–7.

[10] Calvano SE, Xiao WZ, Richards DR, Felciano RM, Baker HV, Cho RJ, et al. A network-based analysis of systemic inflammation in humans. Nature 2005;437(7061):1032–7.

[11] Seok J, Warren H, Cuenca A, Mindrinos M, Baker H, Xu W, et al. Inflammation and host response to injury, Large Scale Collaborative Research Program. Genomic responses in mouse models poorly mimic human inflammatory diseases. Proc Natl Acad Sci USA 2013;110(9):3507–12.
[12] Tegner J, Yeung MKS, Hasty J, Collins JJ. Reverse engineering gene networks: integrating genetic perturbations with dynamical modeling. Proc Natl Acad Sci USA 2003;100(10):5944–9.
[13] Zou M, Conzen SD. A new dynamic Bayesian network (DBN) approach for identifying gene regulatory networks from time course microarray data. Bioinformatics 2005;21(1):71–9.
[14] Santamaria P. Cytokines and chemokines in autoimmune disease: an overview. Adv Exp Med Biol 2003;520:1–7.
[15] Foxwell BMJ, Bondeson J, Brennan F, Feldmann M. Adenoviral transgene delivery provides an approach to identifying important molecular processes in inflammation: evidence for heterogenecity in the requirement for NF kappa B in tumour necrosis factor production. Ann Rheum Dis 2000;59:54–9.
[16] Kitano H, Oda K. Robustness trade-offs and host-microbial symbiosis in the immune system. Mol Syst Biol 2006;2:2006.0022.
[17] Werner SL, Barken D, Hoffmann A. Stimulus specificity of gene expression programs determined by temporal control of IKK activity. Science 2005;309(5742):1857–61.
[18] Muzio M, Polentarutti N, Bosisio D, Prahladan MKP, Mantovani A. Toll-like receptors: a growing family of immune receptors that are differentially expressed and regulated by different leukocytes. J Leukoc Biol 2000;67(4):450–6.
[19] Klipp E, Herwig R, Kowald A, Wierling C, Lehrach H. Systems biology in practice: concepts, implementation and application. Weinheim: John Wiley & Sons; 2005.
[20] Bar-Joseph Z, Gerber GK, Lee TI, Rinaldi NJ, Yoo JY, Robert F, et al. Computational discovery of gene modules and regulatory networks. Nat Biotechnol 2003;21(11):1337–42.
[21] Davidson EH, McCay DR, Hood L. Regulatory gene networks and the properties of the developmental process. Proc Natl Acad Sci USA 2003;100(4):1475–80.
[22] Hood L. Systems biology: integrating technology, biology, and computation. Mech Ageing Dev 2003;124(1):9–16.
[23] Johansson R. System modeling and identification. Englewood Cliffs, NJ: Prentice Hall; 1993.
[24] Wu W-S, Li W-H, Chen B-S. Computational reconstruction of transcriptional regulatory modules of the yeast cell cycle. BMC Bioinformatics 2006;7:421.
[25] Pahl HL. Activators and target genes of Rel/NF-kappa B transcription factors. Oncogene 1999;18 (49):6853–66.
[26] Breitkreutz BJ, Stark C, Tyers M. Osprey: a network visualization system. Genome Biol 2003;4(3): R22.
[27] Oyama N, Iwatsuki K, Homma Y, Kaneko F. Induction of transcription factor AP-2 by inflammatory cytokines in human keratinocytes. J Investig Dermatol 1999;113(4):600–6.
[28] Murakami S, Lefebvre V, de Crombrugghe B. Potent inhibition of the master chondrogenic factor Sox9 gene by interleukin-1 and tumor necrosis factor-alpha. J Biol Chem 2000;275(5):3687–92.
[29] Fukuda K, Yoshida H, Sato T, Furumoto T, Mizutani-Koseki Y, Suzuki Y, et al. Mesenchymal expression of Foxl1, a winged helix transcriptional factor, regulates generation and maintenance of gut-associated lymphoid organs. Dev Biol 2003;255(2):278–89.
[30] Imagawa S, Nakano Y, Obara N, Suzuki N, Doi T, Kodama T, et al. A GATA-specific inhibitor (K-7174) rescues anemia induced by IL-1 beta, TNF-alpha, or L-NMMA. FASEB J 2003;17 (10):1742–4.
[31] Koyano S, Saito Y, Sai K, Kurose K, Ozawa S, Nakajima T, et al. Novel genetic polymorphisms in the NR3C1 (glucocorticoid receptor) gene in a Japanese population. Drug Metab Pharmacokinet 2005;20(1):79–84.
[32] Nakano Y, Imagawa S, Matsumoto K, Stockmann C, Obara N, Suzuki N, et al. Oral administration of K-11706 inhibits GATA binding activity, enhances hypoxia-inducible factor 1 binding activity, and restores indicators in an in vivo mouse model of anemia of chronic disease. Blood 2004;104 (13):4300–7.
[33] Choi SJ, Oba T, Callander NS, Jelinek DF, Roodman GD. AYL-1A and AML-1B regulation of MIP-1 alpha expression in multiple myeloma. Blood 2003;101(10):3778–83.

[34] Hawkins GA, Amelung PJ, Smith RS, Jongepier H, Howard TD, Koppelman GH, et al. Identification of polymorphisms in the human glucocorticoid receptor gene (NR3C1) in a multi-racial asthma case and control screening panel. DNA Seq 2004;15(3):167–73.
[35] Albert R. Scale-free networks in cell biology. J Cell Sci 2005;118(21):4947–57.
[36] Kitano H. Biological robustness. Nat Rev Genet 2004;5(11):826–37.
[37] Boldrick JC, Alizadeh AA, Diehn M, Dudoit S, Liu CL, Belcher CE, et al. Stereotyped and specific gene expression programs in human innate immune responses to bacteria. Proc Natl Acad Sci USA 2002;99(2):972–7.
[38] Le YY, Murphy PM, Wang JM. Formyl-peptide receptors revisited. Trends Immunol 2002;23 (11):541–8.
[39] Leroy O, Gangneux J-P, Montravers P, Mira J-P, Gouin F, Sollet J-P, et al. Epidemiology, management, and risk factors for death of invasive Candida infections in critical care: a multicenter, prospective, observational study in France (2005-2006). Crit Care Med 2009;37(5):1612–18.
[40] Kojic EM, Darouiche RO. Candida infections of medical devices. Clin Microbiol Rev 2004;17 (2):255–67.
[41] Seneviratne CJ, Jin L, Samaranayake LP. Biofilm lifestyle of *Candida*: a mini review. Oral Dis 2008;14(7):582–90.
[42] Olorode OA, Okpokwasili GC. The efficacy of disinfectants on abattoirs' *Candida albicans* isolates in Niger Delta region. Mycoses 2012;55. 323.
[43] Pfaller MA, Diekema DJ. Epidemiology of invasive candidiasis: a persistent public health problem. Clin Microbiol Rev 2007;20(1):133–63.
[44] Lo HJ, Kohler JR, DiDomenico B, Loebenberg D, Cacciapuoti A, Fink GR. Nonfilamentous *C. albicans* mutants are avirulent. Cell 1997;90(5):939–49.
[45] Calderone RA, Fonzi WA. Virulence factors of *Candida albicans*. Trends Microbiol 2001;9 (7):327–35.
[46] Leberer E, Ziegelbauer K, Schmidt A, Harcus D, Dignard D, Ash J, et al. Virulence and hyphal formation of *Candida albicans* require the Ste20p-like protein kinase CaCla4p. Curr Biol 1997;7 (8):539–46.
[47] Jones T, Federspiel NA, Chibana H, Dungan J, Kalman S, Magee BB, et al. The diploid genome sequence of *Candida albicans*. Proc Natl Acad Sci USA 2004;101(19):7329–34.
[48] Goffeau A, Barrell BG, Bussey H, Davis RW, Dujon B, Feldmann H, et al. Life with 6000 genes. Science 1996;274(5287):546.
[49] Ihmels J, Bergmann S, Berman J, Barkai N. Comparative gene expression analysis by a differential clustering approach: application to the *Candida albicans* transcription program. PLoS Genet 2005;1 (3):380–93.
[50] Meeker ND, Trede NS. Immunology and zebrafish: spawning new models of human disease. Dev Comp Immunol 2008;32(7):745–57.
[51] Sullivan C, Kim CH. Zebrafish as a model for infectious disease and immune function. Fish Shellfish Immunol 2008;25(4):341–50.
[52] Amsterdam A, Hopkins N. Mutagenesis strategies in zebrafish for identifying genes involved in development and disease. Trends Genet 2006;22(9):473–8.
[53] Postlethwait J, Amores A, Force A, Yan YL. The zebrafish genome. In: Detrich HW, Westerfield M, Zon LI, editors. Methods in Cell Biology, Vol 60: Zebrafish: Genetics and Genomics. New York, NY: Academic Press; 1999. p. 149–63.
[54] Chao C-C, Hsu P-C, Jen C-F, Chen IH, Wang C-H, Chan H-C, et al. Zebrafish as a model host for *Candida albicans* infection. Infect Immun 2010;78(6):2512–21.
[55] Newman JRS, Ghaemmaghami S, Ihmels J, Breslow DK, Noble M, DeRisi JL, et al. Single-cell proteomic analysis of *S-cerevisiae* reveals the architecture of biological noise. Nature 2006;441 (7095):840–6.
[56] Orntoft TF, Thykjaer T, Waldman FM, Wolf H, Celis JE. Genome-wide study of gene copy numbers, transcripts, and protein levels in pairs of non-invasive and invasive human transitional cell carcinomas. Mol Cell Proteomics 2002;1(1):37–45.
[57] Alon U. An introduction to systems biology: design principles of biological circuits. Boca Raton, FL: Chapman & Hall/CRC; 2007.

[58] Akaike H. New look at statistical-model identification. IEEE Trans Autom Control 1974;AC19 (6):716–23.
[59] Wu W-S, Li W-H, Chen B-S. Identifying regulatory targets of cell cycle transcription factors using gene expression and ChIP-chip data. BMC Bioinformatics 2007;8:188.
[60] Edwards JE, Rotrosen D, Fontaine JW, Haudenschild CC, Diamond RD. Neutrophil-mediated protection of cultured human vascular endothelial-cells from damage by growing *Candida albicans* hyphae. Blood 1987;69(5):1450–7.
[61] Hummert S, Hummert C, Schroeter A, Hube B, Schuster S. Game theoretical modelling of survival strategies of *Candida albicans* inside macrophages. J Theor Biol 2010;264(2): 312–18.
[62] Efron B, Tibshirani R. An introduction to the bootstrap. New York, NY: Chapman & Hall; 1993.
[63] Dyer SA, Dyer JS. Cubic-spline interpolation: Part 1. IEEE Instrum Meas Mag 2001;4(1):44–6.
[64] Esser K. The Mycota: a comprehensive treatise on fungi as experimental systems for basic and applied research. 2nd ed. Berlin; New York: Springer-Verlag; 2004.
[65] Uhl MA, Biery M, Craig N, Johnson AD. Haploinsufficiency-based large-scale forward genetic analysis of filamentous growth in the diploid human fungal pathogen *C. albicans*. EMBO J 2003;22 (11):2668–78.
[66] Yaar L, Mevarech M, Koltin Y. A *Candida albicans* RAS-related gene (CaRSR1) is involved in budding, cell morphogenesis and hypha development. Microbiology 1997;143:3033–44.
[67] Woo M, Lee K, Song K. MYO2 is not essential for viability, but is required for polarized growth and dimorphic switches in *Candida albicans*. FEMS Microbiol Lett 2003;218(1):195–202.
[68] Duenkler A, Wendland J. *Candida albicans* RHO-type GTPase-encoding genes required for polarized cell growth and cell separation. Eukaryot Cell 2007;6(5):844–54.
[69] Barabasi AL, Oltvai ZN. Network biology: understanding the cell's functional organization. Nat Rev Genet 2004;5(2):101–115.
[70] Jeong H, Mason SP, Barabasi AL, Oltvai ZN. Lethality and centrality in protein networks. Nature 2001;411(6833):41–2.
[71] Roig P, Gozalbo D. Depletion of polyubiquitin encoded by the UBI4 gene confers pleiotropic phenotype to *Candida albicans* cells. Fungal Genet Biol 2003;39(1):70–81.
[72] Zou H, Fang H-M, Zhu Y, Wang Y. *Candida albicans* Cyr1, Cap1 and G-actin form a sensor/ effector apparatus for activating cAMP synthesis in hyphal growth. Mol Microbiol 2010;75 (3):579–91.
[73] Reijnst P, Jorde S, Wendland J. *Candida albicans* SH3-domain proteins involved in hyphal growth, cytokinesis, and vacuolar morphology. Curr Genet 2010;56(4):309–19.
[74] Rida PCG, Nishikawa A, Won GY, Dean N. Yeast-to-hyphal transition triggers formin-dependent golgi localization to the growing tip in *Candida albicans*. Mol Biol Cell 2006;17(10):4364–78.
[75] Shapiro RS, Uppuluri P, Zaas AK, Collins C, Senn H, Perfect JR, et al. Hsp90 orchestrates temperature-dependent *Candida albicans* morphogenesis via Ras1-PKA signaling. Curr Biol 2009;19 (8):621–9.
[76] Newport G, Agabian N. KEX2 influences *Candida albicans* proteinase secretion and hyphal formation. J Biol Chem 1997;272(46):28954–61.
[77] Umeyama T, Kaneko A, Nagai Y, Hanaoka N, Tanabe K, Takano Y, et al. *Candida albicans* protein kinase CaHsl1p regulates cell elongation and virulence. Mol Microbiol 2005;55(2):381–95.
[78] Shin DH, Jung S, Park SJ, Kim YJ, Ahn JM, Kim W, et al. Characterization of thiol-specific antioxidant 1 (TSA1) of *Candida albicans*. Yeast 2005;22(11):907–18.
[79] Brand A, Lee K, Veses V, Gow NAR. Calcium homeostasis is required for contact-dependent helical and sinusoidal tip growth in *Candida albicans* hyphae. Mol Microbiol 2009;71(5):1155–64.
[80] Csank C, Schroppel K, Leberer E, Harcus D, Mohamed O, Meloche S, et al. Roles of the *Candida albicans* mitogen-activated protein kinase homolog, Cek1p, in hyphal development and systemic candidiasis. Infect Immun 1998;66(6):2713–21.
[81] Jiang WD, Gerhold D, Kmiec EB, Hauser M, Becker JM, Koltin Y. The topoisomerase I gene from Candida albicans. Microbiology 1997;143:377–86.
[82] Ghiselli G. SMC3 knockdown triggers genomic instability and p53-dependent apoptosis in human and zebrafish cells. Mol Cancer 2006;5:52.

[83] Stockhammer OW, Rauwerda H, Wittink FR, Breit TM, Meijer AH, Spaink HP. Transcriptome analysis of Traf6 function in the innate immune response of zebrafish embryos. Mol Immunol 2010;48(1–3):179–90.
[84] Phelan PE, Mellon MT, Kim CH. Functional characterization of full-length TLR3, IRAK-4, and TRAF6 in zebrafish (*Danio rerio*). Mol Immunol 2005;42(9):1057–71.
[85] Grassme H, Jendrossek V, Gulbins E. Molecular mechanisms of bacteria induced apoptosis. Apoptosis 2001;6(6):441–5.
[86] Buffo J, Herman MA, Soll DR. A characterization of pH-regulated dimorphism in *Candida albicans*. Mycopathologia 1984;85(1–2):21–30.
[87] Weissman Z, Kornitzer D. A family of *Candida* cell surface haem-binding proteins involved in haemin and haemoglobin-iron utilization. Mol Microbiol 2004;53(4):1209–20.
[88] Almeida RS, Brunke S, Albrecht A, Thewes S, Laue M, Edwards Jr. JE, et al. The hyphal-associated adhesin and invasin Als3 of *Candida albicans* mediates iron acquisition from host ferritin. PLoS Pathog 2008;4(11):e1000217.
[89] Fratti RA, Belanger PH, Ghannoum MA, Edwards JE, Filler SG. Endothelial cell injury caused by *Candida albicans* is dependent on iron. Infect Immun 1998;66(1):191–6.
[90] Hudson DA, Sascia OL, Sanders RJ, Norris GE, Edwards PJB, Sullivan PA, et al. Identification of the dialysable serum inducer of germ-tube formation in *Candida albicans*. Microbiology 2004;150:3041–9.
[91] Vidotto V, Accattatis G, Zhang Q, Campanini G, Aoki S. Glucose influence on germ tube production in *Candida albicans*. Mycopathologia 1996;133(3):143–7.
[92] Paranjape V, Datta A. Role of nutritional-status of the cell in pH regulated dimorphism of *Candida albicans*. FEMS Microbiol Lett 1991;80(2–3):333–6.
[93] Wheeler RT, Fink GR. A drug-sensitive genetic network masks fungi from the immune system. PLoS Pathog 2006;2(4). e35-Article No: e35.
[94] Wheeler RT, Kombe D, Agarwala SD, Fink GR. Dynamic, morphotype-specific *Candida albicans* beta-glucan exposure during infection and drug treatment. PLoS Pathog 2008;4(12):e1000227.
[95] Singleton DR, Masuoka J, Hazen KC. Cloning and analysis of a *Candida albicans* gene that affects cell surface hydrophobicity. J Bacteriol 2001;183(12):3582–8.
[96] Singleton DR, Fidel PL, Wozniak KL, Hazen KC. Contribution of cell surface hydrophobicity protein I (Csh1p) to virulence of hydrophobic *Candida albicans* serotype A cells. FEMS Microbiol Lett 2005;244(2):373–7.
[97] Cerenius L, Kawabata S-i, Lee BL, Nonaka M, Soderhall K. Proteolytic cascades and their involvement in invertebrate immunity. Trends Biochem Sci 2010;35(10):575–83.
[98] Ryan O, Shapiro RS, Kurat CF, Mayhew D, Baryshnikova A, Chin B, et al. Global gene deletion analysis exploring yeast filamentous growth. Science 2012;337(6100):1353–6.
[99] Coleman TF, Hulbert LA. A direct active set algorithm for large sparse quadratic programs with simple bounds. Mathematical Programming 1989;45(3):373–406.
[100] Bar-Joseph Z, Gerber G, Gifford DK, Jaakkola TS, Simon I. A new approach to analyzing gene expression time series data. Proceedings of the sixth annual international conference on computational biology: 2002. ACM; 2002. p. 39–48.
[101] De Boor C. A practical guide to splines: with 32 figures. Rev. ed. New York, NY: Springer; 2001.
[102] Jansen CA, Piriou E, De Cuyper IM, van Dort K, Lange JMA, Miedema F, et al. Long-term highly active antiretroviral therapy in chronic HIV-1 infection: evidence for reconstitution of antiviral immunity. Antivir Ther 2006;11(1):105–16.
[103] Perelson AS. Modelling viral and immune system dynamics. Nat Rev Immunol 2002;2(1):28–36.
[104] Mohammadi P, Desfarges S, Bartha I, Joos B, Zangger N, Munoz M, et al. 24 Hours in the life of HIV-1 in a T cell line. PLoS Pathog 2013;9(1):e1003161.
[105] Lu J, Clark AG. Impact of microRNA regulation on variation in human gene expression. Genome Res 2012;22(7):1243–54.
[106] Li CW, Wang WH, Chen BS. Investigating the specific core genetic-and-epigenetic networks of cellular mechanisms involved in human aging in peripheral blood mononuclear cells. Oncotarget 2016;7(8):8556–79.

[107] Keshet I, Yisraeli J, Cedar H. Effect of regional DNA methylation on gene expression. Proc Natl Acad Sci USA 1985;82(9):2560–4.
[108] Li C-W, Chen B-S. Network biomarkers of bladder cancer based on a genome-wide genetic and epigenetic network derived from next-generation sequencing data. Dis Markers 2016;2016:18.
[109] Duskova K, Nagilla P, Le HS, Iyer P, Thalamuthu A, Martinson J, et al. MicroRNA regulation and its effects on cellular transcriptome in Human Immunodeficiency Virus-1 (HIV-1) infected individuals with distinct viral load and CD4 cell counts. BMC Infect Dis 2013;13:250.
[110] Chen BS, Li CW. Constructing an integrated genetic and epigenetic cellular network for whole cellular mechanism using high-throughput next-generation sequencing data. BMC Syst Biol 2016;10(1):18.
[111] Sun GH, Li HT, Wu XW, Covarrubias M, Scherer L, Meinking K, et al. Interplay between HIV-1 infection and host microRNAs. Nucleic Acids Res 2012;40(5):2181–96.
[112] Whisnant AW, Bogerd HP, Flores O, Ho P, Powers JG, Sharova N, et al. In-depth analysis of the interaction of HIV-1 with cellular microRNA biogenesis and effector mechanisms. Mbio 2013;4(2):e000193.
[113] O'Doherty U, Swiggard WJ, Malim MH. Human immunodeficiency virus type 1 spinoculation enhances infection through virus binding. J Virol 2000;74(21):10074–80.
[114] Hsu CW, Juan HF, Huang HC. Characterization of microRNA-regulated protein-protein interaction network. Proteomics 2008;8(10):1975–9.
[115] Liang H, Li WH. MicroRNA regulation of human protein protein interaction network. RNA 2007;13(9):1402–8.
[116] Wang C, Jiang W, Li W, Lian B, Chen X, Hua L, et al. Topological properties of the drug targets regulated by microRNA in human protein-protein interaction network. J Drug Target 2011;19(5):354–64.
[117] Zhang Y, Guo X, Xiong L, Yu L, Li Z, Guo Q, et al. Comprehensive analysis of microRNA-regulated protein interaction network reveals the tumor suppressive role of microRNA-149 in human hepatocellular carcinoma via targeting AKT-mTOR pathway. Mol Cancer 2014;13:253.
[118] Redova M, Svoboda M, Slaby O. MicroRNAs and their target gene networks in renal cell carcinoma. Biochem Biophys Res Commun 2011;405(2):153–6.
[119] Kauder SE, Bosque A, Lindqvist A, Planelles V, Verdin E. Epigenetic regulation of HIV-1 latency by cytosine methylation. PLoS Pathog 2009;5(6):e1000495.
[120] Weber M, Hellmann I, Stadler MB, Ramos L, Paabo S, Rebhan M, et al. Distribution, silencing potential and evolutionary impact of promoter DNA methylation in the human genome. Nat Genet 2007;39(4):457–66.
[121] Valinluck V, Tsai HH, Rogstad DK, Burdzy A, Bird A, Sowers LC. Oxidative damage to methyl-CpG sequences inhibits the binding of the methyl-CpG binding domain (MBD) of methyl-CpG binding protein 2 (MeCP2). Nucleic Acids Res 2004;32(14):4100–8.
[122] Chatr-aryamontri A, Breitkreutz BJ, Oughtred R, Boucher L, Heinicke S, Chen DC, et al. The BioGRID interaction database: 2015 update. Nucleic Acids Res 2015;43(D1):D470–8.
[123] Fu W, Sanders-Beer BE, Katz KS, Maglott DR, Pruitt KD, Ptak RG. Human immunodeficiency virus type 1, human protein interaction database at NCBI. Nucleic Acids Res 2009;37:D417–22.
[124] Hsu PWC, Lin LZ, Hsu SD, Hsu JBK, Huang HD. ViTa: prediction of host microRNAs targets on viruses. Nucleic Acids Res 2007;35:D381–5.
[125] Kozomara A, Griffiths-Jones S. miRBase: annotating high confidence microRNAs using deep sequencing data. Nucleic Acids Res 2014;42(Database issue):D68–73.
[126] Lamb J, Crawford ED, Peck D, Modell JW, Blat IC, Wrobel MJ, et al. The connectivity map: using gene-expression signatures to connect small molecules, genes, and disease. Science 2006;313(5795):1929–35.
[127] Matsunaga A, Hishima T, Tanaka N, Yamasaki M, Yoshida L, Mochizuki M, et al. DNA methylation profiling can classify HIV-associated lymphomas. AIDS 2014;28(4):503–10.
[128] Wang YC, Lin C, Chuang MT, Hsieh WP, Lan CY, Chuang YJ, et al. Interspecies protein-protein interaction network construction for characterization of host-pathogen interactions: a *Candida albicans*-zebrafish interaction study. BMC Syst Biol 2013;7:79.

[129] Coleman TF, Li YY. A reflective Newton method for minimizing a quadratic function subject to bounds on some of the variables. SIAM J Optim 1996;6(4):1040–58.
[130] Lu P, Vogel C, Wang R, Yao X, Marcotte EM. Absolute protein expression profiling estimates the relative contributions of transcriptional and translational regulation. Nat Biotechnol 2007;25 (1):117–24.
[131] Pagano M, Gauvreau K. Principles of biostatistics. 2nd ed. Pacific Grove, CA: Duxbury; 2000.
[132] Brass AL, Dykxhoorn DM, Benita Y, Yan N, Engelman A, Xavier RJ, et al. Identification of host proteins required for HIV infection through a functional genomic screen. Science 2008;319 (5865):921–6.
[133] Konig R, Zhou YY, Elleder D, Diamond TL, Bonamy GMC, Irelan JT, et al. Global analysis of host-pathogen interactions that regulate early-stage HIV-1 replication. Cell 2008;135(1):49–60.
[134] Zhou HL, Xu M, Huang Q, Gates AT, Zhang XHD, Castle JC, et al. Genome-scale RNAi screen for host factors required for HIV replication. Cell Host Microbe 2008;4(5):495–504.
[135] Nakayama-Hosoya K, Ishida T, Youngblood B, Nakamura H, Hosoya N, Koga M, et al. Epigenetic repression of interleukin 2 expression in senescent CD4 + T cells during chronic HIV type 1 infection. J Infect Dis 2015;211(1):28–39.
[136] Warren K, Warrilow D, Meredith L, Harrich D. Reverse transcriptase and cellular factors: regulators of HIV-1 reverse transcription. Viruses 2009;1(3):873–94.
[137] Bourbigot S, Beltz H, Denis J, Morellet N, Roques BP, Mely Y, et al. The C-terminal domain of the HIV-1 regulatory protein Vpr adopts an antiparallel dimeric structure in solution via its leucine-zipper-like domain. Biochem J 2005;387(Pt 2):333–41.
[138] Zhao RY, Li G, Bukrinsky MI. Vpr-host interactions during HIV-1 viral life cycle. J Neuroimmune Pharmacol 2011;6(2):216–29.
[139] Sherer NM, Lehmann MJ, Jimenez-Soto LF, Ingmundson A, Horner SM, Cicchetti G, et al. Visualization of retroviral replication in living cells reveals budding into multivesicular bodies. Traffic 2003;4(11):785–801.
[140] Clarke HJ, Chambers JE, Liniker E, Marciniak SJ. Endoplasmic reticulum stress in malignancy. Cancer Cell 2014;25(5):563–73.
[141] Puoti M, Bruno R, Soriano V, Donato F, Gaeta GB, Quinzan GP, et al. Hepatocellular carcinoma in HIV-infected patients: epidemiological features, clinical presentation and outcome. AIDS 2004;18(17):2285–93.
[142] Weber J. The pathogenesis of HIV-1 infection. Br Med Bull 2001;58:61–72.
[143] Perez CL, Milush JM, Buggert M, Eriksson EM, Larsen MV, Liegler T, et al. Targeting of conserved gag-epitopes in early HIV infection is associated with lower plasma viral load and slower CD4(+) T cell depletion. AIDS Res Hum Retroviruses 2013;29(3):602–12.
[144] Engeland CE, Brown NP, Borner K, Schumann M, Krause E, Kaderali L, et al. Proteome analysis of the HIV-1 Gag interactome. Virology 2014;460–461:194–206.
[145] Lewis B, Whitney S, Hudacik L, Galmin L, Huaman MC, Cristillo AD. Nedd4-mediated increase in HIV-1 Gag and Env proteins and immunity following DNA-vaccination of BALB/c mice. PLoS One 2014;9(3):e91267.
[146] Houzet L, Klase Z, Yeung ML, Wu AN, Le SY, Quinones M, et al. The extent of sequence complementarity correlates with the potency of cellular miRNA-mediated restriction of HIV-1. Nucleic Acids Res 2012;40(22):11684–96.
[147] He GC, Margolis DM. Counterregulation of chromatin deacetylation and histone deacetylase occupancy at the integrated promoter of human immunodeficiency virus type 1 (HIV-1) by the HIV-1 repressor YY1 and HIV-1 activator Tat. Mol Cell Biol 2002;22(9):2965–73.
[148] Spadoni JL, Rucart P, Le Clerc S, van Manen D, Coulonges C, Ulveling D, et al. Identification of genes whose expression profile is associated with non-progression towards AIDS using eQTLs. PLoS One 2015;10(9):e0136989.
[149] Peterlin BM, Trono D. Hide, shield and strike back: how HIV-infected cells avoid immune eradication. Nat Rev Immunol 2003;3(2):97–107.
[150] Xie GQ, Yu ZS, Jia DY, Jiao RJ, Deng WM. E(y)1/TAF9 mediates the transcriptional output of Notch signaling in Drosophila. J Cell Sci 2014;127(17):3830–9.

[151] Mbita Z, Hull R, Dlamini Z. Human immunodeficiency virus-1 (HIV-1)-mediated apoptosis: new therapeutic targets. Viruses 2014;6(8):3181–227.
[152] Hayes AM, Qian SM, Yu LB, Boris-Lawrie K. Tat RNA silencing suppressor activity contributes to perturbation of lymphocyte miRNA by HIV-1. Retrovirology 2011;8:36.
[153] Pedersen IM, Cheng G, Wieland S, Volinia S, Croce CM, Chisari FV, et al. Interferon modulation of cellular microRNAs as an antiviral mechanism. Nature 2007;449(7164). 919-U913.
[154] Santhakumar D, Forster T, Laqtom NN, Fragkoudis R, Dickinson P, Abreu-Goodger C, et al. Combined agonist-antagonist genome-wide functional screening identifies broadly active antiviral microRNAs. Proc Natl Acad Sci USA 2010;107(31):13830–5.
[155] Sanghvi VR, Steel LF. RNA silencing as a cellular defense against HIV-1 infection: progress and issues. FASEB J 2012;26(10):3937–45.
[156] Denkert C, Koch I, von Keyserlingk N, Noske A, Niesporek S, Dietel M, et al. Expression of the ELAV-like protein HuR in human colon cancer: association with tumor stage and cyclooxygenase-2. Modern Pathol 2006;19(9):1261–9.
[157] Kim HC, Choi KC, Choi HK, Kang HB, Kim MJ, Lee YH, et al. HDAC3 selectively represses CREB3-mediated transcription and migration of metastatic breast cancer cells. Cell Mol Life Sci 2010;67(20):3499–510.
[158] Li XR, Chu HJ, Lv T, Wang L, Kong SF, Dai SZ. miR-342-3p suppresses proliferation, migration and invasion by targeting FOXM1 in human cervical cancer. FEBS Lett 2014;588(17):3298–307.
[159] Tai MC, Kajino T, Nakatochi M, Arima C, Shimada Y, Suzuki M, et al. miR-342-3p regulates MYC transcriptional activity via direct repression of E2F1 in human lung cancer. Carcinogenesis 2015;36(12):1464–73.
[160] Vo DT, Abdelmohsen K, Martindale JL, Qiao M, Tominaga K, Burton TL, et al. The oncogenic RNA-binding protein Musashi1 is regulated by HuR via mRNA translation and stability in glioblastoma cells. Mol Cancer Res 2012;10(1):143–55.
[161] Xhemalce B. From histones to RNA: role of methylation in cancer. Brief Funct Genomics 2013;12(3):244–53.
[162] Xu LM, Li LQ, Li J, Li HW, Shen QB, Ping JL, et al. Overexpression of miR-1260b in non-small cell lung cancer is associated with lymph node metastasis. Aging Dis 2015;6(6):478–85.
[163] Yoon SY, Kim JM, Oh JH, Jeon YJ, Lee DS, Kim JH, et al. Gene expression profiling of human HBV- and/or HCV-associated hepatocellular carcinoma cells using expressed sequence tags. Int J Oncol 2006;29(2):315–27.
[164] Arion D, Lewis DA. Altered expression of regulators of the cortical chloride transporters NKCC1 and KCC2 in schizophrenia. Arch Gen Psychiatry 2011;68(1):21–31.
[165] Sola C, Garcia-Ladona FJ, Mengod G, Probst A, Frey P, Palacios JM. Increased levels of the Kunitz protease inhibitor-containing beta APP mRNAs in rat brain following neurotoxic damage. Brain Res Mol Brain Res 1993;17(1–2):41–52.
[166] Rodriguez E, Plaud M, Romeu R, Skolasky R, Melendez L. Late HIV infection modulates the expression and activity of Cathepsin B, and its inhibitors in macrophages: implications in neuropathogenesis. Retrovirology 2010;7:18–19.
[167] Chen Z, Manley JL. Robust mRNA transcription in chicken DT40 cells depleted of TAF(II)31 suggests both functional degeneracy and evolutionary divergence. Mol Cell Biol 2000;20 (14):5064–76.
[168] Frontini M, Soutoglou E, Argentini M, Bole-Feysot C, Jost B, Scheer E, et al. TAF9b (formerly TAF9L) is a bona fide TAF that has unique and overlapping roles with TAF9. Mol Cell Biol 2005;25(11):4638–49.
[169] Dubrow R, Silverberg MJ, Park LS, Crothers K, Justice AC. HIV infection, aging, and immune function: implications for cancer risk and prevention. Curr Opin Oncol 2012;24(5):506–16.
[170] Engels EA, Pfeiffer RM, Goedert JJ, Virgo P, McNeel TS, Scoppa SM, et al. Trends in cancer risk among people with AIDS in the United States 1980-2002. AIDS 2006;20(12):1645–54.
[171] Patel P, Hanson DL, Sullivan PS, Novak RM, Moorman AC, Tong TC, et al. Incidence of types of cancer among HIV-infected persons compared with the general population in the United States, 1992-2003. Ann Intern Med 2008;148(10):728–9.

[172] Formenti SC, Chak L, Gill P, Buess EM, Hill CK. Increased radiosensitivity of normal tissue fibroblasts in patients with acquired-immunodeficiency-syndrome (AIDS) and with kaposis-sarcoma. Int J Radiat Biol 1995;68(4):411−12.

[173] Ghanam RH, Samal AB, Fernandez TF, Saad JS. Role of the HIV-1 matrix protein in Gag intracellular trafficking and targeting to the plasma membrane for virus assembly. Front Microbiol 2012;3:55.

[174] Warrilow D, Tachedjian G, Harrich D. Maturation of the HIV reverse transcription complex: putting the jigsaw together. Rev Med Virol 2009;19(6):324−37.

[175] Giroud C, Chazal N, Gay B, Eldin P, Brun S, Briant L. HIV-1-associated PKA acts as a cofactor for genome reverse transcription. Retrovirology 2013;10:157.

[176] Leng J, Ho HP, Buzon MJ, Pereyra F, Walker BD, Yu XG, et al. A Cell-intrinsic inhibitor of HIV-1 reverse transcription in CD4(+) T cells from elite controllers. Cell Host Microbe 2014;15(6):717−28.

[177] Liang C, Hu J, Russell RS, Kameoka M, Wainberg MA. Spliced human immunodeficiency virus type 1 RNA is reverse transcribed into cDNA within infected cells. AIDS Res Hum Retroviruses 2004;20(2):203−11.

[178] Pak V, Eifler TT, Jager S, Krogan NJ, Fujinaga K, Peterlin BM. CDK11 in TREX/THOC regulates HIV mRNA 3′ end processing. Cell Host Microbe 2015;18(5):560−70.

[179] Berro R, Pedati C, Kehn-Hall K, Wu W, Klase Z, Even Y, et al. CDK13, a new potential human immunodeficiency virus type 1 inhibitory factor regulating viral mRNA splicing. J Virol 2008;82 (14):7155−66.

[180] Aandahl EM, Aukrust P, Skalhegg BS, Muller F, Froland SS, Hansson V, et al. Protein kinase A type I antagonist restores immune responses of T cells from HIV-infected patients. FASEB J 1998;12(10):855−62.

[181] Ellis J, Hotta A, Rastegar M. Retrovirus silencing by an epigenetic TRIM. Cell 2007;131(1):13−14.

[182] Vang T, Liu WH, Delacroix L, Wu SD, Vasile S, Dahl R, et al. LYP inhibits T-cell activation when dissociated from CSK. Nat Chem Biol 2012;8(5):437−46.

[183] Pareek TK, Lam E, Zheng XJ, Askew D, Kulkarni AB, Chance MR, et al. Cyclin-dependent kinase 5 activity is required for T cell activation and induction of experimental autoimmune encephalomyelitis. J Exp Med 2010;207(11):2507−19.

[184] Iordache L, Launay O, Bouchaud O, Jeantils V, Goujard C, Boue F, et al. Autoimmune diseases in HIV-infected patients: 52 cases and literature review. Autoimmun Rev 2014;13(8):850−7.

[185] Xu X, Bieda M, Jin VX, Rabinovich A, Oberley MJ, Green R, et al. A comprehensive ChIP-chip analysis of E2F1, E2F4, and E2F6 in normal and tumor cells reveals interchangeable roles of E2F family members. Genome Res 2007;17(11):1550−61.

[186] Choi S, Kim HR, Leng L, Kang I, Jorgensen WL, Cho CS, et al. Role of macrophage migration inhibitory factor in the regulatory T cell response of tumor-bearing mice. J Immunol 2012;189 (8):3905−13.

[187] Ammar A, Sahraoui Y, Tsapis A, Bertoli AM, Jasmin C, Georgoulias V. Human immunodeficiency virus-infected adherent cell-derived inhibitory factor (P-29) inhibits normal T-cell proliferation through decreased expression of high-affinity interleukin-2 receptors and production of interleukin-2. J Clin Investig 1992;90(1):8−14.

[188] Rowbotham SP, Barki L, Neves-Costa A, Santos F, Dean W, Hawkes N, et al. Maintenance of silent chromatin through replication requires SWI/SNF-like chromatin remodeler SMARCAD1. Mol Cell 2011;42(3):285−96.

[189] Muthumani K, Shedlock DJ, Choo DK, Fagone P, Kawalekar OU, Goodman J, et al. HIV-mediated phosphatidylinositol 3-kinase/serine-threonine kinase activation in APCs leads to programmed death-1 ligand upregulation and suppression of HIV-specific CD8 T cells. J Immunol 2011;187(6):2932−43.

[190] Myers AP. New strategies in endometrial cancer: targeting the PI3K/mTOR pathway—the devil is in the details. Clin Cancer Res 2013;19(19):5264−74.

[191] Christeff N, Gharakhanian S, Thobie N, Rozenbaum W, Nunez EA. Evidence for changes in adrenal and testicular steroids during HIV infection. J Acquir Immune Defic Syndr 1992;5 (8):841−6.

[192] Christeff N, Gherbi N, Mammes O, Dalle MT, Gharakhanian S, Lortholary O, et al. Serum cortisol and DHEA concentrations during HIV infection. Psychoneuroendocrinology 1997;22(Suppl. 1):S11−18.
[193] Christeff N, Melchior JC, de Truchis P, Perronne C, Nunez EA, Gougeon ML. Lipodystrophy defined by a clinical score in HIV-infected men on highly active antiretroviral therapy: correlation between dyslipidaemia and steroid hormone alterations. AIDS 1999;13(16):2251−60.
[194] Christeff N, Melchior JC, Mammes O, Gherbi N, Dalle MT, Nunez EA. Correlation between increased cortisol: DHEA ratio and malnutrition in HIV-positive men. Nutrition 1999;15 (7−8):534−9.
[195] Towers G, Harris J, Lang G, Collins MKL, Latchman DS. Retinoic acid inhibits both the basal activity and phorbol ester-mediated activation of the HIV long terminal repeat promoter. AIDS 1995;9(2):129−36.
[196] Espeseth AS, Fishel R, Hazuda D, Huang Q, Xu M, Yoder K, et al. siRNA screening of a targeted library of DNA repair factors in HIV infection reveals a role for base excision repair in HIV integration. PLoS One 2011;6(3):e17612.
[197] Bushman FD, Malani N, Fernandes J, D'Orso I, Cagney G, Diamond TL, et al. Host cell factors in HIV replication: meta-analysis of genome-wide studies. PLoS Pathog 2009;5(5):e1000437.
[198] Drahovsky D, Lacko I, Wacker A. Enzymatic DNA methylation during repair synthesis in non-proliferating human peripheral lymphocytes. Biochim Biophys Acta 1976;447(2):139−43.
[199] Esteller M, Gaidano G, Goodman SN, Zagonel V, Capello D, Botto B, et al. Hypermethylation of the DNA repair gene O(6)-methylguanine DNA methyltransferase and survival of patients with diffuse large B-cell lymphoma. J Natl Cancer Inst 2002;94(1):26−32.
[200] Baxter J, Sauer S, Peters A, John R, Williams R, Caparros ML, et al. Histone hypomethylation is an indicator of epigenetic plasticity in quiescent lymphocytes. EMBO J 2004;23(22):4462−72.
[201] Zhang Y, Zhao M, Sawalha AH, Richardson B, Lu Q. Impaired DNA methylation and its mechanisms in CD4(+)T cells of systemic lupus erythematosus. J Autoimmun 2013;41:92−9.
[202] Emerson V, Holtkotte D, Pfeiffer T, Wang IH, Schnolzer M, Kempf T, et al. Identification of the cellular prohibitin 1/prohibitin 2 heterodimer as an interaction partner of the c-terminal cytoplasmic domain of the HIV-1 glycoprotein. J Virol 2010;84(3):1355−65.
[203] Kumar A, Zloza A, Moon RT, Watts J, Tenorio AR, Al-Harthi L. Active beta-catenin signaling is an inhibitory pathway for human immunodeficiency virus replication in peripheral blood mononuclear cells. J Virol 2008;82(6):2813−20.
[204] Rajalingam K, Wunder C, Brinkmann V, Churin Y, Hekman M, Sievers C, et al. Prohibitin is required for Ras-induced Raf-MEK-ERK activation and epithelial cell migration. Nat Cell Biol 2005;7(8):837−43.
[205] Benetti L, Calistri A, Ulivieri C, Cabrelle A, Baldari CT, Palu G, et al. Inhibition of ShcA isoforms p46/p52Shc enhances HIV-1 replication in CD4 + T-lymphocytes. J Cell Physiol 2004;199 (1):40−6.
[206] Catrina SB, Lewitt M, Massambu C, Dricu A, Grunler J, Axelson M, et al. Insulin-like growth factor-I receptor activity is essential for Kaposi's sarcoma growth and survival. Br J Cancer 2005;92(8):1467−74.
[207] Woldt E, Matz RL, Terrand J, Mlih M, Gracia C, Foppolo S, et al. Differential signaling by adaptor molecules LRP1 and ShcA regulates adipogenesis by the insulin-like growth factor-1 receptor. J Biol Chem 2011;286(19):16775−82.
[208] Crum NF, Spencer CR, Amling CL. Prostate carcinoma among men with human immunodeficiency virus infection. Cancer 2004;101(2):294−9.
[209] Schlaberg R, Fisher JG, Flamm MJ, Murty VV, Bhagat G, Alobeid B. Chronic myeloid leukemia and HIV-infection. Leuk Lymphoma 2008;49(6):1155−60.
[210] Miller AM, Lundberg K, Ozenci V, Banham AH, Hellstrom M, Egevad L, et al. CD4(+)CD25 (high) T cells are enriched in the tumor and peripheral blood of prostate cancer patients. J Immunol 2006;177(10):7398−405.
[211] Hoxie JA, Alpers JD, Rackowski JL, Huebner K, Haggarty BS, Cedarbaum AJ, et al. Alterations in T4 (CD4) protein and mRNA synthesis in cells infected with HIV. Science 1986;234 (4780):1123−7.

[212] Birge RB, Kalodimos C, Inagaki F, Tanaka S. Crk and CrkL adaptor proteins: networks for physiological and pathological signaling. Cell Commun Signal 2009;7:13.
[213] Minegishi M, Tachibana K, Sato T, Iwata S, Nojima Y, Morimoto C. Structure and function of Cas-L, a 105-kD Crk-associated substrate-related protein that is involved in beta 1 integrin-mediated signaling in lymphocytes. J Exp Med 1996;184(4):1365–75.
[214] Guo M, Shapiro R, Morris GM, Yang XL, Schimmel P. Packaging HIV virion components through dynamic equilibria of a human tRNA synthetase. J Phys Chem B 2010;114(49):16273–9.
[215] Mercenne G, Bernacchi S, Richer D, Bec G, Henriet S, Paillart JC, et al. HIV-1 Vif binds to APOBEC3G mRNA and inhibits its translation. Nucleic Acids Res 2010;38(2):633–46.
[216] Park SG, Schimmel P, Kim S. Aminoacyl tRNA synthetases and their connections to disease. Proc Natl Acad Sci USA 2008;105(32):11043–9.
[217] Folgueira L, Algeciras A, MacMorran WS, Bren GD, Paya CV. The Ras-Raf pathway is activated in human immunodeficiency virus-infected monocytes and participates in the activation of NF-kappa B. J Virol 1996;70(4):2332–8.
[218] Lyman MG, Randall JA, Calton CM, Banfield BW. Localization of ERK/MAP kinase is regulated by the alphaherpesvirus tegument protein Us2. J Virol 2006;80(14):7159–68.
[219] Geng Y, Whoriskey W, Park MY, Bronson RT, Medema RH, Li T, et al. Rescue of cyclin D1 deficiency by knockin cyclin E. Cell 1999;97(6):767–77.
[220] Lukas J, Herzinger T, Hansen K, Moroni MC, Resnitzky D, Helin K, et al. Cyclin E-induced S phase without activation of the pRb/E2F pathway. Genes Dev 1997;11(11):1479–92.
[221] Knudsen ES, Wang JY. Targeting the RB-pathway in cancer therapy. Clin Cancer Res 2010;16(4):1094–9.

CHAPTER 8

Big Regeneration Mechanisms via Systems Biology and Big Database Mining Methods

INTRODUCTION

Mammalian development is a complex multistep process starting with the totipotent embryo containing unlimited developmental potential and terminating in almost 200 differentiated cell types that perform specialized physiological functions. The cells' fate and their developmental potentials are defined by Waddington's epigenetic landscape model, which shows a gradual decrease in the developmental potential of a cell with each differential state. Epigenetic alterations, including DNA methylation and histone modifications, are key factors in the differentiation of stem cells into different tissue subtypes. The cells lose their developmental plasticity with each, irreversible, differentiation step. However, we still find that it is difficult to identify the forces behind the conversion of an almost quiescent, totipotent stem cell into a terminally differentiated, fully functional somatic cell.

Recently, inducing stem cells toward a desired differentiation state has been critical to stem cell-based therapies. Despite successful demonstrations of controlling the differentiation process, the effective use of stem cell-based therapies is suffering from a lack of systematic knowledge regarding the mechanisms of directed differentiation. In this chapter, because of the key role of stem cell in tissue regeneration, we will first construct the dynamic networks of the three differentiation stages of stem cells to reveal essential proteins and functional modules in the differentiation process. Using dynamic modeling and temporal microarray data from the three stages, three dynamic protein–protein interaction (PPI) networks are constructed. The interaction networks derived from the constructed networks are used to delineate the evolution of interaction variations and the underlying mechanisms. Proteins and functional modules in directed differentiation can be investigated based on their proposed relevance scores, i.e., in the differentiation process of stem cells, proteins and functional modules with larger relevance scores become more significant in controlling differentiation and increase the efficiency of stem cell-based therapies.

Big Mechanisms in Systems Biology
DOI: http://dx.doi.org/10.1016/B978-0-12-809479-2.00008-1

Regeneration is one of the most intriguing and fascinating biological phenomena, but the molecular and cellular basis of this process is not yet fully understood. Unlike vertebrates with high regenerative capacity, such as amphibians and fish, we mammals have a restricted ability to regenerate our lost cells, tissues, and organs, in the adult stages. In general, the degree of generative capacity is related to the type of adult stem cells, which differ vastly among species. Therefore, in this chapter (Section 8.1), we explain the molecular mechanism of regeneration in organisms that can invoke special biological programs to replenish lost structures is important for shedding light on the longstanding question of the mechanism of regeneration. In the next part of this chapter, we will investigate the crucial cerebellar mound heading-related pathways and their crosstalk after traumatic brain injury (TBI) in zebrafish. The direct damage and secondary injury in the brain often result in chronic neurological disorders. Due to the multifactorial nature of secondary brain injury and subsequent, complex cellular responses, the underlying mechanisms of regeneration remain unclear. In Section 8.2, we use an adult zebrafish cerebellum injury model to investigate the phenotypes and secondary injury responses for recovery mechanisms in injured brains. Using time-course microarray analysis, a candidate PPI network was refined as a cerebellar wound healing PPI network by dynamic modeling and big data mining. Pathway enrichment and ontological analysis were incorporated into the refined PPI network to highlight the main molecular schemes of cerebellar wound healing. Several significant pathways, involving chemokines, phosphatidylinositol 3-kinases, axon guidance signaling pathways, and their crosstalk through P13K, PAK2, and PLXNA3 were identified as coordinating neurogenesis and angiogenesis, which are essential for the restoration of an injured brain. These findings will provide insight into the molecular restoration mechanisms after TBI, and open up new opportunities to devise treatments for TBI in humans.

Finally, due to some relation between the regeneration in stem cells and the proliferation in cancer cells, we investigate the systematic mechanisms of cancer stem cells. The biological function of normal adult somatic stem cells (SSCs) in their counterpart tissues is to provide a continuous supply of terminally differentiated, fully functional cells in tissues with high cellular turnover. Similar to normal SSCs, cancer stem cells also contain the unique biological characteristics of unlimited proliferation, self-renewal, and differentiation into specialized cell types.

8.1 DYNAMIC SYSTEM MECHANISMS IN THE THREE DIFFERENTIATION STAGES OF STEM CELLS TO REVEAL ESSENTIAL PROTEINS AND FUNCTIONAL MODULES IN THE DIRECTED DIFFERENTIATION PROCESS

Conducting targeted directed differentiation mechanism of stem cells is needed for further stem cell-based therapies. Directed differentiations from pluripotent stem cells into specific types of cells have been successfully demonstrated using extensive experimental protocols. However, effective use of stem cells for cell therapy is suffering from a lack of knowledge about underlying systematic mechanisms of directed differentiations. Dynamic PPI networks of three different differentiation stages were compared through dynamic network modeling to obtain interaction difference networks (IDNs) in the directed differentiation process from human embryonic stem cells (hESCs) into dopamine (DA) neurons. Variations in interaction activity of different differentiation stages can be assessed from IDNs to investigate the underlying systematic differentiation mechanisms. A relevance score to describe essential proteins and functional modules in the directed differentiation is proposed based on the IDNs. In addition to known roles of the Wnt and hedgehog signaling pathways in directed differentiations, the intricate interaction activity variations of essential proteins and functional modules can provide clues in identifying the roles they play in directed differentiation. The plausibility of constructed IDNs was tested by examining known key proteins involved in the directed differentiation process. Further proteins and functional modules essential at different stages of the directed differentiation process were investigated based on variations of their interaction activity in the IDNs. Several essential proteins and functional modules with high relevance scores were investigated for their roles in the directed differentiation process from hESCs to DA neurons, and may serve as a starting point in understanding the systemic mechanisms of directed differentiation as well as in increasing the efficiency of stem cell-based therapies.

Stem cell therapy requires precise control of the differentiation process from pluripotent stem cells to a specific cell type. For use in therapies, stem cells have to proliferate extensively to generate sufficient quantities of the desired cell type, yet growth of the differentiated cells has to be under control to prevent them from becoming a source of cancer. This means that a balance has to be achieved between promoting and restricting the pronounced ability of stem cells to differentiate and proliferate. The embryogenesis is an example of the concept of balance, and the systematic mechanisms of developmental processes have been used to develop experimental protocols. Manipulation of members or regulators of known pathways (e.g., the cell cycle, differentiation) may provide the desired outcomes.

Since 2000, a plethora of elaborate experimental protocols have been employed to induce stem cells toward desired directions of differentiation. In the directed differentiation of human pluripotent cells to ureteric bud kidney progenitor-like cells, a new platform for renal lineage commitment was developed by adding the needed molecules in the kidney embryogenesis [1]. A robust and efficient process to direct differentiation using a series of temporal growth factor manipulations to mimic embryonic intestinal development was established for the directed differentiation of human pluripotent cells into intestinal tissue [2]. In the dual inhibition of SMAD signaling, a possible mechanism was presented and a highly efficient neural conversion performed [3]. A key objective for neural induction in hESCs is the knowledge and capability to initiate signaling pathways [4]. The derivation of floor plate (FP) precursors was induced by a modified dual-SMAD inhibition protocol, progressing to DA neurons with FGF and SHH signals that have the ability to control dopaminergic cell production. Under both LSB and LSB/S/F8 control conditions, based on previous studies of dual-SMAD inhibition of neural induction, a new midbrain DA neuron protocol was developed [5]. Exposure to CHIR99021 (CHIR), a potent GSK3B inhibitor, was found to activate Wnt signaling [6]. Protocols aimed at known and specific pathways or mechanisms to control differentiation work on the assumption that if these mechanisms are invoked, then the desired cell type can be generated. Such protocols have been found to succeed as well as fail. This can be attributed to a lack of knowledge about the systematic mechanisms of directed differentiation.

Directing progenitor cells along a series of molecular events for neuronal differentiation in a systematic manner has proven difficult since the normal developmental process that generates most classes of CNS neurons remains poorly defined [7]. However, in the directed differentiation of mouse induced pluripotent stem cells (iPSCs) to cardiovascular cells, the authors succeeded in inducing the correct direction of differentiation based on knowledge of the systematic differentiation process of cardiovascular cells from iPSCs [8]. Systematic viewpoints of the mechanisms of directed differentiation can thus provide a global perspective on how to balance the promotion and restriction of stem cell growth and how to control the direction of differentiation.

In a modified dual-SMAD protocol study, the authors induced three different outcomes from three different treatments (LSB, LSB/S/F8, and LSB/S/F8/CHIR) and measured gene expression-time profiles for each treatment [5]. Treatments were found to induce different extents of differentiation (i.e., differentiation stages). To discover which parts of the cellular system of iPSCs or hESCs are changed under different treatment conditions (different stages of directed differentiation), dynamic network modeling is introduced in the present study to describe cellular systems of iPSCs. Dynamic network modeling of biological processes has been successful in furthering understanding of these processes from systematic viewpoints [9,10]. Three dynamic PPI networks corresponding to the above three conditions (or differentiation stages) are constructed by identifying the interaction activities among proteins using microarray data in the dynamic network modeling process. The IDNs can then be obtained by comparing the three stage-specific PPI networks. They represent the differentiation stage transitions of the three differentiation conditions. Since a biological function consists of a series of PPIs, a larger interaction activity variation observed in the IDNs may cause a more profound effect on the biological function in the directed differentiation process. Analysis of the variation of interaction activities of a key protein in the networks provides clues about how the outcomes of different differentiation conditions are caused by the interaction activity variations. Further, we define a relevance score for proteins in the networks based on the interaction activity variations of a protein between two different treatment conditions, which represents the importance of the protein in condition transitions. From a higher-level perspective, the directed differentiation process can be viewed as a network consisting of functional modules. Relevance scores of functional modules in the IDNs can be computed by summing the relevance scores of all members in functional modules, and reveal the importance of molecules and functional modules in the directed differentiation process.

8.1.1 Systems biology and big database mining methods for big regeneration mechanism of stem cells

The analysis procedure used in this section to identify essential proteins and functional modules in the directed differentiation process consists of three key steps: (1) microarray data preprocessing, (2) dynamic PPI network construction, and (3) relevance score calculation from IDNs between the different differentiation stages. Details for each steps are described in the following sections.

Microarray data preprocessing

The genome-wide microarray data used in this section are labeled GSE32658 in the GEO database [5]. They consist of measurements of temporal gene expression in the directed differentiation from hESCs to DA neurons in a range of 0–25 days under three different differentiation conditions (LSB, LSB/S/F8, and LSB/S/F8/CHIR), as illustrated in Fig. 8.1A. The LSB, LSB/S/F8, and LSB/S/F8/CHIR conditions respectively represent the first, second, and third differentiation stages from hESCs to DA neurons. By comparing the microarray data from these different conditions, we can investigate the stage-specific directed differentiation information. The genome-wide microarray data (GSE32658) were profiled using an Expression BeadChip Kit (Illumina, San Diego, CA, USA). The dataset contains 8 sampling time points spaced across 25 days (gene expression levels sampled at day 0, 1, 3, 5, 7, 11, 13, 25), and each sample contains three or four biological replications. In this section, we adopted the first 7 time points with three biological replications for each time point. For the LSB condition, the data comprised the first 7 time points of LSB treatment (day 0, 1, 3, 5, 7, 11, 13 with LSB treatment). For the LSB/S/F8 condition, the first 2 time points of LSB treatment and the first 5 time points of LSB/S/F8 treatment were used (day 0, 1 with LSB treatment and day 3, 5, 7, 11, 13 with LSB/S/F8 treatment). For the LSB/S/F8/CHIR condition, data comprised the first 2 time points of LSB treatment, the first time point of LSB/S/F8 treatment, and the first 4 time points of LSB/S/F8/CHIR treatment (day 0, 1 with LSB treatment, day 3 with LSB/S/F8

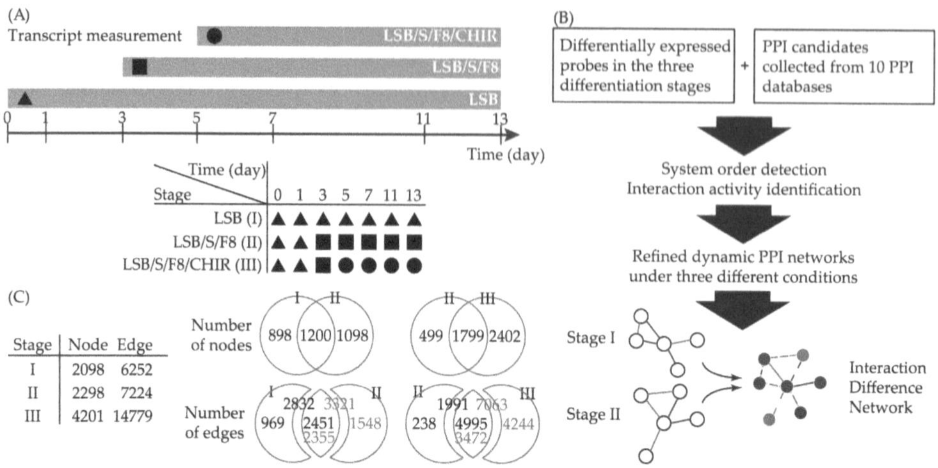

Figure 8.1 Overview of data, flowchart, and summary of the constructed PPI networks in the differentiation process from hESCs to DA neurons. (A) The upper part of the timeline shows time points of microarray data sampling (day 0, 1, 3, 5, 7, 11, 13, 25) under three conditions (LSB: ▲, LSB/S/F8: ■, and LSB/S/F8/CHIR: ●). The lower part of the timeline shows the experimental protocols used in Ref. [5]. The table at the lower right shows the data used to construct the dynamic PPI networks. (B) A brief flowchart of dynamic network construction. (C) A basic summary of the resultant constructed dynamic networks, including the number of nodes and edges of the three networks.

treatment, and day 5, 7, 11, 13 with LSB/S/F8/CHIR treatment) (Fig. 8.1A). Data were normalized relative to controls to avoid masking biological effects [11]. Differentially expressed probes were further filtered from the normalized data using an ANOVA and a significance threshold of FDR-corrected p-values $< 5 \times 10^{-5}$. After filtering, there were 3556, 3731, and 6460 differentially expressed proteins respectively for the LSB, LSB/S/F8, and LSB/S/F8/CHIR conditions, which were used to investigate and compare dynamic interaction networks.

Candidate PPIs from different databases need to be collected before the start of dynamic network modeling to construct PPI networks. These consist of interactions from computational predictions and biological experiments. Here, candidate PPIs were collected from 10 frequently used databases (BIND [12], BioGrid [13], DIP [14], HPRD [15], I2D [16], INTACT [17], MINT [18], PIP [19], Reactome [20], and STRING [21]). Candidate PPIs were then pruned using the microarray data prior to constructing the real dynamic PPI network.

Dynamic PPI network construction

The techniques for dynamic network construction can be found in Ref. [22]. The flowchart for dynamic network construction used in this section is summarized in Fig. 8.1B and briefly described in the following.

A discrete dynamic PPI model was employed in this section to describe PPIs in the network under different differentiation stages. The dynamic model for a target protein p in the candidate PPI network was proposed as follows [22]:

$$z_p[t+1] = z_p[t] + \sum_{q=1}^{Q} b_{pq} z_p[t] z_q[t] + \alpha_p x_p[t] - \beta_p z_p[t] + \omega_p[t+1] \tag{8.1}$$

where $z_p[t]$ is the protein level of p at time t, b_{pq} is the interaction activity of the qth protein and the pth target protein, $z_q[t]$ is the protein level of the qth protein that interacts with the target protein p at time t, α_p is the translation effect from mRNA to protein, $x_p[t]$ is the corresponding mRNA expression level of the target protein p, β_p is the degradation rate of the target protein p, and $\omega_p[t]$ is the stochastic noise. The biological interpretation of Eq. (8.1) is that the protein level of the target protein p at time $t+1$ is associated with the current target protein level, the regulatory interactions with Q interactive proteins, the translation effect from mRNA of target protein p, the degradation effect, and stochastic noise at time $t+1$. Dynamic PPI modeling of p in Eq. (8.1) can be represented as the following regression form:

$$\begin{aligned} z_p[t+1] &= \begin{bmatrix} z_p[t]z_1[t] & \dots & z_p[t]z_Q[t] & x_p[t] & z_p[t] \end{bmatrix} \bullet \begin{bmatrix} b_{p1} \\ \vdots \\ b_{pQ} \\ \alpha_p \\ (1-\beta_p) \end{bmatrix} + \omega_p[t+1] \\ &\triangleq \phi_p[t] \bullet \theta_p + \omega_p[t+1] \end{aligned} \tag{8.2}$$

where $\phi_p[t]$ is the regression data vector and θ_p is the kinetic parameter vector to be estimated. In order to avoid overfitting to a small amount of data in estimating parameter processes, original data points were interpolated to L data points using a cubic spline method [23,24]. Thus, Eq. (8.2) can be presented as the following equation:

$$Z_p = \Phi_p \bullet \theta_p + \Omega_p \tag{8.3}$$

where $Z_p = [z_p[1], \ldots, z_p[L]]^T$, $\Phi_p = [\phi_p[0], \ldots, \phi_p[L-1]]^T$, and $\Omega_p = [\omega_p[1], \ldots, \omega_p[L]]^T$. The parameter estimation problem can then be formulated as the following least square minimization problem:

$$\min_{\theta_p} \frac{1}{2} \left\| Z_p - \Phi_p \bullet \theta_p \right\|_2^2 \tag{8.4}$$

Since large-scale measurement of protein levels are still lacking, mRNA expression profiles were used as substitutes for protein levels when identifying the protein interaction parameters (as seen in Eq. 2.20). Even if the mRNA expression level cannot be completely representative of protein expression level, there are partly positive correlations between them [25,26]. Therefore, the mRNA microarray can represent the trend of the corresponding protein expression level. Once the protein interaction activity b_{pq} is estimated, Akaike's Information Criterion (AIC) (in Eq. 2.23) was applied to eliminate models with false positive protein interactions. Realistic PPI networks were then constructed from the remaining candidate PPI networks.

Relevance score calculation from IDNs under different differentiation stages

The resultant PPI networks at different differentiation stages can be represented as a matrix from previously constructed dynamic PPI networks:

$$\begin{bmatrix} \hat{b}_{11}^{\text{cond.}} & \cdots & \hat{b}_{1Q}^{\text{cond.}} \\ \vdots & \ddots & \vdots \\ \hat{b}_{Q1}^{\text{cond.}} & \cdots & \hat{b}_{QQ}^{\text{cond.}} \end{bmatrix}, \text{cond.} \in \{\text{LSB, LSB/S/F8, LSB/S/F8/CHIR}\} \tag{8.5}$$

where $\hat{b}_{pq}$ is the interaction activity between protein p and protein q in the realistic PPI networks. The interaction difference of two dynamic PPI networks at different differentiation stages is called the IDN in the differentiation process from hESCs into DA neurons and can be expressed as the following interaction difference matrix:

$$D_{\text{cond.2}-\text{cond.1}} = \begin{bmatrix} d_{11} & \cdots & d_{1Q} \\ \vdots & \ddots & \vdots \\ d_{Q1} & \cdots & d_{QQ} \end{bmatrix} = \begin{bmatrix} \hat{b}_{11}^{\text{cond.2}} & \cdots & \hat{b}_{1Q}^{\text{cond.2}} \\ \vdots & \ddots & \vdots \\ \hat{b}_{Q1}^{\text{cond.2}} & \cdots & \hat{b}_{QQ}^{\text{cond.2}} \end{bmatrix} - \begin{bmatrix} \hat{b}_{11}^{\text{cond.1}} & \cdots & \hat{b}_{1Q}^{\text{cond.1}} \\ \vdots & \ddots & \vdots \\ \hat{b}_{Q1}^{\text{cond.1}} & \cdots & \hat{b}_{QQ}^{\text{cond.1}} \end{bmatrix}$$
$$\text{cond.1}, \text{cond.2} \in \{\text{LSB, LSB/S/F8, LSB/S/F8/CHIR}\} \tag{8.6}$$

where d_{pq} is the difference in the interaction activity of protein p and q in the IDN between two different differentiation conditions (or differentiation stages). For essential proteins in the directed differentiation process, two issues are taken into consideration. First, the magnitude of the interaction activity represents the influence of one protein on the other proteins in the realistic PPI network. Second, if a protein plays a critical role in the transition from condition 1 (differentiation stage 1) to condition 2 (differentiation stage 2), the difference in interaction activities will be larger.

Based on these considerations, the relevance score (RS) of a protein in the directed differentiation process can be defined as follows:

$$RS_p = \frac{\sum_{q=1}^{Q} |d_{pq}|}{\text{Degree of protein } p} \tag{8.7}$$

Similarly, the relevance score of the functional module *f*, a set of proteins with a specific function, in the directed differentiation process can also be defined:

$$RS_f = \sum_{p \in f} \frac{\sum_{q=1}^{Q} |d_{pq}|}{\text{Degree of protein } p} \tag{8.8}$$

Classification of a group of proteins in the IDNs into functional modules or pathways can easily be achieved by ontological analyses and relevance scores of the functional modules can be calculated accordingly. Similarly, the importance of a functional module in the directed differentiation can be represented by the relevance score of the functional module.

8.1.2 Results and discussion of differentiation mechanisms of stem cells

Temporal variations in the interaction activities of key proteins in the directed differentiation process

The resultant dynamic PPI networks for three differentiation stages and their nodes and edges are summarized in Fig. 8.1C. A basic question concerns the parts of the dynamic PPI network (system) that are changed with the addition of SHH and FGF8 on the first day and the addition of CHIR on the third day. These parts may play crucial roles in the direction of differentiation. The interaction difference matrices between the LSB/S/F8 and LSB conditions and between the LSB/S/F8/CHIR and LSB/S/F8 conditions ($D_{\text{LSB/S/F8}-\text{LSB}}$ and $D_{\text{LSB/S/F8/CHIR}-\text{LSB/S/F8}}$) were calculated according to Eq. (8.6), allowing the construction of the IDNs (see Fig. A.1). Prior to further analysis of the IDNs in the directed differentiation process, we first inspected key proteins (i.e., OTX2, FOXA2, and LMX1A for DA neuron differentiation from hESCs) and the proteins that interact with them [5].

OTX2 represents the first transcription factor with a proven role in mesencephalic–diencephalic dopaminergic (mdDA) neurogenesis [27] and controls development of several neuronal populations in the midbrain by regulating progenitor identity and neurogenesis [28]. After adding SHH and FGF8, the positive interaction between OTX2 and APP and the negative interaction between OTX2 and FOXA2 emerged and the negative interaction between OTX2 and TLE4 diminished (Fig. 8.2A). Since TLE4 shows negative regulation ability on the transactivating ability of OTX2 [29], reducing the negative effect of TLE4 on OTX2 may enhance the ability of OTX2 to determine cell fate during

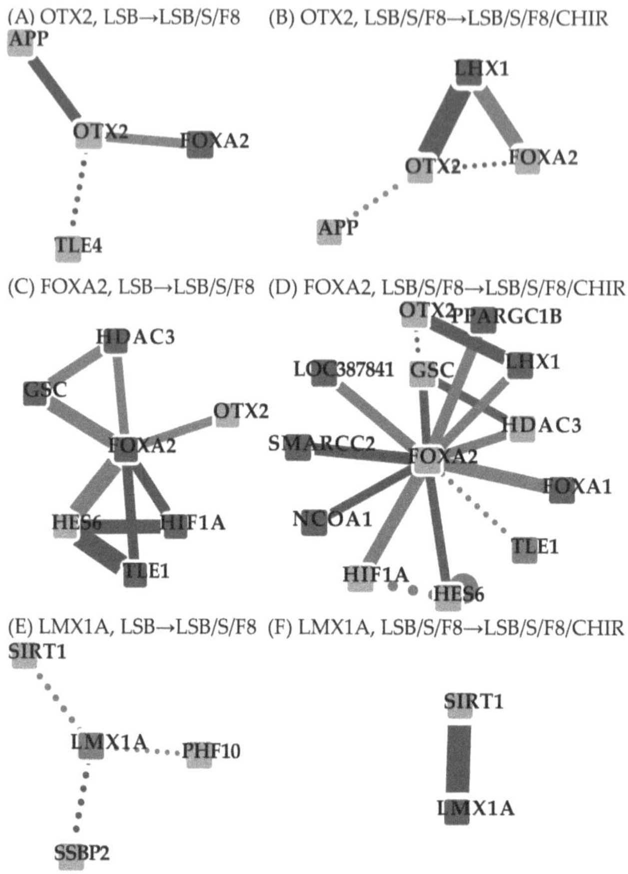

Figure 8.2 The interaction activity variations of known key proteins OTX2, FOXA2, and LMX1A in the IDNs at two differentiation stage transitions. (A) Interaction activity variation of OTX2 from LSB to LSB/S/F8; (B) of OTX2 from LSB/S/F8 to LSB/S/F8/CHIR; (C) of FOXA2 from LSB to LSB/S/F8; (D) of FOXA2 from LSB/S/F8 to LSB/S/F8/CHIR; (E) of LMX1A from LSB to LSB/S/F8; (F) of LMX1A from LSB/S/F8 to LSB/S/F8/CHIR. (Nodes: *Red* indicates disappearance at the previous stage and appearance at the later stage. *Blue* indicates appearance at the previous stage and disappearance at the later stage. *Purple* indicates appearance under both conditions. Edges: *Red* indicates an enhancing interaction (i.e., + → + +, − − → −, 0 → +, and − → 0). *Blue* indicates an attenuating interaction (+ + → +, − → − −, + → 0, and 0 → −). A *solid edge* represents appearance under both conditions. A *dashed edge* represents disappearance under the previous condition and appearance under the later condition. A *dotted edge* represents appearance under the previous condition and disappearance under the later condition. The width of the edge represents the magnitude of the interaction activity variations.)

directed differentiation. After adding CHIR, the positive interaction between OTX2 and APP and the negative interaction between OTX2 and FOXA2 diminished and the positive interaction between OTX2 and LHX1 and the negative interaction between LHX1 and FOXA2 emerged (Fig. 8.2B). The emerging positive interaction between

OTX2 and APP and its following disappearance correlates to the fact that overexpression of the APP gene causes glial differentiation of stem cells and that reduction in APP levels would be useful in promoting neurogenesis [30]. This indicates the necessity of CHIR addition to promote adequate neural differentiation. The connection between OTX2 and FOXA2 is more intricate. After adding SHH and FGF8, a negative interaction emerged; the addition of CHIR then turned the direct negative interaction into indirect interactions through LHX1, which may promote DA differentiation [31].

FOXA2, a transcriptional activator, regulates DA neuronal generation and differentiation [32] and is only significantly differentially expressed after adding SHH and FGF8 (Fig. 8.2C). The negative interaction between FOXA2 and HES6 emerged after adding SHH and FGF8, and FOXA2 exerted indirect positive interactions on HES6 through HIF1A and TLE1. Later, the negative interaction was attenuated by adding CHIR and the indirect positive interactions were diminished (Fig. 8.2D). HES6, a candidate gene for mood disorder susceptibility and antidepressant response, promotes neuronal differentiation [33]. Thus, the interactions between FOXA2 and HES6 may provide a mechanism to control the differentiation directly and indirectly. The positive interaction between FOXA2 and SMARCC2 (also known as BAF170) emerged after adding CHIR. SMARCC2 has been proven to play a role in neurogenesis [34]. The negative interaction between FOXA2 and PPARGC1B (also known as PGC1B), which controls mitochondrial metabolism [35], emerged after adding CHIR. The positive interaction between FOXA2 and NCOA1 (also known as SRC1), emerged after adding CHIR. SRC1 null mice show moderate motor dysfunction and delayed development of cerebellar Purkinje cells [36].

LMX1A, a roof plate marker, indicates the establishment of midbrain DA neuron precursor fate [5] and was diminished after SHH and FGF addition but emerged again after CHIR addition (Fig. 8.2E and F). The interactions between LMX1A and SIRT1 declined throughout the differentiation process. SIRT1 mediates MPP + -induced apoptosis in dopaminergic cells [37]. This may imply that control of apoptosis is exerted at the beginning of the differentiation process and is progressively attenuated.

Most of the interactions among these three known key proteins in the directed differentiation from hESCs to DA neurons were confirmed by these IDNs (Fig. 8.2) and the literature, justifying the reliability of the constructed dynamic PPI network. Based on this finding, we selected possible essential proteins based on their relevance scores and investigated their relationships with temporal variations of protein interaction activity in the directed differentiation process.

Significant proteins with the top 10 relevance scores in the IDNs of stem cells

Following qualitative analysis of the three key proteins of the directed differentiation, we here provide quantitative measurements of the interaction activity variations of the directed differentiation-related proteins at different stages, according to the constructed IDNs. A similar evaluation based on IDNs has given an insight into the pathways and mechanisms of lung carcinogenesis and potential therapeutic targets for lung cancer [38]. Utilizing the identified interaction matrices (Eq. 8.5) from microarray data and the definition of the relevance score of a protein in Eq. (8.7), the importance of a protein at different stages of the directed differentiation process from hESCs to DA neurons can be evaluated. The proteins with the top 10 relevance scores are given in Table 8.1 and the IDNs of these proteins extracted from the whole IDN are depicted in Fig. 8.3.

The complex METTL3-METTL14 showed a negative interaction after adding SHH and FGF8. This complex, a methyltransferase, mediates mammalian nuclear RNA N^6-adenosine methylation [39], which is essential for neural stem cell differentiation [40]. In both of the IDNs ($D_{LSB/S/F8-LSB}$ and $D_{LSB/S/F8/CHIR-LSB/S/F8}$), this complex had a strong interaction with POLR2I, which is a transcription factor and is related to Huntington's disease. Although the role of POLR2I in the directed differentiation from hESCs to DA neurons is not quite clear, the strong interaction may imply the importance of methylation and POLR2I in the differentiation process. Furthermore, the mitochondrial ribosomal protein MRPS34 interacted with a human homolog of the *Drosophila* disk large tumor suppressor protein (hDLG) which may function in the regulation of cell death and development [41]. The differential expression of MRPS34 was also reported in several stem cell lines. EFEMP2 and STYXL1 are related to EGF and the cell cycle, respectively, and their roles in directed differentiation are still not fully explored. The above proteins are common in both IDNs, indicating that methylation, mitochondrial ribosomes, EGF signaling, and cell cycle-related functions are all involved in differentiation stage transitions.

Table 8.1 The 10 proteins with the highest relevance scores in differential process of stem cells

Rank	**LSB→LSB/S/F8**		**LSB/S/F8→LSB/S/F8/CHIR**	
	Relevance score	**Protein**	**Relevance score**	**Protein**
1	33.71857	METTL14	33.71857	METTL14
2	15.33615	METTL3	16.94178	METTL3
3	14.93786	<u>NR2F2</u>	9.465864	<u>GLRX</u>
4	4.348328	POLR2I	3.366032	POLR2I
5	2.916923	<u>CLDN3</u>	2.441295	EFEMP2
6	2.682951	MRPS34	2.041421	MRPS34
7	2.682884	<u>NCOR2</u>	2.001651	<u>GTF2F2</u>
8	2.457388	EFEMP2	1.884138	<u>FAM124B</u>
9	2.227754	<u>CLDN1</u>	1.859985	STYXL1
10	2.141034	STYXL1	1.42109	<u>SEPT10</u>

The cellular roles of underlined proteins have been explored in the directed differentiation process from stem cells to DA neurons and depicted in Fig. 8.3.

In the IDN between LSB/S/F8 and LSB conditions ($D_{LSB/S/F8\text{–}LSB}$; Fig. 8.3A), the ligand-activated transcription factor NR2F2 is involved with OCT4 in mammalian ESC pluripotency [42] and FOXA1 transcription factor networks [43]. Its interaction activity with NCOR2, a nuclear receptor corepressor controlling transcriptional silencing, was reduced after adding SHH and FGF8. By attenuating this interaction, a group of genes can be up-regulated (e.g., HDAC1). In addition, NCOR2 is involved in the Notch signaling pathway, which serves an important role in cell differentiation. Another group emerging in the network was claudin, including CLDN1 and CLDN3. Claudin is related to tight junctions and cell-to-cell adhesion [44]. Cell adhesion controls interactions of stem cells with their niche and signaling environment [45]. One member of the claudin group, CLDN6, was found to participate in the immunological elimination of hESCs and the prevention of cell outgrowth [46]. Hence, CLDN1 and CLDN3 may be involved in controlling the extension of hESCs and should be worthy of further exploration.

In the IDN between LSB/S/F8/CHIR and LSB/S/F8 conditions ($D_{LSB/S/F8/CHIR-LSB/S/F8}$; Fig. 8.3B), there were four specific proteins with large relevance scores: GLRX, GTF2F2, FAM124B, and SEPT10. GLRX is differentially expressed in neuronal differentiation of human retinal pigment epithelial cells and in salamander

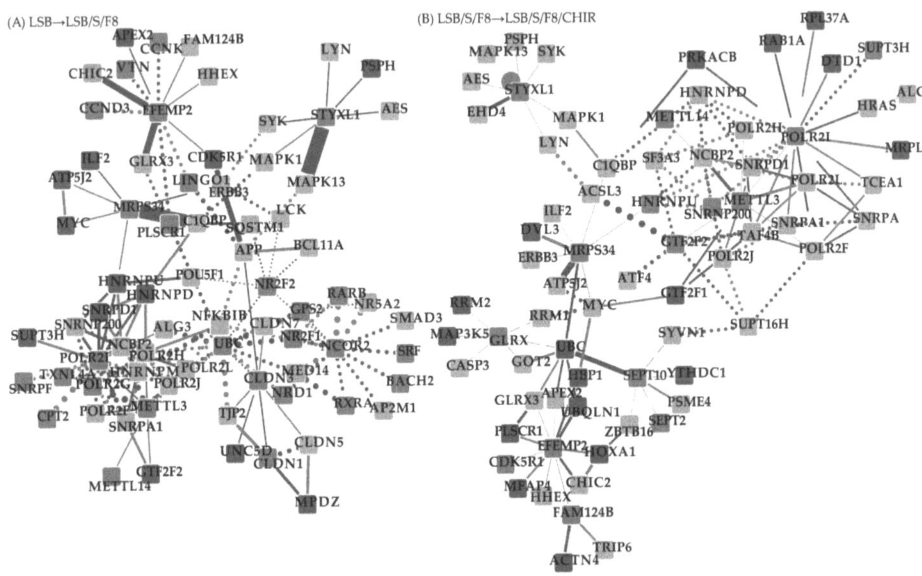

Figure 8.3 Interaction activity variations of the 10 proteins with the highest relevance scores in the IDNs at two differentiation stage transitions. (A) Interaction activity variations of the 10 proteins highest in relevance score from LSB to LSB/S/F8. (B) Interaction activity variations of the 10 proteins highest in relevance score from LSB/S/F8 to LSB/S/F8/CHIR. Node and edge colors and edge line styles have the same interpretation as in Fig. 8.2. A *yellow node* represents a protein with a high relevance score (Table 8.1).

spinal cord regeneration [47] and is involved in the estrogen pathway [48]. GTF2F2 encodes the general transcription factor IIF and is involved in NRF-1 regulated neurite growth [49] and the estrogen pathway [50]. FAM124B is a component of the CHD7- and CHD8-containing complex that is related to neurodevelopmental disorders [51]. Septin10 (SEPT10), an evolutionarily conserved group of GTP-binding and filament-forming proteins, interacts with components of cytoskeletons. Investigation of SEPT2, also a septin, has shown a relationship between septin and neuronal development.

It was observed that some of these proteins (GLRX, GTF2F2, FAM124B, and SEPT10) showed more neuron-specific related functions after the addition of CHIR, while for others (NR2F2, NCOR2, CLDN1, and CLDN3) this was the case after adding SHH and FGF8; the latter group seems to be involved with more general processes of stem cell differentiation. The IDNs also provide a basis for further study. For example, the relations shown as thick edges in the IDNs (Fig. 8.2) imply a greater amount of variation occurring in the stage transitions, which may indicate essential roles in stage transition. Thus, the interaction between METTL3 and METTL14 is one of the most varied, and its role in the directed differentiation of hESCs to DA neurons may be worth further investigation.

Relevance scores of the integrin, FGF, hedgehog, and Wnt signaling pathways in differential mechanism of stem cells

The relevance scores of proteins indicate the essential candidates for important roles in the directed differentiation from hESCs to DA neurons. Complex biological phenomena are however the outcomes of intricate interactions among several functional modules (or signaling pathways) which comprise several proteins with specific functions. To gain a clearer understanding of the mechanisms of directed differentiation of hESCs to DA neurons, an interpretation from a higher-level perspective is therefore indispensable. The method of calculating protein relevance scores can be generalized to a functional module (or a signaling pathway) as shown in Eq. (8.8), and higher-level interpretations of the directed differentiation can be deduced from relevance scores. The prerequisite to calculating relevance scores is to classify proteins into functional modules, which may be achieved through ontological analysis or pathway analysis. Based on the IDNs, ontological analysis yielded several significant enriched functional modules, which are listed with their average interaction variations in Tables 8.2 and 8.3. Based on the rank of relevance scores and p-values, we focused on the integrin, FGF, hedgehog, and Wnt signaling pathways in the following analysis.

The integrin signaling pathway shows the highest rank by relevance score (Tables 8.2 and 8.3) corresponding to differentiation stage transitions both from LSB to LSB/S/F8 and from LSB/S/F8 to LSB/S/F8/CHIR. In the differentiation stage transition from LSB to LSB/S/F8, the IDN has a core network (Fig. 8.4A)

Table 8.2 The 10 functions highest in relevance score from the LSB to LSB/S/F8 condition in the differential process of stem cells

Rank	LSB→LSB/S/F8		
	Relevance score	**Function**	***p*-Value of ontology analysis**
1	7.627697307	Huntington disease	5.55E-03
2	5.389941296	Integrin signaling pathway	1.35E-07
3	4.865942264	Gonadotropin releasing hormone receptor pathway	1.87E-02
4	4.776453757	Angiogenesis	2.76E-10
5	3.318549657	PDGF signaling pathway	1.10E-02
6	3.13212684	Cytoskeletal regulation by Rho GTPase	2.68E-04
7	3.120593534	FGF signaling pathway	1.69E-03
8	3.016342937	EGF receptor signaling pathway	8.17E-05
9	2.417680991	TGF-beta signaling pathway	9.86E-03
10	2.36486466	Parkinson disease	2.78E-02

Table 8.3 The 10 functions highest in relevance score from the LSB/S/F8 to LSB/S/F8/CHIR condition

Rank	LSB/S/F8→LSB/S/F8/CHIR		
	Relevance score	**Function**	***p*-Value of ontology analysis**
1	7.085580216	Gonadotropin releasing hormone receptor pathway	2.84E-04
2	6.595569241	Integrin signaling pathway	3.54E-08
3	4.956190934	FGF signaling pathway	1.68E-06
4	4.634381934	Angiogenesis	6.45E-06
5	4.535296087	EGF receptor signaling pathway	2.35E-06
6	3.702903524	Huntington disease	9.60E-03
7	3.299706184	PDGF signaling pathway	2.75E-02
8	2.664383051	Parkinson disease	8.74E-05
9	2.640768983	TGF-beta signaling pathway	3.05E-02
10	2.61175047	Ras pathway	1.18E-03

consisting of integrins (ITGA2, ITGA3, ITGA6, ITGAV, ITGB1, ITGB2, and ITGB5), collagens (COL2A1, COL5A2, COL6A1, COL9A1, COL9A3, COL11A1, and COL18A1), and laminins (LAMA1, LAMB1, LAMB2, and LAMC2). This core network links to VCL and FYN with enhancing interactions. VCL, a cytoskeletal protein, links to RAS (HRAS) and TGFβ (TGFB1I1). TGFB1I1 then links to PXN, which is related to functions of focal adhesion, and LCK, a well-studied protein that can enhance differentiation of lymphoid cell lines by suppressing

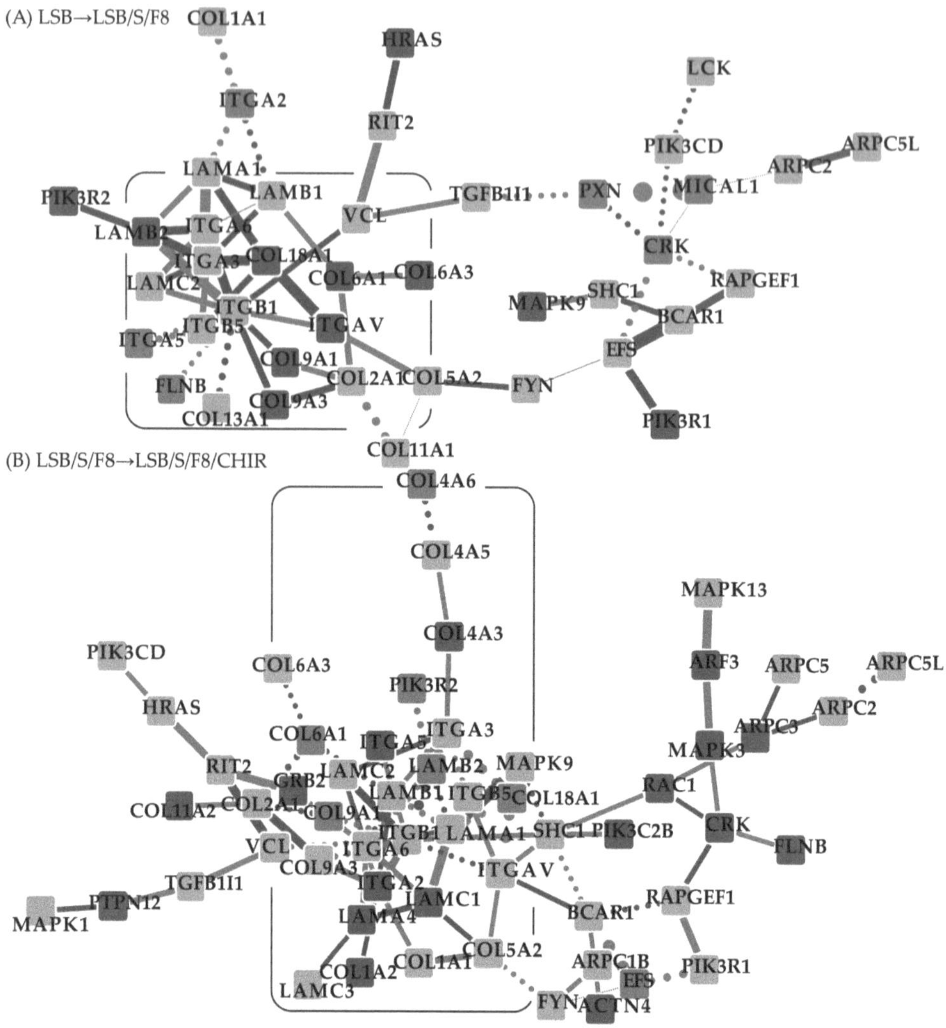

Figure 8.4 The interaction activity variations of the integrin signaling pathway in the directed differentiation process at two differentiation stage transitions. (A) Interaction activity variations from the LSB stage to the LSB/S/F8 stage. (B) Interaction activity variations from the LSB/S/F8 stage to the LSB/S/F8/CHIR stage. Node and edge colors and edge line styles have the same interpretation as in Fig. 8.2.

glucocorticoid sensitivity and apoptosis [52]. FYN, the other protein linking to the core network, has been implicated in the control of cell growth. In addition, some actin-related proteins (ARPC2 and ARPC5L) are included in the core network through PXN. It is notable that many proteins in the integrin signaling pathway also appear in pathways related to cancers. This explains the similarity between cancer cells and stem cells in their almost unlimited differentiation ability, immortality, and the uncontrollable differentiation of induced stem cells. In the differentiation stage transition from LSB/S/F8 to LSB/S/F8/CHIR, the scale of the core network consisting of integrins, collagens, and laminins increased with the inclusion of further instances of these proteins (Fig. 8.4B). The interactions from the core network to VCL and FYN diminished, implying that inhibition of apoptosis was not required. However, requirements for actin polymerization seemed to be increased. More actin-related proteins emerged in this IDN (Fig. 8.4B). In addition, GRB2 emerged in the core network, acting as a suppressant of proliferative signals and triggering programmed cell death, and being attenuated through interactions with RIT2 and ITGA6. Interactions related to cell growth and apoptosis inhibition in the previous stage transition (from LSB to LSB/S/F8) became more negative or vanished in the later stage transition (from LSB/S/F8 to LSB/S/F8/CHIR). In addition, actin polymerization and focal adhesion, which are downstream of the integrin signaling pathways, were more active in the later stage. Those interaction variations indicate the necessity for regulation on cell survivability and neurogenesis.

The FGF signaling pathway also had a high relevance score in the differentiation stage transitions (Tables 8.2 and 8.3). As the hedgehog and Wnt signaling pathways have been implicated in the development of thalamus [53], these two pathways were included in the following analysis together with the FGF signaling pathway, although they are not significantly enriched in the IDNs corresponding to the two differentiation stage transitions. In the transition from LSB to LSB/S/F8, cadherins, FGF, and FGFR comprised a core network (Fig. 8.5A). However, the interactions in the core network (e.g., among CDH1, CDH2, CDH3, CDH7, CDH6, CDH10, and CDH11) were diminished after adding SHH and FGF8. Instead, the interactions between hedgehog and Wnt signaling pathways were increased (e.g., HDAC3 and PPP4C, FZD5 and PTCH1). AES, a molecule in the Wnt signaling pathway, is related to neurogenesis. Its connection to FGF was diminished and a negative interaction with HDAC3 emerged. CSNK2A1 and CSNK2A2, which are related to axon guidance, were also shown to have negative and diminishing interactions with neighboring proteins (Fig. 8.5A). In the transition from LSB/S/F8 to LSB/S/F8/CHIR (Fig. 8.5B), FGF8 lost most of its interactions, and other FGFs and FGFRs connected to form a larger core network, which included cadherins (e.g., CTNNB1, CDH11, CTNNA1, CDH13, PCDH7, CDH3) and axon guidance molecules (CSNK2A1 and CSNK2A2). Some previously noted molecules (AES, CSNK2A1, and CSNK2A2) related to neurogenesis and axon guidance were reconnected to the core network after adding CHIR. The reestablishment of connections between AES, CSNK2A1, and HDAC3 is required for the following differentiation process. Another important interaction occurred between WNT2 and HDAC2. Both of these proteins have been implicated in neuronal differentiation

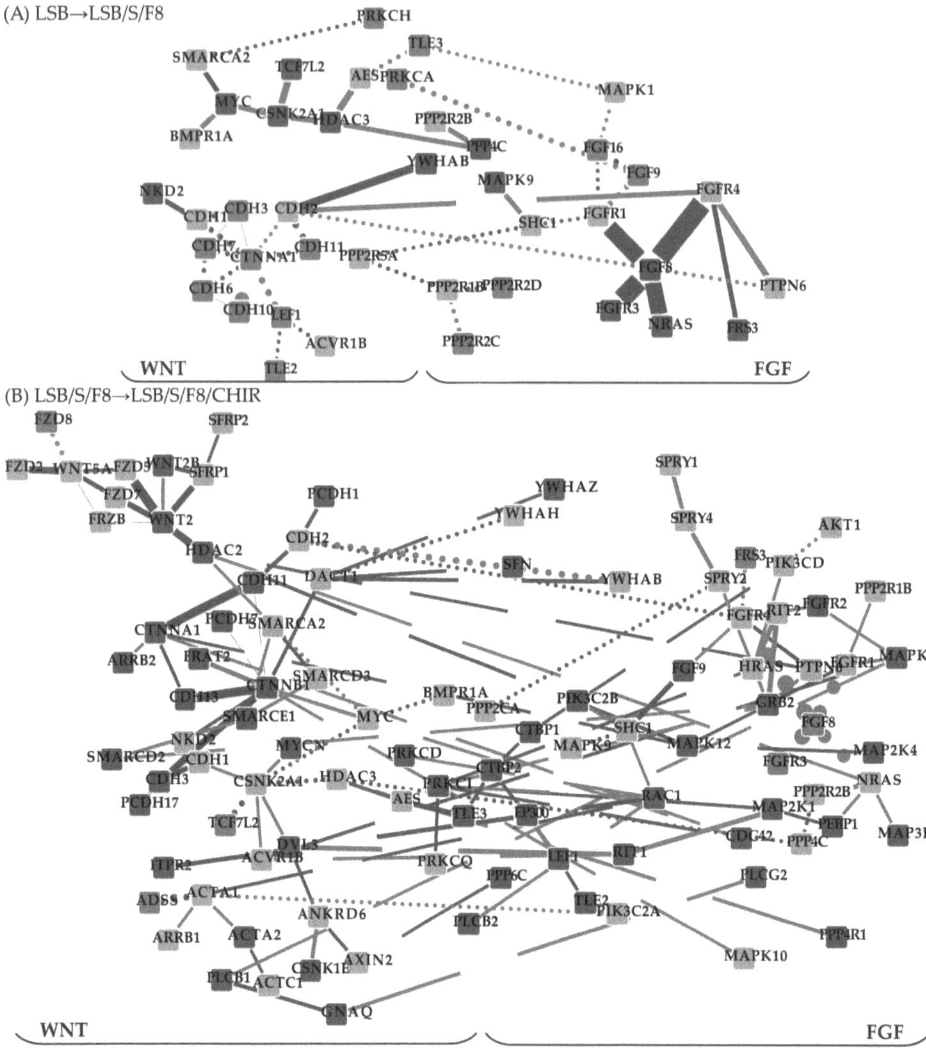

Figure 8.5 The interaction activity variations of the FGF, hedgehog, and Wnt signaling pathway in the directed differentiation process at two differentiation stage transitions. (A) Interaction activity variations from the LSB stage to the LSB/S/F8 stage. (B) Interaction activity variations from the LSB/S/F8 stage to the LSB/S/F8/CHIR stage. Node and edge colors and edge line styles have the same interpretation as in Fig. 8.2. *Yellow nodes* stand for proteins belonging to the FGF signaling pathway.

[54]. GRB2 also emerged in the core network and exerted its influence on cell growth in cooperation with the FGF, Wnt, and hedgehog signaling pathways.

The role of the FGF signaling pathway seemed to change between the early and the later differentiation stage. At the first stage transition, the connections among the

FGF, Wnt, and hedgehog signaling pathways are built up for a general stem cell differentiation scheme. It may be an essential step to recruit them for further neuron differentiation, as the Wnt and hedgehog signaling pathways are well known for being involved in the neural development process. In the later stage transition, the FGF signaling pathway was involved in a more complex interaction network with neural differentiation-related proteins. The integrin signaling pathway also showed a similar trend; i.e., many proteins with higher relevance scores in this pathway (e.g., LCK, FYN, ARP) are involved in a general scheme of stem cell differentiation, while interactions with neurogenesis-related proteins appear later on. These proteins and their interactions may play essential roles in directed differentiation, and require investigation. Although the characteristics of the four discussed signaling pathways are further familiar in the field of differentiation, their roles in directed differentiation are more explicitly indicated by dynamic PPI network-based relevance scores. The diminishing demand for control of cell growth and apoptosis can be observed via the relevance scores. These proposed scores could therefore be a potential method to describe dynamic biological differentiation processes.

8.1.3 Conclusions of systematic differentiation mechanisms of stem cells

Understanding the processes and events occurring during directed differentiations of hESCs provides a basis for controlling the differentiation of hESCs. Employing the microarray data for measuring temporal gene expression profiles at three different differentiation stages [5], differentiation processes and events can be treated as a combination of several essential molecules, functional modules, and their interactions. Timely appearance and disappearance of molecules, interactions among molecules, and functional modules promote the progress of the directed differentiation from hESCs to DA neurons. In this section we have used a computational method to investigate PPI network changes to dissect the directed differentiation process from hESCs to DA neurons and gain an insight into the underlying mechanisms of directed differentiation.

Interactions among proteins have been identified by time profile microarray data and then assembled into dynamic networks for the directed differentiation process through successful dynamic network modeling [9,10,22,38]. Three dynamic PPI networks corresponding to three different differentiation stages of hESCs to DA neurons were constructed. By comparing any two PPI networks at differentiation stages, an interaction difference matrix that can be depicted as an IDN between different differentiation stages was obtained (Figs. 8.2–8.5). The interaction activity variations in the PPIs from one stage to another can be observed. Several known key proteins (OTX2, FOXA2, and LMX1A) in the directed differentiation from hESCs to DA neurons were inspected to justify the reliability of the resultant IDNs in the directed differentiation process. The variations of their interactions indicate the cellular functions promoting the progress of the directed differentiation. The relationships between these interaction variations and the cellular functions in the directed differentiation were verified by literature. According to the interaction activities identified by the dynamic network modeling, the variations in the interaction activity of a protein in the IDN can be calculated and defined as a relevance score. As a biological function consists of a series of PPIs or a larger variation of PPIs, more profound effects are caused by the PPIs. The relevance score is hence representative of the significance of the protein in the differentiation stage transitions. Generalizing the concept from a protein to a cellular function, the significance of a functional module in the differentiation stage transitions can be also evaluated through relevance scores of functional modules.

Trends of interaction variations through the differentiation stage transitions can be observed from the resultant IDN. The transition from LSB to LSB/S/F8 may indicate a general scheme of stem cell differentiation. Evading apoptosis and enhancing survivability of differentiated stem cells emerged as underlying mechanisms of directed differentiation at the initial stage and are expressed as proteins (e.g., MRPS34, EFEMP2, and STYXL1) and functional modules (e.g., LCK and FYN in the integrin signaling and subsiding interactions among cadherins in the FGF signaling). The transition from LSB/S/F8 to LSB/S/F8/CHIR may specify the proteins and functional modules in the directed differentiation mechanism related to neurogenesis and functions of DA neurons (e.g., gonadotropin releasing hormone secretion). The inhibition of apoptosis and the activation of cell growth declined after adding CHIR, a GSK3B inhibitor that also can activate Wnt signaling. The activation of Wnt signaling, although not a significant parameter in the enrichment analysis, was visible in the IDNs from LSB/S/F8 to the LSB/S/F8/CHIR. In addition, the crosstalk among the Wnt, hedgehog, and FGF signaling pathways were increased in the directed differentiation process (Fig. 8.5) and are involved in determining the boundary of brain regions [54].

Several essential proteins and signaling pathways were investigated as underlying mechanisms of the directed differentiation process from hESCs to DA neurons using computational and biological verification. Many of them require experimental verification through their relationship with the directed differentiation of stem cells. For now, a computational basis for understanding the directed differentiation process has been provided and may serve as a starting point in understanding the systematic mechanisms of directed differentiations as well as in increasing the efficiency of stem cell-based therapies.

APPENDIX A: FIGURES

Here we show Fig. A.1.

Figure A.1 The whole interaction difference network from LSB stage to LSB/S/F8 stage and from LSB/S/F8 stage to LSB/S/F8/CHIR stage. (A) The whole interaction difference network from LSB stage to LSB/S/F8 stage. The *blue node* indicates the node exists only in the dynamic PPI network of LSB stage. The *red node* indicates the node exists only in the dynamic PPI network of LSB/S/F8 stage. The *purple node* indicates the node exists in both the dynamic PPI network of LSB and LSB/S/F8 stage. (B) The whole interaction difference network from LSB/S/F8 stage to LSB/S/F8/CHIR stage. The *blue node* indicates the node exists only in the dynamic PPI network of LSB/S/F8 stage. The *red node* indicates the node exists only in the dynamic PPI network of LSB/S/F8/CHIR stage. The *purple node* indicates the node exists in both the dynamic PPI network of LSB/S/F8 and LSB/S/F8/CHIR stage.

8.2 CEREBELLA REGENERATION-RELATED PATHWAYS AND THEIR CROSSTALKS IN MOLECULAR RESTORATION MECHANISMS AFTER TBI IN ZEBRAFISH

After investigating the significant proteins for differentiation mechanisms of stem cells to discuss their regeneration roles in the previous section, we will discuss regeneration-related pathways and the molecular restoration mechanisms of zebrafish after TBI in this following section. Upon injury, the direct damage and the subsequent secondary injury in the brain often result in chronic neurological disorders. Due to multifactorial nature of secondary injury and subsequent complex cellular responses, much of the underlying mechanisms are unclear. This section used an adult zebrafish cerebellum injury model to investigate the phenotypes and the secondary injury responses for recovery mechanisms of injured brain. Using the time-course microarray analysis, a candidate PPI network was refined as cerebellar wound healing PPI network by dynamic modeling and big data mining. Pathway enrichment and ontological analysis were incorporated into the refined network to highlight the main molecular scheme of cerebellar wound healing. Several significant pathways, including chemokine, phosphatidylinositide 3-kinases, and axon guidance signaling pathway and their crosstalks through PI3K, PAK2, and PLXNA3 were identified to coordinate for neurogenesis and angiogenesis, which are essential for the restoration of the injured brain. This finding provides an insight into the molecular restoration mechanisms after TBI, and open up new opportunity to devise the treatment for TBI in human.

TBI, also known as intracranial injury, is a major cause of death and disability worldwide, especially in children and young adults. Recent data have shown that approximately 1.7 million people sustain a TBI annually in the United States [55]. TBI is caused by an impact on the head or a penetrating head injury that disrupts normal brain function. Upon injury, the direct damage and the subsequent secondary injury in the brain often results in chronic neurological disorders. The multifactorial nature of secondary injury after brain injury is characterized by a complex multicellular process that involves apoptosis, inflammation, proliferation of glial cells, and increased progenitor cell activity, which can lead to an increase in neurogenesis [56]. However, much of the underlying molecular restoration mechanisms are still unclear and this leads less effectively therapeutic strategies for TBI in human.

The central nervous system (CNS) is crucial to the most organisms. Once damaged, there are lethal effects if it cannot be healed or regenerated. Recently, advances in neuroscience research have led to the development of innovative therapeutic strategies that aim to regenerate damaged CNS. Tissue regeneration is one of the most interesting biological phenomena. However, the molecular and cellular mechanisms by which regeneration takes place are still unclear [57]. The ability to regenerate lost or damaged parts of the CNS is seen in varying degrees in many organisms. In the past, the adult mammalian CNS was viewed as a system without the capacity for regeneration once damaged [58]. In contrast to mammals with their limited ability to regenerate CNS as adults, *Danio rerio* (zebrafish) can constitutively produce new neurons along the rostrocaudal brain axis throughout its lifespan. Other organs also have an extensive regenerative capacity and respond to injuries such as trauma, lesion, and ischemic episodes by producing new tissues to replace the lost ones. Adult zebrafishes have the extraordinary ability to regenerate their damaged fins, skin, and heart [59,60]. Furthermore, they can regenerate several organs or tissues in the nervous system: spinal cord, photoreceptor, retina, cerebellum, and optic nerve [60–65]. This feature of the adult zebrafish brain relies on the presence of neural stem cell niches that enable stem cells to continuously proliferate, and on a permissive environment for neurogenesis in the brain [66].

Additionally, the zebrafish immune system is also remarkably similar to mammalian immune systems. Overall, the zebrafish genetic map demonstrates a highly conserved synteny, similar to the human genome [67]. To this end, the zebrafish has emerged as a powerful vertebrate model for elucidating the molecular and cellular mechanisms of regeneration [59], and numerous studies have already used this species to study the regeneration of the nervous system [66].

In the present study, the schematic diagram of TBI experiment on zebrafish by stab lesion is established and the gene expression levels are measured through time-course microarray experiments from injury to recovery stage. With the time-course microarray experiments, the molecular mechanisms of the cerebellar wound healing are studied using the dynamic network modeling [22]. The dynamic network model integrates the information from various PPI databases and the time-course microarray data in this section. The resultant PPI network serves as a basis to illustrate the molecular mechanisms of cerebellar wound healing. Meanwhile, the evolution of behavior of zebrafish from injury to recovery is observed under the confocal microscope, video-recorded, and quantitatively analyzed to measure the degree of disability caused by the injuries. This analysis would be helpful to focus on those genes involving the recovery process if a high correlation is shown between the gene expression profile and the zebrafish movement index (ZMI). According to the gene expression profiles and the ZMI, we focus on three groups of genes, i.e., the acutely activated, positively correlated, and negatively correlated to ZMI groups. To extract knowledge from the time-course expression data, several systems biology tools, e.g., STEM [68] and PANTHER [69], were adapted in this section besides the dynamic network modeling. Several significantly enriched pathways in the above three groups are identified, e.g., chemokine signaling pathway (inflammation-related), phosphatidylinositide 3-kinases (PI3K) signaling pathway (cell cycle-related), and axon guidance pathway. Then, using the dynamic cellular PPI network as a backbone, the crosstalks to coordinate with these pathways during the recovery process and the schematic diagram of wound healing-related cellular pathways are also presented in this section. Plausible nodes within the network of signaling pathways of the injured cerebellum were PI3K, PAK2, and PLXNA3 for coordination of neurogenesis and angiogenesis, which are essential for the restoration of injured brain. Moreover, we found these pathways may stimulate subnetworks important for neurogenesis and angiogenesis.

These findings not only confirm that these methods were successful, but also improve the confidence on the wet experiment design employed. Some interactions, such as crosstalks to coordinate among pathways for wound healing, were identified in this chapter, which will aid in deciphering these complex molecular restoration mechanisms in the future. This section provides an enhanced understanding of the molecular restoration mechanisms after a traumatic event to the brain, and open up new opportunity to devise treatment for TBI in human.

8.2.1 Materials and methods of constructing cerebella regeneration PPI network

Stab lesion assay and time-course microarray experiments in zebrafish TBI model

Zebrafish line (Tg(kdr:EGFP)) and wildtype were used to perform the stab lesion assay, microarray, and behavior video tracking experiments. The body length of all zebrafishes used in this section were 2.8–3.8 cm. Before the stab lesion assay, the zebrafishes were kept under a light cycle of 14 h light and 10 h dark at a temperature of 28°C. In the stab lesion assay, 6-month-old adult zebrafish were anesthetized by immersion in the aquarium water in which 200 ppm tricaine (MS222; Sigma, St. Louis, MO, USA) is dissolved for 5 min. A syringe (27G; Thermo Scientific, USA) was punched vertically through the cranial surface into the zebrafish cerebellum to a depth of 1.5 mm (Fig. 8.6). There are three large scales on the skull, which perfectly match with the positions of two optic tectum hemispheres and the cerebellum.

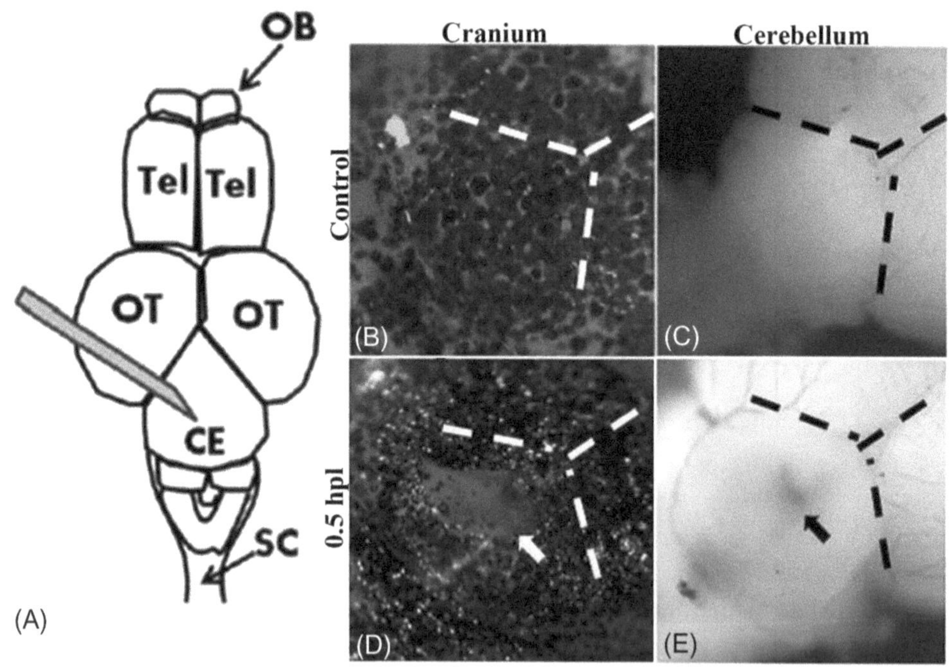

Figure 8.6 The diagram of the stab lesion assay. (A) Schematic diagram of the stab lesion. A 27G syringe was used to create a cranial wound of depth 1.5 mm in the cerebellum (*OB*, olfactory bulb; *Tel*, telencephalon; *OT*, optic tectum; *Ce*, cerebellum; *SC*, spinal cord). (B–E) Bright field images of the cranium (B and D) and exposed brains (C and E) of control fish and after the stab lesion. (B) and (C) show intact cranium and cerebellum of the control (no stab lesion) and (D) and (E) show the injured cranium and cerebellum of an experimental animal at 0.5-h postlesion (hpl). The fresh wound can be seen clearly on both the cranium and the cerebellum (D and E; the *white arrow* shows the lesion site on the cranium and the *black arrow* shows the wound on the exposed cerebellum).

The accumulating pigment around the three scales could also help us locate the lesion site of the wild-type zebrafish's brain. The injury depth is controlled by wrapping the needle with plastic tube and only left 1.5 mm needle tip for stabbing. Also the lesion depth with sagittal brain sections is examined (see Fig. B.1). The injured zebrafish were then put back into fresh aquarium water for recovery. Zebrafish that had apparently ceased bleeding were selected for the subsequent experiments. The control group is not anesthetized. Every fish in the experiment is only anesthetized with tricaine to cause the lesion except the control animals. The tricaine anesthetization is the standard procedure in zebrafish experiments. After lesion, the whole sample preparation process does not use any anesthetic drugs. Before they were sacrificed, we used ice-cold water to immobilize them and quickly dissected the cerebellum out. The dissected cerebellum was collected into a 1.5 mL microcentrifuge tube with 200 μl TRIzol (Invitrogen, Carlsbad, CA, USA) (three cerebellums per tube). After homogenization, we filled each tube with TRIzol up to 1 mL. After lesion, the whole sample preparation process does not use any anesthetic drugs. Therefore, the anesthetic process will not cause much influence on the gene expression profiles. The experimental procedures were approved by the committee for the use of laboratory animals at National Tsing-Hua University (IACUC number: 10140).

For the microarray experiments, zebrafish were euthanized by prolonged immersion in ice-cold water. Then, the zebrafish were dissected quickly to harvest the brains. Each dissection took approximately 4 min. A surgical knife was used to cut out the cerebellums. The separated cerebellums were kept in RNAlater (Ambion, Inc., Austin, TX, USA). We collected the cerebellums at the following time points: Control (no injury), 0.25, 1, 3, 6, 10, 15, 21, and 28 days postlesion (dpl) (the recovery process is also monitored using immunohistochemistry staining; see Fig. B.2). At each time point, the sample consisting of eight cerebellums from male zebrafishes. Microarray analysis of the samples was conducted by WELGENE Biotech Co., Ltd. The quality assurance and quality control data for the samples are shown in Table B.1. The regeneration of cerebellum and the time-course microarray data in the cerebellar wound healing process are shown in Fig. 8.7. The microarray dataset discussed and used to reconstruct the PPI network in this section can be retrieved as Dataset GSE56375 from the NCBI GEO repository website (http://www.ncbi.nlm.nih.gov/geo/).

ZMI experiments

For the ZMI experiments, zebrafish were placed in a glass tank (18 × 12 × 15 cm) filled with 2.5 L aquarium water. For each time point, three male adult zebrafish (with body length ranging from 2.8 to 3.8 cm) were transferred to the observation tank. Behavior was recorded with a video camera (HDR-XR200; Sony, Japan) which was placed above the tank and set to record trails of 5 min duration (Fig. 8.8A–C). The first video of zebrafish behavior was recorded after placing the zebrafish in

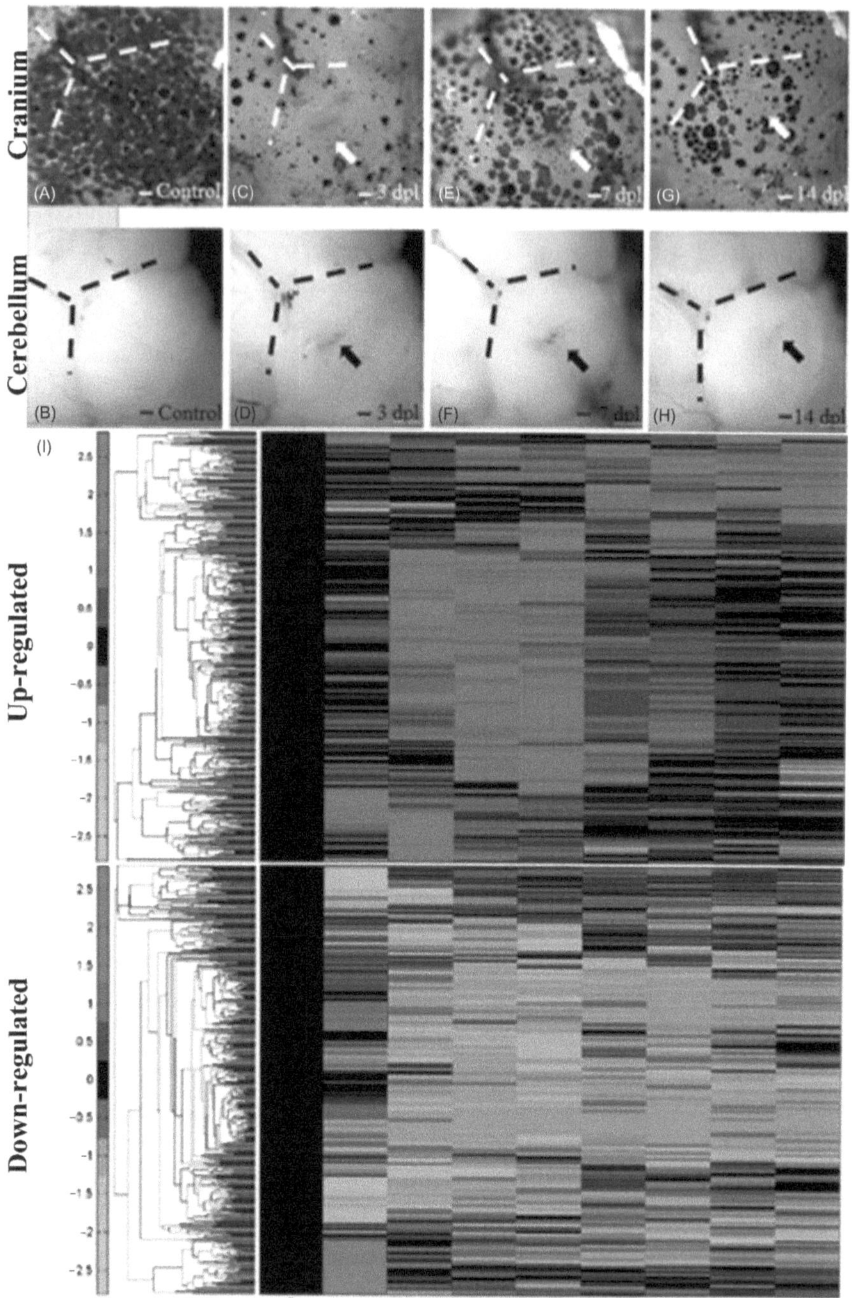

Figure 8.7 Regeneration of cerebellum at different day-post-lesion and the time-course microarray data in cerebellar wound healing process. (A–H) Bright field images of the craniums (A, C, E, and G) and cerebellums (B, D, F, and H) at different time points postlesion. The cranial and cerebellum were intact before stab lesion (A and B). At 3 dpl, the wound can be seen on the cranial and cerebellum. (C and D) At 7 dpl, the cranial is seal and scar is observed in the cerebellum. (E and F) At 14 dpl, the scar was hardly seen. (Scale bar, 100 μm.) (I) The differentially expressed genes (≥1.5- or ≤0.67-fold change) are hierarchically clustered (5839 genes). Columns represent the day-post-lesion (dpl), and the rows represent the genes. The blocks indicate temporally up-regulated and down-regulated genes, respectively. The color bar represents the log value of the ratio relative to the intensity at 0 dpl.

aquarium water for 25 min, and the second video of zebrafish behavior was recorded after a 10 min break. We have chosen three behavioral patterns as criteria for evaluating the movement of zebrafish, including total swimming distance (distance index, DI), single direction turning (turn direction index, TDI), and turning angle (turn angle index, TAI) [70,71]. The formulae of each index are in the following with the notations: C and I stand for control and injured fish, respectively.

$$ZMI = DI + TDI + TAI$$

where

$$DI = \frac{\text{Swimming distance}_C - \text{Swimming distance}_I}{\text{Swimming distance}_C + \text{Swimming distance}_I}$$

$$TDI = \frac{\left|\#\text{of right turn}_I - \#\text{of left turn}_I\right|}{\#\text{of right turn}_I + \#\text{of left turn}_I} - \frac{\left|\#\text{of right turn}_C - \#\text{of left turn}_C\right|}{\#\text{of right turn}_C + \#\text{of left turn}_C}$$

$$TAI = \frac{\left|(\#\text{of fish with turn angle} > 100)_C - (\#\text{of fish with turn angle} > 100)_I\right|}{(\#\text{of fish with turn angle} > 100)_C + (\#\text{of fish with turn angle} > 100)_I}$$

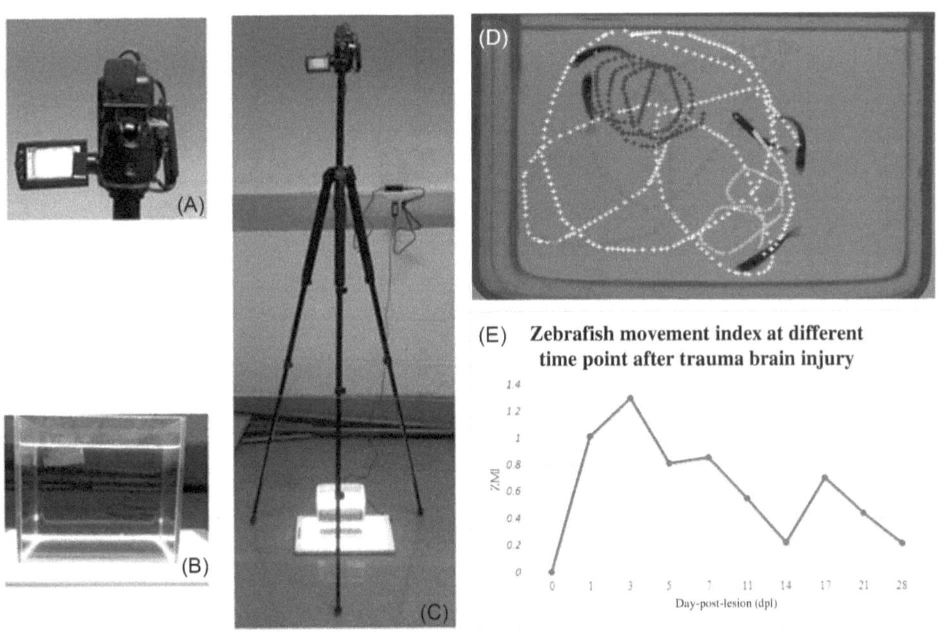

Figure 8.8 Equipment setup for behavior analysis. (A) Close-up views of the video camera. (B) Close-up views of the aquarium. (C) The camera and aquarium setup. (D) Swim sample paths of zebrafish after TBI. (E) Quantitative behavior index (zebrafish movement index, ZMI) of injured zebrafish describes the degree of disability in the behavior of the injured zebrafish. The greater the index value, the greater the degree of disability.

Note that the values of the three indices are between 0 and 1 and would be equal to 0 when the behaviors of injured fish perfectly match to control fish. We used the Multiple Zebrafish Tracking and Behavior Analysis software from Dr. YC Chen's Laboratory (Department of Electrical Engineering, NTHU) to analyze the swimming distance, turn direction, and turn angle for each zebrafish in the video clips. An example of a swim path diagram used for this behavior analysis is shown in Fig. 8.8D.

Big data mining for candidate healing PPI network

Several types of data are integrated to construct the cerebellar wound healing molecular network, including time-course expression profiles (Fig. 8.9), zebrafish and human PPIs (from BioGRID and Reactome) [13,20], zebrafish–human gene ortholog data (ZFIN) [72], and zebrafish gene annotations were retrieved from Gene Ontology (GO) [73]. The PPIs of human and zebrafish collected from the databases are served as a collection of candidates of potential PPIs in the cerebellar wound healing network. However, there is a paucity of information on zebrafish PPIs. Hence, we inferred part of zebrafish PPIs from human PPIs based on ortholog information. Then, we had PPI information (inferred from human + zebrafish) and time-course expression profiles. To set up the protein pool for differentially expressed proteins, these expression profiles were used to select differentially expressed proteins according to fold change. The differentially expressed proteins were identified if their fold changes were ≥ 1.5 or ≤ 0.67 comparing to control time point. These differentially expressed proteins were assumed to be most likely involved in the cerebellar wound healing process. Based on the protein pool and PPI information, a candidate PPI network for the cerebellar wound healing process was constructed.

Dynamic network modeling for cerebellar wound healing PPI network

However, since the candidate PPI network was based on a wide variety of biological experimental conditions and/or ortholog information, it may have contained many interactions irrelevant to the cerebellar wound healing process (i.e., false positive interactions). To develop a realistic PPI network for the cerebellar wound healing process, the candidate PPI network was needed to further prune and then obtain a refined PPI network (see Fig. 8.10 and Appendix B for detailed information) by system identification method and system order detection method using the real microarray data, and the dynamic network model in the following:

$$y_p[t+1] = y_p[t] + \sum_{q=1}^{N} b_{pq} y_p[t] y_q[t] + \alpha_p x_p[t] - \beta_p y_p[t] + \omega_p[t+1]$$

where $y_p[t]$ and $y_q[t]$ represent the protein activity level of the target protein p and the qth protein interacting with p at time t, respectively; b_{pq} denotes the ability of the qth interactive protein to interact with p; α_p denotes the translation effect from mRNA to p;

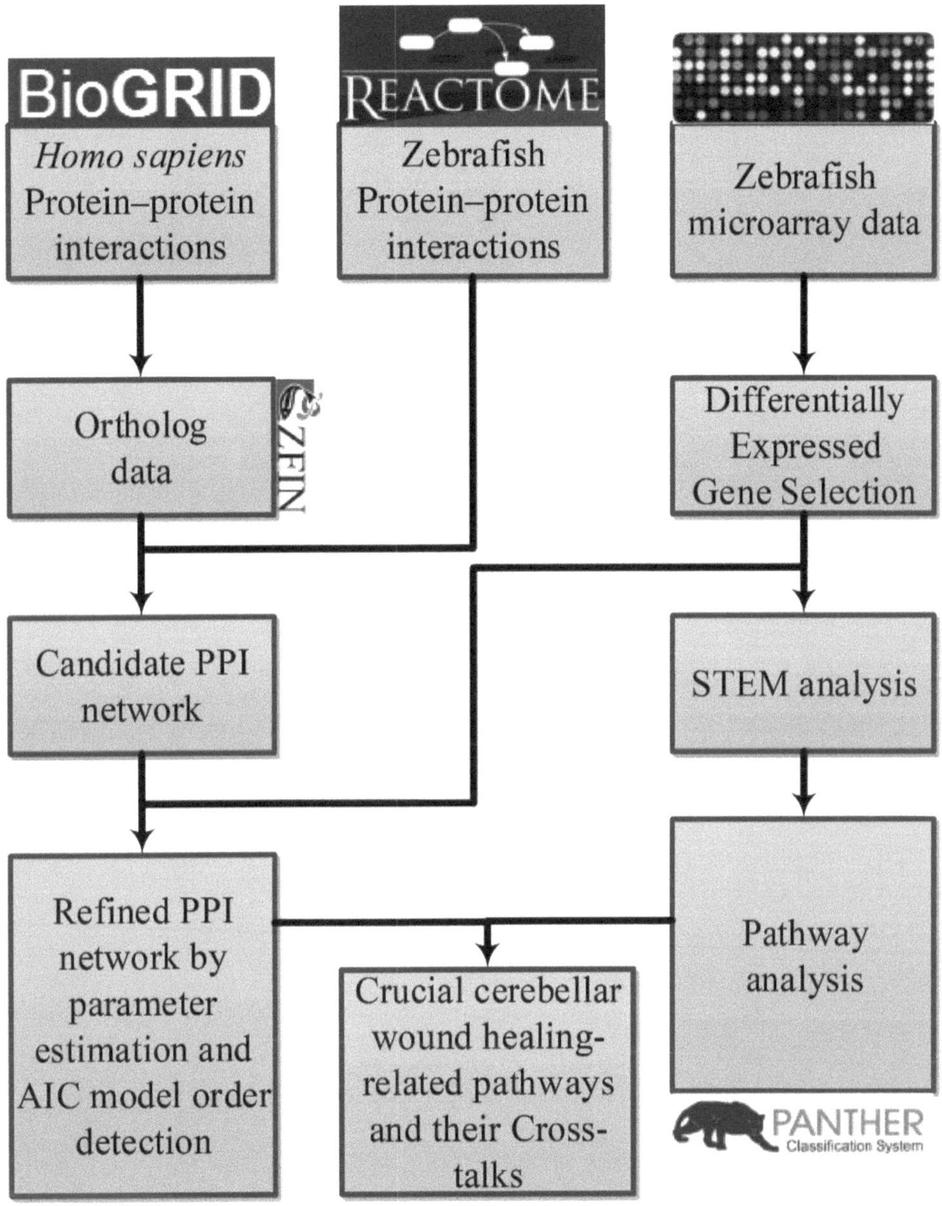

Figure 8.9 Flowchart for constructing the wound healing-related PPI network using a dynamic network model and big data mining. The network is constructed by PPI information from databases mining and systems identification according to the time-course microarray data (as shown in the boxes). Open access systems biology tools STEM and PANTHER are used to identify significantly temporal patterns and enriched cerebellar wound healing-related pathways.

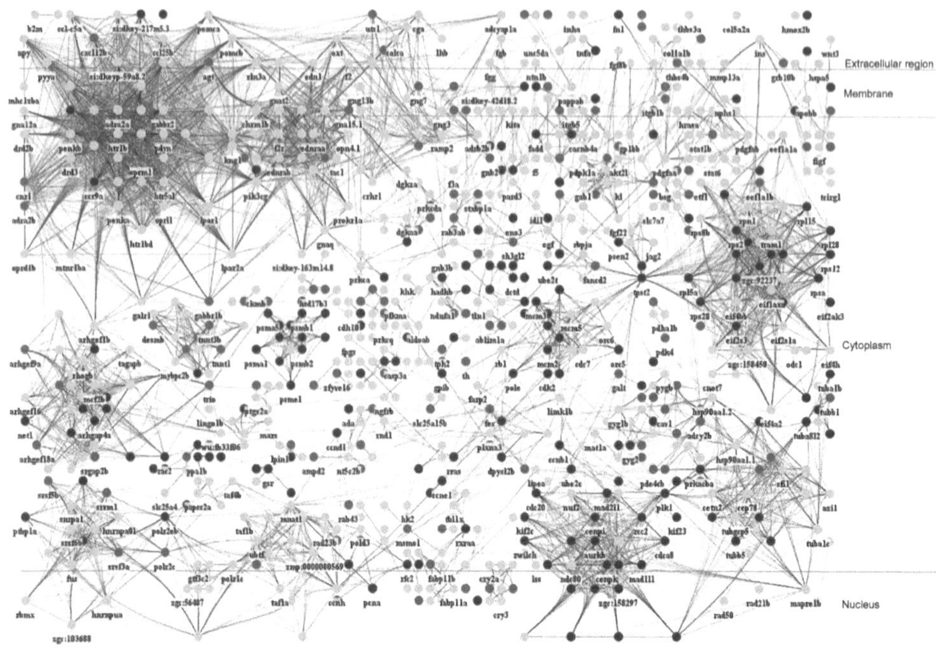

Figure 8.10 The constructed zebrafish cerebellar wound healing-related cellular PPI network by dynamic network modeling via microarray data and big data mining. The cerebellar wound healing-related PPI network of zebrafish contains 5270 PPIs among 802 proteins. The *red, green, and blue nodes* belong to group N, A, and P (see Fig. 8.11 and the context for details). The information to draw the network is summarized in Table B.5.

$x_p[t]$ represents the mRNA expression level of p; β_p indicates the degradation effect of p; and $\omega_p[t+1]$ is stochastic noise.

Rewriting the above dynamic equation with vector notation, we get:

$$y_p[t+1] = \begin{bmatrix} y_p[t]y_1[t] & \ldots & y_p[t]y_N[t] & x_p[t] & y_p[t] \end{bmatrix} \begin{bmatrix} b_{p1} \\ \vdots \\ b_{pN} \\ \alpha_p \\ 1-\beta_p \end{bmatrix} + \omega_p[t]$$
$$= \phi_p[t]\theta_p + \omega_p[t]$$

By collecting all the data from every time point, the interactions between the target protein y_p and its candidates $(y_1,\ldots,y_N)$ can be determined through the system identification and system order detection methods. The technique details are described in Appendix B. To this end, the final PPI network is completed.

Systems biology tools and statistics

After refining the candidate PPI network, two open access programs were used to analyze the time-course expression profile. For identifying the significant patterns of time-course expression profiles, the STEM tool is used to identify significant temporal patterns from time-course expression levels data. The default settings (the default enrichment in STEM is actual size based enrichment, in which the enrichment is computed using the hypergeometric distribution based on the number of genes in the set of interest) are used for the analysis and the significant patterns (with the Bonferoni corrected *p*-value < .05) in Fig. 8.11A are highlighted with color background and ordered by the corrected *p*-value.

For identifying significantly enriched functions in the specific pattern groups identified by STEM, we performed enrichment and ontology analysis through the website of Protein Analysis Through Evolutionary Relationships (PANTHER) [69]. This tool can map gene lists to GO molecular function and biological process categories, as well as PANTHER biological pathways. Furthermore, it can display the results in pathway diagrams to enable visualization of the relationships between genes in known pathways. The PANTHER tool is used to identify the significantly enriched pathways in a group of proteins or genes. The default settings (the binomial test is applied to determine whether there is a statistical overrepresentation or underrepresentation of the

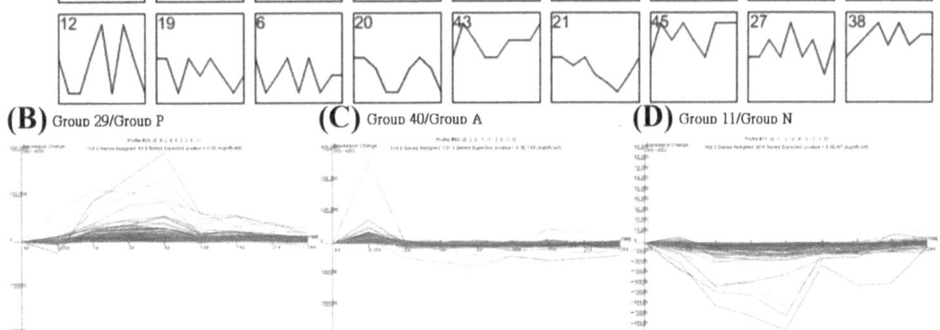

Figure 8.11 Significantly temporal patterns in the wound healing process. (A) Using STEM, significantly temporal patterns are identified (color background) (FDR-corrected *p*-value < .05). According to the STEM results and ZMI, we focused on group 29, 40, and 11, which are positively correlated with ZMI (B), acute response (C), and negatively correlated with ZMI (D).

genes and/or proteins in the test list relative to the reference list provided by PANTHER) are used for the analysis and the Bonferroni correction for multiple testing is used for correcting the original p-value. The significantly enriched pathways are with the corrected p-value $< .05$.

The sign test is applied to justify the strong correlation between the temporal expression profiles and ZMI at the 5% significance level, where the null hypothesis is the probability of having positive correlation is equal to 0 (i.e., the median of the Pearson correlation coefficients between the temporal expression profiles and ZMI is zero) and the alternative hypothesis is the probability of having positive correlation is not equal to 0 (i.e., the median of the Pearson correlation coefficients between the temporal expression profiles and ZMI is not equal to zero).

8.2.2 Results of wound healing mechanisms of TBI in zebrafish

The significantly temporal patterns and PPI network for the cerebellar wound healing process in the zebrafish TBI model

The first principal aim of this section was to construct an integrated cerebellar wound healing-related cellular PPI network. Various types of data are integrated with dynamic network model and microarray data under this framework (Fig. 8.9). There are 802 proteins (nodes) and 5270 interactions (edges) in the final PPI network (i.e., refined PPI network by AIC system order detection; Fig. 8.10). This PPI network serves as the backbone of this further analysis. Information from different systems biology tools can be integrated under this framework. First, the temporal information from the time-course microarray data in Fig. 8.7 are used to build up the PPI network through dynamic network modeling. The time-course microarray data is compared to two sets of published data using zebrafish as a model system to study the regeneration of neural systems (spinal cord and optic nerve) [64,74]. We found this data is highly consistent with the reported findings (the correlation coefficients are about 0.5−0.7) although the experimental conditions (the lesion schemes) and the studying objects (the injured parts of neural systems) are different (see Figs. B.3 and B.4). Thus, this can serve as an independent validation of these results. Next, we use STEM to find significantly temporal patterns in time-course microarray data. Then, several significantly temporal patterns are revealed (Fig. 8.11). The temporal pattern of group 29 (group P hereafter) showed a strong positive (P) correlation with the ZMI (comparing Fig. 8.8E with Fig. 8.11B, the Pearson correlation coefficient is 0.7427 ± 0.1315 with the p-value for sign test is less than .05). The temporal pattern of group 11 (group N hereafter) was the reverse, i.e., there was a highly negative (N) correlation with the ZMI (comparing Fig. 8.8E with Fig. 8.11D, the Pearson correlation coefficient is -0.7050 ± 0.1167 with the p-value for sign test is less than .05). The temporal pattern of group 40 was related to acute (A) responses during the wound healing process (Fig. 8.11C) as the variations of expression levels are aroused earlier than group P and N [75,76].

Second, qualitative functional information were integrated to investigating the molecular systems mechanisms of cerebellar wound healing process. For the genes in group A, some of the most common GO biological processes were related to the regulation of MAP kinase activity, regulation of the MAPK cascade, phosphorylation, and dephosphorylation. For the genes in group P, some of the most common GO biological processes were related to the regulation of the cell cycle, meiosis, DNA replication, and organelle organization. For the genes in group N, some of the most common GO biological processes were related to cation transport, potassium ion transport, ion transmembrane transport, and nervous system development. Further enrichment and ontology analysis to these biological processes of the wound healing will be given in the sequel.

Significant pathways in the wound healing mechanism

The presence of temporally coregulated proteins may imply the involvement of common biological pathways in response to TBI. Therefore, to evaluate the biological functions of temporally coregulated proteins, pathway analysis was conducted using PANTHER. Upon the results, the roles of these significantly enriched pathways of groups (A, N, and P) in the zebrafish TBI model are investigated as follows:

1. Significantly Enriched Pathways of Group A: the Acute Inflammation and Immune Response in the Wound Healing Process

   Examples of the pathways are given in Fig. B.5 which are also primarily involved in the acute inflammation and immune response to TBI (the detail is enlisted in Table B.2).

   a. Endogenous cannabinoid signaling: An endogenous cannabinoid (2-AG) has been implicated to affect the neuroprotective mechanism immediately after TBI [77]. The endo-cannabinoid system is also involved in pain relief, the blocking of working memory, immune responses, etc. The endo-cannabinoid system has therefore been suggested as a potential target for acute therapies for TBI [78].

   b. PI3K/PKB pathway: Phosphoinositide 3-kinase (PI3K) catalyzes the production of phosphatidylinositol-3,4,5-trisphosphate, which is involved in several crucial biological processes, including cell survival, regulation of gene expression and cell metabolism, and cytoskeletal rearrangements. A study has reported that PKB, which belongs to group A, modulates the macrophage inflammatory response to *Francisella* infection and confers a survival advantage in mice [79].

   c. Heterotrimeric G-protein signaling pathway: As signal transducers, G proteins communicate signals from many hormones, neurotransmitters, and other signaling factors [80] occurring ubiquitously in the cerebellar wound healing process.

2. Significantly Enriched Pathways of Group P: Positive Correlation With ZMI in the Wound Healing Process

The primarily involving pathways are given in Fig. B.6 and investigated in the following. (The detail is listed in Table B.3.) They either played important roles in cytoskeleton regulation, angiogenesis and inflammation, or reflected the cellular processes by which behavioral disability increased initially and then decreased.

a. Cell cycle and DNA replication: An ordered series of events led to the replication of cells and a series of integrated protein–protein and protein–DNA interactions and enzymatic reactions, to ensure the accuracy of DNA replication. That may implicate the occurrence of angiogenesis and neurogenesis in the damaged site.
b. Parkinson's disease pathway, Huntington's disease pathway, and Alzheimer's disease–presenilin pathway: Parkinson's, Huntington's, and Alzheimer's diseases are three of the most notorious neurodegenerative diseases. Patients with neurodegenerative diseases suffer from progressive loss of structure or function of neurons, including the death of neurons. Research is currently identifying many similarities at a subcellular level between these diseases. For example, mitochondrial dysfunction and oxidative stress play a causal role in neurodegenerative disease pathogenesis [81]. The appearance of these pathways reveals the relationship between the secondary damage of cerebellar injury and neurodegenerative diseases.
c. Inflammation mediated by chemokine and cytokine signaling pathway, T-cell activation, and B-cell activation: Upon binding to a family of G-protein coupled with seven-transmembrane receptors, chemokines control and manage trafficking and migration of immune cells. This pathway illustrates chemokine-induced adhesion and migration of leukocytes, resulting in infiltration to the tissue and transcriptional activation to enable the recruitment of more leukocytes for protection from infection in the cerebellar wound healing process [82,83]. Inhibiting specific chemokines and receptors could therefore prevent excessive recruitment of leukocytes to effectively control the scale of inflammation. T and B cells are the most abundant lymphocytes. T-cell activation refers to a process in which mature T cells, which express antigen-specific T-cell receptors on their surfaces, recognize their cognate antigens and respond by entering the cell cycle, secreting cytokines or lytic enzymes, and initiating the cell-based functions of the immune system. B-cell activation entails the process by which a pre-B cell matures into a plasma cell. The signaling pathways for this inflammation resolution in the wound healing process are complex.
d. Cytoskeletal regulation by Rho GTPase and Axon guidance: In cerebellum regeneration processes such as wound healing and angiogenesis, many cell types, including neural stem cells, leukocytes, lymphocytes, fibroblasts, neuronal cells, epithelial cells, and endothelial cells, need to change cell morphology for migration from stem cell niche and extravasation from the vascular in the

wound healing process. Changing in cell morphology is a multistep process involving changes in the cytoskeleton, cell–substrate adhesions, and extracellular matrix [84]. The "cytoskeletal regulation by Rho GTPase" pathway is such a change, regulated by Rho GTPase. The integrin signaling pathway is triggered when integrins in the cell membrane bind to extracellular matrix components [85]. Furthermore, axons are guided along specific tracts by attractive and repulsive cues during the establishment of the nervous system network in the cerebellar wound healing process.

3. Significantly Enriched Pathways of Group N: Negative Correlation With ZMI in the Wound Healing Process

   Below are the pathways that were significantly enriched by the genes in group N (Fig. B.7 and Table B.4). These neurotransmitter-related pathways appeared to be inhibited during the wound healing process. They may not be as crucial for regenerability as the above mentioned pathways, but they support the functional recovery of neural transmission in cerebellar wound healing process.

   Some examples of these neurotransmitter-related pathways are: Beta1 adrenergic receptor signaling pathway, Beta2 adrenergic receptor signaling pathway, synaptic vesicle trafficking, muscarinic acetylcholine receptor 2 and 4 signaling pathway, metabotropic glutamate receptor group II pathway, 5-HT1 type receptor mediated signaling pathway, metabotropic glutamate receptor group III pathway, enkephalin release pathway, GABA-B receptor II signaling pathway, Beta3 adrenergic receptor signaling pathway, DA receptor mediated signaling pathway, opioid prodynorphin pathway, opioid proenkephalin pathway, 5-HT4 type receptor mediated signaling pathway, muscarinic acetylcholine receptor 1 and 3 signaling pathway, ionotropic glutamate receptor pathway, and opioid proopiomelanocortin pathway. These pathways are all related to neurotransmitters and receptors in neural transmission and their roles in restoration and neurogenesis of the cerebellar wound healing process need further elaboration.

Crosstalks between the three groups of proteins of PPI network in the cerebellar wound healing mechanism

Moreover, the orchestrated interactions (crosstalks) among pathways are more important than themselves. The crosstalks among signal/regulatory pathways have been discussed to coordinate for efficient inflammatory responses for different stimuli in infectious process [9,10]. Therefore, at first, we examine the crosstalks between the three groups of proteins, extracted from this dynamic PPI network (in Fig. 8.10), in the cerebellar wound healing process.

In groups A and P, i.e., the proteins whose temporary expression profiles are positively correlated with the ZMI and the proteins in the acute phase, there are two major subnetworks (AP1 and AP2) emerging from this PPI network (see Fig. 8.12A; ZFIN

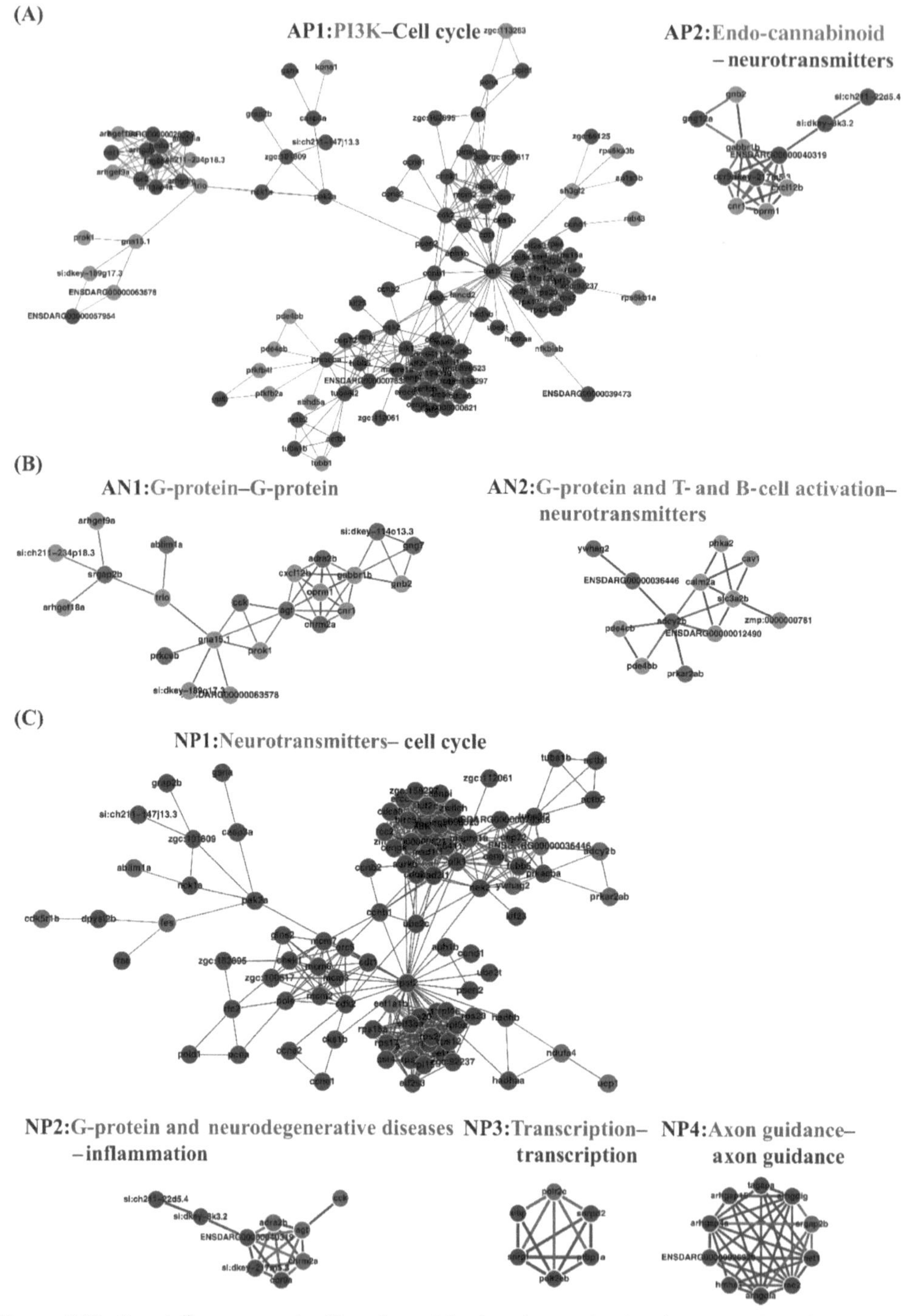

Figure 8.12 Crosstalks among significantly enriched pathways in the three significantly temporal groups of proteins. (A) Crosstalks between groups A and P: Crosstalks between PI3K pathways (enriched in group A) and the cell cycle (enriched in group P) dominated subnetwork AP1. *(Continued)*

symbols for the edges between two groups are given in Table B.5). In the subnetwork AP1, the crosstalks between the PI3K pathway enriched in group A and cell cycle pathway and immune-related pathway enriched in group P are observed. This observation supports our understanding that PI3K pathways are involved in several crucial recovery processes, including cell proliferation and immune response in cerebellar wound healing process. In the subnetwork AP2, the cannabinoid pathway crosstalks with neurotransmitter-related pathways. The cannabinoid pathway has shown its importance in neurogenesis [77,86]. This may imply that neurotransmitter-related pathways should play a role in neurogenesis, which may be regulated by cannabinoid pathway and has not been fully explored [87,88].

In groups A and N, there were also two major subnetworks (AN1 and AN2) (Fig. 8.12B; ZFIN symbols for the edges between two groups are given in Table B.5). The subnetwork, AN1, contains interactions among cellular communication pathways because G-protein pathways are enriched in both groups. In the subnetwork AN2, G-protein pathways and T- and B-cell activation pathways enriched in group A have links to neurotransmitter-related pathways enriched in group N. This crosstalk between immune system and neurotransmitters is an interesting area in cerebellar wound healing process [89–91].

Third, there are four major subnetworks in groups N and P (NP1 to NP4) (Fig. 8.12C; ZFIN symbols for the edges between two groups are given in Table B.5). In the subnetwork NP1, neurotransmitter-related pathways have links to the cell cycle pathway, which constitutes another interesting crosstalks [87,88]. In the subnetwork NP2, inflammation-related pathways of group N have crosstalks with the G-protein and Alzheimer's pathway. Neurodegenerative diseases are often associated with prolonged brain injury. However, the molecular mechanisms by which the prolonged brain injury could induce neurodegenerative diseases are still unclear. Based on this

---

◀ Crosstalks between endo-cannabinoid pathways (enriched in group A) and neurotransmitter-related pathways (enriched in group P) dominated subnetwork AP2. (B) Crosstalks between groups A and N: Crosstalks between G-protein pathways (enriched in proteins from both groups A and N) dominated subnetwork AN1. The crosstalks between G-proteins and T- and B-cell activation pathways (enriched in group A) and neurotransmitter-related pathways (enriched in group N) dominated the subnetwork AN2. (C) Crosstalks between groups N and P: Crosstalks between neurotransmitter-related pathways (enriched in group N) and cell cycle pathways (enriched in group P) dominated subnetwork NP1. Crosstalks between G-protein and neurodegenerative diseases pathways (enriched in group N) and inflammation pathways (enriched in group P) dominated subnetwork NP2. Crosstalks between transcription pathways (enriched in both groups N and P) dominated subnetwork NP3. Crosstalks between axon guidance pathways (enriched in both groups N and P) dominated subnetwork NP4. Red lines (dark gray lines in print versions) indicate the crosstalks between groups and are listed in Table B.5.

network, it suggests that inflammation may be involved in the origination of neurodegenerative diseases. The subnetwork NP3 contains many gene regulatory interactions, e.g., transcriptions, which are enriched in both groups. By examining adjacent nodes of NP3, we find that the subnetwork NP3 links to the cell cycle pathway through nodes that were not included in the three groups. NP3 may play a role in regulating and communicating other genes that were not belong to the three groups. The subnetwork NP4 is dominated by axon guidance pathway, which can be seen as an indication of neurogenesis or functional recovery in cerebellar wound healing process. By examining adjacent nodes of NP4, axon guidance pathway indirectly links to the G-protein pathways through nodes not including in the three groups. In summary, we develop a schematic diagram of crosstalks among the pathways enriched in the three group that may explain the underlying mechanism of the cerebellar wound healing process (Fig. 8.13A).

Generally, the crosstalks exist not only in these three groups but also in the whole PPI network. Next, by examining the crosstalks of the following three pathways (Fig. 8.13B) significantly enriched in the whole network, we find connection among inflammation, neurogenesis, and angiogenesis, which are essential for the restoration from the injured brain. In the chemokine signaling pathway, a pro-inflammatory pathway, chemokine (C−X−C motif) receptor 3 and 4 (CXCR3 and CXCR4) and chemokine (C−C motif) receptor 7 (CCR7) are found in this network and viewed as a start point to activate a series of inflammatory events (e.g., activations of STAT and NF-κB signaling pathways). The PI3K is in the middle of the pathway and also involved in another pathway, PI3K signaling pathway. In the PI3K signaling pathway, several receptor tyrosine kinases (FGFR, NGFR, and PDGFR), insulin receptor substrate 1 (IRS1), cholinergic receptor muscarinic 2 (CHRM2), and growth factor receptor-bound protein 2 (GRB2) are all in this network. The information will be transduced to AKT through PI3K, and then promote cell cycle progression through cyclin D1 (CCND1), which is a sign for new cell generation. More interestingly, the PI3K is also functional in axon guidance pathway through regulating cytoskeleton. In the axon guidance pathway, in particular, the semaphorins-related ones, integrin beta 1 (ITGB1) and plexin A3 (PLXNA3) in this network are receptors for semaphorins and activate downstream regulators: ras-related C3 botulinum toxin substrate 1 (RAC1), dihydropyrimidinase-like 2 (DPYSL2 or CRMP2), and p21 protein (Cdc42/Rac)-activated kinase 2 (PAK2). PAK2 is involved in the vascular stabilization, which may be an important function after angiogenesis. DPYSL2 controls branching guidance and number of neurites. PLXNA3 is related to the morphogenesis of motor neurons and thus could have contributions to the recovery of zebrafish movement. Through this pathway validation, it confirms this network and reveals the relationships among inflammation, neurogenesis, and angiogenesis in cerebellar wound healing process from the aspect of pathways crosstalks.

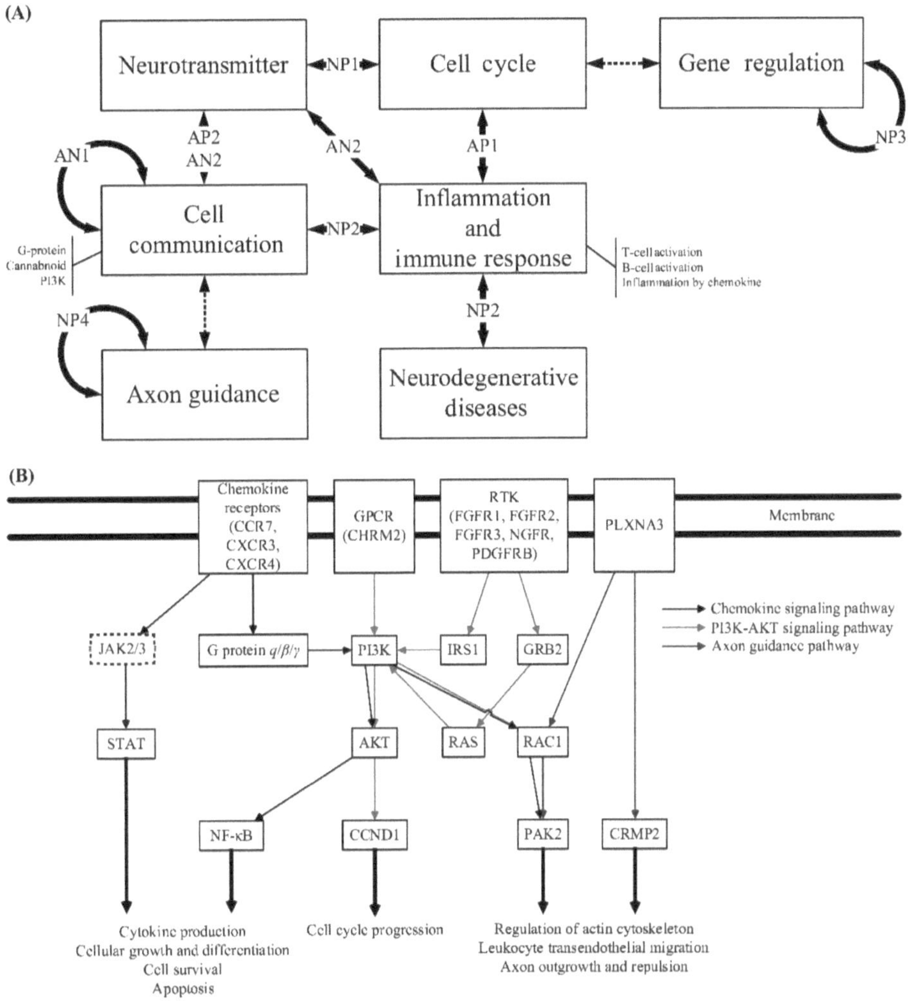

Figure 8.13 Schematic diagram of wound healing-related cellular pathways and their crosstalks among inflammation, neurogenesis, and angiogenesis. (A) Pathways shown in blocks were enriched in the subnetworks (Fig. 8.12). Their crosstalks were observed in the cerebellar wound healing PPI network of zebrafish (*solid arrows* indicate crosstalks involving proteins in groups A, N, and P; *dashed arrows* indicate crosstalks involving proteins not in these groups). To establish cell communication and defense mechanisms (inflammation and immune response) after brain injury, cell communication was related to neurotransmitter-related pathways (through AP2 and AN2), the axon guidance pathway (through NP4 indirectly), and itself (through AN1). Defense mechanisms were related to the cell cycle pathway (through AP1), neurotransmitter-related pathways (through AN2), the cell communication pathway (through NP2), and neurodegenerative diseases (through NP2). Cell cycle pathways also interacted with gene regulation pathways through NP3 indirectly. (B) The crosstalks are found among pathways of inflammation, neurogenesis, and angiogenesis in cerebellar wound healing process. Inflammation is mediated by chemokine signaling pathway. New neural and vascular cells generation is mediated by PI3K-AKT signaling pathway and axon guidance pathway. PI3K in chemokine signaling pathway also regulates the PI3K-AKT signaling pathway and axon guidance pathway. PAK2 in axon guidance pathway also maintains the stability of vascular.

8.2.3 Discussion of cerebella regeneration mechanism in zebrafish

TBI is one of the leading causes of disability in the United States [55]. However, the complex and intricate nature of the brain makes it very difficult to address this problem. To date, there are no FDA-approved pharmacological therapies [92]. Nevertheless, several recent advances have been made, through mass spectrometry, bioinformatics tools, and systems biology, in discovering biomarkers which could serve as potential therapy targets [93].

In this section, we used zebrafish, that can recover from cerebellum injuries, to investigate potential signaling pathways and their crosstalks that may be crucial for TBI recovery and can also constitute useful biomarkers. Based on a set of time-course microarray experiments and the integration of omic data extracted from databases, a zebrafish PPI network of cerebellum wound healing processes was built. By analyzing the temporal expression patterns, three major protein groups (the acute, positively, and negative related groups, i.e., Groups A, P, and N) are identified. Within each group, several significantly enriched pathways are found, and their relationships with the cerebellar wound healing process are confirmed through literature survey. The crosstalks of these significantly enriched pathways are revealed by using the constructed zebrafish PPI network of cerebellum wound healing process (see Fig. 8.13A).

After injury of the cerebellum, cellular communication (i.e., G-protein and endo-cannabinoid pathways) and defense mechanisms (inflammation and immune responses) are provoked to coordinate the subsequent recovery processes. G-protein pathways and endo-cannabinoid pathways interact with neurotransmitter-related pathways (through subnetworks AP2 and AN2), after which those neurotransmitter-related pathways cause the activation of cell cycle pathways (through subnetwork NP1), which is crucial in neurogenesis. Meanwhile, the G-protein pathways also indirectly initiate axon guidance which is maintained by crosstalks between groups N and P (subnetwork NP4). The defense mechanisms, i.e., inflammation and immune responses, interact with cell cycle pathways, and neurotransmitter-related pathways for restoration through subnetworks AP1 and AN2, respectively. Inflammation is interpreted as a necessary and sufficient condition for neurogenesis [94]. Hence, there should be crosstalks between the inflammation and cell cycle pathways for coordination in cerebellar wound healing process. Moreover, inflammation also seems to be connected with neurodegenerative diseases, e.g., Alzheimer's, Parkinson's and Huntington's diseases (through subnetwork NP2), which are common in brain injury patients [95]. The crosstalks between immune responses and neurotransmitter-related pathways, which have also been reported in the literature, may provide an opportunity for TBI therapy. The following pathway validation reconfirms this PPI network and reveals the crosstalks among inflammation neurogenesis, and angiogenesis, which are crucial in the TBI recovery. The PI3K (Fig. 8.13B) adapts the information from chemokines, growth factors, and G-protein coupled receptors and further transduces the information to NF-κB, cyclins, and PAK2. It seemly plays a central role to regulate cytokine production, cell cycle progression, and axon outgrowth. Besides PI3K, we also found some plausible targets for further experimental designs, e.g., the steroid biosynthesis, the hormone biosynthesis, and the notch signaling pathway, which are all enriched in this network.

The systems biology methodology can systematically analyze high-throughput data, and has also been suggested as a promising way to discover biomarkers for TBI [96]. This refined PPI network reveals the crosstalks between several pathways, including some well-known crosstalks to coordinate between the immune response and cell communication processes for regeneration. These results also indicates crosstalks between the immune response process and neurodegenerative diseases, which have been demonstrated by Feala et al. [96]. Neurotransmitters may also be involved in the TBI recovery process through their effect on cell cycle regulations [88], immune responses [97], and cell communication [98]. The roles of the neurotransmitter-related pathways in the TBI process have been less addressed. It is therefore interesting to study the roles of neurotransmitters in neuron regeneration. Finally, the crosstalks and plausible nodes identified in this section will be helpful in deciphering the molecular interactions between these pathways or cells in the wound healing process, and make human CNS regeneration possible in the future.

APPENDIX B: METHODS, TABLES, AND FIGURES

Method B.1 Dynamic model of the wound healing-related cellular PPI network.

In the above candidate PPI network, the candidate interactions could be depicted as a dynamic system, in which the interactive proteins and mRNA are the inputs and the protein activities are the outputs. In particular, given a target protein p that interacts with N proteins in the candidate PPI network, the dynamic model of the protein p is described as follows:

$$y_p[t+1] = y_p[t] + \sum_{q=1}^{N} b_{pq} y_p[t] y_q[t] + \alpha_p x_p[t] - \beta_p y_p[t] + \omega_p[t+1] \qquad \text{(B.1)}$$

where $y_p[t]$ and $y_q[t]$ represent the protein activity level of the target protein p and the qth protein interacting with p at time t, respectively; b_{pq} denotes the ability of the qth interactive protein to interact with p; α_p denotes the translation effect from mRNA to p; $x_p[t]$ represents the mRNA expression level of p; β_p indicates the degradation effect of p; and $\omega_p[t+1]$ is stochastic noise. The rate of PPI is proportional to the product of the concentrations of the two proteins involved [99], i.e., it is proportional to the probability of molecular collisions between two proteins. The interaction is therefore modeled as a nonlinear multiplication scheme. For example, in the PPI network, the phosphorylation of $y_p[t]$ by kinase $y_q[t]$ is proportional to the product of the concentration of kinase $y_q[t]$ and its substrate $y_p[t]$ [99]. The biological meaning of Eq. (B.1) is that the activity level of p at time $t+1$ is attributed to its activity level at time t, plus the effects of interactions with N interactive proteins, plus the translation product from the mRNA, minus the degradation effect, plus any stochastic noise. Because of the undirected nature of protein interactions, there is no direction between interacting proteins in the candidate PPI network. The interaction parameter b_{pq}, translation parameter α_p and decay rate β_p were estimated from the microarray data, as described below.

Method B.2 Interaction parameter identification using the time series microarray data.

The interaction parameters for the dynamic network model were identified using the microarray data collected as described above. The interaction parameters were identified using a least squares parameter estimation formula [100]. Eq. (B.1) can be rewritten as follows:

$$y_p[t+1] = \begin{bmatrix} y_p[t]y_1[t] & \cdots & y_p[t]y_N[t] & x_p[t] & y_p[t] \end{bmatrix} \begin{bmatrix} b_{p1} \\ \vdots \\ b_{pN} \\ \alpha_p \\ 1-\beta_p \end{bmatrix} + \omega_p[t] \tag{B.2}$$

$$= \phi_p[t]\theta_p + \omega_p[t]$$

where $\phi_p[t]$ indicates the regression vector, which can be computed directly from microarray data points, and θ_p is the parameter vector to be estimated. The dataset used for parameter estimation was small, so to avoid overfitting, the cubic spline method [23,24] was used to interpolate extra time points for the gene expression data. For simplicity, at different time points $t = 1, \ldots, L$, Eq. (B.2) can be presented as follows:

$$Y_p = \Phi_p\theta_p + \Omega_p \tag{B.3}$$

where $Y_p = [y_p[2]\ y_p[1] \ldots y_p[L]]^T$, $\Phi_p = [\phi_p[1]\ \phi_p[2] \ldots \phi_p[L-1]]$, and L is the number of data points of microarray data after cubic spline interpolation. The least squares parameter estimation routine can then be formulated as follows:

$$\min_{\theta_p} \left\| Y_p - \Phi_p\theta_p \right\|_2^2 \tag{B.4}$$

Because there was no large-scale measurement of protein activities available, mRNA expression profiles in Fig. 8.7 were used to substitute for protein activity levels when identifying the interaction parameters. Although mRNA expression levels do not reflect protein expression levels exactly, there are partially positive correlations between them [25,26]. The least squares minimization formula (B.4) can be solved using the active set method for quadratic programming in Eq. (2.20) or (2.21) [100]. The PPI parameter b_{pq} was estimated for each protein in the candidate PPI network using the time series microarray data, i.e., through the estimated interaction abilities. However, since the candidate PPI network was constructed using data from a variety of biological experiments under a range of conditions, or inferred from orthology data, it contained many PPIs irrelevant to the wound healing process. The estimated interaction parameters $\hat{b}_{pq}$ were therefore pruned using the model order (true interaction number) detection method, as described below.

Method B.3 Determination of significant interaction pairs.

When the PPI parameters $\hat{b}_{pq}$ were identified, AIC in Eq. (2.23) [101,102] was used to select model order and also identify significant interactions in the cerebellar wound healing-related PPI network, i.e., to determine the number N for target protein p in Eq. (B.1). AIC decreases as the residual error decreases and increases with the number of interactions (model complexity). As the expected residual error decreases with increasing interaction numbers for non-adequate model complexities, AIC should reach a minimum around the correct PPI number [101,102]. Therefore, AIC can be used to detect model order (number of PPIs) based on the protein interaction abilities $\hat{b}_{pq}$ identified above. In this way we used AIC model order detection to prune the candidate PPI network one protein by on protein to obtain the refined PPI network using the time series microarray data and achieve a realistic PPI network.

Staining

Immunohistochemistry staining was performed following the standard procedure. Zebrafish brains were harvested and fixed with 4% paraformaldehyde (Merck) overnight at 4°C. We then washed the zebrafish brains four times for 15 min in PBST. The brains were permeabilized in ice-cold acetone for 90 min at 4°C. They were then washed four times for 15 min with maleic acid buffer. Blocking buffer (2% BSA and 2% goat serum in maleic acid buffer) was used to block the brains for 4 h at room temperature. Primary antibodies, rabbit anti-eGFP (1:600; Novus Biologicals, Littleton, CO, USA) and mouse anti-PCNA (sc-56, 1:50; Santa Cruz Biotechnology, Santa Cruz, CA, USA), were added to the blocking buffer incubate and stored overnight at 4°C. The samples were then washed four times for 30 min with maleic acid buffer. Secondary antibodies, goat antirabbit 488 (1:200; KPL, Gaithersburg, MD, USA) and goat antimouse 549 (1:200, KPL, Gaithersburg, MD, USA), were added to the blocking buffer to be incubated with the brains for 2 h. After that, brains were washed four times with maleic acid buffer for 30 min each time. Finally, brains were stored in FocusClear (CelExplorer Labs, Hsinchu, Taiwan) at 4°C. Fluorescence images were taken using A1R confocal microscope (Nikon, Japan).

Here we show Tables B.1–B.5 and Figs. B.1–B.7.

Table B.1 RNA QA/QC information

Sample name	RNA concentration (ng/μL)	RNA quantity (ng)	OD260/280 ratio	Bioanalyzer chip lane location	28S/18S ration	RIN
6 hpl	227.12	2044.08	2.12	1	1.8	9.4
1 dpl	279.09	2511.81	2.11	2	1.3	N/A
3 dpl	429.08	3861.72	2.12	3	1.0	9.4
6 dpl	308.67	2778.03	2.13	4	1.3	9.7
10 dpl	372.76	3354.84	2.14	5	1.4	9.6
15 dpl	510.94	4598.46	2.13	6	0.9	9.1
21 dpl	354.08	3186.72	2.1	7	1.5	8.6
28 dpl	468.07	4212.63	2.13	8	1.5	9.3
Control	576.71	5190.39	2.17	9	1.4	9.0

Table B.2 The significantly enriched pathways in group A

Pathways	*p*-Value
Endogenous cannabinoid signaling	7.17E-06
PI3 kinase pathway	1.87E-05
Androgen/estrogen/progesterone biosynthesis	1.90E-05
GABA-B receptor II signaling	3.47E-05
Heterotrimeric G-protein signaling pathway-Gq alpha and Go alpha mediated pathway	4.50E-05
Thyrotropin-releasing hormone receptor signaling pathway	2.88E-04
Insulin/IGF pathway-mitogen activated protein kinase kinase/MAP kinase cascade	6.87E-04
Metabotropic glutamate receptor group II pathway	1.26E-03
TCA cycle	1.69E-03
Circadian clock system	3.47E-03

Table B.3 The significantly enriched pathways in group P

Pathways	*p*-Value
Cell cycle	4.94E-08
Parkinson's disease	3.20E-05
Integrin signaling pathway	3.43E-05
De novo pyrimidine deoxyribonucleotide biosynthesis	2.68E-04
Inflammation mediated by chemokine and cytokine signaling pathway	4.15E-04
Axon guidance mediated by semaphorins	6.53E-04
Cytoskeletal regulation by Rho GTPase	8.56E-04
T-cell activation	8.63E-04
p53 pathway	2.02E-03
B-cell activation	3.29E-03
Alzheimer's disease-presenilin pathway	5.94E-03
Huntington's disease	5.98E-03

Table B.4 The significantly enriched pathways in group N

Pathways	*p*-Value
Muscarinic acetylcholine receptor 2 and 4 signaling pathway	3.56E-10
Heterotrimeric G-protein signaling pathway-Gi alpha and Gs alpha mediated pathway	2.84E-08
GABA-B receptor II signaling	9.11E-08
Dopamine receptor mediated signaling pathway	1.39E-07
5HT1 type receptor mediated signaling pathway	2.63E-07
Beta1 adrenergic receptor signaling pathway	3.49E-07
Beta2 adrenergic receptor signaling pathway	3.74E-07
Histamine H2 receptor mediated signaling pathway	1.41E-06
Endothelin signaling pathway	4.66E-06
Enkephalin release	5.30E-06
Metabotropic glutamate receptor group II pathway	8.41E-06
Muscarinic acetylcholine receptor 1 and 3 signaling pathway	2.46E-05
Beta3 adrenergic receptor signaling pathway	5.54E-05
Alpha adrenergic receptor signaling pathway	6.49E-05
Adrenaline and noradrenaline biosynthesis	7.01E-05
Metabotropic glutamate receptor group III pathway	7.30E-05
Opioid prodynorphin pathway	1.07E-04
5-HT4 type receptor mediated signaling pathway	1.07E-04
5-Hydroxytryptamine biosynthesis	1.24E-04
Opioid proopiomelanocortin pathway	1.29E-04
Opioid proenkephalin pathway	1.29E-04
Synaptic vesicle trafficking	1.54E-04
Transcription regulation by bZIP transcription factor	4.92E-04
Heterotrimeric G-protein signaling pathway-Gq alpha and Go alpha mediated pathway	5.89E-04
Nicotine pharmacodynamics pathway	7.76E-04
Alzheimer's disease-amyloid secretase pathway	8.28E-04
Oxytocin receptor mediated signaling pathway	8.28E-04
Thyrotropin-releasing hormone receptor signaling pathway	9.08E-04
Gonadotropin releasing hormone receptor pathway	1.02E-03
5HT2 type receptor mediated signaling pathway	1.15E-03
Parkinson's disease	2.24E-03
Hedgehog signaling pathway	2.83E-03
Apoptosis signaling pathway	2.87E-03
Histamine synthesis	3.99E-03
Inflammation mediated by chemokine and cytokine signaling pathway	4.11E-03
Corticotropin releasing factor receptor signaling pathway	4.41E-03

Table B.5 ZFIN symbols of nodes in the subnetworks

Subnetwork	Group A	Group N	Group A	Group N
AN1	cxcl12b	chrm2a	trio	srgap2b
	prok1	agt	gabbr1b	adra2b
	gabbr1b	chrm2a	gabbr1b	si:dkey-114o13.3
	trio	ablim1a	oprm1	chrm2a
	gabbr1b	gng7	arhgef18a	srgap2b
	cnr1	chrm2a	gna15.1	prkceb
	gna15.1	cck	gnb2	gng7
	si:ch211-234p18.3	srgap2b	oprm1	adra2b
	cxcl12b	adra2b	cnr1	adra2b
	gnb2	si:dkey-114o13.3	gna15.1	agt
	cnr1	agt	prok1	cck
	cxcl12b	agt	gabbr1b	agt
	arhgef9a	srgap2b	oprm1	agt
AN2	pde4bb	adcy2b	calm2a	adcy2b
	pde4cb	adcy2b	slc3a2b	adcy2b
Subnetwork	**Group A**	**Group P**	**Group A**	**Group P**
AP1	trio	rac2	si:ch211-234p18.3	arhgap4a
	sh3gl2	tpst2	arhgef9a	tagapa
	trio	hmha1	sh3gl2	zgc:66125
	trio	arhgap15	tubb1	actb1
	pfkfb4l	prkacba	si:ch211-234p18.3	arhgap15
	pde4bb	prkacba	nfkbiab	tpst2
	fancd2	tpst2	tubb1	tuba8l2
	sh3gl2	ap1s3b	pde4cb	prkacba
	trio	nck1a	trio	pak2a
	arhgef9a	arhgap4a	rab43	ccnd1
	zgc:113263	pcna	trio	tagapa
	tubb1	tuba1b	pfkfb2a	prkacba
	si:ch211-234p18.3	hmha1	rps6kb1a	zgc:92237
	arhgef9a	hmha1	pfkfb2a	gpib
	si:ch211-234p18.3	tagapa	si:ch211-234p18.3	rac2
	arhgef9a	arhgap15	zgc:113263	pold1
	tubb1	actb2	arhgef18a	arhgap15
	arhgef18a	hmha1	arhgef9a	rac2
	arhgef18a	rac2	arhgef18a	arhgap4a
	pfkfb4l	gpib	fancd2	ube2t
	arhgef18a	tagapa	kpna1	casp3a
	trio	arhgap4a		
AP2	cnr1	ccr9a	cxcl12b	si:dkey-217m5.3
	gnb2	gng12a	cnr1	si:dkey-217m5.3
	gabbr1b	si:dkey-217m5.3	gabbr1b	gng12a
	oprm1	si:dkey-217m5.3	oprm1	ccr9a
	gabbr1b	ccr9a	cxcl12b	ccr9a

(*Continued*)

Table B.5 (Continued)

Subnetwork	Group N	Group P	Group N	Group P
NP1	fes	rras	eef1a1b	zgc:92237
	prkar2ab	prkacba	eef1a1b	rpl5a
	adcy2b	prkacba	eef1a1b	rps23
	fes	dpysl2b	eef1a1b	rps17
	cdk5r1b	dpysl2b	eef1a1b	rps28
	ndufa4	hadhaa	eef1a1b	rps20
	ndufa4	hadhb	eef1a1b	rps12
	eef1a1b	eef1g	eef1a1b	rpl4
	ywhag2	cep72	eef1a1b	rps2
	ablim1a	nck1a	ywhag2	prkacba
	fes	pak2a	ywhag2	nek2
	eef1a1b	tpst2	ywhag2	cenpj
	eef1a1b	rpl28	ywhag2	tuba8l2
	eef1a1b	rpl15	ywhag2	tubb5
	eef1a1b	rps15a	ywhag2	mapre1a
	eef1a1b	rpsa	ywhag2	plk1
NP2	adra2b	ccr9a	chrm2a	si:dkey-217m5.3
	agt	ccr9a	adra2b	si:dkey-217m5.3
	agt	si:dkey-217m5.3	chrm2a	ccr9a
NP3	polr2c	polr2eb	polr2c	slbp
	polr2c	snrpf	polr2c	snrpd2
	polr2c	ptbp1a		
NP4	srgap2b	arhgdia	srgap2b	rac2
	srgap2b	arhgap4a	srgap2b	net1
	srgap2b	arhgdig		

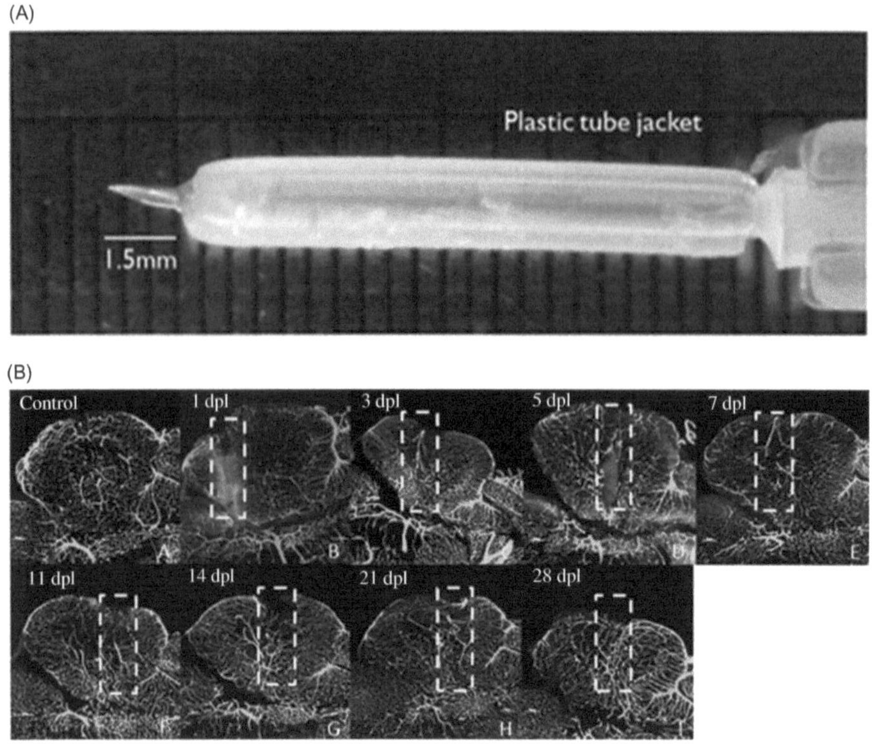

Figure B.1 The wrapped needle and the sagittal brain sections after injury. We controlled the injury depth by wrapping the needle with plastic tube and only left 1.5 mm needle tip for stabbing. Also the lesion depth with sagittal brain sections is examined. The *white-dashed line* surrounds the intact cerebellum, and the *yellow-dashed rectangles* (gray in print versions) in the other pictures point out the lesion depth. Lesion depth caused by modified needle only penetrates the cerebellum region, and does not wound other brain parts. (A–I) represent control, 1-, 3-, 5-, 7-, 11-, 14-, 21-, and 28-dpl cerebellum, respectively.

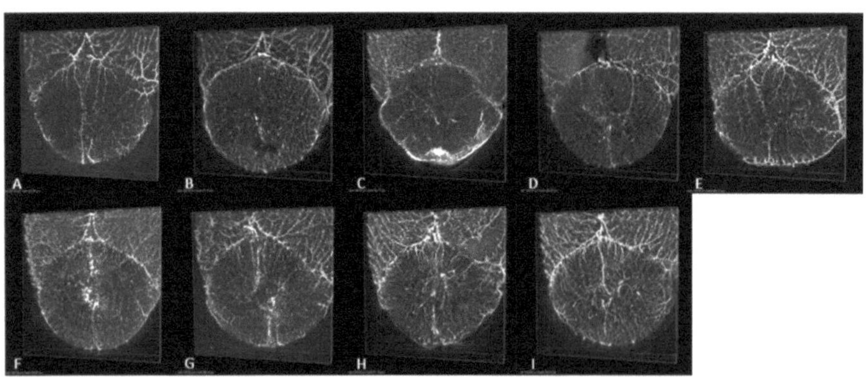

Figure B.2 3D image of blood vessels and proliferating cells during regeneration. (A) 3D image of blood vessels (*green* (light gray in print versions)) and proliferating cells (*red* (dark gray in print versions)) before lesion. (B–I) Represent 1-, 3-, 5-, 7-, 11-, 14-, 21- and 28-dpl cerebellum, respectively. (B) After 1 dpl, a wound can be clearly visible in cerebellum where the signals of blood vessels and proliferating cells are disappear. (C–F) At 3–11 dpl, blood vessels are regrowth in the wound, and proliferating cells are accumulated to the wound. (G) There is a gap in the wound where the blood vessels and proliferating cells are less. (H–I) Blood vessels and proliferating cells fill in the wound and the PCNA signals are decrease if compare to 11 or 14 dpl.

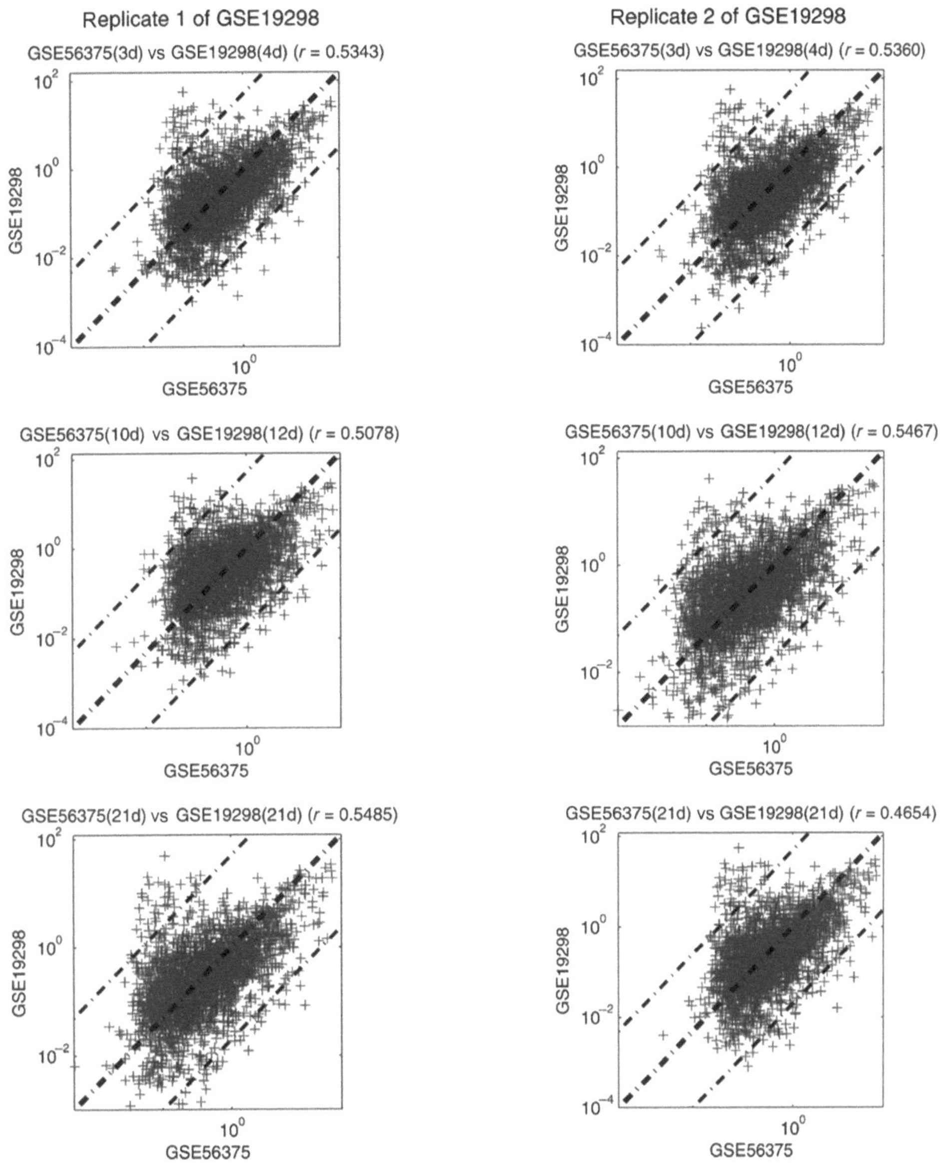

Figure B.3 The published time-course microarray data for validation. The published data of the optic nerve regeneration is GSE19298 [64]. In this section, microarray analysis was performed on total RNA extracted from whole eye following optic nerve crush or sham surgery at defined intervals (4, 12, and 21 days). The *x*-axis is this time-course microarray data (after normalizing). The *y*-axis is the published time-course microarray data (after normalizing). The Pearson correlation coefficient is denoted as *r*.

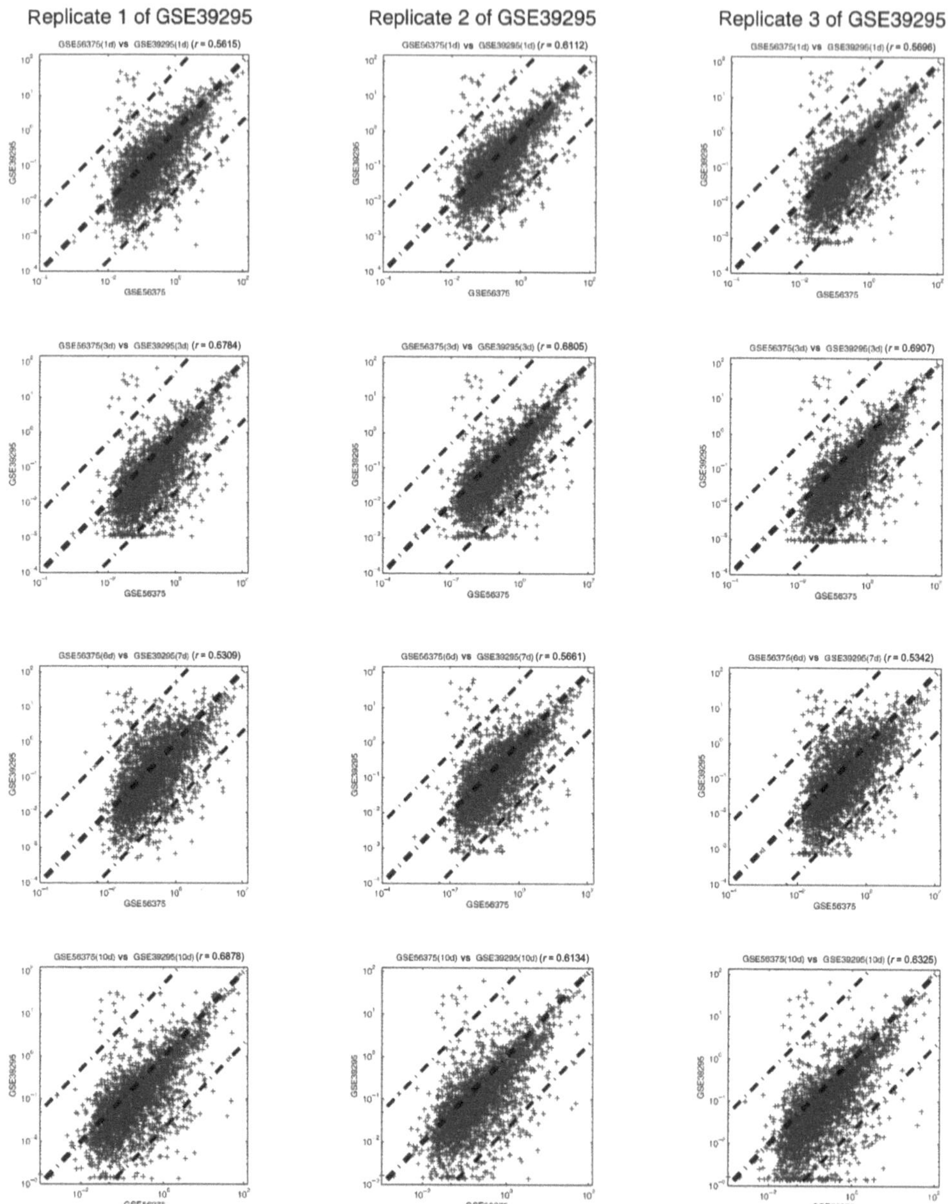

Figure B.4 The published time-course microarray data for validation. The published data of the spinal cord regeneration is GSE39295 [74]. In this section, the spinal cord has been injured by crushing dorso-ventrally for 1 s at the level of 15th/16th vertebrae. Later the wound were sealed by placing a suture. Both spinal cord injured and sham operated fish were allowed to regenerate and the progress of regeneration was observed after 1, 3, 7, 10, and 15 days of injury. Zebrafishes were anesthetized deeply for 5 min in 0.1% tricaine (MS222; Sigma, USA) and approximately 1 mm length of spinal cord both rostrally and caudally from injury epicenter were dissected out from 50 to 60 fishes in each batch and pooled for RNA extraction. The *x*-axis is this time-course microarray data (after normalizing). The *y*-axis is the published time-course microarray data (after normalizing). The Pearson correlation coefficient is denoted as *r*.

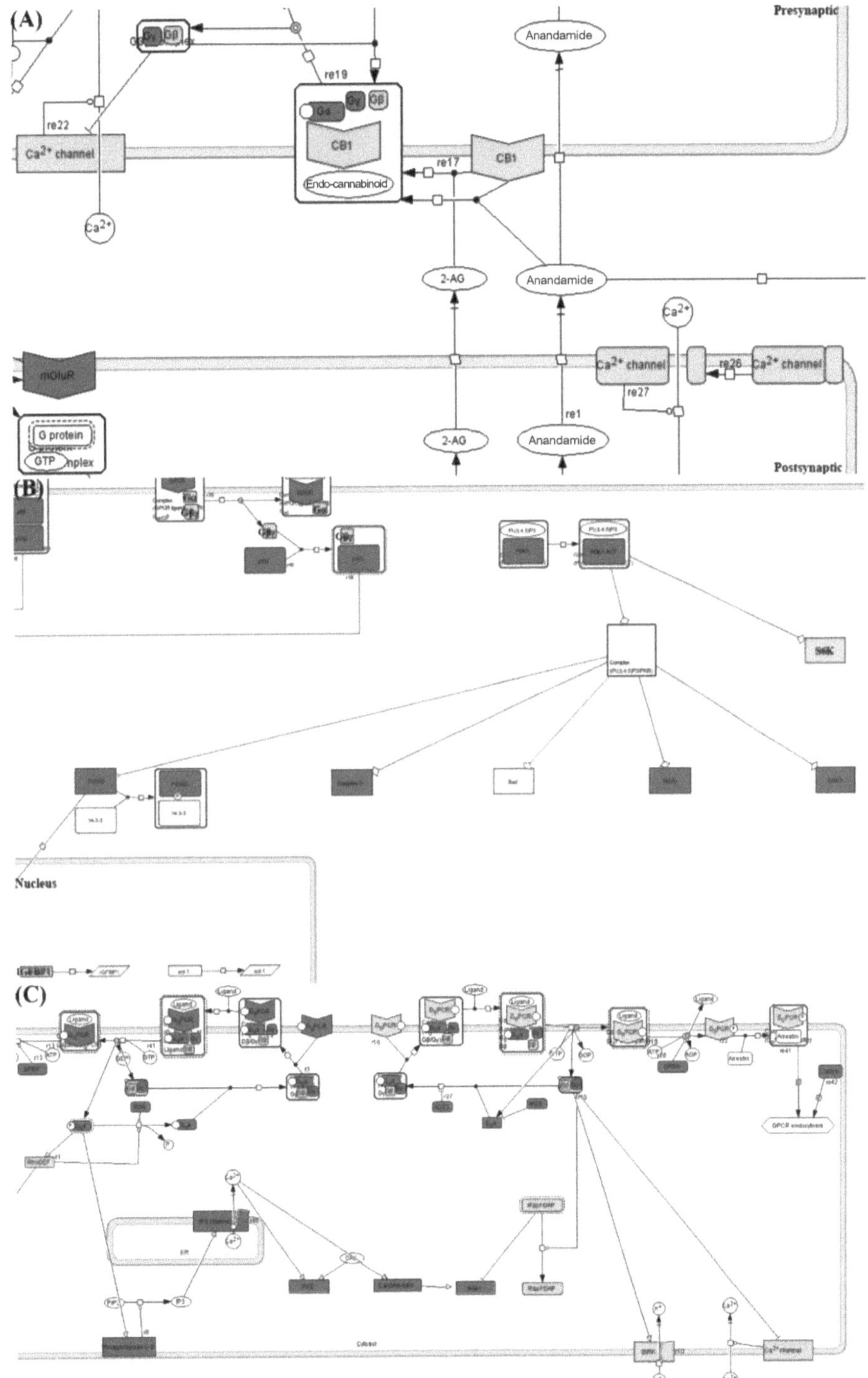

Figure B.5 Examples of the over-represented pathways enriched by group A proteins and involved in the acute inflammation and immune response during the wound healing process. These pathways were primarily involved in the acute inflammation and immune response to TBI. (A) Endogenous cannabinoid signaling pathway; (B) PI3 kinase pathway; (C) Heterotrimeric G-protein signaling pathway-Gq alpha and Go alpha mediated pathway. Proteins in group A are shown in *red* (light gray in print versions).

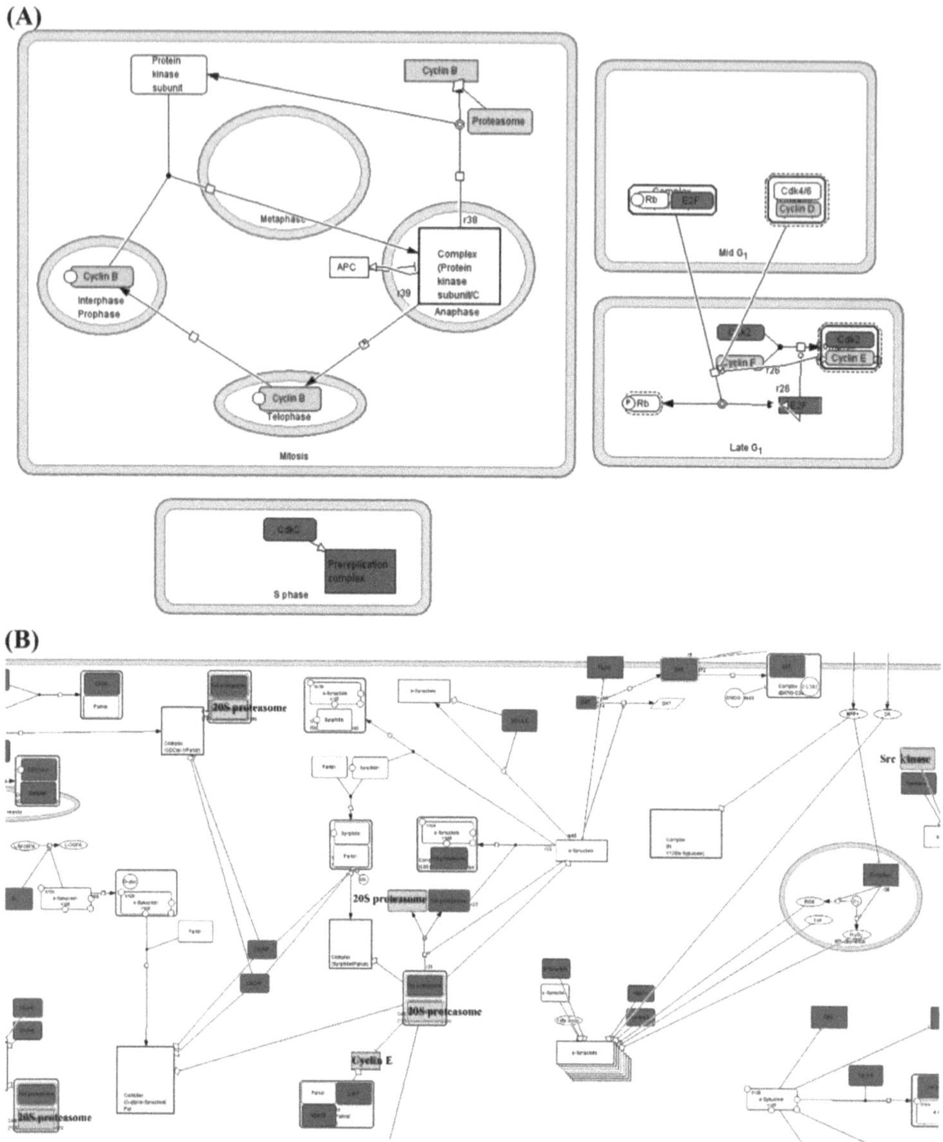

Figure B.6 Examples of the over-represented pathways enriched by proteins from group P and positively correlated with ZMI. These pathways either played an important role in cytoskeleton regulation, angiogenesis and inflammation, or reflected the process in which behavioral disability increased initially and then decreased. (A) Cell cycle pathway; (B) Parkinson's disease pathway. Proteins in group P are shown in *red* (light gray in print versions).

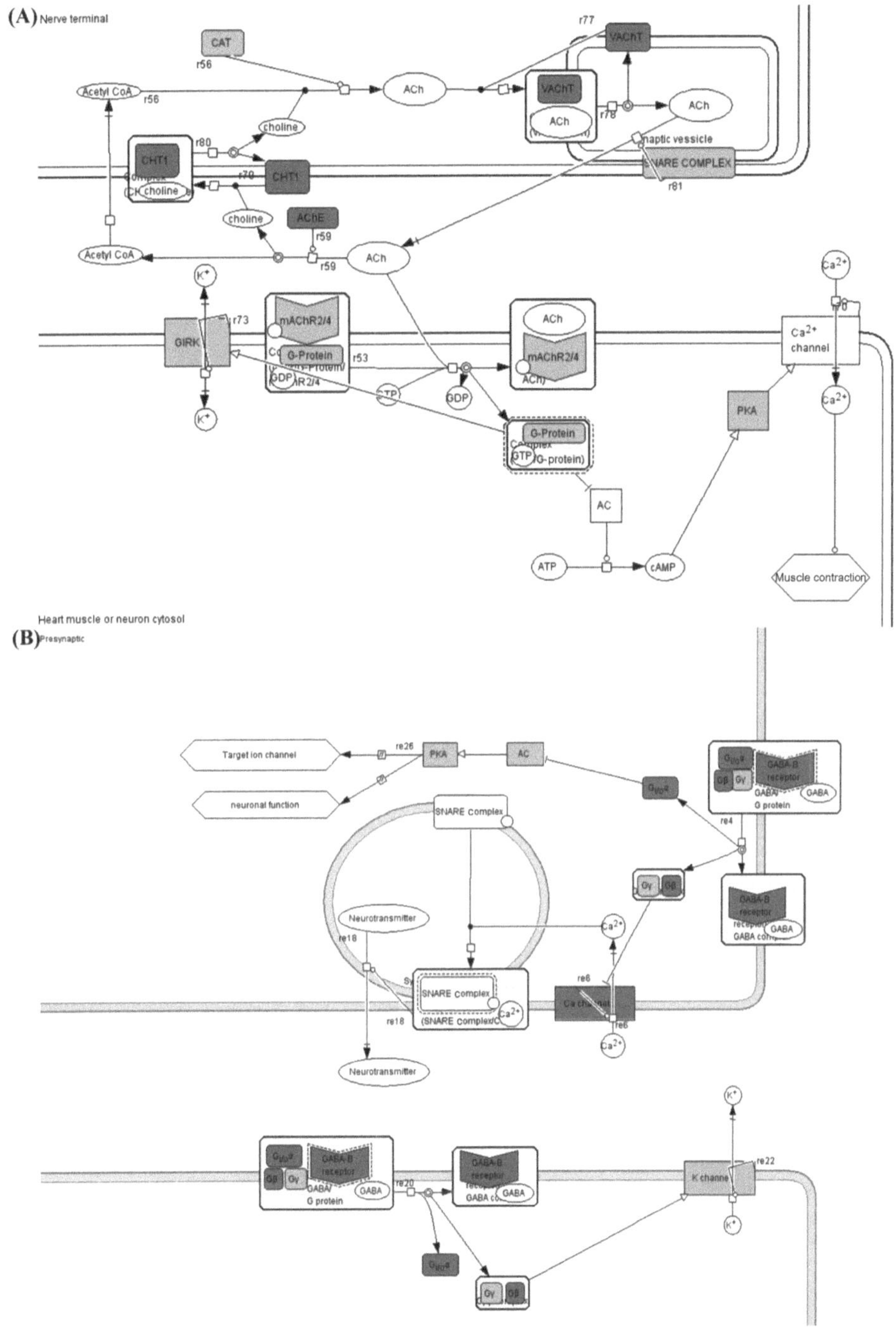

Figure B.7 Examples of the over-represented pathways enriched by proteins from group N and negatively correlated with ZMI. These pathways support the functional recovery of neural transmission. (A) Muscarinic acetylcholine receptor 2 and 4 signaling pathway; (B) GABA-B receptor II signaling pathway. Proteins in group N are shown in *red* (light gray in print versions).

REFERENCES

[1] Xia Y, Nivet E, Sancho-Martinez I, Gallegos T, Suzuki K, Okamura D, et al. Directed differentiation of human pluripotent cells to ureteric bud kidney progenitor-like cells. Nat Cell Biol 2013;15 (12):1507−15.

[2] Spence JR, Mayhew CN, Rankin SA, Kuhar MF, Vallance JE, Tolle K, et al. Directed differentiation of human pluripotent stem cells into intestinal tissue in vitro. Nature 2011;470(7332):105−120.

[3] Chambers SM, Fasano CA, Papapetrou EP, Tomishima M, Sadelain M, Studer L. Highly efficient neural conversion of human ES and iPS cells by dual inhibition of SMAD signaling. Nat Biotechnol 2009;27(3):275−80.

[4] Greber B, Coulon P, Zhang M, Moritz S, Frank S, Mueller-Molina AJ, et al. FGF signalling inhibits neural induction in human embryonic stem cells. EMBO J 2011;30(24):4874−84.

[5] Kriks S, Shim J-W, Piao J, Ganat YM, Wakeman DR, Xie Z, et al. Dopamine neurons derived from human ES cells efficiently engraft in animal models of Parkinson's disease. Nature 2011;480 (7378):547-U177.

[6] Lyashenko N, Winter M, Migliorini D, Biechele T, Moon RT, Hartmann C. Differential requirement for the dual functions of beta-catenin in embryonic stem cell self-renewal and germ layer formation. Nat Cell Biol 2011;13(7):753-U365.

[7] Wichterle H, Lieberam I, Porter JA, Jessell TM. Directed differentiation of embryonic stem cells into motor neurons. Cell 2002;110(3):385−97.

[8] Narazaki G, Uosaki H, Teranishi M, Okita K, Kim B, Matsuoka S, et al. Directed and systematic differentiation of cardiovascular cells from mouse induced pluripotent stem cells. Circulation 2008;118(5):498−506.

[9] Chen B-S, Yang S-K, Lan C-Y, Chuang Y-J. A systems biology approach to construct the gene regulatory network of systemic inflammation via microarray and databases mining. BMC Med Genomics 2008;1:46.

[10] Yang S-K, Wang Y-C, Chao C-C, Chuang Y-J, Lan C-Y, Chen B-S. Dynamic cross-talk analysis among TNF-R, TLR-4 and IL-1R signalings in TNF alpha-induced inflammatory responses. BMC Med Genomics 2010;3:19.

[11] Wu Z, Aryee MJ. Subset quantile normalization using negative control features. J Comput Biol 2010;17(10):1385−95.

[12] Bader GD, Betel D, Hogue CWV. BIND: the Biomolecular Interaction Network Database. Nucleic Acids Res 2003;31(1):248−50.

[13] Stark C, Breitkreutz B-J, Reguly T, Boucher L, Breitkreutz A, Tyers M. BioGRID: a general repository for interaction datasets. Nucleic Acids Res 2006;34:D535−9.

[14] Xenarios I, Rice DW, Salwinski L, Baron MK, Marcotte EM, Eisenberg D. DIP: the Database of Interacting Proteins. Nucleic Acids Res 2000;28(1):289−91.

[15] Peri S, Navarro JD, Amanchy R, Kristiansen TZ, Jonnalagadda CK, Surendranath V, et al. Development of human protein reference database as an initial platform for approaching systems biology in humans. Genome Res 2003;13(10):2363−71.

[16] Brown KR, Jurisica I. Online predicted human interaction database. Bioinformatics 2005;21 (9):2076−82.

[17] Orchard S, Ammari M, Aranda B, Breuza L, Briganti L, Broackes-Carter F, et al. The MIntAct project-IntAct as a common curation platform for 11 molecular interaction databases. Nucleic Acids Res 2014;42(D1):D358−63.

[18] Chatraryamontri A, Ceol A, Palazzi LM, Nardelli G, Schneider MV, Castagnoli L, et al. MINT: the molecular INTeraction database. Nucleic Acids Res 2007;35:D572−4.

[19] McDowall MD, Scott MS, Barton GJ. PIPs: human protein-protein interaction prediction database. Nucleic Acids Res 2009;37:D651−6.

[20] Croft D, Mundo AF, Haw R, Milacic M, Weiser J, Wu G, et al. The Reactome pathway knowledgebase. Nucleic Acids Res 2014;42(D1):D472−7.

[21] Franceschini A, Szklarczyk D, Frankild S, Kuhn M, Simonovic M, Roth A, et al. STRING v9.1: protein-protein interaction networks, with increased coverage and integration. Nucleic Acids Res 2013;41(D1):D808−15.

[22] Wang Y-C, Chen B-S. Integrated cellular network of transcription regulations and protein-protein interactions. BMC Syst Biol 2010;4:20.
[23] Bar-Joseph Z, Gerber G, Gifford DK, Jaakkola TS, Simon I. A new approach to analyzing gene expression time series data. Proceedings of the sixth annual international conference on computational biology: 2002. ACM; 2002. p. 39–48.
[24] De Boor C. with 32 figures, Rev. edn A practical guide to splines. New York, NY: Springer; 2001
[25] Newman JRS, Ghaemmaghami S, Ihmels J, Breslow DK, Noble M, DeRisi JL, et al. Single-cell proteomic analysis of *S. cerevisiae* reveals the architecture of biological noise. Nature 2006;441 (7095):840–6.
[26] Orntoft TF, Thykjaer T, Waldman FM, Wolf H, Celis JE. Genome-wide study of gene copy numbers, transcripts, and protein levels in pairs of non-invasive and invasive human transitional cell carcinomas. Mol Cell Proteomics 2002;1(1):37–45.
[27] Di Salvio M, Di Giovannantonio LG, Omodei D, Acampora D, Simeone A. Otx2 expression is restricted to dopaminergic neurons of the ventral tegmental area in the adult brain. Int J Dev Biol 2010;54(5):939–45.
[28] Vernay B, Koch M, Vaccarino F, Briscoe J, Simeone A, Kageyama R, et al. Otx2 regulates subtype specification and neurogenesis in the midbrain. J Neurosci 2005;25(19):4856–67.
[29] Puelles E, Annino A, Tuorto F, Usiello A, Acampora D, Czerny T, et al. Otx2 regulates the extent, identity and fate of neuronal progenitor domains in the ventral midbrain. Development 2004;131 (9):2037–48.
[30] Sugaya K, Alvarez A, Marutle A, Kwak YD, Choumkina E. Stem cell strategies for Alzheimer's disease therapy. Panminerva Med 2006;48(2):87–96.
[31] Nakatani T, Kumai M, Mizuhara E, Minaki Y, Ono Y. Lmx1a and Lmx1b cooperate with Foxa2 to coordinate the specification of dopaminergic neurons and control of floor plate cell differentiation in the developing mesencephalon. Dev Biol 2010;339(1):101–13.
[32] Kittappa R, Chang WW, Awatramani RB, McKay RDG. The foxa2 gene controls the birth and spontaneous degeneration of dopamine neurons in old age. PLoS Biol 2007;5(12):2875–84.
[33] Glubb DM, Joyce PR, Kennedy MA. Expression and association analyses of promoter variants of the neurogenic gene HES6, a candidate gene for mood disorder susceptibility and antidepressant response. Neurosci Lett 2009;460(2):185–90.
[34] Ahmed M, Xu J, Xu P-X. EYA1 and SIX1 drive the neuronal developmental program in cooperation with the SWI/SNF chromatin-remodeling complex and SOX2 in the mammalian inner ear. Development 2012;139(11):1965–77.
[35] Uldry M, Yang W, St-Pierre J, Lin J, Seale P, Spiegelman BM. Complementary action of the PGC-1 coactivators in mitochondrial biogenesis and brown fat differentiation. Cell Metab 2006;3 (5):333–41.
[36] Nishihara E, Yoshida-Komiya H, Chan CS, Liao L, Davis RL, O'Malley BW, et al. SRC-1 null mice exhibit moderate motor dysfunction and delayed development of cerebellar Purkinje cells. J Neurosci 2003;23(1):213–22.
[37] Park G, Jeong J-W, Kim J-E. SIRT1 deficiency attenuates MPP + -induced apoptosis in dopaminergic cells. FEBS Lett 2011;585(1):219–24.
[38] Wang Y-C, Chen B-S. A network-based biomarker approach for molecular investigation and diagnosis of lung cancer. BMC Med Genomics 2011;4:2.
[39] Liu J, Yue Y, Han D, Wang X, Fu Y, Zhang L, et al. A METTL3-METTL14 complex mediates mammalian nuclear RNA N6-adenosine methylation. Nat Chem Biol 2014;10(2):93–5.
[40] Cortese R, Lewin J, Baeckdahl L, Krispin M, Wasserkort R, Eckhardt F, et al. Genome-wide screen for differential DNA methylation associated with neural cell differentiation in mouse. PLoS One 2011;6(10):e26002.
[41] Tselykh TV, Roos C, Heino TI. The mitochondrial ribosome-specific MrpL55 protein is essential in Drosophila and dynamically required during development. Exp Cell Res 2005;307(2):354–66.
[42] Shi G, Jin Y. Role of Oct4 in maintaining and regaining stem cell pluripotency. Stem Cell Res Ther 2010;1:39.
[43] Paulweber B, Sandhofer F, Levywilson B. The mechanism by which the human apolipoprotein-b gene reducer operates involves blocking of transcriptional activation by hepatocyte nuclear factor-III. Mol Cell Biol 1993;13(3):1534–46.

[44] Anderson JM, Van Itallie CM. Physiology and function of the tight junction. Cold Spring Harb Perspect Biol 2009;1(2):a002584.

[45] Chen S, Lewallen M, Xie T. Adhesion in the stem cell niche: biological roles and regulation. Development 2013;140(2):255–65.

[46] Ben-David U, Nudel N, Benvenisty N. Immunologic and chemical targeting of the tight-junction protein Claudin-6 eliminates tumorigenic human pluripotent stem cells. Nat Commun 2013;4:1992.

[47] Monaghan JR, Walker JA, Page RB, Putta S, Beachy CK, Voss SR. Early gene expression during natural spinal cord regeneration in the salamander *Ambystoma mexicanum*. J Neurochem 2007;101 (1):27–40.

[48] Diwakar L, Kenchappa RS, Annepu J, Saeed U, Sujanitha R, Ravindranath V. Down-regulation of glutaredoxin by estrogen receptor antagonist renders female mice susceptible to excitatory amino acid mediated complex I inhibition in CNS. Brain Res 2006;1125:176–84.

[49] Tong C-W, Wang J-L, Jiang M-S, Hsu C-H, Chang W-T, Huang AM. Novel genes that mediate nuclear respiratory factor 1-regualted neurite outgrowth in neuroblastoma IMR-32 cells. Gene 2013;515(1):62–70.

[50] Marino M, Galluzzo P, Ascenzi P. Estrogen signaling multiple pathways to impact gene transcription. Curr Genomics 2006;7(8):497–508.

[51] Batsukh T, Schulz Y, Wolf S, Rabe TI, Oellerich T, Urlaub H, et al. Identification and characterization of FAM124B as a novel component of a CHD7 and CHD8 containing complex. PLoS One 2012;7(12):e52640.

[52] Harr MW, Caimi PF, McColl KS, Zhong F, Patel SN, Barr PM, et al. Inhibition of Lck enhances glucocorticoid sensitivity and apoptosis in lymphoid cell lines and in chronic lymphocytic leukemia. Cell Death Differ 2010;17(9):1381–91.

[53] Hagemann AIH, Scholpp S. The tale of the three brothers—Shh, Wnt, and Fgf during development of the thalamus. Front Neurosci 2012;6:76.

[54] Grigoryan T, Stein S, Qi J, Wende H, Garratt AN, Nave K-A, et al. Wnt/Rspondin/beta-catenin signals control axonal sorting and lineage progression in Schwann cell development. Proc Natl Acad Sci USA 2013;110(45):18174–9.

[55] Faul M, Xu L, Wald MM, Coronado V, Dellinger AM. Traumatic brain injury in the united states: national estimates of prevalence and incidence, 2002-2006. Injury Prevention 2010;16:A268.

[56] Kizil C, Kaslin J, Kroehne V, Brand M. Adult neurogenesis and brain regeneration in zebrafish. Dev Neurobiol 2012;72(3):429–61.

[57] Tanaka EM, Ferretti P. Considering the evolution of regeneration in the central nervous system. Nat Rev Neurosci 2009;10(10):713–23.

[58] Horner PJ, Gage FH. Regenerating the damaged central nervous system. Nature 2000;407 (6807):963–70.

[59] Antos CL, Tanaka EM. Vertebrates that regenerate as models for guiding stem cells. In: Meshorer E, Plath K, editors. Cell biology of stem cells, vol. 695. Berlin: Springer; 2010. p. 184–214.

[60] Poss KD, Wilson LG, Keating MT. Heart regeneration in zebrafish. Science 2002;298 (5601):2188–90.

[61] Cameron DA, Gentile KL, Middleton FA, Yurco P. Gene expression profiles of intact and regenerating zebrafish retina. Mol Vis 2005;11(93):775–91.

[62] Craig SE, Calinescu AA, Hitchcock PF. Identification of the molecular signatures integral to regenerating photoreceptors in the retina of the zebra fish. J Ocul Biol Dis Infor 2008;1(2–4):73–84.

[63] Guo Y, Ma L, Cristofanilli M, Hart RP, Hao A, Schachner M. Transcription factor Sox11b is involved in spinal cord regeneration in adult zebrafish. Neuroscience 2011;172:329–41.

[64] McCurley AT, Callard GV. Time course analysis of gene expression patterns in zebrafish eye during optic nerve regeneration. J Exp Neurosci 2010;2010(4):17–33.

[65] Qin Z, Barthel LK, Raymond PA. Genetic evidence for shared mechanisms of epimorphic regeneration in zebrafish. Proc Natl Acad Sci USA 2009;106(23):9310–15.

[66] Kishimoto N, Shimizu K, Sawamoto K. Neuronal regeneration in a zebrafish model of adult brain injury. Dis Model Mech 2012;5(2):200–9.

[67] Postlethwait J, Amores A, Force A, Yan YL. The zebrafish genome. In: Detrich HW, Westerfield M, Zon LI, editors. Methods in cell biology, vol. 60: Zebrafish: genetics and genomics. 1999. p. 149–63.
[68] Ernst J, Bar-Joseph Z. STEM: a tool for the analysis of short time series gene expression data. BMC Bioinformatics 2006;7:191.
[69] Mi H, Muruganujan A, Thomas PD. PANTHER in 2013: modeling the evolution of gene function, and other gene attributes, in the context of phylogenetic trees. Nucleic Acids Res 2013;41 (D1):D377–86.
[70] Kalueff AV, Gebhardt M, Stewart AM, Cachat JM, Brimmer M, Chawla JS, et al. Towards a comprehensive catalog of zebrafish behavior 1.0 and beyond. Zebrafish 2013;10(1):70–86.
[71] Levin ED, Cerutti DT. "Behavioral neuroscience of zebrafish," in Methods of Behavior Analysis in Neuroscience, ed. J. J. Buccafusco 2nd Edn. Boca Raton (FL): CRC Press 2009.
[72] Bradford Y, Conlin T, Dunn N, Fashena D, Frazer K, Howe DG, et al. ZFIN: enhancements and updates to the zebrafish model organism database. Nucleic Acids Res 2011;39:D822–9.
[73] Ashburner M, Ball CA, Blake JA, Botstein D, Butler H, Cherry JM, et al. Gene Ontology: tool for the unification of biology. Nat Genet 2000;25(1):25–9.
[74] Hui SP, Sengupta D, Lee SGP, Sen T, Kundu S, Mathavan S, et al. Genome wide expression profiling during spinal cord regeneration identifies comprehensive cellular responses in zebrafish. PLoS One 2014;9(1):e84212.
[75] Singer AJ, Clark RAF. Mechanisms of disease—cutaneous wound healing. N Engl J Med 1999;341 (10):738–46.
[76] Stadelmann WK, Digenis AG, Tobin GR. Physiology and healing dynamics of chronic cutaneous wounds. Am J Surg 1998;176(2A):26S–38S.
[77] Mechoulam R, Panikashvili D, Shohami E. Cannabinoids and brain injury: therapeutic implications. Trends Mol Med 2002;8(2):58–61.
[78] Shohami E, Cohen-Yeshurun A, Magid L, Algali M, Mechoulam R. Endocannabinoids and traumatic brain injury. Br J Pharmacol 2011;163(7):1402–10.
[79] Rajaram MVS, Ganesan LP, Parsa KVL, Butchar JP, Gunn JS, Tridandapani S. Akt/protein kinase B modulates macrophage inflammatory response to *Francisella* infection and confers a survival advantage in mice. J Immunol 2006;177(9):6317–24.
[80] Neves SR, Ram PT, Iyengar R. G protein pathways. Science 2002;296(5573):1636–9.
[81] Lin MT, Beal MF. Mitochondrial dysfunction and oxidative stress in neurodegenerative diseases. Nature 2006;443(7113):787–95.
[82] Lukacs NW. Role of chemokines in the pathogenesis of asthma. Nat Rev Immunol 2001;1(2):108–16.
[83] Vicente-Manzanares M, Sancho D, Yanez-Mo M, Sanchez-Madrid F. The leukocyte cytoskeleton in cell migration and immune interactions. In: Jeon KW, editor. International review of cytology, vol. 216. 2002. p. 233–89.
[84] Ridley AJ. Rho GTPases and cell migration. J Cell Sci 2001;114(15):2713–22.
[85] Giancotti FG, Ruoslahti E. Transduction—integrin signaling. Science 1999;285(5430):1028–32.
[86] Jiang W, Zhang Y, Xiao L, Van Cleemput J, Ji SP, Bai G, et al. Cannabinoids promote embryonic and adult hippocampus neurogenesis and produce anxiolytic- and antidepressant-like effects. J Clin Investig 2005;115(11):3104–16.
[87] Berg DA, Belnoue L, Song H, Simon A. Neurotransmitter-mediated control of neurogenesis in the adult vertebrate brain. Development 2013;140(12):2548–61.
[88] Martins RAP, Pearson RA. Control of cell proliferation by neurotransmitters in the developing vertebrate retina. Brain Res 2008;1192:37–60.
[89] Blalock JE. Production of peptide-hormones and neurotransmitters by the immune-system. Chem Immunol 1992;52:1–24.
[90] Qiu YH, Peng YP, Wang JH. Immunoregulatory role of neurotransmitters. Adv Neuroimmunol 1996;6(3):223–31.
[91] Webster JI, Tonelli L, Sternberg EM. Neuroendocrine regulation of immunity. Annu Rev Immunol 2002;20:125–63.

[92] Guingab-Cagmat JD, Cagmat EB, Hayes RL, Anagli J. Integration of proteomics, bioinformatics, and systems biology in traumatic brain injury biomarker discovery. Front Neurol 2013;4:61.
[93] Neselius S, Brisby H, Theodorsson A, Blennow K, Zetterberg H, Marcusson J. CSF-biomarkers in olympic boxing: diagnosis and effects of repetitive head trauma. PLoS One 2012;7(4):e33606.
[94] Kyritsis N, Kizil C, Zocher S, Kroehne V, Kaslin J, Freudenreich D, et al. Acute Inflammation initiates the regenerative response in the adult zebrafish brain. Science 2012;338(6112):1353–6.
[95] DeKosky ST, Blennow K, Ikonomovic MD, Gandy S. Acute and chronic traumatic encephalopathies: pathogenesis and biomarkers. Nat Rev Neurol 2013;9(4):192–200.
[96] Feala JD, AbdulHameed MDM, Yu C, Dutta B, Yu X, Schmid K, et al. Systems biology approaches for discovering biomarkers for traumatic brain injury. J Neurotrauma 2013;30 (13):1101–16.
[97] Basu S, Dasgupta PS. Dopamine, a neurotransmitter, influences the immune system. J Neuroimmunol 2000;102(2):113–24.
[98] Piomelli D. The molecular logic of endocannabinoid signalling. Nat Rev Neurosci 2003;4 (11):873–84.
[99] Alon U. An introduction to systems biology: design principles of biological circuits. Boca Raton, FL: Chapman & Hall/CRC; 2007.
[100] Coleman TF, Hulbert LA. A direct active set algorithm for large sparse quadratic programs with simple bounds. Math Program 1989;45(3):373–406.
[101] Akaike H. New look at statistical-model identification. IEEE Trans Autom Control 1974;AC19 (6):716–23.
[102] Johansson R. System modeling and identification. Englewood Cliffs, NJ: Prentice Hall; 1993.

CHAPTER 9

Big Tumorigenesis Mechanisms in Systems Cancer Biology via Big Database Mining and Network Modeling

INTRODUCTION

Cancer is a genetic disease. Many screening arrays used to find new cancer genes have been developed, and, as a result, many cancer-related genes have been identified. The Somatic Mutations in Cancer (COSMIC) database contains 487 genes implicated in cancer, i.e., the Cancer Gene Census. Cancer is also a systemic disease. It has become apparent that cancers can be organized into a limited number of biological functions, named the "hallmarks of cancer." There are eight hallmarks related to cell survival, proliferation, and metastatic dissemination, which are the key mechanisms in carcinogenesis. Each function has its own pathway. As more and more genes controlling these functions are identified and arranged into signaling pathways, it appears that these pathways are interconnected through crosstalk at different levels in the biological networks. In this chapter, the hallmarks of cancer can be understood as different ways of looking at the unique process of carcinogenesis. These interconnections and the crosstalk between genes, RNAs, proteins, and other molecules could be combined into an integrated genetic and epigenetic biological network of tumor progression. Therefore, tumorigenesis mechanisms should be investigated from a cancer systems biology perspective, using big database mining, network modeling, and genome-wide identification.

Since it is well established that cancer is a genetic disease and a consequence of the dysregulation of a gene network that controls cell growth and dissemination, the construction of gene regulatory networks (GRNs) is an important, modern approach to understanding cancer. In this chapter (Section 9.1), we first construct and then clarify the dynamic GRN of the cancer cell cycle via database mining and microarray data. The cell cycle is an important clue to understanding the mechanism underlying cancer cell growth. Recently, expression profiles of microarray data of cancer cell cycle have become available showing the dynamic interactions among cancer cell cycle-related

Big Mechanisms in Systems Biology
DOI: http://dx.doi.org/10.1016/B978-0-12-809479-2.00009-3

genes. Therefore, it is useful to construct a dynamic model of the GRN of the cancer cell cycle to gain insight into the control of cell growth and dissemination as a gene regulatory mechanism of cancer cells. In this chapter, we also trace upstream regulatory genes of target genes to infer the regulatory pathways of the gene network by maximum likelihood estimation, and then analyze the regulatory capability and sensitivity of regulatory genes to clarify their roles in the cancer cell cycle. We also present a systematically iterative approach to discern and characterize the transcriptional regulatory network in the HeLa cell cycle from raw repression profiles. We can predict and clarify how the E2F target genes in the G1/S phase are crucial in regulating cell cycle progression and tumorigenesis. From the GRN results and literature confirmation, we infer that the cell cycle genes, MCM4, MCM5, CDC6, CDC25A, UNG, and E2F2, are E2F target genes in the HeLa (human cervical carcinoma) cell cycle.

Since cancer is caused by genetic abnormalities, such as mutations of oncogenes or tumor suppressor genes, which alter downstream signal transduction pathways and protein–protein interactions (PPIs), comparisons of the interactions of proteins in cancerous and normal HeLa cells can shed light on the mechanisms of carcinogenesis. In Section 9.2, we construct a cancer-perturbed protein–protein interaction network (PPIN) from database mining and a dynamic protein interaction model. We construct candidate PPINs involved in the apoptosis of cancerous and normal cells using two human yeast two-hybrid data sets and four online databases. Next, we apply a nonlinear stochastic model, maximum likelihood parameter estimation, and AIC to prune false-positive PPIs in the candidate PPINs from microarray data. Comparisons of the apoptosis networks in HeLa cells and in normal primary lung fibroblasts can provide insights into the mechanism of apoptosis and allow identification of potential drug targets. The potential targets include BCL2, caspase-3, and TP53. The comparison of cancerous and normal cells also allows derivation of several party husband date hubs in the human PPINs involved in caspase activation.

Since lung cancer is the leading cause of cancer deaths and diagnosis from the systems perspective, in Section 9.3, we propose a systems biology approach to integrate microarray gene expression profiles and PPI information from database mining to develop a network-based biomarker for molecular investigation into the mechanism of lung carcinogenesis and diagnosis of lung cancer. The network-based biomarker consists of two protein association networks constructed for cancer samples and noncancer samples. Based on this biomarker, a total of 40 proteins that are significant in lung carcinogenesis are identified and assigned carcinogenesis relevance values (CRVs). In addition, the network-based biomarker, acting as the screening test, is proven effective in diagnosing lung cancer in smokers with clinical signs. Clearly, a network-based biomarker using constructed protein association networks is a useful tool to identify the pathways and mechanisms of lung carcinogenesis, and, more importantly, to provide potential therapeutic targets to combat cancer.

Finally, in Section 9.4, a new method is also proposed for constructing an integrated genetic and epigenetic network (IGEN) to characterize cellular mechanism in bladder carcinogenesis between stage 1 and stage 4 bladder cancer cells by using system model and big database miming through next-generation sequencing (NGS) data. The proposed method not only unravel core network marker but also provide multiple drug design to treat bladder cancer with minimum side effects.

9.1 CONSTRUCTION AND CLARIFICATION OF DYNAMIC NETWORKS OF THE CANCER CELL CYCLE VIA MICROARRAY DATA

The cell cycle is an important clue to unravel the tumorigenesis mechanism of cancer cells. Recently, expression profiles of cDNA microarray data of the cancer cell cycle have become available on the dynamic interactions among cancer cell cycle related genes. Therefore, it is more appealing to construct a dynamic model for GRN of cancer cell cycle to gain more insight into the infrastructure of gene regulatory mechanism of cancer cell via microarray data. Based on the gene regulatory dynamic model and microarray data, we construct the whole dynamic gene regulatory molecular network of cancer cell cycle. In this section, we trace back upstream regulatory genes of a target gene to infer the regulatory pathways of the gene network by maximum likelihood estimation method. Finally, based on the dynamic regulatory network, we analyze the regulatory abilities and sensitivities of regulatory genes to clarify their roles in the mechanism of cancer cell cycle. This section presents a systematically iterative approach to discern and characterize the transcriptional regulatory network in the HeLa cell cycle from the raw expression profiles. The transcription regulatory network in a HeLa cell cycle can also be confirmed by some experimental reviews. Based on the results and some literature reviews, we can predict and clarify the E2F target genes in G1/S phase, which are crucial for regulating cell cycle progression and tumorigenesis. From the results of the network construction and literature confirmation, we infer that *MCM4*, *MCM5*, *CDC6*, *CDC25A*, *UNG*, and *E2F2* are E2F target genes in the HeLa cell cycle.

The losses of cellular regulation give rise to most cases of cancer. In cells, intricate genetic control systems regulate the balance between cell survival and death in response to growth signals, growth-inhibiting signals, and death signals. When some errors occur in the control systems causing cells to proliferate continuously, a tumor comes into being [1]. The proliferation of cancer cell into two individual cells must go through the cell cycle process [2,3]. Cell cycle entails an ordered series of macromolecular events that lead to cell division and the production of two daughter cells. So you can see that the cell cycle is meaningful to the proliferation of cancer cells.

Expression levels of thousands of genes fluctuate throughout the cancer cell cycle [3–5]. Functional genes show periodic transcription to reflect cell growth, DNA synthesis, spindle pole body duplication, and migration through the cell cycle [6]. These processes and their regulation have been extensively investigated at the molecular level [7–11]. Systems biology can be described as "integrative biology" with the ultimate goal of being able to predict de novo biological outcomes given the list of the components involved [12]. Hence it is a coordinated study by: (1) investigating the components of cellular networks and their interactions; (2) applying experimental high-throughput and whole-genome techniques; and (3) integrating computational methods with experimental effort [13]. In this situation, characterization of the genome-wide transcriptional program of the cell division cycle in mammalian cells is a critical step toward understanding the basic cell cycle processes and their roles in cancer. Therefore, it is worth investigating how these periodic patterns are regulated in the GRN of the cancer cell cycle from the systems biology perspective.

Gene expression data of the HeLa cell cycle have been collected [3] and analyzed with many clustering methods to organize which gene is associated with the cell cycle [3,14]. Theoretically, it is possible to reverse-engineer the cell cycle network, if we take cDNA expression levels as the output of gene expression networks, and collect cDNA expression levels of transcription factors as an input. In order to realize how genes are regulated by transcription factors, we must also understand the interactions between target genes and their transcription factors (i.e., which transcription factor(s) binds to which promoters). With all this information and an interactive dynamic model, we get some clues to piece together the gene expression regulatory network in the HeLa cell cycle.

In this section, we attempt to devise an interactive dynamic model to characterize transcriptional regulatory network of the HeLa cell cycle from the cDNA expression data of the human cell cycle in a tumor [3]. Based on this dynamic regulatory network, we not only predict the upstream regulators but also characterize the significance of the regulators depending on quantifying their regulatory abilities based on the corresponding biochemical kinetic parameters. At first, we construct a discrete-time dynamic model system and calculate the system kinetic parameters as the regulatory ability by using the expression data [3] and the system identification method [15]. Second, based on the interactive dynamic model, we detect the transcriptional regulatory function of target genes by the maximum likelihood parameter estimation method. Third, we trace back a group of upstream genes that play a role of transcriptional regulators of target genes in the HeLa cell cycle of *Homo sapiens* via deducing the interactive relationship between the expression profiles of regulators and the detected transcription regulation of specific target genes. The pathway kinetic parameters of transcriptional regulatory network of HeLa cell cycle are also estimated by the cDNA expression profiles of target genes and their upstream regulatory genes. Finally, these upstream regulatory genes are considered as the target genes. By a similar method, we construct their upstream regulatory genes one by one. Iteratively, we can construct the whole GRN of HeLa cell cycle.

We applied this method to a publicly available data set of HeLa cells with microarray experiment on cell cycle [3] to identify the transcriptional regulators of cell cycle and to characterize their regulatory abilities on specific target genes. By means of the quantitative system analysis of the transcriptional regulatory network from HeLa cell cycle genes, several transcription factors were identified and their regulatory abilities were determined. Further, some genes that may be suspected of regulators of HeLa cell cycle are predicted here to be synergistic in harmonizing gene expression. The proposed algorithm provides a novel approach to gain insight into the GRN of HeLa cell by its gene expression data using system identification technique and discrete-time dynamic model. Furthermore, we combine the constructed HeLa cell cycle dynamic network, some experimental reviews and E2F binding site's research [16], to not only confirm the reliability of the dynamic network but also find the E2F target genes in HeLa cell cycle progression. Finally, from the results of this section, we infer that E2F directly regulates *MCM4*, *MCM5*, *CDC6*, *CDC25A*, *UNG*, and *E2F2* in HeLa cell cycle progression.

This approach is so different from the statistical clustering method. It not only provides a suitable interactive dynamic model to decipher the complex signal transduction pathway that regulates gene expressions in HeLa cell cycle, but also predicts some potential regulators that have not been found. We can also quantify the regulatory abilities of the transcription factors by the corresponding kinetic activities to their target genes in the HeLa cell cycle regulatory network. We could construct the cell cycle regulatory network in HeLa cells quantitatively and discuss the sensitivity of regulatory genes to the GRN from the system analysis perspective.

9.1.1 System model and network identification of the HeLa cell cycle

The construction of the transcriptional regulatory network of a HeLa cell cycle can be divided into two steps. First, the transcriptional regulation should be extracted from the gene expression data by the dynamic discrete-time model. Second, the upstream regulators will be traced back by correlating the transcriptional regulation of target genes with the expression profiles of possible regulators of HeLa cell cycle. In this section, 64 transcription regulators, as shown in Table 9.1, are used as candidates of upstream regulators to each target gene. Finally, the kinetic parameters of gene transcriptional regulatory network of HeLa cells are estimated by the cDNA expression profiles of target genes and their upstream regulatory genes.

Dynamic signaling regulatory model

The second-order difference equation is used in the description of dynamic system evolved from the causality of gene regulatory function. Let $X_i(k)$ denote the expression profile of the ith gene at time point k. The following second-order difference equation is proposed to model the cDNA expression level of the ith gene,

$$X_i(k) + a_i X_i(k-1) + b_i X_i(k-2) = G_i(k) + \varepsilon_i(k) \quad (9.1)$$

Table 9.1 CloneIDs of 64 transcription factors in human cell cycle are selected as candidates to regulate downstream target genes

IMAGE:173309(**BCLAF1**)	IMAGE:564803(**FOXM1**) IEA	IMAGE:590774(**MAPK13**) Knebel et al.	IMAGE:202704(**SCML1**) IEA
IMAGE:241474(**BRCA1**) IEA	IMAGE:149809(**GATA2**) IEA	IMAGE:1536451 (**MHC2TA**)	IMAGE:782622 (**SP1**)
IMAGE:815287(**BRD8**) Monden et al.	IMAGE:135688(**GATA2**) IEA	IMAGE:809731(**MNT**) Meroni et al.	IMAGE:80318 (**SP1**)
IMAGE:825210(**C14orf106**) IEA	IMAGE:780958 (**GTF3C4**)	IMAGE:197520(**NCOA3**) IEA	IMAGE:840636 (**SRF**)
IMAGE:726588(**C14orf106**) IEA	IMAGE:291827 (**GTF3C4**)	IMAGE:884438(**NFE2L2**) IEA	IMAGE:545503 (**STAT1**)
IMAGE:1915416(**CDK7**) Shiekhattar et al.	IMAGE:344049(**HCFC1**) Wysocka et al.	IMAGE:1455463 (**NFIC**)	IMAGE:134120 (**STAT5B**)
IMAGE:130242(**CDK7**) Shiekhattar et al.	IMAGE:897806 (**HIF1A**)	IMAGE:265874 (**NFIC**)	IMAGE:712840 (**STAT5B**)
IMAGE:268652(**CDKN1A**) IEA	IMAGE:878184(**HMG20B**) IEA	IMAGE:271198(**NR3C1**) Takahashi et al.	IMAGE:132857 (**STAT5B**)
IMAGE:240367(**CTCF**) Filippova et al.	IMAGE:1842250(**HMGB2**) Shirakawa et al.	IMAGE:245517(**NR5A2**) IEA	IMAGE:272192 (**TCERG1**)
IMAGE:490728 (**DMTF1**)	IMAGE:363103(**HMGB2**) Shirakawa et al.	IMAGE:43229(**PCNA**) IEA	IMAGE:137387 (**TFAP2A**)
IMAGE:487797 (**DR1**)	IMAGE:242952(**ILF2**) Reichman et al.	IMAGE:789182(**PCNA**) IEA	IMAGE:868630(**TGFB1I4**) IEA
IMAGE:566760 (**DR1**)	IMAGE:838829(**JARID1B**) IEA	IMAGE:30114 (**PHTF2**)	IMAGE:366414(**UHRF1**) IEA
IMAGE:884462(**DSCR1**) Fuentes et al.	IMAGE:345056(**KIAA1404**) IEA	IMAGE:1947972(**PKNOX1**) Chen et al.	IMAGE:1550739(**UHRF1**) IEA
IMAGE:236142 (**E2F1**)	IMAGE:510381(**KLF6**) Ratziu et al.	IMAGE:2018976(**PTTG1**) Dominguez et al.	IMAGE:246869(**ZNF207**) Pahl et al.
IMAGE:768260 (**E2F1**)	IMAGE:302549(**KLF9**) Imataka et al.	IMAGE:781089(**PTTG1**) Dominguez et al.	IMAGE:296429(**ZNF24**) IEA
IMAGE:293331 (**E2F2**)	IMAGE:35147(**LASS6**) Banerjee-Basu et al.	IMAGE:845502(**RBPSUH**) Hsieh et al.	IMAGE:280750(**ZNF281**) Law et al.

Gene names are boldly denoted behind or under the CloneIDs in brackets. IEA denotes Inferred from Electronic Annotation in NCBI.

where $G_i(k)$ is the upstream transcriptional regulatory function from regulatory genes to influence the expression profile $X_i(k)$ of the ith gene while a_i and b_i are the parameters that characterize the dynamic inherent property of the gene like degradation and oscillation, and $\varepsilon_i(k)$ is the random noise of current microarray data or the residue of the model. In general, the second-order difference equation has been widely used to model dynamic discrete time systems to efficiently characterize the dynamic properties of damping and resonance of systems in physics and engineering. The reason is that the roots of the characteristic equation of second-order dynamic equation may be a real double root, two real roots or conjugate roots, which could easily describe a system with undamping, overdamping, critical damping, underdamping or oscillation, dependent on the specification of their coefficients [17]. These characteristics cannot be easily described by the first-order dynamic equation. Therefore the second stochastic equation is employed to characterize the biochemical processing of the gene expression.

Evidently, the transcriptional regulatory function $G_i(k)$ controls the synthetic rate of cDNA and the clue of upstream regulatory pathway is involved in $G_i(k)$. Therefore we emphasize how to detect the upstream regulatory function $G_i(k)$ from expression data $X_i(k)$ and this dynamic model equation in Eq. (9.1). In general, it is not easy to extract transcriptional regulatory function $G_i(k)$. In order to extract the input regulatory function $G_i(k)$, we apply the Fourier decomposition method to decompose $G_i(k)$ to generate some harmonic sinusoid functions. When the extraction problem is reduced to a simple parameter estimation problem, $G_i(k)$ can be decomposed by the following Fourier series.

$$G_i(k) \approx \sum_{n=0}^{N} \left[\alpha_n \cos(nk) + \beta_n \sin(nk)\right]. \tag{9.2}$$

Then we need to estimate the parameters of Fourier series, α_n, and β_n, that are the magnitudes of different harmonics of sinusoid functions ($\cos(nt)$ and $\sin(nt)$) for $n = 0, \ldots, N$. The Fourier series is a good tool to synthesize functions with finite energy by harmonic functions in respect of engineering.

Extraction of the transcriptional regulatory function $G_i(k)$

Since $G_i(k)$ has been decomposed, we combine Eqs. (9.1) and (9.2) to get the following dynamic model equation for the expression profile of the ith gene,

$$X_i(k) = -a_i X_i(k-1) - b_i X_i(k-2) + \sum_{n=0}^{N} \left[\alpha_n \cos(nk) + \beta_n \sin(nk)\right] + \varepsilon_i(k). \tag{9.3}$$

In the above dynamic model equation, the parameters a_i, b_i, α_n, and β_n should be estimated by the time profile of expression data of the ith gene in linear scale, i.e., these parameters should be specified so that the simulating output $X_i(k)$ of the dynamic model in Eq. (9.3) should match the expression profile of the ith gene. The maximum likelihood estimation method is employed to estimate these parameters a_i, b_i, α_n, and β_n in Eq. (9.3) from the expression profile $X_i(k)$ in Appendix A.

After the parameters α_n, and β_n of the regulatory function $G_i(k)$ have been estimated, we can present the regulation detection $\hat{G}_i(k)$ as follows,

$$\hat{G}_i(k) = \sum_{n=0}^{N} \left[\hat{\alpha}_n \cos(nk) + \hat{\beta}_n \sin(nk)\right], \tag{9.4}$$

where $\hat{\alpha}_n$, and $\hat{\beta}_n$ are the estimates of α_n, and β_n, respectively.

We know that the input transcriptional regulatory function $G_i(k)$ of the target gene of the HeLa cell cycle is often relative to the bindings of transcription factors or some interactions from the upstream regulators. In the next step, we will trace back the corresponding regulatory genes from the input regulatory function $\hat{G}_i(k)$ of the target gene.

Iterative algorithm for constructing GRN

In biology, the specific biochemical reactions are usually relative to the concentration of specific products. For this purpose, we describe the regulatory function as the following sigmoid function to describe the binding and nonbinding of transcription factors to motif binding sites [7,13]

$$\tilde{X}_j(k) = \frac{1}{1 + e^{-\gamma(X_j(k) - M_j)}}, \tag{9.5}$$

where γ is the transition rate and M_j is the mean expression of the jth regulatory gene's profile $X_j(k)$.

We determine R_i regulatory genes whose regulatory signals $X_j(k)$ $j = 1,\ldots,R$ are the most correlative to the target gene profile $X_i(k)$ of the ith target gene. Then, we could reconstruct the GRN by tracing back the upstream regulators from the extracted regulatory function $\hat{G}_i(k)$, which are contributed by R_i regulatory genes, via the following biochemical kinetic relationship,

$$\hat{G}_i(k) = c_{i0} + \sum_{j \in R_i} c_{ij} \tilde{X}_j(k) + e_i(k), \tag{9.6}$$

where c_{ij} is the pathway kinetic parameters from the regulatory gene j to target gene i, R_i represents the number of the searched upstream regulatory genes selected by the absolute value of correlation coefficient between the target gene expression profile and the regulatory gene expression profile which is more than 0.8 based on the 95% confidence of normalized correlation coefficients of expression profiles of total cell cycle-related genes. In Whitfield et al. [3], the constant c_{i0} is the basal level denoting the regulatory function other than upstream regulatory genes, e.g., due to posttranscriptional regulation, and $e_i(k)$ is the error or the noise of the network model.

Using the maximum likelihood algorithm to estimate the parameters c_{i0} and c_{ij} from $\hat{G}_i(k)$ and $\tilde{X}_j(k)$, the regulation from the upstream regulators is identified as (see Appendix A)

$$\hat{G}_i(k) = \hat{c}_{i0} + \sum_{j \in R_i} \hat{c}_{ij}\tilde{X}_j(k), \tag{9.7}$$

By combining Eq. (9.1) and the above equation, the dynamics of transcriptional regulatory network of the HeLa cell cycle can be represented by the following identified difference equation,

$$X_i(k) = -\hat{a}_i X_i(k-1) - \hat{b}_i X_i(k-2) + \hat{c}_{i0} + \sum_{j \in R_i} \hat{c}_{ij}\tilde{X}_j(k), \tag{9.8}$$

where $i = 1,2,\ldots$ for all profile target genes in HeLa cell cycle.

In fact, Eq. (9.8) contains much information for exploring the regulatory network of each specific target gene of the HeLa cell cycle. The regulatory genes, which belong to a specific set, R_i, represent the potential upstream regulators for target gene i. The estimated chemical kinetic parameter, $\hat{c}_{ij}$, characterizes the type and intensity of the influence of the jth regulatory gene on the ith target gene, in which positive sign indicates activation and a negative sign indicates repression, and the magnitude is defined as the regulatory ability. After the regulatory pathways of the ith target gene is constructed by tracing back their upstream regulatory genes, these R_i upstream regulators are considered as target genes again to trace back their upstream regulatory genes. Iteratively, we can construct the whole GRN of the HeLa cell cycle globally. The goal of reverse engineering GRN is to deduce the possible set of regulators and to identify their associated regulation abilities by the available data set from the dynamic system perspective. For this purpose, we devise a novel algorithm based on the dynamic gene expression model for searching possible upstream regulators and then identifying the relevant regulatory abilities $\hat{c}_{ij}$ according to Eq. (9.8).

9.1.2 Investigating tumorigenesis mechanisms by GRN of cancer cell cycle through microarray data

Data processing and analysis

Data were extracted by superimposing a grid over each array using GenePix 3.0 software (Axon Instruments). Spots of poor quality, determined by visual inspection, were removed from further analysis. Data of HeLa cell collected for each array were stored in the Stanford Microarray Database (SMD) and are available from SMD at http://genome-www5.stanford.edu/ [3,18].

We combine 775 cell cycle-related genes from the human expression of HeLa cell cycle-regulated genes according to the classification by Whitfield et al. (2002) with human expression of cell cycle-regulated genes according to the traditional classification as the target genes. After that, we select 64 transcription factors (Table 9.2) from the 775 cell cycle genes [3].

The raw data were transformed into a linear scale from the original log ratio and applied to this approach. Following the dynamic model in Eq. (9.8), the parameters which characterize the dynamic regulatory mechanism are estimated successfully for each target gene in the pathway. Fig. 9.1 compares the simulation results of the dynamic expression model in Eq. (9.8) with the experimental expression profiles for some important cell cycle-related genes regulated by E2F family [19], such as *CDC25A* [16,20,21], *MCM6* [5,16,22], *E2F1* [19], *CDC6* [16,21], *E2F2* [16,20], *MCM5* [5,16], *MCM4* [5,16], *PCNA* [5,16,20,22], *RFC4* [16,22], and *DHFR* [5]. The extracted regulatory functions $\hat{G}_i(k)$ of these genes which are estimated in Table 9.1 by the maximum likelihood algorithm in Appendix A from their expression profiles are shown in Fig. 9.2. The extracted regulatory function $\hat{G}_i(k)$ in Fig. 9.2 is employed to estimate the kinetic parameters of GRN in Eq. (9.8) by the parameter estimation scheme in Appendix A. This iterative algorithm can find the most likely regulatory genes that may participate in the expression program of HeLa cell cycle genes.

Inference of the regulatory pathway

For illustration purposes, the inferring strategy is applied to the E2F target genes (Bracken et al., 2004) in HeLa cell cycle pathways to recognize their upstream regulatory genes. E2F transcription factors are well studied owing to their importance in both cell cycles [20,23]. Their regulatory abilities are shown in the upstream regulatory functions $\hat{G}_i(k)$ of dynamic equation in Table 9.2. Parameters of regulatory functions $\hat{G}_i(k)$ in Table 9.2 represent the regulatory abilities and sensitivities of the relative transcription factors. It is very exciting that E2F1 and E2F2 are found to be active regulators in most E2F target genes listed in Table 9.2, which agree very well with the previous results [24–26]. The regulatory abilities of the related regulators implying different degrees of influence are converted to the *red-colored lines* (dark gray in print versions) as positive regulations (activations) and the *black-colored lines* as negative regulations (inhibitions) for each target gene. Then, based on the dynamic regulatory equations in Table 9.2, the pathways of E2F target gene in HeLa cell cycle regulatory system are described in Fig. 9.3. The coefficients of these dynamic regulatory equations represent the kinetic activities of regulatory genes. If a regulatory gene is with a large kinetic parameter in the dynamic regulatory equation, it will play an important role in HeLa cell cycle and is more sensitive to the gene expression of target gene.

Table 9.2 Upstream regulatory TFs and their regulatory function $\hat{G}_i(k)$ on E2F target genes in cancer cell cycle

1. $$\hat{G}_{\varpi v25A}(k) = 12.159 \bullet \widetilde{X_{DMTI2}(k)} + 11.804 \bullet \widetilde{X_{I2I2}(k)} - 12.473 \bullet \widetilde{X_{BROAZ}(k)} - 4.7281 \bullet \widetilde{X_{PONA}(k)} - 3.8787 \bullet \widetilde{X_{I2I2}(k)}$$

2. $$\hat{G}_{MCM6}(k) = 4.2865 \bullet X_{PONA}(k) + 4.1629 \bullet \widetilde{X_{I2I2}(k)} + 3.5672 \bullet \widetilde{X_{MNT}(k)} + 3.5642 \bullet \widetilde{X_{DMTT2}(k)} + 2.4202 \bullet \widetilde{X_{I2I2}(k)}$$
$$-12.765 \bullet \widetilde{X_{SP2}(k)}$$

3. $$\hat{G}_{I2I2}(k) = 19.97 \bullet \widetilde{X_{I2I2}(k)} + 3.898 \bullet \widetilde{X_{UHRT2}(k)} - 7.0755 \bullet \widetilde{X_{PONA}(k)}$$

4. $$\hat{G}_{\varpi \infty 6}(k) = 9.2779 \bullet \widetilde{X_{I2I2}(k)} + 8.1237 \bullet \widetilde{X_{I2I2}(k)} + 8.0354 \bullet \widetilde{X_{UHRT2}(k)} - 9.5837 \bullet \widetilde{X_{PONA}(k)}$$

5. $$\hat{G}_{I2I2}(k) = 5.557 \bullet \widetilde{X_{SP2}(k)} + 4.3598 \bullet \widetilde{X_{WRSA2}(k)} + 3.6731 \bullet \widetilde{X_{I2I2}(k)} + 3.0949 \bullet \widetilde{X_{UHRT2}(k)} - 12.012 \bullet \widetilde{X_{BROA2}(k)} - 9.9406 \bullet \widetilde{X_{PONA}(k)}$$
$$- 2.4365 \bullet X_{MNT}(k) - 1.9962 \bullet X_{DMIT2}(k)$$

6. $$\hat{G}_{MCMS}(k) = 6.7316 \bullet \widetilde{X_{I2I2}(k)} - 9.2368 \bullet \widetilde{X_{PONA}(k)}$$

7. $$\hat{G}_{MCMS}(k) = 9.8915 \bullet \widetilde{X_{I2I2}(k)} + 8.8902 \bullet X_{PONA}(k) + 6.3504 \bullet X_{BROA2}(k) + 4.7006 \bullet X_{MNT}(k) + 2.8582 \bullet X_{WJIX2}(k) - 13.12 \bullet X_{SP2}(k)$$
$$- 7.1211 \bullet \widetilde{X_{DMTI2}(k)} - 4.6503 \bullet X_{PONA}(k) - 4.4148 \bullet \widetilde{X_{LASS6}(k)}$$

8. $$\hat{G}_{PONA}(k) = 8.1178 \bullet X_{I2I2}(k)$$

9. $$\hat{G}_{RICA}(k) = 18.628 \bullet \widetilde{X_{DMTT2}(k)} + 15.158 \bullet \widetilde{X_{PONA}(k)} + 14.923 \bullet \widetilde{X_{I2I2}(k)} + 12.721 \bullet \widetilde{X_{BOLAT2}(k)} + 5.6902 \bullet \widetilde{X_{HOIO2}(k)} - 16.276 \bullet \widetilde{X_{SP2}(k)}$$
$$- 11.268 \bullet \widetilde{X_{LASS6}(k)} - 7.872 \bullet \widetilde{X_{I2I2}(k)} - 6.8745 \bullet \widetilde{X_{KLT9}(k)} - 6.1689 \bullet \widetilde{X_{STAT2}(k)} - 2.7893 \bullet \widetilde{X_{BROMZ}(k)} - 2.4802 \bullet \widetilde{X_{WJIx2}(k)} - 2.2907 \bullet \widetilde{X_{PONA}(k)}$$

10. $$\hat{G}_{DHIR}(k) = 10.711 \bullet \widetilde{X_{MNT}(k)} + 6.9038 \bullet \widetilde{X_{PONA}(k)} + 5.6045 \bullet \widetilde{X_{SRT}(k)} + 1.6341 \bullet \widetilde{X_{I2I2}(k)} - 13.422 \bullet \widetilde{X_{WJIX2}(k)} - 1.6894 \bullet \widetilde{X_{BOLAT2}(k)} - 1.5479 \bullet \widetilde{X_{BROA2}(k)}$$

The positive sign implies activations while the negative sign implies inhibitions for each target gene. The magnitudes indicate their regulatory abilities to the downstream target genes.

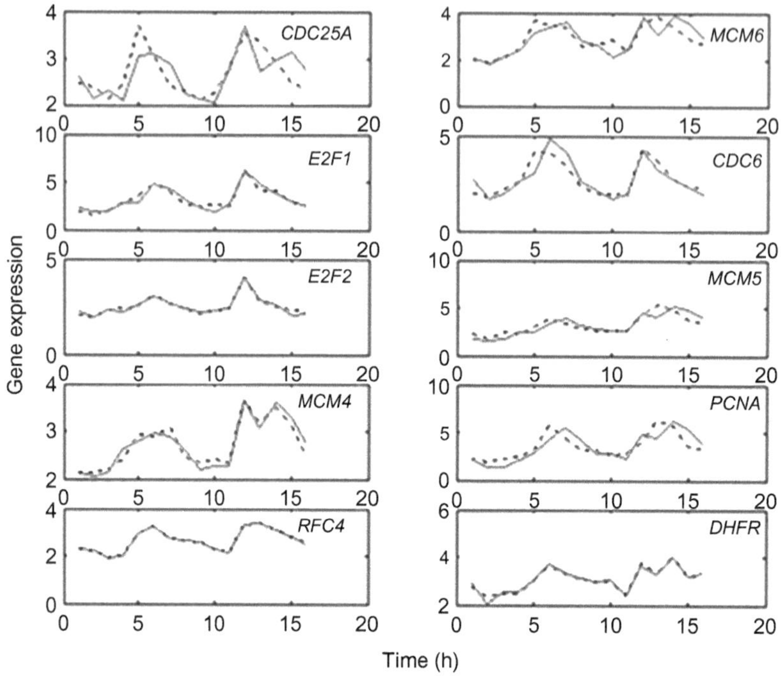

Figure 9.1 The second-order dynamic model fitting of E2F target genes. The expression profile with the corresponding second-order dynamic model fitting for E2F target genes, *CDC25A, MCM6, E2F1, CDC6, E2F2, MCM5, MCM4, PCNA, RFC4,* and *DHFR*. The *red dashed lines* (dark gray in print versions) are the microarray data from Whitfield et al. (2002), and the *blue solid lines* (light gray in print versions) are the estimated dynamic evolution of expression data by the proposed method. Obviously, the proposed second-order dynamic model could efficiently characterize the dynamic properties like damping and resonance of the GRN. The data have been transformed from the log scale in Whitfield et al. to linear scale.

Based on the dynamic regulatory modeling, the 152 E2F target genes are found and shown in Table 9.3. In these 152 E2F target genes, 6 genes match the E2F target genes found by Elkon et al. (2003) from 124 E2F target genes edited by Ren et al. (2002) and 17 genes match the E2F target genes found by Elkon et al. (2003) from 872 periodic genes edited by Whitfield et al. (2002). Ren et al. (2002) have found 124 E2F target genes with E2F binding promoter.

After finishing the construction of the first layer network of E2F target genes [19], we take these upstream regulators as target genes. By using a similar method, we construct the upstream regulatory genes of these target genes. In the second layer network, the regulatory abilities implying different degrees of influence are converted into *pink-colored lines* as positive regulations (activations) and *blue-colored lines* as negative regulations (inhibitions). Then, we combine the first layer network and the

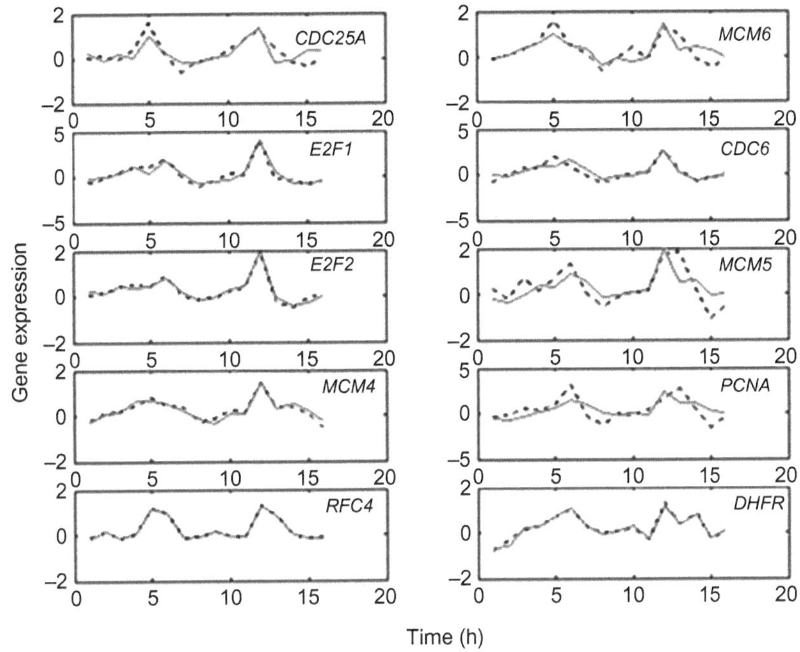

Figure 9.2 The extracted upstream regulatory functions from expression profiles [3] and their fitting by upstream regulatory genes. The upstream transcriptional regulatory functions $\hat{G}(t)$ extracted from expression profiles of corresponding E2F target genes are denoted by the *red dashed lines* (dark gray in print versions), and the *blue solid lines* (light gray in print versions) are fitted by their upstream transcriptional regulatory functions in Eq. (9.6). These results show that the upstream TF regulatory model in Eq. (9.6) could match the regulatory function extracted from microarray data. In the figure, the gene expressions are expressed in linear scale.

second layer network together to form a more complete network to E2F target genes in HeLa cell cycle as described in Fig. 9.4. Iteratively, we can construct the higher layer network to complete the GRN of HeLa cell cycle.

9.1.3 Discussion of GRN of cell cycle for tumorigenesis mechanism of HeLa cells

The losses of cellular regulation give rise to most cases of cancer. The transcription factors are crucial for regulating cell cycle progression and may be related to the development of a cancer. Therefore, to understand these gene regulatory processes, we need to unravel the regulatory mechanisms of these transcription factors in cell cycle. This section presents a systematically iterative approach to discern and characterize the transcriptional regulatory network of 775 cell cycle-related genes from the raw expression profiles of the HeLa cell [3]. Because the transcriptional regulatory

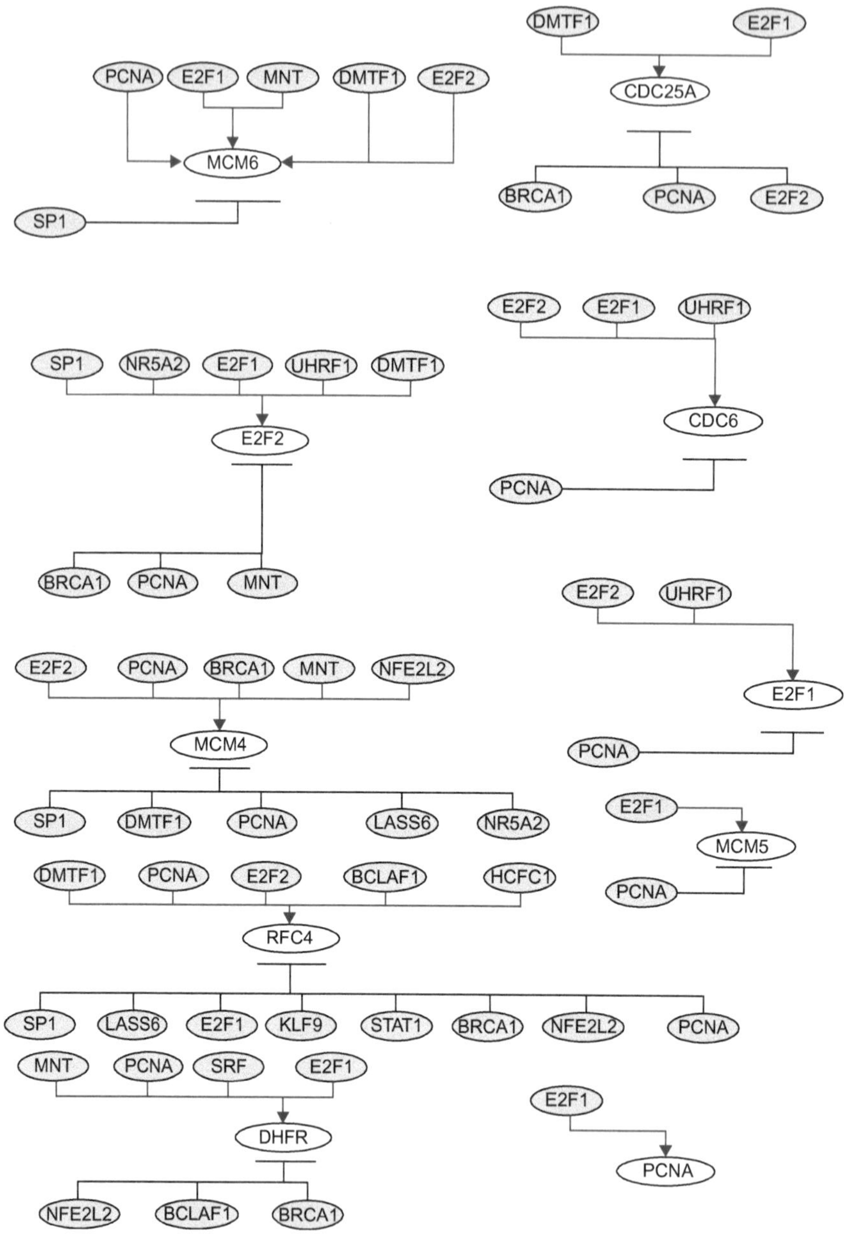

Figure 9.3 The regulatory pathways of E2F target genes in cancer cell cycle based on the dynamic regulatory modeling in Table 9.2. The related genes are represented as *gray ovals* and the regulatory abilities of the related regulators implying different degree of influence are converted into *red-colored lines* (dark gray in print versions) as positive regulations (activations) and *black-colored lines* as negative regulations (inhibitions) for each target gene. These regulatory pathways could be integrated as the regulatory network of cancer cell cycle in Fig. 9.4.

Table 9.3 The 152 E2F target genes in the gene regulatory network based on the dynamic regulatory modeling

BAIAP2	*TOP3A*	*ACYP1*	*BIVM*	*ADCK2*	*OGT*	*C20orf55*
SSR3	*MED31*	*CDC6*[a,b]	*FLJ10154*	*FANCG*	*PILRB*	*MET*
SLBP[b]	*NASP*	*ADCY6*	*MGC15716*	*APEX2*	*FLJ11021*	*NR5A2*
MNT	*CDC25A*[a,b]	*ANKRD10*	*ABCA7*	*LOC197336*	*ASF1B*	*CDKN2D*
LOC221955	*UNG*[a,b]	*BARD1*	*FLJ13912*	*RBBP8*	*SFRS5*[b]	*CENPE*
LOC90110	*MGC3207*	*E2F2*[a]	*Pfs2*	*MCM8*	*C22orf18*	*USP16*
FBXL20	*UBQLN2*	*MCM5*[a,b]	*EGFL5*	*C20orf111*	*LOC58486*	*ARHGAP8*
ANKRD25	*RAMP*	*CCRK*	*NKTR*	*ABCC5*	*MAP3K2*	*PSEN1*
KIAA1529	*MCM6*[a,b]	*TREX1*	*KIAA0092*	*MLF11P*	*EZH2*	*C6*
PTOP	*DMTF1*[b]	*RAB23*[b]	*LOC200895*	*FANCA*	*CENTB5*	*USP6NL*
TRIM26	*KIAA0738*	*FLJ10618*	*HSPB8*	*DNAJC6*	*HIST3H2A*	*LOC51334*
ZMYND19	*PNN*	*OACT1*	*C14orf130*	*RAD18*	*BRCA1*	*AKAP13*
CDCA7	*MAP2K6*	*ORC1L*[b]	*KIAA1586*	*SP1*	*FLJ35740*	*AGPAT3*
CASP8AP2[b]	*GMNN*	*MCM4*[b]	*HELLS*	*USP1*	*HSPC150*	*ASAM*
PLCXD1	*LOC283596*	*MSH2*	*CLSPN*	*MYCBP2*	*ORC3L*	*RAB3A*
ZNF367	*CASP2*	*LCHN*	*RFC4*	*TOPBP1*[b]	*DNAJB4*	*HIST2H2BE*
RNP	*C9or42*	*FLJ20280*	*DONSON*	*DHFR*	*RRM1*	*HIST2H2AA*
USP53	*E2F1*[b]	*PCNA*	*AP4B1*	*DKZP434B168*[b]	*RHOBTB3*	*H2AFY*
SKP2	*CCNE2*[b]	*MBD4*	*MGC2610*	*DLNB14*	*KIAA0841*	*H3F3A*
PDXP	*SCML1*	*TLOC1*	*ATAD2*	*RFC2*	*NFE2L2*	*HIST1H3D*
FLJ20530	*SERPINB3*	*SDC1*	*RPA2*[b]	*NSUN3*	*BBS2*	
PANK2	*UHRF1*	*PRPS2*	*FEN1*	*FKBP6*	*FANCD2*	

The marked letters in superscript are the matched E2F target genes found by Elkon et al. (2003).
[a]E2F target genes found by Elkon et al. (2003) from 124 E2F target genes edited by Ren et al. (2002).
[b]E2F target genes found by Elkon et al. (2003) from 872 periodic genes edited by Whitfield et al. (2002).

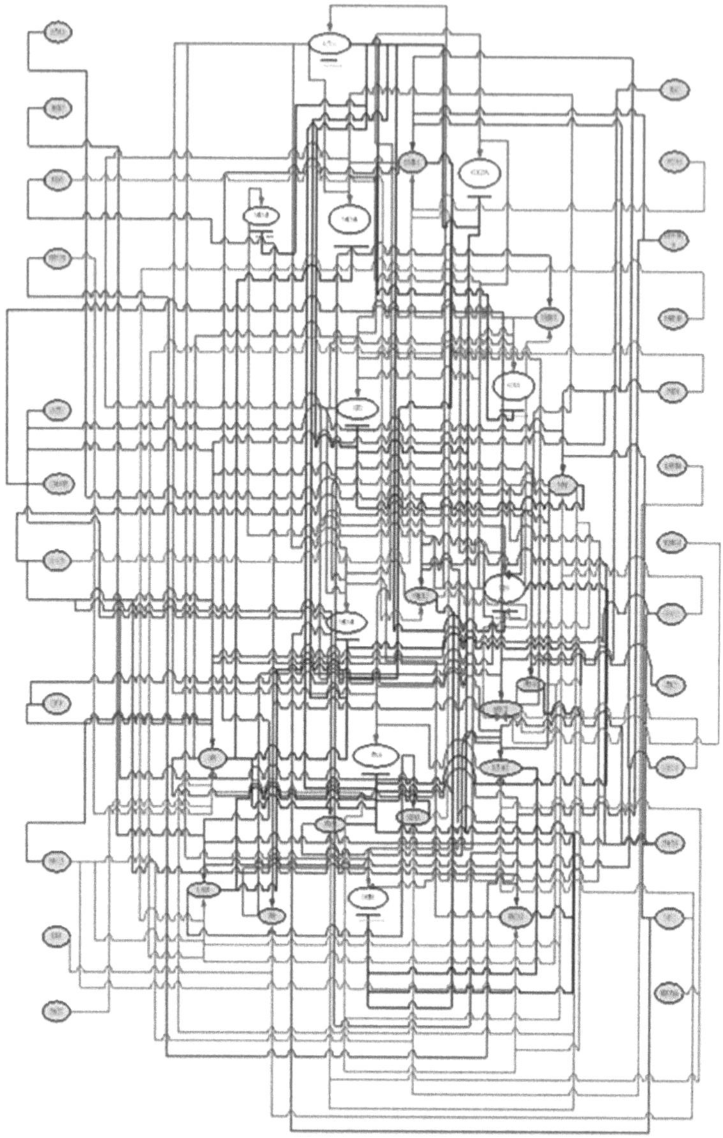

Figure 9.4 The network of E2F target genes in cancer cell cycle based on the dynamic regulatory modeling. The regulatory pathway of the *i*th target gene is constructed by tracing back its upstream regulatory genes. Then, these upstream regulatory genes are considered as target genes again to trace back their upstream regulatory upstream regulatory genes. Iteratively, we can construct the whole GRN of cancer cell cycle globally. In the first layer of the upstream regulators, *red solid lines* represent positive regulation and *black solid lines* represent negative regulation, and in the second layer of the upstream regulators, *pink solid lines* represent positive regulation and *blue solid lines* represent negative regulation. Based on this procedure, this E2F regulatory network is the interaction of regulatory pathways in Fig. 9.3.

network of 775 cell cycle-related genes is very complicated, two miniature GRNs of E2F family during G1/S phase in Fig. 9.4 and the other family during G2/M in Fig. 9.7 are given to illustrate the regulatory mechanism of HeLa cell.

Our approach also offers the following advantages. First, based on the dynamic regulatory model, a GRN of cancer cell could be constructed by the extracted upstream regulatory function through microarray data. Then, the identified regulatory ability for each specific regulator could evaluate the contribution of this regulator; the positive sign stands for activation and the negative sign stands for repression, and the magnitude represents the significance. These advantages of the proposed approach will improve the analysis to cope with rapidly growing microarray data of human. *BRCA1* is one of the important cell cycle-related genes to play as a transcriptional repressor in cell cycle progress [27]. This finding matches the GRN constructed by dynamic regulatory model (shown in Figs. 9.4 and 9.5).

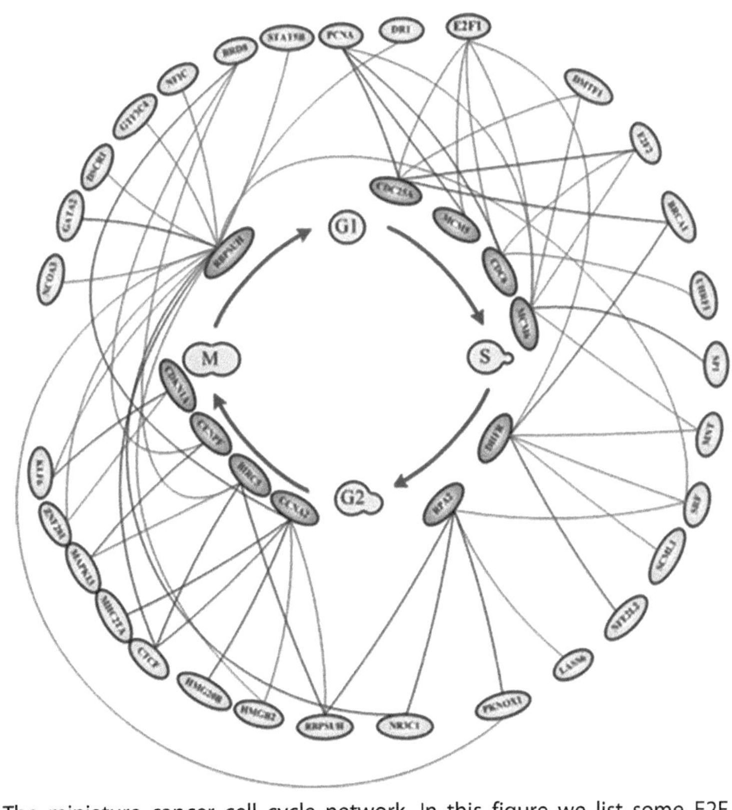

Figure 9.5 The miniature cancer cell cycle network. In this figure we list some E2F target genes and show the role of E2F transcription factors in cancer cell cycle [20]. *Red solid lines* (dark gray in print versions) represent positive regulations and *black solid lines* represent negative regulations.

It is clear that E2F regulates the expression of a host of factors that function during G1/S transition and S phase even the whole cell cycle [19]. E2F is best known for its role in regulating the transcription of genes that positively affect cell cycle progression [28–33]. These results almost match this finding (shown in Figs. 9.4 and 9.5). Hayashi with his collaborator found that E2F1 activates human *MCM8*. In this section, we found that E2F1 activates human *MCM5* and *MCM6* (shown in Figs. 9.4 and 9.5). Elkon et al. (2003) and Ren et al. (2002) have found that human *MCM5* and *MCM6* are two E2F target genes (shown in Table 9.3). As a result of these studies, we infer that human *MCM5* and *MCM6* may be positively regulated directly by E2F1 in HeLa cell cycle. Accumulating evidence indicates that *cdc25A* possesses oncogenic properties. Recently, overexpression of *cdc25A* was found in many breast, head, and neck cancers [34]. *CDC6* plays a critical role in the regulation of the onset of DNA replication in eukaryotic cells and *Cdc6* expression is downregulated in prostate cancer [35]. From the results of Table 9.3, we could infer that E2F directly regulates *MCM4*, *MCM5*, *CDC6*, *CDC25A*, *UNG*, and *E2F2* in HeLa cell cycle progression. In Fig. 9.5, we represent some E2F target genes and show the importance of E2F transcription factor in HeLa cell cycle progression. In Fig. 9.6, we also predict the probable transcription regulations in some target genes, which express in G2/M phase of human cell cycle as shown in the regulatory network of Fig. 9.7. Further, we can construct not only E2F-related regulatory network but also the whole HeLa cell cycle network if we have genome-wide microarray data and CHIP data. However, at present, we still need to get enough evidence and CHIP experiments in the same cellular system to confirm this GRN of HeLa cell cycle [19].

Based on the dynamic regulatory modeling, the 152 E2F target genes are found and shown in Table 9.3. These target genes may be regulated by E2F directly or indirectly. In these 152 E2F target genes, 6 genes match the E2F target genes found by Elkon et al. (2003) from 124 E2F target genes with E2F promoter edited by Ren et al. (2002) and 17 genes match the E2F target genes found by Elkon et al. (2003) from 872 periodic genes in HeLa cell cycle edited by Whitfield et al. (2002).

Finally, in order to validate the proposed approach, an independent validation is also given by randomly reshuffling the time order of microarray experiment but with the same choices of target gene and regulatory genes, to confirm the reliability of the proposed method as shown in Fig. 9.8. As shown previously, BRCA1 plays as a transcriptional repressor in the cell cycle progress [27] (shown in Fig. 9.5). From the shuffling results shown in Fig. 9.8, BRCA1 becomes a transcriptional activator. It is clearly seen that the proposed HeLa cell cycle regulatory pathway in Fig. 9.5 is destroyed by reshuffling the experimental data.

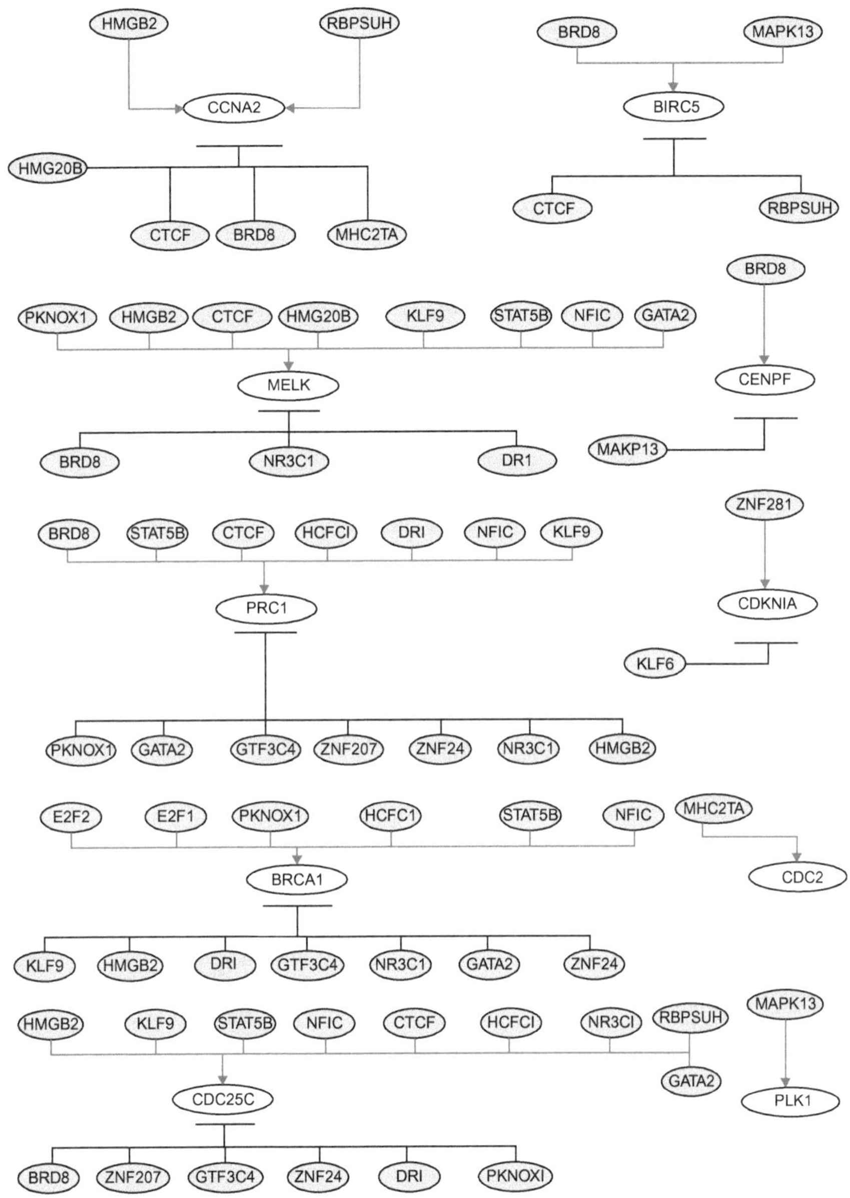

Figure 9.6 The regulatory pathways of target genes, which expressed in G2/M phase of human cell cycle, are based on the dynamic regulatory modeling. The related genes are represented as *gray ovals* and the regulatory abilities of the related regulators implying different degree of influence are converted into *red-colored lines* (dark gray in print versions) as positive regulations (activations) and *black-colored lines* as negative regulations (inhibitions) for each target gene.

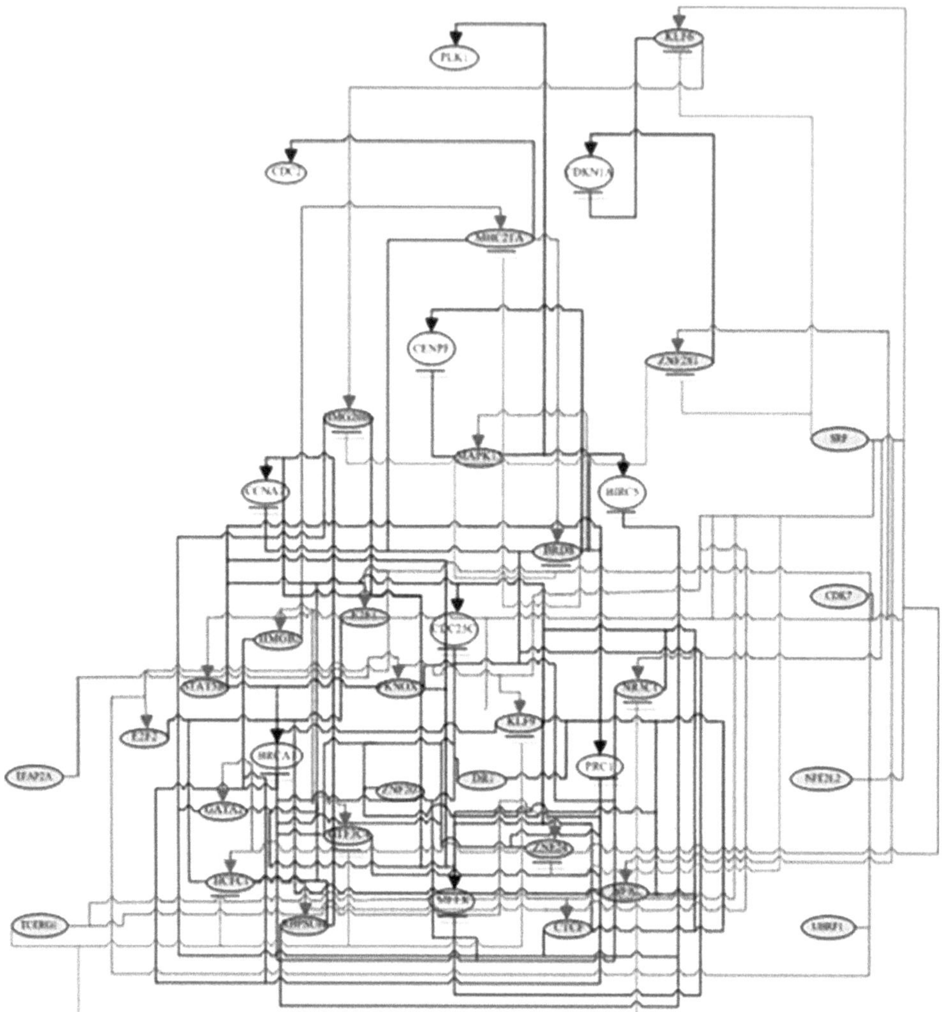

Figure 9.7 The gene regulatory network of in G2/M phase of cancer cell cycle based on the dynamic regulatory modeling and the interactions of regulatory pathways in Fig. 9.6.

APPENDIX A: METHODS

Maximum likelihood estimation of a_i, b_i, α_n, and β_n

The dynamic Eq. (9.3) must match the expression profile at all time points and then is arranged in a vector difference form. Consequently, the vector dynamic form of this equation is applied to m time points of expression profile in order to make the dynamic model work.

$$X_{im} = -a_i X_{im-1} - b_i X_{im-2} + \sum_{n=0}^{N} [\alpha_n \cos(n) + \beta_n \sin(n)] + \varepsilon_{im}, \qquad (A.1)$$

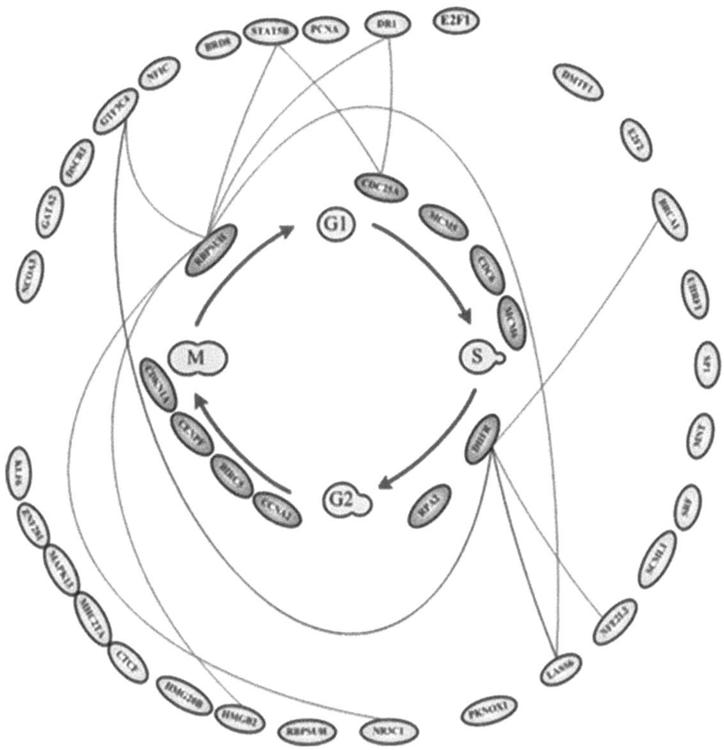

Figure 9.8 The miniature cancer cell cycle network in Fig. 9.5 is repeated as independent validation by randomly reshuffling the time order of microarray experiment but with the same choices of target and regulatory genes. Obviously, the proposed regulatory network in Fig. 9.5 is destroyed by the reshuffling of experimental data. *Red solid lines* (dark gray in print versions) represent positive regulations and *black solid lines* represent negative regulations. For example, BRCA1 in Fig. 9.5 plays as a transcriptional repressor in cell cycle progress [27]. From this figure. BRCA1 becomes a transcriptional activator.

where

$$X_{im} = \begin{bmatrix} X_i(3) \\ \vdots \\ X_i(m) \end{bmatrix}, \quad X_{im-1} = \begin{bmatrix} X_i(2) \\ \vdots \\ X_i(m-1) \end{bmatrix}, \quad X_{im-2} = \begin{bmatrix} X_i(1) \\ \vdots \\ X_i(m-2) \end{bmatrix}$$

$$\cos(n) = \begin{bmatrix} \cos(n3) \\ \vdots \\ \cos(nm) \end{bmatrix}, \quad \sin(n) = \begin{bmatrix} \sin(n3) \\ \vdots \\ \sin(nm) \end{bmatrix}, \quad \varepsilon_{im} = \begin{bmatrix} \varepsilon_i(3) \\ \vdots \\ \varepsilon_i(m) \end{bmatrix}$$

Next, we translate Eq. (A.1) into a matrix form,

$$Y_i = A_i\Phi_i + E_i, \tag{A.2}$$

where $\Phi_i = \begin{bmatrix} a_i \\ b_i \\ \alpha_0 \\ \beta_0 \\ \vdots \\ \alpha_N \\ \beta_N \end{bmatrix}$ and $E_i = \varepsilon_{im}$ are in vector forms, and $A_i = \begin{bmatrix} -X_{im-1} \\ -X_{im-2} \\ \cos(0) \\ \sin(0) \\ \vdots \\ \cos(N) \\ \sin(N) \end{bmatrix}$ is a matrix.

Then we use the maximum likelihood estimation to derive the optimal parameters estimation of $\hat{\Phi}_i$ [15].

We assume that each element in the error vectors, $\varepsilon_i(k)$, $k = \{3,\ldots,m\}$, is an independent random variable with a normal distribution with zero mean and variance σ^2, and we will estimate the parameter $\hat{\Phi}_i$ by the maximum likelihood method.

$$p(Y_i|\Phi_i, \sigma^2) = \frac{1}{\sqrt{2\pi\sigma^2}}\exp\left\{-\frac{[Y_i - A_i\Phi_i]^T[Y_i - A_i\Phi_i]}{2\sigma^2}\right\}, \tag{A.3}$$

The log-likelihood function for given m data points is then described by

$$\log L(\Phi_i, \sigma^2) = -\frac{m-2}{2}\ln\left[2\pi\sigma^2\right] - \frac{1}{2\sigma^2}\sum_{k=1}^{m}[Y_i - A_i\Phi_i]^T[Y_i - A_i\Phi_i], \tag{A.4}$$

Here we expect the log-likelihood function to have the maximum at $\Phi_i = \hat{\Phi}_i$ and $\sigma_i^2 = \hat{\sigma}_i^2$. The necessary condition for maximum likelihood estimates $\hat{\Phi}_i$ and $\hat{\sigma}_i^2$ as follows [15].

$$\begin{aligned} \left.\frac{\partial \log L(\Phi_i, \sigma_i^2)}{\partial \Phi_i}\right|_{\Phi_i = \hat{\Phi}_i} &= 0 \\ \left.\frac{\partial \log L(\Phi_i, \sigma_i^2)}{\partial \sigma_i^2}\right|_{\sigma_i^2 = \hat{\sigma}_i^2} &= 0 \end{aligned}, \tag{A.5}$$

The estimated parameters $\hat{\Phi}_i$ and $\hat{\sigma}_i^2$ are shown below.

$$\hat{\Phi}_i = (A_i^T A_i)^{-1} A_i^T Y_i, \tag{A.6}$$

$$\hat{\sigma}_i^2 = \frac{1}{m-2}(Y_i - A_i\Phi_i)^T(Y_i - A_i\Phi_i), \tag{A.7}$$

Theoretically, E_i is just the noise of the gene expression profile of the microarray chips, but some modeling errors and approximation errors in Eq. (9.2) are also involved in E_i. So taking the modeling error and approximation error in our consideration makes this dynamic model equation more approach the actual situation.

The number of Fourier series N is determined by tradeoff between the computational complexity of parameter estimation in Eq. (A.6) and the accuracy of approximation in Eq. (9.2). According to this expression data, we choose $N = 16$ to make the synthesis of these harmonics be the best approximation to the expression data.

Parameter estimation of c_{i0} and c_{ij}

To estimate the pathway kinetic parameters c_{ij} in Eq. (9.6) by $\hat{G}_i(k)$ and $\tilde{X}_j(k)$ with m time points of upstream regulatory expression profile, Eq. (9.6) is represented with algebraic form as follows:

$$\hat{G}_i = B_i\Omega_i + V_i, \tag{A.8}$$

where $\hat{G}_i = \begin{bmatrix} \hat{G}_i(3) \\ \vdots \\ \hat{G}_i(m) \end{bmatrix}$, $B_i = \begin{bmatrix} 1 & \tilde{X}_1(3) & \cdots & \tilde{X}_{R_i}(3) \\ \vdots & \vdots & \ddots & \vdots \\ 1 & \tilde{X}_1(m) & \cdots & \tilde{X}_{R_i}(m) \end{bmatrix}$, $\Omega_i = \begin{bmatrix} c_{i0} \\ \vdots \\ c_{iR_i} \end{bmatrix}$, and $V_i = \begin{bmatrix} e_i(3) \\ \vdots \\ e_i(m) \end{bmatrix}$.

We assume that each element in the error vectors, $e_i(k)$, $k = \{3,\ldots,m\}$, is an independent random variable with a normal distribution with zero mean and variance $\sigma_{e_i}^2$. Then by the similar procedure, the estimated parameters $\hat{\Omega}_i$ and $\sigma_{e_i}^2$ are shown below.

$$\hat{\Omega}_i = (B_i^T B_i)^{-1} B_i^T \hat{G}_i, \tag{A.9}$$

$$\sigma_{e_i}^2 = \frac{1}{m-2}(\hat{G}_i - B_i\Omega_i)^T(\hat{G}_i - B_i\Omega_i), \tag{A.10}$$

9.2 INVESTIGATING TUMORIGENESIS MECHANISMS BY CANCER-PERTURBED PPINs

After constructing and clarifying the dynamic GRN of cancer cell cycle for target genes to investigate the transcriptional tumorigenesis in HeLa cell in the previous section, we will construct normal and cancer PPINs in this section to investigate the perturbed PPIN in carcinogenesis and identify the potential drug targets in the following section.

Study of interactome, the entire set of molecular interactions within cells, has provided many insights into the etiology and regulation of cancer [36,37]. Tumorigenesis is a multistep process caused by genetic alterations that drive the progressive transformation of normal cells into malignant cells. At the molecular level, genetic mutations, translocations, amplifications, deletions, and viral gene insertions can alter translated proteins and thereby disrupt signal transduction pathways and PPIs that are essential for apoptosis and other important cellular processes [38]. Inactivation of pro-apoptotic proteins or upregulation of antiapoptotic proteins results in unchecked growth of cells and ultimately to cancer [39]. From a systems biology perspective, cancer is mainly caused by malfunctions of perturbed protein interaction networks in the cell [40,41].

Apoptosis is necessary for normal human development and survival, in that cells must die in order to prevent uncontrolled growth [42]. Apoptosis requires activation of multiple pathways via regulated PPIs [43]. Apoptosis is mediated by an intrinsic pathway, which is triggered by "death stimuli" (e.g., DNA damage, oncogene activation) within a cell, or by an extrinsic pathway, which is initiated by binding of an extracellular "death ligand." The extrinsic pathway can link to the intrinsic pathway, which then triggers the release of mitochondria proteins via PPIs [39,43,44]. During apoptosis, several proteins are released from the intermembrane space of the mitochondria into the cytoplasm and these proteins activate initiator caspases and trigger a series of PPIs in the caspase cascade. Evading apoptosis is one of the six acquired capabilities of cancer cells [38], and anticancer treatment using cytotoxic drugs is considered to mediate cell death by activating key elements of the apoptosis program and the cellular stress response [45]. Comprehensive knowledge of PPIs provides a framework for understanding the biology of cancer as an integrated system [46].

Most gene products mediate their functions within complex networks of interconnected macromolecules, forming a dynamic topological interactome [46,47]. High-throughput two-hybrid experiments [48,49] and several online interactome databases, such as BIND [50], HPRD (Human Protein Reference Database) [51], Intact [52], and Himap [53], allow analysis of the global topologies of human PPIs. BIND [54] is a database designed to store full descriptions of interactions, molecular complexes and pathways. HPRD [51] provides detailed data including protein sequences, localization, domains, and motifs, and thousands of PPIs, with other data. Intact [52] contains an enrichment of PPIs, related literature, and experimental detail. Himap [53] combines two data sets of yeast two-hybrid experiments [48,49] to form a human protein reference database [51], with references to functions and predictions.

However, experimental and database approaches often yield "false-positives" [55]. For example, yeast two-hybrid experiments based on transactivation of reporter genes require the presence of auto-activators, where the bait activates gene expression in the absence of any prey [46]. The yeast two-hybrid technique can yield false-positives (spurious interactions detected because of the high-throughput nature of the screening process), and false-negatives (undetected interactions) [54,55]. Computational methods can refine PPINs and result in fewer false-positives [56,57]. Because of the complex nature of interactomes, such as that observed in the apoptosome complex during caspase formation [42,43,56], a nonlinear mathematical model provides better characterization than a linear model [58,59]. In addition, a stochastic model allows consideration of intrinsic and extrinsic molecular "noise" that causes stochastic variations in transcription and translation [58]. In this section, we describe a nonlinear stochastic model that characterizes dynamic PPINs of apoptosis in cancerous and normal cells.

In this section, we built an initial PPIN based on two human yeast two-hybrid data sets [48,49] and four online interactome databases such as BIND [54], HPRD [51], Intact [52], and Himap [53]. Next, we constructed a nonlinear stochastic model of dynamic PPIs to eliminate false-positives from the network by applying a statistical method (Akaike Information Criterion, AIC) in Eq. (2.23) to the high-throughput protein interaction data. We regard all proteins in an organism as a large dynamic interaction system. PPIs are considered nonlinear stochastic processes with several expression profiles of interactive protein partners as input, and the expression profile of a target protein as output. Because of random noise and uncertainties during experiments, we describe PPI with stochastic discrete nonlinear dynamic equations. We considered linear individual (or binary) protein interactions and nonlinear cooperative protein complex interactions, but not DNA–protein or metabolite interactions. First, we constructed PPINs of apoptosis in HeLa cells and normal primary human lung fibroblasts based on microarray data [60]. Next, we obtained the cancer-perturbed PPIN by comparison of apoptosis in normal cells via gain-of-function and loss-of-function networks. Because current drugs designed to induce apoptosis kill cancer cells as well as normal cells, these cancer-perturbed PPINs allow identification of potential selective targets of apoptosis-promoting drugs [40].

9.2.1 Investigating tumorigenesis mechanism by cancer-perturbed PPINs of *apoptosis*

Construction of the cancer-perturbed PPINs of apoptosis

Initially, we selected proteins that are known to have roles in apoptosis and considered them as the "core nodes" of this network. These included *BAX* (BCL2-associated X protein), *BCL2* (B-cell CLL/lymphoma 2), *BID* (BH3 interacting domain death agonist), *CASP3* (caspase-3), *BIRC4* (baculoviral IAP repeat-containing 4), *CASP9* (caspase-9), *CYCS* (cytochrome c, somatic), and *DIABLO* (diablo homolog, Drosophila). Networks, such as ours, that are developed from initially selected genes or proteins as the core nodes are referred to as "BRAC-centered networks" [61]. This initial apoptosis network contained 207 protein nodes and 841 PPI edges.

From Eqs. (B.1) to (B.13) (see Appendix B), we calculated each protein interaction twice, with each partner considered as the target protein. The final results are shown in Fig. 9.9. Fig. 9.9 illustrates the individual and cooperative PPINs of apoptosis in cancerous cells (183 nodes and 552 edges) and normal cells (175 nodes and 547 edges). These networks are easily modeled by undirected graphs, where the nodes are proteins and two nodes are connected by an undirected edge if the corresponding proteins bind one another [62].

In order to identify drug targets for anticancer drugs, it is important to identify cancer-perturbed PPINs to identify drug targets to kill cancer cells [38]. If an interaction is absent in normal cells, but present in cancer cells, we call it "gain-of-function";

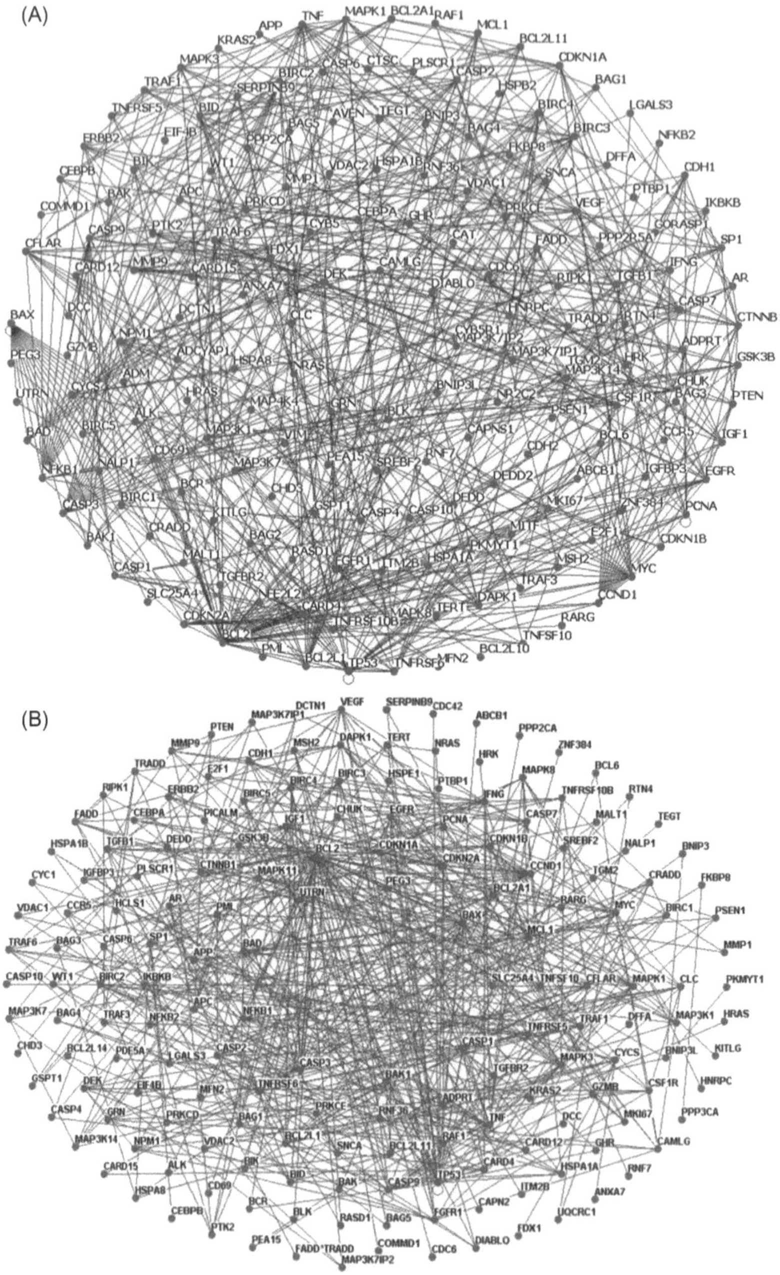

Figure 9.9 Global protein–protein interactions of apoptosis in cancerous and normal cells. (A) Apoptotic protein–protein interaction network in HeLa cells, showing 183 nodes and 552 edges. (B) Apoptotic protein–protein interaction network in normal primary lung fibroblasts, showing 175 nodes and 547 edges. Each interaction was calculated twice and only interactions with two "1" scores after AIC evaluation were considered "true" interactions. All protein–protein interaction networks in this section were constructed with Osprey version 1.2.0.

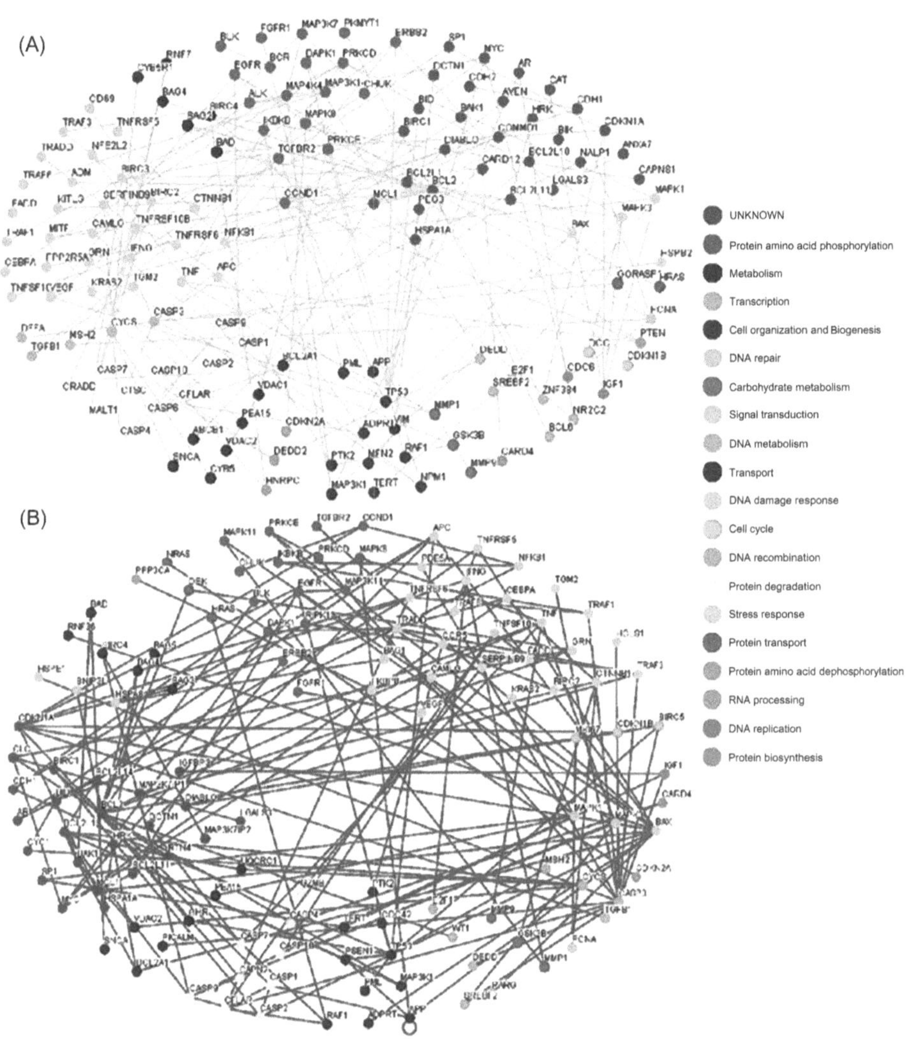

Figure 9.10 Cancer-perturbed protein–protein interactions in the apoptosis network. (A) "Gain-of-function" network, showing 140 nodes and 157 edges. (B) "Loss-of-function" network, showing 126 nodes and 162 edges. Colors of nodes represent GO annotations.

if an interaction is present in normal cells but not in cancerous cells, we term it "loss-of-function." For the 841 interactions that we identified, we classified 157 (18.7%) as gain-of-function and 162 (19.3%) as loss-of-function (Fig. 9.10A and B). This network analysis identified 38% (18.7% + 19.3%) of all PPIs during apoptosis as potential drug targets.

Fig. 9.10 shows nodes that are colored according to protein family, as annotated by the Gene Ontology (GO) hierarchy. Proteins are listed with GO annotations according to the number of perturbed interactions. All protein candidates with more than 5 degrees of perturbations are shown, to illustrate the number of links with perturbed nodes. Gain-of-function proteins with more than 5 degrees of perturbations in Fig. 9.10A include *BCL2*, *CASP3*, *TP53*, *BCL2L1*, *PRKCD*, *MAPK3*, *NFKB1*, *BIRC3*, *CCND1*, and *PCNA*. Loss-of-function proteins with more than 5 degrees of perturbations in Fig. 9.10B include *BCL2*, *BAX*, *CASP3*, *CDKN1A*, *TP53*, *BCL2L1*, *TNF*, *CASP9*, *EGFR*, *MAPK1*, *APC*, *TNFRSF6*, *BAK1*, *MYC*, *CFLAR*, and *APP*. The *BCL2* protein has the highest degree of perturbations (18 and 17) in cancerous and normal cells, respectively.

In order to confirm the topology of these networks (Fig. 9.10), we calculated the false-positive and false-negative rates for the 86 BCL2-interacting proteins in normal cells by use of the HPRD [51] and a literature review. After refinement by these algorithms, we reduced the false-positive rate to 1.16%. However, the false-negative rate remained at 41.87%, indicating incomplete construction of the network from current experiments and databases. Therefore, compensation by basal level k (see Eq. B.1) is important for estimation of model parameters.

Cancer-perturbed apoptosis mechanism at the systems level

In many cancers, pro-apoptotic proteins are inactivated or antiapoptotic proteins are upregulated, leading to unchecked growth and an inability to respond to cellular stresses [39]. These gain- and loss-of-function mutations lead to aberrations in PPINs. An integration of interactome data and genomic data can provide a clearer understanding of the functional relationships that underlie apoptosis and other biological processes [46,57]. The results of this section of the apoptosis mechanism at the systems level and the elucidation of cancer-perturbed PPIN topology are depicted in Fig. 9.10A and B.

Extrinsic pathway, intrinsic pathway, and crosstalk

Members of the death receptor superfamily trigger the extrinsic apoptosis pathway upon recruit of caspase-8 through the adaptor protein FAS (TNFRSF6)-associated death domain (FADD) [42]. Binding of a "death ligand" to a receptor triggers formation of a signaling complex that activates caspase-8 or caspase-10, which then activates caspase-3, and finally promotes cell death [63]. Extracellular and intracellular stress trigger the intrinsic apoptosis pathway (mitochondria pathway), which involves activation of pro-apoptotic members of the Bcl-2 family [43]. Bid, a pro-apoptotic member of the Bcl-2 family, allows crosstalk between the extrinsic and intrinsic pathways. The three subfamilies of Bcl-2-related proteins are the antiapoptotic proteins (e.g., *BCL2* and *BCL2L1*), the pro-apoptotic multidomain proteins (e.g., *BAX* and *BAK*), and pro-apoptotic BH3-only proteins (e.g., *BID* and *BIM*) [44,64].

Of the 19 Bcl-2 proteins or regulators, 11 are present in Fig. 9.10A and B (*BAD*, *BAG4*, *BAK1*, *BAX*, *BCL2*, *BCL2A1*, *BCL2L1*, *BCL2L11*, *BID*, *HRK*, and

MCL1), 4 are present in Fig. 9.10A alone (*BAG2*, *BCL2L10*, *BCL6*, and *BIK*), and 4 are present in Fig. 9.10B alone (*BAG1*, *BAG3*, *BAG5*, and *BCL2L14*). This indicates that there is not a simple dichotomy between gain-of-function and loss-of-function proteins in the cancer interactome. For example, *BAX* has 4 gain-of-function interactions with *BCL2L1*, *TP53*, *MFN2*, and *BCL2L10* but it also has 13 loss-of-function interactions with *RARG*, *MCL1*, *BCL2A1*, *CDKN1A*, *MAPK11*, APP, APC, CASP2, PRKCE, RNF36, ADPRT, TGFBR2, and TNFRSF5 (Fig. 9.10).

Caspases family and caspases regulators

The apoptosis death signal is activated via a series of protease caspases (initiator caspases-2, 8, 9, and 10, and effector caspases-3, 6, and 7) which require activation by proteolysis [43,44]. With the exception of *CASP6*, which is present only in the gain-of-function network, the caspases (including *CASP1*, *CASP2*, *CASP4*, *CASP7*, *CASP9*, and *CASP10*) occur in the gain-of-function and loss-of-function networks. Seven caspase activators, inhibitors, and regulators (including *BAX*, *TP53*, *CFLAR*, *CYCS*, *CARD4*, *BIRC4*, and *DIABLO*) are present in both networks (Fig. 9.10A and B). This reveals the different roles of caspase regulators in cancerous and normal cells.

Regulation of apoptosis at the systems level

Besides Bcl-2 and caspases, proteins involved in apoptosis regulation include *BIRC3*, *PTEN*, *CARD12*, *MAP3K7*, *DEDD2*, *MITF*, *MALT1*, *BCL6*, *NALP1*, *CRADD*, *RTN4*, *PSEN1*, *IGFBP3*, *BNIP3L*, *RARG*, *CFLAR*, *TRAF3*, *TRAF1*, *MCL1*, *CARD4*, *TRAF6*, *VEGF*, *BIRC2*, *FGFR1*, *PEA15*, *DEDD*, *MMP9*, *HRK*, and *TP53* (Fig. 9.10A and B). Thus, we have identified 29 proteins, in addition to members of the Bcl-2 and caspase families, that regulate apoptosis at the systems level.

Apoptosis and cell cycle regulation

Proteins generally function as components of complexes that contain other macromolecules, to carry out specific biological processes. Networks of interactions connect different cellular processes [46]. We have identified 30 proteins in the PPIN of apoptosis that also participate in cell-cycle regulation. These are *PTEN*, *CDC6*, *MFN2*, *PKMYT1*, *DCC*, and *E2F1* in Fig. 9.10A, and *BIRC5*, *NRAS*, *MKI67*, *E1F1*, *PPP3CA*, *BCL2*, *TP53*, *MAPK3*, *CCND1*, *PCNA*, *EGFR*, *MAPK1*, *VEGF*, *BAX*, *CDKN1A*, *TGFB1*, *APC*, *MSH2*, *APP*, *KRAS2*, *HRAS*, *CDKN1B*, *CDKN2A*, and *PML* in Fig. 9.10B.

Although Fig. 9.10A and B describes the perturbation of apoptosis at the systems level, most proteins with high degree of perturbation (Fig. 9.10A) are also included in Fig. 9.10B. In other words, it is not possible to uniquely describe protein hubs as "gain-of-function" or "loss-of-function" hubs. Therefore, we summarized the degree of perturbation to identify the perturbed protein hubs in the network of cancer cells. After identifying targeted proteins as inhibitors or activators, it is necessary to study how a drug target is wired into the control circuitry of a complex cellular network [65]. Next, we show a flowchart for identification and prediction of apoptosis drug targets in cancer drug discovery.

Prediction of apoptosis drug targets using cancer-perturbed networks of apoptosis

Systems-based drug design is a major application of systems biology [40,59,66]. This method constructs disease-perturbed PPINs and identifies potential drug targets by comparison of the networks of normal and abnormal cells. This contrasts with the traditional approach, that reduces cellular processes to their individual components or signal transduction pathways and targets a specific molecule or signaling pathway. A limitation of this traditional approach is that a single molecule or pathway does not adequately describe most biological systems, including those affected in cancer [59]. By comparisons of the PPINs of normal and cancerous cells derived from microarray data, we can identify potential drug targets through a systems-based approach (Fig. 9.11).

Thus, we built initial PPINs from large-scale experiments and databases, and then employed each microarray data set of HeLa cells and primary lung fibroblasts to modify these networks. For network modification, we used a nonlinear stochastic model and AIC in Eq. (2.23). We next compared the networks of cancerous and normal cells, derived gain-of-function and loss-of-function networks, and identified protein hubs with high degree of perturbation as potential drug targets.

Scale-free networks are extremely sensitive to removal of targeted hubs (i.e., attack vulnerability [46,47,67]), so we summed the degree of perturbation (i.e., connectivity) of each node in the cancer-perturbed network to obtain these perturbed hubs. Proteins with sum of degree of perturbation ≥ 8 (Table 9.4) differentiate the cancerous and normal interactomes and are potential drug targets [41,59]. We classified the 17 potential drug targets (Table 9.4) into six categories: (1) Intrinsic pathway: *BCL2*, *BAX*, *BCL2L1*, *BID*, and *CYCS*; (2) Extrinsic pathway: *TNF* and *TNFRSF6*; (3) Common pathway: *CASP3* and *CASP9*; (4) Apoptosis regulators: *TP53*, *MYC*, *CFLAR*, and *EGFR*; (5) Stress-induced signaling: *MAPK1* and *MAPK3*; and (6) Others: *CDKN1A* and *CCND1*. These results indicate that most proteins interact with few partners, whereas hubs interact with many partners, consistent with current views on interactome networks with a scale-free or power law degree distribution [46].

Intrinsic pathway: **BCL2, BAX, BCL2L1,** *and* **CYCS**

Defective apoptosis in human cancers often results from overexpression or inhibition of *BCL2* proteins. These proteins regulate mitochondrial permeability by inhibiting (e.g., *BCL2* and *BCL2L1*) or promoting (e.g., *BAX* and *BID*) the release of

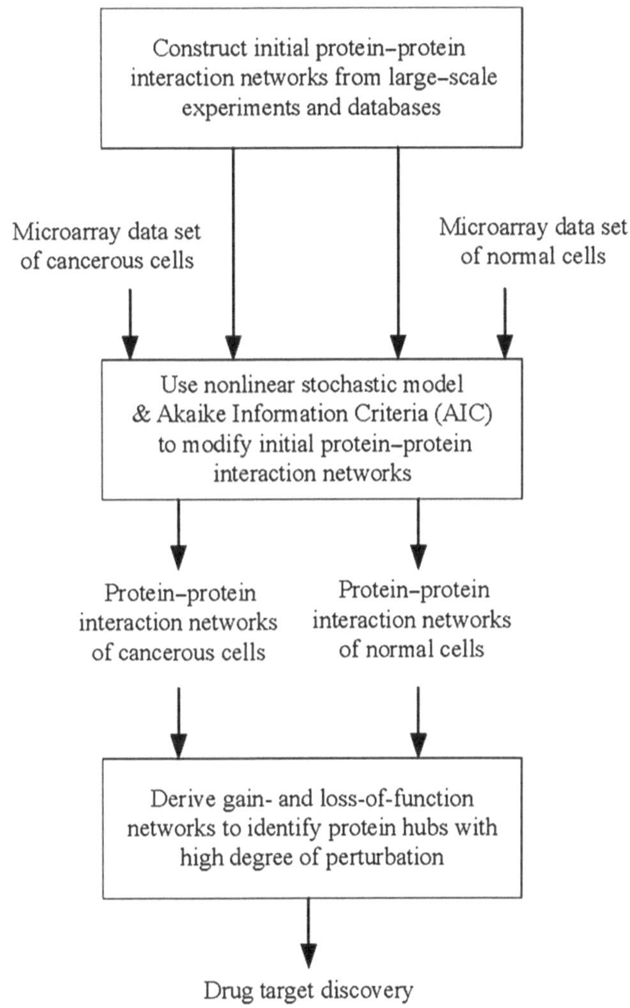

Figure 9.11 Flowchart for identification of potential drug targets in the cancer-perturbed network using microarray data.

cytochrome c (*CYCS*) [68]. *BCL2* and several antiapoptotic relatives, such as *BCL2L1*, associate with the mitochondrial outer membrane and the endoplasmic reticulum (ER) nuclear membrane and maintain the integrity of these membranes. Initiation of apoptosis requires pro-apoptotic family members that closely resemble *BCL2* and distantly related proteins that are related only by the small *BH3* protein-interaction domain [69]. In these results, proteins with gain-of-function interactions with BCL2 include *CCND1*, *BAD*, *MCL1*, *MAPK3*, *ADM*, KITLG, EGFR, *BAG2*, *PKMYT1*, *TP53*, *PCNA*, *MITF*, *ABCB1*, *BCL6*, *ZNF384*, *HRK*, *PPP2R5A*, and *VEGF*. Proteins with loss-of-function interactions with *BCL2* include *CDKN1A*,

Table 9.4 Seventeen potential drug targets ranked by sum of degree of perturbation ≥8

Protein targets	Sum of degree of perturbation	Agents	Literature review
BCL2	35	G3139 (Genasense), ABT-737	[39,68–73]
CASP3	22	Synthetic activators of caspases/ Apoptin/IAP targets surviving	[68,70,71]
BAX	17	Gene therapy through Bax vectors	[71]
TP53	17	ONY-015/INGN201/MDM2 inhibitors	[39,71,73,74]
BCL2L1	13	Antisense BCL-xL	[71]
CDKN1A	13		[75]
TNF	10		[39,71]
EGFR	9	Trastuzumab (Herceptin)	[71]
MAPK1	9	CI-1040/PD0325901/ARRY-142886	[76,77]
MAPK3	9	CI-1040/PD0325901/ARRY-142886	[76,77]
MYC	9	Bortezomib (Proteosome inhibitors)	[78]
BID	8		[71]
CASP9	8	Caspases activators/Apoptin/surviving	[67,70,71]
CCND1	8		[79]
CFLAR	8		[63]
CYCS	8		[73,80]
TNFRSF6	8		[71]

TNF, WT1, BAG4, BCL2L14, DEK, GRN, RAF1, BLK, BAG5, CAPN2, GHR, CDKN1B, RTN4, BNIP3L, MAP3K1, and *CLC.* We predict *BCL2* to be the best potential drug target because this protein best differentiates PPINs of HeLa and normal cells [40,59].

Our analysis agrees with the conclusions of previous studies that showed that *BCL2* protein family members are good targets for cancer therapy. Drugs that target *BCL2* include Genasense [39,71] and *ABT-737* [72]. The activation of Bax can be induced by gene therapy through delivery of Bax vectors, and this approach has been successful in inducing apoptosis in cancer cell lines [39]. Antisense *BCL2L1* (BCL-xL) downregulates the expression of *BCL2* and *BCL2L1*, induces apoptosis, and inhibits growth of several tumor types in vitro and in vivo [71]. Unfortunately, targeting of *BCL2* also causes adverse effects, presumably because many normal cells depend on proteins in the BCL2 family to maintain normal mitochondrial function. Difficulty in using BCL2 antisense DNA or RNA as a delivery system is a problem with Genasense [70]. Cytochrome c, once released into the cytosol, interacts with Apaf-1 and this leads to activation of caspase-9 proenzymes [71]. The chief function of *BCL2* proteins is to regulate the release of cytochrome c from mitochondria. Thus, targeting *CYCS* or *Apaf-1* would also be expected to cause severe adverse effects [73,80].

Extrinsic pathway and crosstalk: **TNF, TNFRSF6,** *and* **BID**

The extrinsic pathway activated by death receptors, such as *Fas* (*TNFRSF6/APO-1/CD95*) and other TNF receptor family members, allows apoptosis to maintain normal tissue homeostasis [42]. Although death receptors of the TNF superfamily members are potential targets for anticancer drugs, toxic side effects have been observed that place limits on their therapeutic use [39]. TNF and Fas (TNFRSF6) were also found to activate non-specific TNF receptors resulting in extensive ischemic and hemorrhagic lesions in several tissues leading to septic shock and fulminating hepatic failure in animal models [71].

A more promising approach involves targeting the *TRAIL* (TNF-Related Apoptosis Inducing Ligand) receptors [39,42,71]. Activation of the *TRAIL* death receptor can kill cancerous cells but not normal cells, whereas monoclonal antibodies against *TRAIL* and recombinant TRAIL ligand can cause *TRAIL* resistance [70]. Thus, administration of such a drug might cause tumors to develop resistance, or cause the death of normal cells. *BID*, a pro-apoptotic Bcl-2 family member, provides crosstalk and integration between the death-receptor and mitochondrial pathways [43,44]. Targeting *BID*, however, is rarely discussed, although such work has a potential for use in combined therapy [71].

Common pathway: **CASP3** *and* **CASP9**

Caspases are the central components of the apoptotic response network. An effector caspase (e.g., caspase-3) is activated by an initiator caspase (e.g., caspase-9) and the initiator caspase is activated via other PPIs [43,44]. Targeting inhibitors of caspases could potentially cause apoptosis of cancerous cells. Synthetic activators of caspases include *Apoptin* and *IAP* [68,71]. Caspase-3 and 9 are subject to inhibition by *IAPs* such as Livin [44]. Like *BCL2* inhibitors, *XIAP* inhibitors must block PPIs. When released from mitochondria, *Smac* binds *XIAP* and inactivates it, triggering apoptosis [70].

Apoptosis regulators: **TP53, MYC, CFLAR,** *and* **EGFR**

One of the most dramatic responses to *p53* is induction of apoptosis and regulation via the intrinsic pathway [74]. Drug trials that target *p53* include gene therapy involving *ONYX-015* and *INGN201* and antisense therapy that targets a protein controlling *p53* activity by Nutlins which blocks *p53/MDM2* interaction [39,71]. The proto-oncogene *c-MYC* encodes a transcription factor that is implicated in various cellular processes, including cell growth, proliferation, loss of differentiation, and apoptosis. The induction of cell-cycle entry sensitizes the cell to apoptosis, so that cell proliferation and apoptotic pathways are coupled [78]. CFLAR (c-FLIP) regulates caspase-8 and *FADD*-like apoptosis. Whereas CFLAR blocks the activation of the initiator caspase-8, XIAP can block the initiation phase (by inhibition of caspase-9) and the execution phase (by blocking caspase-3 and caspase-7) [63]. Some agents, e.g., agent ZD1839 as an EGFR (epidermal growth factor receptor) inhibitor, do not primarily target apoptosis, but indirectly modulate apoptosis [71].

Stress-induced signaling and others: MAPK1, MAPK3, CDKN1A, *and* CCND1

Proteins of the *MAPK* (mitogen-activated protein kinase) family are crucial in many signaling pathways [77]. Three *MEK* (MAPK kinase) inhibitors, *CI-1040*, *PD0325901*, and *ARRY-142886*, are currently in clinical trials for treatment of various cancers [76]. Although some drugs target MAPKs, *MAPK1* (*ERK* or p38), and *MAPK3* (*ERK1*) are not the main drug targets in the apoptotic pathway [71]. CDKN1A (p21) (cyclin-dependent kinase inhibitor-1) plays a role in cell cycle arrest and induction of apoptosis. The activities of cyclin D- and cyclin E-dependent kinases are linked through the Cip/Kip family of Cdk inhibitors, including *p27* and *p21* [75]. *CCND1* is cyclin D1 in the G1/S transition of the cell cycle, and is controlled by the tumor suppressor gene RB through cdk-cyclin D complexes [79].

Prediction of additional drug targets by decreasing the degree of perturbation

In addition to identifying several apoptosis drug targets that have already been identified by other studies, we applied this method to predict additional drug targets by decreasing the perturbation threshold. Thus, if we reduce the sum of degree of perturbation to 7, we predict the following additional targets: *BAK1*, *CASP2*, *BCL2A1*, *IGF1*, *PRKCD*, *NFKB1*, and *PCNA*. *NFKB1* has both anti- and pro-apoptotic functions that are determined by the nature of the death stimulus. The drug PS11445 targets the NFKB1 inhibitor *IKKβ* [71] but the other proteins have not previously been considered as drug targets.

Prediction of new GO annotations of the four proteins: CDKN1A, CCND, PCNA, *and* PRKCD

If we consider all 24 proteins with sum of degree of perturbation ≥ 7, four proteins (*CDKN1A*, *CCND1*, *PRKCD*, and *PCNA*) are also considered to have a role in apoptosis. If the function of any one protein in a network is known, identification of its interacting partners allows prediction of their functions [46]. However, *PCNA* and *PRKCD* are known to have a role in DNA damage-induced apoptosis. *PCNA* (Proliferating Cell Nuclear Antigen) is ubiquitinated and involved in RAD6-dependent DNA repair in response to DNA damage. *PRKCD* (protein kinase C, delta) is also associated with DNA damage-induced apoptosis [81], whereas GO annotations of *PRKCD* do not include apoptosis.

Although other methods (such as identification of protein domains) allow prediction of PPIs in different organisms [82–84], these methods do not allow identification of potential drug targets. This method provides efficient and precise prediction of anticancer drug targets and also specifies these target proteins with detailed GO annotations. This will help researchers identify additional drug targets by examination of other cellular mechanisms involved in cancer. Network modeling has been used to identify genes potentially associated with breast cancer in a BRAC-centered network [61].

However, there are very few time-series microarray databases for cancerous and normal cells, so it is difficult to use this method to compare protein hubs for different types of cancer. Moreover, this method does not address the potential adverse effects of targeting a specific protein and the problem of drug delivery. We expect that more genomic time-series microarray experiments and clinical research will address these limitations in the future.

Caspase activation through static and dynamic hubs

Static network topology is not sufficient to define function, and incorporating time-dependent expression data is important for understanding pathway function [54]. A number of computational approaches have been proposed for predicting PPIs, such as domain–domain interactions [83], the confidence score resulting from sequence similarity and number of edges [36], and integration of genomic data sets [85]. A nonlinear stochastic model can depict PPINs at different times by using linear binary interactions and nonlinear protein complex relationships. Previous studies have used linear stochastic models to describe the multiple feedback loops of p53 [86], but nonlinear effects cannot be depicted by this method. This method also illustrates the dynamic behavior of PPINs, which cannot be examined by probabilistic methods of data integration [85].

Networks consist of party hubs (static hubs) and date hubs (dynamic hubs). Party hubs are found in static complexes with most of their partners present at the same time, whereas date hubs bind their interaction partners at different times or locations [47]. To further investigate dynamic apoptotic properties in human cells, we considered caspases as protein hubs in these dynamic nonlinear stochastic protein–protein models (Fig. 9.12A–D). In order to identify party hubs and date hubs, we first summed the degree of perturbation of each protein as "plus degree of perturbation" and then subtracted the degree of perturbation of each protein as "minus degree of perturbation" at two time periods in cancerous and normal cells. If the plus degree of perturbation of one protein was ≥ 20, we defined this protein as a "hub." If the minus degree of perturbation of the hub was ≤ 3, we defined the hub as a party hub (static hub). If the minus degree of perturbation of the hub was >3, we defined the hub as a date hub.

Fig. 9.12A–D illustrates the time-dependence of the networks, where *bold lines* represent distinct interactions at different times. Caspase signaling results in time-variant PPIs, and dynamic modeling allows specification of the time-dependent interactome. In cancerous cells, the date hubs include *BIRC2*, *CASP2*, and *CASP3*, and the party hubs include *TP53*, *TNF*, *BIRC3*, *BAX*, *CASP1*, and *CASP9*.

In normal cells, date hubs include CASP3 and CASP9 and party hubs include *TNFRSF6*, *TP53*, *BIRC2*, *BIRC3*, *BCL2*, *BAX*, and *CASP1*. Effector caspase-3 is a date hub in both cell types because intrinsic and extrinsic pathways converge on caspase-3. Because date hubs appear to be more important than party hubs [87],

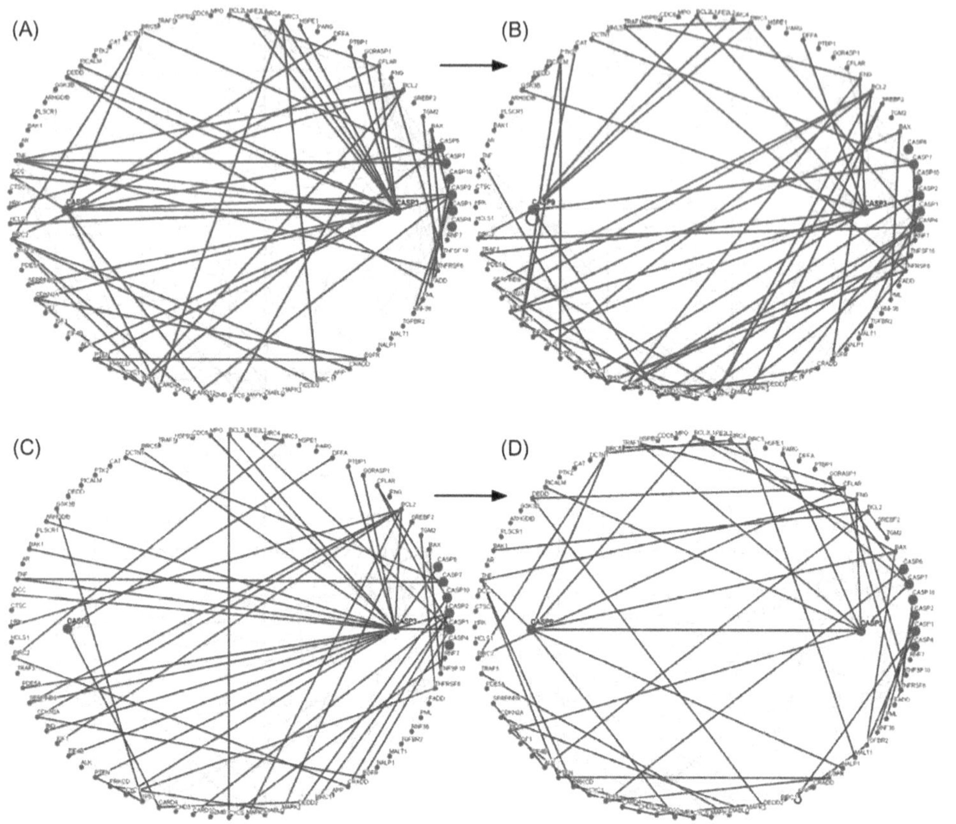

Figure 9.12 Dynamic protein–protein interactions in caspase formation. (A) Protein–protein interactions within the hub caspases of cancer cells during 0–8 h after induction of apoptosis. Distinct interactions at different times are marked with *bold lines.* (B) During 4–30 h after induction of apoptosis in cancer cells. (C) During 0–8 h after induction of apoptosis in normal cells. (D) During 4–36 h after induction of apoptosis in normal cells.

caspase-2 and 9 are important date hubs that differentiate network topologies of cancerous and normal cells. *TP53*, *BIRC3*, *BAX*, and *CASP1* are party hubs in both cell types. Party hubs are found in static complexes where they interact with most of their partners simultaneously [87]. In other words, we believe these four proteins play central roles in functional complexes in both cancerous and normal cells.

9.2.2 Conclusion of tumorigenesis mechanism by cancer-perturbed PPINs of apoptosis

The cancer-perturbed PPINs of apoptosis that we developed here shed light on the mechanism of cancer at the systems level and allow identification of potential drug targets. In this section, we applied nonlinear stochastic modeling to describe individual and cooperative protein interactions. This method is more precise than the linear models used in previous research. We successfully integrated microarray and proteome databases to identify cancer-perturbed PPINs. These predictions of potential drug targets agreed with potential targets identified by other studies and also identified additional targets which may guide the development of new anticancer drugs in future. We also identified static and dynamic hubs in human PPINs, which have heretofore been identified only in yeast.

APPENDIX B: METHODS OF CONSTRUCTING CANCER-PERTURBED PPINs

Construction of initial PPINs

A comprehensive understanding of PPIs in an organism provides a framework for understanding biology as an integrated system, and human perturbed PPINs offer insight into disease mechanisms such as cancer at the systems level [40,46,59]. Before construction of cancer-perturbed PPINs to explore their roles in the mechanism of cancer, PPINs for both cancer and normal cells must be constructed for comparison. The systematic experimental mappings of the human interactome include yeast two-hybrid systems [48,49], which reveal thousands of preys and baits in the protein matrix. Several online databases such as BIND [50], HPRD [51], Intact [52], and Himap [53] provide fundamental global topologies of human PPIs. The combined use of these databases [51−54] assists precise estimation of parameters as the basis of construction of PPINs, because incomplete rough networks can lead to biased or possibly erroneous conclusions [46]. We first constructed an initial PPIN using two yeast-two-hybrid experiments and four online databases.

Studies of large-scale PPIs allow development of protein interaction networks, but all large-scale experiments and databases contain high rates of false-positives [88]. Previous studies have demonstrated that use of multiple functional databases allows better identification of PPIs and leads to better prediction of the function of unknown proteins [85]. Therefore, we integrated different databases and experiments to refine this network [55]. After development of this initial network, we used microarray data to remove false-positive interactions.

Selecting and processing experimental data

We used microarray data [60] to compare gene expression in HeLa cells and normal primary human lung fibroblasts subjected to several stresses (e.g., heat shock from 37°C to 42°C, oxidative stress with menadione or hydrogen peroxide, and ER stress with DTT (dithiothreitol) or tunicamycin). Gene expressions in HeLa cells and normal fibroblasts, both of which were treated with 2.5 mM DTT, were these microarray data sources. Gene expression was recorded at 0−30 h after stress in HeLa cells and 0−36 h in normal cells [60].

Nonlinear stochastic interaction model of initial networks

We used a nonlinear stochastic model to mimic the dynamic interactions among proteins to remove false-positives and refine the network. In recent years, some systems and computational biologists have employed a dynamic perspective to describe biological functions, because of their dynamic nature [7,89,90]. We considered two types of interactions, binary protein interactions and protein complex interactions, but not DNA–protein or metabolite interactions. We denote these basal interactions in this model, whereas $\varepsilon(t)$ represents stochastic molecular events, such as molecular fluctuations of the target protein.

In this section, we define "individual PPIs" as binary PPIs, and "cooperative interactions" as the interaction of a protein complex with the target protein. $x(t)$ denotes the expression profile of target protein x at time point t; a denotes the influence of a target protein at one time point to the target protein at the next time point; b_i denotes individual or binary interaction of protein i with target protein $x(t)$; and c_{ij} denotes the cooperative interaction ability of protein i and protein j on the target protein x. Thus, Fig. B.1 shows three individual or binary PPIs to the target protein $x(t)$ (protein $x_1(t)$, $x_2(t)$, and $x_3(t)$) and one cooperative interaction between protein $x_1(t)$

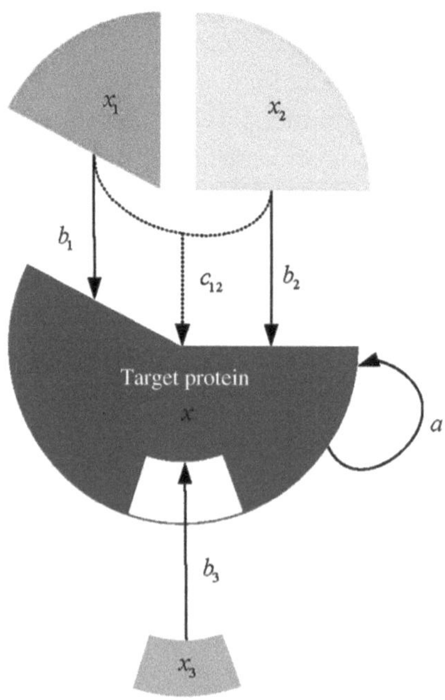

Figure B.1 Graphical representation of individual protein interactions and cooperative protein interactions. This dynamic protein interaction equation includes three individual or binary protein–protein interactions to the target protein $x(t)$ (protein $x_1(t)$, $x_2(t)$, and $x_3(t)$) and one cooperative interaction between protein $x_1(t)$ and $x_2(t)$ to the target protein $x(t)$. a denotes the influence of a target protein at one time point to the target protein at the next time point; b_i denotes individual or binary interaction of protein i with target protein $x(t)$; and c_{ij} denotes the cooperative interaction ability of protein i and protein j on the target protein x.

and $x_2(t)$ to the target protein $x(t)$. By considering basal interactions k and stochastic events $\varepsilon(t)$, we can express the dynamic model of the target protein time profile $x(t)$ as: $x(t) = ax(t) + b_1x_1(t) + b_2x_2(t) + b_3x_3(t) + c_{12}x_{12}(t) + k + \varepsilon(t)$. This equation is based on a previously developed model [91].

Therefore, dynamic interactions between upstream proteins and their targets can be written as:

$$x(t+1) = ax(t) + \sum_{i=1}^{N} b_i x_i(t) + \sum_{i=1}^{N}\sum_{j=1}^{N} c_{ij}x_{ij}(t) + k + \varepsilon(t). \tag{B.1}$$

This equation is based on a previously developed model [13,91,92].

Remark 9.1: The discrete-time dynamic model in Eq. (B.1) is based on the discrete sampling of the following continuous differential model:

$$\dot{x}(t) = -\lambda x(t) + \sum_{i=1}^{N} b_i x_i(t) + \sum_{i=1}^{N}\sum_{j=1}^{N} c_{ij}x_{ij}(t) + k + \varepsilon(t),$$

where $\dot{x}(t) = dx(t)/dt$ denotes the differential of $x(t)$ with respect to the continuous time t, and λ denotes the decay rate of $x(t)$. By unit sampling and with $\dot{x}(t) \approx x(t+1) - x(t)$, we get the following discrete model [15]:

$$x(t+1) - x(t) = -\lambda x(t) + \sum_{i=1}^{N} b_i x_i(t) + \sum_{i=1}^{N}\sum_{j=1}^{N} c_{ij}x_{ij}(t) + k + \varepsilon(t),$$

or

$$\begin{aligned} x(t+1) &= (1-\lambda)x(t) + \sum_{i=1}^{N} b_i x_i(t) + \sum_{i=1}^{N}\sum_{j=1}^{N} c_{ij}x_{ij}(t) + k + \varepsilon(t) \\ &= ax(t) + \sum_{i=1}^{N} b_i x_i(t) + \sum_{i=1}^{N}\sum_{j=1}^{N} c_{ij}x_{ij}(t) + k + \varepsilon(t) \end{aligned},$$

where a denotes the influence of $x(t)$ on $x(t+1)$ and is dependent of the decay rate, λ.

By iteration of Eq. (B.1), we can construct the whole PPIN, which is interconnected through the interactions of $\sum_{i=1}^{N} b_i x_i(t)$ and $\sum_{i=1}^{N}\sum_{j=1}^{N} c_{ij}x_{ij}(t)$ in Eq. (B.1) for all proteins. In Eq. (B.1), $x(t)$ represents the expression profile of the target protein at time point t, which is estimated from mRNA expression profiles via a translational sigmoid function [13,92]:

$$x_{ij}(t) = f(y(t)) = \frac{1}{1 + \exp[-r(y(t) - M)]}. \tag{B.2}$$

In Eq. (B.2), r denotes the transition rate of the sigmoid function, and M denotes the mean level of mRNA expression of the corresponding protein. $\sum_{i=1}^{N} x_i(t)$ denotes all possible individual interactive functions (i.e., all possible binary PPIs) of N interactive protein candidates of the target protein in the initial PPIN, a denotes the influence of the present target protein on the target protein at the next time point, and b_i indicates the individual interactive ability of protein i to target protein $x(t)$. $\sum_{i=1}^{N}\sum_{j=1}^{N} c_{ij}x_{ij}(t)$ denotes all possible interactive functions of cooperative protein partners with the target protein in the initial network, where $x_{ij}(t)$ denotes nonlinear cooperative interaction between protein x_i and protein x_j on the target protein, i.e., $x_{ij}(t) = f(y_i(t)) \cdot f(y_j(t))$, and c_{ij} denotes the cooperative interaction ability of protein i and protein j on the target protein (see Fig. B.1). We obtained all possible cooperative interactions $\sum_{i=1}^{N}\sum_{j=1}^{N} c_{ij}x_{ij}(t)$ from online high-throughput protein interaction databases that contained putative protein complexes within the network. If the multiprotein complex is composed of three or more proteins, we added all combinations of two cooperative proteins from the complex to determine the nonlinear property because one extreme value in the equation will cause significant aberration in other estimated parameters. The basal interaction k in Eq. (B.1) represents unknown PPIs that result from other possible interacting proteins or other influences (e.g., mRNA–protein interactions, protein synthesis). $\varepsilon(t)$ represents random noise arising from model uncertainty and molecular fluctuations of protein interactions with the target protein [58].

With the nonlinear stochastic interaction model (Eq. B.1), we can identify interaction parameters a, b_i, c_{ij}, and k using microarray data. After having identified all the interactions with the target protein, upstream proteins can be considered as target proteins, and by repeating this procedure we can identify all interactions of the initial network. If the estimated interaction parameters $\hat{a}$, $\hat{b}_i$, $\hat{c}_{ij}$, and $\hat{k}$ are not significant according to the AIC, the corresponding interactions will be eliminated from the initial network.

Identification of interactions in the initial PPIN

Before modification of the initial network, we estimated interaction parameters of the initial network by using the maximum likelihood estimation. This represents interactions of all possible protein candidates in the initial network. After further rearrangement, Eq. (B.1) can be rewritten as:

$$x(t+1) = \begin{bmatrix} x(t) & x_1(t) & \dots & x_N(t) & x_{12}(t) & \dots & x_{(N-1)N}(t) & 1 \end{bmatrix} \begin{bmatrix} a \\ b_1 \\ \vdots \\ b_N \\ c_{12} \\ \vdots \\ c_{(N-1)N} \\ k \end{bmatrix} + \varepsilon(t) \qquad \text{(B.3)}$$

$$= \phi(t)\theta + \varepsilon(t)$$

where $\phi(t)$ denotes the regression vector composed of many elements that represent expression levels of protein candidates in the initial network at time point t.

Employing the cubic spline method [7,92] to interpolate microarray data allows us to obtain as many data points as needed. In general, the number of data points should be at least five times the number of parameters to be estimated. The cubic spline method allows us to obtain these values: $x(t)$, $x_i(t_l)$, and $x_j(t_l)$ for $l=1,\ldots,M$, $i=1,\ldots,N$, and $j=1,\ldots,S$, where M denotes the number of microarray data points, N denotes the number of possible protein interactions for the target protein, and S denotes the number of protein complexes in the network. These data points are used as the basis in this regression vector $\phi(t)$. Computation of Eq. (B.3) at different time points allows us to construct the following vector-form equation:

$$\begin{bmatrix} x(t_2) \\ \vdots \\ x(t_M) \end{bmatrix} = \begin{bmatrix} \phi(t_1) \\ \vdots \\ \phi(t_{M-1}) \end{bmatrix} \theta + \begin{bmatrix} \varepsilon(t_1) \\ \vdots \\ \varepsilon(t_{M-1}) \end{bmatrix}. \tag{B.4}$$

For simplicity, we represent this as:

$$X = \Phi\theta + \nu \tag{B.5}$$

In Eq. (B.4), random noise $\varepsilon(t_k)$ is regarded as a white Gaussian noise with zero mean and unknown variance σ^2, i.e., $E[\nu]=0$, and $\Sigma_\nu = E[\nu\nu^{\mathrm{T}}] = \sigma^2 I$. Next, we employed a maximum likelihood estimation method [15] to estimate θ and σ^2 using regression data obtained from the microarray data of the target protein and proteins with which it interacts. Under the assumption that ν is a Gaussian noise vector with $M-1$ elements, its probability density function is given as follows.

$$p(\nu) = ((2\pi)^{M-1}\det\Sigma_\nu)^{-1/2}\exp\left(-\frac{1}{2}\nu^T\Sigma_\nu^{-1}\nu\right). \tag{B.6}$$

Since $\nu = X - \Phi\theta$ (Eq. 9.23), we can rewrite Eq. (B.6) as the following likelihood function:

$$p(\theta,\sigma^2) = (2\pi\sigma^2)^{-(M-1)/2}\exp\left(-\frac{(X-\Phi\theta)^T(X-\Phi\theta)}{2\sigma^2}\right). \tag{B.7}$$

Maximum likelihood parameter estimation involves finding θ and σ^2 to maximize the likelihood function in Eq. (B.7). In order to simplify the computation, it is practical to take the logarithm of Eq. (B.7), which yields the following log-likelihood function:

$$\log L(\theta,\sigma^2) = -\frac{M-1}{2}\log(2\pi\sigma^2) - \frac{1}{2\sigma^2}\sum_{k=1}^{M-1}(x(t_k)-\phi(t_k)\theta)^2. \tag{B.8}$$

In Eq. (B.8), $x(t_k)$ and $\phi(t_k)$ are the kth elements of X and Φ, respectively.

Here, we expect the log-likelihood function to have the maxima at $\theta = \hat{\theta}$ and $\sigma^2 = \hat{\sigma}^2$. The necessary conditions for determining maximum likelihood estimates $\hat{\theta}$ and $\hat{\sigma}^2$ must consist of [15]:

$$\begin{aligned} \frac{\partial \log L(\theta, \sigma^2)}{\partial \theta} &= 0 \\ \frac{\partial \log L(\theta, \sigma^2)}{\partial \sigma^2} &= 0 \end{aligned}. \tag{B.9}$$

After some computational arrangements from Eq. (B.9), the estimated parameters $\hat{\theta}$ and $\hat{\sigma}^2$ are:

$$\hat{\theta} = (\Phi^T \Phi)^{-1} \Phi^T X, \tag{B.10}$$

$$\hat{\sigma}^2 = \frac{1}{M-1} \sum_{k=1}^{M-1} (x(t_k) - \phi(t_k)\hat{\theta})^2 = \frac{1}{M-1} (X - \Phi\hat{\theta})^T (X - \Phi\hat{\theta}). \tag{B.11}$$

After obtaining estimate $\hat{\theta}$, we can rewrite estimated PPI (Eq. B.1) as follows:

$$x(t+1) = \hat{a}x(t) + \sum_{i=1}^{N} \hat{b}_i x_i(t) + \sum_{i=1}^{N} \sum_{j=1}^{N} \hat{c}_{ij} x_{ij}(t) + \hat{k}. \tag{B.12}$$

We quantified the interactions of all candidate proteins by the process described above.

In Eq. (B.12), the estimated parameter $\hat{a}$ denotes the estimation of the residual of target protein, $\hat{b}_i$ denotes individual interactive rate or binary PPI between protein i and the target protein, and $\hat{c}_{ij}$ denotes cooperative interactive rate or protein complex membership between protein i and protein j, i.e., protein complex ij, to the target protein. A positive value implies positive interaction, a negative value implies negative interaction, and the interactions are more likely as the parameters get larger.

Modification of initial PPINs

Although the maximum likelihood estimation method allows quantification of all possible interactions of a target protein, we still do not know the significance of the interaction and whether it can be regarded as a true interaction. Thus, we used a statistical approach that involves model validation to evaluate significance levels and refine the network. In this procedure, we employed the AIC in Eq. (2.23) to validate the model order or the number of model parameters of the network [15].

The AIC considers estimated residual variance and model complexity as one statistic and provides a measure of the information lost when a model is used. In other words, the AIC is an operational way of trading off the complexity of an estimated model against how well the model fits the data. AIC decreases as residual variance $\hat{\sigma}^2$ decreases and increases as the number of parameters p increases.

As the expected residual variance decreases with increasing p for inadequate model complexities, there should be a minimum near the correct number of interaction parameters p. For a protein interaction model with p interaction parameters to fit with data from N samples, the AIC can be rewritten as follows [15]:

$$\mathrm{AIC}(p) = \log\left(\frac{1}{N}(X-\hat{X})^T(X-\hat{X})\right) + \frac{2p}{N}, \tag{B.13}$$

where $\hat{X}$ denotes the estimated expression profile of the target protein, i.e., $\hat{X} = \phi\hat{\theta}$. After the statistical selection of p parameters by minimizing the AIC, we can determine whether the protein interaction is significant or a false-positive. In these results, "1" represents protein i interacting with protein j if the interaction is within p significant interactions, and "0" represents protein i not interacting with protein j if the interaction has less than p significant interactions. It must be considered that PPINs represent mutual binding relationships; if protein i binds to protein j, then protein j also binds to protein i. To determine mutual relationships in the high false-positive yeast two-hybrid experiments, we use the Boolean logical "AND" to determine the interaction maps. Two proteins were considered to interact only if each protein binds to the other protein. If an interaction is detected only in HeLa cells but not in normal cells, we call this a gain-of-function; if the interaction is detected in normal cells but not HeLa cells, we call it a loss-of-function. Table B.1 shows the truth table of all

Table B.1 Truth table of all 16 events in the sample space of comparisons of individual interaction of protein *i* and protein *j* between HeLa and normal cells

AIC detection results in HeLa cells			**AIC detection results in normal cells**			**Comparisons between HeLa and normal cells**
Target protein *i*	**Target protein *j***	**"AND" results**	**Target protein *i***	**Target protein *j***	**"AND" results**	
0	0	0	0	0	0	
0	0	0	0	1	0	
0	0	0	1	0	0	
0	0	0	1	1	1	Loss-of-function
0	1	0	0	0	0	
0	1	0	0	1	0	
0	1	0	1	0	0	
0	1	0	1	1	1	Loss-of-function
1	0	0	0	0	0	
1	0	0	0	1	0	
1	0	0	1	0	0	
1	0	0	1	1	1	Loss-of-function
1	1	1	0	0	0	Gain-of-function
1	1	1	0	1	0	Gain-of-function
1	1	1	1	0	0	Gain-of-function
1	1	1	1	1	1	

16 events in the sample space of comparisons of individual PPIs between HeLa cells and normal cells, as determined with this AIC detection algorithm. Protein complexes are only considered once, because of incomplete information from online databases. We iteratively refined interactions in the initial network using a similar procedure. Finally, this yielded a refined PPIN for HeLa and normal cells.

9.3 A NETWORK-BASED BIOMARKER APPROACH FOR MOLECULAR INVESTIGATION AND DIAGNOSIS OF LUNG CANCER

In Section 9.2, we used cancer-perturbed PPIN to investigate the tumorigenic mechanism of carcinogenesis by genome-wide microarray data and system identification method. In this section, based on significant proteins with high CRVs in cancer PPIN which compared with normal cell PPIN, we will construct network-based biomarker for lung cancer molecular mechanism investigation and therapeutic drug target.

Cancer, the complex disease of uncontrolled cell growth, is one of the leading causes of human death worldwide and the deaths from cancer are projected to continue rising [93,94]. Among all types of cancer, the most commonly diagnosed, as well as the most common cause of cancer deaths, is lung cancer, with a mortality rate as high as 80–85% within 5 years [94,95]. Lung cancer is categorized into two main types: small cell lung carcinoma (SCLC) and nonsmall cell lung carcinoma (NSCLC). NSCLCs are subcategorized into three main subtypes: squamous cell carcinoma, adenocarcinoma, and large cell carcinoma [96]. Previous research has shown that these major histological types of lung cancer are associated with cigarette smoking [97]. In light of this, much research has been devoted to investigating the molecular alterations which ensued from cigarette smoking and the mechanism that links cigarette smoking to lung cancer. Spira et al. used DNA microarray to compare the gene expressions of large-airway epithelial cells from nonsmokers and smokers, and to determine how cigarette smoking alters the transcriptome [98]. Hecht indicated that many tobacco smoke carcinogens, such as polycyclic aromatic hydrocarbons and nicotine-derived nitrosamine ketone, are predominant inducers of lung cancer [99]. Recently, Takahashi et al. showed that induction of IKKβ- and JNK1-dependent inflammation is likely to be an important contributor to the tumor-promoting activity of tobacco smoke [100].

In addition to the investigation on carcinogenesis, many studies identified cancer biomarkers through analysis of genome-wide expression profiles [101,102]. The biomarkers are used either as a diagnostic evaluation to determine the health of a patient with or without the cancer, or as a prognostic indicator to determine the patient's prognosis. Spira et al. used gene expression profiles of samples from lung cancer patients to identify an 80-gene biomarker that distinguished apparent differences between smokers with and without lung cancer [98]. Because there is no effective screening tool for diagnosing lung cancer at an early stage, the 80-gene biomarker could make a beneficial contribution to minimizing high mortality rates by providing a better prognosis [98,103]. However, the biomarker identification method, which strictly uses gene expression profiles, cannot show how the different genes within the biomarker gene set are related to each other, i.e., the biomarkers are not identified using the systems perspective. Further, the gene lists obtained for similarly diagnosed patients by different research groups differ widely and share few common genes [104].

Due to these kinds of limitations and the widely accepted opinion that cancer is a disease of pathways [38,105], both PPI and pathway information are integrated for biomarker identification. Chuang et al. developed a protein-network-based approach that identifies biomarkers not as individual genes but as subnetworks extracted from protein interaction databases. They showed that the subnetwork classification is highly accurate in signature discrimination and provides an accurate account of the network structure [106]. Many other network-based approaches for prioritizing disease genes and protein interaction subnetworks that are discriminative of disease signature have been developed [107−110]. The dynamic structure of the human protein interaction network has recently been examined to aid in predicting breast cancer prognosis, suggesting that network modularity might be a defining feature of tumor phenotype [111].

Network analysis has shown that under different cellular states or in response to diverse stimuli, transcription factors alter their interactions to regulate different genes, thereby rewiring the network [112]. The same situation occurs with protein interaction networks [111,113]. Motivated by the dynamic structure of the human protein interaction network and the observation that interacting proteins tend to result in similar disease phenotypes when dysregulated [114], we developed a computational framework to construct the network-based biomarker for molecular investigation and diagnosis of lung cancer. The network-based biomarker consisted of two protein association networks for cancer and noncancer smokers. Based on the concept of network comparison [115], 40 significant proteins that play potentially important roles in lung carcinogenesis were identified. The network-based biomarker is a useful tool for further distinguishing the presence of cancer in smokers by similarity measurement of molecular patterns. Hopefully, the proposed method can aid in further understanding lung carcinogenesis and providing potential drug targets for humans to combat lung cancer.

9.3.1 Methods of investigating network-based biomarkers of lung cancer

Overview of the network-based biomarker approach for lung cancer investigation

The overall flowchart of the proposed network-based biomarker approach is shown in Fig. 9.13. This goal is to investigate lung cancer by the construction of network-based biomarkers composed of protein association networks for smokers with and without

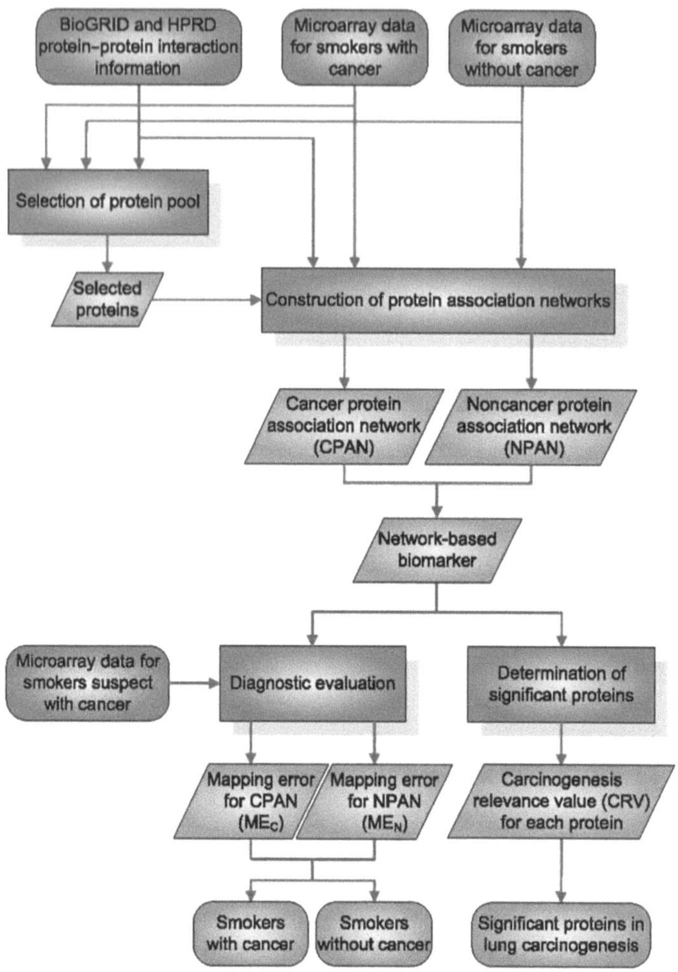

Figure 9.13 The flowchart of constructing the network-based biomarker for lung cancer investigation and diagnosis. The figure indicates the flowchart of the proposed method. *Red* represents the data needed. *Blue* denotes the processing steps of the approach. *Green* represents the processed results of each step and *orange* denotes the overall results from the entire method. In summary, two kinds of data, microarray data and PPI information, are needed for the proposed method. These data are used for protein pool selection, and then the selected proteins and the input data are used for protein association network construction, resulting in cancer protein association network (CPAN) and noncancer protein association network (NPAN). The two constructed protein association networks form the overall network-based biomarker, which can be used for either determination of significant proteins or diagnostic evaluation. With the help of the network-based biomarker, CRV is computed for each protein, and significant proteins in lung carcinogenesis are determined based on the CRVs. These significant proteins provide targets for further characterization. On the other hand, given the microarray data for smokers suspect of cancer, mapping errors for CPAN and NPAN can be computed, respectively, which help diagnose the smokers with cancer or without cancer.

cancer. Microarray gene expression profiles of patient samples and PPI information were integrated for protein selection and network construction. Two protein association networks with quantitative protein association abilities for cancer and noncancer smokers were constructed. By comparing two protein association networks within the network-based biomarker, a CRV was computed to correlate proteins with the level of lung carcinogenesis. A higher score suggests the particular protein plays a more critical role in lung carcinogenesis. A set of significant proteins was selected based on the CRV for each protein and the statistical assessment. Further, using microarray data for smokers suspected of having cancer, it is possible to compute mapping errors for further diagnostic evaluation of smokers with or without cancer.

Data selection and preprocessing

In this section, two kinds of data, microarray gene expression profile and PPI information, were integrated. The microarray data were downloaded from the GEO database http://www.ncbi.nlm.nih.gov/geo/. Spira et al. performed gene expression profiling in histologically normal large-airway epithelial cells obtained by bronchoscopy of current and former smokers. The health of each individual was recorded following the bronchoscopy until a final diagnosis of either presence or absence of lung cancer [98]. Data were collected from a total pool of 187 subjects and divided into primary and prospective data sets: 79 and 73 smokers with and without lung cancer respectively in the primary data set; 18 and 17 smokers with and without lung cancer respectively in the prospective data set. The primary data set was used for network-based biomarker construction and the prospective data set was used for diagnostic evaluation. PPI data were extracted from the Biological General Repository for Interaction Datasets (BioGRID) http://thebiogrid.org/ and the HPRD http://www.hprd.org/. The BioGRID database was developed to house and distribute collections of protein and genetic interactions from major model organism species. BioGRID currently contains over 340,000 interactions derived from both high-throughput studies and conventional focused studies [116]. The HPRD is a database that integrates a wealth of information relevant to the human proteome, including PPIs, posttranslational modifications, disease associations, and tissue expression [51]. Prior to further processing, the gene expression value g_{ij} is normalized to z-transformed scores z_{ij} so that for each gene i the normalized expression value has mean $\mu_i = 0$ and standard deviation $\sigma_i = 1$ over sample j.

Selection of protein pool and construction of network-based biomarker

To integrate the gene expression and PPI information data and construct the network-based biomarker consisting of protein association networks, the expression value of each gene was first overlaid on its corresponding protein. The gene expression for each protein was then used to select differentially expressed proteins using one-way analysis of variance (ANOVA), where the null hypothesis is that the average expression levels for the protein are the same for smokers with and without cancer. The proteins with Bonferroni adjusted p-values less than .05 were selected for the protein pool. Because we used the network-based biomarker as an investigative tool, the differentially expressed proteins without PPI information were excluded from the protein pool, while proteins based on PPI information that were highly connected with pool proteins were also included. In other words, the protein pool consisted of both differentially expressed proteins and the proteins that are highly connected with them. Based on the protein pool and PPI information, a rough PPI network can be easily constructed by linking proteins that share interactions. It is worth noting that since the data for cancer and noncancer samples are limited, the number of proteins selected for rough PPI network construction is also restricted. That is, to avoid overfitting in network construction, the maximum degree of the proteins in the rough PPI network should be less than the cancer/noncancer sample number, thereby restricting the size of the rough PPI network.

Using a simple regression model, the rough PPI network was further refined with the microarray data to highlight the independent protein association for samples with and without lung cancer relative to their respective data sets. For a target protein i in the rough PPI network, the protein was described using the following protein association model:

$$y_i(n) = \sum_{k=1}^{N_i} \alpha_{ik} y_{ik}(n) + \varepsilon_i(n), \tag{9.9}$$

where $y_i(n)$ represents the gene expression level of the target protein i for the sample n; α_{ik} denotes the association ability between the target protein i and its kth interactive protein, which quantifies the expression relation between the interactive proteins and can be identified using this data; $y_{ik}(n)$ indicates the gene expression level of the kth protein that interacts with the target protein i for sample n; N_i represents the number of proteins interacting with the target protein i and can be obtained from the rough PPI network; and $\varepsilon_i(n)$ denotes stochastic noise associated with other factors or model uncertainty. Eq. (9.9) states that, biologically, the expression level of the target protein i is associated with the expression levels of interacting proteins. A protein association model was constructed for each protein in the protein pool.

After the protein association model of the rough PPI network was constructed, the association parameters in Eq. (9.9) were identified using maximum likelihood estimation method [15,92] using microarray data. Since there are two data sets of microarray data (smokers with and without cancer), the association parameters were identified separately for the cancer data set and noncancer data set, resulting in $\alpha_{ik,C}$ and $\alpha_{ik,N}$, respectively. In this case, for each protein in each phenotype, i.e., with cancer or without cancer, a mathematical description was constructed to characterize the respective expression association. Once the association parameters for all proteins in the rough PPI network were identified, the significant protein associations were determined based on the estimated association abilities (α_{ik}'s).

The AIC in Eq. (2.23) [15,117] and a Student's t-test [118] were employed for model order selection and for determining significance of protein. In doing so, the rough PPI network was refined and the protein association networks for smokers with or without cancer were constructed.

Based on the identified protein association abilities, two matrices were established to represent the cancer protein association network (CPAN) and the noncancer protein association network (NPAN).

$$C = \begin{bmatrix} \alpha_{11,C} & \cdots & \alpha_{1K,C} \\ \vdots & \ddots & \vdots \\ \alpha_{K1,C} & \cdots & \alpha_{KK,C} \end{bmatrix}, \quad N = \begin{bmatrix} \alpha_{11,N} & \cdots & \alpha_{1K,N} \\ \vdots & \ddots & \vdots \\ \alpha_{K1,N} & \cdots & \alpha_{KK,N} \end{bmatrix}, \tag{9.10}$$

where $\alpha_{ij,C}$ and $\alpha_{ij,N}$ indicate the quantitative protein association ability between protein i and protein j for CPAN and NPAN, respectively; and K is the number of proteins in the protein association network. For a given protein i and protein j in the protein association network, the association ability α_{ij} quantifies the expression relation between the interactive proteins. If $\alpha_{ij} = 0$, there is no association between protein i and protein j. Further, we said that protein i is associated with protein j which means that the expression level changes of protein i account for the expression level changes of protein j and vice versa. As a consequence, when the estimated protein association ability $\alpha_{ij} \neq \alpha_{ji}$, the one which has larger absolute value would be selected as the association ability between protein i and protein j, i.e., $\alpha_{ij} = \alpha_{ji}$. The resulting CPAN and NPAN constituted the network-based biomarker used for identifying the significant proteins in lung carcinogenesis and diagnostic evaluation.

Determination of significant proteins in lung carcinogenesis via the network-based biomarker

According to Eqs. (9.9) and (9.10), the protein association models for CPAN and NPAN can be represented as the following equations.

$$\begin{aligned} Y_C &= CY_C + E_C, \\ Y_N &= NY_N + E_N, \end{aligned} \tag{9.11}$$

where $Y_C = \begin{bmatrix} y_{1,C}(n) \\ \vdots \\ y_{K,C}(n) \end{bmatrix}$ and $Y_N = \begin{bmatrix} y_{1,N}(n) \\ \vdots \\ y_{K,N}(n) \end{bmatrix}$ denote the vectors of expression levels; and E_C and E_N indicate the noise vectors in cancer and noncancer cases, respectively.

A matrix indicating the difference between two protein association networks is defined as *C-N*.

$$D = \begin{bmatrix} d_{11} & \cdots & d_{1K} \\ \vdots & \ddots & \vdots \\ d_{K1} & \cdots & d_{KK} \end{bmatrix} = \begin{bmatrix} \alpha_{11,C} - \alpha_{11,N} & \cdots & \alpha_{1K,C} - \alpha_{1K,N} \\ \vdots & \ddots & \vdots \\ \alpha_{K1,C} - \alpha_{K1,N} & \cdots & \alpha_{KK,C} - \alpha_{KK,N} \end{bmatrix}, \tag{9.12}$$

where d_{ij} denotes the difference in protein association ability between CPAN and NPAN among protein i and protein j. Using the matrix D to show the difference in network structure between CPAN and NPAN, a CRV was presented to quantify the correlation of each protein with significance of lung carcinogenesis. To identify the significant proteins in lung carcinogenesis, two important issues were taken into consideration. First, the magnitude of the association abilities α_{ij} denotes the significance of one protein to the other one. A higher absolute value of α_{ij} implies that the two proteins are more tightly associated. Second, if a protein plays a more crucial role in lung carcinogenesis, the difference in association numbers linked to the protein for CPAN and NPAN would be larger. For example, if one protein shares a strong association with many proteins in CPAN, but a weaker association (no protein) in NPAN, the protein in question is more likely involved in lung carcinogenesis. As a result, the CRV was determined based on the difference in protein association abilities via the following equation.

$$CRV_i = \sum_{j=1}^{K} \left| d_{ij} \right|. \tag{9.13}$$

For the *i*th protein in the network-based biomarker, the implication of Eq. (9.13) is that the CRV quantifies the extent of protein associations that differentiate CPAN from NPAN.

In addition to the CRV assigned, an empirical p-value was also obtained for each protein to determine the statistical significance of the CRV. To determine the p-value for an observed CRV, a null distribution of CRVs (Fig. 9.14) was generated by repeatedly permuting the network structure of the rough PPI network and computing the CRV for each random network structure. The permutation of the network structure was performed by maintaining network size, i.e., interacted proteins were permuted without altering the total number of protein interactions. The process was repeated 100,000 times and the p-value of the corresponding CRV was estimated as the fraction of random network structures whose CRV is at least as large as the CRV of the real network structure. The CRVs with p-value $\leq .05$ were determined as significant CRVs and the corresponding proteins were identified as significant proteins in lung carcinogenesis.

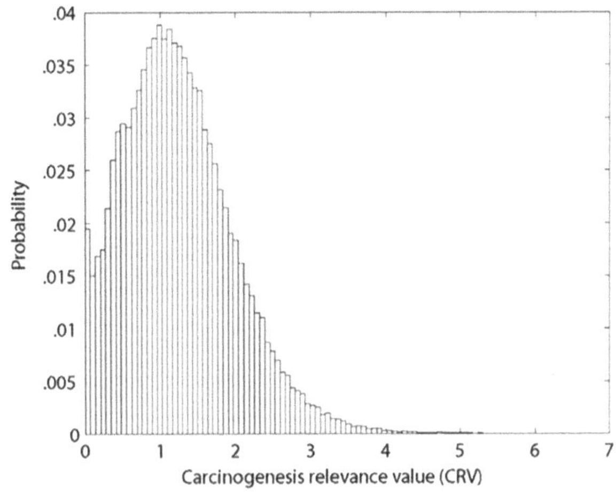

Figure 9.14 Distribution of carcinogenesis relevance values (CRVs) of random networks. The null distribution of CRVs is generated by 100,000 randomly permuted network structures.

Diagnostic evaluation by the network-based biomarker

An important feature of the proposed network-based biomarker approach is its ability to identify significant proteins in lung cancer, and perform accurate diagnosis of smokers suspect of lung cancer. Using the new microarray expression data for smokers, we can classify a sample into smoker with cancer or smoker without cancer based on CPAN and NPAN within the network-based biomarker. The idea comes from the similarity comparison of new sample data between CPAN and NPAN. Specifically, sample data that are more similar to the network structure of CPAN than to NPAN signifies the presence of lung cancer in the smoker, and vice versa. Because the network construction uses more than one sample, the new sample data were mapped to the CPAN and NPAN identified above, with the mapping error employed as the criteria for classification. Assuming we had new sample data $Z = \begin{bmatrix} z_1 \\ \vdots \\ z_K \end{bmatrix}$ from a smoker, based on Eqs. (9.9) and (9.11), the mapping errors for CPAN and NPAN, respectively, are defined as

$$\begin{aligned} ME_C &= \left\| Z - CZ \right\|_2, \\ ME_N &= \left\| Z - NZ \right\|_2, \end{aligned} \tag{9.14}$$

where $\left\| P \right\|_2 = (\sum_{i=1}^{K} p_i^2)^{1/2}$, when $P = \begin{bmatrix} p_1 \\ \vdots \\ p_K \end{bmatrix}$. The mapping errors can be considered as the similarity measurement of the new sample Z to the CPAN and NPAN systems. The smaller the mapping error, the stronger the correlation between the sample data and the protein association network. Consequently, if $ME_C < ME_N$, the new sample Z is more similar to the cancer system and is categorized into smokers with cancer, and vice versa. The criteria of mapping errors have simultaneously taken into account the protein association network structures with quantitative association abilities and the expression levels of the proteins. Further, since the modeling error operates as the criterion of classification, the classification is more dependent on network structure than data alone and thus could also be suitable for classification using independent data. We believe that this kind of classification approach can provide new perspective for diagnostic evaluation.

9.3.2 Investigating network-based biomarker for molecular mechanism of lung cancer

Construction of network-based biomarker and identification of significant proteins in lung carcinogenesis

We applied the proposed network-based biomarker approach for molecular investigation and diagnosis of lung cancer. The primary data set (79 smokers with lung cancer and 73 smokers without lung cancer) of GSE4115 downloaded from the GEO database http://www.ncbi.nlm.nih.gov/geo was used for construction of the network-based biomarker. Based on ANOVA, 199 proteins with PPI information were identified as the differentially expressed proteins and were selected in the protein pool. In addition, the proteins that linked to three differentially expressed proteins in the protein pool according to PPI information were also included in the pool. As a result, the protein pool consisted of 339 proteins. Following this, proteins with PPI information among them were linked together, resulting in the rough PPI network. The expression profiles for smokers with or without cancer and the protein association model (9.9) were further employed to refine the rough PPI network. The CPAN and NPAN, which consisted of 399 and 393 protein associations, respectively, would constitute the network-based biomarker of lung cancer (Fig. 9.15). The difference between CPAN and NPAN is expressed further in Fig. 9.16. According to the CPAN and NPAN with quantitative association abilities, the CRVs for each protein were computed and the significance of these CRVs was determined. Consequently, 40 identified proteins played significant roles in lung carcinogenesis (Tables 9.5 and C.1).

Investigation into the significant proteins in lung carcinogenesis

The 40 identified proteins contributing to lung carcinogenesis can be divided into three categories according to the functional annotations (Table 9.5); the three functional subnetworks are shown in Fig. 9.17. The mechanisms in carcinogenesis using the significant proteins were further investigated through various cell life cycle stages.

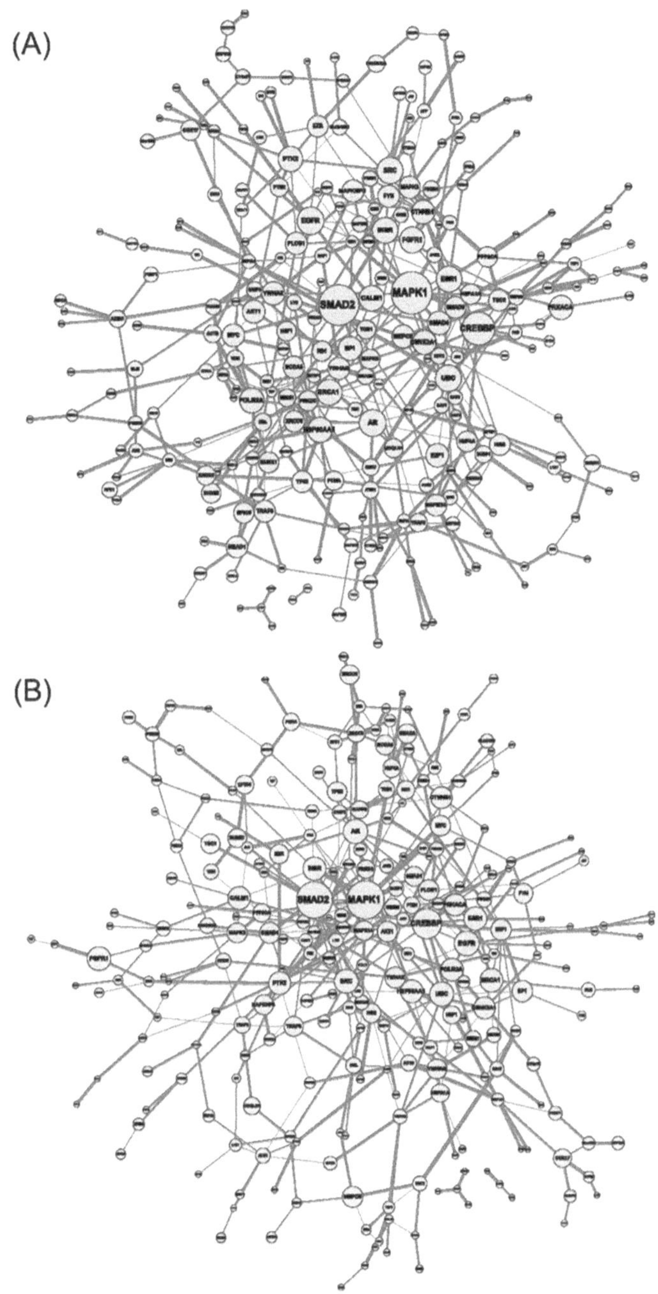

Figure 9.15 The constructed network-based biomarker. (A) Cancer protein association network (CPAN); (B) noncancer protein association network (NPAN). The node size is proportional to the CRV for each protein and the edge width represents the magnitude of the association ability between the two proteins. The figures are created using Cytoscape [119].

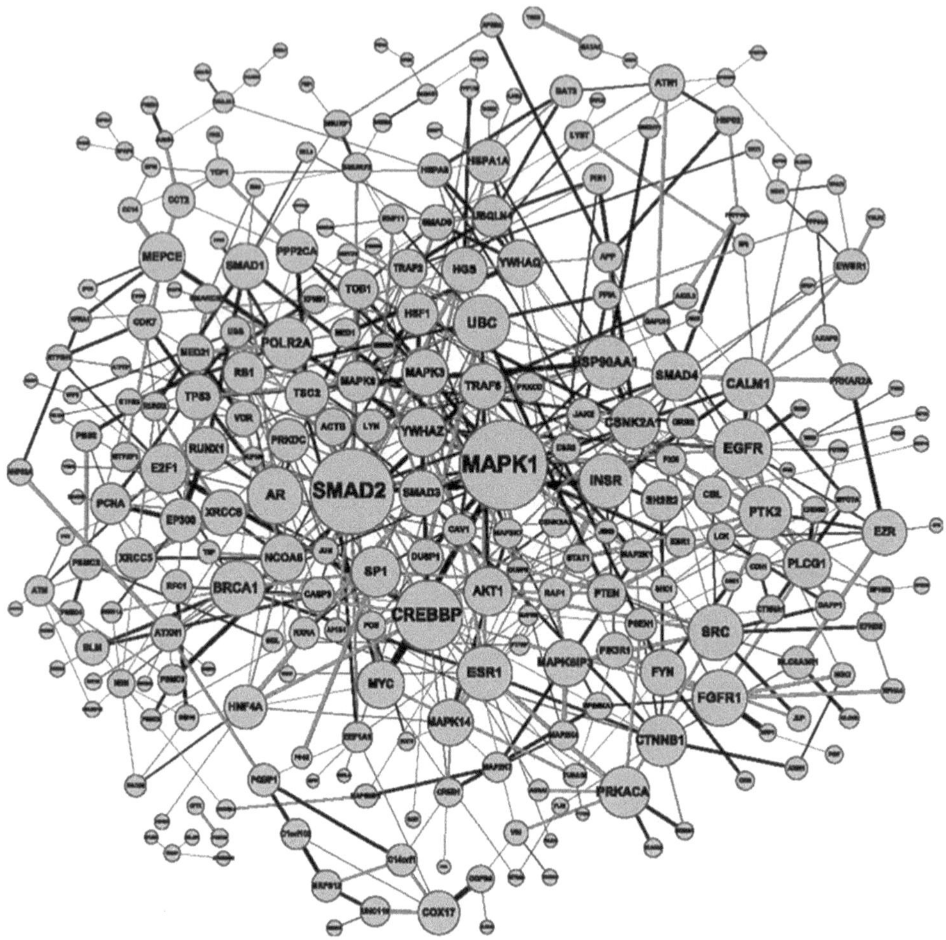

Figure 9.16 The difference between CPAN and NPAN. The node size is proportional to the CRV for each protein and the edge width represents the magnitude of the association ability between the two proteins. *Red* (dark gray in print versions) and *blue* (light gray in print versions) edges indicate the positive and negative values of d_{ij} in Eq. (9.12), respectively. The figure is created using Cytoscape [119].

Cell growth

Cancer is a complex disease of uncontrolled cell growth. Therefore, the proteins responsible for cell growth are likely to play critical roles in lung carcinogenesis. Among the 40 significant proteins identified based on the network-based biomarker, 30 proteins are annotated with cell growth (Table 9.5). The MAPK cascade is a highly conserved module that is identifiable in many cancers. Three MAPK-related proteins annotated with cell growth, *MAPK1* (*ERK2*), *MAPK3* (*ERK1*),

Table 9.5 The identified significant proteins in lung carcinogenesis

Protein symbol[a]	CRV	*p*-Value	Functional annotation[b]			Literature evidence[c]
			Cell growth	Cell survival	Cell migration	
MAPK1	8.3418	< 1e-5	+	+	+	[120,121]
SMAD2	7.7901	< 1e-5	+	+	+	[122]
CREBBP	5.7870	.00002	+			[123]
EGFR	4.3635	.00086	+	+	+	[124–126]
AR	4.0966	.00159	+	+	+	[127]
UBC	4.0331	.00180				
SRC	3.9446	.00218	+	+	+	[128,129]
FGFR1	3.9227	.00237	+		+	[130]
BRCA1	3.9049	.00243	+	+		[131]
ESR1	3.8409	.00295	+	+	+	[132]
INSR	3.7946	.00329	+		+	[133]
PTK2	3.6758	.00432	+	+	+	[129,134]
HSP90AA1	3.6732	.00436	+	+	+	[135]
CALM1	3.6363	.00482		+		
POLR2A	3.5701	.00547				
CSNK2A1	3.4128	.00761	+	+		[136]
PRKACA	3.3688	.00856		+		
CTNNB1	3.2935	.00994	+	+	+	[137]
SP1	3.2397	.01133	+	+		[138]
SMAD4	3.1947	.01266	+	+	+	[122]
E2F1	3.1382	.01407	+	+		[139]
YWHAZ	3.1212	.01467	+			[140]
MEPCE	3.0968	.01545				
AKT1	3.0193	.01857	+	+	+	[141]
PLCG1	2.9654	.02069			+	[142]
MYC	2.8987	.02385	+	+		[143]
MAPK3	2.8545	.02654	+	+	+	[120,144]
NCOA6	2.8132	.02892	+	+		
FYN	2.7833	.03089	+		+	[145]
MAPK8IP3	2.7746	.03141			+	
YWHAQ	2.7582	.03242	+			[146]
TRAF6	2.7150	.03535		+		[147]
SMAD1	2.6940	.03697	+	+	+	[148]
SMAD3	2.6815	.03815	+	+	+	[122]
MAPK14	2.6727	.03894	+	+	+	[149]
TP53	2.6522	.04056	+	+	+	[124,125,150]
XRCC6	2.6270	.04263		+		
EZR	2.6213	.04314			+	[151]
TSC2	2.6116	.04401	+	+	+	[152]
HGS	2.5730	.04744	+			

[a]The full names of these proteins according to UniProt database http://www.uniprot.org/ are listed in Table C.1.
[b]The functional annotations are from the Gene Ontology database http://www.geneontology.org/ and literature.
[c]The literature evidences indicate that overexpression/dysregulation of the specific protein or mutation of the corresponding gene would result in carcinogenesis.

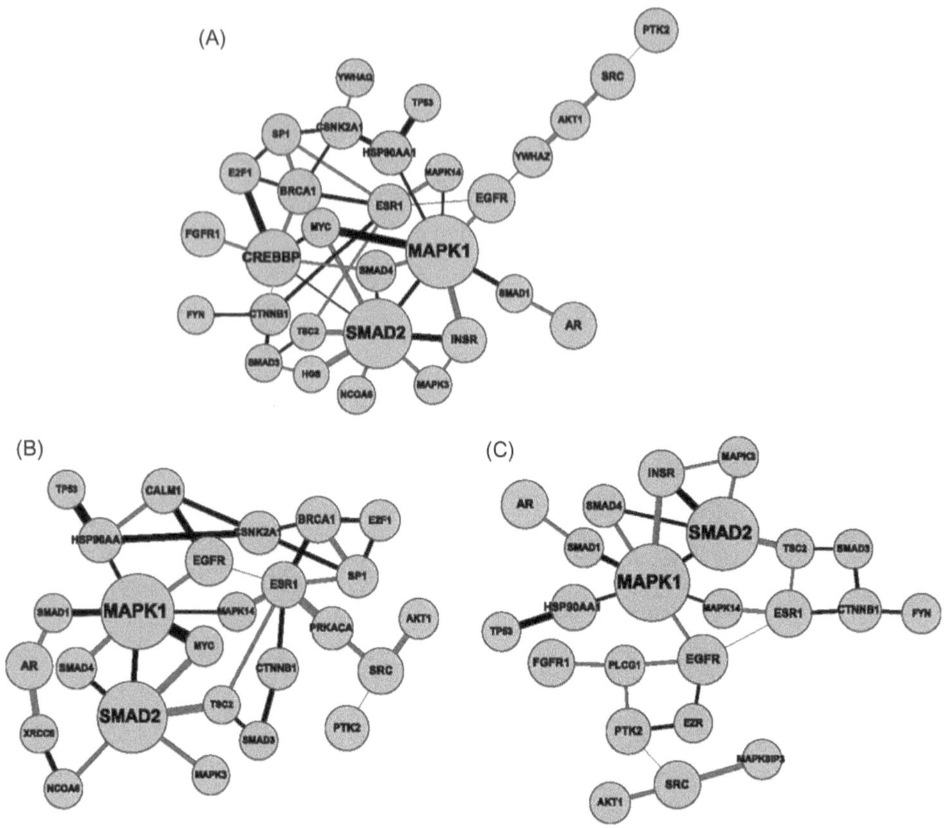

Figure 9.17 The functional subnetworks of significant proteins identified according to the network-based biomarker. (A) Cell growth functional subnetwork; (B) Cell survival functional subnetwork; (C) Cell migration functional subnetwork. All the functional subnetworks are extracted from Fig. 9.16.

and *MAPK14* (*p38α*), were identified as significant proteins in lung carcinogenesis. Elevated expression of activated MAPK1/3 have been observed in *NSCLC* [153] and may play a role in lung metastasis [120]. *MAPK14* (p38α) is well known for its important role in response to inflammation and environmental stress [121]; its protein expression is more than three times lower in human lung tumors compared with normal human lung tissue, suggesting that *MAPK14* may function as a negative regulator of lung carcinogenesis [149]. Four proteins from the SMAD protein family (*SMAD1*, *SMAD2*, *SMAD3*, and *SMAD4*) were also identified as significant proteins in lung carcinogenesis. The SMAD proteins, which consist of three functional classes, are signal transducers and transcriptional modulators [154]. Mutations and altered expression for these four proteins were observed in human cancer [122,148], highlighting the significant, albeit varied, roles of the SMAD protein family in carcinogenesis.

EGFR, *FGFR1*, and *INSR* are receptor tyrosine kinases (RTKs) bound by an epidermal growth factor, a fibroblast growth factor, and insulin, respectively. RTKs have been shown to not only perform a key regulatory role in normal cellular processes, but also become critically involved in the development and progression of human cancers [155]. *EGFR* is one of the most extensively studied proteins in carcinogenesis. It is overexpressed in NSCLC as well as in other common tumors, and its increased expression is associated with aggressive tumor growth and therapy resistance [126]. *EGFR* has also been found to be significantly mutated in lung adenocarcinoma [124,125]. Because of its significance, *EGFR* has become a popular therapeutic target for carcinogenesis. Gefitinib (Iressa) and erlotinib (Tarceva) are two targeted therapies that specifically inhibit *EGFR* tyrosine kinase [125,126]. *FGFR1* and *INSR* are both known to be involved in carcinogenesis and thus become novel, attractive targets for cancer therapeutic strategies [130,133].

Many proteins acting as transcriptional regulators were also identified. *CREBBP* is a transcriptional coactivator downstream of the TGFβ pathway and the mutations and deletions of the *CREBBP* gene are associated with lung cancer [123]. *CTNNB1* (β-catenin) is one of the core components in the Wnt pathway. Mutation of β-catenin, which results in aberrant activation of the Wnt pathway, is a frequent cause of human cancer growth [137]. *E2F1* is one of the significant proteins involved in the cell cycle. Its overexpression has been demonstrated in both *NSCLC* and *SCLC* and is induced by its upstream *RB* protein [139]. *MYC* is a nuclear phosphoprotein and functions as a transcription factor. It controls cell cycle progression by simulating G1/S transition and may result in loss of cell cycle arrest and uncontrolled tumor growth when dysfunctional [143]. Other identified significant proteins such as *AR*, *ESR1*, *SRC*, *FYN*, *YWHAQ*, *YWHAZ*, and *HSP90AA1* were also shown to contribute to the process of carcinogenesis [127,128,132,135,140,145,146].

Cell survival

The ability of tumor cell populations to expand in number is determined not only by the rate of cell proliferation but also by the rate of cell death [38]. In addition, the acquired resistance of programmed cell death, apoptosis, is a hallmark of cancer. Consequently, proteins annotated with cell survival might be important in carcinogenesis. Twenty-seven significant proteins identified by the network-based biomarker were annotated with cell survival. *TP53* (p53) is a well-studied tumor suppressor protein and plays important roles in anticancer mechanisms. Its activation is induced by a number of stress signals such as DNA damage, oxidative stress, and activated oncogenes. Activated *p53* induces cell cycle arrest, apoptosis, and inhibition of angiogenesis and metastasis. Once damaged, tumor suppression is severely reduced, resulting in uncontrolled proliferation of the cell. Due to the importance in carcinogenesis, it is no surprise that *p53* was found to be significantly mutated in lung adenocarcinoma as well as in squamous cell carcinoma and SCLC [124,125,150].

TRAF6 functions as a signal transducer in the NFκB pathway that activates *IKK*, in response to proinflammatory cytokines. The identification of TRAF6 by the proposed network-based biomarker approach reinforces the linkage between inflammation and cancer [147,156]. *BRCA1* is a nuclear phosphoprotein that contributes to genomic stability. The mutant phenotype of *BRCA1* predisposes to breast and ovarian cancer [131]. *SP1* is a transcription factor downstream of the TGFβ pathway and its overexpression contributes to malignant transformation [138]. Other protein kinases, *AKT* and *CSNK2A1* (CK2), were also shown to participate in the carcinogenic process [136,141,157].

Cell migration

With the progression of cancer, the malignant tumor cells acquire the ability to migrate and metastasize to distant sites. As a result, the proteins that are relevant to the cell migration capability are crucial for the carcinogenic process. Twenty-three out of 40 significant proteins were annotated with cell migration. *PTK2* (*FAK*), a protein tyrosine kinase in the RTK pathway, is an important mediator within the cell migration process, as well as in cell proliferation and cell survival. Substantial evidence has shown that activated *PTK2* leads to tumor growth and metastasis [129], and the level of expression is substantially linked to the invasive potential of tumors [134]. High levels of *TSC2* were correlated with increased metastasis and reduced survival in breast cancer patients, revealing a protumorigenic role for *TSC2* [152]. The other two significant proteins, *PLCG1* and *EZR* (ezrin), were demonstrated to play critical roles in the metastatic potential of cancer cells but not in primary tumor growth [142,151].

For 9 out of the 40 significant proteins identified, little is known about their roles in lung cancer (see Table 9.5). UBC is identified as a polyubiquitin precursor. Protein ubiquitination is a fundamental, regulatory posttranslational modification controlling intracellular signaling events. It has been associated with protein degradation, DNA repair, cell cycle regulation, endocytosis, and kinase modification [158]. Dysregulation of ubiquitin-mediated signaling is increasingly implicated in some human diseases. Therefore, UBC may be an important target for further characterization of lung carcinogenesis. *CALM1* is calmodulin, which mediates the control of a large number of enzymes and other proteins by Ca^{2+}. It is an essential regulator of cell cycle progression and cell survival. Further research is needed to examine its relation with carcinogenesis. *PRKACA* is a cAMP-dependent protein kinase. The identification of *PRKACA* as a significant protein implies that cAMP signaling might also be involved in lung carcinogenesis. *MAPK8IP3* functions as a scaffold protein in the RTK pathway; *NCOA6*, *HGS*, and *XRCC6* are annotated with cell growth and/or cell survival. However, until now, no empirical evidence has linked their relevance to carcinogenesis, which makes them potential targets for further investigation into lung carcinogenesis.

Diagnostic evaluation of smokers suspect of lung cancer using the network-based biomarker

The network-based biomarker was constructed based on the primary data set of GSE4115. An independent data set (the prospective data set of GSE4115, 18 smokers with lung cancer and 17 smokers without lung cancer) was then used to evaluate the diagnostic performance of the proposed network-based biomarker. Among the 35 samples, 26 were accurately classified, resulting in an accuracy of 74.29%. The sensitivity and specificity of the proposed approach were also evaluated. The network-based molecular diagnosis can identify smokers with or without cancer, using a high level of sensitivity of 83.33%, and a moderate specificity of 64.71%: this enables the proposed network-based biomarker to effectively diagnose the smokers with lung cancer. Further, this approach enables the network-based biomarker to act as a screening test, which, with the aid of other clinical diagnostic tools, both accelerates the process and improves the sensitivity of diagnosis.

The cause of the moderate specificity was further investigated. The difference between the specificity value and the sensitivity value can be attributed to a misclassification of a number of smokers without cancer in the cancer category. The misclassification may be due to similarities in molecular pattern, i.e., the gene expression profiles of smokers with cancer are similar to those of smokers without cancer. In order to validate the hypothesis, Pearson correlation coefficients [118] of gene expression profiles for both smokers with and without cancer were calculated. The mean correlation coefficient of smokers with cancer was 0.9616, whereas the mean correlation coefficient of smokers without cancer was 0.9441. In addition, the mean correlation coefficient of smokers both with and without cancer (pooled) was as high as 0.9437, suggesting that the molecular patterns shared among smokers with and without cancer are indeed highly similar. Because of the highly similar molecular patterns and the fact that cigarette smoking is the main instigator of lung cancer, it is likely smokers without cancer could one day develop lung cancer.

In order to validate the predictive performance of the proposed network-based biomarker, several comparisons were made. First, we tested the predictive performance without the information of PPIs. The 199 differentially expressed proteins selected by ANOVA using the primary data set were used for classification of the prospective data set. A simple hierarchical clustering was performed (see Fig. C.1), illustrating that the use of gene expression alone cannot accurately classify the prospective data set (65.71% accuracy). The comparison shows that the integration of gene expression profiles and protein interaction information can improve lung cancer diagnosis. Second, we compared the predictive performance of the proposed method with randomly selected networks. The average accuracy for 100,000 randomly selected networks of 339 proteins was 48.44%, highlighting the significance of the proposed network-based biomarker. Third, in addition to evaluating the predictive performance using an independent data set, a fivefold cross-validation was applied to the primary data set plus the prospective data set. The accuracy of the cross-validation is similar to the initial accuracy of 74.29% computed using the independent data set, illustrating the robustness and reproducibility of the proposed network-based biomarker approach.

9.3.3 Discussion of network-based biomarker of lung cancer

Cancer is a complex disease and carcinogenesis in humans is a multistep process that transforms normal cells into malignant derivatives. Many researchers are investigating the underlying mechanisms that prompt the uncontrolled cell proliferation and metastasis. They have successfully identified some key components of the various steps in the carcinogenesis and some therapeutic interventions have been developed to at least slow down the carcinogenic process. However, because of the complexity, the therapy that targets some specific molecules is only partially effective and tumor-specific. Therefore, investigation of the carcinogenesis from the systems perspective is inevitable. On the other hand, biomarker identification for cancer diagnosis has been a primary research focus in the biomedical field since the use of biomarkers could provide early detection of cancer. As a result, in this section, a network-based biomarker approach has been proposed to simultaneously account for molecular investigation and diagnosis. The proposed approach was applied on the sample data obtained from smokers with and without lung cancer and 40 significant proteins were identified in lung carcinogenesis. The network-based biomarker considers not only differentially expressed proteins but also the protein association network structure. This allows an accurate identification of proteins with low discriminative potentials if such proteins were associated with many other significant proteins [106]. This property is important for the identification of significant proteins in lung carcinogenesis and provides a mechanistic insight into the process. From the mechanism investigation of the 40 significant proteins identified using the network-based biomarker, we found that the significant proteins identified are involved in the pathways that are responsible for cellular processes, including proliferation, differentiation, apoptosis, and metastasis. More importantly, from the results presented, we found that dysregulated signals exist in multiple pathways. There are two possible explanations for the result: the genetic mutations are accumulated respectively to components of different pathways, or the aberrant signals affect different pathways through crosstalk mechanisms. Further investigation is needed to address these two hypotheses. In addition to the investigation of significant proteins, the network-based biomarker can be used as a type of screening test with high sensitivity. Using the same data (the prospective data set of GSE4115), conventional bronchoscopy was shown to be 44% sensitive to cancer detection [98], which is only half of the proposed network-based biomarker. This further reinforces the clinical utility of the network-based biomarker.

Although the proposed method is shown to be useful, some limitations exist and the need for further improvements remains. In the proposed network-based biomarker approach, gene expression profiles were overlaid to the corresponding proteins for further analysis. However, levels of mRNA do not always correlate with protein levels and do not provide information on posttranslational modification such as phosphorylation that may be critical in regulating protein activity [125]. Consequently, emerging high-throughput proteomic techniques such as protein microarrays would benefit this method by significantly improving the detection performance over mRNA microarray data. In addition, if the genome-wide gene expression levels and protein expression levels can be obtained simultaneously, we are then able to construct the integrated cellular networks of transcription regulations and protein interactions which provide a more integrated network-based biomarker [159]. The PPI data from public databases also plays important roles in the proposed method. Nevertheless, there is a large variation in the coverage of protein interaction data across the interaction databases [160]. Therefore, HPRD and BioGRID databases were integrated for the PPI information in this section. We believe that the increased quality and coverage of protein interaction data would enhance the proposed network-based biomarker approach for characterization of lung carcinogenesis. Another limitation of the proposed method was the restriction in the size of the protein association network from the sample size available due to the avoidance of overfitting in the network construction. This results in the exclusion of some well-studied proteins that are relevant to the lung carcinogenic process in the network-based biomarker, including *KRAS*, *MET*, and *PI3KCA*. To overcome the problem, more samples are needed. We believe that improvements to diagnostic evaluation using the network-based biomarker lie in the expansion of the constructed protein association networks. Many other groups identified the discriminative subnetworks using different methods, especially graph theory-based methods [161,162]. For example, Tian et al. proposed a hypergraph-based iterative learning algorithm for subnetwork identification, which minimizes a cost function under a unified regularization framework [162]. These graph-based methods can also be incorporated to improve the significant protein selection in the proposed method. Further, in this section, the samples for gene expression profiling are simply divided into two groups: smokers with cancer, and without cancer. With more sample data, particularly cancer stage-specific samples, we can determine how the network evolves and changes during cancer progression using the proposed method.

Our network-based biomarker provides both a systematic insight into the lung carcinogenic process and a good method for identifying significant proteins, categorized as lung cancer-related proteins and many others that have not been previously reported. These proteins not only provide new targets for further research into understanding the mechanisms of lung carcinogenesis, but are also potential targets for therapeutic interventions. The main challenge of cancer research is to find an effective therapeutic approach that specifically kills malignant cells. Conventional chemotherapy acts by killing all rapidly dividing cells, resulting in toxic effects and damage to normal tissues [163]. With the advances in understanding the mechanisms of the carcinogenic process, the so-called targeted therapy, which is more effective and less harmful to normal cells, is developed to inhibit the specific molecules that play crucial roles in tumor growth. The significant proteins identified by the proposed network-based biomarker provide suitable molecules to be targeted. For example, gefitinib (Iressa) and erlotinib (Tarceva) are two tyrosine kinase inhibitors that specifically target EGFR. Despite their effectiveness, there are still patients that do not respond well to these drugs [164]. One explanation is that single-target agents are likely to result in network compensation and drug resistance [165]. As a result, multitarget therapeutic interventions that affect multiple targets simultaneously may be required for effective control against cancer. Because biological systems are unable to perform optimally under the influence of two or more simultaneously administered drugs, multitarget therapeutics can prove effective as they may be less vulnerable to adaptive resistance from the human body [166]. With the help of significant proteins identified by the proposed network-based biomarker approach and the pathway information, it is possible the multitarget therapeutic interventions that act on different critical pathways in lung carcinogenesis can be developed.

9.3.4 Conclusions of significant molecular mechanisms by network-based biomarkers

Lung cancer is the leading cause of cancer deaths worldwide. Understanding the causes and the underlying mechanisms can help fight the disease. In this section, a network-based biomarker approach, which integrated gene expression profiles and protein interaction information, was developed for molecular investigation and diagnosis for lung cancer. From a systems perspective, the constructed network-based biomarker further evaluated the lung carcinogenic process by use of significant protein identification and diagnostic evaluation. The diagnostic results indicate that the network-based biomarker is sensitive to the diagnosis of smokers with lung cancer and can be used as one kind of screening test. More importantly, the significant proteins identified by the network-based biomarker give mechanistic insights into the carcinogenic process and provide potential therapeutic targets to combat cancer.

APPENDIX C: TABLES AND FIGURES

Here we show Table C.1 and Fig. C.1.

Table C.1 The full names of the significant proteins identified by a network-based biomarker approach for molecular investigation and diagnosis of lung cancer

MAPK1	Mitogen-activated protein kinase 1	*E2F1*	Transcription factor E2F1
SMAD2	Mothers against decapentaplegic homolog 2	*YWHAZ*	14-3-3 Protein zeta/delta
CREBBP	CREB-binding protein	*MEPCE*	7SK snRNA methylphosphate capping enzyme
EGFR	Epidermal growth factor receptor	*AKT1*	RAC-alpha serine/threonine-protein kinase
AR	Androgen receptor	*PLCG1*	1-phosphatidylinositol-4,5-bisphosphate phosphodiesterase gamma-1
UBC	Polyubiquitin-C	*MYC*	Myc proto-oncogene protein
SRC	Proto-oncogene tyrosine-protein kinase Src	*MAPK3*	Mitogen-activated protein kinase 3
FGFR1	Basic fibroblast growth factor receptor 1	*NCOA6*	Nuclear receptor coactivator 6
BRCA1	Breast cancer type 1 susceptibility protein	*FYN*	Tyrosine-protein kinase Fyn
ESR1	Estrogen receptor	*MAPK8IP3*	C-Jun-amino-terminal kinase-interacting protein 3
INSR	Insulin receptor	*YWHAQ*	14-3-3 Protein theta
PTK2	Focal adhesion kinase 1	*TRAF6*	TNF receptor-associated factor 6
HSP90AA1	Heat shock protein HSP 90-alpha	*SMAD1*	Mothers against decapentaplegic homolog 1
CALM1	Calmodulin	*SMAD3*	Mothers against decapentaplegic homolog 3
POLR2A	DNA-directed RNA polymerase II subunit RPB1	*MAPK14*	Mitogen-activated protein kinase 14
CSNK2A1	Casein kinase II subunit alpha	*TP53*	Cellular tumor antigen p53
PRKACA	cAMP-dependent protein kinase catalytic subunit alpha	*XRCC6*	X-ray repair cross-complementing protein 6
CTNNB1	Catenin beta-1	*EZR*	Ezrin
SP1	Transcription factor Sp1	*TSC2*	Tuberin
SMAD4	Mothers against decapentaplegic homolog 4		

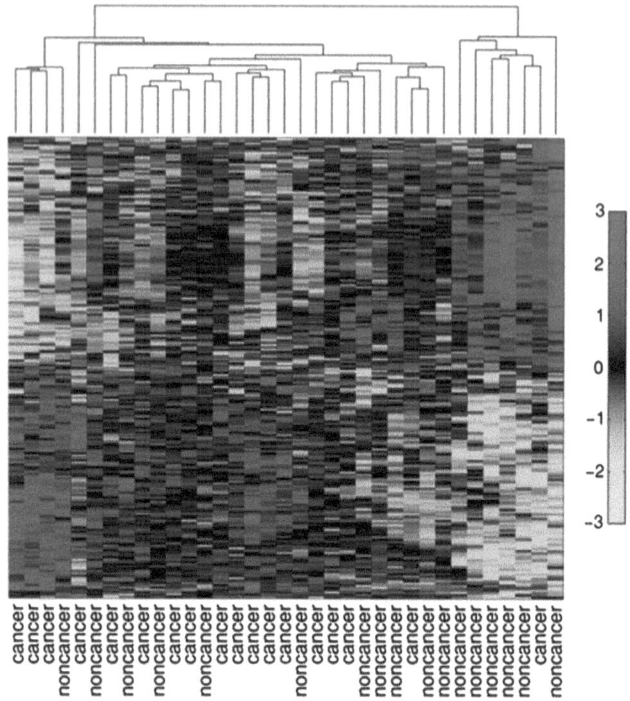

Figure C.1 The hierarchical clustering for 199 differentially expressed proteins of the prospective data set.

9.4 NETWORK BIOMARKERS OF BLADDER CANCER BASED ON A GENOME-WIDE GENETIC AND EPIGENETIC NETWORK DERIVED FROM NGS DATA

In the previous sections, we have constructed the GRN and PPIN to investigate the tumorigenesis mechanisms from microarray data. However, tumorigenesis is involved in more complex genetic regulations and epigenetic interactions. In this section, we will construct an IGEN to combine gene regulation, protein interaction, microRNA (miRNA) regulation and DNA methylation from NGS data to investigate the tumorigenesis mechanism from the more whole perspective.

Epigenetic regulation and miRNA regulation are associated with carcinogenesis and the development of cancer. The available omics data, including those from NGS, genome-wide methylation profiling, candidate IGEN analysis, and drug response genome-wide microarray analysis, allow us to elucidate the mechanisms behind bladder carcinogenesis. This information enables us to design multiple drug combinations for treating bladder cancer with minimal side effects. Currently, it is still difficult to construct an accurate IGEN of bladder cancer cells on a genome-wide scale that includes protein interactions, transcription and miRNA regulation, and DNA methylation. According to the omics data on human cells, we constructed an IGEN system based on three coupling regression models that characterize PPINs, GRNs, miRNA regulatory networks (MRNs), and epigenetic regulatory networks. By applying principal genome-wide network projection (PGNP) to IGEN analysis, we identified the core network biomarkers of bladder carcinogenesis to investigate its underlying mechanisms. We found that the major factors, including downregulated miR1-2, and aging- and smoking-related proteins, lead to the progression from normal bladder cells to stage 1 bladder cancer cells through SUMOylation, ubiquitination, and proteasome (SUP) signaling pathway and ER signaling pathway. The smoking-related protein HSP90AA1 and DNA methylation of *ECT2* mediate metastasis in bladder cancer. The activated DNA repair of bladder cancer cells leads to metastasis of those cells owing to their immortality.

We found that when the accumulated genetic mutations and epigenetic alterations lead to the dysregulation of the tumor necrosis factor (TNF) pathway in inflammation, the accumulated misfolded proteins in the ER pathway may induce cell proliferation in stage 1 bladder cancer. Regulation of the ER and TNF pathways is then adaptive to the accumulated genetic mutations and epigenetic alterations through the SUP pathway. The progression of DNA repair and cell proliferation in stage 1 bladder cancer ultimately results not only in the derepression of miR200a and miR200b but also the regulation of the TNF pathway to metastasis-related genes or proteins, cell proliferation, and DNA repair in stage 4 bladder cancer. Therefore, we designed a multiple drug combination comprising gefitinib, estradiol, yohimbine, and fulvestrant for treating stage 1 bladder cancer with minimal side effects, and another multiple drug combination comprising gefitinib, estradiol, chlorpromazine, and LY294002 for treating stage 4 bladder cancer with minimal side effects. An accurate genome-wide IGEN would allow us not only to elucidate bladder carcinogenic mechanisms, but also to improve drug safety and efficacy in the treatment of bladder cancer.

Bladder cancer is still one of the most common cancers worldwide. Single gene markers have been proposed for improving cancer treatment [167]. However, single gene markers cannot overcome treatment side effects because the markers are not implicated in genome-wide networks, and the analysis of a genome-wide network is a complicated issue from a systems biology perspective. The rapid development of molecular biology techniques has produced a great deal of high-throughput experimental data, including genome-wide microarray data, genome-wide methylation profiles, NGS data, miRNA profiles, genetic sequences, protein abundance data, and drug response genome-wide microarray data. These kinds of omics data provide an opportunity to design multiple drug combinations for the treatment of bladder cancer by applying the network biomarkers identified by systems biology.

To date, genetic regulation systems, including PPINs and GRNs, have been applied to analyze the functional mechanisms behind human aging and cancer [168]. We now know that epigenetic alterations are much more rapid and adaptive with regards to influencing genome-wide gene expression than genetic changes [169]. Rapid and slow response mechanisms, i.e., epigenetic alterations and genetic changes, respectively, coordinate an efficient and robust system. Epigenetic regulation, including DNA methylation and histone modification, results in potentially reversible alterations in gene expression that do not involve permanent changes to the DNA sequence. miRNAs that are influenced by aberrant epigenetic regulation also mediate the regulation of gene expression [170]. It has been found that DNA methylation directly affects the binding affinities of miRNAs, RNA polymerase, and transcription factors (TFs) [171], and indirectly influences PPIs [172]. Methylation analysis of human genomic DNA in 12 tissues revealed that DNA methylation profiles are tissue-specific [173]. Therefore, omics data and systems biology methods are required to unravel the mechanisms underlying carcinogenesis from the complex molecular biology and design anticancer drugs for the treatment of bladder cancer.

The Human Genome Project (HGP) has identified 30,000-40,000 genes in human DNA, including miRNAs. The genes, proteins, and their associations, miRNA regulation, and DNA methylation constitute the IGEN, which coordinates cellular responses. PPIN in human lung cancer [174] and GRN in human aging [175] of the genes with significant expression differences between cancer cells (or aged people) and normal cells (or young people) have been identified for the extraction of the core network biomarkers according to the estimated association abilities between TFs (or upstream proteins) and target genes (or target proteins). Aging is associated with cancer [176]. The association abilities estimated by the network models assume that the binding affinities of TFs (or upstream proteins) to target genes (or proteins) are the same. According to a recent study in primary human somatic and germline cells [171], the impact of the binding affinities of miRNAs, RNA polymerase, and TFs on gene expression is mediated by DNA methylation. According to the available genome-wide methylation profiles and NGS data for bladder cancer in The Cancer Genome Atlas (TCGA), DNA methylation and miRNA regulation can be also characterized by the GRN model to identify the genome-wide IGEN. In this section, we identified the IGENs in normal bladder cells and bladder cancer cells, then investigated the impact of epigenetic regulation and miRNA regulation on bladder carcinogenesis by comparing the IGEN in normal bladder cells with that in bladder cancer cells.

Although a genome-wide IGEN can be identified based on well-defined system identification techniques [168,174], the means by which the core network biomarkers are extracted from the identified genome-wide network is still an important issue. The total association capabilities of a single node can affect the contribution it makes to its neighbors. However, the genome-wide IGEN including transcriptional gene regulations, miRNA regulations, and PPIs constitutes a genome-wide network structure. The contribution made by one node to its neighbors is not sufficient to explain its impact on a genome-wide scale network of bladder cells. In this section, we applied a PGNP based on principal component analysis (PCA) to identify core network biomarkers in bladder carcinogenesis, with the objective of extracting the most significant part from a genome-wide network structure. Because the drug response genome-wide microarray data are now available [177], we analyzed the drug response microarray data of the core network biomarkers to design multiple drug combinations with minimal side effects for bladder cancer treatment. Therefore, the identified core network biomarkers could provide an opportunity to design such drug combinations for bladder cancer treatment. Furthermore, it has been reported that aging (over 45 years old) and smoking are two major risk factors for bladder carcinogenesis [178]. Therefore, we used the core network biomarkers to elucidate the cellular mechanisms by which aging and smoking elevate bladder cancer risk through epigenetic regulation, miRNA regulation, and signaling pathways.

According to the strategy shown in the flowchart (Fig. 9.18), we integrated omics data, including genome-wide methylation profiles, NGS expression data, miRNA profiles in TCGA, drug response genome-wide microarray data in The Connectivity Map (CMAP) [177], drug–gene interaction data in the Drug Gene Interaction Database (DGIdb) [179], miRNA–target gene association data in TargetScan [180], PPIs in BioGRID, transcription regulations in the Human Transcriptional Regulation Interactions database (HTRIdb) [181], the Integrated Transcription Factor Platform (ITFP) [182], and the TRANSFAC [183], biological processes and pathways in a GO database, the National Center for Biotechnology Information (NCBI) Entrez Gene database, and the Kyoto Encyclopedia of Genes and Genomes (KEGG) pathway database [184]. We used miRNA–target gene association data, PPIs, and transcription regulations to build the candidate IGEN for general molecular mechanisms. We then constructed a regression IGEN model to characterize the molecular mechanisms including miRNA regulation, PPIs, transcription regulation and DNA methylation in cells. To prune the false-positive connections in the candidate IGEN and identify the model parameters of the IGEN in the real human bladder cells, we used methylation profiles, NGS expression data, and miRNA profiles in normal bladder cells and stage 1 and stage 4 bladder cancer cells. We then applied the constrained least squares method and the AIC in Eq. 2.23 [15], a system order detection method, to prune the false-positive connections for obtaining the real IGENs in the three stages of human bladder carcinogenesis. The three genome-wide real IGENs in normal bladder cells and stage 1 and stage 4 bladder cancer cells were then projected into the three core networks of the three stages of bladder carcinogenesis, respectively. Because the core networks contain the identified signal transduction pathways, i.e., the receptors and TFs of the core network can be directly or indirectly connected by the core

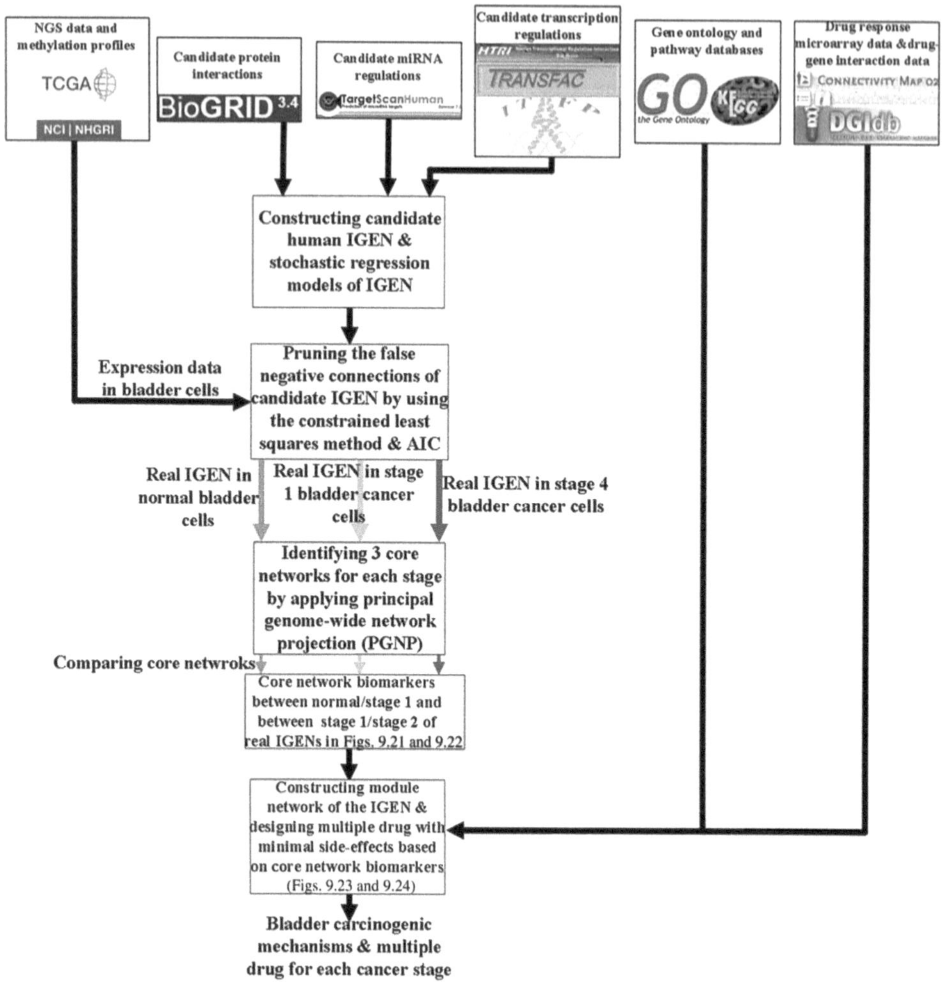

Figure 9.18 Flowchart of the proposed method for constructing the core network biomarkers and identifying bladder carcinogenesis mechanisms.

proteins/TFs, the proteins/TFs and the corresponding genes that participate in the identified signaling pathways of the core networks are considered as the core network biomarkers for normal and cancerous cells, respectively. The miRNAs with very different connections in regulating the genes of the core network biomarkers between two cells are also involved in the core network biomarkers. By comparing the identified connections of the IGENs, we investigated how the connection changes of the core network biomarkers from normal bladder cells to stage 1 bladder cancer cells, and from stage 1 bladder cancer cells to stage 4 bladder cancer cells contribute to bladder carcinogenesis.

We also investigated how the module network of the core network biomarkers, including the KEGG pathways and biological processes, participates in bladder carcinogenesis. According to the information on the biological processes and signaling pathways in the GO database, the NCBI Entrez Gene database, and the KEGG pathway database, the roles of the TFs/proteins in the core network biomarkers are projected into three pathways: the SUP pathway; the TNF signaling pathway; and the ER signaling pathway. The roles of the downstream genes in the core network biomarkers are projected into three biological processes: cell proliferation, DNA repair, and metastasis. The module network, including the KEGG pathways, TFs, miRNAs, and biological processes, are connected according to the three identified IGENs in the three types of bladder cell. By comparing the connection changes of the module networks from normal bladder cells to stage 1 bladder cancer cells, and from stage 1 bladder cancer cells to stage 4 bladder cancer cells, we ultimately unraveled the cellular mechanisms behind bladder carcinogenesis and proposed two multiple drug combinations for treating stage 1 and stage 4 bladder cancers, respectively.

Additionally, to determine how the two major risk factors, aging and smoking, influence bladder carcinogenesis, we not only highlighted the significantly expressed genes between smokers and nonsmokers, but also the significantly expressed genes between young ($\leq$45 years old) and old ($>$45 years old) people in the core network biomarkers of bladder carcinogenesis. Finally, we investigated the carcinogenic mechanism of human bladder cells by which the identified major factors, including downregulated miR1-2, aging, and smoking, lead to the progression from normal bladder cells to stage 1 bladder cancer cells through the SUP and ER signaling pathways. The smoking-related protein HSP90AA1 and DNA methylation of *ECT2* mediate the progression from stage 1 bladder cancer cells to metastasis in stage 4 bladder cancer. Activated DNA repair and accumulated epigenetic alterations lead to the phenotypic changes of bladder cells from normal to cancerous, and from cancerous to metastatic cells owing to the immortality of cancer cells. Based on the core network biomarkers in bladder carcinogenesis, a multiple drug combination comprising gefitinib, estradiol, yohimbine, and fulvestrant was designed for treating stage 1 bladder cancer with minimal side effects, while a multiple drug combination comprising gefitinib, estradiol, chlorpromazine, and LY294002 was designed for treating stage 4 bladder cancer with minimal side effects.

9.4.1 IGEN system, system identification, and PGNP

According to the flowchart in Fig. 9.18, we constructed a candidate human IGEN by mining large databases, including BioGRID, TargetScan, HTRIdb, TRANSFAC, and ITFP. However, many false-positive and insignificant connections existed in the candidate human IGEN for normal and cancerous bladder cells. Using the NGS expression data, miRNA profiles, and the methylation profiles of normal and cancerous bladder cells in TCGA, we identified the association parameters of the network connections. We also applied AIC to detect the systems order, i.e., the number of connections, and to delete the insignificant connections that were out of system order to prune the false-positive connections in the candidate IGEN and obtain the two real IGENs for normal and cancerous bladder cells, respectively. By applying PGNP to the two real IGENs in normal and cancerous cells, we first identified the core proteins/TFs that played a major role in the principal networks of the IGENs, constituting the core IGENs in normal and cancerous cells. To determine how the signaling cascades from the core receptor proteins to the core TFs participate in bladder carcinogenesis, the core proteins, which mediate the signal transductions from the core receptor proteins to the core TFs, and their corresponding genes were considered the core network biomarkers of the normal and cancerous cells. The miRNAs with very different connections in regulating the genes of the core network biomarkers between normal and cancerous cells were also involved in the core network biomarkers. Finally, by comparing the connection changes of the core network biomarkers from normal cells to stage 1 cancer cells, and from stage 1 cancer cells to stage 4 cancer cells, we investigated the cellular mechanisms of bladder carcinogenesis.

Data preprocessing of omics data

We downloaded the genome-wide mRNA and miRNA NGS data and the methylation profiles from TCGA, including 17 samples for normal bladder cells, 348 samples for stage 1 bladder cancer cells, and 56 samples for stage 4, i.e., metastatic stage, bladder cancer cells. The data also contained 6 samples for young ($\leq$45 years old) people, 477 samples for old ($>$45 years old) people, 98 samples for nonsmokers, and 323 samples for smokers. We used ANOVA to identify significant differences in gene expression between smokers and nonsmokers, and between young and old people (p-value $< .05$). We used the gene symbols of the human gene information data downloaded from the NCBI FTP site as standard human gene names to integrate the omics data, including NGS data, methylation profiles, drug response genome-wide microarray data in CMAP, drug–gene interaction DGIdb data, miRNA–target gene association data in TargetScan, PPIs in BioGRID, transcription regulations in HTRIdb, and ITFP and TRANSFAC data. We also used the GO database, the NCBI Entrez Gene database, and the KEGG pathway database to find the biological processes and pathways of each gene. We used Matlab's text-file and string manipulation tools for text mining.

Construction of the stochastic regression models for the IGEN system

The goal of the stochastic regression model is to characterize molecular mechanisms, including PPIs, transcription regulations, miRNA regulations, and epigenetic regulations via DNA methylation, by NGS data through detecting false-positives of candidate IGENs in human cells. For the stochastic regression model of the gene regulatory subnetwork in the candidate human IGEN, including transcription regulations, miRNA regulations, and epigenetic regulations via DNA methylation, we identified the regulation capabilities of TFs and miRNAs in the GRN of the candidate IGEN. For the expression levels of the ith gene, its DNA methylation and its jth TF/protein and lth miRNA in the nth sample are denoted by $x_i(n)$, $m_i(n)$, $y_j(n)$, and $s_l(n)$, respectively. Then, the stochastic regression model of GRN is described by the following stochastic regression equation:

$$x_i(n) = \sum_{\substack{j \equiv \Omega_i \\ j \neq i}} a_{ij} M_i(n) y_j(n) + \sum_{l \equiv \delta_i} c_{li} M_i(n) x_i(n) s_l(n) + b_i M_i(n) + \nu_i(n), \quad (9.15)$$
$$\text{for } i = 1, \ldots, K, \text{ and } n = 1, \ldots, N$$

where the repression ability from the lth miRNA to the ith gene $c_{li} \leq 0$; the basal level of the ith gene expression $b_i \geq 0$; $\Omega_i \subset \Omega \equiv \{1,\ldots,K\}$; $\delta_i \subset \delta \equiv \{1,\ldots,L\}$; $M_i(n) = 1/\left[1 + \left(m_i(n)/0.5\right)^2\right]$; Ω_i and δ_i denote the candidate regulations based on the databases of transcription regulation and miRNA-target association, respectively; a_{ij} indicates the regulatory ability from the jth TF $y_j(n)$ to the ith gene; $\nu_i(n)$ represents the stochastic noise due to the modeling residue and fluctuation in the ith gene; and K, L, and N are the total number of TFs, miRNAs and data samples in the omics data, respectively. $M_i(n)$ denotes the effect of methylation $m_i(n)$ on the binding affinity of TFs, miRNAs, or RNA polymerase on the ith gene which also represents the impact of DNA methylation of the ith gene on the binding affinities of miRNAs, RNA polymerase and TFs in the gene expression process. The effect of DNA methylation on binding affinities $M_i(n)$, for $i = 1,\ldots,K$, is ranged between 0.2 and 1, while the expression range of the genome-wide DNA methylation $m_i(n)$, for $i = 1,\ldots,K$, is between 0 and 1. If DNA methylation of the ith gene is close to 1, the effect on the binding affinity to the ith gene is close to 0.2, which implicates the impact of DNA methylation on the binding affinities of miRNAs, RNA polymerase and TFs to be like an inhibitor. The ith mRNA expression results from transcription regulations $\sum_{\substack{j \equiv \Omega_i \\ j \neq i}} a_{ij} M_i(n) y_j(n)$, miRNA repressions $\sum_{l \equiv \delta_i} c_{li} M_i(n) x_i(n) s_l(n)$, the mRNA basal expression $b_i M_i(n)$, and the stochastic noise due to measurement and random fluctuations $\nu_i(n)$. In the model (9.15), the TF regulations, miRNA regulations and basal levels are all influenced by the DNA methylation $m_i(n)$ on the ith gene.

For the stochastic regression model of the miRNA regulatory subnetwork in the candidate IGEN, the expression levels of the lth miRNA and its ith target gene in the

nth sample, denoted by $s_l(n)$ and $x_i(n)$, respectively, could be described by the stochastic regression model of MRN as the following stochastic regression equation:

$$s_l(n) = \sum_{i \equiv \delta_l} c_{li} M_i(n) x_i(n) s_l(n) + M_l(n) z_l + e_l(n), \quad \text{for } l = 1, \ldots, L, \text{ and } n = 1, \ldots, N \tag{9.16}$$

where the repression ability of the lth miRNA to the ith gene $c_{li} \leq 0$; the basal level of the lth miRNA expression $z_l \geq 0$; $\delta_l \subset \delta \equiv \{1, \ldots, L\}$; δ_l denotes the candidate regulations of the lth miRNA based on the database of miRNA–target gene association; $e_l(n)$ represents the stochastic noise owing to the modeling residue and fluctuation in the lth miRNA. The lth miRNA expression in Eq. (9.16) results from miRNA–gene interactions $\sum_{i \equiv \delta_l} c_{li} M_i(n) x_i(n) s_l(n)$, the miRNA basal expression z_l, and the stochastic noise $e_l(n)$.

For the stochastic regression model of the PPI subnetwork in the candidate IGEN, the expression level of the jth protein and its kth connecting protein in the nth sample, denoted by $y_j(n)$ and $y_k(n)$, respectively, could be described by the stochastic regression model of PPIN as the following stochastic regression equation:

$$y_j(n) = \sum_{\substack{k \equiv \Omega_j \\ k \neq j}} d_{jk} y_k(n) y_j(n) + h_j + w_j(n), \quad \text{for } j = 1, \ldots, K, \text{ and } n = 1, \ldots, N \tag{9.17}$$

where the basal level of the jth protein expression $h_j \geq 0$; $\Omega_j \subset \Omega \equiv \{1, \ldots, K\}$; Ω_j denotes the candidate interactions of the jth protein based on the PPI database; d_{jk} indicates the interaction ability of the kth protein to the jth protein; and $w_j(n)$ represents the stochastic noise owing to the modeling residue and fluctuation in the jth protein. The jth protein expression in Eq. (9.17) results from the rate of formation of the protein complex $y_k(n) y_j(n)$ proportional to the product of the concentration of each protein $\sum_{\substack{k \equiv \Omega_j \\ k \neq j}} d_{jk} y_k(n) y_j(n)$ [185,186], the protein basal expression h_j, and the stochastic noise $w_j(n)$.

We proposed general stochastic regression models to characterize cellular mechanisms, including genetic and epigenetic regulations, in human cells. A number of parameters, including the TF regulatory ability a_{ij}, the miRNA repression ability c_{li}, and the protein interaction ability d_{jk}, needed to be estimated, and were determined using the databases of PPI, miRNA–target gene association, and transcription regulation.

Identification of the TF regulatory ability a_{ij}, the miRNA repression ability c_{li}, and the protein interaction ability d_{jk}, and their statistical significance testing

We used the mRNA and miRNA expression data from the NGS as the expression levels for $x_i(n)$ and $s_l(n)$ respectively, and used DNA methylation profiles as the expression level of $m_i(n)$ to identify the model parameters a_{ij}, c_{li}, d_{jk}, b_i, z_l, and h_j in Eqs. (9.15–9.17). Because large-scale measurement of protein activities has yet to be realized and 73% of the variance in protein abundance can be explained by mRNA abundance [187], mRNA expression profiles were always used to substitute for the protein expression profiles. Therefore, we also applied mRNA expression levels in the NGS data as the expression levels of $y_j(n)$ to identify the parameters in Eqs. (9.15–9.17). If the simultaneously measured genome-wide protein expression data and the mRNA expression data in each bladder cancer stage are available, the general models in Eqs. (9.15–9.17) can also be applied to identify the real IGEN of the cancer more precisely. Because the parameters in Eq. (9.15) have certain constraints, such as the nonpositive miRNA repressions and nonnegative basal levels, the regulatory parameters were identified by solving the constrained least square parameter estimation problem in the following.

In order to identify the parameters in Eq. (9.15), the stochastic regression model of GRN was rewritten as the following linear regression form:

$$x_i(n) = \left[M_i(n)y_1(n) \quad \ldots \quad M_i(n)x_i(n)y_K(n) \quad M_i(n)x_i(n)s_1(n) \quad \ldots \quad M_i(n)x_i(n)s_L(n) \quad M_i(n)\right] \begin{bmatrix} a_{i1} \\ \vdots \\ a_{iK} \\ c_{1i} \\ \vdots \\ c_{Li} \\ b_i \end{bmatrix} + \nu_i(n) = \phi_i(n)\theta_i^1 + \nu_i(n), \quad \text{for } i=1,\ldots,K, \text{ and } n=1,\ldots,N, \tag{9.18}$$

where $\phi_i(n)$ denotes the regression vector and θ_i^1 is the parameter vector of target gene i to be estimated. $x_i(n)$ and $\phi_i(n)$ are available in the omics data.

The regression model (9.18) at different data samples can be rearranged as follows.

$$\begin{bmatrix} x_i(1) \\ \vdots \\ x_i(N) \end{bmatrix} = \begin{bmatrix} \phi_i(1) \\ \vdots \\ \phi_i(N) \end{bmatrix} \theta_i^1 + \begin{bmatrix} \nu_i(1) \\ \vdots \\ \nu_i(N) \end{bmatrix} \tag{9.19}$$

where N denotes the number of data samples in the NGS data of a bladder cancer stage.

For simplicity, we define the notations X_i, Φ_i, and V_i to represent Eq. (9.19) as follows:

$$X_i = \Phi_i\theta_i^1 + V_i \tag{9.20}$$

The constrained least squares parameter estimation problem of θ_i^1 is formulated as follows:

$$\min_{\theta_i^1} \left\| \Phi_i \theta_i^1 - X_i \right\|_2^2 \text{ subject to } \begin{bmatrix} \overbrace{\begin{matrix} 0 & \cdots & \cdots & 0 \\ \vdots & \ddots & \ddots & \vdots \\ \vdots & & \ddots & \vdots \\ 0 & \cdots & \cdots & 0 \end{matrix}}^{K} & \overbrace{\begin{matrix} 1 & 0 & \cdots & 0 \\ 0 & \ddots & \ddots & \vdots \\ \vdots & \ddots & 1 & 0 \\ 0 & \cdots & 0 & -1 \end{matrix}}^{L+1} \end{bmatrix} \theta_i^1 \leq \begin{bmatrix} 0 \\ \vdots \\ \vdots \\ 0 \end{bmatrix} \tag{9.21}$$

This gives the constraints to force the miRNA repression c_{li} to be always nonpositive, and the basal level b_i to be always nonnegative in Eq. (9.15), i.e., $c_{li} \leq 0$ and $b_i \geq 0$.The constrained least square problem was solved in Matlab using the active set method for quadratic programming [188,189].

Similarly, the stochastic regression model of the miRNA regulatory subnetwork in Eq. (9.16) was rewritten in the following regression form.

$$\begin{aligned} s_l(n) &= \begin{bmatrix} M_1(n)x_1(n)s_l(n) & \ldots & M_K(n)x_K(n)s_l(n) & M_l(n) \end{bmatrix} \begin{bmatrix} c_{l1} \\ \vdots \\ c_{lK} \\ z_l \end{bmatrix} + e_l(n) \\ &= \vartheta_l(n)\theta_l^2 + e_l(n), \quad \text{for } i = 1, \ldots, K, \text{ and } n = 1, \ldots, N, \end{aligned} \tag{9.22}$$

where $\vartheta_l(t)$ indicates the regression vector and θ_l^2 is the parameter vector to be estimated.

For simplicity, we define the notations S_l, Ψ_l, and E_l to represent Eq. (9.22) as follows:

$$S_l = \Psi_l \theta_l^2 + E_l \tag{9.23}$$

The parameter identification problem is then formulated as follows:

$$\min_{\theta_i^2} \left\| \Psi_l \theta_l^2 - S_l \right\|_2^2 \text{ subject to } \begin{bmatrix} \overbrace{\begin{matrix} 1 & 0 & \cdots \\ 0 & \ddots & \ddots \\ \vdots & \ddots & 1 \\ 0 & \cdots & 0 \end{matrix}}^{K} & \begin{matrix} 0 \\ \vdots \\ 0 \\ -1 \end{matrix} \end{bmatrix} \theta_l^2 \leq \begin{bmatrix} 0 \\ \vdots \\ \vdots \\ 0 \end{bmatrix} \tag{9.24}$$

This gives the constraint to force the miRNA repression c_{li} to be always nonpositive, and the basal level z_l to be always nonnegative in Eq. (9.16), i.e., $c_{li} \leq 0$, and $z_l \geq 0$. Finally, the protein model (9.17) uses the same methods as above to make sure $h_j \geq 0$.

Furthermore, in order to extract the core network biomarkers from normal and cancerous cells, we first used NGS data and methylation profiles in the normal and stage 1 and 4 bladder cancer cells to identify an IGEN for normal bladder cells and a general IGEN for bladder cancer cells. The two identified IGENs were used to extract the core network biomarkers in bladder carcinogenesis. We then used the association parameters in the general IGEN of bladder cancer cells as the initial condition of the constrained least square parameter estimation and applied the data on stage 1 and 4 bladder cancer cells to identify the IGENs for stages 1 and stage 4, respectively. According to the three identified IGENs in normal bladder cells, and the stage 1 and 4 bladder cancer cells, we determined the cellular mechanisms of the core network biomarkers in bladder carcinogenesis.

By applying the Student's t-test to the parameter estimation method [190], the p-values for the estimated parameters, including the TF regulatory ability a_{ij}, the miRNA repression ability c_{li}, and the protein interaction ability d_{jk}, were calculated to determine the significance of the parameters. Additionally, to determine the significance of expression level and DNA methylation profile of a gene/miRNA between normal bladder cells and cancerous bladder cells, we applied one-way ANOVA to calculate the p-value.

After the parameter identification problem had been solved, we identified the IGEN for each bladder cell type. For example, we identified the regulatory parameter $a_{\mathrm{RPS20,JUN}} = 0.26$ from the TF JUN to the target gene RPS20 (p-value $< .02$) in stage 4 bladder cancer cells, the interaction parameter $d_{\mathrm{HUWE1,ADRM1}} = 1.2$ between the two proteins ADRM1 and HUWE1 (p-value $< .005$) in stage 1 bladder cancer cells, and the coupling rate $c_{\mathrm{RPS20,MIR155}} = -1.2$ between the miRNA miR155 and the mRNA RPS20 in stage 4 bladder cancer cells (p-value $< .07$).

Principal genome-wide network projection (PGNP)

After the identification of the IGENs in normal and cancer cells, we extracted the core network biomarkers of the IGENs based on the perspectives of the functional modules and pathways to reveal the cellular mechanisms behind bladder carcinogenesis. To extract the core network biomarkers, including the core proteins, their corresponding genes, and their upstream miRNAs, from an IGEN on a genome-wide scale, we first decomposed the combined network matrix of the IGEN to left- and right-singular vectors and singular values based on singular value decomposition. The top left- and right-singular vectors with the top singular values constitute the principal network of the IGEN. The projection distance of each gene/protein/miRNA to these top singular vectors represents the significance of this gene/protein/miRNA in the IGEN. The genes/proteins/miRNAs with the top projection distance ultimately constitute the core network biomarkers of the IGEN. Let the combined network matrix of the TF regulatory ability a_{ij}, the miRNA repression ability c_{li}, and the protein interaction ability d_{jk} of the IGEN in Eqs. (9.15–9.17) be represented by:

$$A = \begin{bmatrix} a_{11} & \cdots & a_{1K} \\ \vdots & \ddots & \vdots \\ a_{K1} & \cdots & a_{KK} \\ c_{11} & \cdots & c_{1K} \\ \vdots & \ddots & \vdots \\ c_{L1} & \cdots & c_{LK} \\ d_{11} & \cdots & d_{1K} \\ \vdots & \ddots & \vdots \\ d_{K1} & \cdots & d_{KK} \end{bmatrix} \tag{9.25}$$

By applying PGNP, the matrix A is then be decomposed as follows:

$$A = UDV^T = \begin{bmatrix} u_1 & \ldots & u_K \end{bmatrix} \begin{bmatrix} d_1 & 0 & 0 \\ 0 & \ddots & 0 \\ 0 & 0 & d_K \end{bmatrix} \begin{bmatrix} v_1 & \ldots & v_K \end{bmatrix}^T = \sum_{i=1}^{K} u_i d_i v_i^T \tag{9.26}$$

where u_i, $v_i \in \mathfrak{R}^K$ are the ith left- and right-singular vectors of A, respectively. The diagonal entries of D are the K singular values of A in descending order, $d_1 \geq \ldots \geq d_K$.

The eigenexpression fraction (E_m) is defined as

$$E_m = \frac{d_m^2}{\sum\limits_{m=1}^{K} d_m^2} \tag{9.27}$$

We choose the top M singular vectors of V such that $\sum_{m=1}^{M} E_m \geq 0.85$, with the minimal M, so that the top M principal components contain 85% of the IGEN from an energy point of view. The principal projections of A to the top M singular vectors of V, or similarities, is defined as follows:

$$S(k,m) = a_k \bullet v_m{}^T \quad \text{for } k = 1, \ldots, (2K+L), \text{ and } m = 1, \ldots, M \tag{9.28}$$

where a_k and v_m^T denote the kth row vector of A, and the mth singular vector of V, respectively. Furthermore, we defined the 2-norm distance from the target genes, miRNAs, and proteins/TFs to the top M singular vectors, respectively, as follows:

$$D(k) = \left[\sum_{m=1}^{M} [S(k,m)]^2 \right]^{1/2}, \quad \text{for } k = 1, \ldots, (2K+L), \tag{9.29}$$

where $D(k)$ for $k = 1,\ldots, K$, for $k = K+1,\ldots, K+L$, and for $k = K+L+1,\ldots, 2K+L$, are the 2-norm distances from the target genes, miRNAs, and proteins/TFs to the top M singular vectors, respectively. According to $D(k)$ for $k = K+L+1,\ldots,2K+L$, we can identify the core proteins/TFs that play a major role in the principal networks of the IGENs, constituting the core IGENs in normal and cancer cells. The identified core proteins/TFs contain receptors that mediate the signaling cascades connected to core TFs. The core proteins, which participate in signal transduction from core receptors to core TFs, and their corresponding genes were considered the core network biomarkers for normal and cancerous cells. The miRNAs with very different connections in regulating the genes of the core network biomarkers between two cells were also involved in the core network biomarkers.

Design of a multiple drug combination with minimal side effects for the treatment of bladder cancer

To design a multiple drug combination with minimal side effects for the treatment of bladder cancer based on the core network biomarkers of the IGEN, we considered two databases, CMAP and DGIdb. CMAP contains the genome-wide microarray data in response to 1327 drugs in five cell lines, while DGIdb comprises a drug–gene interaction database. Multiple drug therapy induces a genome-wide response. The strategy of multiple drug screening is that the multiple drugs should inhibit the highly expressed genes, activate the reduced expression of the genes, and not influence the nondifferentially expressed genes in the core network biomarkers of bladder cancer cells compared with normal bladder cells. The binding protein of the designed multiple drug combination can also be obtained using the DGIdb. The strategy leads to improved drug safety and efficacy in the treatment of bladder cancer

9.4.2 Results and discussion

Construction of IGENs

We first used NGS expression data and methylation profiles in normal bladder cells and stage 1 and 4 bladder cancer cells to identify a real IGEN for normal bladder cells and a general real IGEN for bladder cancer cells (see Section 9.4.1). By applying PGNP to the real IGEN of the normal bladder cells and the general real IGEN of the bladder cancer cells, we then obtained 115 core proteins/TFs for the core IGEN of the normal bladder cells, and 138 core proteins/TFs for the core IGEN of the bladder cancer cells. To determine how the signaling cascades from the core receptor proteins to the core TFs participate in bladder carcinogenesis, the core proteins, which mediate the signal transductions from core receptor proteins to core TFs, and their corresponding genes were considered the core network biomarkers. The miRNAs with a high number of different connections regulating the genes of the core network biomarkers between normal and cancerous cells were also involved in the core network biomarkers. Moreover, to identify the mechanism of carcinogenesis from stage 1 to stage 4 bladder cancer, we used the identified parameters of the models (9.15–9.17) in the general IGEN of bladder cancer cells as the initial condition of the constrained least square parameter estimation. We then applied the data for stage 1 and 4 bladder cancer cells to obtain the two real IGENs for stage 1 and 4 bladder cancer, respectively. Furthermore, we analyzed the connection changes of the core network biomarkers between normal bladder cells and stage 1 bladder cancer cells (Fig. 9.19), and between stage 1 and 4 bladder cancer cells (Fig. 9.20) to determine the mechanisms of bladder carcinogenesis, and accordingly design multiple drug combinations for treating bladder cancer with minimal side effects.

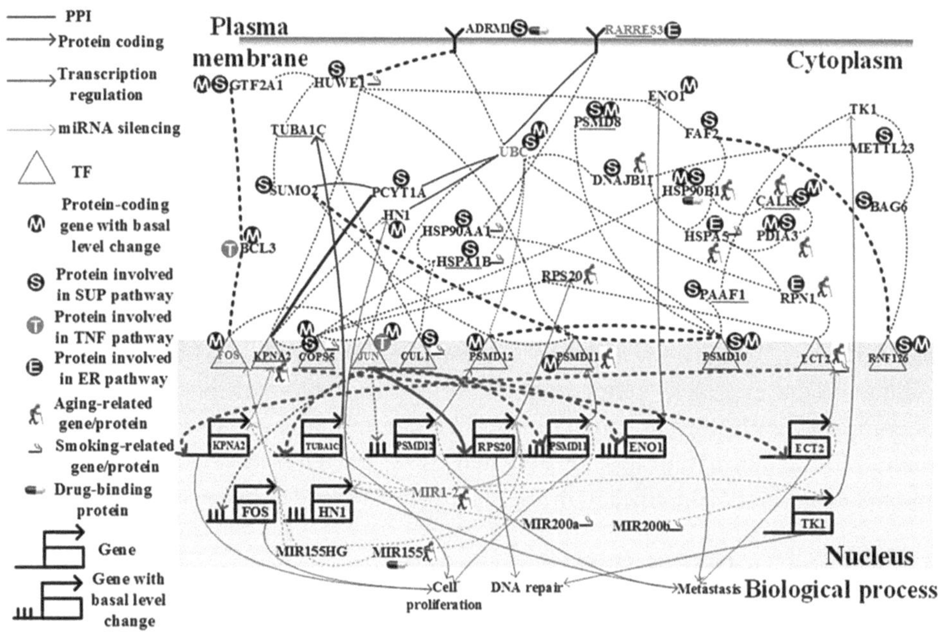

Figure 9.19 Comparison of genetic and epigenetic alterations and connection changes in the core network biomarkers of bladder carcinogenesis between normal bladder cells and stage 1 bladder cancer cells. *Red* (dark gray in print versions), *blue* (light gray in print versions), and *black* gene/miRNA symbols represent the highly expressed genes, the suppressed genes, and the nondifferentially expressed genes in stage 1 bladder cancer cells, respectively, compared with normal bladder cells. *Dashed and solid lines* denote the identified connections in normal and cancerous cells, respectively. The identified connections of the core network biomarkers do not only exist in normal bladder cells. *Bold lines* indicate the high regulatory or interaction parameters, i.e., a_{ij}, c_{li}, and d_{jk}, identified in the stochastic regression models (9.15–9.17) of the IGEN. The *underlined proteins*, including RARRES3, TUBA1C, PSMD8, HSPA1B, RPS20, CALR, PAAF1, and KPNA2, were the identified core network biomarkers. The major factors, including downregulated miR1-2, the aging-related proteins, HSP90B1, CALR, HSPA5, PDIA3, RPN1, and ECT2, the smoking-related proteins, HUWE1, HSPA5, and ECT2, and the epigenetic regulation of *ENO1*, *HSP90B1*, *CALR* and *PDIA3* lead to the progression from normal bladder cells to stage 1 bladder cancer cells through the SUP and ER signaling pathways.

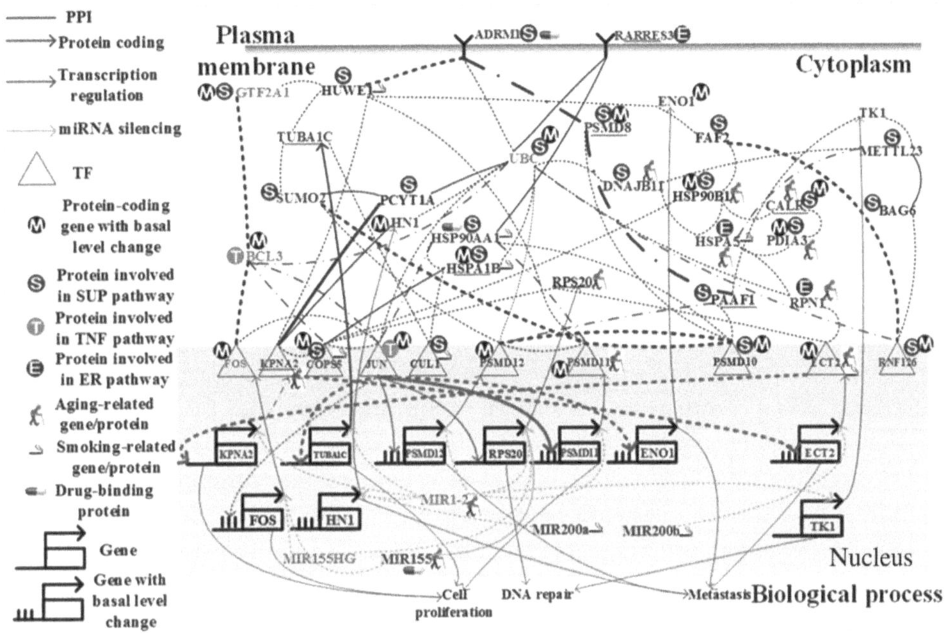

Figure 9.20 Comparison of genetic and epigenetic alterations and connection changes in the core network biomarkers of bladder carcinogenesis between stage 1 and stage 4 bladder cancer cells. *Red* (dark gray in print versions), *blue* (light gray in print versions), and *black* gene/miRNA symbols represent the highly expressed genes, the suppressed genes, and the nondifferentially expressed genes in stage 4 bladder cancer cells, respectively, compared with stage 1 bladder cancer cells. *Dashed*, *dash-dot*, and *solid lines* denote the identified connections in stage 1 cancer cells, stage 4 cancer cells, and in both stage 1 and 4 cancer cells, respectively. *Bold lines* indicate the high regulatory or interaction parameters, i.e., a_{ij}, c_{li}, and d_{jk}, identified in the stochastic regression models (9.15–9.17) of the IGEN. The *underlined proteins* RARRES3, TUBA1C, PSMD8, HSPA1B, RPS20, CALR, PAAF1, and KPNA2 were the identified core network biomarkers. The smoking-related protein HSP90AA1 and DNA methylation of *ECT2* mediate metastasis of bladder cancer.

To investigate the impact of the major risk factors, aging and smoking, on the core network biomarkers of bladder carcinogenesis, we highlighted the significantly expressed genes between young and old people and between nonsmokers and smokers in the core network biomarkers (p-value < .05). Additionally, the genes with changes in the basal level of Eq. (9.15) between normal bladder cells and stage 1 bladder cancer cells, and between stage 1 and 4 bladder cancer cells were also highlighted in the core network biomarkers of Figs. 9.19 and 9.20, respectively. The basal level change of a gene between two cell types has been implicated in the epigenetic regulation of gene expression. The expression of a gene that exhibits a basal level change and a significant change (p-value < .05) of its methylation profile between the two bladder cell types is probably regulated by DNA methylation in bladder carcinogenesis.

Projection of the core network biomarkers into biological processes and signaling pathways to investigate carcinogenic mechanisms of bladder cancer

According to the information of the biological processes and signaling pathways in the GO and KEGG pathway databases, the roles of the genes in the core network biomarkers (Figs. 9.19 and 9.20) are projected into three pathways: the SUP, TNF signaling, and ER signaling pathways; and three biological processes: cell proliferation, DNA repair, and metastasis.

It has been reported that the SUP pathway is associated with increased proliferation in urinary bladder carcinogenesis [191]. HuaChanSu (HCS), a class of toxic steroids, has been used to show that the TNF pathway mediates the inhibition of cell proliferation in bladder cancer [192]. Moreover, the viability of human bladder cancer cells is reduced by using cantharidin, a natural toxin, through the ER pathway [193]. Therefore, the proteins of the core network biomarkers participating in the SUP, TNF, and ER signaling pathways play an important role in bladder carcinogenesis. We then determined how the core network biomarkers mediate bladder carcinogenesis through the influences of aging, smoking, epigenetic regulation, and miRNA regulation.

The role of the SUP pathway is to degrade misfolded proteins, influence PPIs, translocate proteins, and stabilize protein structure. Owing to the accumulation of genetic mutations and epigenetic alterations in cancer cells, the SUP pathway plays a crucial role in the maintenance of many important cellular processes in cancer cells. The repressed activity of ubiquitin C (UBC), which encodes the polyubiquitin precursor, influences degradation and translation of several proteins in stage 1 and stage 4 bladder cancer cells. For example, the repression of UBC affects the signal transduction of RARRES3, a tumor suppressor, in bladder carcinogenesis. To maintain the cellular functions of cancer cells, the regulation of the SUP pathway adapts to the accumulated genetic mutations and epigenetic alterations.

In normal cells, the TNF pathway is critical for inducing inflammation, which can cause cell death. Accumulated DNA damage, epigenetic alterations, or stresses can induce the TNF pathway, and the pathway then triggers cell death. JUN, one of the TFs in the TNF pathway, plays an important role in promoting the invasion and migration of bladder cancer cells [194]. We determined that the repressed expression of JUN in stage 1 bladder cancer cells leads to cancer cell immortality and causes accumulated genetic mutations and epigenetic alterations. Additionally, the results revealed that JUN was activated in stage 4 bladder cancer cells to mediate metastasis. The role of JUN in the metastasis of bladder cancer cells can also be supported [194]. It has also been reported that the TNF pathway acts as a switch between inflammation and cancer [195]. Moreover, downregulated BCL3, which participates in the TNF pathway in the adipose tissue of the bladder wall, leads to reduced inflammation in bladder carcinogenesis [196].

The ER pathway participates in the regulation of protein folding, protein synthesis, and posttranslational modifications [197]. Misfolded proteins, arising from genetic mutations, epigenetic alterations, or stresses, induce the ER pathway to restore cellular homeostasis in normal cells. Owing to the immortal nature of cancer cells, the accumulated genetic mutations and epigenetic alterations in bladder cancer cells can activate most of the genes that contribute to the ER pathway in bladder carcinogenesis (Figs. 9.19 and 9.20). In the ER pathway, only RARRES3, a tumor suppressor gene, was downregulated in stage 1 bladder cancer cells.

The impact of aging, smoking, and miRNA and epigenetic regulation on bladder carcinogenesis through the core network biomarkers

Major factors, including downregulated miR1-2 and aging- and smoking-related proteins, may lead to the progression from normal bladder cells to stage 1 bladder cancer cells through the SUP and ER signaling pathways.

It has been reported that aging and smoking are the major factors that accumulate genetic and epigenetic alterations, and ultimately induce bladder carcinogenesis. In Fig. 9.19, our results reveal that ADRM1 regulates KPNA2, which promotes proliferation, and is mediated by the aging-related proteins, HSP90B1, CALR, HSPA5, PDIA3, RPN1, and ECT2, the smoking-related proteins, HUWE1, HSPA5, and ECT2, and the epigenetic regulation of ENO1, HSP90B1, CALR, and PDIA3, through the SUP and ER signaling pathways. ADRM1 knockdown leads to a reduction of cancer cell proliferation and has been found in gastric [198], ovarian [199], liver [200], colorectal cancers [201], and acute leukemia [202]. Therefore, the results support the hypothesis that aging is the most important factor in inducing bladder carcinogenesis through the SUP pathway. Additionally, our results (Fig. 9.19) show that the inhibited aging-related miRNA miR1-2 in stage 1 bladder cancer cells leads to miR1-2 dysregulation of genes including *KPNA2*, *TUBA1C*, *HN1*, *PSMD11*, *PSMD12*, and *TK1*, which influence cell proliferation, DNA repair, and metastasis. miR1-2 has also been identified as a tumor suppressor in bladder cancer cells [203].

miR1-2 and miR200b mediate the reduction of cell proliferation and metastasis through KPNA2 and ECT2, respectively

The receptor ADRM1 signal triggers the signaling cascade from the smoking-related protein HUWE1 to the aging-related proteins HSP90B1 and RPS20 and the smoking-related TF COPS5. The TF COPS5 upregulates the metastasis-associated gene *ECT2*, which is suppressed by miR200b in stage 1 bladder cancer cells. The results show the cross-regulation between the transcription of the smoking-related protein COPS5 and the aging-related protein ECT2. The aging-related miRNA miR1-2 and the smoking-related miRNA miR200b act as a switch to depress the proliferation-associated protein KPNA2 in stage 1 and stage 4 bladder cancer cells

(Figs. 9.19 and 9.20), and the metastasis-associated gene *ECT2* in stage 4 bladder cancer cells (Fig. 9.20), respectively.

The smoking-related protein HSP90AA1 and DNA methylation of ECT2 mediate the metastasis of bladder cancer

Our results reveal that receptor RARRES3 signaling triggers the activated TF JUN mediated by the smoking-related protein HSP90AA1, and JUN then activates the metastasis-associated gene *PSMD12* in stage 4 bladder cancer cells (Fig. 9.20). Receptor ADRM1 signaling also triggers the metastasis-associated protein PSMD12 through the proteins PSMD8 and PAAF1, and epigenetic regulation in stage 4 bladder cancer cells. This shows that metastasis-associated *ECT2* is activated by epigenetic regulation in stage 4 bladder cancer cells. Receptor RARRES3 signaling also triggers the aging-related and proliferation-associated TF PSMD11 through the smoking-related protein HSP90AA1 in stage 4 bladder cancer cells. The activated TF JUN also regulates the proliferation-associated gene *PSMD11* and the DNA repair-associated gene *RPS20*. There is evidence that curcumin (diferuloylmethane) can suppress tumor initiation, promotion, and metastasis. Curcumin can also inhibit the expression of JUN [204]. Additionally, the RNAi-induced induction of *ECT2* suppresses cell migration, invasion, and metastasis [205]. Our results indicate that the upregulation of *ECT2* in stage 4 bladder cancer cells is regulated by epigenetic regulation of *ECT2* expression. This is also supported by the significant change in the DNA methylation profiles in *ECT2* between normal bladder cells and stage 4 bladder cancer cells (p-value $< .007$).

Functional module network analysis in bladder carcinogenesis

The activated DNA repair of bladder cancer cells leads to metastasis owing to the immortality of cancer cells.

According to the modular information in the GO database and the KEGG pathway database, the genes/proteins in the core network biomarkers (Figs. 9.21 and 9.22) are projected into three pathways: the SUP pathway, the TNF signaling pathway, and the ER signaling pathway; and three biological processes: cell proliferation, DNA repair, and metastasis. The module networks in Figs. 9.21 and 9.22 show that the activated TFs KPNA2, COPS5, PSMD12, and ECT2 play an important role in mediating the signal transduction of the SUP and ER pathways to activate cell proliferation and metastasis in stage 1 bladder cancer. The metastasis of the stage 1 bladder cancer is repressed by the activated miRNAs miR200a and miR200b, as shown in Fig. 9.21. The activated signal transduction from SUP and ER pathways also triggers DNA repair through the epigenetically regulated TFs PSMD11 and RNF126.

Additionally, the activated TFs PSMD11, RNF126, and JUN mediate the signal transduction from SUP, TNF, and ER pathways to trigger cell proliferation, DNA repair, and metastasis in stage 4 bladder cancer, as shown in Fig. 9.22. Although

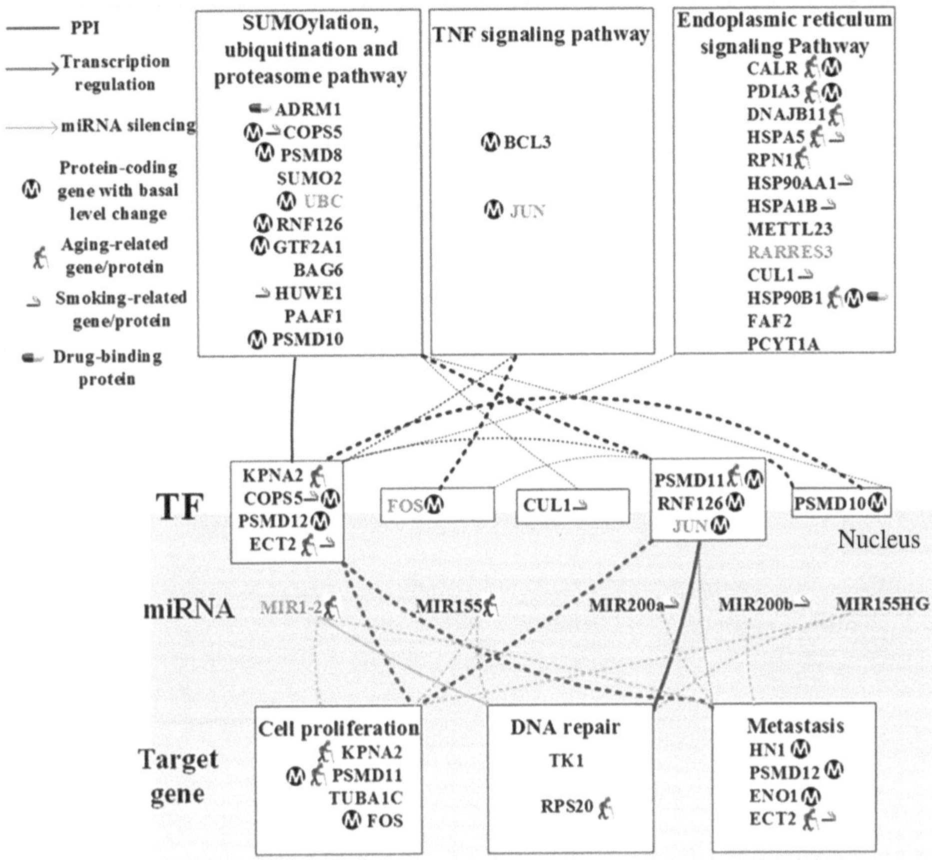

Figure 9.21 Module network of the core network biomarkers in Fig. 9.19 for investigating the bladder carcinogenic mechanisms from normal bladder cells to stage 1 bladder cancer cells. The notations of gene/miRNA symbols and line styles are the same as those in Fig. 9.19. The activated TFs KPNA2, COPS5, PSMD12, and ECT2 play an important role in mediating the signal transduction of the SUP and ER pathways to activate cell proliferation and metastasis in stage 1 bladder cancer. The metastasis of the stage 1 bladder cancer is repressed by the activated miRNAs miR200a and miR200b.

miR155 is activated in stage 1/4 bladder cancer, miR155 suppresses FOS and RPS20 in stage 1 bladder cancer, and miR155 only suppresses the DNA repair-associated gene *RPS20* in stage 4 bladder cancer. Furthermore, we suggest that DNA repair may play a critical role in repairing DNA damage, which results from genetic and epigenetic alterations, leading to phenotypic change of the bladder cells from normal cells to stage 1 cancer cells, and from stage 1 cancer cells to metastatic cancer cells.

In summary, aging and epigenetic regulation dominate bladder carcinogenesis through CALR, PDIA3, DNAJB11, HSPA5, RPN1, HSP90B1, KPNA2, ECT2, and PSMD11; and through COPS5, PSMD8, RNF126, CALR, PDIA3, HSP90B1,

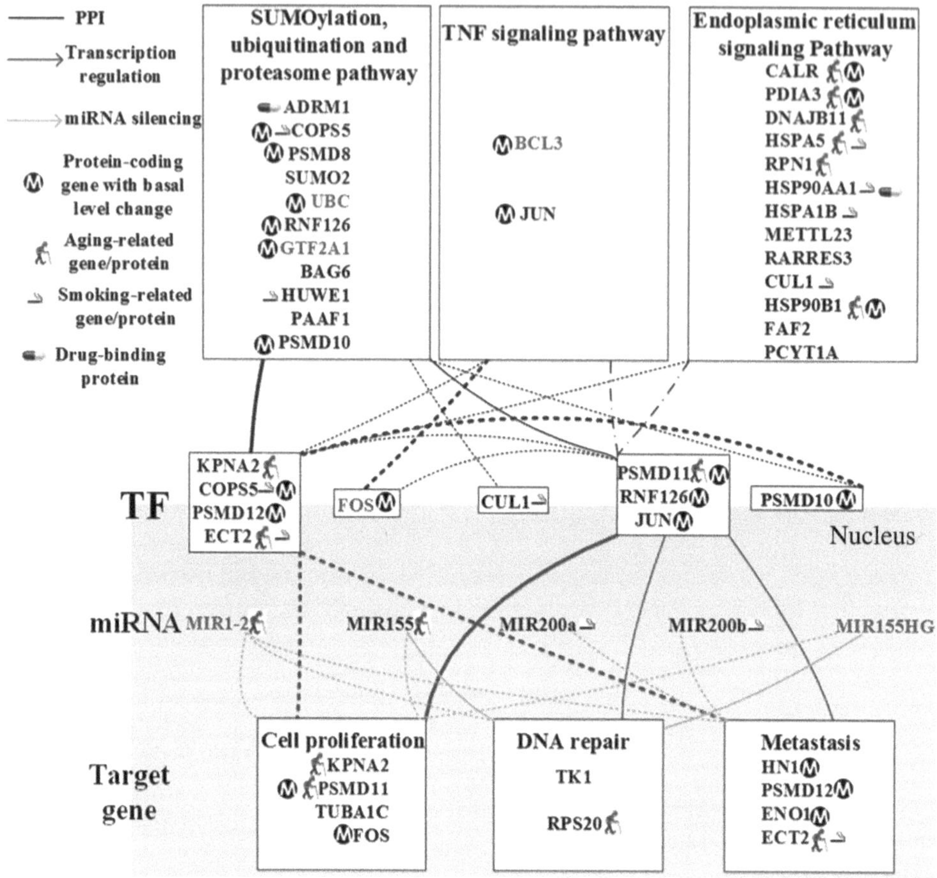

Figure 9.22 Module network of the core network biomarkers in Fig. 9.20 to investigate the bladder carcinogenic mechanisms from stage 1 to stage 4 bladder cancer cells. The notations of gene/miRNA symbols and line styles are the same as those in Fig. 9.20. The activated DNA repair of bladder cancer cells leads to metastasis owing to the immortality of cancer cells. The activated JUN in the TNF pathway induces cell proliferation, DNA repair, and metastasis in stage 4 bladder cancer cells.

PSMD12, PSMD11, JUN, HN1, and ENO1, respectively. Smoking promotes bladder carcinogenesis especially in metastasis. Finally, the cellular mechanisms from normal to stage 1 bladder cancer cells, and from stage 1 to stage 4 bladder cancer cells are summarized in Fig. 9.23A and B, respectively. When the accumulated genetic mutations and epigenetic alterations lead to the dysregulation of the TNF pathway in inflammation, the accumulated misfolded proteins in the ER pathway induce cell proliferation in stage 1 bladder cancer (Fig. 9.23A). Regulation of the ER and TNF pathways adapts to the accumulated genetic mutations and epigenetic alterations through the SUP pathway. The progression of DNA repair and cell proliferation in stage 1 bladder cancer ultimately results not only in the repression of miR200a and miR200b during

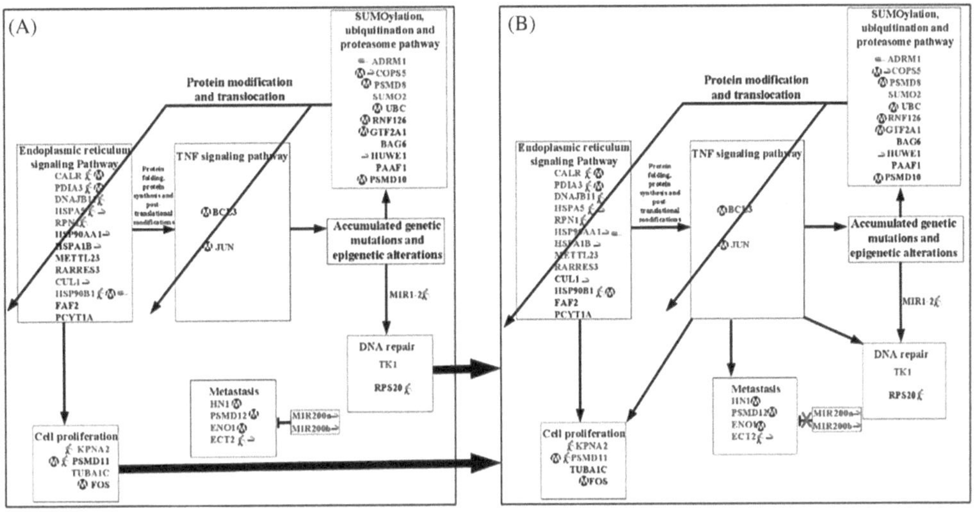

Figure 9.23 The carcinogenic mechanisms from normal to stage 1 bladder cancer cells (A), and from stage 1 to stage 4 bladder cancer cells (B). When the accumulated genetic mutations and epigenetic alterations lead to the dysregulation of the TNF pathway in inflammation, the accumulated misfolded proteins in the ER pathway induce cell proliferation in stage 1 bladder cancer (A). The regulations of ER and TNF pathways are adaptive to the accumulated genetic mutations and epigenetic alterations through the SUP pathway. The progression of DNA repair and cell proliferation in stage 1 bladder cancer ultimately results not only in the repression of miR200a and miR200b during metastasis, but also the regulation of the TNF pathway to metastasis, cell proliferation, and DNA repair in stage 4 bladder cancer (B).

metastasis, but also the regulation of the TNF pathway to metastasis, cell proliferation, and DNA repair in stage 4 bladder cancer (Fig. 9.23B).

Two separate drug combinations for treating stage 1 and stage 4 bladder cancer cells with minimal side effects

The design of a multiple drug combination for treating stage 1 bladder cancer depends on a strategy of inhibiting the highly expressed genes *ADRM1, COPS5, PSMD8, SUMO2, CALR, PDIA3, DNAJB11, HSPA5, RPN1, CUL1, HSP90B1, KPNA2, PSMD12, ECT2, TK1, TUBA1C, HN1,* and *ENO1*; activating the suppressed genes *UBC, JUN, RARRES3,* and *FOS*; and suppressing the drug's effect on the nondifferentially expressed genes *BAG6, HUWE1, PAAF1, PSMD10, FAF2, PCYT1A,* and *PSMD10*. According to the drug design strategy (see Section 9.4.1), a multiple drug combination comprising gefitinib, estradiol, yohimbine, and fulvestrant was obtained for treating stage 1 bladder cancer.

The design of a multiple drug combination for treating stage 4 bladder cancer depends on a strategy of inhibiting the highly expressed genes *ADRM1, COPS5, PSMD8, SUMO2, RNF126, CALR, PDIA3, DNAJB11, HSPA5, RPN1, HSP90AA1,*

HSPA1B, *METTL23*, *RARRES3*, *KPNA2*, *PSMD12*, *ECT2*, *JUN*, *TK1*, *TUBA1C*, *HN1*, and *ENO1*; activating the suppressed genes *BCL3*, *FOS*, *UBC*, and *GTF2A1*; and suppressing the drug's effect on the nondifferentially expressed genes, which are the same as those in stage 1 bladder cancer. We obtained a multiple drug combination comprising gefitinib, estradiol, chlorpromazine, and LY294002 for treating stage 4 bladder cancer (Table 9.6). According to the information in DGIdb, miR155, the HSP90 protein family, ADRM1, and estrogen receptor are the direct targets of the multiple drug combination comprising gefitinib, estradiol, yohimbine, and fulvestrant in stage 1 bladder cancer, respectively (Figs. 9.19 and 9.21), while the same proteins are also the direct targets of the multiple drug combination comprising gefitinib, estradiol, chlorpromazine, and LY294002 in stage 4 bladder cancer, respectively (Figs. 9.20 and 9.22). Moreover, the analysis of drug response genome-wide microarray data reveals that high doses of yohimbine can activate BAG6 in stage 1 bladder cancer, while high doses of chlorpromazine can activate HSPA5 and JUN in stage 4 bladder cancer. Therefore, low-dose yohimbine and low-dose chlorpromazine could avoid side effects in the treatment of stage 1 and stage 4 bladder cancer cells, respectively. Ultimately, we designed one specific drug combination for treating stage 1 bladder cancer and another specific drug combination for treating stage 4 bladder cancer with minimal side effects (Table 9.6).

9.4.3 Conclusion

In this section, we proposed a new method for constructing an IGEN for characterizing cellular mechanisms in bladder carcinogenesis by using system regression modeling and large-scale database mining. We then applied PGNP to obtain the core network biomarkers of the IGEN. By comparing the connection changes of the core network biomarkers between normal bladder cells and stage 1 bladder cancer cells, and between stage 1 and stage 4 bladder cancer cells, we investigated the progression mechanisms of bladder carcinogenesis. Database mining provided all possible candidates for genetic and miRNA regulations and protein interactions in IGEN. We used AIC and statistical assessment to prune the false-positive regulations and interactions by applying the regression coupling model to NGS data and methylation profiles. We compared the connection differences in the core network biomarkers between different cellular types to explore bladder carcinogenic mechanisms. According to the comparison of the connection changes in the core network biomarkers between normal cells and stage 1 cancer cells, and between stage 1 and stage 4 cancer cells, we investigated how the genetic and epigenetic regulations, miRNA regulations, and aging-related and smoking-related genes affect the biological functions that lead to bladder carcinogenesis. According to gene expression changes in the core network biomarkers between normal bladder cells and stage 1 bladder cancer cells, and between stage 1 and stage 4 bladder cancer cells, we then identified two separate drug combinations

Table 9.6 The multiple drug design strategy and potential multiple drug combination for stage 1 and 4 cancers

	Stage 1 bladder cancer	Stage 4 bladder cancer
The highly expressed genes for potential inhibition strategy of multiple drug design	*ADRM1*, *COPS5*, *PSMD8*, *SUMO2*, *CALR*, *PDIA3*, *DNAJB11*, *HSPA5*, *RPN1*, *CUL1*, *HSP90B1*, *KPNA2*, *PSMD12*, *ECT2*, *TK1*, *TUBA1C*, *HN1*, and *ENO1*	*ADRM1*, *COPS5*, *PSMD8*, *SUMO2*, *RNF126*, *CALR*, *PDIA3*, *DNAJB11*, *HSPA5*, *RPN1*, *HSP90AA1*, *HSPA1B*, *METTL23*, *RARRES3*, *KPNA2*, *PSMD12*, *ECT2*, *JUN*, *TK1*, *TUBA1C*, *HN1*, and *ENO1*
The suppressed genes for potential activation strategy of multiple drug design	*UBC*, *JUN*, *RARRES3*, and *FOS*	*BCL3*, *FOS*, *UBC*, and *GTF2A1*
The nondifferentially expressed genes to avoid side effect of multiple drug design	*BAG6*, *HUWE1*, *PAAF1*, *PSMD10*, *FAF2*, *PCYT1A*, and *PSMD10*	*BAG6*, *HUWE1*, *PAAF1*, *PSMD10*, *FAF2*, *PCYT1A*, and *PSMD10*
The potential multiple drug combination	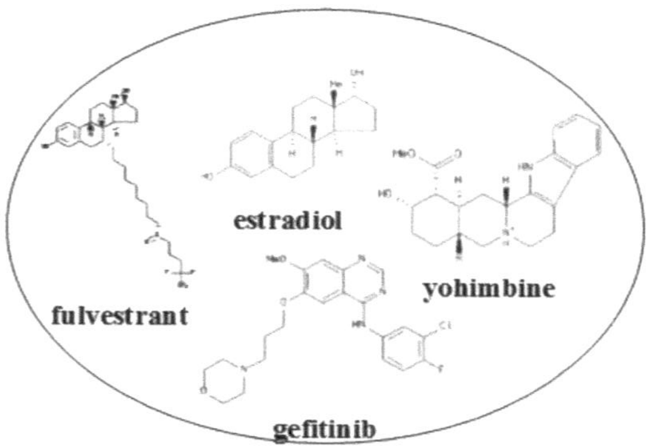	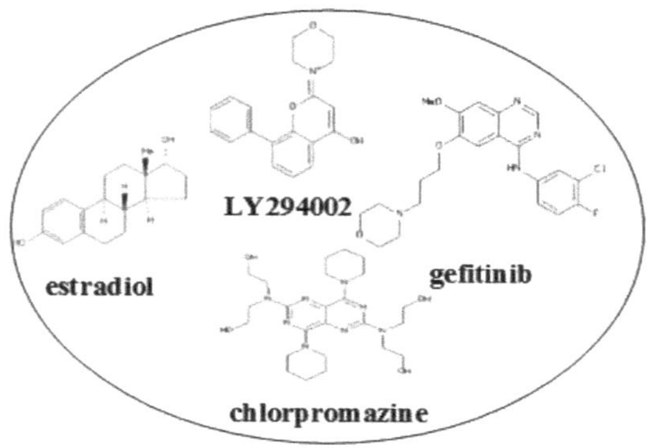

for treating stage 1 and 4 bladder cancer cells. Therefore, the proposed IGEN construction method and PGNP allow us not only to unravel the carcinogenic mechanisms from genetic and epigenetic alterations of the core network biomarkers, but also to design multiple drug combinations for treating bladder cancer with minimal side effects.

REFERENCES

[1] Badawi R, Birns J, Watson T, Kalra L. Growth factors and their relationship to neoplastic and paraneoplastic disease. Eur J Intern Med 2005;16(2):83–94.
[2] Valsesia-Wittmann S, Magdeleine M, Dupasquier S, Garin E, Jallas AC, Combaret V, et al. Oncogenic cooperation between H-Twist and N-Myc overrides failsafe programs in cancer cells. Cancer Cell 2004;6(6):625–30.
[3] Whitfield ML, Sherlock G, Saldanha AJ, Murray JI, Ball CA, Alexander KE, et al. Identification of genes periodically expressed in the human cell cycle and their expression in tumors. Mol Biol Cell 2002;13(6):1977–2000.
[4] Cho RJ, Huang MX, Campbell MJ, Dong HL, Steinmetz L, Sapinoso L, et al. Transcriptional regulation and function during the human cell cycle. Nat Genet 2001;27(1):48–54.
[5] Ishida S, Huang E, Zuzan H, Spang R, Leone G, West M, et al. Role for E2F in control of both DNA replication and mitotic functions as revealed from DNA microarray analysis. Mol Cell Biol 2001;21(14):4684–99.
[6] Cho RJ, Campbell MJ, Winzeler EA, Steinmetz L, Conway A, Wodicka L, et al. A genome-wide transcriptional analysis of the mitotic cell cycle. Mol Cell 1998;2(1):65–73.
[7] Chen HC, Lee HC, Lin TY, Li WH, Chen BS. Quantitative characterization of the transcriptional regulatory network in the yeast cell cycle. Bioinformatics 2004;20(12):1914–27.
[8] Hinchcliffe EH, Sluder G. "It takes two to tango": understanding how centrosome duplication is regulated throughout the cell cycle. Genes Dev 2001;15(10):1167–81.
[9] Nurse P. A long twentieth century of the cell cycle and beyond. Cell 2000;100(1):71–8.
[10] Shah JV, Cleveland DW. Waiting for anaphase: Mad2 and the spindle assembly checkpoint. Cell 2000;103(7):997–1000.
[11] Stillman B. Cell cycle control of DNA replication. Science 1996;274(5293):1659–64.
[12] Liu ET. Systems biology, integrative biology, predictive biology. Cell 2005;121(4):505–6.
[13] Klipp E, Herwig R, Kowald A, Wierling C, Lehrach H. Systems biology in practice: concepts, implementation and application. Weinheim, Germany: John Wiley & Sons; 2005.
[14] Cho YS, Kim MK, Tan LZ, Srivastava R, Agrawal S, Cho-Chung YS. Protein kinase A RI alpha antisense inhibition of PC3M prostate cancer cell growth: Bcl-2 hyperphosphorylation, Bax up-regulation, and Bad-hypophosphorylation. Clin Cancer Res 2002;8(2):607–14.
[15] Johansson R. System modeling and identification. Englewood Cliffs, NJ: Prentice Hall; 1993.
[16] Ren B, Cam H, Takahashi Y, Volkert T, Terragni J, Young RA, et al. E2F integrates cell cycle progression with DNA repair, replication, and G(2)/M checkpoints. Genes Dev 2002;16(2):245–56.
[17] Kreyszig E. Advanced engineering mathematics. 7th ed. New York, NY: Wiley; 1993.
[18] Sherlock G, Hernandez-Boussard T, Kasarskis A, Binkley G, Matese JC, Dwight SS, et al. The Stanford Microarray Database. Nucleic Acids Res 2001;29(1):152–5.
[19] Bracken AP, Ciro M, Cocito A, Helin K. E2F target genes: unraveling the biology. Trends Biochem Sci 2004;29(8):409–17.
[20] Muller H, Helin K. The E2F transcription factors: key regulators of cell proliferation. Biochim Biophys Acta-Rev Cancer 2000;1470(1):M1–12.
[21] Stanelle J, Stiewe T, Theseling CC, Peter M, Putzer BM. Gene expression changes in response to E2F1 activation. Nucleic Acids Res 2002;30(8):1859–67.
[22] Polager S, Kalma Y, Berkovich E, Ginsberg D. E2Fs up-regulate expression of genes involved in DNA replication, DNA repair and mitosis. Oncogene 2002;21(3):437–46.

[23] Nevins JR. The Rb/E2F pathway and cancer. Hum Mol Genet 2001;10(7):699–703.
[24] Cam H, Dynlacht BD. Emerging roles for E2F: beyond the G1/S transition and DNA replication. Cancer Cell 2003;3(4):311–16.
[25] Iveyhoyle M, Conroy R, Huber HE, Goodhart PJ, Oliff A, Heimbrook DC. Cloning and characterization of E2F-2, a novel protein with the biochemical-properties of transcription factor-E2F. Mol Cell Biol 1993;13(12):7802–12.
[26] Lang SE, McMahon SB, Cole MD, Hearing P. E2F transcriptional activation requires TRRAP and GCN5 cofactors. J Biol Chem 2001;276(35):32627–34.
[27] Kennedy RD, Gorski JJ, Quinn JE, Stewart GE, James CR, Moore S, et al. BRCA1 and c-Myc associate to transcriptionally repress psoriasin, a DNA damage-inducible gene. Cancer Res 2005;65 (22):10265–72.
[28] Di Stefano L, Jensen MR, Helin K. E2F7, a novel E2F featuring DP-independent repression of a subset of E2F-regulated genes. EMBO J 2003;22(23):6289–98.
[29] Hayashi R, Goto Y, Haga A, Kobayashi D, Ikeda R, Yoshida K. Comparative genomics on MCM8 orthologous genes reveals the transcriptional regulation by transcription factor E2F. Gene 2006;367:126–34.
[30] Ortega S, Malumbres M, Barbacid M. Cyclin D-dependent kinases, INK4 inhibitors and cancer. Biochim Biophys Acta-Rev Cancer 2002;1602(1):73–87.
[31] Sherr CJ, Roberts JM. CDK inhibitors: positive and negative regulators of G(1)-phase progression. Genes Dev 1999;13(12):1501–12.
[32] Stott FJ, Bates S, James MC, McConnell BB, Starborg M, Brookes S, et al. The alternative product from the human CDKN2A locus, p14(ARF), participates in a regulatory feedback loop with p53 and MDM2. EMBO J 1998;17(17):5001–14.
[33] Trimarchi JM, Lees JA. Sibling rivalry in the E2F family. Nat Rev Mol Cell Biol 2002;3(1):11–20.
[34] Wu W, Fan YH, Kemp BL, Walsh G, Mao L. Overexpression of cdc25A and cdc25B is frequent in primary non-small cell lung cancer but is not associated with overexpression of c-myc. Cancer Res 1998;58(18):4082–5.
[35] Robles LD, Frost AR, Davila M, Hutson AD, Grizzle WE, Chakrabarti R. Down-regulation of Cdc6, a cell cycle regulatory gene, in prostate cancer. J Biol Chem 2002;277(28):25431–8.
[36] Jonsson PF, Bates PA. Global topological features of cancer proteins in the human interactome. Bioinformatics 2006;22(18):2291–7.
[37] Rhodes DR, Kalyana-Sundaram S, Mahavisno V, Barrette TR, Ghosh D, Chinnaiyan AM. Mining for regulatory programs in the cancer transcriptome. Nat Genet 2005;37(6):579–83.
[38] Hanahan D, Weinberg RA. The hallmarks of cancer. Cell 2000;100(1):57–70.
[39] Fesik SW. Promoting apoptosis as a strategy for cancer drug discovery. Nat Rev Cancer 2005;5 (11):876–85.
[40] Hood L, Heath JR, Phelps ME, Lin BY. Systems biology and new technologies enable predictive and preventative medicine. Science 2004;306(5696):640–3.
[41] Kitano H. Innovation—a robustness-based approach to systems-oriented drug design. Nat Rev Drug Discov 2007;6(3):202–10.
[42] Danial NN, Korsmeyer SJ. Cell death: critical control points. Cell 2004;116(2):205–19.
[43] Hengartner MO. The biochemistry of apoptosis. Nature 2000;407(6805):770–6.
[44] Riedl SJ, Shi YG. Molecular mechanisms of caspase regulation during apoptosis. Nat Rev Mol Cell Biol 2004;5(11):897–907.
[45] Herr I, Debatin KM. Cellular stress response and apoptosis in cancer therapy. Blood 2001;98 (9):2603–14.
[46] Cusick ME, Klitgord N, Vidal M, Hill DE. Interactome: gateway into systems biology. Hum Mol Genet 2005;14:R171–81.
[47] Han JDJ, Bertin N, Hao T, Goldberg DS, Berriz GF, Zhang LV, et al. Evidence for dynamically organized modularity in the yeast protein-protein interaction network. Nature 2004;430 (6995):88–93.
[48] Rual JF, Venkatesan K, Hao T, Hirozane-Kishikawa T, Dricot A, Li N, et al. Towards a proteome-scale map of the human protein-protein interaction network. Nature 2005;437(7062):1173–8.

[49] Stelzl U, Worm U, Lalowski M, Haenig C, Brembeck FH, Goehler H, et al. A human protein-protein interaction network: a resource for annotating the proteome. Cell 2005;122(6):957–68.
[50] Bader GD, Betel D, Hogue CWV. BIND: the Biomolecular Interaction Network Database. Nucleic Acids Res 2003;31(1):248–50.
[51] Peri S, Navarro JD, Amanchy R, Kristiansen TZ, Jonnalagadda CK, Surendranath V, et al. Development of human protein reference database as an initial platform for approaching systems biology in humans. Genome Res 2003;13(10):2363–71.
[52] Hermjakob H, Montecchi-Palazzi L, Lewington C, Mudali S, Kerrien S, Orchard S, et al. IntAct: an open source molecular interaction database. Nucleic Acids Res 2004;32:D452–5.
[53] Rhodes DR, Chinnaiyan AM. Integrative analysis of the cancer transcriptome. Nat Genet 2005;37: S31–7.
[54] Bader JS, Chaudhuri A, Rothberg JM, Chant J. Gaining confidence in high-throughput protein interaction networks. Nat Biotechnol 2004;22(1):78–85.
[55] Carter GW. Inferring network interactions within a cell. Brief in Bioinformatics 2005;6(4):380–9.
[56] Chiang J-H, Chao S-Y. Modeling human cancer-related regulatory modules by GA-RNN hybrid algorithms. BMC Bioinformatics 2007;8:91.
[57] Hornberg JJ, Bruggeman FJ, Westerhoff HV, Lankelma J. Cancer: a systems biology disease. Biosystems 2006;83(2–3):81–90.
[58] Chen BS, Wang YC. On the attenuation and amplification of molecular noise in genetic regulatory networks. BMC Bioinformatics 2006;7:52.
[59] Chen B-S, Li C-W. Analysing microarray data in drug discovery using systems biology. Expert Opin Drug Discov 2007;2(5):755–68.
[60] Murray JI, Whitfield ML, Trinklein ND, Myers RM, Brown PO, Botstein D. Diverse and specific gene expression responses to stresses in cultured human cells. Mol Biol Cell 2004;15(5):2361–74.
[61] Pujana MA, Han JD, Starita LM, Stevens KN, Tewari M, Ahn JS, et al. Network modeling links breast cancer susceptibility and centrosome dysfunction. Nat Genet 2007;39(11):1338–49.
[62] Aittokallio T, Schwikowski B. Graph-based methods for analysing networks in cell biology. Brief Bioinformatics 2006;7(3):243–55.
[63] Riedl SJ, Salvesen GS. The apoptosome: signalling platform of cell death. Nat Rev Mol Cell Biol 2007;8(5):405–13.
[64] Corey SJ, Minden MD, Barber DL, Kantarjian H, Wang JCY, Schimmer AD. Myelodysplastic syndromes: the complexity of stem-cell diseases. Nat Rev Cancer 2007;7(2):118–29.
[65] Araujo RP, Liotta LA, Petricoin EF. Proteins, drug targets and the mechanisms they control: the simple truth about complex networks. Nat Rev Drug Discov 2007;6(11):871–80.
[66] Hood L, Perlmutter RM. The impact of systems approaches on biological problems in drug discovery. Nat Biotechnol 2004;22(10):1215–17.
[67] Barabasi AL, Oltvai ZN. Network biology: understanding the cell's functional organization. Nat Rev Genet 2004;5(2):101–115.
[68] Andersen MH, Becker JC, Straten PT. Regulators of apoptosis: suitable targets for immune therapy of cancer. Nat Rev Drug Discov 2005;4(5):399–409.
[69] Cory S, Adams JM. The BCL2 family: regulators of the cellular life-or-death switch. Nat Rev Cancer 2002;2(9):647–56.
[70] Garber K. New apoptosis drugs face critical test. Nat Biotechnol 2005;23(4):409–11.
[71] Ghobrial IM, Witzig TE, Adjei AA. Targeting apoptosis pathways in cancer therapy. CA Cancer J Clin 2005;55(3):178–94.
[72] Oltersdorf T, Elmore SW, Shoemaker AR, Armstrong RC, Augeri DJ, Belli BA, et al. An inhibitor of Bcl-2 family proteins induces regression of solid tumours. Nature 2005;435(7042):677–81.
[73] Reed JC. Apoptosis-based therapies. Nat Rev Drug Discov 2002;1(2):111–21.
[74] Vousden KH, Lane DP. p53 in health and disease. Nat Rev Mol Cell Biol 2007;8(4):275–83.
[75] Sherr CJ, McCormick F. The RB and p53 pathways in cancer. Cancer Cell 2002;2(2):103–12.
[76] Sebolt-Leopold JS, Herrera R. Targeting the mitogen-activated protein kinase cascade to treat cancer. Nat Rev Cancer 2004;4(12):937–47.

[77] Wada T, Penninger JM. Mitogen-activated protein kinases in apoptosis regulation. Oncogene 2004;23(16):2838–49.
[78] Pelengaris S, Khan M, Evan G. c-MYC: more than just a matter of life and death. Nat Rev Cancer 2002;2(10):764–76.
[79] Lewin B. Genes VIII. Oxford; New York: Oxford University Press; 2004.
[80] Lehninger AL, Nelson DL, Cox MM. Lehninger principles of biochemistry. 4th ed. New York, NY: W.H. Freeman; 2005.
[81] Basu A. Involvement of protein kinase C-delta in DNA damage-induced apoptosis. J Cell Mol Med 2003;7(4):341–50.
[82] Ben-Hur A, Noble WS. Kernel methods for predicting protein-protein interactions. Bioinformatics 2005;21:I38–46.
[83] Liu Y, Liu NJ, Zhao HY. Inferring protein-protein interactions through high-throughput interaction data from diverse organisms. Bioinformatics 2005;21(15):3279–85.
[84] Martin S, Roe D, Faulon JL. Predicting protein-protein interactions using signature products. Bioinformatics 2005;21(2):218–26.
[85] Troyanskaya OG. Putting microarrays in a context: integrated analysis of diverse biological data. Brief Bioinformatics 2005;6(1):34–43.
[86] Chu LH, Chen BS. Comparisons of robustness and sensitivity between cancer and normal cells by microarray data. Cancer Informatics 2008;6:165–81.
[87] Ekman D, Light S, Bjorklund AK, Elofsson A. What properties characterize the hub proteins of the protein-protein interaction network of *Saccharomyces cerevisiae*? Genome Biol 2006;7(6):R45.
[88] Gandhi TKB, Zhong J, Mathivanan S, Karthick L, Chandrika KN, Mohan SS, et al. Analysis of the human protein interactome and comparison with yeast, worm and fly interaction datasets. Nat Genet 2006;38(3):285–93.
[89] Hood L. Systems biology: integrating technology, biology, and computation. Mech Ageing Dev 2003;124(1):9–16.
[90] Lin LH, Lee HC, Li WH, Chen BS. Dynamic modeling of cis-regulatory circuits and gene expression prediction via cross-gene identification. BMC Bioinformatics 2005;6:258.
[91] Alon U. An introduction to systems biology: design principles of biological circuits. Boca Raton, FL: Chapman & Hall/CRC; 2007.
[92] Chang Y-H, Wang Y-C, Chen B-S. Identification of transcription factor cooperativity via stochastic system model. Bioinformatics 2006;22(18):2276–82.
[93] Kanavos P. The rising burden of cancer in the developing world. Ann Oncol 2006;17:15–23.
[94] Parkin DM, Bray F, Ferlay J, Pisani P. Global cancer statistics, 2002. CA Cancer J Clin 2005;55(2):74–108.
[95] Spira A, Beane JE, Shah V, Steiling K, Liu G, Schembri F, et al. Airway epithelial gene expression in the diagnostic evaluation of smokers with suspect lung cancer. Nat Med 2007;13(3):361–6.
[96] Brambilla E, Travis WD, Colby TV, Corrin B, Shimosato Y. The new World Health Organization classification of lung tumours. Eur Respir J 2001;18(6):1059–68.
[97] Khuder SA. Effect of cigarette smoking on major histological types of lung cancer: a meta-analysis. Lung Cancer 2001;31(2–3):139–48.
[98] Spira A, Beane J, Shah V, Liu G, Schembri F, Yang XM, et al. Effects of cigarette smoke on the human airway epithelial cell transcriptome. Proc Natl Acad Sci USA 2004;101(27):10143–8.
[99] Hecht SS. Tobacco smoke carcinogens and lung cancer. J Nat Cancer Inst 1999;91(14):1194–210.
[100] Takahashi H, Ogata H, Nishigaki R, Broide DH, Karin M. Tobacco smoke promotes lung tumorigenesis by triggering IKK beta- and JNK1-dependent inflammation. Cancer Cell 2010;17(1):89–97.
[101] Alizadeh AA, Eisen MB, Davis RE, Ma C, Lossos IS, Rosenwald A, et al. Distinct types of diffuse large B-cell lymphoma identified by gene expression profiling. Nature 2000;403(6769):503–11.
[102] Golub TR, Slonim DK, Tamayo P, Huard C, Gaasenbeek M, Mesirov JP, et al. Molecular classification of cancer: class discovery and class prediction by gene expression monitoring. Science 1999;286(5439):531–7.

[103] Hirsch FR, Merrick DT, Franklin WA. Role of biomarkers for early detection of lung cancer and chemoprevention. Eur Respir J 2002;19(6):1151–8.

[104] Ein-Dor L, Zuk O, Domany E. Thousands of samples are needed to generate a robust gene list for predicting outcome in cancer. Proc Natl Acad Sci USA 2006;103(15):5923–8.

[105] Vogelstein B, Kinzler KW. Cancer genes and the pathways they control. Nat Med 2004;10 (8):789–99.

[106] Chuang H-Y, Lee E, Liu Y-T, Lee D, Ideker T. Network-based classification of breast cancer metastasis. Mol Syst Biol 2007;3:140.

[107] Chen J, Aronow BJ, Jegga AG. Disease candidate gene identification and prioritization using protein interaction networks. BMC Bioinformatics 2009;10:73.

[108] Nibbe RK, Koyutuerk M, Chance MR. An integrative-omics approach to identify functional sub-networks in human colorectal cancer. PLoS Comput Biol 2010;6(1):e1000639.

[109] Nibbe RK, Markowitz S, Myeroff L, Ewing R, Chance MR. Discovery and scoring of protein interaction subnetworks discriminative of late stage human colon cancer. Mol Cell Proteomics 2009;8(4):827–45.

[110] Vanunu O, Magger O, Ruppin E, Shlomi T, Sharan R. Associating genes and protein complexes with disease via network propagation. PLoS Comput Biol 2010;6(1):e1000641.

[111] Taylor IW, Linding R, Warde-Farley D, Liu Y, Pesquita C, Faria D, et al. Dynamic modularity in protein interaction networks predicts breast cancer outcome. Nat Biotechnol 2009;27 (2):199–204.

[112] Luscombe NM, Babu MM, Yu HY, Snyder M, Teichmann SA, Gerstein M. Genomic analysis of regulatory network dynamics reveals large topological changes. Nature 2004;431(7006):308–12.

[113] Yang SK, Wang YC, Chao CC, Chuang YJ, Lan CY, Chen BS. Dynamic cross-talk analysis among TNF-R, TLR-4 and IL-1R signalings in TNF alpha-induced inflammatory responses. BMC Med Genomics 2010;3:19.

[114] Oti M, Snel B, Huynen MA, Brunner HG. Predicting disease genes using protein-protein interactions. J Med Genet 2006;43(8):691–8.

[115] Wang Y-C, Lan C-Y, Hsieh W-P, Murillo LA, Agabian N, Chen B-S. Global screening of potential *Candida albicans* biofilm-related transcription factors via network comparison. BMC Bioinformatics 2010;11:53.

[116] Stark C, Breitkreutz B-J, Reguly T, Boucher L, Breitkreutz A, Tyers M. BioGRID: a general repository for interaction datasets. Nucleic Acids Res 2006;34:D535–9.

[117] Akaike H. Citation classic—a new look at the statistical-model identification. Curr Contents/Eng Technol Appl Sci 1981;51 22-22.

[118] Pagano M, Gauvreau K. Principles of biostatistics. 2nd ed. Pacific Grove, CA: Duxbury; 2000.

[119] Shannon P, Markiel A, Ozier O, Baliga NS, Wang JT, Ramage D, et al. Cytoscape: a software environment for integrated models of biomolecular interaction networks. Genome Res 2003;13 (11):2498–504.

[120] Hu K, Gan Y-H, Li S-L, Gao Y, Wu D-C, Wang C-Y, et al. Relationship of activated extracellular signal-regulated kinase 1/2 with lung metastasis in salivary adenoid cystic carcinoma. Oncol Rep 2009;21(1):137–43.

[121] Kyriakis JM, Avruch J. Mammalian mitogen-activated protein kinase signal transduction pathways activated by stress and inflammation. Physiol Rev 2001;81(2):807–69.

[122] Bierie B, Moses HL. TGF-beta and cancer. Cytokine Growth Factor Rev 2006;17(1–2):29–40.

[123] Kishimoto M, Kohno T, Okudela K, Otsuka A, Sasaki H, Tanabe C, et al. Mutations and deletions of the CBP gene in human lung cancer. Clin Cancer Res 2005;11(2):512–19.

[124] Ding L, Getz G, Wheeler DA, Mardis ER, McLellan MD, Cibulskis K, et al. Somatic mutations affect key pathways in lung adenocarcinoma. Nature 2008;455(7216):1069–75.

[125] Herbst RS, Heymach JV, Lippman SM. Molecular origins of cancer: lung cancer. N Engl J Med 2008;359(13):1367–80.

[126] Langer CJ. Emerging role of epidermal growth factor receptor inhibition in therapy for advanced malignancy: focus on NSCLC. Int J Radiat Oncol Biol Phys 2004;58(3):991–1002.

[127] Vander Griend DJ, D'Antonio J, Gurel B, Antony L, DeMarzo AM, Isaacs JT. Cell-autonomous intracellular androgen receptor signaling drives the growth of human prostate cancer initiating cells. Prostate 2010;70(1):90–9.
[128] Byers LA, Sen B, Saigal B, Diao L, Wang J, Nanjundan M, et al. Reciprocal regulation of c-Src and STAT3 in non-small cell lung cancer. Clin Cancer Res 2009;15(22):6852–61.
[129] Mitra SK, Schlaepfer DD. Integrin-regulated FAK-Src signaling in normal and cancer cells. Curr Opin Cell Biol 2006;18(5):516–23.
[130] Behrens C, Lin HY, Lee JJ, Raso MG, Hong WK, Wistuba II, et al. Immunohistochemical expression of basic fibroblast growth factor and fibroblast growth factor receptors 1 and 2 in the pathogenesis of lung cancer. Clin Cancer Res 2008;14(19):6014–22.
[131] Welcsh PL, Owens KN, King MC. Insights into the functions of BRCA1 and BRCA2. Trends Genet 2000;16(2):69–74.
[132] Henderson BE, Feigelson HS. Hormonal carcinogenesis. Carcinogenesis 2000;21(3):427–33.
[133] Cox ME, Gleave ME, Zakikhani M, Bell RH, Piura E, Vickers E, et al. Insulin receptor expression by human prostate cancers. Prostate 2009;69(1):33–40.
[134] McLean GW, Carragher NO, Avizienyte E, Evans J, Brunton VG, Frame MC. The role of focal-adhesion kinase in cancer—a new therapeutic opportunity. Nat Rev Cancer 2005;5(7): 505–15.
[135] Gallegos Ruiz MI, Floor K, Roepman P, Rodriguez JA, Meijer GA, Mooi WJ, et al. Integration of gene dosage and gene expression in non-small cell lung cancer, identification of HSP90 as potential target. PLoS One 2008;3(3):e0001722.
[136] Wang HM, Davis A, Yu SH, Ahmed K. Response of cancer cells to molecular interruption of the CK2 signal. Mol Cell Biochem 2001;227(1–2):167–74.
[137] Barker N, Clevers H. Mining the Wnt pathway for cancer therapeutics. Nat Rev Drug Discov 2006;5(12):997–1014.
[138] Deniaud E, Baguet J, Mathieu AL, Pages G, Marvel J, Leverrier Y. Overexpression of Sp1 transcription factor induces apoptosis. Oncogene 2006;25(53):7096–105.
[139] Imai MA, Oda Y, Oda M, Nakanishi I, Kawahara E. Overexpression of E2F1 associated with LOH at RB locus and hyperphosphorylation of RB in non-small cell lung carcinoma. J Cancer Res Clin Oncol 2004;130(6):320–6.
[140] Li Z, Zhao J, Du Y, Park HR, Sun S-Y, Bernal-Mizrachi L, et al. Down-regulation of 14-3-3 zeta suppresses anchorage-independent growth of lung cancer cells through anoikis activation. Proc Natl Acad Sci USA 2008;105(1):162–7.
[141] West KA, Linnoila IR, Belinsky SA, Harris CC, Dennis PA. Tobacco carcinogen-induced cellular transformation increases activation of the phosphatidylinositol 3′-Kinase/Akt pathway in vitro and in vivo. Cancer Res 2004;64(2):446–51.
[142] Sala G, Dituri F, Raimondi C, Previdi S, Maffucci T, Mazzoletti M, et al. Phospholipase Cgamma1 is required for metastasis development and progression. Cancer Res 2008;68 (24):10187–96.
[143] Zajac-Kaye M. Myc oncogene: a key component in cell cycle regulation and its implication for lung cancer. Lung Cancer 2001;34:S43–6.
[144] Vicent S, Lopez-Picazo JM, Toledo G, Lozano MD, Torre W, Garcia-Corchon C, et al. ERK1/2 is activated in non-small-cell lung cancer and associated with advanced tumours. Br J Cancer 2004;90(5):1047–52.
[145] Chen Z-Y, Cai L, Bie P, Wang S-G, Jiang Y, Dong J-H, et al. Roles of Fyn in pancreatic cancer metastasis. J Gastroenterol Hepatol 2010;25(2):293–301.
[146] Wang B, Liu K, Lin H-Y, Bellam N, Ling S, Lin W-C. 14-3-3 tau regulates ubiquitin-independent proteasomal degradation of p21, a novel mechanism of p21 downregulation in breast cancer. Mol Cell Biol 2010;30(6):1508–27.
[147] Inoue J-i, Gohda J, Akiyama T, Semba K. NF-kappa B activation in development and progression of cancer. Cancer Sci 2007;98(3):268–74.
[148] Langenfeld EM, Kong Y, Langenfeld J. Bone morphogenetic protein 2 stimulation of tumor growth involves the activation of Smad-1/5. Oncogene 2006;25(5):685–92.

[149] Ventura JJ, Tenbaum S, Perdiguero E, Huth M, Guerra C, Barbacid M, et al. p38 alpha MAP kinase is essential in lung stem and progenitor cell proliferation and differentiation. Nat Genet 2007;39(6):750–8.
[150] Hollstein M, Sidransky D, Vogelstein B, Harris CC. p53 Mutations in human cancers. Science 1991;253(5015):49–53.
[151] Deng X, Tannehill-Gregg SH, Nadella MVP, He G, Levine A, Cao Y, et al. Parathyroid hormone-related protein and ezrin are up-regulated in human lung cancer bone metastases. Clin Exp Metastasis 2007;24(2):107–19.
[152] Liu H, Radisky DC, Nelson CM, Zhang H, Fata JE, Roth RA, et al. Mechanism of Akt1 inhibition of breast cancer cell invasion reveals a protumorigenic role for TSC2. Proc Natl Acad Sci USA 2006;103(11):4134–9.
[153] Klinakis A, Lobry C, Abdel-Wahab O, Oh P, Haeno H, Buonamici S, et al. A novel tumour-suppressor function for the Notch pathway in myeloid leukaemia. Nature 2011;473(7346):230–3.
[154] Shi YG, Massague J. Mechanisms of TGF-beta signaling from cell membrane to the nucleus. Cell 2003;113(6):685–700.
[155] Zwick E, Bange J, Ullrich A. Receptor tyrosine kinase signalling as a target for cancer intervention strategies. Endocr Relat Cancer 2001;8(3):161–73.
[156] Coussens LM, Werb Z. Inflammation and cancer. Nature 2002;420(6917):860–7.
[157] Potter CJ, Pedraza LG, Xu T. Akt regulates growth by directly phosphorylating Tsc2. Nat Cell Biol 2002;4(9):658–65.
[158] Komander D. The emerging complexity of protein ubiquitination. Biochem Soc Trans 2009;37:937–53.
[159] Wang YC, Chen BS. Integrated cellular network of transcription regulations and protein-protein interactions. BMC Syst Biol 2010;4:20.
[160] Mathivanan S, Periaswamy B, Gandhi TKB, Kandasamy K, Suresh S, Mohmood R, et al. An evaluation of human protein-protein interaction data in the public domain. BMC Bioinformatics 2006;7:S19.
[161] Rapaport F, Zinovyev A, Dutreix M, Barillot E, Vert J-P. Classification of microarray data using gene networks. BMC Bioinformatics 2007;8:35.
[162] Tian Z, Hwang T, Kuang R. A hypergraph-based learning algorithm for classifying gene expression and arrayCGH data with prior knowledge. Bioinformatics 2009;25(21):2831–8.
[163] Petrelli A, Giordano S. From single- to multi-target drugs in cancer therapy: when aspecificity becomes an advantage. Curr Med Chem 2008;15(5):422–32.
[164] Kris MG, Natale RB, Herbst RS, Lynch TJ, Prager D, Belani CP, et al. Efficacy of gefitinib, an inhibitor of the epidermal growth factor receptor tyrosine kinase, in symptomatic patients with non-small cell lung cancer—a randomized trial. JAMA 2003;290(16):2149–58.
[165] Erler JT, Linding R. Network-based drugs and biomarkers. J Pathol 2010;220(2):290–6.
[166] Zimmermann GR, Lehar J, Keith CT. Multi-target therapeutics: when the whole is greater than the sum of the parts. Drug Discov Today 2007;12(1–2):34–42.
[167] Schroeder GL, Lorenzo-Gomez MF, Hautmann SH, Friedrich MG, Ekici S, Huland H, et al. A side by side comparison of cytology and biomarkers for bladder cancer detection. J Urol 2004;172(3):1123–6.
[168] Li CW, Chen BS. Identifying functional mechanisms of gene and protein regulatory networks in response to a broader range of environmental stresses. Comp Funct Genomics 2010;408705.
[169] Chernov AV, Reyes L, Xu Z, Gonzalez B, Golovko G, Peterson S, et al. Mycoplasma CG- and GATC-specific DNA methyltransferases selectively and efficiently methylate the host genome and alter the epigenetic landscape in human cells. Epigenetics 2015;10(4):303–18.
[170] Bandres E, Agirre X, Bitarte N, Ramirez N, Zarate R, Roman-Gomez J, et al. Epigenetic regulation of microRNA expression in colorectal cancer. Int J Cancer 2009;125(11):2737–43.
[171] Weber M, Hellmann I, Stadler MB, Ramos L, Paabo S, Rebhan M, et al. Distribution, silencing potential and evolutionary impact of promoter DNA methylation in the human genome. Nat Genet 2007;39(4):457–66.

[172] Valinluck V, Tsai HH, Rogstad DK, Burdzy A, Bird A, Sowers LC. Oxidative damage to methyl-CpG sequences inhibits the binding of the methyl-CpG binding domain (MBD) of methyl-CpG binding protein 2 (MeCP2). Nucleic Acids Res 2004;32(14):4100–8.
[173] Ghosh S, Yates AJ, Fruhwald MC, Miecznikowski JC, Plass C, Smiraglia D. Tissue specific DNA methylation of CpG islands in normal human adult somatic tissues distinguishes neural from non-neural tissues. Epigenetics 2010;5(6):527–38.
[174] Wang YC, Chen BS. A network-based biomarker approach for molecular investigation and diagnosis of lung cancer. BMC Med Genomics 2011;4:2.
[175] Tu CT, Chen BS. On the increase in network robustness and decrease in network response ability during the aging process: a systems biology approach via microarray data. IEEE-ACM Trans Comput Biol Bioinform 2013;10(2):468–80.
[176] Blagosklonny MV. Prevention of cancer by inhibiting aging. Cancer Biol Ther 2008;7(10):1520–4.
[177] Lamb J, Crawford ED, Peck D, Modell JW, Blat IC, Wrobel MJ, et al. The connectivity map: using gene-expression signatures to connect small molecules, genes, and disease. Science 2006;313 (5795):1929–35.
[178] Fink SEK, Pahernik S, Hallscheidt P, Zeier M. Urothelial carcinoma. Onkologe 2015;21(8):739–45.
[179] Wagner AH, Coffman AC, Ainscough BJ, Spies NC, Skidmore ZL, Campbell KM, et al. DGIdb 2.0: mining clinically relevant drug-gene interactions. Nucleic Acids Res 2015.
[180] Bartel DP. MicroRNAs: target recognition and regulatory functions. Cell 2009;136(2):215–33.
[181] Bovolenta LA, Acencio ML, Lemke N. HTRIdb: an open-access database for experimentally verified human transcriptional regulation interactions. BMC Genomics 2012;13:405.
[182] Zheng G, Tu K, Yang Q, Xiong Y, Wei C, Xie L, et al. ITFP: an integrated platform of mammalian transcription factors. Bioinformatics 2008;24(20):2416–17.
[183] Matys V, Fricke E, Geffers R, Gossling E, Haubrock M, Hehl R, et al. TRANSFAC: transcriptional regulation, from patterns to profiles. Nucleic Acids Res 2003;31(1):374–8.
[184] Kanehisa M, Goto S. KEGG: kyoto encyclopedia of genes and genomes. Nucleic Acids Res 2000;28(1):27–30.
[185] Ferreira AEN, Freire AMJP, Voit EO. A quantitative model of the generation of N-epsilon-(carboxymethyl)lysine in the Maillard reaction between collagen and glucose. Biochem J 2003;376:109–21.
[186] Voit EO, Ferreira AEN. Computational analysis of biochemical systems: a practical guide for biochemists and molecular biologists. Cambridge; New York: Cambridge University Press; 2000.
[187] Lu P, Vogel C, Wang R, Yao X, Marcotte EM. Absolute protein expression profiling estimates the relative contributions of transcriptional and translational regulation. Nat Biotechnol 2007;25 (1):117–24.
[188] Coleman TF, Hulbert LA. A direct active set algorithm for large sparse quadratic programs with simple bounds. Math Program 1989;45(3):373–406.
[189] Gill PE, Murray W, Wright MH. Practical optimization. London; New York: Academic Press; 1981.
[190] Seber GAF, Lee AJ. Linear regression analysis. 2nd ed. Hoboken, NJ: Wiley-Interscience; 2003.
[191] Romanenko A, Kakehashi A, Morimura K, Wanibuchi H, Wei M, Vozianov A, et al. Urinary bladder carcinogenesis induced by chronic exposure to persistent low-dose ionizing radiation after Chernobyl accident. Carcinogenesis 2009;30(11):1821–31.
[192] Yang T, Shi RL, Chang L, Tang K, Chen K, Yu G, et al. Huachansu suppresses human bladder cancer cell growth through the Fas/Fasl and TNF- alpha/TNFR1 pathway in vitro and in vivo. J Exp Clin Cancer Res 2015;34:21.
[193] Su C-C, Liu S-H, Lee K-I, Huang K-T, Lu T-H, Fang K-M, et al. Cantharidin induces apoptosis through the calcium/PKC-regulated endoplasmic reticulum stress pathway in human bladder cancer cells. Am J Chin Med 2015;43(03):581–600.

[194] Lee EJ, Lee SJ, Kim S, Cho SC, Choi YH, Kim WJ, et al. Interleukin-5 enhances the migration and invasion of bladder cancer cells via ERK1/2-mediated MMP-9/NF-kappa B/AP-1 pathway: involvement of the p21WAF1 expression. Cell Signal 2013;25(10):2025–38.

[195] Sethi G, Sung B, Aggarwal BB. TNF: a master switch for inflammation to cancer. Front Biosci 2008;13:5094–107.

[196] O'Rourke RW, Metcalf MD, White AE, Madala A, Winters BR, Maizlin II, et al. Depot-specific differences in inflammatory mediators and a role for NK cells and IFN-gamma in inflammation in human adipose tissue. Int J Obes 2009;33(9):978–90.

[197] Martinon F. Targeting endoplasmic reticulum signaling pathways in cancer. Acta Oncol 2012;51 (7):822–30.

[198] Jang SH, Park JW, Kim HR, Seong JK, Kim HK. ADRM1 gene amplification is a candidate driver for metastatic gastric cancers. Clin Exp Metastasis 2014;31(6):727–33.

[199] Fejzo MS, Anderson L, von Euw EM, Kalous O, Avliyakulov NK, Haykinson MJ, et al. Amplification target ADRM1: role as an oncogene and therapeutic target for ovarian cancer. Int J Mol Sci 2013;14(2):3094–109.

[200] Yang X, Miao XY, Wen Y, Hu JX, Dai WD, Yin BL. A possible connection between adhesion regulating molecule 1 overexpression and nuclear factor kappa B activity in hepatocarcinogenesis. Oncol Rep 2012;28(1):283–90.

[201] Chen W, Hu XT, Shi QL, Zhang FB, He C. Knockdown of the novel proteasome subunit Adrm1 located on the 20q13 amplicon inhibits colorectal cancer cell migration, survival and tumorigenicity. Oncol Rep 2009;21(2):531–7.

[202] Zheng X, Guo Y, Chen Y, Chen M, Lin Z, Wu Y, et al. Knockdown of adhesion-regulating molecule 1 inhibits proliferation in HL60 Cells. Acta Haematol 2015;134(2):88–100.

[203] Yamasaki T, Yoshino H, Enokida H, Hidaka H, Chiyomaru T, Nohata N, et al. Novel molecular targets regulated by tumor suppressors microRNA-1 and microRNA-133a in bladder cancer. Int J Oncol 2012;40(6):1821–30.

[204] Aggarwal BB, Kumar A, Bharti AC. Anticancer potential of curcumin: preclinical and clinical studies. Anticancer Res 2003;23(1A):363–98.

[205] Xie J, Lei PF, Hu YH. Small interfering RNA-induced inhibition of epithelial cell transforming sequence 2 suppresses the proliferation, migration and invasion of osteosarcoma cells. Exp Ther Med 2015;9(5):1881–6.

[206] Hocking RR. A Biometrics invited paper. The analysis and selection of variables in linear regression. Biometrics 1976;1–49.

CHAPTER 10

Big Evolutionary Mechanisms of Network Robustness and Signaling Transductivity in Aging and Carcinogenic Process by System Modeling and Database Mining

INTRODUCTION

Living organisms are complex systems characterized by emergent properties. One ubiquitous property in complex systems is robustness. Robustness means that living organisms are capable of continuing to perform their functions despite significant variations both in the environment and in their physiological organizations, including their genetic backgrounds. Robustness is a nontrivial biological mechanism that is difficult to study using a purely reductionist approach. The mechanism underlying robustness is distributed in all biological systems, and guarantees organization of the whole system and cooperation between various parts. From the point of view of mathematical modeling and computational systems biology, the robustness mechanism is a design principle in modeling, i.e., a system model characterized by robust mechanisms is likely to correspond more closely to biological reality.

In the cancer field, questions about the mechanism of biological robustness become crucially relevant. There are two important questions: How do cancer cells manage to appear and invade normal tissues? Furthermore, why is cancer so difficult to treat? That is, cancer as a disease is robust, because it represents a significant system change that appears during the organism's life, and it is able to withstand therapeutic interventions. The theory of biological and cancer robustness mechanisms is still in its infancy. Many hypotheses on the cancer robustness mechanism are discussed, yet there are still no efficient methods to measure the network robustness of cancer. As cancer is a kind of aging disease and the network response ability is a complementary system characteristic of network robustness, i.e., a network with large network robustness must have small network response abilities, and vice versa. In Section 10.1, we propose new methods to measure network robustness and response ability in aging and carcinogenesis, based on dynamic system modeling and microarray data.

Big Mechanisms in Systems Biology
DOI: http://dx.doi.org/10.1016/B978-0-12-809479-2.00010-X

Beginning with the following two definitions, "robustness: the network mechanism to maintain systematic performance in the face of intrinsic perturbations," and "response ability: the network mechanism to respond to external stimuli or transduce the stimuli to downstream regulators (the transductivity)," we present two novel systematic measurement methods, Network Robustness Measurement (NRM) and Response Ability Measurement (RAM) in this chapter to estimate the network robustness and response ability of a gene regulatory network (GRN) or protein–protein interaction network (PPIN) based on the dynamic network model constructed from the microarray data. We demonstrate the efficiency of NRM and RAM in analyzing GRNs and PPINs, respectively, by considering aging- and cancer-related microarray datasets. When applied to an aging-related GRN, the results indicate that such a network is more robust to intrinsic network perturbations due to the elderly tolerating more accumulated genetic and epigenetic variations than the young, and that the elderly are therefore less responsive to external stimuli. When applied to a PPIN of fibroblasts and HeLa cells, we observed that the network response ability of cancer cells is lower than that of healthy cells. Accordingly, we propose that generalized NRM and RAM methods represent effective tools for exploring and analyzing different system-level dynamic mechanisms using microarray data. Making use of such properties can facilitate prediction and application and provide useful information on clinical strategy and drug target selection and design specifications from a systems biology perspective.

In general, many DNA mutations and epigenetic alterations that have accumulated at oncogenes and tumor-suppressor genes are the cause of cancer. Because the differential expression of these oncogenes and tumor suppressors is primarily due to their signal transduction pathways (STPs), the signal transductivities of pathways in cancer cells need to be investigated systematically with microarray data and then compared with those in noncancer cells. In Section 10.2, based on systems analysis using microarray data from leukemic cells and nonleukemic normal cells, we report that leukemic cells with greater network robustness than nonleukemia normal cells have been shown to tolerate more intrinsic genetic mutations and epigenetic alterations, especially in acute myeloid leukemia (AML). The transductivity of the transcription factors (TFs) STATs is enhanced in leukemic cells relative to normal cells, while the transductivities of other TFs, including p53, C/EBPα, and CSLs, are attenuated in cancer cells. Furthermore, the mechanisms of transductivity and sensitivity of proteins in coupled STPs of leukemic and nonleukemic normal cells could also reveal their roles in leukemia carcinogenesis. Based on a transductivity analysis, we trace the main cause of TF dysfunctions, detect mutant proteins such as DVL, PSE2, and ERK, and discuss the role of crosstalk between multiple STPs in the leukemogenic process. Since TFs downstream of STPs also play an important role in regulation of genes related to cellular functions in the nucleus, the network robustness of the GRN consisting of these TFs in leukemic cells is also estimated from microarray data and compared with nonleukemia normal cells to investigate network robustness in the leukemic process.

Cancer is driven by somatic evolution cell lineages. Molecular biologists have long recognized carcinogenesis as an evolutionary process that involves natural selection. In Section 10.3, the evolution of somatic cancer cell lineage during carcinogenesis was modeled as an equilibrium point (i.e., phenotype of attractor) shifting process of a nonlinear stochastic evolutionary biological network. This process is subject to intrinsic random network fluctuations because of somatic genetic and epigenetic variations as well as extrinsic disturbances because of carcinogens and stresses. In Section 10.3, the nature selection scheme of an evolutionary biological network of carcinogenesis was derived from a robust negative feedback scheme based on the nonlinear stochastic Nash game strategy.

10.1 NEW MEASUREMENT METHODS OF NETWORK ROBUSTNESS AND RESPONSE ABILITY IN AGING AND CARCINOGENIC PROCESS VIA MICROARRAY DATA AND DYNAMIC SYSTEM MODEL

"Robustness," the network mechanism to maintain systematic performance in the face of intrinsic perturbations, and "response ability," the network mechanism to respond to external stimuli or transduce them to downstream regulators, are two important complementary system mechanisms that must be considered when discussing biological system performance. However, at present, these features cannot be measured directly for all network components in an experimental procedure. Therefore, we present two novel systematic measurement methods—NRM and RAM—to estimate the network robustness and response ability of a GRN or PPIN based on the dynamic network model constructed by the corresponding microarray data. We demonstrate the efficiency of NRM and RAM in analyzing GRNs and PPINs, respectively, by considering aging- and cancer-related datasets. When applied to an aging-related GRN, these results indicate that such a network is more robust to intrinsic network perturbations in the elderly than in the young, and is therefore less responsive to external stimuli. When applied to a PPIN of fibroblast and HeLa cells, we observe that the network of cancer cells possesses better robustness than that of normal cells. Moreover, the response ability of the PPIN calculated from the cancer cells is lower than that from healthy cells. Accordingly, we propose that generalized NRM and RAM methods represent effective tools for exploring and analyzing different systems-level dynamical properties via microarray data. Making use of such properties can facilitate prediction and application, providing useful information on clinical strategy, drug target selection, and design specifications of synthetic biology from a systems biology perspective.

Biological processes such as development, differentiation, tumorigenesis, and aging are increasingly being described in terms of temporal ordering of highly orchestrated transcriptional programs [1]. Although these processes can be analyzed using high-throughput gene expression microarray data at multiple time points, the computational methods available only identify which genes vary significantly, and how they vary across some or all of the time points measured [2,3]. Despite the technological limitations, system-level mechanisms such as network robustness increasingly attracted serious scientific interest in systems biology. The robustness mechanism—the system ability to maintain continued performance against intrinsic network perturbations and uncertainty—has long been recognized as a key mechanism of living systems and has been reviewed extensively elsewhere [4,5]. Moreover, this fundamental and ubiquitously observed system-level mechanism cannot be understood by focusing on the individual components, even though individual components of a system may or may not be robust themselves.

The term robustness is encountered widely in very different scientific fields, from engineering and control theory to dynamic systems [4] and biology [6–8]. It is important to note that robustness describes a relative property, not an absolute one, because no system can maintain stability in all functions when it encounters a perturbation. In other words, the robustness mechanism is not immutable. If a system cannot alter itself, it cannot adapt to intrinsic network perturbations or changes to its internal environment. Most robust biological systems possess a set of certain mechanisms to achieve this: in particular, both positive and negative feedback control are ubiquitous. Negative feedback, the main mechanism used to achieve a robust response to network perturbations, promotes the restoration of an initial condition of a system. Positive feedback amplifies sensitivity to changes and perturbations, a necessary feature for a cell that needs to make decisions robustly [9,10]. Therefore, feedback control is coupled with redundancy to enhance robustness. A signal transduction system is an additional network system that can be considered: it also plays an important role in responding to upstream signals or transducing environmental stimuli to downstream TFs in intracellular communication. From the signal energy point of view, the network response ability can be considered as system gain. Nevertheless, there is still no systematic method to measure the response ability of a network system at present.

In the last decade, with advances in experimental techniques, many researchers have utilized high-throughput data from DNA microarrays, yeast two-hybrid assays, co-immunoprecipitation, and ChIP-on-chip to study many kinds of biomolecular networks. To this end, these kinds of data are often integrated to construct GRNs and PPINs. These molecular networks have been demonstrated to have great potential for revealing the basic functions and essential mechanisms of various biological phenomena; this is accomplished by understanding biological systems on a system-wide, holistic level rather than in terms of their individual components [11].

Although the concept of robustness trade-offs has been proposed [12], few studies have quantitatively and systematically calculated the robustness and response ability of a network system. The qualitative notion that a biological system that is highly tolerant to perturbations is robust is largely uncontested; yet, there is no generally accepted approach using microarray data that can analyze network robustness quantitatively and systematically from a systems biology perspective. Moreover, in a GRN or PPIN, each component can transduce the effects of external stimuli downstream, but the response ability of the network from the global point of view cannot be experimentally measured for all network components. To solve this problem, we propose estimating the response ability of a GRN or PPIN via microarray data computations. In this section, we present a new measurement method, the NRM, to estimate the relative robustness of a GRN or PPIN at different time stages of biological processes and under distinct biological conditions, checking their ability to tolerate intrinsic perturbations and uncertainties based on Lyapunov stability theory. Moreover, we also propose another novel measurement method, the RAM, to estimate the ability of a network system to respond to and then transduce external stimuli to downstream regulators.

In addition, we illustrate these approaches by applying them to two distinct network systems: a GRN and a PPIN. First, the NRM and RAM are applied to an aging-related GRN constructed via microarray data. We show that a GRN is more robust to intrinsic perturbations in the elderly than in the young, and as such possesses poor response ability to external stimuli. We then applied these methods to PPINs of fibroblast and HeLa cells under oxidative stress conditions. We observe that the PPIN of cancer cells possesses better robustness than that of normal cells under oxidative stress. Moreover, the response ability of the PPIN calculated from the cancer cells is lower than that of the healthy cells. Since "robustness" and "response ability" are two complementary antagonistic mechanisms of a network system in the context of system performance (i.e., a more robust system will be less responsive, and vice versa [13]), measurements of these systematic mechanisms could provide verification of the system characteristics of biological networks constructed from microarray data. Measuring the properties of such trade-offs between the robustness and response mechanisms of a network system may provide insight into systemic mechanism changes that occur under different biological conditions. These generalized NRM and RAM methods represent effective tools for exploring and analyzing different systems-level dynamical properties. Making use of such properties can facilitate not only predictions and applications in gene therapy and drug target selection from the perspective of systems biology, but also help calibrate design specifications of robust synthetic biological networks for practical applications [14–16]. Therefore, the proposed methods may have great potential for analyzing and designing the next generation of synthetic GRNs or PPINs.

10.1.1 Materials and methods of measuring the network robustness and response ability

Since we provide test cases of measuring the robustness and response ability of a GRN and a PPIN in the Section 10.1.2, here we will focus on how to construct dynamic system models for a GRN and a PPIN from microarray data and other databases. Then using these dynamic models, we will propose two methods to measure their robustness and response ability.

Constructing a dynamic model for a GRN

It is assumed that the network of genes of interest consists of Z genes, and the time-series microarray data used for constructing the GRN contains K time points. In order to take into account the delayed effects of molecular trafficking and cellular signal transduction, the GRN can be represented by the following linear discrete-time dynamic system:

$$X[t+1] = AX[t] + H + n[t],$$

$$A = \begin{bmatrix} \alpha_{1,1} & \cdots & \alpha_{1,Z} \\ \vdots & \ddots & \vdots \\ \alpha_{Z,1} & \cdots & \alpha_{Z,Z} \end{bmatrix}, H = \begin{bmatrix} h_1 \\ \vdots \\ h_Z \end{bmatrix}, \tag{10.1}$$

where the state vector $X[t] = [X_1[t]...X_Z[t]]^T$ stands for the discrete-time mRNA expression levels of total Z genes at times $t = 1, \ldots, K$. The system matrix A denotes the gene interactions in the gene network: i.e., $\alpha_{i,j}$ denotes the interaction from gene j to gene i when $i \neq j$, $\forall i,j = 1, \ldots, Z$; the constant vector H represents the basal level: i.e., h_i denotes the basal level of the ith gene. $n[t]$ denotes the model residue and measurement noise. If the GRN is nonlinear, then the linear gene network in Eq. (10.1) can be considered as the linearized system at the operation point of interest and H will be related to the location of the operation point of linearization: e.g., for a nonlinear gene network $X[t+1] = f(X[t]) + n[t]$ for some nonlinear interaction $f(X[t])$ the linearized gene network around equilibrium point of interest X_0 can be represented as $X[t+1] = AX[t] + C + \Delta f(X[t]) + n[t]$, where $A = \frac{\partial f(X[t])}{\partial X}\Big|_{X[t]=X_0}$ and $\Delta f(X[t])$ denotes the high order nonlinear term. In this case, $H = C + \Delta f(X[t])$ represents the basal level of the linearized system at this equilibrium point of interest. A case of a more general, nonlinear GRN will be discussed later.

Remark 10.1: The network parameters of system matrix A and basal level H of GRN in Eq. (10.1) are identified by microarray data via recursive least-squares estimation method [17] in Appendix A.2.

In the following sections, we would like to measure robustness and response ability of a PPIN for a given biological phenomenon. Therefore, we need to construct the rough PPIN by data mining and then prune it using microarray data to obtain the refined PPIN for the phenomenon.

Constructing the rough (or candidate) PPIN

In order to obtain a PPIN for a biological phenomenon and investigate its network robustness and response ability, we first need to reconstruct the rough PPIN from microarray data and protein–protein interaction (PPI) information: in this case, these data was taken from STRING (Search Tool for the Retrieval of Interacting Genes/Proteins). STRING, a database of known and predicted protein interactions, comprises direct (physical) and indirect (functional) associations mined from four sources: genomic context, high-throughput experiments, conserved co-expression, and previous knowledge. The database currently covers 5,214,234 proteins from 1133 organisms. As a preliminary measure, we first perform a Bonferroni correction to select those genes with significantly differential expression (the corrected p-value $= 0.01/n$, where $n = 25{,}802$ denotes the number of genes) from microarray data for inclusion in the target pool. Bonferroni correction, one of the strictest and most conservative multiple testing corrections, adjusts the individual p-value for each gene to keep the overall error rate less than or equal to the user-specified p-cutoff value when performing a statistical test on a group of genes [18,19]. Subsequently, this target pool and the PPI database (STRING) are integrated to indicate the possible PPIs between two selected candidates, and the rough PPIN can then be constructed based on the selected candidates in the protein pool and the potential PPIs among them.

Pruning the rough (or candidate) PPIN via a dynamic model

Modern large-scale studies of PPIs permit the development of protein interaction networks, but all large-scale experiments and databases to date have returned high rates of false-positives [20]. Previous evidence has clearly demonstrated that utilizing multiple functional databases not only allows for better identification of PPIs, but also leads to better prediction of the function of unknown proteins [21]. The rough PPIN constructed here comprehensively details the protein interactions possible under all kinds of biological situations and experimental conditions. As such, these interactions must be further narrowed down by microarray data so that they represent only the protein interactions appropriate for the conditions under consideration. Therefore, we use a dynamic PPI model and model order selection method, Akaike Information Criterion (AIC) in Eq. (2.23) [22], together to prune the rough PPIN using time-series microarray data, deleting false positive PPIs in this PPIN. Here, a dynamic PPI model for a rough PPIN can be represented as follows:

$$\omega_i[t+1] = a_{ii}\omega_i[t] + \sum_{j=1}^{Q} b_{ij}\omega_j[t] + k_i + v_i[t], \tag{10.2}$$

where $\omega_i[t]$ denotes the expression profile of the ith target protein $\omega_i[t]$ in the rough PPIN at time point t; $v_i[t]$ denotes the model residue and measurement noise; a_{ii} denotes the influence of the ith target protein on itself at the next time point; b_{ij} denotes the individual interactive ability of protein j in the rough PPIN with target protein $\omega_i[t]$ when $i \neq j$, $\forall i, j = 1, \ldots, Q$ and the basal interaction k_i represents unknown PPIs or other influences such as mRNA–protein interactions or protein synthesis. Therefore, the PPIN with dynamic protein interactions in Eq. (10.2) can be represented by

$$W[t+1] = RW[t] + K + V[t],$$

where

$$W[t] = \begin{bmatrix} \omega_1[t] \\ \omega_2[t] \\ \vdots \\ \omega_Q[t] \end{bmatrix}, \; R = \begin{bmatrix} a_{11} & b_{12} & \cdots & b_{1Q} \\ b_{21} & a_{22} & \ddots & \vdots \\ \vdots & \ddots & \ddots & b_{(Q-1)Q} \\ b_{Q1} & \cdots & b_{Q(Q-1)} & a_{QQ} \end{bmatrix}, \; V[t] = \begin{bmatrix} v_1[t] \\ v_2[t] \\ \vdots \\ v_Q[t] \end{bmatrix}, \; K = \begin{bmatrix} k_1 \\ k_2 \\ \vdots \\ k_Q \end{bmatrix}, \tag{10.3}$$

R denotes the interactive matrix of the PPIN, K denotes the basal levels of the PPIs, and $V[t]$ denotes the modeling and measurement errors.

Remark 10.2: The protein interaction model in Eq. (10.2) can be used to describe any target protein in the rough PPIN. Therefore, the whole PPIN can be described by the protein interaction model described in Eq. (10.3).

In Eq. (10.2), $\omega_i[t]$ denotes the expression profile of the ith target protein at time point t, which is estimated from mRNA expression profiles through the following translational sigmoid function [23]:

$$\omega_i[t] = f_i(x_i[t]) = \frac{1}{1 + \exp[-r(x_i[t] - M_i)]}, \tag{10.4}$$

where r denotes the transition rate of the sigmoid function, M_i denotes the mean mRNA expression level of the corresponding protein i, and $x_i[t]$ represents the corresponding mRNA expression profiles of the ith target gene in the rough PPIN.

Remark 10.3: The network parameters R and K of PPIN in Eq. (10.3) are also identified by the recursive least square estimation method. Further the AIC in Eq. (2.23) is employed to detect the number Q of protein interactions in interactive matrix R in Eq. (10.3), i.e., use AIC to prune the false positive protein interactions in R (see Appendix A.2).

NRM of a dynamic network system

After constructing the dynamic models of the GRN in Eq. (10.1) and the PPIN in Eq. (10.3) from time-series microarray data by the recursive least-squares estimation method as detailed above, we will measure their network robustness based on these dynamic models. The steady state X_s of the dynamic system in Eq. (10.1) is obtained as:

$$X_s = AX_s + H, \quad \text{or} \quad X_s = (I-A)^{-1}H, \tag{10.5}$$

To simplify analysis of "robustness" of the steady state (phenotype) of the GRN, the origin of the dynamic system is shifted to the steady state X_s (i.e., $X[t] = \tilde{X}[t] + X_s$), achieving the following shifted dynamic system by subtracting Eq. (10.1) from (10.5) [24].

$$\tilde{X}[t+1] = A\tilde{X}[t], \tag{10.6}$$

Therefore, the robustness of the network's steady state becomes the robustness of the shifted network in Eq. (10.6) at the origin $\tilde{X}[t] \equiv 0$. In the shifted dynamic system in Eq. (10.6), if the eigenvalues of A are all inside the unit circle $|Z| \leq 1$ in the complex $\mathbb{Z}$ domain, then $\tilde{X}[t] \to 0$, and $X[t] \to X_s$, or approaches the steady state asymptotically: i.e., the phenotype will still be maintained after a small perturbation $\tilde{X}[t]$. Since robustness of the dynamic system in Eq. (10.1) at the steady state X_s is equivalent to robustness of the dynamic system in Eq. (10.6) at the origin $\tilde{X}[t] \equiv 0$ the shifted dynamic system in Eq. (10.6) will simplify the robustness analysis procedure of a network system [25].

Remark 10.4: The steady state W_s of the PPIN in Eq. (10.3) is obtained as:

$$W_s = RW_s + K, \quad \text{or} \quad W_s = (I-R)^{-1}K$$

The origin of the dynamic PPIN is shifted to the steady state W_s (i.e., $W[t] = \tilde{W}[t] + W_s$), achieving the following shifted dynamic system [24].

$$\tilde{W}[t+1] = R\tilde{W}[t], \tag{10.7}$$

Therefore, the robustness of the PPIN phenotype becomes the robustness of the above shifted PPIN at the origin $\tilde{W}[t] \equiv 0$. We then use this shifted PPIN in Eq. (10.7) to estimate its robustness and response ability in the same way as the GRN in Eq. (10.6) by following the same NRM and RAM methods. Bearing this in mind, we only discuss the NRM and RAM of the GRN in Eq. (10.6). The measurements of NRM and RAM of a PPIN follow the same procedure.

The robustness of the shifted network in Eq. (10.6) is a measure of its ability to tolerate intrinsic molecular-level perturbation [13]. If the linear network system suffers from these kinds of network perturbations, whether due to process noise, molecular fluctuations or genetic mutations in the aging or carcinogenic process, the interactive matrix A is perturbed as $A(1+\eta)$, where η denotes the ratio of intrinsic perturbation, and the corresponding additional system perturbation $\Delta A = \eta A$ [13]. Then the network system with intrinsic perturbation considerations factored in can be represented by:

$$\tilde{X}[t+1] = A(1+\eta)\tilde{X}[t], \tag{10.8}$$

A higher η value means more intrinsic network perturbations in aging or carcinogenic process, and the larger the η value that a network can tolerate, the greater robustness the network possesses. Since state-dependent intrinsic perturbations can influence the stability of a network system, we must further discuss the robustness of the perturbative network system in Eq. (10.8). According to quadratic stability theory [25,26], if the Lyapunov equation of energy function for the perturbative network system in Eq. (10.8) is chosen as $V(\tilde{X}[t]) = \tilde{X}^T[t]P\tilde{X}[t]$ for a positive symmetric matrix $P = P^T > 0$, then the system in Eq. (10.8) is quadratically stable if $\Delta V(\tilde{X}[t]) = V(\tilde{X}[t+1]) - V(\tilde{X}[t]) \leq 0$, i.e., if the energy of the GRN is not increased by intrinsic perturbations. Based on this idea, we are able to obtain the following robust stability principle for a GRN of interest that is subject to intrinsic perturbations.

Proposition 10.1: Suppose a network system suffers from intrinsic network perturbations due to genetic variations in aging or carcinogenic process as described in Eq. (10.8). The perturbative network is robustly stable if the following inequality has a positive definite solution $P = P^T > 0$:

$$[A^T(1+\eta)]P[A(1+\eta)] \leq P \tag{10.9}$$

Proof: See Appendix A.3.

Remark 10.5:

1. If a network system is free of intrinsic perturbations as in Eq. (10.6) (i.e., $\eta = 0$), the stable condition is reduced to the following matrix inequality, $A^TPA \leq P$, which has a positive definite solution $P > 0$ [13]. In order to guarantee the stability of a perturbation-free network system, eigenvalues of the system interaction matrix A should be inside the unit circle $|Z| \leq 1$ in the complex $\mathbb{Z}$ complex domain: i.e., $A^TA \leq 1$. From Eq. (10.9), it can be seen that when η is increased gradually,

eigenvalues of system interactive matrix A of the perturbation system in Eq. (10.8) should be nearer the origin of the complex domain $\mathbb{Z}$ than those from the perturbation-free system in Eq. (10.6) in order to tolerate more intrinsic perturbation η. Some eigenvalues of system interactive matrix A near the unit circle are more easily perturbed outside the unit circle by intrinsic fluctuations, making the perturbative network system unstable. Therefore, the distance between eigenvalue locations of A to the unit circle represents a measurement of robustness for a linear network system.

2. Based on the above analysis, the robustness η^o of this network system is the maximum perturbation tolerance allowed by the network, and it can be measured by solving the following constrained optimization:

$$\begin{gathered}\eta^o = \max \eta \\ \text{subject to } P > 0 \text{ and Eq. (10.9)}\end{gathered} \tag{10.10}$$

i.e., the maximum perturbation ratio η^o tolerable by the network system while maintaining stability. The constrained optimization problem in Eq. (10.10) can be solved by increasing η until no positive solution $P > 0$ exists in Eq. (10.9): i.e., to the highest possible η without violating the robust stability in Eq. (10.9). A positive definite solution $P > 0$ in Eq. (10.9) can be easily obtained by using the Linear Matrix Inequality (LMI) Toolbox of Matlab.

We can solve the above constrained optimization problem to measure the robustness of the network system simply by increasing η until no positive definite solution $P > 0$ can be found in Eq. (10.9), and then the maximal perturbation η^o tolerated by the network system represents the robustness under intrinsic perturbations. Hence, we are able to obtain the relative robustnesses of network systems we would like to investigate under many different biological circumstances. When comparing η^o for each network, the higher the value of η^o, the greater the robustness estimate for its respective network.

RAM of a dynamic network system

In this section, we turn to how to estimate the response ability of a network system to external stimuli after its robustness has been calculated. Assuming a network system responds to external stimuli $U[t]$, which might include upstream regulatory signals and external signals from factors such as carcinogens, oxidative stress, and ambient pro-inflammatory molecules, the dynamic network (10.6) is then modified as follows [27]:

$$\begin{aligned}\tilde{X}[t+1] &= A\tilde{X}[t] + U[t], \\ Y[t] &= C\tilde{X}[t],\end{aligned} \tag{10.11}$$

where $U[t] = [u_1[t] \ldots u_Z[t]]^T$ represents external stimuli and $Y[t]$ denotes the output signal responses of the network system of interest. For example, if the output signal response of all genes within the network system is analyzed, then $C = I$, an identity matrix. If only the response of the last gene to $U[t]$ is analyzed, the matrix will be $C = \begin{bmatrix} 0 & \cdots & 0 & 0 \\ \vdots & \ddots & \vdots & \vdots \\ 0 & \cdots & 0 & 0 \\ 0 & \cdots & 0 & 1 \end{bmatrix}$, i.e., all elements of C are zero except for the last element.

In a network system, each component may amplify or attenuate the effect of external stimuli when communicating downstream. The effect of input signals $U[t]$ on output signals $Y[t]$ is less than or equal to a positive value δ, if the following inequality [25,26]:

$$\frac{\sum_{t=0}^{\infty} Y^T[t]Y[t]}{\sum_{t=0}^{\infty} U^T[t]U[t]} \leq \delta^2, \quad \text{or} \quad \sum_{t=0}^{\infty} Y^T[t]Y[t] \leq \delta^2 \sum_{t=0}^{\infty} U^T[t]U[t], \tag{10.12}$$

holds for all possible bounded input signals $U[t] \neq 0$ and $\tilde{X}[0] = 0$. The physical meaning attributable to Eq. (10.12) is that the effect of all possible external stimuli $U[t]$ on $Y[t]$ is less than or equal to δ^2 from the energy point of view: i.e., δ denotes the upper bound of the effect of $U[t]$ on $Y[t]$. If $\tilde{X}[0] \neq 0$, then inequality Eq. (10.12) should be modified as the following:

$$\sum_{t=0}^{\infty} Y^T[t]Y[t] \leq \tilde{X}^T[0]P\tilde{X}[0] + \delta^2 \sum_{t=0}^{\infty} U^T[t]U[t], \tag{10.13}$$

to account for the effect of $\tilde{X}[0]$ on $Y[t]$. δ^o, the smallest upper bound of δ in Eq. (10.12), is called the "network response ability" of the system to all bounded stimulus signals $U[t]$. From δ^o, we obtain a more systematic insight into the ability of a network to respond to external stimuli, which is useful when considering signal transduction or intracellular communication. Via this method, we can measure and compare network response abilities δ^o of various network systems under different biological conditions and states.

Proposition 10.2: The response of the network system in Eq. (10.11) has an upper bound δ in Eq. (10.12), if there exists a positive definite $P = P^T > 0$ solution to the following LMI [28]:

$$\begin{bmatrix} A^T PA + C^T C - P & A^T P \\ PA & P - \delta^2 I \end{bmatrix} \leq 0, \tag{10.14}$$

i.e., if the above LMI holds for some $P > 0$, then the effect of $U[t]$ on $Y[t]$ must be less than or equal to δ: i.e., Eq. (10.12) or (10.13) hold for the network system in Eq. (10.11).

Proof: See Appendix A.4.

Remark 10.6:

1. The response ability δ^{o} of a network system to external stimuli can be obtained by minimizing its upper bound δ, achieved by solving the following constrained optimization problem:

$$\begin{gathered}\delta^{\text{o}} = \min_{P} \delta \\ \text{subject to } \delta > 0, P > 0, \text{ and Eq. (10.14)}\end{gathered}, \tag{10.15}$$

   This constrained optimization can be solved by decreasing the upper bound δ in Eq. (10.12) until no $P > 0$ exists in Eq. (10.14) using Matlab LMI toolbox, which is based on Newton's search method [27].
2. δ^{o}, the smallest upper bound of δ in Eq. (10.12), is called the network response ability of a network system to external stimuli. If $\delta^{\text{o}} < 1$, it means external stimuli are attenuated (buffered) by the network system; if $\delta^{\text{o}} > 1$, external stimuli are amplified.
3. Since the output/input signal energy ratio (i.e., the response) in Eq. (10.12) or (10.13) is considered for all possible external stimuli, the measure of network response ability in Eqs. (10.14) and (10.15) is dependent more on system matrix A than on the external stimuli. This is analogous to the way a lowpass filter is dependent more upon the characteristics of the filter than on noise. Therefore, it is possible to measure the network response ability of a network system from its system characteristics without knowledge of what the external stimuli are, with the caveat that these signals must be bounded. The system matrix A can be estimated from time-series microarray data as in the previous sections. Therefore, employing the Matlab LMI toolbox, we can use Eq. (10.15) to measure network response abilities using microarray data under different specific experimental conditions.

Extension of network robustness and RAMs to nonlinear network systems

Suppose the GRN or PPIN could be constructed by microarray data for the following nonlinear network system:

$$X[t+1] = f(X[t]), \tag{10.16}$$

Suppose we are interested in the phenomenon around the equilibrium point X_e, i.e., $X_e = f(X_e)$ and let $\tilde{X}[t] = X[t] - X_e$. Then, we get the shifted nonlinear network system as:

$$\tilde{X}[t+1] = f(\tilde{X}[t] + X_e) - f(X_e) \triangleq \tilde{f}(\tilde{X}[t]), \tag{10.17}$$

In this system, $\tilde{X}[t] \equiv 0$ is the equilibrium point of interest. Suppose the nonlinear network in Eq. (10.17) could be globally linearized as described in Ref. [27] for all $\tilde{X}[t] = X[t] - X_e$ as:

$$\left.\frac{\partial \tilde{f}(\tilde{X}[t])}{\partial X}\right|_{\tilde{X}[t]} \in \Omega, \tag{10.18}$$

where $\Omega \in \mathbb{R}^{n \times n}$ denotes the system parameter set of these globally linearized systems. Suppose Ω could be described by a list of its vertices via the following convex hull:

$$\Omega \in Co(A_0, \ldots, A_m) \tag{10.19}$$

where Co denotes linear combination of matrices. By the global linearization technique described in Ref. [27], all the globally linearized systems in Ω of the nonlinear system in Eq. (10.16) or (10.17) could be interpolated by the m linear system $\tilde{X}[t+1] = A_i\tilde{X}[t]$, for $i = 1, \ldots, m$, at the vertices of convex hull of the polytope:

$$\tilde{X}[t+1] = \tilde{f}(\tilde{X}[t]) = \sum_{i=1}^{m} h_i(\tilde{X})A_i\tilde{X}[t], \tag{10.20}$$

where the interpolation functions $h_i(\tilde{X})$, for $i = 1, \ldots, m$, have the following properties:

$$0 \leq h_i(\tilde{X}) \leq 1 \text{ and } \sum_{i=1}^{m} h_i(\tilde{X}) = 1, \tag{10.21}$$

i.e., the trajectory of the nonlinear system in Eq. (10.16) or (10.17) could be interpolated by the trajectories of m linear systems in Eq. (10.20) via the nonlinear interpolation functions $h_i(\tilde{X})$, for $i = 1, \ldots, m$.

Suppose the nonlinear network system suffers from the following intrinsic network perturbations:

$$\tilde{X}[t+1] = \sum_{i=1}^{m} h_i(\tilde{X})A_i(1+\eta)\tilde{X}[t], \tag{10.22}$$

Based on the robust stability principle, we deduce the following.

Proposition 10.3: Suppose the nonlinear network system suffers from intrinsic network perturbations as in Eq. (10.22). The perturbative nonlinear network system is robustly stable if the following LMIs have a positive definition $P = P^T > 0$:

$$[A_i^T(1+\eta)]P[A_i(1+\eta)] \leq P, \quad \text{for } i = 1, \ldots, m \tag{10.23}$$

Proof: See Appendix A.5.

The physical meaning of the LMIs in Eq. (10.23) is that the perturbation η is tolerable within all linearized network systems. Therefore, based on the global linearization method, the robustness η° of the nonlinear network system in Eq. (10.16) can be measured by solving the following constrained optimization:

$$\begin{aligned} &\eta^o = \max \eta \\ &\text{subject to } P > 0, \text{ and LMIs in Eq. (10.23)} \end{aligned} \tag{10.24}$$

i.e., the maximum perturbation η°, which is tolerable by all linearized network systems, is a global extension of Eq. (10.15) from a linear network system to a nonlinear network system.

Similarly, if the shifted nonlinear network system with output $Y[t]$ suffers from external stimuli as in:

$$\begin{aligned}\tilde{X}[t+1] &= \tilde{f}(\tilde{X}[t]) + U[t] \\ Y[t] &= C\tilde{X}[t]\end{aligned} \tag{10.25}$$

then, based on the global linearization technique detailed in Ref. [27], the shifted nonlinear network system of Eq. (10.25) with output $Y[t]$ can be represented by the following interpolated system:

$$\begin{aligned}\tilde{X}[t+1] &= \sum_{i=1}^{m} h_i(\tilde{X})A_i\tilde{X}[t] + U[t] \\ Y[t] &= C\tilde{X}[t]\end{aligned} \tag{10.26}$$

Then we get the following result:

Proposition 10.4: The network response of the nonlinear network system in Eq. (10.26) has an upper bound δ in Eq. (10.12), if there exists a positive definite solution $P = P^T > 0$ to the following LMIs:

$$\begin{bmatrix} A_i^T PA_i + C^T C - P & A_i^T P \\ PA_i & P - \delta^2 I \end{bmatrix} \leq 0, \quad \text{for } i = 1, \ldots, m, \tag{10.27}$$

i.e., if the above LMIs hold for some $P > 0$ then the effect of $U[t]$ on $Y[t]$ must be less than or equal to δ. That is, Eq. (10.12) or (10.13) holds for the nonlinear network system (10.25) or (10.26).

Proof: See Appendix A.6.

Therefore, based on the global linearization method, the network response ability δ^{o} of the nonlinear network system in Eq. (10.25) or (10.26) can be measured by solving the following constrained optimization:

$$\begin{gathered}\delta^{\text{o}} = \min_{P} \delta \\ \text{subject to } \delta > 0, P > 0, \text{ and LMIs in Eq. (10.27)}\end{gathered} \tag{10.28}$$

Remark 10.7: If the nonlinear network system is constructed by the system identification method from microarray or other experimental data, then the network robustness η^{o} and response ability δ^{o} can be measured by Eqs. (10.24) and (10.28), respectively. However, to approximate the nonlinear network system via interpolation, these systems must undergo the global linearization in Eqs. (10.18) and (10.19) to find the vertices of the convex hull of the polytope in order to interpolate this linearized system at the vertices (as in the system in Eq. 10.26). This is generally a very complex task. If we only want to measure the network robustness and response ability of a network system near an equilibrium point (e.g., phenotype), then the measurement methods for network robustness and response ability by Eqs. (10.10) and (10.15) based on a linearized system at its equilibrium point are sufficient.

In the following section, we give several practical examples based on linear network systems for measuring the evolution of network robustness and response ability of a GRN and PPIN in the aging process and carcinogenesis constructed by time-series microarray data.

10.1.2 Evolution of network robustness and response ability in aging and carcinogenic process

In this section, we have presented two novel computational methods, NRM and RAM, which estimate the robustness and response ability of network systems in aging carcinogenic process using time-series microarray data. Briefly, the basic methodologies consist of integrating time-series microarray data with either experimental literature or PPI information to construct the dynamic system of a GRN or PPIN. Both of these systems correspond to real phenomena in living organisms and can be used to further evaluate network robustness and response ability (see Fig. 10.1). Details have been described in the preceding section. These methods are equally applicable to and useful for GRNs and PPINs. In order to demonstrate the potential applications of NRM and RAM for estimating system characteristics evolved under different biological conditions, we elaborate on the usage of the above methods in two separate cases and explain the network evolution within the following section.

Case 1: Evolution of the network robustness and response ability of a GRN at different stages of life

Aging is an extremely complex and system-level process, and it has attracted much attention in medical research, especially since chronic diseases are quite prevalent in elderly populations. These illnesses may be the result of both gene mutations that lead to intrinsic perturbations and environmental changes that stimulate signaling in the body. In addition, aging may result in reduced responses to environmental stimuli such as carcinogens, oxidative stress, and pro-inflammatory molecules. Therefore, measuring robustness to intrinsic perturbations and response ability to external stimuli of an aging-related GRN could provide key insights into the system-level mechanisms involved in aging.

Construction of a GRN associated with aging-related pathophysiological phenotypes

Combing data from several references in the literature, we construct a GRN with multiple regulatory loops that are highly associated with aging-related pathophysiological phenotypes, consisting of the following 16 genes: *FOXOs, NF-κB, p53, SIRT1,*

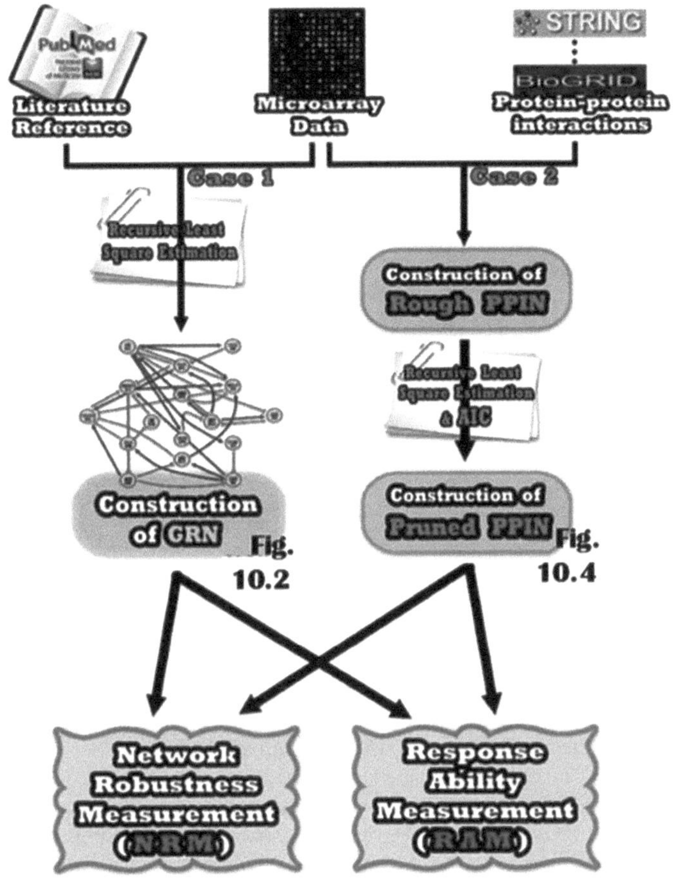

Figure 10.1 Flowchart of the proposed methods to estimate network robustness and response ability. This flowchart delineates the process used to construct the gene regulatory network (GRN) and protein–protein interaction network (PPIN), and the subsequent estimation of network robustness and response ability by the NRM and RAM methods, respectively. The flowchart on the left represents the Case 1 study analysis of the GRN; the flowchart on the right represents the Case 2 study, where methods are applied to a PPIN.

HIC1, Mdm2, Arf1, PTEN, PI3K, Akt, JNK, IKKs, IB, BTG3, E2F, and *ATM* (Fig. 10.2). All genes of interest are critical for regulating cellular functions related to longevity, including detoxification of reactive oxygen species (ROS), cell cycle arrest, repair of damaged DNA, apoptosis, and pro-inflammation, as age increases [29,30]. The GRN can be represented by the linear discrete-time dynamic system shown in Table 10.1.

In this case, we want to estimate the network robustness and response ability of this GRN, and also to investigate whether they might be altered with increasing age.

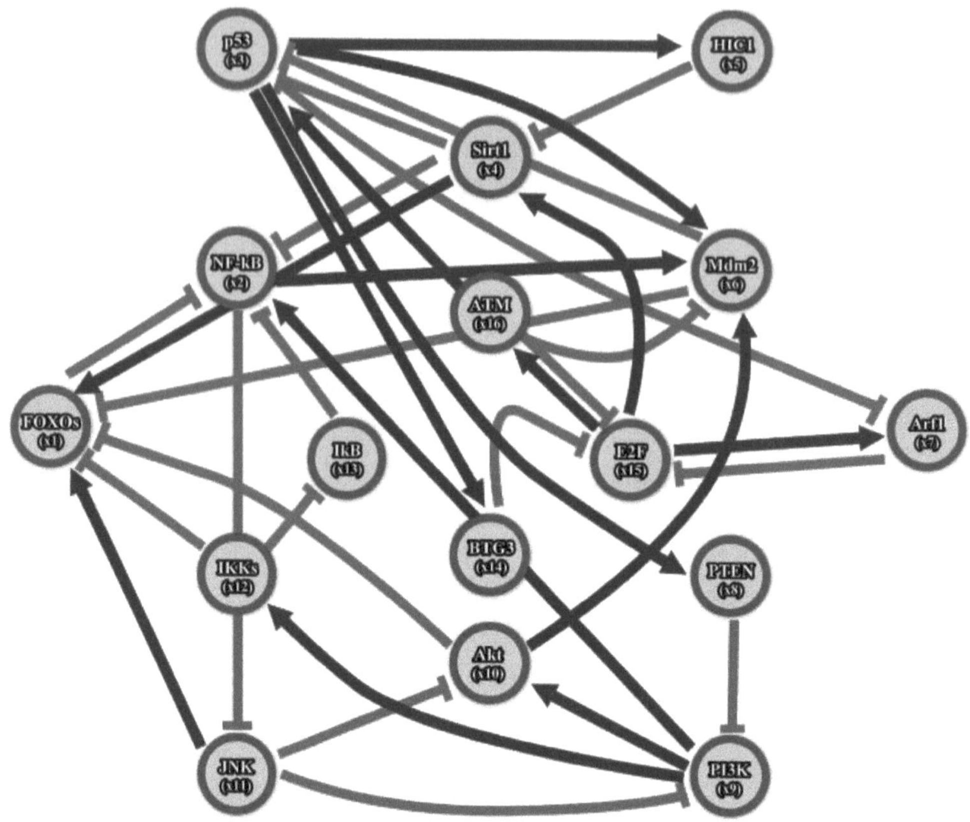

Figure 10.2 Multiple regulatory loops of a GRN associated with aging-related pathophysiological phenotypes. This network includes the following 16 genes: *FOXOs, NF-κB, p53, SIRT1, HIC1, Mdm2, Arf1, PTEN, PI3K, Akt, JNK, IKKs, IκB, BTG3, E2F,* and *ATM*. *Blue arrows* (light gray in print versions) indicate activation; *blunt red arrows* (dark gray in print versions) indicate suppression.

Table 10.1 Ordinary differential equations of gene regulatory networks in Fig. 10.2 for 16 genes associated with aging-related pathophysiological phenotypes

Gene		Dynamic equations of 16 genes of interest
x_1	*FOXOs*	$x_1[t+1] = \alpha_{1,1}x_1[t] + \alpha_{1,4}x_4[t] + \alpha_{1,6}x_6[t] + \alpha_{1,10}x_{10}[t] + \alpha_{1,11}x_{11}[t] + \alpha_{1,12}x_{12}[t] + h_1$
x_2	*NF-κB*	$x_2[t+1] = \alpha_{2,1}x_1[t] + \alpha_{2,2}x_2[t] + \alpha_{2,4}x_4[t] + \alpha_{2,9}x_9[t] + \alpha_{2,13}x_{13}[t] + h_2$
x_3	*p53*	$x_3[t+1] = \alpha_{3,3}x_3[t] + \alpha_{3,4}x_4[t] + \alpha_{3,6}x_6[t] + \alpha_{3,16}x_{16}[t] + h_3$
x_4	*SIRT1*	$x_4[t+1] = \alpha_{4,4}x_4[t] + \alpha_{4,5}x_5[t] + \alpha_{4,15}x_{15}[t] + h_4$
x_5	*HIC1*	$x_5[t+1] = \alpha_{5,3}x_3[t] + \alpha_{5,5}x_5[t] + h_5$
x_6	*Mdm2*	$x_6[t+1] = \alpha_{6,2}x_2[t] + \alpha_{6,3}x_3[t] + \alpha_{6,6}x_6[t] + \alpha_{6,7}x_7[t] + \alpha_{6,10}x_{10}[t] + \alpha_{6,16}x_{16}[t] + h_6$
x_7	*Arf1*	$x_7[t+1] = \alpha_{7,3}x_3[t] + \alpha_{7,7}x_7[t] + \alpha_{7,15}x_{15}[t] + h_7$
x_8	*PTEN*	$x_8[t+1] = \alpha_{8,3}x_3[t] + \alpha_{8,8}x_8[t] + h_8$
x_9	*PI3K*	$x_9[t+1] = \alpha_{9,8}x_8[t] + \alpha_{9,9}x_9[t] + \alpha_{9,11x11}[t] + h_9$
x_{10}	*Akt*	$x_{10}[t+1] = \alpha_{10,9}x_9[t] + \alpha_{10,10}x_{10}[t] + \alpha_{10,11}x_{11}[t] + h_{10}$
x_{11}	*JNK*	$x_{11}[t+1] = \alpha_{11,2}x_2[t] + \alpha_{11,11}x_{11}[t] + h_{11}$
x_{12}	*IKKs*	$x_{12}[t+1] = \alpha_{12,9}x_9[t] + \alpha_{12,12}x_{12}[t] + h_{12}$
x_{13}	*IκB*	$x_{13}[t+1] = \alpha_{13,12}x_{12}[t] + \alpha_{13,13}x_{13}[t] + h_{13}$
x_{14}	*BTG3*	$x_{14}[t+1] = \alpha_{13,3}x_3[t] + \alpha_{14,14}x_{14}[t] + h_{14}$
x_{15}	*E2F*	$x_{15}[t+1] = \alpha_{15,7}x_7[t] + \alpha_{15,14}x_{14}[t] + \alpha_{15,15}x_{15}[t] + \alpha_{15,16}x_{16}[t] + h_{15}$
x_{16}	*ATM*	$x_{16}[t+1] = \alpha_{16,15}x_{15}[t] + \alpha_{16,16}x_{16}[t] + h_{16}$

Furthermore, those altered system characteristics associated with a high incidence of aging-related disease, such as cancer, are prime targets for discussion. The high-throughput microarray data assembled in Ref. [31], which details the age-related effects on gene expression in the thymus and spinal cord dissected from male C57BL/6 mice of ages 1, 6, 16, and 24 months, are used as these gene expression profiles. In order to make this data more useful for identifying the system parameters of this dynamic GRN, we first rescale the original gene expression time profiles. For young mice, those aged 17 and 19 days are categorized as 3-week-old animals; 40, 43, and 44 days as 6-week-old animals; 174, 180, and 186 days as 25-week-old animals; and, 193 and 194 days as 28-week-old animals. For aged mice, those aged 476, 481, and 485 days are categorized as 68-week-old animals; 495 and 498 days as 71-week-old animals; 714 and 719 days as 102-week-old animals; and, 730, 733, and 743 days as 105-week-old animals. (All animals had been placed on ad libitum diets.) All of the above expression level data are calculated in log 2 scale, and then the Z-score method is used to normalize the gene array expression data as has been previously described [32].

We construct multiple regulatory loops of a GRN containing 16 genes (Fig. 10.2) associated with aging-related pathophysiological phenotypes. Using the recursive least-squares parameter estimation method [17] and the microarray data from the thymus and spinal cord [31], a dynamic model of the 16 genes within the GRN in Eq. (10.1) is constructed (shown in Table 10.1) with estimated parameters at different stages of life (summarized in Table 10.2). With Eq. (10.1) and Tables 10.1 and 10.2, we calculate 16 eigenvalues for interactive matrix A at the young and aged stages. The result (Fig. 10.3) shows some eigenvalues of the interactive matrix A clustered together in similar regions near the point $Z=1$, demonstrating that some modes of the same network will more readily approach some constant steady state than others [12,33].

Evolution of network robustness and response mechanisms of GRNs at different stages of life by NRM and RAM methods

It has previously been shown that gene networks associated with aging-related pathophysiological phenotypes suffer from intrinsic perturbations mainly due to noise processes like molecular fluctuation or genetic mutation. The robustness of the GRN is defined as its ability to tolerate intrinsic perturbations and plays a principal role as a fail-safe mechanism during evolutionary processes [12]. Therefore, comparisons of robustness η^{o} for different GRNs pertaining to young and aged stages may quantitatively describe their relative ability to tolerate these intrinsic network perturbations in the aging process. By using this NRM method, we evaluate the network robustness as shown in Table 10.3 to represent the maximum intensity of intrinsic network perturbations that the network system can tolerate in aging process. In other words, the

Table 10.2 Estimated parameters of gene regulatory networks in Table 10.1 with 16 genes in the thymus (A) and spinal cord (B) at the young and aged stages

Parameters of young stage				Parameters of aged stage			
(A) Thymus							
$\alpha_{1,1}=0.407$	$\alpha_{1,4}=0.183$	$\alpha_{1,6}=-0.213$	$\alpha_{1,10}=-0.031$	$\alpha_{1,1}=0.258$	$\alpha_{1,4}=0.143$	$\alpha_{1,6}=-0.023$	$\alpha_{1,10}=-0.089$
$\alpha_{1,11}=0.060$	$\alpha_{1,12}=-0.408$	$\alpha_{2,1}=0.099$	$\alpha_{2,2}=0.964$	$\alpha_{1,11}=0.033$	$\alpha_{1,12}=-0.401$	$\alpha_{2,1}=0.186$	$\alpha_{2,2}=0.525$
$\alpha_{2,4}=-0.130$	$\alpha_{2,9}=0.036$	$\alpha_{2,13}=0.126$	$\alpha_{3,3}=0.990$	$\alpha_{2,4}=0.408$	$\alpha_{2,9}=0.182$	$\alpha_{2,13}=-0.122$	$\alpha_{3,3}=1.002$
$\alpha_{3,4}=-0.071$	$\alpha_{3,6}=-0.027$	$\alpha_{3,16}=-0.013$	$\alpha_{4,4}=1.050$	$\alpha_{3,4}=0.055$	$\alpha_{3,6}=0.109$	$\alpha_{3,16}=0.100$	$\alpha_{4,4}=0.978$
$\alpha_{4,5}=0.026$	$\alpha_{4,15}=0.004$	$\alpha_{5,3}=-0.001$	$\alpha_{5,5}=0.998$	$\alpha_{4,5}=0.036$	$\alpha_{4,15}=0.04$	$\alpha_{5,3}=0.019$	$\alpha_{5,5}=0.978$
$\alpha_{6,2}=0.413$	$\alpha_{6,3}=0.125$	$\alpha_{6,6}=0.424$	$\alpha_{6,7}=0.120$	$\alpha_{6,2}=-0.087$	$\alpha_{6,3}=-0.008$	$\alpha_{6,6}=0.199$	$\alpha_{6,7}=-0.174$
$\alpha_{6,10}=0.200$	$\alpha_{6,16}=0.065$	$\alpha_{7,3}=0.028$	$\alpha_{7,7}=0.937$	$\alpha_{6,10}=0.070$	$\alpha_{6,16}=-0.292$	$\alpha_{7,3}=-0.093$	$\alpha_{7,7}=1.096$
$\alpha_{7,15}=-0.006$	$\alpha_{8,3}=0.008$	$\alpha_{8,8}=1.106$	$\alpha_{9,8}=-0.011$	$\alpha_{7,15}=-0.107$	$\alpha_{8,3}=0.05$	$\alpha_{8,8}=1.001$	$\alpha_{9,8}=-0.027$
$\alpha_{9,9}=0.985$	$\alpha_{9,11}=-0.001$	$\alpha_{10,9}=0.009$	$\alpha_{10,10}=1.068$	$\alpha_{9,9}=0.982$	$\alpha_{9,11}=0.014$	$\alpha_{10,9}=-0.02$	$\alpha_{10,10}=0.998$
$\alpha_{10,11}=-0.024$	$\alpha_{11,2}=0.021$	$\alpha_{11,11}=1.015$	$\alpha_{12,9}=0.012$	$\alpha_{10,11}=0.007$	$\alpha_{11,2}=0.003$	$\alpha_{11,11}=0.994$	$\alpha_{12,9}=-0.003$
$\alpha_{12,12}=1.019$	$\alpha_{13,12}=0.009$	$\alpha_{13,13}=0.998$	$\alpha_{14,3}=-0.003$	$\alpha_{12,12}=1.028$	$\alpha_{13,12}=0.029$	$\alpha_{13,13}=1.009$	$\alpha_{14,3}=-0.092$
$\alpha_{14,14}=1.030$	$\alpha_{15,7}=0.176$	$\alpha_{15,14}=-0.0799$	$\alpha_{15,15}=0.964$	$\alpha_{14,14}=0.965$	$\alpha_{15,7}=0.366$	$\alpha_{15,14}=0.253$	$\alpha_{15,15}=0.454$
$\alpha_{15,16}=-0.061$	$\alpha_{16,15}=-0.010$	$\alpha_{16,16}=1.014$		$\alpha_{15,16}=-0.045$	$\alpha_{16,15}=0.025$	$\alpha_{16,16}=0.997$	
(B) Spinal Cord							
$\alpha_{1,1}=0.388$	$\alpha_{1,4}=0.082$	$\alpha_{1,6}=0.004$	$\alpha_{1,10}=-0.161$	$\alpha_{1,1}=0.886$	$\alpha_{1,4}=0.044$	$\alpha_{1,6}=0.085$	$\alpha_{1,10}=0.043$
$\alpha_{1,11}=0.001$	$\alpha_{1,12}=-0.462$	$\alpha_{2,1}=-0.022$	$\alpha_{2,2}=0.920$	$\alpha_{1,11}=-0.159$	$\alpha_{1,12}=-0.101$	$\alpha_{2,1}=0.186$	$\alpha_{2,2}=0.835$
$\alpha_{2,4}=-0.013$	$\alpha_{2,9}=0.120$	$\alpha_{2,13}=-0.267$	$\alpha_{3,3}=0.954$	$\alpha_{2,4}=0.069$	$\alpha_{2,9}=0.034$	$\alpha_{2,13}=0.233$	$\alpha_{3,3}=0.869$
$\alpha_{3,4}=0.005$	$\alpha_{3,6}=0.268$	$\alpha_{3,16}=0.051$	$\alpha_{4,4}=0.979$	$\alpha_{3,4}=-0.004$	$\alpha_{3,6}=0.003$	$\alpha_{3,16}=-0.195$	$\alpha_{4,4}=0.911$
$\alpha_{4,5}=-0.043$	$\alpha_{4,15}=0.129$	$\alpha_{5,3}=0.080$	$\alpha_{5,5}=0.974$	$\alpha_{4,5}=0.242$	$\alpha_{4,15}=-0.364$	$\alpha_{5,3}=-0.044$	$\alpha_{5,5}=0.998$
$\alpha_{6,2}=-0.074$	$\alpha_{6,3}=0.084$	$\alpha_{6,6}=0.049$	$\alpha_{6,7}=-0.003$	$\alpha_{6,2}=0.140$	$\alpha_{6,3}=-0.013$	$\alpha_{6,6}=0.825$	$\alpha_{6,7}=-0.289$
$\alpha_{6,10}=-0.152$	$\alpha_{6,16}=-0.102$	$\alpha_{7,3}=0.187$	$\alpha_{7,7}=1.040$	$\alpha_{6,10}=0.137$	$\alpha_{6,16}=0.045$	$\alpha_{7,3}=0.003$	$\alpha_{7,7}=0.991$
$\alpha_{7,15}=-0.157$	$\alpha_{8,3}=0.037$	$\alpha_{8,8}=1.009$	$\alpha_{9,8}=-0.006$	$\alpha_{7,15}=-0.007$	$\alpha_{8,3}=-0.019$	$\alpha_{8,8}=0.998$	$\alpha_{9,8}=-0.091$
$\alpha_{9,9}=1.000$	$\alpha_{9,11}=-0.008$	$\alpha_{10,9}=0.009$	$\alpha_{10,10}=0.981$	$\alpha_{9,9}=0.816$	$\alpha_{9,11}=0.126$	$\alpha_{10,9}=0.009$	$\alpha_{10,10}=0.980$
$\alpha_{10,11}=0.034$	$\alpha_{11,2}=0.013$	$\alpha_{11,11}=1.033$	$\alpha_{12,9}=0.018$	$\alpha_{10,11}=0.001$	$\alpha_{11,2}=0.004$	$\alpha_{11,11}=0.992$	$\alpha_{12,9}=-0.057$
$\alpha_{12,12}=1.023$	$\alpha_{13,12}=-0.002$	$\alpha_{13,13}=1.015$	$\alpha_{14,3}=0.596$	$\alpha_{12,12}=0.972$	$\alpha_{13,12}=0.007$	$\alpha_{13,13}=1.011$	$\alpha_{14,3}=-0.011$
$\alpha_{14,14}=0.878$	$\alpha_{15,7}=-0.064$	$\alpha_{15,14}=0.148$	$\alpha_{15,15}=0.681$	$\alpha_{14,14}=1.002$	$\alpha_{15,7}=-0.352$	$\alpha_{15,14}=0.130$	$\alpha_{15,15}=0.667$
$\alpha_{15,16}=0.256$	$\alpha_{16,15}=-0.021$	$\alpha_{16,16}=1.009$		$\alpha_{15,16}=-0.205$	$\alpha_{16,15}=-0.030$	$\alpha_{16,16}=1.009$	

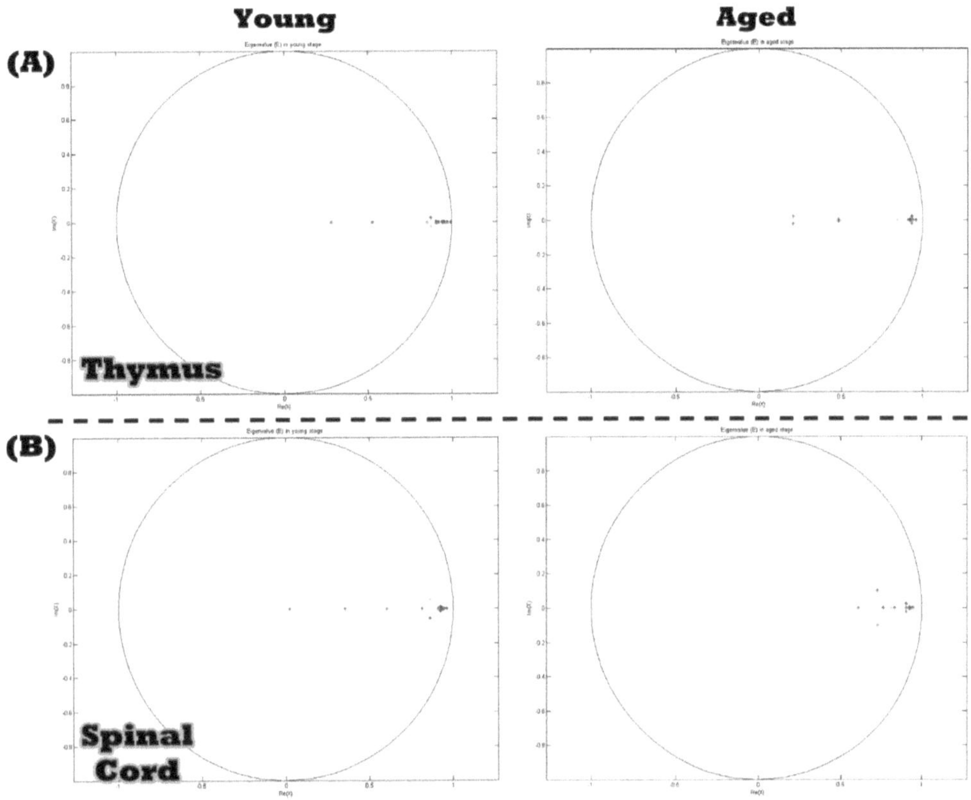

Figure 10.3 The evolution of the 16 eigenvalues of different tissues at the young and aged stages, respectively. Some eigenvalues of the interactive matrix *A* for the thymus (A) and spinal cord (B) are located together at similar regions near the unit circle $|Z| = 1$.

Table 10.3 Evolution of network robustness (η^o) and response ability (δ^o) in aging and carcinogenic process

(A)		Young	Aged
Thymus	η^o	0.2233	0.3852
	δ^o	1.177	0.9362
Spinal cord	η^o	0.236	0.6417
	δ^o	1.1936	0.9073
(B)		**Fibroblast**	**HeLa**
Menadione treatment	η^o	0.125	1.0016
	δ^o	1.1524	0.6159

Network robustness (η^o) and response ability (δ^o) of the GRN with 16 genes across different tissues at different stages of life are shown in (A). In addition, the network robustness (η^o) and response ability (δ^o) of PPINs evaluated in normal fibroblast and HeLa cancer cells under menadione treatment are indicated in (B).

higher the η^o calculated from a given GRN, the more robust the GRN is. Obviously, from Table 10.3, it is seen that the evolutionary network robustness increases in all three tissues because the GRN needs to tolerate the accumulated genetic network variations to maintain its network robust stability in the aging process.

Concerning the RAM, if a GRN also responds to external stimuli $U[t]$, such as upstream signals or biological molecules outside the network, as in Eq. (10.11), the network response ability measured should be based on the output/input energy ratio within a time interval as detailed in Eq. (10.12). Supposing the ratio of the effect of input signals $U[t]$ on $Y[t]$ is upper-bounded by a positive value δ, then we can calculate the smallest value δ^o in Eq. (10.12) (i.e., the minimum upper bound), and call it the "network response ability" to external stimuli. This method, RAM, is based on system gain theory, and incorporates principles behind the external input signal in system theory perspectives [27]. In general, a GRN with greater robustness always has lower response ability, and vice versa. This is an inherent characteristic of physical systems [27]. The trade-off between robustness and response ability is demonstrably visible in Table 10.3 and constitutes secondary proof of this proposed method.

Comparing the GRNs of the thymus and spinal cord data between the "young" group and the "aged" group, we show that both tissues evaluated have a higher robustness η^o value in the aged group than the young group. These network robustness evolution results are shown in Table 10.3. The aging-related GRN examined in the first case study is less robust to intrinsic perturbations in the young group than in the aged group. In other words, the GRN at the aged stage can function in spite of the accumulation of DNA damage or genetic mutations accompanying the aging process. In addition, response ability (δ^o) of the GRN calculated from the aged group is lower than that from the young group (Table 10.3A). In order to tolerate and survive the accumulated intrinsic perturbations (e.g., genetic mutations) associated with the increase in age, the elderly GRN becomes more robust than the young GRN. Therefore, elderly GRNs are less able to respond to external stimuli or transduce them downstream. Hence, the response ability protecting organismal function against external stimuli (e.g., pro-inflammatory molecules) may become worse in the elderly GRN. To confirm the power of the proposed methods, the proposed network robustness and response ability is validated by computer simulation in Figs. A.1–A.3.

In addition, a more robust GRN could harbor more accumulated genetic mutations, which might provide more evolutionary paths to other GRN phenotypes via random drift, and even lead to aging-related diseases like cancers, metabolic disorders, and arthritis [34,35]. On the other hand, the young aging-related GRN is less robust, with a greater response to external stimuli. From the above data, we can conclude that the GRNs of the young group more efficiently respond to external stimuli. These values show that while the GRN of the young group is less robust to intrinsic network perturbations, external stimuli elicit a strong response from it, suggesting that gene expression may be readily reprogrammed to mediate downstream genes or regulators. This also allows for modulation of gene expression in response to external stimuli such as exposure to oxidative stress, carcinogens, and pro-inflammatory molecules.

Case 2: Evolution of the robustness and response ability of a PPIN from normal to cancer cells

Robustness, an intrinsic and systematic property of complex biological systems, enables living organisms to maintain their functions in the face of various network perturbations to assure their survival [5,12,36]. Moreover, most genes code for sensors, actuators, and the complex regulatory networks that control them, and thus confer robustness to the cell under various challenging circumstances rather than providing merely the basic functionality required for survival under a single, ideal environment. Normal cells, as well as cancer cells, utilize the same processes. At present, although researchers have contended that cancer is a highly robust disease, and clinical observations suggest that inefficiencies in cancer therapy may be due to the according drug resistance, a useful computational method to evaluate the network robustness from microarray data and compare it between cancer and normal cells remains lacking. Therefore, in the second case, the NRM method we have proposed is applied to the evolution of the PPIN robustness of HeLa cells versus normal fibroblasts under menadione treatment (i.e., exposed to oxidative stress) from corresponding microarray data and protein interaction information. The evolution of network response ability is also evaluated and discussed.

Constructing refined PPINs of normal fibroblast cells and HeLa cancer cells under menadione treatment

We use microarray data downloaded from the Gene Expression Omnibus (GEO) database on the NCBI website in this section to investigate the network robustness and response ability of a PPIN in normal cells versus cancer cells under oxidative stress [37]. Normal human diploid lung fibroblasts and HeLa S3 cervical carcinoma cells are treated with 10 μM menadione bisulfate for 0.5, 1, 3, 4, 6, 8, 12, 24, and 36 h and 0.5, 1, 2, 4, 8, 12, 24, and 32 h, respectively. For the oxidative stress condition, the temporal responses at several time points over an interval of 24–36 h are measured, allowing stress effects to be observed over an interval longer than a normal cell cycle duration.

As shown in the flowchart in Fig. 10.1, the microarray data are integrated with database information concerning protein interactions as the input for the proposed method (see Section 10.1.1). The candidate protein interactions are taken from all experimental conditions in these data to construct the rough PPIN. To proceed further, the corresponding microarray data are used to identify the parameters in

Eq. (10.7) via the recursive least-squares estimation method [17], and then to prune down the rough PPINs based on these identified parameters, thus obtaining the refined PPINs of both fibroblast and HeLa cells under menadione treatment via the AIC (shown in Fig. 10.4) [22]. After obtaining the refined PPINs, we can see there are still 161 candidate proteins with 371 interactions in the fibroblast PPIN and 222 candidate proteins with 607 interactions in the HeLa PPIN under oxidative stress conditions. Furthermore, the respective system interactive matrices R of the fibroblast and HeLa cell PPINs can be acquired and then applied to the sequential evaluation of network robustness by Eq. (10.10) and response ability by Eq. (10.15) with A being replaced by R.

Evolution of network robustness and response ability from the PPINs of normal fibroblasts to HeLa cells

We compare the PPINs of fibroblasts and HeLa cells while both are under an oxidative stress insult and show that the PPIN of HeLa cells possesses a higher robustness value η^{o} than the PPIN of fibroblasts using this NRM method. These results are shown in Table 10.3B. The PPIN is more robust to such network perturbation in the cancer cells than in the normal cells. Moreover, the response ability (δ^{o}) of the PPIN calculated from the HeLa cells is lower than that from the fibroblasts (Table 10.3B). Therefore, the PPIN of cancer cells is less efficient in transducing or responding to external stimuli when compared with that of normal cells. The analytic results support previous contentions that cancer cells are more robust than normal cells [38]. Accordingly, compelling evidence has previously suggested that the level of ROS is higher in many types of cancer cells than in their normal counterparts [39–41]. Although a moderate increase in ROS can promote cell proliferation and differentiation, excessive amounts can cause oxidative damage to DNA, protein, and lipids. However, during tumorigenesis or advanced malignant transformation, oncogenic signals can elevate cellular levels of ROS to the extent that numerous genes are altered, accumulating changes via both genetic and epigenetic mechanisms [42,43]. Therefore, cancer cells themselves actively generate high levels of endogenous ROS and are persistently exposed to them, stimulating cell proliferation through redox-sensitive TFs and promoting antioxidant adaptive mechanisms to minimize oxidative damage [44,45]. Therefore, the intrinsic oxidative stress generated during tumor initiation or malignant transformation may exert selective pressure on the cell population and enrich those capable of stress adaptation: i.e., cancer cells. These cells that survive oxidative stress may have adapted mechanisms to counteract the cytotoxic effects of increased ROS and to promote pathways essential to cellular survival [45]. All of these alterations confer cellular resistance and system robustness to subsequent intrinsic or extrinsic oxidative stress, even at high levels. In clinical cancer therapy, one chemotherapeutic strategy is to elevate excess cellular oxidative stress in order to deplete the antioxidant system and overwhelm the cancer system's ability to neutralize and eliminate such insults, inducing apoptosis and destroying the cancer

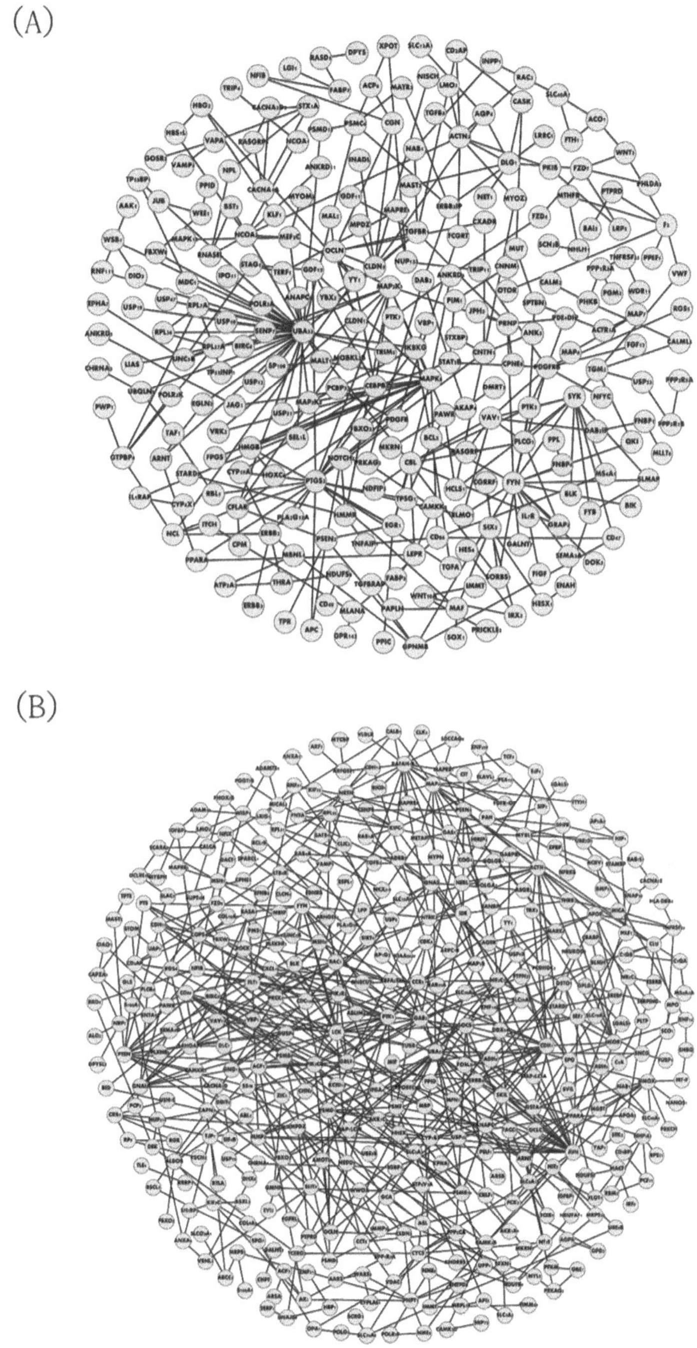

Figure 10.4 Refined protein–protein interaction network in fibroblast and HeLa cells. The figure shows the refined PPIN with effective protein interactions under oxidative stress in fibroblast (A) and HeLa cells (B). A dynamic PPI model and AIC are used together to prune the rough candidate PPIN using the time-series microarray data to delete unrealistic and false positive PPIs in the PPIN.

network system. However, the efficiency of such chemotherapy or radiotherapy is still unsatisfactory. Moreover, previous evidence has suggested that tumors are highly robust and maintain their proliferative potential against a wide range of anticancer therapies. Heterogeneities have been regarded as a vital property of cancer cells because they facilitate robustness through redundancy, and so subsystems killed by chemotherapy can be functionally replaced to ensure tumor proliferation and survival [38]. Hence, the notion of the PPIN of HeLa cells having higher robustness, as evaluated by the proposed NRM method, is consistent with previous contentions and experimental observations. Utilizing microarray data in system-level computations, we have sufficiently demonstrated that cancer cells possess better network robustness and less response ability in the carcinogenic process.

10.1.3 Conclusions of the evolutionary network robustness and response mechanisms in aging and cancer cells

Our proposed methods, NRM and RAM, consist of multiple analytical steps, from constructing dynamic GRNs and PPINs via microarray data to estimating network robustness and response ability to understand their evolutionary mechanism based on their dynamic network systems. However, each individual step can be employed separately, depending on the specific application in question. For example, in the first step we select a set of differentially expressed genes in aging (Case 1) or cancer (Case 2) as these target genes. However, another user could replace this set of genes with other sets of genes of interest to facilitate targeted studies over a wide range of issues. In addition, the proposed systematic methods can evaluate the evolutionary network robustness and response mechanisms not only for GRNs (Case 1) but also for PPINs (Case 2) depending on the needs of the user. The generalized NRM and RAM methods are effective tools for exploring and analyzing different systems-level evolutionary properties. Making use of such evolutionary network robustness and response mechanisms can facilitate biomedical predictions and applications, providing useful information for clinical strategy, drug target selection, and design specifications of synthetic biology from the perspective of systems biology [10,14–16].

APPENDIX A: METHODS AND FIGURES

Appendix A.1: Parameter identification of a GRN by microarray data

The focus of this section is whether the robustness and response ability of a GRN would be distinct and dependent on dynamic behavior under different biological conditions or stages. For this reason, the application of the proposed measurement method for robustness and response ability is highly dependent on the GRN dynamic model, which can be constructed from the time profiles of microarray data at these different stages. In general, for the parameter estimation method for the GRN dynamic model, the amount of microarray data for parameter estimation must be much (i.e., at least three times) larger than the number of parameters to be estimated [17], or it will lead to overfitting due to the data regression vector becoming singular in the parameter estimation process. That is, the overfitting due to the shortage of data points will lead to a large distortion in the parameter estimation of the GRN dynamic model. Therefore, we use the recursive least square estimation method in Eq. (2.21) [17] to estimate the interactive matrix A and basal level H in Eq. (10.1) from the time-series microarray data. Based on the least square parameter estimation in Eq. (2.21), let us denote $V(\hat{\theta}) = \sum_{t=1}^{M} (X(t+1) - \hat{A}X(t) - \hat{H})^T (X(t+1) - \hat{A}X(t) - \hat{H}) = \sum_{t=1}^{M} \varepsilon^T(t)\varepsilon(t)$, where $\hat{\theta} = \begin{bmatrix} \hat{A} \\ \hat{H} \end{bmatrix}$ denotes the estimate of parameters $\theta = \begin{bmatrix} A \\ H \end{bmatrix}$, and M is the amount of microarray data for parameter estimation. Using the recursive least square parameter estimation algorithm, we can estimate $\hat{\theta}$ from the microarray data. In order to improve the parameter estimate when only small amounts of microarray data are available, the recursive least square algorithm in Eq. (2.21) is performed iteratively for several rounds, with the estimated parameters of each round as the initial values of the next round, repeated until the estimated parameters converge. After the recursive least-squares parameter estimation of the final round, the covariance of estimation error $\varepsilon(t)$ is given by $Cov(\varepsilon) = \frac{2}{M-2} V(\hat{\theta})$ and the covariance matrix of $\hat{\theta}$ is given by $Cov(\hat{\theta}) = Cov(\varepsilon)\left[\sum_{t=1}^{M} X^T(t)X(t) + I\right]^{-1}$. These parameter estimation methods for constructing GRNs are termed "reverse engineering methods" in systems biology [33]. Based on the dynamic model described above and on microarray data, the system parameters A for GRNs can be estimated for use in NRM and RAM methods.

Appendix A.2: Network parameter identification of PPIN and pruning of false positive protein interactions

After constructing the discrete dynamic model (10.2) for all possible proteins in the rough PPIN, we use the least-squares estimation method to estimate the interaction parameters a_{ii}, b_{ij}, and k_i in Eq. (10.2) from time-series microarray data $x_i[t]$. Hence, we can rewrite Eq. (10.2) in the following regression form:

$$\omega_i[t+1] = [\omega_i[t]\omega_1[t]\ldots\omega_Q[t]\ 1] \cdot \begin{bmatrix} a_{ii} \\ b_{i1} \\ \vdots \\ b_{iQ} \\ k_i \end{bmatrix} \equiv \phi_i[t] \cdot \theta_i + v_i[t] \tag{A.1}$$

where Q denotes the number of possible interactions with the target protein and $\phi_i[t]$ stands for the regression vector obtained from time-series microarray data in the above process. θ_i is the parameter vector of the PPIN to be estimated.

By computing Eq. (A.1) at different time points, we can construct the following vector-form equation:

$$\begin{bmatrix} \omega_i[t_2] \\ \omega_i[t_3] \\ \vdots \\ \omega_i[t_M] \end{bmatrix} = \begin{bmatrix} \phi_i[t_1] \\ \phi_i[t_2] \\ \vdots \\ \phi_i[t_M - 1] \end{bmatrix} \theta_i + \begin{bmatrix} \nu_i[t_1] \\ \nu_i[t_2] \\ \vdots \\ \nu_i[t_M - 1] \end{bmatrix} \tag{A.2}$$

where M denotes the number of microarray data points. These data points are utilized as the basis of this regression vector $\phi_i[t]$. For simplicity, we further represent the above equation as follows:

$$W_i = \Phi_i \cdot \theta_i + V_i \tag{A.3}$$

We then use the recursive least square estimation method in Eq. (2.21) [17] to estimate θ_i using regression data acquired from microarray data of the ith target protein and its corresponding interacting proteins.

After obtaining the estimated parameter $\hat{\theta}_i$ by the recursive least square estimation method, the estimated PPI in Eq. (10.2) can be written as follows:

$$\omega_i[t+1] = \hat{a}_{ii}\omega_i[t] + \sum_{j=1}^{Q} \hat{b}_{ij}\omega_j[t] + \hat{k}_i \tag{A.4}$$

In Eq. (A.4), the parameter $\hat{a}_{ii}$ represents the estimation of the transition rate of the target protein, and $\hat{b}_{ij}$ denotes the individual interactive rate between protein j and the target protein i. A positive value implies positive interaction, while a negative value implies negative interaction, and both interactions become more likely as the parameters get larger.

After identifying the parameters of the dynamic model in Eq. (10.2) through several rounds of the recursive least-squares estimation method, the reality of the interaction is still unknown, as is whether it can be considered true. Therefore, in order to determine whether a PPI (i.e., $\hat{b}_{ij}$ in Eq. A.4) is realistic or simply a false positive in the rough PPIN, we propose a statistical approach based on model order detection for evaluating the reality of interaction parameters in the rough PPIN and pruning from it the false positive PPIs. Here, we employ AIC to validate the model order, or the number of PPI parameters of the network. The AIC, a method for model order selection, attempts to include both the estimated residual variance and model complexity in one statistic. AIC decreases as residual variance decreases, and increases as the number of parameters increases. As the expected residual variance decreases with increasing parameter numbers for excessive model complexity, a minimum should appear near the correct parameter number [17,22]. For a protein interaction dynamic model with Q interaction parameters to fit with data from N samples, the AIC in Eq. (2.23) can be rewritten as follows:

$$AIC(Q) = \log\left(\frac{1}{N}(W_i - \hat{W}_i)^T(W_i - \hat{W}_i)\right) + \frac{2Q}{N} \tag{A.5}$$

where $\hat{W}_i$ denotes the estimated expression profile of the ith target protein: i.e., $\hat{W}_i = \Phi_i \cdot \hat{\theta}_i$. After the statistical selection of Q interaction parameters derived from minimizing the AIC, we can determine whether the PPI is realistic or false positive for target protein i. After a number test of interaction parameters in $\hat{\theta}_i$ by using the AIC system order test for all target proteins, we can construct a realistic dynamic PPIN in Eq. (10.3) from microarray data.

Appendix A.3: Proof of proposition 10.1

Let us denote the Lyapunov (energy) function of perturbative GRN in Eq. (10.8) as $V(\tilde{X}[t]) = \tilde{X}^T[t]P\tilde{X}[t]$, for some symmetric positive definite matrix $P = P^T > 0$.

$$\begin{aligned}\Delta V(\tilde{X}[t]) &= V(\tilde{X}[t+1]) - V(\tilde{X}[t]) \\ &= \tilde{X}^T[t+1]P\tilde{X}[t+1] - \tilde{X}^T[t]P\tilde{X}[t] \\ &= \tilde{X}^T[t][A(1+\eta)]^T P[A(1+\eta)]\tilde{X}[t] - \tilde{X}^T[t]P\tilde{X}[t] \\ &= \tilde{X}^T[t]\{[A(1+\eta)]^T P[A(1+\eta)] - P\}\tilde{X}[t] \\ &\text{If } [A(1+\eta)]^T P[A(1+\eta)] - P \le 0, \text{ then } \Delta V(\tilde{X}[t]) \le 0\end{aligned}$$

Based on quadratic stability [24,25], if the LMI in Eq. (10.9) holds, then the perturbative network is robustly stable.

Appendix A.4: Proof of proposition 10.2

Let us denote the Lyapunov function of GRN under external stimulation signals in Eq. (10.11) as $V(\tilde{X}[t]) = \tilde{X}^T[t]P\tilde{X}[t]$, for some symmetric positive definite matrix $P = P^T > 0$. We have

$$\begin{aligned}V(\tilde{X}[t+1]) &= \tilde{X}^T[t+1]P\tilde{X}[t+1] \\ &= \{A\tilde{X}[t] + U[t]\}^T P\{A\tilde{X}[t] + U[t]\} \\ &= \tilde{X}^T[t]A^T PA\tilde{X}[t] + \tilde{X}^T[t]A^T PU[t] + U^T[t]PA\tilde{X}[t] + U^T[t]PU[t] \\ &= \tilde{X}^T[t]A^T PA\tilde{X}[t] + \tilde{X}^T[t]A^T PU[t] + U^T[t]PA\tilde{X}[t] + U^T[t]PU[t] \\ &\quad + Y^T[t]Y[t] - \delta^2 U^T[t]U[t] - Y^T[t]Y[t] + \delta^2 U^T[t]U[t] - V(\tilde{X}[t]) + V(\tilde{X}[t]) \\ &= \tilde{X}^T[t]A^T PA\tilde{X}[t] + \tilde{X}^T[t]A^T PU[t] + U^T[t]PA\tilde{X}[t] + U^T[t]PU[t] \\ &\quad + \tilde{X}^T[t]C^T C\tilde{X}[t] - \delta^2 U^T[t]U[t] - \tilde{X}^T[t]P\tilde{X}[t] + V(\tilde{X}[t]) - Y^T[t]Y[t] \\ &\quad + \delta^2 U^T[t]U[t]\end{aligned}$$

If the following inequality holds:

$$\begin{aligned}&\tilde{X}^T[t]A^TPA\tilde{X}[t] + \tilde{X}^T[t]A^TPU[t] + U^T[t]PA\tilde{X}[t] + U^T[t]PU[t]\\&+ \tilde{X}^T[t]C^TC\tilde{X}[t] - \delta^2U^T[t]U[t] - \tilde{X}^T[t]P\tilde{X}[t] \le 0\end{aligned} \tag{A.6}$$

then, we get $V(\tilde{X}[t+1]) \le V(\tilde{X}[t]) - Y^T[t]Y[t] + \delta^2U^T[t]U[t]$.

So we sum the above inequality from 0 to ∞ to get

$$\begin{aligned}&V(\tilde{X}[\infty]) \le V(\tilde{X}[0]) - \sum_{t=0}^{\infty}Y^T[t]Y[t] + \delta^2\sum_{t=0}^{\infty}U^T[t]U[t]\\&\Rightarrow V(\tilde{X}[\infty]) + \sum_{t=0}^{\infty}Y^T[t]Y[t] \le V(\tilde{X}[0]) + \delta^2\sum_{t=0}^{\infty}U^T[t]U[t]\\&\Rightarrow \sum_{t=0}^{\infty}Y^T[t]Y[t] \le V(\tilde{X}[\infty]) + \sum_{t=0}^{\infty}Y^T[t]Y[t] \le V(\tilde{X}[0]) + \delta^2\sum_{t=0}^{\infty}U^T[t]U[t]\end{aligned} \tag{A.7}$$

$$\Rightarrow \sum_{t=0}^{\infty}Y^T[t]Y[t] \le \delta^2\sum_{t=0}^{\infty}U^T[t]U[t] \text{ or } \frac{\sum_{t=0}^{\infty}Y^T[t]Y[t]}{\sum_{t=0}^{\infty}U^T[t]U[t]} \le \delta^2,$$ if $V(\tilde{X}[0]) = 0$ in the case $\tilde{X}[0] = 0$.

If the $V(\tilde{X}[0]) \ne 0$ in the case $\tilde{X}[0] \ne 0$, then the inequality in Eq. (A.7) should be modified as:

$$\sum_{t=0}^{\infty}Y^T[t]Y[t] \le \tilde{X}^T[0]P\tilde{X}[0] + \delta^2\sum_{t=0}^{\infty}U^T[t]U[t] \tag{A.8}$$

The above results hold only when the inequality Eq. (A.6) holds, i.e., the following inequality holds:

$$\begin{aligned}&\tilde{X}^T[t]A^TPA\tilde{X}[t] + \tilde{X}^T[t]A^TPU[t] + U^T[t]PA\tilde{X}[t] + U^T[t]PU[t]\\&+ \tilde{X}^T[t]C^TC\tilde{X}[t] - \delta^2U^T[t]U[t] - \tilde{X}^T[t]P\tilde{X}[t] \le 0\end{aligned}$$

The above inequality can be equivalent to the following inequality:

$$\begin{bmatrix}\tilde{X}\\U\end{bmatrix}^T\begin{bmatrix}A^TPA + C^TC - P & A^TP\\PA & P - \delta^2I\end{bmatrix}\begin{bmatrix}\tilde{X}\\U\end{bmatrix} \le 0 \tag{A.9}$$

The inequality Eq. (A.9) holds for all $\tilde{X}[t]$ and $U[t]$ if and only if the following LMI holds:

$$\begin{bmatrix}A^TPA + C^TC - P & A^TP\\PA & P - \delta^2I\end{bmatrix} \le 0$$

Appendix A.5: Proof of proposition 10.3

Following the proof in Appendix A.3, let us choose Lyapunov function as $V(\tilde{X}[t]) = \tilde{X}^T[t]P\tilde{X}[t]$ for some symmetric positive definite matrix $P = P^T > 0$. By the fact Eqs. (10.21) and (10.22), we get

$$
\begin{aligned}
&\Delta V(\tilde{X}[t+1]) \\
&\quad = V(\tilde{X}[t+1]) - V(\tilde{X}[t]) \\
&\quad = \tilde{X}^T P\tilde{X}[t+1] - \tilde{X}^T[t]P\tilde{X}[t] \\
&\quad = \left(\sum_{i=1}^{m} h_i(\tilde{X})[A_i(1+\eta)]\tilde{X}[t]\right)^T P\left(\sum_{j=1}^{m} h_j(\tilde{X})[A_j(1+\eta)]\tilde{X}[t]\right) - \tilde{X}^T[t]P\tilde{X}[t] \\
&\quad = \sum_{i=1}^{m}\sum_{j=1}^{m} h_i(\tilde{X})h_j(\tilde{X})\tilde{X}^T[t][A_i(1+\eta)]^T P[A_j(1+\eta)]\tilde{X}[t] - \tilde{X}^T[t]P\tilde{X}[t] \\
&\leq \sum_{i=1}^{m}\sum_{j=1}^{m} h_i(\tilde{X})h_j(\tilde{X})\tilde{X}^T[t]\left(\frac{1}{2}[A_i(1+\eta)]^T P[A_i(1+\eta)] + \frac{1}{2}[A_j(1+\eta)]^T P\left[A_j(1+\eta)\right]\right)\tilde{X}[t] \\
&\quad - \tilde{X}^T[t]P\tilde{X}[t] \\
&= \frac{1}{2}\sum_{i=1}^{m} h_i(\tilde{X})\tilde{X}^T[t][A_i(1+\eta)]^T P[A_i(1+\eta)]\tilde{X}[t] \\
&\quad + \frac{1}{2}\sum_{j=1}^{m} h_j(\tilde{X})\tilde{X}^T[t][A_j(1+\eta)]^T P\left[A_j(1+\eta)\right]\tilde{X}[t] - \tilde{X}^T[t]P\tilde{X}[t] \\
&= \sum_{i=1}^{m} h_i(\tilde{X})\tilde{X}^T[t][A_i(1+\eta)]^T P[A_i(1+\eta)]\tilde{X}[t] - \tilde{X}^T[t]P\tilde{X}[t] \\
&= \sum_{i=1}^{m} h_i(\tilde{X})\tilde{X}^T[t]\left([A_i(1+\eta)]^T P[A_i(1+\eta)] - P\right)\tilde{X}[t]
\end{aligned}
$$

If $[A_i(1+\eta)]^T P[A_i(1+\eta)] - P \leq 0$, then $\Delta V(\tilde{X}[t]) \leq 0$, for $i = 1,2,\ldots m$

Based on quadratic stability [24,25], if the LMI in Eq. (10.23) holds, then the perturbative nonlinear network system is robustly stable.

Appendix A.6: Proof of proposition 10.4

Following the proof in Appendix A.4, we choose Lyapunov $V(\tilde{X}[t]) = \tilde{X}^T[t]P\tilde{X}[t]$, for some symmetric positive definite matrix $P = P^T > 0$. We have

$$
\begin{aligned}
&V(\tilde{X}[t+1])\\
&= \tilde{X}^T[t+1]P\tilde{X}[t+1]\\
&= \left\{\sum_{i=1}^{m} h_i(\tilde{X})A_i\tilde{X}[t] + U[t]\right\}^T P\left\{\sum_{j=1}^{m} h_j(\tilde{X})A_j\tilde{X}[t] + U[t]\right\}\\
&= \sum_{i=1}^{m}\sum_{j=1}^{m} h_i(\tilde{X})h_j(\tilde{X})[A_i\tilde{X}[t] + U[t]]^T P[A_j\tilde{X}[t] + U[t]]\\
&\le \sum_{i=1}^{m}\sum_{j=1}^{m} h_i(\tilde{X})h_j(\tilde{X})\left\{\frac{1}{2}[A_i\tilde{X}[t] + U[t]]^T P\left[A_i\tilde{X}[t] + U[t]\right]\right.\\
&\qquad\left. + \frac{1}{2}[A_j\tilde{X}[t] + U[t]]^T P\left[A_j\tilde{X}[t] + U[t]\right]\right\}\\
&= \frac{1}{2}\sum_{i=1}^{m} h_i(\tilde{X})[A_i\tilde{X}[t] + U[t]]^T P\left[A_i\tilde{X}[t] + U[t]\right] + \frac{1}{2}\sum_{j=1}^{m} h_j(\tilde{X})[A_j\tilde{X}[t] + U[t]]^T P\left[A_j\tilde{X}[t] + U[t]\right]\\
&= \sum_{i=1}^{m} h_i(\tilde{X})[A_i\tilde{X}[t] + U[t]]^T P[A_i\tilde{X}[t] + U[t]]\\
&= \sum_{i=1}^{m} h_i(\tilde{X})\left\{\tilde{X}^T[t]A_i^T PA_i\tilde{X}[t] + \tilde{X}^T[t]A_i^T PU[t] + U^T[t]PA_i\tilde{X}[t] + U^T[t]PU[t]\right\}\\
&= \sum_{i=1}^{m} h_i(\tilde{X})\left\{\begin{aligned}&\tilde{X}^T[t]A_i^T PA_i\tilde{X}[t] + \tilde{X}^T[t]A_i^T PU[t] + U^T[t]PA_i\tilde{X}[t] + U^T[t]PU[t]\\ &+ Y^T[t]Y[t] - \delta^2 U^T[t]U[t] - Y^T[t]Y[t] + \delta^2 U^T[t]U[t] - V(\tilde{X}[t]) + V(\tilde{X}[t])\end{aligned}\right\}\\
&= \sum_{i=1}^{m} h_i(\tilde{X})\left\{\begin{aligned}&\tilde{X}^T[t]A_i^T PA_i\tilde{X}[t] + \tilde{X}^T[t]A_i^T PU[t] + U^T[t]PA_i\tilde{X}[t] + U^T[t]PU[t]\\ &+ \tilde{X}^T[t]C^T C\tilde{X}[t] - \delta^2 U^T[t]U[t] - \tilde{X}^T[t]P\tilde{X}[t] + V(\tilde{X}[t]) - Y^T[t]Y[t]\\ &+ \delta^2 U^T[t]U[t]\end{aligned}\right\}
\end{aligned}
$$

If the following inequality holds:

$$\begin{aligned}&\tilde{X}^T[t]A_i^TPA_i\tilde{X}[t]+\tilde{X}^T[t]A_i^TPU[t]+U^T[t]PA_i\tilde{X}[t]+U^T[t]PU[t]\\&+\tilde{X}^T[t]C^TC\tilde{X}[t]-\delta^2U^T[t]U[t]-\tilde{X}^T[t]P\tilde{X}[t]\le 0\end{aligned} \tag{A.10}$$

then, we get $V(\tilde{X}[t+1])\le V(\tilde{X}[t])-Y^T[t]Y[t]+\delta^2U^T[t]U[t]$.

So we sum the above inequality from 0 to ∞ to get

$$\begin{aligned}&V(\tilde{X}[\infty])\le V(\tilde{X}[0])-\sum_{t=0}^{\infty}Y^T[t]Y[t]+\delta^2\sum_{t=0}^{\infty}U^T[t]U[t]\\&\Rightarrow V(\tilde{X}[\infty])+\sum_{t=0}^{\infty}Y^T[t]Y[t]\le V(\tilde{X}[0])+\delta^2\sum_{t=0}^{\infty}U^T[t]U[t]\\&\Rightarrow\sum_{t=0}^{\infty}Y^T[t]Y[t]\le V(\tilde{X}[\infty])+\sum_{t=0}^{\infty}Y^T[t]Y[t]\le V(\tilde{X}[0])+\delta^2\sum_{t=0}^{\infty}U^T[t]U[t]\end{aligned} \tag{A.11}$$

$$\Rightarrow\sum_{t=0}^{\infty}Y^T[t]Y[t]\le\delta^2\sum_{t=0}^{\infty}U^T[t]U[t]\ \text{or}\ \frac{\sum_{t=0}^{\infty}Y^T[t]Y[t]}{\sum_{t=0}^{\infty}U^T[t]U[t]}\le\delta^2,\quad \text{if}\ V(\tilde{X}[0])=0\ \text{in the}$$

case $\tilde{X}[0]=0$

If the $V(\tilde{X}[0])\neq 0$ in the case $\tilde{X}[0]\neq 0$, then the inequality in Eq. (A.11) should be modified as

$$\sum_{t=0}^{\infty}Y^T[t]Y[t]\le\tilde{X}^T[0]P\tilde{X}[0]+\delta^2\sum_{t=0}^{\infty}U^T[t]U[t] \tag{A.12}$$

The above results hold only when the inequality Eq. (A.10) holds, i.e., the following inequality holds:

$$\begin{aligned}&\tilde{X}^T[t]A_i^TPA_i\tilde{X}[t]+\tilde{X}^T[t]A_i^TPU[t]+U^T[t]PA_i\tilde{X}[t]+U^T[t]PU[t]\\&+\tilde{X}^T[t]C^TC\tilde{X}[t]-\delta^2U^T[t]U[t]-\tilde{X}^T[t]P\tilde{X}[t]\le 0\end{aligned}$$

The above inequality is equivalent to the following inequality

$$\begin{bmatrix}\tilde{X}\\U\end{bmatrix}^T\begin{bmatrix}A_i^TPA_i+C^TC-P & A_i^TP\\ PA_i & P-\delta^2I\end{bmatrix}\begin{bmatrix}\tilde{X}\\U\end{bmatrix}\le 0 \tag{A.13}$$

The inequality Eq. (A.13) for all $\tilde{X}[t]$ and $U[t]$ if and only if the following LMIs hold:

$$\begin{bmatrix}A_i^TPA_i+C^TC-P & A_i^TP\\ PA_i & P-\delta^2I\end{bmatrix}\le 0,\quad \text{for } i=1,\ldots,m$$

Appendix A.7: Computer simulation of the evolution of network robustness and response ability in the aging process

In order to confirm the results of proposed approaches, two simulations are given in illustration, one for the confirmation of network robustness and another for the confirmation of network response ability as follows:

Simulation 10.1: From the network robustness in Table 10.3, the network robustness of aging-related GRN in thymus tissue is $\eta^{o} = 0.2233$ at the young stage and is $\eta^{o} = 0.3852$ at the aged stage, i.e., the maximum intrinsic perturbations η in Eq. (10.13) could be tolerated by GRNs in young thymus and aged thymus are 0.2233 and 0.3852, respectively. Therefore, according to the proposed network robustness, if an intrinsic perturbation $\eta = 0.3$, it could not be tolerated by GRN in young thymus, but could be tolerated by GRN in aged thymus. In the simulation results in Figs. A.1 and A.2, the gene expressions of two GRNs without intrinsic perturbation are shown in Figs. A.1A and A.2A, respectively. However, under the intrinsic perturbation $\eta = 0.3$, it is seen that the some gene expressions of GRN in young thymus become divergent (i.e., Sirt1, Mdm2, Akt, and ATM in Fig. A.1B) while all gene expressions of GRN in aged thymus are still robust in Fig. A.2B. These simulation results obviously confirm the proposed network robustness of aging-related GRNs by the proposed method.

Simulation 10.2: Similarly, from the evolution of network response ability results in Table 10.3, it is seen that the network response ability of young thymus is larger than that of aged thymus. In order to confirm these network evolution results, the response abilities to an unit step signal for 16 genes of GRNs in young and aged thymus are obtained in Fig. A.3A based on computer simulation, respectively. In general, the gene response abilities of most genes in young thymus are larger than aged thymus. From Fig. A.3B, it is seen that the network response ability of GRN in young thymus is larger than that in aged thymus. These simulation results also confirm the evolution of network response ability of aging-related GRNs.

Here we show Figs. A.1–A.3.

10.2 EVOLUTION OF SIGNAL TRANSDUCTIVITIES OF COUPLED SIGNAL PATHWAYS IN THE CARCINOGENIC PROCESS

Recently, biologists found that carcinogenesis are mainly caused by accumulated dysfunctions in STPs due to genetic mutations in somatic cells. After investigating the evolution of network robustness and response ability in cancer and aging process, we will investigate the evolution of the transductivity of STPs to get more insight into the carcinogenesis of leukemia by microarray data in this section.

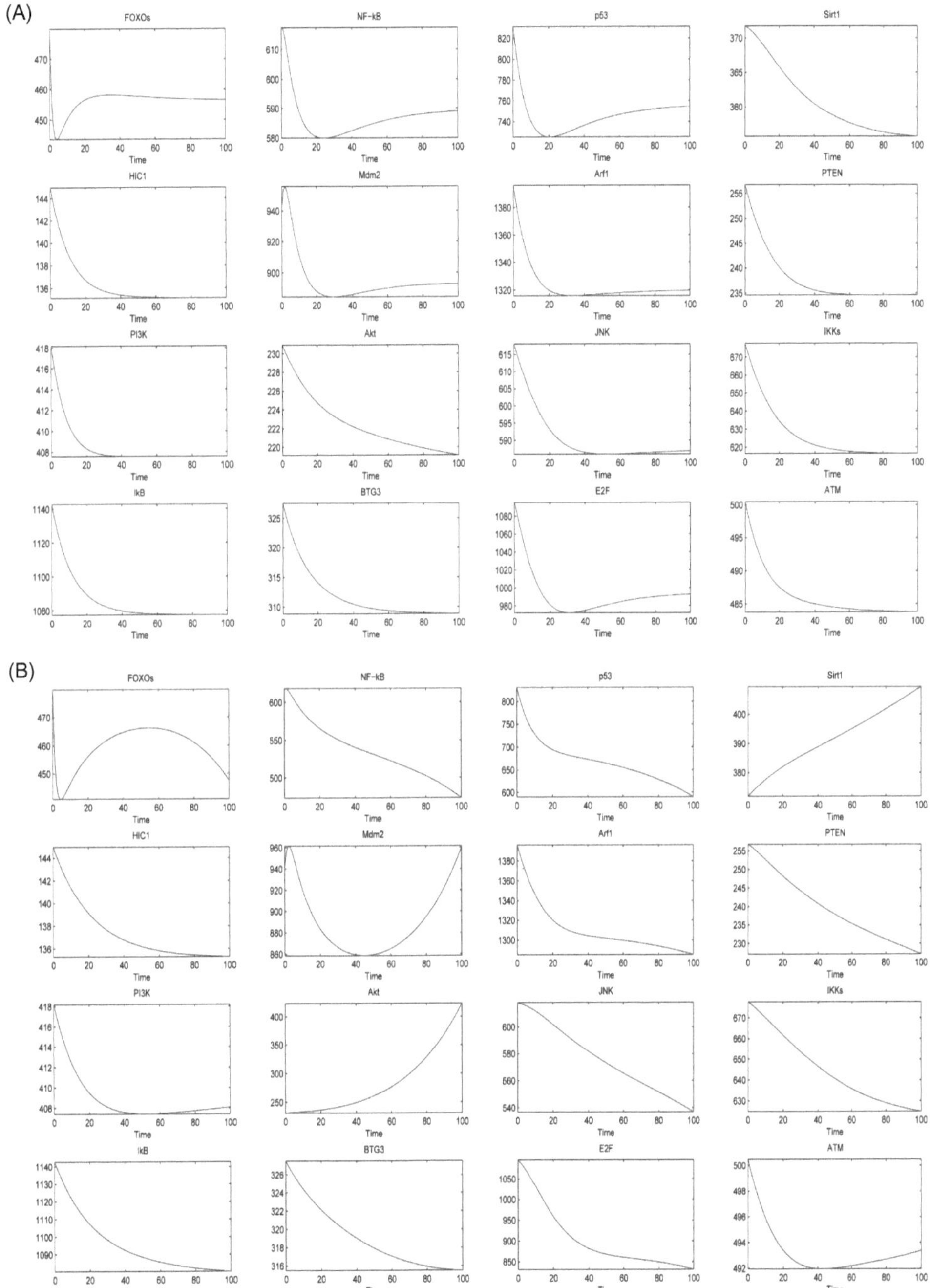

Figure A.1 Gene expressions of aging-related GRN in thymus at young stage without intrinsic perturbation in (A) and under intrinsic network perturbation $\eta = 0.3 > \eta^{o} = 0.2233$ in (B).

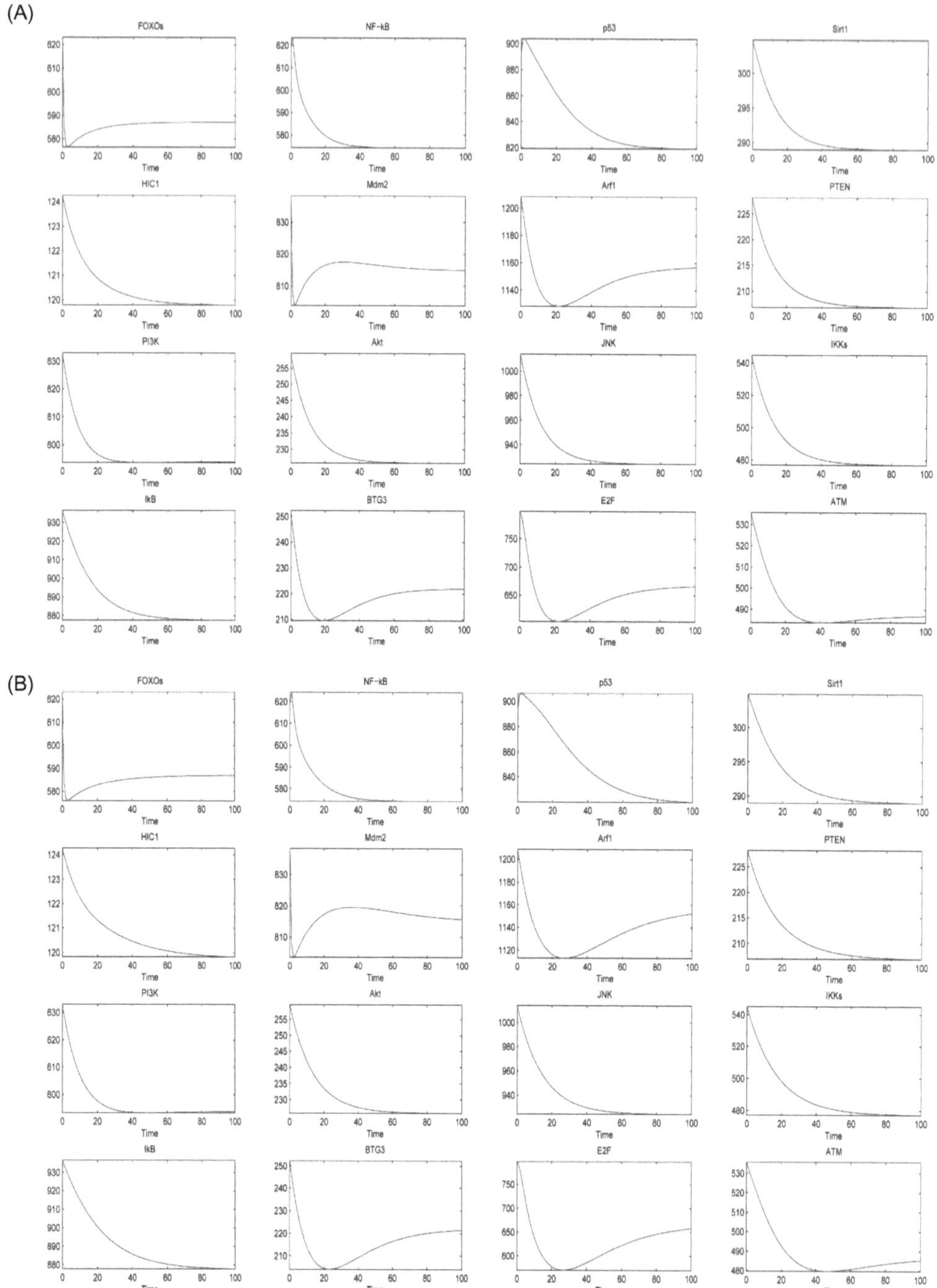

Figure A.2 Gene expressions of aging-related GRN in thymus at aged stage without intrinsic network perturbation in (A) and under intrinsic perturbation $\eta = 0.3 < \eta^{\circ} = 0.3852$ in (B).

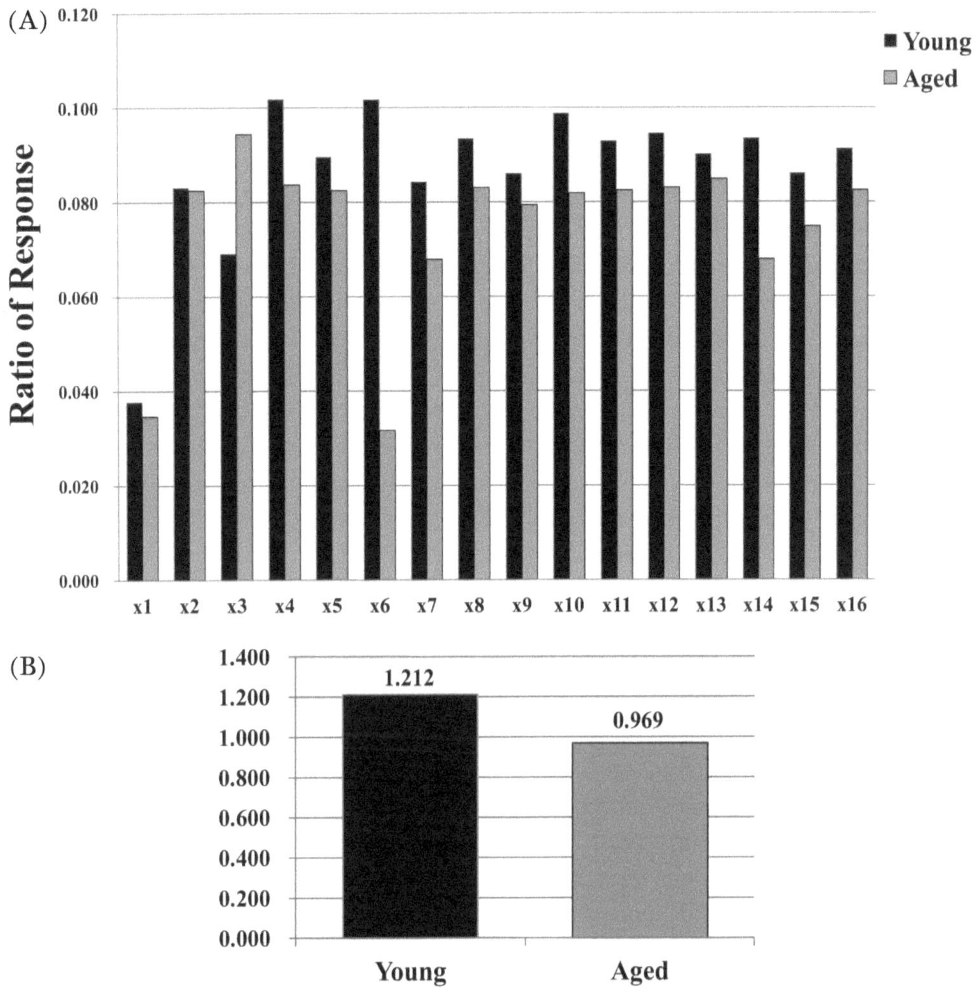

Figure A.3 Evolution of gene and network response ability under external input of unit step signal. (A) Gene response abilities of 16 genes in GRNs of young and aged thymus, respectively. Obviously, the response abilities of most genes of young thymus are larger than those of aged thymus. (B) Network response abilities in GRNs of young and aged thymus, respectively. Obviously, the network response ability of young thymus is larger than the aged thymus. The response abilities in this simulation are larger than the estimated response abilities in Table 10.3 because the estimated response ability δ^o in Eq. (10.15) is the lower bound value for all possible input signals. However, the simulated evolution results in this case is only based on the unit step input signal and therefore cannot achieve their lower bounds.

Mutation accumulation in oncogenes and tumor-suppressor genes is the main cause of cancer. Because differential expression of these genes is primarily observed in STPs, the signal transduction ability or signal transductivity of these pathways in cancer cells should be investigated systematically by obtaining microarray data and then by comparing these data with those resulting from noncancer cells. Systems analysis of microarray data from a population of samples in leukemic and nonleukemic cells showed that the transductivity of TFs, signal transducer and activator of transcriptions (STAT) was enhanced while transductivities of other TFs, such as p53, C/EBPα, and CSLs, were attenuated in leukemic cells. Further, the transductivity evolution levels of proteins in coupling STPs of leukemic and nonleukemic cells highlighted their roles in leukemogenesis. Transductivity analysis showed that dysfunctions in TFs mutations in genes encoding proteins such as DVL, PSE2, and ERK and crosstalk between multiple STPs were responsible for leukemogenesis. Most importantly, transductivity analysis identified proteins responsible for progression of myelodysplastic syndrome (MDS) to AML and other leukemic subtypes. Because TFs that act downstream of STPs play important roles in regulating the genes responsible for various cellular functions, the network robustness of GRNs involving these TFs was estimated using microarray sample data from leukemic cells and was compared with the robustness of GRNs in nonleukemic cells to determine the roles of GRNs in leukemogenesis. We observed that, compared with the GRNs of nonleukemic cells, the GRNs of leukemic cells, particularly AML cells, were more robust and could tolerate a greater number of intrinsic mutations, resulting in evolutionary heterogeneities.

Leukemias such as AML and chronic lymphocytic leukemia (CLL) are hematological cancers that are classified based on the proliferation rate of abnormal blood cells and their origin. Gene profiling analyses have shown that leukemic stem cells (LSCs) possess all the fundamental properties of hematopoietic stem cells (HSCs), such as quiescence, self-renewal, and differentiation [46]. The self-renewal property of LSCs allows them to escape differentiation, maintain the diseased state, and resist chemotherapy. Moreover, abnormal blood cells that do not undergo differentiation show aberrant proliferation and survival and replace normal blood cells, leading to the dysregulation of essential mechanisms leukemogenesis in the body [47].

In nonleukemic cells, STPs perceive and transduce extracellular signals through inter- and intracellular communication and influence TFs to perform specific functions. A delicate balance between apoptosis, proliferation, and cell survival is essential for proper functioning of the body. According to the 2-hit hypothesis, leukemogenesis is a multistep process induced by at least 2 classes of mutations that cooperate with each other [48]. One type of mutation deregulates STPs involved in the proliferation of hematopoietic cells and the other type of mutation affects TFs that are primarily involved in hematopoietic differentiation. It has been proposed that mutations resulting from chromatin translocation can occur in receptors, upstream proteins, or TFs involved in STPs, and may dysregulate STPs, thus affecting various proteins or TFs through crosstalk among these pathways. These mutations play a critical role in oncogenesis and leukemogenesis [46,48,49]. For example, mutations can occur in genes encoding cytokines, growth factors, and interleukins that trigger various STPs, including the JAK/STAT, Raf/MEK/ERK, NF-κB, PI3 kinase/PTEN/Akt/mTOR pathways. Leukemic cells are insensitive to apoptotic pathways induced by tumor necrosis factor (TNF) signaling [50,51]. Further, the self-renewal property of LSCs is coordinated by signals from self-renewal pathways, including PI3 kinase, Wnt, Notch, hedgehog, and transforming growth factor beta (TGFβ) superfamily signaling pathways [46,47,52].

Multiple signaling cascades are triggered in response to different molecular signals that ultimately influence downstream TFs through phosphorylation, ubiquitylation, and acetylation. Because TFs move between the cell cytoplasm and nucleus, corresponding cellular processes such as apoptosis, differentiation, growth arrest, DNA repair, and stress resistance are coordinated by transcriptional regulation [53]. In addition, pluripotency and self-renewal of stem cells are controlled by transcriptional regulation [54,55]. Thus, transcriptional regulation plays a crucial role in the fate of leukemic cells, including the minor population of LSCs, during myelopoiesis and leukemogenesis. Many studies have proposed that dysregulated transcription of various genes is responsible for the pathogenesis of leukemia [46,50,56–59]. For example, signal transducer and activator of transcription 3 (STAT3) transcriptionally regulates myeloid cell leukemia-1 (Mcl-1) and promotes aberrant cell proliferation [59]. Mutation in the gene encoding CCAATT/enhancer-binding protein alpha (C/EBPα) dysregulates cell differentiation, leading to leukemia [57]. In addition, a subclass of forkhead family of TFs (FOXO3) is inhibited by a dysregulated PI3 kinase/PTEN/Akt/mTOR pathway, which in turn prevents apoptosis [50]. Inappropriate NOTCH activation induces CSLs and NOTCH-dependent genes to maintain leukemic cell growth [46,56]. Therefore, construction of a GRN and investigation of its robustness allows identification of the ability of the GRN to tolerate intrinsic perturbations such as mutations or deregulation by other TFs under normal and leukemic conditions.

STPs are complex information processors because kinases and other proteins transduce molecular signals through these STPs. STPs transmit external molecular signals such as hormones, pheromones, and environmental stress through different signaling molecules with different transduction abilities to regulate the activities of downstream TFs. From a systematic viewpoint, transduction ability (termed "transductivity" in this section) of a biological network is a measure of the input/output signal energy ratio of a TF in response to external stimuli. During information processing, signals transduced by proteins in STPs are either amplified or attenuated. From the perspective of energy, transductivity is viewed as a system gain of signals in STPs. On a molecular level, information transmitted by signaling molecules is amplified or attenuated in STPs through enzyme catalysis, molecular modification, and activation or inhibition. In general, most genetic mutations in cancer cells occur in response to signals from STPs that lead to mechanism change in signal transductivity. Transductivity of a specific TF or protein varies in leukemic and nonleukemic cells under different environmental conditions to maintain specific functions and mechanisms. Therefore, investigation of transductivities of different proteins in the STPs of leukemic and nonleukemic cells is important from the viewpoint of systems biology. The robustness of

GRNs to intrinsic variations and transductivity of molecular signals are ubiquitous systematic mechanisms of biological networks to maintain their phenotypes. Therefore, it is important to understand how proteins transduce a signal and regulate the activities of downstream TFs by using a systems biology approach involving system theory and models of STPs and GRNs rather than by performing case studies.

Many studies have focused on leukemogenesis by examining PPIs and network information. For example, Lin et al. performed yeast 2-hybrid screening and Gal4-dependent transcription assay to identify the functions of and interactions between promyelocytic leukemia zinc finger protein and zinc finger protein 179 by determining their transcriptional activity and cellular localization [60]. Patel et al. used co-immunoprecipitation and antibody assays to identify new binding partners of $p210^{BCR-ABL1}$ that play a major role in chronic myeloid leukemia [61]. Majetl et al. proposed a detailed annotated pathway involving 3005 differentially expressed genes identified by comparing gene expression profiles of normal HSCs and LSCs from patients with AML [62]. Zhang et al. constructed an asynchronous Boolean model to systematically analyze long-term survival by studying the activation of normal cytotoxic T lymphocytes (CTL) and deregulation of survival signaling in leukemic T-cell large granular lymphocytes (T-LGL) [51]. They predicted key mediators and pathways associated with survival and found that IL-15 and PDGF were present in all the pathways and mediated other signaling abnormalities. Saadatpour et al. used dynamic and structural perturbation analyses along with the asynchronous Boolean model and survival signaling pathway proposed by Zhang et al. to show the hyperactivity of SMAD and identify 19 potential therapeutic targets, of which 68% were verified in T-LGL leukemia [51,63]. Both these studies identified new therapeutic targets by producing perturbations on nodes (proteins) and determining stationary states (fixed points) through the asynchronous Boolean model. Although this method computes rapidly, it generally does not provide precise information (either 0 or 1) for smaller signaling networks and interactions between proteins. Cheng et al. measured the gene expression profiles of F3-treated THP-1 cells to construct plausible GRNs [64]. The GRNs were estimated using BSIP, a web server for identifying parameters based on time-course microarray analysis, QPCR assays, and assumption of S-system mathematical model. They studied and identified GRNs through which F3 extracted from *Ganoderma lucidum* polysaccharides induced apoptosis in human leukemic THP-1 cells.

The robustness mechanism of GRNs is their systematic ability to tolerate intrinsic network variations such as genetic mutations and to withstand environmental disturbances to maintain their phenotype. Transductivity is a measure of the protein sensitivities of STPs in response to external stimuli such as environmental stress. Recent studies have discussed the robustness and transductivities of biological networks based on dynamic system models [65–67] and have measured these for genetic and protein networks by using time-course microarray analysis. Chu and Chen used both time-course microarray analysis and linear discrete stochastic model under various stress conditions to measure and compare the robustness and sensitivity of feedback loops of *p53* by employing the system control theory [66]. They found that cancer is a robust but fragile system that resists apoptosis but is sensitive to cell cycle progression and metastasis. This result may be used to design and develop robustness-based anticancer drugs. Chen and Wu used time-course microarray data with a nonlinear signal transduction system to describe the dynamic behavior of STPs [65]. They proposed a measurement to quantify the transductivity of each molecule to transduce signals from the system gain viewpoint by solving a LMI-constrained optimization problem. The characteristics of signal transduction were determined using two small STPs. The measurements obtained were viewed as useful tools to characterize and design a synthetic signaling pathway. Tu and Chen observed and compared the gene response ability and robustness of aging-related GRNs by using a discrete-time linear dynamical system in the previous section [67]. They used a quantitative and systematic method to show that pathophysiological phenotypes increased network robustness and decreased network response ability with an increase in aging process.

Large amounts of single time point microarray data are available from various populations. These microarray sample data are valuable for analyzing network robustness and transductivity to obtain systematic insights into genetic networks from a systems biology viewpoint. Use of the data to measure of the robustness and transductivities of GRNs and STPs, respectively, through their system models is difficult by employing conventional methods. Therefore, more efforts are needed to measure network robustness and transductivity by using these microarray data. The present study aimed to develop new measurements of the network robustness and transductivities of GRNs and multiple STPs, respectively, by employing linear regression model and patient microarray sample data from a systems gain perspective. Analysis of the robustness of GRNs showed that leukemic cells, particularly AML and MDS cells, showed better evolution because of their higher robustness to harbor intrinsic mutations than nonleukemic cells. Transductivity mechanism analysis of multiple coupling STPs in leukemic and nonleukemic cells was performed to investigate the pathophysiological phenotypes contributing to leukemogenesis. Transductivity mechanism analysis determined the cellular functions of some TFs contributing to the development of AML and MDS. Comparison of apoptotic and differentiation phenotypes between leukemic and nonleukemic cells could be used to determine the main cause of the progression of MDS to AML. Transductivity mechanism analysis also highlighted the role of crosstalk between multiple STPs in AML and MDS cells. We found that in AML cells, signals from the PI3K-AKT pathway were attenuated by RAS-mediated crosstalk between the PI3K-AKT and classical MAPK pathways. This decreased the antiapoptotic effects of the PI3K-AKT pathway in AML cells. Moreover, the effects of the Wnt pathway were attenuated by GSK3β-mediated crosstalk between the PI3K-AKT and canonical Wnt pathways, particularly in MDS cells. This decreased the difference in accumulated β-catenin between MDS and nonleukemic cells compared with that between AML and nonleukemic cells. We speculated that signals from the crosstalk blocked GSK3β in

MDS cells. In addition, we predicted some mutations that dysregulate STPs and result in AML or MDS. These analyses provided further insights for developing targeted therapies from a systems biology perspective.

10.2.1 Materials and methods of network transductivity and robustness mechanisms of leukemias

Data selection and preprocessing

Four types of data were integrated for constructing cellular networks. We used microarray raw data without normalization from the GEO database (accession no. GSE13159). To technically support clinical diagnosis of hematological malignancies, Haferlach et al. [68] obtained gene expression profiles of 2096 patients from 11 laboratories to identify biomarkers and validate their clinical accuracy. Of these 2096 patients, 2022 were classified into 17 leukemic subtypes and 74 were classified as being nonleukemic. To prevent overfitting and inaccuracy in identifying model parameters, we selected seven leukemic subtypes that had a greater number of samples than the number of samples for the nonleukemic subtype. In addition, while constructing the models of integrated cellular network of coupling STPs and GRNs, we overlaid the expression value of each gene on its corresponding protein to integrate gene expression and PPI information.

Rewiring the integrated cellular network of STPs and GRN associated with pathophysiological phenotypes of leukemia

Several studies [46,48,49,52,69–71] have constructed STPs that participate in leukemogenesis and have used the Kyoto Encyclopedia of Genes and Genomes (KEGG) database [72,73] to partially identify the relationships among kinases in these STPs. Leukemogenic STPs include the PI3K-AKT, classical MAPK, JAK-STAT, TGFβ, apoptosis, Notch, hedgehog, and canonical Wnt pathways (Fig. 10.5). In leukemic cells, signal transduction mediated by binding of ligands to STP receptors results in their abnormal activation because of genetic mutations in genes encoding receptors, kinases, and phosphatases involved in these STPs [49]. Crosstalk, which is a convergence between various pathways, can also influence and dysregulate STPs because of the effects of upstream mutations [46,49]. For example, GSK3β is crucial for both the Wnt and PI3K-AKT pathways. Aberrant or constitutive activation of these pathways can trigger dysfunctions by activating or repressing TFs. The resultant events facilitate cell proliferation, prevent apoptosis, and block hematopoietic cell differentiation. Aberrant activation of pathways such as the Wnt and Notch pathways favors self-renewal of LSCs, which is crucial for myeloid malignancy [46]. In addition, genetic mutations not only enhance tumor malignancy but also resist the effects of cancer therapies [46,47,49]. Collectively, dysfunctioning of proliferation, prevention of apoptosis, blockage of differentiation,

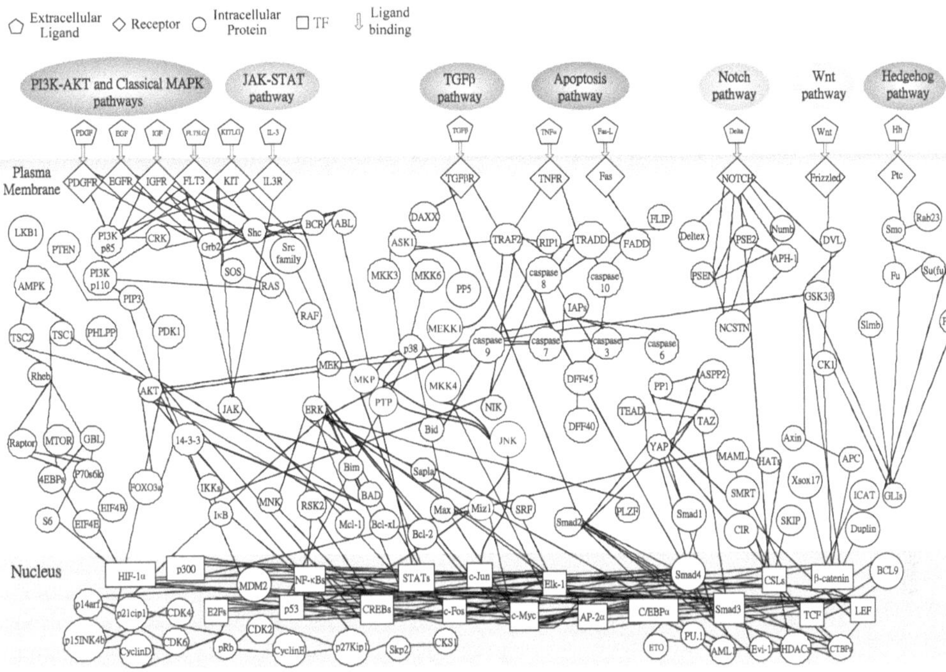

Figure 10.5 Integrated cellular interactions network of eight leukemogenesis-related coupling STPs. There are 159 groups of proteins and 28 groups of TFs. 18 TFs at the downstream of 8 STPs are determined by gene regulations from TRANSFAC database and are viewed as 18 genes of TFs of GRN in Fig. 10.6. The *solid lines* between TFs in nucleus denote GRN as shown in Fig. 10.6 while the others in cytoplasm denote PPIs. Other graphical symbols are defined and shown on top of this figure. The leukemogenesis-related STPs comprising of PI3K-AKT, classical MAPK, JAK-STAT, TGFβ, Apoptosis, Notch, Hedgehog and the canonical Wnt pathways are constructed by the literatures [46,48,49,52,69–71] and KEGG database [72,73]. Different STPs are coupled by the crosstalk between pathways. As the genetic mutations or dysregulations occur at TFs or upstream proteins in the STPs, the accumulated influence from mutations may lead to dysfunction of TFs through the propagation and crosstalk in coupling STPs.

and maintenance of LSCs are involved in the pathophysiology of leukemia. To investigate whether signal transductivity of pathophysiological STPs is distorted in leukemic cells and whether proteins in STPs receive exact information from the coupling pathways, we constructed a system model of coupling STPs containing 159 groups of proteins and 28 groups of TFs (Fig. 10.5) [46,48,49,52,62,69,71]. We focused on the transductivity of downstream TFs p53, C/EBPα, ETO, STATs (STAT1, STAT3, and STAT5), AP-2α, c-Jun, CSLs (RBP-JK and RBPJL), p300, CTBP, β-catenin, and FOXO3a to discuss whether the signal transductivity of pathophysiological STPs is distorted or not (see Section 10.2.3). Transcriptional regulation of GRNs was determined using the TF database

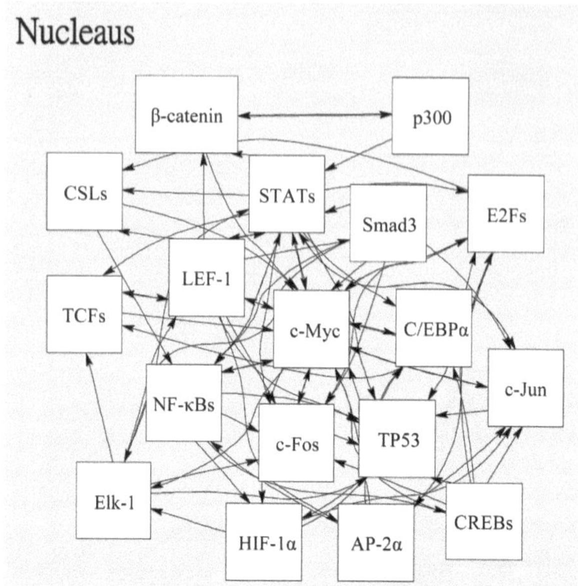

Figure 10.6 Multiple loops of GRN associated with leukemogenesis-related pathophysiological phenotypes. This is copied from the GRN of leukemogenesis-related TFs in nucleus, which consist of 18 genes of TFs at the downstream of coupling STPs in Fig. 10.5. Each TF regulates other genes and acts as a gene of TF regulated by other TFs. The 18 leukemogenesis-related genes of TFs are determined by TRANSFAC database.

(TRANSFAC; http://www.gene-regulation.de/) [74]. We selected genes encoding 18 TFs that regulated the transcription of 28 TFs to construct a GRN and to calculate its robustness in the 7 leukemic subtypes and 1 nonleukemic subtype (Fig. 10.6). These 18 genes included AP2-2α, β-catenin, C/EBPα, p300, c-Myc, CREBs (ATF4, CREB1, Luman, OASIS, CREB3L2, CREB-H, AibZIP, and CRE-BPa), CSLs, E2Fs (E2F1, E2F2, and E2F3), Elk-1, c-Fos, HIF-1α, c-Jun, LEF-1, NF-κBs (NF-κB1, NF-κB1-p50, NF-κB2-p52, NF-κB2, NF-κB2-p52, c-Rel, RelA-p65, and RelB), Smad3, STATs, TCFs (TCF-1, TCF-3, and TCF-4), and p53. The robustness of the GRN is discussed in Section 10.2.3.

Construction of GRN and calculation of the network robustness of GRN

1. Construction of GRN by Microarray Data From a Population of Samples

For the purpose of the system identification of GRN from a population of leukemia samples, we use a simple regression model, GRN model, to depict transcription regulations in GRN (Fig. 10.6). It is assumed that the GRN consists of M genes of TFs ($M = 18$, in Fig. 10.6). We use microarray data from a population of leukemia samples with one time-point microarray data for each individual patient to construct GRN. The samples of microarray data are derived from K patients. Let the state variable of the jth gene of TF in GRN be denoted as $y_j(k)$. In the steady state case, the expressions of GRN can be modeled as the following linear regression equation.

$$\begin{aligned} y_j(k) = {} & c_{j,1}y_1(k) + \cdots + c_{j,j-1}y_{j-1}(k) + c_{j,j+1}y_{j+1}(k) + \cdots \\ & + c_{j,M}y_M(k) + h_j + w_j(k), \quad \text{for } j = 1, \ldots, M \text{ and } k = 1, \ldots, K \end{aligned} \tag{10.29}$$

where $y_j(k)$ indicates the expression level of the jth gene of TF in the GRN at the kth clinical sample; $c_{j,j-1}$ is transcription regulatory ability from $(j-1)$th TF to the jth gene of TF. If $c_{j,j-1}$ is positive, it means the TF $j-1$ performs an activating regulation on gene j. Otherwise it is an inhibitive regulation; h_j denotes the basal level of the jth gene expression; $w_j(k)$ represents model uncertainties of the jth gene. Note that $c_{j,j}$ representing the transcription regulatory ability from the jth TF to its gene is set to be zero in this method. The model (10.29) biologically states that the gene expression $y_j(k)$ of the target gene j is associated with the expressions of its regulatory TFs.

2. Identification of Parameters in GRN Model by Microarray Data From a Population of Samples

To identify the regulatory parameters of Eq. (10.29), i.e., $\theta_j(k)$, we formulate the linear regression models in Eq. (10.29) as the following forms

$$\begin{aligned} y_j(k) &= \begin{bmatrix} y_1(k) & \ldots & y_{j-1}(k) & y_{j+1}(k) & \ldots & y_M(k) & 1 \end{bmatrix} \cdot \begin{bmatrix} c_{j,1} \\ \vdots \\ c_{j,j-1} \\ c_{j,j+1} \\ \vdots \\ c_{j,M} \\ h_j \end{bmatrix} + w_j(k) \\ &= \phi_j^T(k)\theta_j(k) + w_j(k), \quad \text{for } j = 1, \ldots, M \text{ and } k = 1, \ldots, K \end{aligned} \tag{10.30}$$

Because microarray data from a population of samples is used to identify the parameters of the gene regulatory models, we use recursive least-squares identification to estimate the parameter $\theta_j(k)$ one sample by one sample. The recursive identification algorithm for estimate $\theta_j(k)$ of the regulatory parameters of the jth gene in Eq. (10.30) is given as follows [17]:

$$\begin{aligned}
&\theta_j(k) = \theta_j(k-1) + P_j(k)\phi_j(k)\varepsilon_j(k) \\
&\varepsilon_j(k) = y_j(k) - \phi_j^T(k)\theta_j(k-1) \\
&P_j(k) = P_j(k-1) - \frac{P_j(k-1)\phi_j(k)\phi_j^T(k)P_j(k-1)}{1+\phi_j^T(k)P_j(k-1)\phi_j(k)}, \quad P_j(0) \text{ given,} \\
&\qquad \text{for } j = 1, \ldots, M \text{ and } k = 1, \ldots, K
\end{aligned} \tag{10.31}$$

where $\theta_j(k)$ is the parameter estimate and $\varepsilon_j(k)$ is the prediction error at the kth step. The matrix $P_j(k)$ constitutes an estimate of the parameter estimation error covariance at recursive step k. Initially, $P_j(0)$ is chosen as $diag(10, \ldots, 10)_{M\times M}$ for GRN model because of large estimation error at the initial estimate $\theta_j(0)$. The dimension of $P_j(0)$ is determined by $(M+1)$ regulatory parameters for each protein in GRN model needed to be estimated. We identify the regulatory parameters of linear regression model one gene by one gene (i.e., $j = 1, \ldots, M$ in Eq. 10.30) to obtain the whole regulatory parameters of GRN.

Remark 10.8: For more precise parameter estimation in Eq. (10.31), the result of signal parameter estimate $\theta_j(k)$ after K iterations could be considered as the initial parameter estimate $\theta_j(0)$ in Eq. (10.31) for another round of recursive parameter estimation.

By using the recursive identification algorithm in Eq. (10.31), the parameter estimate $\theta_j(k)$ of transcription regulatory abilities for each gene in GRN can be updated as the new samples of microarray data are measured. Therefore, the recursive identification algorithm in Eq. (10.31) is suitable for identifying the parameters of the regression model for GRN by microarray data from a population of samples.

3. Estimation of the Network Robustness of GRN

The network robustness in GRN is a measurement of the system performance of biological networks to tolerate intrinsic network variations like genetic mutation to maintain their phenotype. From an engineering perspective, network robustness and transductivity are two important and complementary system characteristics to discuss system performance, i.e., a more robust system will be with a less transductivity, and vice versa [67]. To systematically measure the network robustness in GRN, the one gene regulatory model of GRN in Eq. (10.29) must be augmented for M genes as the following multivariate regulatory model.

$$y = \hat{C}y + \hat{H} + W \tag{10.32}$$

where

$$y = \begin{bmatrix} y_1(k) & \dots & y_M(k) \end{bmatrix}^T; \hat{H} = \begin{bmatrix} \hat{h}_1 & \dots & \hat{h}_M \end{bmatrix}^T; \; W = \begin{bmatrix} w_1(k) & \dots & w_M(k) \end{bmatrix}^T$$

$$\text{and } \hat{C} = \begin{bmatrix} 0 & \hat{c}_{1,2} & \dots & \hat{c}_{1,j-1} & \hat{c}_{1,j} & \hat{c}_{1,j+1} & \dots & \hat{c}_{1,M} \\ \hat{c}_{2,1} & 0 & \ddots & \dots & \dots & \dots & \dots & \hat{c}_{2,M} \\ \vdots & \ddots & \ddots & \ddots & \dots & \dots & \dots & \vdots \\ \hat{c}_{j-1,1} & \dots & \ddots & 0 & \hat{c}_{j-1,j} & \dots & \dots & \hat{c}_{j-1,M} \\ \hat{c}_{j,1} & \dots & \dots & \hat{c}_{j,j-1} & 0 & \hat{c}_{j,j+1} & \dots & \hat{c}_{j,M} \\ \hat{c}_{j+1,1} & \dots & \dots & \dots & \hat{c}_{j+1,j} & 0 & \ddots & \hat{c}_{j+1,M} \\ \vdots & \dots & \dots & \dots & \dots & \ddots & \ddots & \vdots \\ \hat{c}_{M,1} & \dots & \dots & \dots & \dots & \dots & \ddots & 0 \end{bmatrix}.$$

Let us denote y as the expression vector of M genes in GRN; $\hat{C}$ represents the system matrix of transcription regulatory ability in GRN; $\hat{H}$ represents the vector of basal levels for M genes; $\hat{C}$ and $\hat{H}$ consist of the parameters identified by the recursive identification algorithm in Eq. (10.31) through K samples of microarray data; W represents the vector of model uncertainties for M genes. Eq. (10.32) can be rearranged as follows:

$$(I_M - \hat{C})y = \hat{H} + W \tag{10.33}$$

And the gene expression vector y in GRN can be represented by

$$y = (I_M - \hat{C})^{-1}[\hat{H} + W] \tag{10.34}$$

where $(I_M - \hat{C})$ represents the transcription regulatory matrix in GRN and I_M denotes identity matrix with dimension M.

Proposition 10.5: The network robustness ξ of GRN in a leukemia or nonleukemic subtype is estimated as follows:

$$\xi = \min_j \sigma_j(I_M - \hat{C}) \triangleq \sigma_{\min}(I_M - \hat{C}) \tag{10.35}$$

where σ_j denotes the jth singular value of $(I_M - \hat{C})$ and $\sigma_{\min}$ denotes the minimum singular value.

Remark 10.9: Let us decompose $(I_M - \hat{C})$ by the following singular value decomposition [17].

$$(I_M - \hat{C}) \triangleq \tilde{C} = \sum_{j=1}^{M} \sigma_j u_j^T v_j = U^T \Sigma V = U^T \bullet diag(\sigma_{\max}(\tilde{C}), \ \ldots \ , \sigma_{\min}(\tilde{C})) \bullet V \tag{10.36}$$

where σ_j denotes the jth singular value and is ordered decreasingly. That is, $\sigma_1(\tilde{C}) = \sigma_{\max}(\tilde{C}) \geq \sigma_2(\tilde{C}) \geq \ldots \geq \sigma_M(\tilde{C}) = \sigma_{\min}(\tilde{C})$. If all singular values are free of zero, the inverse of $\tilde{C}$ exists. That is, the steady state of this GRN model in Eq. (10.34) exists. If at least one singular value of $\tilde{C}$ is perturbed to 0, the inverse of $\tilde{C}$ doesn't exist.

***Proof of Proposition 10.5*:**
Suppose the regulatory matrix $\tilde{C}$ suffers from the following intrinsic parameter variation ΔC due to genetic variation, splicing and other uncertainties in the carcinogenic process as follows:

$$y = (\hat{C} - \Delta C)y + \hat{H} + W \tag{10.37}$$

In this situation, Eq. (10.33) could be changed to

$$(I_M - \hat{C} + \Delta C)y = \hat{H} + W \tag{10.38}$$

If the GRN can tolerate ΔC, then the inverse of perturbative regulatory matrix $(I_M - \hat{C} + \Delta C)$ exists and the gene expression vector y of perturbative GRN can be represented by

$$y = (I_M - \hat{C} + \Delta C)^{-1}[\hat{H} + W] \tag{10.39}$$

Let Eq. (10.38) be modified as

$$(I_M - \hat{C})(I_M + (I_M - \hat{C})^{-1}\Delta C)y = \hat{H} + W \tag{10.40}$$

If the following robustness condition holds [75]

$$\left\|(I_M - \hat{C})^{-1}\right\|_2 \left\|\Delta C\right\|_2 < 1 \quad \text{or} \quad \left\|\tilde{C}^{-1}\right\|_2 \left\|\Delta C\right\|_2 < 1 \tag{10.41}$$

then $I_M + (I_M - \hat{C})^{-1}\Delta C$ is always nonsingular, i.e., the inverse of $(I_M - \hat{C} + \Delta C)$ exists. The physical meaning is that the perturbative GRN can tolerate the variation and the steady state of model (10.39) exists. According to Ref. [76], it can be shown that

$$\left\|(I_M - \hat{C})^{-1}\right\|_2 = \frac{1}{\sigma_{\min}(I_M - \hat{C})} \quad \text{or} \quad \left\|\tilde{C}^{-1}\right\|_2 = \frac{1}{\sigma_{\min}(\tilde{C})} \tag{10.42}$$

where $\sigma_{\min}(\tilde{C}) = \min_j \sigma_j(\tilde{C})$, i.e., the minimum singular value of the transduction regulatory matrix $\tilde{C} = I_M - \hat{C}$. Therefore, from Eqs. (10.41) and (10.42), the robustness condition can be rewritten as:

$$\left\|\Delta C\right\|_2 < \sigma_{\min}(I_M - \hat{C}) \quad \text{or} \quad \left\|\Delta C\right\|_2 < \sigma_{\min}(\tilde{C}) \tag{10.43}$$

i.e., the network robustness ξ is equal to the largest tolerance of perturbation $\sigma_{\min}(I_M - \hat{C})$ or $\sigma_{\min}(\tilde{C})$.

Remark 10.10: Let us decompose the transcription regulatory matrix $I_M - \hat{C} + \Delta C$ of perturbative GRN under intrinsic perturbation by the singular value decomposition. If the intrinsic parameter variation ΔC is specified as $\Delta C = -\sigma_{\min}(I_M - \hat{C})I_M$, which violates the robustness condition in Eq. (10.43). Then from Eq. (10.36), the inverse of $I_M - \hat{C} + \Delta C$ will cease to exist, i.e.,

$$(I_M - \hat{C} + \Delta C) = \tilde{C} + \Delta C = U^T \bullet diag(\sigma_{\max}(\tilde{C}), \quad \ldots \quad, 0) \bullet V \tag{10.44}$$

It means that the model of GRN can't tolerate the intrinsic parameter variation and the phenotype $y = (I_M - \hat{C} + \Delta C)^{-1}[\hat{H} + W]$ doesn't exist again. Hence, the robustness measurement ξ of the GRN is defined as

$$\xi = \sigma_{\min}(I_M - \hat{C}) \tag{10.45}$$

4. Construction of Coupling STPs and Calculation of Transductivity, Transductivity Evolution Level in STP

a. *Construction of Coupling STPs by Microarray Data from a Population of Samples*

For the purpose of the system identification of STPs from a population of leukemia samples, we use the other simple regression model, STP model, to depict PPIs and the bindings of ligands to receptors in the coupling STPs (Fig. 10.5). It is assumed that the coupling STPs consist of N proteins ($N = 159$, in Fig. 10.5). We use microarray data from a population of leukemia samples with one time-point microarray data for each individual patient to construct the STPs. The samples of microarray data are derived from K patients. Let the state variable of the ith protein and its ligand in the STPs be denoted as $x_i(k)$ and $u_i(k)$, respectively. In the steady state case, the PPI model for the STPs in Fig. 10.5 can be modified as the following linear regression model.

$$\begin{aligned} x_i(k) = {} & a_{i,1}x_i(k) + \cdots + a_{i,i-1}x_{i-1}(k) + a_{i,i+1}x_{i+1}(k) + \cdots \\ & + a_{i,N}x_N(k) + b_i u_i(k) + d_i + v_i(k), \quad \text{for } i = 1, \ldots, N \text{ and } k = 1, \ldots, K \end{aligned} \tag{10.46}$$

where $x_i(k)$ and $u_i(k)$ indicate respectively the expression levels of the ith protein and its ligand in the STPs at the kth clinical sample; $a_{i,i-1}$ and b_i respectively represent the interaction abilities between the ith and $(i-1)$th proteins and the binding ability between the ith protein and its ligand; d_i denotes basal levels of the ith expression levels; $v_i(k)$ represents stochastic noises associated with other factors of the ith protein; If the ith protein doesn't serve as a receptor, b_i is set to be zero. Note that the interaction ability $a_{i,i}$ of the ith proteins itself is set to be zero in this method. The model in Eq. (10.46) states that, biologically, the expression level $x_i(k)$ of the target protein i is associated with the expression levels of its ligand and interactive proteins. It is worthy to note that the interaction parameters $a_{i,j}$, d_i, and b_i of regression model (10.46) in the STPs are identified systematically according to microarray data in the following. Note that $i \neq j$, for $j = 1, \ldots, N$.

5. Identification of Parameters in STP Model by Microarray Data from a Population of Samples

To identify the protein interaction parameters of Eq. (10.46), i.e., $\theta_i(k)$, we formulate the models in Eq. (10.46) as the following forms

$$x_i(k) = \begin{bmatrix} x_1(k) & \cdots & x_{i-1}(k) & x_{i+1}(k) & \cdots & x_N(k) & u_i(k) & 1 \end{bmatrix} \bullet \begin{bmatrix} a_{i,1} \\ \vdots \\ a_{i,i-1} \\ a_{i,i+1} \\ \vdots \\ a_{i,N} \\ b_i \\ d_i \end{bmatrix} + v_i(k)$$

$$= \phi_i^T(k)\theta_i(k) + v_i(k), \quad \text{for } i = 1, \ldots, N \text{ and } k = 1, \ldots, K \tag{10.47}$$

Because microarray data from a population of samples is used to identify the parameters of the protein interaction models of the STPs in Fig. 10.5, we use recursive least-squares identification algorithm in Eq. (10.31) to estimate the interactive parameter $\theta_i(k)$ of linear interaction model (10.46) of the ith protein in the STPs. We identify the interaction parameters of each protein of STPs with other proteins and ligands in Fig. 10.5 by the recursive least square algorithm in Eq. (10.31) one protein by one protein. Then we could estimate the whole interactive parameters of the STPs in Fig. 10.5. After whole PPI abilities of leukemogenesis-related STPs in Fig. 10.5 are constructed by microarray data, we will investigate the transductivity and transductivity evolution level of leukemogenesis-related STPs in the following section.

a. *Investigation of the Transductivity and Transductivity Evolution Level of Leukemogenesis-related STPs*

The transduction ability of a protein in STP, termed transductivity, measures the system response to environmental changes like environmental stress or the ability to transduce from the external molecular signals to other interactive proteins. To systematically measure the transductivity in the STPs in Fig. 10.5, we augment the PPI model of STPs in Eq. (10.46) for N proteins as the following multivariate regression model.

$$x = \hat{A}x + \hat{B}u + \hat{D} + V \tag{10.48}$$

where

$$x = [x_1 \quad \dots \quad x_N]^T;\ \hat{B} = diag(\hat{b}_1, \quad \dots, \hat{b}_N);\ u = [u_1 \quad \dots \quad u_N]^T;$$
$$\hat{D} = [\hat{d}_1 \quad \dots \quad \hat{d}_N]^T;\ V = [v_1 \quad \dots \quad v_N]^T$$

and

$$\hat{A} = \begin{bmatrix} 0 & \hat{a}_{1,2} & \cdots & \hat{a}_{1,i-1} & \hat{a}_{1,i} & \hat{a}_{1,i+1} & \cdots & \hat{a}_{1,N} \\ \hat{a}_{2,1} & 0 & \ddots & \cdots & \cdots & \cdots & \cdots & \hat{a}_{2,N} \\ \vdots & \ddots & \ddots & \ddots & \cdots & \cdots & \cdots & \vdots \\ \hat{a}_{i-1,1} & \cdots & \ddots & 0 & \hat{a}_{i-1,i} & \cdots & \cdots & \hat{a}_{i-1,N} \\ \hat{a}_{i,1} & \cdots & \cdots & \hat{a}_{i,i-1} & 0 & \hat{a}_{i,i+1} & \cdots & \hat{a}_{i,N} \\ \hat{a}_{i+1,1} & \cdots & \cdots & \cdots & \hat{a}_{i+1,i} & 0 & \ddots & \hat{a}_{i+1,N} \\ \vdots & \cdots & \cdots & \cdots & \cdots & \ddots & \ddots & \vdots \\ \hat{a}_{N,1} & \cdots & \cdots & \cdots & \cdots & \cdots & \ddots & 0 \end{bmatrix}$$

where x indicates the expression vector for N proteins in the STPs; $\hat{A}$ represents the system matrix of PPIs in the STPs; $\hat{B}$ is a diagonal matrix and represents the system matrix of binding abilities between the receptor proteins and their ligands in the STPs. u denotes expression vector of ligands in the STPs as input signals; $\hat{D}$ represents the vector of basal levels for N proteins; and $\hat{A}$, $\hat{B}$, and $\hat{D}$ consist of the interaction parameters identified by the recursive identification algorithm; V represents the vector of stochastic noises associated with other interaction factors for N proteins. The model in Eq. (10.48) can be rearranged as follows:

$$(I_N - \hat{A})x = \hat{B}u + \hat{D} + V \tag{10.49}$$

Since we want to measure the transductivity of protein of interest in the STPs, we denote the expression level of the protein of interest as the output signal.

Therefore, the expression vector of all proteins in STPs and the output signal of a chosen protein z of interest in the STPs can be represented by

$$\begin{cases} x = (I_N - \hat{A})^{-1}\hat{B}u + (I_N - \hat{A})^{-1}[\hat{D} + V] \\ z = Gx \end{cases} \tag{10.50}$$

where G is used to choose the protein of interest and to measure its transductivity. If we are interested in the i target protein in the STPs and would like to measure its transductivity, all elements of the vector G are zero except for a single one at the ith component. For example, if we want to measure the transductivity of the first protein, G is set to be $\begin{bmatrix} 1 & 0 & \dots & 0 \end{bmatrix}$. Further, the output signal z of a chosen protein of interest can be represented by

$$z = Gx = G(I_N - \hat{A})^{-1}\hat{B}u + G(I_N - \hat{A})^{-1}[\hat{D} + V] \tag{10.51}$$

where $G(I_N - \hat{A})^{-1}\hat{B}$ denotes the transductive function from external signals u to the chosen protein z.

We first define transductivity ρ from input signals u to each protein at a leukemic subtype in the STP model as follows [27,65]:

$$\begin{aligned} \rho(\text{protein, leukemic subtype}) &= \sup_{u \in l_2} \frac{\|z\|_2}{\|u\|_2} = \|G(I_N - \hat{A})^{-1}\hat{B}\|_2 \\ &= \sigma_{\max}(G(I_N - \hat{A})^{-1}\hat{B}) \end{aligned} \tag{10.52}$$

where u and z denotes the input signals and the output signal of the STP. For an input signal vector $u = \begin{bmatrix} u_1 & \dots & u_N \end{bmatrix}^T$, the l_2 norm for u is defined by $\|u\|_2 = \sqrt{u_1{}^2 + \dots + u_N{}^2}$. l_2 denotes the set of all bounded signals with $\|u\|_2 < \infty$. The l_2 induced matrix norm in Eq. (10.52) denotes the system gain from input signal u to the output signal z and is defined as $\rho = \sup_{u \in l_2} \|z\|_2 / \|u\|_2$. That means the transductivity ρ in Eq. (10.52) is derived from the worst-case output/input ratio $\|z\|_2 / \|u\|_2$ for all possible bounded inputs from the system gain perspective. $\sigma_{\max}(G(I_N - \hat{A})^{-1}\hat{B})$ denotes the largest singular value of $(G(I_N - \hat{A})^{-1}\hat{B})$. The physical meaning of transductivity is the maximal energy ratio of the output signal to all possible bounded input signals. The reason for using the maximal energy ratio of the output signals to all input signals is that the input signals may vary with environmental conditions, causing the energy ratio to change. Thus, transductivity should be evaluated according to the maximal effect of all possible external signals on the output levels of expression. It exhibits a greater dependency on the systematic characteristics of the steady system in accordance with system gain [27].

In this section, the transductivity of a protein in the STPs implies its abilities to transduce signals from all possible external signals to the downstream interactive proteins.

We use the transductivity to discuss how much a protein transduces signals in the STPs in response to all possible input signals and further orchestrates the program to associate with the cellular functions about apoptosis, cell survival, cell cycle progression, differentiation, cell proliferation, and detoxification of ROS. To compare the transductivities of a protein in leukemic cells with nonleukemic cells in the STPs, the transductivity evolution levels of different leukemic subtypes are also measured to investigate the transductivity changes of these proteins in the STPs during the leukemogenic process. If the transductivity of the protein p53 in the STPs of the MDS subtype is denoted by $\rho(\text{p53}, \text{MDS})$, the evolution level of transductivity evolution level $\Delta\rho$ of a protein between each leukemic subtype and nonleukemic subtype is denoted as follows:

$$\Delta\rho(\text{protein}, \text{subtype}) = \frac{\rho(\text{protein}, \text{subtype}) - \rho(\text{protein}, \text{nonleukemia})}{\rho(\text{protein}, \text{subtype}) + \rho(\text{protein}, \text{nonleukemia})} \tag{10.53}$$

Let us illustrate how to use the measurements of original transductivity ρ and transductivity evolution level $\Delta\rho$. The transductivity ρ of a specific TF in the STPs at each subtype implies its ability to transduce signals from all external signals to the downstream genes at each subtype cell. If ρ of the specific TF is larger than 1, then all external signals will be amplified through the TF. That means with small amount of ligands, the TF can easily transduce the information to the downstream genes for transcription regulation. If the ρ of the specific TF is less than 1, then all external signals will be attenuated or buffered through the TF. That means the TF will require more amount of ligands to transduce the signal to the downstream gene. Also, the TF with larger ρ can transduce more signals to respond to all external signals. Based on $\Delta\rho$, we can realize whether TF is more sensitive to all external signals than nonleukemic subtype. If the $\Delta\rho$ of specific TF is sufficient positive/negative around ± 1, then the TF is more sensitive in transductivity with different tendencies such as gain/loss of functions to leukemogenesis. Further, the resultant signal transductive programs of cellular functions about apoptosis, cell cycle progression, and differentiation help us get insight into the pathophysiological phenotypes in leukemic cells. Consequently, investigating $\Delta\rho$ of specific TF between leukemic subtype and nonleukemic subtype reveals how much the TF functions in the leukemic cells change in response to all external signals through the crosstalk in coupling STPs.

10.2.2 Evolution mechanisms in network robustness and signal transductivity in the carcinogenic process of leukemias

More network robustness of GRN in seven leukemic subtypes than in nonleukemic subtype

The measurement of network robustness (see Section 10.2.1) can reveal the extent how GRN in Fig. 10.5 tolerates the intrinsic perturbation due to noise processing, molecular

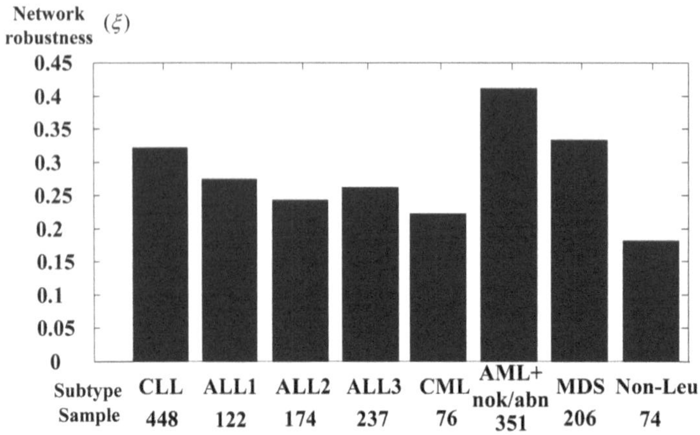

Abbreviations	Full Name
CLL	Chronic lymphocytic leukemia
ALL1	c-ALL/pre-B-ALL with t(9;22)
ALL2	T-cell acute lymphoblastic leukemia
ALL3	c-ALL/pre-B-ALL without t(9;22)
CML	Chronic myelogenous leukemia
AML+nok/abn	Acute myeloid leukemia with normal karyotype+other abnormalities
MDS	Myelodysplastic syndrome
Non-Leu	Non-leukemia and healthy bone marrow
AML+nok/abn +t(8;21)	acute myeloid leukemia with normal karyotype+other abnormalities+t(8;21)

Figure 10.7 The evolution of network robustness (ξ) of the GRN in Fig. 10.6 with 18 TFs across 7 leukemic subtypes and one nonleukemic subtype. The nonleukemic subtype is with the smallest robustness (0.18) by comparison to other leukemic subtypes. It means that nonleukemic subtype is less tolerable to intrinsic perturbation such as mutation or chromosomal translocation and more responsible to external molecular signals than leukemic subtypes. Among these leukemic subtypes, the GRNs of AML and MDS with larger robustness than other leukemic subtypes can give harbor to a diversity of intrinsic mutations and further develop a variety of evolutionary process than other leukemic subtypes.

fluctuations, or genetic mutations. From the NRM of the GRN across seven leukemic subtypes in Fig. 10.7, the nonleukemic subtype is with the smallest robustness (0.18) in comparison to other leukemic subtypes. It means that nonleukemic subtype is less tolerable to intrinsic perturbation such as mutation or chromosomal translocation and more responsible to external molecular signals than leukemic subtypes. Among these leukemic subtypes, the GRNs of AML and MDS subtypes are with larger robustness than other leukemic subtypes, ALL1, CML, CLL, and so on. It implies that the GRNs of AML and MDS subtypes can withstand more intrinsic perturbation than other leukemic subtypes so that they harbor more genetic mutations to develop a diversity of heterogeneity for evolution to compete and eventually dominate either nonleukemic or other leukemic subtypes. Thus, we further focus on the transductivity evolution levels of downstream 28 TFs in the coupling STPs of AML and MDS subtypes.

Transductivity evolution levels of downstream 28 TFs in the coupling STPs of AML, MDS, and nonleukemic subtypes

The transductivity evolution levels $\Delta\rho$, the changes of transductivity between leukemic and nonleukemic subtypes (see Section 10.2.1), and the corresponding transductivities ρ of 28 TFs at AML and MDS subtypes are all shown in Fig. 10.8.

The transductivities ρ for 28 TFs in AML, MDS and non-leukemia

TF	AML	MDS	Non-Leu
CSLs	0.2267	0.4392	0.0059
p300	0.0065	0.0173	0.0002
CTBPs	0.1233	0.0184	0.2262
AP-2α	0.0016	0.0150	0.0000
STATs	0.8370	1.5071	0.1456
c-Jun	0.0001	0.0016	0.0001
β-catenin	0.0659	0.0197	0.0158
LEF-1	0.0537	0.0037	0.2228
TCFs	0.0096	0.0029	0.0192
c-Myc	0.0009	0.0035	0.0169
c-Fos	0.0000	0.0005	0.0000
CREBs	0.0033	0.0015	0.0198
Elk-1	0.0034	0.0124	0.0004
p53	0.0026	0.0028	0.0187
FOXO3a	0.0000	0.0254	0.0042
NF-κBs	0.0000	0.0360	0.0025
HIF-1α	0.0000	0.0000	0.0000
E2Fs	0.0000	0.0000	0.0001
pRb	0.0004	0.0012	0.0004
C/EBPα	0.0014	0.0361	0.0002
AML1	0.0134	0.0068	0.0347
ETO	0.0059	0.0235	0.0062
PLZF	0.0012	0.0009	0.0000
PU.1	0.0045	0.0003	0.0040
Smad2	0.0297	0.0428	0.0362
Smad3	0.1089	0.0696	0.4576
Smad4	0.1090	0.0608	0.0146
GLIs(Ci)	0.0132	0.0001	0.0804

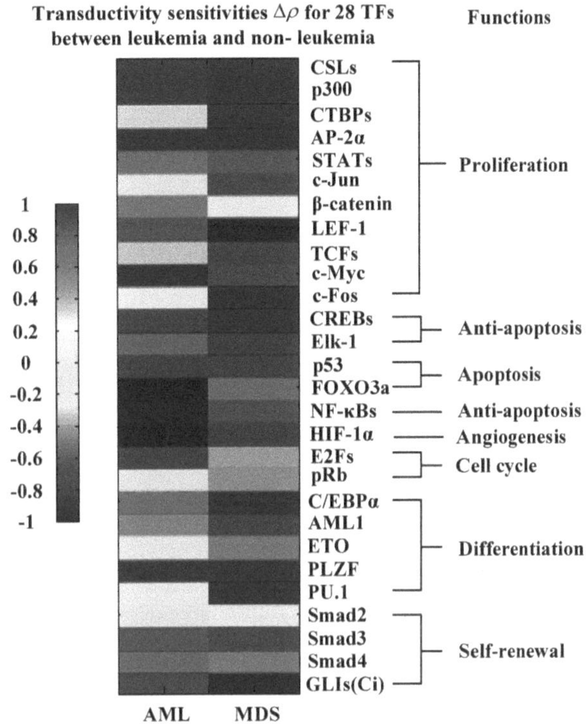

Figure 10.8 The transductivity evolution levels ($\Delta\rho$) and the transductivities (ρ) for 28 TFs in the STPs at AML, MDS, and nonleukemic subtype. Both the transductivity and transductivity evolution level are sorted by rows of cellular functions contributing to leukemia. For example, STATs, c-Jun, β-catenin, and AP-2α are sorted together at the cellular function of cell proliferation. The color bar denotes the extent and development direction of transductivity evolution in leukemic subtypes when compared with nonleukemic subtype from *green* to *blue* or from *green* to *red*. If the TF can transduces more signals and is more sensitive in leukemic subtype than nonleukemic subtype, the color more approaches to *red*. If the TF transduces less signals and is more sensitive in leukemic subtype than nonleukemic subtype, the color more approaches to *blue*. Besides, the dysfunction of a TF happens when the TF is sensitive to all external signals with different tendencies to transduce signals, which further cause gain of functions (*red*) or loss of functions (*blue*). We choose TFs, CSLs, p300, CTBPs, AP-2α, STATs, c-Jun, c-Fos, β-catenin, p53, FOXO3a, C/EBPα, AML1, and ETO to discuss their dysfunctions because we prefer to reveal which TF would be more sensitive to external signals from all receptors at AML and MDS than nonleukemia. Some TFs, such as CSLs, p300, STATs, c-Jun, β-catenin, C/EBPα, and AP-2α, are dysregulated and gain more functions at AML and MDS.p53, CTBPs and AML1 lose some functions at AML and MDS. Notably, the other TFs, c-Jun, c-Fos, and FOXO3a, seem to gain more functions at MDS than nonleukemia and AML through their contrast color of transductivity evolution levels. That is consistent with the fact that AML blocks in differentiation and has more antiapoptotic phenotypes, whereas MDS differentiates normally and has pro-apoptotic or antiapoptotic phenotypes since MDS consists of early MDS and advanced MDS. Besides, the corresponding transductivities reveal which TF can amplify or buffer or attenuate signals. It is worth to note that only STATs can amplify and buffer signals from external signals in multiple STPs at AML and MDS, respectively.

The overall transductivity evolution levels $\Delta\rho$ of 28 TFs at 7 leukemic subtypes and 1 nonleukemic subtype can refer to Fig. 10.8. AML patients whose incidence increases with age account for 80% of all adult leukemia and tumor recurs frequently after therapies. And MDS is partly related to AML, possibly as a pre-AML or preexisting evolution of AML. Thus, we focus on the transductivity evolution level $\Delta\rho$ of these TFs at AML and MDS. To obviously observe whether the signal transductivity of pathophysiological STPs is distorted or not, transductivity evolution levels $\Delta\rho$ and transductivities ρ of 28 TFs are sorted by rows according to the clustering of cellular functions contributing to leukemia in Fig. 10.8.

For example, STATs, c-Jun, β-catenin, and AP-2α are sorted together at the cellular function of cell proliferation. The color bar in Fig. 10.8 denotes the extent and development direction of transductivity evolution level in leukemic subtypes when compared with nonleukemic subtype from *green* to *blue* or from *green* to *red*. If the TF can transduce more signals and is more sensitive in leukemic subtype than nonleukemic subtype, the color more approaches to *red*. If the TF transduces less signals and is more sensitive in leukemic subtype than nonleukemic subtype, the color more approaches to *blue*. Besides, the dysfunction of a TF happens when the TF is sensitive to all external signals with different tendencies to transduce signals, which further cause gain of functions (*red*) or loss of functions (*blue*). When these TFs are dysregulated by accumulating the influences from genetic mutations in oncogenes, tumor-suppress genes and upstream proteins in the coupling STPs, their aberrant transductivity would contribute to abnormal transcription regulations and generate dysfunctions in comparison with nonleukemic subtype.

We select the TFs, CSLs, p300, CTBPs, AP-2α, STATs, c-Jun, c-Fos, β-catenin, p53, FOXO3a, C/EBPα, AML1, and ETO, to discuss their dysfunctions in leukemia because we can reveal that these TFs, which are more sensitive to external signals than nonleukemia, can gain or lose some functions to cause dysfunctions. The transductivities of the TFs in Fig. 10.8 point out which TF can amplify, buffer or attenuate the signals from all ligands at AML or MDS. Here, we observe that only STATs has more transductivity near or larger than one to buffer and amplify signals at AML and MDS. When the specific TF with sufficient positive/negative transductivity evolution level around ± 1, the TF is more sensitive in transductivity with different tendencies such as gain/loss of functions to leukemogenesis. Biologically, if the specific TF transduces much more or less signals from external signals in leukemic cells than nonleukemic cells, the TF is more sensitive to influence interactive proteins. Some TFs, such as CSLs, p300, STATs, c-Jun, β-catenin, C/EBPα, and AP-2α, are dysregulated and gain more functions at AML and MDS.p53, CTBPs, and AML1 lose some functions at AML and MDS. Notably, the other TFs, c-Jun, c-Fos, and FOXO3a, seem to gain more functions at MDS than nonleukemia and AML (Fig. 10.8). These results are consistent with the fact that AML blocks in differentiation and has more antiapoptotic phenotypes, whereas MDS differentiates normally and has pro-apoptotic or antiapoptotic phenotypes since MDS consists of early MDS and advanced MDS. All the above transductivity evolution levels can reveal the phenotypes for MDS progression to AML or the different phenotypes between them.

10.2.3 Discussion of evolution mechanisms of network robustness and transductivity in the carcinogenic process

Evolution analysis of network robustness of GRN in seven leukemic subtypes and one nonleukemic subtype

In general, a GRN with larger network robustness is less responsible to external molecular signals [67]. It was also found that GRN in cancer is more robust to tolerate intrinsic network perturbation [66]. In Fig. 10.7, we estimate the network robustness of GRN in Fig. 10.6 at seven leukemic subtypes and one nonleukemic subtype. The reason why all leukemic subtypes can tolerate more intrinsic perturbations than nonleukemia is that leukemia is a heterogeneous group of blood malignancies which results from chromosomal translocation and mutation accumulated at oncogene, tumor-suppressor gene, and receptor. Therefore, GRNs in all leukemic subtypes need larger network robustness to tolerate more intrinsic network perturbations to maintain their functions [49,67]. Obviously, the network robustness of AML (0.42) is the largest robustness to intrinsic network perturbation than other subtypes. It indicates that AML better tolerates intrinsic genetic mutation, noise processing, and molecular fluctuation and less responds to external molecular signals than the other leukemic subtypes. Since the genetic and molecular diversities observed at AML have made the development of therapies very difficult and cumbersome, the subtype of AML accounts for 80% adult leukemia with low survival rate and the incidence increases with age [77,78]. Thus, the largest network robustness at AML is attributed to the diversity or heterogeneity of mutations involved in leukemogenic process. Biologically, to tolerate more network perturbations or fluctuations resulted from diversity of genetic mutations, the GRN subtype of AML provides the largest network robustness to tolerate intrinsic genetic mutations, process noise and molecular fluctuations, and further develop a variety of heterogeneities for evolutionary process. On the other hand, they are less responsible or not be adaptable to external molecular signals or stresses than the others leukemic subtypes and nonleukemic subtype. The biological system bearing a diversity of mutations finally breaks down and leads to incidences of AML. Consequently, if a GRN could provide the larger robustness, the network could tolerate more intrinsic network perturbation due to the accumulated genetic variations, which might lead to dysfunctions, and vice versa.

Transductivity evolution analysis of downstream TFs in the coupling STPs of AML, MDS, and nonleukemic subtypes

1. p53

   The p53 responds to many stresses and the DNA damage to repress tumor cell growth as a key guardian of genome [67]. It prevents cells from tumorigenesis by keeping an moderate program about generating cell cycle arrest to reduce error rate and promote DNA repair and triggering apoptosis to eliminate abnormal cell [79]. p53 is crucial for cell fate through many transcriptions of genes. When low or repairable levels of DNA damage accumulate, the transcription of p21 cyclin-dependent kinase inhibitory protein-1 (p21Cip1) would induce G1 checkpoint of cell cycle to affect cell cycle progression and conduct DNA repair effectively [80]. When DNA damage is high, the apoptotic process is performed by transcriptions of pro-apoptotic genes such as Noxa, Puma, and Bax, which result in cytochrome c release [80]. Besides, deficiency of p21 would decrease the capacity of self-renewal in HSCs. Additionally, DNA repair process would be enhanced through transcriptions of p48 and p53R2 (a ribonucleotide reductase subunit) [79].

   As shown in Figs. 10.8 and 10.9A, p53 is more sensitive and transduces less signals in leukemic subtypes than nonleukemic subtype, and probably tends to perform insufficient functions or lose some functions about apoptosis, cell cycle arrest and DNA repair upon the external signals. Besides, loss of function in p53 could lead to genomic instability and promote tumorigenesis. p53 is frequently found mutant in leukemia and its functionality could govern the susceptibility to many kinds of therapies [81]. Since the predicted mutations influence on the upstream proteins and accumulate these influences to cause the dysregulation in PI3K-AKT and p38 MAPK pathways, the appropriate functionality of p53 is disrupted. Collectively, the transductivity analysis of p53 in Figs. 10.10 and 10.11 is consistent with its cellular function during leukemogenesis [71,82].

2. C/EBPα, AML1, and ETO

   The C/EBPα is mutated approximately in 10% of AML patients, and two major mutant types increase cell proliferation of premalignant HSCs and deregulate the differentiation programs of normal cells in hematopoiesis to further lead to leukemia [57]. Also, the FLT3 mutation detected in 30% of AML patients was found to inhibit the C/EBPα through ERK1/2 and results in the differentiation blockade of leukemic blasts [83]. C/EBPα coordinates the differentiation program with cell cycle arrest to protect normal myeloid cells from malignant transformation. The differentiation program is dependent on repression of E2Fs [84]. E2Fs activates the transcription of c-Myc to perform the block of differentiation. When E2Fs is inhibited by C/EBPα, the downregulation of c-Myc leads to differentiation. Loss of this repression by C/EBPα contributes to the block of differentiation during leukemogenesis. Besides, C/EBPα inhibits CDK2 and CDK4 to arrest cell proliferation

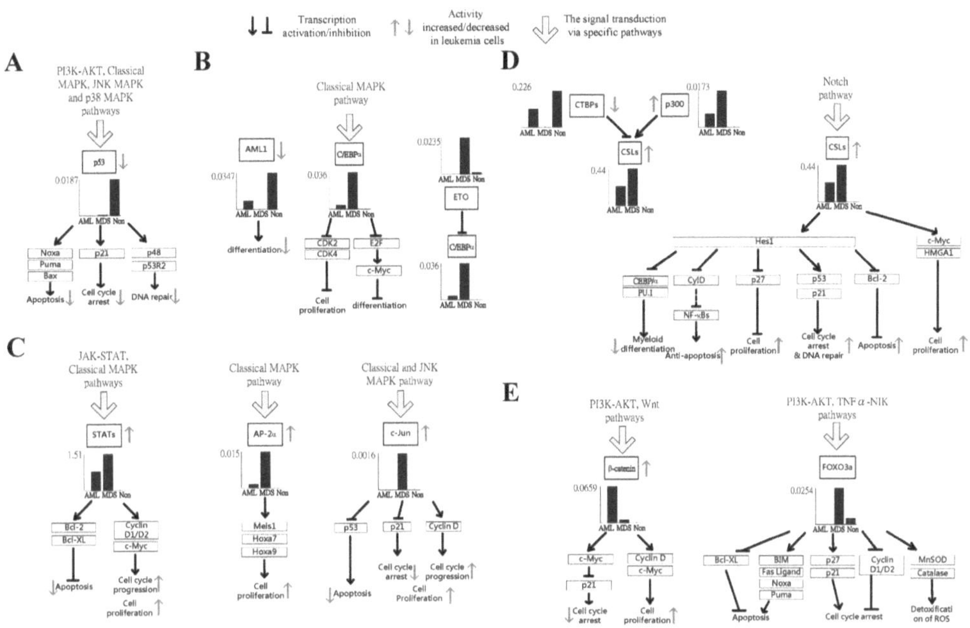

Figure 10.9 The effect of evolutionary transductivity of individual TF of p53, C/EBPα, STATs, CSLs, β-catenin, and FOXO3a on their downstream regulators and cellular functions during leukemogenic process. Some major downstream regulators of these TFs are shown to mediate distinct cellular functions. *Arrows in black* indicate upregulation between TFs and their downstream regulators; *Blunt arrows in black* indicate downregulation between TFs and their downstream regulators. *Arrows in blue* (gray in print versions) represent the change of transductivity and development of cellular functions from nonleukemic subtype to AML or/and MDS. (A) p53 is more sensitive and transduces less signals at leukemic subtypes than nonleukemic subtype, and probably tends to perform insufficient functions or lose some functions about apoptosis, cell cycle arrest, and DNA repair upon the external signals. (B) C/EBPα tends to lose much more functions about differentiation and arrest of cell proliferation at AML than MDS and nonleukemic subtype. It gains more functions of differentiation at MDS than AML and nonleukemic subtype. Importantly, the dysfunction of differentiation attributes to the influence from the mutations predicted in AML1. Although more functions about differentiation and cell proliferation arrest in C/EBPα at AML and MDS subtypes than nonleukemic subtype, genetic mutation influences AML1 at AML and MDS subtype, leading to much more function of myeloid differentiation arrest at AML than MDS and nonleukemic subtype. (C) STATs tends to gain much more functions including prevention from apoptosis, facilitating cell cycle progression and cell proliferation at leukemic subtypes than nonleukemic subtype, which finally result in the dysfunction of STATs in leukemia. AP-2α gains more function of cell proliferation at AML and MDS than nonleukemic subtype. Also, c-Jun and c-Fos tend to gain more functions about cell proliferation and cell survival. (D) CSLs tends to gain more about cell proliferation, cell survival, cell cycle arrest and blockade of differentiation at leukemic subtypes than nonleukemic subtype. Here, we use the transductivity pattern of p300 and CTBPs to represent one transductivity pattern of coactivators and corepressors, respectively. (E) β-catenin tends to gain more function of cell proliferation and possibly could not maintain the quiescence of HSC at leukemic subtypes than nonleukemic subtype. Finally, FOXO3a tends to gain some functions about apoptosis, cell cycle arrest and detoxification of ROS at MDS than AML. It tends to lose some functions about apoptosis, cell cycle arrest and detoxification of ROS at AML than MDS.

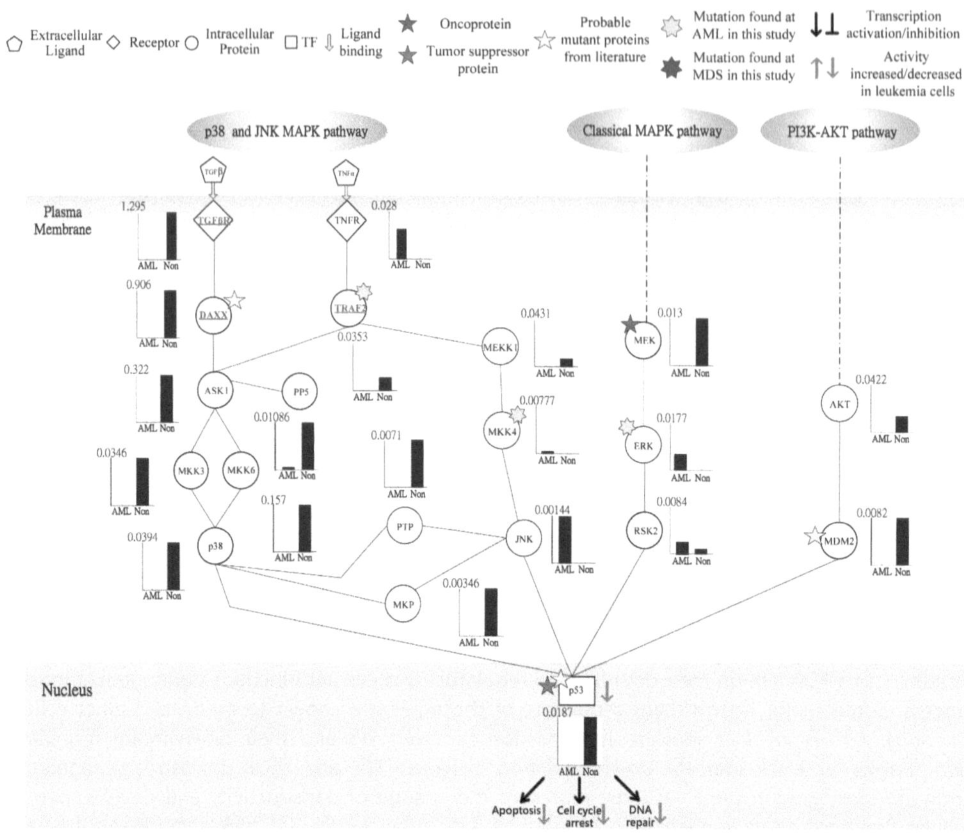

Figure 10.10 Transductivity evolution of the proteins in p38 MAPK, JNK MAPK, and PI3K-AKT coupling pathways contributing to the loss of transductivity of TF p53 at AML subtype. The *solid line in black* represents the PPIs in cytoplasm. Also, the *dot and dash line in black* represents the PPIs omitted here but shown in Fig. 10.12. The proteins shown in *red* symbols with underline in the STPs are analyzed as the main causes of dysfunction of TF. Some genetic mutations in proteins are predicted at AML and MDS subtypes through this transductivity analysis and show them in *cyan* and *blue* seven pointed stars, respectively. Five pointed stars in *yellow*, *red*, and *green* represent mutations, possible functions of oncoprotein and tumor-suppressor protein from literatures [49,86–90], respectively. Parts of predicted mutations from this analysis are verified in literatures when the seven pointed stars meet *yellow* five pointed stars. Some transductivity analyses are concluded as follows: The TF p53 interacts with the p38 MAPK, the classical MAPK, the JNK MAPK and PI3K-AKT pathways. Probable mutations are predicted in ERK, TRAF2, and MKK4 through this analysis. ERK reversibly transduces more apoptotic signals at AML subtype than nonleukemic subtype, which is unfavorable to leukemic cell growth. The mutation predicted in TRAF2 favors to less transductivity in their coupling protein, ASK1, at AML subtype than nonleukemic subtype. The ASK1 in the p38 MAPK pathway is dominated by DAXX. It is worthy to note that the apoptotic signals would be attenuated in RSK2 at AML subtype to favor to leukemic cell growth whereas enhanced in RSK2 at nonleukemic subtype to strengthen the ability of defensive mechanism in comparison with the transductivity of ERK in the classical MAPK pathway. The complex of ubiquitin ligase MDM2/p53 marks p53 so that it is degraded, thus leading to antiapoptotic effects. MDM2 leads to less antiapoptotic effects at AML subtype. Consequently, the loss of function of p53 and dysregulation in p38 MAPK pathway are mainly attributed to DAXX and TGFβ*R* either at AML subtype or MDS subtype.

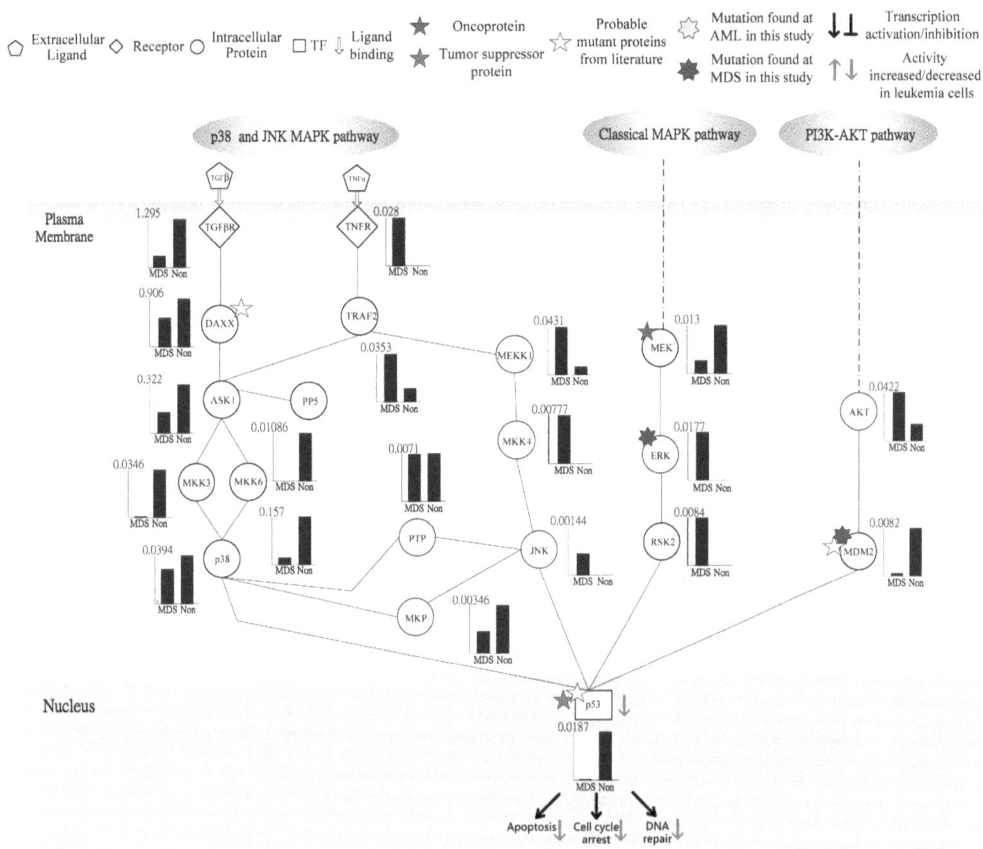

Figure 10.11 Transductivity evolution of the proteins in p38 MAPK, JNK MAPK, and PI3K-AKT coupling pathways contributing to the loss of transductivity of TF p53 at MDS subtype. The *dot and dash line in black* represents the PPIs omitted here but shown in Fig. 10.13. Some explanations of graphical symbols, lines, and transductivity analyses are similar to the results shown in Fig. 10.10. There is a speculation that if we can induce the ability of RSK2 to transduce more apoptotic signals at AML, the progression from MDS to AML could be prevented. Consequently, the progression from MDS to AML attributes to RSK2.

[85]. When C/EBPα is repressed, the cell proliferates in the absence of arrests associated with CDK2 and CDK4. Additionally, the fusion protein, AML1-ETO, is frequently observed in leukemic cells. AML1-ETO inhibits either C/EBPα, or the AML1-targeted gene which is crucial for x blasts [83]. C/EBPα coordinates the differentiation program with cell cycle arrest to protect normal myeloid cells from malignant transformation. The differentiation program is dependent on repression of E2Fs [84]. E2Fs activates the transcription of c-Myc to perform the block of differentiation. When E2Fs is inhibited by C/EBPα, the downregulation of c-Myc leads to differentiation. Loss of this repression by C/EBPα contributes to the block of

differentiation during leukemogenesis. Besides, C/EBPα inhibits CDK2 and CDK4 to arrest cell proliferation [85]. When C/EBPα is repressed, the cell proliferates in the absence of arrests associated with CDK2 and CDK4. Additionally, the fusion protein, AML1-ETO, is frequently observed in leukemic cells. AML1-ETO inhibits either C/EBPα, or the AML1-targeted gene which is crucial for myeloid differentiation and the development of HSCs [86].

As shown in Figs. 10.8 and 10.9B, C/EBPα transduces less signals at AML whereas transduces more signals at MDS than nonleukemic subtype. It implies that C/EBPα tends to gain more function of differentiation at MDS than AML and nonleukemic subtype. It is observed that AML1 performs insufficient functions for myeloid differentiation and development of HSCs since AML1 transduces less signals at AML and MDS than nonleukemic subtype to influence the resultant events of AML1 target genes. If AML1 and ETO form the fusion protein AML1-ETO, it is also thought that AML1-ETO performs insufficient functions to arrest the myeloid differentiation from C/EBPα at AML. From the observation of only ETO which transduces more signals at MDS than nonleukemia and AML, more inhibition from ETO at MDS and loss of function of AML1 at AML result in the insufficient function about the differentiation and arrest of cell proliferation at MDS and AML subtypes, respectively. In total, the deregulated programs of differentiation contributing to leukemogenic process seem to lose more functions about differentiation and evading arrest of cell proliferation at AML and gain more functions of differentiation at MDS than nonleukemic subtype and AML. It has been reported that AML would block in differentiation, whereas MDS would differentiate normally and then fail to achieve the next stage and finally die [91]. The accumulated influences from mutations in the AML1 and ETO may result in the deregulated programs of differentiation contributing to leukemogenic process [48,83].

3. STATs

Abnormal and constitutive inducements of STATs including STAT1, STAT3, STAT5A, and STAT5B have been demonstrated in many leukemic subtypes. With the stimulations of ligands through the JAK-STAT and classical MAPK coupling pathways, the downstream STATs translocates from cytoplasm into nucleus and accumulates more rapidly than ones which are not induced [92]. Enhanced STATs is crucial for programs about prevention from apoptosis, cell cycle progression, and cell proliferation through the target genes Bcl-2, Bcl-XL, cyclin D, and c-Myc. The antiapoptotic proteins Bcl-2 and Bcl-XL escape from apoptosis which is dependent on various stimuli, such as p53 and anticancer agents in malignant cells [78]. As Bcl-2 cooperates with c-Myc to control the programs of cell proliferation and apoptosis, the c-Myc escapes from its induction of apoptosis and promotes cell proliferation [93]. The cyclin D, important for G0/G1 phase of cell cycle, promotes cell cycle progression [78].

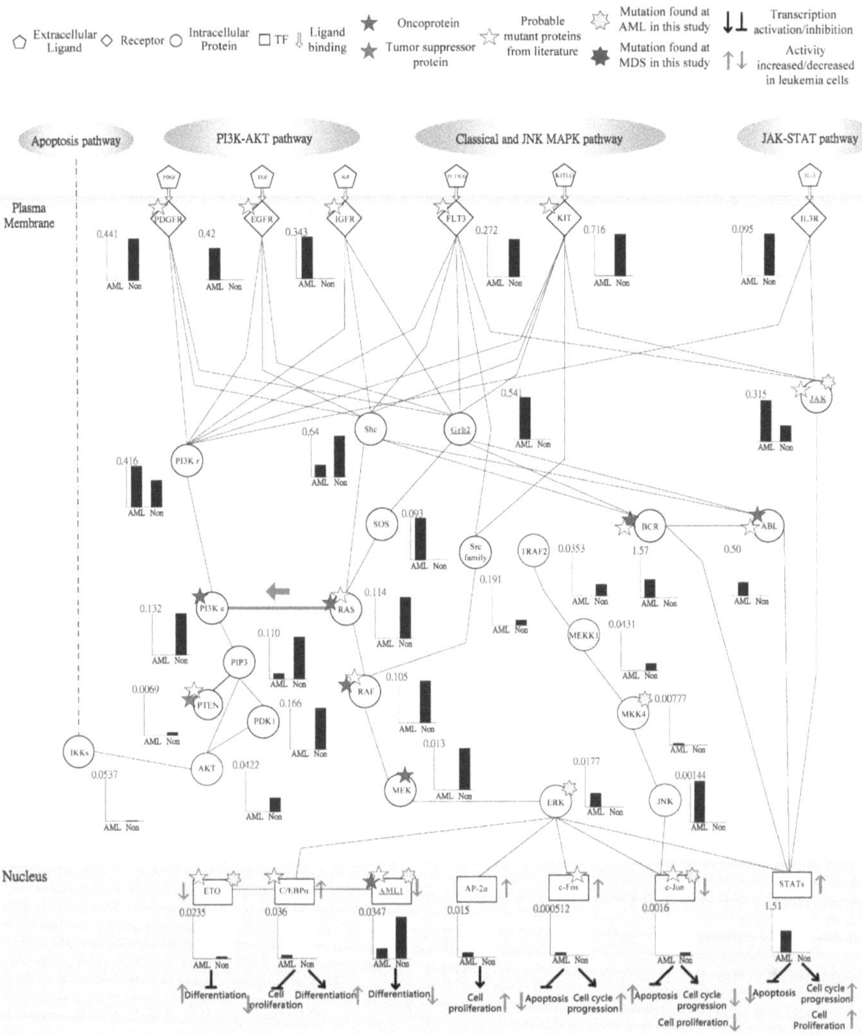

Figure 10.12 Transductivity evolution of the proteins in JAK-STAT and MAPK coupling pathways contributing to the gain of transductivity of STATs and C/EBPα at AML subtype. The *dot and dash line in black* represents the PPIs omitted here but shown in Fig. 10.14. *The bold solid line with an arrow in pink* demonstrates the direction that signals transduce through the crosstalk between two pathways according to this transductivity analysis. Some explanations of graphical symbols and lines are similar to the results shown in Fig. 10.10. Some TFs, STATs, AP-2α, c-Jun, c-Fos, C/EBPα, AML1, and ETO, are associated with the JAK-STAT, JNK MAPK, and classical MAPK coupling pathways. Only STATs interact with both JAK-STAT and classical MAPK coupling pathways, and the other TFs interact with MAPK pathway. We predict JAK, ERK, AML1, ETO are influenced by their mutations at AML subtype. The cell proliferative signal of RAS is determined by Shc which is mainly affected by PDGFR, FLT3 and KIT. It is worth noting that the signal of RAS can affect the signal of PI3Kc, attenuating the survival signals especially at AML subtype through the crosstalk between PI3K-AKT and classical MAPK pathways at AML subtype. The gain of function of STATs, which interacts with classical MAPK and JAK-STAT pathways, mainly attributes to BCR, ABL and especially mutant JAK rather than the mutations predicted in ERK. BCR and ABL are dominated by signals from Grb2, EGFR, and IGFR among multiple receptors. Also, the gain of functions of c-Fos and AP-2α, which interacts with classical MAPK pathway, mainly attributes to the influence from the mutations predicted in ERK. The loss of functions of c-Jun, which interacts with the classical MAPK and JNK MAPK pathway, mainly attributes to the mutations predicted in c-Jun. The loss of function of differentiation in C/EBPα, AML1 and ETO attribute to the influence from the mutations predicted in AML1.

As shown in Figs. 10.8 and 10.9C, STATs is more sensitive and transduces more signals at AML and MDS than nonleukemic subtype. It indicates that STATs tends to gain much more functions including prevention from apoptosis, facilitating cell cycle progression and cell proliferation, which finally result in the dysfunction of STATs. The dysfunction is consistent with the fact that how STATs contributes to leukemia [78,94]. To better understand the mechanism of this dysfunction, we focus on the influences resulting from dysregulated tyrosine kinases, such as breakpoint cluster region (BCR)-abelson (ABL), epidermal growth factor receptor (EGFR), mutant FMS related tyrosine kinase3 (FLT3), JAK family and the classical MAPK pathway.

4. AP-2α and c-Jun

The TF AP-2α serves as a tumor suppressor in various types of cancer but also serves as a tumor promoter in certain cancer, such as AML, a subtype of leukemia [95]. Although the decreased expression of AP-2α promotes cancer progression and predicts poor prognosis in other cancers, its low expression in cytoplasm accompanied with high expression in nucleus enhances tumor malignancy in certain cancers. Overexpression of Hoxa7 or Hoxa9 and Hox cofactor Meis1 induces short-latency AML in mice. Besides, it has been shown that a simultaneous upregulation of AP-2α and its target genes, Hoxa7, Hoxa9, and Meis1, are involved in the survival and proliferation of AML cells. Thus, AP-2α is required for the cell proliferation and survival at AML. Besides, the TF c-Jun has been demonstrated to play the tumor promoting role and be overexpressed at AML [96]. It activates the transcription of cyclin D to promote the G1 phase of cell cycle and inhibits p21 to reduce the arrest of cell cycle, and finally accelerates cell proliferation. It also antagonizes the p53-dependent apoptosis and protect the cell from cell death in leukemia. Additionally, the upregulation of c-Jun seems to downregulate C/EBPα to block the differentiation in leukemic cells.

As shown in Figs. 10.8 and 10.9C, AP-2α is more sensitive and transduces more signals at AML and MDS than a nonleukemic subtype. It implies that AP-2α gains more function of cell proliferation upon the external signals at AML and MDS than nonleukemia. c-Jun is sensitive and tends to gain more functions about cell proliferation and prevention from cell death at MDS whereas it is almost insensitive and tends to lose more functions at AML than nonleukemic subtype. Such transductivity analyses of AP-2α and c-Jun are consistent with the dysfunctions which contribute to leukemogenesis [95,96]. Although c-Jun interacts with the classical MAPK and JNK MAPK pathway, the dysfunction of c-Jun seems to attribute to the genetic mutation in c-Jun based on the observation of transductivity patterns. Since AP-2α is phosphorylated and induced by RAF/MEK/ERK cascades in the classical MAPK pathway [97,98], the dysfunctions of AP-2α attribute to the accumulation of the impact from predicted mutation in proteins such as RAS and RAF.

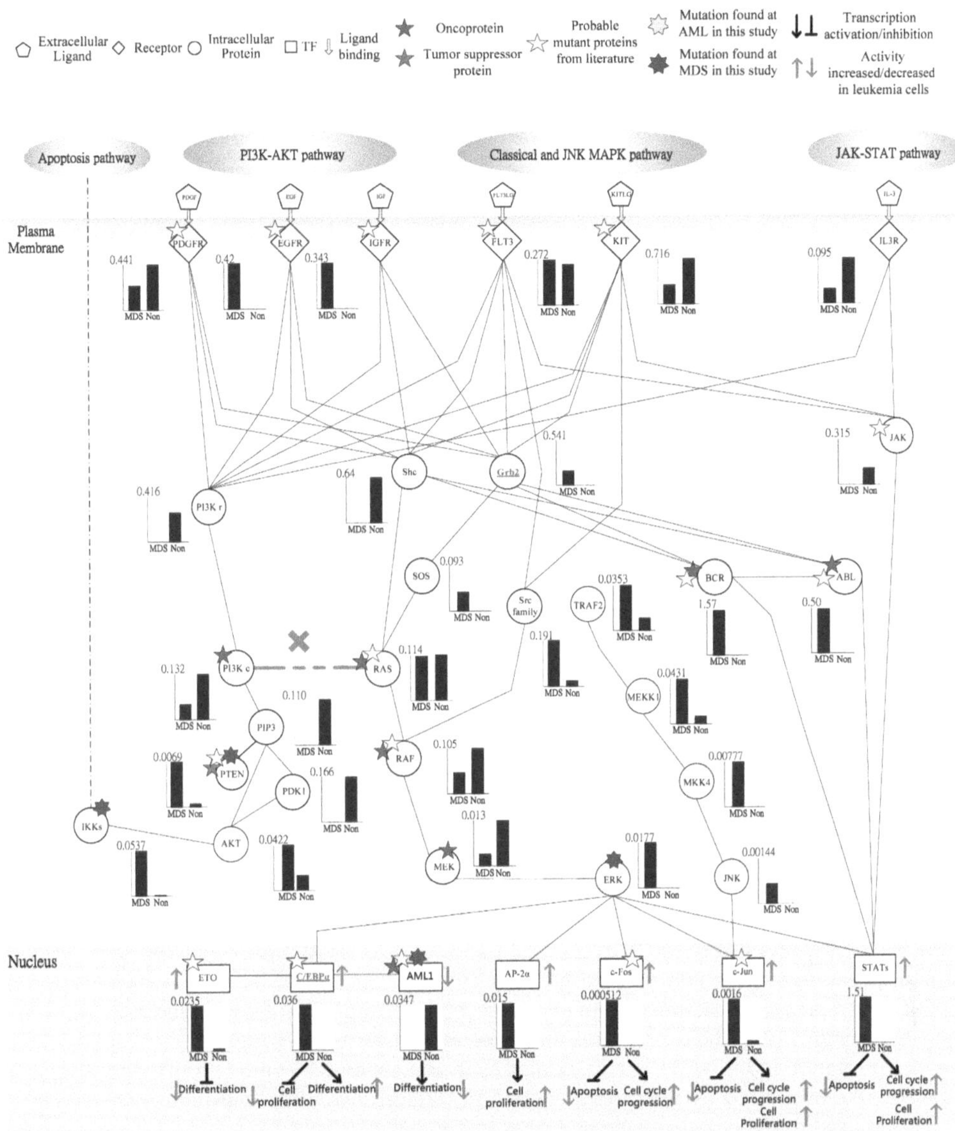

Figure 10.13 Transductivity evolution of the proteins in JAK-STAT and MAPK coupling pathways contributing to the gain of transductivity of STATs and C/EBPα at MDS subtype. The *dot and dash line in black* represents the PPIs omitted here but shown in Fig. 10.15. *The bold dashed line with a cross in pink* represents that no signals transduce through the crosstalk between two pathways. Some explanations of graphical symbols and lines are similar to the results shown in Fig. 10.10. The gain of function of STATs mainly attributes to BCR and ABL dominated by signals from Grb2, EGFR, IGFR, and FLT3 at MDS subtype. Through the observation of transductivity patterns of Shc, SOS, RAS, and PI3Kr at MDS subtype, there is no signal to transduce and affect the transductivity of each other through the crosstalk between RAS and PI3Kr. RAF, ERK, AML1, and PTEN are influenced by their mutations predicted at MDS subtype. Notably, RAF is affected by the mutant RAF and further attenuates the proliferative effect from RAS at MDS subtype. Although there is less proliferative effect at MDS than nonleukemic subtype, the influence from mutations predicted in ERK substantially enhances the proliferative effect at MDS subtype, which will cause the dysfunctions of c-Jun and AP-2α. Consequently, the gain of function of differentiation at MDS than nonleukemic subtype attribute to C/EBPα. The progression from MDS to AML is determined by C/EBPα and the mutations predicted in JAK, ERK, and AML1.

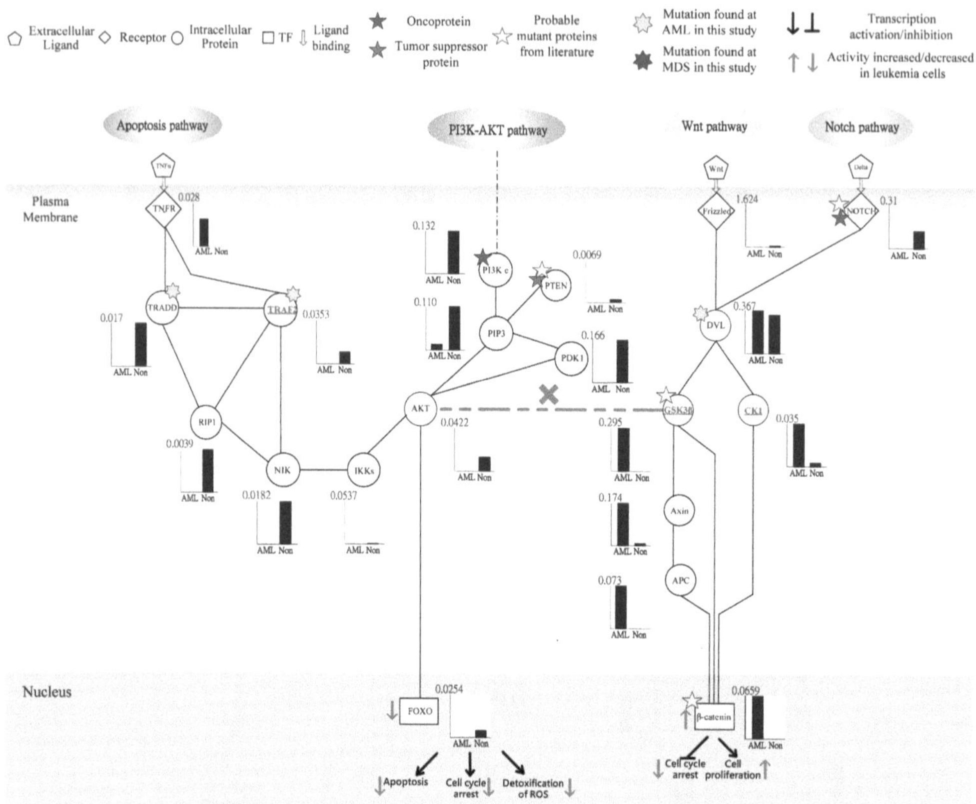

Figure 10.14 Transductivity evolution of the proteins in the canonical Wnt, PI3K-AKT, and apoptosis coupling pathways associated with β-catenin and FOXO3a at AML subtype. The *dot and dash line in black* represents the PPIs omitted here but shown in Fig. 10.12. *The bold dashed line with a cross in pink* represents that no signals transduce through the crosstalk between two pathways. Some explanations of graphical symbols and lines are similar to the results shown in Fig. 10.10. Two TFs, β-catenin and FOXO3a, are associated with the canonical Wnt, PI3K-AKT, and Apoptosis pathways. DVL influenced by predicted mutation in DVL can reversibly transduce more signals to cause more accumulation of β-catenin at AML subtype than nonleukemic subtype. There is no signal transduction through the crosstalk between AKT and GSK3β since mutations predicted in TRADD and TRAF2 reversibly attenuate the respondence for stress or apoptotic signal from TNFR at AML subtype. It is worth noting that the transductivity for degrading the destructive complex can be substantially attenuated at the nonleukemic subtype but attenuated a little at AML subtype when the signals transduce to GSK3β and CK1. It implies that GSK3β and CK1 themselves serve as a brake to stop from signal for the degradation of destructive complex at nonleukemic subtype but may not work well at AML subtype. Note also that both the transductivities of IKKs at AML and nonleukemic subtypes are substantially attenuated so that no signal transduction from IKKs to AKT and both AML and nonleukemic subtypes cannot respond the apoptotic signal. Also, the loss of function in FOXO3a at AML subtype attributes to the signal in the PI3K-AKT pathway. Because the difference of functions of FOXO3a between AML and nonleukemic subtype is small, the less antiapoptotic effect resulting from FOXO3a may not significantly impact on the dysfunction of antiapoptosis at AML subtype.

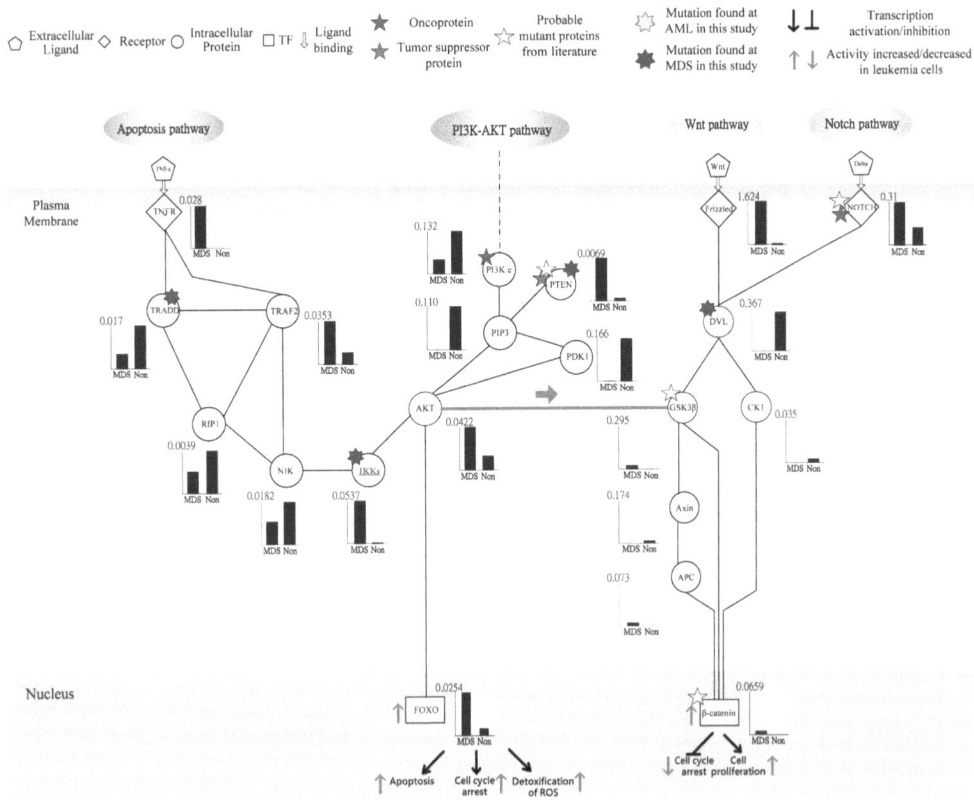

Figure 10.15 Transductivity evolution of the proteins in the canonical Wnt, PI3K-AKT, and apoptosis coupling pathways associated with β-catenin and FOXO3a at MDS subtype. The *dot and dash line in black* represents the PPIs omitted here but shown in Fig. 10.13. *The bold solid line with an arrow in pink* demonstrates the direction that signals transduce through the crosstalk between two pathways according to this transductivity analysis. Some explanations of graphical symbols and lines are similar to the results shown in Fig. 10.10. Two TFs, β-catenin and FOXO3a, are associated with the canonical Wnt, PI3K-AKT, and Apoptosis pathways. DVL influenced by predicted mutation in DVL causes less proliferative effect through the less degradation of destructive complex at MDS subtype. We can observe that the transductivity of GSK3β in Wnt pathway can be influenced and attenuated by the signal from PI3K-AKT pathway through the crosstalk between these pathways at MDS subtype. As the signal is influenced by possible mutations in IKKs and transduces to GSK3β, GSK3β can reversibly transduce more signals to cause more proliferative effect at MDS subtype than nonleukemic subtype. Thus, the gain of function in β-catenin attributes to GSK3β influenced by the signal of AKT through the crosstalk. And the gain of function of FOXO3a mainly attributes to the apoptotic signal of possible mutation in IKKs. Consequently, the progression from MDS to AML attributes to the mutations predicted in GSK3β, CK1 and the mutations predicted in IKKs.

5. CSLs

The aberrant inducement of the nuclear TF CSLs is involved in abnormal or mutant Notch receptor-mediated signaling. When the intracellular Notch (ICN) is released, it shuttles to the nucleus and induces the transcriptions of Notch target genes through the binding with CSLs, resulting in transcriptions of Notch target genes, such as Hairy-Enhancer of Split 1 (Hes1), c-Myc, and High mobility group A1 (HGMA1). In myeloid leukemia, it has been reported that transcription of Hes1 influences myeloid differentiation, cell cycle arrest and apoptosis through interactions with C/EBPα, PU.1, p53, p21, and Bcl-2 [99,100]. Hes1 suppresses C/EBPα, PU.1, and Bcl-2 to drive block of differentiation and apoptosis. Also, Hes1 induces p53 and p21 to arrest the AML growth. On the other hand, it has been suggested that transcription regulations of Hes1, c-Myc, and HMGA1 affect cell proliferation and antiapoptosis through interactions with p27Kip1 and the deubiquitinase CylD in lymphoblastic leukemia [101,102]. Hes1 represses p27Kip1 to promote cell proliferation and down-regulates CylD to allow the survival of quiescent cells by the inhibition of the IκB. Besides, the transcription regulations of c-Myc and HGMA1 promote the cell proliferation. The Notch signaling pathway regulates cell fate and differentiation processes during leukemogenesis. In detail, the Notch signaling has been reported that it serves as both oncogenic and tumor-suppressor functions in the development of myeloid and lymphoblastic leukemia [46,103]. For example, the activation of Notch signaling promotes cell growth in lymphoblastic leukemia but sometimes inhibits in myeloid leukemic cells.

As shown in Figs. 10.8, 10.9D, 10.16, and 10.17 CSLs is more sensitive and can transduce more signals at AML and MDS than nonleukemic subtype. It implies that CSLs tends to gain more or perform redundant functions about cell proliferation, cell survival, cell cycle arrest, and blockade of differentiation. Since the transcriptions of downstream genes are associated with the Notch signaling and the crosstalk with Wnt signaling in the BM niche [46], the dysfunction induced in this pathway is the result of mutant Notch signaling and aberrant crosstalk.

6. β-Catenin

The unphosphorylated β-catenin which accumulates and translocates to the nucleus is important for the maintenance of LSC to promote myeloid malignancies [46]. It has been shown that abnormal inducement of β-catenin in Wnt pathway contributes to the impaired myelomonocytic differentiation in normal progenitors. Also, aberrant inducement of β-catenin leads to the enhanced LSC growth and self-renewal through transcription regulations of the cell cycle regulators, c-Myc and cyclin D. The cyclin D promotes cell cycle progression and cooperates with c-Myc to further exert the proliferative effect. Besides, the maintenance of HSC quiescence, resting state in G0 phase, is determined by cyclin-dependent kinase inhibitor p21 without repression of up-regulated c-Myc [52]. Additionally, both of c-Myc and cyclin D facilitate the cell proliferation in the cell cycle.

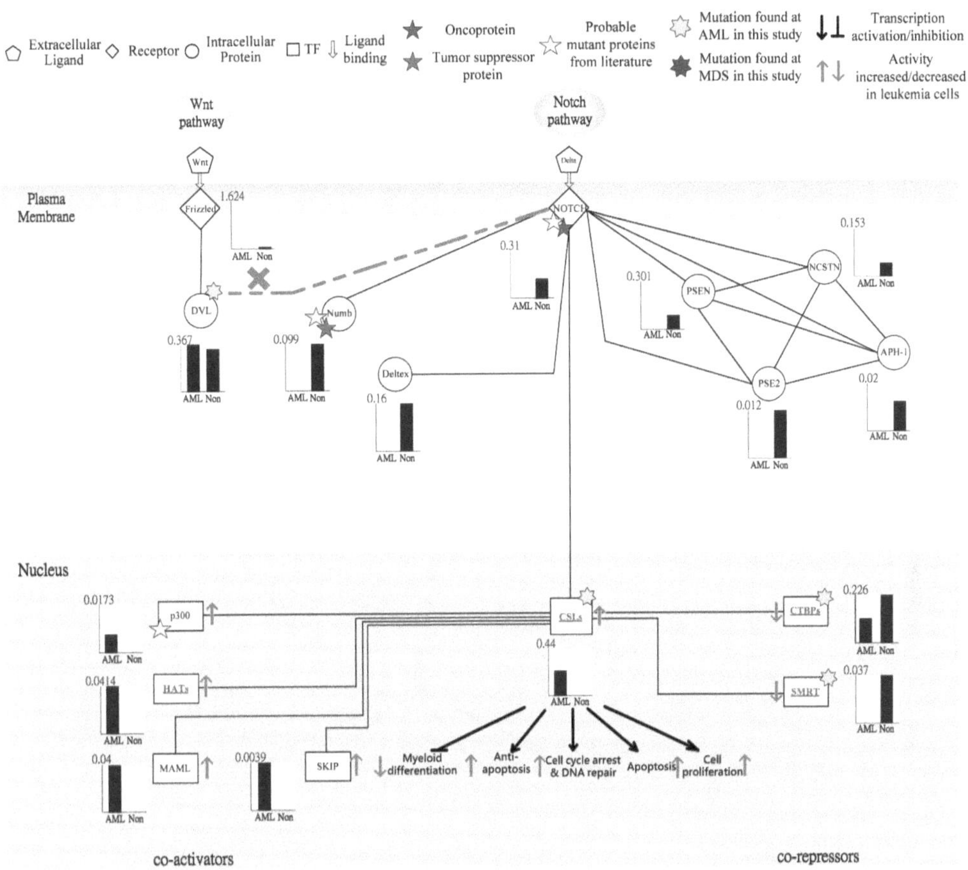

Figure 10.16 Transductivity evolution of the proteins in Notch pathways contributing to the gain of transductivity of TF CSLs at AML subtype. *The bold dashed line with a cross in pink* represents that no signals transduce through the crosstalk between two pathways. Some explanations of graphical symbols and lines are similar to the results shown in Fig. 10.10. NOTCH receptor protein can transduce less signals and lose some functions at AML subtype. The proteolytic cleavages mediated by the PSEN, NCSTN, PSE2, and APH-1 inadequately influence on NOTCH receptor and almost no ICN is released from NOTCH receptor to induce the transcription regulations of CSLs at AML subtype. The loss of function of cell differentiation can be observed from the less transductivity of Numb at AML subtype. The mutations in DVL, CSLs, CTBPs, and SMRT are predicted due to changes of their transductivity patterns at AML subtype. DVL doesn't affect NOTCH receptor through the crosstalk due to the different transductivity patterns of DVL and NOTCH receptor. And the more transductivity of HATs at AML than nonleukemic subtype can contribute to the enhancement of transcription regulation of CSLs. The gain of function of CSLs attributes to the mutations predicted in CSLs, CTBPs and SMRT at AML subtype.

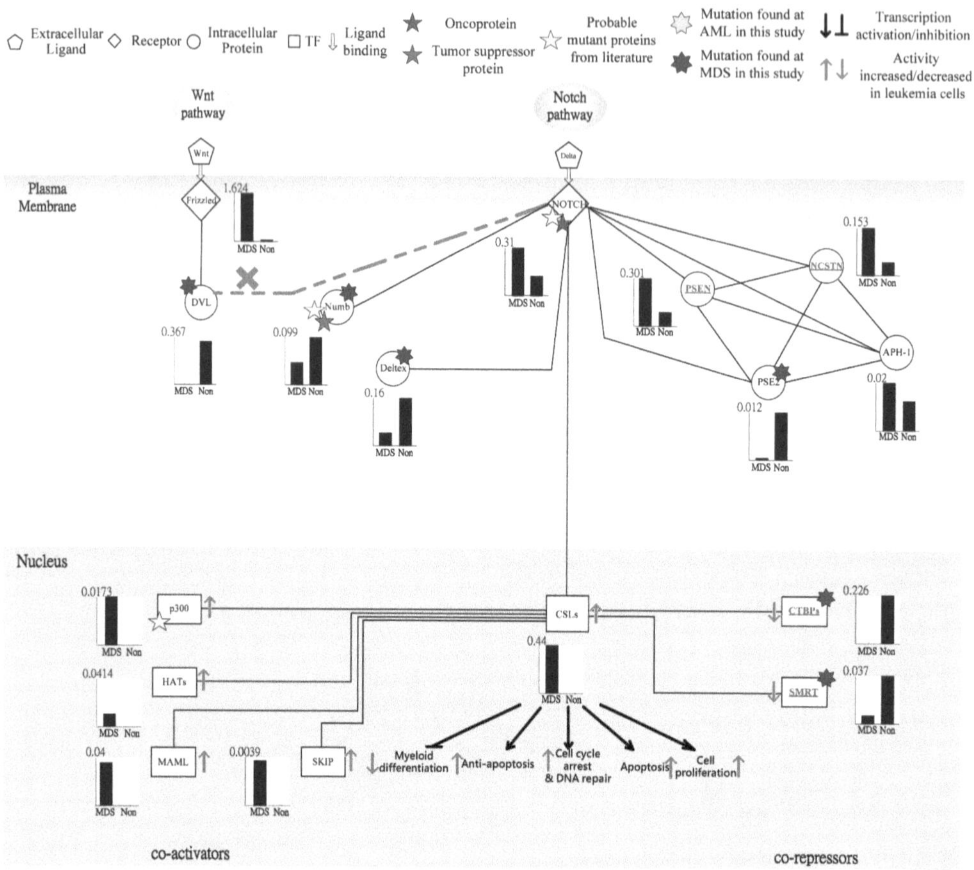

Figure 10.17 Transductivity evolution of the proteins in Notch pathways contributing to the gain of transductivity of TF CSLs in MDS subtype. *The bold dashed line with a cross in pink* represents that no signals transduce through the crosstalk between two pathways. Some explanations of graphical symbols and lines are similar to the results shown in Fig. 10.10. NOTCH receptor protein can transduce more signals and gain much more functions at MDS subtype. More ICN is released by more proteolytic cleavages from PSEN, NCSTN, and APH-1, with larger transductivities at MDS than nonleukemic subtype. Mutations are predicted at DVL, Numb, Deltex, PSE2, CTBPs, and SMRT due to changes of their transductivity patterns at MDS subtype. The transductivity pattern of Numb also implies less cell differentiation at MDS subtype than nonleukemic subtype. Importantly, the observation that Numb transduces less signals at AML than MDS is consistent with the fact that decreasing level of Numb leads to the progression from MDS with low risk to AML. The mutations predicted in CTBPs and SMRT influence their transductivity patterns contributing to less attenuation of barrier to transcription regulations of CSLs. Thus, the gain of function of CSLs attributes to the upstream proteins PSEN, NCSTN and the mutations predicted in CTBPs and SMRT at MDS subtype. The progression from MDS to AML is determined by PSEN and NCSTN.

As shown in Figs. 10.8 and 10.9E, β-catenin is more sensitive and can transduce more signals at AML than a nonleukemic subtype. It suggests that β-catenin tends to gain more function of cell proliferation and possibly could not maintain the quiescence of HSC. This transductivity analysis of β-catenin is consistent with the fact how β-catenin contributes to leukemogenesis [46,104]. Because GSK3β is probably repressed by either DVL in Wnt pathway or AKT in PI3K-AKT pathway, the dysfunction of cell proliferation is attributed to mutations or dysregulation of the Wnt canonical pathway best characterized in the hematopoietic system and PI3K-AKT pathway.

7. FOXO3a
FOXO3a, a member of the forkhead TFs, antagonizes the cell proliferation and survival in the absence of active AKT. The inducement of FOXO3a mediates some functions of apoptosis, cell cycle arrest, differentiation, and detoxification of ROS through the transcription regulations of downstream genes. The transcription regulations of the pro-apoptotic genes, BIM, Noxa, Puma, and Fas ligands, result in apoptosis [49]. Also, the triggered FOXO3a represses the transcription regulation of antiapoptotic Bcl-XL to induce cell death. The regulations of Cdk inhibitor p27Kip1 and p21 could arrest the cell cycle of G1 phase [105]. The inhibited regulations of cyclin D1 and D2 mediate the cell cycle arrest [106]. Besides, the upregulation of p21 blocks the differentiation. Additionally, the function of detoxification of ROS is involved in the transcription regulations of the genes which encode free radical scavenging enzymes, such as Mn superoxide dismutase (MnSOD) and catalase [106]. Thus, FOXO3a serves as an important sensor to repair the damage due to ROS. If the FOXO3a is down-regulated by the elevated level of AKT, the poor induction about cell cycle arrest, apoptosis, and detoxification of ROS would favor the survival of abnormal cells and accumulation of damage. It has been reported that the maintenance of small subset of AML cells, leukemia-initiating cells (LICs), requires the elevated FOXO3a which is associated with low level of AKT [107]. Besides, the myeloid leukemia cells suffer differentiation through the downregulation of p21 associated with the depletion of FOXO3a. Collectively, FOXO3a acts as tumor-suppressor and converse roles in leukemogenesis.
As shown in Figs. 10.8 and 10.9E, FOXO3a transduces less signals and is more sensitive at AML than nonleukemia. It implies that FOXO3a tends to lose some functions linked to apoptosis, cell cycle arrest and detoxification of ROS at AML. On the other hand, FOXO3a transduces more signals and is more sensitive at MDS than nonleukemia. It indicates that FOXO3a tends to gain some functions of apoptosis, cell cycle arrest, and detoxification of ROS at MDS, the pre-AML. We speculate that FOXO3a maintains the function for immature state of LICs in a part of patients with MDS. Thus, the disease progression from MDS to AML is observed or speculated from the above analysis. Since the PI3K-AKT pathway and

the apoptosis pathway from TNFR to IKKs affect directly and indirectly FOXO3a, respectively, the dysfunctions favoring to leukemogenesis are caused by accumulated and propagated influences from mutations or abnormalities in these deregulated pathways.

Transductivity evolution analysis of proteins in STPs contributing to leukemogenesis at AML, MDS, and nonleukemic cells

To investigate the dysregulation of cellular processes, including apoptosis, cell cycle progression, DNA repair, differentiation, cell proliferation, and self-renewal of stem cells, in AML and MDS cells, we determined transductivities of proteins in STPs contributing to leukemogenesis by using four examples, including the transductivities of p53, STATs and C/EBPα, CSLs, and β-catenin and FOXO3a, from two aspects. First, we determined the transductivity of each TF to show its dysfunction. Second, we determined the transductivities of proteins in STPs to understand how their dysregulation resulted in the dysfunctioning of TFs. Transductivity analysis showed that dysfunctioning of TFs occurred because of the effects of, or perturbations from, mutations in genes encoding these TFs or other upstream proteins. Mutations in these genes were predicted by comparing the changes in the transductivity patterns of the corresponding proteins and their upstream proteins in leukemic and nonleukemic cells. A sudden change in transductivity pattern was observed when transductivity of a protein was higher and that of its upstream protein was lower in AML cells compared with nonleukemic cells. These results suggested that changes in transductivity patterns were influenced by mutations in genes encoding proteins involved in STPs (Fig. 10.18). Biologically, sudden changes in the transductivity patterns of proteins are attributed to mutations in genes encoding these proteins in leukemic cells compared with those in nonleukemic cells. In other words, if a protein involved in a coupling STP is mutated, its transductivity pattern is changed in that STP and influences other downstream proteins, resulting in the dysregulation of that STP and dysfunctioning of the downstream proteins. Figs. 10.9–10.16 show some predicted mutations in genes encoding proteins with different transductivities in AML and MDS cells (indicated by *cyan* and *blue* seven pointed stars, respectively). Five pointed stars in *yellow*, *red*, and *green* represent possible mutations and functions of oncoproteins and tumor suppressor.

1. Transductivity Evolution Analysis of Proteins in the p38 MAPK, JNK MAPK, Classical MAPK, and PI3K-AKT Coupling Pathways That Contribute to the Loss of Transductivity of p53 at AML and MDS Cells

The *dot and dashed black lines* represent PPIs omitted from Fig. 10.5. The bold solid line with a *pink arrow* indicates the direction in which signals are transduced through crosstalk between two pathways according to this transductivity analysis. The bold dashed line with a *pink cross* indicates the absence of signal transduction through

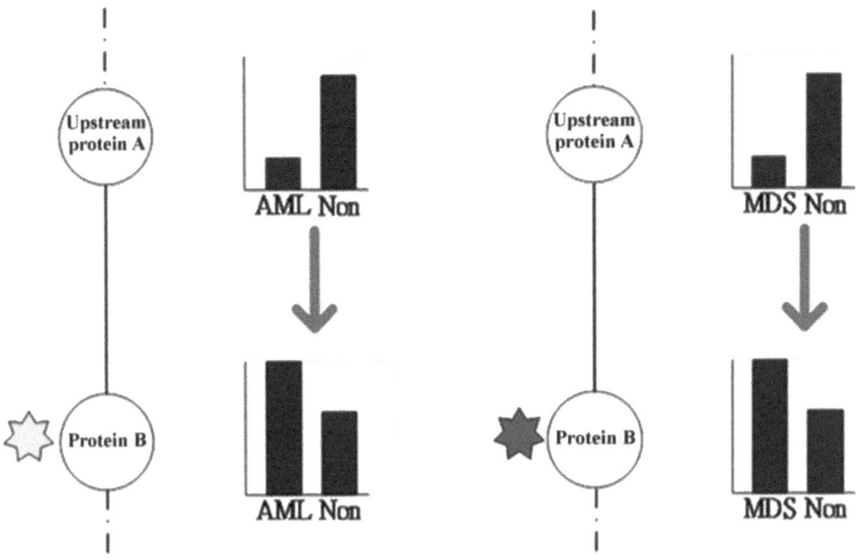

Figure 10.18 The illustration of the evolutionary transductivity patterns due to the effect of genetic mutations. The mutations in proteins are speculated and predicted when the transductivity pattern changes between protein and its upstream protein through the comparison of their transductivities in leukemic subtypes and nonleukemic subtype, i.e., we can observe a sudden change between transductivity patterns when the transductivity of a protein is larger at AML subtype and the transductivity of its upstream protein is less at AML subtype than at nonleukemic subtype. Biologically, the sudden change of transductivity pattern represents the effect of genetic mutations on the transductivity of protein in leukemic subtypes in comparison with nonleukemic subtype. If a genetic mutation occurs at a protein in STPs, then its transductivity pattern will be suddenly changed in the STP to influence the downstream proteins for further reactions such as dysregulation of STP and dysfunction of a TF.

crosstalk between two pathways. Proteins highlighted by *red* symbols with an underline are the main causes of dysfunctions in TFs.

Figs. 10.9, 10.12, and 10.13 show that p53 interacts with the p38 MAPK, JNK MAPK, classical MAPK, and PI3K-AKT pathways. The transductivities of p53 in AML, MDS, and nonleukemia cells were 0.00263, 0.00283, and 0.01875, respectively. Further, the lower transductivity of p53 in leukemic cells than in nonleukemic cells implied that p53, a tumor-suppressor protein, transduced signals with more difficultly in leukemic cells than in nonleukemic cells. We speculated that dysfunctioning of p53 that lead to abnormal activities and loss of function occurred more frequently in AML cells. This is consistent with the finding that p53, a caretaker and gatekeeper protein, orchestrates apoptosis, cell cycle arrest, and DNA repair without any control in leukemic cells. Dysfunctioning of p53 resulted in genomic instability and deregulation of cell growth and apoptosis.

The results of transductivity analysis (Figs. 10.12 and 10.13) showed that the loss of function of p53 was mainly attributed to proteins in the p38 MAPK pathway in AML and MDS cells. Moreover, progression of MDS to AML was determined by RSK2 and a mutation predicted in the gene encoding ERK. Next, we discuss the causes of dysregulation of STPs and dysfunctioning of p53. Transductivity analysis (Figs. 10.12 and 10.13) showed that transductivities of p53 in both AML and MDS cells were lower than its transductivity in nonleukemic cells. Some proteins omitted in the classical MAPK and PI3K-AKT pathways are shown in Figs. 10.14 and 10.15. Transductivities of proteins in the p38 MAPK pathway triggered by TGFβ1, TGFβ2, TGFβ3, and TNFα, except TNFR, were lower in AML cells than in nonleukemic cells (Fig. 10.10). Negligible changes were observed in the transductivity patterns of TNFR and TRAF2 in AML cells, indicating that the transductivity pattern of TRAF2 could be influenced by a predicted mutation in its gene. This change made the transductivity pattern of TRAF2 similar to that of DAXX and decreased the transductivity of its coupling protein ASK1 in AML cells compared with that in nonleukemic cells. Because the p38 MAPK pathway plays an important role in response to stress signals and in cellular functions such as cell cycle arrest, DNA repair, and apoptosis, decreased transductivities of proteins in this pathway in AML cells compared with those in nonleukemic cells indicated a loss of their cellular functions. Further, the transductivity pattern of MEKK1 was more similar to that of its upstream protein TRAF2 than to its downstream protein MKK4 and MKK4 is thus speculated to be a mutant due to the change of transductivity patterns between MEKK1 and MKK4. The transductivity pattern of JNK was similar to that of MKK4 in AML cells; however, this transductivity of p53 was mainly influenced by p38 and MDM2. Larger differences in PTP and MKP between AML and nonleukemic cells indicated higher transductivity of JNK in AML cells than in nonleukemic cells. Although the responses of JNK to stress signals could induce an apoptotic effect through p53, which prevented the growth of leukemic cells, the effects of p38 and the probable mutation in MDM2 became the main cause of p53 dysfunction.

The transductivity patterns of MEK and its upstream proteins RAS and RAF were similar to those of p53 in the classical MAPK pathway (Figs. B.3 and B.4), which responds to stress signals and triggers a defensive mechanism in cells similar to the p38 MAPK pathway. Changes were observed in the transductivity pattern of ERK (Fig. B.4), which were thought to be influenced by a mutation in its gene. Mutant ERK reversibly transduced more apoptotic signals in AML cells than in nonleukemic cells, which in turn inhibited leukemic cell growth. Comparison of transductivities of ERK and RSK2 (Fig. 10.10) in AML and nonleukemic cells showed that the transductivity of RSK2 was inhibited in AML cells but was enhanced in nonleukemic cells. A smaller difference in the transductivities of JNK, RSK2, and MDM2 between AML and nonleukemic cells indicated that the JNK and classical MAPK pathways may not prominently influence p53 compared with the p38 MAPK and PI3K-AKT pathways in AML cells (Fig. 10.10). Although the classical MAPK pathway may not prominently influence p53 in AML cells, it is important to note the transductivity pattern of RSK2. RSK2 attenuated apoptotic signals in AML cells to favor leukemic cell growth and increased them in nonleukemic cells to strengthen the defense mechanism compared with ERK in the classical MAPK pathway. In addition, p53 was degraded by the interaction of ubiquitin ligase MDM2, which in turn inhibited apoptosis. The transductivity patterns of AKT and MDM2 indicated that AKT decreased the induction of MDM2-mediated ubiquitination and the degradation of p53 in AML cells compared with nonleukemic cells (Fig. 10.10). This decreased the antiapoptotic effect in AML cells. Although the PI3K-AKT pathway decreased the antiapoptotic effect in AML cells compared with that in nonleukemic cells (which may not decrease the apoptotic function of p53), the loss of apoptotic function of p53 in AML cells could not be determined using this pathway. This is because the difference in the antiapoptotic effects of the PI3K-AKT pathway was lower than the difference in the apoptotic effects of the p38 MAPK pathway between AML and nonleukemic cells, implying that p38 in the p38 MAPK pathway dominated multiple interactions in other pathways because of the loss of function of p53. Because TRAF2, which showed a lower difference in transductivities between AML and nonleukemic cells (Fig. 10.10), had very little effect on ASK1, upstream proteins in the p38 MAPK pathway, such as DAXX and TGFβR, which showed the highest differences of transductivities between AML and nonleukemic cells, could be the main cause of dysregulation of the p38 MAPK pathway.

In contrast, the transductivities of proteins in the p38 pathways, except TNFR and TRAF2, were lower in MDS cells than in nonleukemic cells (Fig. 10.11). Changes were observed in the transductivity pattern of ASK1; however, these changes were not attributed to mutations in its gene but were attributed to the transductivity pattern of DAXX. Transduction of apoptotic signals was lower in MDS cells than in nonleukemic cells. Higher transductivity of TRAF2 induced MKK4 in the JNK MAPK pathway in MDS cells than in nonleukemic cells. This was not attributed to mutations because no change was observed in the transductivity patterns of MEKK1 and MKK4. Although MKK4 was not predicted to be mutated in MDS cells, the transductivity of MKK4 in MDS cells was higher than that in nonleukemic cells, which was similar to that observed in AML cells (Figs. 10.12 and 10.13). This suggested that MKK4 plays an important role in enhancing apoptotic signals in AML or MDS cells compared to nonleukemic cells. However, the increased transductivities of PTP and MKP inhibited the apoptotic signal of JNK in MDS cells. In addition, the transductivity pattern of MEK was opposite to that of ERK and was thought to be influenced by a mutation in the gene encoding ERK involved in the classical MAPK pathway. In MDS cells, stress signals were reversibly increased by ERK, which further affected its downstream protein RSK2 to exert increased apoptotic effects. The higher transductivity of AKT in the PI3K-AKT pathway increased MDM2-mediated ubiquitination and resulted in higher antiapoptotic effects in MDS cells than in nonleukemic cells. The effect was resulted from a mutation predicted in its gene due to changes in the transductivity patterns of AKT and MDM2. Signal transduction to MDM2 inhibited p53 degradation by MDM2-mediated ubiquitination, which in turn decreased the antiapoptotic effects in MDS cells compared with those in nonleukemic cells. Comparison of differences in the antiapoptotic effects of the PI3K-AKT pathway and apoptotic effects of the classical MAPK, JNK MAPK, and p38 pathways in MDS and nonleukemic cells showed that the antiapoptotic effects of the p38 pathway were lower than the apoptotic effects of the p38 pathway. This indicated that the p38 MAPK pathway contributed to the loss of function of p53 via multiple pathways in MDS cells. Notably, transductivity of p53 resulted in increased apoptosis, cell cycle arrest, and DNA repair in MDS cells than in AML cells. These functions of p53 in MDS cells compared with those in AML cells were mainly attributed to RSK because RSK2 induced higher apoptotic effects in MDS cells than in AML cells. This suggested that the ability of RSK2 to transduce higher apoptotic signals in AML cells could prevent the progression of MDS to AML.

2. Transductivity Evolution Analysis of the Proteins in JAK-STAT and MAPK Coupling Pathways That Contribute to the Gain of Transductivities of STATs and C/EBPα at AML and MDS

Some TFs such as STATs, AP-2α, c-Fos, c-Jun, C/EPBα, AML1, and ETO in the JAK-STAT and classical MAPK coupling pathways are discussed in this section (Figs. 10.12 and 10.13). Transductivity analysis showed that only STATs interacted with both the JAK-STAT and classical MAPK coupling pathways, c-Jun interacted with both the classical MAPK and JNK MAPK coupling pathways, and other TFs interacted with the classical MAPK pathway. First, we discuss the transductivity of STATs.

Transductivities of STATs in AML, MDS, and nonleukemic cells were 0.837029, 1.507073, and 0.1456, respectively. Transductivities in leukemic cells were much higher than those in nonleukemic cells, indicating that STATs significantly buffered signals from all external stimuli in AML cells and enhance the signals in MDS cells. Moreover, STATs attenuated large amount of energy of the signals from all external signals in nonleukemic cells, indicating that STATs were more sensitive to all external stimuli in leukemic cells than in nonleukemic cells even if the external stimuli were small. Higher signal transductivity of STATs inhibited apoptosis and promoted cell cycle progression and cell proliferation in AML and MDS cells. This is consistent with the important role of STATs in leukemogenesis. Other TFs such as AP-2α and c-Fos attenuated external signals because their transductivities were <1. They showed the same transductivity pattern in both AML and MDS cells and higher transductivities in nonleukemic cells, indicating that AP-2α and c-Fos inhibited apoptosis and promoted cell cycle progression and cell proliferation in leukemic cells compared to nonleukemic cells, eventually leading to leukemogenesis. Although c-Jun did not inhibit apoptosis and did not induce cell proliferation in AML, transductivity of c-Jun may play an important role in the progression of MDS to AML. It is because, in contrast, c-Jun inhibited apoptosis and induced cell proliferation in MDS cells compared to nonleukemic cells. Bernasconi also showed that the apoptosis phenotype in MDS cells shifted to be antiapoptosis phenotype in AML cells [108].

Transductivities of C/EBPα were <1 in AML, MDS, and nonleukemic cells, indicating that C/EBPα attenuated all external stimuli (Figs. 10.12 and 10.13). C/EBPα was involved in cell differentiation and proliferation. Because the functions of C/EBPα were influenced by other TFs such as ETO and AML1, which inhibited differentiation, we have discussed the transductivities of these three TFs to explain the functions of C/EBPα. Although C/EBPα increased the differentiation of leukemic cells compared to nonleukemic cells, the transductivities of AML1 in AML and MDS cells were lower than those in nonleukemic cells, indicating an insufficient activity of AML1 target genes during myeloid differentiation in AML and MDS cells. Transductivity patterns of C/EBPα and AML1 indicated that C/EBPα gain more function of differentiation at MDS cells than those at AML cells. It has been proposed that AML cells cannot undergo differentiation while MDS cells differentiate normally at first and then fail to differentiate and finally disappear [91].

Next, we discuss the transductivities of proteins involved in STPs and in crosstalks between catalytic PI3K (p110 subunit; called PI3Kc) and RAS (Figs. 10.12 and 10.13). It is important to note that RAS signaling affected PI3Kc signaling and attenuated survival signals, particularly in AML cells, through crosstalk between the PI3K-AKT and classical MAPK pathways (Fig. 10.12). We also found that the transductivity patterns of JAK, ERK, MKK4, c-Jun, AML1, and ETO and those of ERK, AML1, and PTEN were influenced by mutations in the corresponding genes in AML and MDS cells, respectively. Moreover, progression from MDS to AML was determined by C/EBPα and by mutations predicted in genes encoding JAK, ERK, and AML1. STATs, which interacted with the classical MAPK and JAK-STAT pathways, showed gain of function in AML cells. This was mainly attributed to mutations predicted in genes encoding BCR, ABL, and JAK. BCR and ABL mainly transmit signals from Grb2, EGFR, and IGFR. Changes in the transductivity patterns of FLT3, KIT, IL3R, and JAK suggested that JAK was influenced by mutations in its gene. Further, mutations in JAK affected its downstream TF STATs. Because the ability of ERK to promote cell proliferation of AML and nonleukemic cells was much lower than that of BCR,

ABL, and JAK, dysfunction of STATs was determined by mutations predicted in the genes encoding BCR, ABL, and JAK rather than by mutations predicted in the gene encoding ERK. Moreover, the gain of function of c-Fos and AP-2α, which interacted with the classical MAPK pathway, was mainly attributed to mutations predicted in the gene encoding ERK. Changes in the transductivity patterns of MEK and ERK suggested that the transductivity pattern of ERK was influenced by mutations predicted in the gene encoding ERK. The cell proliferation signal of RAS was determined using Shc, which was mainly affected by PDGFR, FLT3, and KIT. Moreover, mutations in the gene encoding MKK4 may have favored the proliferative effect of JNK in AML cells. Further, c-Jun is predicted to be mutated because of the changes in the transductivity patterns of c-Jun, ERK, and JNK. It was observed that mutations predicted in genes encoding ERK and JNK favored the higher proliferative effect of c-Jun in AML cells than in nonleukemic cells. However, the proliferative effect in AML cells was attenuated by mutations predicted in the gene encoding c-Jun, suggesting that c-Jun did not contribute to leukemogenesis in AML cells. Transductivities of regulatory PI3K (called PI3Kr) were determined using EGFR and IGFR. Changes in the transductivity patterns of PI3Kr and PI3Kc were not attributed to mutations in genes encoding PI3Kc because the signals were transduced by crosstalk between RAS and PI3Kc. RAS signaling influenced PI3Kc signaling and decreased the survival signal in AML cells. Further, tumor-suppressor protein PTEN dephosphorylated PIP3 to PIP2 and attenuated the transduction of the survival signal by the PI3K-AKT pathway. Thus, PTEN served as a brake to attenuate the transductivities of PIP3 in AML and nonleukemic cells, thus attenuating the transduction of the survival signal by AKT than by PIP3. Dysfunctioning of differentiation was attributed to mutations predicted in the gene encoding AML1. Changes in the transductivity patterns of AML1, C/EBPα, and ETO were observed. Changes in the transductivity patterns of AML1 and ETO were suggested to be influenced by mutations in their corresponding genes. The transductivity of C/EBPα in AML cells was influenced by a signal from ERK, which in turn was influenced by mutations predicted in the gene encoding ERK. Although C/EBPα increased the differentiation and cell proliferation arrest of AML cells compared to nonleukemic cells, mutations in the gene encoding AML1 further inhibited the myeloid differentiation of AML cells. Transductivity analysis showed that transductivities of AML1, ETO, and C/EBPα inhibited the differentiation of AML cells compared with that of nonleukemic cells.

In contrast, the gain of function of STATs was mainly attributed to BCR and ABL that were predominantly activated by signals from Grb2, EGFR, IGFR, and FLT3 in MDS cells (Fig. 10.13). In addition, the gain of functions of c-Fos and AP-2α was attributed to mutations predicted in the gene encoding ERK. Further, the gain of function of c-Jun was attributed to a mutation predicted in the gene encoding ERK and increased proliferative signals from TRAF2 in the JNK MAPK pathway. The transductivity patterns of Shc, SOS, RAS, and PI3Kr in MDS cells were not affected by the crosstalk between RAS and PI3Kr. Further, the transductivity pattern of RAS was affected by signals from SOS and Grb2 in MDS cells. Notably, the transductivity patterns of RAF and RAS showed that RAS-induced proliferation of MDS cells was attenuated by RAF. Although RAF and MEK decreased the proliferation of MDS cells compared with that of nonleukemic cells, mutations predicted in the gene encoding ERK substantially increased the proliferation of MDS cells, which was the main cause of dysfunctioning of c-Fos, c-Jun, and AP-2α. The transductivities of PTEN and AML1 were influenced by mutations predicted in their corresponding genes. Transductivity analysis of PIP3, PTEN, and AKT indicated that the transductivity of PTEN was influenced by a mutation in its corresponding gene that possibly destroyed the integrity of the genome and was the first step of malignant transformation. PTEN transduced more signals and possibly increased the attenuation of the survival signal in MDS cells compared with that in nonleukemic cells. Transductivity of AKT was mainly influenced by mutations predicted in genes encoding IKKs involved in the apoptotic pathway (Fig. 10.13). Therefore, AKT transduced more apoptotic signals in MDS cells than in nonleukemic cells. C/EBPα-induced differentiation of MDS cells was attributed to signals from ERK. Although ETO transduced more signals to inhibit the differentiation of MDS cells, myeloid differentiation arrest of MDS cells was inhibited by a mutation in the gene encoding AML1, which hampered its transductivity. Therefore, inhibition of AML cell differentiation and promotion of MDS cell differentiation compared with that of nonleukemic cells were attributed to AML1 and C/EBPα, respectively. Progression of MDS to AML was determined by mutations predicted in genes encoding C/EBPα, JAK, ERK, and AML1.

3. Transductivity Evolution Analysis of the Proteins in Notch Pathway That Contribute to the Gain of Transductivity of CSLs at AML and MDS Cells

This section discusses the transductivities of CSLs, coactivators, and corepressors that act downstream of the Notch pathway (Figs. 10.14 and 10.15). Coactivators included in this analysis were p300, HATs, MAML, and SKIP, and corepressors were CTBP and SMRT. These TFs attenuated signals from all external stimuli because their transductivities were <1. The transductivities of CSLs and coactivators were higher in leukemic cells than in nonleukemic cells, whereas the transductivities of corepressors were higher in nonleukemic cells than in leukemic cells. Interaction of CSLs with coactivators and corepressors had positive and negative effects, respectively, on its activities, indicating that transductivities of CSLs in AML and MDS cells were enhanced according to its transductivity patterns. Further, gain of function of CSLs was observed in the two leukemic cell types.

Pathways and transductivity patterns can be used to briefly explain biological PPIs (Figs. 10.14 and 10.15). The Notch pathway was triggered when the NOTCH receptor interacted with ligands DLL-1, DLL-3, and DLL-4 released by neighboring cells. The gamma-secretase complex containing PSEN, NCSTN, PSE2, and APH-1 mediated proteolytic cleavages to release ICN, which sent a signal to CSLs in the nucleus. Transcriptional regulation of CSLs occurred after corepressors CTBP and SMRT were translocated and coactivators p300, HATs, MAML, and SKIP were recruited by the binding of ICN to CSLs. DVL attenuated ICN through crosstalk between the Wnt and Notch pathways. Further, Numb, a tumor-suppressor protein, interacted with NOTCH1; the upregulation of Numb was associated with cell differentiation. An analysis of transductivity patterns indicated that the gain of function of CSLs may be attributed to mutations in genes encoding CTBP and SMRT in AML cells (Fig. 10.16). The transductivity pattern of NOTCH in AML cells was similar to that of the gamma-secretase complex, Numb, and corepressors (except DVL and CSLs), and coactivators. This implied that in AML cells, gamma-secretase complex-mediated proteolytic cleavage inadequately influenced the NOTCH receptor, resulting in a decreased transduction of signals in response to external stimuli. This also indicated that decreased differentiation of AML cells was attributed to the decreased transductivity of Numb. Mutations in genes encoding DVL, CSLs, CTBP, and SMRT could be predicted by changes in the transductivity patterns of these proteins in AML cells. If the signal of DVL transduced NOTCH receptor, the more attenuation of ICN could be speculated. However, the different transductivity patterns of DVL and NOTCH receptor suggested that DVL did not affect the NOTCH receptor through crosstalk between Wnt and Notch pathways. In fact, the transductivity pattern of CSLs was influenced by a mutated NOTCH receptor that increased the transcriptional regulation of CSLs, with higher transductivity in AML cells than in nonleukemic cells. The transductivity patterns of CTBP and SMRT showed that the transductivities of corepressors involved in the transcriptional regulation of CSLs were attenuated in AML cells compared with those in MDS cells and that this was mainly attributed to mutations predicted in their corresponding genes. Moreover, higher transductivity of HATs in AML cells than in nonleukemic cells increased the transcriptional regulation of CSLs. In addition, increased binding of ICN to CSLs increased the transcriptional regulation of CSLs through the higher transductivities of coactivators in AML cells than in nonleukemic cells. Thus, the gain of function of CSLs was attributed to mutations predicted in genes encoding CSLs, CTBP, and SMRT in AML cells.

Further, increased proteolytic cleavage by the gamma-secretase complex because of increased transductivities of PSEN, NCSTN, and APH-1 released more ICN in MDS cells than in nonleukemic cells (Fig. 10.17). Changes in the transductivity patterns of DVL, Numb, Deltex, PSE2, CTBP, and SMRT were attributed to mutations predicted in the corresponding genes in MDS cells. The transductivity pattern of Numb was influenced by a mutation in its corresponding gene, indicating decreased differentiation of MDS cells. Importantly, the observation that Numb transduced higher number of signals in MDS cells than in AML cells was consistent with the finding that increase in the level of Numb was associated with a high risk of progression of MDS to AML. In addition, mutations predicted in genes encoding CTBP and SMRT influenced their transductivity patterns, which in turn decreased the transductivities of corepressors as barriers of transcriptional regulation of CSLs in MDS cells compared to nonleukemic cells. Thus, the gain of function of CSLs was attributed to upstream proteins PSEN and NCSTN and mutations predicted in genes encoding CTBP and SMRT in MDS cells. Progression of MDS to AML was determined using the transductivity patterns of PSEN and NCSTN.

In this section, CSLs showed higher transductivity in both myeloid leukemia and lymphoblastic leukemia cells than in nonleukemic cells (Fig. 10.8). This implied that CSLs in leukemic cells probably transduced more signals in response to external stimuli and showed gain of function compared with CSLs in nonleukemic cells. Because the gain of function of CSLs contributed to tumor suppression by inducing apoptosis, DNA repair, and cell cycle arrest in addition to its oncogenic functions, its contribution to leukemogenesis remained ambiguous. Thus, transductivity analysis provided three results (Figs. 10.14 and 10.15): (1) NOTCH receptor transduced few signals and lost some function in AML cells but transduced more signals and gained more function in MDS cells. This result supported the therapeutic potential of inducing NOTCH receptor in AML cells [99]; (2) The transductivity pattern in NOTCH and the mutation-affected transductivity patterns of CTBP and SMRT were the main causes of the gain of function of CSLs; and (3) Progression of MDS to AML was determined by changes in the transductivities of PSEN and NCSTN. The speculated mutations may provide inaccurate results if they occur in the NOTCH receptor because no data are available on the comparison of transductivity patterns of extracellular proteins.

4. Transductivity Evolution Analysis of Proteins Involved in the Wnt, PI3K-AKT, and Apoptosis Coupling Pathways Associated With β-catenin and FOXO3a in AML, MDS, and Nonleukemic Cells

This section discusses the transductivities of 2 TFs β-catenin and FOXO3a at downstream of the canonical Wnt, PI3K-AKT, and apoptosis pathways (Figs. 10.16 and 10.17). Transductivities of β-catenin were 0.065852, 0.019664, and 0.015767 and those of FOXO3a were 0.0000179, 0.025403, and 0.004194 in AML, MDS, and nonleukemic cells, respectively. This indicated that β-catenin and FOXO3a attenuated signals from all external stimuli in these three cell types. The transductivity pattern of β-catenin (Figs. 10.16 and 10.17) indicated that it transduced more signals in response to external stimuli in leukemic cells, especially AML cells, than in nonleukemic cells. This indicated that β-catenin showed gain of function that favored leukemogenesis of AML cells compared to other cells. The gain of functions included induction of stem cells to promote their departure from the state of quiescence, to increase their self-renewal ability, and to decrease their apoptosis. FOXO3a transduced more signals in MDS cells but few signals in AML cells than in nonleukemic cells. This indicated that FOXO3a gained more functions in MDS cells and lost some functions in AML cells compared to nonleukemic cells. This also indicated the presence of heterogeneous phenotypes in MDS and AML cells. FOXO3a orchestrated apoptosis, cell cycle arrest, and detoxification of ROS. Gain of functions of FOXO3a allowed it to transduce more signals in response to external disruptive stimuli, induce cell cycle arrest and apoptosis, and repair abnormal cells and vice versa. This in turn increased the apoptosis of MDS cells and decreased the apoptosis of AML cells. The observation of heterogeneous pathophysiological phenotypes of MDS and AML cells is consistent with that reported in previous clinical studies [91]. In addition, comparison of transductivities of coupling STPs indicated that AML cells showed better survival than nonleukemic cells (Figs. 10.16 and 10.17). The dysfunctions observed in AML and MDS cells and their main causes are discussed in subsequent sections.

The loss of functions of FOXO3a in AML cells was attributed to a signal from the PI3K-AKT pathway because signals from other pathways were not detected (Fig. 10.14). Further, the gain of function of FOXO3a in MDS cells was attributed to the survival signal from the PI3K-AKT pathway and apoptotic signal from the apoptosis pathway (Fig. 10.15). Transductivity of GSK3β in the Wnt pathway was influenced and attenuated by a signal from the PI3K-AKT pathway through crosstalk between these pathways in MDS cells. Some omitted PPIs in the PI3K-AKT pathway are shown in Figs. 10.12 and 10.13. Biologically, Wnt proteins bind to their receptors, i.e., Frizzled family proteins, and DVL phosphorylates and degrades the destructive complex to produce β-catenin in the nucleus for transcriptional regulation. Transductivity analysis in the present study showed that the destructive complex containing GSK3β, CK1, Axin, and APC antagonized and stabilized downstream β-catenin to reduce the self-renewal of LSCs. In AML cells, the gain of function of β-catenin was influenced by mutant DVL (Fig. 10.14). Mutant DVL affected the transductivity pattern of DVL and reversibly transduced more signals, resulting in the accumulation of β-catenin in AML cells compared with that in nonleukemic cells. Comparison of the transductivity patterns of DVL, GSK3β, and AKT showed that signal transduction did not occur through the crosstalk between AKT and GSK3β. Transduction of signals to GSK3β and CK1 substantially attenuated the degradation of the destructive complex in nonleukemic cells but only slightly attenuated this degradation in AML cells. This implied that GSK3β and CK1 served as brakes to attenuate the degradation of the destructive complex in nonleukemic cells but may not be active in AML cells. Moreover, this modified signal further influenced Axin, APC, and β-catenin through the transduction of signals from GSK3β and CK1. This indicated that GSK3β and CK1 deregulated and increased cell proliferation in AML cells by accumulating β-catenin. The transductivity patterns of TRADD and TRAF2 in the apoptotic and PI3K-AKT pathways were influenced by mutations predicted in their corresponding genes (Fig. 10.14). These mutations might reversibly attenuate response to stress or apoptotic signals from TNFR in AML cells. Importantly, transductivities of IKKs in both AML and nonleukemic cells were substantially attenuated (0.000228 and 0.001056, respectively), which in turn attenuated signal transduction from IKKs to AKT, thus inhibiting the apoptosis of both AML and nonleukemic cells. The loss of function of FOXO3a in AML cells was attributed to the attenuation of a signal from the PI3K-AKT pathway. Because the difference in the function of FOXO3a in AML and nonleukemic cells was small, it was suggested that decreased antiapoptotic effect of FOXO3a was not responsible for the significant dysfunctioning of apoptosis in AML cells.

In contrast, the gain of function of FOXO3a in MDS cells was attributed to the survival signal from the PI3K-AKT pathway and apoptotic signal from the apoptosis pathway (Fig. 10.15). In contrast to the attenuation of antiapoptotic signal in AML cells, apoptotic signal from a possible IKK mutant in MDS cells influenced the transductivities of AKT, FOXO3a, and GSK3β through crosstalk between the apoptosis and PI3K-AKT pathways. No change was observed in the transductivity pattern of TRAF2. Therefore, more signals were transduced from TNFR to TRAF2 in MDS cells than in nonleukemic cells. Although the transductivity of TRAF2 in MDS cells was higher than that in nonleukemic cells, the transductivity of RIP1 was prominently influenced by a mutation predicted in the gene encoding TRADD. In the presence of mutant IKKs, signals from NIK were reversibly transduced, resulting in higher transductivity in MDS cells than in nonleukemic cells. This suggested that mutations predicted in genes encoding IKKs in MDS cells could not antagonize the apoptotic signals from NIK in AML cells. Collectively, the gain of function of FOXO3a was mainly attributed to apoptotic signals from mutant IKKs. This suggested that FOXO3a resulted in significant dysfunctioning of apoptosis in MDS cells. The transductivity pattern of DVL was affected by a mutation predicted in its corresponding gene in MDS cells. Mutant DVL transduced few signals in MDS cells, resulting in decreased proliferation because of decreased degradation of the destructive complex. Signal transduction from mutant IKKs to GSK3β caused GSK3β to reversibly transduce more signals, resulting in increased proliferation of MDS cells compared to nonleukemic cells. Because almost no differences were observed in the transductivities of Axin, APC, and CK1 between MDS and nonleukemic cells, the gain of function of β-catenin was attributed to GSK3β that was influenced by a signal from AKT through crosstalk between the canonical Wnt, Notch, PI3K-AKT, and apoptosis pathways. Therefore, progression of MDS to AML was attributed to GSK3β and CK1 and mutations predicted in IKKs.

Strategies for developing anticancer therapies based on the results of evolutionary transductivity analysis

AML, the most common subtype of leukemia, accounts for approximately 80% cases of adult leukemia and is associated with a low survival rate [77,78]. After treatment, prognosis of patients at AML is poorer than that of patients with other leukemic subtypes because of genetic and molecular diversities, which consequently result in the recurrence of AML [109]. Genetic mutations and chromosomal translocations observed in AML might disrupt the transcriptional regulation of differentiation, apoptosis, and proliferation of hematopoietic cells to promote malignancy [86] and result in poor response to therapeutic agents. Multiple biological factors, including aberrant STPs, survival of LSCs, multidrug resistance proteins, microenvironmental interactions, and immune tolerance, make treatment of AML difficult, eventually leading to relapse [50].

Targeted therapy focuses on a single protein or proteins in few aberrant STPs without considering the impact on the network. This strategy may fail when other proteins stimulate the target proteins, resulting in malignant transformation through interactions within the network because of PPIs and GRNs. Therefore, to provide a novel therapeutic approach, we proposed a strategy involving determination of drug targets from a systematic viewpoint by using the results of transductivity analysis. Transductivity analyses of proteins involved in multiple STPs that favor the development of LSCs can easily determine the roles of these proteins in transducing signals in response to external stimuli (Fig. 10.5). Changes in transductivity patterns can further identify proteins or pathways responsible for the dysfunction of TFs in multiple STPs to provide potential drug targets. Dysfunctioning of TFs during leukemogenesis occurs because of two reasons: (1) Some proteins involved in STPs either transduce few signals in response to stress or apoptotic signals, resulting in the disruption of defense mechanism, (e.g., DAXX and TGFβR) or transduce more signals in response to external stimuli, resulting in cell proliferation (e.g., Grb2 and GSK3β) and (2) Some proteins in STPs are affected by genetic mutations, resulting in the dysregulation of signal transduction by these STPs (e.g., ERK, JAK, and Numb indicated in *cyan* seven pointed stars in Figs. 10.10–10.17). We believe that these proteins are can be used for designing single drug targets for treating leukemogenesis.

Recent studies have shown that cancer has more redundant robustness because cancer cells buffer various genetic mutations that accumulate during carcinogenesis, resulting in diverse evolution and elimination of normal cells [110]. In addition, clinical treatments are not very effective against relapsed leukemia or drug-resistant leukemia [50]. Generally, dual/multiple mutations do not occur simultaneously; moreover, mutated proteins are resistant to therapies. Therefore, we proposed another strategy involving multiple drug targets to prevent drug resistance associated with mutant proteins and to allow effective treatment. Mutant proteins identified through transductivity analysis in the present study can be used as drug targets for more effective therapy. For example, targeting Grb2 and GSK3β by using their respective pharmacological inhibitors (peptidimer-c [111] and SB216763/SB415286 [112]) in patients with AML can attenuate abnormal proliferation. We believe that therapy based on transductivity analysis can provide higher efficiency from a systematic viewpoint. Other potential drug targets for treating AML and MDS and preventing the progression of MDS to AML are listed in Table 10.4. Explanations regarding these proteins as potential drug targets are provided in Section 10.2.3 and are shown in Figs. 10.10–10.17.

Based on the transductivity analysis, we find a list of proteins as signatures to identify clinically which people suffer from MDS, AML, and further diagnose people who are with disease in MDS and may worsen then in progress to AML. These proteins are determined by transductivity analysis at (see Section 10.2.3) and their transductivities at nonleukemic, MDS and AML subtypes are listed in Table 10.4. Based on transductivity analysis, dysfunctions of which TFs attribute to the proteins are also listed in Table 10.4. The cause why proteins contribute to leukemogenesis at MDS, AML and progression from MDS to AML may be due to five reasons:

1. Proteins may transduce less signals to respond to either the stress or apoptotic signals to cause the deficiency of defensive mechanism like DNA repair.
2. Proteins may transduce more signals in some pathways to facilitate cell proliferation and antiapoptosis.
3. Protein may be affected by their genetic mutations to dysregulate the signal transduction of STPs.
4. Protein may transduce less signals in the pathways to facilitate cell differentiation.
5. Protein may transduce more signals in the pathways to facilitate cell differentiation.

Thus, some proteins among these proteins can be regarded as potential drug targets and the corresponding treatments for therapies are determined. Since TFs are associated with lots of downstream transcriptions for essential mechanism, the TFs cannot be regarded as potential drug targets, resulting in no treatments for therapies. We provide the fold changes and variances of each signature to verify the reliability of signatures based on the transductivity analysis. The fold change is calculated as the ratio of the mean of leukemic subtype to nonleukemic subtype. And the variance is calculated by patients at nonleukemic, MDS, AML subtypes. The fold change, which is equal to or larger than one, is shown in bold and represents the differentially expressed protein between leukemic subtype and nonleukemic subtype. The signatures in

Table 10.4 The signatures for the identifications of the progression from MDS to AML and specific subtypes based on transductivity analysis

(A) Protein signatures for identification of MDS

Protein signature	Transductivity		Suggested treatment	A ratio of mean expression in MDS to Non-Leu	A variance of expression in patients		Functions in leukemogenesis (dysfunction of TF)
	Non-Leu	MDS			MDS	Non-Leu	
C/EBP-α	0.0002	0.0361		0.7726	848,251.61	19,217,040.95	5
PSEN	0.0987	0.3008	Inhibitor	**1.0179**	248,411.94	167,573.03	2 (CSLs)
NCSTN	0.0416	0.1530	Inhibitor	**1.2328**	24,603.49	18,137.11	2 (CSLs)
IKKs	0.0011	0.0537	Inhibitor	**1.1028**	12,665.02	7749.33	2, 3 (FOXO3a, β-catenin)

(B) Proteins as signatures for identification of AML

Protein signature	Transductivity		Suggested treatment	A ratio of mean expression in MDS to Non-Leu	A variance of expression in patients		Functions in leukemogenesis (dysfunction of TF)
	Non-Leu	AML		AML	AML	Non-Leu	
JAK	0.1532	0.3145	Inhibitor	0.7109	4,898,376.06	2,360,573.35	2, 3 (STATs)
TRAF2	0.0111	0.0020	Agonist	**1.2894**	619,492.22	111,782.08	1, 3 (p53)
CK1	0.0106	0.0351	Inhibitor	**1.2629**	59,237.12	8404.98	2 (β-catenin)
AML1	0.0347	0.0134		0.9662	10,077.42	4985.50	3, 4 (C/EBP-α)

(C) Proteins as signatures for identification of progression from MDS to AML

Protein signature	Transductivity		Suggested treatment	A ratio of mean expression in MDS to Non-Leu		A variance of expression in patients			Signature with asterisk meets multistep theory of carcinogenesis from MDS to AML	Functions in leukemogenesis (dysfunction of TF)
	MDS	AML		MDS	AML	Non-Leu	MDS	AML		
HATs	0.011	0.041	Inhibitor	**1.053**	0.950	368,974.9	790,829.9	1,209,579.2	*	2 (CSLs)
p38	0.028	0.001	Agonist	**1.001**	0.989	227,582.0	215,526.5	246,799.2	*	1 (p53)
DAXX	0.542	0.004	Agonist	**1.005**	0.676	217,649.0	282,772.2	132,876.4		1 (p53)
Numb	0.049	0.004	Agonist	**1.064**	**1.227**	6602.0	7642.0	16,924.1		3, 4
SMRT	0.007	0.001	Agonist	**1.084**	**1.169**	5876.6	9131.0	10,391.0	*	2, 3 (CSLs)
GSK3β	0.059	0.295	Inhibitor	**1.031**	0.957	9139.6	13,001.6	10,144.9		2 (β-catenin)
TGFβR	0.331	0.050	Agonist	0.810	0.663	27,157.9	16,646.1	8484.0		1 (p53)
Grb2	0.181	0.541	Inhibitor	0.873	0.996	2152.2	1702.7	3478.1		2 (STATs)

Functions in leukemogenesis: (1) Proteins may transduce less signals to respond to either the stress or apoptotic signals to cause the deficiency of defensive mechanism like DNA repair. (2) Proteins may transduce more signals in some pathways facilitate cell proliferation and antiapoptosis. (3) Protein may be affected by their genetic mutations to dysregulate the signal transduction of STPs. (4) Protein may transduce less signals in the pathways to facilitate cell differentiation. (5) Proteins may transduce more signals to facilitate cell differentiation.

The fold change, which is equal to or larger than one, is shown in bold and represents the differentially expressed protein between leukemic subtype and nonleukemic subtype.

Table 10.4A–C are arranged in descending order of variances for patients at MDS, AML, and AML subtypes, respectively. Consequently, some of these signatures with fold changes larger than one and bigger variances can identify people, who suffer from progression from MDS to AML and specific subtypes from the viewpoint of not only the differentially expressed proteins but also the diversity of characteristics or mutations of patients. Since the progression from MDS to AML is an example of the multistep (evolutionary) theory of carcinogenesis, we further check whether mutations are increasingly accumulated at upstream proteins of each signature in the pathway at MDS in comparison with AML. In Table 10.4C, if the signature at MDS in comparison with AML meets the multistep theory of carcinogenesis based on the change of transductivity patterns, a star would be put on the signature.

Signatures for identifying the progression of MDS to AML and other leukemic subtypes

MDS is partially related to AML, possibly pre-AML or a preexisting evolution of AML. According to the 2008 revision of the World Health Organization classification of myeloid neoplasms and acute leukemia, AML is characterized by 20% more blasts in the bone marrow compared with MDS. Because both MDS and AML are characterized by chromosomal abnormalities, absence of apoptosis, and abnormal differentiation and proliferation of immature blasts, it is difficult to distinguish between these two leukemic subtypes. The American Cancer Society has estimated that at least one-fifth of people with leukemia are not diagnosed because they show vague symptoms. Only the number of blasts that affects the function of mature cells in the bone marrow can be used to briefly differentiate between the two leukemic subtypes. In addition, one goal of therapy for MDS is to decrease the progression to AML. Therefore, it is important to identify proteins as signatures to diagnose patients with MDS who may progress to AML.

Transductivity evolution analysis identified proteins contributing to leukemogenesis of MDS and AML and progression of MDS to AML (Table 10.4A–C). We determined the transductivities of these proteins in nonleukemic, MDS, and AML cells to identify targets for therapy. For example, p38 is a signature for the progression of MDS to AML (Table 10.4C). Transductivity analysis showed that p38 attenuated p53-induced apoptosis. Low transductivity of p38 in AML cells compared with that in MDS cells indicated that it transduced few signals in response to stress or apoptotic signals, thus affecting defensive mechanisms such as DNA repair. Therefore, if a patient has MDS and may progress to AML, as determined by the transductivity pattern of p38, he or she can be treated using a p38 agonist to relieve discomfort and pain. In contrast, higher transductivity of GSK3β in AML cells than in MDS cells indicated that it transduced more signals in some pathways to facilitate cell proliferation and prevent apoptosis. Thus, the therapy to prevent the progression of MDS to AML would be to use GSK3β inhibitors. Transductivities of other proteins involved in leukemogenesis along with their signatures for identifying progression of MDS to AML and corresponding treatments are listed in Table 10.4C. This table also lists TFs responsible for dysfunctioning of proteins. Because TFs are involved in the transcriptional regulation of various genes and are essential for cellular functions, they cannot be used as potential drug targets (Table 10.4A and B). Proteins may contribute to leukemogenesis of MDS and AML and progression of MDS to AML through any of the following five ways: (1) Proteins may transduce few signals in response to stress or apoptotic signals to disrupt defensive mechanisms such as DNA repair; (2) Proteins may transduce more signals via some pathways to facilitate cell proliferation and prevent apoptosis; (3) Proteins affected by genetic mutations may dysregulate signal transduction of multiple STPs; (4) Proteins may transduce few signals to facilitate cell differentiation; (5) Proteins may transduce more signals to facilitate cell differentiation.

Further, we determined fold changes and variances in the signature of each protein to verify the reliabilities of these signatures based on transductivity analysis. Fold change was calculated as a ratio of mean protein signatures in MDS or AML cells to the mean protein signature in nonleukemic cells, and variance was calculated using each protein signature in patients at nonleukemic, MDS and AML subtypes. Fold change of ≥ 1 is shown in bold and represents differentially expressed proteins in leukemic and nonleukemic cells. Signatures in Table 10.4A–C are arranged in the descending order of variance for each protein signature in patients at nonleukemic, MDS and AML subtypes. Some of these signatures with fold changes of >1 and higher variances identified patients with MDS who may progressed to AML and patients with specific leukemic subtypes based on differentially expressed proteins and diverse characteristics or mutations. Because the progression of MDS to AML is an example of the multistep theory of carcinogenesis, we further checked whether mutations were increasingly accumulated in genes encoding proteins that act upstream of each STP in both MDS and AML cells. According to Table 10.4C, if the signature of a protein in MDS and AML cells met the multistep theory of carcinogenesis based on the changes in its transductivity patterns, a star would be put on the signature. Thus, we identified proteins as signatures to diagnose patients with MDS and AML and patients with MDS who may progress to AML (Table 10.4A–C).

10.2.4 Conclusion of evolution mechanisms of network robustness and signal transductivity in the carcinogenesis of leukemic cells

In this section, we developed an efficient method to systematically investigate the transductivities and network robustness of multiple coupling STPs and their downstream GRNs, respectively, in leukemic and nonleukemic cells by using microarray sample data instead of conventional methods such as time-course data and single pathway analysis. A flowchart for investigating both transductivities and network robustness for STPs and GRNs, respectively, is shown in Fig. 10.19. Coupling STPs included in this analysis were the PI3K-AKT, MAPK, JAK-STAT, TGFβ, apoptosis, Notch, hedgehog, and canonical Wnt pathways, along with their 28 downstream TFs. We also discussed the transductivities and transductivity evolution levels of TFs such as p53, STATs, AP-2α, c-Jun, C/EBPα, AML1, ETO, CSLs, p300, CTBP, β-catenin, and FOXO3a and their upstream proteins in AML and MDS cells because the GRNs

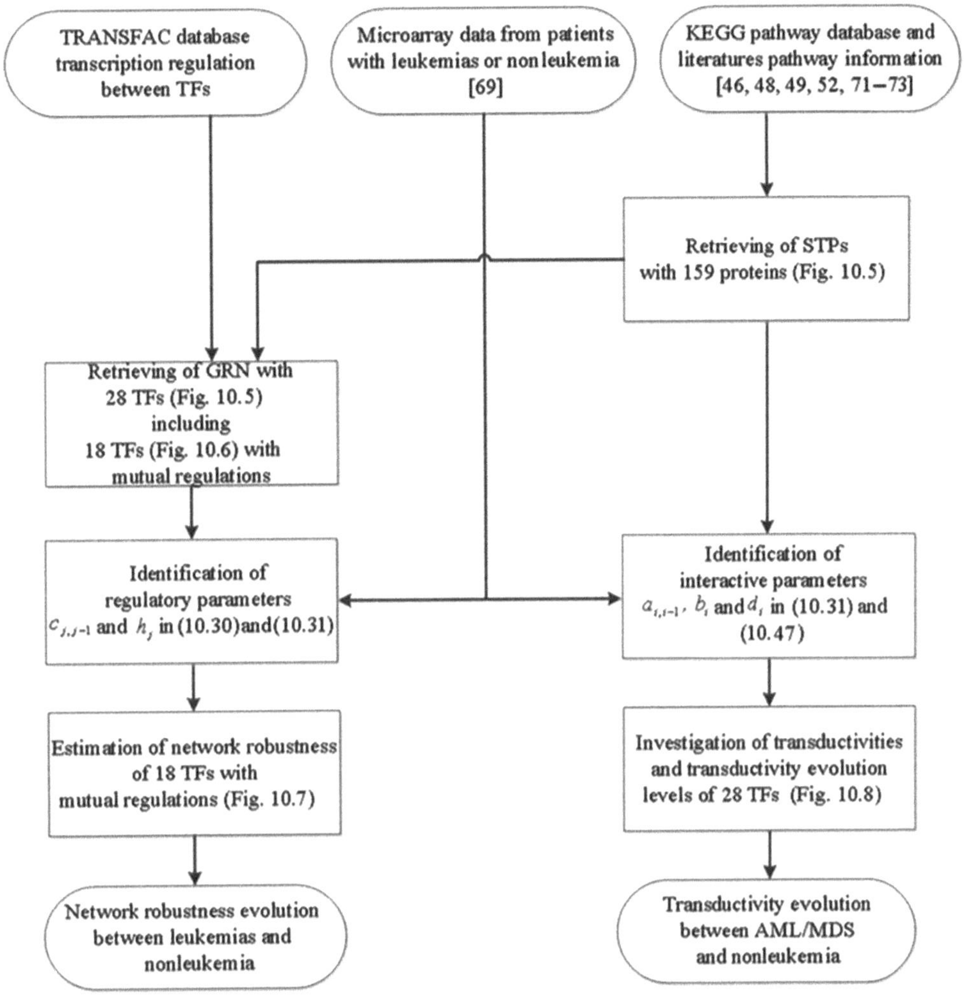

Figure 10.19 Flowchart of the proposed method to estimate the evolution of transductivity of each protein in coupling STPs in Fig. 10.5 and the evolution of network robustness of GRN in Fig. 10.6. To clarify the relation between the integrated cellular interaction network of STPs and GRN, we use the blocks of STPs and GRN to show this method.

of these TFs in these two leukemic subtypes show the highest robustness and harbor many mutations, resulting in their diverse evolution. Transductivity analyses of these TFs showed that STATs enhanced external signals while other TFs attenuated the external signals. Further, transductivity evolution levels of these TFs established cellular functions that lead to the pathophysiological phenotypes of leukemogenesis. Transductivity analysis also showed that possible mutations in genes encoding these TFs may affect other proteins through crosstalk between coupling STPs, resulting in the dysregulation of these STPs or their downstream TFs. We found that RAS signaling affected PI3Kc through crosstalk between the PI3K-AKT and classical MAPK pathways in AML cells and that AKT signaling affected GSK3β signaling through crosstalk between the PI3K-AKT and Wnt pathways in MDS cells. We also identified the two main causes of dysfunctions in TFs in AML and MDS cells. In this section, mutations in genes were predicted by changes in the transductivity patterns of their encoded proteins such as DVL, Deltex, PSE2, CSLs, CTBP, SMRT, ERK, TRADD, TRAF2, Numb, PTEN, JAK, AML1, ETO, and MDM2. Some of these predicted mutations such as those in genes encoding Numb, PTEN, JAK, AML1, ETO, and MDM2 were consistent with those reported in previous studies [49,86–90]. In addition, we proposed two drug designing strategies by using single and multiple drug targets, such as Grb2 and GSK3β, against leukemogenesis. We showed that proteins listed in Table 10.4 were signatures that can be used for diagnosing patients with MDS and AML and for identifying patients with MDS who may progress to AML. These signatures based on transductivity analyses can not only identify differentially expressed proteins but also identify diverse mutations that are characteristic to patients with leukemia. We also constructed a GRN of 18 genes encoding TFs that were determined using gene regulations obtained from the TRANSFAC database [74]. We observed that GRNs in all leukemic cells tolerated more intrinsic mutations than those in nonleukemic cells. GRNs in AML cells showed the highest network robustness compared with those in other leukemic cells, with various mutations and low response to external molecular signals or stress. This indicated a diverse evolution of leukemogenesis, which hampered the treatment of AML with various drugs.

Patients with AML and some patients with MDS are characterized by chromosomal abnormalities, absence of apoptosis, and abnormal cell proliferation of immature blasts compared with nonleukemic cells [108]. Moreover, AML and MDS cells do not undergo differentiation and show increased attenuation of apoptosis, indicating that these cells differentiate normally at first, fail to differentiate further, and eventually die [91], thus showing both pro-apoptotic and antiapoptotic phenotypes. However, some observations regarding predicted mutations and cellular functions based on transductivity evolution levels of proteins in AML and MDS cells are not consistent with the above findings. This may be because of the following reasons: First, if the number of leukemic/nonleukemic samples is less than the maximum interaction number of proteins in the model of STPs in Eq. (10.46), overfitting may occur, resulting in the inaccurate estimation of the recursive parameter of PPIs in Eq. (10.31). Second, measurement of transductivity may be inaccurate because of the diversity of samples from different patients. Third, the effect of receptors that were not considered for STPs included in this section may provide different results for transductivity analysis. Fourth, mutations may randomly occur in samples from different patients, and there is no comparison of transductivity patterns between possible mutant receptors and their upstream proteins to determine the mutations in these receptors. Fifth, MDS may include early and advanced cell types with different phenotypes. Furthermore, some patients with MDS may not progress to AML. Therefore, this future work will involve developing a better method by using microarray data to measure the transductivity and network robustness of STPs and GRNs, respectively, to eliminate inaccuracy arising from the above causes.

APPENDIX B: FIGURES

Here we show Figs. B.1–B.5.

10.3 NONLINEAR STOCHASTIC GAME STRATEGY FOR EVOLUTION MECHANISMS OF ORGAN CARCINOGENESIS UNDER A NATURAL SELECTION SCHEME

Molecular biologists have long recognized carcinogenesis as an evolutionary process that involves natural selection. Cancer is driven by the somatic evolution of cell lineages. After investigating the network evolution by the change of robustness and transductivity in aging and carcinogenic process in the Section 10.2, we will discuss the evolutionary strategy of organ carcinogenesis. In this section, the evolution of somatic cancer cell lineages during carcinogenesis was modeled as an equilibrium point (i.e., phenotype of attractor) shifting, the process of a nonlinear stochastic evolutionary biological network. This process is subject to intrinsic random fluctuations

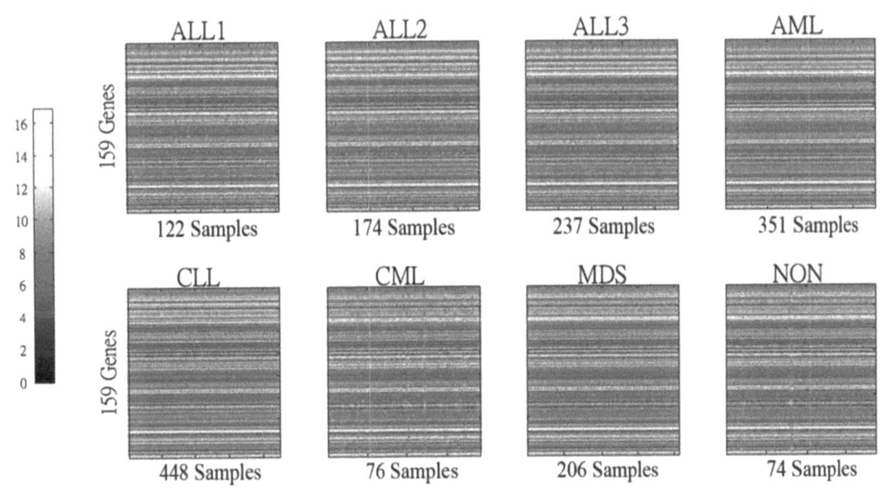

Figure B.1 The heat map of the binary logarithm of the original microarray data without normalization.

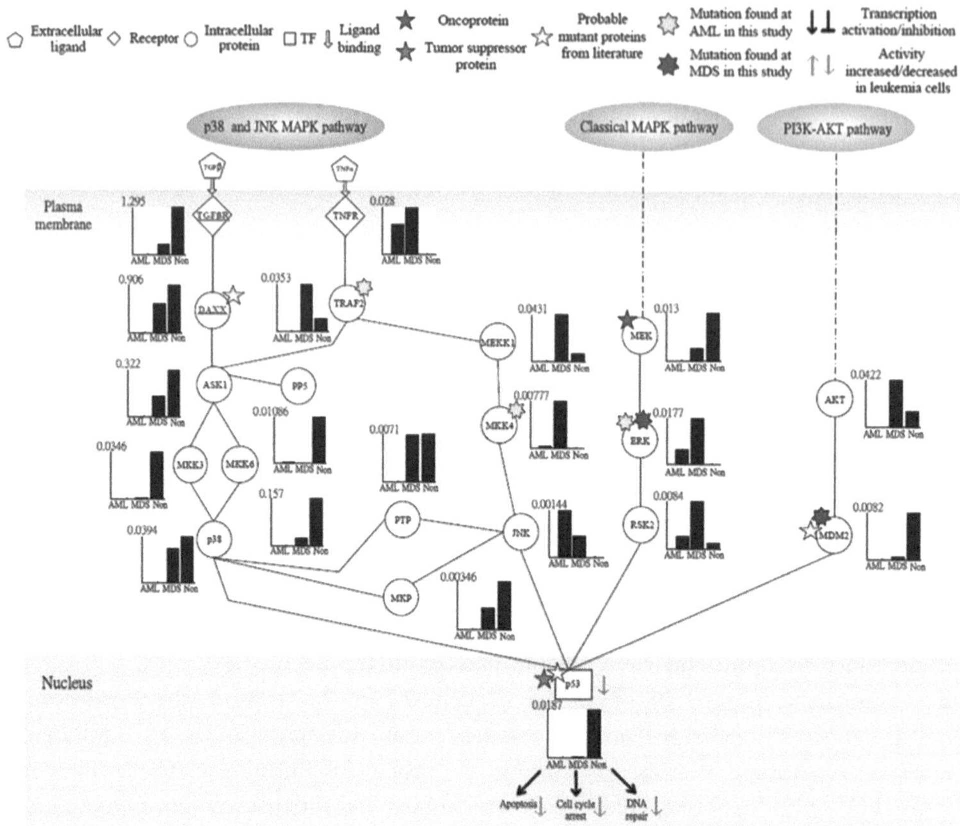

Figure B.2 Transductivity evolution of the proteins in p38 MAPK coupling pathways at AML and MDS subtypes.

due to somatic genetic and epigenetic variations, as well as extrinsic disturbances due to carcinogens and stressors.

In order to maintain the normal function (i.e., phenotype) of an evolutionary biological network subjected to random intrinsic fluctuations and extrinsic disturbances, a network robustness scheme that incorporates natural selection needs to be developed. This can be accomplished by selecting certain genetic and epigenetic variations to modify the network structure to attenuate intrinsic fluctuations efficiently, and to resist extrinsic disturbances in order to maintain the phenotype of the evolutionary biological network at an equilibrium point (attractor). However, during carcinogenesis, the remaining (or neutral) genetic and epigenetic variations accumulate, and the extrinsic disturbances become too large to maintain the normal phenotype at the desired equilibrium point for the nonlinear evolutionary biological network. The network is thus shifted to a cancer phenotype at a new equilibrium point that begins a new evolutionary process.

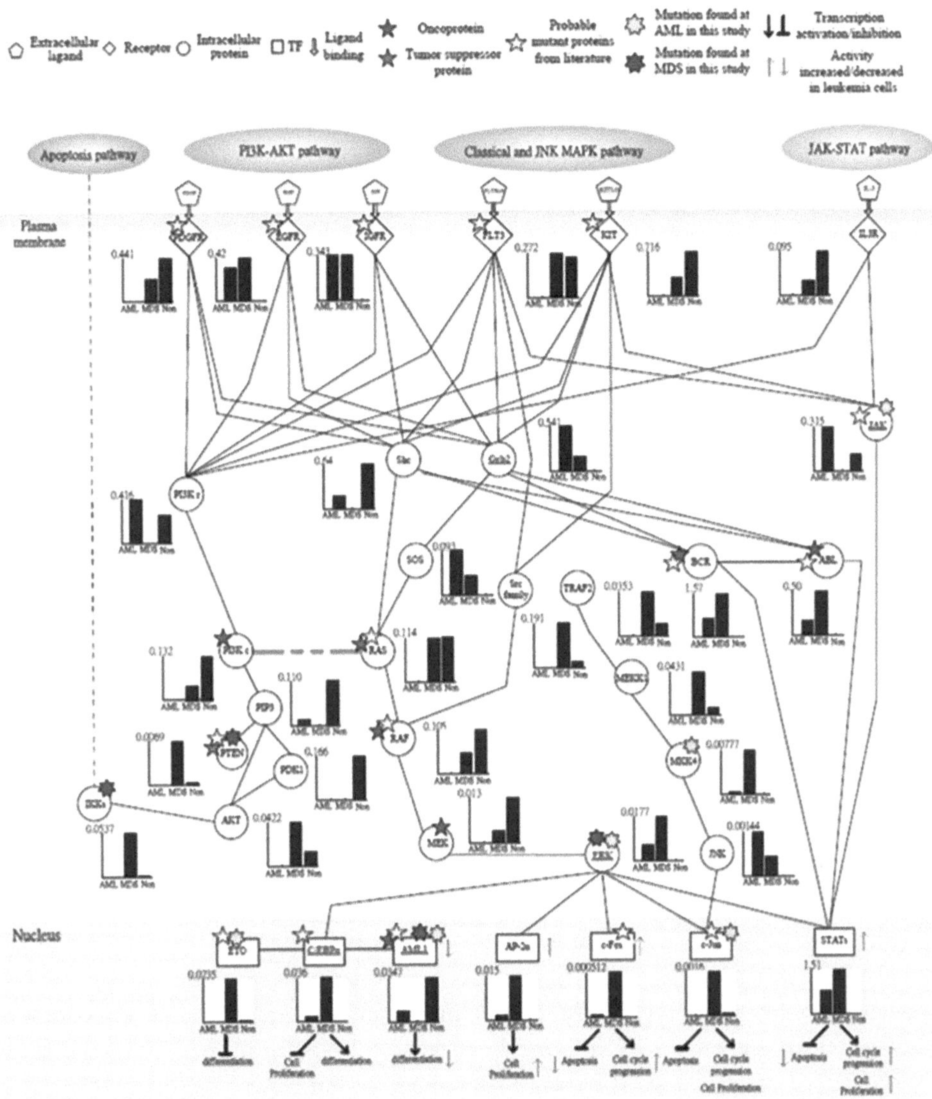

Figure B.3 Transductivity evolution of the proteins in JAK-STAT and MAPK coupling pathways at AML and MDS subtypes.

In this section, the natural selection scheme of an evolutionary biological network of carcinogenesis was derived from a robust negative feedback scheme based on the nonlinear stochastic Nash game strategy. The evolvability and phenotypic robustness criterion of the evolutionary cancer network were also estimated by solving a Hamilton–Jacobi inequality (HJI)-constrained optimization problem. The simulation

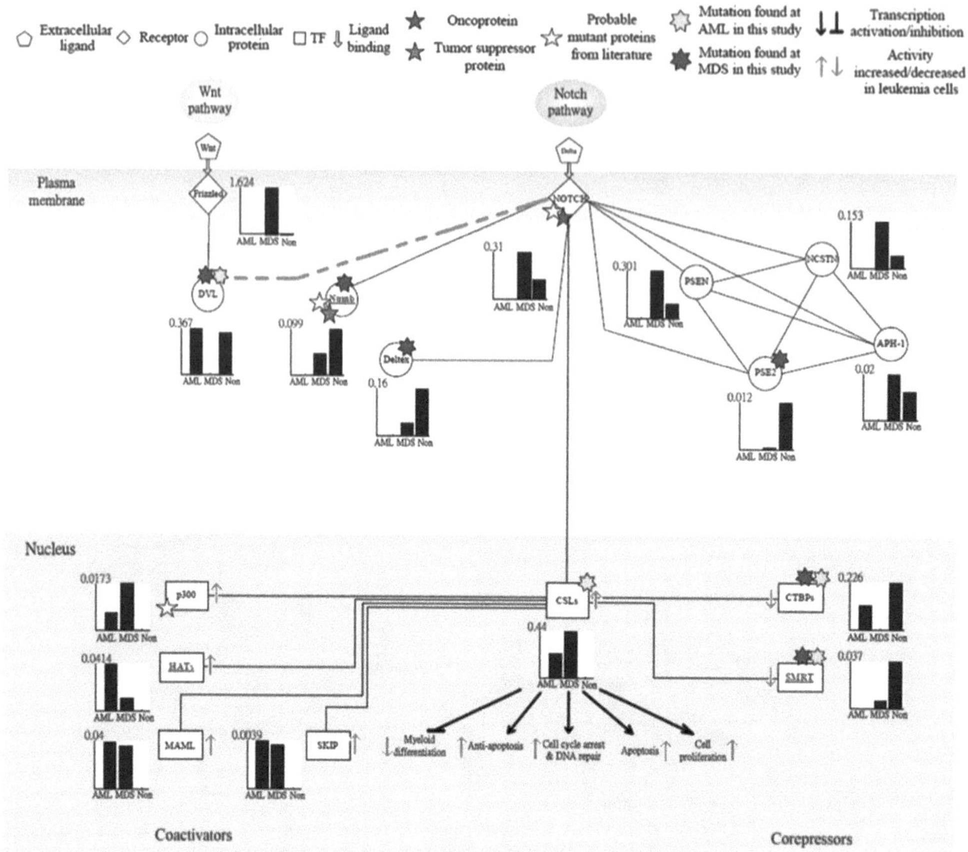

Figure B.4 Transductivity evolution of the proteins in Notch pathways at AML and MDS subtypes.

revealed that the phenotypic shift of the lung cancer-associated cell network takes 54.5 years from a normal state to stage I cancer, 1.5 years from stage I to stage II cancer, and 2.5 years from stage II to stage III cancer, with a reasonable match for the statistical result of average age of lung cancer. These results suggest that a robust negative feedback scheme, based on a stochastic evolutionary game strategy, plays a critical role in an evolutionary biological network of carcinogenesis under a natural selection scheme.

10.3.1 Introduction to evolutionary somatic cellular network in the organ carcinogenesis and aging process

Biological systems are subject to intrinsic random genetic mutations, epigenetic alterations and noises arising from internal and external stochastic fluctuations [15,113].

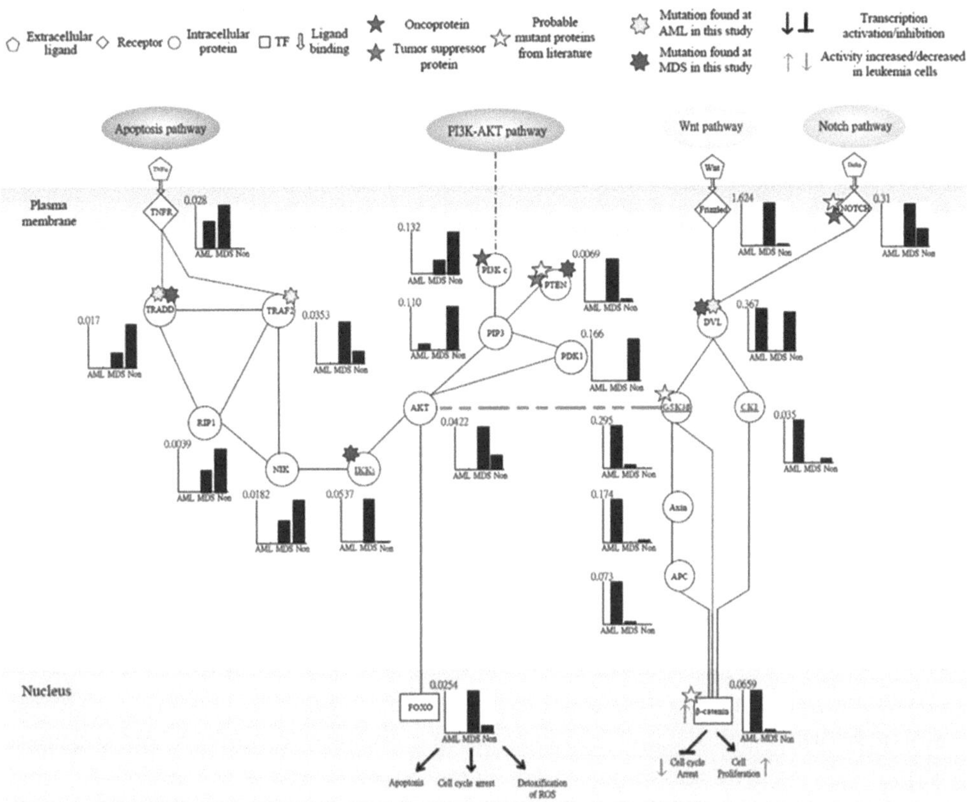

Figure B.5 Transductivity evolution of the proteins in the canonical Wnt, PI3K-AKT, and apoptosis pathways at AML/MDS.

However, it is very important that the potential perturbative effects of stochastic noise [114], genetic mutations, and epigenetic alterations are buffered at the systemic phenotypic level. Indeed, biological systems are expected to display phenotypes that are robust to these genetic and epigenetic variations and noise [115–119]. However, phenotypic changes in evolutionary fitness pressure require that biological systems are able to adapt by producing heritable phenotypic variants. This ability has been termed as evolvability. Recently, a paradoxical question about the intuitively expected trade-off between robustness and evolvability has attracted much attention [120–123]: how can biological systems generate phenotypes that are resistant to genetic mutations, epigenetic alterations and noise, yet also retain their ability to evolve as a result of these effects? One answer is that the ability of a phenotype to resist mutations might enable genetic diversity, which could then be translated to phenotypic diversity as a result of subsequent genetic and epigenetic alterations, or environmental changes in selection pressure [124] through what has been termed the neutral network [119,125–127].

The onset of cancer is characterized by an accumulation of genetic mutations and epigenetic alterations that are caused by different stresses, including tobacco, chemical agents, radiation, and viruses. These mutations and alterations typically modify the structures of DNA and chromatin, and consequently alter the gene products or the regulation of gene expression [128]. Molecular biologists have long recognized carcinogenesis as an evolutionary process that involves natural selection. Indeed, cancer is driven by the somatic evolution of cell lineages [129–142]. In this work, the somatic evolution of cell lineages in an organ was modeled as a nonlinear stochastic biological network to study natural selection in evolutionary carcinogenesis. The evolutionary biological network of an organ is driven by intrinsic random fluctuations due to genetic and epigenetic variations and external disturbances attributed to carcinogens and other stressors. We therefore modeled the intrinsic random fluctuations due to genetic and epigenetic variations as a Poisson counting process, and the external disturbances were described by an uncertain signal. Therefore, an organ with different cell species undergoing carcinogenesis was modeled by a nonlinear stochastic system with an intrinsic Poisson counting process and external random disturbances [143,144].

A natural selection scheme develops a robust property to efficiently attenuate intrinsic fluctuations and withstand external disturbances at the beginning of carcinogenesis. This maintains the normal function (i.e., phenotype) of an evolutionary biological network under intrinsic random fluctuations due to genetic and epigenetic variations and external disturbances attributed to environmental stresses. A robust feedback scheme [120–122] may exist in natural selection to maintain the healthy phenotype of a biological network at the desired equilibrium point x_e (attractor; Fig. 10.20). Because the environmental disturbances and the intrinsic random Poisson fluctuations are unpredictable, the minimax evolution game strategy was employed.

This strategy utilized a robust optimal feedback scheme that minimized the worst-case effects of external disturbances and intrinsic fluctuations on the robust stability of the phenotype at an equilibrium point favored by natural selection. Therefore, the strategy of natural selection on the evolutionary biological network during carcinogenesis was formulated as a nonlinear stochastic evolutionary game problem. Based on the stochastic Nash game strategy, effective feedback loops could be developed by natural selection for a cancer-associated biological network. This strategy can significantly improve the network's robustness to genetic and epigenetic variations and extrinsic disturbances, as well as maintain the organ's normal phenotype. However, neutral genetic mutations and epigenetic alterations gradually accumulated, which lead to the decline of robust stability and result in aging, and eventually overrode the network's robustness developed through natural selection. The robust stability of the biological network thereby deteriorated, which lead to a shift from the healthy phenotypic state at the equilibrium point x_e, to the cancerous phenotypic state at another equilibrium

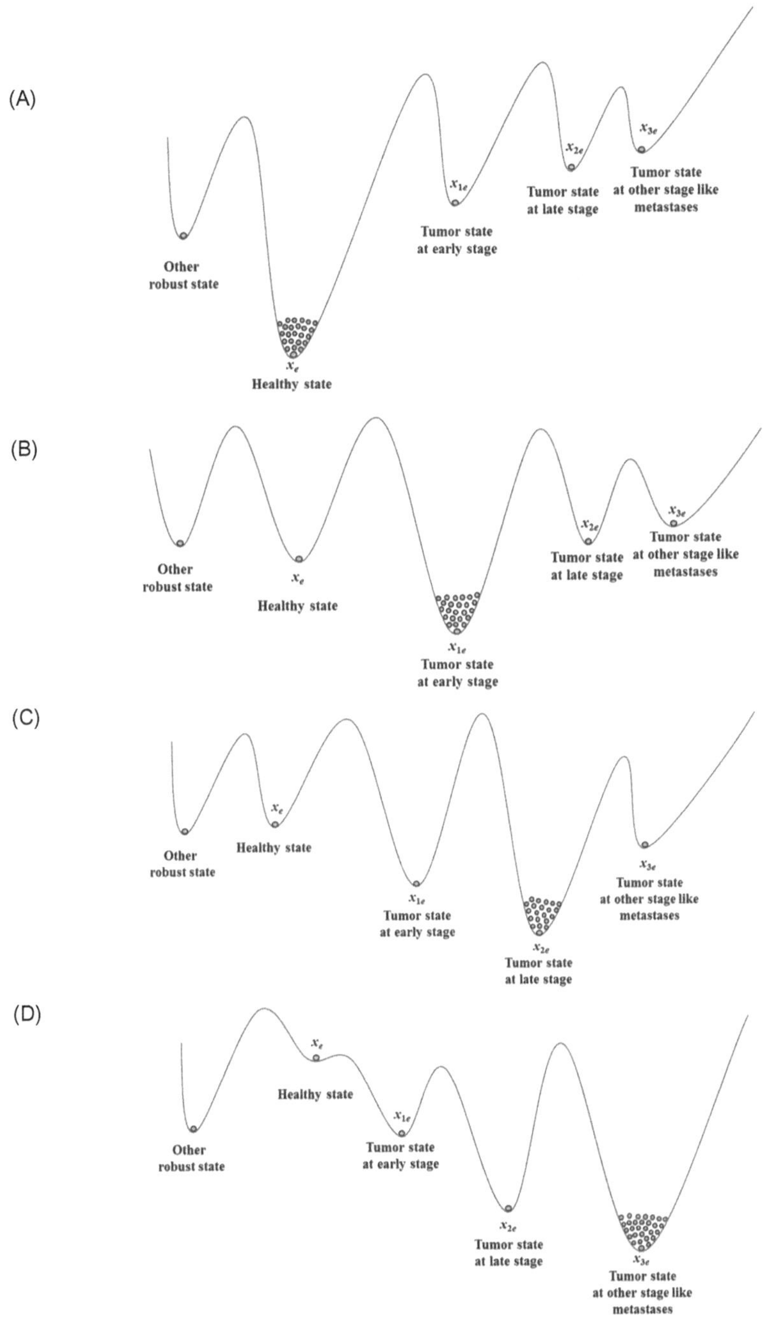

Figure 10.20 Four equilibrium points in the phenotypic landscape of the nonlinear stochastic biological network of organ in Eq. (10.63) or (10.64) in carcinogenesis. The *dots* in the basin of the equilibrium points denote the accumulation of Poisson phenotypic fluctuations (population) of biological networks due to genetic and epigenetic changes in carcinogenesis. The vertical states illustrate the relative stability of robust states, healthy and tumors of different stages at different

(Continued)

point x_{1e}. Hence, cancer initiated another new evolutionary process (Fig. 10.20). As time increased, the abovementioned shift of phenotypic state always occurred when the accumulated genetic mutations and epigenetic alterations were large enough. Therefore, cancer can be considered an aging disease.

In this study, based on the stochastic Nash game strategy for the robust stabilization of evolutionary nonlinear biological networks in carcinogenesis [145–150], both natural selection and external disturbances affected network evolution. Natural selection minimized network evolution (or maximized network fitness), while external disturbances, including genetic mutations and epigenetic alterations, maximized network evolution (or minimized network fitness). We solved a HJI-constrained optimization problem using the stochastic Nash game strategy to obtain a robust evolutionary feedback scheme for natural selection during carcinogenesis. Furthermore, the phenotype robustness criterion was also derived from the stochastic Nash game strategy to obtain insights into the evolvability of the cancer biological network. In order to simplify the procedure, the global linearization method was employed to interpolate several local linear stochastic systems and to approximate the nonlinear stochastic system of an organ undergoing carcinogenesis. Therefore, the HJI-constrained optimization problem could be replaced by a LMI-constrained optimization problem, which was solved easily using the LMI Control Toolbox in MATLAB [27,151,152].

Finally, an *in silico* example of evolutionary oncogenesis in lung cancer has been provided to illustrate natural selection based on the stochastic Nash game strategy. The nonlinear stochastic biological network of lung cancer-related tissue contains parenchymal cells, stromal cells, vascular endothelial cells (VECs) and immature VECs. The network was constructed using Poisson genetic and epigenetic variations and environmental

---

◀ equilibrium points, in the multiple dimensional states space. (A) At the beginning, the phenotypic trait of biological network is near the equilibrium point x_e at healthy state in aging process. (B) By suffering the accumulation of neutral phenotypic variations and the influence of environmental stresses in carcinogenesis, the robust stability of healthy state cannot be maintained through the evolutionary strategy $u(t)$ with the effect of natural selection and then the phenotypic trait may be shifted from x_e to x_{1e} of tumor state at the early stage to start another evolutionary process (carcinogenesis). (C) Similarly, if the accumulated neutral phenotypic variations are large enough, and environmental disturbance $v(t)$ are strong enough, the evolutionary strategy $u(t)$ cannot confer enough robust network stability to maintain the network local stability at x_{1e}. Then, the phenotypic trait of biological network is shifted from cancer stage I at the equilibrium point x_{1e} to cancer stage II at the equilibrium point x_{2e} to begin another stage of evolutionary processes (carcinogenesis). (D) Finally, if the accumulation of neutral phenotypic variations is large enough to violate the phenotypic robustness criterion at x_{2e}, cancer is transited from cancer stage II to cancer stage III. Since the equilibrium points x_e in (B) and (C) are still locally stable, they are curable (treatable) i.e., the curable stages in (B) and (C) may be reversible to (A) by adequate surgical or medical treatment (one kind of enhanced negative feedback). If the healthy state x_e is incurable (saddle point) as shown in (D), then it is impossible that this process is revertible to (A).

disturbances. By using a robust feedback scheme with the natural selection property based on the proposed nonlinear stochastic Nash game strategy, we found that the tissue maintained a normal phenotype for approximately 54.5 years in the aging process under genetic and epigenetic variations and environmental disturbances. When the robustness of the tissue network was violated by the accumulated neutral genetic and epigenetic variations and environmental disturbances, the normal phenotype of the tissue shifted from its original equilibrium point to another equilibrium point with a new, cancerous phenotype. This new phenotype [135,153] took approximately 1.5 years to progress from stage I to stage II cancer, and about 2.5 years to progress from stage II to stage III cancer.

10.3.2 Stochastic biological network of an organ in carcinogenesis

First, we considered a simplified analysis by using the linear biological network of a tissue or organ in carcinogenesis as the following random dynamic process.

$$\frac{dx(t)}{dt} = \left(A + \sum_{k=1}^{N(t)} A_k p(t - t_k)\right) x(t) + Bv(t) \tag{10.54}$$

where $x(t) = [x_1(t)\ldots x_n(t)]^T$ denotes the masses at time t of n cell species in the biological network of a tissue or organ (in the liquid tumor case i.e., lymphomas and leukemias, the masses of cell species are changed to the densities of cancer cells in blood or lymph), $v(t)$ denotes the environmental disturbances (signals or mutagens) due to different kinds of damages or carcinogens. $p(t)$ is a Poisson point process and A_k denotes the effects of random genetic mutations and epigenetic alterations at $t = t_k$ on the phenotype of the tissue. $N(t)$ denotes a Poisson counting process of phenotypic variations in tissue level due to random cellular genetic and epigenetic variations [128] in carcinogenesis as follows:

$$N(t) = \sum_{k=1}^{N(t)} p(t - t_k) = \sum_{k=1}^{N(t)} 1 \tag{10.55}$$

with mean λt and variance λt [144], and the weighting Poisson counting process

$$\sum_{k=1}^{N(t)} A_k p(t - t_k) = \sum_{k=1}^{N(t)} A_k, \quad \text{for } t \geq 0 \tag{10.56}$$

denotes the total phenotype effect of random genetic and epigenetic variations on system matrix A. Poisson point processes are useful for modeling a wide range of event-driven phenomena involving the occurrence of events at random points in time, such as emission of subatomic particle from radioactive material, the detection of photons by the receiver of a weak optical signal, genetic mutations, neural discharges, lightning discharges, seismic events and the arrival of customers always used to model a dynamic system exposed to sudden, infrequent, highly localized changes [144,154]. Therefore, it is suitable to model discontinuous genetic and epigenetic variations of biological network in the carcinogenesis process. The system matrix A denotes the interactions among these n cell species in the tissue or organ, and B denotes the coupling matrix between environment disturbances (signals) and cell species in the organ, i.e., it denotes the effect of environmental signals on the biological network of organ, where

$$A = \begin{bmatrix} a_{11} & a_{12} & \cdots & a_{1n} \\ a_{21} & a_{22} & \cdots & a_{2n} \\ \vdots & \vdots & \ddots & \vdots \\ a_{n1} & a_{n2} & \cdots & a_{nn} \end{bmatrix}; \; A_k = \begin{bmatrix} a_{k11} & a_{k12} & \cdots & a_{k1n} \\ a_{k21} & a_{k22} & \cdots & a_{k2n} \\ \vdots & \vdots & \ddots & \vdots \\ a_{kn1} & a_{kn2} & \cdots & a_{knn} \end{bmatrix};$$
$$B = \begin{bmatrix} b_{11} & b_{12} & \cdots & b_{1m} \\ b_{21} & b_{22} & \cdots & b_{2m} \\ \vdots & \vdots & \ddots & \vdots \\ b_{n1} & b_{n2} & \cdots & b_{nm} \end{bmatrix}; \; v(t) = \begin{bmatrix} v_1(t) \\ v_2(t) \\ \vdots \\ v_m(t) \end{bmatrix}; \tag{10.57}$$

$a_{ij} = 0$ if there exists no interaction between cell species i and cell species j; $a_{kij} = 0$ if the kth genetic or epigenetic variation has no phenotype effect on a_{ij}; and $b_{ij} = 0$ if the disturbance $v_j(t)$ has no effect on the cell species i.

In general, the biological processes in the cancer are time-varying across time, it should be replaced by $A(t)$ and $B(t)$ in Eq. (10.54). In this study, due to discontinuous time-varying parameter variations, $A(t)$ is replaced by $A + \sum_{k=1}^{N(t)} A_k p(t - t_k)$ in Eq. (10.54). $B(t)$ is replaced by time-invariant B but with varying part merged to uncertain disturbance $v(t)$. In this study, $v(t)$ is any bounded environmental stress. Therefore, in this study, we are with time-invariant system model A and B, but with random process to represent the time-varying process across time.

In order to operate successfully in the organ within the context of the organism, all cells in the organ must be integrated into an architecture as system matrix A in the biological network of the organ in Eq. (10.54).

From Eq. (10.54), we get a linear stochastic evolutionary biological network in carcinogenesis as

$$\frac{dx(t)}{dt} = Ax(t) + \sum_{k=1}^{N(t)} A_k x(t) p(t - t_k) + Bv(t) \tag{10.58}$$

Since the weighted Poisson counting process $\sum_{k=1}^{N(t)} A_k p(t - t_k)$ in Eq. (10.54) or (10.58) is an accumulating process of phenotypic changes of biological network in tissue level due to random genetic and epigenetic variations in cellular level, the stochastic biological network in Eq. (10.58) can represent a population of biological networks over all possible Poisson genetic and epigenetic variations in the cancer evolutionary process, i.e., one possible Poisson genetic or epigenetic variation represents one possible biological network in a population of the organs in carcinogenesis. The accumulated phenotypic variations $\sum_{k=1}^{N(t)} A_k x(t)p(t - t_k)$ in Eq. (10.58) may lead to permanent changes in the phenotypic trait of the organ. $v(t)$ denotes the environmental stresses due to environmental changes or any agent capable of causing cancer. Therefore, the phenotypic trait $x(t)$ of biological network will be perturbed away from the equilibrium point $x_e = 0$. Recently, as seen in Ref. [155], the lifetime risk of cancers of many different types is shown to be strongly correlated (0.81) with the total number of divisions of the normal self-renewing cells maintaining that tissue's homeostasis, i.e., to be strongly related to the accumulated Poisson phenotypic changes $\sum_{k=1}^{N(t)} A_k x(t)p(t - t_k)$ in the aging process.

Under the phenotypic variations and environmental disturbances in Eq. (10.58), in order to maintain its normal function and homeostasis, i.e., to operate successfully around the equilibrium point x_e, the natural selection strategy of biological network in Eq. (10.58) needs to select some adequate phenotypic variations to improve the biological network robustness to resist the effects of genetic and epigenetic variations and environment changes in carcinogenesis.

Suppose that the natural selection strategy of evolutionary biology network in the organ is to select the following phenotypic variations to improve the network robustness in carcinogenesis

$$u(t) = \sum_{k=1}^{N_1(t)} A_k x(t)p(t - t_k') \tag{10.59}$$

and the remaining (or neutral) phenotype variations are

$$R(t) = \sum_{k=1}^{N_2(t)} A_k x(t)p(t - t_k^*) = \sum_{k=1}^{N(t)} A_k x(t)p(t - t_k) - \sum_{k=1}^{N_1(t)} A_k x(t)p(t - t_k') \tag{10.60}$$

which have not been selected by natural selection and are accumulated in the biological network. The temporal parameters t_k' and t_k^* were defined as $t_{N_1(t)}' \geq \ldots \geq t_1'$ and $t_{N_2(t)}^* \geq \ldots \geq t_1^*$, respectively. It is worth noting that the t_k' is not equal to t_k^*.

From Eqs. (10.58) – (10.60), it is seen that

$$N(t) = N_1(t) + N_2(t), \quad \sum_{k=1}^{N(t)} A_k x(t) p(t - t_k) = u(t) + R(t) \tag{10.61}$$

where the Poisson counting processes $1 \sim N_1(t)$ and $1 \sim N_2(t)$ have means $\lambda_1 t$ and $\lambda_2 t$, respectively; and $\lambda = \lambda_1 + \lambda_2$.

Therefore, the stochastic evolutionary biological network of organ under somatic genetic and epigenetic variations, environmental disturbances and natural selection in Eq. (10.58) can be represented as follows:

$$\begin{aligned} \frac{dx(t)}{dt} &= Ax(t) + \sum_{k=1}^{N_1(t)} A_k x(t) p(t - t_k') + \sum_{k=1}^{N_2(t)} A_k x(t) p(t - t_k^*) + Bv(t) \\ &= Ax(t) + u(t) + \sum_{k=1}^{N_2(t)} A_k x(t) p(t - t_k^*) + Bv(t) \end{aligned} \tag{10.62}$$

where $u(t) = \sum_{k=1}^{N_1(t)} A_k x(t) p(t - t_k')$ denotes the evolutionary strategy of biological network by natural selection, i.e., how to select adequate phenotypic variations to improve the network robustness to phenotypic variations and environmental disturbance $v(t)$ and to finally maintain the normal function of biological network (or organ) in the network evolutionary process (carcinogenesis) was considered.

In the real organ, the stochastic evolutionary biological networks for n cell species, $x_1, \ldots, x_n$, in carcinogenesis are always nonlinear and the linear stochastic evolution network in Eq. (10.58) is thus modified as

$$\frac{dx(t)}{dt} = f(x) + \sum_{k=1}^{N(t)} f_k(x) p(t - t_k) + Bv(t) \tag{10.63}$$

where $f(x)$ denotes the nonlinear interactive vector among n different cell species in the biological network, and $f_k(x)$, for $k = 1, \ldots, N(t)$, denote the nonlinear phenotypic variations due to Poisson genetic and epigenetic variations in carcinogenesis.

Suppose that the phenotypic variations in Eq. (10.63) could be separated into the part of evolutionary strategy $u(t) = \sum_{k=1}^{N_1(t)} f_k(x) p(t - t_k')$ by natural selection and the remaining neutral phenotypic variations $\sum_{k=1}^{N_2(t)} f_k(x) p(t - t_k^*)$, which are accumulated as random dots in the basin of healthy state x_e in Fig. 10.20A. The above nonlinear stochastic evolutionary biological network in carcinogenesis was then modified as

$$\frac{dx(t)}{dt} = f(x) + u(t) + \sum_{k=1}^{N_2(t)} f_k(x) p(t - t_k^*) + Bv(t) \tag{10.64}$$

where

$$\begin{aligned}\sum_{k=1}^{N(t)} f_k(x)p(t-t_k) &= \sum_{k=1}^{N_1(t)} f_k(x)p(t-t_k^{'}) + \sum_{k=1}^{N_2(t)} f_k(x)p(t-t_k^{*}) \\ &= u(t) + \sum_{k=1}^{N_2(t)} f_k(x)p(t-t_k^{*})\end{aligned} \tag{10.65}$$

The nonlinear stochastic biological system in Eq. (10.64) represents an evolutionary organ under evolutionary strategy $u(t) = \sum_{k=1}^{N_1(t)} f_k(x)p(t-t_k^{'})$ by natural selection, and the neutral phenotypic variations $\sum_{k=1}^{N_2(t)} f_k(x)p(t-t_k^{*})$ and environmental disturbance $Bv(t)$ are accumulated in the basin of the healthy state at the equilibrium point x_e in Fig. 10.20A during the carcinogenesis.

The nonlinear stochastic biological network in Eqs. (10.63) and (10.64) has many equilibrium points (phenotypic traits) (Fig. 10.20). Suppose that the phenotypic trait of the biological network near the stable equilibrium point x_e is the healthy state, which is the phenotypic trait with normal function of interest by natural selection. For the convenience of analysis, the origin of nonlinear stochastic biological network in Eq. (10.64) is shifted to the equilibrium point (phenotypic trait)x_e. In this case, if the shifted nonlinear stochastic network is robustly stable at the origin, the equilibrium point (phenotype trait) x_e of interest is also robustly stable. Let us denote the phenotypic deviation around the phenotype at x_e as $\tilde{x}(t) = x(t) - x_e$ so that the following shifted nonlinear stochastic biological network of organ is obtained as

$$\begin{aligned}\frac{d\tilde{x}(t)}{dt} &= f(\tilde{x}) + \sum_{k=1}^{N(t)} f_k(\tilde{x})p(t-t_k) + Bv(t) \\ &= f(\tilde{x}) + u(t) + \sum_{k=1}^{N_2(t)} f_k(\tilde{x})p(t-t_k^{*}) + Bv(t)\end{aligned} \tag{10.66}$$

where the notations $f(\tilde{x}+x_e)$ and $f_k(\tilde{x}+x_e)$ are respectively simplified by $f(\tilde{x})$ and $f_k(\tilde{x}).u(t) = \sum_{k=1}^{N_1(t)} f_k(\tilde{x})p(t-t_k^{'})$ denotes the evolutionary strategy of nonlinear stochastic biological network by considering natural selection to maintain the normal function of evolutionary biological network.

In this situation, the origin $\tilde{x}(t) = 0$ of the shifted nonlinear stochastic biological network in Eq. (10.66) is the equilibrium point x_e of the original biological network in Eq. (10.64).

Remark 10.11:

1. The linear stochastic biological network of organ in Eq. (10.62) was considered as the linearized system of Eq. (10.66) at the equilibrium point $\tilde{x}=0$ (or at equilibrium point x_e in the nonlinear stochastic biological network in Eq. (10.64) with $A=\frac{\partial f(\tilde{x})}{\partial \tilde{x}}|_{\tilde{x}=0}$ and $A_k=\frac{\partial f_k(\tilde{x})}{\partial \tilde{x}}|_{\tilde{x}=0}$ since $x(t)=x_e$ (i.e., $\tilde{x}(t)=x(t)-x_e=0$) is the only equilibrium point in the linear biological system in Eq. (10.58) or (10.62).
2. The nonlinear stochastic system in Eq. (10.66) was considered as a Darwinian dynamic equation that satisfies Darwin's conditions of variability, heritability and the strategy to survive [148]. However, the heritability in this study implies the heritability of somatic cell evolution in carcinogenesis.

10.3.3 Natural selection in carcinogenesis

In carcinogenesis, the evolutionary biological network of an organ was modeled as the nonlinear stochastic system in Eq. (10.66). Under the effects of random genetic and epigenetic variations and environmental disturbance $v(t)$, in order to maintain the normal function of evolutionary biological network to be in the healthy state at the equilibrium point x_e in Fig. 10.20, an evolutionary strategy $u(t)=\sum_{k=1}^{N_1(t)} f_k(\tilde{x})p(t-t_k^{'})$ must be selected by natural selection to improve the robust stability of the healthy state or the equilibrium point at $x(t)=x_e$ (or $\tilde{x}(t)=0$) of evolutionary biological network. According to the robust stochastic stabilization strategy, the evolutionary strategy needs to make the system state (phenotype) of evolutionary biological network deviate from the healthy state as small as possible with a parsimonious effort of evolutionary strategy $u(t)$ selected by natural selection in response to environmental disturbances, genetic and epigenetic variations and initial state. Because the environmental disturbance $v(t)$ and neutral somatic genetic and epigenetic variations are unpredictable, the evolution level e of the biological network due to the worst-case effect of all possible bounded disturbances, neutral somatic genetic and epigenetic variations, and initial conditions phenotype deviation and strategy effort was defined as [125,127]

$$e=\max_{\substack{v(t)\in L_2[0,t_p]\\ R(t)=\sum_{k=1}^{N_2(t)} f_k(\tilde{x})p(t-t_k^*)}} \frac{E\int_0^{t_p}\left\{\tilde{x}^T(t)Q\tilde{x}(t)+u^T(t)u(t)\right\}dt}{E\{\tilde{x}^T(0)\tilde{x}(0)\}+E\int_0^{t_p}\left\{\sum_{k=1}^{N_2(t)} f_k^T(\tilde{x})f_k(\tilde{x})p(t-t_k^*)+v^T(t)v(t)\right\}dt} \tag{10.67}$$

where $L_2[0, t_p]$ denotes the set of all possible bounded environmental disturbances, i.e., $v(t) \in L_2[0, t_p]$, if $\int_0^{t_p} v^T(t)v(t)dt < \infty$. $E\{*\}$ denotes the expectation of $*$. The integration $E\int_0^{t_p}\{\tilde{x}^T(t)Q\tilde{x}(t) + u^T(t)u(t)\}dt$ in the numerator denotes the performance index of natural selection strategy of biological network in carcinogenesis. The positive symmetric weighting matrix Q is a trade-off between the state deviation $\tilde{x}(t)$ and the effort $u(t)$, i.e., a large Q will lead to a small state deviation in $\tilde{x}(t)$ but with large effort $u(t)$, and vice versa. If Q is selected as diagonal matrix $Q = diag[q_1, \ldots, q_i, \ldots, q_n]$, in the situation, the selection of a larger q_i will make the corresponding ith state $x_i(t)$ be not easier to deviate from the equilibrium point, i.e., a less $\tilde{x}_i(t)$ in carcinogenesis, and vice versa.

The physical meaning of evolution performance in Eq. (10.67) is that the worst-case effect of the initial condition, all possible environmental disturbance $v(t)$ and neutral genetic and epigenetic variations on the robust stability performance of $\tilde{x}(t)$ and control strategy effort of $u(t)$ was considered as the evolution level of the biological network because the environmental disturbances and neutral genetic and epigenetic variations are unpredictable in carcinogenesis. The deviation $\tilde{x}(t)$ and evolutionary strategy $u(t)$ are simultaneously considered in the evolutionary performance in Eq. (10.67) because natural selection employs a parsimonious strategy to achieve the robust stability with a less effort.

The phenotypic fitness (or phenotypic robustness) of the biologic network around the healthy phenotypic trait at the equilibrium point x_e is inversed to the evolution level of biological network [150], i.e.,

$$f \propto \frac{1}{e} \tag{10.68}$$

In other words, a biological network with a low evolution level e in carcinogenesis will lead to a high fitness. Therefore, the maximization of the biological network fitness to the phenotypic trait at equilibrium point x_e under the effect of genetic and epigenetic variations and environmental disturbances is equivalent to the minimization of evolution level as

$$\max f \sim \min e \tag{10.69}$$

Therefore, by combining Eqs. (10.67) and (10.69), the evolutionary strategy $u(t)$ for biological network suffers from effect of natural selection to tolerate neutral phenotypic variations and resist environmental disturbances and to finally maintain the normal phenotype of biological network in carcinogenesis. It was to maximize the fitness of the biological network to the healthy state x_e as the following minimax evolutionary game problem

$$e_0 = \min_{u(t) \in \sum_{k=1}^{N_1(t)} f_k(\tilde{x})p(t-t_k')} = \min_{u(t) \in \sum_{k=1}^{N_1(t)} f_k(\tilde{x})p(t-t_k')} \left[\max_{v(t) \in L_2[0,t_p]} \left(\frac{E\int_0^{t_p} \{\tilde{x}^T(t)Q\tilde{x}(t) + u^T(t)u(t)\}dt}{E\{\tilde{x}^T(0)\tilde{x}(0)\} + E\int_0^{t_p}\left\{\sum_{k=1}^{N_2(t)} f_k^T(\tilde{x})f_k(\tilde{x})p(t-t_k^*)\}dt + E\int_0^{t_p}\{v^T(t)v(t)\right\}dt} \right) \right] \tag{10.70}$$

where $u(t)$ was selected as $u(t) = \sum_{k=1}^{N_1(t)} f_k(\tilde{x})p(t-t_k')$ from the total phenotypic variations $\sum_{k=1}^{N(t)} f_k(\tilde{x})p(t-t_k)$ to maximize the fitness (or minimize the evolution level) of biological network in carcinogenesis. Since the initial phenotypic deviation $\tilde{x}(0)$ also affects $\tilde{x}(t)$ and $u(t)$, its effect on the evolution level was also considered in the denominator of Eq. (10.70). e_0 in Eq. (10.70) was called the network evolvability of biological network with the minimax (game) strategy $u(t)$ by considering natural selection, i.e., the Nash stochastic game is performed for biological network of the organ by natural selection to resist the unpredictable environmental disturbances and tolerate the neutral phenotypic variations due to somatic genetic and epigenetic variations in carcinogenesis. Therefore, the maximum phenotypic robustness (fitness maximization) strategy of biological network by natural selection in carcinogenesis was formulated as a stochastic Nash game problem in Eq. (10.70) for nonlinear stochastic biological network of Eq. (10.66), i.e., how to select $u(t) = \sum_{k=1}^{N_1(t)} f_k(\tilde{x})p(t-t_k')$ from the total phenotypic variations $\sum_{k=1}^{N(t)} f_k(\tilde{x})p(t-t_k)$ due to the genetic and epigenetic variations in carcinogenesis to solve the minimum evolution (or maximum fitness) problem of the biological network at healthy state x_e was considered.

The evolution game in Eq. (10.70), has two players $u(t)$ and $v(t)$. The environmental disturbance $v(t)$ and neutral phenotype variations $\sum_{k=1}^{N_2(t)} f_k(\tilde{x})p(t-t_k^*)$ will maximize their effect on the deviation of phenotype $\tilde{x}(t)$ and the evolution strategy effort $u(t)$ of the biological network from its equilibrium point x_e (i.e., the worst-case effect of all possible disturbance $v(t) \in L_2[0, t_p]$ and neutral phenotypic variations $\sum_{k=1}^{N_2(t)} f_k(\tilde{x})p(t-t_k^*)$ were considered in the evolution game) and the other player $u(t)$ selected phenotypic variations to significantly improve the network robustness (fitness) of the phenotype x_e by minimizing the worst effect of all possible neutral phenotypic variations, environmental disturbances and initial condition on the phenotype deviation and evolution strategy effort in carcinogenesis.

In general, it is very difficult to solve the stochastic Nash game problem in Eq. (10.70) directly for the evolutionary strategy of nonlinear stochastic biological network of organs in carcinogenesis. In this study, a suboptimal method was introduced to solve the stochastic Nash game problem by minimizing the upper bound e of e_0 indirectly as follows:

$$\min_{u(t)\in\sum_{k=1}^{N_1(t)} f_k(\tilde{x})p(t-t'_k)} \left[\max_{v(t)\in L_2[0,t_p]} \left(\frac{E\int_0^{t_p}\left\{\tilde{x}^T(t)Q\tilde{x}(t)+u^T(t)u(t)\right\}dt}{E\{\tilde{x}^T(0)\tilde{x}(0)\}+E\int_0^{t_p}\sum_{k=1}^{N_2(t)}\left\{f_k^T(\tilde{x})f_k(\tilde{x})p(t-t_k^*)\right\}dt+E\int_0^{t_p}\left\{v^T(t)v(t)\right\}dt} \right) \right] \le e \tag{10.71}$$

The upper bound e will be given beforehand to solve the suboptimal Nash game problem in Eq. (10.71) at first and e will be decreased as small as possible to approach to the solution e_0 of the stochastic Nash game problem in Eq. (10.70).

The suboptimal stochastic Nash game problem in Eq. (10.71) is equivalent to the following constrained stochastic Nash game problem [145,146,149]

$$\min_{u(t)\in\sum_{k=1}^{N_1(t)} f_k(\tilde{x})p(t-t'_k)} \max_{v(t)\in L_2[0,t_p]} E\int_0^{t_p}\{\tilde{x}^T(t)Q\tilde{x}(t)+u^T(t)u(t)-ev^T(t)v(t) - e\sum_{k=1}^{N_2(t)} f_k^T(\tilde{x})f_k(\tilde{x})p(t-t_k^*)\}dt \le eE\{\tilde{x}^T(0)\tilde{x}(0)\} \tag{10.72}$$

Let us denote

$$J=E\int_0^{t_p}\left\{\tilde{x}^T(t)Q\tilde{x}(t)+u^T(t)u(t)-ev^T(t)v(t) - e\sum_{k=1}^{N_2(t)} f_k^T(\tilde{x})f_k(\tilde{x})p(t-t_k^*)\right\}dt \tag{10.73}$$

Therefore, we developed two steps to solve the constrained Nash stochastic game in Eq. (10.71) or (10.72). The first step was to solve the following stochastic Nash game problem

$$
\begin{aligned}
J_0 &= \min_{u(t)\in \sum_{k=1}^{N_1(t)} f_k(\tilde{x})p(t-t'_k)} \; \max_{v(t)\in L_2[0,t_p]} J \\
&= \min_{u(t)\in \sum_{k=1}^{N_1(t)} f_k(\tilde{x})p(t-t'_k)} \; \max_{v(t)\in L_2[0,t_p]} E\int_0^{t_p} \Bigg\{ \tilde{x}^T(t)Q\tilde{x}(t) + u^T(t)u(t) \\
&\quad - ev^T(t)v(t) - e\sum_{k=1}^{N_2(t)} f_k{}^T(\tilde{x})f_k(\tilde{x})p(t-t_k^*) \Bigg\} dt
\end{aligned} \tag{10.74}
$$

The second step is to solve the following constraint problem

$$
J_0 \leq eE\{\tilde{x}^T(0)\tilde{x}(0)\} \tag{10.75}
$$

Minimax Nash game strategy of linear evolutionary biological network in carcinogenesis

For the convenience of illustration, we first solve the evolutionary game strategy of linear biological network in carcinogenesis. By solving the constrained stochastic Nash game problem in Eqs. (10.74) and (10.75) (or Eq. (10.72)) for the natural selection strategy of the linear evolutionary stochastic biological network in Eq. (10.62) with the effect of the natural selection in carcinogenesis, we get the following result.

Proposition 10.6: For the linear stochastic evolutionary biological network Eq. (10.62) in carcinogenesis, the constrained stochastic Nash game problem in Eq. (10.71) or (10.72) can be solved by the following evolutionary game strategy $u^*(t)$ and worst-case environmental disturbance $v^*(t)$

$$
u^*(t) = \sum_{k=1}^{N_1(t)} A_k x(t)p(t-t'_k) = -Px(t) \tag{10.76}
$$

$$
v^*(t) = \frac{1}{e}B^T Px(t) \tag{10.77}
$$

where the positive definite matrix $P>0$ is the solution of the following Riccati-like inequalities

$$PA + A^T P - PP + Q + \sum_{k=1}^{N_{2p}} \lambda_2 (PA_k + A_k{}^T P + A_k{}^T PA_k - eA_k{}^T A_k) + \frac{1}{e} PBB^T P \leq 0 \tag{10.78}$$

and

$$P \leq eI \tag{10.79}$$

where $N_{2p} \triangleq N_2(t_p)$, and $1 \sim N_{2p}$ represent the total number of selected phenotypic variations and neutral phenotypic variations to the present time t_p, respectively.

Proof: See Appendix C.1.

Proposition 10.6 is the solution of the suboptimal Nash stochastic game in Eq. (10.71) or (10.72). We need to minimize the upper bound e to approach the network evolvability e_0 in Eq. (10.70) of the linear biological network (10.62) in carcinogenesis. The stochastic Nash game in Eq. (10.70) for network evolvability e_0 of the biological network in Eq. (10.62) needs to solve the following constrained optimization problem.

$$e_0 = \min_{P>0} e \quad \text{subject to Eqs. (10.78) and (10.79)} \tag{10.80}$$

Remark 10.12:

1. The solution of the above network evolvability e_0 of linear biological network (10.62) in carcinogenesis can be obtained by decreasing e until no positive solution P in Eq. (10.78) within $0 < P \leq eI$ in Eq. (10.80).

   After solving e_0 from Eq. (10.80), the Riccati-like inequality in Eq. (10.78) becomes

$$PA + A^T P - PP + Q + \sum_{k=1}^{N_{2p}} \lambda_2 (PA_k + A_k{}^T P + A_k{}^T PA_k - e_0 A_k{}^T A_k) + \frac{1}{e_0} PBB^T P \leq 0 \tag{10.81}$$

   which can be considered as the phenotypic robustness criterion of stochastic evolutionary biological network in Eq. (10.62) because the inequality in Eq. (10.81) is required for the robust stability of the evolutionary biological network under genetic and epigenetic variations and environmental disturbances in carcinogenesis.

2. The term $\sum_{k=1}^{N_{2p}} \lambda_2(PA_k + A_k{}^T P + A_k{}^T PA_k - e_0 A_k{}^T A_k)$ in the phenotypic robustness criterion (10.81) is the effect of neutral phenotypic variations that are not selected by natural selection but are accumulated randomly at the biological network in carcinogenesis.

After solving e_0 and $P>0$ from the constrained optimization problem in Eq. (10.80), the solution of stochastic Nash game strategy in Eq. (10.70) for biological network of organ by natural selection in carcinogenesis is obtained by modifying Eqs. (10.76) and (10.77) as

$$u^*(t) = \sum_{k=1}^{N_1(t)} A_k x(t) p(t - t_k^{'}) = -Px(t) \tag{10.82}$$

$$v^*(t) = \frac{1}{e_0} B^T Px(t) \tag{10.83}$$

where $P>0$ is solution of Eq. (10.81). The minimax strategy $u^*(t)$ by considering natural selection is to select an adequate amount of phenotypic variations $\sum_{k=1}^{N_1(t)} A_k p(t - t_k^{'}) = \sum_{k=1}^{N_1(t)} A_k$ to approach $-P$ to enhance the robust stability of biological system and minimize the evolution level e (or maximize the phenotypic fitness of the biological network of organ via $f_0 = \frac{1}{e_0}$). With the negative feedback $-Px$ due to Nash game evolution strategy by natural selection, the phenotypic robustness criterion in Eq. (10.81) was rearranged as

$$\begin{aligned} &P\left(A - \frac{1}{2}P\right) + \left(A - \frac{1}{2}P\right)^T P + Q \\ &+ \sum_{k=1}^{N_{2p}} \lambda_2(PA_k + A_k{}^T P + A_k{}^T PA_k - eA_k{}^T A_k) + \frac{1}{e_0} PBB^T P \le 0 \end{aligned} \tag{10.84}$$

The eigenvalues of $\left(A - \frac{1}{2}P\right)$ are farther to the left-hand side of the complex s-domain than those of A, such that the network robustness of biological network is improved by negative feedback loops of the minimax Nash game strategy $u^*(t) = \sum_{k=1}^{N_1(t)} A_k x(t) p(t - t_k^{'}) = -Px(t)$ in carcinogenesis.

Further, the phenotypic robustness criterion in Eq. (10.84) of evolutionary biological network in carcinogenesis was reformulated as

$$\underbrace{Q+\frac{1}{e_0}PBB^TP}_{\text{Environmental robustness}} + \underbrace{\sum_{k=1}^{N_{2p}}\lambda_2(PA_k+A_k{}^TP+A_k{}^TPA_k-e_0A_k{}^TA_k)}_{\text{Phenotypic robustness}} \le \underbrace{-[P\left(A-\frac{1}{2}P\right)+\left(A-\frac{1}{2}P\right)^TP]}_{\text{Enhanced network robustness}} \quad (10.85)$$

The physical meaning of network phenotypic robustness criterion in Eq. (10.85) is that if the enhanced network robustness of the biological network through negative feedback loops of the minimax evolutionary strategy in Eq. (10.70) can confer the environmental robustness to resist the environmental disturbances and phenotypic robustness to tolerate the neutral Poisson phenotypic variations, the phenotypic trait of biological network of organ is maintained in carcinogenesis. Otherwise, the phenotypic trait of biological network of organ may not be maintained under environmental disturbances and random Poisson genetic and epigenetic variations in carcinogenesis [147]. According to the statistical results [155], the lifetime risk of many different cancers is strongly correlated (0.81) with the total number of divisions of the normal self-renewing cells maintaining that tissue's homeostasis. Also, the violation of phenotypic robustness criterion in Eq. (10.85) can be explained by the number of cell divisions, and a third of cancer risk among tissues are attributable to environmental factors. Namely, the cause of the violation of phenotypic robustness criterion in Eq. (10.85) during carcinogenesis is attributed 1/3 risk (i.e., $Q+\frac{1}{e_0}PBB^TP$) to environmental factor and 2/3 risk to intrinsic phenotypic factor (i.e., $\sum_{k=1}^{N_{2p}}\lambda_2(PA_k+A_k{}^TP+A_k{}^TPA_k-e_0A_k{}^TA_k)$) in Eq. (10.85).

Through the minimax game strategy of biological network of organ in carcinogenesis, the accumulated phenotypic variations $\sum_{k=1}^{N_1(t)}A_kp(t-t_k^{'})=\sum_{k=1}^{N_1(t_p)}A_kp(t_p-t_k^{'})=\sum_{k=1}^{N_{1p}}A_k$, where $N_{1p}\triangleq N_1(t_p)$, are chosen by natural selection to the amount of $-P$ to construct negative feedback loops $-Px(t)$ in Eq. (10.82). This is why many feedback loops, and the epigenetic methylations and miRNA regulations in cancer are always found to be with negative feedback (inhibition). The stochastic biological network of organ in Eq. (10.62) was thus of the following form

$$\frac{dx(t)}{dt}=(A-P)x(t)+\sum_{k=1}^{N_2(t)}A_kx(t)p(t-t_k^*)+Bv(t) \quad (10.86)$$

Because the eigenvalues of $(A - P)$ of biological network are in the far left-hand side of complex s-domain, the biological network of organ is more robust by the minimax game strategy of natural selection to resist environmental disturbances and to tolerate neutral phenotypic variations.

3. The minimax game strategy by natural selection can provide negative feedback loop $-Px(t)$ for a greater network stability robustness improvement to buffer more neutral phenotypic variations $\sum_{k=1}^{N_2(t)} A_k x(t)p(t-t_k^*)$ in carcinogenesis. However, when the neutral phenotypic variations due to Poisson genetic and epigenetic variations are accumulated in carcinogenesis, i.e., the term $\sum_{k=1}^{N_{2p}} \lambda_2(PA_k + A_k{}^T P + A_k{}^T PA_k - eA_k{}^T A_k)$ is large enough, and the network phenotypic robustness criterion of biological network in Eq. (10.85) is violated eventually, the biological network will become unstable at the healthy state, $x(t) = x_e$. Finally, the organ will become cancer.

Minimax game strategy of nonlinear biological network in carcinogenesis

After solving the minimax game strategy of linear biological network of organ by natural selection in carcinogenesis, we will solve the minimax game problem in Eq. (10.70) for the evolutionary strategy of nonlinear biological network in Eq. (10.66) by natural selection in carcinogenesis. Firstly, we solved the suboptimal Nash game problem in Eq. (10.71), or equivalently in Eqs. (10.74) and (10.75), for nonlinear stochastic biological network in Eq. (10.70) operating near the healthy state x_e in Fig. 10.20A.

Proposition 10.7: For the nonlinear stochastic biological network (10.66) in carcinogenesis, the suboptimal stochastic Nash game problem in Eq. (10.71) or (10.74) and (10.75) can be solved by the following evolutionary game strategy $u^*(t)$ and worst-case environmental disturbance $v^*(t)$

$$u^*(t) = \sum_{k=1}^{N_1(t)} f_k(\tilde{x}(t))p(t-t_k) = -\frac{1}{2}\frac{\partial V(\tilde{x}(t))}{\partial \tilde{x}} \tag{10.87}$$

$$v^*(t) = \frac{1}{2e} B^T \frac{\partial V(\tilde{x}(t))}{\partial \tilde{x}} \tag{10.88}$$

where the Lyapunov function $V(\tilde{x}(t)) > 0$ is the solution of the following constrained HJI

$$\begin{aligned}
&\left(\frac{\partial V(\tilde{x}(t))}{\partial \tilde{x}}\right)^T f(\tilde{x}) + \sum_{k=1}^{N_2(t)} (V(\tilde{x} + f_k(\tilde{x})p(t-t_k^*)) - V(\tilde{x})) \\
&- \frac{1}{4}\left(\frac{\partial V(\tilde{x}(t))}{\partial \tilde{x}}\right)^T \left(\frac{\partial V(\tilde{x}(t))}{\partial \tilde{x}}\right) + \frac{1}{4e}\left(\frac{\partial V(\tilde{x}(t))}{\partial \tilde{x}}\right)^T B^T B \left(\frac{\partial V(\tilde{x}(t))}{\partial \tilde{x}}\right) \\
&+ \tilde{x}^T(t)Q\tilde{x}(t) - e\sum_{k=1}^{N_2(t)} f_k{}^T(\tilde{x})f_k(\tilde{x})p(t-t_k^*) \le 0
\end{aligned} \tag{10.89}$$

with

$$E\{V(\tilde{x}(t))\} \leq eE\{\tilde{x}^T(0)\tilde{x}(0)\} \tag{10.90}$$

Proof: See Appendix C.2.

After solving the above suboptimal Nash game problem in Eq. (10.71) or (10.74) and (10.75), for the stochastic biological network in Eq. (10.66). The minimax evolutionary strategy by natural selection in Eq. (10.70) was solved by minimizing the upper bound e as follows:

$$\begin{gathered} e_0 = \min_{V(\tilde{x}(t))>0} e \\ \text{subject to Eqs. (10.89) and (10.90)} \end{gathered} \tag{10.91}$$

If $V(\tilde{x})$ and e_0 are obtained by solving the HJI-constrained optimization in Eq. (10.91), the network robust stability criterion in Eqs. (10.89) and (10.90) in carcinogenesis was modified as

$$\begin{aligned} &\left(\frac{\partial V(\tilde{x}(t))}{\partial \tilde{x}}\right)^T f(\tilde{x}(t)) + \sum_{k=1}^{N_2(t)} (V(\tilde{x}(t) + f_k(\tilde{x}(t))p(t-t_k^*)) - V(\tilde{x}(t))) \\ &- \frac{1}{4}\left(\frac{\partial V(\tilde{x}(t))}{\partial \tilde{x}}\right)^T \left(\frac{\partial V(\tilde{x}(t))}{\partial \tilde{x}}\right) + \frac{1}{4e_0}\left(\frac{\partial V(\tilde{x}(t))}{\partial \tilde{x}}\right)^T B^T B \left(\frac{\partial V(\tilde{x}(t))}{\partial \tilde{x}}\right) \\ &+ \tilde{x}^T(t)Q\tilde{x}(t) - e_0 \sum_{k=1}^{N_2(t)} f_k{}^T(\tilde{x})f_k(\tilde{x})p(t-t_k^*) \leq 0 \end{aligned} \tag{10.92}$$

with $V(\tilde{x}(t)) \leq e_0\tilde{x}^T(0)\tilde{x}(0)$. Since the HJI in Eq. (10.92) is necessary for the phenotype at $\tilde{x}(t) = 0$ or $x(t) = x_e$ to be robust, it can be considered as the phenotypic trait robustness criterion of nonlinear stochastic network in Eq. (10.66) at the phenotypic state x_e in carcinogenesis.

In this situation the evolutional game strategy $u^*(t)$ in Eq. (10.87) and the worst-case environmental disturbance $v^*(t)$ in Eq. (10.88) were modified, respectively, as follows:

$$u^*(t) = \sum_{k=1}^{N_1(t)} f_k(\tilde{x}(t))p(t-t_k') = -\frac{1}{2}\frac{\partial V(\tilde{x}(t))}{\partial \tilde{x}} \tag{10.93}$$

$$v^*(t) = \frac{1}{2e_0}B^T\frac{\partial V(\tilde{x}(t))}{\partial \tilde{x}} \tag{10.94}$$

where the Lyapunov function $V(\tilde{x}(t))$ is the solution of the network robust stability criterion in Eq. (10.92). The minimax strategy of $u^*(t)$ is to select $\sum_{k=1}^{N_1(t)} f_k(\tilde{x}(t))p(t-t_k')$

to approach the nonlinear negative feedback $-\frac{\partial V(\tilde{x}(t))}{\partial \tilde{x}}$ to minimize the evolution level e, and v^* denotes the worst-case disturbance to maximize evolution level e.

The robust stability criterion in Eq. (10.92) was rewritten as the following phenotypic robustness criterion in carcinogenesis.

$$\begin{aligned}
&\underbrace{\sum_{k=1}^{N_2(t)}\left(V\left(\tilde{x}(t)+f_k(\tilde{x}(t))p(t-t_k^*)\right)-V(\tilde{x}(t))\right)-e_0\sum_{k=1}^{N_2(t)}f_k{}^T(\tilde{x})f_k(\tilde{x})p(t-t_k^*)}_{\text{Phenotypic robustness}}\\
&\qquad\underbrace{+\tilde{x}^T(t)Q\tilde{x}(t)+\frac{1}{4e_0}\left(\frac{\partial V(\tilde{x}(t))}{\partial \tilde{x}}\right)^T B^T B\left(\frac{\partial V(\tilde{x}(t))}{\partial \tilde{x}}\right)}_{\text{Environmental robustness}}\\
&\qquad\underbrace{\leq -\frac{1}{2}\left(\frac{\partial V(\tilde{x}(t))}{\partial \tilde{x}}\right)^T\left(f(\tilde{x}(t))-\frac{1}{4}\left(\frac{\partial V(\tilde{x}(t))}{\partial \tilde{x}}\right)\right)}_{\text{Enhanced network robustness}}\\
&\qquad\underbrace{-\frac{1}{2}\left(f(\tilde{x}(t))-\frac{1}{4}\left(\frac{\partial V(\tilde{x}(t))}{\partial \tilde{x}}\right)\right)^T\left(\frac{\partial V(\tilde{x}(t))}{\partial \tilde{x}}\right)}_{\text{Enhanced network robustness}}
\end{aligned} \tag{10.95}$$

The first two terms are due to neutral phenotypic variations from the genetic and epigenetic variations in carcinogenesis, the last two terms on the left-hand side of Eq. (10.95) are by the worst-case environmental disturbance, and the terms on the right-hand side of Eq. (10.95) are due to the negative feedback through evolutionary minimax game strategy $u^*(t)=-\frac{1}{2}\frac{\partial V(\tilde{x}(t))}{\partial \tilde{x}}$. The biological meaning of phenotypic robustness criterion in Eq. (10.95) is that if the enhanced network robustness by evolutionary minimax strategy $u^*(t)$ can confer phenotypic robustness to tolerate the neutral phenotypic variations $\sum_{k=1}^{N_2(t)} f_k(\tilde{x}(t))p(t-t_k^*)$ and environmental disturbances, the healthy state (or phenotypic trait) x_e was be maintained in carcinogenesis. By substituting the minimax game evolutionary strategy $u^*(t)$ in Eq. (10.95) into the nonlinear stochastic biological network in Eq. (10.66), we get

$$\frac{d\tilde{x}(t)}{dt}=\left(f(\tilde{x}(t))-\frac{1}{2}\left(\frac{\partial V(\tilde{x}(t))}{\partial \tilde{x}}\right)\right)+\sum_{k=1}^{N_2(t)} f_k(\tilde{x}(t))p(t-t_k^*)+Bv(t) \tag{10.96}$$

From Eq. (10.96), it is seen that $\left(f(\tilde{x}(t)) - \frac{1}{2}\left(\frac{\partial V(\tilde{x}(t))}{\partial \tilde{x}}\right)\right)$ can make the nonlinear stochastic biological network of organ more robustly stable than $f(\tilde{x}(t))$ due to the negative feedback loops of minimax game evolutionary strategy by considering natural selection in carcinogenesis.

If the network robustness on the right-hand side of Eq. (10.95) is improved by the minimax game evolutionary strategy, from the landscape of phenotype in Fig. 10.20A, the healthy phenotypic trait is with deeper basin and steep cliff at the equilibrium point x_e, and the nonlinear biological network can harbor much more neutral phenotypic variations $\sum_{k=1}^{N_2(t)} f_k(\tilde{x}(t))p(t-t_k^*)$, i.e., much more dots in the basin of x_e. However, as time increases, the neutral phenotypic variations are accumulated to a sufficiently large amount in carcinogenesis that the enhanced network robustness on the right-hand side of Eq. (10.95) cannot provide enough phenotypic robustness to tolerate these accumulated neutral phenotypic variations and resist environment disturbances, and eventually the phenotypic robustness criterion in Eq. (10.95) is violated. This is why cancer is an aging-associated disease and the accumulating process of network phenotypic variations can be considered as an aging process of the tissue. In this situation, these accumulated neutral phenotypic variations may provide raw materials for biological network evolution to cause a possible phenotypic transition to the early state tumor state at the equilibrium point x_{e1}, i.e., a phenotypic transition may occur from the basin x_e to another basin x_{e1} in Fig. 10.20B to start another cancer phenotypic evolutionary process at the equilibrium point x_{e1} in carcinogenesis.

Remark 10.13: For the early stage of tumor at the equilibrium point x_{e1} of stage I in Fig. 10.20B, by the similar analysis in the above, we could shift the origin of nonlinear stochastic biological network of organ to x_{e1} as follows:

$$\frac{d\tilde{x}'(t)}{dt} = f(\tilde{x}'(t)) + u(t) + \sum_{k=1}^{N_2'(t)} f_k(\tilde{x}'(t))p(t-t_k^*) + Bv'(t) \tag{10.97}$$

where $\tilde{x}'(t) = x(t) - x_{e1}$.

As the minimax game evolutionary strategy in Eq. (10.93) is chosen for nonlinear stochastic network in Eq. (10.96) by natural selection in the healthy stage of carcinogenesis. The biological network of organ with early cancer stage is of the following form through the evolutionary game strategy at x_{e1}:

$$\frac{d\tilde{x}'(t)}{dt} = \left(f(\tilde{x}'(t)) - \frac{1}{2}\left(\frac{\partial V(\tilde{x}'(t))}{\partial \tilde{x}'}\right)\right) + \sum_{k=1}^{N_2'(t)} f_k(\tilde{x}'(t))p(t-t_k^*) + Bv'(t) \tag{10.98}$$

As the neutral phenotypic variations $\sum_{k=1}^{N_2'(t)} f_k(\tilde{x}'(t))p(t-t_k^*)$ are accumulated to a sufficient large amount in the early stage of carcinogenesis so that the basin of x_{e1} could not tolerate these neutral phenotypic variations and has not enough ability to resist environmental disturbances simultaneously, the biological network transits from the equilibrium point x_{e1} to the equilibrium point x_{e2} of another phenotype of cancer to begin the second stage of carcinogenesis in Fig. 10.20C. Finally, the phenotype of cancer stage II will be transited to the phenotype of cancer stage III in Fig. 10.20D when the phenotypic robustness criterion at x_{2e} is violated by Poisson random genetic and epigenetic variations in carcinogenesis. In the late stage of carcinogenesis, the network evolutionary strategy that maintains network robustness can be hijacked by cancer biological network to maintain dysfunction at the other equilibrium point in Fig. 10.20, as occurs in cancer drug resistance [136,137].

Remark 10.14:

1. The feedback loop $-\frac{1}{2}\left(\frac{\partial V(\tilde{x}(t))}{\partial \tilde{x}}\right) = \sum_{k=1}^{N_1(t)} f_k(\tilde{x}(t))p(t-t_k')$ provides the orderly tissue structure of organ by considering natural selection. However, the accumulated neutral phenotypic variations $\sum_{k=1}^{N_2(t)} f_k(\tilde{x}(t))p(t-t_k^*)$ are the source of random and complex tissue structure of cancer and provide raw material to increase the heterogeneity of tumor cells for a new evolutionary possibility to cause a transition from one cancer state to another cancer state in carcinogenesis.
2. In Fig. 10.20B and C, the cancer states at x_{1e} and x_{2e} are still curable, i.e., the carcinogenesis is still reversible to Fig. 10.20A by adequate surgical or medical treatment (one kind of enhanced negative feedback). If the healthy state is incurable (saddle point) as shown in Fig. 10.20D, then it is impossible that this process is reversible to the healthy state in Fig. 10.20A.

The stochastic evolutionary game of nonlinear cancer-associated network in carcinogenesis

In general, it is very difficult to solve the HJI-constrained optimization problem in Eq. (10.91) for the evolution game strategy $u^*(t)$ of the evolutionary biological network in Eq. (10.64) in carcinogenesis. At present, there is no efficient method to solve the nonlinear partial differential HJI in Eq. (10.92) either analytically or numerically. Therefore, in order to simplify the stochastic evolutionary game strategy, the global linearization technique [27] is employed to interpolate the nonlinear stochastic biological network in Eq. (10.66) with a set of local stochastic linearized biological networks like Eq. (10.62) at different operation points. Using the global linearization technique [27], we suppose that all the global linearization of the nonlinear stochastic biological network in Eq. (10.66) are bounded by the following L vertices as [27]:

$$\begin{bmatrix} \frac{\partial f(\tilde{x})}{\partial \tilde{x}} \\ \frac{\partial f_1(\tilde{x})}{\partial \tilde{x}} \\ \vdots \\ \frac{\partial f_{N(t)}(\tilde{x})}{\partial \tilde{x}} \end{bmatrix} = C_0 \left[\begin{bmatrix} A_1 \\ A_{11} \\ \vdots \\ A_{N(t)1} \end{bmatrix} \cdots \begin{bmatrix} A_i \\ A_{1i} \\ \vdots \\ A_{N(t)i} \end{bmatrix} \cdots \begin{bmatrix} A_L \\ A_{1L} \\ \vdots \\ A_{N(t)L} \end{bmatrix} \right], \forall \tilde{x}(t) \tag{10.99}$$

where C_0 denotes the convex hull of the polytope with L vertices. That is, if the local linearized biological systems at all $\tilde{x}(t)$ are inside the convex hull C_0, the trajectory $\tilde{x}(t)$ of the nonlinear evolutionary biological network in Eq. (10.66) can be represented by the convex combination of the stable trajectories of the following L local linearized biological networks at L vertices of the polytope in Eq. (10.99)

$$\frac{d\tilde{x}(t)}{dt} = A_i\tilde{x}(t) + Bv(t) + u(t) + \sum_{k=1}^{N_2(t)} A_{ki}p(t - t_k^*)\tilde{x}(t), \quad \text{for } i = 1, \ldots, L \tag{10.100}$$

Based on the global linearization theory [27], if Eq. (10.99) holds, every trajectory of nonlinear stochastic evolutionary biological network in Eq. (10.66) can be represented by a convex combination of L local linear biological networks in Eq. (10.100), i.e.,

$$\frac{d\tilde{x}(t)}{dt} = \sum_{i=1}^{L} \alpha_i(\tilde{x}) \left(A_i\tilde{x}(t) + Bv(t) + u(t) + \sum_{k=1}^{N_2(t)} A_{ki}p(t - t_k^*)\tilde{x}(t) \right) \tag{10.101}$$

where the interpolation functions $\alpha_i(\tilde{x})$ satisfy the following constraints:$0 \leq \alpha_i(\tilde{x}) \leq 1$ and $\sum_{i=1}^{L} \alpha_i(\tilde{x}) = 1$. The physical meaning of Eq. (10.101) is that the evolutionary trajectory of the nonlinear stochastic biological network in Eq. (10.66) can be represented by the evolutionary trajectory of the interpolated biological network in Eq. (10.101). In Eq. (10.101), the evolutionary trajectory $u(t) = \sum_{k=1}^{N_1(t)} f_k(\tilde{x})p(t - t_k^{'})$ in Eq. (10.66) can be represented as $u(t) = \sum_{i=1}^{L} \sum_{k=1}^{N_1(t)} \alpha_i(\tilde{x})A_{ki}p(t - t_k^{'})\tilde{x}(t)$ by the global linearization technique. For the Nash game evolutionary strategy problem in Eq. (10.70) in the carcinogenesis of the nonlinear stochastic network in Eq. (10.101), we get the following results for Nash game problem in Eq. (10.72), or (10.74), and (10.75).

Proposition 10.8: For the interpolated evolutionary biological network (10.101) in carcinogenesis, the stochastic evolutionary Nash game problem in Eq. (10.72) is solved by the following evolutionary strategy $u^*(t)$ and the worst-case environmental disturbance $v^*(t)$

$$u^*(t) = \sum_{i=1}^{L}\sum_{k=1}^{N_1(t)} \alpha_i(\tilde{x})A_{ki}p(t-t_k')\tilde{x}(t) = -P\tilde{x} \tag{10.102}$$

$$v^*(t) = \frac{1}{e}B^T P\tilde{x}(t) \tag{10.103}$$

where the positive definite matrix P is the solution of the following Riccati-like inequalities

$$PA_i + A_i^T P - PP + Q + \sum_{k=1}^{N_{2p}} \lambda_2(PA_{ki} + A_{ki}^T P + A_{ki}^T PA_{ki} - eA_{ki}^T A_{ki}) + \frac{1}{e}PBB^T P \le 0, \quad i = 1, \ldots, L \tag{10.104}$$

$$0 < P < eI \tag{10.105}$$

Proof: See Appendix C.3.

Remark 10.15:

1. In comparison with Proposition 10.6 of a stochastic linear biological network in carcinogenesis in Eq. (10.62), the Riccati-like inequality in Eq. (10.78) is based on the local linearized network at $\tilde{x}(t) = 0$, while the L Riccati-like inequalities in Eq. (10.104) are based on L local linearized biological networks in Eq. (10.101). Hence, the result of the Nash evolutionary game strategy in Proposition 10.8 is more suitable for nonlinear stochastic biological system in carcinogenesis.
2. The Riccati-like inequalities in Eq. (10.104) can be considered as the local linearization of HJI in Eq. (10.89) at the L vertices of polytope in Eq. (10.99).

For solving P in Eq. (10.104), the L Riccati-like inequalities in Eq. (10.104) can be transformed to the following equivalent LMIs by the Schur complement transformation method [27]

$$\begin{bmatrix} PA_i + A_i^T P + Q + \sum_{k=1}^{N_{2p}} \lambda_2(PA_{ki} + A_{ki}^T P + A_{ki}^T PA_{ki} - eA_{ki}^T A_{ki}) & P \\ P & -\left(\frac{1}{e}BB^T - I\right)^{-1} \end{bmatrix} \le 0,$$

$$\text{for } i = 1, \ldots, L \tag{10.106}$$

These LMIs in Eq. (10.106) can be easily solved by using the LMI toolbox in Matlab. However, e in Eqs. (10.104) and (10.105) is only the upper bound of the network evolvability e_0. The network evolvability e_0 of the nonlinear biological network under evolutionary game strategy can be obtained by solving the following LMIs-constrained optimization problem

$$\begin{gathered} e_0 = \min_{P>0} e \\ \text{subject to Eqs. (10.105) and (10.106)} \end{gathered} \tag{10.107}$$

Remark 10.16:

1. The e_0 in Eq. (10.107) could be solved by decreasing e until Eq. (10.106) has no positive solution for $0 < P \leq eI$ with the help of LMI toolbox in Matlab based on Newton's search method [27].

   After solving e_0 in Eq. (10.107) the Riccati-like inequalities for the network robustness criteria of local linearized biological networks in Eq. (10.104) become

$$\begin{aligned} &PA_i + A_i{}^T P - PP + Q + \sum_{k=1}^{N_{2p}} \lambda_2 (PA_{ki} + A_{ki}{}^T P + A_{ki}{}^T PA_{ki} - e_0 A_{ki}{}^T A_{ki}) \\ &+ \frac{1}{e_0} PBB^T P \leq 0, \quad i = 1, \ldots, L \end{aligned} \tag{10.108}$$

2. After solving P and e_0 from Eq. (10.107), the evolutionary game strategy $u^*(t) = -P\tilde{x}(t)$, and the worst-case disturbance $v^*(t) = \frac{1}{e_0} B^T P\tilde{x}(t)$ are solved for the nonlinear stochastic network in Eq. (10.66).

   The Riccati-like inequalities for local network robustness criteria in Eq. (10.104) can be rearranged as

$$\begin{aligned} &P\left(A_i - \frac{1}{2}P\right) + \left(A_i - \frac{1}{2}P\right)^T P + Q \\ &+ \sum_{k=1}^{N_{2p}} \lambda_2 (PA_{ki} + A_{ki}{}^T P + A_{ki}{}^T PA_{ki} - e_0 A_{ki}{}^T A_{ki}) \\ &+ \frac{1}{e_0} PBB^T P \leq 0, \quad \text{for } i = 1, \ldots, L \end{aligned} \tag{10.109}$$

   Obviously, every interaction matrix A_i of local linearized biological networks is shifted by $-\frac{1}{2}P$ so that the eigenvalues of $(A_i - \frac{1}{2}P)$ are on the farther left-hand side of the s-complex domain, and thus the evolutionary biological network is with more network robustness to tolerate local genetic and epigenetic variations and resist environmental disturbances in carcinogenesis.

3. The minimax game evolutionary strategy $u^*(t)$ is to select the total amount of phenotypic variations $\sum_{i=1}^{L}\sum_{k=1}^{N_1(t)}\alpha_i(\tilde{x}(t))A_{ki}p(t-t_k^{'}) = -P$. This is so as to select adequate genetic and epigenetic variations to form negative feedback loops $-P\tilde{x}(t)$ in the nonlinear biological network, which can resist environmental disturbances and buffer neutral genetic and epigenetic variations in carcinogenesis. Therefore, the increase of inhibitive epigenetic methylation and miRNA regulation can be observed [128] in carcinogenesis [156].

The phenotypic robustness criterion in Eq. (10.109) can be rearranged as

$$\underbrace{Q+\frac{1}{e_0}PBB^TP}_{\text{Local environmental robustness}} \quad +\underbrace{\sum_{k=1}^{N_{2p}}\lambda_2(PA_{ki}+A_{ki}{}^TP+A_{ki}{}^TPA_{ki}-e_0A_{ki}{}^TA_{ki})}_{\text{Local phenotypic robustness}}$$
$$\leq \underbrace{-\left[P\left(A_i-\frac{1}{2}P\right)+\left(A_i-\frac{1}{2}P\right)^TP\right]}_{\text{Enhanced local network robustness}}, \quad \text{for } i=1,\ldots,L \tag{10.110}$$

The biological meaning in Eq. (10.110) is that if the enhanced local network robustness of each local linearized biological network by using evolutionary game strategy $u^*(t)$ under natural selection can confer both local phenotypic robustness to buffer neutral local genetic and epigenetic variations and local environmental robustness and resist the local environmental disturbances in carcinogenesis, the phenotype of nonlinear stochastic network with Nash game evolutionary strategy can be maintained in the basin of the equilibrium point in carcinogenesis.

4. From the phenotype robustness criterion in Eq. (10.110) in the cancer evolutionary process, it is more appealing to make the left-hand side of Eq. (10.110) as small as possible and to make the right-hand side of Eq. (10.110) as large as possible. In this situation, the coupling matrix B between biological network, environment disturbances and phenotypic effects A_{ki} (due to genetic mutations and epigenetic alterations) should be as small as possible. In order to maintain network robustness, there always exist some membrane or transparent membrane to make B smaller to isolate the biological network from environmental disturbance, and some redundant and modular structures make the phenotypic effect A_{ki} of genetic mutations and epigenetic alterations as small as possible. Further, the negative feedback loop $-P\tilde{x}(t)$ mediated by evolutionary strategy $u^*(t)$ makes the right-hand side of Eq. (10.110) as large as possible. These networks are also hijacked by cancer to maintain dysfunction in the late stage of carcinogenesis. This always occurs in cancer drug resistance [136,137].

5. Because the Nash game evolutionary strategy can improve significantly the local network robustness of each local linearized biological network, the nonlinear stochastic biological network will harbor more neutral genetic and epigenetic variations. However, as the harbored neutral genetic and epigenetic variations are accumulated to the extent that the third term in Eq. (10.110) becomes very large, and the phenotypic robustness criterion of biological network in Eq. (10.110) cannot be guaranteed in carcinogenesis, the phenotype of biological network might be shifted to cancer state at another equilibrium point in Fig. 10.20 with the help of the environmental disturbances and start another period of cancer network evolution at the other equilibrium point favored by natural selection.

10.3.4 *In silico* example

Considering the dynamic evolutionary game strategy in the oncogenesis of lung cancer, some cell lineages related to lung cancer are described as follows. The tissue contains parenchymal cells and stromal cells. The stroma typically contains VECs, pericytes, and smooth muscle cells in its blood vessels [138,139,141,142]. Along with fibroblasts and other cell types, parenchymal cells tend to exhibit considerable phenotypic variations due to genetic and epigenetic variations during oncogenesis [135].

Let $x_1(t)$ and $x_2(t)$ be the mass of parenchyma cells with phenotype 1 and 2, respectively. Further, let $y(t)$ be the mass of immature VECs within the tumor and $z(t)$ be the mass of VECs, which can be represented by the length of existing micro-vessel in 1 g of undiseased tissue (i.e., one unit is equal to microvessel length in 1 gram of undiseased tissue). The dynamic system of cells related to lung cancer is thus described as follows [135]:

$$\begin{aligned}
\frac{dx_1(t)}{dt} &= \Phi_1(v)x_1(t) - 0.01x_1(t)\\
\frac{dx_2(t)}{dt} &= \Phi_2(v)x_2(t) - 0.01x_2(t)\\
\frac{dy(t)}{dt} &= (0.06H(x_1, x_2, z) - 0.04)y(t)\\
\frac{dz(t)}{dt} &= 3y(t) - 0.004m(t)z(t)
\end{aligned} \tag{10.111}$$

where $h_1(m) = 0.28c(m)e^{-0.06c(m)}$, $h_2(m) = 0.4c(m)e^{-0.06c(m)}$, $m(t) = z(t)/[x_1(t) + x_2(t)]$, $c(m) = 95m(t)/[1.375 + m(t)]$, and $H(x_1, x_2, z) = [x_1(t)h_1(m) + x_2(t)h_2(m)]/[x_1(t) + x_2(t)]$.

The variable $m(t)$ represents tumor vascularization (perfusion) in microvessel units per gram of parenchyma ($m(t)$, to a physiologist, is proportional to tumor microvessel length density). The functions $\Phi_i(m)$ express per capita growth rate of cell type i as a function of blood supply and are represented by the following equations, respectively [135,142]

$$\Phi_1(m) = \frac{0.05c^2(m)}{0.76^2 + c^2(m)} - 0.05\left(1 - \frac{c^2(m)}{0.4^2 + c^2(m)}\right) \tag{10.112}$$

and

$$\Phi_2(m) = \frac{0.05c^2(m)}{0.53^2 + c^2(m)} - 0.03\left(1 - \frac{c^2(m)}{0.1^2 + c^2(m)}\right) \tag{10.113}$$

In carcinogenesis, the tissue dynamic system in Eq. (10.111) suffers from phenotypic variations due to genetic and epigenetic variations and environmental disturbances as the following nonlinear stochastic Poisson system

$$\begin{bmatrix} \frac{dx_1(t)}{dt} \\ \frac{dx_2(t)}{dt} \\ \frac{dy(t)}{dt} \\ \frac{dz(t)}{dt} \end{bmatrix} = f(x) + \sum_{k=1}^{N_1(t)} f_k(x)p(t - t_k^{'}) + \sum_{k=1}^{N_2(t)} f_k(x)p(t - t_k^{*}) + v(t) \tag{10.114}$$

where $x(t) = \begin{bmatrix} x_1(t) \\ x_2(t) \\ y(t) \\ z(t) \end{bmatrix}$, $f(x) = \begin{bmatrix} \Phi_1(m)x_1(t) - 0.01x_1(t) \\ \Phi_2(m)x_2(t) - 0.01x_2(t) \\ (0.06H(x_1, x_2, z) - 0.04)y(t) \\ 3y(t) - 0.004m(t)z(t) \end{bmatrix}$, $f_k(x) = 0.5f(x)$ and

the mean of Poisson process is imposed as $\lambda_2 = 0.05$.

The equilibrium point x_e of the nominal lung tissue system is at $x_e = [x_{e1}, x_{e2}, x_{e3}, x_{e4}]^T = [6.19 \times 10^{-11}\ \ 2.20 \times 10^{-3}\ \ 0\ \ 9.88 \times 10^{-6}]^T$ of the nominal system in Eq. (10.111) simulated from the initial condition $x(0) \triangleq [10^{-2}\ \ -1.20 \times 10^{-2}\ \ 0\ \ 1.16 \times 10^{-7}]^T$. For the simplicity, the state of the lung tissue in Eq. (10.114) is shifted to the form $\tilde{x}(t) = x(t) - x_e$. Thus, the new equilibrium point of the phenotype concerned is at $\tilde{x}_e = O_{4\times 1}$, through the following *in silico* example.

Based on the above evolutionary game strategy, the phenotypic variations $\sum_{k=1}^{N_1(t)} f_k(\tilde{x})p(t - t_k^{'})$ are selected by considering natural selection to construct negative feedback loops to modify the lung cancer-associated cell system in Eq. (10.111), in order to resist environmental disturbances and to tolerate the effect due to the neutral phenotypic variations $\sum_{k=1}^{N_2(t)} f_k(\tilde{x})p(t-t_k^*)$. According to the global linearization scheme in Eqs. (10.99–10.101), the nonlinear stochastic Poisson system in Eq. (10.114) can be approximated by interpolating the following l local linear stochastic Poisson systems

$$\begin{bmatrix} \frac{d\tilde{x}_1(t)}{dt} \\ \frac{d\tilde{x}_2(t)}{dt} \\ \frac{d\tilde{x}_3(t)}{dt} \\ \frac{d\tilde{x}_4(t)}{dt} \end{bmatrix} = \sum_{i=1}^{l} \phi_i(\tilde{x}) \left(A_i\tilde{x}(t) + \sum_{k=1}^{N_1(t)} A_{ki}\tilde{x}(t)p(t-t_k^{'}) + \sum_{k=1}^{N_2(t)} A_{ki}\tilde{x}(t)p(t-t_k^*) \right) + v(t) \tag{10.115}$$

where A_i are given in Appendix C.4, $A_{ki}=0.5A_i$ and $l=5$.

Based on the evolutionary game strategy in Proposition 10.8, natural selection is to select phenotypic variations to construct the following negative feedback loops to improve the system robustness

$$u^*(t) = \sum_{i=1}^{l}\sum_{k=1}^{N_1(t)} \phi_i(\tilde{x}(t))A_{ki}\tilde{x}(t)p(t-t_k^{'}) = -P\tilde{x}(t)$$

where the symmetric matrix P is the solution of the constrained optimization problem in Eq. (10.107) as follows:

$$P = 10^{-4} \times \begin{bmatrix} 1.59 & -0.07 & -0.69 & -0.86 \\ * & 23.34 & 7.37 & -4.55 \\ * & * & 191.87 & -41.90 \\ * & * & * & 11.97 \end{bmatrix}$$

We also find the evolvability $e_0=0.4688$ and fitness $f_0=2.1331$ for the lung tissue system with the phenotype at the equilibrium point $x_e=[6.19\times10^{-11}\ 2.20\times10^{-3}\ 0\ 9.88\times10^{-6}]^T$ in Eq. (10.114). From the computer simulation in Fig. 10.21, it is seen that the phenotype of the lung tissue is maintained at equilibrium point x_e under the Poisson genetic and epigenetic variations and environmental disturbances in the period of time $t=0\sim19906$ (days) ≈ 54.5 (years), which can be considered as aging process of tissue.

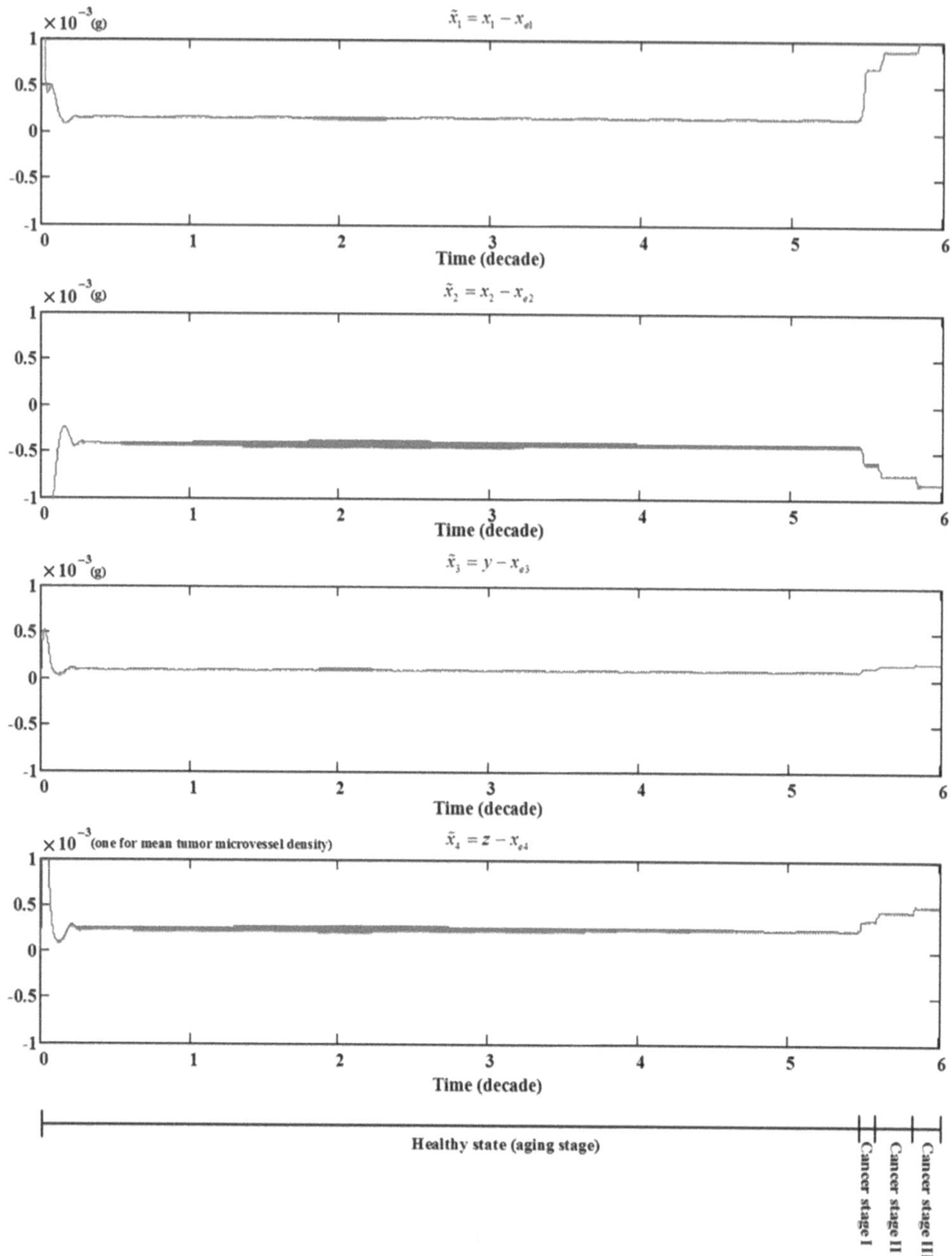

Figure 10.21 The trajectories of the nonlinear stochastic biological network of lung cancer cells in carcinogenesis with states $\tilde{x} = [x_1 - x_{e1}, x_2 - x_{e2}, y - x_{e3}, z - x_{e4}]^T$, suffering from Poisson genetic and epigenetic variations with Poisson counting process as shown in Fig. 10.22 and environmental disturbance $v(t)$. Even the network robustness is enhanced by evolutionary game strategy; (*Continued*)

The neutral phenotypic variations $\sum_{k=1}^{N_2(t)} f_k(\tilde{x})p(t - t_k^*)$ are accumulated large enough as shown in Fig. 10.22 so that the phenotypic robustness criterion in Eq. (10.95) or (10.110) of nonlinear stochastic biological system in Eq. (10.114) or (10.115) is eventually violated in carcinogenesis. The network robust stability is thus broken down and the phenotypic trait of the lung tissue is shifted to another

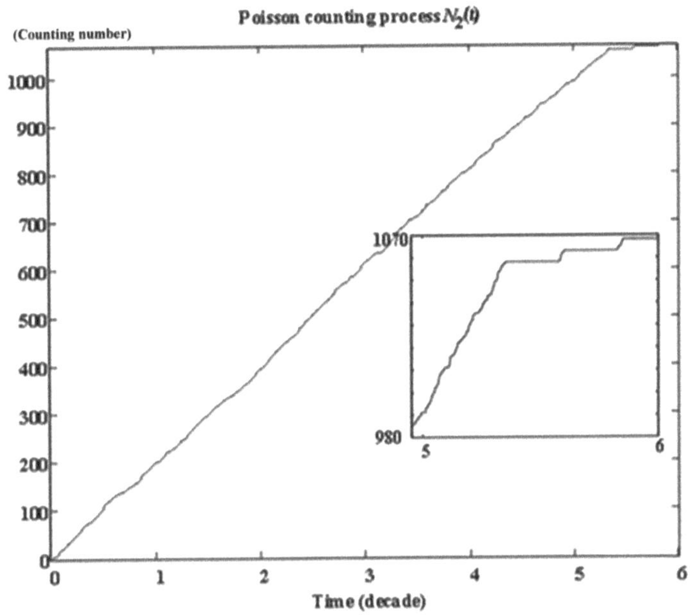

Figure 10.22 The Poisson counting process of the neutral genetic and epigenetic variations $N_2(t)$ which are accumulated in carcinogenesis of lung cancer cell lineages in the example. The vertical axis denotes the counting number of phenotypic changes of Poisson counting process $N_2(t)$ in carcinogenesis.

◀ however, it is seen that the phenotype will shift to another phenotype when the neutral phenotype variations are accumulated to a sufficiently large amount (aging process) such that the phenotypic robustness criterion in Eq. (10.95) or (10.110) is violated (i.e., the normal phenotype of the lung cancer-associated network is shifted from the equilibrium point at $x_e = [6.19 \times 10^{-11}\ 2.20 \times 10^{-3}\ 0\ 9.88 \times 10^{-6}]^T$ to the phenotype of first stage lung cancer state at the equilibrium point $x_{1e} = \tilde{x}_{1e} + x_e = [1.46 \times 10^{-4}\ 1.80 \times 10^{-3}\ 1.02 \times 10^{-4}\ 2.75 \times 10^{-4}]^T$ at 54.5 years of age, and then the second stage lung cancer state at the equilibrium point $x_{2e} = \tilde{x}_{2e} + x_e = [7.09 \times 10^{-4}\ 1.60 \times 10^{-3}\ 1.30 \times 10^{-4}\ 3.60 \times 10^{-4}]^T$ at 58.5 years of age and so on). Obviously, the cancer network evolution seems to increase its step in the late stages of carcinogenesis. From the literature in Ref. [157], lung cancer occurs mostly in men over 50 years, often at age of 60 ~ 75 years. Moreover, the units of state x_1, x_2, and y are gram, and the unit of state z represents the length of existing microvessel in 1 g of undiseased tissue (i.e., one unit is equal to microvessel length in 1 g of undiseased tissue).

equilibrium point $x_{2e} = \tilde{x}_{2e} + x_e = [7.09 \times 10^{-4}\ 1.60 \times 10^{-3}\ 1.30 \times 10^{-4}\ 3.60 \times 10^{-4}]^T$, which can be considered as the phenotype at the first stage of lung cancer, to start another evolutionary process in carcinogenesis. The simulation result in Fig. 10.21 indicates that the phenotypic shift of the lung cancer-associated cell network takes 54.5 years from the normal state to stage I cancer, 1.5 years from stage I to stage II cancer, and 2.5 years from stage II to III cancer. For clarity, the evolutionary simulation in the phase plane is given in Fig. 10.23 to illustrate the transition of lung cancer from the healthy state at $x_e = [6.19 \times 10^{-11}\ 2.20 \times 10^{-3}\ 0\ 9.88 \times 10^{-6}]^T$ to stage I cancer at $x_{1e} = [1.46 \times 10^{-4}\ 1.80 \times 10^{-3}\ 1.02 \times 10^{-4}\ 2.75 \times 10^{-4}]^T$, then to stage II cancer at $x_{2e} = [7.09 \times 10^{-4}\ 1.60 \times 10^{-3}\ 1.30 \times 10^{-4}\ 3.60 \times 10^{-4}]^T$, and finally to stage III cancer at $x_{3e} = [1.01 \times 10^{-3}\ 1.35 \times 10^{-3}\ 1.80 \times 10^{-4}\ 5.12 \times 10^{-4}]^T$. Based on the results of simulation, we found that the average age of lung cancer is more than 50 years, the transition from stage I cancer to stage II cancer is about 1.5 years, and the period from stage II to stage III lung cancer is about 2.5 years. The literature[157] seems to give a statistical support to these results.

10.3.5 Discussion

For ecologists and evolutionary biologists, natural selection and evolution are usually viewed as the domains of peppered moths and finches, driven to adapt by predators and competition. A long time ago, few evolutionary biologists could conceive that their field of biology would have a pivotal role in the understanding and combat of complex diseases such as cancer [158]. Recently, molecular biologists have acknowledged carcinogenesis as an evolutionary process involving natural selection that buffers neutral phenotypic variations and withstands environmental disturbances in the aging process [159]. In order to maintain the phenotypic stability of interest by natural selection, a stochastic Nash game strategy was developed using natural selection to improve the stability of a cancer-associated biological network. This strategy minimized the worst-case effects of uncertain neutral genetic and epigenetic variations and environmental disturbances on the variations from its normal carcinogenic phenotype. With the stochastic Nash game strategy, the robust phenotypic stability criterion for evolutionary biological networks in carcinogenesis is found in Eq. (10.95) or (10.110). This criterion states that the phenotypic robustness plus the environmental robustness should be less than or equal to the enhanced network robustness. Essentially, if the enhanced network robustness by the stochastic Nash game strategy can confer both phenotypic robustness to buffer neutral Poisson and epigenetic and genetic variations, and environmental robustness to withstand environmental disturbances in carcinogenesis, by considering natural selection, the network phenotype of the cancer-associated biological network can be robustly stable in the basin around the equilibrium point x_e. In the evolutionary period, the linear network interaction matrix A is

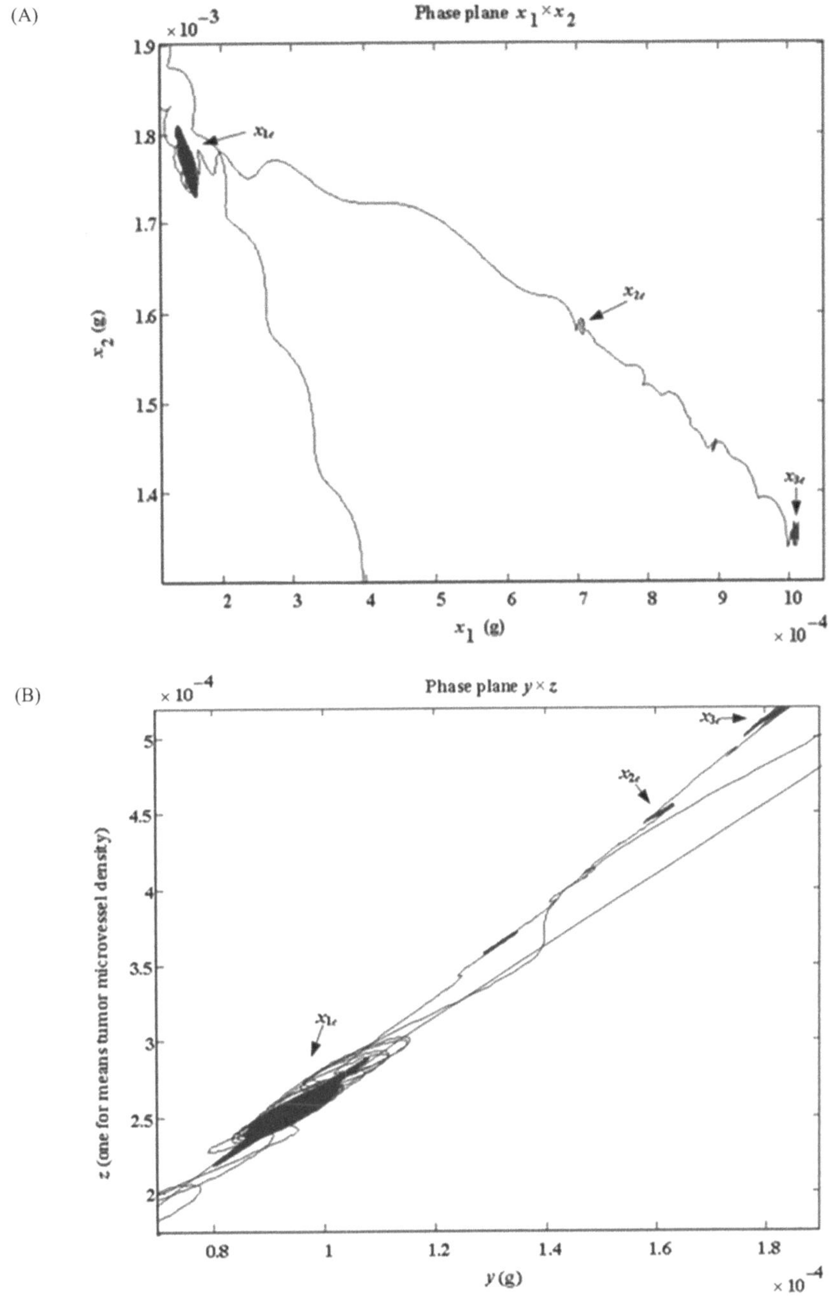

Figure 10.23 The evolutionary simulation in phase plane for the biological network of *in silico* example in carcinogenesis. This evolutionary simulation in phase plane could support the phenotypic transition of lung cancer from the healthy state at the equilibrium point $x_e = [6.19 \times 10^{-11}\ 2.20 \times 10^{-3}\ 0\ 9.88 \times 10^{-6}]^T$, to the cancer stage I state at the equilibrium point $x_{1e} = [1.46 \times 10^{-4}\ 1.80 \times 10^{-3}\ 1.02 \times 10^{-4}\ 2.75 \times 10^{-4}]^T$, then to the cancer stage II state at the equilibrium point $x_{2e} = [7.09 \times 10^{-4}\ 1.60 \times 10^{-3}\ 1.30 \times 10^{-4}\ 3.60 \times 10^{-4}]^T$, and finally to the cancer stage III state at the equilibrium point $x_{3e} = [1.01 \times 10^{-3}\ 1.35 \times 10^{-3}\ 1.80 \times 10^{-4}\ 5.12 \times 10^{-4}]^T$. Moreover, the units of state x_1, x_2, and y are gram, and the unit of state z represents the length of existing microvessel in 1 g of undiseased tissue (i.e., one unit is equal to microvessel length in 1 g of undiseased tissue).

modified by $(A - \frac{1}{2}P)$ in Eq. (10.84) (or $f(\tilde{x}(t))$ is modified by $f(\tilde{x}(t)) - \frac{1}{4}\frac{\partial V(\tilde{x}(t))}{\partial \tilde{x}}$ in Eq. (10.95) for a nonlinear network case) through negative feedback loops of the stochastic Nash game strategy. This could then decrease the evolvability e_0 of the biological network in carcinogenesis or increase the fitness f_0 of the biological network in Eq. (10.68). In this situation, the phenotype is more robust due to the larger and deeper basin in the phenotypic landscape, as presented in Fig. 10.20. However, when large amounts of neutral genetic and epigenetic variations are continuously accumulated in a larger and deep basin in a period of carcinogenesis, they may provide more raw material for new evolutionary heterogenic possibilities through random genetic drift [135,136,160]. Therefore, the stochastic Nash game strategy may improve the network robustness in order to maintain the phenotypic trait in an evolutionary period and may eventually improve the evolution of the cancer biological network, with a phenotypic shift from the normal phenotype at the equilibrium point x_e to the cancer phenotype at the equilibrium point x_{1e}. This evolutionary phenomenon has been illustrated by the *in silico* example of the evolutionary network of lung cancer cells in Fig. 10.21.

From the phenotypic robustness criterion in Eq. (10.95) or (10.110), the results reveal two ways to improve the phenotypic robustness of the evolutionary biological network in carcinogenesis. One way is to improve the stability and robustness of the network by making the right-hand side of Eq. (10.95) or (10.110) as large a value as possible so that the phenotypic robustness criterion always holds. The other way to improve the stability and robustness of the network is to reduce the effect of neutral genetic variations A_{ki} and the coupling B of environmental disturbances $v(t)$ on the biological network. This has the effect of making the left-hand side of Eq. (10.95) or (10.110) as small as possible. The evolutionary game strategy involves selecting certain somatic phenotypic variations to develop negative feedback loops $-\frac{1}{2}\frac{\partial V(\tilde{x}(t))}{\partial \tilde{x}}$ in Eq. (10.93) as the first way to improve the robustness of nonlinear biological networks and thus maintain their normal phenotypic function. This is why many negative feedback loops and inhibitive epigenetic methylations and miRNA regulations in biological networks have been found to improve the robustness of the network and prevent a phenotypic shift to the cancerous state. Furthermore, there are numerous redundant, membranous, modular, and complex structures in cancer biological networks, which are helpful in attenuating the effect of intrinsic neutral genetic and epigenetic variations and environmental disturbances on the left-hand side of Eq. (10.95) or (10.110). This is the second way to maintain the phenotype of the biological network in carcinogenesis. After enduring several evolutionary stages, the biological network phenotype is shifted to late-stage cancer at certain equilibrium points. These robust network strategies are always hijacked by cancer in order to resist anticancer drugs. This is why many anticancer treatments have limited success and cures remain elusive [136,137]. Therefore, based on the phenotype robustness criterion of the evolutionary network in Eq. (10.95) or (10.110), we could gain deep insight into the development of complex negative feedback, and redundant, modular, membranous, and scale-free structures of cancer tissue by natural selection during carcinogenesis.

Recently, it has been reported that the phenotype of a synthetic gene network always evolves due to genetic mutations from one generation to another generation, thus making it difficult to properly design a robust synthetic gene network [153,160–163]. Therefore, the robust phenotypic criterion in Eqs. (10.95) and (10.110), based on the proposed stochastic Nash game strategy, may be a suitable remedy for the limitations of conventional synthetic design methods.

10.3.6 Conclusion

In this section, evolutionary biological networks with unpredictable somatic genetic variations and uncertain environmental disturbances in carcinogenesis were modeled as nonlinear Poisson dynamic systems. An evolutionary Nash game strategy was developed to select certain phenotypic somatic variations to construct negative feedback loops. This strategy buffered neutral genetic and epigenetic variations and unpredictable environmental disturbances to efficiently improve network robustness and to avoid shifting to the cancerous state, thus maintaining the phenotype favored by natural selection during carcinogenesis. The phenotypic robustness criterion for each cancer-related state at an equilibrium point of the evolutionary biological network during carcinogenesis was also derived from this study. We determined that the evolutionary Nash game strategy was able to select certain phenotypic variations due to somatic Poisson genetic and epigenetic variations. Negative feedback loops were constructed to make the basin of the phenotype equilibrium point significantly deeper and wider in the phenotypic landscape (Fig. 10.20), so that the biological network could buffer more neutral genetic and epigenetic variations and withstand additional environmental disturbances during aging and carcinogenesis. However, as enough random neutral genetic and epigenetic variations accumulated over a long period of time, they provided the raw heterogenic material for new adaptation and evolutionary innovation. This enabled the shift to a new phenotype at a neighboring equilibrium point to commence another period of network evolution during carcinogenesis. Taken together, these results demonstrate why cancer is a disease that is associated with aging.

The global linearization technique was also employed to simplify the solution procedure for the nonlinear biological network in carcinogenesis. In this manner, a difficult HJI-constrained optimization problem for the evolutionary Nash game strategy could be simplified to an equivalent LMI-constrained optimization problem. Therefore, the trade-off between evolvability, genetic and epigenetic robustness, and network phenotypic robustness, can be easily discussed in terms of the local stability robustness and local environmental disturbance-filtering ability of a set of local linear evolutionary biological networks. In the example of a dynamic system of cells related to lung cancer, the results of the simulation support the proposed game strategy of an evolutionary biological network during carcinogenesis.

APPENDIX C

Appendix C.1: Proof of proposition 10.6

For the linear stochastic biological network in Eq. (10.62), since $x_e = 0$, $\tilde{x}(t) = x(t) - x_e = x(t)$. From Eq. (10.73), we get

$$J = E\int_0^{t_p}\left\{x^T(t)Qx(t) + u^T(t)u(t) - ev^T(t)v(t) - e\sum_{k=1}^{N_2(t)} x^T(t){A_k}^T A_k x(t)p(t-t_k^*)\right\}dt$$

$$= E\{x^T(0)Px(0)\} - E\{x^T(t_p)Px(t_p)\} + E\int_0^{t_p}\left\{x^T(t)Qx(t) + u^T(t)u(t) - ev^T(t)v(t)\right.$$

$$\left. - e\sum_{k=1}^{N_2(t)} x^T(t){A_k}^T A_k x(t)p(t-t_k^*) + \frac{d}{dt}x(t)^T Px(t)\right\}dt$$

$$= E\{x^T(0)Px(0)\} - E\{x^T(t_p)Px(t_p)\} + E\int_0^{t_p}\left\{x^T(t)Qx(t) + u^T(t)u(t) - ev^T(t)v(t)\right.$$

$$+ \left(x^T(t)A^T + u^T(t) + v^T(t)B^T\right)Px(t) + x^T(t)P(Ax(t) + u(t) + Bv(t))$$

$$+ \left(\sum_{k=1}^{N_2(t)}\left(\left(x^T(t) + x^T(t){A_k}^T p(t-t_k^*)\right)P\left(x(t) + A_k x(t)p(t-t_k^*)\right) - x^T(t)Px(t)\right)\right)$$

$$\left. - e\sum_{k=1}^{N_2(t)} x^T(t){A_k}^T A_k x(t)p(t-t_k^*)\right\}dt$$

$$= E\{x^T(0)Px(0)\} - E\{x^T(t_p)Px(t_p)\} + E\int_0^{t_p}\left\{x^T(t)Qx(t) + u^T(t)u(t) - ev^T(t)v(t)\right.$$

$$+ x^T(t)A^T Px(t) + u^T(t)Px(t) + v^T(t)B^T Px(t) + x^T(t)PAx(t) + x^T(t)Pu(t)$$

$$+ x^T(t)PBv(t) + \sum_{k=1}^{N_2(t)} x^T(t)Px(t) + \sum_{k=1}^{N_2(t)} x^T(t){A_k}^T Px(t)p(t-t_k^*)$$

$$+ \sum_{k=1}^{N_2(t)} x^T(t)PA_k x(t)p(t-t_k^*) + \sum_{k=1}^{N_2(t)} x^T(t){A_k}^T PA_k x(t)p(t-t_k^*) - \sum_{k=1}^{N_2(t)} x^T(t)Px(t)$$

$$\left. - e\sum_{k=1}^{N_2(t)} x^T(t){A_k}^T A_k x(t)p(t-t_k^*)\right\}dt$$

$$= E\{x^T(0)Px(0)\} - E\{x^T(t_p)Px(t_p)\} + E\int_0^{t_p}\Big\{x^T(t)Qx(t) + u^T(t)u(t)$$
$$- ev^T(t)v(t) + x^T(t)A^TPx(t) + u^T(t)Px(t) + v^T(t)B^TPx(t) + x^T(t)PAx(t)$$
$$+ x^T(t)Pu(t) + x^T(t)PBv(t) + \sum_{k=1}^{N_2(t)} x^T(t)A^T{}_kPx(t)p(t-t_k^*)$$
$$+ \sum_{k=1}^{N_2(t)} x^T(t)PA_kx(t)p(t-t_k^*) + \sum_{k=1}^{N_2(t)} x^T(t)A_k{}^TPA_kx(t)p(t-t_k^*)$$
$$- e\sum_{k=1}^{N_2(t)} x^T(t)A_k{}^TA_kx(t)p(t-t_k^*)\Big\}dt$$
$$= E\{x^T(0)Px(0)\} - E\{x^T(t_p)Px(t_p)\} + E\int_0^{t_p}\Big\{x^T(t)Qx(t) + x^T(t)A^TPx(t)$$
$$+ x^T(t)PAx(t) + \sum_{k=1}^{N_2(t)} x^T(t)A_k{}^TPx(t)p(t-t_k^*) + \sum_{k=1}^{N_2(t)} x^T(t)PA_kx(t)p(t-t_k^*)$$
$$+ \sum_{k=1}^{N_2(t)} x^T(t)A_k{}^TPA_kx(t)p(t-t_k^*) - e\sum_{k=1}^{N_2(t)} x^T(t)A_k{}^TA_kx(t)p(t-t_k^*)$$
$$+ (u(t)+Px(t))^T(u(t)+Px(t)) - x^T(t)PPx(t)$$
$$- \left(\sqrt{e}v(t) - \tfrac{1}{\sqrt{e}}B^TPx(t)\right)^T\left(\sqrt{e}v(t) - \frac{1}{\sqrt{e}}B^TPx(t)\right) + \frac{1}{e}x^T(t)PBB^TPx(t)\Big\}dt$$

Then the stochastic Nash game in Eq. (10.74) become

$$J_0 = \min_{u\in\sum_{k=1}^{N_1(t)}A_kx(t)p(t-t_k')} \max_{v(t)} J$$
$$= E\{x(0)^TPx(0)\} - E\{x(t_p)^TPx(t_p)\} + E\int_0^{t_r}\Big\{x(t)^T(PA + A^TP - PP + Q$$
$$+ \sum_{k=1}^{N_2(t)} \lambda_2(PA_k + A_k{}^TP + A_k{}^TPA_k - eA_k{}^TA_k) + \frac{1}{e}PBB^TP)x(t)\Big\}dt$$

By the Riccati-like inequality in Eq. (10.78), we get J_0 has the upper bound as follows:

$$J_0 \leq E\{x^T(0)Px(0)\}$$

By the second step of stochastic Nash game in Eq. (10.75), we get $P \leq eI$ which is the inequality in Eq. (10.79).

Appendix C.2: Proof of proposition 10.7

For the nonlinear stochastic biological network in Eq. (10.66), we get

$$J = E\int_0^{t_p}\left\{\tilde{x}^T(t)Q\tilde{x}(t) + u^T(t)u(t) - ev^T(t)v(t) - e\sum_{k=1}^{N_2(t)} f_k{}^T(\tilde{x})f_k(\tilde{x})p(t-t_k^*)\right\}dt \quad \text{(C.1)}$$

For some Lyapunov positive function $V(\tilde{x}(t))$, we get the following equivalent equation

$$\begin{aligned} J = {} & E\{V(\tilde{x}(0))\} - E\{V(\tilde{x}(t_p))\} + E\int_0^{t_p}\{\tilde{x}^T(t)Q\tilde{x}(t) + u^T(t)u(t) - ev^T(t)v(t) \\ & - e\sum_{k=1}^{N_2(t)} f_k{}^T(\tilde{x})f_k(\tilde{x})p(t-t_k^*)\}dt + E\int_0^{t_p}\{dV(\tilde{x}(t))\} \end{aligned} \quad \text{(C.2)}$$

By the Itô's formula of Poisson process in Eq. (10.66), we get

$$\begin{aligned} & E\int_0^{t_p}\{dV(\tilde{x}(t))\} \\ & = E\int_0^{t_p}\left\{\left(\frac{\partial V(\tilde{x}(t))}{\partial \tilde{x}}\right)^T (f(\tilde{x}) + u(t) + Bv(t) + \sum_{k=1}^{N_2(t)}\left(V\left(\tilde{x} + f_k(\tilde{x})p(t-t_k^*)\right) - V(\tilde{x})\right)\right\}dt \end{aligned} \quad \text{(C.3)}$$

Substituting Eq. (C.3) into (C.2), we get

$$\begin{aligned} J = {} & E\{V(\tilde{x}(0))\} - E\{V(\tilde{x}(t_p))\} + E\int_0^{t_p}\left\{\left(\tilde{x}^T(t)Q\tilde{x}(t) + u^T(t)u(t)\right.\right. \\ & \left. - ev^T(t)v(t) - e\sum_{k=1}^{N_2(t)} f_k{}^T(\tilde{x})f_k(\tilde{x})p(t-t_k^*)\right) + \left(\frac{\partial V(\tilde{x}(t))}{\partial \tilde{x}}\right)^T \left(f(\tilde{x}) + u(t) + Bv(t)\right. \\ & \left.\left. + \sum_{k=1}^{N_2(t)}\left(V\left(\tilde{x} + f_k(\tilde{x})p(t-t_k^*)\right) - V(\tilde{x})\right)\right)\right\}dt \end{aligned} \quad \text{(C.4)}$$

By the fact that

$$
\begin{aligned}
& u^T(t)u(t) + \left(\frac{\partial V(\tilde{x}(t))}{\partial \tilde{x}}\right)^T u(t) \\
& = \left(u(t)+\frac{1}{2}\left(\frac{\partial V(\tilde{x}(t))}{\partial \tilde{x}}\right)\right)^T \left(u(t)+\frac{1}{2}\left(\frac{\partial V(\tilde{x}(t))}{\partial \tilde{x}}\right)\right) - \frac{1}{4}\left(\frac{\partial V(\tilde{x}(t))}{\partial \tilde{x}}\right)^T \\
& \times \left(\frac{\partial V(\tilde{x}(t))}{\partial \tilde{x}}\right) - ev(t)v(t) + \left(\frac{\partial V(\tilde{x}(t))}{\partial \tilde{x}}\right)^T Bv(t) \qquad \text{(C.5)} \\
& = -\left(\sqrt{e}v(t)-\frac{1}{2\sqrt{e}}B^T\left(\frac{\partial V(\tilde{x}(t))}{\partial \tilde{x}}\right)\right)^T \left(\sqrt{e}v(t)-\frac{1}{2\sqrt{e}}B^T\left(\frac{\partial V(\tilde{x}(t))}{\partial \tilde{x}}\right)\right) \\
& + \frac{1}{4e}\left(\frac{\partial V(\tilde{x}(t))}{\partial \tilde{x}}\right)^T BB^T \left(\frac{\partial V(\tilde{x}(t))}{\partial \tilde{x}}\right)
\end{aligned}
$$

Substituting Eq. (C.5) into (C.7), we get

$$
\begin{aligned}
J = {} & E\{V(\tilde{x}(0))\} - E\{V(\tilde{x}(t_p))\} + E\int_0^{t_p} \left\{ \tilde{x}^T(t)Q\tilde{x}(t) + \left(\frac{\partial V(\tilde{x}(t))}{\partial \tilde{x}}\right)^T f(\tilde{x}) \right. \\
& + \sum_{k=1}^{N_2(t)} \left(V\left(\tilde{x} + f_k(\tilde{x})p(t-t_k^*)\right) - V(\tilde{x})\right) - \frac{1}{4}\left(\frac{\partial V(\tilde{x}(t))}{\partial \tilde{x}}\right)^T \left(\frac{\partial V(\tilde{x}(t))}{\partial \tilde{x}}\right) \\
& + \left(u(t)+\frac{1}{2}\left(\frac{\partial V(\tilde{x}(t))}{\partial \tilde{x}}\right)\right)^T \left(u(t)+\frac{1}{2}\left(\frac{\partial V(\tilde{x}(t))}{\partial \tilde{x}}\right)\right) \qquad \text{(C.6)} \\
& - \left(\sqrt{e}v(t))-\frac{1}{2\sqrt{e}}B^T\left(\frac{\partial V(\tilde{x}(t))}{\partial \tilde{x}}\right)\right)^T \left(\sqrt{e}v(t))-\frac{1}{2\sqrt{e}}B^T\left(\frac{\partial V(\tilde{x}(t))}{\partial \tilde{x}}\right)\right) \\
& \left. + \frac{1}{4e}\left(\frac{\partial V(\tilde{x}(t))}{\partial \tilde{x}}\right)^T BB^T \left(\frac{\partial V(\tilde{x}(t))}{\partial \tilde{x}}\right) - e\sum_{k=1}^{N_2(t)} f_k{}^T(\tilde{x})f_k(\tilde{x})p(t-t_k^*) \right\} dt
\end{aligned}
$$

Then

$$\begin{aligned}
\min_{u(t)}\max_{v(t)} J &= E\{V(\tilde{x}(0))\} - E\{V(\tilde{x}(t_p))\} + \min_{u(t)}\max_{v(t)} E\int_0^{t_p}\Bigg\{\tilde{x}^T(t)Q\tilde{x}(t) \\
&+ \left(\frac{\partial V(\tilde{x}(t))}{\partial \tilde{x}}\right)^T f(\tilde{x}) + \sum_{k=1}^{N_2(t)}\left(V\left(\tilde{x}+f_k(\tilde{x})p(t-t_k^*)\right) - V(\tilde{x})\right) \\
&- \frac{1}{4}\left(\frac{\partial V(\tilde{x}(t))}{\partial \tilde{x}}\right)^T\left(\frac{\partial V(\tilde{x}(t))}{\partial \tilde{x}}\right) \\
&+ \left(u(t)-\frac{1}{2}\left(\frac{\partial V(\tilde{x}(t))}{\partial \tilde{x}}\right)\right)^T\left(u(t)-\frac{1}{2}\left(\frac{\partial V(\tilde{x}(t))}{\partial \tilde{x}}\right)\right) \\
&- \left(\sqrt{e}v(t)-\frac{1}{2\sqrt{e}}B^T\left(\frac{\partial V(\tilde{x}(t))}{\partial \tilde{x}}\right)\right)^T\left(\sqrt{e}v(t)-\frac{1}{2\sqrt{e}}B^T\left(\frac{\partial V(\tilde{x}(t))}{\partial \tilde{x}}\right)\right) \qquad (C.7) \\
&+ \frac{1}{4e}\left(\frac{\partial V(\tilde{x}(t))}{\partial \tilde{x}}\right)^T BB^T\left(\frac{\partial V(\tilde{x}(t))}{\partial \tilde{x}}\right) - e\sum_{k=1}^{N_2(t)} f_k^{\,T}(\tilde{x})f_k(\tilde{x})p(t-t_k^*)\Bigg\}dt \\
&= E\{V(\tilde{x}(0))\} - E\{V(\tilde{x}(t_p))\} + E\int_0^{t_p}\Bigg\{\tilde{x}^T(t)Q\tilde{x}(t) + \left(\frac{\partial V(\tilde{x}(t))}{\partial \tilde{x}}\right)^T f(\tilde{x}) \\
&+ \sum_{k=1}^{N_2(t)}\left(V\left(\tilde{x}+f_k(\tilde{x})p(t-t_k^*)\right) - V(\tilde{x})\right) - \frac{1}{4}\left(\frac{\partial V(\tilde{x}(t))}{\partial \tilde{x}}\right)^T\left(\frac{\partial V(\tilde{x}(t))}{\partial \tilde{x}}\right) \\
&+ \frac{1}{4e}\left(\frac{\partial V(\tilde{x}(t))}{\partial \tilde{x}}\right)^T BB^T\left(\frac{\partial V(\tilde{x}(t))}{\partial \tilde{x}}\right) - e\sum_{k=1}^{N_2(t)} f_k^{\,T}(\tilde{x})f_k(\tilde{x})p(t-t_k^*)\Bigg\}dt
\end{aligned}$$

By the HJI in Eq. (10.89) and the fact $E\{V(\tilde{x}(t_p))\} \geq 0$, we get

$$\min_{u(t)}\max_{v(t)} J \leq E\{V(\tilde{x}(0))\} \leq eE\{\tilde{x}^T(0)\tilde{x}(0)\} \qquad (C.8)$$

which is the inequality Eq. (10.90).

Appendix C.3: Proof of proposition 10.8

Since we replace the nonlinear stochastic network in Eq. (10.66) by the interpolated local linear biological network in Eq. (10.101), the HJI in Proposition 10.7 can be replaced with

$$\begin{aligned}&\left(\frac{\partial V(\tilde{x}(t))}{\partial \tilde{x}}\right)\left(\sum_{i=1}^{L}\alpha_i(\tilde{x})A_i\tilde{x}(t)\right)+\tilde{x}^T(t)Q\tilde{x}(t)\\&-\frac{1}{4}\left(\frac{\partial V(\tilde{x}(t))}{\partial \tilde{x}}\right)^T\left(\frac{\partial V(\tilde{x}(t))}{\partial \tilde{x}}\right)+\frac{1}{4e}\left(\frac{\partial V(\tilde{x}(t))}{\partial \tilde{x}}\right)^T BB^T\left(\frac{\partial V(\tilde{x}(t))}{\partial \tilde{x}}\right)\\&+\sum_{k=1}^{N_2(t)}\left(V\left(\tilde{x}+\sum_{i=1}^{L}\alpha_i(\tilde{x})A_{ki}\tilde{x}(t)p(t-t_k^*)\right)-V(\tilde{x})\right)\\&-e\sum_{k=1}^{N_2(t)}\sum_{i=1}^{L}\sum_{j=1}^{L}\alpha_i(\tilde{x})\alpha_j(\tilde{x})A_{ki}A_{kj}p(t-t_k^*)\leq 0\end{aligned}\tag{C.9}$$

with $V(\tilde{x}(0))\leq e\tilde{x}^T(0)\tilde{x}(0)$. If we choose $V(\tilde{x}(t))=\tilde{x}^T(t)P\tilde{x}(t)$, by the fact $\sum_{i=1}^{L}\alpha_i(\tilde{x})=1$ and $\frac{\partial V(\tilde{x}(t))}{\partial \tilde{x}}=2P\tilde{x}(t)$, we get

$$\begin{aligned}&\sum_{i=1}^{L}\alpha_i(\tilde{x})\tilde{x}^T(t)\Bigg(PA_i+A_i{}^TP-PP+Q+\frac{1}{e}PBB^TP\\&+\sum_{k=1}^{N_{2p}}\lambda_2\left(PA_{ki}+A_{ki}{}^TP+A_{ki}{}^TPA_{ki}-eA_{ki}{}^TA_{ki}\right)\Bigg)\tilde{x}(t)\leq 0\end{aligned}\tag{C.10}$$

The above inequality hold if the Riccati-like inequalities in Eq. (10.104) hold. Further

$$u^*(t)=-\frac{1}{2}\frac{\partial V(\tilde{x}(t))}{\partial \tilde{x}}=-P\tilde{x}(t),v^*(t)=\frac{1}{e}B^TPx(t)\tag{C.11}$$

which are the solution of Nash game evolutionary strategy

$$u^*(t)=\sum_{i=1}^{L}\sum_{k=1}^{N_1(t)}\alpha_i(\tilde{x})A_ip(t-t_k^{'})=-P\tilde{x}(t),v^*(t)=\frac{1}{e}B^TP\tilde{x}(t)\tag{C.12}$$

Appendix C.4: Parameters A_i of global linearization scheme *In silico* Example

$$A_1 = \begin{bmatrix} -0.0534782 & -0.0254467 & 0.0520144 & 0.0003126 \\ 0.0005168 & -0.0038916 & -0.0135070 & -0.0001828 \\ 0.0000006 & 0.0000004 & -0.0237363 & -4.9803948 \times 10^{-9} \\ -0.0000367 & -0.0000331 & -0.0000629 & -0.0237372 \end{bmatrix}$$

$$A_2 = \begin{bmatrix} -0.0596919 & 0.0036681 & 0.0000219 & 0.5821825 \\ -0.0001722 & -0.0417919 & -0.0003613 & -0.2061930 \\ -0.0003118 & -0.0006342 & -0.0395600 & 0.6118400 \\ 1.2427067 \times 10^{-9} & -2.6932285 \times 10^{-9} & 3.0000000 & -0.0000029 \end{bmatrix}$$

$$A_3 = \begin{bmatrix} -0.0599859 & 0.0000641 & 0.0001830 & 0.3217975 \\ -0.0000170 & -0.0395879 & -0.0000082 & -0.0324759 \\ -0.0000328 & 0.0001248 & -0.0398643 & 0.1280925 \\ 2.0485833 \times 10^{-10} & -2.7951458 \times 10^{-10} & 3.0000000 & -0.0000008 \end{bmatrix}$$

$$A_4 = \begin{bmatrix} -0.0599820 & 0.0004016 & -0.0000326 & 0.2173435 \\ -0.0000171 & -0.0400054 & 0.0000678 & -0.0178187 \\ -0.0000197 & 0.0000895 & -0.0398498 & 0.0719197 \\ 1.1501642 \times 10^{-10} & -1.9750672 \times 10^{-9} & 3.0000000 & -0.0000005 \end{bmatrix}$$

$$A_5 = \begin{bmatrix} -0.0600004 & 0.0005292 & 0.0002048 & 0.1781146 \\ -0.0000175 & -0.0401022 & 0.0000980 & -0.0078165 \\ -0.0000255 & 0.0000218 & -0.0397513 & 0.0722756 \\ 1.7267656 \times 10^{-10} & -1.3279752 \times 10^{-9} & 3.0000000 & -0.0000005 \end{bmatrix}$$

REFERENCES

[1] Mandel EM, Grosschedl R. Transcription control of early B cell differentiation. Curr Opin Immunol 2010;22(2):161–7.

[2] Storey JD, Xiao WZ, Leek JT, Tompkins RG, Davis RW. Significance analysis of time course microarray experiments. Proc Natl Acad Sci USA 2005;102(36):12837–42.

[3] Zhu DX, Hero AO, Cheng H, Khanna R, Swaroop A. Network constrained clustering for gene microarray data. Bioinformatics 2005;21(21):4014–20.

[4] Chen BS, Lee TS, Feng JH. A nonlinear H-infinity control design in robotic systems under parameter perturbation and external disturbance. Int J Control 1994;59(2):439–61.
[5] Stelling J, Sauer U, Szallasi Z, Doyle FJ, Doyle J. Robustness of cellular functions. Cell 2004; 118(6):675–85.
[6] Hopfield JJ. Kinetic proofreading: a new mechanism for reducing errors in biosynthetic processes requiring high specificity. Proc Natl Acad Sci USA 1974;71(10):4135–9.
[7] Pastor-Satorras R, Vespignani A. Evolution and structure of the Internet: a statistical physics approach. Cambridge; New York: Cambridge University Press; 2004.
[8] Yan J, Magnasco MO, Marko JF. Kinetic proofreading can explain the supression of supercoiling of circular DNA molecules by type-II topoisomerases. Phys Rev E 2001;63(3).
[9] Angeli D, Ferrell JE, Sontag ED. Detection of multistability, bifurcations, and hysteresis in a large class of biological positive-feed back systems. Proc Natl Acad Sci USA 2004;101(7): 1822–7.
[10] Hasty J, McMillen D, Collins JJ. Engineered gene circuits. Nature 2002;420(6912):224–30.
[11] Zhang SH, Jin GX, Zhang XS, Chen LN. Discovering functions and revealing mechanisms at molecular level from biological networks. Proteomics 2007;7(16):2856–69.
[12] Kitano H. Biological robustness. Nat Rev Genet 2004;5(11):826–37.
[13] Doyle JC, Francis BA, Tannenbaum A. Feedback control theory. New York, NY: Macmillan Pub. Co.; 1992.
[14] Chen BS, Wu CH. A systematic design method for robust synthetic biology to satisfy design specifications. BMC Syst Biol 2009;3:66.
[15] Kaern M, Blake WJ, Collins JJ. The engineering of gene regulatory networks. Annu Rev Biomed Eng 2003;5:179–206.
[16] Murphy KF, Balazsi G, Collins JJ. Combinatorial promoter design for engineering noisy gene expression. Proc Natl Acad Sci USA 2007;104(31):12726–31.
[17] Johansson R. System modeling and identification. Englewood Cliffs, NJ: Prentice Hall; 1993.
[18] Bland JM, Altman DG. Multiple significance tests—the Bonferroni Method. Br Med J 1995;310 (6973):170.
[19] Noble WS. How does multiple testing correction work? Nat Biotechnol 2009;27(12):1135–7.
[20] Gandhi TKB, Zhong J, Mathivanan S, Karthick L, Chandrika KN, Mohan SS, et al. Analysis of the human protein interactome and comparison with yeast, worm and fly interaction datasets. Nat Genet 2006;38(3):285–93.
[21] Troyanskaya OG. Putting microarrays in a context: integrated analysis of diverse biological data. Brief Bioinform 2005;6(1):34–43.
[22] Akaike H. Citation classic—a new look at the statistical-model identification. Curr Contents/Eng Technol Appl Sci 1981;51:22.
[23] Klipp E. Systems biology in practice: concepts, implementation and application. Weinheim: Wiley-VCH; 2005.
[24] Kuo BC, Golnaraghi MF. Automatic control systems. 8th ed. New York, NY: John Wiley & Sons; 2003.
[25] Chen BS, Zhang WH. Stochastic H(2)/H(infinity) control with state-dependent noise. IEEE Trans Autom Control 2004;49(1):45–57.
[26] Zhang WH, Chen BS. State feedback H(infinity) control for a class of nonlinear stochastic systems. SIAM J Control Optim 2006;44(6):1973–91.
[27] Boyd SP. Linear matrix inequalities in system and control theory. Philadelphia, PA: Society for Industrial and Applied Mathematics; 1994.
[28] Lin CL, Lai CC, Huang TH. A neural network for linear matrix inequality problems. IEEE Trans Neural Networks 2000;11(5):1078–92.
[29] Chung HY, Sung B, Jung KJ, Zou Y, Yu BP. The molecular inflammatory process in aging. Antioxid Redox Signal 2006;8(3–4):572–81.
[30] Judge S, Jang YM, Smith A, Hagen T, Leeuwenburgh C. Age-associated increases in oxidative stress and antioxidant enzyme activities in cardiac interfibrillar mitochondria: implications for the mitochondrial theory of aging. FASEB J 2005;19(1):419–21.

[31] Zahn JM, Poosala S, Owen AB, Ingram DK, Lustig A, Carter A, et al. AGEMAP: a gene expression database for aging in mice. PLoS Genet 2007;3(11):2326–37.
[32] Cheadle C, Vawter MP, Freed WJ, Becker KG. Analysis of microarray data using Z score transformation. J Mol Diagn 2003;5(2):73–81.
[33] Tegner J, Yeung MKS, Hasty J, Collins JJ. Reverse engineering gene networks: integrating genetic perturbations with dynamical modeling. Proc Natl Acad Sci USA 2003;100(10):5944–9.
[34] de Visser J, Hermisson J, Wagner GP, Meyers LA, Bagheri HC, Blanchard JL, et al. Perspective: evolution and detection of genetic robustness. Evolution 2003;57(9):1959–72.
[35] Lenski RE, Barrick JE, Ofria C. Balancing robustness and evolvability. PLoS Biol 2006;4(12): 2190–2.
[36] Csete ME, Doyle JC. Reverse engineering of biological complexity. Science 2002;295(5560): 1664–9.
[37] Murray JI, Whitfield ML, Trinklein ND, Myers RM, Brown PO, Botstein D. Diverse and specific gene expression responses to stresses in cultured human cells. Mol Biol Cell 2004;15(5):2361–74.
[38] Kitano H. Opinion—cancer as a robust system: implications for anticancer therapy. Nat Rev Cancer 2004;4(3):227–35.
[39] Kawanishi S, Hiraku Y, Pinlaor S, Ma N. Oxidative and nitrative DNA damage in animals and patients with inflammatory diseases in relation to inflammation-related carcinogenesis. Biol Chem 2006;387(4):365–72.
[40] Szatrowski TP, Nathan CF. Production of large amounts of hydrogen-peroxide by human tumor-cells. Cancer Res 1991;51(3):794–8.
[41] Toyokuni S, Okamoto K, Yodoi J, Hiai H. Persistent oxidative stress in cancer. FEBS Lett 1995; 358(1):1–3.
[42] Schneider BL, Kulesz-Martin M. Destructive cycles: the role of genomic instability and adaptation in carcinogenesis. Carcinogenesis 2004;25(11):2033–44.
[43] Wallace DC. The mitochondrial genome in human adaptive radiation and disease: on the road to therapeutics and performance enhancement. Gene 2005;354:169–80.
[44] Irmak MB, Ince G, Ozturk M, Cetin-Atalay R. Acquired tolerance of hepatocellular carcinoma cells to selenium deficiency: a selective survival mechanism? Cancer Res 2003;63(20):6707–15.
[45] Trachootham D, Alexandre J, Huang P. Targeting cancer cells by ROS-mediated mechanisms: a radical therapeutic approach? Nat Rev Drug Discov 2009;8(7):579–91.
[46] Sands WA, Copland M, Wheadon H. Targeting self-renewal pathways in myeloid malignancies. Cell Commun Signal 2013;11:33.
[47] Pattabiraman DR, Weinberg RA. Tackling the cancer stem cells—what challenges do they pose? Nat Rev Drug Discov 2014;13(7):497–512.
[48] Takahashi S. Downstream molecular pathways of FLT3 in the pathogenesis of acute myeloid leukemia: biology and therapeutic implications. J Hematol Oncol 2011;4:13.
[49] Steelman LS, Abrams SL, Whelan J, Bertrand FE, Ludwig DE, Basecke J, et al. Contributions of the Raf/MEK/ERK, PI3K/PTEN/Akt/mTOR and Jak/STAT pathways to leukemia. Leukemia 2008;22(4):686–707.
[50] Koschmieder SK, Krug U. Myeloid leukemia—basic mechanisms of leukemogenesis. InTech; 2011.
[51] Zhang RR, Shah MV, Yang J, Nyland SB, Liu X, Yun JK, et al. Network model of survival signaling in large granular lymphocyte leukemia. Proc Natl Acad Sci USA 2008;105(42):16308–13.
[52] Rizo A, Vellenga E, de Haan G, Schuringa JJ. Signaling pathways in self-renewing hematopoietic and leukemic stem cells: do all stem cells need a niche? Hum Mol Genet 2006;15:R210–19.
[53] Meshkini A, Yazdanparast R. Regulation of p38, PKC/Foxo3a/p73 signaling network by GTP during erythroid differentiation in chronic myelogenous leukemia. Cell Biochem Biophys 2013;67(2):675–87.
[54] Palomero T, Ferrando AA. Genomic tools for dissecting oncogenic transcriptional networks in human leukemia. Leukemia 2009;23(7):1236–42.
[55] Sanda T, Lawton LN, Barrasa MI, Fan ZP, Kohlhammer H, Gutierrez A, et al. Core transcriptional regulatory circuit controlled by the TAL1 complex in human t cell acute lymphoblastic leukemia. Cancer Cell 2012;22(2):209–21.

[56] Moellering RE, Cornejo M, Davis TN, Del Bianco C, Aster JC, Blacklow SC, et al. Direct inhibition of the NOTCH transcription factor complex (vol. 462, p. 182, 2009). Nature 2010;463 (7279):384.
[57] Shima Y, Kitabayashi I. Deregulated transcription factors in leukemia. Int J Hematol 2011;94(2): 134–41.
[58] Tenen DG, Hromas R, Licht JD, Zhang DE. Transcription factors, normal myeloid development, and leukemia. Blood 1997;90(2):489–519.
[59] Teramo A, Gattazzo C, Passeri F, Lico A, Tasca G, Cabrelle A, et al. Intrinsic and extrinsic mechanisms contribute to maintain the JAK/STAT pathway aberrantly activated in T-type large granular lymphocyte leukemia. Blood 2013;121(19):3843–54.
[60] Lin DY, Huang CC, Hsieh YT, Lin HC, Pao PC, Tsou JH, et al. Analysis of the interaction between Zinc finger protein 179 (Znf179) and promyelocytic leukemia zinc finger (Plzf). J Biomed Sci 2013;20:98.
[61] Patel H, Nteliopoulos G, Nikolakopoulou Z, Jackson A, Gordon MY. Antibody arrays identify protein-protein interactions in chronic myeloid leukaemia. Br J Haematol 2011;152(5):611–14.
[62] Majetl R, Becker MW, Tian Q, Lee TLM, Yan XW, Liu R, et al. Dysregulated gene expression networks in human acute myelogenous leukemia stem cells. Proc Natl Acad Sci USA 2009;106(9): 3396–401.
[63] Saadatpour A, Wang RS, Liao AJ, Liu X, Loughran TP, Albert I, et al. Dynamical and structural analysis of a t cell survival network identifies novel candidate therapeutic targets for large granular lymphocyte leukemia. PLoS Comput Biol 2011;7(11):e1002267.
[64] Cheng KC, Huang HC, Chen JH, Hsu JW, Cheng HC, Ou CH, et al. *Ganoderma lucidum* polysaccharides in human monocytic leukemia cells: from gene expression to network construction. BMC Genomics 2007;8:411.
[65] Chen BS, Wu CC. On the calculation of signal transduction ability of signaling transduction pathways in intracellular communication: systematic approach. Bioinformatics 2012;28(12): 1604–11.
[66] Chu LH, Chen BS. Comparisons of robustness and sensitivity between cancer and normal cells by microarray data. Cancer Inform 2008;6:165–81.
[67] Tu CT, Chen BS. On the increase in network robustness and decrease in network response ability during the aging process: a systems biology approach via microarray data. IEEE-ACM Trans Comput Biol Bioinformatics 2013;10(2):468–80.
[68] Haferlach T, Kohlmann A, Wieczorek L, Basso G, Kronnie GT, Bene MC, et al. Clinical utility of microarray-based gene expression profiling in the diagnosis and subclassification of leukemia: report from the International Microarray Innovations in Leukemia Study Group. J Clin Oncol 2010;28(15): 2529–37.
[69] Etet PFS, Vecchio L, Kamdje AHN. Signaling pathways in chronic myeloid leukemia and leukemic stem cell maintenance: key role of stromal microenvironment. Cell Signal 2012;24(9):1883–8.
[70] Martelli AM, Evangelisti C, Chappell W, Abrams SL, Basecke J, Stivala F, et al. Targeting the translational apparatus to improve leukemia therapy: roles of the PI3K/PTEN/Akt/mTOR pathway. Leukemia 2011;25(7):1064–79.
[71] Steelman LS, Franklin RA, Abrams SL, Chappell W, Kempf CR, Basecke J, et al. Roles of the Ras/Raf/MEK/ERK pathway in leukemia therapy. Leukemia 2011;25(7):1080–94.
[72] Kanehisa M, Goto S. KEGG: Kyoto Encyclopedia of Genes and Genomes. Nucleic Acids Res 2000;28(1):27–30.
[73] Kanehisa M, Goto S, Sato Y, Kawashima M, Furumichi M, Tanabe M. Data, information, knowledge and principle: back to metabolism in KEGG. Nucleic Acids Res 2014;42(D1):D199–205.
[74] Matys V, Kel-Margoulis OV, Fricke E, Liebich I, Land S, Barre-Dirrie A, et al. TRANSFAC (R) and its module TRANSCompel (R): transcriptional gene regulation in eukaryotes. Nucleic Acids Res 2006;34:D108–10.
[75] Noble B, Daniel JW. Applied linear algebra. 2nd ed. Englewood Cliffs, NJ: Prentice Hall; 1977.
[76] Press WH. Numerical recipes software (firm): numerical recipes in the art of scientific computing. 2nd ed. Cambridge, England; New York, NY: Cambridge University Press; 1992.

[77] Frohling S, Scholl C, Gilliland DG, Levine RL. Genetics of myeloid malignancies: pathogenetic and clinical implications. J Clin Oncol 2005;23(26):6285–95.
[78] Lee HJ, Daver N, Kantarjian HM, Verstovsek S, Ravandi F. The role of JAK pathway dysregulation in the pathogenesis and treatment of acute myeloid leukemia. Clin Cancer Res 2013;19(2): 327–35.
[79] Harris SL, Levine AJ. The p53 pathway: positive and negative feedback loops. Oncogene 2005;24(17): 2899–908.
[80] Nii T, Marumoto T, Tani K. Roles of p53 in various biological aspects of hematopoietic stem cells. J Biomed Biotechnol 2012;2012:903435.
[81] McCubrey JA, Steelman LS, Chappell WH, Abrams SL, Franklin RA, Montalto G, et al. Ras/Raf/ MEK/ERK and PI3K/PTEN/Akt/mTOR cascade inhibitors: how mutations can result in therapy resistance and how to overcome resistance. Oncotarget 2012;3(10):1068–111.
[82] Zhao Z, Zuber J, Diaz-Flores E, Lintault L, Kogan SC, Shannon K, et al. p53 loss promotes acute myeloid leukemia by enabling aberrant self-renewal. Genes Dev 2010;24(13):1389–402.
[83] Radomska HS, Basseres DS, Zheng R, Zhang P, Dayaram T, Yamamoto Y, et al. Block of C/EBP alpha function by phosphorylation in acute myeloid leukemia with FLT3 activating mutations. J Exp Med 2006;203(2):371–81.
[84] Nerlov C. C/EBP alpha mutations in acute myeloid leukaemias. Nat Rev Cancer 2004;4(5): 394–400.
[85] Cleaves R, Wang QF, Friedman AD. C/EBP alpha p30, a myeloid leukemia oncoprotein, limits G-CSF receptor expression but not terminal granulopoiesis via site-selective inhibition of C/EBP DNA binding. Oncogene 2004;23(3):716–25.
[86] Elsaesser A, Franzen M, Kohlmann A, Weisser M, Schnittger S, Schoch C, et al. Proto-oncogene c-jun expression is induced by AML1-ETO in a JNK-dependent manner: possible role in the pathogenesis of acute myeloid leukemia. Blood 2002;100(11):536a.
[87] Lindstrom MS, Jin AW, Deisenroth C, Wolf GW, Zhang YP. Cancer-associated mutations in the MDM2 zinc finger domain disrupt ribosomal protein interaction and attenuate MDM2-induced p53 degradation. Mol Cell Biol 2007;27(3):1056–68.
[88] Murati A, Brecqueville M, Devillier R, Mozziconacci MJ, Gelsi-Boyer V, Birnbaum D. Myeloid malignancies: mutations, models and management. BMC Cancer 2012;12:304.
[89] Oberender C, Kaeda J, Pawlaczyk-Peter B, Daniel P, Arnold R, Dorken B, et al. Sequencing of NUMB transcripts in chronic myeloid leukemia detects two single nucleotide polymorphisms. Leuk Lymphoma 2013;54(2):421–2.
[90] Pereira JKN, Traina F, Machado-Neto JA, Duarte ADS, Lopes MR, Saad STO, et al. Distinct expression profiles of MSI2 and NUMB genes in myelodysplastic syndromes and acute myeloid leukemia patients. Leuk Res 2012;36(10):1300–3.
[91] Corey SJ, Minden MD, Barber DL, Kantarjian H, Wang JCY, Schimmer AD. Myelodysplastic syndromes: the complexity of stem-cell diseases. Nat Rev Cancer 2007;7(2):118–29.
[92] Benekli M, Baumann H, Wetzler M. Targeting signal transducer and activator of transcription signaling pathway in leukemias. J Clin Oncol 2009;27(26):4422–32.
[93] Yang E, Korsmeyer SJ. Molecular thanatopsis: a discourse on the BCL2 family and cell death. Blood 1996;88(2):386–401.
[94] Brzezianska E, Domanska D. Review: the JAK/STAT protein activation—role in cancer development and targeted therapy. Current Signal Transduction Therapy 2012;7(3):187–201.
[95] Ding XF, Yang ZJ, Zhou FL, Wang FM, Li XX, Chen C, et al. Transcription factor AP-2 alpha regulates acute myeloid leukemia cell proliferation by influencing Hoxa gene expression. Int J Biochem Cell Biol 2013;45(8):1647–56.
[96] Kollmann K, Heller G, Ott RG, Scheicher R, Zebedin-Brandl E, Schneckenleithner C, et al. c-JUN promotes BCR-ABL-induced lymphoid leukemia by inhibiting methylation of the 5′ region of Cdk6. Blood 2011;117(15):4065–75.
[97] Xing Y, Hogge DE. Combined inhibition of the phosphoinosityl-3-kinase (PI3Kinase) P110 delta subunit and mitogen-extracellular activated protein kinase (MEKinase) shows synergistic cytotoxicity against human acute myeloid leukemia progenitors. Leukemia Res 2013;37(6):697–704.

[98] Abdullah M, Seman Z. Role of signaling pathways in acute myeloid leukemia. InTech, 2011.

[99] Kannan S, Sutphin RM, Hall MG, Golfman LS, Fang W, Nolo RM, et al. Notch activation inhibits AML growth and survival: a potential therapeutic approach. J Exp Med 2013;210(2):321–37.

[100] Chen PM, Yen CC, Wang WS, Lin YJ, Chu CJ, Chiou TJ, et al. Down-regulation of Notch-1 expression decreases PU.1-mediated myeloid differentiation signaling in acute myeloid leukemia. Int J Oncol 2008;32(6):1335–41.

[101] Weng AP, Millholland JM, Yashiro-Ohtani Y, Arcangeli ML, Lau A, Wai C, et al. c-Myc is an important direct target of Notch1 in T-cell acute lymphoblastic leukemia/lymphoma. Genes Dev 2006;20(15):2096–109.

[102] Pear WS. Hooked on Hes: a T-ALL of Addiction. Immunity 2010;33(5):645–7.

[103] Klinakis A, Lobry C, Abdel-Wahab O, Oh P, Haeno H, Buonamici S, et al. A novel tumour-suppressor function for the Notch pathway in myeloid leukaemia. Nature 2011;473(7346):230–3.

[104] Simon M, Grandage VL, Linch DC, Khwaja A. Constitutive activation of the Wnt/ss-catenin signalling pathway in acute myeloid leukaemia. Oncogene 2005;24(14):2410–20.

[105] Park S, Chapuis N, Tamburini J, Bardet V, Cornillet-Lefebvre P, Willems L, et al. Role of the PI3K/AKT and mTOR signaling pathways in acute myeloid leukemia. Haematologica 2010;95(5):819–28.

[106] Greer EL, Brunet A. FOXO transcription factors at the interface between longevity and tumor suppression. Oncogene 2005;24(50):7410–25.

[107] Sykes SM, Lane SW, Bullinger L, Kalaitzidis D, Yusuf R, Saez B, et al. AKT/FOXO signaling enforces reversible differentiation blockade in myeloid leukemias. Cell 2011;147(1):247.

[108] Bernasconi P. Molecular pathways in myelodysplastic syndromes and acute myeloid leukemia: relationships and distinctions—a review. Br J Haematol 2008;142(5):695–708.

[109] Sykes SM, Lane SW, Bullinger L, Kalaitzidis D, Yusuf R, Saez B, et al. AKT/FOXO signaling enforces reversible differentiation blockade in myeloid leukemias. Cell 2011;146(5):697–708.

[110] Lynch M, Latta L, Hicks J, Giorgianni M. Mutation, selection, and the maintenance of life-history variation in a natural population. Evolution 1998;52(3):727–33.

[111] Ye YB, Lin JY, Chen Q, Liu F, Chen HJ, Li JY, et al. The cytotoxicity of a Grb2-SH3 inhibitor in Bcr-Abl positive K562 cells. Biochem Pharmacol 2008;75(11):2080–91.

[112] Meijer L, Flajolet M, Greengard P. Pharmacological inhibitors of glycogen synthase kinase 3. Trends Pharmacol Sci 2004;25(9):471–80.

[113] Kaern M, Elston TC, Blake WJ, Collins JJ. Stochasticity in gene expression: from theories to phenotypes. Nat Rev Genet 2005;6(6):451–64.

[114] Chu LH, Chen BS. Construction of a cancer-perturbed protein-protein interaction network for discovery of apoptosis drug targets. BMC Syst Biol 2008;2:56.

[115] Ao P. Laws in Darwinian evolutionary theory. Phys Life Rev 2005;2(2):117–56.

[116] Carter AJ, Hermisson J, Hansen TF. The role of epistatic gene interactions in the response to selection and the evolution of evolvability. Theor Popul Biol 2005;68(3):179–96.

[117] Bloom JD, Labthavikul ST, Otey CR, Arnold FH. Protein stability promotes evolvability. Proc Natl Acad Sci USA 2006;103(15):5869–74.

[118] Aldana M, Balleza E, Kauffman S, Resendiz O. Robustness and evolvability in genetic regulatory networks. J Theor Biol 2007;245(3):433–48.

[119] Felix MA, Wagner A. Robustness and evolution: concepts, insights and challenges from a developmental model system. Heredity (Edinb) 2008;100(2):132–40.

[120] Chen BS, Lin YP. A unifying mathematical framework for genetic robustness, environmental robustness, network robustness and their trade-off on phenotype robustness in biological networks. Part I: Gene regulatory networks in systems and evolutionary biology. Evol Bioinform Online 2013;9:43–68.

[121] Chen BS, Lin YP. A unifying mathematical framework for genetic robustness, environmental robustness, network robustness and their tradeoff on phenotype robustness in biological networks. Part II: Ecological networks. Evol Bioinform Online 2013;9:69–85.

[122] Chen BS, Lin YP. A unifying mathematical framework for genetic robustness, environmental robustness, network robustness and their trade-offs on phenotype robustness in biological networks. Part III: Synthetic gene networks in synthetic biology. Evol Bioinform Online 2013;9:87–109.

[123] Kitano H. Biological robustness. Nat Rev Genet 2004;5(11):826–37.
[124] Wagner A. Robustness and evolvability: a paradox resolved. Proc R Soc B Biol Sci 2008;275 (1630):91–100.
[125] Chen BS, Wu WS, Wu WS, Li WH. On the adaptive design rules of biochemical networks in evolution. Evol Bioinform Online 2007;3:27–39.
[126] Draghi JA, Parsons TL, Wagner GP, Plotkin JB. Mutational robustness can facilitate adaptation. Nature 2010;463(7279):353–5.
[127] Chen B-S, Lin Y-P. On the interplay between the evolvability and network robustness in an evolutionary biological network: a systems biology approach. Evol Bioinform Online 2011;7:201.
[128] Barillot E. Computational systems biology of cancer. Boca Raton, FL: CRC Press; 2013.
[129] Wong YH, Chen RH, Chen BS. Core and specific network markers of carcinogenesis from multiple cancer samples. J Theor Biol 2014;362:17–34.
[130] Tawfik DS. Messy biology and the origins of evolutionary innovations. Nat Chem Biol 2010;6 (10):692–6.
[131] Ao P, Galas D, Hood L, Zhu X. Cancer as robust intrinsic state of endogenous molecular-cellular network shaped by evolution. Med Hypotheses 2008;70(3):678–84.
[132] Weinberg RA. The biology of cancer. Concord, CA: Garland Publishing; 2006.
[133] Murgia C, Pritchard JK, Kim SY, Fassati A, Weiss RA. Clonal origin and evolution of a transmissible cancer. Cell 2006;126(3):477–87.
[134] Vogelstein B, Kinzler KW. Cancer genes and the pathways they control. Nat Med 2004;10(8): 789–99.
[135] Nagy JD. Competition and natural selection in a mathematical model of cancer. Bull Math Biol 2004;66(4):663–87.
[136] Kitano H. Cancer as a robust system: implications for anticancer therapy. Nat Rev Cancer 2004; 4(3):227–35.
[137] Kitano H. Cancer robustness: tumour tactics. Nature 2003;426(6963):125.
[138] Lobov IB, Brooks PC, Lang RA. Angiopoietin-2 displays VEGF-dependent modulation of capillary structure and endothelial cell survival in vivo. Proc Natl Acad Sci USA 2002;99 (17):11205–10.
[139] Yancopoulos GD, Davis S, Gale NW, Rudge JS, Wiegand SJ, Holash J. Vascular-specific growth factors and blood vessel formation. Nature 2000;407(6801):242–8.
[140] Greaves MF. Cancer: the evolutionary legacy. Oxford; New York, NY: Oxford University Press; 2000.
[141] Carmeliet P, Jain RK. Angiogenesis in cancer and other diseases. Nature 2000;407(6801):249–57.
[142] Weibel ER. The pathway for oxygen: structure and function in the mammalian respiratory system. Cambridge, MA: Harvard University Press; 1984.
[143] Chen BS, Ho SJ. The stochastic evolutionary game for a population of biological networks under natural selection. Evol Bioinform Online 2014;10:17–38.
[144] Gardner WA. Introduction to Random Processes: with applications to signals and systems, vol. 31. New York, NY: McGraw-Hill; 1990.
[145] Chen B-S, Uang H-J, Tseng C-S. Robust tracking enhancement of robot systems including motor dynamics: a fuzzy-based dynamic game approach. IEEE Trans Fuzzy Syst 1998;6(4):538–52.
[146] Chen B-S, Tseng C-S, Uang H-J. Fuzzy differential games for nonlinear stochastic systems: suboptimal approach. IEEE Trans Fuzzy Syst 2002;10(2):222–33.
[147] de Visser JA, Hermisson J, Wagner GP, Ancel Meyers L, Bagheri-Chaichian H, Blanchard JL, et al. Perspective: evolution and detection of genetic robustness. Evolution 2003;57(9): 1959–72.
[148] Vincent TL, Brown JS. Evolutionary game theory, natural selection, and Darwinian dynamics. Cambridge; New York, NY: Cambridge University Press; 2005.
[149] Chen BS, Chang CH, Lee HC. Robust synthetic biology design: stochastic game theory approach. Bioinformatics 2009;25(14):1822–30.
[150] Chen BS, Hsu CY, Liou JJ. Robust design of biological circuits: evolutionary systems biology approach. J Biomed Biotechnol 2011;2011:304236.

[151] Chen B-S, Zhang W. Stochastic H_2/H_∞ control with state-dependent noise. IEEE Trans Autom Control, 2004;49(1):45–57.
[152] Zhang W, Chen B-S. State feedback H_∞ control for a class of nonlinear stochastic systems. SIAM J Control Optim 2006;44(6):1973–91.
[153] Wang YC, Chen BS. A network-based biomarker approach for molecular investigation and diagnosis of lung cancer. BMC Med Genomics 2011;4:2.
[154] Chen W-H, Chen B-S. Robust stabilization design for nonlinear stochastic system with Poisson noise via fuzzy interpolation method. Fuzzy Sets Syst 2013;217(0):41–61.
[155] Tomasetti C, Vogelstein B. Variation in cancer risk among tissues can be explained by the number of stem cell divisions. Science 2015;347(6217):78–81.
[156] Shukla S, Meeran SM. Epigenetics of cancer stem cells: pathways and therapeutics. Biochim Biophys Acta 2014;1840(12):3494–502.
[157] Kozielski J, Kaczmarczyk G, Porębska I, Szmygin-Milanowska K, Gołecki M. Lung cancer in patients under the age of 40 years. Contemp Oncol (Pozn) 2012;16:413–15.
[158] Crespi B, Summers K. Evolutionary biology of cancer. Trends Ecol Evol 2005;20(10):545–52.
[159] Campisi J. Aging, tumor suppression and cancer: high wire-act!. Mech Ageing Dev 2005;126(1): 51–8.
[160] Waddington CH. Canalization of development and the inheritance of acquired characters. Nature 1942;150:563–5.
[161] Voit EO, Ferreira AEN. Computational analysis of biochemical systems: a practical guide for biochemists and molecular biologists. Cambridge; New York, NY: Cambridge University Press; 2000.
[162] Hasty J, Dolnik M, Rottschafer V, Collins JJ. Synthetic gene network for entraining and amplifying cellular oscillations. Phys Rev Lett 2002;88(14):148101.
[163] Isaacs FJ, Hasty J, Cantor CR, Collins JJ. Prediction and measurement of an autoregulatory genetic module. Proc Natl Acad Sci USA 2003;100(13):7714–19.

CHAPTER 11

Big Mechanisms of Aging via System Identification and Big Database Mining

INTRODUCTION

Aging, an extremely complex and system-level process, has attracted much attention in medical research, especially because chronic diseases are quite prevalent in the elderly population. These may be the result of both gene mutations that lead to intrinsic perturbations and environmental changes that may stimulate signaling in the body. Therefore, analysis of network robustness to tolerate intrinsic perturbations and network response ability of gene networks to respond to external stimuli during the aging process may provide insight into the systematic changes caused by aging. This chapter first proposes novel methods to estimate network robustness and measure network response ability of gene regulatory networks (GRNs) by using their corresponding microarray data in the aging process in Section 11.1. Then, the chapter reveals that an aging-related gene network is more robust to intrinsic perturbations in the elderly than the young, and therefore is less responsive to external stimuli. Finally, the chapter indicates that the response abilities of individual genes, especially *FOXOs*, *NF-κB* and *p53*, are significantly different in the young versus the aged subjects. These observations are consistent with experimental findings in the aged population, e.g., elevated incidence of tumorigenesis and diminished resistance to oxidative stress. The proposed method can also be used for exploring and analyzing the dynamic properties of other biological processes via corresponding microarray data to provide useful information on clinical strategy and drug target selection.

Complex biological processes such as aging do not result from an isolated event in a single protein, chemical enzyme, or individual cell type. Rather, aging process occurs as a coordinated result of interacting cell types and tissues, as well as genetic and epigenetic alterations in gene regulation and expression, miRNA regulation, DNA methylation, signal pathways, and biological networks. Therefore, the chapter will investigate specific core genetic and epigenetic network (GEN) for cellular mechanism of human aging from next generation sequencing (NGS) data and database mining in Section 11.2. Further, we will obtain core proteins, transcription factors (TFs), target genes and microRNAs (miRNAs) in the core GENs between young and elderly and between old women and men to provide a platform for useful medicine therapy to facilitate the development of new antiaging drugs.

Big Mechanisms in Systems Biology
DOI: http://dx.doi.org/10.1016/B978-0-12-809479-2.00011-1

11.1 ON THE SYSTEMATIC MECHANISM OF GRN IN THE AGING PROCESS: A SYSTEMS BIOLOGY APPROACH VIA MICROARRAY DATA

Although the average lifespan of humans has increased in recent times, the process of aging still occurs. Aging is an incessant postmaturational process that results in a reduced response to environmental stimuli [1]. Decades of research have demonstrated that higher incidence of many pathophysiological phenotypes, such as cancer, inflammation, diabetes, cardiovascular diseases, and neurodegenerative disorders, are associated with the process of aging [2–6]. The gradual decline in cellular function associated with aging and its related pathophysiology is the result of cumulative changes in expression or activity of a number of genes. Therefore, the dynamic GRN plays a key role in the systems modeling of interactions and regulations among genes in the aging process.

Recent experimental evidence has demonstrated that the efficiency of the p53 pathway declines with age, predicting increased rates of mutation. These mutations may be the result of decreased DNA repair and fixation of mutations caused by a decline in p53-mediated apoptosis [7]. The p53-regulatory loop is considered to be a key guardian of the genome, linking DNA damage and stress responses. This leads to either cell cycle arrest for carrying out DNA repair or induction of cell apoptosis and promotion of irreversible growth arrest (senescence) in cases where DNA damage is nonrepairable [8,9]. In recent years, cancer has been regarded as a disease related to aging. An overwhelming number of cancers occur in the elderly, resulting from an accumulation of genetic mutations. The causes of these mutations include irregularities in the cell cycle and apoptosis of cells subject to significant stress or DNA damage [10,11].

Recent studies have indicated that the inflammatory process is involved in aging and age-related diseases. This includes increased production of COX-derived reactive species, including reactive oxygen species (ROS), reactive nitrogen species, and reactive lipid species, as well as increased gene expression of pro-inflammatory molecules such as TNF-α, IL-1β, IL-6, inducible nitric oxide synthase (iNOS), and COX-2 [12,13]. One of the major pathogenesis regulators involved in many inflammatory diseases is the NF-κB regulatory loop. In response to internal or external stress stimuli, the free NF-κB complex is translocated to the nucleus, where it promotes expression of pro-inflammatory genes that are important for activation of immune responses, including genes encoding cytokines, chemokines, and adhesion molecules, as well as enzymes and molecules with microbicidal activity [14,15]. On the other hand, the activation of NF-κB also prevents cells from apoptosis [16–18] by either inducing the expression of antiapoptotic proteins, including BCL-XL [19], FLICE-like inhibitory protein (FLIP) [20,21] and members of the inhibitor of apoptosis (IAP) family [22,23], or inhibiting the function of c-Jun N-terminal kinase (JNK), a well-known kinase that regulates apoptotic signals [24,25]. Because activated NF-κB can both inhibit apoptosis and stimulate production of pro-inflammatory cytokines, it further promotes proliferation of the cells in which these processes have been initiated. Hence, tissue exposed to a chronic inflammation or to a pro-inflammatory environment may be more susceptible to carcinogenic processes [26].

With age, there is generally an increase in oxidative stress [27], which is always accompanied by the upregulation of NF-κB activity. This heightened stress plays a pathophysiological role in chronic diseases that are highly prevalent in the elderly. Although the causes of increased oxidative stress in aging have not been explicitly identified, excessive production of ROS causing elevated oxidative stress has been associated with several metabolic factors, including hyperglycemia, hyperlipidemia, and high levels of advanced glycation end products. Hence, maintaining the ability to detoxify ROS is of vital importance during the aging process. Calorie restriction (CR) has been shown to delay age-related biological changes and suppress a number of age-associated pathological abnormalities in both genders and across mammalian and nonmammalian species [28]. One consistent finding in CR animals is reduced oxidative stress and the result of well-maintained redox status, which greatly impacts the regulation of many redox-sensitive transcriptomes and expression of associated genes.

Our current understanding of the biological phenomenon of aging and how it affects GRNs and signaling pathways is based on bio-experimental results delineated in the literature. Aging is a complex and system-level process. With increasing age, there may be a reduced defensive response to external environmental stimuli such as carcinogens, oxidative stress, and pro-inflammatory molecules. However, at present, quantitative and systematic methods for discussing pathophysiological phenotypes related to aging are still lacking. From an engineering perspective [29], the robustness to tolerate intrinsic system variations and the ability to respond to as well as transduce external stimuli downstream (i.e., response ability) are two important and complementary system characteristics to discuss system performance; i.e., a more robust system will be less responsive, and vice versa [30]. Therefore, the main focus of this section is to develop new measurements to estimate whether the system robustness and response ability of a GRN would be changed with increasing age, and to investigate whether the changes to system characteristics are associated with high incidence of aging-related diseases, such as cancer. Under this premise, analysis of aging-related GRNs via microarray data across various tissues at different stages of life is a suitable methodology to adopt to gain a greater understanding of the systematic changes accompanied by increased age.

The robustness of a GRN is defined by the level of tolerance to intrinsic perturbations. The robustness of a GRN also plays a principal role in the fail-safe mechanism of evolution. Although the concept of robustness trade-offs has been proposed

[31–34], few studies have quantitatively and systematically calculated the robustness of GRNs associated with aging-related pathophysiological phenotypes. Recently, systems biology approaches have been widely applied to analyze biological systems from the dynamic network point of view [31,35,36]. In this section, we use microarray data to develop a dynamic model of a GRN with multiple regulatory loops involved in the aging process and the associated pathophysiological phenotypes. Subsequently, we employ this model to quantify the relative robustness of each GRN with respect to the young, aged, and CR stages subject to intrinsic perturbations based on Lyapunov stability theory. Furthermore, we develop a gene response ability measurement (GRAM) method for estimating the ability of a gene network to transduce external stimuli to its downstream targets from the systematic perspective. The gene response ability to amplify or attenuate external stimulation signals, either from upstream regulators or environmental stimuli, can be estimated for each GRN via corresponding microarray data with respect to different stages of life across various tissues. Based on dynamic modeling through microarray data, we have found that an aging-related GRN increases network robustness and decreases network response ability during the aging process. Since "network robustness" and "network response ability" are two complementary antagonistic characteristics of a network system [30], the measurements of both of these variables could be used to double-check and confirm the system characteristics of aging-related GRNs constructed by microarray data. Measuring the trade-offs between network robustness and network response ability of aging-related GRNs may provide insight into systematic changes that occur during the aging process and potentially have applications for gene therapy and drug target selection.

11.1.1 Methods to construct dynamic model of aging-related GRNs

From the systems biology perspective, the systematic changes in pathophysiological phenotypes accompanied by an increase in age should be related to the dynamic process of aging-related GRNs. The network robustness of a GRN to tolerate intrinsic fluctuations and the network response ability of a GRN to respond to environmental stimuli are the two most important systematic characteristics of a dynamic network to be evaluated if we are to gain more insight into systematic properties of the aging-related gene network. However, the network robustness and network response ability of GRNs from the global system point of view are not experimentally measurable through all the network components. Therefore, to analyze GRNs involved in the aging process as a dynamic system, we first need to construct and compare the dynamic GRNs accomplished by microarray data across different tissues at different stages of life [37]. Then, the constructed aging-related GRNs at different life stages are used to estimate their network robustness to tolerate intrinsic perturbations and their network response abilities to respond to external signals from upstream regulation and from environmental stimuli. Finally, we analyze network robustness and response abilities of these gene networks to gain greater insight into the dynamics of aging and investigate how these changes in systematic properties accompanied by an increase in age affect the GRN as a whole.

Data selection and process

We use previously reported microarray data [37], which profiles the effects of aging on gene expression in the thymus, spinal cord, and eye tissues dissected from male C57BL/6 mice of ages 1, 6, 16, and 24 months with five mice per age cohort, as our mRNA expression profiles. The DNA microarray comprises 16,896 cDNA probes corresponding to 8932 genes. In order to make good use of these data for identifying the system parameters of our dynamic model of GRN, we rescale the original gene expression time profiles for 3, 6, 25, 28, 68, 71, 102, and 105 weeks, with two or three mice per age cohort (duplicate or triplicate repeats). The detailed categories are as follows: (1) young mice, with ages of 17 and 19 days termed as 3-week-old animals, 40, 43, and 44 days as 6-week-old animals, 174, 180, and 186 days as 25-week-old animals, 193 and 194 days as 28-week-old animals, placed on ad libitum diets; (2) aged mice, with ages of 476, 481, and 485 days termed as 68-week-old animals, 495 and 498 days as 71-week-old animals, 714 and 719 days as 102-week-old animals, and 730, 733, and 743 days as 105-week-old animals, placed on ad libitum diets; (3) CR mice, with ages of 472 and 474 days termed as 68-week-old animals, 482, 485, and 495 days as 71-week-old animals, and 673, 673, and 673 days as 96-week-old animals, fed on 60% caloric-restricted diets beginning at 24 weeks of age. Therefore, there are four time points for both the young and aged groups, and three time points for the CR group. All of the above data are calculated on a log 2 scale (expression level), and then the *Z*-score method used to normalize the gene array expression data has been previously described [38].

Construction of a dynamic model for aging-related GRNs

To investigate whether the ability of pathophysiological phenotypes, including detoxification of ROS, inflammation, apoptosis, cell cycle arrest, and cell growth, would be unregulated with increasing age, we focus on FOXOs-, NF-κB-, and p53-regulatory loops highly involved in these corresponding pathophysiological phenotypes. There are several highly related genes within these regulatory loops; hence, we further use one-way analysis of variance (ANOVA; p-value $< .05$) to select 16 genes with differential expression across young, aged, and CR groups: *FOXOs*, *NF-κB*, *p53*, *SIRT1*, *HIC1*, *Mdm2*, *Arf1*, *PTEN*, *PI3K*, *Akt*, *JNK*, *IKKs*, *IκB*, *BTG3*, *E2F*, and *ATM*.

For constructing aging-related GRNs, it is assumed that the network of genes of interest consists of N genes ($N = 16$, in Fig. 11.1) and time-series microarray data used for constructing the gene network contains K time points. In order to simplify analysis and take into account delayed effects of molecular trafficking and cellular

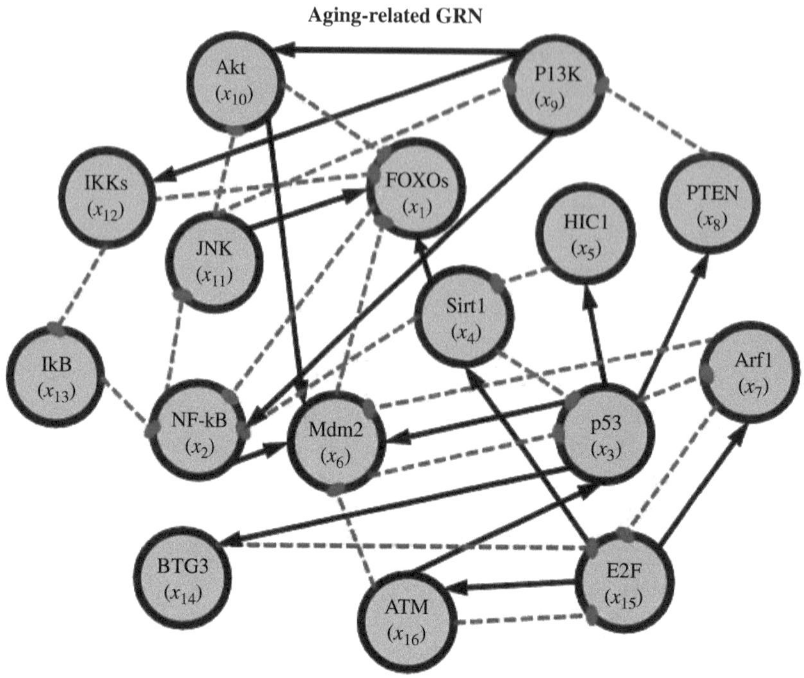

Figure 11.1 Multiple loops of gene regulatory network associated with aging-related pathophysiological phenotypes. This aging-related gene regulatory network includes 16 genes: *FOXOs, NF-κB, p53, SIRT1, HIC1, Mdm2, Arf1, PTEN, PI3K, Akt, JNK, IKKs, IκB, BTG3, E2F,* and *ATM.*

signal transduction, the GRN of aging we want to investigate (shown in Table 11.1) could be represented by the following linear discrete-time dynamic system:

$$X[t+1] = AX[t] + H$$

$$A = \begin{bmatrix} \alpha_{1,1} & \alpha_{1,2} & \cdots & \cdots & \alpha_{1,N} \\ \alpha_{2,1} & \ddots & & & \\ \vdots & & \alpha_{i,j} & & \\ \vdots & & & \ddots & \\ \alpha_{N,1} & & & & \alpha_{N,N} \end{bmatrix}, \quad H = \begin{bmatrix} h_1 \\ h_2 \\ \vdots \\ \vdots \\ h_N \end{bmatrix} \tag{11.1}$$

where the state vector $X[t] = [x_1[t] \ldots x_N[t]]^T$ stands for the discrete-time mRNA expression levels of total N genes at times $t = 1, \ldots, K$. The system matrix A denotes the gene interactions in the gene network, i.e., $\alpha_{i,j}$ denotes the interaction from gene j to gene i when $i \neq j$ $\forall i,j = 1, \ldots, N$ ($N = 16$ for the aging-related GRNs in Fig. 11.1); the constant vector H represents the basal level, i.e., h_i denotes the basal level of the ith gene. If the GRN is nonlinear, then the linear gene network in

Table 11.1 Ordinary difference equations of the gene regulatory network in Fig. 11.1 with 16 genes associated with aging-related pathophysiological phenotypes

	Gene	Dynamic equations of 16 genes of interest
x_1	*FOXOs*	$x_1[t+1]=\alpha_{1,1}x_1[t]+\alpha_{1,4}x_4[t]+\alpha_{1,6}x_6[t]+\alpha_{1,10}x_{10}[t]+\alpha_{1,11}x_{11}[t]+\alpha_{1,12}x_{12}[t]+h_1$
x_2	*NF-κB*	$x_2[t+1]=\alpha_{2,1}x_1[t]+\alpha_{2,2}x_2[t]+\alpha_{2,4}x_4[t]+\alpha_{2,9}x_9[t]+\alpha_{2,13}x_{13}[t]+h_2$
x_3	*p53*	$x_3[t+1]=\alpha_{3,3}x_3[t]+\alpha_{3,4}x_4[t]+\alpha_{3,6}x_6[t]+\alpha_{3,16}x_{16}[t]+h_3$
x_4	*SIRT1*	$x_4[t+1]=\alpha_{4,4}x_4[t]+\alpha_{4,5}x_5[t]+\alpha_{4,15}x_{15}[t]+h_4$
x_5	*HIC1*	$x_5[t+1]=\alpha_{5,3}x_3[t]+\alpha_{5,5}x_5[t]+h_5$
x_6	*Mdm2*	$x_6[t+1]=\alpha_{6,2}x_2[t]+\alpha_{6,3}x_3[t]+\alpha_{6,6}x_6[t]+\alpha_{6,7}x_7[t]+\alpha_{6,10}x_{10}[t]$
x_7	*Arf1*	$x_7[t+1]=\alpha_{7,3}x_3[t]+\alpha_{7,7}x_7[t]+\alpha_{7,15}x_{15}[t]+h_7$
x_8	*PTEN*	$x_8[t+1]=\alpha_{8,3}x_3[t]+\alpha_{8,8}x_8[t]+h_8$
x_9	*PI3K*	$x_9[t+1]=\alpha_{9,8}x_8[t]+\alpha_{9,9}x_9[t]+\alpha_{9,11}x_{11}[t]+h_9$
x_{10}	*Akt*	$x_{10}[t+1]=\alpha_{10,9}x_9[t]+\alpha_{10,10}x_{10}[t]+\alpha_{10,11}x_{11}[t]+h_{10}$
x_{11}	*JNK*	$x_{11}[t+1]=\alpha_{11,2}x_2[t]+\alpha_{11,11}x_{11}[t]+h_{11}$
x_{12}	*IKKs*	$x_{12}[t+1]=\alpha_{12,9}x_9[t]+\alpha_{12,12}x_{12}[t]+h_{12}$
x_{13}	*IκB*	$x_{13}[t+1]=\alpha_{13,12}x_{12}[t]+\alpha_{13,13}x_{13}[t]+h_{13}$
x_{14}	*BTG3*	$x_{14}[t+1]=\alpha_{14,3}x_3[t]+\alpha_{14,14}x_{14}[t]+h_{14}$
x_{15}	*E2F*	$x_{15}[t+1]=\alpha_{15,7}x_7[t]+\alpha_{15,14}x_{14}[t]+\alpha_{15,15}x_{15}[t]+\alpha_{15,16}x_{16}[t]+h_{15}$
x_{16}	*ATM*	$x_{16}[t+1]=\alpha_{16,15}x_{15}[t]+\alpha_{16,16}x_{16}[t]+h_{16}$

Eq. (11.1) can be considered as the linearized system around an equilibrium point (i.e., phenotype) of interest and H is related to the location of the equilibrium point of linearization, i.e., for a nonlinear gene network $X[t+1]=f(X[t])$ for some nonlinear interaction $f(X)$ the linearized gene network around an equilibrium point X_0 of interest can be represented as $X[t+1]=AX[t]+C+\Delta f(X)$ where $A=\partial f(X[t])/\partial X[t]\big|_{X[t]=X_0}$ and $C=f(X_0)$ and $\Delta f(X)$ denotes the high order nonlinear term. In this case, $H=C+\Delta f(X)$ can represent the basal level of linearized system at this equilibrium point of interest.

Parameter estimation of GRN by microarray data

Our proposed focus is to investigate whether the robustness and response ability of GRN at different stages of life would be distinct. For this reason, the application of the proposed method is dependent on the time profiles of the microarray data. In other words, our approach will perform better when more time points of microarray data are available to allow thorough exploration of the characteristic changes of a dynamic GRN. Therefore, using the cubic spline method to interpolate microarray data makes it possible to obtain as many data points as needed to avoid overfitting in the parameter estimation process of GRN [39–41]. We then use the least squares parameter estimation method [42] to estimate the interactive matrix A and basal level H in Eq. (11.1) from the time-series microarray data. Based on the least square parameter estimation in Ref. [42], let us denote the total quadratic estimation error as $V(\hat{\theta}) = \sum_{t=1}^{M} \left(X(t+1) - \hat{A}X(t) - \hat{H}\right)^T \left(X(t+1) - \hat{A}X(t) - \hat{H}\right) = \sum_{t=1}^{M} \varepsilon(t)^T \varepsilon(t)$, where $\hat{\theta} = \begin{bmatrix} \hat{A} \\ \hat{H} \end{bmatrix}$ denotes the estimate of parameter $\theta = \begin{bmatrix} A \\ H \end{bmatrix}$ and M is the number of microarray data for parameter estimation. The least squares parameter estimation method [42] requires selecting $\hat{\theta}$ so as to minimize the total quadratic estimation error, $V(\hat{\theta})$. After the least squares parameter estimation by Eq. (2.20) or Eq. (2.21), the covariance of estimation error $\varepsilon(t)$ is given by $Cov(\varepsilon) = 2/M - 2V(\hat{\theta})$ and the covariance matrix of $\hat{\theta}$ is given by $Cov(\hat{\theta}) = Cov(\varepsilon)\left[\sum_{t=1}^{M} X^T(t)X(t) + I\right]^{-1}$. These parameter estimation methods for constructing GRNs are called the "reverse engineering method" in systems biology [43,44]. Based on the dynamic model described earlier and parameter estimation methods via microarray data across different tissues from young, aged, and CR mice, the system parameters A for aging-related GRNs in Fig. 11.1 and Table 11.1 are estimated for young, aged, and CR mice (Table 11.2). These parameters are used to measure network robustness and response ability in the following section.

Relative network robustness of aging-related GRNs at young, aged, and CR stages under intrinsic perturbations

After constructing the dynamic models of the GRN in Eq. (11.1) by time-series microarray data in the "Construction of a dynamic model for aging-related GRNs" section, we measure their network robustness based on their dynamic models. The steady state X_S of the dynamic system in Eq. (11.1) is obtained as:

$$X_s = \hat{A}X_s + \hat{H}, \quad \text{or} \quad X_s = (I - \hat{A})^{-1}\hat{H} \tag{11.2}$$

To simplify analysis of "robustness" of the steady state (phenotype), the origin of the dynamic system is shifted to the steady state X_S, i.e., $X[t] = \tilde{X}[t] + X_s$, allowing the following shifted dynamic system to be achieved by subtracting Eq. (11.1) from (11.2) [45]

$$\tilde{X}[t+1] = \hat{A}\tilde{X}[t] \tag{11.3}$$

Therefore, the robustness of the phenotype (i.e., the steady state) of the gene network becomes the robustness of the shifted gene network in Eq. (11.3) at the origin $\tilde{X}[t] \equiv 0$. In the shifted dynamic system in Eq. (11.3), if the eigenvalues of $\hat{A}$ are all inside the unit circle $|Z| \leq 1$ in the complex Z domain, then $\tilde{X}[t] \to 0$, and $X[t] \to X_S$, i.e., approaches the steady state asymptotically. Since the network robustness of the dynamic system in Eq. (11.1) at the steady state X_S is equivalent to the network robustness of the dynamic system in Eq. (11.3) at the origin $\tilde{X}[t] \equiv 0$, the shifted dynamic system in Eq. (11.3) will simplify the network robustness analysis procedure for GRNs [45–47].

Table 11.2 Estimated parameters of gene regulatory networks in Table 11.1 with 16 gene across thymus, spinal cord, and eye tissues at time young, aged, and calorie-restricted (CR) stages

Estimated parameters from microarray data

Thymus	
Young	$\alpha_{1,1}=0.406$ $\alpha_{1,4}=0.181$ $\alpha_{1,6}=-0.210$ $\alpha_{1,10}=-0.030$ $\alpha_{1,11}=0.060$ $\alpha_{1,12}=-0.401$ $\alpha_{2,1}=0.098$ $\alpha_{2,2}=0.949$ $\alpha_{2,4}=-0.128$ $\alpha_{2,9}=0.036$ $\alpha_{2,13}=-0.124$ $\alpha_{3,3}=0.982$ $\alpha_{3,4}=0.070$ $\alpha_{3,6}=0.027$ $\alpha_{3,16}=0.013$ $\alpha_{4,4}=1.048$ $\alpha_{4,5}=0.026$ $\alpha_{4,15}=0.004$ $\alpha_{5,3}=0.001$ $\alpha_{5,5}=0.992$ $\alpha_{6,2}=0.408$ $\alpha_{6,3}=0.123$ $\alpha_{6,6}=0.422$ $\alpha_{6,7}=0.119$ $\alpha_{6,10}=0.199$ $\alpha_{6,16}=0.064$ $\alpha_{7,3}=0.028$ $\alpha_{7,7}=0.933$ $\alpha_{7,15}=-0.006$ $\alpha_{8,3}=0.008$ $\alpha_{8,8}=1.015$ $\alpha_{9,8}=-0.011$ $\alpha_{9,9}=0.974$ $\alpha_{9,11}=0.001$ $\alpha_{10,9}=0.009$ $\alpha_{10,10}=1.067$ $\alpha_{10,11}=-0.024$ $\alpha_{11,2}=0.021$ $\alpha_{11,11}=1.001$ $\alpha_{12,9}=-0.012$ $\alpha_{12,12}=1.016$ $\alpha_{13,12}=0.009$ $\alpha_{13,13}=0.986$ $\alpha_{14,3}=-0.003$ $\alpha_{14,14}=1.020$ $\alpha_{15,7}=0.173$ $\alpha_{15,14}=-0.079$ $\alpha_{15,15}=0.958$ $\alpha_{15,16}=-0.061$ $\alpha_{16,15}=0.010$ $\alpha_{16,16}=1.014$
Aged	$\alpha_{1,1}=0.254$ $\alpha_{1,4}=0.143$ $\alpha_{1,6}=-0.023$ $\alpha_{1,10}=-0.089$ $\alpha_{1,11}=0.032$ $\alpha_{1,12}=-0.398$ $\alpha_{2,1}=0.184$ $\alpha_{2,2}=0.520$ $\alpha_{2,4}=-0.402$ $\alpha_{2,9}=0.181$ $\alpha_{2,13}=-0.120$ $\alpha_{3,3}=0.986$ $\alpha_{3,4}=0.055$ $\alpha_{3,6}=0.109$ $\alpha_{3,16}=0.099$ $\alpha_{4,4}=0.976$ $\alpha_{4,5}=0.036$ $\alpha_{4,15}=0.004$ $\alpha_{5,3}=0.019$ $\alpha_{5,5}=0.977$ $\alpha_{6,2}=-0.087$ $\alpha_{6,3}=0.008$ $\alpha_{6,6}=0.196$ $\alpha_{6,7}=0.174$ $\alpha_{6,10}=0.070$ $\alpha_{6,16}=0.291$ $\alpha_{7,3}=0.092$ $\alpha_{7,7}=1.090$ $\alpha_{7,15}=-0.106$ $\alpha_{8,3}=0.005$ $\alpha_{8,8}=0.999$ $\alpha_{9,8}=-0.027$ $\alpha_{9,9}=0.975$ $\alpha_{9,11}=0.014$ $\alpha_{10,9}=0.002$ $\alpha_{10,10}=0.987$ $\alpha_{10,11}=-0.007$ $\alpha_{11,2}=0.003$ $\alpha_{11,11}=0.994$ $\alpha_{12,9}=-0.003$ $\alpha_{12,12}=1.009$ $\alpha_{13,12}=0.029$ $\alpha_{13,13}=0.994$ $\alpha_{14,3}=-0.091$ $\alpha_{14,14}=0.955$ $\alpha_{15,7}=0.364$ $\alpha_{15,14}=-0.248$ $\alpha_{15,15}=0.449$ $\alpha_{15,16}=-0.045$ $\alpha_{16,15}=0.025$ $\alpha_{16,16}=0.979$
CR	$\alpha_{1,1}=0.132$ $\alpha_{1,4}=0.231$ $\alpha_{1,6}=-0.124$ $\alpha_{1,10}=-0.063$ $\alpha_{1,11}=0.040$ $\alpha_{1,12}=-0.194$ $\alpha_{2,1}=0.132$ $\alpha_{2,2}=0.583$ $\alpha_{2,4}=-0.246$ $\alpha_{2,9}=0.155$ $\alpha_{2,13}=-0.045$ $\alpha_{3,3}=0.114$ $\alpha_{3,4}=0.023$ $\alpha_{3,6}=0.170$ $\alpha_{3,16}=0.219$ $\alpha_{4,4}=0.920$ $\alpha_{4,5}=0.162$ $\alpha_{4,15}=0.019$ $\alpha_{5,3}=0.036$ $\alpha_{5,5}=0.979$ $\alpha_{6,2}=0.248$ $\alpha_{6,3}=0.170$ $\alpha_{6,6}=0.513$ $\alpha_{6,7}=0.236$ $\alpha_{6,10}=0.122$ $\alpha_{6,16}=0.299$ $\alpha_{7,3}=0.030$ $\alpha_{7,7}=0.991$ $\alpha_{7,15}=-0.001$ $\alpha_{8,3}=0.106$ $\alpha_{8,8}=1.022$ $\alpha_{9,8}=-0.220$ $\alpha_{9,9}=0.436$ $\alpha_{9,11}=0.252$ $\alpha_{10,9}=0.099$ $\alpha_{10,10}=0.970$ $\alpha_{10,11}=-0.087$ $\alpha_{11,2}=0.026$ $\alpha_{11,11}=0.997$ $\alpha_{12,9}=-0.089$ $\alpha_{12,12}=0.954$ $\alpha_{13,12}=0.039$ $\alpha_{13,13}=1.045$ $\alpha_{14,3}=-0.088$ $\alpha_{14,14}=0.977$ $\alpha_{15,7}=0.138$ $\alpha_{15,14}=-0.161$ $\alpha_{15,15}=0.926$ $\alpha_{15,16}=-0.185$ $\alpha_{16,15}=0.018$ $\alpha_{16,16}=1.088$
Spinal cord	
Young	$\alpha_{1,1}=0.382$ $\alpha_{1,4}=0.081$ $\alpha_{1,6}=0.004$ $\alpha_{1,10}=-0.161$ $\alpha_{1,11}=0.001$ $\alpha_{1,12}=-0.453$ $\alpha_{2,1}=0.022$ $\alpha_{2,2}=0.915$ $\alpha_{2,4}=0.013$ $\alpha_{2,9}=0.119$ $\alpha_{2,13}=-0.263$ $\alpha_{3,3}=0.941$ $\alpha_{3,4}=-0.005$ $\alpha_{3,6}=0.267$ $\alpha_{3,16}=0.050$ $\alpha_{4,4}=0.973$ $\alpha_{4,5}=0.043$ $\alpha_{4,15}=-0.128$ $\alpha_{5,3}=0.079$ $\alpha_{5,5}=0.970$ $\alpha_{6,2}=-0.073$ $\alpha_{6,3}=-0.084$ $\alpha_{6,6}=0.048$ $\alpha_{6,7}=-0.030$ $\alpha_{6,10}=0.149$ $\alpha_{6,16}=0.102$ $\alpha_{7,3}=0.186$ $\alpha_{7,7}=1.021$ $\alpha_{7,15}=0.156$ $\alpha_{8,3}=-0.037$ $\alpha_{8,8}=1.004$ $\alpha_{9,8}=-0.006$ $\alpha_{9,9}=0.992$ $\alpha_{9,11}=-0.008$ $\alpha_{10,9}=0.009$ $\alpha_{10,10}=0.972$ $\alpha_{10,11}=0.034$ $\alpha_{11,2}=0.013$ $\alpha_{11,11}=1.017$ $\alpha_{12,9}=0.018$ $\alpha_{12,12}=1.008$ $\alpha_{13,12}=0.002$ $\alpha_{13,13}=1.001$ $\alpha_{14,3}=-0.591$ $\alpha_{14,14}=0875$ $\alpha_{15,7}=-0.064$ $\alpha_{15,14}=0.146$ $\alpha_{15,15}=0.668$ $\alpha_{15,16}=0.252$ $\alpha_{16,15}=-0.021$ $\alpha_{16,16}=1.008$

(Continued)

Table 11.2 (Continued)

Estimated parameters from microarray data

Aged	$\alpha_{1,1}=0.880$ $\alpha_{1,4}=0.043$ $\alpha_{1,6}=0.086$ $\alpha_{1,10}=-0.042$ $\alpha_{1,11}=0.158$ $\alpha_{1,12}=-0.100$ $\alpha_{2,1}=0.186$ $\alpha_{2,2}=0.829$ $\alpha_{2,4}=0.068$ $\alpha_{2,9}=0.033$ $\alpha_{2,13}=-0.231$ $\alpha_{3,3}=0.860$ $\alpha_{3,4}=-0.004$ $\alpha_{3,6}=0.003$ $\alpha_{3,16}=0.194$ $\alpha_{4,4}=0.900$ $\alpha_{4,5}=0.239$ $\alpha_{4,15}=-0.359$ $\alpha_{5,3}=0.043$ $\alpha_{5,5}=0.991$ $\alpha_{6,2}=-0.139$ $\alpha_{6,3}=-0.014$ $\alpha_{6,6}=0.810$ $\alpha_{6,7}=-0.285$ $\alpha_{6,10}=0.136$ $\alpha_{6,16}=0.047$ $\alpha_{7,3}=0.003$ $\alpha_{7,7}=0.990$ $\alpha_{7,15}=0.007$ $\alpha_{8,3}=-0.019$ $\alpha_{8,8}=0.998$ $\alpha_{9,8}=-0.091$ $\alpha_{9,9}=0.812$ $\alpha_{9,11}=-0.124$ $\alpha_{10,9}=0.009$ $\alpha_{10,10}=0.969$ $\alpha_{10,11}=0.001$ $\alpha_{11,2}=0.004$ $\alpha_{11,11}=0.987$ $\alpha_{12,9}=0.056$ $\alpha_{12,12}=0.970$ $\alpha_{13,12}=0.007$ $\alpha_{13,13}=1.021$ $\alpha_{14,3}=-0.011$ $\alpha_{14,14}=0.999$ $\alpha_{15,7}=-0.347$ $\alpha_{15,14}=0.128$ $\alpha_{15,15}=0.659$ $\alpha_{15,16}=0.203$ $\alpha_{16,15}=-0.030$ $\alpha_{16,16}=1.013$
CR	$\alpha_{1,1}=0.134$ $\alpha_{1,4}=0.143$ $\alpha_{1,6}=0.153$ $\alpha_{1,10}=-0.157$ $\alpha_{1,11}=0.043$ $\alpha_{1,12}=-0.054$ $\alpha_{2,1}=0.074$ $\alpha_{2,2}=0.802$ $\alpha_{2,4}=0.319$ $\alpha_{2,9}=0.125$ $\alpha_{2,13}=-0.132$ $\alpha_{3,3}=0.669$ $\alpha_{3,4}=-0.087$ $\alpha_{3,6}=0.380$ $\alpha_{3,16}=0.231$ $\alpha_{4,4}=0.972$ $\alpha_{4,5}=0.017$ $\alpha_{4,15}=-0.191$ $\alpha_{5,3}=0.025$ $\alpha_{5,5}=0.934$ $\alpha_{6,2}=-0.065$ $\alpha_{6,3}=-0.374$ $\alpha_{6,6}=0.472$ $\alpha_{6,7}=-0.054$ $\alpha_{6,10}=0.224$ $\alpha_{6,16}=0.130$ $\alpha_{7,3}=0.201$ $\alpha_{7,7}=0.604$ $\alpha_{7,15}=0.057$ $\alpha_{8,3}=-0.024$ $\alpha_{8,8}=1.117$ $\alpha_{9,8}=-0.061$ $\alpha_{9,9}=0.971$ $\alpha_{9,11}=-0.049$ $\alpha_{10,9}=0.041$ $\alpha_{10,10}=0.722$ $\alpha_{10,11}=0.445$ $\alpha_{11,2}=0.001$ $\alpha_{11,11}=1.004$ $\alpha_{12,9}=0.009$ $\alpha_{12,12}=1.005$ $\alpha_{13,12}=0.005$ $\alpha_{13,13}=0.987$ $\alpha_{14,3}=-0.065$ $\alpha_{14,14}=0.944$ $\alpha_{15,7}=-0.202$ $\alpha_{15,14}=0.108$ $\alpha_{15,15}=0.367$ $\alpha_{15,16}=0.218$ $\alpha_{16,15}=-0.132$ $\alpha_{16,16}=0.964$
Eye	
Young	$\alpha_{1,1}=0.108$ $\alpha_{1,4}=0.036$ $\alpha_{1,6}=0.141$ $\alpha_{1,10}=-0.103$ $\alpha_{1,11}=0.130$ $\alpha_{1,12}=-0.113$ $\alpha_{2,1}=0.328$ $\alpha_{2,2}=1.050$ $\alpha_{2,4}=-0.030$ $\alpha_{2,9}=0.035$ $\alpha_{2,13}=-0.071$ $\alpha_{3,3}=0.341$ $\alpha_{3,4}=-0.069$ $\alpha_{3,6}=-0.153$ $\alpha_{3,16}=0.401$ $\alpha_{4,4}=0.999$ $\alpha_{4,5}=0.044$ $\alpha_{4,15}=0.062$ $\alpha_{5,3}=0.258$ $\alpha_{5,5}=0.917$ $\alpha_{6,2}=-0.250$ $\alpha_{6,3}=-0.038$ $\alpha_{6,6}=0.548$ $\alpha_{6,7}=0.116$ $\alpha_{6,10}=-0.371$ $\alpha_{6,16}=-0.112$ $\alpha_{7,3}=0.023$ $\alpha_{7,7}=0.986$ $\alpha_{7,15}=-0.017$ $\alpha_{8,3}=0.046$ $\alpha_{8,8}=1.012$ $\alpha_{9,8}=-0.012$ $\alpha_{9,9}=1.039$ $\alpha_{9,11}=0.006$ $\alpha_{10,9}=0.418$ $\alpha_{10,10}=1.110$ $\alpha_{10,11}=0.074$ $\alpha_{11,2}=0.057$ $\alpha_{11,11}=1.072$ $\alpha_{12,9}=0.015$ $\alpha_{12,12}=0.975$ $\alpha_{13,12}=0.013$ $\alpha_{13,13}=1.008$ $\alpha_{14,3}=-0.132$ $\alpha_{14,14}=1.005$ $\alpha_{15,7}=0.099$ $\alpha_{15,14}=0.005$ $\alpha_{15,15}=0.980$ $\alpha_{15,16}=-0.183$ $\alpha_{16,15}=0.046$ $\alpha_{16,16}=0.880$
Aged	$\alpha_{1,1}=0.547$ $\alpha_{1,4}=0.095$ $\alpha_{1,6}=0.347$ $\alpha_{1,10}=-0.164$ $\alpha_{1,11}=0.143$ $\alpha_{1,12}=-0.135$ $\alpha_{2,1}=0.046$ $\alpha_{2,2}=0.752$ $\alpha_{2,4}=-0.188$ $\alpha_{2,9}=0.133$ $\alpha_{2,13}=-0.190$ $\alpha_{3,3}=0.581$ $\alpha_{3,4}=-0.078$ $\alpha_{3,6}=-0.291$ $\alpha_{3,16}=0.077$ $\alpha_{4,4}=0.747$ $\alpha_{4,5}=0.149$ $\alpha_{4,15}=0.657$ $\alpha_{5,3}=0.026$ $\alpha_{5,5}=1.017$ $\alpha_{6,2}=-0.259$ $\alpha_{6,3}=-0.095$ $\alpha_{6,6}=0.694$ $\alpha_{6,7}=0.128$ $\alpha_{6,10}=-0.291$ $\alpha_{6,16}=-0.215$ $\alpha_{7,3}=0.083$ $\alpha_{7,7}=0.930$ $\alpha_{7,15}=-0.007$ $\alpha_{8,3}=0.165$ $\alpha_{8,8}=0.946$ $\alpha_{9,8}=-0.015$ $\alpha_{9,9}=1.018$ $\alpha_{9,11}=0.010$ $\alpha_{10,9}=0.002$ $\alpha_{10,10}=0.967$ $\alpha_{10,11}=0.019$ $\alpha_{11,12}=0.028$ $\alpha_{11,11}=1.019$ $\alpha_{12,9}=0.012$ $\alpha_{12,12}=1.021$ $\alpha_{13,12}=0.033$ $\alpha_{13,13}=1.000$ $\alpha_{14,3}=-0.110$ $\alpha_{14,14}=1.005$ $\alpha_{15,7}=0.226$ $\alpha_{15,14}=0.083$ $\alpha_{15,15}=0.922$ $\alpha_{15,16}=-0.109$ $\alpha_{16,15}=0.222$ $\alpha_{16,16}=0.959$

The network robustness of the shifted gene network in Eq. (11.3) is measured by its ability to tolerate intrinsic molecular perturbations [30]. If the linear GRN suffers from intrinsic molecular perturbations mainly due to process noise, thermal fluctuations, or genetic mutations, the interactive matrix $\hat{A}$ is perturbed as $\hat{A}(1+\eta)$, where η denotes the ratio of intrinsic network perturbation: i.e., the corresponding additional system perturbation $\Delta\hat{A}=\eta\hat{A}$ [30]. Then, a GRN with intrinsic network perturbation can be represented by

$$\tilde{X}[t+1]=\hat{A}(1+\eta)\tilde{X}[t]=(\hat{A}+\Delta\hat{A})\tilde{X}[t] \tag{11.4}$$

The higher the η value, the more numerous the intrinsic perturbations and the larger the η value that a gene network can tolerate, and therefore the greater the robustness of the network. Since the state-dependent intrinsic network perturbations can influence the stability of GRNs, more effort is needed to discuss the robustness of perturbative GRN in Eq. (11.4). According to the quadratic stability theory [46,47], if the Lyapunov equation of energy function for the perturbative gene network in Eq. (11.4) is chosen as $V(\tilde{X}[t])=\tilde{X}^T[t]P\tilde{X}[t]$, for a positive symmetric matrix $P=P^T>0$, then the perturbative GRN in Eq. (11.4) is robustly stable if $\Delta V(\tilde{X}[t])=V(\tilde{X}[t+1])-V(\tilde{X}[t])\leq 0$: i.e., the energy of the gene network is not increased by intrinsic network perturbations. Based on this idea, we are able to obtain the following robust stability principle for aging-related GRN subjected to intrinsic perturbations.

Suppose an aging-related GRN suffers from intrinsic network perturbations as described in Eq. (11.4): then, from Proposition 10.1 the perturbative network is robustly stable in the aging process if the following inequality has a positive definite solution $P=P^T>0$:

$$[\hat{A}(1+\eta)]^T P[\hat{A}(1+\eta)]\leq P \tag{11.5}$$

Remark 11.1:

1. If a GRN is free of intrinsic network perturbations as in Eq. (11.3), i.e., $\eta=0$, the stable condition is reduced to the matrix inequality $\hat{A}^T P\hat{A}\leq P$, which has a positive definite solution $P>0$ [30]. In order to guarantee the stability of a perturbation-free GRN, eigenvalues of the system interaction matrix $\hat{A}$ should be inside the unit circle $|Z|\leq 1$ in the Z complex domain: i.e., $\hat{A}^T\hat{A}<1$. From Eq. (11.5), it can be seen that the eigenvalues of system interactive matrix $\hat{A}$ of the perturbation system in Eq. (11.4) should be nearer the origin of the complex domain Z than those of $\hat{A}$ of the perturbation-free system in Eq. (11.3) when η is increased gradually. Some eigenvalues of system interactive matrix $\hat{A}$ near the unit circle are more easily perturbed outside the unit circle by intrinsic parameter fluctuations, which would make the perturbative GRN become unstable. Therefore, the distance between eigenvalue locations of $\hat{A}$ to the unit circle can be considered as a measurement of robustness for a linear GRN.

2. Based on the above analysis, the network robustness η° of the GRN is the maximum perturbation tolerance by the network without violating the network stability, and can be measured by solving the following constrained optimization:

$$\begin{aligned} &\eta^{\circ} = \max \eta \\ &\text{subject to } P > 0 \text{ and Eq. (11.5)} \end{aligned} \tag{11.6}$$

i.e., the maximum perturbation ratio η° tolerable by the GRN such that stability is still maintained. The constrained optimization problem in Eq. (11.6) can be solved by increasing η up until no positive solution $P > 0$ exists in Eq. (11.5): i.e., to the highest value possible without violating the robust stability in Eq. (11.5). A positive definite solution $P > 0$ in Eq. (11.5) can be obtained by the linear matrix inequality (LMI) Toolbox of Matlab.

We can solve the above constrained optimization problem to measure the network robustness of aging-related GRNs only by increasing η until no positive definite solution $P > 0$ is found in Eq. (11.5), and then the maximal network perturbation η° that could be tolerated by the gene network represents the network robustness of the gene network under intrinsic perturbations. Hence, we are able to obtain relative robustness of GRNs at different stages of life. In comparing the values of η° for each network, higher values of η° imply a greater network robustness estimate for different aging-related GRNs during the aging process. The relative network robustness η° to tolerate normalized network perturbations for the young, aged, and CR groups' GRNs with the system interaction matrices in Table 11.2 are shown in Table 11.3. GRNs for the aged group are more robust than for the young and CR groups in different tissues, as may be seen in Table 11.3. Further detailed discussion will be given in Section 11.1.3.

Table 11.3 The network robustness (η°) and the network response ability (δ°) of the gene regulatory network with 16 genes across different tissues at the young, aged, and CR stages

Tissue	Young	Aged	CR
Thymus			
η°	0.2310	0.3750	0.2050
δ°	1.1653	0.9463	1.2279
Spinal cord			
η°	0.2410	0.6270	0.1400
δ°	1.1630	0.9367	1.2645
Eye			
η°	0.1600	0.2910	
δ°	1.2376	0.8798	

After the network robustness of aging-related GRNs is measured based on the microarray-identified system model for tolerance to intrinsic biochemical parametric fluctuations, we then examine the response abilities of genes to external stimuli. Assuming the GRN responds to external stimuli $U[t]$, including upstream regulatory signals and external signals outside the network, such as hormones, carcinogens, oxidative stress, or ambient biomedical molecules that could bring about an effect on the network, the dynamic Eq. (11.3) could be modified as follows [29]:

$$\begin{aligned}\tilde{X}[t+1] &= \hat{A}\tilde{X}[t] + U[t]\\ Y[t] &= C\tilde{X}[t]\end{aligned} \tag{11.7}$$

where $U[t] = [u_1[t] \ \ldots \ u_N[t]]^T$ represents external stimuli and $Y[t]$ denotes output signal responses of some genes of interest. For example, if we want to analyze the output signal response of gene i to external stimuli $U[t]$, then $C = diag(0,\ldots,0,1,0,\ldots,0)$, i.e., all elements of C are zero except 1 at the ith diagonal component, e.g., $C = diag(0,0,1,0,\ldots,0)$ for discussing the gene response of the p53 gene. If the network response ability of the entire GRN to external stimuli is analyzed, then $C = I$, which is an identity matrix.

In a GRN, each component may amplify or attenuate the effect of external stimuli to its downstream regulators. The effect of input signals $U[t]$ on output signals $Y[t]$ is less than or equal to a positive value δ, if the following inequality holds [46,47]:

$$\frac{\sum_{t=0}^{\infty} Y^T[t]Y[t]}{\sum_{t=0}^{\infty} U^T[t]U[t]} \leq \delta^2 \quad \text{or} \quad \sum_{t=0}^{\infty} Y^T[t]Y[t] \leq \delta^2 \sum_{t=0}^{\infty} U^T[t]U[t] \tag{11.8}$$

for all possible bounded input signals $U[t] \neq 0$ and $\tilde{X}[0] = 0$. The physical meaning described in Eq. (11.8) is that the effect of all possible external stimuli $U[t]$ on $Y[t]$ is less than or equal to δ^2 from the energy point of view: i.e., δ denotes the upper bound of the effect of $U[t]$ on $Y[t]$. If $\tilde{X}[0] \neq 0$, then the inequality in Eq. (11.8) should be modified as [46,47]:

$$\sum_{t=0}^{\infty} Y^T[t]Y[t] \leq \tilde{X}^T[0]P\tilde{X}[0] + \delta^2 \sum_{t=0}^{\infty} U^T[t]U[t] \tag{11.9}$$

for all bounded input signals $U[t]$. δ^o, the smallest upper bound of δ in Eq. (11.8), is called the "gene response ability" of the GRN in response to all bounded stimulation signals $U[t]$. The gene response ability δ^o permits us to obtain a more systematic insight into the ability of gene response to external stimuli for individual genes or the entire GRN. According to the GRAM method, we can measure and compare gene response ability δ^o of the aging-related GRNs at the young, aged, and CR stages of life.

According to Proposition 10.2, the gene response of the aging-related GRN in Eq. (11.7) has an upper bound δ in Eq. (11.8), if there exists a positive definite $P = P^T > 0$ solution to the following LMI [48]:

$$\begin{bmatrix} \hat{A}^T P \hat{A} + C^T C - P & \hat{A}^T P \\ P\hat{A} & P - \delta^2 I \end{bmatrix} \leq 0 \tag{11.10}$$

i.e., if the above LMI holds for some $P > 0$, then the effect of $U[t]$ on $Y[t]$ must be less than or equal to δ: i.e., Eq. (11.8) or (11.9) holds for the aging-related GRN in Eq. (11.7).

Remark 11.2: According to the gene response ability δ^{o} (i.e., the minimal δ) of the GRN to external stimuli, the gene response ability of the GRN can be obtained by solving the following constrained optimization problem:

$$\begin{gathered} \delta^{\mathrm{o}} = \min_{P} \delta \\ \text{subject to } \delta > 0, P > 0, \text{and Eq. (11.10)} \end{gathered} \tag{11.11}$$

This constrained optimization can be solved by decreasing the upper bound δ in Eq. (11.8) until no $P > 0$ exists in Eq. (11.10) using Matlab LMI toolbox [29]. The response ability δ^{o} of GRNs is estimated by Eq. (11.11) for the young, aged, and CR groups separately, and the results are shown in Table 11.3. Further detailed discussion of δ^{o} in aging process will be given in Section 11.1.3.

1. δ^{o}, the smallest upper bound of δ in Eq. (11.8), is called the gene response ability of the GRN to external stimuli. If $\delta^{\mathrm{o}} < 1$, external stimuli are attenuated (buffered) by the GRN. If $\delta^{\mathrm{o}} > 1$, external stimuli are amplified by the GRN.
2. Since the output/input signal energy ratio (i.e., the response) in Eq. (11.8) or (11.9) is considered for all possible external stimuli, the measure of gene response ability in Eqs. (11.10) and (11.11) is more dependent on system matrix A than on the external stimuli, in the same way a lowpass filter more dependent on the system response characteristics of the filter than on noise itself. Therefore, it is possible to measure the gene response ability of a GRN from system characteristics of the network without knowledge of what the external stimuli are, with the exception that these signals should be bounded. The system matrix A could be estimated by time-series microarray data in the previous sections. Therefore, via the help of the Matlab LMI toolbox, we use Eq. (11.11) to measure the system response abilities of the aging-related gene networks across different tissues at the young, aged, and CR stages via microarray data.

11.1.2 Discussion of big cellular mechanisms in aging-related networks

Construction of multiple regulatory loops of gene networks associated with aging-related pathophysiological phenotypes

We combine several references from the literature to construct multiple regulatory loops of gene networks associated with aging-related pathophysiological phenotypes for 16 genes: namely, *FOXOs, NF-κB, p53, SIRT1, HIC1, Mdm2, Arf1, PTEN, PI3K, Akt, JNK, IKKs, IκB, BTG3, E2F,* and *ATM* (Fig. 11.1). All genes selected are considered critical for regulating cellular functions during the aging process, including detoxification of ROS, cell cycle arrest, repair of damaged DNA, apoptosis and pro-inflammation [13,28]. The complicated regulation mechanisms of the genes selected above are described in detail as follows: Akt-mediated phosphorylation of FOXOs leads to the release of FOXOs from DNA, followed by sequestration in the cytoplasm and inhibition of FOXO-mediated transcription of genes involved in cell cycle arrest, apoptosis, and insulin sensitivity [49]. The IKKs also antagonize the FOXOs by interacting with, phosphorylating, and targeting FOXOs for ubiquitin-mediated degradation [50]. Moreover, IKKs also promote the degradation of the IκB protein, which inhibits nuclear translocation and activation of *NF-κB*. NF-κB, a major regulator of gene expression programs associated with mammalian aging, is primarily involved in transcription of genes contributing to innate and adaptive immunity, apoptosis, and production of pro-inflammatory molecules. Excess NF-κB activity can also induce insulin resistance, muscle atrophy, and neurotoxicity in Alzheimer's disease, three prevalent age-associated morbidities [51,52]. PTEN can interfere with NF-κB activity via inhibition of *PI3K*. Furthermore, NF-κB upregulates the expression of *Mdm2*, an ubiquitin-ligase and inhibitor of tumor suppressor p53 that induces cell cycle arrest and apoptosis [53]. Although the activated *Mdm2* can target *p53* by both inhibiting its activity as a TF and reinforcing its degradation rate, *Mdm2* itself can also be transcriptionally activated by p53 to form a negative feedback loop, sustaining very low levels of p53 activity in the absence of cellular damage. Upon significant stress or DNA damage, p53 and Mdm2 are phosphorylated by ATM/ATR and other kinases in a manner that prevents Mdm2 from interacting with p53, thus resulting in p53 stabilization [54,55]. Transcription of the *Arf1* gene, a tumor suppressor encoded by the *INK4a/ARF* locus, is positively regulated by β-catenin and negatively regulated by p53. Furthermore, Arf protein binds to Mdm2 protein and inhibits its ubiquitin ligase activity, thus increasing the level of p53 protein [56]. The *Arf* gene is also a transcriptional target of E2F1, and it was originally assumed that Arf would be a critical mediator of E2F1-induced, p53-dependent apoptosis. E2F1-induced apoptosis requires ATM and other members of the ATM network, including NBS1 and Chk2. In the absence of ATM, E2F1 can still increase p53 protein levels through upregulation of Arf [57]. BTG3 expression is increased through activated p53, which then acts to keep the activity of E2F1 at bay and to facilitate the proper execution of checkpoint function. Moreover, upon oncogenic stress, p53 accumulation due to activated E2F1 and increased ARF production would lead to increased BTG3 expression, which in turn would inhibit E2F1, forming a regulatory loop.

In some cell types, p53 can positively regulate *PTEN*, inhibiting PI3K and thus attenuating phosphorylation of Akt. While low Akt activity not only promotes FOXOs activity and expression of target genes, it also prevents Mdm2 from interacting with p53 resulting in activation of p53 [58]. This connection forms a positive feedback loop for enhancing p53 activity and decreasing Akt activity. In addition, the longevity-factor, SIRT1, regulates the function of *FOXOs* and *NF-κB* by protein deacetylation. Interestingly, SIRT1 inhibits the function of *NF-κB*, but simultaneously acts as an activator of *FOXOs* to enhance nuclear translocation and (likely) regulate gene-specific transcription [59]. The roles of SIRT1 in regulating cellular responses to DNA damage involve a complex feedback loop encompassing HIC1 and p53: SIRT1-mediated deacetylation of p53 prevents p53-dependent transactivation of p21 and BAX after DNA damage, and SIRT1 is able to deacetylate all major p53 acetylation sites [60]. Furthermore, the transcriptional repressor hypermethylated in cancer 1 (HIC1) also negatively regulates *SIRT1* transcription. HIC1, a C-terminal binding protein 1 (CTBP1), and SIRT1 form a corepressor complex that binds enhancer elements upstream of the *SIRT1* promoter and inhibits *SIRT1* expression [61].

On the network robustness and response ability of GRNs at the young, aged, and CR stages

We construct multiple regulatory loops of gene network with 16 genes (Fig. 11.1) associated with aging-related pathophysiological phenotypes. Using the least squares parameter estimation [42] and microarray data covering thymus, spinal cord, and eye tissues [37], a dynamic model of the 16 genes within the gene network in Eq. (11.1) is constructed (shown in Table 11.1) with estimated parameters across different tissues at different stages of life (summarized in Table 11.2). With Eq. (11.1) and Tables 11.1 and 11.2, we calculate 16 eigenvalues of interactive matrix A with regard to the different tissues at the young, aged, and CR stages. Fig. 11.2 shows that some eigenvalues of the interactive matrix A are located together at similar regions near the unit circle $|Z| = 1$, demonstrating some modes of the network will more readily approach some constant steady state.

Trade-off between network robustness and network response ability during the aging process: Network robustness increases and network response ability decreases in the aging process

It has previously been shown that gene networks associated with aging-related pathophysiological phenotypes suffer intrinsic perturbations mainly due to process noise, molecular fluctuations, or genetic mutations during aging [52]. The network robustness of a GRN is defined as its ability to tolerate intrinsic perturbations, and it plays a

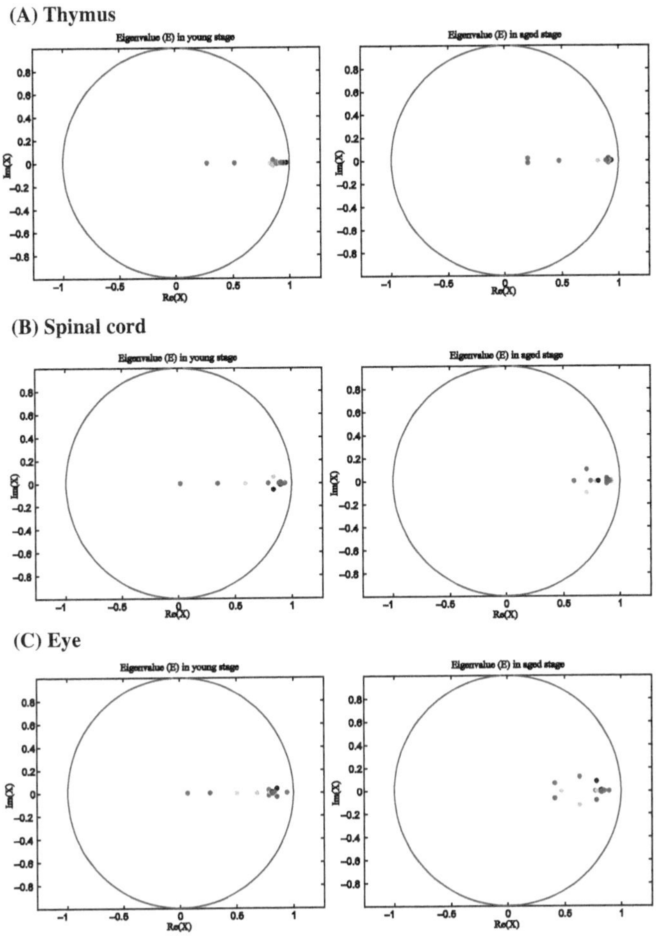

Figure 11.2 The locations of the 16 eigenvalues of different tissues at the young, aged, and calorie-restricted (CR) stages in (A) thymus, (B) spinal cord, and (C) eye. For lack of CR data in eye, we cannot obtain the eigenvalues of the corresponding aging-related network in CR case.

principal role in the fail-safe mechanism of the evolutionary process [31]. Therefore, comparison of network robustness η° for different gene networks pertaining to the young, aged, and CR stages shown in Table 11.3 may clearly depict their relative ability to tolerate intrinsic perturbation. We calculate the robustness in Eq. (11.6) based on Lyapunov stability theory, representing the maximal intensity of intrinsic perturbations that the system can tolerate. In other words, the higher the η° we calculate from a given gene network, the more robust the gene network is. Additionally, if a gene network also responds to external stimuli $U[t]$ in Eq. (11.7)—e.g., upstream signals or biomedical molecules outside the network—the network response ability measured should be based on the output/input energy ratio with $C = I$ in Eq. (11.7) and within the time interval in Eq. (11.8). Suppose the effect of input signals $U[t]$ on $Y[t]$ is bounded by a positive value δ (i.e., an upper bound). Then, we calculate the smallest value δ^o in Eq. (11.8) (i.e., the minimum upper bound), called the "network response ability" to external stimuli. This measurement is based on the system gain theory to the external input signal from the system theory perspective [29]. In general, a GRN with greater robustness always has lower response ability, and vice versa. This is an inherent characteristic of physical systems [29]. The trade-off between network robustness and network response ability can be confirmed in Table 11.3 and can be considered as secondary proof of the proposed method.

We compare the "young" group with the "aged" group for the thymus, spinal cord, and eye tissues and find that the older group of all tissues analyzed has a higher network robustness η^o value than the "young" group and the "CR" group. These results are shown in Table 11.3. Based on the analytical methodology described earlier, the higher the value of η^o, the more intrinsic network perturbations a gene network can tolerate, and thus the more robust the network. Therefore, the aging-related gene network examined in this section is less robust to intrinsic network perturbations in the young group than in the aged group, i.e., aging is a network robustness increasing process. In addition, the gene response ability (δ^o) of the entire gene network calculated from the "aged" group is lower than from the "young" group, i.e., aging is a network response ability decreasing process. This ability is reversed in the "CR" group (Table 11.3). In order to tolerate the accumulated intrinsic network parameter fluctuations due to genetic mutations, the elderly GRN becomes more robust than the young GRN. Therefore, elderly GRNs are less responsive to external stimuli due to their robustness. Hence, the network protective ability against external stimuli becomes worse in gene networks of the elderly. However, a more robust GRN could harbor more accumulated genetic mutations, which might provide more evolutionary paths to other phenotypes of the gene network by random drift [62,63] and might lead to some aging-related diseases like cancers, metabolic disorders, and arthritis. On the other hand, the young aging-related GRNs are less robust, with greater response to the external stimuli. From the above data, we conclude that the gene network of the young group is more efficient in their response to external stimuli. Thus, we have developed a systematic model that shows the GRN of a given "young" group to be less robust to intrinsic perturbations but more responsive toward external stimuli than the comparable "aged" group, meaning gene expression could be readily reprogrammed to mediate downstream genes or regulators. This also allows for modulation of gene expression in response to external stimuli, such as exposure to oxidative stress, carcinogens, and pro-inflammatory molecules.

Based on the trade-off between network robustness and network response ability of gene networks in Table 11.3 in the aging process, we can make a network robustness-based hypothesis on network aging. In the aging process, the network robustness of the GRN is gradually increased in order to tolerate the accumulated genetic variations, and simultaneously, the antagonistic network response ability to external stimuli is gradually decreased. Because the power of natural selection fades with age, it is possible for hazardous late-acting genes to exist due to the accumulation of genetic variations. Observations to support these hypotheses have been found in late-life mortality plateaus [64] and in *Drosophila melanogaster* testing [65]. In order to buffer the effects of these accumulated genetic variations, gene networks need more robustness to retain their functions. However, these robust gene networks with large amounts of harbored genetic variations at the late stage of life might provide the gene networks with more paths for other phenotypes by random drift. This is the cause of aging-related diseases like atherosclerosis, prostate hypertrophy, Alzheimer's disease, and cancers. Furthermore, a large response ability to external stimuli, such as to growth factors at a young stage, may be useful for growth and reproduction. With increasing age, the decrease in the network response ability to external stimuli, such as free radicals, inflammation, infection, and pathogenic invasion, results in failure to trigger protective mechanisms and might be the cause of aging.

Gene response ability of individual genes in the aging-related regulatory network in the young, aged, and CR groups

After analyzing network robustness and network response ability of the entire GRN associated with aging, we focus on the response abilities of individual genes with $C = diag(0,\ldots,0,1,0,\ldots,0)$ in Eq. (11.7) for the ith gene to be analyzed, i.e., all are zero except the ith diagonal term. The purpose of this analysis is to gain a greater understanding of the protective and/or defensive genetic mechanisms inherent to the aging-related GRN. If a gene has greater response ability to external stimuli, the gene should be more efficient in the transduction of active signals to downstream genes for further reaction. For this reason, we measure downstream transduction ability of each individual gene (called individual gene response ability) for a number of genes in the aging-related gene network (shown in Fig. 11.1) in the "young," "aged," and "CR" groups. Response abilities of these individual genes across different tissues are summarized in Fig. 11.3. Each subgraph represents response abilities of the individual genes with respect to the young, aged, and CR groups in different tissues. We find that these genes examined have different gene response abilities at different stages of life, with the most conspicuous difference and parallel analytic results among different tissues for (1) the gene *FOXOs* (Fig. 11.3A) in the thymus ($\delta^{o}_{\text{FOXOs\_young}} = 0.4599$, $\delta^{o}_{\text{FOXOs\_aged}} = 0.2330$, $\delta^{o}_{\text{FOXOs\_CR}} = 0.1887$), spinal cord ($\delta^{o}_{\text{FOXOs\_young}} = 0.6004$, $\delta^{o}_{\text{FOXOs\_aged}} = 0.4347$, $\delta^{o}_{\text{FOXOs\_CR}} = 1.0769$), and eye ($\delta^{o}_{\text{FOXOs\_young}} = 0.1797$, $\delta^{o}_{\text{FOXOs\_aged}} = 0.1171$); (2) the gene *NF-κB* (Fig. 11.3B) in the thymus ($\delta^{o}_{\text{NF-}\kappa B\text{\_young}} = 0.1486$, $\delta^{o}_{\text{NF-}\kappa B\text{\_aged}} = 0.3373$, $\delta^{o}_{\text{NF-}\kappa B\text{\_CR}} = 0.3094$), spinal cord ($\delta^{o}_{\text{NF-}\kappa B\text{\_young}} = 0.4020$, $\delta^{o}_{\text{NF-}\kappa B\text{\_aged}} = 0.4758$, $\delta^{o}_{\text{NF-}\kappa B\text{\_CR}} = 0.3667$),

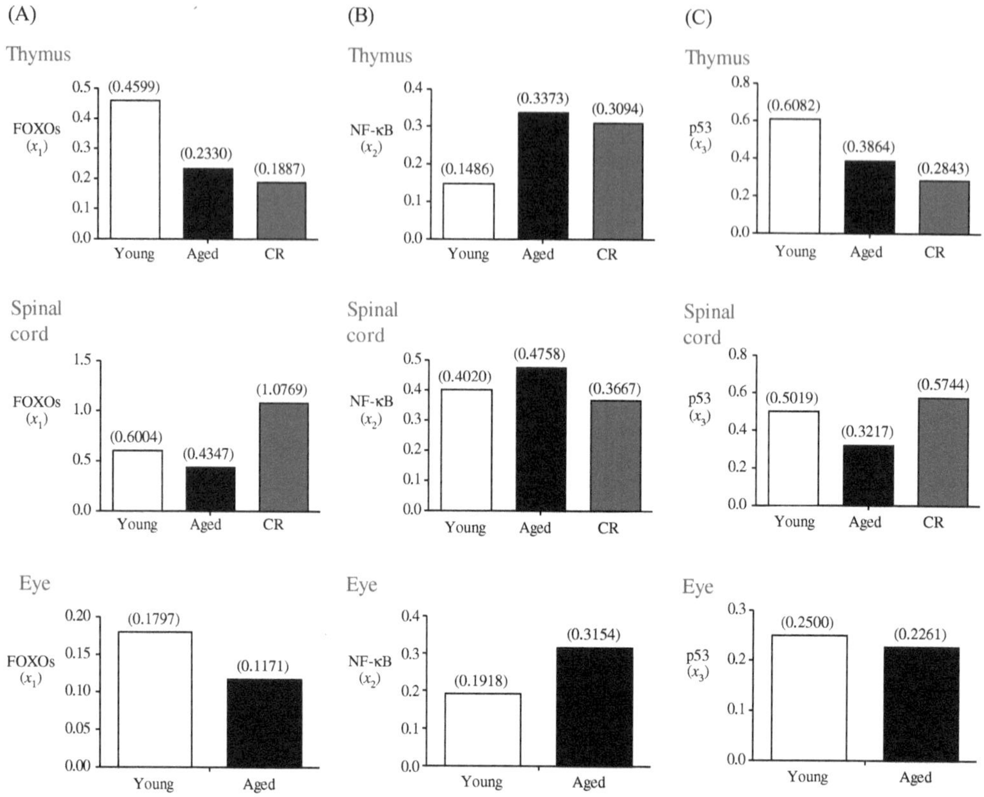

Figure 11.3 Comparison of individual gene response abilities (δ^o) of all genes in the aging-related gene regulatory network in Fig. 11.1 across different tissues at the young, aged, and calorie-restricted (CR) stages. The most conspicuous difference and parallel analytic results among different tissues are the genes *FOXOs* (A), *NF-κB* (B), and *p53* (C), which will be further discussed in Fig. 11.4.

and eye ($\delta^o_{\text{NF-}\kappa B\_\text{young}} = 0.1918$, $\delta^o_{\text{NF-}\kappa B\_\text{aged}} = 0.3154$); (3) the gene *p53* (Fig. 11.3C) in the thymus ($\delta^o_{\text{p53\_young}} = 0.6082$, $\delta^o_{\text{p53\_aged}} = 0.3864$, $\delta^o_{\text{p53\_CR}} = 0.2843$), spinal cord ($\delta^o_{\text{p53\_young}} = 0.5019$, $\delta^o_{\text{p53\_aged}} = 0.3217$, $\delta^o_{\text{p53\_CR}} = 0.5744$), and eye ($\delta^o_{\text{p53\_young}} = 0.2500$, $\delta^o_{\text{p53\_aged}} = 0.2261$). Further comparisons and discussion are given in the following subsections.

Comparison of individual gene response abilities for *FOXOs* among the young, aged, and CR groups

The free radical theory of aging postulates a direct causal relationship between aging and accumulation of damage due to ROS [66]. The role of FOXOs in controlling cellular oxidative stress and lifespan regulation notably substantiates this theory. According to our measurement of individual gene response ability for *FOXOs*

(Fig. 11.3A), we have obtained a conspicuously lower gene response ability for the "aged" group ($\delta^{o}_{\mathrm{FOXOs\_aged}} = 0.2330$ in the thymus, $\delta^{o}_{\mathrm{FOXOs\_aged}} = 0.4347$ in the spinal cord, $\delta^{o}_{\mathrm{FOXOs\_aged}} = 0.1171$ in the eye) versus the "young" group ($\delta^{o}_{\mathrm{FOXOs\_young}} = 0.4599$ in the thymus, $\delta^{o}_{\mathrm{FOXOs\_young}} = 0.6004$ in the spinal cord, $\delta^{o}_{\mathrm{FOXOs\_young}} = 0.1797$ in the eye). This may imply that FOXOs has less efficient transduction for orchestrating the programs of gene expression that regulate apoptosis, cell cycle arrest, and stress resistance during aging.

Changes in several key FOXOs target genes mediate the ability of *FOXOs* to induce cell death, including: upregulation of BIM, which is a pro-apoptotic BCL-2 family member; downregulation of BCL-XL, which is an antiapoptotic BCL-2 family member [19]; and upregulation of genes that encode death cytokines, such as Fas ligand and TRAIL [67] (see Fig. 11.4A). In addition to upregulating cell apoptosis, activated FOXOs upregulates its downstream genes that encode the Cdk inhibitor p27KIP1, p21, and the Rb family member p130 to mediate G1 arrest as well as to promote cell cycle arrest by repressing the expression of two positive cell cycle regulators, cyclin D1 and D2 [68]. Several FOXOs target genes, such as Growth Arrest and DNA Damage 45 (GADD45) and cyclin G2, have been shown to mediate the effect of FOXOs on the G2/M boundary to stimulate DNA repair or apoptosis and to inhibit cell proliferation [69]. Hence, the lower gene response ability of *FOXOs* that

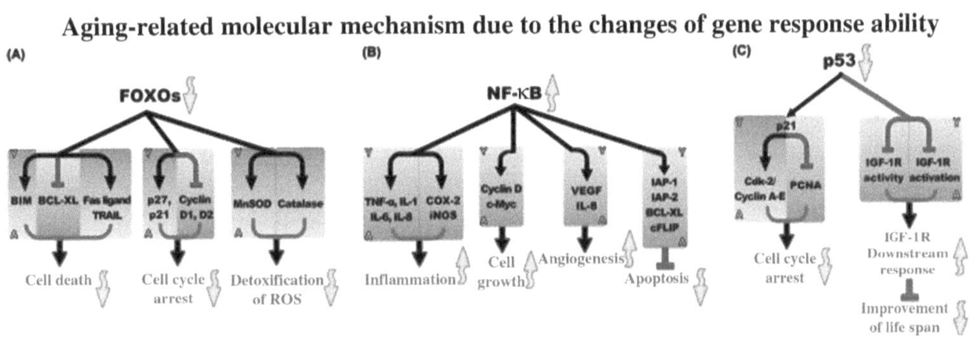

Figure 11.4 The effect of changing gene response ability of an individual gene on its downstream regulators and cellular functions during aging processes. Some major downstream regulators of *FOXOs* (A), *NF-κB* (B), and *p53* (C) are shown to mediate distinct downstream cellular functions. *Arrows in black* indicate activation; *blunt arrows in red* indicate suppression. *Wavy arrows in yellow* represent the changes to gene response ability and the downstream cellular functions such as cell death, cell cycle arrest, detoxification of ROS, etc. The *upward arrows* indicate an increase and the *downward arrows* indicate a decrease. The gradual changes between *light blue* and *dark blue* in rectangle indicate the dynamic changing process of the gene response ability on its downstream regulators from the young stage to aged stage. All detailed regulatory changes occurring due to the change of gene response ability are described in the Results and Discussion in section 11.1.2. (*Y*, young stage; *A*, aged stage.)

we estimate during the aging process may imply that poor efficiency for induction of cell cycle arrest and apoptosis not only results in a decreased ability of cells to arrest at key checkpoints, leading to tumor development, but also leads to the survival of abnormal or mutated cells that would normally die, resulting in tumorigenesis. Moreover, a lower gene response ability of *FOXOs* in response to cumulative oxidative stress does not allow time for repair of damaged DNA and detoxification of cells from ROS. Adequate induction of cell apoptosis for eliminating damaged or abnormal cells may also not exist in the aged GRNs due to the reduced response ability of *FOXOs*, which may lead to a reduced transduction of cell damage information to the downstream corresponding regulators of cell death. Therefore, many damaged or abnormal cells accumulate in the body, resulting in aging.

In addition, active *FOXOs* permit detoxification of ROS, thus increasing cellular resistance by upregulating downstream genes that encode free radical scavenging enzymes, such as Mn superoxide dismutase (MnSOD), which converts superoxide byproducts to hydrogen peroxide and oxygen, and catalase, which transforms hydrogen peroxide into water and oxygen [68,70]. Thus, *FOXOs* control two aspects of cellular resistance to stress: repair of damage resulting from ROS and detoxification of ROS (see Fig. 11.4A). Stress resistance is highly coupled with organismal longevity, raising the possibility of a stress resistance mechanism by which *FOXOs* extends lifespan. Accordingly, a poor response ability of *FOXOs* implies deterioration of longevity and tumor suppression.

Individual gene response ability of *NF-κB* in the young, aged, and CR groups

Several studies have shown oxidative stress to be a major factor in the aging process. *NF-κB*, a pleiotropic and ubiquitous regulator, is well known to be sensitive to various external stimuli, particularly oxidative stress. Activation of *NF-κB* mediates expression of many genes that are central to the production of inflammatory cytokines and tumorigenesis; these include interleukins (IL-1β, IL-2, and IL-6), tumor necrosis factors (TNF-α), adhesion molecules (ICAM-1, V-CAM, and E-selectin), enzymes (iNOS and COX-2), and gene products that negatively regulate apoptosis (IAP-1, IAP-2, Bcl-xL, and cFLIP). Based on our estimation of individual gene response ability for *NF-κB* (illustrated in Fig. 11.3B), its gene response abilities in the "aged" group ($\delta^{o}_{\text{NF-}\kappa B\_\text{aged}} = 0.3373$ in the thymus, $\delta^{o}_{\text{NF-}\kappa B\_\text{aged}} = 0.4758$ in the spinal cord, $\delta^{o}_{\text{NF-}\kappa B\_\text{aged}} = 0.3154$ in the eye) are much higher than the corresponding values for the "young" group ($\delta^{o}_{\text{NF-}\kappa B\_\text{young}} = 0.1486$ in the thymus, $\delta^{o}_{\text{NF-}\kappa B\_\text{young}} = 0.4020$ in the spinal cord, $\delta^{o}_{\text{NF-}\kappa B\_\text{young}} = 0.1918$ in the eye). Furthermore, this gene response ability is reversible by CR ($\delta^{o}_{\text{NF-}\kappa B\_\text{CR}} = 0.3094$ in the thymus, $\delta^{o}_{\text{NF-}\kappa B\_\text{CR}} = 0.3667$ in the spinal cord) when compared with the response values for the "aged" group. Lowered gene response ability in the young group indicates less transduction to the downstream regulators of NF-κB signaling. This observation may conform with evidence corroborating the idea that NF-κB activation from external stimuli is an inducible but transient event, where the IκB gene is first activated and then NF-κB is transported from the nucleus to the cytoplasm [71]. Additionally, in multiple tissues, CR could directly or indirectly increase gene expression of Nfkbia, an inhibitor of *NF-κB*, thus eliminating the downstream responses originating from *NF-κB* [72].

In comparison with the young group however, the higher gene response ability of *NF-κB* in the aged group is consistent with previous studies. Much evidence in the literature has shown that *NF-κB* can be activated during the aging process and be considered as a signaling pathway to regulate several downstream gene expressions. Moreover, the binding activity of *NF-κB* is highly increased during the aging process, a fact that has been also observed in multiple tissues, such as the brain, liver, and kidney. Accordingly, the elevated gene response ability of *NF-κB* may imply a general mechanism with increasing age [73–76]. Thus, *NF-κB* may lose its transient activation nature and become constitutively activated during the aging process [77]. Higher gene response ability also indicates a better efficiency of signal transduction to its downstream regulators, which may be associated with the increase in inflammatory cytokines that contribute to the progression of many degenerative diseases and tumorigenesis. Higher gene response ability of *NF-κB* may efficiently induce expression of certain genes that encode several inflammatory cytokines, including IL-1, IL-6, IL-8, and TNF-α, as well as the inflammatory enzymes cyclooxygenase-2 (COX-2) and iNOS [12,13]. It is important to note that these inflammatory proteins contain redox-sensitive NF-κB-specific DNA binding sites in the promoter regions and that their production is greatly influenced by the oxidative stress that persists during aging [78] (see Fig. 11.4B). On the other hand, the activated *NF-κB* may induce several genes, so it not only acts as a potent *NF-κB* activator to form an auto-activating loop that sustains its downstream regulation, but also promotes cell proliferation via growth factors such as TNF-α, IL-1, and IL-6. In addition to the growth factors, certain cell-cycle regulatory proteins, such as cyclin D and cMyc, two NF-κB downstream regulators, have an important role in cell growth and proliferation [79], and key angiogenesis factors, such as vascular endothelial growth factor (VEGF) and interleukin-8 [80], are directly or indirectly enhanced by *NF-κB* activation. Several gene products, including IAP-1, IAP-2, BCL-XL, and cFLIP, that negatively regulate apoptosis are mediated by *NF-κB* activation as well [81]. Therefore, the elevated *NF-κB* gene response ability we estimate for the aged group seems to mimic the gain of function in *NF-κB*, which not only brings about dysregulation of proliferation and induction of antiapoptosis, but also facilitates cancer maintenance and invasion, and may be involved in the increased incidence of tumorigenesis in old age.

Comparison of individual gene response abilities of the p53 gene among the young, aged, and CR groups

Throughout life, all organisms incessantly suffer from various stresses, which induce counteracting cellular responses to prevent further deterioration. p53 plays a central role in integrating environmental stresses and cellular responses [82]. Significant stress and DNA damage may cause either cell cycle arrest so that DNA repair may be performed or the induction of apoptosis and promotion of cellular senescence [8,9]. According to GRAM measurements, the response ability of the *p53* gene (shown in Fig. 11.3C) is conspicuously lower in the aged group ($\delta^{o}_{p53\_aged} = 0.3864$ in the thymus, $\delta^{o}_{p53\_aged} = 0.3217$ in the spinal cord, $\delta^{o}_{p53\_aged} = 0.2261$ in the eye) than in the young group ($\delta^{o}_{p53\_young} = 0.6082$ in the thymus, $\delta^{o}_{p53\_young} = 0.5019$ in the spinal cord, $\delta^{o}_{p53\_young} = 0.2500$ in the eye), manifesting that p53 in older organisms has a poor ability to efficiently transduce active regulatory signals to downstream regulators in response to external stimuli.

The function of p53, particularly in tumor suppression, involves transactivation or transrepression of a large number of genes involved in cell cycle regulation and apoptosis, such as the cyclin-dependent kinase (Cdk) inhibitor p21 and the insulin-like growth factor receptor IGF-1R. p53 appears to be essential for upregulation of *p21* following DNA damage [83], and *p21* subsequently encodes a checkpoint protein that interacts with cell cycle proteins such as Cdks and blocks activity of cyclin/Cdk complexes. p21 not only binds to Cdk2/cyclin A–E complexes and prevents exit from G1 into S [84], but can also pause the cell cycle during S-phase by inhibiting proliferating cell nuclear antigen, which is required for replication of DNA and DNA repair [85] (see Fig. 11.4C). The response ability of the *p53* gene that we estimate for the aged group may imply less expression of *p21* and thus abnormal regulation of the cell cycle, which may affect the efficiency of cells to repair damaged DNA. Loss of *p53* function suggests that not only tumor risk may increase, but also that tissue homeostasis may be impaired during the aging process.

The level of *IGF-1R* is controlled by several tumor suppressors, including WT-1, BRCA1, and p53 [86]. p53 suppresses *IGF-1R* activity by reducing endogenous levels of *IGF-1R* mRNA [87] and inhibiting ligand-induced activation of IGF-1R by stimulating transcription of IGF-binding protein-3 (IGFBP-3). Loss of p53 function due to DNA mutations has been associated with over-expression of *IGF-1R* and an increased risk of tumorigenesis. High serum levels of the IGF-1 ligand have been correlated to increased risk for prostate and breast cancers in human population studies [88]. In addition, previous evidence suggests that DNA mutations that cause a decline in IGF-1R levels or block signal transduction by this receptor also significantly improve lifespan in the nematode *Caenorhabditis elegans* and the fruit fly *D. melanogaster*. A similar result has been obtained from a cohort study of elderly Dutch citizens, revealing that genetic variations causing reduced insulin/IGF-1 signaling are beneficial for survival in old age [89]. Therefore, the attenuated response ability of *p53* that we estimate from the aged group may be the main cause of loss of *p53* function, resulting in increased responsiveness of *IGF-1R*. This may also explain the increased risk of tumorigenesis and other associated deleterious effects on longevity during the aging process.

11.1.3 Conclusion

We have constructed an aging-related GRN from microarray data and developed some quantitative and systematic methods to measure robustness and gene response ability for young, aged, and CR groups. We show that the GRN of an aged group is more robust than that of a young group, thus allowing for greater tolerance to the accumulation of a large amount of intrinsic perturbations due to damage and mutation. By comparing network response abilities of young and aged groups across different tissues, we have found that the aging-related GRN of younger organisms is more responsive to external stimuli than the corresponding GRN of older organisms. Furthermore, an improvement to network response ability caused by CR is also demonstrated. In general, we have found that an aging-related GRN with higher network robustness has lower network response ability, and vice versa: i.e., there is a trade-off between network robustness and network response ability in the aging process. These antagonistic characteristics may be confirmed by system theory and provide secondary proof for our calculated results. From these observations, we speculate the strategy of the evolutionary process may be that GRNs with lower robust network stability have greater network response ability to external stimuli during the growing process. This is inversed during the aging process, where increased network robustness functions as a conservative fail-safe mechanism to buffer more genetic variations. However, this network mechanism may lead to diminished responses to some harmful stressors like carcinogens, oxidative stresses, and pro-inflammatory molecules. Hence, downstream genes may experience loss of function, prohibiting efficient activation to respond to these harmful stressors, such that some diseases like cancers, metabolic disorders, and arthritis may occur more easily with increasing age. By using the proposed measurement methods, it may become feasible to estimate changes in network robustness and network response ability through an organism's lifespan. Therefore, changes in aging phenotypes may be thoroughly explained or inferred by these systematic characteristic changes to gene networks from the systems point of view. In addition, the quantitative and systematic measurement methods proposed in this section also allow for system analysis of GRNs for other biological processes. In the near future, these procedures may have potential applications in informing methodologies of clinical strategy and drug target selection from the perspective of systems biology.

11.2 INVESTIGATING SPECIFIC CORE GEN FOR CELLULAR MECHANISMS OF HUMAN AGING VIA NGS DATA

Since aging is a very complex and system-level process, it is involved in cellular processes like gene regulation, protein interaction, miRNA regulation, and DNA methylation. After investigating the big mechanisms of aging by GRN through microarray data, we will investigate the core GEN by NGS data to investigate more cellular mechanisms of human aging to provide the development of new antiaging drug in this section.

Aging is an inevitable part of life for humans, and slowing down the aging process has become a main focus of human endeavor. Therefore, understanding the underlying mechanisms behind aging can help us retard the process. Here, we applied a systems biology approach to construct protein—protein interaction networks (PPINs), GRNs, and epigenetic networks, i.e., GENs, of elderly individuals and young controls. We then compared these GENs to extract mechanisms associated with human aging by using their corresponding microarray data in peripheral blood mononuclear cells, miRNA data, and database mining. In addition, we discuss the different aging mechanisms in old men and women identified from peripheral blood mononuclear cells, and investigate the core genetic and epigenetic mechanisms involved in the human aging process.

The core GENs of elderly individuals and young controls were obtained by applying principal network projection (PNP) to GENs based on Principal Component Analysis (PCA). Through comparison of the core GENs, common core GEN, and specific core GENs of young and elderly individuals we provide insight into the human aging mechanism. Similarly, the common and specific core GENs of the elderly were also obtained to investigate the distinctive aging mechanism between old men and women. Taken together, our results provide a platform for drug therapy to retard human aging.

Complex biological processes such as aging do not result from an isolated event in a single protein, chemical, enzyme, or individual cell type. Rather, such processes occur as the coordinated result of interacting cell types and tissues, as well as alterations in gene regulation and expression, signaling pathways, and biological networks. It is the combined and synergistic effects of the individual parts that lead to aging [90]. Aging is a complex process comprising a wide variety of interconnected features and effects, such as a progressive functional decline, gradual deterioration of physiological function, and decrease in fertility and viability. Deterioration of physical health is the principal factor associated with aging [91]. Although aging is inevitable, understanding the underlying molecular mechanisms can help to retard the process. This is a key goal of aging research, in addition to designing drug targets to combat the aging process.

Several aging-related studies have been conducted in a variety of animals in the last three decades. For example, basic research conducted in model organisms, particularly *C. elegans*, identified a number of single-gene mutations that confer increased lifespan phenotypes. In addition, genome-wide expression studies have revealed distinct expression profiles associated with aging [92–94]. Jurk et al. demonstrated that chronic, and progressive low-grade inflammation induced by knockout of the nfkb1 subunit of the TF NF-κB induces premature aging in mice [95]. However, this study merely demonstrates the effect of a single event, which does not adequately explain the underlying aging mechanisms.

miRNAs are endogenous small RNAs that regulate gene expression mainly at the posttranscriptional level, and have gained increasing attention in the last decade [96]. A single miRNA can target multiple genes, and multiple miRNAs can also target the same gene [97]. The miRNAs involved in aging and lifespan determination have been studied extensively in the last several years. For instance, Liang et al. demonstrated upregulation of miRNA expression with concomitant inverse downregulation of target genes with aging [98]. Additionally, Yang et al. demonstrated in *C. elegans*, that a mir-34 mutant led to an extended lifespan [99]. Understanding how miRNAs associate with aging-related biological processes and pathways can provide deeper insight into the mechanisms of aging.

DNA methylation patterns are shaped by two opposing processes of adding and removing a methyl group at position five of cytosine in DNA [100]. Age-associated alterations in DNA methylation are commonly grouped into a phenomenon known as "epigenetic drift," which is characterized by gradual extensive demethylation of the genome and hypermethylation of a number of promoter-associated CpG islands [101]. Hypermethylation of CpG islands in gene promoters induces gene silencing and alters gene expression. Aging induced DNA methylation has been studied extensively in the last several years. Meaghan et al. found that healthy human aging occurs throughout the lifetime. The authors discussed the dynamics of DNA methylation as well as how the interactions between genomics, the environment, and epigenomics influence aging rates [102].

Protein–protein interactions (PPIs) are crucial for signal transduction in biological processes. Therefore, compiling PPIs provides new insight into protein function [103]. PPIs have revealed global topological and dynamic features related to well-understood biological properties. This indicates that studying PPIs will allow further understanding of disease mechanisms at a systematic level [104]. Wang et al. identified *Candida albicans*-zebrafish interspecies PPIs and used this information to highlight the association between *C. albicans* pathogenesis and the zebrafish redox process, indicating that redox status is critical in the battle between the host and pathogen [105]. Therefore, PPIs can be an effective way to study the complex biological processes involved in aging.

TFs are the main regulators of DNA transcription. Thus, knowing the genes that are targeted by a specific TF is of utmost importance for understanding developmental processes, the cellular stress response, or disease etiology [106]. Recently, Lee et al. constructed a network of TF-gene interactions in *Saccharomyces cerevisiae* [107]. This network of regulator-gene interactions describes potential pathways that yeast cells can use to regulate global gene expression programs. Additionally, Tu et al. indicated that the network robustness and response ability of dynamic GRNs play a key role in the aging process [108]. Accordingly, in the current study, PPIs, GRNs, and miRNA regulation were used to construct GENs in order to understand the molecular mechanisms of human aging.

PCA is a multivariate technique that analyzes a data table in which observations are described by several intercorrelated quantitative dependent variables. The goal is to extract the most important information from the data table. PCA is probably the most popular multivariate statistical technique used in all scientific disciplines. Here, the PNP method based on PCA was utilized to obtain the core GENs for young and old women and men.

We analyzed the GENs of human aging using a regression model and big data mining. Based on the PPI candidate, TF-gene regulation, and miRNA regulation databases, as well as the gene expression profiles of young and old peripheral blood mononuclear cell samples, two GENs with quantitative regulatory abilities for young and old individuals were constructed. By applying the PNP method, we obtained the core proteins, TFs, target genes, and miRNAs for each of the core GENs in young and old individuals. The intersection of young and old GENs is called the common core GEN, and the distinction between young and old GENs is referred to as the specific core GEN. We garnered insight into the mechanism of human aging by investigating the specific core GEN of old individuals. We observed that these specific core proteins, TFs, target genes, miRNAs, and DNA methylated proteins are involved in aging-associated cellular mechanisms, suggesting that drug therapy to delay human aging can be designed against these targets. In addition, we also investigated gender differences in human aging based on specific core GENs as well as information obtained from the analysis of peripheral blood mononuclear cells.

Overall, this study provides new age-related molecular drug targets as well as unravels the molecular mechanisms of gender-specific changes in the human aging process.

Analysis of the specific core GEN of old individuals demonstrated that inhibition of three key genes (*FLNB*, *CDK4*, and *ZNF274*) mitigated the dysregulation of MAPK, T-cell receptor, and neurotrophin signaling pathways as well as aberrant cell cycle and apoptosis regulation. The specific core GEN of old men and revealed differences in biological regulation during aging. In women, inhibition of *TAOK3* and *TRAF6* genes circumvented dysregulation of the MAPK and Toll-like receptor (TLR) signaling pathways as well as immune, proliferative, and metabolic dysfunction. However, in men, inhibition of *STMN1* and *LRRFIP2* could not overcome dysregulation of the MAPK and Wnt signaling pathways or aberrant cell cycle and apoptosis regulation, thus resulting in cancer. Based on these findings, we conclude that the MAPK signaling pathway plays the most important role in the human aging process.

11.2.1 Methods to construct GENs and core GEN of human aging

Overview of the construction of common and specific core GENs of human aging

Fig. 11.5 shows an outline of the process for constructing the common and specific core GENs of human aging. Microarray data of young and old samples, miRNA data from peripheral blood mononuclear cells, as well as the miRNA, methylation, TF-gene, and BioGRID databases were searched in order to construct candidate GENs.

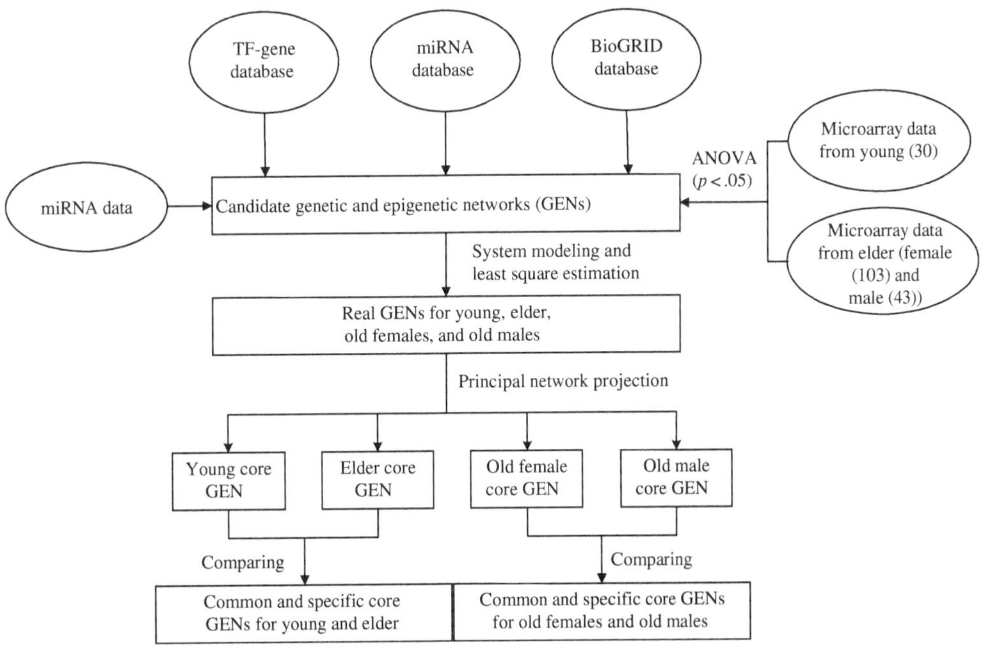

Figure 11.5 Flowchart for constructing the common and specific core GENs for human aging. Microarray data of young and elderly samples, miRNA data, and miRNA, TF-gene, and BioGRID databases were searched to construct candidate GENs, which consisted of interactive candidate GRN, PPIN, and miRNA regulation network. False positives of candidate GENs were then pruned for constructing real GENs of young and old men and women by system modeling and least square estimation. The core GENs were obtained by PNP. The intersection of young and elderly core GENs is called a common core GEN. The distinction between young and elderly core GENs are called young and/or elderly specific core GENs. We investigated different molecular mechanisms between the young and elderly by their specific core GENs. We further investigated the gender-specific aging mechanisms in old women and men.

These GENs consisted of interactive candidate GRN, PPI, and miRNA regulation networks. In order to avoid false positives, we applied system modeling and least square estimation [42]. The core GENs for young and old men and women were obtained by PNP. The intersection of young and old core GENs is called a common core GEN, whereas the distinction between young and old core GENs is referred to as the young and/or old specific core GENs. We investigated the gender-specific aging mechanisms in young and old individuals based on their specific core GENs as well as data obtained from analyzing peripheral blood mononuclear cells from the respective individuals.

Data selection and preprocessing

Microarray datasets were obtained from the NCBI Gene Expression Omnibus (GEO). A study by Marttila et al. demonstrated gender-specific changes during aging of the human immune system based on transcriptomic analysis of peripheral blood mononuclear cells derived from elderly individuals and young controls [109]. Using the GEO, we obtained the data from this study for young and elderly individuals (GSE65219) [109]. The sample numbers for the young (19–30 years of age) and elderly (90–99 years of age) groups were 30 and 146, respectively. In addition, 103 and 43 of the 146 samples were old women and men, respectively.

The PPI candidate dataset for *Homo sapiens* was obtained from the Biological General Repository for Interaction Database (BioGRID). BioGRID is a freely accessible database of physical and genetic interactions available at http://www.thebiogrid.org [110]. More than 700,000 interactions are available from major model organism species and from more than 50,000 publications.

A TF-gene regulation dataset is available at The Human Transcriptional Regulation Interactions database (HTRIdb) [111] and Integrated Transcription Factor Platform (ITFP) [112]. HTRIdb has been populated with 284 TFs totaling 51,871 regulations. In addition, ITFP has been populated with 4015 TFs and 69,496 regulations.

The miRNA regulation dataset was collected from starBase v2.0 (http://starbase.sysu.edu.cn/) and miRTarBase (http://mirtarbase.mbc.nctu.edu.tw/). The miRTarBase database provides the most current and comprehensive information of experimentally validated miRNA-target interactions [113]. All the above large databases were used as candidates for identifying GENs of young and old male and female samples.

Selection of a protein pool and construction of the candidate GENs for young and old men and women

In order to construct the candidate GENs, which are integrated through gene expression with microarray data for young and elderly samples, miRNA data, and the miRNA, TF-gene, and BioGRID databases were used based on the following two steps to select the protein pool. First, gene expression values were overlaid onto the corresponding proteins. Then, a one-way ANOVA was performed to select differentially expressed proteins ($p < .05$). The null hypothesis (H_0) assumes that the average protein expression levels of young and elderly samples are the same. Proteins with p-values $< .05$ were selected for the protein pool to construct the candidate GENs containing candidate GRN, PPIN, and the miRNA regulatory network. A total of 2306 differentially expressed proteins between young and elderly samples were selected as candidate GENs. Therefore, the candidate network consisted of 4581 PPIs, and 4400 TF, and 53,099 miRNAs regulations.

The regression model of the GENs for young and old men and women

A candidate GEN contains false positive interactions and regulations because miRNA, TF-gene, and BioGRID databases include all experimental conditions and various experimental environments. The false positives of a candidate GEN should therefore be pruned by using real microarray and miRNA data through the system identification and system order detection method. Two regression models, the protein association model and gene regulatory model, are used to characterize the GEN in cells.

The PPIs of the protein i in the candidate PPIN can be described by the following protein association model:

$$y_i(n) = \sum_{j=1, i\neq j}^{N_i} a_{ij} y_i(n) y_j(n) + h_i + v_i(n), \quad \text{for } i = 1, 2, \ldots, N \tag{11.12}$$

where $y_i(n)$ represents the mRNA expression level of the protein i for the sample n, $y_j(n)$ denotes the mRNA expression level of the jth protein interacting with protein i for the sample n, a_{ij} represents the association ability between protein i and its interactive protein j, N_i indicates the number of proteins interacting with protein i in the candidate PPIN, h_i is the basal expression level of protein i, and $v_i(n)$ characterizes the stochastic noise of the ith protein for sample n. The model demonstrates that protein i is affected by N_i proteins, the basal expression level, and stochastic noise. The PPI association equation of protein i in Eq. (11.12) can be represented by following regression form:

$$y_i(n) = \begin{bmatrix} y_i(n)y_1(n) & \ldots & y_i(n)y_{j-1}(n) & y_i(n)y_{j+1}(n) & \ldots & y_i(n)y_{N_i}(n) & 1 \end{bmatrix}$$
$$\times \begin{bmatrix} a_{i1} \\ \vdots \\ a_{ij-1} \\ a_{ij+1} \\ \vdots \\ a_{iN_i} \\ h_i \end{bmatrix} + v_i(n) = \Phi_i(n)\Theta_i + v_i(n), \quad \text{for } i = 1, 2, \ldots, N \tag{11.13}$$

In addition, the TF and miRNA regulation of the target gene i can be described using the following gene regulatory model:

$$x_i(n) = \sum_{j=1, i \neq j}^{N_i} b_{ij} y_j(n) - \sum_{l=1}^{O_i} c_{il} mir_l(n) + k_i + w_i(n), \quad \text{for } i = 1, 2, \ldots, N \tag{11.14}$$

where $x_i(n)$ represents the mRNA expression level of target gene i for sample n, $y_j(n)$ denotes the mRNA expression level of the jth TF binding to the target gene i for sample n, b_{ij} indicates the regulatory ability of the jth TF binding to the ith target gene, $mir_l(n)$ represents the expression level of the lth miRNA interacting with target gene i for sample n, c_{il} denotes the regulatory ability of the lth miRNA to the ith target gene, N_i and O_i indicate the number of TFs and miRNAs binding to target gene i in the candidate GRN, respectively, k_i is the basal expression level of the ith gene, and $w_i(n)$ characterizes the stochastic noise of gene i for sample n. The model demonstrates that target gene i is affected by N_i TFs, O_i miRNAs, the basal expression level, and stochastic noise. Expression of the ith gene in Eq. (11.14) can be represented by following regression form:

$$x_i(n) = [\, y_1(n) \;\; \ldots \;\; y_{j-1}(n) \;\; y_{j+1}(n) \;\; \ldots \;\; y_{N_i}(n) \;\; -mir_1(n) \;\; \ldots \;\; -mir_{O_i}(n) \;\; 1 \,] \times \begin{bmatrix} b_{i1} \\ \vdots \\ b_{ij-1} \\ b_{ij+1} \\ \vdots \\ b_{iN_i} \\ c_{i1} \\ \vdots \\ c_{iO_i} \\ k_i \end{bmatrix} + w_i(n) = \phi_i(n)\theta_i + w_i(n), \quad \text{for } i = 1, 2, \ldots, N \tag{11.15}$$

Note that $\Phi_i(n)$ in Eq. (11.13) and $\phi_i(n)$ in Eq. (11.15) denote the regression vector, which can be obtained from microarray and miRNA expression data. In addition, Θ_i in Eq. (11.13) and θ_i in Eq. (11.15) represent the regulatory parameter vector of the ith target gene to be estimated by the least squares parameter estimation method in Eq. (2.20) or Eq. (2.21) through the real expression of genes and miRNAs.

Construction of GENs of human aging using the system identification and system order detection methods

After constructing the regression models of the PPIN and GRN in GENs, we identified the association ability parameters a_{ij} in Eq. (11.12), the TF regulatory ability parameters b_{ij}, and the miRNA regulatory ability c_{il} in Eq. (11.14) using the least squares parameter estimation method in Eq. (2.20) with microarray and miRNA data. All proteins in the candidate PPIN and genes in the candidate GRN were identified one protein/gene by one protein/gene. Next, the Akaike information criterion (AIC) in Eq. (2.23) [42] and Student's t-test [114] were used for determining regression model orders (i.e., N_i and O_i) and pruning false positive PPIs (i.e., a_{ij}) as well as TF and miRNA regulations (i.e., b_{ij} and c_{il}, respectively) out of the system model order. Specifically, the insignificant parameters, a_{ij}, b_{ij}, and c_{il}, out of the regression model order were deleted from candidate GENs to obtain the real GENs.

Finally, the GENs of young and elderly individuals were constructed (Figs. 11.6A and 11.7A), as well as those for old women and men (Figs. 11.8A and 11.9A). The functional networks of these GENs (Figs. 11.6B, 11.7B, 11.8C, and 11.9B) were used to illustrate the important functional interactions of the cellular mechanisms in young and old women and men, respectively. However, further analysis is required to fully understand these GENs. Therefore, we used the PNP method based on PCA to obtain the core GEN from the corresponding GEN.

Constructing core GENs using the PNP method

After using the AIC order detection model and Student's t-test to prune candidate GENs to obtain the real GENs of human aging ($p < .05$), we then obtained the PPIN matrix A and GRN matrix B as follows:

$$A = \begin{bmatrix} a_{1,1} & \cdots & a_{1,j} & \cdots & a_{1,N} \\ \vdots & \ddots & \vdots & \ddots & \vdots \\ a_{i,1} & \cdots & a_{i,j} & \cdots & a_{i,N} \\ \vdots & \ddots & \vdots & \ddots & \vdots \\ a_{N,1} & \cdots & a_{N,j} & \cdots & a_{N,N} \end{bmatrix},$$

$$B = \begin{bmatrix} b_{1,1} & \cdots & b_{1,j} & \cdots & b_{1,N} & -c_{1,1} & \cdots & -c_{1,l} & \cdots & -c_{1,O} \\ \vdots & \ddots & \vdots & \ddots & \vdots & \vdots & \ddots & \vdots & \ddots & \vdots \\ b_{i,1} & \cdots & b_{i,j} & \cdots & b_{i,N} & -c_{i,1} & \cdots & -c_{i,l} & \cdots & -c_{i,O} \\ \vdots & \ddots & \vdots & \ddots & \vdots & \vdots & \ddots & \vdots & \ddots & \vdots \\ b_{N,1} & \cdots & b_{N,j} & \cdots & b_{N,N} & -c_{N,1} & \cdots & -c_{N,l} & \cdots & -c_{N,O} \end{bmatrix}$$

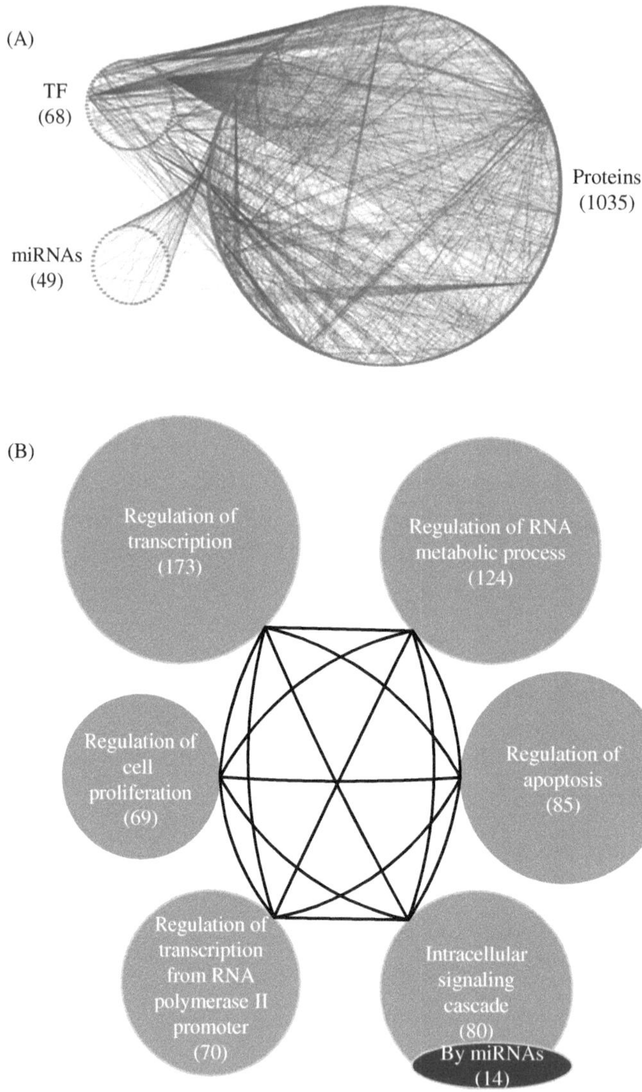

Figure 11.6 (A) The GEN of young individuals and (B) the associated functional network. (A) The upper left circle represents the TFs (68), the lower left circle represents the miRNAs (49), and the right circle represents the proteins (1035) for the GEN of young individuals. Links within the "proteins" circle represent PPIs, links between the "TFs" circle and the "proteins" circle represent regulation of TFs on target genes, and links between the "miRNAs" circle and the "proteins" circle represent regulation of miRNAs on target genes in the GEN of young individuals. (B) In the functional network of the GEN of young individuals, the first six enriched GO terms (biological processes) are regulation of transcription, regulation of RNA metabolic process, regulation of apoptosis, intracellular signaling cascade, regulation of transcription from RNA polymerase II promoter, and regulation of cell proliferation. The numbers in parentheses represents the number of proteins involved in each function.

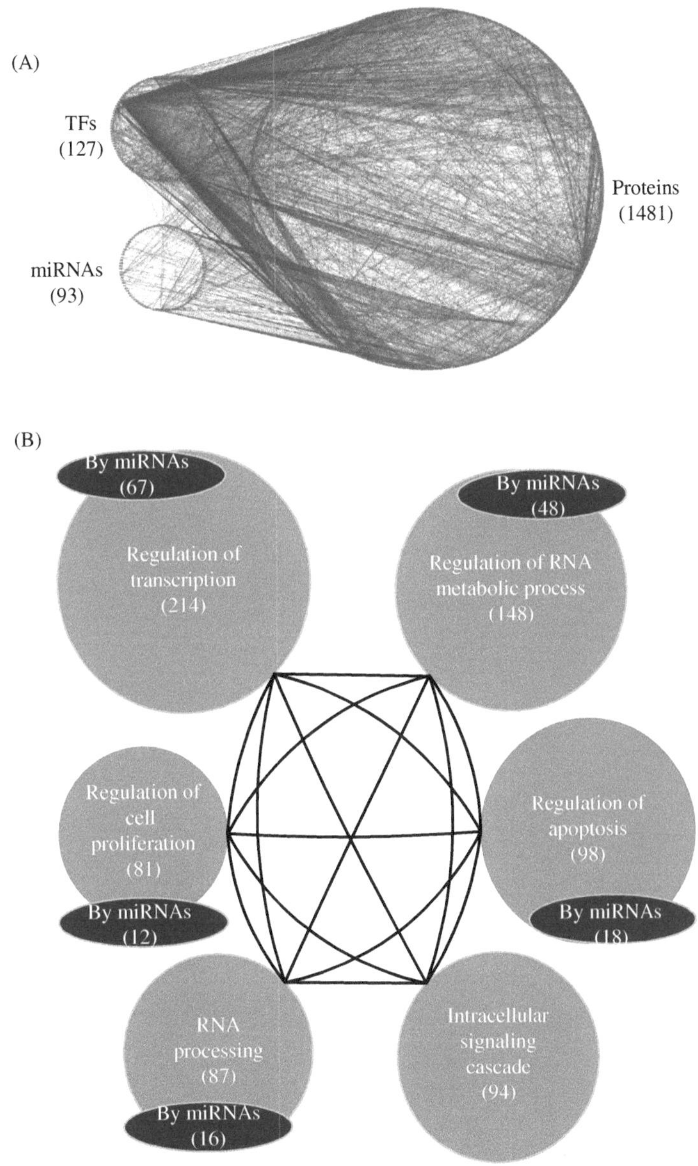

Figure 11.7 (A) The GEN of elderly individuals and (B) the associated functional network. (A) The upper left circle represents the TFs (127), the lower left circle represents the miRNAs (93), and the right circle represents the proteins (1481) for the GEN of elderly individuals. Links within the "proteins" circle represent PPIs, links between the "TFs" circle and the "proteins" circle represent regulation of TFs on target genes, and links between the "miRNAs" circle and the "proteins" circle represent regulation of miRNAs on target genes in the GEN of elderly individuals. (B) In the functional network of the GEN of elderly individuals, the first six enriched GO terms (biological processes) are regulation of transcription, regulation of RNA metabolic process, regulation of apoptosis, intracellular signaling cascade, RNA processing, and regulation of cell proliferation. The numbers in parentheses represent the number of proteins involved in each function.

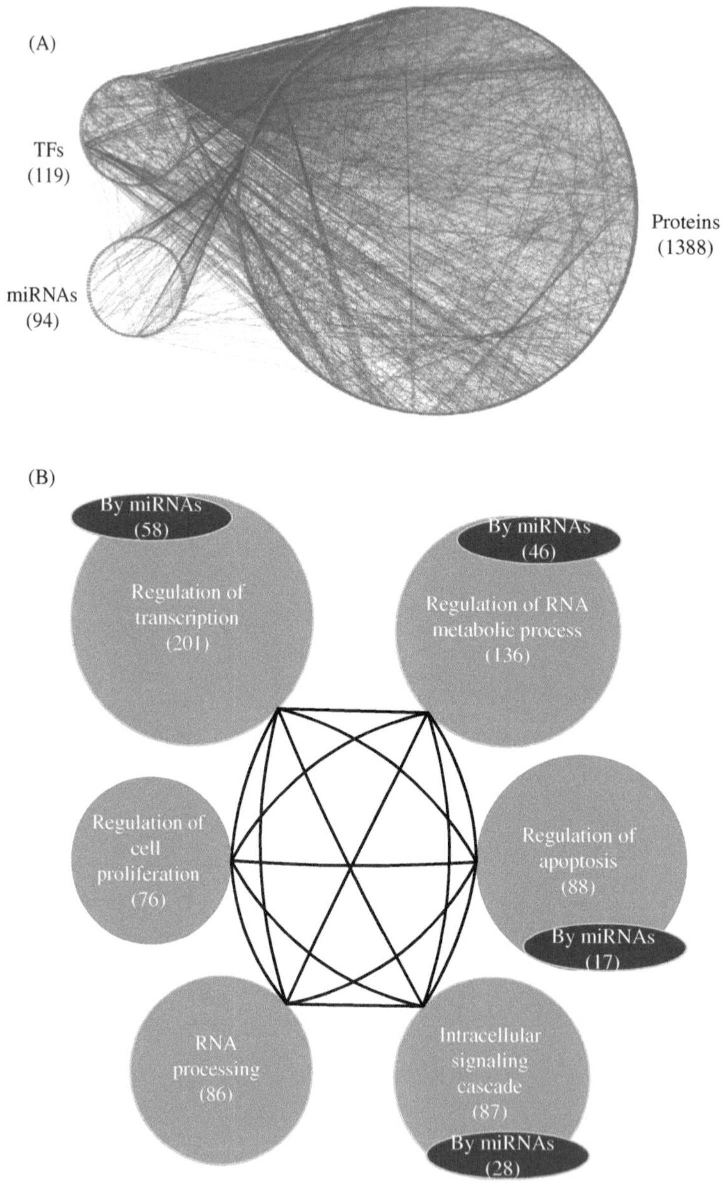

Figure 11.8 (A) The GEN of old women and (B) the associated functional network. (A) The upper left circle represents the TFs (119), the lower left circle represents the miRNAs (94), and the big right circle represents the proteins (1388) for the GEN of old women. Links within the "proteins" circle represent PPIs, links between the "TFs" circle and the "proteins" circle represent regulation of TFs on target genes, and links between the "miRNAs" circle and the "proteins" circle represent regulation of miRNAs on target genes in the GEN of old women. (B) In the functional network of the GEN of old women, the first six enriched GO terms (biological processes) are regulation of transcription, regulation of RNA metabolic process, regulation of apoptosis, intracellular signaling cascade, RNA processing, and regulation of cell proliferation. The number in parentheses represents the number of proteins involved in each function.

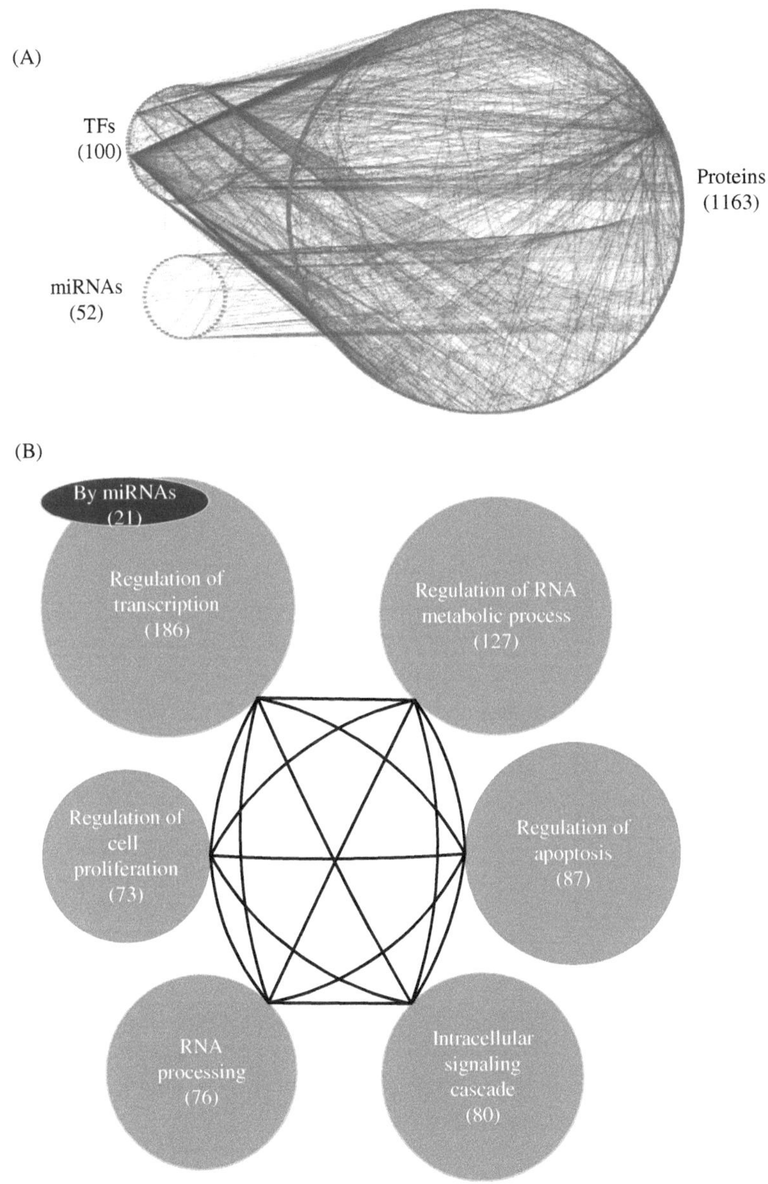

Figure 11.9 (A) The GEN of old men and (B) the associated functional network. (A) The upper left circle represents the TFs (100), the lower left circle represents the miRNAs (52), and the right circle represents the proteins (1163) for the GEN of old men. Links within the "proteins" circle represent PPIs, links between the "TFs" circle and the "proteins" circle represent regulation of TFs on target genes, and links between the "miRNAs" circle and the "proteins" circle represent regulation of miRNAs on target genes in the GEN of old men. (B) In the functional network of the GEN of old men, the first six enriched GO terms (biological processes) for are regulation of transcription, regulation of RNA metabolic process, regulation of apoptosis, intracellular signaling cascade, RNA processing, and regulation of cell proliferation. The number in parentheses represents the number of proteins involved in each function.

where some $a_{i,j}$, $b_{i,j}$, and $c_{i,l}$ remained while others pruned using the AIC system order detection method in the previous section were padded with zero. We then combined A and B as network matrix H of GENs as follows:

$$H = \begin{bmatrix} a_{1,1} & \cdots & a_{1,j} & \cdots & a_{1,N} & b_{1,1} & \cdots & b_{1,j} & \cdots & b_{1,N} & -c_{1,1} & \cdots & -c_{1,l} & \cdots & -c_{1,O} \\ \vdots & \ddots & \vdots & \ddots & \vdots & \vdots & \ddots & \vdots & \ddots & \vdots & \vdots & \ddots & \vdots & \ddots & \vdots \\ a_{i,1} & \cdots & a_{i,j} & \cdots & a_{i,N} & b_{i,1} & \cdots & b_{i,j} & \cdots & b_{i,N} & -c_{i,1} & \cdots & -c_{i,l} & \cdots & -c_{i,O} \\ \vdots & \ddots & \vdots & \ddots & \vdots & \vdots & \ddots & \vdots & \ddots & \vdots & \vdots & \ddots & \vdots & \ddots & \vdots \\ a_{N,1} & \cdots & a_{N,j} & \cdots & a_{N,N} & b_{N,1} & \cdots & b_{N,j} & \cdots & b_{N,N} & -c_{N,1} & \cdots & -c_{N,l} & \cdots & -c_{N,O} \end{bmatrix}^T$$

PCA is based on the following singular value decomposition,

$$H = U \times D \times V^T \tag{11.16}$$

where $U \in \mathfrak{R}^{(2N+O)\times N}$, $V \in \mathfrak{R}^{N \times N}$, and h_k denote the kth row vectors of H for $k = 1,\ldots,(2N+O)$, i.e., $H = [h_1, \ldots, h_k, \ldots, h_{2N+O}]$, and the m column vectors of V are denoted as v_m, for $m = 1,\ldots,N$, which are defined as the right-singular vectors of H. The diagonal entries of the diagonal matrix, $D = diag(d_1, \ldots, d_m, \ldots, d_N)$, are the N singular values of H in descending order, i.e., $d_1 \geq d_2 \geq \ldots \geq d_N$. Note that, $diag(d_1, d_2)$ denotes the diagonal matrix of d_1 and d_2, such that $diag(d_1, d_2) = \begin{bmatrix} d_1 & 0 \\ 0 & d_2 \end{bmatrix}$. The Eigen expression fraction (E_m) is defined as

$$E_m = \frac{{d_m}^2}{\sum\limits_{m=1}^{N} {d_m}^2} \tag{11.17}$$

We selected the top M singular vectors v_m of V such that $\sum_{m=1}^{M} E_m \geq 0.85$ with the minimal M, so that the M principal components contain 85% of the GEN from an energy point of view. The projections of H to the top M singular vectors v_m of V, or the similarities, are defined respectively as follows:

$$S(k, m) = h_k \times {v_m}^T, \quad \text{for } k = 1, \ldots, (2N+O) \text{ and } m = 1, \ldots, M \tag{11.18}$$

We further defined the 2-norm distance from each gene, protein, and miRNA in GEN to the top M singular vectors as:

$$D(k) = \left[\sum_{m=1}^{M} [S(k, m)]^2\right]^{1/2}, \quad \text{for } k = 1, \ldots, (2N+O) \tag{11.19}$$

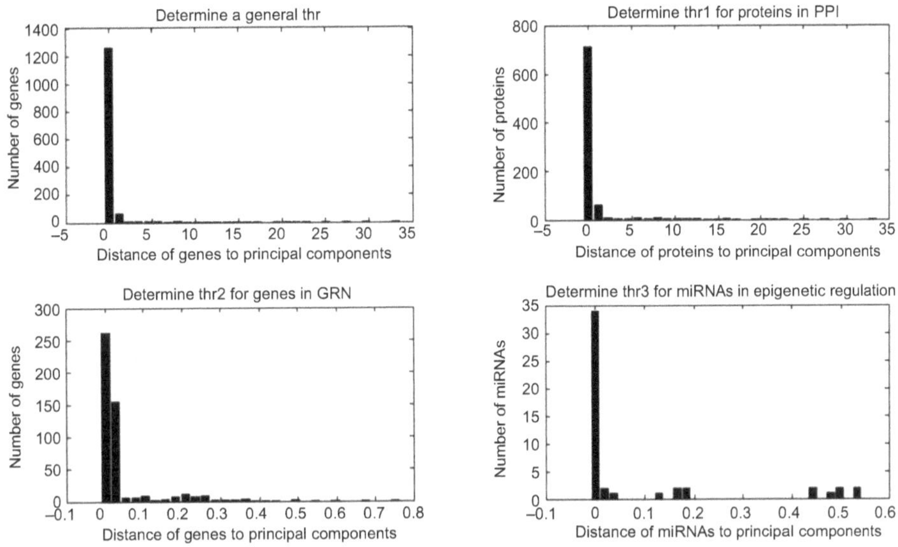

Figure 11.10 The th1, th2, and th3 for proteins, genes, and miRNAs for the PNP from the young GEN in Fig. 11.6A to the young core GEN in Fig. 11.12. In the young GEN, the threshold value $th1$ for proteins and TFs in PPI is 4.0, the threshold value $th2$ for target genes in GRN is 0.015, and the threshold value $th3$ for miRNAs in epigenetic regulation is 0.01. The vertical axis represents the number of proteins, genes, or miRNAs, whereas the horizontal axis represents the distance of proteins, genes, or miRNAs to principal components.

If $D(k)$ is close to zero, it implies that the kth gene, protein, and miRNA is independent of the top M singular vectors. Thus, we determine three thresholds, $th1$, $th2$, and $th3$ as shown in Figs. 11.10 and 11.11 for young and elderly individuals, to respectively identify the core genes, $D(k) \geq th1$ for $k = 1,\ldots,N$, the core proteins, $D(k) \geq th2$ for $k = N+1,\ldots,2N$, and the core miRNAs $D(k) \geq th3$ for $k = 2N+1,\ldots,2N+O$. These were used to obtain the core GENs as shown in Figs. 11.12 and 11.13, which have the principal network structures of the GENs for young and elderly individuals, respectively.

Using the PNP method based on PCA, we obtained the core genes, proteins, miRNAs, and their corresponding interactions and regulatory abilities in core GENs for young and elderly individuals. Comparing these core GENs of young and elderly individuals, part of the intersection is called the common core GEN and part of the distinction sets are called the specific core GENs. In the following section, we further explore the aging mechanisms by investigating the common core GEN and specific core GENs of elderly individuals and young controls.

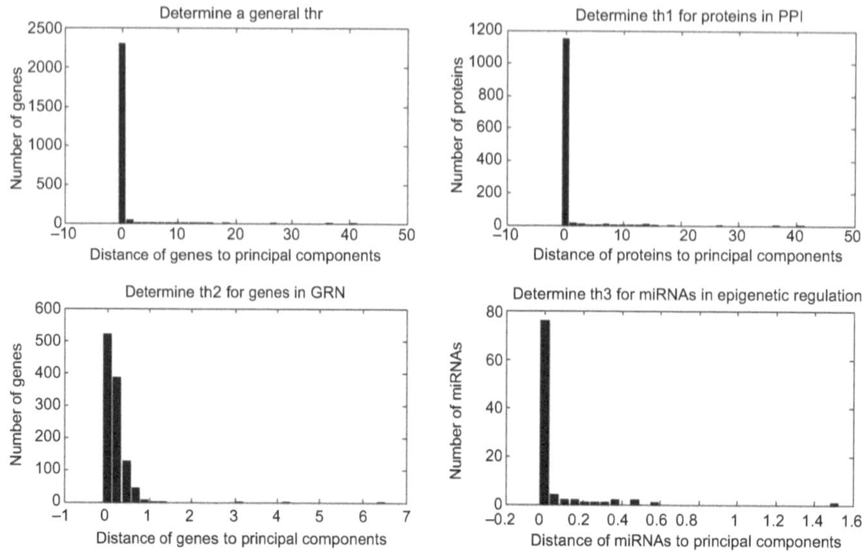

Figure 11.11 The th1, th2, and th3 for proteins, genes, and miRNAs for the PNP from the elderly GEN in Fig. 11.7A to the elderly core GEN in Fig. 11.13. In the elderly GEN, the threshold value *th*1 for proteins and TFs in PPI is 0.6, the threshold value *th*2 for target genes in GRN is 0.35, and the threshold value *th*3 for miRNAs in epigenetic regulation is 0.03. The vertical axis represents the number of proteins, genes, or miRNAs, whereas the horizontal axis represents the distance of proteins, genes, or miRNAs to principal components.

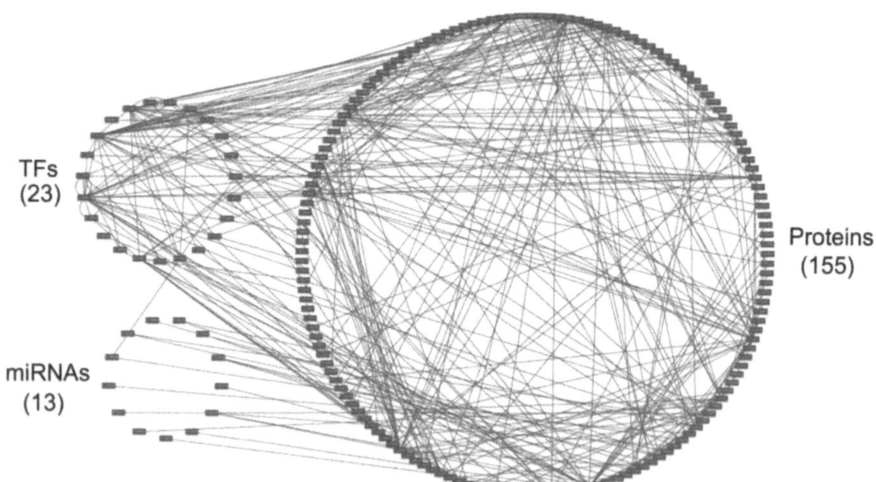

Figure 11.12 The core GEN of young individuals from Fig. 11.6A, as determined by the PNP method. The upper left circle represents the TFs (23), the lower left circle represents the miRNAs (13), and the right circle represents the proteins (155) for the core GEN of young individuals. Links within the "proteins" circle represent the significant PPIs, links between the "TFs" circle and the "proteins" circle represent significant regulation of TFs on target genes, and links between the "miRNAs" circle and the "proteins" circle represent significant regulation of miRNAs on target genes in the core GEN of young individuals.

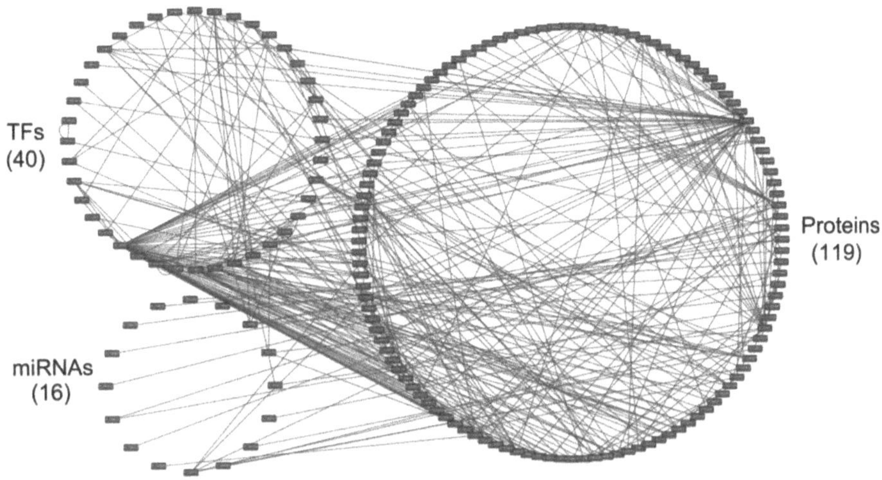

Figure 11.13 The core GEN of elderly individuals from Fig. 11.7a, as determined by the PNP method. The upper left circle represents the TFs (40), the lower left circle represents the miRNAs (16), and the right circle represents the proteins (119) for the core GEN of elderly individuals. Links within the "proteins" circle represent the significant PPIs, links between the "TFs" circle and the "proteins" circle represent significant regulation of TFs on target genes, and links between the "miRNAs" circle and the "proteins" circle represent significant regulation of miRNAs on target genes in the core GEN of elderly individuals.

11.2.2 Results

Molecular mechanisms of human aging in the specific core GENs of elderly men and women

This study is focused on investigating the different molecular mechanisms between young and elderly individuals as well as gender-specific distinctive molecular mechanisms between old individuals in the human aging process. Using the PNP of GENs, we obtained four core GENs for young and old men and women, as shown in Figs. 11.12, 11.13, 11.18, and 11.19, respectively. By intersecting and distinguishing these core proteins, TFs, target genes and miRNAs between young and old core GENs, we identified one common core GEN and two specific core GENs for young and old individuals. Similarly, we also acquired one common core GEN and two specific core GENs for old women and men. In this section, we further discuss the molecular mechanisms of these specific core GENs.

Functional network and pathway analysis of proteins in the GENs for young and old men and women

Based on the GEN of young and elderly individuals (Figs. 11.6A and 11.7A), we observed that the number of nodes and links in the elderly is higher than in the young. This illustrates that the GEN of the elderly may contain more pathways.

In order to overcome genetic mutations in the aging process, more genetic and epigenetic regulatory mechanisms and pathways are developed to maintain normal cellular functions. This could account for why the GEN of the elderly is more complex than that of the young [108,115]. Similarly, the GEN of old women contained more nodes and links (Fig. 11.8A) than that of old men (Fig. 11.9A), illustrating that the GEN of old women may contain more pathways.

The GEN of old women is developed to eliminate the dysfunction of accumulated genetic mutations in order to maintain normal cellular functions during the aging process, which reflects the increased longevity of old women compared with old men.

We investigated the cellular enrichment functions of these proteins in the GENs of young and old men and women. Using the website tool DAVID (https://david.ncifcrf.gov/) and gene ontology analysis, we identified the functional networks of these GENs (Figs. 11.6B, 11.7B, 11.8B, and 11.9B) to illustrate the important functional interactions of cellular mechanisms in young and old men and women, respectively. As shown in Fig. 11.6B, the first six enriched GO terms (biological processes) for GENs of young individuals are regulation of transcription, regulation of RNA metabolic process, regulation of apoptosis, intracellular signaling cascade, regulation of transcription from RNA polymerase II promoter, and regulation of cell proliferation. The first six enriched GO terms for GENs of elderly individuals are regulation of transcription, regulation of RNA metabolic process, regulation of apoptosis, intracellular signaling cascade, RNA processing, and regulation of cell proliferation (Fig. 11.7B). Although the first six important functions are similar between the cohorts, the number of proteins involved in each function is higher in the elderly than in the young. Furthermore, there are more regulatory miRNAs in the GEN of elderly than in young individuals.

The analysis also demonstrated that the first six enriched GO terms for GENs of old men and women are also similar (Figs. 11.8B and 11.9B). They include regulation of transcription, regulation of RNA metabolic process, regulation of apoptosis, intracellular signaling cascade, RNA processing, and regulation of cell proliferation. Although they share the same biological processes, the number of proteins involved in each function is higher in old women than in old men, and there are more regulatory miRNAs in old women than in old men. In addition, it has been suggested that the human mutation rate is much higher in men than in women, and this increases with paternal age [116,117]. Accumulation of a large number of abnormal or mutated cells will lead to cancer, which is an aging disease. In the GEN of old women, there is more complexity, which is developed to reduce the dysfunction of accumulated genetic mutations in the aging process. This is reflected in the increased longevity of old women.

The bioinformatics database DAVID, which automatically outputs the results from Kyoto Encyclopedia of Genes and Genomes (KEGG) pathway analysis, was used for

Table 11.4 Pathway analysis of proteins in the GEN of elderly individuals

KEGG pathway	Numbers	*p*-Value
Pathways in cancer	42	1.7E-4
Neurotrophin signaling pathway	19	2.4E-3
MAPK signaling pathway	32	3.5E-3
NOD-like receptor signaling pathway	12	3.5E-3
Cytosolic DNA-sensing pathway	10	1.4E-2
B-cell receptor signaling pathway	12	1.5E-2
ErbB signaling pathway	13	1.8E-2
Toll-like receptor signaling pathway	14	2.4E-2

In the GEN of elderly individuals, the first eight enriched KEGG pathways are pathways in cancer, the neurotrophin signaling pathway, MAPK signaling pathway, NOD-like receptor signaling pathway, cytosolic DNA-sensing pathway, B-cell receptor signaling pathway, ErbB signaling pathway, and Toll-like receptor signaling pathway.

the pathway analysis of these proteins. The results of the pathway analysis on the GEN of the elderly are summarized in Table 11.4. The first eight enriched KEGG pathways are cancer, neurotrophin signaling, MAPK signaling, NOD-like receptor signaling, cytosolic DNA-sensing, B-cell receptor signaling, ErbB signaling, and TLR signaling pathways. It is possible that the accumulated genetic mutations led to dysregulation of these pathways.

Pathway analysis of core proteins and the impact of miRNAs and DNA methylation on the elder specific core GEN

For the PNP process, a threshold value is necessary in order to obtain core proteins, TFs, and their target genes, as well as core miRNAs of the core GEN. In the GEN of young individuals, the threshold value *th*1 for proteins and TFs in PPI was 4.0, the threshold value *th*2 for their target genes in GRN was 0.015, and the threshold value *th*3 for miRNAs in epigenetic regulation was 0.01 (Fig. 11.10). In the GEN of the elderly, the threshold value *th*1 for proteins and TFs in PPI was 0.6, the threshold value *th*2 for their target genes in GRN was 0.35 and the threshold value *th*3 for miRNAs in epigenetic regulation was 0.03 (Fig. 11.11).

The PNP based on the assigned thresholds was used to generate the core GENs of young and old individuals (Figs. 11.11 and 11.12). By intersecting and distinguishing these core proteins, TFs and target genes, and miRNAs between the young and core GENs, we identified the common core GEN between the young and elderly, as well as the young and elderly specific core GENs (Fig. 11.14). Several genes demonstrated differences in basal levels (i.e., k_i in Eq. 11.14 are different) between young and elderly individuals. This is mainly due to DNA methylation of corresponding genes, which could potentially be developed to eliminate accumulated genetic mutations and maintain normal cellular functions.

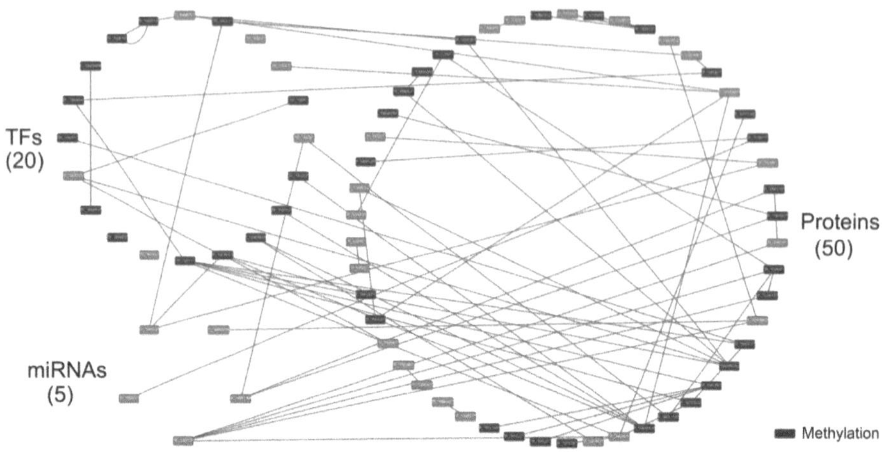

Figure 11.14 The specific core GEN of elderly individuals, as determined by the distinction between the core GEN of the young and elderly. The upper left circle represents the TFs (20), the lower left circle represents the miRNAs (5), and the right circle represents the proteins (50) for the specific core GEN of elderly individuals. The *red nodes* (dark gray in print versions) represent DNA methylation of genes. Links within the "proteins" circle represent the significant PPIs, links between the "TFs" circle and the "proteins" circle represent significant regulation of TFs on target genes, and links between the "miRNAs" circle and the "proteins" circle represent significant regulation of miRNAs on target genes in the specific core GEN of elderly individuals.

Table 11.5 Pathway analysis of core proteins in the specific core GEN of elderly individuals during the aging process

KEGG pathways	Numbers	*p*-Value
Toll-like receptor signaling pathway	5	9.1E-2
T-cell receptor signaling pathway	5	1.1E-1
Fc epsilon RI signaling pathway	3	2.0E-1
Neurotrophin signaling pathway	4	3.8E-1
Wnt signaling pathway	4	4.8E-1
MAPK signaling pathway	5	5.5E-1

In the specific core GEN of elderly individuals, the first six enriched KEGG pathways are the Toll-like receptor signaling pathway, T-cell receptor signaling pathway, Fc epsilon RI signaling pathway, neurotrophin signaling pathway, Wnt signaling pathway, and MAPK signaling pathway.

In order to investigate the mechanisms that cause aging, we analyzed the elderly specific core GEN. The bioinformatics database, DAVID, was used to examine the function and related pathways of these core specific proteins (Table 11.5). The first six enriched KEGG pathways are the TLR signaling, T-cell receptor signaling, Fc epsilon RI signaling, neurotrophin signaling, Wnt signaling, and MAPK signaling pathways. We considered that these six pathways might play important roles in the aging process.

Table 11.6 Core protein-coding genes regulated by core miRNAs in the specific core GEN of elderly individuals and the related pathways

miRNAs	Genes	Pathways
let-7d	*CDK4*	p53 signaling pathway PI3K-Akt signaling pathway T-cell receptor signaling pathway
mir-223	*FLNB*	MAPK signaling pathway
mir-130a	*ZNF274*	Neurotrophin signaling pathway

In the specific core GEN of elderly individuals, *CDK4* is inhibited by let-7d, *FLNB* is inhibited by mir-223, and ZNF274 is inhibited by mir-130a. Moreover, CDK4 is involved in the p53 signaling pathway, PI3K-Akt signaling pathway, and T-cell receptor signaling pathway. FLNB is involved in the MAPK signaling pathway, and *ZNF274* is involved in neurotrophin signaling pathway.

In addition, from the identified specific core GEN of elderly individuals (Fig. 11.14), we observed that there are several core genes that are inhibited by core miRNAs. *CDK4, APPBP2, ZNF10, CDK13, and AKTIP* are inhibited by let-7d, APPBP2 is inhibited by mir-200a, *KLC1, FLNB, MYBL1, and GCN1L1* are inhibited by mir-223, *FBXW7, ZNF274,* and *IRF8* are inhibited by mir-130a, and *FUBP1* is inhibited by mir-25. Next, we therefore investigated the mechanism by which these core protein-coding genes are regulated by core miRNAs and the related pathways (Table 11.6). We observed that CDK4 is involved in the p53 signaling, PI3K-Akt signaling, and T-cell receptor signaling pathways, FLNB is involved in the MAPK signaling pathway, and ZNF274 is involved in neurotrophin signaling pathway.

The results of the pathway analysis demonstrate the importance of these core miRNAs in the elderly specific core GEN, as a significant number of mutations in these pathways often lead to cellular dysfunction. We observed that there are three core miRNAs, mir-223, let-7d, and mir-130a that regulate *FLNB*, *CDK4*, and *ZNF274*, respectively (Fig. 11.15). Those genes are involved in the MAPK signaling, T-cell receptor signaling, and neurotrophin signaling pathways, respectively. Because these three miRNAs and genes only appear in the elderly specific core GEN, we speculate that they play critical roles in the aging process.

Pathway analysis of core proteins and the impact of miRNAs and DNA methylation on the specific core GEN of old women

As previously mentioned, in the PNP process it is necessary to select threshold values in order to obtain core proteins, TFs and their target genes, and core miRNAs of the core GEN. In the GEN of old women, the threshold value *th*1 for proteins and TFs in PPI was 0.5, the threshold value *th*2 for their target genes in GRN was 0.4, and the threshold value *th*3 for miRNAs in epigenetic regulation was 0.1 (Fig. 11.16). In the GEN of old men, the threshold value *th*1 for proteins and TFs in PPI was 0.8,

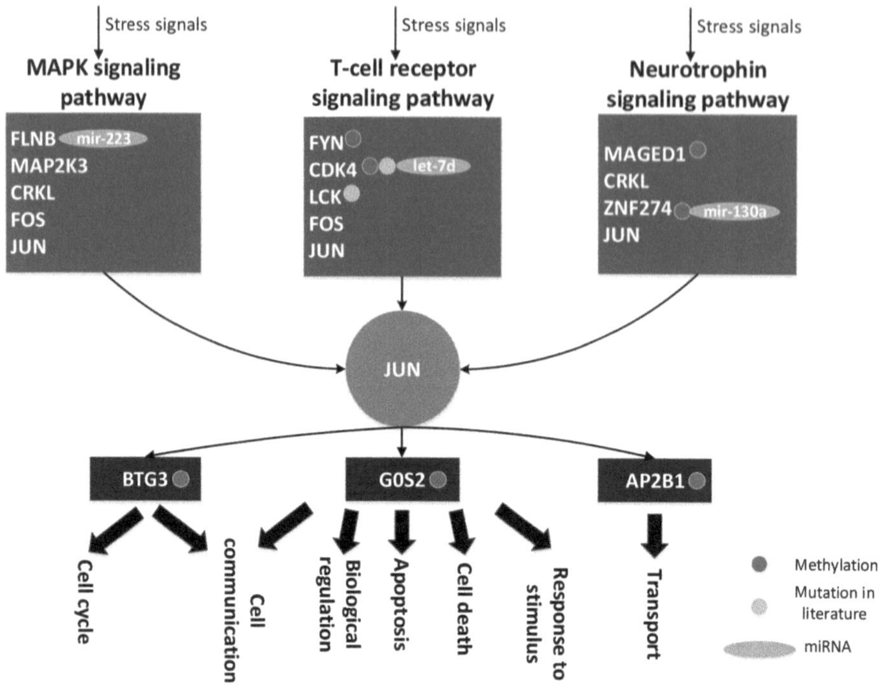

Figure 11.15 The role of miRNA regulation and DNA methylation on the human aging mechanism based on the specific core GEN of the elderly. During the aging process, FLNB, MAP2K3, CRKL, FOS, and JUN play key roles in the MAPK signaling pathway. In addition, FYN, CDK4, LCK, FOS, and JUN are involved in T-cell receptor signaling pathway, and MAGED1, CRKL, ZNF274, and JUN are involved in neurotrophin signaling pathway. In order to overcome the cellular dysfunction caused by accumulated genetic mutations during the aging process, the transcription factor JUN can be activated by a wide variety of stress signals, including the MAPK signaling, T-cell receptor signaling, and neurotrophin signaling pathways. This results in various cellular responses, including cell cycle, cell communication, apoptosis, cell death, biological regulation, cell transport, and response to stimulus. In this figure, the *light blue blocks* represent the proteins involved in the indicated pathways, the *dark green node* represents the TF the *dark blue blocks* represent the TFs target genes, the *light green nodes* represent the miRNAs, the *red nodes* represent DNA methylation of genes, the *yellow nodes* represent the mutation of genes, arrows represent activation of proteins, and link represents interaction between these two proteins.

the threshold value *th*2 for their target genes in GRN was 0.003, and the threshold value *th*3 for miRNAs in epigenetic regulation was 0.003 (Fig. 11.17).

Using the PNP method based on thresholds in Figs. 11.16 and 11.17, we obtained the old male and female core GENs (Figs. 11.18 and 11.19). By intersecting and distinguishing these core proteins, TFs and target genes, and miRNAs between old female and male core GENs, we identified the common core GEN between old women and men, the old female-specific core GEN (Fig. 11.20), and the old male specific core GEN (Fig. 11.21). We observed that there were greater differences in the

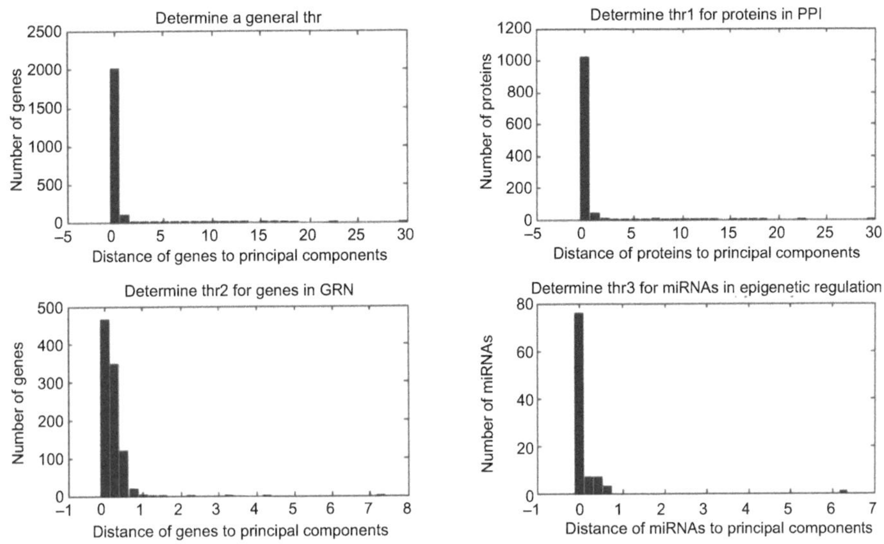

Figure 11.16 The th1, th2, and th3 for proteins, genes, and miRNAs for the PNP from the old female GEN in Fig. 11.8A to the old female core GEN in Fig. 11.18. In the old female GEN, the threshold value *th*1 for proteins and TFs in PPI is 0.5, the threshold value *th*2 for target genes in GRN is 0.4, and the threshold value *th*3 for miRNAs in epigenetic regulation is 0.1. The vertical axis represents the number of proteins, genes, or miRNAs whereas the horizontal axis represents the distance from proteins, genes, or miRNAs to principal components.

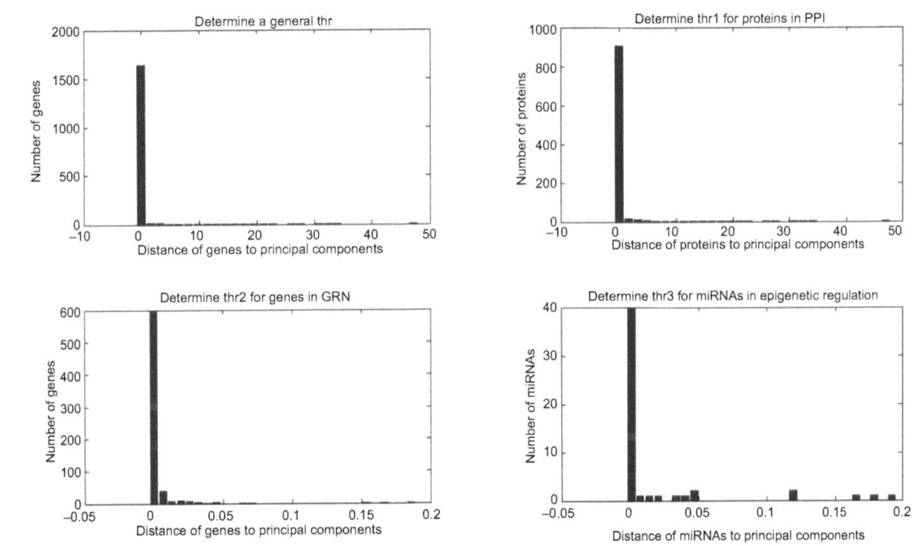

Figure 11.17 The th1, th2, and th3 for proteins, genes, and miRNAs for the PNP from the old male GEN in Fig. 11.9A to the old male core GEN in Fig. 11.19. In the old male GEN, the threshold value *th*1 for proteins and TFs in PPI is 0.8, the threshold value *th*2 for target genes in GRN is 0.003, and the threshold value *th*3 for miRNAs in epigenetic regulation is 0.003. The vertical axis represents the number of proteins, genes, or miRNAs whereas the horizontal axis represents the distance from proteins, genes, or miRNAs to principal components.

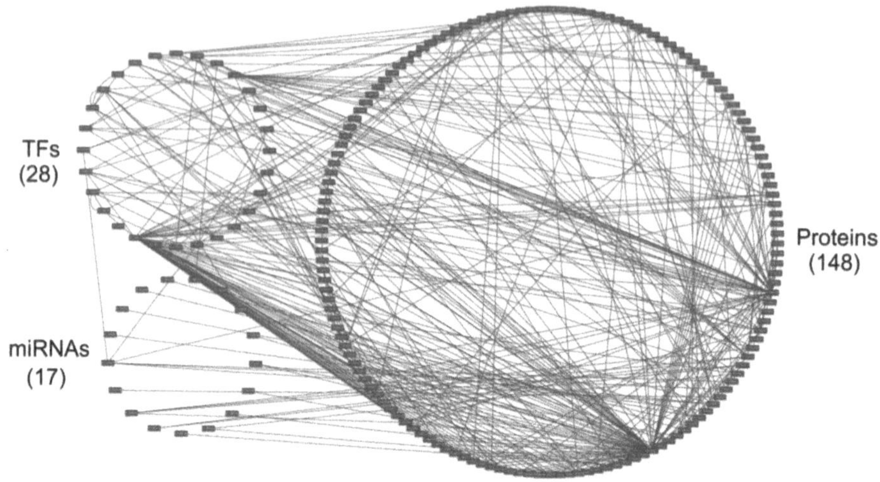

Figure 11.18 The core GEN of old women from Fig. 11.8A, as determined by the PNP method. The upper left circle represents the TFs (28), the lower left circle represents the miRNAs (17), and the right circle represents the proteins (148) for the core GEN of old women. Links within the "proteins" circle represent the significant PPIs, links between the "TFs" circle and the "proteins" circle represent significant regulation of TFs on target genes, and links between the "miRNAs" circle and the "proteins" circle represent significant regulation of miRNAs on target genes in the core GEN of old women.

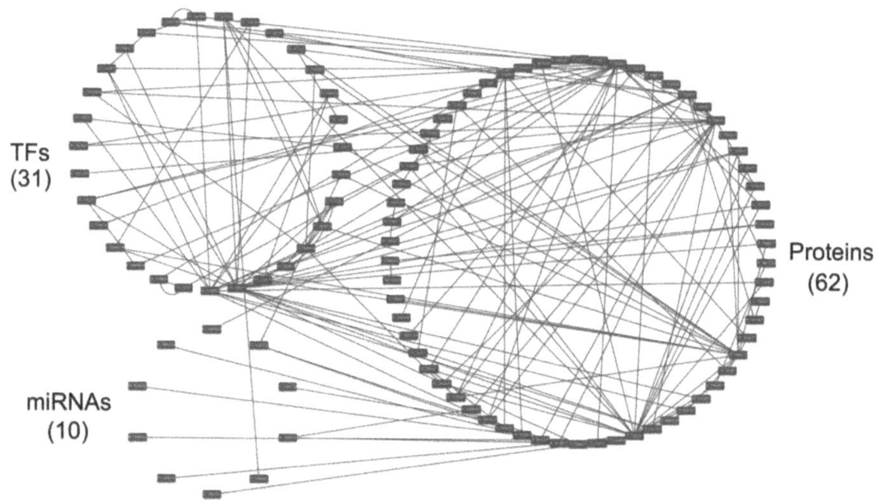

Figure 11.19 The core GEN of old men from Fig. 11.9A, as determined by the PNP method. The upper left circle represents the TFs (31), the lower left circle represents the miRNAs (10), and the right circle represents the proteins (62) for the core GEN of old men. Links within the "proteins" circle represent the significant PPIs, links between the "TFs" circle and the "proteins" circle represent significant regulation of TFs on target genes and links between the "miRNAs" circle and the "proteins" circle represent significant regulation of miRNAs on target genes in the core GEN of old men.

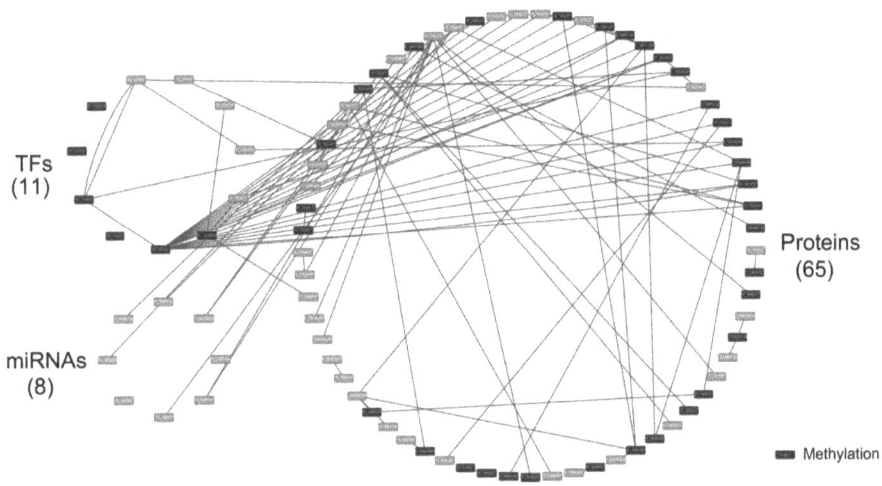

Figure 11.20 The specific core GEN of old women as determined by the distinction between the core GEN of old women and men. The upper left circle represents the TFs (11), the lower left circle represents the miRNAs (8), and the right circle represents the proteins (65) for the specific core GEN of old women. The *red nodes* (dark gray in print versions) represent DNA methylation of genes. Links within the "proteins" circle represent the significant PPIs, links between the "TFs" circle and the "proteins" circle represent significant regulation of TFs on target genes and links between the "miRNAs" circle and the "proteins" circle represent significant regulation of miRNAs on target genes in the specific core GEN of old women.

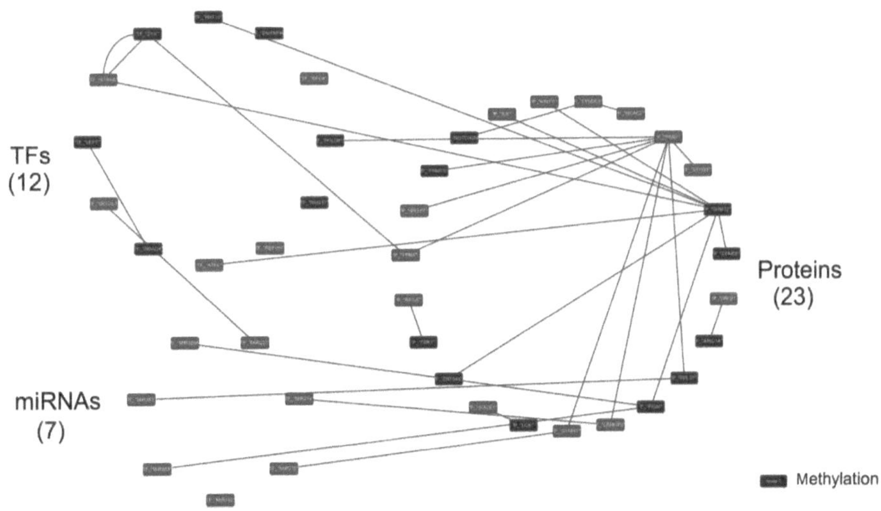

Figure 11.21 The specific core GEN of old men as determined by the distinction between the core GEN of old women and old men. The upper left circle represents the TFs (12), the lower left circle represents the miRNAs (7), and the right circle represents the proteins (23) for the specific core GEN of old men. The *red nodes* (dark gray in print versions) represent DNA methylation of genes. Links within the "proteins" circle represent the significant PPIs, links between the "TFs" circle and the "proteins" circle represent significant regulation of TFs on target genes and links between the "miRNAs" circle and the "proteins" circle represent significant regulation of miRNAs on target genes in the specific core GEN of old men.

basal levels of genes (i.e., k_i in Eq. 11.14 are different) in the specific core GEN of old women than in that of old men, which are mainly due to DNA methylation of corresponding genes. We speculated that old women probably developed these methylated genes in an effort to eliminate the dysfunction of accumulated genetic mutations in the aging process. We analyzed the old female and male specific core GENs to determine the distinctive mechanisms between old women and men that contribute to the gender-specific differences in the aging process.

The results of the pathway analysis of the specific core GEN of old women are summarized in Table 11.7. The first four enriched KEGG pathways are the TLR, MAPK, NOD-like receptor, and neurotrophin signaling pathways. This indicates that these pathways are critical to the aging process in women.

In addition, we found that there are many core genes that are inhibited by the core miRNAs (Fig. 11.20). *TRAF6 and TAOK3* are inhibited by mir-373, *GYK* and *APPBP2* are inhibited by mir-200a, *TRAF6, TAOK3,* and *ZNF302* are inhibited by mir-141, *CBX2* is inhibited by mir-1, *APBB1* is inhibited by mir-148b, *POLR3D* is inhibited by mir-374b, *NAE1* is inhibited by mir-503, and *GTF2I* is inhibited by mir-98.

Next, we investigated the pathways and mechanisms involved in the regulation of these core protein-coding genes by core miRNAs. The results are summarized in Table 11.8. TRAF6 is involved in the MAPK signaling, NF-κB signaling, TLR signaling, NOD-like receptor signaling, RIG-I-like receptor signaling, and neurotrophin signaling pathways. In addition, TAOK3 is involved in the MAPK signaling pathway, POLR3D is involved in metabolic and cytosolic DNA-sensing pathways, and GTF2I is involved in the cGMP-PKG signaling pathway. Based on these findings, it is possible that the observed negative regulation of miRNAs were developed as a protective mechanism to ensure normal cellular functions.

Pathway analyses demonstrated that there are two core miRNAs, mir-373 and mir-141 (Fig. 11.22). *TAOK3* and *TRAF6* are inhibited by mir-373, and are involved in the MAPK signaling pathway. However, *TRAF6* is also inhibited by mir-141, and is also involved in the TLR signaling pathway. Because these two miRNAs and genes only appear in the old female-specific core GEN, it is likely that these two miRNAs, genes, and pathways play a critical role in the female-specific aging process.

Table 11.7 Pathway analysis of core proteins in the specific core GEN of old women during the aging process

KEGG pathways	Numbers	*p*-Value
Toll-like receptor signaling pathway	5	3.8E-2
MAPK signaling pathway	8	4.9E-2
NOD-like receptor signaling pathway	3	1.8E-1
Neurotrophin signaling pathway	4	2.0E-1

In the specific core GEN of old women, the first four enriched KEGG pathways are the Toll-like receptor signaling pathway, MAPK signaling pathway, NOD-like receptor signaling pathway, and neurotrophin signaling pathway.

Table 11.8 Core protein-coding genes regulated by core miRNAs in the specific core GEN of old women and the related pathways

miRNAs	Genes	Pathways
mir-141 mir-373	*TRAF6*	MAPK signaling pathway NF-κB signaling pathway Toll-like receptor signaling pathway NOD-like receptor signaling pathway RIG-I-like receptor signaling pathway Neurotrophin signaling pathway
mir-141 mir-373	*TAOK3*	MAPK signaling pathway
mir-374b	*POLR3D*	Metabolic pathways Cytosolic DNA-sensing pathway
mir-98	*GTF2I*	cGMP-PKG signaling pathway

In the specific core GEN of old women, *TRAF6* and *TAOK3* are inhibited by mir-141 and mir-373, *POLR3D* is inhibited by mir-374b, and *GTF2I* is inhibited by mir-98. Moreover, TRAF6 is involved in the MAPK signaling pathway, NF-κB signaling pathway, Toll-like receptor signaling pathway, NOD-like receptor signaling pathway, RIG-I-like receptor signaling pathway, and neurotrophin signaling pathway. TAOK3 is involved in the MAPK signaling pathway, POLR3D is involved in metabolic pathways and the cytosolic DNA-sensing pathway, and GTF2I is involved in the cGMP-PKG signaling pathway.

Pathway analysis of core proteins and the impact of miRNAs and DNA methylation on the specific core GEN of old men

Using the bioinformatics database, DAVID, we investigated the molecular pathways related to the core proteins in the old male specific core GEN (Table 11.9). It is noteworthy that only one enriched KEGG pathway is related to cancer.

In addition, from our specific core GEN of old men (Fig. 11.21), we observed that there are many core genes that are inhibited by the core miRNA. For instance, *TUBG1* is inhibited by mir-152, *STMN1* is inhibited by mir-210, *LRRFIP2* and *UNG* are inhibited by mir-214, *GCN1L1* is inhibited by mir-221, *RPL37* is inhibited by mir-381, and *PIGN* is inhibited by mir-320a and mir-653. In Table 11.10, we show the specific pathways that these core protein-coding genes are involved in. STMN1 is involved in the MAPK signaling pathway, and LRRFIP2 is involved in the Wnt signaling pathway.

The results of the pathway analysis indicate no intersection between the two findings (Fig. 11.23). Therefore, we can only investigate how these two core protein-coding genes are regulated by core miRNAs in the old male specific core GEN and the associated pathways. We observed that there are two core miRNAs, mir-210 and mir-214, that regulate *STMN1* and *LRRFIP2*, respectively which are involved in the MAPK and Wnt signaling pathways, respectively. These two miRNAs and genes only appear in the old male specific core GEN indicating that they are important for the male-specific aging process.

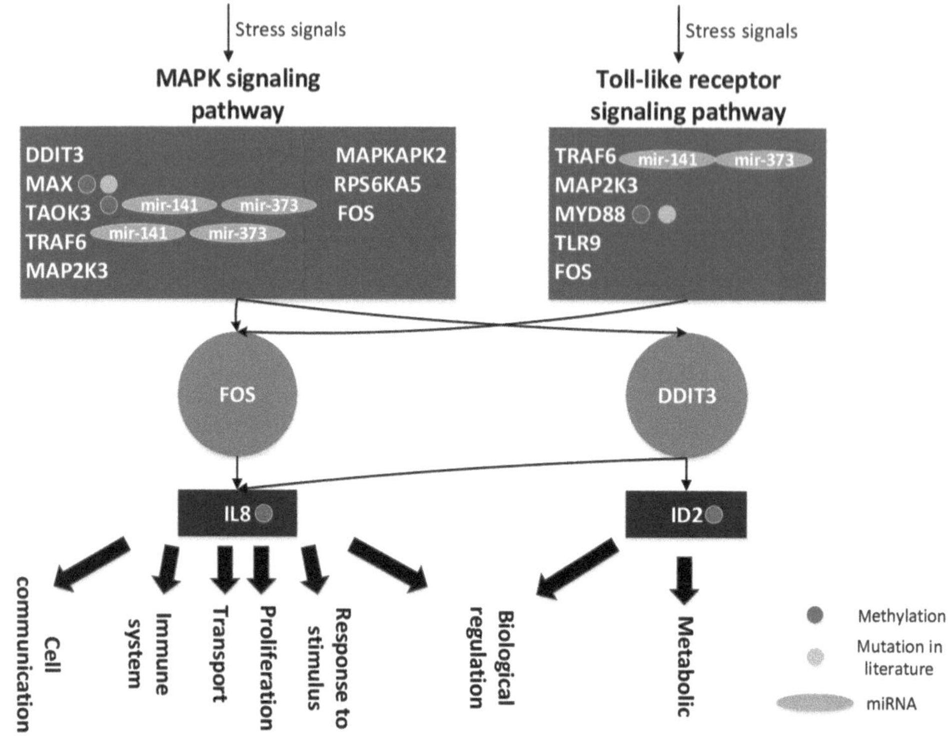

Figure 11.22 The role of miRNA regulation and DNA methylation on gender-specific aging mechanisms based on the specific core GEN of old women. In the aging process, DDIT3, MAX, TAOK3, TRAF6, MAP2K3, MAPKAPK2, RPS6KA5, and FOS are involved in the MAPK signaling pathway and TRAF6, MAP2K3, MYD88, TLR9, and FOS are involved in the Toll-like receptor signaling pathway. In order to overcome the dysfunction caused by accumulated genetic mutations during the aging process, the transcription factor FOS and DDIT3 can be activated by a wide variety of stress signals, including the MAPK signaling and Toll-like receptor signaling pathways. This results in various cellular responses, including immune system activation, cell proliferation, metabolism, cell communication, cell transport, biological regulation, and response to stimulus. In this figure, the *light blue blocks* represent the proteins involved in the indicated pathways, the *deep green nodes* represent the TFs, the *deep blue blocks* represent the TFs target genes, the *light green nodes* represent the miRNAs, the *red nodes* represent DNA methylation of genes, the *yellow nodes* represent the mutation of genes, arrows represent activation of proteins, and the link represents interaction between these two proteins.

Table 11.9 Pathway analysis of core proteins in the specific core GEN of old men during the aging process

KEGG pathways	Numbers	*p*-Value
Pathway in cancer	3	3.0E-1

In the specific core GEN of old men, only one KEGG pathway is enriched in the aging process.

Table 11.10 Core protein-coding genes regulated by core miRNAs in the specific core GEN of old men and the related pathways

miRNAs	Genes	Pathways
mir-210	*STMN1*	MAPK signaling pathway
mir-214	*LRRFIP2*	Wnt signaling pathway

In the specific core GEN of old men, *STMN1* is inhibited by mir-210, and *LRRFIP2* is inhibited by mir-214. In addition, STMN1 is involved in the MAPK signaling pathway, and LRRFIP2 is involved in the Wnt signaling pathway.

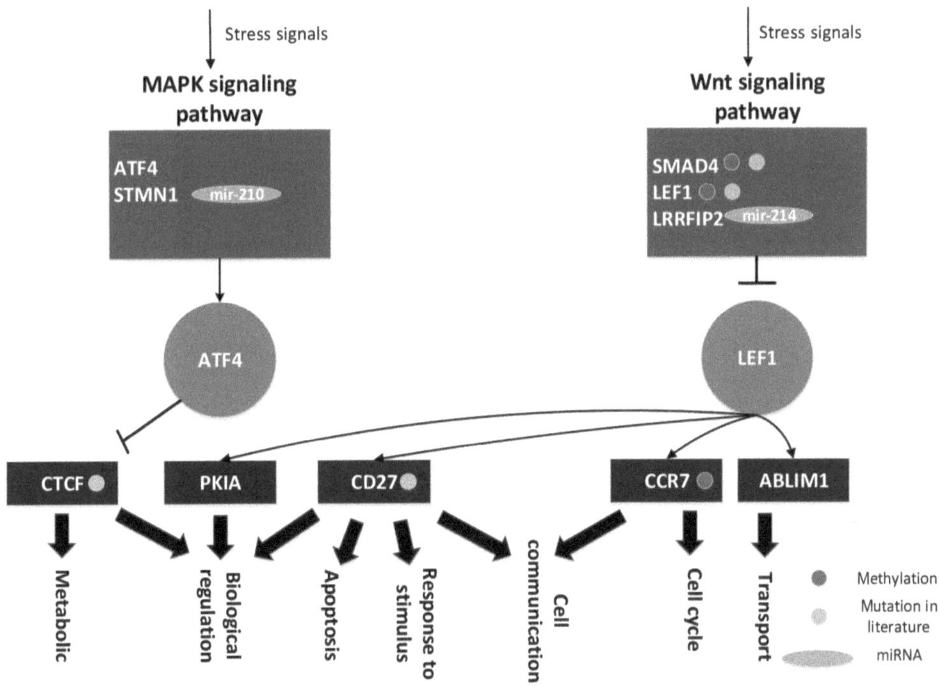

Figure 11.23 The role of miRNA regulation and DNA methylation in gender-specific aging mechanisms based on the specific core GEN of old men. During the aging process, ATF4 and STMN1 are involved in the MAPK signaling pathway, and SMAD4, LEF1, and LRRFIP2 play key roles in the Wnt signaling pathway. Since these genetic and epigenetic regulation observed cannot overcome the dysfunction due to accumulated genetic mutations during the aging process, the transcription factor ATF4 can be activated and LEF1 can be inhibited by a wide variety of stress signals, including the MAPK and Wnt signaling pathways. This results in deregulation of various cellular responses, including cell cycle, apoptosis, metabolism, cell communication, cell transport, biological regulation, and response to stimulus. In this figure, the *light blue blocks* represent the proteins involved in the indicated pathways, the *deep green nodes* represent the TFs, the *deep blue blocks* represent the TFs target genes, the *light green nodes* represent the miRNAs, the *red nodes* represent DNA methylation of genes, the *yellow nodes* represent the mutation of genes, arrows represent activation of proteins, and the link represents interaction between these two proteins.

11.2.3 Discussion

The role of miRNA regulation and DNA methylation in human aging based on the specific core GEN of elderly individuals

Aging is a complex process driven by diverse molecular pathways. Based on the analysis of the specific core GEN of elderly individuals, we demonstrate that FLNB, MAP2K3, CRKL, FOS, and JUN are involved in the MAPK signaling pathway (Fig. 11.15). This pathway is activated by diverse extracellular and intracellular stimuli including peptide growth factors, cytokines, hormones, and various cellular stressors such as oxidative and endoplasmic reticulum stress. It has been suggested that the MAPK pathway regulates a variety of cellular activities including proliferation, differentiation, survival, and death. Dysregulation of the MAPK signaling pathway has been implicated in the development of many human diseases and aging [118].

In addition, we found that FYN, CDK4, LCK, FOS, and JUN are involved in the T-cell receptor signaling pathway. This signaling pathway, in response to antigen recognition, plays a central role in the adaptive immune response [119]. T-cell activation by foreign antigens induces antigen specific T-cell clonal expansion and differentiation, and this response is regulated by signal transduction pathways initiated by antigen receptors and costimulatory molecules [120].

Additionally, MAGED1, CRKL, ZNF274, and JUN are involved in the neurotrophin signaling pathway. This pathway is activated by neurotrophins binding to cognate receptors, which have been shown to regulate almost all aspects of neuronal development and function, including precursor proliferation and commitment, cell survival, axon and dendrite growth, membrane trafficking, synapse formation and function, as well as glial differentiation and interactions with neurons [121].

Humans possess the capability to respond to a variety of external signals through cell surface receptors. In Fig. 11.15, extracellular and intracellular stimuli through the MAPK signaling, T-cell receptor signaling, and neurotrophin signaling pathways lead to activation of JUN, and result in various cellular responses through transcriptional regulation of its target genes. The TF JUN can activate many target genes, including *BTG3* ($b_{ij} = 0.150252$), *G0S2* ($b_{ij} = 0.246038$), and *AP2B1* ($b_{ij} = 0.043467$). BTG3 is involved in cell cycle and cell communication, G0S2 is involved in apoptosis, cell communication, cell death, biological regulation and response to stimuli, and AP2B1 is involved in cell transport.

We speculate that increased genetic and epigenetic regulatory mechanisms and pathways are developed during the aging process in order to maintain normal cellular function and overcome accumulated genetic mutations. Accumulation of a large number of abnormal or mutated cells is part of the aging process. Mutation of CDK4 was found to be associated with a variety of cancers [122] and mutation of LCK was found to be associated with immunodeficiency [123]. We observed that expression of these two genes vary significantly between young and old individuals ($p < .05$), which supports the above findings.

In addition, we found seven specific genes that demonstrate differing basal expression levels (i.e., k_i in Eq. 11.14 are different) between young and elderly individuals, mainly due to DNA methylation. They are *FYN, CDK4, MAGED1, ZNF274, BTG3, G0S2,* and *AP2B*. Analysis of the MethHC database (a database of DNA methylation and gene expression in Human Cancer, http://methhc.mbc.nctu.edu.tw/php/index.php) [124] demonstrated that DNA methylation of *FYN, CDK4, MAGED1, ZNF274, BTG3, G0S2,* and *AP2B1* varies significantly ($p < .05$) between tumor and normal samples, which support our findings since cancer is an aging disease. In addition, *FLNB* is inhibited by mir-223 ($c_{il} = -0.21954$), *CDK4* is inhibited by miRNA let-7d ($c_{il} = -0.50305$), and *ZNF274* is inhibited by mir-130a ($c_{il} = -0.10597$). These results were observed only in the specific core GEN of elderly individuals. The differences between the young and elderly specific core GENs might be responsible for human aging mechanisms. In addition, it has been reported that *FOS* and *JUN* are aging-related human genes (Fig. 11.15 and GenAge: The Ageing Gene Database, http://genomics.senescence.info/genes/) [125], which also support our results.

In order to overcome the accumulated mutation of genes in the aging process, the TF JUN can be activated by a wide variety of stress signals in the aging process, including the MAPK signaling, T-cell receptor signaling, and neurotrophin signaling pathways. This results in various cellular responses, including cell cycle, cell communication, apoptosis, cell death, biological regulation, cell transport, and response to stimulus. If not, aging is accompanied by a decline in the function of the immune system, which increases susceptibility to infections. In addition, dysfunction of the cell cycle results in a decreased ability of cells to arrest at key checkpoints and also leads to the survival of abnormal or mutated cells that would normally die, resulting in tumorigenesis. Accumulation of a large number of abnormal or mutated cells leads to aging. Furthermore, a defective apoptotic processes results in a wide variety of diseases, such as cancer. In short, aging is characterized by deterioration in the maintenance of homeostatic processes over time [115], leading to functional decline and increased risk for disease and death. Therefore, drugs designed to target these DNA methylated genes (*BTG3, G0S2, and AP2B1*) or genes inhibited by miRNAs, mir-223, let-7d, and mir-130a (*FLNB, CDK4, and ZNF274*) may delay aging.

The role of miRNA regulation and DNA methylation on human gender aging mechanisms based on the specific core GEN of old women

Analysis of gender-specific aging mechanisms based on the specific core GEN of old women demonstrated that DDIT3, MAX, TAOK3, TRAF6, MAP2K3, MAPKAPK2, RPS6KA5, and FOS are involved in the MAPK signaling pathway (Fig. 11.22). The MAPK family plays an important role in complex cellular programs such as proliferation, differentiation, development, transformation, and apoptosis [126].

In addition, TRAF6, MAP2K3, MYD88, TLR9, and FOS are involved in the TLR signaling pathway. TLRs play an essential role in the activation of innate immunity by recognizing specific patterns of microbial components [127]. Stimulation of different TLRs induces distinct patterns of gene expression, which not only lead to activation of innate immunity but also instruct the development of antigen-specific acquired immunity [128].

Extracellular and intracellular stimuli through the MAPK signaling and TLR signaling pathways lead to activation of FOS and DDIT3, and result in various cellular responses through transcriptional regulation of its target genes (Fig. 11.22). The TF FOS can activate the *IL8* gene ($b_{ij} = 1.136866$). IL8 is involved in a number of cellular functions of the immune system, cell proliferation, cell communication, cell transport, biological regulation, and the response to various stimuli. In addition, the TF DDIT3 can activate many target genes, including *IL8* ($b_{ij} = 1.143526$) and *ID2* ($b_{ij} = 0.458447$). The functions of ID2 include metabolism and biological regulations.

Accumulation of mutations play a key role in the aging process, and it was shown that mutation of *MAX* is associated with hereditary pheochromocytoma [129], whereas mutation of MYD88 was found to be associated with many hematological malignancies [130]. We found that the expression of these two genes vary significantly between young and old individuals ($p < .05$), which supports the above results.

In addition, we found that five specific genes, *MAX, TAOK3, MYD88, IL8*, and *ID2*, have varying basal expression levels (i.e., k_i in Eq. 11.14 are different) between young and old women, which is mainly due to the DNA methylation of corresponding genes. It has been reported that DNA methylation and the expression levels of these five genes vary significantly between tumor and normal samples ($p < .05$; via MethHC) [124], supports our results. In addition, *TAOK3* and *TRAF6* are inhibited by mir-141 ($c_{il} = -0.2235$-*TAOK3*; $c_{il} = -0.36144$-*TRAF6*) and mir-373 ($c_{il} = -0.10454$-*TAOK3*; $c_{il} = -0.12645$-*TRAF6*). Mutations in these genes, inhibition of miRNAs, and DNA methylation only appear in the specific core GEN of old women. In addition, it has been reported that *DDIT3, MAX*, and *FOS* are aging-related human genes (Fig. 11.22; GenAge database). Other aging-related genes were not identified in our analyses.

The TFs FOS and DDIT3 can be activated by a wide variety of stress signals in the female aging process, including the MAPK signaling and TLR signaling pathways, resulting in various cellular responses, including immune activation, cell proliferation, metabolism, cell communication, cell transport, biological regulation, and response to various stimuli. If these TFs are not activated during the aging process, the innate immune system becomes dysregulated, which is characterized by persistent inflammatory responses that involve multiple immune and nonimmune cell types. This leads to proliferation of abnormal or mutated cells and the increased incidence of tumorigenesis. Many tumors accumulate in the body during aging. In addition, many metabolic pathways decline with age, and this is a major consequence of the aging process, which induces susceptibility to a number of diseases. Therefore, drugs designed to target *IL8, ID2, TRAF6, or TAOK3* may improve the aging process in women.

The role of miRNA regulation and DNA methylation in the human aging mechanism based on the specific core GEN of old men

Analysis of the specific core GEN of old men revealed that ATF4 and STMN1 are involved in the MAPK signaling pathway (Fig. 11.23). The MAPK signal pathway regulates a wide variety of cellular functions in response to cellular stresses [131]. In addition, SMAD4, LEF1, and LRRFIP2 are involved in the Wnt signaling pathway. It has been suggested that the Wnt signaling pathway has important functions in stem cell biology, cardiac development and differentiation, angiogenesis, cardiac hypertrophy, cardiac failure, and aging [132]. Aberrant Wnt signaling underlies a wide range of pathologies in humans [133].

We show that extracellular and intracellular stimuli through the MAPK signaling and Wnt signaling pathways lead to activation of ATF4 and inhibition of LEF1, resulting in various cellular responses through transcriptional regulation of target genes (Fig. 11.23). The TF ATF4 can inhibit the *CTCF* gene ($b_{ij} = -0.57079$), which plays critical roles in metabolism and biological regulation. In addition, the TF LEF1 can activate many target genes, including *ABLIM1* ($b_{ij} = 0.659135$), *CCR7* ($b_{ij} = 0.833707$), *CD27* ($b_{ij} = 0.850484$), and *PKIA* ($b_{ij} = 0.485246$). ABLIM1 is important for cell transport, CCR7 is involved in cell cycle and cell communication, CD27 plays a role in apoptosis, cell communication, biological regulation, and response to stimulus, and PKIA is necessary for biological regulation.

As previously mentioned, accumulation of genetic mutations is highly correlated with the aging process, and it has been shown that mutation of *SMAD4* is associated with pancreatic cancer [134], juvenile polyposis syndrome [135], and hereditary hemorrhagic telangiectasia syndrome [136]. Moreover, mutation of *LEF1* was found to be associated with somatic sebaceous tumors [137], whereas mutation of *CTCF* is associated with invasive breast cancers [138], prostate cancers [139], and Wilms' tumors [139]. Mutation of *CD27* is associated with lymphoproliferative syndrome 2 [140]. Interestingly, the expression levels of these four genes differ significantly ($p < .05$) between young and old men in our data, which supports the above findings.

In addition, we found that three specific genes have differing basal level expression (i.e., k_i in Eq. 11.14 are different) between young and old men, which is mainly due to DNA methylation of corresponding genes. These genes are *SMAD4, LEF1*, and *CCR7*. It has been reported that DNA methylation of *SMAD4, LEF1,* and *CCR7* demonstrate varying expression levels ($p < .05$) between tumor and normal samples (MethHC database) [124], which support our results. Additionally, *STMN1* is inhibited by mir-210 ($c_{il} = -0.25071$) and *LRRFIP2* is inhibited by the mir-214

($c_{il} = -0.48908$). These genetic mutations, miRNA inhibitions, and DNA methylation of genes only appear in the old male specific core GEN. These differences might account for the aging mechanisms in old men. It has been suggested that the mutation rate is much higher in men than in women, and increases with age. The specific core GEN of old men is less developed to eliminate dysfunctions due to accumulated genetic mutations in the aging process compared to old women, which reflects the shorter lifespan of old men. It has been reported that *CD27* is an aging-related human gene (GenAge).

Since these genetic and epigenetic regulations cannot overcome the accumulated genetic mutations in the aging process, the TFs ATF4 can be activated and LEF1 can be inhibited by a wide variety of stress signals in the male-specific aging process. This includes the MAPK signaling and Wnt signaling pathways, which result in dysregulation of various cellular responses, including cell cycle, apoptosis, metabolism, cell communication, cell transport, biological regulation, and the response to various stimuli. Furthermore, apoptosis enables an organism to eliminate unwanted and defective cells through an orderly process of cellular disintegration. Improper regulation of apoptosis contributes to disorders such as cancer, viral infection, and autoimmune diseases. Moreover, the cell cycle represents a series of events responsible for cell duplication. Dysregulation of the cell cycle leads cancer development. Medical statistics indicate that men are more likely than women to get cancer. This may explain why men have a lower average life expectancy compared to women. Therefore, designing drugs to the DNA methylated genes, *CTCF, CD27,* and *CCR7*, or the genes inhibited by miRNAs, mir-210 and mir-214, *STMN1* and *LRRFIP2*, may improve the male-specific aging process.

The overall mechanism of human aging

Aging is characterized by a general decline in cellular function, which ultimately affects homeostasis of the entire body [115]. Aging is an inevitable part of life and comes with all sorts of physical and mental ailments, including common metabolism, inflammation, immune decline, and cancer [141]. Extracellular and intracellular activation of various pathways, such as the MAPK signaling, T-cell receptor signaling, neurotrophin signaling, TLR signaling, and Wnt signaling pathways lead to various cellular responses. Genetic mutations lead to dysregulation of pathways and dysfunction of various cellular responses necessary during the aging process. In order to overcome these dysfunctions, genetic and epigenetic regulation is enhanced to maintain normal cellular functions. In addition, it has been suggested that the human mutation rate is much higher in men than in women and increases with paternal age. Accumulation of a large number of abnormal or mutated cells will lead to cancer, which is an aging disease. For old women, GEN is more developed than that of old men, which reflects the increased longevity of old women.

11.2.4 Conclusion

In this section, to investigate human aging mechanisms from peripheral blood mononuclear cells, the GENs of young and old men and women were constructed based on their corresponding microarray data, miRNA data, methylation data, and database mining via the least squares parameter estimation method, AIC order detection model, and the Student's *t*-test. The core GENs were obtained using the PNP method based on PCA. However, the common core and specific core GENs were acquired using the intersection and distinction of core proteins, TFs, target genes, and miRNAs in core GENs between the young and elderly, and between old women and men. We found that in the specific core GEN of elderly individuals, *FLNB, CDK4,* and *ZNF274* are inhibited by mir-223, let-7d, and mir-130a, respectively, and DNA methylation of *FYN, CDK4, MAGED1,* and *ZNF274* in order to overcome dysregulation of the MAPK signaling, T-cell receptor signaling, and neurotrophin signaling pathways, as well as deregulated cell cycle and apoptosis processes. The specific core GEN of old women demonstrated that *TAOK3* and *TRAF6* are inhibited by mir-141 and mir-373, respectively, and DNA methylation of *MAX, TAOK3,* and *MYD88* in order to overcome dysregulation of the MAPK signaling and TLR signaling pathways as well as dysfunctions of the immune system, proliferation, and metabolism. The specific core GEN of old men showed that *STMN1* and *LRRFIP2* are inhibited by mir-210 and mir-214, respectively, and DNA methylation of *SMAD4* and *LEF1*, which may lead to dysregulation of the MAPK signaling and Wnt signaling pathways as well as deregulation of the cell cycle and apoptosis, thus resulting in cancer. The results obtained from the present study provides a platform for useful medical therapy to facilitate the development of new antiaging drugs.

REFERENCES

[1] Troen BR. The biology of aging. Mt Sinai J Med 2003;70(1):3–22.
[2] Chobotova K. Aging and cancer: converging routes to disease prevention. Integr Cancer Ther 2009;8(2):115–22.
[3] Campisi J, Yaswen P. Aging and cancer cell biology, 2009. Aging Cell 2009;8(3):221–5.
[4] Ueki K. The mechanisms whereby insulin/IGF-1 signaling regulates aging and longevity. Nippon Rinsho 2009;67(7):1315–20.
[5] Bruunsgaard H. The clinical impact of systemic low-level inflammation in elderly populations. With special reference to cardiovascular disease, dementia and mortality. Dan Med Bull 2006;53 (3):285–309.
[6] Vasto S, Candore G, Listi F, Balistreri CR, Colonna-Romano G, Malavolta M, et al. Inflammation, genes and zinc in Alzheimer's disease. Brain Res Rev 2008;58(1):96–105.
[7] Feng Z, Hu W, Teresky AK, Hernando E, Cordon-Cardo C, Levine AJ. Declining p53 function in the aging process: a possible mechanism for the increased tumor incidence in older populations. Proc Natl Acad Sci USA 2007;104(42):16633–8.
[8] Tyson JJ. Another turn for p53. Mol Syst Biol 2006;2 0032.
[9] Aranda-Anzaldo A, Dent MA. Reassessing the role of p53 in cancer and ageing from an evolutionary perspective. Mech Ageing Dev 2007;128(4):293–302.

[10] Saunders LR, Verdin E. Sirtuins: critical regulators at the crossroads between cancer and aging. Oncogene 2007;26(37):5489−504.
[11] Sherr CJ. Tumor surveillance via the ARF-p53 pathway. Genes Dev 1998;12(19):2984−91.
[12] Kim HJ, Jung KJ, Yu BP, Cho CG, Choi JS, Chung HY. Modulation of redox-sensitive transcription factors by calorie restriction during aging. Mech Ageing Dev 2002;123(12):1589−95.
[13] Chung HY, Sung B, Jung KJ, Zou Y, Yu BP. The molecular inflammatory process in aging. Antioxid Redox Signal 2006;8(3−4):572−81.
[14] Lenardo MJ, Baltimore D. NF-kappa B: a pleiotropic mediator of inducible and tissue-specific gene control. Cell 1989;58(2):227−9.
[15] Ghosh S, May MJ, Kopp EB. NF-kappa B and Rel proteins: evolutionarily conserved mediators of immune responses. Annu Rev Immunol 1998;16:225−60.
[16] Liu ZG, Hsu H, Goeddel DV, Karin M. Dissection of TNF receptor 1 effector functions: JNK activation is not linked to apoptosis while NF-kappaB activation prevents cell death. Cell 1996;87(3):565−76.
[17] Van Antwerp DJ, Martin SJ, Kafri T, Green DR, Verma IM. Suppression of TNF-alpha-induced apoptosis by NF-kappaB. Science 1996;274(5288):787−9.
[18] Wang CY, Mayo MW, Baldwin Jr AS. TNF- and cancer therapy-induced apoptosis: potentiation by inhibition of NF-kappaB. Science 1996;274(5288):784−7.
[19] Tsukahara T, Kannagi M, Ohashi T, Kato H, Arai M, Nunez G, et al. Induction of Bcl-x(L) expression by human T-cell leukemia virus type 1 Tax through NF-kappaB in apoptosis-resistant T-cell transfectants with Tax. J Virol 1999;73(10):7981−7.
[20] Kreuz S, Siegmund D, Scheurich P, Wajant H. NF-kappaB inducers upregulate cFLIP, a cycloheximide-sensitive inhibitor of death receptor signaling. Mol Cell Biol 2001;21(12):3964−73.
[21] Micheau O, Lens S, Gaide O, Alevizopoulos K, Tschopp J. NF-kappaB signals induce the expression of c-FLIP. Mol Cell Biol 2001;21(16):5299−305.
[22] Wang CY, Mayo MW, Korneluk RG, Goeddel DV, Baldwin Jr AS. NF-kappaB antiapoptosis: induction of TRAF1 and TRAF2 and c-IAP1 and c-IAP2 to suppress caspase-8 activation. Science 1998;281(5383):1680−3.
[23] Stehlik C, de Martin R, Kumabashiri I, Schmid JA, Binder BR, Lipp J. Nuclear factor (NF)-kappaB-regulated X-chromosome-linked iap gene expression protects endothelial cells from tumor necrosis factor alpha-induced apoptosis. J Exp Med 1998;188(1):211−16.
[24] Papa S, Zazzeroni F, Pham CG, Bubici C, Franzoso G. Linking JNK signaling to NF-kappaB: a key to survival. J Cell Sci 2004;117(Pt 22):5197−208.
[25] Nakano H, Nakajima A, Sakon-Komazawa S, Piao JH, Xue X, Okumura K. Reactive oxygen species mediate crosstalk between NF-kappaB and JNK. Cell Death Differ 2006;13(5):730−7.
[26] Modugno F, Ness RB, Chen C, Weiss NS. Inflammation and endometrial cancer: a hypothesis. Cancer Epidemiol Biomarkers Prev 2005;14(12):2840−7.
[27] Judge S, Jang YM, Smith A, Hagen T, Leeuwenburgh C. Age-associated increases in oxidative stress and antioxidant enzyme activities in cardiac interfibrillar mitochondria: implications for the mitochondrial theory of aging. FASEB J 2005;19(3):419−21.
[28] Yu BP. Calorie restriction as a potent anti-aging intervention: suppression of oxidative stress. Aging interventions and therapies. Singapore: World Scientific Publishing; 2005.
[29] Boyd SP. Linear matrix inequalities in system and control theory. Philadelphia, PA: Society for Industrial and Applied Mathematics; 1994.
[30] Doyle JC, Tannenbaum A. Feedback control theory. New York, NY: Macmillan Publishing Company; 1992.
[31] Kitano H. Biological robustness. Nat Rev Genet 2004;5(11):826−37.
[32] Krantz M, Ahmadpour D, Ottosson LG, Warringer J, Waltermann C, Nordlander B, et al. Robustness and fragility in the yeast high osmolarity glycerol (HOG) signal-transduction pathway. Mol Syst Biol 2009;5:281.
[33] Csete ME, Doyle JC. Reverse engineering of biological complexity. Science 2002;295(5560):1664−9.

[34] Savageau MA. Demand theory of gene regulation. I. Quantitative development of the theory. Genetics 1998;149(4):1665–76.
[35] Alon U. Biological networks: the tinkerer as an engineer. Science 2003;301(5641):1866–7.
[36] Tu CT, Chen BS. New measurement methods of network robustness and response ability via microarray data. PLoS One 2013;8(1):e55230.
[37] Zahn JM, Poosala S, Owen AB, Ingram DK, Lustig A, Carter A, et al. AGEMAP: a gene expression database for aging in mice. PLoS Genet 2007;3(11):e201.
[38] Cheadle C, Vawter MP, Freed WJ, Becker KG. Analysis of microarray data using Z score transformation. J Mol Diagn 2003;5(2):73–81.
[39] Lin LH, Lee HC, Li WH, Chen BS. Dynamic modeling of cis-regulatory circuits and gene expression prediction via cross-gene identification. BMC Bioinformatics 2005;6:258.
[40] Chen HC, Lee HC, Lin TY, Li WH, Chen BS. Quantitative characterization of the transcriptional regulatory network in the yeast cell cycle. Bioinformatics 2004;20(12):1914–27.
[41] Chen BS, Yang SK, Lan CY, Chuang YJ. A systems biology approach to construct the gene regulatory network of systemic inflammation via microarray and databases mining. BMC Med Genomics 2008;1:46.
[42] Johansson R. System modeling and identification. Englewood Cliffs, NJ: Prentice Hall; 1993.
[43] Tegner J, Yeung MK, Hasty J, Collins JJ. Reverse engineering gene networks: integrating genetic perturbations with dynamical modeling. Proc Natl Acad Sci USA 2003;100(10): 5944–9.
[44] D'Haeseleer P, Liang S, Somogyi R. Genetic network inference: from co-expression clustering to reverse engineering. Bioinformatics 2000;16(8):707–26.
[45] Kuo BC, Golnaraghi F. Automatic control systems. New York, NY: John Wiley & Sons Inc.; 2009.
[46] Chen BS, Zhang W. Stochastic H_2/H_∞ control with state-dependent noise. IEEE Trans Autom Control 2004;49(1):45–57.
[47] Zhang W, Chen BS. State feedback H_∞ control for a class of nonlinear stochastic systems. SIAM J Control Optim 2006;44(6):1973–91.
[48] Lin CL, Lai CC, Huang TH. A neural network for linear matrix inequality problems. IEEE Trans Neural Network 2000;11(5):1078–92.
[49] Burgering BM, Medema RH. Decisions on life and death: FOXO Forkhead transcription factors are in command when PKB/Akt is off duty. J Leukoc Biol 2003;73(6):689–701.
[50] Hu MC, Lee DF, Xia W, Golfman LS, Ou-Yang F, Yang JY, et al. IkappaB kinase promotes tumorigenesis through inhibition of forkhead FOXO3a. Cell 2004;117(2):225–37.
[51] Cai D, Frantz JD, Tawa Jr. NE, Melendez PA, Oh BC, Lidov HG, et al. IKKbeta/NF-kappaB activation causes severe muscle wasting in mice. Cell 2004;119(2):285–98.
[52] Chen J, Zhou Y, Mueller-Steiner S, Chen LF, Kwon H, Yi S, et al. SIRT1 protects against microglia-dependent amyloid-beta toxicity through inhibiting NF-kappaB signaling. J Biol Chem 2005;280(48):40364–74.
[53] Tergaonkar V, Pando M, Vafa O, Wahl G, Verma I. p53 stabilization is decreased upon NFkappaB activation: a role for NFkappaB in acquisition of resistance to chemotherapy. Cancer Cell 2002;1 (5):493–503.
[54] Toledo F, Wahl GM. Regulating the p53 pathway: in vitro hypotheses, in vivo veritas. Nat Rev Cancer 2006;6(12):909–23.
[55] Harris SL, Levine AJ. The p53 pathway: positive and negative feedback loops. Oncogene 2005;24 (17):2899–908.
[56] Sherr CJ. Divorcing ARF and p53: an unsettled case. Nat Rev Cancer 2006;6(9):663–73.
[57] Powers JT, Hong S, Mayhew CN, Rogers PM, Knudsen ES, Johnson DG. E2F1 uses the ATM signaling pathway to induce p53 and Chk2 phosphorylation and apoptosis. Mol Cancer Res 2004;2 (4):203–14.
[58] Zhou BP, Liao Y, Xia W, Zou Y, Spohn B, Hung MC. HER-2/neu induces p53 ubiquitination via Akt-mediated MDM2 phosphorylation. Nat Cell Biol 2001;3(11):973–82.
[59] Giannakou ME, Partridge L. The interaction between FOXO and SIRT1: tipping the balance towards survival. Trends Cell Biol 2004;14(8):408–12.

[60] Luo J, Nikolaev AY, Imai S, Chen D, Su F, Shiloh A, et al. Negative control of p53 by Sir2alpha promotes cell survival under stress. Cell 2001;107(2):137–48.
[61] Chen WY, Wang DH, Yen RC, Luo J, Gu W, Baylin SB. Tumor suppressor HIC1 directly regulates SIRT1 to modulate p53-dependent DNA-damage responses. Cell 2005;123(3):437–48.
[62] Lenski RE, Barrick JE, Ofria C. Balancing robustness and evolvability. PLoS Biol 2006;4(12):e428.
[63] de Visser JA, Hermisson J, Wagner GP, Ancel Meyers L, Bagheri-Chaichian H, Blanchard JL, et al. Perspective: evolution and detection of genetic robustness. Evolution 2003;57(9):1959–72.
[64] Charlesworth B. Patterns of age-specific means and genetic variances of mortality rates predicted by the mutation-accumulation theory of ageing. J Theor Biol 2001;210(1):47–65.
[65] Reynolds RM, Temiyasathit S, Reedy MM, Ruedi EA, Drnevich JM, Leips J, et al. Age specificity of inbreeding load in *Drosophila melanogaster* and implications for the evolution of late-life mortality plateaus. Genetics 2007;177(1):587–95.
[66] Harman D. The aging process. Proc Natl Acad Sci USA 1981;78(11):7124–8.
[67] Modur V, Nagarajan R, Evers BM, Milbrandt J. FOXO proteins regulate tumor necrosis factor-related apoptosis inducing ligand expression. Implications for PTEN mutation in prostate cancer. J Biol Chem 2002;277(49):47928–37.
[68] Ramaswamy S, Nakamura N, Sansal I, Bergeron L, Sellers WR. A novel mechanism of gene regulation and tumor suppression by the transcription factor FKHR. Cancer Cell 2002;2(1):81–91.
[69] Furukawa-Hibi Y, Yoshida-Araki K, Ohta T, Ikeda K, Motoyama N. FOXO forkhead transcription factors induce G(2)-M checkpoint in response to oxidative stress. J Biol Chem 2002;277 (30):26729–32.
[70] Kops GJ, Dansen TB, Polderman PE, Saarloos I, Wirtz KW, Coffer PJ, et al. Forkhead transcription factor FOXO3a protects quiescent cells from oxidative stress. Nature 2002;419(6904):316–21.
[71] Sethi G, Sung B, Aggarwal BB. Nuclear factor-kappaB activation: from bench to bedside. Exp Biol Med (Maywood) 2008;233(1):21–31.
[72] Swindell WR. Genes regulated by caloric restriction have unique roles within transcriptional networks. Mech Ageing Dev 2008;129(10):580–92.
[73] Helenius M, Hanninen M, Lehtinen SK, Salminen A. Aging-induced up-regulation of nuclear binding activities of oxidative stress responsive NF-kB transcription factor in mouse cardiac muscle. J Mol Cell Cardiol 1996;28(3):487–98.
[74] Donato AJ, Eskurza I, Silver AE, Levy AS, Pierce GL, Gates PE, et al. Direct evidence of endothelial oxidative stress with aging in humans: relation to impaired endothelium-dependent dilation and upregulation of nuclear factor-kappaB. Circ Res 2007;100(11):1659–66.
[75] Helenius M, Kyrylenko S, Vehvilainen P, Salminen A. Characterization of aging-associated up-regulation of constitutive nuclear factor-kappa B binding activity. Antioxid Redox Signal 2001;3 (1):147–56.
[76] Csiszar A, Wang M, Lakatta EG, Ungvari Z. Inflammation and endothelial dysfunction during aging: role of NF-kappaB. J Appl Physiol 2008;105(4):1333–41.
[77] Gosselin K, Abbadie C. Involvement of Rel/NF-kappa B transcription factors in senescence. Exp Gerontol 2003;38(11–12):1271–83.
[78] Jobin C, Bradham CA, Russo MP, Juma B, Narula AS, Brenner DA, et al. Curcumin blocks cytokine-mediated NF-kappa B activation and proinflammatory gene expression by inhibiting inhibitory factor I-kappa B kinase activity. J Immunol 1999;163(6):3474–83.
[79] Guttridge DC, Albanese C, Reuther JY, Pestell RG, Baldwin Jr AS. NF-kappaB controls cell growth and differentiation through transcriptional regulation of cyclin D1. Mol Cell Biol 1999;19 (8):5785–99.
[80] Richmond A. Nf-kappa B, chemokine gene transcription and tumour growth. Nat Rev Immunol 2002;2(9):664–74.
[81] Mukhopadhyay A, Banerjee S, Stafford LJ, Xia C, Liu M, Aggarwal BB. Curcumin-induced suppression of cell proliferation correlates with down-regulation of cyclin D1 expression and CDK4-mediated retinoblastoma protein phosphorylation. Oncogene 2002;21(57):8852–61.
[82] Vousden KH, Lu X. Live or let die: the cell's response to p53. Nat Rev Cancer 2002;2 (8):594–604.

[83] Macleod KF, Sherry N, Hannon G, Beach D, Tokino T, Kinzler K, et al. p53-dependent and independent expression of p21 during cell growth, differentiation, and DNA damage. Genes Dev 1995;9(8):935–44.
[84] Sherr CJ, Roberts JM. CDK inhibitors: positive and negative regulators of G1-phase progression. Genes Dev 1999;13(12):1501–12.
[85] Gartel AL, Radhakrishnan SK. Lost in transcription: p21 repression, mechanisms, and consequences. Cancer Res 2005;65(10):3980–5.
[86] Abramovitch S, Glaser T, Ouchi T, Werner H. BRCA1-Sp1 interactions in transcriptional regulation of the IGF-IR gene. FEBS Lett 2003;541(1–3):149–54.
[87] Werner H, Karnieli E, Rauscher FJ, LeRoith D. Wild-type and mutant p53 differentially regulate transcription of the insulin-like growth factor I receptor gene. Proc Natl Acad Sci USA 1996;93(16):8318–23.
[88] Werner H, Shalita-Chesner M, Abramovitch S, Idelman G, Shaharabani-Gargir L, Glaser T. Regulation of the insulin-like growth factor-I receptor gene by oncogenes and antioncogenes: implications in human cancer. Mol Genet Metab 2000;71(1–2):315–20.
[89] van Heemst D, Beekman M, Mooijaart SP, Heijmans BT, Brandt BW, Zwaan BJ, et al. Reduced insulin/IGF-1 signalling and human longevity. Aging Cell 2005;4(2):79–85.
[90] Poland GA, Ovsyannikova IG, Kennedy RB, Lambert ND, Kirkland JL. A systems biology approach to the effect of aging, immunosenescence and vaccine response. Curr Opin Immunol 2014;29:62–8.
[91] Jonker MJ, Melis JP, Kuiper RV, van der Hoeven TV, Wackers PF, Robinson J, et al. Life spanning murine gene expression profiles in relation to chronological and pathological aging in multiple organs. Aging Cell 2013;12(5):901–9.
[92] Lund J, Tedesco P, Duke K, Wang J, Kim SK, Johnson TE. Transcriptional profile of aging in *C. elegans*. Curr Biol 2002;12(18):1566–73.
[93] McCarroll SA, Murphy CT, Zou S, Pletcher SD, Chin CS, Jan YN, et al. Comparing genomic expression patterns across species identifies shared transcriptional profile in aging. Nat Genet 2004;36(2):197–204.
[94] Zahn JM, Sonu R, Vogel H, Crane E, Mazan-Mamczarz K, Rabkin R, et al. Transcriptional profiling of aging in human muscle reveals a common aging signature. PLoS Genet 2006;2(7):e115.
[95] Jurk D, Wilson C, Passos JF, Oakley F, Correia-Melo C, Greaves L, et al. Chronic inflammation induces telomere dysfunction and accelerates ageing in mice. Nat Commun 2014;2:4172.
[96] Inukai S, Slack F. MicroRNAs and the genetic network in aging. J Mol Biol 2013;425(19):3601–8.
[97] Krek A, Grun D, Poy MN, Wolf R, Rosenberg L, Epstein EJ, et al. Combinatorial microRNA target predictions. Nat Genet 2005;37(5):495–500.
[98] Liang R, Bates DJ, Wang E. Epigenetic control of microRNA expression and aging. Curr Genomics 2009;10(3):184–93.
[99] Yang J, Chen D, He Y, Melendez A, Feng Z, Hong Q, et al. MiR-34 modulates *Caenorhabditis elegans* lifespan via repressing the autophagy gene atg9. Age (Dordr) 2013;35(1):11–22.
[100] Jung M, Pfeifer GP. Aging and DNA methylation. BMC Biol 2015;13:7.
[101] Zampieri M, Ciccarone F, Calabrese R, Franceschi C, Burkle A, Caiafa P. Reconfiguration of DNA methylation in aging. Mech Ageing Dev 2015;151:60–70.
[102] Jones MJ, Goodman SJ, Kobor MS. DNA methylation and healthy human aging. Aging Cell 2015;14(6):924–32.
[103] Stelzl U, Worm U, Lalowski M, Haenig C, Brembeck FH, Goehler H, et al. A human protein-protein interaction network: a resource for annotating the proteome. Cell 2005;122(6):957–68.
[104] Rual JF, Venkatesan K, Hao T, Hirozane-Kishikawa T, Dricot A, Li N, et al. Towards a proteome-scale map of the human protein-protein interaction network. Nature 2005;437(7062):1173–8.
[105] Wang YC, Lin C, Chuang MT, Hsieh WP, Lan CY, Chuang YJ, et al. Interspecies protein-protein interaction network construction for characterization of host-pathogen interactions: a *Candida albicans*-zebrafish interaction study. BMC Syst Biol 2013;7:79.

[106] Sikora-Wohlfeld W, Ackermann M, Christodoulou EG, Singaravelu K, Beyer A. Assessing computational methods for transcription factor target gene identification based on ChIP-seq data. PLoS Comput Biol 2013;9(11):e1003342.

[107] Lee TI, Rinaldi NJ, Robert F, Odom DT, Bar-Joseph Z, Gerber GK, et al. Transcriptional regulatory networks in *Saccharomyces cerevisiae*. Science 2002;298(5594):799–804.

[108] Tu CT, Chen BS. On the increase in network robustness and decrease in network response ability during the aging process: a systems biology approach via microarray data. IEEE/ACM Trans Comput Biol Bioinform 2013;10(2):468–80.

[109] Marttila S, Jylhava J, Nevalainen T, Nykter M, Jylha M, Hervonen A, et al. Transcriptional analysis reveals gender-specific changes in the aging of the human immune system. PLoS One 2013;8(6):e66229.

[110] Stark C, Breitkreutz BJ, Reguly T, Boucher L, Breitkreutz A, Tyers M. BioGRID: a general repository for interaction datasets. Nucleic Acids Res 2006;34(Database issue):D535–539.

[111] Bovolenta LA, Acencio ML, Lemke N. HTRIdb: an open-access database for experimentally verified human transcriptional regulation interactions. BMC Genomics 2012;13:405.

[112] Zheng G, Tu K, Yang Q, Xiong Y, Wei C, Xie L, et al. ITFP: an integrated platform of mammalian transcription factors. Bioinformatics 2008;24(20):2416–17.

[113] Hsu SD, Tseng YT, Shrestha S, Lin YL, Khaleel A, Chou CH, et al. miRTarBase update 2014: an information resource for experimentally validated miRNA-target interactions. Nucleic Acids Res 2014;42(Database issue):D78–85.

[114] Pagano M, Gauvreau K. Principles of biostatistics. 2nd ed. Pacific Grove, CA: Duxbury; 2000.

[115] Chen BS, Tsai KW, Li CW. Using nonlinear stochastic evolutionary game strategy to model an evolutionary biological network of organ carcinogenesis under a natural selection scheme. Evol Bioinform 2015;11:155–78.

[116] Crow JF. The high spontaneous mutation rate: is it a health risk? Proc Natl Acad Sci USA 1997;94(16):8380–6.

[117] Thomas GH. High male:female ratio of germ-line mutations: an alternative explanation for postulated gestational lethality in males in X-linked dominant disorders. Am J Hum Genet 1996;58(6):1364–8.

[118] Kim EK, Choi EJ. Pathological roles of MAPK signaling pathways in human diseases. Biochim Biophys Acta 2010;1802(4):396–405.

[119] Huse M. The T-cell-receptor signaling network. J Cell Sci 2009;122(Pt 9):1269–73.

[120] Cantrell DA. T-cell antigen receptor signal transduction. Immunology 2002;105(4):369–74.

[121] Reichardt LF. Neurotrophin-regulated signalling pathways. Philos Trans R Soc Lond B Biol Sci 2006;361(1473):1545–64.

[122] Rane SG, Cosenza SC, Mettus RV, Reddy EP. Germ line transmission of the Cdk4(R24C) mutation facilitates tumorigenesis and escape from cellular senescence. Mol Cell Biol 2002;22(2):644–56.

[123] Hauck F, Randriamampita C, Martin E, Gerart S, Lambert N, Lim A, et al. Primary T-cell immunodeficiency with immunodysregulation caused by autosomal recessive LCK deficiency. J Allergy Clin Immunol 2012;130(5):1144–52.e1111.

[124] Huang WY, Hsu SD, Huang HY, Sun YM, Chou CH, Weng SL, et al. MethHC: a database of DNA methylation and gene expression in human cancer. Nucleic Acids Res 2015;43(Database issue):D856–861.

[125] de Magalhaes JP, Toussaint O. GenAge: a genomic and proteomic network map of human ageing. FEBS Lett 2004;571(1–3):243–7.

[126] Zhang W, Liu HT. MAPK signal pathways in the regulation of cell proliferation in mammalian cells. Cell Res 2002;12(1):9–18.

[127] Takeda K, Akira S. TLR signaling pathways. Semin Immunol 2004;16(1):3–9.

[128] Akira S, Takeda K. Toll-like receptor signalling. Nat Rev Immunol 2004;4(7):499–511.

[129] Burnichon N, Cascon A, Schiavi F, Morales NP, Comino-Mendez I, Abermil N, et al. MAX mutations cause hereditary and sporadic pheochromocytoma and paraganglioma. Clin Cancer Res 2012;18(10):2828–37.

[130] Je EM, Kim SS, Yoo NJ, Lee SH. Mutational and expressional analyses of MYD88 gene in common solid cancers. Tumori 2012;98(5):663–9.
[131] Tanoue T, Nishida E. Docking interactions in the mitogen-activated protein kinase cascades. Pharmacol Ther 2002;93(2–3):193–202.
[132] Rao TP, Kuhl M. An updated overview on Wnt signaling pathways: a prelude for more. Circ Res 2010;106(12):1798–806.
[133] Clevers H, Nusse R. Wnt/beta-catenin signaling and disease. Cell 2012;149(6):1192–205.
[134] Blackford A, Serrano OK, Wolfgang CL, Parmigiani G, Jones S, Zhang X, et al. SMAD4 gene mutations are associated with poor prognosis in pancreatic cancer. Clin Cancer Res 2009;15(14):4674–9.
[135] Houlston R, Bevan S, Williams A, Young J, Dunlop M, Rozen P, et al. Mutations in DPC4 (SMAD4) cause juvenile polyposis syndrome, but only account for a minority of cases. Hum Mol Genet 1998;7(12):1907–12.
[136] Iyer NK, Burke CA, Leach BH, Parambil JG. SMAD4 mutation and the combined syndrome of juvenile polyposis syndrome and hereditary haemorrhagic telangiectasia. Thorax 2010;65(8):745–6.
[137] Takeda H, Lyle S, Lazar AJ, Zouboulis CC, Smyth I, Watt FM. Human sebaceous tumors harbor inactivating mutations in LEF1. Nat Med 2006;12(4):395–7.
[138] Aulmann S, Blaker H, Penzel R, Rieker RJ, Otto HF, Sinn HP. CTCF gene mutations in invasive ductal breast cancer. Breast Cancer Res Treat 2003;80(3):347–52.
[139] Filippova GN, Qi CF, Ulmer JE, Moore JM, Ward MD, Hu YJ, et al. Tumor-associated zinc finger mutations in the CTCF transcription factor selectively alter tts DNA-binding specificity. Cancer Res 2002;62(1):48–52.
[140] Salzer E, Daschkey S, Choo S, Gombert M, Santos-Valente E, Ginzel S, et al. Combined immunodeficiency with life-threatening EBV-associated lymphoproliferative disorder in patients lacking functional CD27. Haematologica 2013;98(3):473–8.
[141] Houtkooper RH, Argmann C, Houten SM, Canto C, Jeninga EH, Andreux PA, et al. The metabolic footprint of aging in mice. Sci Rep 2011;1:134.

CHAPTER 12

Big Drug Design Mechanisms via Systems Biology and Big Database Mining

INTRODUCTION

In the past three decades, medicine has sought a "magic bullet" that specifically targets a single disease-causing molecule and that could cure cancer, diabetes, and other complex diseases. Although some drugs have proven successful, many others have been found to be ineffective or the cause of significant side effects. This disappointing outcome highlights the limitations of the single target-drug paradigm that is considered to be the cause of stagnation in productivity in the pharmaceutical industry. However, compound efficacy and safety in humans, rather than target selection, are the criteria that determine which drug candidates enter the clinic. In Section 12.1, drug discovery using big database mining and systems biology for drug development is introduced. Systems biology approaches to disease begin with the idea that disease-perturbed regulatory networks differ from their normal counterparts. Microarray data analyses reveal global changes in gene or protein expression in response to genetic and environmental changes, and, accordingly, are well suited to construct normal, disease-perturbed, and drug-affected networks that are useful for drug discovery in the pharmaceutical industry. The integration of modeling, microarray data, database mining, and systems biology approaches will allow for a true breakthrough in silico absorption, distribution, metabolism, excretion, and toxicity assessments in drug design. Therefore, drug discovery through big database mining and systems biology by means of microarray analyses can change the approach to drug design and significantly reduce the time and cost of new drug development.

Because of the importance of cancer in public health, considerable efforts are invested in understanding tumorigenesis, in order to develop new therapeutic molecules and to ameliorate disease. Historically, the first therapeutic molecules for cancer treatment targeted fast replicating cells with no specificity for a particular target gene or protein and exerting their action on DNA, as nitrogen mustards did. The discovery of the so-called oncogenes and tumor suppressor genes in the 1980s, followed by the identification of their signaling pathways revived the old theory of a "magic bullet." According to this paradigm, specific receptors associated with cancer cells can be

Big Mechanisms in Systems Biology
DOI: http://dx.doi.org/10.1016/B978-0-12-809479-2.00012-3

targeted with a drug that binds to the receptor in order to kill cancer cells. However, cancers are complex and multidimensional in nature, involving many molecules acting in interconnected pathways in a large and complex network. Targeting only one molecule is likely to open many bypass roads in the regulatory network and feedback loops that will compensate for the effects of the therapeutic molecule. To remedy this, in Section 12.2, we introduce a systems biology method to find specific, core network markers of cancers for multiple drug targets. This encourages the use of combinations of specific drugs or multitarget inhibitors. Since the behavior of complex biological networks involves crosstalk and feedback loops, systematic analysis of the robustness and transductivity of complex networks is necessary to find the weak points. This will be essential to searching for drug targets and designing new drugs.

In Section 12.3, based on the multiple drug targets of the specific network markers investigated in Section 12.2, a method for systems drug design is introduced. A systems drug design for multiple targets is developed and several methods to discover drug targets for multitarget cocktail therapies are introduced.

12.1 OVERVIEW OF DRUG DISCOVERY USING SYSTEMS BIOLOGY

12.1.1 Introduction

Many of the drug candidates that fail in clinical trials are withdrawn because of unforeseen effects of human metabolism, such as toxicity and unfavorable pharmacokinetic profiles. Early preclinical estimation of such compounds is important but not yet possible. Currently available preclinical testing systems are split into three categories: (1) structure–activity relationship (SAR) computational models based on compound structure; (2) "pattern" databases of tissue or organ response to drug, compiled from high-throughput experiments; and (3) "systems biology" databases of metabolic pathways, genes, and regulatory networks. Compounds at different stages of preclinical testing also undergo in silico ADME/TOX evolution, varying from straightforward solubility and reactivity filtering to the advanced modeling for brain–blood barrier permeability and compound metabolism. The ultimate goal is to be able to predict human metabolism, toxicity and pharmacokinetics based solely on the structure formula of novel compounds. Such a preclinical testing system would allow the elimination of unfavorable compounds early in development, at the level of lead compound section and, ultimately, initial discovery chemistry used in bioarrays.

It is believed that the creation of comprehensive and reliable in silico systems for the evaluation of human metabolism and relevant toxicity is technically feasible and that such systems will be developed in near future. Such systems include components as follows: (1) there is the computational modeling of metabolism and toxicity based on chemical structure of lead compounds. These models use vast set of experimental data (toxicity end points, metabolic products and intermediates) for elucidation of empirical rules describing the response to xenobiotics. These data are used for building SARs within groups of similar compounds and mathematical model (such as neural networks and nonlinear regressions) are built for effective separation of compounds. (2) There are powerful high-throughput experimental technologies that catalog whole genomes, tissues, and organ responses to external stimuli. (3) There are databases of cellular functional pathways and enzymatic reactions, assembled from experimental literature. These systems biology databases cover information such as metabolic pathways and networks, eukaryotic signal transduction, transcriptional regulation, membrane receptors, and protein interactions.

Because a metabolism, toxicity and pharmacokinetic profile is a system-level problem, the solution will require interaction of these technologies into one comprehensive environment, connecting the relevant chemistry with key cellular metabolic pathways. In general, none of these testing technologies is sufficient for the task of predicting ADME/TOX behavior based solely on chemical structure. For example, empirical structure-based models do not deal with the underlying mechanisms of toxicity and mechanism. Genomics-based testing systems are based on the assumption that novel compounds have similar toxicity to a reference compound with a similar profile; in reality, this is true only for structurally similar molecules and the techniques therefore have limited predictive value. Furthermore, the pathways do not reflect spatial or temporal patterns of pathway activity in response to lead compounds. Systems biology has come to mean a more integrated approach to drug discovery where the effect of a lead compound on whole pathways and networks of pathways, not just the effects on isolated proteins, are evaluated process. What is needed is a platform that integrates different types of in silico modeling with high-throughput molecular assays and that enables the construction of system-level picture of the behavior of a drug.

The completion of listing the human genome in 2003 marked the beginning of a new era of biomedical research. In this new era, scientists are expected to answer at least the following questions: (1) What are the functions of the 25,000 human genes identified or predicted? What are the genes that are related to a particular disease? (2) How do these genes control these metabolic or developmental processes? In other words, what are the roles of these genes in various regulatory or protein interaction networks? (3) What novel diagnostic techniques can be derived from knowledge derived from (1) and (2)? (4) What novel therapies and new drugs can be developed based on knowledge from (1) and (2)? Obviously, these questions focus on systems biology-based drug discovery.

The answers to these questions are not only of scientific significance but will also lead to revolutionary developments of molecular diagnostics and therapy and drug discovery. Next-generation sequencing (NGS) data and mRNA time profiles were monitored for all genes in the genome through high-throughput microarray data, and these data could be integrated with protein–protein and protein–DNA interaction data to construct a cellular network from systematic perspective. Recently, there are several network construction methods through microarray data using the Bayesian method to dynamic models. The system module provided new insights into the control of a metabolic process and its interactions with other cellular processes. It also suggested several concepts from systems approaches to human disease.

As protein and gene regulatory networks (GRNs) dynamically changed upon intrinsic and environmental perturbations, the dynamic progress of disease should similarly be reflected in time profile changes from the normal state to various stages of disease-perturbed networks [1–5]. For example, cancer arises from multiple spontaneous and/or inherited mutations functioning in networks that control central cellular events. It is becoming clear that the evolving states of prostate cancer are reflected in dynamically changing expression patterns of the genes and proteins within the diseased cell. Given enough mRNA microarray data, one can presumably identify types of a particular cancer, the various stages in the progression of each disease type, the partition of disease into categories defined by critical therapeutic targets, and the measurement of how drugs alter the disease pattern. The changes in the tissue-specific marker may identify critical points within these perturbed networks that may be affected by drugs, either to convert the diseased network back toward normally or to permit the specific killing of the diseased cell.

The pharmaceutical industry embraced DNA microarray technology when it was first introduced in the mid-1990s [6,7]. This can be explained by the gene-consuming nature of the modern drug discovery process. The cumulative attrition rate of this process is so high that a large number of genes have to get into the discovery pipeline in order to introduce a single new chemical entity on the drug market. Typically, a small molecule discovery pipeline consists of target identification and validation, lead identification, lead optimization, preclinical development, and three phases of clinical development. However, the cumulative success rate amounts to 1.7%. In other words, for every 60 drug targets identified, only one new chemical entity enters the drug market. It is therefore not surprising that high-throughput research techniques are highly welcomed by pharmaceutic industry.

One area where the use of microarrays has an enormous impact is in disease classification and prediction of clinical outcome [8–10]. Usually, the expression of a single marker is not predictive for prognosis of a complex disease, but predictive power greatly improves when a set of genes or proteins is analyzed. In drug target discovery, microarray are used to identify genes and/or proteins that are expressed under relevant (pathological) conditions [8,10]. The dynamics of transcriptome are often studied to unravel signal transduction pathways or gene function. It has been conventionally shown that the dynamics of gene expression contain information about gene function.

12.1.2 Systems biology-based drug discovery

Systems biology tries to understand how a biological system functions, including its dynamics and its robustness against disturbances, and also the interrelationships among different parts of the system. As recent reviews indicate [11–14], systems biology approaches offer novel strategies to shorten the cumbersome path from identified target to an approved drug. Systems biology provides in silico models for cost-effective decision making during drug development. Biological networks such as metabolic networks are probably the best known examples of biological systems. The development of systems biology stems from recent advances in large-scale and high-throughput technologies such as genomic sequencing, microarrays, NGS, and CHIP-chip. Due to these novel technologies, scientists have been able to collect data samples at a rate that was beyond imagination a decade ago. The huge amounts of data produced cannot be understood without a network or system viewpoint and without highly sophisticated computational analyses. As a result, systems biology has become a frontier of modern biology research [7–9,15–18]. Accordingly, new methodologies must be developed to describe and unravel how a biological system works rather than how an individual component in the system functions. In this regard, the biological network is the most widely used approach to unravel all components (parts) of a biological system, to describe the interrelationships among different parts of system (especially the upstream–downstream relationships) and to quantify the dynamics and robustness of the systems. At a high level, several networks can be connected to model the interactions between systems. These discussions lead to challenge of how a quantitative model of a biological system can be systematically derived from the huge amount of data, especially to construct a disease-perturbed dynamic model for drug discovery.

From the perspective of system science, a comprehensive description of a system include two major components: network structure and system dynamics. A good understanding about network structure and dynamic behavior of a biological system then allows us to design novel approaches to adjust the system in case it fails to perform properly due to either malfunction of its components or interference of the outside world. In other words, we can design novel therapies and new drugs based on this understanding of network structure and system dynamics. Therefore, the advances in systems biology research may bring breakthrough of novel therapies and drugs.

A quantitative model of a biological system is to derive a network structure from microarray data or NGS data. The network structure specifies how the components in the biological system are linked together as well as how they interact with each other. Since there are many types of biological systems, there are many kinds of biological networks, including GRNs, protein–protein interaction networks (PPINs), biochemical networks, and genetic interaction networks. The first two types of network currently receive most attention. We shall therefore concentrate on these two types of networks. A GRN is composed of nodes representing genes or DNA segments, with input being proteins such as transcription factors and outputs being the profiles of gene expression. Uncovering regulatory networks helps us to understand the complex cellular process. Cellular elements, including physiological, differentiation, and development events, involve the interplay of the regulatory network.

PPINs refer to the interactions among protein molecules and the study of these interactions from the perspective of biochemistry and networks. Proteins may combine into macromolecular complexes to execute essential tasks such as replication, transcription, protein transport, or metabolic reaction catalysis. In particular, a protein may carry another protein, or a protein may interact briefly with another protein just to modify it. The interactions among proteins are important for many biological functions. Identifying protein complexes and how they share components is an essential step in describing cellular biology on the molecular basis. Further, we need to study system dynamics for GRNs and PPINs. System dynamics aim to answer the following questions: (1) How does a network respond to input stimuli, i.e., pathogens or compounds? (2) How stable is it in the presence of noise and/or disturbance. i.e., xenobiotics? (3) How robustly can it endure any malfunction of a component? (4) How is control realized? These questions are essential for drug discovery from the systems biology perspective.

A good understanding about the network structure and dynamic behavior of a biological system allows us to design novel approaches to adjust the biological system in case it fails to perform properly due to either malfunction of its components or interference from the outside world. In other words, we can design a more robust biological system, or novel therapies and/or new drugs accordingly. In all aspects, systems biology is the natural path for future development of biomedical science. The ultimate objective of biomedical research is to achieve a comprehensive understanding of how these cells, these organs, and even our body function. From the scientific point of view, the most comprehensive and precise description of a system is a quantitative model that accurately predicts how a biological system will behave under all kinds of conditions. Nevertheless, a quantitative system model must be constructed based on a sufficient quantity of observation. Otherwise, the system model would not be realistic. In summary, based on analyzing microarray data and database mining, we can construct a more realistic disease-perturbed biologic network. Then, based on a disease-perturbed biologic network to analyze robustness and sensitivity of system properties, we could have more insight into the mechanisms of toxicity and drug metabolism.

For the past half-century, biologists have been uncovering details of countless molecular events. Furthermore, with largest large-scale and high-throughput technologies and big database mining schemes, scientists have been able to collect data samples at a rate that was beyond imagination a decade age. Therefore, we are currently at the turning point of future development in biomedical sciences and drug design engineering. The development of systems biology not only will revolutionize the understanding of how these cells, organs, and bodies operate, but also will open the door for development of new therapies and design of new drugs. With a realistic quantitative model of a biological system, we can figure out the most effective approach to fix the biological system, in case it fails to perform properly due to either malfunction of its components or xenobiotics from the outside world. For example, a good understanding of a pathway that leads to development of cancer cells may enable one to invent a novel therapy that downregulates or blocks the pathway. In addition, a good understanding of how an antigen interacts with proteins in our body will provide important clues about how to design a more effective drug.

12.1.3 Drug discovery through comparing biological networks estimated by microarray data and big database mining

In this section, we describe computational methods for identifying and validating drug targets based on biological networks estimated from microarray data. Recently, microarray technology has produced a large volume of genome-wide gene expression data and protein microarray data under various experimental conditions such as gene disruptions, gene overexpressions, stresses, toxicities, cancer cells, and host–pathogen interactions. Based on this new data production, there have been considerable attempts to estimate gene networks from microarray data and several computational methods have been proposed together with mathematical models for gene clusters, gene networks, and biochemical networks such as Boolean networks [19,20], dynamic networks [21–24], and Bayesian networks [25–27].

For drug discovery, we need to use the normal and disease-perturbed microarray data to construct normal and disease-perturbed biological networks. By comparing the difference between normal and disease-perturbed biological networks, we could screen the targets of drug. Based on database mining, we can target a three-dimensional structure and pharmacophore model such as ligand-based binding method and structure-based binding method, leading to compounds to drug candidates developed by docking methods. After the drug candidates are selected, the validations of drug action on target genes and antitarget-mediated side effects could also be obtained through biological networks estimated by microarray data, e.g., the antitarget and ADME/TOX screening for bioactivity profiling and early stage risk assessment of potential side effects and toxicity using microarray data and database mining. The validation of drug action and side effects could result in feedback for optimizing lead compounds to clinical candidates. Based on the analyses above, the analyses of microarray data for construction of GRNs and PPINs is the first target for drug discovery from the systems biology. These will be discussed below.

Gene clusters (modules identification)

The first phase of microarray data analyses is gene classification. Groups of genes that exhibit similar expression patterns are considered likely to be coregulated [5,28–36]. The expectation is that the clusters identified will point to relevant sites of expression, functional, and regulatory networks, and normal or pathological functions in organs and whole organisms or discover disease subtype (e.g., cancer diagnosis and drug discovery).

The next phase of microarray data analyses requires systematic retrieval of functional information to identify the general biological process associated with lists of selected genes. Unveiling the biological relevance of gene expression profiles is initially performed by reading preexisting biochemical literature. In the last few years, there has been a lot of interest in analytical literature- and text-mining tools that help find the most relevant information. Frequently, information retrieval incorporates keyword-based approaches. Gene Ontology (GO) has been recently developed for annotating genes in various organisms. It constitutes a standardized nomenclature for expressing and sharing biological concepts, and formalizing knowledge about biological processes, molecular functions, and cellular components. This opens new perspectives in functional annotation to quantify the difference between terms associated with gene clusters by scoring them according to the functional coherence of corresponding gene groups. Recent studies reported that GO functional groups can be used as a framework for numerical clustering to create more meaning clusters.

At present, most computational analyses of microarray data rely on clustering methods. However, these methods only broadly identify biological functions into subgroups and hide many possible mechanisms. Additionally, a clustering approach is incapable of constructing the signal pathways and quantifying the systematic features of the gene regulatory circuits such as stability, robustness, and sensitivity, which are useful for drug design from systems biology perspective.

Dynamic network reconstruction via microarray analyses

Biological phenomena at different organismic levels have revealed some sophisticated systematic architectures of cellular and physiological activities. These systematic architecture were built upon the biochemical processes before the emergence of proteome and transcriptome. Under the molecular machinery, the biochemical processes are mostly interpreted as frameworks of connectivity between biochemical compounds and proteins, which are synthesized from genes to function as transcription factors binding to regulatory sites of other genes, as enzymes catalyzing metabolic reactions, or as components of signal transduction pathways. This implies that, in order to understand the molecular mechanism of genes in the control of intracellular or intercellular processes, the scope should be broadened from DNA sequences coding for proteins to the systems of GRNs that determine which genes are expressed, when and where in the organism, and to what extent. In the field of engineering, the systematic architecture and dynamic model is used to investigate the characteristics of signal regulatory networks. Similarly, learning how to construct the dynamic model of a biological network via microarray data and database mining is important for systems biology, and might be a key to the door of systems biology-based drug discovery.

A statistical model of Bayesian network [25–27,37–41] was proposed to model GRNs. The technique uses a probabilistic score to evaluate the networks which respect to the expression data and searches for the network with the optimal score. The dynamic Bayesian network [25–27,38,39,41] was proposed to learn the network structure and parameters by maximizing the posterior probability via Bayes rule of a priori probability and marginal likelihood. Another algorithm of Boolean network [19,20,42] can also be employed to model the dynamic evolution of gene expression, where the state of a gene can be simplified to being either active (on, 1) or inactive (off, 0). The probabilistic nature of Bayesian networks is capable of handling noise inherent in both the biological processes and the microarray experiments. This makes Bayesian networks superior to Boolean networks, which is deterministic in nature. The validity of dynamic Bayesian networks is evaluated by the sensitivity:specificity score ratio, which depends on the training size, and the degree of accuracy of prior assumption. Recently, dynamic Bayesian networks have been employed for identifying and validating drug target genes from microarray gene expression data. The gene networks are estimated from disruptant microarray data and drug response microarray data by the Bayesian network model and then are used to identify sets of genes or gene regulatory pathway affected by the drug via virtual gene technique data and database mining for identifying drug-affected genes and exploring druggable genes.

Recently, the dynamic system approaches have been employed to model how a target gene's expression profile is regulated by its upstream regulatory genes from the system causality point of view. These derived regulatory functions are closely related to the causal upstream information of the network and will create a basis for inferring the regulatory network through microarray analysis and system identification from the systems biology point of view. By comparing the normal and drug-affected GRNs constructed by the corresponding microarray data, we could identify gene regulatory pathways affected by drug. Then we could validate drug action on target genes and antitarget-mediated side effects through the kinetic parameters of the dynamic drug-affected gene network, i.e., the antitarget and ADME/TOX screening for bioactivity profiling and early stage risk assessment of potential side effect and toxicity could be performed by dynamic regulatory network via microarray data (Fig. 12.1). The validation of drug action and side effect could be used for optimization of lead compounds to chemical candidates.

However, due to underlying network complexity and large amount of microarray data, we could not identify the GRN of human at one time. Therefore, singular value decomposition or principle component analysis or some intelligent methods are employed to treat the reconstruction of GRNs by neglecting some weak linking

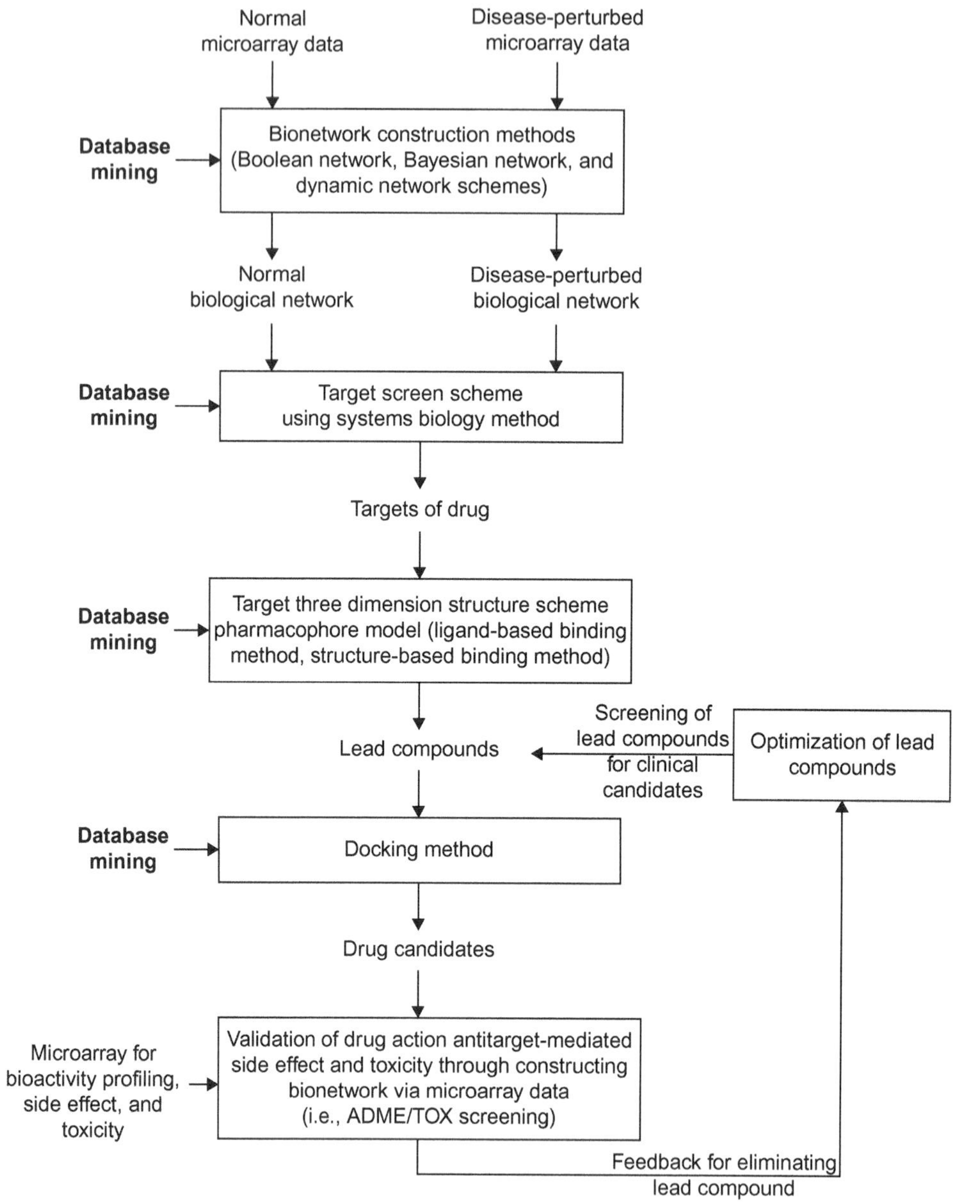

Figure 12.1 Flow chart for drug discovery using systems biology based on microarray data analyses and database mining.

paths. Recently, an iterative construction method was proposed to overcome the difficulty [43]. Using the dynamic equation and microarray data, the regulation function from the upstream regulators is estimated for each target gene. From the estimated regulatory function, we deduce for each target gene its upstream regulators by tracing back upstream signal pathways. Then, upstream regulatory genes are taken as target genes to trace back their upstream regulatory genes. Iteratively, we can construct the whole regulatory network to the genome-wide using the dynamic regulatory model and microarray data from the systems biology point of view (Fig. 12.2). Therefore, the dynamic GRN via microarray data has much potential to drug target discovery and validation.

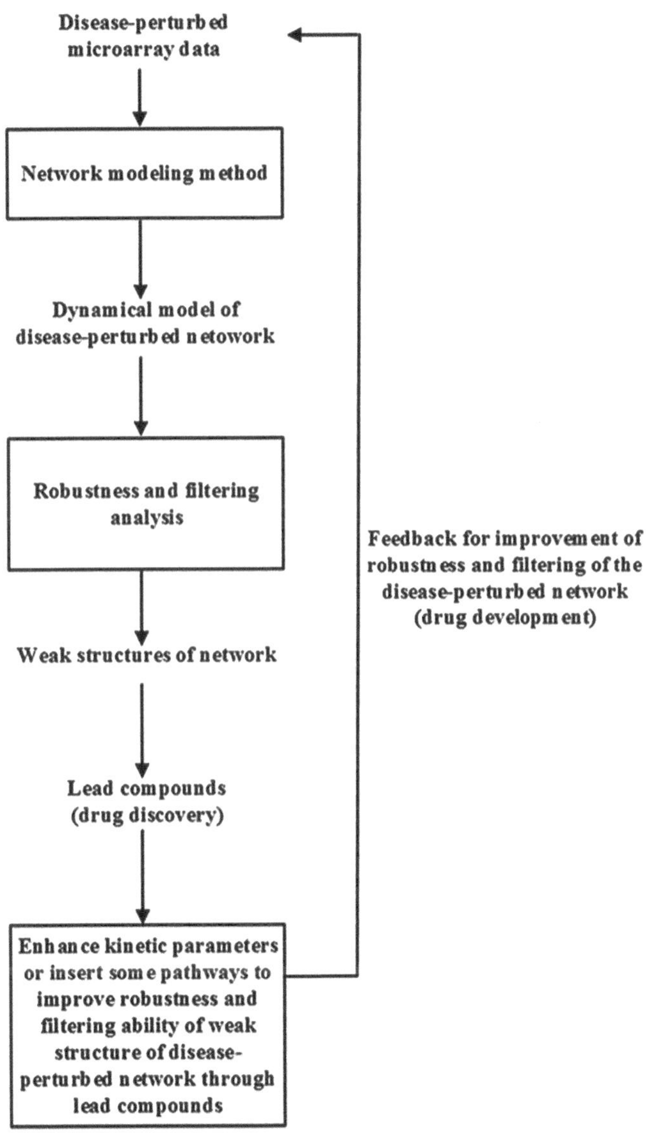

Figure 12.2 Drug discovery based on engineered robust gene circuit design of dynamic gene network constructed by microarray data and system model.

In the next step, we need to relate gene expression patterns with physiology, pathology, or drug design, through the functional network, and regulatory mechanisms in which the relevant genes are involved. Due to underlying network complexity and combinatorial expression in the number of potential network structures, available methods often decompose networks into basic functional and structure units. For instance, biological pathways are networks of relationships between biological entities, which were established steadily through a century of biochemistry and molecular biology. With the advent of genome-scale biology, systematic cataloging and storage of pathways has recently gained increasing importance. A wide variety of resources and tools exists to display pathway information, enabling presentation of microarray data in the framework of documented biological pathways.

As actual knowledge about biological networks is often incomplete, an exact solution will not be determined, but a solution space can be generated, containing all possible cellular functions that are consistent with the information assembled and the biological context. This constitutes a starting point for inferring new knowledge by integrating other "omic" space, based on data to contribute to a systems approach toward understanding biological complexity and to the development of drug designs for novel therapeutics to tackle human diseases.

Drug target discovery often starts with a transcriptomics study to identify genes that show altered expression during, for instance, a disease condition. Similarly, proteomics monitors alterations in the protein content of samples. The lack of correlation between mRNA and protein level is not surprising because both are controlled by individual rates of synthesis and turnover. However, the timing in changes of levels do correlate. This is consistent with the notion that the majority of changes in protein levels result from a change in mRNA level.

12.1.4 Conclusion and discussion

Systems biology describes the computational systems and models that deal with the analyses of vast experimental data, especially microarray data. These computational systems and models have multiple applications in drug discovery, including the validation of drug targets via cross-referencing of experimental data, pharmacogenomics, polymorphism haplotyping, and the understanding of drug metabolism and toxicity.

For drug discovery, tremendous challenges are posited to systems biologists because of the complex nature of biological systems and because of uncertain and noisy nature of large-scale data. Drug design challenges to overcome are being able to construct biological system models based on GRNs and PPINs via microarray data.

Additionally, there are several challenges arising from microarray data analyses such as Chip-Chip data and other data:

1. Design of advanced multivariate analyses algorithms—A microarray experiment typically includes thousands of genes and proteins. Therefore, if the drug discovery is to identify those genes involved in a pathway that leads to the development of cancer cells (or other disease) from microarray data, then we are dealing with a dataset in a vector space with thousands of dimensions. Therefore, design of novel and advanced multivariate analyses algorithms is of great significance.
2. Design of advanced data clustering (unsupervised learning) algorithms—Data clustering algorithms have been widely exploited in microarray data analyses for identifying functionally correlated genes. No data clustering algorithm can provide a complete view of the picture for drug discovery, so design of a novel and advanced data clustering algorithm is an important work.
3. Causality analyses—Causal inference is a critical step in constructing GRNs such as cellular signaling pathways, which need complete determination of interactions among genes, proteins, and metabolites. Ultimately, we are interested in the flow of causality. Such knowledge of the flow information through a cell in response to lead compound help us to understand the underlying mechanism of complex cellular process under lead compounds. Many different models have been suggested and applied to the issue, including differential equations, Boolean networks, and Bayesian networks. They all have strengths and limitations for drug discovery. Recent studies have aimed to infer global networks directly from available whole genome microarray data. Nevertheless, existing approaches often require a large number of time-course data or suffer high computational complexity.
4. Engineering principles—Some theoretical engineering principles that have been developed in the context of human engineered systems are useful in understanding biological functions for drug discovery. For example, dynamic system theory, feedback control design theory, game theory and learning algorithms have been useful for synthetic biology, signaling and communication, and robust circuit design in systems biology, which are useful for drug design.

In addition to the topics addressed earlier for drug discovery, another category of algorithms that have been widely exploited in bioinformatics applications is the data clarification (supervised learning) algorithm. With respect to construction of biological networks, the data clarification algorithm has been exploited in sequences and structure analyses, which could provide scientists with initial clues of the function of a gene or a protein in the drug discovery procedure. Nevertheless, as the sizes of contemporary bioinformatics databases continue to grow at exponential rates, the efficiency of the data clarification algorithm has become a major concern in drug discovery procedure. In this regard, the support vector machine (SVM), which is the prevailing supervised learning algorithm employed in bioinformatics applications, suffers a major deficiency due to its high time complexity at the model selection stage. Therefore, design of highly efficient data classification algorithms to cope with the ever-growing bioinformatics databases is of great importance in the drug design procedure.

Concerning analyses of system dynamics of a biological network, the major challenge arises from the fact that many biological systems are nonlinear. Although nonlinear systems are not new to scientists, recent developments in statistical learning have laid a solid foundation for design of more advanced schemes to model nonlinear systems. For example, in recent years, a novel nonlinear regression algorithm called support vector regression (SVR) has been developed by exploiting the fundamental principles of SVM. SVR delivers higher accuracy than traditional nonlinear regression algorithms. Furthermore, it has systematic procedures to fine-tune parameter settings in the SVR algorithm. Therefore, it is conceivable that the SVR algorithm will provide the driving force for advances in modeling nonlinear biochemical systems. With a good understanding about the network structure and dynamic behavior of a biological system, we can proceed to design novel remedies to correct the system in case of malfunction either due to failure of its components or xenobiotics. In other words, we can design novel therapies and/or new drugs accordingly.

The medicine of today is reactive, with focus on developing the therapies for preexisting diseases, typically late in their progression. In the next decades, medicine will move toward predictive and preventive modes. Preventive medicine will follow as disease-perturbed networks can be used to identify drug targets first for therapy, and later for prevention. Pharmacological intervention will focus on preventing disease-mediated transition, as well as reversing or terminating those that have occurred. This will require building a fundamental understanding of systems biology that underlies normal biological and pathological processes, and the development of new microarray analyses methods for systems biology will be required to achieve this goal. Predictive and preventive medicine will lead naturally to personalized medicine that will revolutionize health care. Drug companies will have the opportunity for more effective means of drug discovery guided by systems biology through microarray data analyses and big database mining schemes.

After the description of drug discovery therapy using systems biology method in this section, an example is given in Section 12.2 to illustrate systems drug design for multiple-target therapy of four cancers.

12.2 INVESTIGATING CORE AND SPECIFIC NETWORK MARKERS OF CANCERS FOR MULTIPLE DRUG TARGETS

Cancer is the leading cause of death worldwide and is generally caused by mutations in multiple proteins or the dysregulation of pathways. Understanding the causes and the underlying carcinogenic mechanisms can help fight this disease. In this section, a systems biology approach was used to construct the PPINs of four cancers and the noncancers by their corresponding microarray data, protein–protein interaction (PPI) modeling and database-mining. By comparing PPINs between cancer and noncancer samples to find significant proteins with large PPI changes during carcinogenesis process, core and specific network markers were identified by the intersection and difference of significant proteins, respectively, with carcinogenesis relevance values (CRVs) for each cancer. A total of 28 significant proteins were identified as core network markers in the carcinogenesis of four types of cancer, 2 of which are novel cancer-related proteins (e.g., UBC and PSMA3). Moreover, seven crucial common pathways were found among these cancers based on their core network markers, and some specific pathways were particularly prominent based on the specific network markers of different cancers (e.g., the RIG-I-like receptor (RLR) pathway in bladder cancer, the proteasome pathway and TCR pathway in liver cancer, and the HR pathway in lung cancer). Additional validation of these network markers using the literature and new tested datasets could strengthen these findings and confirm the proposed method. From these core and specific network markers, we could not only gain an insight into crucial common and specific pathways in the carcinogenesis, but also obtain a high promising PPI target for multiple drug design of cancer therapy in Section 12.2.1.

Cancer is the leading cause of one in eight deaths worldwide [44]. It is a large group of over 100 distinct diseases with various risk factors and its development is a multistep process, including the dysfunction of genes that regulate the proliferation, division, and death of cells, as well as the tissue microenvironment. Defects and glitches in these gene-encoded proteins lead to a series of acquired functional capabilities that allow undisciplined growth and invasion of abnormal cells into other tissues [45]. Thus, evidence suggests that most cancers are not caused by one single factor or event; instead, a normal cell develops through a sequence of premalignant states into an invasive cancer. To determine the molecular mechanisms of cancer, the detection of potential cancer in early stages before tumors are visible is practicable.

The identification of biomarkers that can distinguish normal tissue from cancerous tissue is one of the main goals of cancer research. Much effort has been made to identify the causes of cancer, which can help our understanding of cancer and assist in the design of drugs for effective therapies. For example, comparisons within a single cancer type or a cancer family have revealed hidden oncogenic mechanisms that have been successfully applied to establish therapeutic strategies for individual types of cancer [46–49]. Additionally, several bioinformatics methods have been developed and applied to detect differentially expressed genes between normal and cancer samples [50–52], and these genes have been determined to be related to cancers and can be used as biomarkers [43]. Unfortunately, comparing data from single types of cancer has little success at recognizing compatible changes among the distinct types of malignant tumors (cancers) [53], and, many of the differentially expressed genes detected in one experiment do not always correspond to another experiment for the same cancer, especially for complex cancers [43]. While complex cancers are usually caused by the dysregulation of functional networks that consist of a set of multiregulated genes [43,54–56], how to identify a carcinogenic mechanism that is related to cancer development becomes an important research topic, and the first step is to discover common and specific features among various cancer tissues. Once we identify the different features, we can determine appropriate therapeutic targets and then target at risk patients for further medication and treatment [57].

Many systems biology approaches have been applied in various areas, such as reconstruction and design of genetic network [58–67]. Despite extensive research, there are still large gaps and controversies in the understanding of the fundamental molecular mechanisms of cancer in various tissues. Complex diseases are well recognized by the deregulation of biological systems or molecular networks. Generally, molecule regulations and interactions vary at different times and in different tissues—their changes are causally related to cancer progression. Therefore, molecules that are found to interact with different members of the molecular interaction network can be connected with the relevant cancer [68]. Moreover, PPINs also provide valuable information for understanding cellular functions and biological processes [69,70]. Due to the substantial increase in human protein interaction data, PPIN analysis is used to unravel the molecular mechanisms of disease, and, in particular, to analyze cancer mechanisms [71–74]. For example, the cancer-perturbed PPINs of apoptosis shed light on the mechanism of cancer at the systems level and allow the identification of potential novel drug targets [75]. Therefore, it is critical to identify relevant PPINs to investigate cancer development processes. Recently, Wang et al. proposed a network approach through the construction of protein association networks to successfully shed light upon the pathways and mechanisms in the lung carcinogenic process and to provide potential therapeutic targets for novel drug design [74].

We analyzed various cancers, specifically, bladder, colorectal, liver, and lung cancers [76–78], through regression modeling, microarray data, maximum likelihood parameter estimation, and big database mining. Based on their PPI information and the gene expression data from normal and patient samples, two PPINs with quantitative protein association abilities for cancer and the surrounding noncancerous tissue of each cancer were constructed, respectively. The cancer and noncancer PPINs for each cancer were then compared with respect to their network structure and their protein association abilities for obtaining sets of significant proteins with large PPI variations from noncancer to cancer, which play important roles in carcinogenesis for each cancer type. The intersections of the four sets of significant proteins in each cancer were considered to be core network markers of these four cancers, while the distinctive significant proteins unique to one cancer type were considered to be a specific network marker for each cancer. Therefore, a network marker of each cancer can be divided into a core network marker and a specific network marker. We found that there were 28 significant proteins that were core network markers of bladder, colorectal, liver, and lung cancers; the correlated pathways and molecular mechanism are further discussed. These significant proteins are involved in cancer-related pathways and play important roles in carcinogenesis. Finding the core network marker is a novel approach to investigate the common network mechanism of carcinogenesis from the perspective of multiple cancers during the process of oncogenesis. After establishing the core network marker from each cancer network marker, the specific network marker of each cancer was determined. From the specific network markers of each cancer, we can investigate specific pathways or molecular mechanisms for gaining an insight into the specific carcinogenesis of each cancer. Furthermore, these analyses demonstrate the efficiency of this method and the validity of the identified network markers, which are consistent with results reported in the literature and further validate these results.

12.2.1 Materials and methods

Overview of the construction process of core and specific network markers

A flowchart representing the construction of core and specific network markers for carcinogenesis is shown in Fig. 12.3. In summary, three kinds of data, specifically: (1) microarray data of cancer and noncancer samples of bladder, colorectal, liver, and lung cancers; (2) data from the GO database; and (3) data from the PPI databases were required to construct the PPIN for each type of cancer and noncancer. These data were used for PPI pool selection and then the selected PPIs and the microarray data were used for PPIN construction, resulting in a cancer PPIN (CPPIN) and a noncancer PPIN (NPPIN) by regression modeling and the maximum likelihood parameter estimation method. The two constructed CPPINs and NPPINs were compared to obtain the sets of significant proteins with large PPI changes between two PPIs for each cancer based on the CRV for each protein via the statistical assessment. From the intersection of these significant proteins, we were able to determine the core network markers of these four cancers. Distinctive from core network marker for each cancer, the specific network marker was obtained by extracting each unique network marker for each of the four cancers.

Data selection and preprocessing

The microarray gene expression datasets were obtained from the NCBI Gene Expression Omnibus (GEO). In this section, we chose four different types of cancer datasets as research objects, including bladder cancer (GSE13507) [79], colorectal cancer

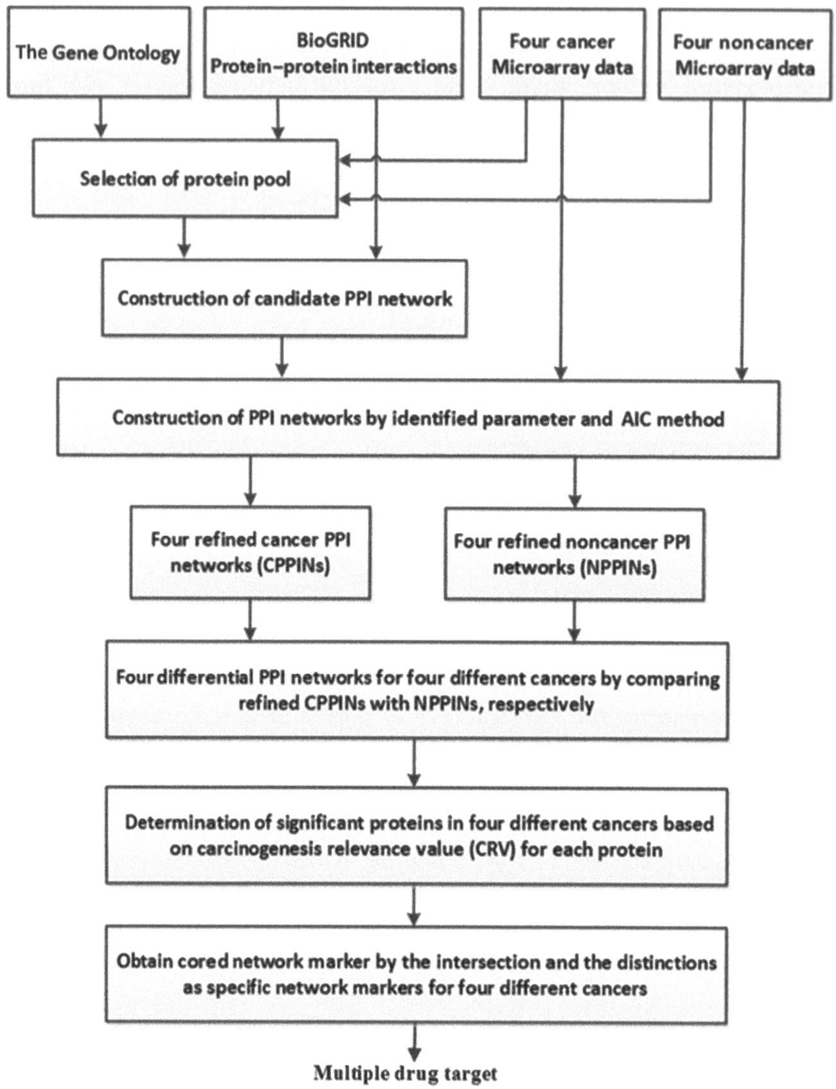

Figure 12.3 The flowchart of constructing the core network marker and the specific network marker for cancers investigation. In summary, three kinds of data, microarray data, GO database, and PPI information, are needed for the PPIN construction method. These data are used for pool selection, and then the selected proteins and the microarray data are used for the contribution of protein–protein interaction network (PPIN) by maximum likelihood estimation and model order detection method, resulting in cancer PPIN (CPPIN) and noncancer PPIN (NPPIN). The two constructed PPINs can be used for the determination of significant proteins of tumorigenesis by the difference between two PPI matrices of two constructed PPINs. With the help of the differential PPI matrix (network) between CPPIN and NPPIN, carcinogenesis relevance value (CRV) is computed for each protein, and significant proteins in carcinogenesis are determined based on *p*-value of the CRVs of these proteins in the differential PPI matrix between CPPIN and NPPIN. These significant proteins are obtained for four cancers. The intersection of these significant proteins and their PPIs in four cancers is featured as the core network marker of four cancers and the distinctive proteins and their PPIs of each cancer are featured as specific network for each cancer.

(GSE41258) [80], liver cancer (GSE14520) [81,82], and lung cancer (GSE19188) [83]. The data were selected such that both cancer and noncancer samples were contained in the same dataset. Due to the discrepancies in gene expression that are intrinsic to cell culture and fixation [84,85], only data derived from nonprocessed primary biopsies could be used [57]. Therefore, each dataset contains primary tumor samples from patients and adjacent nontumor tissue samples from the same cancer patients, which could be considered as control samples. By defining an early phenotype of cancer, cancers are classified into four stages according to their degree of invasion and migration using the TNM staging system defined by the American Joint Committee on Cancer (AJCC) and the International Union Against Cancer (UICC). In general, stage I and stage II describe early stage cancers that have higher curability under medical treatment, while stage III and stage IV are late stages. Based on this concept, only stage I and stage II were taken into consideration [83] to investigate the early processes of cancer development. To make all the data compatible and reliable, Pearson correlation coefficients of the gene expression profiles for cancer and noncancer were calculated. The data marked as "included in analysis: NO" or data that were not required were eliminated. One aspect that should be noted to avoid overfitting in network construction relates to the fact that the maximum degree of the proteins in the PPIN should be less than the cancer/noncancer sample number [74]. Among these datasets, we shall have a greater number of samples in cancer and noncancer for overcoming the restriction in the size of network from the sample size. Therefore, four types of cancer were analyzed by microarray datasets of tumor and nontumor samples, of which the sample numbers are listed respectively as follows (tumor/nontumor samples): bladder cancer (106/58), colorectal cancer (72/33), liver cancer (119/152), and lung cancer (NSCLC) (71/65) (Table 12.1). Prior to further analysis, the gene expression value, h_{ij}, was normalized to z-transformed scores, g_{ij}, for each gene, i, and then the normalized expression value had a mean $\mu_i = 0$ and standard deviation $\sigma_i = 1$ over sample j [43,74].

The PPI data in *Homo sapiens* were extracted from the Biological General Repository for Interaction Database (BioGRID, downloaded in October 2012). The BioGRID is an open-access archive of genetic and protein interactions that are curated from the primary biomedical literature for all major model organisms. As of September 2012, BioGRID houses more than 500,000 manually annotated interactions from more than 30 model organisms [198]. The gene annotations of *H. sapiens* were extracted from the GO database, a widely used gene annotation database that is systematically constructed based on biological function-depicting terms; further, these terms are able to annotate genes or proteins. The GO database currently covers over 11 million gene products from over 350,000 species [199]. Each of the above databases was mined for PPINs of the four types of cancer and noncancer by their corresponding microarray data. These cancer and noncancer PPINs were then compared to obtain network markers for each cancer.

Table 12.1 The identified 28 significant proteins of core network marker in common carcinogenesis

Protein symbol	Cancer type				Functional annotation		
	Bladder	Colorectal	Liver	Lung	Cell growth	Cell survival	Immunity
BRCA1		[86,87]		[88]	+	+	
CDK2	[89]	[90]	[91]	[92]	+	+	
CEBPB		[93]	[94]	[95,96]	+	+	+
CREBBP	[97]	[98]	[99]	[100,101]	+		
CTNNB1	[102]	[103,104]	[105,106]	[103]	+	+	
CUL1	[107]	[108]	[109]	[110,111]	+		
CUL3	[112]	[113,114]	[115]	[116]	+		
EP300	[117]	[98]	[118]	[101,119]	+		
ER1	[120]	[121]	[122,123]	[124]	+	+	
HDAC1	[125]	[98,126]	[127,128]	[129,130]	+		
HDAC2	[131]	[126]	[128]	[132]	+		
HDAC4	[133]	[126,134]	[135]	[130]	+		
IRAK1				[136]	+	+	+
ISG15	[137,138]	[138,139]	[140]	[141,142]	+	+	+
KIAA0101		[143]	[144]	[145]		+	
MDM2	[146,147]	[148]	[149]	[150,151]	+	+	
MYC	[152,153]	[154]	[155]	[156]	+	+	
PCNA	[157]	[158,159]	[91,160]	[161]	+		
PRKDC [162]	[163]	[164,165]	[166]	[167,168]		+	+
PSMA3							
RB1	[169]	[90,158]	[170,171]	[172,173]	+	+	
SRC	[174,175]	[176]	[177]	[178]	+	+	+
TERF1		[179]	[180]	[181]		+	
TP53	[157,182]	[182,183]	[160,182]	[182,184]			
TRAF2	[185]	[186]	[187]	[188]			
UBC							
XRCC6	[189]	[190]	[191]	[192,193]			
YWHAZ	[45]	[194]	[195]	[196]			

These 28 significant proteins are the intersection of the identified significant proteins of four cancers in Table A.3. The literature evidences indicate that overexpression/deregulation of the specific protein or mutation of the corresponding gene would result in carcinogenesis. The full names of these proteins are listed in Table A.7. The functional annotations are from the Gene Ontology database [197].

Selection of protein pool and identification of the PPINs for cancerous and noncancerous cells

To integrate gene expression with PPI data to construct the required PPIN, we set up a protein pool, which has differentially expressed proteins. The gene expression values were overlaid with the corresponding proteins and were assumed to correlate with protein expression levels. The expression of each protein was analyzed by using one-way analysis of variance (ANOVA) to select proteins with differential expression levels. This method allowed determination of significant differences between cancer and noncancer sets to find proteins to be significant of carcinogenesis. The null hypothesis (H_0) here is based on the assumption that the mean protein expression levels of cancer and noncancer sets are the same. Adjustment with Bonferroni [200], a type of multiple testing, was used to detect and correct proteins with discrepancy. Proteins with a *p*-value of less than .01 were included in the protein pool. However, if the proteins in the protein pool did not have PPI information, they were eliminated. In addition, proteins not already in the protein pool were included if their PPI information has determined that they had a tight relationship with proteins already in the pool. As a result, the protein pool contained proteins that had certain differences in expression level and proteins that had tight relationships with the aforementioned proteins. In this case, the protein pool consisted of 573 proteins in bladder cancer, 1237 proteins in colorectal cancer, 2942 proteins in liver cancer, and 3404 proteins in lung cancer.

Furthermore, to make the protein pool more relevant, we used the GO database to identify proteins relating to cancer. The selected cancer-related functions included cell cycle arrest, apoptosis, cell proliferation, and inflammation, among others. To ensure that the selected proteins were as closely related to cancer as possible, the selected GO terms corresponding to the aforementioned cancer functions belonged to a rather lower level; detailed information is listed in Table A.2. As a result, 3029 proteins were selected to be in the functional pool. Combining the protein pool with the functional pool, we obtained a new pool consisting of 1753 proteins that were not only significantly expressed, but also had cellular functions in cancer. On the strength of the significant and functional pool and the PPI information, candidate PPINs for different cancers were constructed for each cancer and noncancer by linking the proteins that interact among them, i.e., the proteins that had PPI information through the pool were linked together, resulting in the candidate PPINs.

As the candidate PPIN included all possible PPIs under various environments, different organisms, and experimental conditions, the candidate PPINs were further confirmed by microarray data to identify appropriate PPIs according to the biological processes that are relevant to cancer. To remove false positive PPIs from each candidate PPIN for different biological conditions, we used a PPI model, the maximum likelihood parameter estimation method, and a model order detection method together to prune each candidate PPIN by the corresponding microarray data to approach the actual PPIN. Here, the PPIs of a target protein *i* in the candidate PPIN can be depicted by the following protein association model:

$$x_i[n] = \sum_{j=1}^{M_i} \alpha_{ij} x_j[n] + \omega_i[n] \qquad (12.1)$$

where $x_i[n]$ represents the expression level of the target protein i for the sample n; $x_j[n]$ represents the expression level of the jth protein interacting with the target protein i for the sample n; α_{ij} denotes the association ability between the target protein i and its jth interactive protein; M_i represents the number of proteins interacting with the target protein i; and $\omega_i[n]$ represents the stochastic noise due to other factors or model uncertainty. The biological meaning of Eq. (12.1) is that the expression level of the target protein i is associated with the expression levels of the proteins interacting with it. Consequently, a protein association (interaction) model for each protein in the protein pool can be built.

After constructing Eq. (12.1) for the PPI model of each protein in the candidate PPIN, we used the maximum likelihood estimation method [201] to identify the association parameters in Eq. (12.1) by microarray data as follows (see Appendix A.1):

$$x_i(n) = \sum_{j=1}^{M_i} \hat{\alpha}_{ij} x_j(n) + w_i(n) \tag{12.2}$$

where $\hat{\alpha}_{ij}$ is identified by the maximum likelihood estimation method.

Once the association parameters for all proteins in the candidate PPIN were identified for each protein, the true protein associations were determined by pruning the false positive PPIs. Akaike Information Criterion (AIC) [201] and a Student's t-test [202] were employed to achieve model order selection for the pruning of false positive protein associations in $\hat{\alpha}_{ij}$ (see Appendix A.2).

Determination of significant proteins and their network structures in the carcinogenesis of four types of cancers

After the AIC order detection and use of the Student's t-test to determine p-values of $\hat{\alpha}_{ij}$, the false positive PPIs $\hat{\alpha}_{ij}$ in Eq. (12.2) were pruned away and only significant PPIs were refined as follows:

$$x_i(n) = \sum_{j=1}^{M_i'} \hat{\alpha}_{ij} x_j(n) + w_i'(n), \quad i = 1, \ldots, M \tag{12.3}$$

where $M_i' \leq M_i$ denotes the number of true PPIs, with the target protein i, i.e., a number of $M_i - M_i'$ (or false positives) are pruned in the PPIs of target protein i. One protein by one protein (i.e., $i = 1, 2, \ldots, M$ for all proteins in refined PPIN in Eq. 12.3) results in refined PPIN

$$X(n) = AX(n) + w(n) \tag{12.4}$$

where

$$X(n) = \begin{bmatrix} x_1(n) \\ x_2(n) \\ \vdots \\ x_M(n) \end{bmatrix}, \quad A = \begin{bmatrix} \hat{\alpha}_{11} & \cdots & \hat{\alpha}_{1M} \\ \vdots & \ddots & \vdots \\ \hat{\alpha}_{M1} & \cdots & \hat{\alpha}_{MM} \end{bmatrix}, \quad w(n) = \begin{bmatrix} w'_1(n) \\ w'_2(n) \\ \vdots \\ w'_M(n) \end{bmatrix},$$

and the interaction matrix A denotes the PPIs.

If there is no PPI between protein i and protein j or it is pruned by AIC order detection due to the false positive PPIs in the refined PPIN, then $\hat{\alpha}_{ij} = 0$. In general, $\hat{\alpha}_{ij} = \hat{\alpha}_{ji}$, but if this is not the case, the larger one was chosen as $\hat{\alpha}_{ij} = \hat{\alpha}_{ji}$ to avoid the situation $\hat{\alpha}_{ij} \neq \hat{\alpha}_{ji}$. The above PPIN construction method was employed to construct the refined PPINs for cancer and noncancer cells of bladder, colorectal, liver, and lung cancer, respectively, as shown in Fig. A.2.

The interaction matrices A of refined PPINs in Eq. (12.4) for cancer and noncancer cells of the four cancers were constructed, respectively, as follows:

$$A_C^k = \begin{bmatrix} \hat{\alpha}_{11,C}^k & \cdots & \hat{\alpha}_{1M,C}^k \\ \vdots & \ddots & \vdots \\ \hat{\alpha}_{M1,C}^k & \cdots & \hat{\alpha}_{MM,C}^k \end{bmatrix}, \quad \text{and} \quad A_N^k = \begin{bmatrix} \hat{\alpha}_{11,N}^k & \cdots & \hat{\alpha}_{1M,N}^k \\ \vdots & \ddots & \vdots \\ \hat{\alpha}_{M1,N}^k & \cdots & \hat{\alpha}_{MM,N}^k \end{bmatrix} \tag{12.5}$$

where k = bladder cancer, colorectal cancer, liver cancer, and lung cancer; A_C^k and A_N^k denote the interaction matrices of refined PPIN of the kth cancer and noncancer, respectively; M is the number of proteins in the refined PPIN. Therefore, the protein association model for CPPIN and NPPIN in the kth cancer and noncancer can be represented by the following equations according to Eqs. (12.4) and (12.5):

$$\begin{aligned} x_C^k(n) &= A_C^k x_C(n) + w_C^k(n) \\ x_N^k(n) &= A_N^k x_N(n) + w_N^k(n) \end{aligned} \tag{12.6}$$

where k = bladder cancer, colorectal cancer, liver cancer, and lung cancer; $x_C^k(n) = [x_{1C}^k x_{2C}^k \ldots x_{MC}^k]^T$ and $x_N^k(n) = [x_{1N}^k x_{2N}^k \ldots x_{MN}^k]^T$ denote the vectors of expression levels; $w_C^K(n)$ and $w_N^K(n)$ indicate the noise vectors of PPINs in the kth cancer and noncancer cells, respectively.

The different matrix $A_C^k - A_N^k$ of differential PPIN between CPPIN and NPPIN in the kth cancer is defined as:

$$D^k = \begin{bmatrix} d_{11}^k & \cdots & d_{1M}^k \\ \vdots & \ddots & \vdots \\ d_{M1}^k & \cdots & d_{MM}^k \end{bmatrix} = \begin{bmatrix} \hat{\alpha}^k{}_{11,C} - \hat{\alpha}^k{}_{11,N} & \cdots & \hat{\alpha}^k{}_{1M,C} - \hat{\alpha}^k{}_{1M,N} \\ \vdots & \ddots & \vdots \\ \hat{\alpha}^k{}_{M1,C} - \hat{\alpha}^k{}_{M1,N} & \cdots & \hat{\alpha}_{MM,C}^k - \hat{\alpha}_{MM,N}^k \end{bmatrix} \tag{12.7}$$

where k = bladder cancer, colorectal cancer, liver cancer, and lung cancer; d_{ij}^k denotes the PPI variation between the ith protein and the jth protein of differential PPIN by comparing CPPIN with NPPIN in the kth type of cancer; the matrix D^k indicates the difference in network structure between CPPIN and NPPIN in the kth type of cancer. In order to investigate carcinogenesis from the difference matrix D^k between CPPIN and NPPIN of the kth cancer in Eq. (12.7), a score named the CRV was presented to quantify the correlation of PPI variations of each protein in D^k with the significance of carcinogenesis as follows [74]:

$$CRV^k = \begin{bmatrix} CRV_1^k \\ \vdots \\ CRV_i^k \\ \vdots \\ CRV_M^k \end{bmatrix} \tag{12.8}$$

where $CRV_i^k = \sum_{j=1}^{M} \left| d_{ij}^k \right|$, and k = bladder cancer, colorectal cancer, liver cancer, and lung cancer.

The CRV_i^k in Eq. (12.8) quantifies the differential extent of PPI variations of the ith protein (the absolute sum of the ith row of D^k in Eq. (12.7)) and CRV^k can differentiate CPPIN from NPPIN in the kth cancer. In other words, the CRV_i^k in Eq. (12.8) could calculate the total PPI variations of the ith protein by comparing the network structure differences between the cancer and noncancer networks, which can be used to check which proteins are involved with the kth cancer.

In order to investigate which proteins are more likely involved in the kth cancer, we needed to calculate the corresponding empirical p-value to determine the statistical significance of CRV_i^k. To determine the observed p-value of each CRV_i^k, we repeatedly permuted the network structure of the candidate PPIN of the kth type of cancer as a random network of the kth cancer. Each protein in the random network of the kth type of cancer will have its own CRV to generate a distribution of CRV_i^k for k = bladder, colorectal, liver, and lung cancer (Fig. A.1). Although there was random disarrangement of the network structure, the linkages of each protein were maintained, i.e., the proteins with which a particular protein interacted were permuted without changing the total number of protein interactions. This procedure was repeated 100,000 times and the corresponding p-value was calculated as the fraction of random network structure in which CRV_i^k is at least as large as the CRV of the real network structure. According to the distributions of CRV_i^k of random networks in Fig. A.1, the CRV_i^k in Eq. (12.8) with a p-value of less than or equal to .01 was regarded as a significant CRV and the corresponding protein was determined to be a significant protein in the carcinogenesis of the kth cancer: a protein with a p-value $>.01$ was removed from the list of significant proteins in carcinogenesis (in other words, if the p-value of $CRV_i^k > .01$, then the ith protein was removed from CRV^k in Eq. (12.8) and the remainders in CRV^k with p-values of CRVs less than .01 were significant proteins of the kth cancer). Based on the p-value of the CRVs for all proteins ($i = 1, 2, \ldots, M$) and the four types of cancer (k = bladder, colorectal, liver, and lung cancer), we generated four lists of significant proteins (Table A.3) for the cancers according to the CRV and the statistical assessment of each significant protein in CRV^k in Eq. (12.8). As shown in Table A.3, we found 107 significant proteins in bladder cancer, 110 significant proteins in liver cancer, 60 significant proteins in colorectal cancer, and 86 proteins in lung cancer. These proteins have significant PPI changes between the CPPIN and NPPIN in the carcinogenic process for their corresponding cancer and we suspect that they may play important roles in carcinogenesis, warranting further investigation.

The intersection of these significant proteins in the four cancers and their PPIs is known as the core network markers, while the differences of these significant proteins are the unique significant proteins of each cancer and their PPIs in each of the cancers are known as the specific network markers for each cancer. We found that there were 28 significant proteins that could be classified as a core network marker and 26, 4, 24, and 13 significant proteins that were specific network markers of bladder, colorectal, liver, and lung cancer, respectively. The core network and specific network markers for the cancers are shown in Fig. 12.4. This insight into the carcinogenic mechanisms of common core and specific network markers in different cancers will be discussed in detail in the following section.

Pathway analysis

Much valuable information can be found in the known pathways, which are useful for most "normal" biological phenomena. All of the known pathways are the result of repeated testing and verification and the entire pathway network has given definitions for most links. Therefore, the identified proteins in the above network markers were mapped onto the known pathway networks (e.g., the Kyoto Encyclopedia of Genes and Genomes (KEGG) or Protein ANalysis THrough Evolutionary Relationships (PANTHER) pathway) to investigate significant pathways with the network marker and to explore the relationships between these pathways and cancer. It supports the view that systems biology can help identify significant network biomarkers in both normal and cancerous pathways that play a central role in the pathogenesis of cancer.

Together with comprehensive pathway databases such as the KEGG, we used a series of bioinformatics pathway analysis tools to identify cancer-relevant pathways from the significant proteins in network markers for each cancer [203]. KEGG includes manually curated biological pathways that cover three main categories: systems information (e.g., human diseases, drugs), genomics information (e.g., gene catalogs, sequence similarities), and chemical information (e.g., metabolites, biochemical

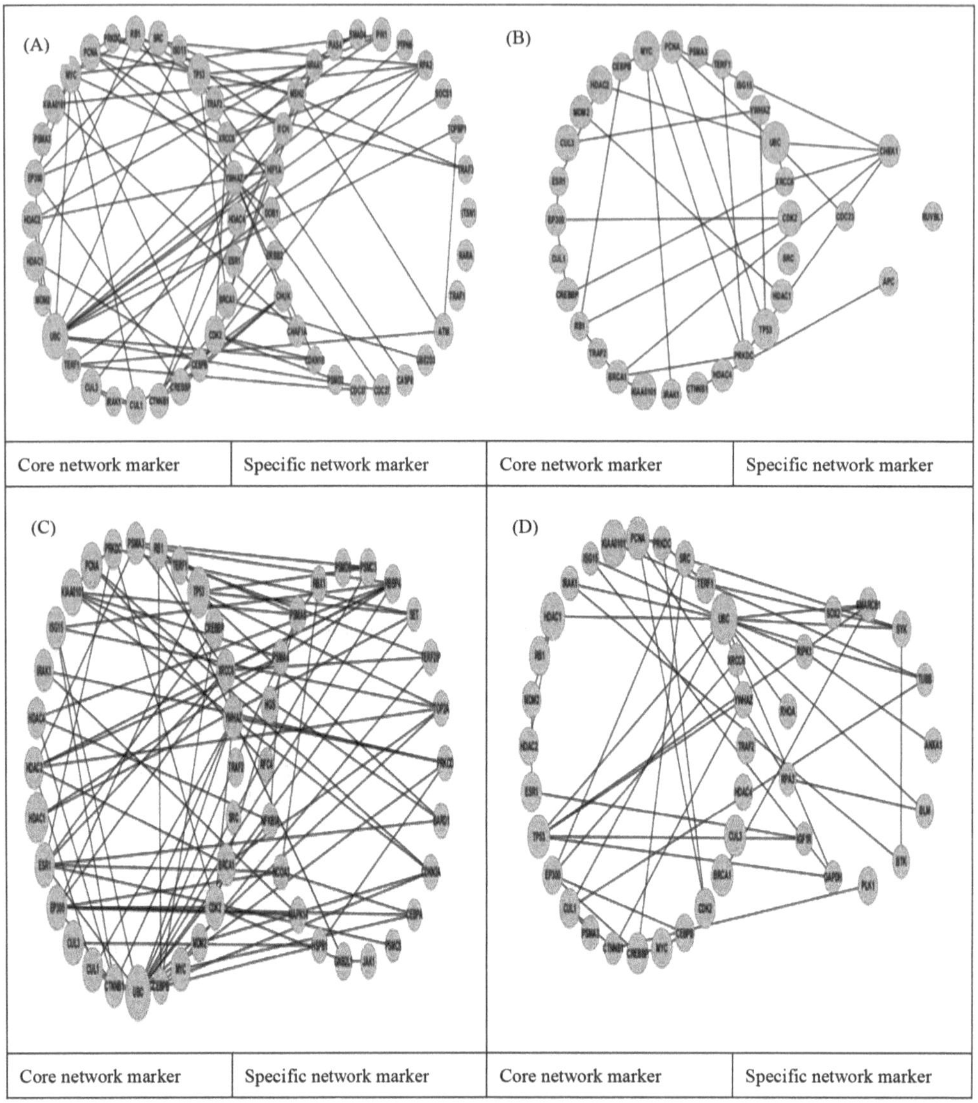

Figure 12.4 The core network marker and specific network marker for four types of cancer. Network markers of (A) bladder cancer, (B) colorectal cancer, (C) liver cancer, and (D) lung cancer, respectively. The left parts represent the core network marker, and right parts represent the specific network markers of each cancer. Further, the number of core network marker is 28 proteins, and protein numbers of specific network marker with respect to four cancers are 26, 4, 24, and 13 in bladder, colorectal, liver, and lung cancer, respectively. The node size is proportional to the CRV for each protein in each cancer from Table A.3. All the network markers (subnetworks) are extracted from Fig. A.2.

reactions). At present, KEGG contains 134,511 distinct pathways generated from 391 original reference pathways [204]. Therefore, to investigate the pathways involved in carcinogenesis, the bioinformatics database DAVID, which automatically outputs the results from KEGG pathway analysis [205], was used for the pathway analysis of significant proteins in the core and specific network marker sets.

Analysis of the networks and pathways of cancer and benign tumors

Depending on whether or not they can spread by invasion and metastasis, tumors are classified into benign or malignant tumors. Benign tumors are tumors that cannot spread by invasion or metastasis; hence, they only grow locally. Malignant tumors are tumors that are capable of spreading by invasion and metastasis. By definition, the term "cancer" applies only to malignant tumors [206] (see Fig. 12.3). So, in most cases in the field of cancer research, if a tumor is diagnosed as "benign," it will be no different from normal tissue, and there will be no further investigation. We know there are some studies (not most) that focus on the difference between benign tumors and malignant tumors; however, it is not the main purpose of this section. We focus on finding the significant difference of PPI between "cancer" and normal tissue, as network markers for the purpose to discover new therapy target in this section. We begin trying to analyze the difference between cancer and benign tumor. We received the original microarray dataset from NCBI GEO database. There are a lot of data combining the cancer and noncancer samples at the same time, but very few of them contain benign tumor samples. We find that GSE27554 and GSE27409 both contain lung cancer and benign lung tumors. The numbers of samples in these two platforms are very small, so we need to combine them together. After combining, there are still only 20 and 10 samples for cancerous and benign tumors, respectively. Too few samples in this case will make this model being overestimated (i.e., overfitting in parameter estimation). However, we have applied this model on these samples, and the results will be shown next.

12.2.2 Common core network markers of four types of cancer for multiple drug target

Network mechanisms of carcinogenesis based on the common core network markers and specific network markers in four types of cancer

From the differential networks between CPPIN and NPPIN of four types of cancer and through the p-value of CRV for significant proteins, we found the core network markers and specific network markers for different cancers by the intersection and difference of significant proteins in different cancers, respectively. In this section, we aimed to investigate the common carcinogenic mechanisms of the core network markers and distinctive carcinogenic mechanisms by investigating the specific network markers in different cancers.

Validation of the feasibility of the proposed method and a novel predictor of core network markers for four types of cancer

A total of 28 proteins in the core network marker were identified to be significant proteins in common carcinogenesis using the GEO database (Table 12.1); thus, the core network marker is a common feature that contains 28 significant proteins related to carcinogenesis. Before applying the proposed method to construct a global view of the core network markers, we first validated the feasibility of the proposed method. To verify the accuracy, we investigated whether these significant proteins were already known to be related to carcinogenesis. Table 12.1 lists the relevance of these 28 proteins to carcinogenesis. Experimental studies that were used to investigate the overexpression or dysregulation of these significant proteins or mutations of the corresponding gene have confirmed that these proteins are involved in carcinogenesis at the early stages.

As shown in Table 12.1, 21 of the 28 identified proteins have been verified in all four types of cancer previously, suggesting that they play a role in early carcinogenesis. The sensitivity is 75.0%, which shows that the common core network markers can be successfully identified by the proposed method, thus validating the feasibility. With the exception of the above proteins, five have not been shown to exist in all four cancers, namely, BRCA1, CEBPB, IRAK1, KIAA0101, and TERF1. This may be due to the experimental constraints or the relationships between these five proteins and the four cancers, which researchers have not yet investigated fully. However, these five proteins significantly influence other cancer types. For instance, the mutant phenotype of BRCA1 has been found in breast and ovarian cancer [207]. Thus, we suppose these as novel predictors of the four cancers.

For 2 out of the 28 significant proteins identified as core network markers, namely, UBC and PSMA3, little is known about their roles in cancer (Table 12.1). UBC is classified as a polyubiquitin precursor protein. Protein ubiquitination is a posttranslational modification (PTM) related to a wide variety of cellular processes. It is associated with protein degradation, antigen processing, DNA repair, cell cycle regulation and division, differentiation and development, kinase modification, endocytosis, the immune response and inflammation, and regulation of other cell signaling pathways [208]. Functionally, dysregulation of ubiquitin-related signaling has been progressively implicated in human diseases [138]. Therefore, UBC should be a key protein for further characterization in carcinogenesis. With respect to PSMA3, 53 articles refer to this gene in PubMed. Functionally, it has been proposed to participate in some metabolic pathways (proteasome or proteasome complex) and processes (interspecies interaction between organisms, proteolysis involved in cellular protein catabolic processes, ubiquitin-dependent protein catabolic processes) [209,210]. Due to its important functions, PSMA3 may also be involved in carcinogenesis despite a lack of evidence indicating its relevance to carcinogenesis; this consequently makes it a novel target for further investigation.

Although the two significant proteins described earlier were found to be core network markers in the four types of cancers examined, they have not been shown to be molecular markers of cancer in any previous literature. However, according to the proposed network marker method based on microarray data of multiple cancers, they were found to be significant proteins in the core network marker. There may be novel molecular markers that are worthy of further experimental investigation.

Investigation of the significant proteins in the core network marker in carcinogenesis

These 28 common proteins in the core network marker seem to be an important network feature in tumorigenesis. The mutations of these genes promote carcinogenesis since many of these translated proteins help regulate cell birth or cell death processes by apoptosis while others participate in repairing damaged DNA. Carcinogenesis changes many fundamental properties of cells, allowing these cells to evade the control of normal growth and confer the full cancer phenotype [211]. In addition, these mutated cells that are destined to become a tumor also have to be able to grow prosperously in a chronic inflammatory microenvironment, escape immune recognition, and suppress immune reactivity. Both the genetic and molecular definition of these three immune biomarkers of cancer allow the development of specific countermeasures to reverse the immune-suppressive effect in favor of the immune system, and, eventually, the patient [212]. Therefore, these 28 significant proteins can be classified into three categories according to their functional annotations (Tables 12.1 and A.7). The mechanisms employed by the 28 significant proteins in the core network marker group to promote carcinogenesis were further investigated from the perspective of cell growth, cell survival, immunity, and proteolysis.

Cell growth

Cancer is a genetic disease that arises from an acquisition of aberrant functions in critical genes that regulate normal cell growth and proliferation [213]. Uncontrolled cell proliferation is the hallmark of malignant tumors [161]. The unregulated constitutive activity of certain proteins and their overproduction promote cell proliferation and transformation, thereby contributing to carcinogenesis. Thus, the proteins responsible for cell growth also tend to play important roles in carcinogenesis [74]. Among these 28 significant proteins identified, 21 of them have functions relating to cell growth, reinforcing its significance. CUL1 and CUL3 belong to the CULLIN family that provides a scaffold for ubiquitin ligases, plays an important role in protein degradation and ubiquitination, and maintains the balance between normal cell growth and uncontrolled proliferation. Therefore, dysregulation of the activity of CULLIN has been shown to promote oncogenesis. At present, CUL1 is the most representative of the CULLIN family and it plays an elementary role in embryonic development as well as contributes to normal and malignant cellular processes. Additionally, loss of

Cul3 expression was detected in human cancers and correlated directly with tumor dedifferentiation. There is an increasing evidence that CUL1 and CUL3 will be targets of new therapeutics against cancer and other human diseases.

Many proteins serving as transcriptional regulators are also recognized by this method. Both histone deacetylase (HDAC) and histone acetyltransferase (HAT) are histone-modifying enzymes that regulate the acetylation level of histones and transcription factors, which play a critical role in the control of cell survival. In turn, these activities are regulated by a variety of signal transduction pathways. These proteins have been shown to regulate the expression and activity of numerous proteins involved in both cancer initiation and cancer progression. Three of HDAC-associated molecules are HDAC1, HDAC2, and HDAC4, and two of HAT-associated molecules are CREB-binding protein (CBP) and protein p300, which were all identified to be significant proteins in carcinogenesis. These proteins interact with oncogene and antioncogene products, thus modifying or mediating their transcriptional functions related to cell differentiation and cell proliferation [214–216]. This result suggests that monitoring these proteins may help to predict the prognosis of cancer patients.

CTNNB1 (β-catenin), a multifunctional protein, is one of the essential components of the Wnt pathway and the elevation of its protein levels is a hallmark of Wnt pathway activation. Mutation of β-catenin, which is a frequent cause of human cancers, blocks its phosphorylation and targeted degradation, resulting in the accumulation of active transcriptional activators in the nucleus and abnormal activation of the Wnt pathway. MYC is a nuclear phosphorylation protein that serves as a transcription factor. It plays a role in several fundamental functions of cell biology, including the regulation of cell growth and proliferation, metabolism, differentiation, apoptosis, and angiogenesis. Malfunction of MYC, which causes MYC to be constitutively expressed leading to a lack of cell cycle arrest and uncontrolled cancer growth, has also been found in many types of cancers [217]. Other critical proteins that were identified, such as CDK2, ESR1, MDM2, PCNA, SRC, and YWHAZ, also participate in the process of carcinogenesis.

Cell survival

The expanding ability of tumor cell populations is determined by the proportion of cell proliferation versus cell death. The acquired resistance to programmed cell death (PCD), also known as apoptosis, is a critical biomarker of cancer [213]. Therefore, proteins that are involved in cell survival may be significant in carcinogenesis. Eighteen key proteins in the core network marker have roles in cell survival. RB (retinoblastoma-associated) and TP53 are two tumor suppressor proteins; they act as two control center nodes within two significant complementary cellular regulatory networks that determine whether cells proliferate or, in contrast, stimulate proteins that can activate both cell senescence and apoptotic pathways [218]. Changing of the

RB protein significantly impacts cells and the whole body [219]. Therefore, cancer cells with defects in the RB pathway thus lose their functions as a pivotal gatekeeper of cell cycle progression; hence, RB absence induces persistent cell proliferation [218]. TP53 (p53) is a tumor suppressor protein that has been extensively studied and plays key roles in anticancer mechanisms. The p53 signaling pathway is triggered to respond to various external stress signals, allowing p53 to organize transcription behaviors that collaborate to suppress tumor growth. Activated p53 can induce cell cycle suspension, stimulate the apoptotic process, and inhibit metastasis and angiogenesis. Mutation in the p53 protein itself or perturbation of p53 pathways signaling will cause the functional loss of p53; indeed, it is a generic phenomenon in most human cancers [220]. This makes p53 an extremely attractive target for tumor therapy. Due to their key roles in carcinogenesis, it is not surprising that p53 and RB were both identified to be significant proteins in the core network marker group. Other proteins related to DNA repair mechanisms, such as PRKDC and XRCC6, were also shown to be involved in the carcinogenic process. A lot of research have discussed that p53 is the target for cancer therapy [221,222]. It is not only a therapy for itself, but also act as PPI target. The authors mentioned about these facts again and again as following: (1) The MDM2-p53 and CDK-pRB PPI complexes play significant roles in neutralizing tumor suppressive function. (2) The MDM2-p53 and Myc-Max PPI complex play important roles in evading growth suppression and cell death, promoting genomic instability and cancer metabolism. (3) The important thing is that several MDM2-p53 PPI inhibitors have entered clinical trials. (4) The MDM2-p53 PPI inhibitor, RO5503781, now is in a Phase I trial in advanced malignant patients [221].

Immunity

The immune response has key functions in cancer as it plays decisive roles at different stages of tumor development. Consequently, proteins related to immunity are important in carcinogenesis and are appealing targets in cancer therapy and prevention. Six out of 28 significant proteins were involved in immunity. TRAF2, a member of the TRAF family, is involved in a variety of intracellular signal transduction pathways, the most important of which is the NF-κB pathway. Furthermore, it is an unique adaptor protein that mediates several signaling pathways leading to protection against apoptosis. Alteration of the expression of TRAF2 protein was observed in human cancers, indicating a significant role in carcinogenesis. ISG15 is an ubiquitin-like protein (UBL) that acts as a signaling messenger that governs several cellular functions, such as cell proliferation, apoptosis, and the cell cycle. It shares several common properties with other proteins and is regulated by specific signal transduction pathways that play a role in innate immunity. ISG15 is overexpressed and conjugated to many proteins in tumors and is presumably implicated in tumorigenesis through the modification of several oncogenes and tumor suppressors [223]. The identification of TRAF2 and ISG15 in the core network marker group reinforces the link between immunity and cancer.

Proteolysis

We know that aggressiveness and invasivity are the main characteristics of cancer. These results show that these four proteins (CUL3, MDM2, BRCA1, CUL1) are related to the Ubiqutined proteolysis pathway. It is also known that cell cycle, proteolysis, and cancer are highly related, and we saw that CUL1 and MDM2 also appear in the cell cycle pathway [224]. Yamasaki et al. also discussed about the topic whether ubiquitin ligases could be targeted for the therapy of human cancer or not. From traditional viewpoint of small molecular inhibitors, the ubiquitin ligases have not only lack of a canonical active enzymatic site for small molecule to dock, but also involve in the PPI mode. However, small molecule inhibitors of Mdm2 have been developed, such as Nutlins [224]. Moreover, from today's viewpoint, PPIs could be as a new target for cancer therapy, i.e., as we have described, people can use small molecules to interfere the PPIs [221,222]. These results can explain the aggressiveness and invasivity characteristics of cancers very well: (1) As we described earlier, the 28 proteins are the network marker of these four cancers. By the pathway analysis of David database, we know that there are four proteins (CUL3, MDM2, BRAC1, and CUL1) related to the has04120 pathway, i.e., Ubiquitin-mediated proteolysis (see Rank 6 of Table A.5). The ratio (4/28) is high. This result may prove that protease activity is a typical feature of cancer. In this section, we emphasize that these 28 proteins are responsible for the housekeeping mechanisms of cancer. We saw 4 of these 28 proteins are related to the Ubiquitin-mediated proteolysis pathway. (2) There is only PSMA3 in these 28 proteins related to the map03050, i.e., Proteasome pathway (see Rank 8 of Table A.5). These results show that 5 of these 28 core proteins are related to proteolysis and all of them are appeared in the core network marker. We also find that there is not any protein in the specific network being related to proteolysis. Both findings suggest that proteolysis may be one of the housekeeping mechanisms during various cancer developments.

Investigating common pathways that underlie the core network markers in cancer

The proteins in the core network marker are altered in these four cancer types, resulting in cancer initiation and tumor development. This suggests that there may be some common key pathways that exist in the initial stage of these cancers. Even small changes in the expression level of a single protein can affect a wide range of signaling pathways that are involved in various biological functions. In this section, we attempted to characterize the core network markers. To do this, we focused on whether the identified core network markers (28 proteins) were related to the known KEGG pathways and whether they operated through these pathways.

As shown in Tables 12.2 and A.4, there are seven pathways that the 28 significant proteins of the core network marker group are involved in, with a p-value $\leq .05$. Firstly, the cell cycle is composed of two consecutive periods; they are characterized by DNA replication and sequential differentiation and segregation of replicated chromosomes into two separate daughter cells. Both positive-acting and negative-acting proteins control the cells' entry and advancement through the cell cycle, which is composed of four distinct phases: G1, S (synthesis), G2, and M (mitosis) [225]. Dysregulation of the cell cycle governs deviant cell proliferation in cancer. Losing the ability to control cell cycle checkpoints induces abnormal genetic instability. It may be due to the result of the activation of tumorigenic mutations. They have been recognized in various tumors at different levels in the mitogenic signal transduction pathways: (1) ligands

Table 12.2 The pathways analysis for specific network marker for each carcinogenesis of four cancers

(A) Bladder cancer	Colorectal cancer	Liver cancer	Lung cancer
hsa04622:RIG-I-like receptor signaling pathway		hsa03050: Proteasome hsa04660: T-cell receptor signaling pathway	hsa03440: Homologous recombination

(B) Pathway title	Cancer type			
	Bladder	Colorectal	Liver	Lung
Cell cycle	+	+	+	+
Wnt signaling pathway	+	+	+	+
Notch signaling pathway	+	+	+	+
Adherens junction	+	+	+	+
TGF-beta signaling pathway	+	+	+	+
Ubiquitin-mediated proteolysis	+	+	+	+
p53 signaling pathway	+	+	+	+
RIG-I-like receptor signaling pathway	+			
Proteasome			+	
T-cell receptor signaling pathway			+	
Homologous recombination				+

The significant pathways via DAVID Bioinformatics database are selected for specific network marker for each carcinogenesis of four cancers. The significantly enriched pathways for each cancer results from Table A.5 are given with p-value $\leq .05$. Thus, we can find out different pathways by comparing significant pathways of specific network markers with those of the core network markers for each cancer. (A) The table is simplified to indicate only different pathways of each cancer. The significant pathways of specific network marker in colorectal cancer are also enriched in core network marker. Therefore, there is not different pathway in specific network marker of colorectal cancer. (B) Combination of pathways of core and specific network markers for each cancer from Tables A.4 and 12.2A, respectively.

and receptors (receptor mutations of HER2/neu [ErB2] or the amplification of the *HER2* gene); (2) downstream signal transduction networks (Raf/Ras/MAPK or PI3K-AKT-mTOR); and (3) regulatory genes of the cell cycle (cyclin D1/CDK4, CDK6, and cyclin E/CDK2) [226]. Increasing evidence convincingly shows aberrant expression of cell cycle regulators in multiple cancers. Here, 13 of the core network markers identified are involved in the cell cycle pathway (Fig. 12.5(A)).

The Wnt/β-catenin signaling pathways (Fig. 12.5B) are a branch of a function network; they are a bundle of signaling pathways composed of proteins that transduce signals from the outside of a cell through the receptors of the cell surface to the inside of the cell. They contribute significantly to the developmental process, particularly to direct cell attachment and proliferation. Under normal conditions, the Wnt signaling pathway is critical for healthy and normal development, while in adult cells, as switching between the cell proliferation and invasion modes. Therefore, the Wnt signaling pathway participates in each of the stages of malignant cancer development and clearly contributes to human tumor progression. Other identified common pathways such as the Notch signaling pathway, adherens junctions, the TGF-β signaling pathway, ubiquitin-mediated proteolysis, and the p53 signaling pathway are also associated with cancer [211,227–230].

Pathway analysis of specific network markers of each cancer

The functional annotations of significant proteins common to the cancer pathways identified via the KEGG database show that the core network markers can take part in multiple cancer processes. Therefore, the core network marker indicates that there may be a basic or common initiator of carcinogenesis, i.e., the housekeeping mechanism of cancers. Next, we aimed to understand the differences between each cancer and from that in core network markers by investigating each specific network marker. The pathway analysis of each specific network marker was also conducted using KEGG (Table A.5). In the following sections, we focus on different parts of the prominent pathways of specific network markers in each cancer type, as shown in Table 12.2.

Investigating the mechanisms underlying the specific network markers in bladder cancer

In Table 12.2, the RLR signaling pathway was the only pathway specifically elevated in bladder cancer relative to the other three cancers (Fig. 12.6A). RLRs are a family of intracellular pattern-recognition receptors. They can sense nucleic acids derived from viruses and stimulate antiviral innate immune responses. RLR proteins are expressed in both immune and nonimmune cells. By recognizing these nucleic acid species, RLRs recruit particular intracellular adaptor proteins to initiate signaling pathways that culminate in the activation of NF-κB, MAP kinases, and IRFs.

(A)

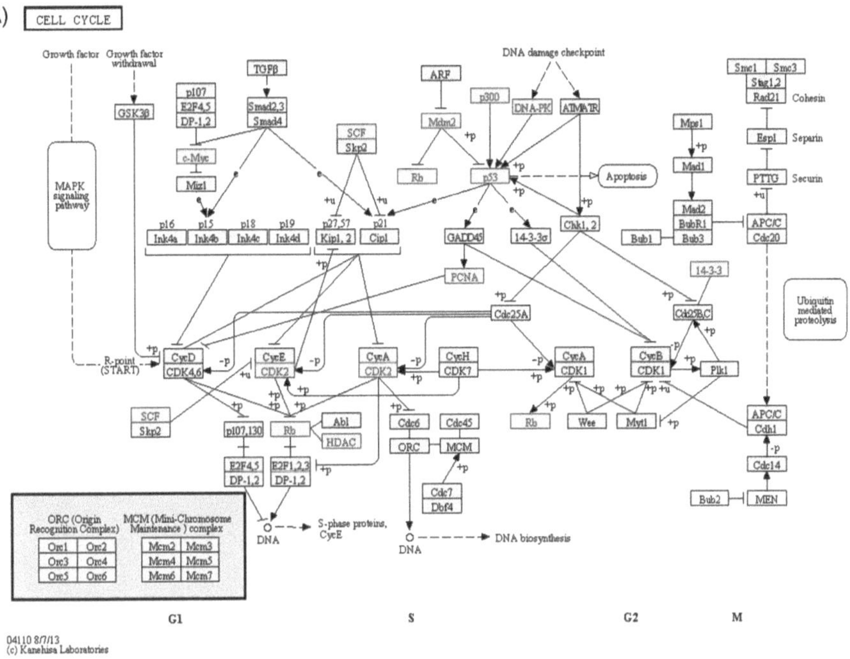

(B)

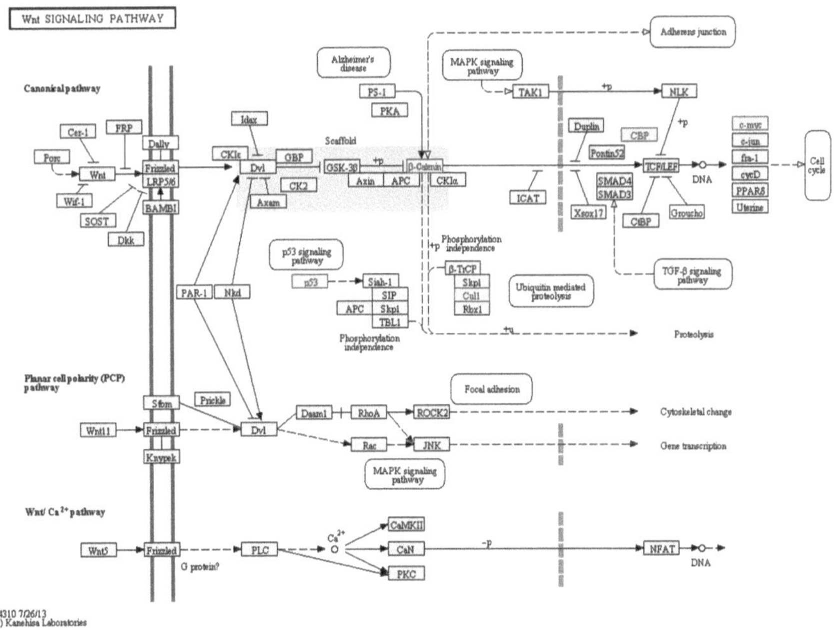

Figure 12.5 Overview of significant pathways in core network marker. (A) The proteins in the core network marker are enriched in "hsa04110: Cell cycle" (Rank 1 in Table A.4). (B) Proteins in the core network marker are enriched in "hsa04310: Wnt signaling pathway" (Rank 2 in Table A.4). The list of 28 significant proteins in core network marker is important for four types of cancer. Among these KEGG pathways via DAVID tool (Table A.4) showing a significant association with common proteins, two mainly entitled pathways (Rank 1 and Rank 2) are illustrated, respectively. The proteins in core network marker highlighted by *pink* (dark gray in print versions) show potentially targets in the pathways.

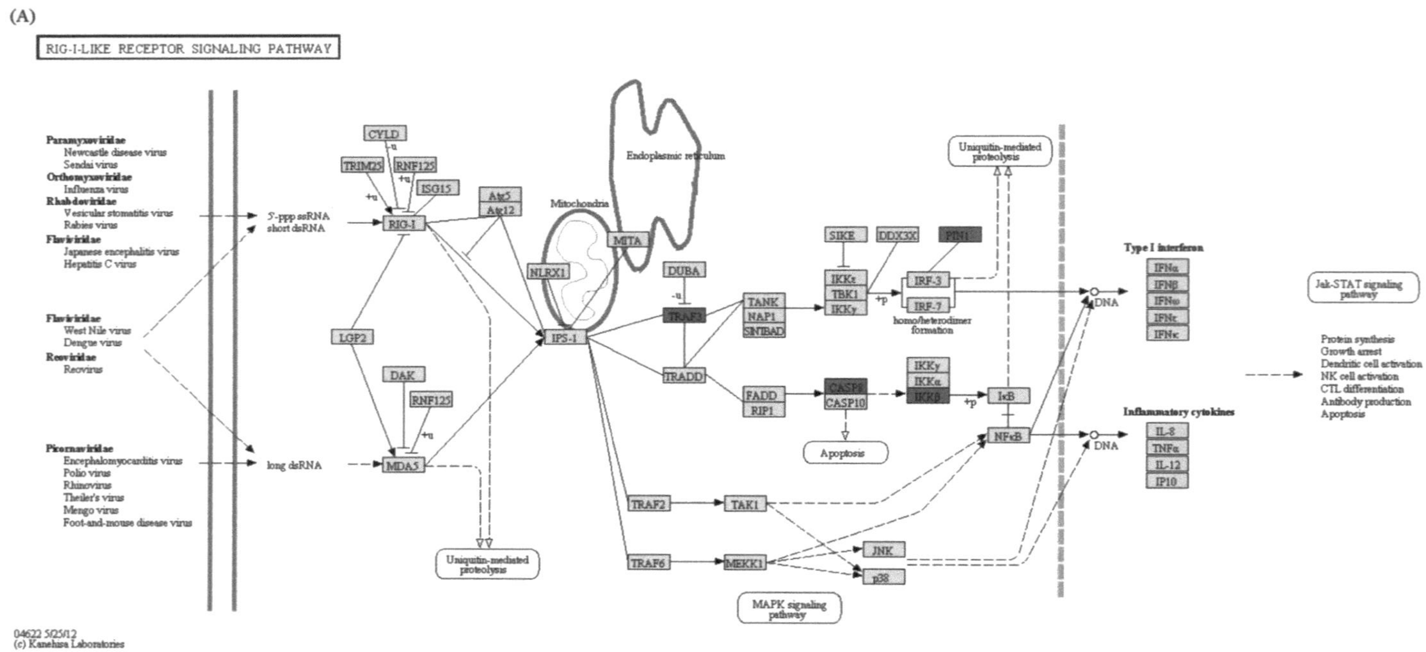

Figure 12.6 Overview of significant pathways in specific network marker of each cancer. Among these KEGG pathways via DAVID tool (Table 12.2) showing a significant association with specific proteins of each cancer, these molecular pathways entitled with *p*-value $\leq .05$ and different from other cancers and common significant pathways are illustrated, respectively. It shows that these specific pathways are identified to play an important role in the carcinogenesis mechanism of specific cancer. The proteins in specific network markers of each cancer highlighted by *pink* (dark gray in print versions) show potentially targets in the pathways. (A) Proteins in the specific network marker of bladder cancer are enriched in "hsa04622: RIG-I-like receptor signaling pathway." (B) Proteins in the specific network marker of liver cancer are largest enriched in "hsa03050: Proteasome" (Table A.5, Liver Cancer). (C) Proteins in the specific network marker of liver cancer are second enriched in "hsa04660: T-cell receptor signaling pathway" (Table A.5, Liver Cancer). (D) Proteins in the specific network marker of lung cancer are enriched in "hsa03440: Homologous recombination."

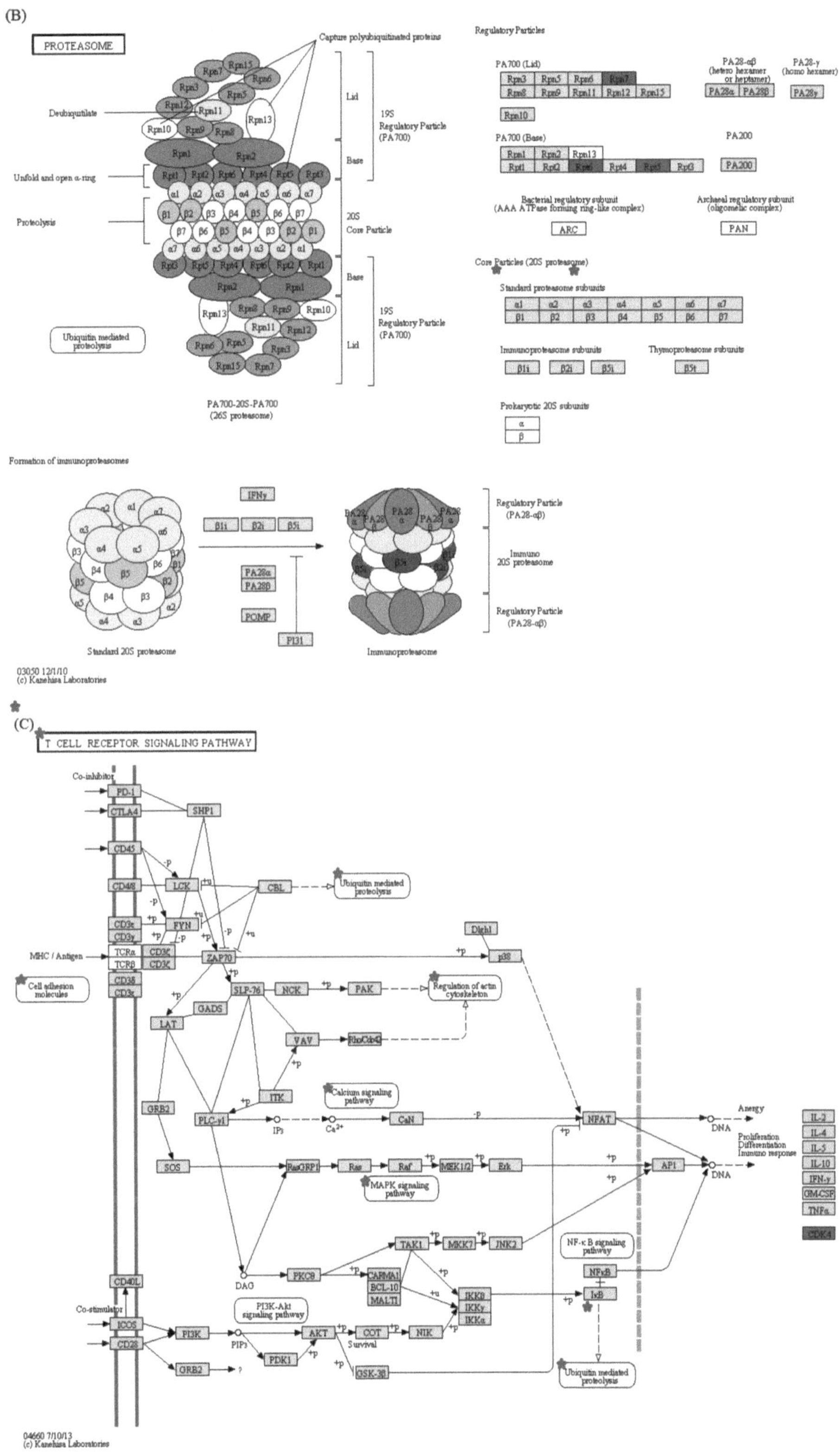

Figure 12.6 (Continued)

(D)

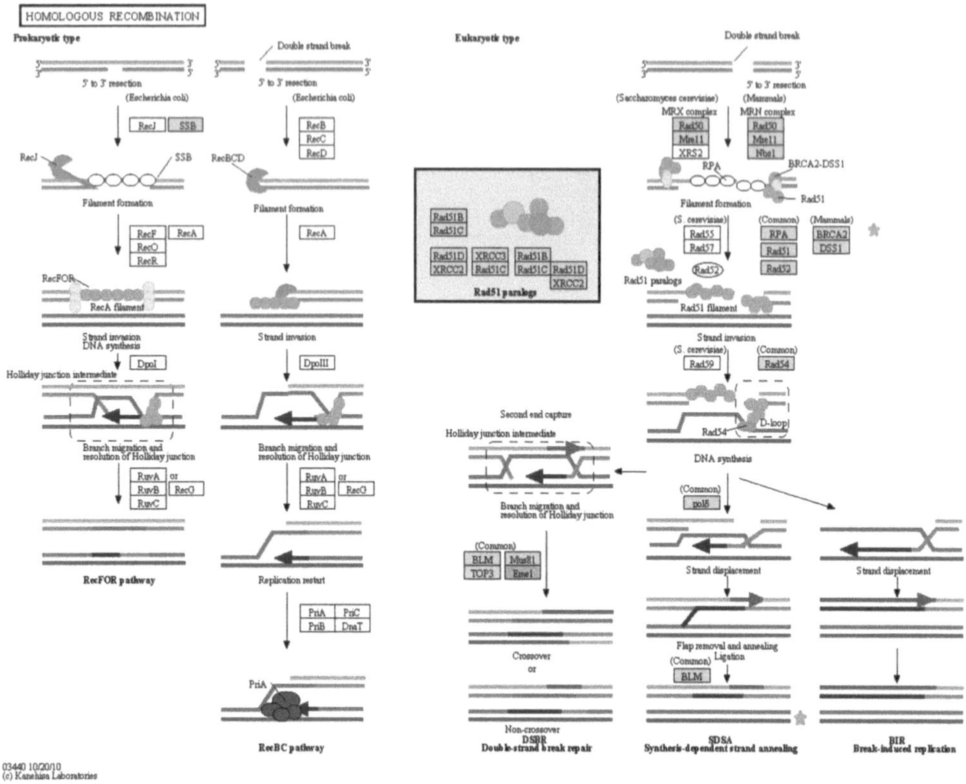

Figure 12.6 (Continued)

These proteins control the transcription of genes encoding type I interferon and other inflammatory cytokines that are important for killing viruses [231]. Currently, there are 53 companies with partners developing 33 RLRs signaling pathway-targeting drugs in 106 developmental cancer projects [232]. Although there are no studies indicating that bladder cancer is directly connected with this pathway, we can infer that the identified RLR signaling pathway may be an important mechanism in bladder cancer. Therefore, RLRs merit further study due to both the importance of this pathway and these results.

On the other hand, we found that four identified proteins, CASP8, CHUK, TRAF3, and PIN1, in the specific network marker of bladder cancer, are involved in the RLR signaling pathway. CASP8 is a caspase protein and most likely acts upon caspase 3.

CASP8 participates in PCD induced by the *Fas* gene and diverse apoptotic stimuli. CHUK, also known as IKK-α or IKK1, is a protein kinase and it is part of the IκB kinase complex that plays a critical role in regulating the NF-κB transcription factor.

However, CHUK has many other additional cellular targets; it is thought to operate independently from the NF-κB pathway to regulate epidermal differentiation. TRAF3 is a member of the TNF receptor-associated factor (TRAF) protein family. It regulates the pathways leading to the activation of MAP kinases and NF-κB, as well as having an important role in the regulation of B-cell survival. Furthermore, the TRAF3 signaling pathway also leads to the production of interferons and cytokines. PIN1 is an enzyme that regulates the outcome of proline-directed kinase signaling, and, consequently, regulates both cell proliferation and survival. Recent data also indicates that PIN1 plays a key role in immune response. In particular, the upregulation of PIN1 may result in the development of certain cancers and the inhibition of PIN1 may have therapeutic implications for immune disorders and even cancer.

Investigating the mechanisms underlying specific network markers in liver cancer

The pathways that are enhanced in the specific network marker group are different from the pathways of core network markers (Table 12.2). The proteasome is the largest pathway that is most prominent in liver cancer (Fig. 12.6B). The proteasome acts as a protein-destroying machine in many crucial cellular functions, such as cell cycle control, cell differentiation, signaling pathways, antigen processing for proper immunological processing, response to cellular stress and inflammation, and the apoptotic process. It can degrade various cellular proteins in a quick and timely manner. Most substrate proteins are posttranslationally modified by ubiquitin before their degradation by the proteasome. The proteasome is a large protein complex in the cytosol and it is responsible for degrading proteins that have been tagged for destruction by ubiquitin or by some other methods. In this section, there were five specific network markers identified in liver cancer that are related to the proteasome, i.e., PSMC5, PSMA6, PSMC3, PSMA4, and PSMD6. Most proteins are degraded via the proteasome-dependent and ubiquitin-independent pathways that are involved in growth control processes that, when disrupted, can promote oncogenesis. This is clear for both uninfected and virus-infected cells and tumors [233]. Since the proteasome can degrade many cellular proteins, several with regulatory functions, it is not surprising that proteasome inhibitors affect many biological processes and can prevent cancer. Proteasome inhibition has already proven to be a novel and promising strategy for the treatment of cancer. Additionally, it has been reported that proteasome inhibitor treatment is used in alcoholic liver disease [234] and in rat liver cells [235]. These findings suggest that the proteasome pathway has a significant relationship with liver cancer and may be involved in the early carcinogenic process.

Another identified pathway with a p-value $\leq .05$ is the T-cell receptor (TCR) signaling pathway (Fig. 12.6C). Activation of T lymphocytes plays a critical role in the efficient response of the immune system. The activation of the TCR signaling pathway can be induced by the interaction of the TCR and its coreceptor with the MHC/peptide complex. This process involves costimulatory molecules and signals mediated by cytokine receptors, which eventually lead to a series of signaling cascades [236]. These cascades comprise an array of protein-tyrosine kinases, phosphatases, GTP-binding proteins, and adaptor proteins that regulate generic and specialized functions, leading to T-cell proliferation, cytokine production, and differentiation into effector cells. Recent studies show that the tumor-specific T-cell response is compromised in both primary and secondary liver cancer [237]. There are also some studies documenting that examination of the TCR signaling pathway has also yielded new therapeutic targets for the next generation of anticancer drugs that allow T cells to kill tumors unimpeded [238]. In addition, there are many new and innovative treatments for liver cancer that are currently being tested and considered, including the possible use of T cells to boost the body's efficiency in combating liver cancer on its own. These results show the relationship between the TCR signaling pathway and liver cancer, indicating that it has an important role in liver cancer.

In this section, three specific protein markers of liver cancer, including MAPK14, NFKBIA, and CDK4, were identified in TCR signaling. The mitogen-activated protein kinase (MAPK) cascade is a highly conserved module that is relevant to many cancers. One MAPK-related protein, MAPK14 (p38α), has been well-studied for its key roles in the response to both inflammation and environmental stress [74]. NFKBIA (NF-κB inhibitor α, or IκBα) is a member of the NF-κB family of cellular proteins that inhibit the NF-κB transcription factor. It is mutated in some Hodgkin's lymphoma (HL) cells; such mutations inactivate the IκBα protein leading to the chronic activation of NF-κB in lymphoma tumor cells. This activity contributes to the malignant state of these tumor cells. CDK4 is a member of the Ser/Thr protein kinase family and is important for cell cycle G1 phase progression. Mutations in CDK4, as well as in its related proteins including D-type cyclins p16(INK4a) and Rb, were all found to be associated with tumorigenesis in a variety of cancers.

Investigating the mechanisms underlying the specific network markers in lung cancer

Homologous recombination (HR) is a pathway that is enriched in specific network markers as compared to core network markers in lung cancer (Fig. 12.6D). HR comprises a series of interrelated pathways and plays a crucial role in the repair of DNA double-strand breaks (DSBs) where the DSBs are lethal lesions. It occurs in the late S/G2 phase and possibly generates single-stranded regions of DNA, followed by strand invasion, DNA synthesis, formation of Holliday junctions, branch migration, and resolution. If there are a large number of HR proteins, most cancers will possibly have lost a subpathway of HR. It may cause genetic instability and induce further genetic changes, inducing cancer growth [239]. In addition, some studies have found that lung tumors, compared with normal tissues, exhibit significant overexpression of DNA repair genes that are associated with HR pathways. Transcriptional overexpression of some of these genes showed strong correlation with more aggressive clinical behavior. It may explain the low level NSCLC response to combination chemotherapy [240]. For instance, etoposide (VP-16) is usually used to treat small cell lung cancer and it has been found that large quantities of the HR protein RAD51 correlate with increased repair of DNA damage induced by etoposide and drug resistance in lung cancer cells [239]. Therefore, we are not surprised to identify this pathway in lung cancer.

In these results, there are two specific network markers in lung cancer, namely BLM and RPA3, that map onto this pathway. BLM participates in DNA replication and repair. Cancer results from genetic changes that allow cells to divide in an uncontrolled way. Altered BLM protein activity may also lead to an increase in cell death, resulting in slow growth in affected individuals. Another protein, RPA3, is also required for DNA recombination, repair, and replication.

Investigating the mechanisms underlying the specific network markers in colorectal cancer

The number of proteins that are specific network markers in colorectal cancer is too small (only four proteins) to allow mapping of the significant pathways (p-value $\leq .05$) by this analysis tool. Therefore, this will not be further discussed with respect to pathways that are enhanced in colorectal cancer relative to the other three cancers (Table 12.2).

However, the proteins in the specific network marker of colorectal cancer were further investigated at a molecular level as follows. We took four proteins mapping to the KEGG database separately (Table A.6) and then found that two proteins, APC and RUVBL1, both of which are involved in the Wnt signaling pathway, one of common pathways. The Wnt signaling pathway is well-studied in colorectal cancer. APC is also well-studied and plays a tumor-suppressor role in colorectal cancer. It is clear that the first step in colon carcinogenesis involves the loss of a functional *APC* gene, resulting in the formation of colon polyps (precancerous growths) on the inside of the colon wall [241]. Indeed, several studies have shown that APC inactivation in vivo in mice is sufficient to initiate tumor development [229,242,243]. The APC protein does so by preventing the Wnt signal-transduction pathway from activating the expression of proto-oncogenes, including the *c-myc* gene. The absence of functional APC protein leads to inappropriate production of Myc, resulting in a proliferation rate that is greater than that of normal polyps [211]. Another protein, RUVBL1, is an ATPase associated with chromatin-remodeling complexes and has important roles in transcriptional regulation, the DNA damage response, telomerase activity, snoRNP assembly, cellular transformation, and cancer metastasis. It has been shown that EHF-mediated RUVBL1 expression allows colon tumor cells to avoid apoptosis mediated by p53. Thus, RUVBL1 may be a novel molecular target for the treatment of colon tumors [244]. In summary, they may have malignancy-specific roles in colorectal cancer by regulating an aberrant program.

Strengthening the validation of the core and specific network markers by outside groups

In this section, to further strengthen these identified result and approach, we performed a simple validation by outside groups. We applied both core and specific network markers from the above four cancer datasets to two other independent hold-out datasets of lung cancer (GSE18842 [245] and GSE10072 [246]) from the Affymetrix platform. Only the early stage of cancer was considered. Note that all of these tested datasets were not used in "Investigating the mechanisms underlying the specific network markers in colorectal cancer" section, thereby allowing the evaluation of these results objectively. The validation was carried out with this flowchart again by tested datasets, and then we compared the new results with the previously identified core and specific network markers, individually. The percentages of overlapping proteins can be used to validate these results as shown in Table 12.3. The results show that

Table 12.3 The validation for core network marker and specific network markers

(A)

				Specific network marker for each cancer			
GEO	Cancer type	Platform	Cored marker	Bladder	Colorectal	Liver	Lung
			(28)	(26)	(4)	(24)	(13)
GSE18842	Lung	GPL570	**25 (89.3%)**	6 (23.1%)	2 (50.0%)	5 (20.8%)	**8 (61.5%)**
GSE10072	Lung	GPL96	**24 (85.7%)**	3 (11.5%)	1 (25.0%)	7 (29.2%)	**6 (46.2%)**

(B)

				Specific network marker for each cancer			
GEO	Cancer type	Platform	Cored marker	Bladder	Colorectal	Liver	Lung
			(28)	(26)	(4)	(24)	(13)
GSE6344	Kidney	GPL96	**23 (82.1%)**	4 (15.4%)	0 (0%)	7 (29.2%)	2 (15.4%)
GSE6956	Prostate	GPL571	**22 (78.6%)**	2 (7.7%)	1 (25.0%)	6 (25.0%)	2 (15.4%)

To test the identified result and approach, a simple validation is used, showing that percentages of mapping proteins by outside groups. The tested datasets are downloaded from GEO databases, which are searched from the Affymetrix platform and are different from above datasets in Table A.1. Within grid for specific network marker of each cancer in table, marked number and ratio (%) implies numbers of mapping protein in tested result and percentages with protein numbers of identified core and specific network marker, respectively. (A) Tested results of two independent datasets of lung cancer which is the one of above four cancers. (B) Tested results of two new types of cancers different from above four cancers. Numbers in bold denote the better results in the validation.

core network markers achieved higher scores on all lung cancer datasets. However, the specific network markers of lung cancer were not effectively identified, although they are still better than the results for other three cancers, implying that the specific network marker for lung cancer is more specific to lung cancer than other cancers. It is worth noting that since the number of specific network markers in colorectal cancer was too small, the results are not statistically robust. Kidney cancer and prostate cancer selected from the GSE6344 [247] and GSE6956 [248] datasets, two additional cancer types, were also examined for these network markers. Among the 28 proteins in core network marker group, 23 and 22 were mapped in kidney and prostate cancer, respectively, resulting in a comparable testing score of 82.1% and 78.6%. This shows that this core network markers are also relevant to other cancers. Furthermore, the testing scores for the specific network markers in the kidney and prostate cancer datasets were all below 25%, thus implying the specific network markers are specific as compared to other cancers.

Some other extra preliminary results

1. Here we just show the preliminary results, and we'll develop it to an independent topic in the future. Fig. A.3 A1 and A2 shows the cancer-PPIN and benign tumor-PPIN (BTPPIN) for lung cancer and benign lung tumor networks.
2. Identifying the PPI interface to develop small molecule inhibitors has become a new direction for cancer target therapy. This section is a prior work to analyze the carcinogenesis mechanism from early to late stage bladder cancer by network-based biomarker approach. Other research does not distinct these two stage cancers to get more insight into carcinogenesis of bladder cancer so far. The network marker and their related pathways identified in the early stage bladder cancer most are related to ordinary cancer mechanisms, which just show a highly active state of the early stage and cannot find more novel results. However, from identified two specific significant pathways found in late stage bladder cancer, i.e., ribosome pathway and spliceosome pathway, it shows a novel result to be the target for cancer therapy. Fig. A.3 B1–B4 are the results for early and late stage bladder cancer. We can see the evolution of network marker.

12.2.3 Discussion

Cancer is a complex disease; carcinogenesis in humans is a multistep process that transforms normal cells into malignant derivatives. Investigating carcinogenesis from a systems perspective is inevitable. Some novel studies about some of the most deadly cancers, such as leukemia [249] and endometritis [250], have indicated that, while the carcinogenic genes are similar, invasiveness varies in different parts of the body. The bioinformatics analyses of microarray datasets publicly available for bladder, colorectal, liver, and lung cancers have provided an opportunity to investigate the hypothesis in which there are foundational molecular mechanisms for all types of organ-based cancers. The proposed method considers not only differentially expressed proteins but also the protein association network structure. This property is important for the discovery of significant proteins in carcinogenesis and provides some insights into the mechanism responsible for carcinogenesis. We identified 28 significant proteins as core network markers, many of which are involved in fundamental cellular processes that are responsible for proliferation, differentiation, apoptosis, and immunity. In normal cells, these cellular processes are precisely regulated to achieve appropriate responses. However, in carcinogenesis, mutations can damage the key components. Therefore, these mutations dysregulate cellular processes, resulting in aberrant expression of the significant proteins and the transformed malignancy. We also found that this dysregulation exists in multiple cellular processes and pathways. In addition, further analysis of the common and specific proteins in cancer networks may be important to determine the mechanisms leading to carcinogenesis and may contribute to the identification of susceptibility proteins for each cancer. In the development of carcinogenesis, common genetic variations of core network markers suggest a nonspecific influence because they operate in the general pathways without specificity. Therefore, the core network markers are common to many cancers, while specific network markers can be regarded as distinctive for each cancer. In these results, we found that many biological pathways are closely related to carcinogenesis. For example, the critical proliferation- and survival-related pathways in almost each type of cancer are abnormal in terms of the common part, while some cancer cells express specific abnormal pathways in terms of distinctive parts [251]. Thus, this hypothesis agrees with reports in the literature.

In this section, we use the so-called systems biology approach to study the cancer's network marker. Firstly, microarray data, PPI databases and PPI interaction model in Eq. (12.1) are employed to construct the PPINs of normal and cancer cells by the maximum likelihood parameter estimation method (see Appendix A.1). The AIC system order detection method (Appendix A.2) is employed to prune the false positive PPIs to obtain the real PPINs of normal and cancer cells, i.e., we use the so-called reverse engineering method to construct PPINs of normal and cancer cells. Then the differential PPIN between cancer PPIN and normal PPIN is obtained in Eq. (12.7) to investigate PPI variations of each protein in the differential PPIN due to the carcinogenesis. Finally, the CRV based on PPI variations is also proposed to evaluate the significance of carcinogenesis for each protein of differential PPIN. Proteins with significant CRV (p-value $<.01$) are considered significant proteins of the cancer. The 28 significant proteins in Table 12.1 are the intersection of these significant proteins of four cancers, and these 28 significant proteins and their PPIs construct core network markers of cancers in Fig. 12.4. The distinctive significant proteins outside of these 28 significant proteins form the specific network marker of each cancer. Finally, based on these core network markers and specific networks, we investigate the mechanism of carcinogenesis process with the help of databases (e.g., the GO database, DAVID and KEGG pathway database) and try to find a multiple network target therapy of cancer. Unlike the conventional theoretical methods, which always give a single mathematical model for cancer network for a more detailed theoretical analysis, this study is a systems biology approach to cancer network markers based on real microarray data through the use of so-called reverse engineering, theoretical statistical methods and data mining method from combination of big databases.

Recently, a PPI target seems to have become a paradigm for the drug discovery of cancer therapy and precision medicine [1,2]. Unlike the conventional drug design focusing on the inhibition of a single protein, usually an enzyme or receptor, the small-molecule inhibition of direct PPIs that mediate many important biological processes is an emerging and challenging scheme in drug design, especially for cancer. Extensive biological and clinical investigations have led to the identification of PPI hubs and nodes that are critical for the acquisition and maintenance of characteristics for cell transformation in cancer. These kinds of cancer-enabling PPIs have become significant therapeutic target in anticancer strategies. The technological progresses in the discovery of PPI modulator and the validation of clinical PPI-targeting agents, makes it possible to have an anticancer process by interfering with PPI [221,222].

Therefore, future research directed at PPI target discovery, PPI interface characterization, and PPI-focused chemical libraries is expected to accelerate the development of next generation of PPI-based anticancer agents. However, the PPINs of cancer are very complex and are some different for various cancers. In the situation, we focus on the PPIN markers, whose PPIs are all with significant CRV, to estimate the PPI changes during carcinogenesis process, to exploit the important targets and their PPI interface characterization for different cancers. Therefore we could not only gain an insight into the crucial common pathways in the carcinogenesis, but also could obtain a highly promising PPI target for cancers. The specific network marker could be considered a specific PPI target of each cancer. If we could develop some combining anticancer strategies to target PPIs in both core network marker and specific network marker simultaneously in future, it may provide emerging opportunities for anticancer.

The conventional methods to find hubs or significant proteins in PPIN of cancer are to find those proteins with the most links to other proteins. In this section, we compared cancer and normal PPINs, which are obtained by parameter estimation method and AIC order detection schemes from microarray data of patient and normal people, respectively, to obtain the PPI differential network to represent the PPI variations during the tumorigenesis process. Further, we also developed the CRV of each protein in Eq. (12.8) based on the total variations of PPI interaction parameters with other proteins in the PPI differential network to validate the significant PPI changes during tumorigenesis process. Finally, with the intersection and distinction of these significant proteins with high CRV, we obtained the core network marker for four cancers and specific network marker for each cancer. These core network makers and specific network markers are novel and could provide efficient PPI target, for a small-molecule drug to bind to destroy tumors. All of the calculation and estimation process are based on theoretical calculation and estimation theory. We know maximum likelihood parameter estimation and AIC model selection method are famous and widely used system identification methods from experimental data. During these estimation and learning processes, the model can learn the most possible PPIN from large amount of microarray data and big databases to interpret the hidden biological mechanisms.

Our theoretical background is one kind of the reconstruction of genetic and PPINs from microarray data and big databases. Hidde de Jong's review article [252] gives a summary of these theories and modeling methods, in particular directed graphs, Bayesian networks, Boolean networks and their generalizations, ordinary and partial differential equations, qualitative differential equations, stochastic equations, and rule-based formalisms. To summarize briefly, these theories include two main parts: one is the original raw data where the biological genetic or PPI mechanisms are hidden in them, such as microarray data or vast literature; another is the theoretical methods or mathematical modeling to reveal the hidden mechanisms in the original raw data. So, this work is not only empirical, it is also as important as these works summarized in this section. Further, we use more novel theoretical idea of AIC to do model order selection to reduce the size of candidate PPIN to more real network by pruning the false positive PPIs.

Our results are only the beginning, as the significant proteins identified prompt the next stage of experimental research and therapeutic targeting, which will confirm the role of these network markers in multiorgan cancers. Furthermore, the proteins found to interact with different partners in the network could correlate with the relevant disease [68]. Increasing evidence shows that the interaction and network of proteins play important roles and are thus worth investigating with respect to the molecular mechanisms of cancer. Therefore, it is necessary and important to introduce this new concept into cancer research in order to improve diagnosis, therapies, and the prognosis of diseases [253]. From the results presented, we found that the edges (interactions) and magnitude of nodes of the core network markers for each cancer were also different. For instance, mutations of TP53, a well-studied protein that was found to be a core network marker, have been reported to occur in almost every type of cancer at rates ranging from 10% to 100%. Interestingly, the abnormalities in TP53 were shown to occur at different phases of the multistep process of malignant transformation, suggesting that it contributes differently to tumorigenesis in each cancer type. TP53 has been shown to interact with multiple proteins in the current study.

In liver cancer, it is thought that TP53 is eliminated together with the RB and C/EBPα tumor suppressors following the increase in gankyrin at the early stage of tumorigenesis [253]. Although there are fewer studies, TP53 is also associated with YWHAZ of cancer. It has been reported that YWHAZ downregulates TP53 in mammary epithelial cells [254], suggesting that this link between TP53 and YWHAZ may also exist in colorectal cancer. In addition, TP53 plays different roles in multiple cancers, resulting in different pathophysiological phenotypes. The increasing evidence of aberrant TP53 function has already led to the recognition of some important molecules with the potential for clinical treatment [220]. Therefore, analysis of above types of PPIs can provide important data at different levels of the PPINs in cancer for further study.

Furthermore, it is worth noting that some limitations exist in the proposed method. Firstly, gene expression profiles were overlaid with the corresponding proteins for further analysis. However, levels of mRNA do not always correlate with protein levels and do not provide information on PTMs, such as phosphorylation, which may be critical for regulating protein activity. Secondly, there is a large variation in the coverage of protein interaction data across the interaction databases. In this section, we only used the BioGRID database, but increased PPI information may be used to enhance the result. Thirdly, since these four cancers have to be compared to each other simultaneously, the smallest network will restrict the number of proteins ultimately obtained. Moreover, the size of the protein association network is restricted by the sample size available to avoid overfitting in network construction. This results in the exclusion of some well-studied proteins that are relevant to carcinogenesis in these results of core network markers, including KRAS, MET, and PI3KCA. In addition, to obtain more valuable and realistic results and to reach our objective, we adopted a more rigorous statistic (p-value $\leq .01$), only chose the intersecting and distinct proteins, and neglected the overlapping proteins between cancers. Therefore, some representative proteins in certain cancers were not included in these results. Although the so-called specific network markers are more specific to its cancer than other specific network markers of other cancers, they are not clearly representative of a certain type of cancer. To overcome this problem and improve the results, we could increase the numbers of different types of cancer or subtypes of the same cancer in different facts, relax restrictions, and unify the protein numbers of all cancers for comparison. Another limitation of the proposed method was related to the addition of a functional pool before the construction of a candidate PPIN. Although the selected proteins from the final pool are both significantly expressed and are functionally related to cancer, some proteins with other functions are thus not retained, and may be discarded unnecessarily. These "lost proteins" may be involved in cancer, but it is not yet known. In spite of this, the results in this section are useful.

Our results give systematic insights into the network markers of carcinogenic processes and identify significant proteins relevant to them. The significant proteins of the core network marker group consist of those that are cancer-related and some others that have not been previously reported. These significant proteins not only provide new targets for further studies to understand the molecular mechanisms in carcinogenesis but also are potential targets for therapeutic interventions. The main challenge of cancer research is to find effective therapeutic approaches that can specifically kill malignant cells. With the advances in understanding the mechanisms involved in the carcinogenic process, targeted therapy, which is more effective and less harmful to normal cells, has been developed to inhibit the specific molecules that play crucial roles in tumor growth. The significant proteins identified by the proposed network-based biomarker approach have the potential to be novel targets for cancer treatment. In addition, identifying the type of cancer is crucial for cancer diagnosis. Theoretically, considering the application from a clinical viewpoint (e.g., tissue or blood sample testing), it is possible to diagnose cancer at an early stage by analyzing the core network markers (28 proteins), while confirming the type of cancer with the specific network markers for each cancer. To do this, we used a simple verification to test whether the identified core and specific network markers in the tested datasets were able to simultaneously change. However, further investigation is needed to reinforce these results. Furthermore, pending further validation in the future, these core and specific network markers identified by this approach may prove to be biomarkers for drug responses and prognosis in various cancer types.

In this section, we only focused on the core network marker and specific network markers at the initial stage of four types of cancer to investigate the carcinogenesis mechanism of initial stage of cancers. In the future, the core network marker and specific network markers of the late stage of four cancers will be constructed by the late stage microarray data. Then the evolution of core network markers and specific network markers in response to mutation selection on a cancer cell during carcinogenesis process could be investigated by comparing network markers between different stages. These evolution network markers can gain more insight into the evolution mechanisms during the whole carcinogenesis process.

12.2.4 Conclusion

In this section, to investigate multiple carcinogenesis, the PPINs of four cancers and noncancers were constructed using corresponding microarray data and database-mining via AIC order detection and statistical t-tests (i.e., CPPINs and NPPINs). The core network markers and specific network markers were obtained, respectively, by the intersection and difference of proteins of four sets of significant proteins based on the structure changes of protein associations that differentiate CPPIN from

NPPIN (i.e., CRV). We found that the core network with multiple cancers consists of a common 28 proteins but with some differences of PPIs in each cancer, while specific networks consist of different proteins for each cancer. A total of 2 of the 28 significant proteins in the core network marker group (i.e., UBC and PSMA3) still require further study as not much is known about them. We found that there were seven pathways (cell cycle, Wnt signaling pathway, Notch signaling pathway, TGF-β signaling pathway, and p53 signaling pathway) that were particularly prominent in the 28 significant proteins of the core network, which are considered to be the housekeeping mechanism of multiple cancers (e.g., proliferation, differentiation, apoptosis, and immunity). From the key pathways, with respect to specific network markers, we found that the RLR signaling pathway was specific to bladder cancer, the proteasome and TCR signaling pathways were specific to liver cancer, and that the HR pathway was specific for lung cancer. These specific pathways play an important role in the carcinogenesis mechanism of these specific cancers. After validation from outside groups, the proposed core network markers were able to determine the core carcinogenesis mechanism of multiple cancers and the specific network markers seemed specific for different cancers. The proposed method can be used to provide practical medical therapy and facilitate the development of new anticancer drugs in this section.

APPENDIX A: METHODS, TABLES, AND FIGURES

Appendix A.1: Parameter identification of regression model in Eq. (12.1) by maximum likelihood method

Eq. (12.1) can be written as the following requiring form

$$x_i[n] = \begin{bmatrix} x_1[n] & \ldots & x_{Mi}[n] \end{bmatrix} \begin{bmatrix} \alpha_{i1} \\ \vdots \\ \alpha_{iM_i} \end{bmatrix} + \omega_i[n] = \phi_i[n] \cdot \theta_i + \omega_i[n], \tag{A.1}$$

where $\phi_i[n]$ denotes the regression vector which can be obtained from microarray data, θ_i is the parameter vector to be estimated. Suppose that there are m samples, then it is easy to acquired values of $\{x_i[n]\ \phi_i[n]\}$ for $n \in \{1, \ldots m\}$. In this case, Eq. (A.1) for different samples can be represented as the following vector form:

$$\begin{bmatrix} x_i[1] \\ \vdots \\ x_i[m] \end{bmatrix} = \begin{bmatrix} \phi_i[1] \\ \vdots \\ \phi_i[m] \end{bmatrix} \cdot \theta_i + \begin{bmatrix} \omega_i[1] \\ \vdots \\ \omega_i[m] \end{bmatrix}, \tag{A.2}$$

where $\phi_i[m] = [x_i(m) \ldots x_{M_i}(n)], \theta_i = [\alpha_{i1} \ldots \alpha_{iM_i}]^T$

For simplicity, it can be represented as follows:

$$X_i = \Phi_i \cdot \theta_i + e_i, \tag{A.3}$$

where $e_i = [w_1(1)\ldots w_i(m)]^T$.

In Eq. (A.3), the noise e_i for different samples was regarded as independent random variables of normal distribution with zero mean and unknown variance σ_i^2, i.e., $E\{e_i\} = 0$, and $\sum_i = E\{e_i e_i{}^T\} = \sigma_i^2 I$, where I is the identity matrix. The probability density function of e_i is given as follows.

$$p(e_i) = \frac{1}{((2\pi)^m \det\Sigma_i)^{1/2}} \exp\left(-\frac{1}{2} e_i^T \sum\nolimits_i^{-1} e_i\right), \tag{A.4}$$

From Eq. (A.4), we can obtain the likelihood function

$$L(\theta_i, \sigma_i^2) = p(\theta_i, \sigma_i^2) = \frac{1}{(2\pi\sigma_i^2)^{m/2}} \exp\left(-\frac{(X_i - \Phi_i\theta_i)^T (X_i - \Phi_i\theta_i)}{2\sigma_i^2}\right), \tag{A.5}$$

Maximum likelihood estimation method aims at finding θ_i and σ_i^2 to maximize the likelihood function in Eq. (A.5). In order to simplify the computation, it is practical to take the logarithm of the likelihood function, which yields the following log-likelihood function:

$$\log L(\theta_i, \sigma_i^2) = -\frac{m}{2}\log(2\pi\sigma_i^2) - \frac{1}{2\sigma_i^2}\sum_{n=1}^{m} [y_i[n] - \phi_i[n] \cdot \theta_i]^2, \tag{A.6}$$

where $x_i[n]$ and $\phi_i[n]$ are the nth element of X_i and Φ_i in Eq. (A.3), respectively.

By the maximum likelihood parameter estimation method, we expect the log-likelihood function to have the maximum at $\theta_i = \hat{\theta}_i$ and $\sigma_i^2 = \hat{\sigma}_i^2$. The necessary conditions for the maximum likelihood estimates $\hat{\theta}_i$ and $\hat{\sigma}_i^2$ must conform to the following two equations:

$$\begin{aligned} \frac{\partial \log L(\theta_i, \sigma_i^2)}{\partial \theta_i} &= 0, \\ \frac{\partial \log L(\theta_i, \sigma_i^2)}{\partial \sigma_i^2} &= 0. \end{aligned} \tag{A.7}$$

The estimated parameters $\hat{\theta}_i$ and $\hat{\sigma}_i^2$ are shown as

$$\hat{\theta}_i = (\Phi_i^T \Phi_i)^{-1} \Phi_i^T Y_i, \tag{A.8}$$

$$\hat{\sigma}_i^2 = \frac{1}{m}\sum_{n=1}^{m} [x_i[n] - \phi_i[n] \cdot \hat{\theta}_i]^2 = \frac{1}{m}(X_i - \Phi_i \cdot \hat{\theta}_i)^T (X_i - \Phi_i \cdot \hat{\theta}_i), \tag{A.9}$$

where X_i and Φ_i can be obtained from the microarray in the rough PPIN. Since there are two data sets of microarray data, two association parameters for cancer and noncancer were separately identified.

Appendix A.2: Determination of significant protein associations by AIC and student's *t*-test

When association parameters of all the proteins in rough PPIN were identified as Eq. (12.2), significant protein associations were determined by parameter estimates $\hat{\alpha}_{ij}$ of their association abilities. In order to determine whether the association was significant or not, AIC in Eq. (2.23) and Student's *t*-test, which is used to calculate the *p*-values of the association abilities, are employed to detect the system model order (or the number of model parameters) and determine the significance of these model parameters. The AIC, a method for model order detection, attempts to include both the estimated residual variance and model complexity in one statistic. AIC decreases as residual variance decreases, and increases as the number of parameters increases. As the expected residual variance decreases with increasing parameter numbers for excessive model complexity, a minimum should appear near the correct parameter number. Thus, the AIC criterion, in which estimated parameters were obtained above, was used to select model structure. Due to computation efficiency, it is impractical to compute the AIC statistics for all possible regression models. Here, the stepwise regression method which combines forward selection method and backward elimination method was applied to compute the AIC statistics. Once the estimated association parameters were examined using the AIC model detection criteria, the student's *t*-test was employed to calculate the *p*-values for the association abilities under the null hypothesis $H_0{:}\alpha_{ij} = 0$ to determine the significant protein associations. The *p*-values computed were then adjusted by Bonferroni correction to avoid a lot of spurious positives. The associations which adjusted *p*-value $\leq .05$ were determined as significant associations and were preserved in the protein association network.

Briefly, we use the AIC method to obtain how many system orders, which mean the numbers of interactions, in the dynamic system of the association abilities (model). We use the above maximum likelihood estimate method to identify the parameter $\hat{\alpha}_{ij}$ and then employ AIC and student *t*-test to calculate *p*-values of association abilities for determining the significant PPIs for the target protein *i* by pruning the insignificant PPIs.

Here we show Tables A.1–A.9 and Figs. A.1–A.3.

Table A.1 Descriptive information on datasets extracted from the GEO database used in this section

Cancer type	GEO accession number	Tumor (*N*)	Adjacent normal (*N*)	Platform
Bladder cancer	GSE13507	106	58	GPL6102
Colorectal cancer	GSE41258	72	33	GPL96
Liver cancer	GSE14520	119	152	GPL3921
Lung cancer	GSE19188	71	65	GPL570

Descriptive information on datasets extracted from the GEO database is used in this section. Cases are grouped by control and cancer tissues came from human patients of early stage. For each dataset, sample sizes of control and cancer tissues are provided.

Table A.2 Descriptive information on datasets extracted from the Gene Ontology database used for cancer network construction in this section

GO term	Biological process	Number
Cell cycle arrest		
GO:0000075	Cell cycle checkpoint	302
GO:0071157	Negative regulation of cell cycle arrest	24
GO:0071158	Positive regulation of cell cycle arrest	87
Apoptotic process		
GO:0043066	Negative regulation of apoptotic process	841
GO:0043065	Positive regulation of apoptotic process	603
GO:2001233	Regulation of apoptotic signaling pathway	74
Cell proliferation		
GO:0008285	Negative regulation of cell proliferation	826
GO:0008284	Positive regulation of cell proliferation	930
Cell division		
GO:0051782	Negative regulation of cell division	10
GO:0051781	Positive regulation of cell division	84
Wnt receptor signaling pathway		
GO:0090090	Negative regulation of canonical Wnt receptor signaling pathway	106
GO:0090263	Positive regulation of canonical Wnt receptor signaling pathway	60
Cell differentiation		
GO:0030855	Epithelial cell differentiation	442
GO:0048863	Stem cell differentiation	160
GO:0048762	Mesenchymal cell differentiation	184
NF-κB transcription factor activity		
GO:0051092	Positive regulation of NF-κB transcription factor activity	115
DNA repair		
GO:0006281	DNA repair	762
Inflammatory response		
GO:0002544	Chronic inflammatory response	31
GO:0002534	Cytokine production involved in inflammatory response	12
GO:0002437	Inflammatory response to antigenic stimulus	45
GO:0002532	Production of molecular mediator involved in inflammatory response	42
GO:0050727	Regulation of inflammatory response	277
GO:0002246	Wound healing involved in inflammatory response	7
Immune response		
GO:0002418	Immune response to tumor cell	11
GO:0045087	Innate immune response	778
GO:0002281	Macrophage activation involved in immune response	19
GO:0050776	Regulation of immune response	916
GO:0002286	T-cell activation involved in immune response	95

Detailed information from the GO database is used in this section to identify proteins relating to cancer function. To make the chosen proteins as related to cancer as possible, the selected GO terms corresponding to the function belong to rather lower level. Here, we selected 3029 proteins to the functional pool.

Table A.3 The identified significant proteins in bladder, colorectal, liver, and lung carcinogenesis

Bladder cancer ($N = 107$)			Liver cancer ($N = 110$)		
CRV	Name	*p*-Value	CRV	Name	*p*-Value
158.5321	UBC[a]	<1E-5	137.284	UBC[a]	<1E-5
19.5883	TP53[a]	<1E-5	19.5736	HDAC1[a]	<1E-5
19.2879	HDAC1[a]	<1E-5	18.7422	TP53[a]	<1E-5
16.7366	CDK2[a]	<1E-5	16.7663	KIAA0101[a]	<1E-5
15.8188	KIAA0101[a]	<1E-5	15.8533	SUMO1	<1E-5
13.3217	PCNA[a]	<1E-5	15.1438	PCNA[a]	<1E-5
13.2669	CUL1[a]	<1E-5	14.0690	CDK2[a]	<1E-5
13.0423	MYC[a]	<1E-5	13.2218	EP300[a]	<1E-5
13.0117	CUL3[a]	<1E-5	12.9519	CUL3[a]	<1E-5
12.1078	EP300[a]	<1E-5	12.6421	YWHAZ[a]	<1E-5
12.0995	CREBBP[a]	<1E-5	11.2802	CUL1[a]	<1E-5
11.6611	VHL	<1E-5	10.8312	CUL5	<1E-5
11.6233	CUL2	<1E-5	10.7821	MYC[a]	<1E-5
11.5863	BRCA1[a]	<1E-5	10.6794	AR	<1E-5
11.0938	HDAC2[a]	<1E-5	10.3758	ESR1[a]	<1E-5
9.2003	TRAF2[a]	<1E-5	10.2716	CUL2	<1E-5
8.8026	SRC[a]	<1E-5	9.7752	HDAC2[a]	.00002
8.7995	YWHAZ[a]	<1E-5	9.5993	CTNNB1[a]	.00002
8.6685	ATM[b]	<1E-5	9.5856	CREBBP[a]	.00002
8.6624	UCHL5	<1E-5	9.5585	XRCC6[a]	.00002
8.5157	RELA	<1E-5	9.1997	HDAC6	.00004
7.9985	CSNK2A1	<1E-5	8.7723	RELA	.00007
7.6796	CTNNB1[a]	<1E-5	7.7978	PSMA3[a]	.00010
7.6209	SMARCA4	<1E-5	7.7449	SMARCA4	.00010
7.1345	CUL5	.00001	7.0852	PARP1	.00016
7.0345	AKT1	.00001	6.8526	CUL4A	.00017
7.0177	SMAD3	.00001	6.8236	BRCA1[a]	.00017
6.8922	RB1[a]	.00002	6.7171	PIK3R1	.00019
6.7340	IKBKG	.00003	6.6842	PSMA2	.00022
6.6287	RAD21	.00004	6.6460	IRAK1[a]	.00023
6.5794	SIRT1	.00004	6.4883	TRIM28[b]	.00027
6.5444	RAD23A	.00004	6.4146	PPP2CA	.00031
6.4468	RPA1	.00005	6.3572	UCHL5	.00031
6.2665	CAV1	.00009	6.2256	FYN	.00033
6.1783	PML	.00010	6.0239	ISG15[a]	.00038
5.9832	PIN1[b]	.00014	6.0180	PTK2	.00038
5.8659	HDAC4[a]	.00016	5.9417	EGFR	.00041
5.8136	PSMD10	.00017	5.9391	MAPK1	.00041
5.8022	PSMA3[a]	.00018	5.9369	PRKDC[a]	.00041
5.7800	PARP1	.00020	5.9326	IKBKG	.00042

(*Continued*)

Table A.3 (Continued)

Bladder cancer ($N = 107$)			Liver cancer ($N = 110$)		
CRV	Name	*p*-Value	CRV	Name	*p*-Value
5.7285	TRAF6	.00022	5.8956	SMAD2	.00044
5.6189	ESR1[a]	.00026	5.8397	HDAC4[a]	.00048
5.5193	CDKN1A	.00035	5.8021	SOCS3	.00048
5.5065	MAPK1	.00036	5.7348	PSMC2	.00053
5.4992	PTPN11	.00037	5.6979	VHL	.00055
5.4808	PTK2	.00038	5.6600	POLR2A	.00057
5.4642	TERF1[a]	.00038	5.5696	BUB1B	.00064
5.4465	FYN	.00038	5.5623	NFKB1	.00064
5.4175	CBL	.00043	5.5531	RB1[a]	.00065
5.3860	MCM2	.00049	5.5034	ARRB2	.00071
5.3247	MAPK3	.00061	5.4753	CDKN1A	.00071
5.2871	XRCC6[a]	.00063	5.4193	RUVBL2	.00075
5.2666	HIF1A[b]	.00068	5.3650	CDK9	.00080
5.2443	ZBTB16	.00072	5.3402	CDK4[b]	.00082
5.2303	CEBPB[a]	.00075	5.2323	RAD23A	.00099
5.1927	EGFR	.00081	5.1866	PSMA1	.00104
5.1852	MSH2[b]	.00081	5.1507	CDC20	.00109
5.1728	E2F1	.00085	5.1194	CAV1	.00115
5.1682	NFKB1	.00087	5.1193	CCNB1	.00115
5.1451	RUVBL2	.00094	5.0936	YY1	.00121
5.1106	ITCH[b]	.00100	5.0894	AKT1	.00122
5.0836	STUB1	.00108	5.0858	CUL4B	.00124
5.0181	MGMT	.00124	5.0219	TRAF6	.00137
4.9555	DDB1[b]	.00138	5.0135	HDAC3	.00137
4.9001	SHC1	.00157	4.9742	RPA1	.00143
4.8586	PTPN6[b]	.00164	4.9434	RBBP4[b]	.00152
4.8562	HDAC6	.00165	4.9046	TERF1[a]	.00163
4.8206	CDC27[b]	.00182	4.8937	NR3C1	.00166
4.8138	CHAF1A[b]	.00182	4.8138	PTPN11	.00189
4.7516	RARA[b]	.00211	4.7783	PSMC3[b]	.00203
4.6807	JUP	.00250	4.7703	TRAF2[a]	.00207
4.6605	CHUK[b]	.00263	4.7530	MDM2[a]	.00213
4.6426	NR4A1[b]	.00277	4.7299	GNB2L1[b]	.00223
4.6161	CASP8[b]	.00298	4.7228	CDT1	.00224
4.6009	PSMB5	.00308	4.7213	RAF1	.00224
4.5775	MCM7	.00321	4.6929	HGS[b]	.00232
4.5727	UBE2D1	.00325	4.5666	LYN	.00292
4.5647	MDM2[a]	.00331	4.5359	PSMD6[b]	.00304
4.5400	GSK3B	.00347	4.5338	PSMA6[b]	.00304
4.5387	CDC37[b]	.00347	4.5098	STUB1	.00324

(*Continued*)

Table A.3 (Continued)

Bladder cancer ($N = 107$)			Liver cancer ($N = 110$)		
CRV	Name	*p*-Value	CRV	Name	*p*-Value
4.5346	PIAS4[b]	.00348	4.4813	RBX1[b]	.00345
4.5274	UBE2D3[b]	.00353	4.4330	CEBPB[a]	.00390
4.5167	CDK9	.00361	4.4270	SNCA	.00394
4.5072	RAF1	.00363	4.4188	MGMT	.00395
4.4953	RPA2[b]	.00381	4.4114	CCNA2	.00398
4.4856	ISG15[a]	.00389	4.3524	UBE2D1	.00443
4.4437	ERBB2[b]	.00436	4.3404	PSMD10	.00453
4.4333	BCL2	.00444	4.3377	CDKN2A[b]	.00453
4.4200	IKBKB	.00460	4.3368	BARD1[b]	.00454
4.4138	SOCS1[b]	.00465	4.2604	CCND1	.00514
4.3966	SMAD4[b]	.00479	4.2135	MAPK14[b]	.00566
4.3747	ITSN1[b]	.00500	4.2124	TOP2A[b]	.00567
4.3516	POLR2A	.00519	4.2020	JAK1[b]	.00579
4.3468	CCND1	.00526	4.1882	JUP	.00595
4.3101	CSNK2B	.00571	4.1824	HSPB1[b]	.00602
4.2893	PSMD2[b]	.00606	4.1402	NFKBIA[b]	.00646
4.2145	JUN	.00710	4.1298	PRKCD[b]	.00656
4.1837	AR	.00755	4.1223	NCOA3[b]	.00665
4.1501	TRAF1[b]	.00801	4.1125	TERF2IP[b]	.00678
4.1157	IRAK1[a]	.00861	4.1122	PSMA4[b]	.00678
4.0798	SOCS3	.00912	4.0767	SRC[a]	.00737
4.0781	PRKDC[a]	.00912	4.0556	RFC4[b]	.00764
4.0759	TRAF3[b]	.00914	4.0447	MAD2L1	.00781
4.0575	CDKN1B[b]	.00950	4.0104	PSMC5[b]	.00835
4.0569	TOPBP1[b]	.00951	4.0023	SET[b]	.00855
4.0514	PIK3R1	.00965	3.9986	JUN	.00858
4.0396	YY1	.00991	3.9947	E2F1	.00862
			3.9564	CSNK2B	.00930
			3.9551	SMARCA2	.00932
			3.9283	CEBPA[b]	.00983

Colorectal cancer ($N = 60$)			Lung cancer ($N = 86$)		
CRV	Name	*p*-Value	CRV	Name	*p*-Value
80.3851	UBC[a]	<1E-5	120.1210	UBC[a]	<1E-5
25.9329	TP53[a]	<1E-5	20.2633	KIAA0101[a]	<1E-5
20.0595	MYC[a]	<1E-5	19.9208	HDAC1[a]	<1E-5
19.1403	KIAA0101[a]	<1E-5	15.1811	TP53[a]	<1E-5
18.9463	HDAC2[a]	<1E-5	14.8108	PCNA[a]	<1E-5
16.3169	CUL3[a]	.00004	13.2794	SUMO1	.00002
15.2868	SUMO1	.00005	12.5188	BRCA1[a]	.00003
14.4739	CDK2[a]	.00008	11.5249	CUL3[a]	.00006

(Continued)

Table A.3 (Continued)

Colorectal cancer ($N = 60$)			Lung cancer ($N = 86$)		
CRV	Name	*p*-Value	CRV	Name	*p*-Value
11.7823	HDAC1[a]	.00017	11.2892	CREBBP[a]	.00006
11.5906	CREBBP[a]	.00018	1.3383	SIRT1	.00011
11.5648	MDM2[a]	.00020	9.9420	CDK2[a]	.00014
10.8636	BRCA1[a]	.00026	9.5803	NFKB1	.00015
10.4098	CUL4A	.00031	9.4508	MYC[a]	.00017
9.88493	CHEK1[b]	.00036	8.7261	ARRB2	.00024
9.7845	HDAC4[a]	.00036	8.6696	ESR1[a]	.00025
9.6282	PCNA[a]	.00040	8.5670	MAPK1	.00026
9.3040	CTNNB1[a]	.00043	8.5044	CUL1[a]	.00026
9.1547	CAV1	.00049	8.3205	RB1[a]	.00029
9.1407	SRC[a]	.00049	8.0708	TERF1[a]	.00036
9.0806	BCL2L1	.00051	8.0389	CUL5	.00038
8.7090	ZBTB16	.00060	7.8990	AKT1	.00040
8.5589	PRKDC[a]	.00065	7.7916	HDAC2[a]	.00042
8.3187	EP300[a]	.00075	7.3717	BCL2	.00059
8.1656	RUVBL2	.00078	7.3549	CBL	.00059
8.0875	PSMA3[a]	.00083	7.2717	CEBPB[a]	.00061
7.9038	YWHAZ[a]	.00089	7.2591	EGFR	.00061
7.5644	NR3C1	.00117	7.1278	EP300[a]	.00068
7.5019	HDAC3	.00121	7.0299	PLK1[b]	.00072
7.4694	SNCA	.00122	6.9997	PARP1	.00075
7.3650	TRAF2[a]	.00135	6.9776	RELA	.00076
7.1217	XRCC5	.00162	6.9598	STUB1	.00077
7.0177	IRAK1[a]	.00177	6.8477	MCM7	.00083
6.9782	BUB1B	.00180	6.8167	IKBKB	.00084
6.8718	RAD21	.00197	6.5798	YWHAZ[a]	.00098
6.8303	LYN	.00202	6.5564	SMAD2	.00098
6.8046	UCHL5	.00207	6.4431	PML	.00104
6.5630	IKBKG	.00257	6.4281	TRAF2[a]	.00106
6.5303	GSK3B	.00274	6.3976	RAD23A	.00112
6.3873	ESR1[a]	.00312	6.3707	SRC[a]	.00116
6.3738	SMAD2	.00316	6.2816	MDM2[a]	.00123
6.2595	TERF1[a]	.00353	6.2299	SYK[b]	.00131
6.2559	CCNB1	.00353	6.2225	HDAC4[a]	.00131
6.2297	APC[b]	.00360	6.2008	MAPK3	.00135
6.2042	CDC23[b]	.00366	6.1279	NR3C1	.00137
6.1770	FYN	.00377	6.0414	RPA1	.00150
6.0231	XRCC6[a]	.00436	5.9577	SMARCA2	.00164
6.0129	RUVBL1[b]	.00438	5.9449	RHOA[b]	.00166
5.9763	RB1[a]	.00456	5.9140	MAD2L1	.00172
5.9162	SMARCA4	.00489	5.7953	CUL4B	.00201

(Continued)

Table A.3 (Continued)

Colorectal cancer (*N* = 60)			Lung cancer (*N* = 86)		
CRV	**Name**	***p*-Value**	**CRV**	**Name**	***p*-Value**
5.8124	ISG15[a]	.00537	5.7088	RAD21	.00218
5.7154	JUN	.00589	5.6736	PRKDC[a]	.00226
5.6948	PSMA1	.00597	5.6706	CCNA2	.00226
5.5716	DAXX	.00676	5.6438	GSK3B	.00233
5.5283	CUL1[a]	.00703	5.5661	XRCC5	.00263
5.5263	MAPK3	.00706	5.5173	DAXX	.00279
5.5114	PSMB5	.00719	5.4934	HDAC3	.00286
5.5094	PPP2CA	.00720	5.4764	PSMA2	.00293
5.4262	CEBPB[a]	.00788	5.4590	SHC1	.00301
5.3316	PML	.00876	5.4418	CUL2	.00310
5.2106	CDKN1A	.00993	5.3539	BLM[b]	.00329
			5.3437	RIPK1[b]	.00333
			5.3230	TRAF6	.00337
			5.3163	SMARCB1[b]	.00340
			5.2831	PSMA3[a]	.00351
			5.2059	BUB1B	.00395
			5.1777	IRAK1[a]	.00409
			5.1574	CSNK2A1	.00420
			5.1245	SMAD3	.00433
			5.0643	ANXA1[b]	.00468
			5.0284	CUL4A	.00487
			5.0036	CDC20	.00502
			4.9761	IGF1R[b]	.00516
			4.9430	ISG15[a]	.00538
			4.9186	ZBTB16	.00560
			4.8891	MCM2	.00577
			4.8671	XRCC6[a]	.00597
			4.8075	BCL2L1	.00659
			4.8015	CTNNB1[a]	.00663
			4.7498	PSMC2	.00705
			4.7441	GAPDH[b]	.00712
			4.6974	BTK[b]	.00756
			4.6282	RPA3[b]	.00827
			4.6223	CDT1	.00831
			4.6200	TUBB[b]	.00834
			4.5938	SOX2[b]	.00868
			4.5918	AR	.00870

[a]The common proteins of four types of cancer.
[b]The distinctive proteins from other three cancers.

Table A.4 The pathways analysis for 28 common significant proteins in carcinogenesis

Rank	Term	Count	Symbol	*p*-Value
1	hsa04110: Cell cycle	13	YWHAZ, CREBBP, TP53, PRKDC, RB1, CDK2, HDAC2, EP300, HDAC1, PCNA, MDM2, MYC, CUL1	1.50E-14
2	hsa04310: Wnt signaling pathway	6	EP300, CREBBP, TP53, MYC, CUL1, CTNNB1	3.78E-04
3	hsa04330: Notch signaling pathway	4	EP300, HDAC2, HDAC1, CREBBP	.001008
4	hsa04520: Adherens junction	4	EP300, CREBBP, SRC, CTNNB1	.004180
5	hsa04350: TGF-beta signaling pathway	4	EP300, CREBBP, MYC, CUL1	.005889
6	hsa04120: Ubiquitin-mediated proteolysis	4	CUL3, MDM2, BRCA1, CUL1	.020254
7	hsa04115: p53 signaling pathway	3	TP53, MDM2, CDK2	.034267
8	map03050: Proteasome pathway	1	PSMA3	[a]

The significant pathways via DAVID Bioinformatics database are selected for the 28 significant proteins in carcinogenesis. Furthermore, the result excludes the disease pathway.
[a]There is only PSMA3 related to the proteasome pathway, so its *p*-value is low. This pathway is important for cancer, so we list it here.

Table A.5 The complete pathways analysis for specific network marker for each carcinogenesis of four cancers in Table 12.2

Term	Count	Symbol	*p*-Value
Bladder cancer			
hsa04120: Ubiquitin-mediated proteolysis	6	UBE2D3, PIAS4, DDB1, SOCS1, ITCH, CDC27	1.90E-04
hsa04622: RIG-I-like receptor signaling pathway	4	CASP8, CHUK, TRAF3, PIN1	.002897
hsa04110: Cell cycle	4	CDKN1B, SMAD4, CDC27, ATM	.013957
hsa04520: Adherens junction	3	PTPN6, ERBB2, SMAD4	.039471
hsa04210: Apoptosis	3	CASP8, CHUK, ATM	.059239
hsa04620: Toll-like receptor signaling pathway	3	CASP8, CHUK, TRAF3	.064224
hsa03430: Mismatch repair	2	RPA2, MSH2	.090981
Liver cancer			
hsa03050: Proteasome	5	PSMC5, PSMA6, PSMC3, PSMA4, PSMD6	8.12E-06
hsa04660: T-cell receptor signaling pathway	3	MAPK14, NFKBIA, CDK4	.039196
hsa04722: Neurotrophin signaling pathway	3	MAPK14, NFKBIA, PRKCD	.050351
hsa04110: Cell cycle	3	CDKN2A, CDK4, RBX1	.051083
Lung cancer			
hsa03440: Homologous recombination	2	BLM, RPA3	.048517
hsa05130: Pathogenic *Escherichia coli* infection	2	TUBB, RHOA	.096550

The significant pathways via DAVID Bioinformatics database are selected specific network marker for each carcinogenesis of four cancers. Only colorectal cancer is not shown in the table, because protein numbers of colorectal cancer are too small to output the results from KEGG pathway analysis. Shaded background indicates *p*-value >.05.

Table A.6 The pathways analysis for specific network marker in colorectal cancer

Colorectal cancer

Term	Count	Symbol
ko04310: Wnt signaling pathway	2	APC, RUVBL1
ko04810: Regulation of actin cytoskeleton	1	APC
ko04390: Hippo signaling pathway	1	APC
ko05166: HTLV-I infection	1	APC

The pathways by giving four proteins of specific network marker are searched against KEGG pathway maps and found proteins are marked in result. Shaded background indicate no consider in this section.

Table A.7 Full names of the significant proteins in Table 12.1

Symbol	Full names	Symbol	Full names
BRCA1	Breast cancer type 1 susceptibility protein	KIAA0101	PCNA-associated factor
CDK2	Cyclin-dependent kinase 2	MDM2	E3 ubiquitin-protein ligase Mdm2
CEBPB	CCAAT/enhancer-binding protein beta	MYC	Myc proto-oncogene protein
CREBBP	CREB-binding protein	PCNA	Proliferating cell nuclear antigen
CTNNB1	Catenin beta-1	PRKDC	DNA-dependent protein kinase catalytic subunit
CUL1	Cullin-1	PSMA3	Proteasome subunit alpha type-3
CUL3	Cullin-3	RB1	Retinoblastoma-associated protein
EP300	E1A-binding protein p300	SRC	Proto-oncogene tyrosine-protein kinase Src
ESR1	Estrogen receptor	TERF1	Telomeric repeat-binding factor 1
HDAC1	Histone deacetylase 1	TP53	Cellular tumor antigen p53
HDAC2	Histone deacetylase 2	TRAF2	TNF receptor-associated factor 2
HDAC4	Histone deacetylase 4	UBC	Polyubiquitin-C
IRAK1	Interleukin-1 receptor-associated kinase 1	XRCC6	X-ray repair cross-complementing protein 6
ISG15	Ubiquitin-like protein ISG15	YWHAZ	14-3-3 protein zeta/delta

Table A.8 The identified significant proteins in benign lung tumor

Benign lung cancer ($N = 2$)

CRV	Name	*p*-Value
4.927824	CBR3	<1E-5
4.426606	TFF3	<1E-5

Table A.9 The pathways analysis for two significant proteins in benign lung tumor

Benign lung tumor

Term	Count	Symbol	*p*-Value
Amino sugar and nucleotide sugar metabolism	1	CRB3	*

The asterisk denotes non p-value.

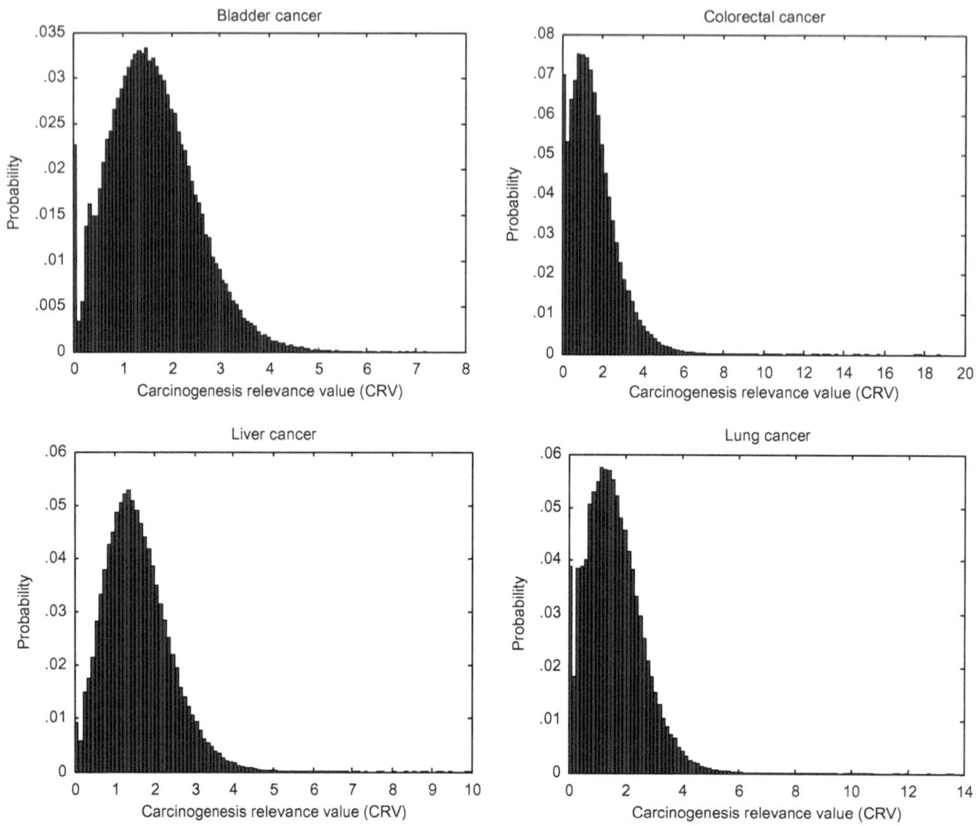

Figure A.1 Distribution of carcinogenesis relevance values (CRVs) of random candidate PPIN networks for different cancers. The null distributions of CRVs for k = bladder, colorectal, liver, and lung cancer are generated by 100,000 randomly permuted network structures.

12.3 SYSTEMS DRUG DESIGN MECHANISMS FOR MULTIPLE DRUG TARGETS

After investigating 28 significant proteins as core network markers for multiple drug targets in the process, we will use a system drug design mechanism to select multiple drug targets for multiple drug design to provide multiple-target cocktail therapy. Computer-aided drug design (CADD) has a long history of being applied to discover new molecules to treat various cancers, but it has always been focused on single targets. The development of systems biology has let scientists reveal more hidden mechanisms of cancers, but attempts to apply systems biology to cancer therapies remain at preliminary stages. Section 12.2 successfully developed various systems biology models for several cancers. Based on these results, we present the first attempt to combine a multiple-target therapy with systems biology.

In Section 12.2, we identified 28 significant proteins, i.e., common core network markers, of four types of cancers as house-keeping proteins of these cancers. In this section, we ranked these proteins by summing their CRVs across the four cancers, and then performed docking and pharmacophore modeling to undertake virtual screening on the NCI database for anticancer drugs. We also performed pathway analysis on these proteins using Panther and MetaCore to reveal more systematic mechanisms of these cancer house-keeping proteins.

We designed several approaches to discover targets for multitarget cocktail therapies. In the first one, we identified the top 20 drugs for each of the 28 cancer house-keeping proteins, and analyzed the docking pose to further understand the interaction mechanisms of these drugs. After screening for duplicates, we found that 13 of these drugs could target 11 proteins, simultaneously. In the second approach, we chose the top five pharmacophore and used them in a virtual screening against the Life-Chemical library for anticancer drugs. Based on these results, wet-lab bio-scientists could freely investigate combinations of these drugs for multiple-target therapy for cancers, in contrast to the traditional single target therapy. Combination of systems biology with CADD could help us develop novel drug cocktails with multiple targets. We believe this will enhance the efficiency of therapeutic practice and lead to new directions for cancer therapy.

Cancer is the leading cause of human death worldwide. It is a complex disease, and people have tried to reveal its underlying mechanisms to guide the development of novel therapy strategies. In the last two decades, cancer researchers have generated an abundance of knowledge about cancer, and revealed the etiology of various cancers at DNA, RNA, and protein levels [255]. Weinberg et al. summarized the first and next generation of cancer hallmarks to expand the current understanding of the basic mechanisms of cancer [213,218]. Recently, due to the scale up in high-throughput data, availability of integrated OMICS data and various advanced statistical analysis methods, many novel systems biology approaches have been employed to reveal the deeper underlying systematic mechanisms of various cancers [23,256,257].

The traditional CADD focuses on a single target for therapy, such as Src, Fak, and EGFR in the case of cancer [197,258]. CADD has a long history and many successful examples. Researchers have used virtual screening with de novo methods to develop small molecules that in most cases inhibit these targets (although sometime they are agonists), and accordingly reduce the expression of these proteins to killing cancer cells. CADD methods can be divided into structure-based and ligand-based methods [259]. Methods in the former category analyze both the structures of the target protein and the small molecule inhibitors to design drugs: examples include the docking method and molecular dynamics simulations. On the other hand, methods in the latter category use only the structures of small molecule inhibitors (drugs) to do statistical calculations to determine the relationship between drugs IC_{50} and their corresponding molecular properties: examples include HypoGen pharmacophore modeling, COMFA (and COMSIA) methods [260], and many other machine learning and regression methods [261,262].

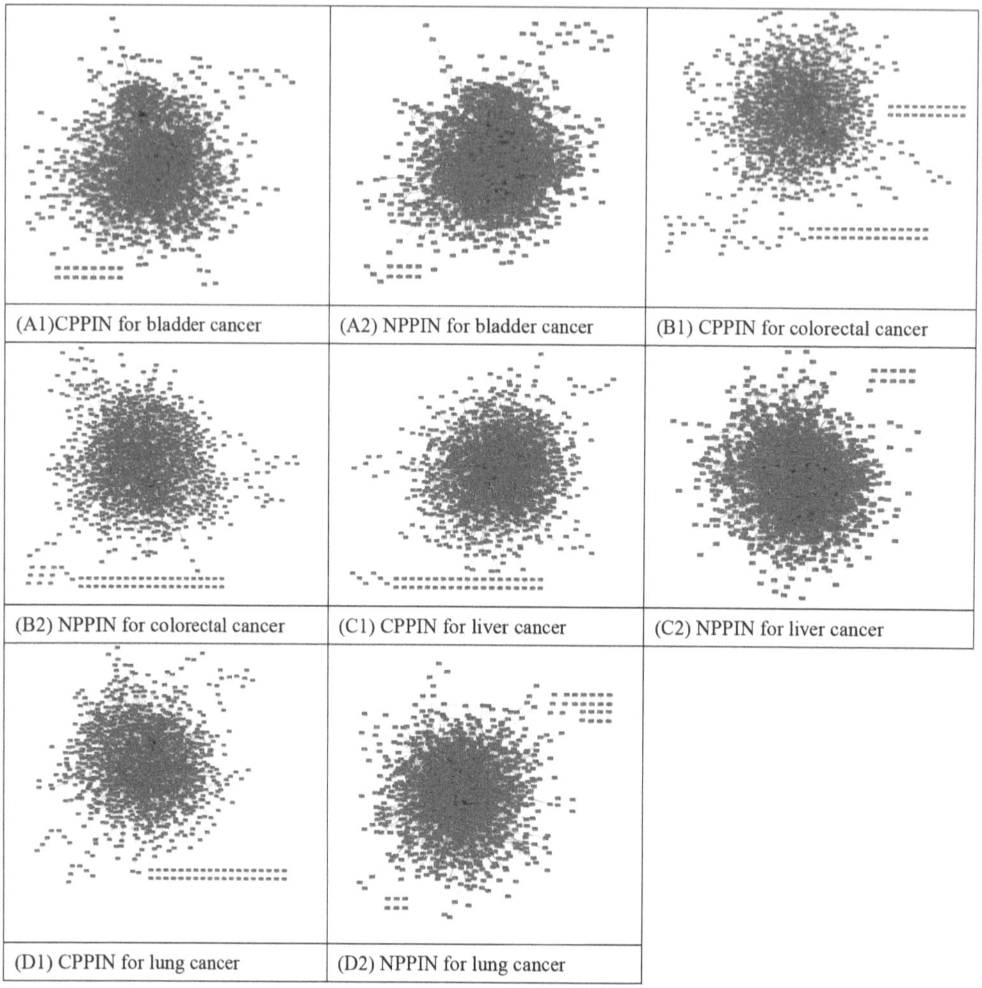

Figure A.2 The constructed cancer PPIN (CPPIN) and noncancer PPIN (NPPIN) for (A1 and A2) bladder, (B1 and B2) colorectal, (C1 and C2) liver, and (D1 and D2) lung cancer. The protein association numbers of CPPIN and NPPIN with respect to four cancers are listed below (CPPIN/NPPIN): bladder cancer (2612/2220), colorectal cancer (1411/1609), liver cancer (2148/3286), and lung cancer (1836/2426). The figures are created using Cytoscape.

One of the main differences between Western and Chinese medical philosophy is that the former always focuses on single target, while the latter always focuses on multiple targets simultaneously [263]. Systems biology constructs the regulation relationships within genetic, metabolic, and PPINs. These biological networks are highly complex, so robustness and sensitivity are their key system-level features [264,265]. The intertwined nature of these networks shows us that inhibiting a single protein directly is not the only way to depress its expression. Systems biology will help us to

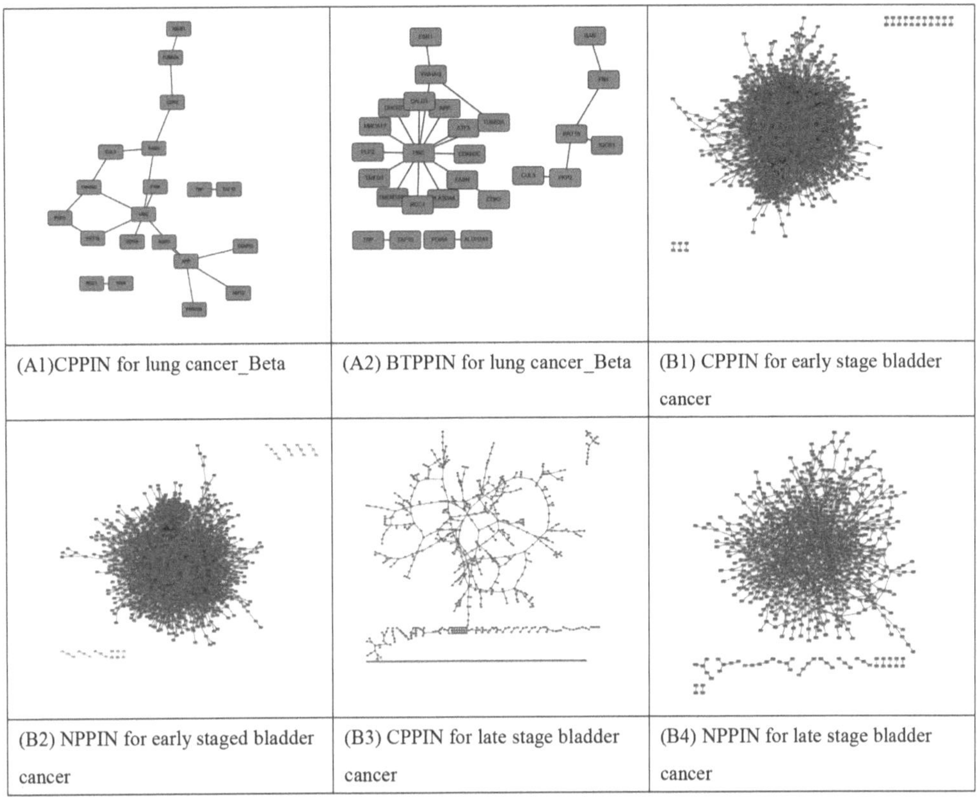

(A1)CPPIN for lung cancer_Beta
(A2) BTPPIN for lung cancer_Beta
(B1) CPPIN for early stage bladder cancer
(B2) NPPIN for early staged bladder cancer
(B3) CPPIN for late stage bladder cancer
(B4) NPPIN for late stage bladder cancer

Figure A.3 Three set of network including (A) lung cancer and benign lung tumor (B) CPPIN and NPPIN of early and late stage bladder cancer. We can apply this method to various cancer dataset. (A1) and (A2) show the cancer-PPIN and begin tumor-PPIN(BTPPIN) for lung cancer and benign lung tumor network. (B1)–(B4) are the results of CPPIN and NPPIN for early and late stage bladder cancer, it help us see the evolution of network.

identify several protein targets to be inhibited, simultaneously: due to their network behavior, this multitarget approach will procedure the same or better effect, than focusing on a single target protein. Also, recent research has demonstrated that to inhibit PPIs is another novel anticancer strategy [221]. Inspired by the above ideas, based on the results of our previous systems biology studies [266], we have developed a novel multitarget cocktail therapy to focus on common core network markers of four different cancers. The strategy is different from Dr. Ho's famous cocktail therapy for AIDS [267], which is not target therapy. This method is to apply traditional CADD methods simultaneously to multiple drug targets. We regard this as a great advance in novel anticancer strategies.

In Section 12.2, we analyzed various cancers, specifically, bladder, colorectal, liver, and lung cancers, through regression modeling, microarray data, maximum likelihood parameter estimations, and big database mining. Based on known PPI information and gene expression data from normal and patient samples, we built a cancer PPIN [266]. Here, we focus on not only the PPIN of single cancer types but also common core network markers of four different cancers through the intersection of their respective PPINs. Cancer is a complex disease, and so we tried to reveal the house-keeping proteins significant in different cancers, i.e., the common core network markers shared by the different cancers. As the first trial to develop multiple-target therapy, cancer house-keeping proteins (common core network markers) may be a good choice. In this section, we also analyze the sensitivity and robustness of the common core network markers, provide more clues to develop novel multiple targets therapy strategy.

12.3.1 Materials and methods for multiple drug design

Identify the core network markers as multiple drug targets of four cancers (28 proteins)—a brief review of these previous methods

In Section 12.2 [266], we used a systems biology approach to study network markers of various cancers. Firstly, microarray data, PPI databases, and PPI interaction models were employed to construct the PPINs of normal and cancer cells by the maximum likelihood parameter estimation method (see Appendix A.1). The AIC system order detection method (see Appendix A.2) is employed to prune the false positive PPIs to obtain the real PPINs of normal and cancer cells, i.e., we use the so-called reverse engineering method to construct PPINs of normal and cancer cells. Then, the differential PPIN, obtained by contrasting the cancer PPIN and normal PPIN, was used to investigate PPI variations of each protein in the differential PPIN due to the carcinogenesis. Finally, the CRV in Eq. (12.8) based on PPI variations is also proposed to evaluate the significance of carcinogenesis for each protein of differential PPIN. Proteins with significant CRV (p-value $<.01$) are considered as significant proteins of the cancer. The 28 significant proteins in Fig. 12.7 (see also Table 12.5) are the intersection of these significant proteins of four cancers, and these 28 significant proteins and their PPIs construct a common core network marker of cancers. The distinctive significant proteins outside of these 28 significant proteins form the specific network marker of each cancer. Finally, based on the common core network marker and specific network marker, we investigate the mechanism of carcinogenesis process with the help of databases (e.g., GO database [268], DAVID database [205,269], and KEGG pathway database [270,271]) and try to find multiple drug target therapy of cancer. Unlike the conventional theoretical methods, which always give a single mathematical model for cancer network for a more detailed theoretical analysis, this chapter is through a systems biology approach to cancer network markers based on real microarray data through the so-called reverse engineering, theoretical statistical method and data mining method in combination with big databases. These are the novelty and significance of this section.

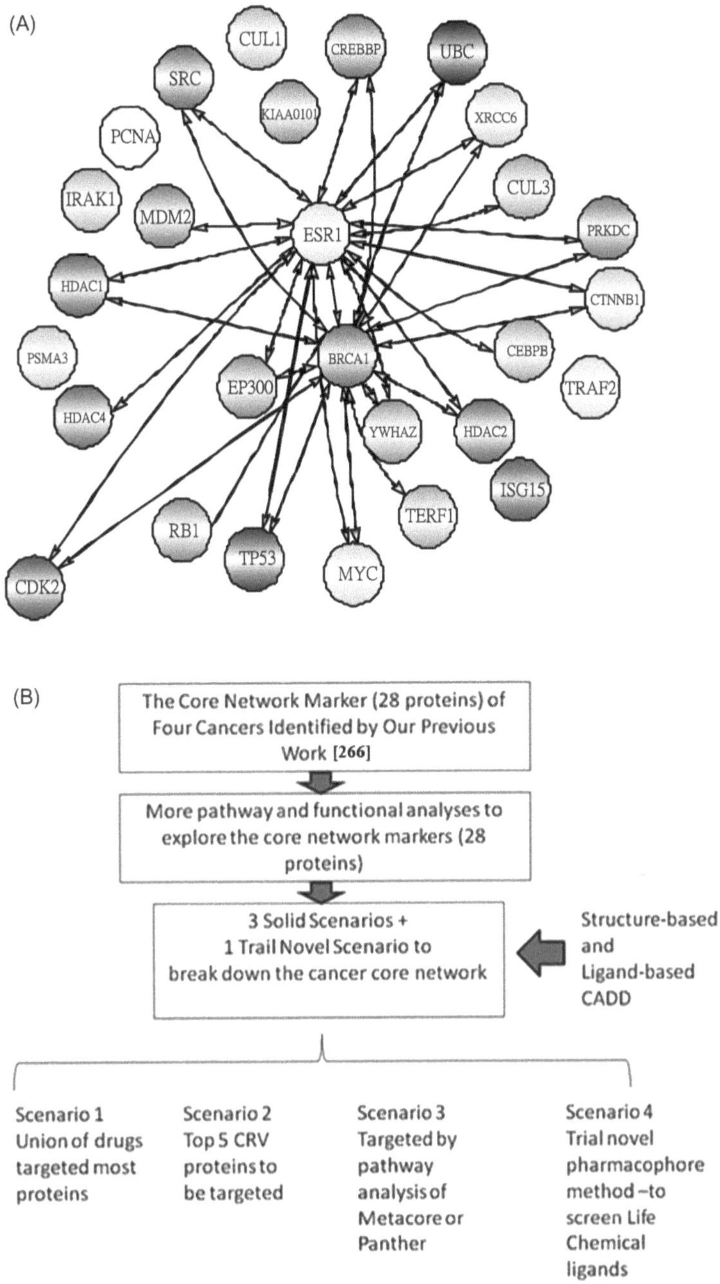

Figure 12.7 (A) Core network marker to be selected as multiple drug targets of four cancers. This is the PPIN of the 28 proteins of the core network marker of four cancers. We see both ESR1 and BRAC1 are most highly connected with other proteins. (B) System Flowchart. We use the CADD method to target the 28 core proteins obtained by previous systems biology method. We use both structure-based and ligand-based CADD with pathway and functional analyses to develop four scenarios for selecting multiple drug targets to break down the core network.

Theoretical biological background

Recently, the PPI target seems to have become a paradigm for the drug discovery of cancer therapy and precision medicine [221,222]. Unlike the conventional drug design focusing on the inhibition of a single protein, usually an enzyme or receptor, small-molecule inhibition of direct PPIs that mediate many important biological processes is an emerging and challenging scheme in drug design, especially for cancer. Extensive biological and clinical investigations have led to the identification of PPI hubs and nodes that are critical for the acquisition and maintenance of characteristics for cell transformation in cancer. Such cancer-enabling PPIs have become a significant therapeutic target in the anticancer strategies. The technological progresses in the discovery of PPI modulators and the validation of clinical PPI-targeting agents make anticancer by interfering PPI become true [221,222].

Therefore, future studies directed at the PPI target discovery, PPI interface characterization, and PPI-focused chemical libraries are expected to accelerate the development of next generation of PPI-based anticancer agents. However, the PPINs of cancer are very complex and different for various cancers. In the situation, we focus on the common PPIN markers and their PPIs are all with significant CRV, which is considered to estimate the PPI changes during carcinogenesis process, to exploit the important targets and their PPI interface characteristics as the house-keeping proteins for different cancers. This allows us to not only gain insight into crucial common pathways of different cancers in carcinogenesis, but also obtain highly promising PPI targets for each cancer. Combining anticancer strategies to target PPIs in both common core network markers and specific network markers simultaneously in the future may provide emerging opportunities for anticancer therapeutics.

Theoretical mathematical and statistical method

The conventional methods to find hubs or significant proteins in PPIN of cancer are to find those proteins with the most links to other proteins. In this section, we compare cancer and normal PPINs, which are obtained by parameter estimation method and AIC system order detection scheme from microarray data of patient and normal people, respectively, to obtain the PPI differential network to represent the PPI variations of PPIN during the tumorigenesis process. Further, we also develop the CRV of each protein in Eq. (12.8) based on the total variations of PPI interaction abilities with other proteins in the PPI differential network to validate the significant PPI changes during tumorigenesis process. Finally, with the intersection and distinction of these significant proteins with high CRV, we obtain the core network marker for four cancers and specific network marker for each cancer. These core network markers and specific network markers are novel and could provide efficient PPI targets, for a small-molecule drug to bind to destroy tumors. All of the calculation and estimation process are based on theoretical calculation and estimation theory through real microarray data.

We know the maximum likelihood parameter estimation method and AIC model order selection method are famous and widely-used system identification methods from experimental data. During these estimation and learning processes, the PPI interaction mathematical model can learn the most possible PPIN for cancers and normal patients from large amount of microarray data and big databases to interpret the hidden biological mechanisms. The above paragraphs are adapted from our previous study [266].

Several scenarios of selecting multiple drug targets to break down the cancer core network

In this section, we used several different approaches to select multiple drug targets to break down the cancer core network. This is the key feature of this research, which confers its novelty. In contrast to traditional single target therapy, we used a multiple-target cocktail therapy to attack common core network markers, i.e., to inhibit several key proteins simultaneously instead of inhibiting only one single target. We assume that precisely targeting the core network would reduce the activity of the cancer. Based on this assumption, we believe that attacking core network markers simultaneously will be more efficient than attacking only one target. The first important problem is how to choose the group of key proteins as multiple drug targets. The first two novel approaches shown below to break down the cancer core network are based on the systems biology approach employed in our previous investigations [266]. We developed the third approach by performing pathway analysis using Panther [272] and MetaCore (GeneGo Inc.) (see Appendix B.1). In addition, we used a "confusion pharmacophore model" to develop the fourth novel trial approach.

In the first approach, we chose the compounds with highest docking scores for each of the 28 significant proteins. Then we removed the redundant compounds, and analyzed which compounds can target more than two proteins simultaneously. The basic strategy is to minimize the number of compounds necessary to target in order to break down the core network. Since there are 28 proteins in this core network, even choosing only one ligand to inhibit each of these 28 proteins would necessitate 28 ligands, which may be unsuitable for wet-lab validation.

In the second approach, we used the summation of the CRVs (Table 12.4) as the criterion to choose the key proteins to inhibit. In our previous study, we showed that the CRV quantifies the extent of a protein's association with other proteins, and thus it is optimal to target proteins with the highest summation of CRVs. This strategy depends on the cancer being targeted; in the present study, we focused on the common core network markers of the four cancers, so we chose the highest summation of CRVs across the four cancers.

In the third approach, we employed many new and valuable pathway analyses. Wet-lab cancer researchers can choose the pathway they want to focus on as the

Table 12.4 (A) CRVs of the 28 core proteins of each of the four cancers. (B) Top five ligands for the top five CRV proteins

(A) Protein name	CRV of bladder cancer	CRV of liver cancer	CRV of colorectal cancer	CRV of lung cancer	CRV SUM of four cancers	Protein name	CRV SUM (sorted)
BRCA1	11.59	6.82	10.86	12.52	41.79	UBC	496.32
CDK2	16.74	14.07	14.47	9.94	55.22	TP53	79.44
CEBPB	5.23	4.43	5.43	7.27	22.36	KIAA0101	71.99
CREBBP	12.10	9.59	11.59	11.19	44.46	HDAC1	70.56
CTNNB1	7.68	9.60	9.30	4.80	31.38	CDK2	55.22
CUL1	13.27	11.38	5.53	8.50	38.58	CUL3	53.81
CUL3	13.01	12.95	16.32	11.52	53.81	MYC	53.33
EP300	12.11	13.22	8.32	7.13	40.78	PCNA	52.90
ESR1	5.62	10.38	6.39	8.67	31.05	HDAC2	47.61
HDAC1	19.29	19.57	11.78	19.92	70.56	CREBBP	44.46
HDAC2	11.09	9.78	18.95	7.79	47.61	BRCA1	41.79
HDAC4	5.87	5.84	9.78	6.22	27.71	EP300	40.78
IRAK1	4.12	6.65	7.02	5.18	22.96	CUL1	38.58
ISG15	4.49	6.023	5.81	4.94	21.26	YWHAZ	35.93
KIAA0101	15.82	16.77	19.14	20.26	71.99	CTNNB1	31.38
MDM2	4.56	4.75	11.56	6.28	27.16	ESR1	31.05
MYC	13.04	10.78	20.06	9.45	53.33	SRC	28.39
PCNA	13.32	15.14	9.63	14.81	52.90	TRAF2	27.76
PRKDC	4.08	5.94	8.56	5.67	24.25	HDAC4	27.71
PSMA3	5.80	7.80	8.09	5.28	26.97	MDM2	27.16
RB1	6.89	5.55	5.98	8.32	26.74	PSMA3	26.97
SRC	8.80	4.08	9.14	6.37	28.39	RB1	26.74
TERF1	5.46	4.90	6.26	8.07	24.70	XRCC6	25.74
TP53	19.59	18.74	25.93	15.18	79.44	TERF1	24.70
TRAF2	9.20	4.77	7.37	6.43	27.76	PRKDC	24.25
UBC	158.53	137.28	80.39	120.12	496.32	IRAK1	22.96
XRCC6	5.29	9.56	6.02	4.87	25.74	CEBPB	22.36
YWHAZ	8.80	12.64	7.90	6.58	35.93	ISG15	21.26

(B) UBC(4FJV)		TP53(1TSR)		KIAA0101		HDAC1(4BKX)		CDK2(3QQK)	
NCI drug	LibDock score	NCI drug	LibDock score	NCI drug	LibDock score	NCI drug	LibDock score	NCI drug	LibDock score
719481	122.89	673172	126.11	655102	165.72	627865	134.69	680359	108.34
633409	114.19	695409	125.82	669588	164.04	625439	134.10	678636	101.88
672968	111.87	682236	117.73	407811	158.57	647638	125.51	669299	101.27
734999	111.58	695405	115.63	704564	148.68	2426	124.72	679065	101.01
688121	111.50	667504	114.33	698687	147.05	707841	117.08	376791	100.92

therapy target. There are many different combinations dependent on different conditions of cancers, so we do not list all the possibilities. Here, we demonstrate a single example, and others can develop their own scenarios following this example with the help of the docking results (see Appendix B.2).

The fourth novel approach is to develop confusion and individual pharmacophore models of core network markers (28 proteins) to undertake virtual screening using the compounds stored in the Life Chemical database.

We highlight the different approaches in this section, because they form the crux of this study. In the following sections, we describe the related methods necessary for each of the approaches.

More pathway and functional analyses

PANTHER

In our previous studies, we performed pathway analysis using the DAVID database. Here, we expand on the previous analysis by performing more pathway and functional analyses in order to develop a more efficient strategy for multiple-target therapy. This approach will allow us to accumulate more information on possible therapy strategies. The 28 proteins comprising the cancer core network markers were analyzed for their molecular functions, molecular processes, and subcellular localizations by the PANTHER classification system (http://www.pantherdb.org) [272]. PANTHER was designed to classify proteins (and their genes) in order to facilitate high-throughput analysis. Proteins have been classified according to:

1. Family and subfamily: families are groups of evolutionarily related proteins; subfamilies are related proteins that also have the same function.
2. Molecular function: the function of the protein by itself or with directly interacting proteins at a biochemical level, e.g., a protein kinase.
3. Biological process: the function of the protein in the context of a larger network of proteins that interact to accomplish a process at the level of the cell or organism, e.g., mitosis.
4. Pathway: similar to biological process, but a pathway also explicitly specifies the relationships between the interacting molecules.

MetaCore

MetaCore includes a manually annotated database of gene interactions and metabolic reactions obtained from the scientific literature, including the most newly updated ones. The enrichment analysis of the biological process was based on the hypergeometric distribution algorithm and relevant pathway maps. The mathematical foundation of MetaCore is shown in Appendix B.1.

More pathway and functional analyses are fundamental to learning more about the hidden mechanisms of cancer networks. So, to interpret the results for the third approach in a meaningful way, this topic must be explained and described beforehand. The novelty of this section is in its combination of systems biology with

CADD. This research opens many new directions for multiple drug design to target multiple drug targets to break down the core network biomarkers; we describe only a portion of these results in this section, for illustrative purposes. These results clearly show that these approaches show great promise for future research to target special pathways through the design of drugs having multiple targets.

Protein and ligands structures

The following section illustrates the CADD strategy applied to target the 28 common core network proteins. The first three approaches need both the 3D structures of proteins and ligands, while the fourth pharmacophore approach only needs the structures of the ligands. The first thing we need is thus the 3D structures of the proteins. At this stage, 24 of them are available in the well-known PDB database, and we can download them directly, while there are four proteins (CEBPB, IRAK1, KIAA0101, and PSMA3) whose 3D structures have not been solved at this stage (Table 12.5). We used the NCI (46872 ligands) and Life Chemical (31742 ligands) drug libraries with the filter anticancer. All drug structures were prepared and minimized by DS3.5. People seldom do the virtual screening on multiple targets since it is computationally intense. We performed this work on an IBM server with 160 cores and 1 TB of memory.

Homology model and binding site prediction

Proteins do not always have 3D structures available in the PDB database. Since only amino acid sequence data without corresponding structures were available for four of the proteins targeted in this section (CEBPB, IRAK1, KIAA0101, and PSMA3) we used homology modeling to predict the structures for these proteins. Among many famous software and webservers, the I-Tasser webserver developed by Zhang et al. [273] is the most well known homology modeling webserver. We used this server to perform homology modeling of these proteins of interest that lacked available 3D structures. For some well-studied protein targets, structures with embedded ligands are available, so the exact binding site for the docking experiments is known. However, binding site information was unavailable for most proteins, so we used the COACH webserver [274] developed by Zhang et al. to predict the binding sites.

Docking

We use LibDock Score of DS 3.5 to perform docking simulation and then rank the docking results, and finally we choose the top 20 compounds for further analysis, such as the construction of pharmacophore models.

Table 12.5 Structure information of 28 core proteins

No	Protein	PDB ID	Resolution (Å)	Chains	Length	Ligand	Full name
1	BRCA1	4IFI	2.20	A	214	BAAT peptide	Breast cancer type 1 susceptibility protein
2	CDK2	3QQK	1.86	A	306	X02	Cyclin-dependent kinase 2
3	CEBPB	[a]					CCAAT/enhancer-binding protein beta
4	CREBBP	4A9K	1.81	A, B	119	TYL	CREB-binding protein
5	CTNNB1	3TX7	2.76	A	527	P6L	Catenin (cadherin-associated protein), beta 1
6	CUL1	4F52	3.00	A, C	282	n.a.	Cullin-1
7	CUL3	4EOZ	2.40	B, D	346	n.a.	Cullin-3
8	EP300	4BHW	2.80	A, B	578	01K	E1A binding protein p300
9	ESR1	1UOM	2.28	A	254	PTI	Estrogen receptor 1
10	HDAC1	4BKX	3.00	B	482	n.a.	Histone deacetylase 1
11	HDAC2	4LY1	1.57	A, B, C	369	20Y	Histone deacetylase 2
12	HDAC4	2VQM	1.80	A	413	HA3	Histone deacetylase 4
13	IRAK1	[a]					Interleukin-1 receptor-associated kinase 1
14	ISG15	3SDL	2.29	C, D	164	n.a.	ISG15 ubiquitin-like modifier
15	KIAA0101	[a]					KIAA0101
16	MDM2	4MDN	1.90	A	94	Y30	Mouse double minute 2 homolog
17	MYC	1NKP	1.80	A, D	88	n.a.	V-myc myelocytomatosis viral oncogene homolog
18	PCNA	3WGW	2.80	A, B	261	T2B	Proliferating cell nuclear antigen
19	PRKDC	3KGV	6.60	A–F, O, P–T, X–Y	4128	n.a.	Protein kinase, DNA-activated, catalytic polypeptide
20	PSMA3	[a]					Proteasome subunit alpha type-3
21	RB1	3POM	2.50	A, B	352	n.a.	Retinoblastoma 1
22	SRC	2SRC	1.50	A	452	ANP	Proto-oncogene tyrosine-protein kinase Src
23	TERF1	3BQO	2.00	A	211	TIN2 peptide	Telomeric repeat-binding factor 1
24	TP53	1TSR	2.20	A, B, C	219	n.a.	Tumor protein p53
25	TRAF2	1D0A	2.00	A–F	168	OX40L peptide	TNF receptor-associated factor 2
26	UBC	4FJV	2.05	B, D	86	n.a.	Ubiquitin C
27	XRCC6	1JEQ	2.70	A	609	n.a.	X-ray repair cross-complementing 6
28	YWHAZ	4HKC	2.20	A	250	Alpha-4 peptide	14-3-3 Protein zeta/delta

[a]3D structures prepared by homology model, I-Tasser.

Pharmacophore model

1. **HypoGen Method and Virtual Screening**

   There are three stages in the generation of the HypoGen model: constructive phase, subtractive phase, and optimization phase. (1) Constructive phase: Active ligands within a given range of the maximum activity are chosen initially. The two most active ligands are used to enumerate the pharmacophores. (2) Subtractive phase: After the constructive phase, a database with a large number of pharmacophore structures is generated. The purpose of the subtractive phase is to identify the inactive pharmacophores and eliminate them from the database. (3) Optimization phase: The simulated annealing method is used to modify the scoring function of each hypothesis to be tested. After optimization, the HypoGen method generates the top 10 hypotheses. This study is the first attempt to combine systems biology with CADD. Pharmacophore modeling is one of the most powerful approaches in CADD. Here, we developed a novel prototype pharmacophore model, which is different from the traditional one. As mentioned earlier, this is a novel idea, and the pharmacophore model is just a prototype: it will require further modification in future studies. In traditional pharmacophore modeling, the IC_{50} value for each drug under investigation is necessary, but this model does not require them. Instead, we linearly transformed LibDock scores to generate putative IC_{50} values, and used these values to build the pharmacophore. We combined the top three ligands for each of the 28 proteins to make a ligand pool, and then used this ligand pool to construct a common pharmacophore; this is qualitatively different from building 28 individual pharmacophores.

2. **PharmGist**

   PharmGist is another method to construct ligand-based pharmacophores. Because it does not require IC_{50} values of ligands [275], it is easier to construct pharmacophores by PharmGist than by the HypoGen model. For each protein, we constructed one PharmGist model: these models can be used to do virtual screening in the future.

Cocktail multiple-target strategy: a novel model combined with systems biology and CADD

In summary, we used systems biology to construct the common core network marker of four different cancers, which contains 28 proteins as the housekeeping proteins shared by the different cancers. Then we used the docking and pharmacophore methods to perform virtual screening on these 28 core proteins and get the top compounds for each protein. In contrast to traditional single target methods, we suggest using a combination of the top compounds to perform cell proliferation experiments. We have provided an example for each of these first three approaches. This section provides a novel direction for multiple drug target therapy for cancers.

Wet-lab biomedical scientists can combine the top ligands for each of the 28 core proteins based on their experimental conditions and the pathways that they want to focus on. As an early stage project, we have taken a conservative approach, only using the NCI anticancer drug library. As these drugs have proven anticancer activity, this cocktail of multiple drugs can be expected to slightly or moderately enhance the therapeutic effect under the right wet-lab conditions. Collaboration with and input from wet-lab experts would permit the modification and optimization of this model, and the scope could expand to include other drug libraries in the virtual screening.

12.3.2 Results and discussion of multiple drug target

Review of the results of previous methods

Our previous study identified 28 core proteins in the common core network marker for four cancers [266]. These are the proteins intersecting between the four cancers' PPINs with the top CRVs. Fig. 12.7 shows the PPI information of each common core network marker. As stated previously, these 28 core proteins could be responsible for the housekeeping mechanisms of these four cancers, so using the minimum possible number of ligands to target the maximum possible number of the 28 proteins could be the most efficient strategy to attack the cancers.

Several scenarios to break the cancer core network marker

1. We chose the top 20 compounds with the highest docking scores for each of the 28 significant proteins. Of a total of 560 ligands, we found there were 13 ligands that target 2 or 3 proteins simultaneously (Table 12.6A) after redundancy analysis. These 13 ligands combined 11 target proteins (Table 12.6B). We used the 13 ligands as the minimum package for breaking the core network by inhibiting these 11 proteins. While this does not target all 28 proteins, the main purpose here was to minimize the number of ligands at this first stage trial. (As we have said, this is the first trial of this primary model.) Of course, according to these docking results of these 28 proteins, there are thousands of possible drug combinations that can be used based on different analysis methods as per a given researchers scope. It is quite possible to develop combinations to target all the 28 proteins based on these docking results.
2. In our previous research, we observed the CRVs of the 28 core proteins for each of the four cancers (Table 12.4A). In this study, we summed up the 4 CRVs for each of the 28 proteins. After ranking these summations, we listed the top 5 CRV summations (Table 12.4(B)), and suggested them to be the therapeutic targets for breaking cancer core network marker. According to our previous systems biology analysis, we believe that inhibiting proteins with the highest CRVs is a highly efficient way to attack cancers. As the first trial, the combination of proteins with the top 5 CRVs should simply be taken as a proof of concept: one can develop combinations according to one's particular needs based on these CRVs.

Table 12.6 (A) NCI drugs target more than one protein. (B) The proposed 13 drugs totally target the following 11 proteins

(A) No.	NCI drug	Protein name	LibDock score
1	625439	KIAA0101	145.708
	625439	HDAC1(4BKX)	134.097
	625439	HDAC4(2VQM)	149.324
2	645378	CEBPB(2E_42)	158.637
	645378	KIAA0101	140.103
3	668448	BRCA1(4IFI)	192.069
	668448	PSMA3	166.516
4	668577	MYC(1NKP)	125.603
	668577	PSMA3	180.11
5	682094	TERF1(3BQO)	123.287
	682094	YWHAZ(4HKC)	182.534
6	687363	MYC(1NKP)	145.138
	687363	YWHAZ(4HKC)	184.649
7	695175	TRAF2(1D0A)	101.267
	695175	PSMA3	148.83
8	695409	TP53(1TSR)	125.821
	695409	PSMA3	164.906
9	704565	KIAA0101	137.38
	704565	HDAC4(2VQM)	153.564
10	719660	MYC(1NKP)	134.021
	719660	YWHAZ(4HKC)	200.903
11	724305	CEBPB(2E_42)	167.724
	724305	HDAC4(2VQM)	141.195
12	726771	BRCA1(4IFI)	142.956
	726771	HDAC4(2VQM)	148.366
13	742856	MYC(1NKP)	140.093
	742856	TP53(1TSR)	100.787

(B) No.	Protein name
1	BRCA1(4IFI)
2	CEBPB(2E42)
3	ISG15(3SDL)
4	MDM2(4MDN)
5	PCNA(3WGW)
6	PRKDC(3KGV)
7	PSMA3
8	SRC(2SRC)
9	TERF1(3BQO)
10	TP53(1TSR)
11	XRCC6(1JEQ)

3. We have also listed pathway analyses using Panther and MetaCore on the core network proteins. Other scientists can use this information to choose the pathways that they want to target, and then choose the best combination of drugs from these virtual screening results. For instance, from the pathway analysis results using Panther (see Table 12.7), researchers interested in targeting the p53 pathway can choose between EP300, CREBBP, HDAC1, HDA2, TRAF2, CDK2, MDM2, and TP53 as the therapy targets. For the above scenario, researchers can choose the top drugs. MetaCore gives us more information for the purpose of target therapy. The top three modules mapped to these 28 core proteins are listed in Table 12.8. Taking the first map/module (DNA damage_ATM/ATR regulation of G1/S checkpoint) as an example, we see CDK2, p53, Ubiquitin, Brcal, c-Myc, PCNA, and MDM2 are related to this module. As above, we can use the drugs in Appendix B.1 to choose the optimal combination of drugs to attack this module.

Table 12.7 The pathway analysis results of Panther

Rank	Pathway title	Count	Gene
1	p53 pathway	8	*EP300, CREBBP, HDAC1, HDA2, TRAF2, CDK2, MDM2, TP53*
2	Wnt signaling pathway	7	*EP300, CREBBP, HDAC1, MYC, HDAC2, CTNN1, TP53*
3	p53 pathway feedback loops 2	6	*MYC, RB1, CTNNB1, CDK2, MDM2, TP53*
4	Parkinson disease	4	*YWHAZ, CUL1, PSMA3, SRC*
4	Gonadotropin releasing hormone receptor pathway	4	*EP300, CREBBP, CTNNB1, SRC*
5	Huntington disease	3	*EP300, CREBBP, TP53*
6	Apoptosis signaling pathway	2	*TRAF2, TP53*
6	Angiogenesis	2	*CTNNB1, SRC*
6	P53 pathway feedback loops 1	2	*MDM2, TP53*
6	Cadherin signaling pathway	2	*CTNNB1, SRC*
6	Transcription regulation by bZIP transcription factor	2	*EP300, CREBBP*
6	TGF-beta signaling pathway	2	*EP300, CREBBP*
7	Interleukin signaling pathway	1	*MYC*
7	Alzheimer disease-presenilin pathway	1	*CTNNB1*
7	Integrin signaling pathway	1	*SRC*
7	Insulin/IGF pathway-protein kinase B signaling cascade	1	*MDM2*
7	Hypoxia response via HIF activation	1	*CREBBP*
7	Ubiquitin proteasome pathway	1	*MDM2*
7	p53 pathway by glucose deprivation	1	*TP53*

Please see their statistical distribution in Fig. 12.8.

Table 12.8 Top three modules/maps given by MetaCore

No.	Processes	Map of core proteins	*Z*-Score	*G*-Score
1	DNA damage_ATM/ATR regulation of G1/S checkpoint: negative regulation of transcription, DNA-dependent (60.0%), negative regulation of macromolecule metabolic process (70.0%), negative regulation of gene expression (62.0%), negative regulation of RNA metabolic process (60.0%), negative regulation of cellular metabolic process (68.0%)	CDK2, p53, ubiquitin, Brca1, c-Myc, PCNA, MDM2	107.78	107.78
2	Transcription_P53 signaling pathway: positive regulation of transcription from RNA polymerase II promoter (80.0%), positive regulation of macromolecule biosynthetic process (86.7%), positive regulation of nucleobase-containing compound metabolic process (86.7%), regulation of transcription from RNA polymerase II promoter (86.7%), positive regulation of nitrogen compound metabolic process (86.7%)	CDK2, p53, CBP, p300, Rb protein, MDM2, Beta-catenin	67.47	67.47
3	DNA damage_Brca1 as a transcription regulator: cellular response to interferon-gamma (36.7%), interferon-gamma-mediated signaling pathway (32.7%), response to interferon-gamma (36.7%), T-cell costimulation (30.6%), lymphocyte costimulation (30.6%)	p53, ESR1 (nuclear), Brca1, c-Myc, PCNA, Rb protein	60.47	60.47

More pathway and functional analysis

PANTHER

The results of PANTHER are listed in Fig. 12.8 and Table 12.7. We see the 28 core network markers hit many important cancer-related pathways, such as p53, Wnt signaling, p53 feedback loops 2, and apoptosis. As we have described earlier, this study is a prototype model. Inhibition of the right proteins hitting the key pathways is an

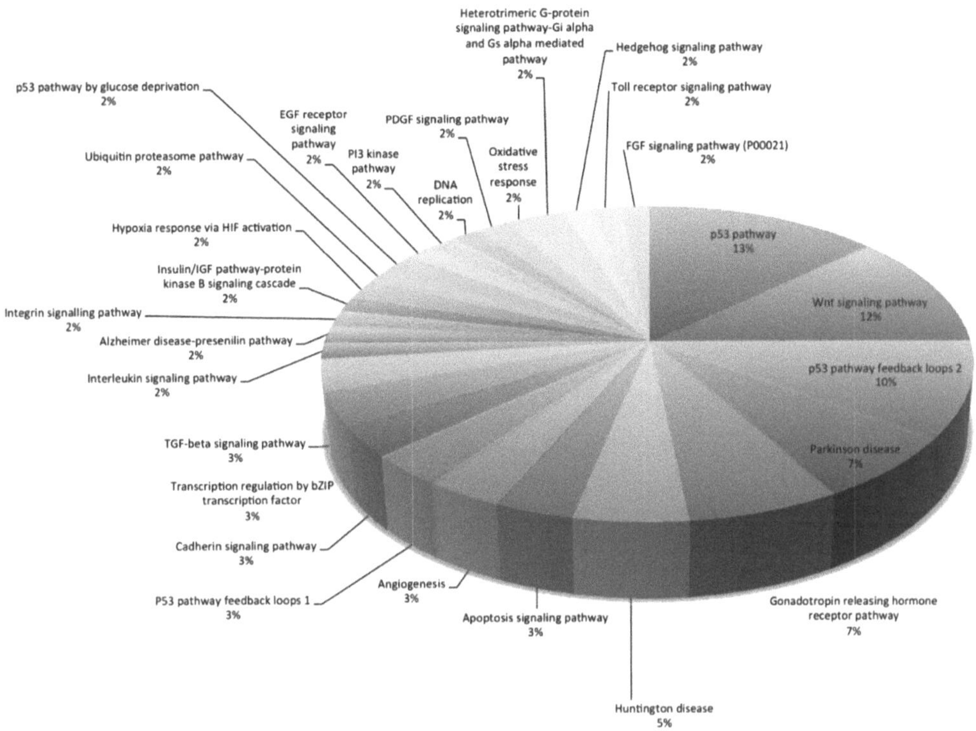

Figure 12.8 Panther. We see the 28 proteins of core network marker can hit many important cancer-related pathways, such as p53, Wnt signaling, p53 feedback loops 2, and apoptosis. As we have described earlier, this section is only a prototype model. By using the Panther analysis, we know which proteins should be inhibited that hit the key pathways, which is one of the important strategies in a real clinical situation. This result offers another reference for doctors to design the best combination of multiple inhibitors. We will do more deep analysis in the future if more clinical doctors could offer the clinical data to us. The preliminary results also could help us exclude the cancer-unrelated pathways at the first stage, such as Huntington disease and Alzheimer.

important strategic consideration in real clinical situations. These results offer another reference for doctors to design the best combination of multiple inhibitors. In the future, using clinical data from doctors will help us perform deeper analysis. These preliminary results also could help us exclude pathways unrelated to cancer at the first stage, such as those related to Huntington's disease, and Alzheimer's disease.

MetaCore

The results of MetaCore analysis are described below. We have described the top three maps/modules (Table 12.8).

1. DNA damage_ATM/ATR regulation of G1/S checkpoint: The highest scoring map (i.e., the map with the lowest *p*-value) is based on the enrichment distribution sorted by the "Statistically significant maps" set. Experimental data from all files is

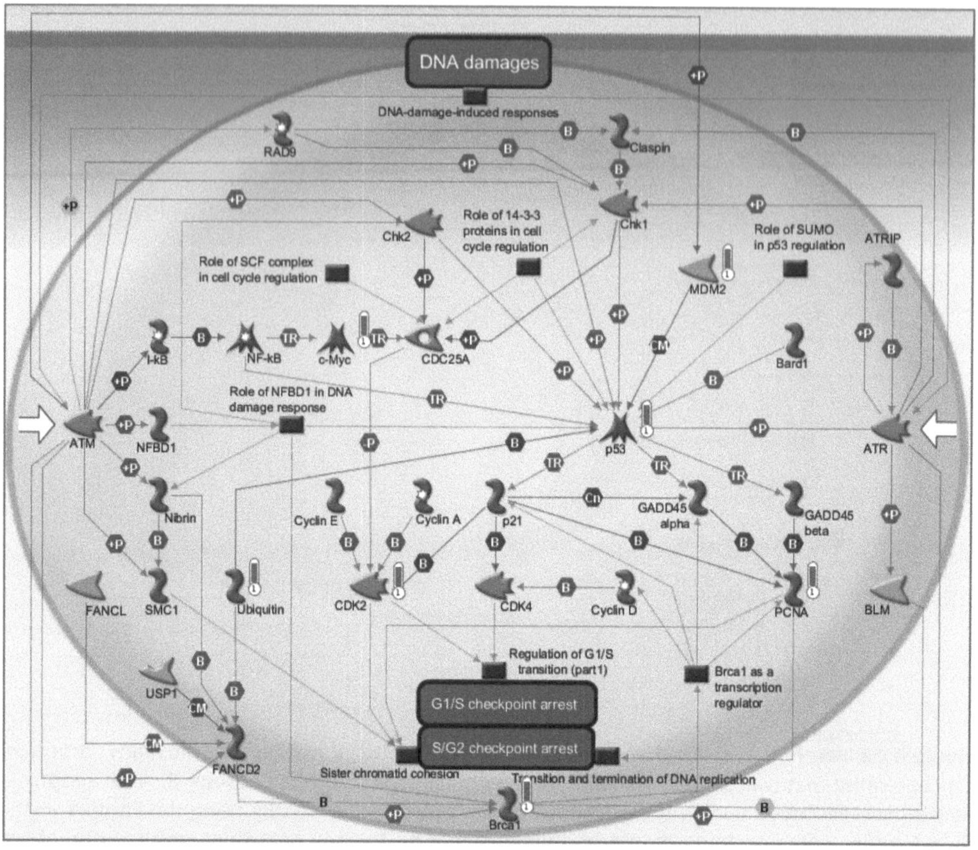

Figure 12.9 DNA damage ATM/ATR regulation of G1/S checkpoint. The top scored map (map with the lowest *p*-value) is based on the enrichment distribution sorted by "Statistically significant Maps" set. Experimental data from all files is linked to and visualized on the maps as thermometer-like figures. Upward thermometers have *red color* to indicate upregulated signals and downward (*blue*) ones indicate downregulated expression levels of the genes.

linked to and visualized on the maps as thermometer-like figures. Upward thermometers are *red* and indicate upregulated signals, and downward (*blue*) ones indicate downregulated expression levels of the genes (Fig. 12.9).

2. The second highest scoring map (i.e., the map with the second lowest *p*-value) is based on the enrichment distribution sorted by the "Statistically significant maps" set. Experimental data from all files is linked to and visualized on the maps as thermometer-like figures. Upward thermometers are *red color* and indicate upregulated signals, and downward (*blue*) ones indicate downregulated expression levels of the genes (Fig. 12.10).

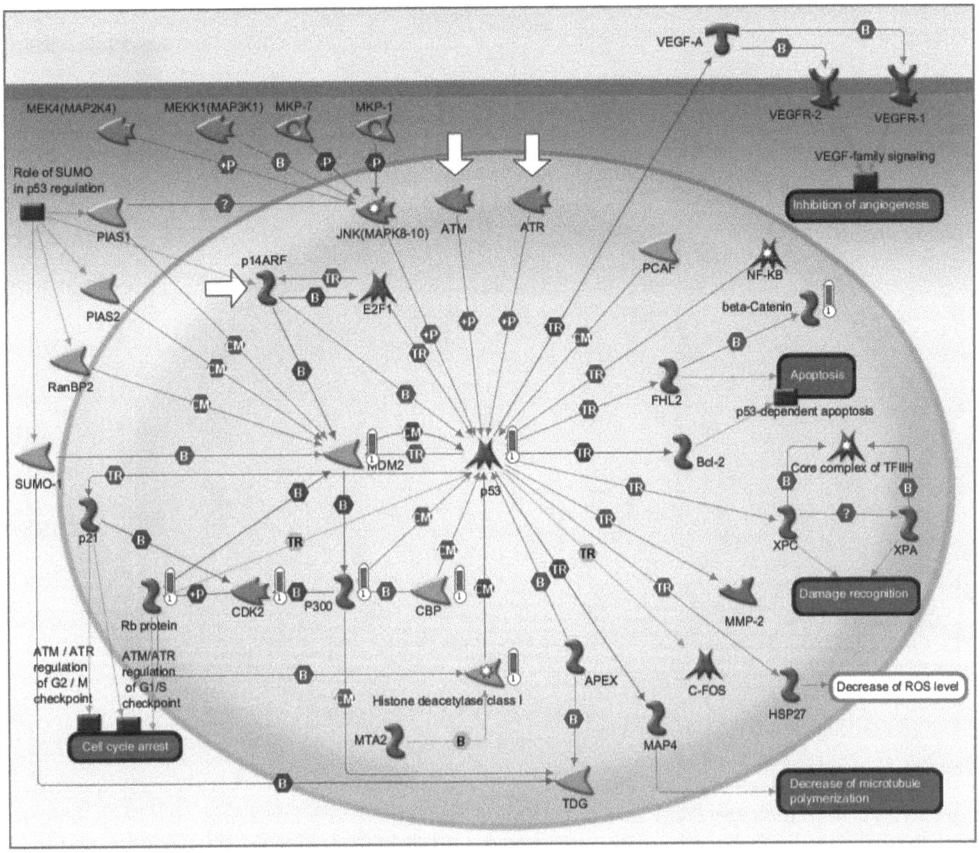

Figure 12.10 Transcription p53 signaling pathway. The second scored map (map with the second lowest *p*-value) is based on the enrichment distribution sorted by "Statistically significant Maps" set. Experimental data from all files is linked to and visualized on the maps as thermometer-like figures. Upward thermometers have *red color* to indicate upregulated signals and downward (*blue*) ones indicate downregulated expression levels of the genes.

3. The third highest scoring map (i.e., the map with the third lowest *p*-value) is based on the enrichment distribution sorted by the "Statistically significant maps" set. Experimental data from all files is linked to and visualized on the maps as thermometer-like figures. Upward thermometers are *red* and indicate upregulated signals, and downward (*blue*) ones indicate downregulated expression levels of the genes (Fig. 12.11).

Statistical analysis results of the three maps are shown in Table 12.8. These results also help us choose more suitable targets and exclude less suitable ones.

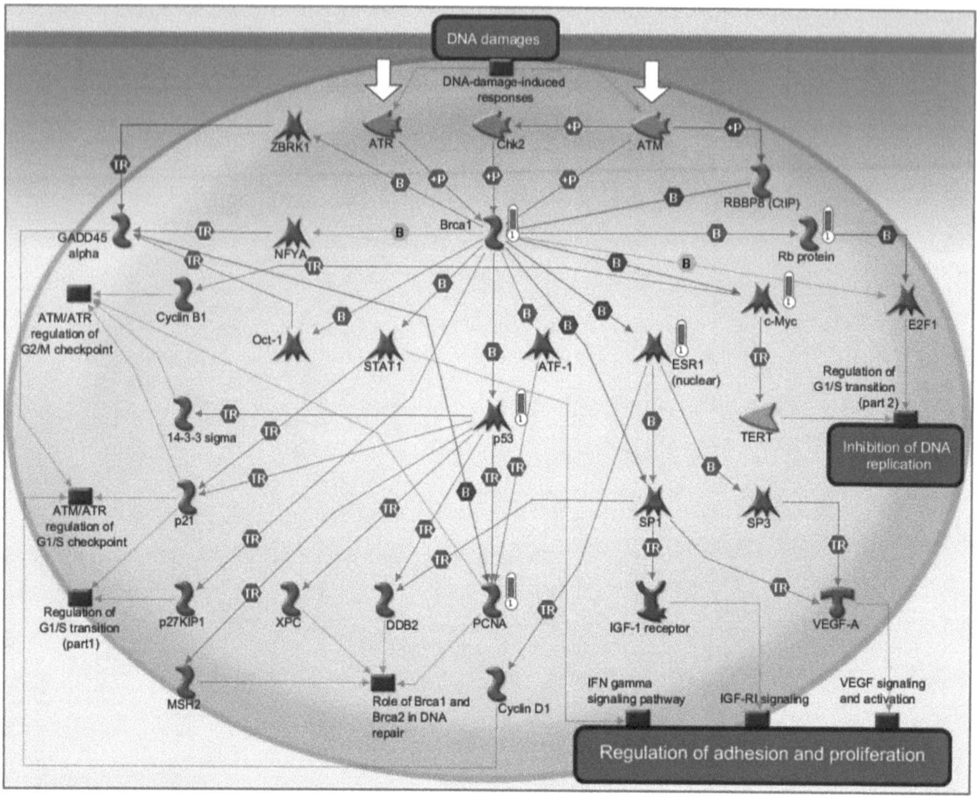

Figure 12.11 DNA damage_Brca1 as a transcription regulator. The third scored map (map with the third lowest *p*-value) is based on the enrichment distribution sorted by "Statistically significant Maps" set. Experimental data from all files is linked to and visualized on the maps as thermometer-like figures. Upward thermometers have *red color* to indicate upregulated signals and downward (*blue*) ones indicate downregulated expression levels of the genes.

Protein and ligands structures

Table 12.5 shows detailed information of these 28 proteins including protein name, PDB ID, resolution, chains, length, ligands, and their full names. Most of these 3D structures were downloaded from the PDB database. 3D structures of CEBPB, IRAK1, KIAA0101, and PRKDC were constructed by homology modeling using I-TASSER. We can see that many protein structures also contained a ligand in a bound conformation. We used the locations of these ligands within the target protein as the docking site. Binding sites of proteins without bound ligands were predicted using COACH.

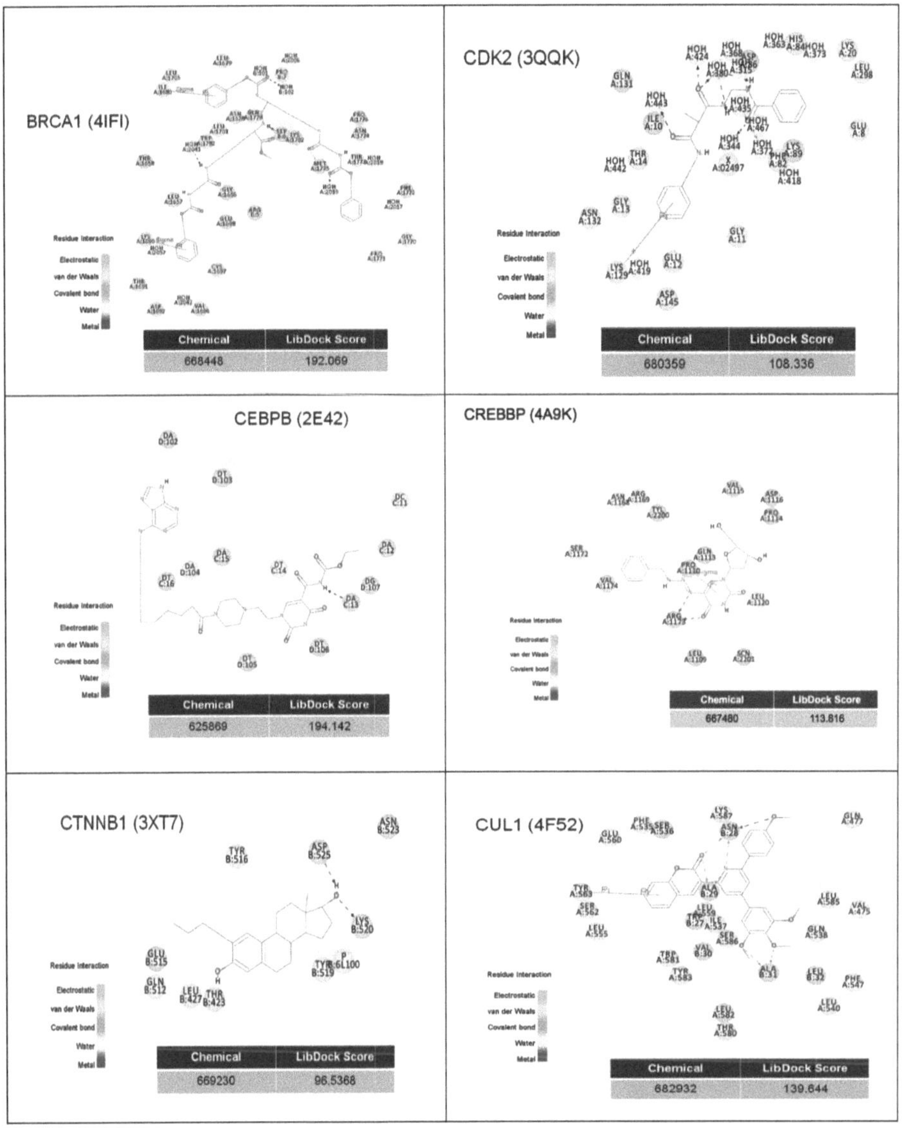

Figure 12.12 The docking poses analysis shows the key residues of the proteins which interact with the ligands. The analysis reveals more mechanisms to help us design de novo drugs.

Docking

Fig. 12.12 shows the docking pose of the top compounds (i.e., those with the highest docking score). Their docking scores ranged from 95 to 241. The analysis also shows the key residues in ligand binding. This section is the first to perform high-throughput multiple-protein docking, and this work requires large computational

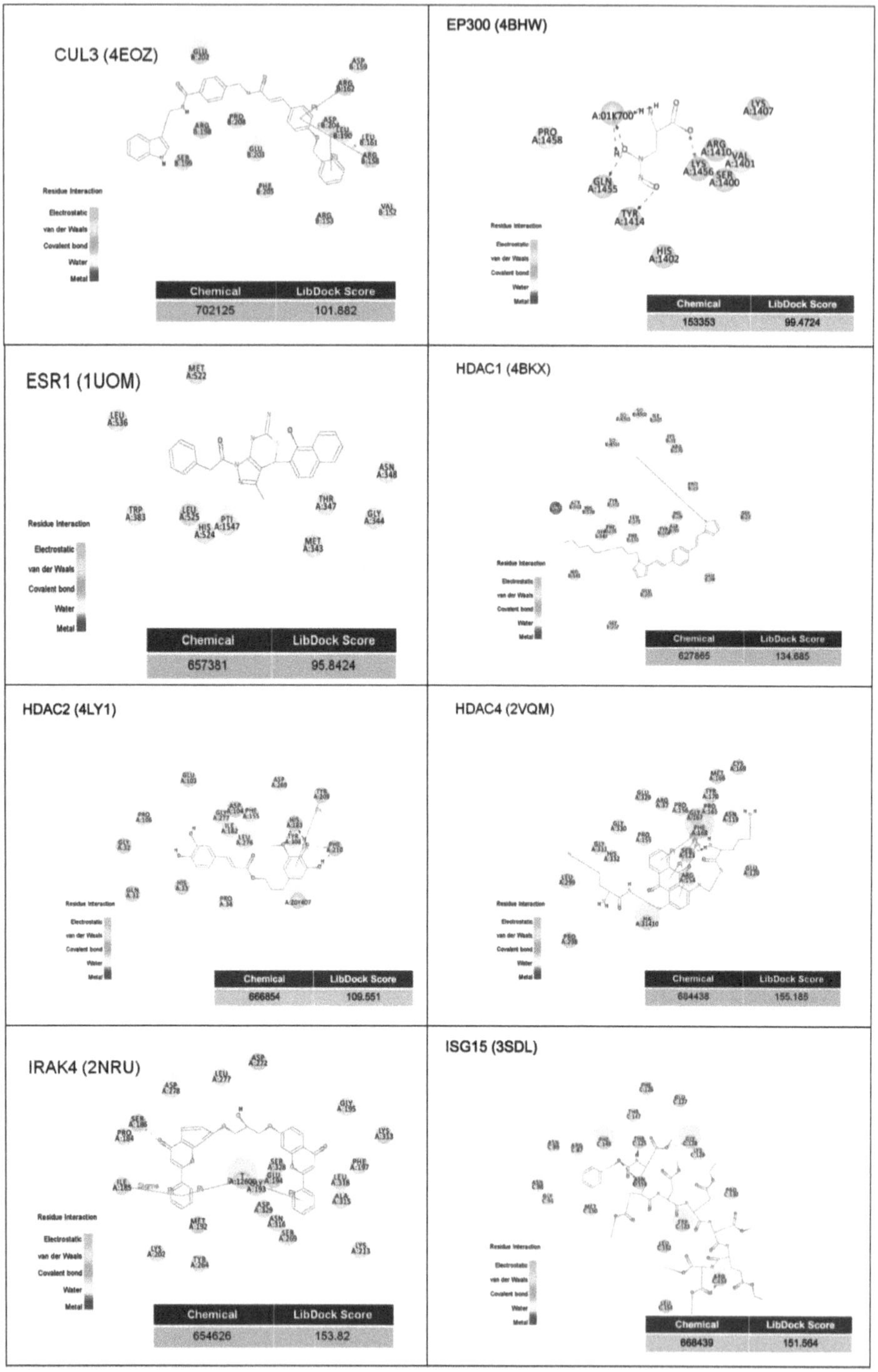

Figure 12.12 (Continued)

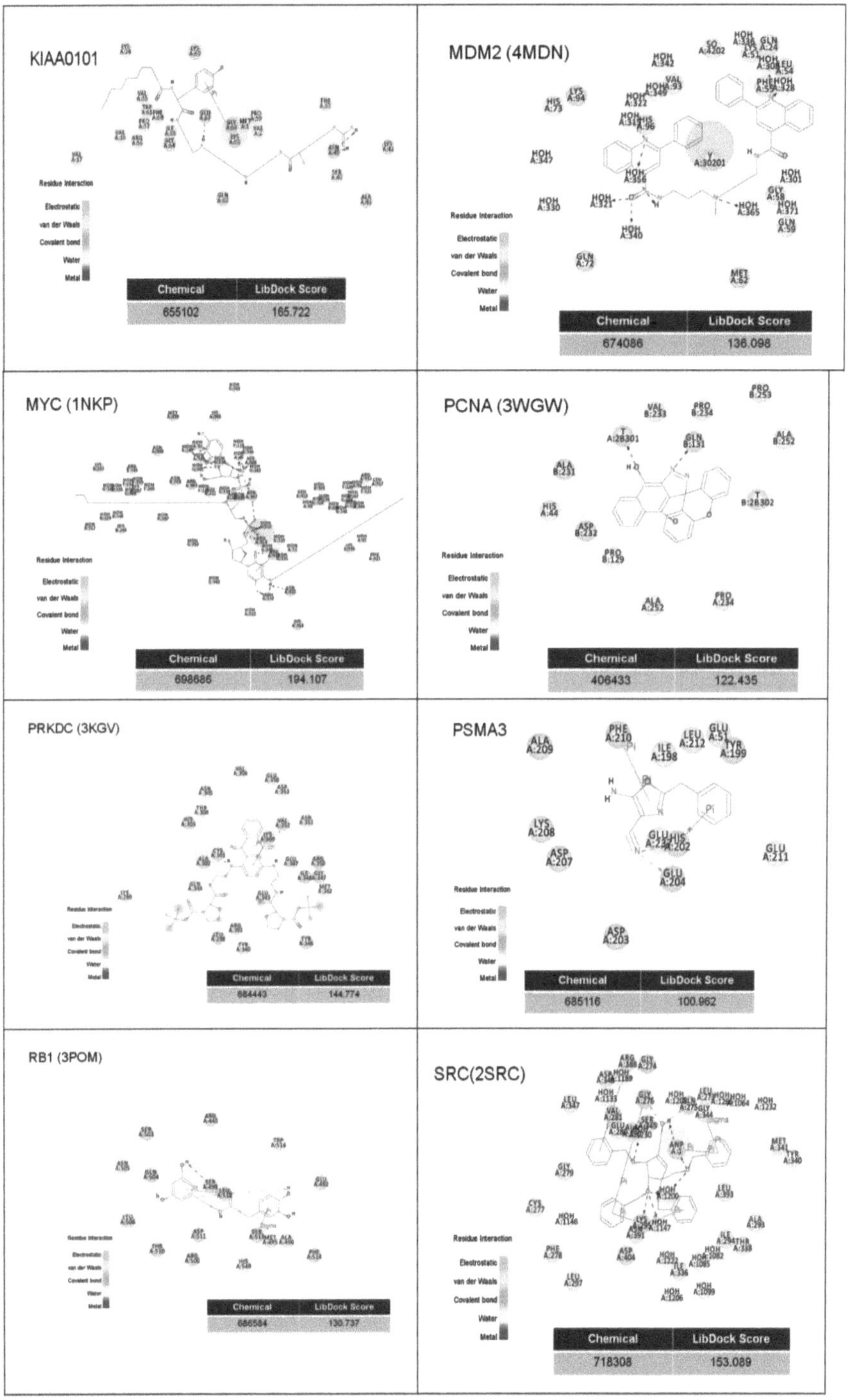

Figure 12.12 (Continued)

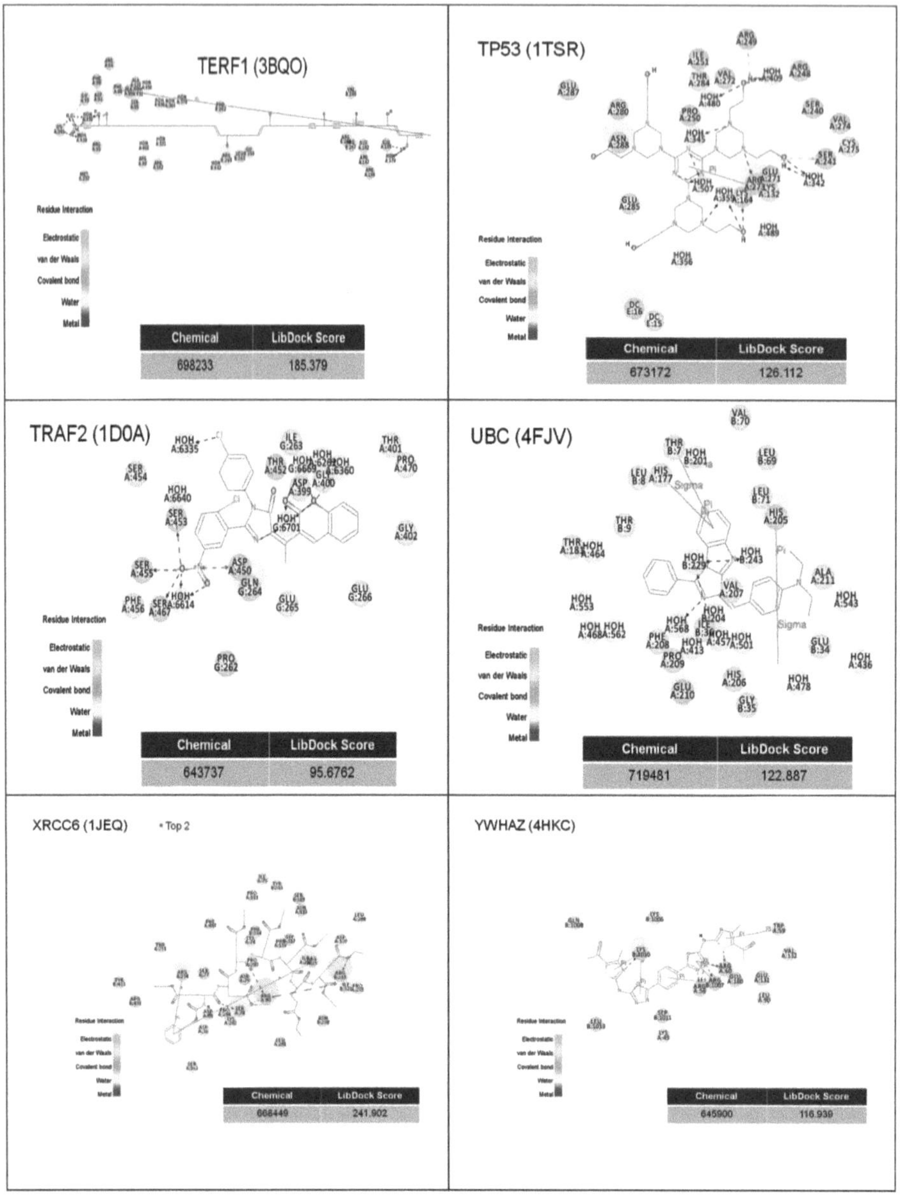

Figure 12.12 (Continued)

resources. Due to space limitations, we have not listed another table for these key residues. The information could be useful for future drug design and these top compounds could be considered as control ligands.

Pharmacophore

3D pharmacophore modeling is another powerful method to perform virtual screening on large ligand databases. It is as powerful as the docking method, and is always more efficient than docking methods. For example, the large ligands database ZINC [276] has more than 200 million ligands. By calibrating the parameters of the pharmacophore virtual screening, it is possible to virtually screen the whole ZINC database using this IBM server with 160 cores. We have even performed virtual screening for Src inhibitors. However, constructing a proper HypoGen model is also very time-consuming. As an initial study using a novel approach, we used relatively unrestricted conditions to set up a rough HypoGen model. It would be beneficial to do virtual screening on another ligand database in the future.

Fig. 12.13 lists the total 54 ligands used to construct the HypoGen pharmacophore model. We present the structure of the pharmacophore in Fig. 12.14; NSC-625869 refers to the ligand in the NCI library of anticancer drugs that was screened and matched by this pharmacophore. The numerical results of the HypoGen model are shown in Table 12.9. After having constructed this pharmacophore, we were able to use it to screen against other ligand databases, such as the Life Chemical database. Fig. 12.15 and Table 12.10 show the compounds in the Life Chemical database screened and matched by this pharmacophore.

Compared to the HypoGen model, the pharmacophore model derived by PharmGist is easier to obtain because PharmGist does not require the IC_{50} values of the compounds for computation. We have listed the results in Fig. 12.16 and Table 12.11.

Novel model combined with both systems biology and CADD

Our method has reversed the normal procedure of CADD with the help of systems biology. In the traditional CADD method, a single validated target, such as Src, EGFR, and Fak, is used as the protein target. The first step is usually virtual screening by docking or the pharmacophore method. The second step is to validate the top compounds from virtual screening with ELISA or Western blot wet-lab experiments. If they are found to bind to the target proteins, then researchers could employ these ligands in cell proliferation experiments. For this novel approach, we choose the top compounds from several targets simultaneously, and then perform the cell proliferations experiment at the beginning. This hypothesis is that these

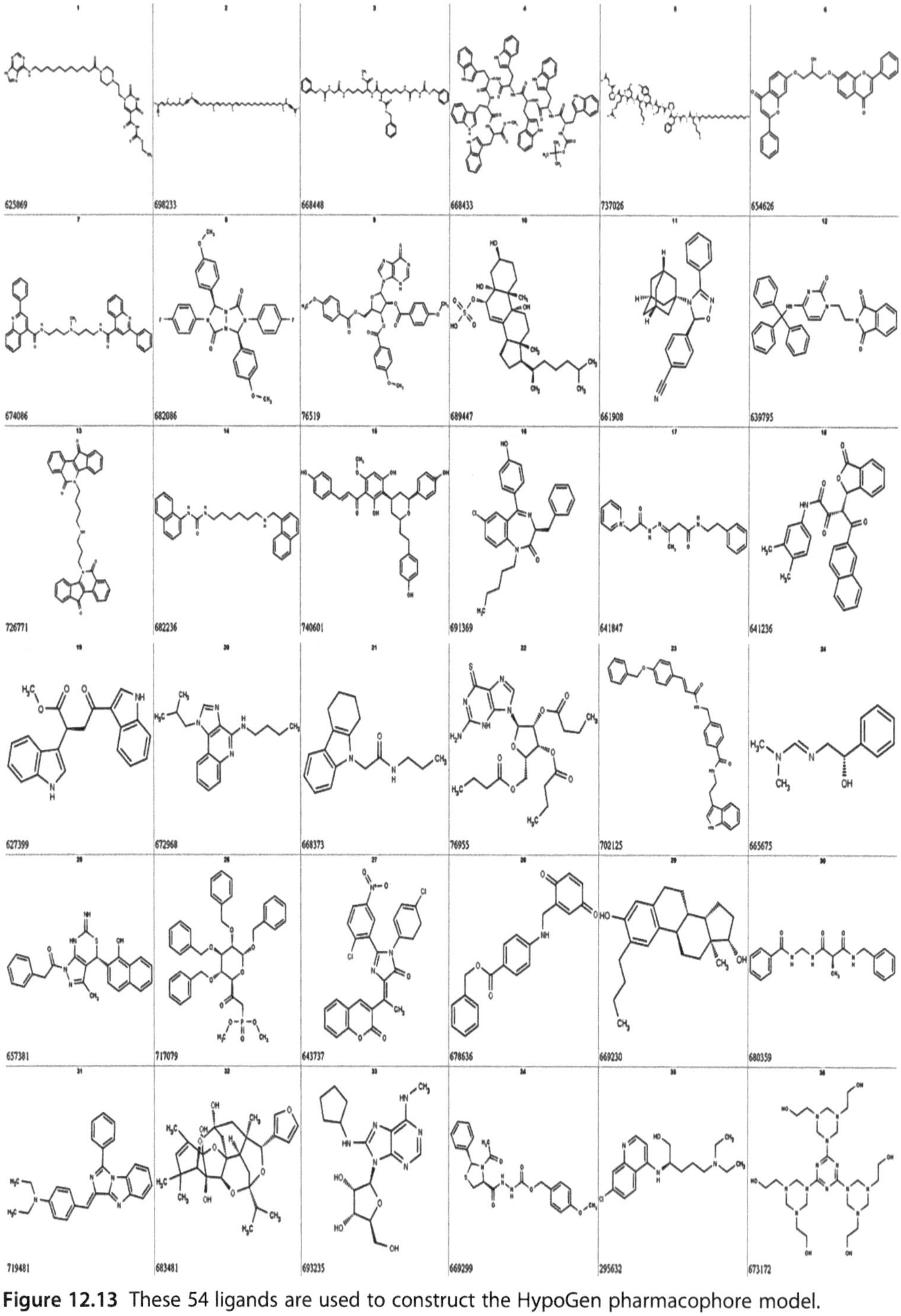

Figure 12.13 These 54 ligands are used to construct the HypoGen pharmacophore model.

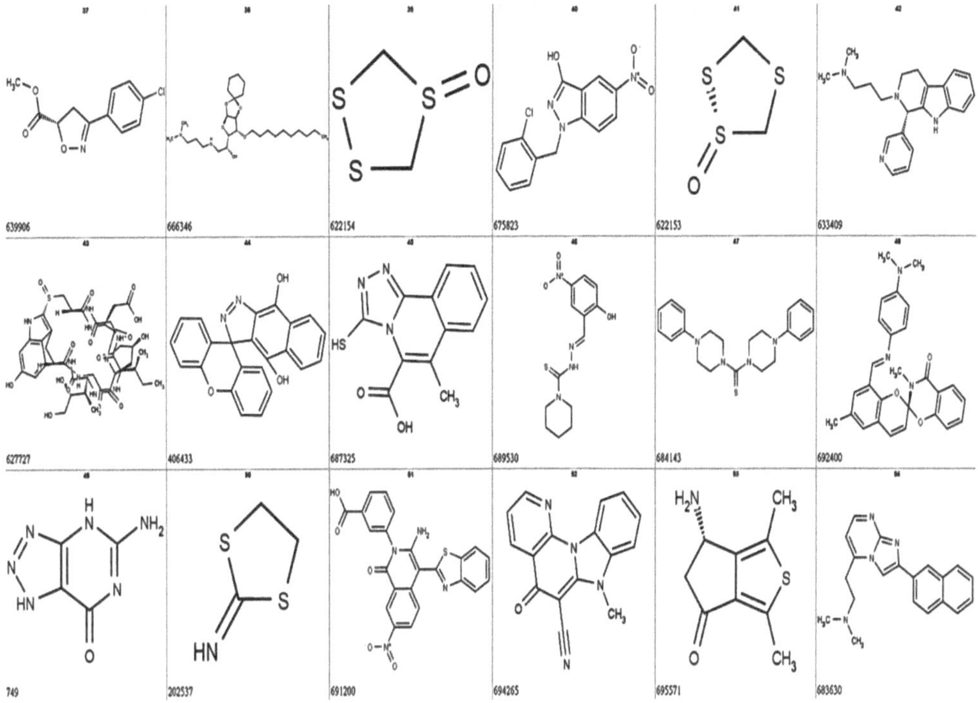

Figure 12.13 (Continued)

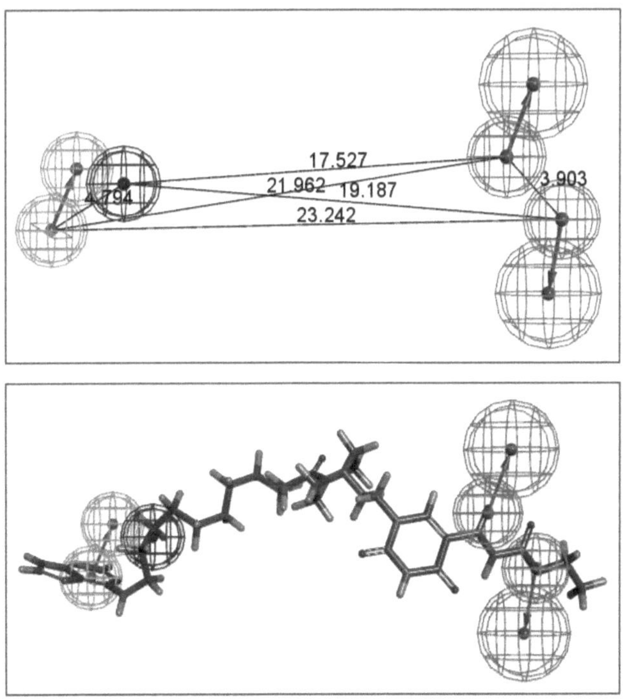

Figure 12.14 The HypoGen pharmacophore model built by DS3.5. We show the distances between the pharmacophores. No. 625869 is the drug in NCI anticancer library screened and matched by this pharmacophore.

Table 12.9 (A) Top pharmacophore description. (B) Top 10 pharmacophore descriptions. (C) Experimental and estimation values of IC_{50} for each of the 54 ligands

(A) Category	Number of ligands	IC_{50} range (nM)	Feature	Correlation
28 targets	54	2–50,000	HBA, HBA, HY, RA	0.92

(B) Hypothesis	Total cost	Δ Cost	RMS	Correlation (*r*)	Features
1	211.414	33.787	0.58	0.924995	HBA, HBA, HY, RA
2	211.745	33.456	0.60	0.920410	HBA, HBA, HY, RA
3	212.295	32.906	0.61	0.918758	HBA, HBA, HY, RA
4	212.530	32.671	0.61	0.917975	HBA, HBA, HY, RA
5	213.564	31.637	0.64	0.908625	HBA, HBA, HY, RA
6	213.870	31.331	0.65	0.905497	HBA, HBA, HY, RA
7	214.412	30.789	0.67	0.900153	HBA, HBA, HY, RA
8	215.039	30.162	0.69	0.893115	HBA, HBA, HY, RA
9	215.282	29.919	0.69	0.891929	HBA, HBA, HY, RA
10	215.758	29.443	0.71	0.885327	HBA, HBA, HY, RA

Null cost = 245.201; fixed cost = 202.014; configuration cost = 19.2545; Δ Cost = Null cost − Total cost. All costs are in units of bits. HBA, hydrogen bond acceptor; HY, hydrophobic; RA, ring aromatic.

(C) NCI drug	IC_{50} (nM)		Error	NCI drug	IC_{50} (nM)		Error
	Exp.	Est.			Exp.	Est.	
625869	2.0	2.4	1.2	669299	4792.2	5656.4	1.2
668448	66.3	142.2	2.1	641847	5026.9	4480.2	−1.1
698233	256.5	97.5	−2.6	669230	5282.5	4302.8	−1.2
668433	380.2	229.7	−1.7	657381	5358.4	4451.3	−1.2
737026	810.9	366.4	−2.2	643737	5376.8	4463.6	−1.2
654626	1377.0	1868.5	1.4	691369	5509.1	4822.7	−1.1
726771	1877.1	4272.0	2.3	627399	5527.5	4285.1	−1.3
717079	2067.7	4413.1	2.1	627727	5892.9	14154.0	2.4
674086	2234.0	4138.9	1.9	693235	6107.2	5310.1	−1.2
666346	2270.0	5579.5	2.5	675823	6377.0	5049.0	−1.3
740601	2627.9	4280.5	1.6	665675	6591.5	4380.4	−1.5
639795	2742.6	4358.1	1.6	295632	7254.4	4402.1	−1.6
673172	2823.0	4721.0	1.7	668373	7390.1	4343.3	−1.7
641236	2827.4	4418.8	1.6	749	7534.5	21062.2	2.8
76955	2836.7	4936.6	1.7	689447	8003.5	5397.9	−1.5
719481	3033.7	6903.6	2.3	202537	12700.6	21711.5	1.7
406433	3064.1	12371.9	4.0	639906	13142.7	5880.0	−2.2
683481	3344.0	4540.1	1.4	622153	14753.4	5742.3	−2.6
682236	3395.0	4677.1	1.4	622154	15831.2	4771.3	−3.3
661908	3469.7	4310.7	1.2	692400	50000.0	13059.1	−3.8
633409	3661.4	4382.2	1.2	687325	50000.0	14121.3	−3.5
672968	3845.3	4363.1	1.1	689530	50000.0	17997.5	−2.8
680359	4140.3	4290.9	1.0	691200	50000.0	21974.7	−2.3
76519	4334.6	3133.5	−1.4	684143	50000.0	32999.6	−1.5
702125	4732.2	4288.4	−1.1	683630	50000.0	40400.2	−1.2
678636	4732.9	12631.9	2.7	694265	50000.0	51746.7	1.0
682086	4770.1	4479.7	−1.1	695571	50000.0	61122.2	1.2

Exp., experimental value; Est., estimation value by pharmacophore; error = Est./Exp.

Mapping	ID number	Fit value	Estimate IC_{50} (nM)
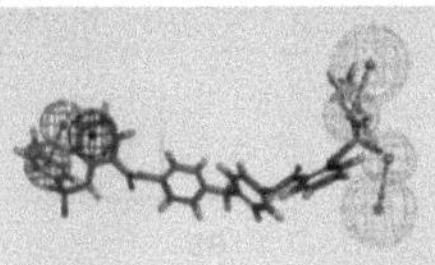	F0325-0146	6.580	4.599
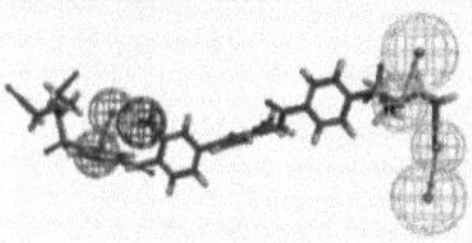	F0289-0199	6.446	6.266
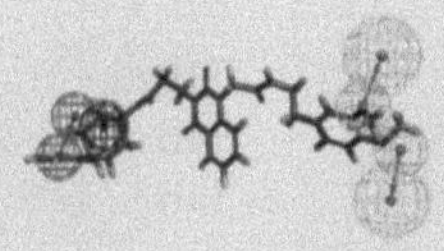	F0922-0370	5.935	20.327
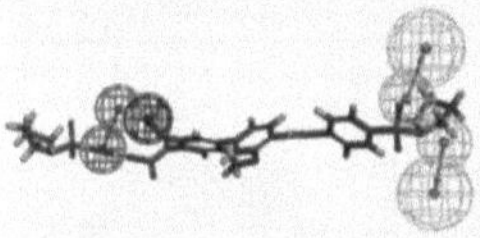	F0012-0228	5.690	35.775
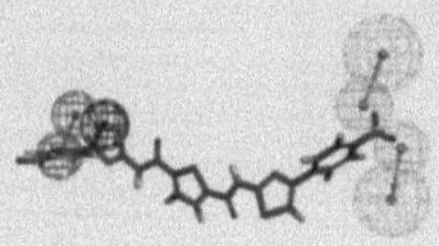	F0496-0019	5.623	41.719

Figure 12.15 These are the top ligands of Life-Chemical drug library screened by this HypoGen pharmacophore model. Please see Table 12.8 for the detailed descriptions of these mapping results.

Table 12.10 The Life-Chemical ligands are virtually screened by the HypoGen model

ID No.	Fit value	Estimate	HBA_1	HBA_2	HY_3	RA_4
F0325-0146	6.58	4.60	1	1	1	1
F0289-0199	6.45	6.27	1	1	1	1
F0922-0370	5.94	20.33	1	1	1	1
F0012-0228	5.69	35.78	1	1	1	1
F0496-0019	5.62	41.72	1	1	1	1
F0737-0405	5.55	49.32	1	1	1	1
F0382-0020	5.51	51.02	1	1	1	1
F0325-0148	5.12	132.56	1	1	1	1
F0737-0312	5.00	173.44	1	1	1	1
F0725-0356	4.99	179.14	1	1	1	1
F0301-0263	4.92	210.04	1	1	1	1
F0473-0314	4.91	216.74	1	1	1	1
F0922-0913	4.87	235.21	1	1	1	1
F1601-0068	4.58	456.06	1	1	1	1
F0463-0195	4.51	542.61	1	1	1	1
F0866-0317	4.26	956.83	1	1	1	1
F0537-0936	4.18	1163.54	1	1	1	1
F0180-0144	3.97	1880.45	1	1	1	1
F0922-0900	3.96	1930.82	1	1	1	1

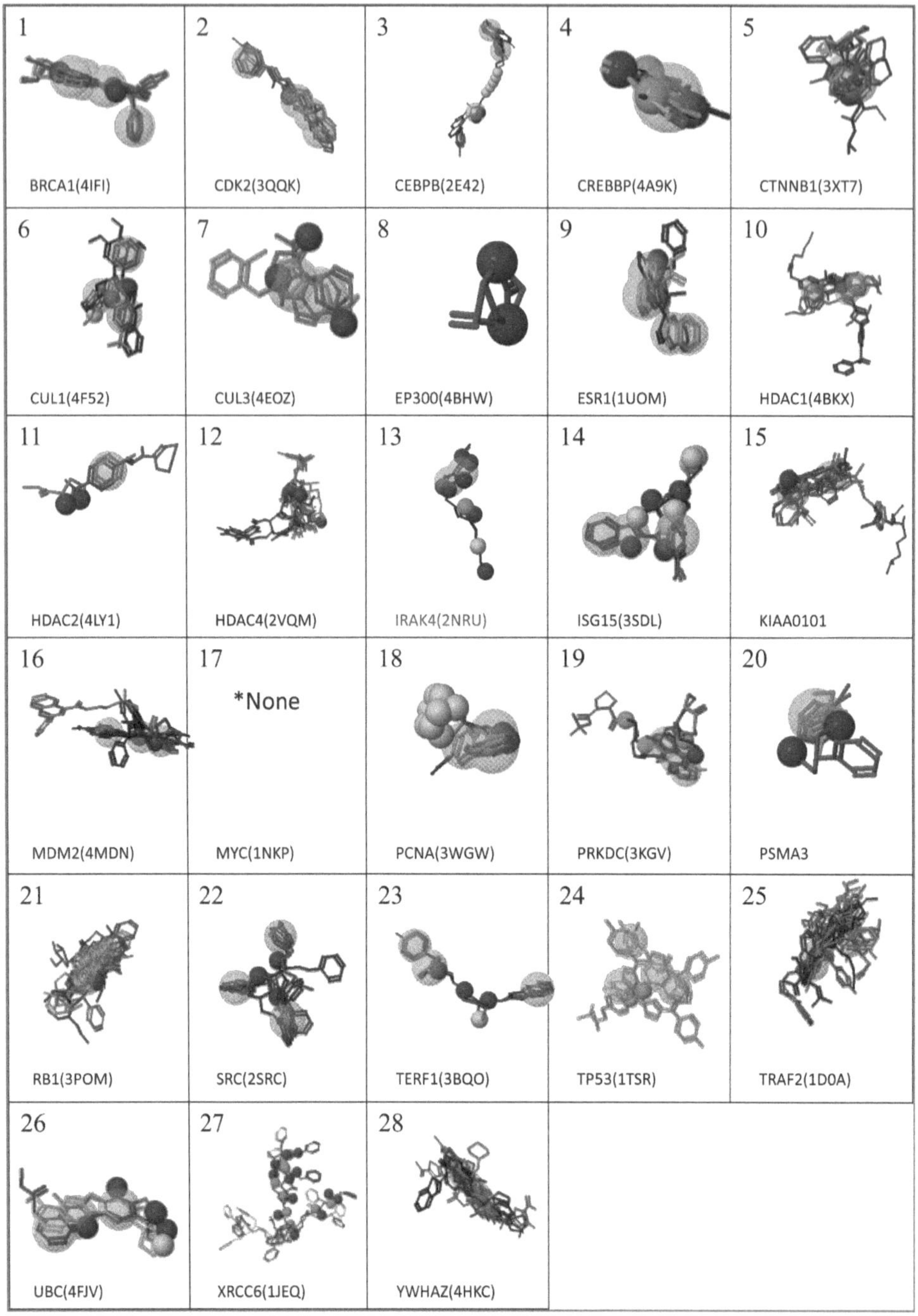

Figure 12.16 The pharmacophores for each of the 28 significant proteins built by PharmGist. Detail descriptions of these pharmacophores, please see Table 12.11. *17:MYC(1NKP)'s pharmacophore is failed to generated.

Table 12.11 The pharmacophores descriptions of PharmGist

No.	Protein	PDB	Score	Features	Spatial features	R	H	D	A	N	P	Chemicals	Fig.
1	BRCA1	4IFI	39.2	7	7	4	0	0	2	0	1	3	9.1
2	CDK2	3QQK	33.1	5	5	4	0	0	1	0	0	3	9.2
3	CEBPB	2E42	40.4	12	12	3	5	0	4	0	0	3	9.3
4	CREBBP	4A9K	29.9	5	5	1	0	2	2	0	0	9	9.4
5	CTNNB1	3TX7	23.7	3	3	2	0	0	1	0	0	7	9.5
6	CUL1	4F52	37.5	7	7	4	1	0	2	0	0	3	9.6
7	CUL3	4EOZ	25.7	5	5	2	0	0	3	0	0	3	9.7
8	EP300	4BHW	3.0	2	2	0	0	0	2	0	0	2	9.8
9	ESR1	1UOM	33.1	5	5	4	0	0	1	0	0	3	9.9
10	HDAC1	4BKX	26.5	4	4	3	0	0	0	0	1	4	9.10
11	HDAC2	4LY1	6.0	3	3	1	0	0	2	0	0	2	9.11
12	HDAC4	2VQM	34.0	7	5	0	0	2	5	0	0	3	9.12
13	IRAK4	2NRU	44.5	13	11	2	1	3	6	0	1	3	9.13
14	ISG15	3SDL	42.7	16	14	3	7	2	3	0	1	3	9.14
15	KIAA0101		26.5	4	4	2	0	1	1	0	0	3	9.15
16	MDM2	4MDN	36.0	5	5	3	0	0	2	0	0	6	9.16
17	MYC	1NKP											9.17
18	PCNA	3WGW	36.0	14	14	3	9	0	2	0	0	3	9.18
19	PRKDC	3KGV	36.7	7	7	3	0	3	1	0	0	3	9.19
20	PSMA3		14.7	3	3	1	0	0	2	0	0	3	9.20
21	RB1	3POM	36.4	4	4	3	0	0	1	0	0	9	9.21
22	SRC	2SRC	35.7	6	6	3	0	0	3	0	0	4	9.22
23	TERF1	3BQO	36.5	7	7	3	1	0	3	0	0	4	9.23
24	TP53	1TSR	33.8	6	6	4	1	1	0	0	0	3	9.24
25	TRAF2	1D0A	30.9	3	3	2	0	0	1	0	0	16	9.25
26	UBC	4FJV	41.6	10	9	3	1	1	5	0	0	3	9.26
27	XRCC6	1JEQ	141.1	22	22	0	0	10	12	0	0	3	9.27
28	YWHAZ	4HKC	34.9	3	3	3	0	0	0	0	0	12	9.28

Please see their pharmacophore diagrams in Fig. 12.15.

multiple drug cocktails would destroy the core network, although we have not experimentally validated whether each top compound binds to its corresponding target or not.

12.3.3 Conclusions

In this section, we combined systems biology with traditional CADD to design drug cocktails to inhibit the 28 proteins that comprise the common core network markers of four cancers. The results of our previous study have led us to believe that those proteins likely represent housekeeping proteins of these four cancers. Wet-lab researchers could use the cocktails of multiple drugs indicated by this analysis as objects of study in experiments on treating the four cancers. Moreover, with the help of sensitivity analysis, we also found the most likely multiple drug targets for each individual cancer.

APPENDIX B: METHOD AND TABLE

Appendix B.1: Mathematical foundation of MetaCore

1. *Shortest paths*:

   Builds a network consisting of shortest paths between pairs of initial objects in each direction, using the standard Dijkstra's shortest paths algorithm. There are from and to lists of objects; by default, the whole list of seed nodes is taken for each of them. Canonical pathways are considered by this algorithm as a single step, and are used for network building only if the list of objects used for network building contains both from and "to" items for a given pathway. The Z-score, G-score, and p-value are three different scoring functions used to rank the small networks created by the network building algorithm "Analyze Network." When viewing the most relevant networks list, one can sort the networks by a desired score, the actual size of the network, or the target size of the network by clicking on a column header.
2. *Z-score*:

   Each subnetwork is associated with a Z-score which ranks the subnetworks according to saturation with the objects from the initial list of seed nodes. The Z-score ranks the subnetworks of the Analyze Network algorithm with regard to their Saturation with genes subjected to analysis in the experiment. A high Z-score means the network is highly saturated with these genes.

Z-score formula is given by

$$(r - nR/N) \Big/ \sqrt{n(R/N)(1 - R/N)(1 - (n-1)/(N-1))}$$

where

N—total number of nodes in MetaCore database;

R—number of the network objects corresponding to the genes and proteins in your list;

n—total number of nodes in each small network generated from your list;

r—number of nodes with data in each small network generated from your list.

3. *G-score*:

The *G*-score modifies the *Z*-score based on the number of Canonical Pathways used to build the network. If a network has a high *G*-score, it is saturated with expressed genes (from *Z*-score) and it contains many Canonical Pathways. Sorting the table by this value essentially enables you to sort the table by two factors at once.

4. *p-Value*:

The *p*-values throughout MetaCore—for maps, networks, and processes—are all calculated using the same basic formula for hypergeometric distribution. The *p*-value essentially represents the probability for a particular mapping of an experiment to a map (or network, or process, etc.) to arise by chance, considering the numbers of genes in experiment versus the number of genes in the map (resp. network, process, etc.) within the "full set" of all genes on maps (resp. networks, processes, etc.).

***p*-Value formula:**

$$p\text{-Value} = (R!n!(N-R)!(N-n)!/N!) \cdot \sum_{i=\max(r,R+n-N)}^{\min(n,R)} 1/(i!(R-i)!(n-i)!(N-R-n+i)!)$$

where

N—total number of nodes in MetaCore database;

R—number of the network objects corresponding to the genes and proteins in your list;

n—total number of nodes in each small network generated from your list;

r—number of nodes with data in each small network generated from your list.

Appendix B.2: Docking results: Top 20 ligands for each of the 28 proteins

Here we show Table B.1.

Table B.1 Top 20 ligands for the 28 proteins

BRCA1 (4IFI)		CDK2 (3QQK)		CEBPB (2E42)		CREBBP (4A9K)		CTNNB1 (3XT7)		CUL1 (4F52)		CUL3 (4EOZ)	
NCI drug	LibDock score	NCI drug	LibDock score	NCI drug	LibDock score	NCI drug	LibDock score	NCI drug	LibDock score	NCI drug	LibDock score	NCI drug	LibDock score
668448	192.07	680359	108.34	625869	194.14	49714	88.69	669230	96.54	682932	139.64	702125	101.88
726771	142.96	678636	101.88	690021	171.24	668373	80.38	665675	85.82	666346	135.44	675823	87.41
639795	127.39	669299	101.27	690246	168.34	749	79.47	295632	81.26	616358	130.95	689447	76.65
697231	121.82	679065	101.01	724305	167.72	675278	78.13	690569	80.97	713370	130.83	680304	73.65
629023	121.09	376791	100.92	722651	167.12	674553	77.17	516286	80.67	682363	130.03	673517	72.20
617007	120.47	667673	98.24	700431	166.67	680827	77.14	162907	80.54	736612	124.15	697400	69.87
661197	118.59	687005	97.37	719325	165.99	690575	75.21	639398	74.46	716163	120.72	302988	67.42
670653	116.92	370589	96.34	668604	164.56	619158	75.05	623238	73.95	722175	119.85	684059	66.85
677390	116.56	668394	95.64	651241	164.16	22709	74.88	668375	73.83	717217	115.27	669691	65.00
661832	116.02	57975	95.37	719276	162.46	686663	73.57	666127	73.37	35489	114.48	31664	64.69
726260	115.75	667671	94.85	669663	160.59	632947	73.25	125214	73.27	663289	114.35	9908	64.23
633643	115.34	664145	94.08	697189	160.29	715694	72.87	684753	72.86	679514	113.93	665663	63.87
656722	114.91	668455	94	700443	160.11	1847	72.75	3039	72.64	630363	110.62	689741	62.07
740001	114.42	685267	93.97	371682	158.80	625159	72.44	329138	71.94	701735	109.54	701992	62.02
667476	113.58	361672	93.43	748494	158.73	675585	71.64	715452	71.90	617824	108.94	2023	61.85
658140	112.83	687003	93.37	645378	158.64	37049	70.96	610187	71.29	29638	108.22	677782	61.48
709899	112.25	684968	92.73	685968	158.45	694099	70.83	656987	70.52	689190	105.36	644582	61.47
694154	111.93	676392	91.70	697160	158.44	694100	70.83	610186	70.47	617970	104.38	620111	60.80
717077	111.32	631964	91.29	98535	158.04	1390	70.64	717122	70.18	623055	100.18	688119	60.59
629021	111.02	693324	91.19	10010	158.03	693886	70.45	698948	69.69	652565	98.22	740122	60.05

EP300(4BHW)		ESR1(1UOM)		HDAC1(4BKX)		HDAC2(4LY1)		HDAC4(2VQM)		IRAK4(2NRU)		ISG15(3SDL)	
NCI drug	LibDock score	NCI drug	LibDock score	NCI drug	LibDock score	NCI drug	LibDock score	NCI drug	LibDock score	NCI drug	LibDock score	NCI drug	LibDock score
153353	99.47	657381	95.84	627865	134.69	666854	109.55	684438	155.19	654626	153.82	76519	106.13
9345	98.54	627399	94.33	625439	134.10	697191	103.81	699408	154.59	697155	125.34	682086	101.50
146876	88.20	693235	89.50	647638	125.51	701745	103.66	704565	153.56	60673	119.34	641847	98.95
17391	86.81	632624	88.80	2426	124.72	700011	98.38	673354	152.44	693118	113.55	682087	96.54
643467	83.15	645318	88.28	707841	117.08	696561	97.97	691703	150.68	735798	111.15	102815	93.55
74518	82.93	702754	87.62	633063	113.58	529861	96.29	625439	149.32	677934	108.87	679185	86.13
521778	81.26	646922	87.39	645389	113.54	698037	96.07	668434	148.45	670053	107.44	698173	83.87
626779	78.10	691839	87.20	265875	113.19	742553	93.32	726771	148.37	726442	106.94	711014	79.25
126766	75.40	700991	87.07	734996	112.45	706418	92.29	699222	146.25	157036	104.91	698168	78.93
163326	71.36	715174	86.20	637131	112.40	624664	90.89	618571	144.19	37881	104.76	735034	78.41
110344	59.45	674102	86.20	740107	112.17	697236	90.49	718726	142.11	735213	103.88	673343	77.92
202537	56.54	677995	86.13	645392	111.94	122870	90.27	653419	141.80	676423	102.07	711012	77.67
2654	55.73	723380	86.08	666063	111.78	721045	90.18	724305	141.20	703068	101.71	742410	77.44
609696	53.57	13378	86.04	635565	107.59	670681	89.88	715563	140.30	708572	101.68	711010	76.74
622153	52.26	637882	86.01	703134	99.42	280594	88.95	686479	139.76	669655	100.90	72861	76.70
622154	51.68	706028	85.76	705953	99.26	650907	88.93	676817	138.37	703065	100.80	640674	75.72
155080	50.66	681075	85.52	121129	98.97	696560	88.90	681462	137.76	703063	100.73	632935	74.49
744479	41.06	367238	85.29	668021	98.82	691823	88.88	735846	137.43	703067	100.18	665775	74.10
617069	31.74	609965	84.72	624177	9355	716908	88.84	643589	136.64	698983	99.42	711018	73.97
716706	27.74	710269	84.59	652603	9272	659999	88.55	743122	136.64	657377	99.25	711026	73.61
KIAA0101		**MDM2 (4MDN)**		**MYC (1NKP)**		**PCNA (3WGW)**		**PRKDC (3KGV)**		**PSMA3**		**RB1 (3POM)**	
NCI drug	LibDock score	NCI drug	LibDock score	NCI drug	LibDock score	NCI drug	LibDock score	NCI drug	LibDock score	NCI drug	LibDock score	NCI drug	LibDock score
655102	165.72	674086	136.10	698686	194.11	406433	122.44	684443	144.77	668577	180.11	686584	130.74
669588	164.04	76955	125.90	668433	181.27	683481	118.43	713288	137.91	717708	172.98	682292	121.51
407811	158.57	709896	118.47	737026	168.29	661908	116.71	710559	135.82	668448	166.52	154020	119.13
704564	148.68	688217	116.34	646614	155.61	626160	115.55	660224	129.37	695409	164.91	664154	118.97
698687	147.05	724910	116.15	350895	147.53	351521	113.94	639966	126.24	727367	161.91	657190	118.09
664983	147	62485	115.32	687363	145.14	685837	113.93	617309	124.85	653260	161.66	631392	115.87
625451	146.74	676436	112.72	742856	140.09	707572	112.97	725739	122.53	729648	160.32	93033	115.66
625439	145.71	671153	111.63	737025	135.85	61900	112.61	617613	122.32	671864	160.07	350378	115.18
623527	144.82	652623	106.66	722654	134.50	714364	111.81	602077	121.59	697156	158.81	675972	114.65
683897	142.76	716911	105.40	719660	134.02	706157	111.76	625885	121.37	661423	157.04	735341	114.25

(Continued)

Table B.1 (Continued)

KIAA0101		MDM2 (4MDN)		MYC (1NKP)		PCNA (3WGW)		PRKDC (3KGV)		PSMA3		RB1 (3POM)	
NCI drug	LibDock score	NCI drug	LibDock score	NCI drug	LibDock score	NCI drug	LibDock score	NCI drug	LibDock score	NCI drug	LibDock score	NCI drug	LibDock score
713288	141.28	724562	105.20	668432	131.50	627732	111.44	668579	119.73	720458	156.30	700411	113.87
645378	140.10	720388	104.21	677662	129.50	692402	110.59	655900	119.68	65104	155.24	734791	113.54
697157	139.49	626315	104.09	653867	128.86	631391	110.40	726365	118.21	629733	153.84	692602	113.51
704565	137.38	658958	102.92	691531	126.62	169517	110.39	638037	117.44	683781	152.79	20534	113.48
668891	136.34	695619	102.82	668577	125.60	661902	110.24	639018	117.17	668873	149.83	691581	112.42
706419	136.00	59349	102.71	708447	125.33	653264	109.89	647426	116.74	239400	149.49	666125	112.34
674709	135.98	665904	100.19	613528	125.01	657989	109.79	56614	115.36	695175	148.83	308848	112.19
695800	135.26	740033	98.88	676779	124.62	631307	109.12	650721	114.74	668871	147.28	183519	112.17
715546	133.38	741107	98.56	627966	122.23	661909	108.86	722664	114.52	618486	146.72	627221	112.07
707826	133.33	645673	97.79	633660	121.78	710117	108.48	747167	114.52	743438	146.21	340555	112.03
SRC (2SRC)		**TERF1 (3BQO)**		**TP53 (1TSR)**		**TRAF2 (1D0A)**		**UBC (4FJV)**		**XRCC6 (1JEQ)**		**YWHAZ (4HKC)**	
NCI drug	**LibDock score**	**NCI drug**	**LibDock score**	**NCI drug**	**LibDock score**	**NCI drug**	**LibDock score**	**NCI drug**	**LibDock score**	**NCI drug**	**LibDock score**	**NCI drug**	**LibDock score**
718308	153.09	698233	185.38	673172	126.11	653000	127.24	719481	122.89	668437	304.62	698229	213.08
718013	143.95	740601	129.25	695409	125.82	686501	125.10	633409	114.19	668449	241.90	725986	207.04
717079	139.21	641236	126.04	682236	117.73	715469	113.53	672968	111.87	668447	225.63	719660	200.90
718165	138.31	682094	123.29	695405	115.63	692357	111.72	734999	111.58	668571	219.59	684034	195.22
694702	138.11	661210	121.90	667504	114.33	712006	111.45	688121	111.50	745943	212.87	705931	194.33
717101	136.96	740555	121.09	682235	107.12	652584	110.44	633410	111.34	737027	211.33	722308	193.58
699416	136.53	642077	119.01	698974	106.27	670859	107.91	715227	110.41	672428	209.12	623553	193.25
624953	134.05	686379	118.35	649748	105.65	652583	107.68	672967	110.11	676934	206.55	630197	192.76
681461	132.45	669392	118.30	725903	104.68	731648	105	688122	108.74	668436	204.28	697179	190.80
301477	131.73	735031	118.19	682576	102.01	745501	103.33	637660	108.61	682348	204.08	687369	189.45
665308	131.55	715583	115.34	649749	101.87	728316	102.95	639174	108.33	668450	203.96	619027	186.67
681322	131.41	715585	115.33	742856	100.79	676857	102.07	703028	106.90	746847	199.55	698230	185.97
633395	131.14	306231	114.93	726149	100.55	642924	101.55	703032	106.41	633663	199.31	695401	185.90
123127	131.13	677586	114.86	649551	100.45	695175	101.27	176323	106.39	707766	196.19	642601	184.83
704581	130.41	697230	114.81	721386	97.67	732429	100.45	716853	105.48	684748	194.12	687363	184.65
682935	130.24	715588	114.64	742858	97.30	276361	99.09	1760	105.12	668438	191.73	625849	184.48
631687	129.23	715587	113.82	622170	96.78	730045	98.72	732175	104.87	699190	189.46	618938	183.84
661213	128.91	655438	113.48	700214	96.62	625349	98.41	249910	104.71	698684	188.82	730631	183.83
672445	128.72	641223	112.79	734002	96.46	668579	98.18	13597	104.20	668439	187.45	680418	183.13
652617	128.67	677089	112.11	628445	96.05	743407	97.99	11930	104.03	651806	184.19	682094	182.53

REFERENCES

[1] Grant GM, Fortney A, Gorreta F, Estep M, Del Giacco L, Van Meter A, et al. Microarrays in cancer research. Anticancer Res 2004;24(2A):441−8.
[2] Huang Y, Sadee W. Drug sensitivity and resistance genes in cancer chemotherapy: a chemogenomics approach. Drug Discov Today 2003;8(8):356−63.
[3] Li CW, Chu YH, Chen BS. Construction and clarification of dynamic gene regulatory network of cancer cell cycle via microarray data. Cancer Inform 2006;2:223−41.
[4] Mocellin S, Provenzano M, Rossi CR, Pilati P, Nitti D, Lise M. DNA array-based gene profiling: from surgical specimen to the molecular portrait of cancer. Ann Surg 2005;241(1):16−26.
[5] Warner GC, Reis PP, Jurisica I, Sultan M, Arora S, Macmillan C, et al. Molecular classification of oral cancer by cDNA microarrays identifies overexpressed genes correlated with nodal metastasis. Int J Cancer 2004;110(6):857−68.
[6] Hood L. Systems biology: integrating technology, biology, and computation. Mech Ageing Dev 2003;124(1):9−16.
[7] Hood L, Heath JR, Phelps ME, Lin B. Systems biology and new technologies enable predictive and preventative medicine. Science 2004;306(5696):640−3.
[8] Chen BS, Wang YC. On the attenuation and amplification of molecular noise in genetic regulatory networks. BMC Bioinformatics 2006;7:52.
[9] Chen BS, Wang YC, Wu WS, Li WH. A new measure of the robustness of biochemical networks. Bioinformatics 2005;21(11):2698−705.
[10] Hasty J, McMillen D, Collins JJ. Engineered gene circuits. Nature 2002;420(6912):224−30.
[11] Bugrim A, Nikolskaya T, Nikolsky Y. Early prediction of drug metabolism and toxicity: systems biology approach and modeling. Drug Discov Today 2004;9(3):127−35.
[12] Butcher EC. Can cell systems biology rescue drug discovery? Nat Rev Drug Discov 2005;4(6):461−7.
[13] Butcher EC, Berg EL, Kunkel EJ. Systems biology in drug discovery. Nat Biotechnol 2004;22(10):1253−9.
[14] Ma H, Horiuchi KY. Chemical microarray: a new tool for drug screening and discovery. Drug Discov Today 2006;11(13−14):661−8.
[15] Isaacs FJ, Hasty J, Cantor CR, Collins JJ. Prediction and measurement of an autoregulatory genetic module. Proc Natl Acad Sci USA 2003;100(13):7714−19.
[16] Kitano H. Computational systems biology. Nature 2002;420(6912):206−10.
[17] Kitano H. Systems biology: a brief overview. Science 2002;295(5560):1662−4.
[18] Kitano H. Biological robustness. Nat Rev Genet 2004;5(11):826−37.
[19] Akutsu T, Miyano S, Kuhara S. Algorithms for identifying Boolean networks and related biological networks based on matrix multiplication and fingerprint function. J Comput Biol 2000;7(3-4):331−43.
[20] Holter NS, Maritan A, Cieplak M, Fedoroff NV, Banavar JR. Dynamic modeling of gene expression data. Proc Natl Acad Sci USA 2001;98(4):1693−8.
[21] Chang WC, Li CW, Chen BS. Quantitative inference of dynamic regulatory pathways via microarray data. BMC Bioinformatics 2005;6:44.
[22] Chang YH, Wang YC, Chen BS. Identification of transcription factor cooperativity via stochastic system model. Bioinformatics 2006;22(18):2276−82.
[23] Chen BS, Li CW. Analysing microarray data in drug discovery using systems biology. Expert Opin Drug Discov 2007;2(5):755−68.
[24] Chen HC, Lee HC, Lin TY, Li WH, Chen BS. Quantitative characterization of the transcriptional regulatory network in the yeast cell cycle. Bioinformatics 2004;20(12):1914−27.
[25] Kim S, Imoto S, Miyano S. Dynamic Bayesian network and nonparametric regression for nonlinear modeling of gene networks from time series gene expression data. Biosystems 2004;75(1−3):57−65.
[26] Tamada Y, Imoto S, Tashiro K, Kuhara S, Miyano S. Identifying drug active pathways from gene networks estimated by gene expression data. Genome Inform 2005;16(1):182−91.

[27] Zou M, Conzen SD. A new dynamic Bayesian network (DBN) approach for identifying gene regulatory networks from time course microarray data. Bioinformatics 2005;21(1):71–9.
[28] Brown MP, Grundy WN, Lin D, Cristianini N, Sugnet CW, Furey TS, et al. Knowledge-based analysis of microarray gene expression data by using support vector machines. Proc Natl Acad Sci USA 2000;97(1):262–7.
[29] Bussemaker HJ, Li H, Siggia ED. Regulatory element detection using correlation with expression. Nat Genet 2001;27(2):167–71.
[30] Galm U, Shen B. Expression of biosynthetic gene clusters in heterologous hosts for natural product production and combinatorial biosynthesis. Expert Opin Drug Discov 2006;1(5):409–37.
[31] He F, Zeng AP. In search of functional association from time-series microarray data based on the change trend and level of gene expression. BMC Bioinformatics 2006;7:69.
[32] Ji X, Li-Ling J, Sun Z. Mining gene expression data using a novel approach based on hidden Markov models. FEBS Lett 2003;542(1-3):125–31.
[33] Jonsson PF, Cavanna T, Zicha D, Bates PA. Cluster analysis of networks generated through homology: automatic identification of important protein communities involved in cancer metastasis. BMC Bioinformatics 2006;7:2.
[34] Mirza A, Wu Q, Wang L, McClanahan T, Bishop WR, Gheyas F, et al. Global transcriptional program of p53 target genes during the process of apoptosis and cell cycle progression. Oncogene 2003;22(23):3645–54.
[35] Spellman PT, Sherlock G, Zhang MQ, Iyer VR, Anders K, Eisen MB, et al. Comprehensive identification of cell cycle-regulated genes of the yeast *Saccharomyces cerevisiae* by microarray hybridization. Mol Biol Cell 1998;9(12):3273–97.
[36] Tamayo P, Slonim D, Mesirov J, Zhu Q, Kitareewan S, Dmitrovsky E, et al. Interpreting patterns of gene expression with self-organizing maps: methods and application to hematopoietic differentiation. Proc Natl Acad Sci USA 1999;96(6):2907–12.
[37] Barenco M, Tomescu D, Brewer D, Callard R, Stark J, Hubank M. Ranked prediction of p53 targets using hidden variable dynamic modeling. Genome Biol 2006;7(3):R25.
[38] Chakraborty S, Ghosh M, Maiti T, Tewari A. Bayesian neural networks for bivariate binary data: an application to prostate cancer study. Stat Med 2005;24(23):3645–62.
[39] Dojer N, Gambin A, Mizera A, Wilczynski B, Tiuryn J. Applying dynamic Bayesian networks to perturbed gene expression data. BMC Bioinformatics 2006;7:249.
[40] Imoto S, Tamada Y, Savoie CJ, Miyano S. Analysis of gene networks for drug target discovery and validation. Methods Mol Biol 2006;360:33–56.
[41] Liang Y, Kelemen AG. Hierarchical Bayesian neural network for gene expression temporal patterns. Stat Appl Genet Mol Biol 2004;3 Article 20.
[42] Savoie CJ, Aburatani S, Watanabe S, Eguchi Y, Muta S, Imoto S, et al. Use of gene networks from full genome microarray libraries to identify functionally relevant drug-affected genes and gene regulation cascades. DNA Res 2003;10(1):19–25.
[43] Liu KQ, Liu ZP, Hao JK, Chen L, Zhao XM. Identifying dysregulated pathways in cancers from pathway interaction networks. BMC Bioinformatics 2012;13:126.
[44] Stratton MR, Campbell PJ, Futreal PA. The cancer genome. Nature 2009;458(7239):719–24.
[45] Neal CL, Yu D. 14-3-3zeta as a prognostic marker and therapeutic target for cancer. Expert Opin Ther Targets 2010;14(12):1343–54.
[46] Cowling VH, Cole MD. Mechanism of transcriptional activation by the Myc oncoproteins. Semin Cancer Biol 2006;16(4):242–52.
[47] Pacal M, Bremner R. Insights from animal models on the origins and progression of retinoblastoma. Curr Mol Med 2006;6(7):759–81.
[48] Contessa JN, Hampton J, Lammering G, Mikkelsen RB, Dent P, Valerie K, et al. Ionizing radiation activates Erb-B receptor dependent Akt and p70 S6 kinase signaling in carcinoma cells. Oncogene 2002;21(25):4032–41.
[49] Codegoni AM, Nicoletti MI, Buraggi G, Valoti G, Giavazzi R, D'Incalci M, et al. Molecular characterisation of a panel of human ovarian carcinoma xenografts. Eur J Cancer 1998;34(9):1432–8.

[50] Golub TR, Slonim DK, Tamayo P, Huard C, Gaasenbeek M, Mesirov JP, et al. Molecular classification of cancer: class discovery and class prediction by gene expression monitoring. Science 1999;286(5439):531–7.
[51] Thomas JG, Olson JM, Tapscott SJ, Zhao LP. An efficient and robust statistical modeling approach to discover differentially expressed genes using genomic expression profiles. Genome Res 2001;11 (7):1227–36.
[52] Han H, Bearss DJ, Browne LW, Calaluce R, Nagle RB, Von Hoff DD. Identification of differentially expressed genes in pancreatic cancer cells using cDNA microarray. Cancer Res 2002;62 (10):2890–6.
[53] Uramoto H, Sugio K, Oyama T, Nakata S, Ono K, Nozoe T, et al. Expression of the p53 family in lung cancer. Anticancer Res 2006;26(3A):1785–90.
[54] Peltonen L, McKusick VA. Genomics and medicine. Dissecting human disease in the postgenomic era. Science 2001;291(5507):1224–9.
[55] Glazier AM, Nadeau JH, Aitman TJ. Finding genes that underlie complex traits. Science 2002;298 (5602):2345–9.
[56] Merikangas KR, Low NC, Hardy J. Commentary: understanding sources of complexity in chronic diseases—the importance of integration of genetics and epidemiology. Int J Epidemiol 2006;35 (3):590–2 discussion 593–6.
[57] Rybaczyk LA, Bashaw MJ, Pathak DR, Huang K. An indicator of cancer: downregulation of monoamine oxidase-A in multiple organs and species. BMC Genomics 2008;9:134.
[58] Lin CL, Liu YW, Chuang CH. Control design for signal transduction networks. Bioinform Biol Insights 2009;3:1–14.
[59] Lin CL, Chuang CH. Review of control theory and dynamics in systems biology. Int J Syst Synth Biol 2010;1(1):39–61.
[60] Chuang C-H, Lin C-L. On robust state estimation of gene networks. Biomed Eng Comput Biol 2010;2:23–36.
[61] Wu CC, Huang HC, Juan HF, Chen ST. GeneNetwork: an interactive tool for reconstruction of genetic networks using microarray data. Bioinformatics 2004;20(18):3691–3.
[62] Lin CC, Lee CH, Fuh CS, Juan HF, Huang HC. Link clustering reveals structural characteristics and biological contexts in signed molecular networks. PLoS One 2013;8(6):e67089.
[63] Lin LL, Wang YH, Lai CY, Chau CL, Su GC, Yang CY, et al. Systems biology of meridians, acupoints, and chinese herbs in disease. Evid Based Complement Alternat Med 2012;2012: 372670.
[64] Lin LL, Huang HC, Juan HF. Revealing the molecular mechanism of gastric cancer marker annexin A4 in cancer cell proliferation using exon arrays. PLoS One 2012;7(9):e44615.
[65] Wu WS. Different functional gene clusters in yeast have different spatial distributions of the transcription factor binding sites. Bioinform Biol Insights 2011;5:1–11.
[66] Yang TH, Wu WS. Identifying biologically interpretable transcription factor knockout targets by jointly analyzing the transcription factor knockout microarray and the ChIP-chip data. BMC Syst Biol 2012;6:102.
[67] Wu WS, Li WH. Systematic identification of yeast cell cycle transcription factors using multiple data sources. BMC Bioinformatics 2008;9:522.
[68] Liu X, Liu ZP, Zhao XM, Chen L. Identifying disease genes and module biomarkers by differential interactions. J Am Med Inform Assoc 2012;19(2):241–8.
[69] Lin CC, Juan HF, Hsiang JT, Hwang YC, Mori H, Huang HC. Essential core of protein-protein interaction network in *Escherichia coli*. J Proteome Res 2009;8(4):1925–31.
[70] Lee TY, Horng JT, Juan HF, Huang HD, Wu LC, Tsai MF, et al. An agent-based system to discover protein-protein interactions, identify protein complexes and proteins with multiple peptide mass fingerprints. J Comput Chem 2006;27(9):1020–32.
[71] Kar G, Gursoy A, Keskin O. Human cancer protein-protein interaction network: a structural perspective. PLoS Comput Biol 2009;5(12):e1000601.
[72] Liu R, Aihara K, Chen L. Dynamical network biomarkers for identifying critical transitions and their driving networks of biologic processes. Quant Biol 2013;1(2):105–14.

[73] Liu X, Liu R, Zhao XM, Chen L. Detecting early-warning signals of type 1 diabetes and its leading biomolecular networks by dynamical network biomarkers. BMC Med Genomics 2013;6(Suppl. 2):S8.
[74] Wang YC, Chen BS. A network-based biomarker approach for molecular investigation and diagnosis of lung cancer. BMC Med Genomics 2011;4:2.
[75] Chu LH, Chen BS. Construction of a cancer-perturbed protein-protein interaction network for discovery of apoptosis drug targets. BMC Syst Biol 2008;2:56.
[76] Chen CH, Chuang SM, Yang MF, Liao JW, Yu SL, Chen JJ. A novel function of YWHAZ/beta-catenin axis in promoting epithelial-mesenchymal transition and lung cancer metastasis. Mol Cancer Res 2012;10(10):1319–31.
[77] Wang CC, Lin SY, Lai YH, Liu YJ, Hsu YL, Chen JJ. Dimethyl sulfoxide promotes the multiple functions of the tumor suppressor HLJ1 through activator protein-1 activation in NSCLC cells. PLoS One 2012;7(4):e33772.
[78] Lin SY, Hsueh CM, Yu SL, Su CC, Shum WY, Yeh KC, et al. HLJ1 is a novel caspase-3 substrate and its expression enhances UV-induced apoptosis in non-small cell lung carcinoma. Nucleic Acids Res 2010;38(18):6148–58.
[79] Kim WJ, Kim EJ, Kim SK, Kim YJ, Ha YS, Jeong P, et al. Predictive value of progression-related gene classifier in primary non-muscle invasive bladder cancer. Mol Cancer 2010;9:3.
[80] Sheffer M, Bacolod MD, Zuk O, Giardina SF, Pincas H, Barany F, et al. Association of survival and disease progression with chromosomal instability: a genomic exploration of colorectal cancer. Proc Natl Acad Sci USA 2009;106(17):7131–6.
[81] Roessler S, Long EL, Budhu A, Chen Y, Zhao X, Ji J, et al. Integrative genomic identification of genes on 8p associated with hepatocellular carcinoma progression and patient survival. Gastroenterology 2012;142(4) 957–66 e912.
[82] Roessler S, Jia HL, Budhu A, Forgues M, Ye QH, Lee JS, et al. A unique metastasis gene signature enables prediction of tumor relapse in early-stage hepatocellular carcinoma patients. Cancer Res 2010;70(24):10202–12.
[83] Hou J, Aerts J, den Hamer B, van Ijcken W, den Bakker M, Riegman P, et al. Gene expression-based classification of non-small cell lung carcinomas and survival prediction. PLoS One 2010;5(4):e10312.
[84] Vogl A, Sartorius U, Vogt T, Roesch A, Landthaler M, Stolz W, et al. Gene expression profile changes between melanoma metastases and their daughter cell lines: implication for vaccination protocols. J Invest Dermatol 2005;124(2):401–4.
[85] Medeiros F, Rigl CT, Anderson GG, Becker SH, Halling KC. Tissue handling for genome-wide expression analysis: a review of the issues, evidence, and opportunities. Arch Pathol Lab Med 2007;131(12):1805–16.
[86] Suchy J, Cybulski C, Gorski B, Huzarski T, Byrski T, Debniak T, et al. BRCA1 mutations and colorectal cancer in Poland. Fam Cancer 2010;9(4):541–4.
[87] Grabsch H, Dattani M, Barker L, Maughan N, Maude K, Hansen O, et al. Expression of DNA double-strand break repair proteins ATM and BRCA1 predicts survival in colorectal cancer. Clin Cancer Res 2006;12(5):1494–500.
[88] Gachechiladze M, Skarda J. The role of BRCA1 in non-small cell lung cancer. Biomed Pap Med Fac Univ Palacky Olomouc Czech Repub 2012;156(3):200–3.
[89] Huang WW, Yang JS, Pai SJ, Wu PP, Chang SJ, Chueh FS, et al. Bufalin induces G(0)/G(1) phase arrest through inhibiting the levels of cyclin D, cyclin E, CDK2 and CDK4, and triggers apoptosis via mitochondrial signaling pathway in T24 human bladder cancer cells. Mutat Res 2012;732(1-2):26–33.
[90] Yamamoto H, Monden T, Ikeda K, Izawa H, Fukuda K, Fukunaga M, et al. Coexpression of cdk2/cdc2 and retinoblastoma gene products in colorectal cancer. Br J Cancer 1995;71(6):1231–6.
[91] Li KK, Ng IO, Fan ST, Albrecht JH, Yamashita K, Poon RY. Activation of cyclin-dependent kinases CDC2 and CDK2 in hepatocellular carcinoma. Liver 2002;22(3):259–68.
[92] Kawana H, Tamaru J, Tanaka T, Hirai A, Saito Y, Kitagawa M, et al. Role of p27Kip1 and cyclin-dependent kinase 2 in the proliferation of non-small cell lung cancer. Am J Pathol 1998;153(2):505–13.

[93] Sun L, Fu BB, Liu DG. Systemic delivery of full-length C/EBP beta/liposome complex suppresses growth of human colon cancer in nude mice. Cell Res 2005;15(10):770–6.
[94] Flodby P, Liao DZ, Blanck A, Xanthopoulos KG, Hallstrom IP. Expression of the liver-enriched transcription factors C/EBP alpha, C/EBP beta, HNF-1, and HNF-4 in preneoplastic nodules and hepatocellular carcinoma in rat liver. Mol Carcinog 1995;12(2):103–9.
[95] Xi S, Yang M, Tao Y, Xu H, Shan J, Inchauste S, et al. Cigarette smoke induces C/EBP-beta-mediated activation of miR-31 in normal human respiratory epithelia and lung cancer cells. PLoS One 2010;5(10):e13764.
[96] Ikezoe T, Yang Y, Saitoh T, Heber D, McKenna R, Taguchi H, et al. PC-SPES down-regulates COX-2 via inhibition of NF-kappaB and C/EBPbeta in non-small cell lung cancer cells. Int J Oncol 2006;29(2):453–61.
[97] Boorjian SA, Heemers HV, Frank I, Farmer SA, Schmidt LJ, Sebo TJ, et al. Expression and significance of androgen receptor coactivators in urothelial carcinoma of the bladder. Endocr Relat Cancer 2009;16(1):123–37.
[98] Ishihama K, Yamakawa M, Semba S, Takeda H, Kawata S, Kimura S, et al. Expression of HDAC1 and CBP/p300 in human colorectal carcinomas. J Clin Pathol 2007;60(11):1205–10.
[99] Dimova EY, Jakubowska MM, Kietzmann T. CREB binding to the hypoxia-inducible factor-1 responsive elements in the plasminogen activator inhibitor-1 promoter mediates the glucagon effect. Thromb Haemost 2007;98(2):296–303.
[100] Kishimoto M, Kohno T, Okudela K, Otsuka A, Sasaki H, Tanabe C, et al. Mutations and deletions of the CBP gene in human lung cancer. Clin Cancer Res 2005;11(2 Pt 1):512–19.
[101] Karamouzis MV, Konstantinopoulos PA, Papavassiliou AG. Roles of CREB-binding protein (CBP)/p300 in respiratory epithelium tumorigenesis. Cell Res 2007;17(4):324–32.
[102] Hirata H, Hinoda Y, Ueno K, Shahryari V, Tabatabai ZL, Dahiya R. MicroRNA-1826 targets VEGFC, beta-catenin (CTNNB1) and MEK1 (MAP2K1) in human bladder cancer. Carcinogenesis 2012;33(1):41–8.
[103] Barker N, Clevers H. Mining the Wnt pathway for cancer therapeutics. Nat Rev Drug Discov 2006;5(12):997–1014.
[104] Maier TJ, Janssen A, Schmidt R, Geisslinger G, Grosch S. Targeting the beta-catenin/APC pathway: a novel mechanism to explain the cyclooxygenase-2-independent anticarcinogenic effects of celecoxib in human colon carcinoma cells. FASEB J 2005;19(10):1353–5.
[105] Fatima S, Lee NP, Luk JM. Dickkopfs and Wnt/beta-catenin signalling in liver cancer. World J Clin Oncol 2011;2(8):311–25.
[106] Ozen C, Yildiz G, Dagcan A, Cevik D, Ors A, Keles U, et al. Genetics and epigenetics of liver cancer. N Biotechnol 2013;30(4):381–4.
[107] Wang W, Liu Z, Qu P, Zhou Z, Zeng Y, Fan J, et al. Knockdown of regulator of cullins-1 (ROC1) expression induces bladder cancer cell cycle arrest at the G2 phase and senescence. PLoS One 2013;8(5):e62734.
[108] Korzeniewski N, Zheng L, Cuevas R, Parry J, Chatterjee P, Anderton B, et al. Cullin 1 functions as a centrosomal suppressor of centriole multiplication by regulating polo-like kinase 4 protein levels. Cancer Res 2009;69(16):6668–75.
[109] Lee J, Zhou P. Cullins and cancer. Genes Cancer 2010;1(7):690–9.
[110] Salon C, Brambilla E, Brambilla C, Lantuejoul S, Gazzeri S, Eymin B. Altered pattern of Cul-1 protein expression and neddylation in human lung tumours: relationships with CAND1 and cyclin E protein levels. J Pathol 2007;213(3):303–10.
[111] Peng Z, Wang H, Shan C. Expression of ubiquitin and cullin-1 and its clinicopathological significance in benign and malignant lesions of the lung. Zhong Nan Da Xue Xue Bao Yi Xue Ban 2009;34(3):204–9.
[112] Grau L, Luque-Garcia JL, Gonzalez-Peramato P, Theodorescu D, Palou J, Fernandez-Gomez JM, et al. A quantitative proteomic analysis uncovers the relevance of CUL3 in bladder cancer aggressiveness. PLoS One 2013;8(1):e53328.
[113] Lee DF, Kuo HP, Liu M, Chou CK, Xia W, Du Y, et al. KEAP1 E3 ligase-mediated downregulation of NF-kappaB signaling by targeting IKKbeta. Mol Cell 2009;36(1):131–40.

[114] Du M, Sansores-Garcia L, Zu Z, Wu KK. Cloning and expression analysis of a novel salicylate suppressible gene, Hs-CUL-3, a member of cullin/Cdc53 family. J Biol Chem 1998;273(38):24289–92.
[115] Kossatz U, Breuhahn K, Wolf B, Hardtke-Wolenski M, Wilkens L, Steinemann D, et al. The cyclin E regulator cullin 3 prevents mouse hepatic progenitor cells from becoming tumor-initiating cells. J Clin Invest 2010;120(11):3820–33.
[116] Thu KL, Pikor LA, Chari R, Wilson IM, Macaulay CE, English JC, et al. Genetic disruption of KEAP1/CUL3 E3 ubiquitin ligase complex components is a key mechanism of NF-kappaB pathway activation in lung cancer. J Thorac Oncol 2011;6(9):1521–9.
[117] Kim WJ, Lee JW, Quan C, Youn HJ, Kim HM, Bae SC. Nicotinamide inhibits growth of carcinogen induced mouse bladder tumor and human bladder tumor xenograft through up-regulation of RUNX3 and p300. J Urol 2011;185(6):2366–75.
[118] Li M, Luo RZ, Chen JW, Cao Y, Lu JB, He JH, et al. High expression of transcriptional coactivator p300 correlates with aggressive features and poor prognosis of hepatocellular carcinoma. J Transl Med 2011;9:5.
[119] Hou X, Li Y, Luo RZ, Fu JH, He JH, Zhang LJ, et al. High expression of the transcriptional co-activator p300 predicts poor survival in resectable non-small cell lung cancers. Eur J Surg Oncol 2012;38(6):523–30.
[120] Teng J, Wang ZY, Jarrard DF, Bjorling DE. Roles of estrogen receptor alpha and beta in modulating urothelial cell proliferation. Endocr Relat Cancer 2008;15(1):351–64.
[121] Nussler NC, Reinbacher K, Shanny N, Schirmeier A, Glanemann M, Neuhaus P, et al. Sex-specific differences in the expression levels of estrogen receptor subtypes in colorectal cancer. Gend Med 2008;5(3):209–17.
[122] Freise C, Ruehl M, Seehofer D, Hoyer J, Somasundaram R. The inhibitor of Ca(2 +)-dependent K (+) channels TRAM-34 blocks growth of hepatocellular carcinoma cells via downregulation of estrogen receptor alpha mRNA and nuclear factor-kappaB. Invest New Drugs 2013;31(2):452–7.
[123] Liu WH, Yeh SH, Lu CC, Yu SL, Chen HY, Lin CY, et al. MicroRNA-18a prevents estrogen receptor-alpha expression, promoting proliferation of hepatocellular carcinoma cells. Gastroenterology 2009;136(2):683–93.
[124] Henderson BE, Feigelson HS. Hormonal carcinogenesis. Carcinogenesis 2000;21(3):427–33.
[125] Ozawa A, Tanji N, Kikugawa T, Sasaki T, Yanagihara Y, Miura N, et al. Inhibition of bladder tumour growth by histone deacetylase inhibitor. BJU Int 2010;105(8):1181–6.
[126] Mariadason JM. HDACs and HDAC inhibitors in colon cancer. Epigenetics 2008;3(1):28–37.
[127] Xie HJ, Noh JH, Kim JK, Jung KH, Eun JW, Bae HJ, et al. HDAC1 inactivation induces mitotic defect and caspase-independent autophagic cell death in liver cancer. PLoS One 2012;7(4):e34265.
[128] Quint K, Agaimy A, Di Fazio P, Montalbano R, Steindorf C, Jung R, et al. Clinical significance of histone deacetylases 1, 2, 3, and 7: HDAC2 is an independent predictor of survival in HCC. Virchows Arch 2011;459(2):129–39.
[129] Sasaki H, Moriyama S, Nakashima Y, Kobayashi Y, Kiriyama M, Fukai I, et al. Histone deacetylase 1 mRNA expression in lung cancer. Lung Cancer 2004;46(2):171–8.
[130] Weichert W. HDAC expression and clinical prognosis in human malignancies. Cancer Lett 2009;280(2):168–76.
[131] Niegisch G, Knievel J, Koch A, Hader C, Fischer U, Albers P, et al. Changes in histone deacetylase (HDAC) expression patterns and activity of HDAC inhibitors in urothelial cancers. Urol Oncol 2012;31(8):1770–9.
[132] Jung KH, Noh JH, Kim JK, Eun JW, Bae HJ, Xie HJ, et al. HDAC2 overexpression confers oncogenic potential to human lung cancer cells by deregulating expression of apoptosis and cell cycle proteins. J Cell Biochem 2012;113(6):2167–77.
[133] Xu XS, Wang L, Abrams J, Wang G. Histone deacetylases (HDACs) in XPC gene silencing and bladder cancer. J Hematol Oncol 2011;4:17.
[134] Wilson AJ, Byun DS, Nasser S, Murray LB, Ayyanar K, Arango D, et al. HDAC4 promotes growth of colon cancer cells via repression of p21. Mol Biol Cell 2008;19(10):4062–75.

[135] Zhang J, Yang Y, Yang T, Liu Y, Li A, Fu S, et al. microRNA-22, downregulated in hepatocellular carcinoma and correlated with prognosis, suppresses cell proliferation and tumourigenicity. Br J Cancer 2010;103(8):1215–20.

[136] Behrens C, Feng L, Kadara H, Kim HJ, Lee JJ, Mehran R, et al. Expression of interleukin-1 receptor-associated kinase-1 in non-small cell lung carcinoma and preneoplastic lesions. Clin Cancer Res 2010;16(1):34–44.

[137] Andersen JB, Aaboe M, Borden EC, Goloubeva OG, Hassel BA, Orntoft TF. Stage-associated overexpression of the ubiquitin-like protein, ISG15, in bladder cancer. Br J Cancer 2006;94(10):1465–71.

[138] Hoeller D, Hecker CM, Dikic I. Ubiquitin and ubiquitin-like proteins in cancer pathogenesis. Nat Rev Cancer 2006;6(10):776–88.

[139] Desai SD, Haas AL, Wood LM, Tsai YC, Pestka S, Rubin EH, et al. Elevated expression of ISG15 in tumor cells interferes with the ubiquitin/26S proteasome pathway. Cancer Res 2006;66(2):921–8.

[140] Breuhahn K, Vreden S, Haddad R, Beckebaum S, Stippel D, Flemming P, et al. Molecular profiling of human hepatocellular carcinoma defines mutually exclusive interferon regulation and insulin-like growth factor II overexpression. Cancer Res 2004;64(17):6058–64.

[141] Feng Q, Sekula D, Guo Y, Liu X, Black CC, Galimberti F, et al. UBE1L causes lung cancer growth suppression by targeting cyclin D1. Mol Cancer Ther 2008;7(12):3780–8.

[142] Bektas N, Noetzel E, Veeck J, Press MF, Kristiansen G, Naami A, et al. The ubiquitin-like molecule interferon-stimulated gene 15 (ISG15) is a potential prognostic marker in human breast cancer. Breast Cancer Res 2008;10(4):R58.

[143] Simpson F, Lammerts van Bueren K, Butterfield N, Bennetts JS, Bowles J, Adolphe C, et al. The PCNA-associated factor KIAA0101/p15(PAF) binds the potential tumor suppressor product p33ING1b. Exp Cell Res 2006;312(1):73–85.

[144] Liu L, Chen X, Xie S, Zhang C, Qiu Z, Zhu F. Variant 1 of KIAA0101, overexpressed in hepatocellular carcinoma, prevents doxorubicin-induced apoptosis by inhibiting p53 activation. Hepatology 2012;56(5):1760–9.

[145] Kato T, Daigo Y, Aragaki M, Ishikawa K, Sato M, Kaji M. Overexpression of KIAA0101 predicts poor prognosis in primary lung cancer patients. Lung Cancer 2012;75(1):110–18.

[146] Simon R, Struckmann K, Schraml P, Wagner U, Forster T, Moch H, et al. Amplification pattern of 12q13-q15 genes (MDM2, CDK4, GLI) in urinary bladder cancer. Oncogene 2002;21(16):2476–83.

[147] Lianes P, Orlow I, Zhang ZF, Oliva MR, Sarkis AS, Reuter VE, et al. Altered patterns of MDM2 and TP53 expression in human bladder cancer. J Natl Cancer Inst 1994;86(17):1325–30.

[148] Ray RM, Bhattacharya S, Johnson LR. Mdm2 inhibition induces apoptosis in p53 deficient human colon cancer cells by activating p73- and E2F1-mediated expression of PUMA and Siva-1. Apoptosis 2011;16(1):35–44.

[149] Endo K, Ueda T, Ohta T, Terada T. Protein expression of MDM2 and its clinicopathological relationships in human hepatocellular carcinoma. Liver 2000;20(3):209–15.

[150] Higashiyama M, Doi O, Kodama K, Yokouchi H, Kasugai T, Ishiguro S, et al. MDM2 gene amplification and expression in non-small-cell lung cancer: immunohistochemical expression of its protein is a favourable prognostic marker in patients without p53 protein accumulation. Br J Cancer 1997;75(9):1302–8.

[151] Yu H, Zou Y, Jiang L, Yin Q, He X, Chen L, et al. Induction of apoptosis in non-small cell lung cancer by downregulation of MDM2 using pH-responsive PMPC-b-PDPA/siRNA complex nanoparticles. Biomaterials 2013;34(11):2738–47.

[152] Schmitz-Drager BJ, Schulz WA, Jurgens B, Gerharz CD, van Roeyen CR, Bultel H, et al. c-myc in bladder cancer. Clinical findings and analysis of mechanism. Urol Res 1997;25(Suppl. 1):S45–49.

[153] Tang Y, Simoneau AR, Liao WX, Yi G, Hope C, Liu F, et al. WIF1, a Wnt pathway inhibitor, regulates SKP2 and c-myc expression leading to G1 arrest and growth inhibition of human invasive urinary bladder cancer cells. Mol Cancer Ther 2009;8(2):458–68.

[154] Smith DR, Myint T, Goh HS. Over-expression of the c-myc proto-oncogene in colorectal carcinoma. Br J Cancer 1993;68(2):407–13.

[155] Shachaf CM, Kopelman AM, Arvanitis C, Karlsson A, Beer S, Mandl S, et al. MYC inactivation uncovers pluripotent differentiation and tumour dormancy in hepatocellular cancer. Nature 2004;431(7012):1112–17.

[156] Zajac-Kaye M. Myc oncogene: a key component in cell cycle regulation and its implication for lung cancer. Lung Cancer 2001;34(Suppl. 2):S43–46.

[157] Inagaki T, Ebisuno S, Uekado Y, Hirano A, Hiroi A, Shinka T, et al. PCNA and p53 in urinary bladder cancer: correlation with histological findings and prognosis. Int J Urol 1997;4(2):172–7.

[158] McKay JA, Douglas JJ, Ross VG, Curran S, Loane JF, Ahmed FY, et al. Analysis of key cell-cycle checkpoint proteins in colorectal tumours. J Pathol 2002;196(4):386–93.

[159] Qasim BJ, Ali HH, Hussein AG. Immunohistochemical expression of PCNA and CD34 in colorectal adenomas and carcinomas using specified automated cellular image analysis system: a clinicopathologic study. Saudi J Gastroenterol 2012;18(4):268–76.

[160] Stroescu C, Dragnea A, Ivanov B, Pechianu C, Herlea V, Sgarbura O, et al. Expression of p53, Bcl-2, VEGF, Ki67 and PCNA and prognostic significance in hepatocellular carcinoma. J Gastrointestin Liver Dis 2008;17(4):411–17.

[161] Nguyen VN, Mirejovsky P, Mirejovsky T, Melinova L, Mandys V. Expression of cyclin D1, Ki-67 and PCNA in non-small cell lung cancer: prognostic significance and comparison with p53 and bcl-2. Acta Histochem 2000;102(3):323–38.

[162] Hsu FM, Zhang S, Chen BP. Role of DNA-dependent protein kinase catalytic subunit in cancer development and treatment. Transl Cancer Res 2012;1(1):22–34.

[163] Wang SY, Peng L, Li CP, Li AP, Zhou JW, Zhang ZD, et al. Genetic variants of the XRCC7 gene involved in DNA repair and risk of human bladder cancer. Int J Urol 2008;15(6):534–9.

[164] Hosoi Y, Watanabe T, Nakagawa K, Matsumoto Y, Enomoto A, Morita A, et al. Up-regulation of DNA-dependent protein kinase activity and Sp1 in colorectal cancer. Int J Oncol 2004;25(2):461–8.

[165] Lu Y, Zhang HL, Li YZ, Zhao P. [Clinicopathological significance of expressions of DNA dependent protein kinase catalytic subunit and P16 in colorectal carcinoma]. Zhonghua Yi Xue Za Zhi 2008;88(29):2025–9.

[166] Long XD, Yao JG, Huang YZ, Huang XY, Ban FZ, Yao LM, et al. DNA repair gene XRCC7 polymorphisms (rs#7003908 and rs#10109984) and hepatocellular carcinoma related to AFB1 exposure among Guangxi population, China. Hepatol Res 2011;41(11):1085–93.

[167] Xing J, Wu X, Vaporciyan AA, Spitz MR, Gu J. Prognostic significance of ataxia-telangiectasia mutated, DNA-dependent protein kinase catalytic subunit, and Ku heterodimeric regulatory complex 86-kD subunit expression in patients with nonsmall cell lung cancer. Cancer 2008;112 (12):2756–64.

[168] Hao Y, Hu X, Liu Y, Jin B, Wen H, Hou K, et al. [The expression of ERCC1, DNA-PKcs protein and the relation to prognosis in non-small cell lung cancer]. Zhongguo Fei Ai Za Zhi 2008;11(2):226–30.

[169] Chatterjee SJ, George B, Goebell PJ, Alavi-Tafreshi M, Shi SR, Fung YK, et al. Hyperphosphorylation of pRb: a mechanism for RB tumour suppressor pathway inactivation in bladder cancer. J Pathol 2004;203(3):762–70.

[170] Llovet JM, Bruix J. Molecular targeted therapies in hepatocellular carcinoma. Hepatology 2008;48 (4):1312–27.

[171] Azechi H, Nishida N, Fukuda Y, Nishimura T, Minata M, Katsuma H, et al. Disruption of the p16/cyclin D1/retinoblastoma protein pathway in the majority of human hepatocellular carcinomas. Oncology 2001;60(4):346–54.

[172] Tanaka H, Fujii Y, Hirabayashi H, Miyoshi S, Sakaguchi M, Yoon HE, et al. Disruption of the RB pathway and cell-proliferative activity in non-small-cell lung cancers. Int J Cancer 1998;79 (2):111–15.

[173] Kitamura H, Yazawa T, Sato H, Okudela K, Shimoyamada H. Small cell lung cancer: significance of RB alterations and TTF-1 expression in its carcinogenesis, phenotype, and biology. Endocr Pathol 2009;20(2):101–7.

[174] Kihira S, Yoshida J, Kawada Y, Hitomi Y, Asada T, Hisatomi R, et al. Membrane microdomain-associated uroplakin IIIa contributes to Src-dependent mechanisms of anti-apoptotic proliferation in human bladder carcinoma cells. Biol Open 2012;1(10):1024–34.
[175] Saini S, Arora S, Majid S, Shahryari V, Chen Y, Deng G, et al. Curcumin modulates microRNA-203-mediated regulation of the Src-Akt axis in bladder cancer. Cancer Prev Res (Phila) 2011;4(10):1698–709.
[176] Lieu C, Kopetz S. The SRC family of protein tyrosine kinases: a new and promising target for colorectal cancer therapy. Clin Colorectal Cancer 2010;9(2):89–94.
[177] Ito Y, Kawakatsu H, Takeda T, Sakon M, Nagano H, Sakai T, et al. Activation of c-Src gene product in hepatocellular carcinoma is highly correlated with the indices of early stage phenotype. J Hepatol 2001;35(1):68–73.
[178] Byers LA, Sen B, Saigal B, Diao L, Wang J, Nanjundan M, et al. Reciprocal regulation of c-Src and STAT3 in non-small cell lung cancer. Clin Cancer Res 2009;15(22):6852–61.
[179] Garcia-Aranda C, de Juan C, Diaz-Lopez A, Sanchez-Pernaute A, Torres AJ, Diaz-Rubio E, et al. Correlations of telomere length, telomerase activity, and telomeric-repeat binding factor 1 expression in colorectal carcinoma. Cancer 2006;106(3):541–51.
[180] Igarashi M, Suda T, Hara H, Takimoto M, Nomoto M, Takahashi T, et al. Interferon can block telomere erosion and in rare cases result in hepatocellular carcinoma development with telomeric repeat binding factor 1 overexpression in chronic hepatitis C. Clin Cancer Res 2003;9(14):5264–70.
[181] Hu J, Sun L, Zhang C, Zhou X. Expression of telomeric repeat binding factor 1 in non-small cell lung cancer. J Surg Oncol 2006;93(1):62–7.
[182] Hollstein M, Sidransky D, Vogelstein B, Harris CC. p53 mutations in human cancers. Science 1991;253(5015):49–53.
[183] Iacopetta B. TP53 mutation in colorectal cancer. Hum Mutat 2003;21(3):271–6.
[184] Singhal S, Vachani A, Antin-Ozerkis D, Kaiser LR, Albelda SM. Prognostic implications of cell cycle, apoptosis, and angiogenesis biomarkers in non-small cell lung cancer: a review. Clin Cancer Res 2005;11(11):3974–86.
[185] Zheng QY, Li PP, Jin FS, Yao C, Zhang GH, Zang T, et al. Ursolic acid induces ER stress response to activate ASK1-JNK signaling and induce apoptosis in human bladder cancer T24 cells. Cell Signal 2013;25(1):206–13.
[186] Dai S, Jiang L, Wang G, Zhou X, Wei X, Cheng H, et al. HSP70 interacts with TRAF2 and differentially regulates TNFalpha signalling in human colon cancer cells. J Cell Mol Med 2010;14(3):710–25.
[187] Luedde T, Schwabe RF. NF-kappaB in the liver--linking injury, fibrosis and hepatocellular carcinoma. Nat Rev Gastroenterol Hepatol 2011;8(2):108–18.
[188] Zheng M, Morgan-Lappe SE, Yang J, Bockbrader KM, Pamarthy D, Thomas D, et al. Growth inhibition and radiosensitization of glioblastoma and lung cancer cells by small interfering RNA silencing of tumor necrosis factor receptor-associated factor 2. Cancer Res 2008;68(18):7570–8.
[189] Pucci S, Mazzarelli P, Rabitti C, Giai M, Gallucci M, Flammia G, et al. Tumor specific modulation of KU70/80 DNA binding activity in breast and bladder human tumor biopsies. Oncogene 2001;20(6):739–47.
[190] Mazzarelli P, Parrella P, Seripa D, Signori E, Perrone G, Rabitti C, et al. DNA end binding activity and Ku70/80 heterodimer expression in human colorectal tumor. World J Gastroenterol 2005;11(42):6694–700.
[191] Hsu CM, Yang MD, Chang WS, Jeng LB, Lee MH, Lu MC, et al. The contribution of XRCC6/Ku70 to hepatocellular carcinoma in Taiwan. Anticancer Res 2013;33(2):529–35.
[192] Busser B, Sancey L, Josserand V, Niang C, Khochbin S, Favrot MC, et al. Amphiregulin promotes resistance to gefitinib in nonsmall cell lung cancer cells by regulating Ku70 acetylation. Mol Ther 2010;18(3):536–43.
[193] Bau DT, Tsai CW, Wu CN. Role of the XRCC5/XRCC6 dimer in carcinogenesis and pharmacogenomics. Pharmacogenomics 2011;12(4):515–34.

[194] Lu B, Xu J, Zhu Y, Zhang H, Lai M. Systemic analysis of the differential gene expression profile in a colonic adenoma-normal SSH library. Clin Chim Acta 2007;378(1−2):42−7.
[195] Choi JE, Hur W, Jung CK, Piao LS, Lyoo K, Hong SW, et al. Silencing of 14-3-3zeta over-expression in hepatocellular carcinoma inhibits tumor growth and enhances chemosensitivity to cis-diammined dichloridoplatium. Cancer Lett 2011;303(2):99−107.
[196] Li Z, Zhao J, Du Y, Park HR, Sun SY, Bernal-Mizrachi L, et al. Down-regulation of 14-3-3zeta suppresses anchorage-independent growth of lung cancer cells through anoikis activation. Proc Natl Acad Sci USA 2008;105(1):162−7.
[197] Gupta A, Bhunia S, Balaramnavar V, Saxena A. Pharmacophore modelling, molecular docking and virtual screening for EGFR (HER 1) tyrosine kinase inhibitors. SAR QSAR Environ Res 2011;22(3−4):239−63.
[198] Chatr-Aryamontri A, Breitkreutz BJ, Heinicke S, Boucher L, Winter A, Stark C, et al. The BioGRID interaction database: 2013 update. Nucleic Acids Res 2013;41(Database issue): D816−823.
[199] Gene Ontology C. The Gene Ontology: enhancements for 2011. Nucleic Acids Res 2012;40 (Database issue):D559−564.
[200] Bland JM, Altman DG. Multiple significance tests: the Bonferroni method. BMJ 1995;310 (6973):170.
[201] Johansson R. System modeling and identification. Englewood Cliffs, NJ: Prentice Hall; 1993.
[202] Pagano M, Gauvreau K. Principles of biostatistics. Pacific Grove, CA: Duxbury; 2000.
[203] Kanehisa M. Molecular network analysis of diseases and drugs in KEGG. Methods Mol Biol 2013;939:263−75.
[204] Satoh J. Molecular network of microRNA targets in Alzheimer's disease brains. Exp Neurol 2012;235(2):436−46.
[205] Huang da W, Sherman BT, Lempicki RA. Systematic and integrative analysis of large gene lists using DAVID bioinformatics resources. Nat Protoc 2009;4(1):44−57.
[206] National Cancer Institute. Understanding cancer. <http://www.cancer.gov/about-cancer/understanding/what-is-cancer>.
[207] Welcsh PL, Owens KN, King MC. Insights into the functions of BRCA1 and BRCA2. Trends Genet 2000;16(2):69−74.
[208] Komander D. The emerging complexity of protein ubiquitination. Biochem Soc Trans 2009;37 (Pt 5):937−53.
[209] Livinskaya VA, Ivanov VA, Fedorova OA, Vorontsova DN, Barlev NA, Nikiforov AA. Polyclonal antibodies against human proteasome subunits PSMA3, PSMA5, and PSMB5. Hybridoma (Larchmt) 2012;31(4):272−8.
[210] Fedorova OA, Moiseeva TN, Nikiforov AA, Tsimokha AS, Livinskaya VA, Hodson M, et al. Proteomic analysis of the 20S proteasome (PSMA3)-interacting proteins reveals a functional link between the proteasome and mRNA metabolism. Biochem Biophys Res Commun 2011;416 (3-4):258−65.
[211] Lodish H, Berk A, Kaiser C, Krieger M, Bretscher A, Ploegh H. Molecular cell biology. New York: W. H. Freeman; 6th edition; 2013.
[212] Cavallo F, De Giovanni C, Nanni P, Forni G, Lollini PL. 2011: The immune hallmarks of cancer. Cancer Immunol Immunother 2011;60(3):319−26.
[213] Hanahan D, Weinberg RA. The hallmarks of cancer. Cell 2000;100(1):57−70.
[214] Osada H, Tatematsu Y, Sugito N, Horio Y, Takahashi T. Histone modification in the TGFbetaRII gene promoter and its significance for responsiveness to HDAC inhibitor in lung cancer cell lines. Mol Carcinog 2005;44(4):233−41.
[215] Bi G, Jiang G. The molecular mechanism of HDAC inhibitors in anticancer effects. Cell Mol Immunol 2006;3(4):285−90.
[216] Iljin K, Wolf M, Edgren H, Gupta S, Kilpinen S, Skotheim RI, et al. TMPRSS2 fusions with oncogenic ETS factors in prostate cancer involve unbalanced genomic rearrangements and are associated with HDAC1 and epigenetic reprogramming. Cancer Res 2006;66(21): 10242−6.

[217] Watters AD, Latif Z, Forsyth A, Dunn I, Underwood MA, Grigor KM, et al. Genetic aberrations of c-myc and CCND1 in the development of invasive bladder cancer. Br J Cancer 2002;87 (6):654–8.
[218] Hanahan D, Weinberg RA. Hallmarks of cancer: the next generation. Cell 2011;144(5):646–74.
[219] Ali MJ, Parsam VL, Honavar SG, Kannabiran C, Vemuganti GK, Reddy VAP. RB1 gene mutations in retinoblastoma and its clinical correlation. Saudi J Ophthalmol 2010;24(4): 119–23.
[220] Muller PA, Vousden KH. p53 mutations in cancer. Nat Cell Biol 2013;15(1):2–8.
[221] Ivanov AA, Khuri FR, Fu HA. Targeting protein-protein interactions as an anticancer strategy. Trends Pharmacol Sci 2013;34(7):393–400.
[222] White AW, Westwell AD, Brahemi G. Protein-protein interactions as targets for small-molecule therapeutics in cancer. Expert Rev Mol Med 2008;10:e8.
[223] Wood LM, Pan ZK, Seavey MM, Muthukumaran G, Paterson Y. The ubiquitin-like protein, ISG15, is a novel tumor-associated antigen for cancer immunotherapy. Cancer Immunol Immunother 2012;61(5):689–700.
[224] Yamasaki L, Pagano M. Cell cycle, proteolysis and cancer. Curr Opin Cell Biol 2004;16(6): 623–8.
[225] Camacho J. Molecular oncology: principles and recent advances. Bentham Science Publishers: Mexico city, Mexico; 2012.
[226] Williams GH, Stoeber K. The cell cycle and cancer. J Pathol 2012;226(2):352–64.
[227] Allenspach EJ, Maillard I, Aster JC, Pear WS. Notch signaling in cancer. Cancer Biol Ther 2002; 1(5):466–76.
[228] Bolos V, Grego-Bessa J, de la Pompa JL. Notch signaling in development and cancer. Endocr Rev 2007;28(3):339–63.
[229] Schneikert J, Behrens J. The canonical Wnt signalling pathway and its APC partner in colon cancer development. Gut 2007;56(3):417–25.
[230] Derynck R, Akhurst RJ, Balmain A. TGF-beta signaling in tumor suppression and cancer progression. Nat Genet 2001;29(2):117–29.
[231] Kawai T, Akira S. Toll-like receptor and RIG-I-like receptor signaling. Ann N Y Acad Sci 2008;1143:1–20.
[232] Bioseeker. RIG-I-like receptor signaling pathway in cancer drug pipeline update 2013; April 2013.
[233] Hwang J, Winkler L, Kalejta RF. Ubiquitin-independent proteasomal degradation during oncogenic viral infections. Biochim Biophys Acta 2011;1816(2):147–57.
[234] Bardag-Gorce F. Proteasome inhibitor treatment in alcoholic liver disease. World J Gastroenterol 2011;17(20):2558–62.
[235] Levine L. Proteasome inhibitors: their effects on arachidonic acid release from cells in culture and arachidonic acid metabolism in rat liver cells. BMC Pharmacol 2004;4:15.
[236] Dai P, Liu X, Li QW. [Function of the Lck and Fyn in T cell development]. Yi Chuan 2012;34 (3):289–95.
[237] Pedroza-Gonzalez A, Verhoef C, Ijzermans JN, Peppelenbosch MP, Kwekkeboom J, Verheij J, et al. Activated tumor-infiltrating CD4 + regulatory T cells restrain antitumor immunity in patients with primary or metastatic liver cancer. Hepatology 2013;57(1):183–94.
[238] Cronin SJ, Penninger JM. From T-cell activation signals to signaling control of anti-cancer immunity. Immunol Rev 2007;220:151–68.
[239] Helleday T. Homologous recombination in cancer development, treatment and development of drug resistance. Carcinogenesis 2010;31(6):955–60.
[240] Saviozzi S, Ceppi P, Novello S, Ghio P, Lo Iacono M, Borasio P, et al. Non-small cell lung cancer exhibits transcript overexpression of genes associated with homologous recombination and DNA replication pathways. Cancer Res 2009;69(8):3390–6.
[241] Lodish H. Molecular cell biology. New York, NY: W. H. Freeman and Company; 2000.
[242] Polakis P. Wnt signaling in cancer. Cold Spring Harb Perspect Biol 2012;4(5).

[243] Markowitz SD, Bertagnolli MM. Molecular origins of cancer: molecular basis of colorectal cancer. N Engl J Med 2009;361(25):2449–60.
[244] Taniue K, Oda T, Hayashi T, Okuno M, Akiyama T. A member of the ETS family, EHF, and the ATPase RUVBL1 inhibit p53-mediated apoptosis. EMBO Rep 2011;12(7):682–9.
[245] Sanchez-Palencia A, Gomez-Morales M, Gomez-Capilla JA, Pedraza V, Boyero L, Rosell R, et al. Gene expression profiling reveals novel biomarkers in nonsmall cell lung cancer. Int J Cancer 2011;129(2):355–64.
[246] Landi MT, Dracheva T, Rotunno M, Figueroa JD, Liu H, Dasgupta A, et al. Gene expression signature of cigarette smoking and its role in lung adenocarcinoma development and survival. PLoS One 2008;3(2):e1651.
[247] Tun HW, Marlow LA, von Roemeling CA, Cooper SJ, Kreinest P, Wu K, et al. Pathway signature and cellular differentiation in clear cell renal cell carcinoma. PLoS One 2010;5(5): e10696.
[248] Wallace TA, Prueitt RL, Yi M, Howe TM, Gillespie JW, Yfantis HG, et al. Tumor immunobiological differences in prostate cancer between African-American and European-American men. Cancer Res 2008;68(3):927–36.
[249] Cancer Genome Atlas Research Network. Genomic and epigenomic landscapes of adult de novo acute myeloid leukemia. N Engl J Med 2013;368(22):2059–74.
[250] Cancer Genome Atlas Research Network, Kandoth C, Schultz N, Cherniack AD, Akbani R, Liu Y, et al. Integrated genomic characterization of endometrial carcinoma. Nature 2013;497 (7447):67–73.
[251] Guo H, Zhu YP, Li D, He FC. [Identification, modeling and simulation of key pathways underlying certain cancers]. Yi Chuan 2011;33(8):809–19.
[252] de Jong H. Modeling and simulation of genetic regulatory systems: a literature review. J Comput Biol 2002;9(1):67–103.
[253] Wu D, Rice CM, Wang X. Cancer bioinformatics: a new approach to systems clinical medicine. BMC Bioinformatics 2012;13:71.
[254] Rivlin N, Brosh R, Oren M, Rotter V. Mutations in the p53 tumor suppressor gene: important milestones at the various steps of tumorigenesis. Genes Cancer 2011;2(4):466–74.
[255] Danes CG, Wyszomierski SL, Lu J, Neal CL, Yang W, Yu D. 14-3-3 zeta down-regulates p53 in mammary epithelial cells and confers luminal filling. Cancer Res 2008;68(6):1760–7.
[256] Sager R. Expression genetics in cancer: shifting the focus from DNA to RNA. Proc Natl Acad Sci USA 1997;94(3):952–5.
[257] Nibbe RK, Koyuturk M, Chance MR. An integrative -omics approach to identify functional sub-networks in human colorectal cancer. PLoS Comput Biol 2010;6(1):e1000639.
[258] Zhang XW, Wei D, Yap Y, Li L, Guo SY, Chen F. Mass spectrometry-based "omics" technologies in cancer diagnostics. Mass Spectrom Rev 2007;26(3):403–31.
[259] Randhawa V, Bagler G. Identification of SRC as a potent drug target for asthma, using an integrative approach of protein interactome analysis and in silico drug discovery. OMICS 2012;16(10): 513–26.
[260] Huang HJ, Lee KJ, Yu HW, Chen CY, Hsu CH, Chen HY, et al. Structure-based and ligand-based drug design for HER 2 receptor. J Biomol Struct Dyn 2010;28(1):23–37.
[261] Pirhadi S, Ghasemi JB. 3D-QSAR analysis of human immunodeficiency virus entry-1 inhibitors by CoMFA and CoMSIA. Eur J Med Chem 2010;45(11):4897–903.
[262] Fernandez M, Caballero J, Fernandez L, Sarai A. Genetic algorithm optimization in drug design QSAR: Bayesian-regularized genetic neural networks (BRGNN) and genetic algorithm-optimized support vectors machines (GA-SVM). Mol Divers 2011;15(1):269–89.
[263] Kinnings SL, Liu N, Tonge PJ, Jackson RM, Xie L, Bourne PE. A machine learning-based method to improve docking scoring functions and its application to drug repurposing. J Chem Inf Model 2011;51(2):408–19.
[264] Zhao XF, Li Q, Bian LJ, Zheng XH, Zheng JB, Zhang YY, et al. Using immobilized G-protein coupled receptors to screen bioactive traditional Chinese medicine compounds with multiple targets. J Pharm Biomed Anal 2012;70:549–52.

[265] Chen BS, Hsu CY, Liou JJ. Robust design of biological circuits: evolutionary systems biology approach. J Biomed Biotechnol 2011;2011:304236.
[266] Tu CT, Chen BS. New measurement methods of network robustness and response ability via microarray data. PLoS One 2013;8(1):e55230.
[267] Wong YH, Chen RH, Chen BS. Core and specific network markers of carcinogenesis from multiple cancer samples. J Theor Biol 2014;362:17–34.
[268] Lamptey P, Cates Jr W. Opinion: an ounce of prevention is worth a million lives. AIDSCaptions 1997;.
[269] Ashburner M, Ball CA, Blake JA, Botstein D, Butler H, Cherry JM, et al. Gene ontology: tool for the unification of biology. The Gene Ontology Consortium. Nat Genet 2000;25(1):25–9.
[270] Huang DW, Sherman BT, Tan Q, Kir J, Liu D, Bryant D, et al. DAVID Bioinformatics Resources: expanded annotation database and novel algorithms to better extract biology from large gene lists. Nucleic Acids Res 2007;35(Web Server issue):W169–175.
[271] Kanehisa M, Goto S. KEGG: Kyoto Encyclopedia of Genes and Genomes. Nucleic Acids Res 2000;28(1):27–30.
[272] Kanehisa M, Goto S, Sato Y, Kawashima M, Furumichi M, Tanabe M. Data, information, knowledge and principle: back to metabolism in KEGG. Nucleic Acids Res 2014;42(Database issue): D199–205.
[273] Mi H, Muruganujan A, Thomas PD. PANTHER in 2013: modeling the evolution of gene function, and other gene attributes, in the context of phylogenetic trees. Nucleic Acids Res 2013;41 (Database issue):D377–386.
[274] Zhang Y. I-TASSER server for protein 3D structure prediction. BMC Bioinformatics 2008;9:40.
[275] Yang JY, Roy A, Zhang Y. Protein-ligand binding site recognition using complementary binding-specific substructure comparison and sequence profile alignment. Bioinformatics 2013;29 (20):2588–95.
[276] Schneidman-Duhovny D, Dror O, Inbar Y, Nussinov R, Wolfson HJ. PharmaGist: a webserver for ligand-based pharmacophore detection. Nucleic Acids Res 2008;36:W223–8.
[277] Irwin JJ, Sterling T, Mysinger MM, Bolstad ES, Coleman RG. ZINC: a free tool to discover chemistry for biology. J Chem Inf Model 2012;52(7):1757–68.

INDEX

Note: Page numbers followed by "*f*" and "*t*" refer to figures and tables, respectively.

A

E

F

I

N

R

T

U

V

W

X

Y

Z

Kolibri 360 GmbH
Ferdinand-Jühlke-Straße 7
99095 Erfurt, Deutschland
produktsicherheit@kolibri360.de